P9-AFF-666

Biotechnology

Second Edition

Volume 2

Genetic Fundamentals and Genetic Engineering

Biotechnology
Second Edition

Fundamentals

Volume 1
Biological Fundamentals

Volume 2
Genetic Fundamentals and
Genetic Engineering

Volume 3
Bioprocessing

Volume 4
Measuring, Modelling, and Control

Products

Volume 5
Genetically Engineered Proteins and
Monoclonal Antibodies

Volume 6
Products of Primary Metabolism

Volume 7
Products of Secondary Metabolism

Volume 8
Biotransformations

Special Topics

Volume 9
Enzymes, Biomass, Food and Feed

Volume 10
Special Processes

Volume 11
Environmental Processes

Volume 12
Modern Biotechnology:
Legal, Economic and
Social Dimensions

Distribution:

VCH, P. O. Box 101161, D-6940 Weinheim (Federal Republic of Germany)

Switzerland: VCH, P. O. Box, CH-4020 Basel (Switzerland)

United Kingdom and Ireland: VCH (UK) Ltd., 8 Wellington Court, Cambridge CB1 1HZ (England)

USA and Canada: VCH, 220 East 23rd Street, New York, NY 10010–4606 (USA)

ISBN 3-527-28312-9 (VCH, Weinheim)
Set ISBN 3-527-28310-2 (VCH, Weinheim)

ISBN 1-56081-152-8 (VCH, New York)
Set ISBN 1-56081-602-3 (VCH, New York)

A Multi-Volume Comprehensive Treatise

Biotechnology

Second, Completely Revised Edition

Edited by
H.-J. Rehm and G. Reed
in cooperation with
A. Pühler and P. Stadler

Volume 2

Genetic Fundamentals and Genetic Engineering

Edited by
A. Pühler

Weinheim · New York · Basel · Cambridge

Series Editors:
Prof. Dr. H.-J. Rehm
Institut für Mikrobiologie
Universität Münster
Corrensstraße 3
D-4400 Münster

Dr. G. Reed
2131 N. Summit Ave.
Apartment #304
Milwaukee, WI 53202-1347
USA

Prof. Dr. A. Pühler
Biologie VI (Genetik)
Universität Bielefeld
P.O. Box 100131
D-4800 Bielefeld 1

Dr. P. J. W. Stadler
Bayer AG
Verfahrensentwicklung Biochemie
Leitung
Friedrich-Ebert-Straße 217
D-5600 Wuppertal 1

Volume Editor:
Prof. Dr. A. Pühler
Biologie VI (Genetik)
Universität Bielefeld
P.O. Box 100131
D-4800 Bielefeld 1

Published jointly by
VCH Verlagsgesellschaft mbH, Weinheim (Federal Republic of Germany)
VCH Publishers Inc., New York, NY (USA)

Editorial Director: Dr. Hans-Joachim Kraus
Editorial Manager: Christa Maria Schultz
Copy Editor: Karin Dembowsky
Production Director: Maximilian Montkowski
Production Manager: Dipl. Wirt.-Ing. (FH) Hans-Jochen Schmitt

Library of Congress Card No.: applied for

British Library Cataloguing-in-Publication Data:
A catalogue record for this book is available from the British Library

Die Deutsche Bibliothek – CIP-Einheitsaufnahme
Biotechnology : a multi volume comprehensive treatise / ed. by
H.-J. Rehm and G. Reed. In cooperation with A. Pühler and P.
Stadler. – 2., completely rev. ed. – Weinheim; New York;
Basel; Cambridge: VCH.
NE: Rehm, Hans J. [Hrsg.]

2., completely rev. ed.
Vol. 2. Genetic fundamentals and genetic engineering / ed. by A. Pühler.
– 1993
ISBN 3-527-28312-9 (Weinheim)
ISBN 1-56081-152-8 (New York)
NE: Pühler, Alfred [Hrsg.]

Printed on acid-free and low-chlorine paper.

Composition and Printing: Zechnersche Buchdruckerei, D-6720 Speyer.
Bookbinding: Klambt-Druck GmbH, D-6720 Speyer
Printed in the Federal Republic of Germany

Preface

In recognition of the enormous advances in biotechnology in recent years, we are pleased to present this Second Edition of "Biotechnology" relatively soon after the introduction of the First Edition of this multi-volume comprehensive treatise. Since this series was extremely well accepted by the scientific community, we have maintained the overall goal of creating a number of volumes, each devoted to a certain topic, which provide scientists in academia, industry, and public institutions with a well-balanced and comprehensive overview of this growing field. We have fully revised the Second Edition and expanded it from ten to twelve volumes in order to take all recent developments into account.

These twelve volumes are organized into three sections. The first four volumes consider the fundamentals of biotechnology from biological, biochemical, molecular biological, and chemical engineering perspectives. The next four volumes are devoted to products of industrial relevance. Special attention is given here to products derived from genetically engineered microorganisms and mammalian cells. The last four volumes are dedicated to the description of special topics.

The new "Biotechnology" is a reference work, a comprehensive description of the state-of-the-art, and a guide to the original literature. It is specifically directed to microbiologists, biochemists, molecular biologists, bioengineers, chemical engineers, and food and pharmaceutical chemists working in industry, at universities or at public institutions.

A carefully selected and distinguished Scientific Advisory Board stands behind the series. Its members come from key institutions representing scientific input from about twenty countries.

The volume editors and the authors of the individual chapters have been chosen for their recognized expertise and their contributions to the various fields of biotechnology. Their willingness to impart this knowledge to their colleagues forms the basis of "Biotechnology" and is gratefully acknowledged. Moreover, this work could not have been brought to fruition without the foresight and the constant and diligent support of the publisher. We are grateful to VCH for publishing "Biotechnology" with their customary excellence. Special thanks are due Dr. Hans-Joachim Kraus and Christa Schultz, without whose constant efforts the series could not be published. Finally, the editors wish to thank the members of the Scientific Advisory Board for their encouragement, their helpful suggestions, and their constructive criticism.

H.-J. Rehm
G. Reed
A. Pühler
P. Stadler

Scientific Advisory Board

Contents

Contributors

Prof. Dr. Jozef Anné
Rega Instituut
Katholieke Universiteit Leuven
Minderbroedersstraat 10
B-3000 Leuven
Belgium
Chapter 4

Prof. Dr. Gottfried Brem
Lehrstuhl für Molekulare Tierzuchtforschung
Universität München
Veterinärstraße 13
D-8000 München 22
Federal Republic of Germany
Chapter 18

Dr. Anneliese Crueger
Bayer AG
Verfahrensentwicklung Biochemie
Friedrich-Ebert-Straße 217
D-5600 Wuppertal 1
Federal Republic of Germany
Chapter 1

Prof. Dr. Joachim W. Engels
Institut für Organische Chemie
Universität Frankfurt
Niederurseler Hang
D-6000 Frankfurt am Main 50
Federal Republic of Germany
Chapter 9

Prof. Dr. mult. Karl Esser
Lehrstuhl für Allgemeine Botanik
Ruhr-Universität Bochum
P.O. Box 102148
D-4630 Bochum 1
Federal Republic of Germany
Chapter 3

Prof. Dr. Werner Frommer
Claudiusweg 17
D-5600 Wuppertal 1
Federal Republic of Germany
Chapter 19

Prof. Dr. Hermann Geldermann
Institut für Tierhaltung und Tierzüchtung
Universität Stuttgart
Garbenstraße 17
D-7000 Stuttgart 70
Federal Republic of Germany
Chapter 5

Prof. Dr. Friedrich Götz
Lehrstuhl für Mikrobielle Genetik
Universität Tübingen
Auf der Morgenstelle 28
D-7400 Tübingen
Federal Republic of Germany
Chapter 6

Dr. Hansjörg Hauser
Gesellschaft für Biotechnologische Forschung mbH
Mascheroder Weg 1
D-3300 Braunschweig
Federal Republic of Germany
Chapter 17

Prof. Dr. Bruce W. Holloway
Department of Genetics and Developmental Biology
Monash University
Wellington Road
Clayton, Victoria, 3168
Australia
Chapter 2

Prof. Dr. Günter Kahl
Fachbereich Biologie
Universität Frankfurt
Siesmayerstraße 70
D-6000 Frankfurt am Main 1
Federal Republic of Germany
Chapter 15

Dr. Jörn Kalinowski
Biologie VI (Genetik)
Universität Bielefeld
P.O. Box 100131
D-4800 Bielefeld 1
Federal Republic of Germany
Chapter 12

Prof. Dr. Ralf Mattes
Institut für Industrielle Genetik
Universität Stuttgart
Azenbergstraße 18
D-7000 Stuttgart 1
Federal Republic of Germany
Chapter 7

Dr. Günther Muth
Institut für Angewandte Molekularbiologie der Universität des Saarlandes
Im Stadtwald
D-6600 Saarbrücken
Federal Republic of Germany
Chapter 12

Priv.-Doz. Dr. Ursula B. Priefer
Biologie VI (Genetik)
Universität Bielefeld
P.O. Box 100131
D-4800 Bielefeld 1
Federal Republic of Germany
Chapter 11

Dr. Johan Robben
Faculteit Landbouwwetenschappen
Laboratorium voor Gentechnologie
Katholieke Universiteit Leuven
Willem de Croylaan, 42
B-3001 Leuven
Belgium
Chapter 8

Dr. Helmut Schwab
Institut für Biotechnologie
Arbeitsgruppe Genetik
Technische Universität Graz
Petersgasse 12
A-8010 Graz
Austria
Chapter 10

Dr. Reinhard Simon
TÜV Südwestdeutschland e.V.
Fachgruppe Biologische Sicherheit
Robert-Bunsen-Straße 1
D-7800 Freiburg i.Br.
Federal Republic of Germany
Chapter 19

Dipl.-Chem. Belinda Sprunkel
Institut für Organische Chemie
Universität Frankfurt
Niederurseler Hang
D-6000 Frankfurt am Main 50
Federal Republic of Germany
Chapter 9

Prof. Dr. Ulf Stahl
Fachbereich 13
Institut für Biotechnologie
Technische Universität Berlin
Gustav-Meyer-Allee 25
D-1000 Berlin 65
Federal Republic of Germany
Chapter 3

Dr. Peter E. Sudbery
Department of Molecular Biology and Biotechnology
The University of Sheffield
P.O. Box 594
Sheffield S10 2UH
United Kingdom
Chapter 13

Prof. Dr. Geoffrey Turner
Department of Molecular Biology and Biotechnology
The University of Sheffield
P.O. Box 594
Sheffield S10 2UH
United Kingdom
Chapter 14

Dr. Eugen Uhlmann
Institut für Organische Chemie
Universität Frankfurt
Niederurseler Hang
D-6000 Frankfurt am Main 50
Federal Republic of Germany
Chapter 9

Dr. Peter Verhasselt
Faculteit Landbouwwetenschappen
Laboratorium voor Gentechnologie
Katholieke Universiteit Leuven
Willem de Croylaan, 42
B-3001 Leuven
Belgium
Chapter 8

Marleen Voet
Faculteit Landbouwwetenschappen
Laboratorium voor Gentechnologie
Katholieke Universiteit Leuven
Willem de Croylaan, 42
B-3001 Leuven
Belgium
Chapter 8

Prof. Dr. Guido Volckaert
Faculteit Landbouwwetenschappen
Laboratorium voor Gentechnologie
Katholieke Universiteit Leuven
Willem de Croylaan, 42
B-3001 Leuven
Belgium
Chapter 8

Prof. Dr. W. Eberhard Weber
Institut für Angewandte Genetik
Universität Hannover
Herrenhäuser Straße 2
D-3000 Hannover
Federal Republic of Germany
Chapter 5

Dr. Kurt Weising
Fachbereich Biologie
Universität Frankfurt
Siesmayerstraße 70
D-6000 Frankfurt am Main 1
Federal Republic of Germany
Chapter 15

Prof. Dr. Lothar Willmitzer
Institut für Genbiologische Forschung Berlin GmbH
Ihnestraße 63
D-1000 Berlin 33
Federal Republic of Germany
Chapter 16

Dr. Manfred Wirth
Gesellschaft für Biotechnologische Forschung mbH
Mascheroder Weg 1
D-3300 Braunschweig
Federal Republic of Germany
Chapter 17

Prof. Dr. Wolfgang Wohlleben
Institut für Angewandte Molekularbiologie der Universität des Saarlandes
Im Stadtwald
D-6600 Saarbrücken
Federal Republic of Germany
Chapter 12

Prof. Dr. Günter Wricke
Institut für Angewandte Genetik
Universität Hannover
Herrenhäuser Straße 2
D-3000 Hannover
Federal Republic of Germany
Chapter 5

Introduction

ALFRED PÜHLER

Bielefeld, Federal Republic of Germany

The immense progress made in genetics during the last two decades has had a tremendous effect on biotechnology. In contrast to former times when strains could be modified by classical mutagenesis only, it is now possible to design organisms of prokaryotic and eukaryotic origin which conduct specific production processes. This is achieved by applying newly developed techniques which combine biochemistry, molecular biology and genetics, and from the new discipline of so-called genetic engineering. Every scientist involved in biotechnology whether working as a chemist, biologist or engineer, should be aware of these powerful techniques. For this reason, the special volume on "Genetic Fundamentals and Genetic Engineering" was conceived to give an introduction to classical genetics, provide information on molecular genetics and present the most recent progress made in the field of genetic engineering using microorganisms, plants and animals. It should be pointed out that this volume was not designed as a textbook of classical and molecular genetics, but was written to emphasize the biotechnological aspects thereof. Examples of industrial and agricultural applications were therefore added to each chapter. This is, consequently, the first book which deals comprehensively with all aspects of genetics in biotechnology and will be of interest to advanced students as well as to scientists employed in institutes or industrial companies.

In *Part I* the classical genetic techniques relevant to biotechnology are presented. Besides mutagenesis which still plays an important role in strain improvement, the genetic exchange processes for prokaryotes and lower eukaryotes are extensively discussed. Genetic exchange processes remain powerful tools to genetically modify Gram-negative and Gram-positive bacteria as well as lower eukaryotes such as yeasts and filamentous fungi. One chapter deals specifically with the cell fusion technique which permits the exchange of genetic information between organisms which normally do not exchange genetic material. A further chapter is devoted to gene mapping techniques applicable to animals and plants. These techniques are of increasing importance for the construction of defined organisms which can be employed in biotechnological processes in the future.

Part II highlights recent advances in molecular genetics which have revolutionized our knowledge in the field. It starts with a chapter on the structure and function of DNA where the diversity of nucleic acid structures is presented. A subsequent chapter, which accentuates microorganisms used predominantly in industrial production processes, deals specifically with the principles of gene expression in prokaryotes. The chapter on DNA sequencing introduces a completely new development in the field of genetics. For the first time, automation and robotization combined with computer analysis will accelerate the acquisition of knowledge of basic genetic information. The chapter on DNA synthesis is of similar importance, since newly synthesized oligonucleotides have already found wide-spread application in biology and medicine. Such oligonucleotides, which are used as hybridization probes or as primers in the polymerase chain reaction, play a key role in site-specific mutagenesis and the biosynthesis of synthetic genes.

Part III deals with the genetic engineering of microorganisms. Techniques, first developed for the model organism *Escherichia coli,* were then extended to other Gram-negative bacteria. Gene transfer techniques for these organisms are involved in the use of broad-host-range plasmids. The genetic engineering techniques for Gram-positive bacteria pertain predominantly to Bacilli, Streptomycetes and Corynebacteria, with specific emphasis on cloning vectors and DNA transfer methods. With respect to the genetic engineering of eukaryotic microorganisms, the methods developed for yeasts and filamentous fungi are described including integration vectors and yeast artificial chromosomes.

Parts IV and V concentrate on plants and animals, since these multicellular organisms require special genetic engineering techniques. The methods developed to genetically engineer *plant* cells are extensively described and include all possible DNA transfer techniques. In a further chapter, the regeneration to intact plants and the controlled expression of foreign genes are summarized. This is followed by a comprehensive list of transgenic plants which have been modified to enhance their agronomic or industrial applicability. Genetic engineering techniques pertaining to *animals* are similarly discussed. The first chapter of *Part V* presents a complete survey of all genetic engineering aspects of animal cells including cloning vectors, gene transfer techniques and expression of foreign genes. The subsequent chapter concentrates on transgenic animals, discussing the current knowledge of transgenic mice, livestock, chicken and fish and also mentions the problems and future possibilities of such transgenic organisms.

In the last section, *Part VI*, biosafety concepts in modern biotechnology with particular emphasis on the contained use of mircoorganisms are summarized, demonstrating that production using genetically modified organisms can be carried out safely using the established systems.

Barcelona, September 1992 A. Pühler

I. Classical Genetics

1 Mutagenesis

ANNELIESE CRUEGER

Wuppertal, Federal Republic of Germany

1 Introduction

Mutagenesis means the induction of inheritable changes - mutations - into the genetic material of any organism. These mutations are caused by changes in the genotype and can be detected as a modified phenotype of the organism (mutant). Mutations occur *in vivo* spontaneously or after induction by mutagens, e.g., radiation or chemical agents. Mutations can also be induced *in vitro* by the use of genetic engineering techniques.

Mutagenesis is important in many aspects. The occurrence of mutagenesis is a basic requirement of evolution. Moreover, in microbial strain improvement and in plant cultivation mutagenesis is used for breeding new varieties. In food and feed as well as in the environment mutagens are a source of danger due to the fact that many mutagens also act as carcinogens and/or teratogens.

Ever since mutation was emphasized by DE VRIES at the beginning of this century as a fundamental genetic process, it was quickly recognized that mutation constitutes the raw material for evolution. The first of several advances in the analysis of mutational processes was the discovery by MULLER in 1927 that mutations could be induced in *Drosophila* by using X-ray radiation. The first success in mutagenesis by chemical agents in 1941 was the use of mustard gas in *Drosophila* (AUERBACH and ROBSON, 1947). A prerequisite for the deeper understanding of mutagenesis on a molecular basis was the discovery of the essential role of the nucleic acids in heredity, fist demonstrated by AVERY et al. (1944) and then analyzed by WILKINS, CRICK, and WATSON (1953) (see WILKINS et al., 1953). Since then much of our understanding of mutational processes has arisen from studies of prokaryotic microorganisms, particularly bacteriophages due to some advantages, such as their relative simplicity of composition, genetic homogenity, haploidy, rapid mode of growth, and the opportunity for examining rare events within huge populations.

The types of DNA damage caused by different mutagens and the action of the cellular DNA repair pathways on these damages could be elucidated mainly in *Escherichia coli*. More recent years have seen an extension of these results into the realm of eukaryotic and multicellular organisms. Finally, the development of gene technology has led to revolutionary new methods of direct mutagenesis which make it possible to isolate mutants of specific genes of interest.

The topic of this chapter has been reviewed by AUERBACH (1976), DRAKE (1970, 1989), CLARKE (1975, 1976), JACOBSON (1981), SARGENTINI and SMITH (1985), and LINDAHL (1990).

2 Taxonomy of Mutational Lesions

For an understanding of the details of the mutation process it will be convenient to begin with considering the elementary types of mutational changes. Mutations may arise spontaneously or after induction by radiation or chemical agents as a result of structural changes in the genome. These changes can be distinguished initially according to whether they produce large or small alterations in the genome, although a certain amount of overlap does connect the different classes:

Genome mutations may cause changes in the number of chromosome sets.

Chromosome mutations may change the order of genes within the chromosome.

Gene or *point mutations* may result from changes in the base sequence in a gene.

Premutational lesions or mutations may arise through (1) reactions between mutagen and DNA, (2) use of mutagen altered precursors or base analogs in DNA replication or repair, (3) errors in DNA replication, recombination or repair, or (4) indirectly through errors in transcription and translation.

2.1 Genome Mutations

Among chromosomal variations the easiest to observe are those that involve changes in the number of chromosomes of a cell. These ploidy mutations just occur in eukaryotes. Two types are distinguished: Variations with changes in the number of entire sets of chromosomes (*euploidy*), where the term *polyploidy* refers to any organism in which the number of complete chromosome sets exceeds that of the diploid. The second type of variations has changes in the number of only one or a few single chromosomes within the set (*aneuploidy*).

In the case of euploidy, haploid or monoploid animals or plants can be induced by various means but usually develop abnormally. One reason for this is that due to the absence of homologous pairing of chromosomes in meiosis, where chromosomes are randomly distributed in the daughter cells, gametes with less than the complete chromosome set are produced.

Polyploidy mutations occur spontaneously by fusion of gametes with non-reduced chromosome sets yielding triploids or tetraploids or may be induced by plant growth promoters (e.g., indole-3-acetic acid) or colchicine. Colchicine and the synthetic equivalent colcemid interfere with the formation of spindle fibers in mitosis. The metaphase chromosomes remain scattered in the cytoplasm and do not move to a metaphasic plate yielding tetraploid cells. Triploids ($3n$) and tetraploids ($4n$) often are more vigorous than the diploid varieties. The triploids are usually sterile because of defective gamete formation. By repeating the colchicine treatment higher ploidies are available, but usually are not viable.

Aneuploidies with missing or additional chromosomes in the genome are a very heterogeneous group of variations. Well investigated are trisomies where an additional chromosome fits into the diploid chromosome set ($2n+1$). Trisomy is mainly caused by non-disjunction during mitosis or meiosis. In humans the trisomy for chromosome 21 (Down syndrome, previously Mongolian idiocy) which occurs with a frequency of one out of 600 births is well known. Other autosomal trisomies in humans are the Pätau's syndrome (chromosome 13) and Edwards' syndrome (chromosome 18). Well known as aneuploidies in human sex chromosomes are the Turner's syndrome where one X chromosome is missing (X0) and the Klinefelter's syndrome where the individuals have the full male XY complement set but are polysomic for one or more additional X chromosomes (XXY, XXXY, etc.) (BORGAONKAR, 1980).

2.2 Chromosome Mutations

A change of information occurs through a change in chromosome structure by loss, addition, or rearrangement of particular sections of the chromosome or its cell division subunit, the chromatid. The number of chromosomes usually remains the same. Chromosome mutations are of minor importance for prokaryotic organisms due to their single chromosome. In eukaryotes chromosome mutations are caused by breaks in the interphase chromosomes. According to the circumstances three types of structural changes may be observed: (1) Loss of a chromosome region. Terminal losses formerly were called deletions, intercalary losses deficiencies, but nowadays the term "deficiencies" is used for all these events. (2) Addition of a chromosomal section (duplication). (3) Rearrangement of a chromosomal sequence either within the same chromosome resulting in a change in gene order within the chromosome (inversion), or between different chromosomes by the transfer of a section of one chromosome to a non-homologous chromosome (translocation).

If only one chromosome is affected by a break, deficiencies or inversions are generated (intrachromosomal chromosome mutations); with two or even higher numbers of chromosomes involved duplications or translocations result from the rearrangement process.

Chromosome mutations can be induced by chemical mutagens, such as sulfur and nitrogen mustards, nitrosomethylurea, azaserine, mitomycin C, or by radiation such as UV-light or X-rays. All types of chromosome mutations have been described in plants, animals, and humans.

2.3 Gene Mutations

Gene mutations have come to mean the process by which new alleles of a gene are produced. In the case of *macrolesions,* a larger number of nucleotides is concerned. By replication errors inversions, deficiencies and insertions may occur which involve in some cases large sequences of a gene. In *microlesions* or *point mutations* one single base is substituted by another or one or a few nucleotides are deleted or inserted. Point mutations are divided into two classes: base pair substitutions and frameshift mutations.

Base pair substitutions affect only one base pair at a time. A wild-type allele may be replaced by another base in the mutant allele. A *transition* refers to the exchange of a purine with another purine or a pyrimidine with another pyrimidine:

AT ⇌ GC and CG ⇌ TA

A *transversion* is the substitution of a pyrimidine for a purine or *vice versa*:

AT ⇌ TA; AT ⇌ CG; TA ⇌ GC and CG ⇌ GC

Base pair substitutions produce characteristic effects upon proteins. When they occur within a gene, they produce either amino acid substitutions (missense mutations), or chain terminating codons (UAA, UAG, or UGA; nonsense mutations). One characteristic of point mutants is that they can revert.

Frameshift mutations consist of deletions or additions of small numbers of base pairs, thus altering the reading frame in the following transcription and translation process. From the mutational site onwards, each codon will be different, and as a result, the amino acids will be exchanged within the limitation of codon degeneracy.

Radiation and a multitude of chemical agents mainly induce point mutations at low doses of mutagen. The mutational changes were brought about either by errors in replication or repair by incorporating the wrong base into the polynucleotide chain, or by a direct chemical interaction with a base.

3 Spontaneous Mutations

All living organisms suffer from spontaneous mutations. The mechanisms responsible for the generation of spontaneous mutations have been broadly investigated in microorganisms, but remain to be established in prokaryotes as well as in eukaryotes (SARGENTINI and SMITH, 1985). These mechanisms may be classified into exogenous processes caused by chemical and physical agents, such as environmental substances (e.g., insecticides, herbicides, fungicides, metals in water and soil) or radiation. Endogenous processes, which are now believed to be more significant, include intracellular production of mutagens (e.g., free radicals), or errors during replication and recombination of DNA. In addition mutator genes or transposable elements are involved. The *dam* mutation in *Escherichia coli* also causes increased mutation rates (see mismatch repair, Sect. 7.2.3). All premutational lesions produced by the various sources of spontaneous mutation first pass through DNA-repair processes (see Sect. 7). Following the work of SPEYER (1965), evidence has been established that many spontaneous mutations in *E. coli* and its phages arise by mistakes made by post-replicative mismatch repair. The overall error rate measured during replication of the *E. coli* chromosome is only one mistake in 10^8 to 10^{10} nucleotides polymerized, yet theoretical and experimental evidence (TOPAL and FRESCO, 1976) suggests an error rate of 10^{-4} to 10^{-5} during the incorporation of deoxynucleotide triphosphates by the DNA polymerases. FERSHT et al. (1982) found misinsertion frequencies in *E. coli* for GT, CA, GA, and CT mispairings of 8×10^{-5}, 2×10^{-5}, 1×10^{-5} and 2×10^{-7}, respectively. The observed increase in fidelity is caused by a polymerase/exonuclease complex with editing or proofreading activities. This system is thought to follow the replication fork, scrutinizing the newly replicated DNA for mismatches resulting from errors of DNA replication and correcting them partially (BRUTLAG and KORNBERG, 1972).

SCHAAPER and DUNN (1987) investigated DNA sequence changes in 487 spontaneous mutations in the N-terminal part of the *lacI* gene in *mutH, mutL,* and *mutS* strains of *E. coli* with high spontaneous mutation rates due to a deficiency in the post-replicative mismatch repair thus revealing the original pattern of spontaneous DNA replication errors: The spectra consist of base substitutions (75%), where transitions (both AT→GC and

GC→AT) are favored over transversions (96% versus 4%), and single-base deficiencies (25%). In yeast the results suggest that many spontaneous mutations arise by mutagenic repair of spontaneous lesions (HASTINGS et al., 1976). Spontaneous mutations in higher eukaryotes may occur in the germ line (generative mutations), whereby the mutated gene is transferred to the descendents, a prerequisite of evolutionary development, or they occur in somatic cells (somatic mutations).

Spontaneous mutations in somatic cells are rare; 10^{-9} to 10^{-12} mutations per nucleotide per cell have been estimated (DRAKE, 1969), but often their detection is very difficult due to the diploid status of the cell. Spontaneous mutations have been implicated as a causative factor in aging (HARMAN, 1981), carcinogenesis (AMES, 1983), and cell injury (KLEBANOFF, 1988). LOEB and CHENG (1990) have suggested endogenous factors as possible sources for spontaneous mutations in somatic cells, as replication errors, depurination of DNA, and damage of DNA by the generation of free radicals. Depurination, depyrimidination, and deamination account for the introduction of about 10^4 premutational lesions per day in the DNA of each human cell. Similar levels of damage are expected to be introduced by DNA alkylation by the methyl group donor *S*-adenosylmethionine (LINDAHL, 1990).

4 Mutation Rates

Mutation rate means the frequency of mutation per gene locus and generation. Spontaneous mutation rates in different organisms are given in Tab. 1. Among eukaryotes the mutation rates are appreciably higher due to the fact that in more complex organisms the generation time includes many successive cell divisions. The spontaneous mutation rate may vary in different genes of the same genome.

The measured mutation rates do not represent the true number of mutational events in the genome of an organism due to the fact that changes in the nucleotide sequence do not necessarily always lead to detectable phenotypic changes of the organisms. Some mutations are 'silent': They may affect regions of the protein that do not lead to functional changes, or the nucleotide substitutions may not lead to amino acid exchanges, because of codon degeneracy.

Apart from silent mutations, mutant effects may be absent due to reversion events. While

Tab. 1. Spontaneous Mutation Rates of Specific Loci for Various Organisms (GLASS and RITTERHOFF, 1956; STADLER, 1942; SCHLAGER and DICKIE, 1967; VOGEL and RATHENBERG, 1975)

Organisms	Trait	Mutation Rates
T4 bacteriophage	to rapid lysis (r^+ r)	7×10^{-5}
	to new host range (h^+ h)	1×10^{-7}
Escherichia coli K12	$leu^- \rightarrow leu^+$	7×10^{-10}
	$arg^- \rightarrow arg^+$	4×10^{-9}
	$trp^- \rightarrow trp^+$	6×10^{-8}
	$ara^+ \rightarrow ara^-$	2×10^{-6}
Drosophila melanogaster (males)	e^+ to ebony	2×10^{-5}
	ey^+ to eyeless	6×10^{-5}
	y^+ to yellow	1×10^{-4}
Corn	C to colorless	2×10^{-6}
	Pr to purple	1×10^{-5}
Mouse	b^+ to brown	4×10^{-6}
	c^+ to albino	1×10^{-5}
Humans	Huntington's chorea	5×10^{-6}
	Intestinal polyposis	1×10^{-5}
	Neurofibromatosis	1×10^{-4}

true reversions of forward mutations resulting in the original genotype are very rare events, second-site reversions are more frequent because of the presence of a second mutant called suppressor. *Intragenic suppressors* compensate for the effect of a mutation in the same gene in which the forward mutation is located, e.g., frameshifts or amino acid substitutions whose effects on the protein structure compensate for the first mutational change. *Intergenic suppressors* adjust the effect of mutations in other genes by modifying the environment, thereby enabling the mutant protein to function. An example is the extraordinary sensitivity to zinc ions of the *td-24* tryptophan synthase mutation in *Neurospora,* where the second mutation lowered the zinc concentration in the cell (SUSKIND and KUREK, 1959).

Due to specific requirements for back mutations their mutation rate is just 10% of forward mutations. Deeper insights in mutation rates in non-replicating and in replicating systems have been given by DRAKE (1970).

5 Mutational Spectra

The analysis of a large collection of point mutations by fine-scale mapping or sequencing shows that the mutations are not randomly distributed. This phenomenon of non-random mutational spectra was first described by BENZER (1961) and has been documented for several genes in a variety of organisms (GUTZ, 1961; ISHIKAWA, 1967; CRAWFORD et al., 1970; GORDON et al., 1990). By analyzing the two cistrons *rIIA* and *rIIB* of the bacteriophage T4, BENZER found a total of 1609 mutants, distributed among 250 sites. 815 of these mutants fall into just two sites (*r*17, with 517; and *r*131, with 298). The sites with large numbers of mutations are known as *hot spots.* The pattern of hot spots in a given organism differs considering spontaneous or induced mutations, where individual mutagens induce different mutational spectra and hot spots (e.g., FREESE, 1959; LOPRIENO, 1967; DHILLON and DHILLON, 1974). Non-random mutational spectra clearly cannot be explained by a single cause. In silent sites the low number of mutations is due to a combination of a failure to detect mutations through tolerance of the gene product, and through degeneracy of the code. Concerning hot spots, the mutative rate of a given site depends not only upon the base pair configuration at that site, but also on the nature of nearby base pairs (for review see CLARKE and JOHNSTON, 1976).

The base sequence as cause for the occurrence of hot spots was demonstrated in the *lacI* gene of *E. coli,* where gly codons with their sequence 5′-GGN-3′ predominate as potential targets for MNNG (N-methyl-N′-nitro-N-nitrosoguanidine)-directed mutation (GORDON et al., 1990). The *NarI* sequence (GGCGCC) in the *E. coli* genome is especially susceptible to -2 frameshift mutations induced by N-acetoxy-N-2-acetylaminofluorene, which covalently binds to the C(8) position of guanine residues (KOEHL et al., 1990; BELGUISE-VALLADIER and FUCHS, 1991) (see frameshift mutagens, Sect. 8.2.2).

The importance of neighboring base effects in the siting of hot spots now seems clear, but the precise way in which the effect is mediated can at present only be guessed. It could operate not only at the level of the initial reactivity of the site leading to a premutational lesion, but also subsequently at the level of DNA repair, recombination, or replication.

In some cases rare or modified bases may be involved. In *E. coli,* 5-methylcytosine frequently occurs as modified base arising from cytosine by the action of a methylase. 5-Methylcytosine is deaminated with high frequencies resulting in a thymine thus producing a GC→AT transition.

6 Mutator Genes

Heritable increases in the rate of spontaneous mutations in other genes are usually attributed to the so-called mutator genes. First evidence was given by the analysis of the Florida strain of *Drosophila melanogaster* showing a high mutability (1.09%) caused by a recessive mutator gene on the 2nd chromosome (DEMEREC, 1937). Since then, further mutator genes have been found in *Drosophila* (GREEN,

1976), in other eukaryotes, such as maize (RHOADES, 1941; MCCLINTOCK, 1951; BRINK, 1973) or *Oenothera* (SEARS and SOKALSKI, 1991), and in a variety of prokaryotes. On the basis of our current knowledge of mutagenesis, the following mechanisms of mutator gene action could be envisaged: an alteration of polymerases resulting in replication errors in normal nucleotide sequences; modification of one or more of the different bases in DNA resulting in replication errors; production of mutagenic base analogs leading to incorporation or replication errors. It is not known whether all of these mechanisms are operative in the cell.

Mutator genes were found in three laboratory strains of *E. coli* (TREFFERS et al., 1954; GOLDSTEIN and SMOOT, 1955; SIEGEL and BRYSON, 1964), and the genetic location of *mutT,* the Treffers mutator gene of K12, *mutS1* of *E. coli* B, and *ast* of the Harvard strain were determined (SKAAR, 1956; SIEGEL and BRYSON, 1967; ZAMENHOF, 1966). The *mutT* locus increases the mutation frequency of a number of bacterial markers from about 10^{-8} to about 10^{-5}. This type of mutational change has been identified by YANOFSKY et al. (1966) as an AT→CG transversion, occurring with strong unidirectional preference. The *mutS1* locus of *E. coli* B has been found to induce AT⇌GC transitions (SIEGEL and BRYSON, 1967). In the meantime, mutations in at least nine loci are known to cause mutator phenotypes in *E. coli* (HORIUCHI et al., 1989), with mutator *mutD5* as the most potent mutator known so far. The extreme mutability of this *E. coli* strain results, in addition to a proofreading defect of exonuclease from DNA polymerase III, from a defect in post-replicational DNA mismatch repair (SCHAAPER and RADMAN, 1989).

The plasmid pKM101 in *E. coli* exhibits the mutator effect exhibits (WALKER, 1984). The gene products of the plasmid operon, namely the MucA and MucB proteins, seem to be identical with the UmuDC proteins of the error-prone SOS repair (see Sect. 7.2.4).

7 Repair Mechanisms

Mutagenesis is a complex process. Premutational lesions of DNA occur spontaneously or through the action of mutagenic agents; only part of these result in stable mutants during subsequent replication. The following structural changes occur in DNA:

- Pyrimidine dimers, in which two adjacent pyrimidines on a DNA strand are coupled by additional covalent bonds and thus lose their ability to pair.
- Crosslinks between the complementary DNA strands, which prevent their separation in replication.
- Intercalation of mutagenic agents into the DNA, causing frameshift mutations.
- Single-strand breaks.
- Double-strand breaks.

These premutational structural changes can lead to mutations directly by causing pairing errors in replication or indirectly by error-prone repair in the next round of DNA replication.

Repair systems play a significant role in the mutation process. As a result of repair, potentially lethal changes in the DNA may be eliminated. If the repair systems function in an error-free manner, potentially mutagenic lesions are eliminated before they can be converted into final mutations. A number of repair systems have thus far be discerned in microorganisms and in higher organisms. In the following, the most important of these repair mechanisms are discussed. Genes involved in the various repair processes in *E. coli* are compilated in Tab. 2.

7.1 Photo-Reactivation

Far ultraviolet radiation (220–300 nm) affects DNA in a number of ways, whereby pyrimidine dimers are the major photo-products, a state in which two adjacent pyrimidine molecules are chemically joined to *cis, syn*-cyclobutane pyrimidine dimers (Fig. 1), so that replication of the DNA cannot occur. When such an ultraviolet-irradiated population is subsequently exposed to near UV or visible light (300–450 nm), the survival rate increases and

Tab. 2. Genes Involved in *E. coli* DNA Repair Pathways

Gene	Gene Product	Function	References
Photo-Reactivation (error-free)			SETLOW and SETLOW (1963)
phr	Photolyase	Photo-reactivation	
Excision repair			GROSSMAN and YEUNG (1990)
Nucleotide excision repair (error-free)			
uvrA *uvrB* *uvrC* }	UvrABC endonuclease complex	Excision of pyrimidine dimers and refilling of gaps	
uvrD	DNA helicase II	Unwinding of DNA	
Long-patch excision repair (error-prone)			
recA *lexA* }	see SOS repair		
Base excision repair (error-free)			LINDAHL (1990)
xth	Exonuclease III	Repair of AP sites: cleavage of C3′-O-P bond 5′ to AP sites	
nfo	Endonuclease IV	Hydrolysis of C3′-O-P bond 5′ to AP site	
nth	Endonuclease III	Cleavage 3′ to Ap sites, glycosylase activity	
fpg	Fapy-DNA glycosylase	Cleavage 3′ to AP sites, glycosylase activity	
tag	DNA glycosylase I	Removal of N^3-methyladenine	RIAZUDDIN and LINDAHL (1978); WALLACE (1988)
Postreplication Repair			WANG and SMITH (1981, 1983, 1985); YOUNGS and SMITH (1976)
Postreplicative recombination repair (error-free)			
recA	44 kDa RecA protein	DNA strand exchange in recombination and repair; proteolytic activity	
recB	Exonuclease V subunit }	Repair of double-strand breaks	
recC	Exonuclease V subunit }		
recF	unknown	Repair of daughter strand gaps opposite the lesion	
sbcB	DNA exonuclease I	Conversion of strand gaps to DNA double-strand breaks	
Mismatch repair (error-free)			GRILLEY et al. (1990)
dam	Methylase	Identification of newly synthesized DNA strand by recognizing a GATC sequence	
mutH *mutL* *mutS*	unknown	Mediate mismatch recognition	
mutU (*uvrD*)	DNA helicase II	Unwinding of DNA	

Tab. 2. Continued

Gene	Gene Product	Function	References
SOS Repair (error-prone)			WALKER (1984);
recA	44 kDa Protein	DNA strand exchange in recombination and repair; proteolytic activity	RUPP et al. (1971)
lexA	22 kDa Repressor	Control of genes involved in SOS response	
ssb	Single-strand DNA Binding protein	Involved in induction of SOS rsponse	WHITTIER and CHASE (1981)
din	unknown	Involved in SOS response	FRAM et al. (1986);
umuC	unknown	Involved in the induction of the repair pathway(s) associated with the SOS response	SHINAGAWA et al. (1988)
umuD	unknown		
Repair of Alkylations, Adaptive Response (error-free)			VOLKER (1988);
ada	39 kDa O^6-Methyl guanosine DNA methyltransferase	Transfer of methyl groups, in methylated form transcriptional promotor	TAKANO et al. (1991)
alkA	3-Methyladenine DNA glycosylase II	Excision of alkylated bases	
alkB	unknown	unknown	
aidB	unknown	unknown	
ogt	19 kDa O^6-Methyl guanosine DNA methyltransferase	Transfer of methyl groups	POTTER et al. (1987); MARGISON et al. (1990)
Repair of DNA Double-Strand-Breaks (error-prone)			PICKSLEY et al. (1984)
recA *lexB*	see SOS repair		
recN	unknown	unknown	

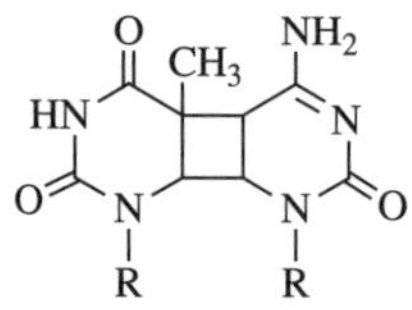

Fig. 1. Thymine-cytosine-cyclobutane dimer, the photo-product formed as a result of ultraviolet radiation.

the frequency of mutation decreases (KELNER, 1949; DULBECCO, 1949). To date, photo-reactivation has been found in over 50 organisms including prokaryotic and eukaryotic microorganisms, archaebacteria, and multicellular plants and animals. As demonstrated by SETLOW and SETLOW (1963), the photo-reactivation primarily relies upon a specific enzyme, a DNA photolyase. In *Escherichia coli* the photolyase is controlled by *phr;* mutations in this gene block the photo-repair. DNA photolyases bind in a dark reaction to the pyrimidine dimers and utilize light energy to break the carbon–carbon bonds in the cyclobutane ring of the dimer (Fig. 2). UV-induced crosslinks can also be photo-reactivated.

All characterized photolyases (*Anacystis nidulans, Streptomyces griseus, Escherichia coli, Saccharomyces cerevisiae* and *Scenedesmus acutus*) contain two chromophores: e.g., *E. coli* and *S. cerevisiae* contain 1,5-dihydroflavin ($FADH_2$) and 5,10-methenyltetrahydrofolate in 1:1 stoichiometry with the apoenzyme (55–66 kDa, JOHNSON et al., 1988). Photolyases of other organisms contain in addition to $FADH_2$

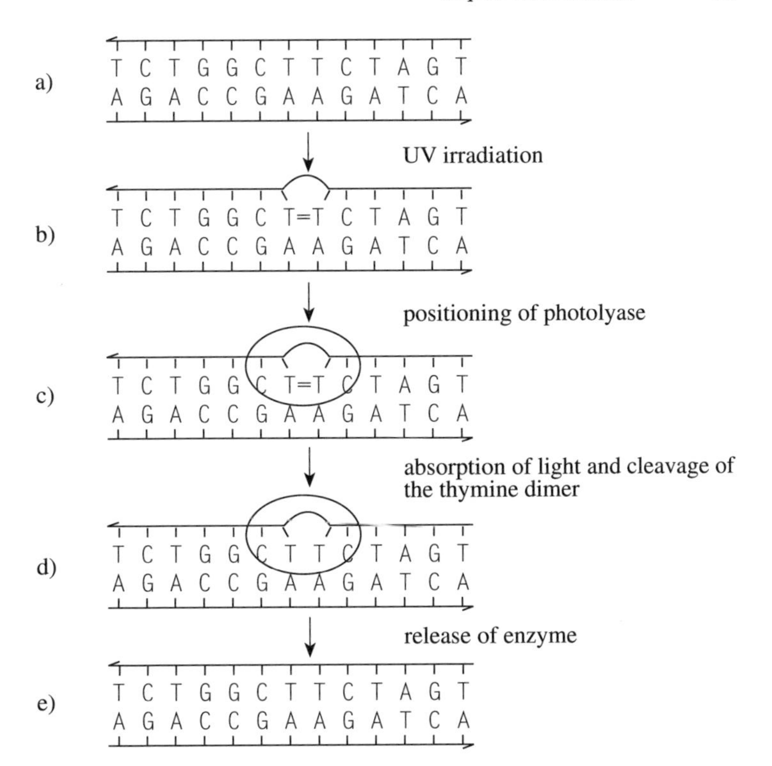

Fig. 2. Photo-reactivation. Repair mechanism for the elimination of thymine dimers: An intact DNA double-strand (a) exhibits after UV irradiation a pyrimidine dimer (b); photolyase locates itself at the point of DNA lesion (c); after absorption of visible light in the blue end of the spectrum the photolyase is enabled to split the dimer (d), resulting in the original DNA strand (e).

an unusual derivative of 8-hydroxy-5-deazaflavin (EKER et al., 1988). *E. coli* contains 10–20 molecules of photolyase per cell, and yeast 75 - 300, indicating the high efficiency at recognizing pyrimidine dimers. In addition to the cyclobutane ring, the unique structure imposed by the dimer upon the sugar–phosphate backbone and attached bases is required by the enzyme.

The wavelengths utilized for photo-reactivation are $\lambda_{max} = 365–405$ nm or 435–445 nm depending on the enzyme. In the folate class of photolyases the folate chromophore functions as the primary absorber of light energy for the use by the flavin chromophore, which donates an electron to the dimer to form a pyrimidine dimer anion (SANCAR and SANCAR, 1988). This system is unstable and will spontaneously collapse to two free pyrimidines plus an electron. A review of physical properties and action mechanisms of dimer recognition and photolysis has been given by SANCAR (1990).

Photo-reactivation functions in an error-free manner and thus does not allow mutations to occur.

7.2 Excision Repair

In contrast to photo-reactivation, which is possible with single-stranded DNA, the complementary strand is required for excision repair. In a dark reaction, damages to the DNA are recognized by specific enzymes. Excision repair is the most important repair event in all groups of organisms.

7.2.1 Nucleotide Excision Repair

This mechanism of excision repair for UV-induced lesions is different from that for lesions induced by alkylation or deamination. By nucleotide excision repair defective nucleotides are cut out and replaced, according to the mechanism illustrated in Fig. 3. For this mechanism to operate, the normal DNA replication process is not required. For reviews of the repair mechanism in *E. coli* see GROSSMAN and YEUNG (1990), in mammalian cells HOIJMAKERS et al. (1990).

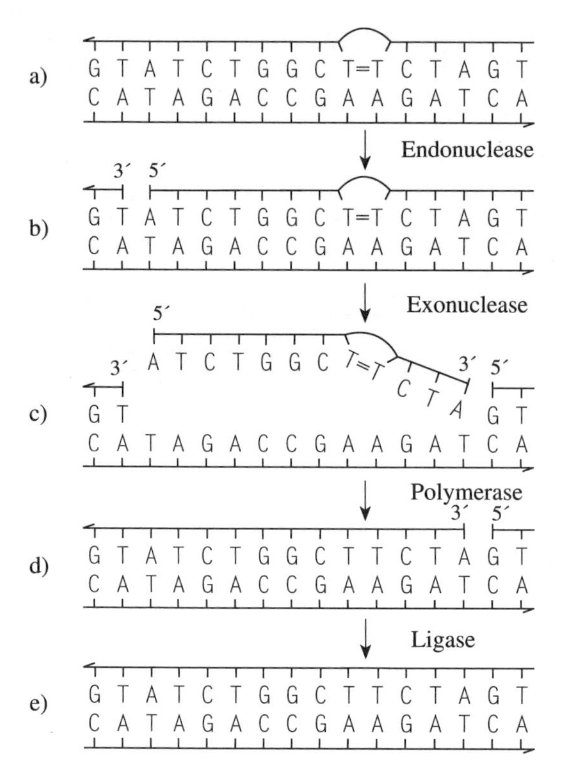

Fig. 3. Nucleotide excision repair of UV-induced lesions: A pyrimidine dimer (a) is repaired by incision of the damaged strand by the UvrABC endonuclease system (b), excision of the lesion by a 5′-exonuclease (c). After further enlargement of the gap by the exonuclease the gap is filled by DNA polymerase (d) and closed by a polynucleotide ligase (e).

Four steps are involved in nucleotide excision repair in *E. coli*: incision, excision, resynthesis, and ligation. In incision the UvrABC endonuclease system together with ATP is involved. This enzyme system is a product of three *E. coli uvr* genes (*ucrA, uvrB,* and *uvrC*). The initial DNA interaction is with UvrA, which is dimerized in the presence of ATP, while nucleoprotein formation seems to take place on undamaged regions of DNA by $(UvrA)_2$. In the presence of UvrB and driven by ATP binding, topological unwinding of DNA to a single helical turn takes place. The $Uvr(A)_2B$ complex translocates to a damaged site of DNA provided by the UvrB-associated ATPase. The incision reaction can only be started, when UvrC is binding to the $Uvr(A)_2B$ nucleoprotein complex. DNA helicase (UvrD) is necessary for turnover. Incision is a dual event in which breakage occurs 7 nucleotides 5′ and 3 to 4 nucleotides 3′ to the same damaged site, this step being without ATP requirement. With the help of a 5′-exonuclease the resulting gap is expanded to approximately 30 nucleotides. The missing nucleotides are filled by DNA polymerase I (*polA*) starting from the 3′ end and are connected by a polynucleotide ligase (*lig*). The excision repair is almost error-free, and mutation is thus avoided. While the majority of UV lesions are repaired by the constitutive UvrABC endonuclease complex, a few are processed by the long patch excision repair (Cooper, 1982). This *recA-lexA*-dependent inducible function requires DNA replication and seems to be mutagenic, since large gaps of up to 1500 nucleotides are cut out. It has been suggested that long patch repair is a form of the SOS repair (see Sect. 7.2.4).

7.2.2 Base Excision Repair

Excision repair of modified bases operates in a different manner. At an apurinic/apyrimidinic (AP) sites, alkylated or deaminated bases are recognized and repaired by different enzyme systems.

AP endonucleases and AP lyases
Most spontaneous DNA lesions are AP sites, which are repaired by enzymatic cleavage of the phosphodiester backbone of DNA. This excision repair mechanism was first described in *E. coli* and in mammalian cells (Verly and Rassart, 1975; Ljungquist et al., 1974), but is now found in virtually every organism. Two types of enzymes have been described. Enzymes hydrolyzing the C3′-O-P bond 5′ to the AP sites are *AP endonucleases* (Lindahl, 1990; Laval et al., 1990; Doetsch and Cunningham, 1990).

An alternative minor slow working pathway involves the breaking of the C3′-O-P bond 3′ to the AP sites by *AP lyases* by a *β*-elimination mechanism (Bailly and Verly, 1989; Lindahl, 1990; Bailly et al., 1989).

Escherichia coli possesses at least four AP site nicking enzymes. The *xth* and *nfo* gene products are classified as endonucleases. The product of the *xth* gene (WEISS and GROSSMAN, 1987) is the major AP endonuclease, but it has also the property to act as 3′-5′-exonuclease on double-stranded DNA and is often designated as exonuclease III. The *nfo* gene product is the endonuclease IV, a 31.6 kDa monomeric metalloenzyme. The *nth* gene product is an endonuclease III, the *fpg* product a Fapy-DNA glycosylase (Fapy: 2,6-diamino-4-hydroxy-5-N-methylformamidopyrimidine). Both lyases are endowed with DNA glycosylase activities acting on ring-modified pyrimidines (*nth*), or on imidazole ring-opened purines (*fpg*) (SANCAR and SANCAR, 1988; O'CONNOR and LAVAL, 1989).

Incision on the 5′ side of the AP site generates a 5′ end with a base-less deoxyribophosphate residue, which is removed by a 52 kDa 5′dRpase (DNA deoxyribophosphodiesterase) requiring Mg^{2+} as a cofactor (FRANKLIN and LINDAHL, 1988). The produced gap is filled by action of exonuclease, DNA polymerase I, and ligase.

DNA glycosylases
This group of enzymes which has been reviewed in detail by WALLACE (1988), LINDAHL (1990), and SAKUMI and SEKIGUCHI (1990), acts on alkylated and deaminated bases. *E. coli* has six different known enzymes of this type, which eliminate the following altered bases from DNA: uracil, 3-methyladenine, hypoxanthine, formamidopyrimidine, and pyrimidine hydrates and related compounds including ring-fragmented pyrimidines. In human cells an additional enzyme removes 5-hydroxymethyluracil. Non-repairable by these enzymes: xanthine, 1-methyladenine, 3-methylthymine, 3-methylcytosine, which spontaneously are produced by *S*-adenosyl-L-methionine as intracellular methyl group donor. 7-Methylguanine is tolerated by the cells and therefore not mutagenic. The N-glycosylases split the bond between base and deoxyribose so that an AP site is formed which is repaired as described.

7.2.3 Post-Replication Repair

Post-replicative recombination repair
UV-induced pyrimidine dimers which have survived photo-reactivation and excision repair hinder base pairing. A gap in the daughter strand of approximately 1000 nucleotides is formed during replication. According to Fig. 4, which is partially hypothetical, these gaps are filled with material from the parent strands through recombination processes by the action of the *recA* gene protein, which is responsible for the exchange of DNA strands. The repair of the parent strands occurs through repair replication with the daughter strands as a matrix by means of DNA polymerase I (HOWARD-FLANDERS, 1973; SMITH, 1978). Post-replication repair works error-free. The *recA* gene and other *rec* genes are involved in different recombination repair pathways; at least five separate genetically controlled pathways are known in *E. coli* (YOUNGS and SMITH, 1976).

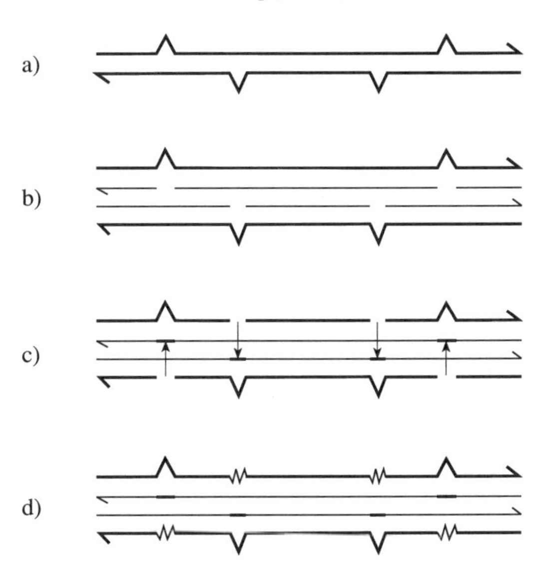

Fig. 4. Model of post-replicative repair. (a) DNA double-strand with premutative lesion (⋀); (b) gaps in daughter strands which have resulted during replication; (c) and (d) hypothetical exchange of parent strand DNA and replication repair (SMITH, 1978).

Two major pathways of UV-radiation damaged DNA are the *recAB* pathway and the *recF* pathway (ROTHMAN et al., 1975; WANG and SMITH, 1981). These two pathways are involved in the repair of different DNA lesions (WANG and SMITH, 1983). The *recF*-dependent pathway is largely responsible for the repair of DNA daughter-strand gaps which are produced in nascent DNA when the replication complex passes the UV-radiation induced pyrimidine dimers (KATO, 1977; WANG and SMITH, 1983). On the other hand, the *recBC*-dependent pathway, where RecBC are the two subunits of exonuclease V, is mainly responsible for the repair of DNA double-strand breaks that arise from unrepaired DNA daughter-strand gaps. The *sbcB* gene product, DNA exonuclease I, is involved in this reaction. It is required for the conversion of a DNA daughter-strand gap to a DNA double-strand break, which is then accessible to the *recBC* pathway (WANG and SMITH, 1985).

Several functions of recombination repair are analogous to conjugational recombination, but are not totally identical. The existence of a network is to be expected where nucleases, polymerases, and other enzymes of recombination cooperate together with specific factors to form the repair system which is not yet fully understood.

Post-replicative DNA mismatch repair

Another form of post-replication repair is the so-called mismatch repair, a system that follows the replication fork during DNA replication, scrutinizing the newly synthesized DNA for mismatches resulting from errors of DNA replication. This repair mechanism has been demonstrated to exist in bacteriophages, bacteria, fungi, and higher eukaryotes, and is able to catch and correct 99% of the mistakes made during normal replication.

Bacteria have two classes of mismatch correction.

(1) A long-patch repair system that processes a variety of base pairing errors, where excision repair is involved excising sequences of several kilobases.

(2) One or more short-patch repair systems correct a particular type of mismatch usually involving only a few nucleotides by excision repair processes (GRILLEY et al., 1990; RADMAN and WAGNER, 1986; MODRICH, 1989).

After replication the newly synthesized DNA strand can be distinguished by not yet being methylated. The *dam* gene product is a methylase which recognizes the non-methylated A, in a GATC sequence thereby identifying the newly synthesized DNA strand. Base pairing errors in this strand can now be repaired by the long-patch repair system. The gene products of *mutH, mutL,* and *mutS* mediate the mismatch recognition, and the *mutU(uvrD)* gene product, DNA helicase II, acts in unwinding the DNA duplex. Inactivation of any of these loci by mutation confers high spontaneous mutability, indicating that the products of these genes contribute to genetic stability.

One of the short-patch repair systems in *E. coli* is the VSP mismatch repair (VSP: very short patch), which is believed to correct GT mismatches that arise through spontaneous deamination of 5-methylcytosine to the correct GC base pair. VSP mismatch repair is dependent on *mutL* and *mutS* gene products (LIEB, 1987; ZELL and FRITZ, 1987).

7.2.4 SOS Repair

Photoreactivation, excision repair, and post-replication repair generally operate as error-free mechanisms. Error-prone – and therefore mutation-inducing – repair pathways also exist, of which the best known is the SOS-repair system of *E. coli* (for reviews see WITKIN, 1976; LITTLE and MOUNT, 1982; WALKER, 1984; for *Streptomyces coelicolor,* see MISURACA et al., 1991). Overlapping gaps, such as the ones which can result in errors in both complementary strands in the replication of a DNA, would be lethal for the cell, but are filled by the SOS repair despite the absence of DNA template. Thus the chemical structure of DNA is reconstructed, but the heredity information is defective; as a result, SOS repair very likely results in mutations. In contrast to the constitutive repair systems thus far described, the SOS repair activity is inducible, being repressed in untreated wild-type cells. Together with repair other phenomena are induced, the so-called SOS response: (1) Weigle reactivation, the error-prone reactivation of bacteriophage DNA (WEIGLE, 1953); (2) DNA error-prone repair leading to mutagenesis (for a first model see RADMAN, 1975); (3) inhibition of cell division; (4) prophage induction in *E. coli.*

In *E. coli*, the SOS responses including the SOS repair are regulated by *recA* and *lexA* genes. The SOS genes are under the negative control of the *lexA* gene which produces a repressor for *recA* and at least 17 chromosomal genes (FRAM et al., 1986), five of which are the so-called *din* (damage inducible) genes. When SOS response is induced by some effectors, which are produced as a consequence of DNA damage or of the blockage of DNA replication (perhaps an ssDNA region), the RecA protein acquires protease activity. It cleaves the LexA repressor, and within a few minutes leads to derepression of the SOS loci (SLITALY and LITTLE, 1987) including the *din* genes and *umuDC* which in *E. coli* is essential for efficient SOS mutagenesis (SHINAGAWA et al., 1988). UmuDC may enhance the ability of a stalled polymerase to replicate past DNA blocking lesions. In addition, a massive synthesis of RecA protein takes place to ensure a quick degradation of LexA repressor protein. SOS mutagenesis is believed to be carried out by the polymerase III holoenzyme which has an intrinsic $3' \rightarrow 5'$ exonuclease activity. This enzyme is responsible for proofreading of newly synthesized DNA. A specific subunit (the product of the *dnaQ/mutD* gene), causing this exonuclease activity, is inhibited under SOS conditions, thus lowering the proofreading of the polymerase III. This allows the enzyme to bypass the lesions by insertion of a 'wrong' nucleotide opposite the lesion leading to SOS targeted mutations (SHWARTZ et al., 1988). After repair, the RecA protein loses its proteolytic activity, LexA protein accumulates and switches off the SOS genes. At present there are no experiments that directly demonstrate the existence of a similar inducible error-prone repair system in eukaryotes.

7.2.5 Repair of Alkylations, Adaptive Response

Alkylations of DNA, mainly at the O^6 position of guanine, are important premutational events in all organisms (LOVELESS, 1969). To counteract such effects, many organisms from bacteria to mammalians possess DNA methyltransferases, which repair these lesions by transferring the alkyl groups to their own molecule (SEKIGUCHI and NAKABEPPU, 1987; LINDAHL et al., 1988; KARRAN et al., 1990; REBECK et al., 1989).

In *E. coli* at least two repairing methyltransferases are produced (TAKANO et al., 1991). One system, the *adaptive response* (for review see WALKER, 1984; LINDAHL et al., 1988; VOLKERT, 1988), is inducible and has been found to occur after lengthy treatment with sublethal concentrations of alkylating agents. Four genes arranged in three transcriptional units constitute the adaptive response: the *ada-alkB* operon, *alkA,* and *aidB*. The Ada protein is a 39 kDa O^6-methylguanosine-DNA-methyltransferase which attacks O^6-methylguanine, O^4-methylthymine, and methylphosphotriester residues. *alkA* (VOLKERT, 1988) specifies 3-methyladenine DNA glycosylase II, an enzyme that is involved in the excision of 3-methyladenine as well as 3-methylguanine, 7-methylguanine, 7-methyladenine, O^2-methylthymine, and O^2-methylcytosine. A second DNA glycosylase, the *tag* gene product 3-methyladenine DNA glycosylase I, which removes N^3-methyladenine from alkylated DNA (see excision repair, DNA glycosylases, Sect. 7.2.2), is constitutively expressed (RIAZUDDIN and LINDAHL, 1978) and is not under regulation of the Ada protein. The functions of the *alkB* and *aidB* gene products are not yet known.

The exact mechanism of the Ada methyltransferase is unclear. Methyl groups from methylphosphotriester residues are transferred from modified DNA to the cysteine-69 residue of Ada. Methyl groups from O^6-methylguanine or O^4-methylthymine are transferred to the cysteine-321 residue of the enzyme which is irreversibly inactivated by these reactions thus reducing the numbers of lesions that can be repaired. When the methylphosphotriester acceptor site is occupied, the methylated form of the Ada protein serves as a transcriptional promotor for the *ada* gene itself and the other genes (*alkA, alkB, aidB*) of the *ada* regulon (VOLKERT and NGUYEN, 1984; TEO et al., 1986). By this mechanism, the induced *E. coli* cell contains more than 3000 molecules of the Ada protein, while in the uninduced cell between 13–60 molecules are present (ROBINS and CAIRNS, 1979). The 3-methyladenine DNA glycosylase II content in the induced state is increased by a factor of 20.

The adaptive response works almost without error, but the capacity of this repair system is limited. It enables the bacteria to resist the killing and mutagenic effect of further acute damage, but alkylated bases accumulate when larger mutagen doses are used, and with O^6-methylguanine the mutation frequency is directly proportional to the O^6-methylguanine concentrations.

A second methyltransferase is encoded by the *ogt* gene in *E. coli* (POTTER et al., 1987). The 19 kDa Ogt protein repairs O^6-methylguanine and O^4-methylthymine and has a striking homology with the C-terminal sequence of the Ada protein, but is not under its regulation.

Similar phenomena of adaptive response in mammalian cells after treatment with various alkylating agents (FRIEDBERG, 1985), or in response to ionizing radiation (WOLFF and WIENCKE, 1988; WOJCIK and TUSCHL, 1990; WANG et al., 1991) have been demonstrated. A pretreatment with low mutagen doses results in a reduction of chromosomal aberrations, double-strand DNA exchanges (sister chromatid exchanges) or micronuclei induced by a subsequent treatment with a higher dose.

8 Induced Mutagenesis

Many mutagens induce more than one type of potentially mutagenic lesion. Thus, they frequently cause mutation directly as a result of pairing errors and indirectly as a result of errors during the repair process. In the following, the most commonly used mutagens are listed, together with their molecular reaction mechanisms.

8.1 Physical Mutagenesis

Mutagenesis can be induced in microorganisms and in cells of higher eukaryotes by physical agents such as heat, freezing and thawing, or by radiation.

8.1.1 Heat

According to AGER and HAYNES (1990), heat acts specifically at GC base pairs causing GC to AT transitions via deamination of cytosine to form uracil, or GC to CG transversions via transmigration of the guanine N-9 N-glycosidic bond to the C-2 amino group which allows base pairing with guanine. In phagc T4 the mutation rate is 4×10^{-8} per GC base pair per day (DRAKE and BALTZ, 1976).

8.1.2 Freezing and Thawing

Freezing and thawing has been proposed by a number of workers to be mutagenic by introducing single- and double-strand breaks, which are repaired by a *rec*-dependent pathway (CALCOTT and THOMAS, 1981). In *E. coli*, freeze drying was also found to be mutagenic (TANAKA et al., 1979).

8.1.3 Far Ultraviolet Radiation

The wavelengths effective for mutagenesis by far UV are between 200–300 nm with an optimum at 254 nm, which is the absorption maximum of DNA. The most important DNA lesions are cyclobutane–pyrimidine dimers (see Fig. 1) – T^T, C^T, C^C – in a ratio of 2:1:1, formed between adjacent pyrimidines or between pyrimidines of complementary strands resulting in crosslinks. Other lesions have been reported, such as hydration by insertion of a water molecule into the C=C double bond of cytosine, or pyrimidine(6-4)pyrimidone photoproducts (GLICKMAN et al., 1986). Both pyrimidine dimers and (6-4) lesions are mutagenic. Few, if any, DNA single- or double-strand breaks are produced directly by UV radiation (RAHN and PATRICK, 1976). Single-strand breaks occur during excision repair. Following high fluxes of UV radiation (e.g., 3–4 J/m² in *E. coli*), the number of double-strand breaks correlates with the number of unrepaired DNA daughter-strand gaps.

Far UV radiation mainly induces base substitutions (60 to 75 %), followed by single-base frameshifts, deficiencies, and duplications. In UV-induced mutations in the *lacI* gene of *E.*

coli among the base substitutions, both transitions (72.5%) and transversions (27.5%) were observed. Transitions of GC→AT (60%) were the largest group of base substitutions (SCHAAPER et al., 1987).

As in other bacterial and rodent cells, in human lymphoblastoid cells (line TK 6) a similar pattern of transitions and transversions has been found (KEOHAVONG et al., 1991).

During the repair of UV lesions, up to 1000 pyrimidine dimers per genome can be repaired within 20 minutes, and with the exception of adaptive response, all repair systems are involved (KATO, 1977). After UV radiation of *Saccharomyces cerevisiae* the induction of 'petite' or *rho*$^-$ mutation (RANT and SIMPSON, 1955) is due to the presence of pyrimidine dimers in the mitochondrial DNA (mtDNA). Excision repair has been shown not to take place in mtDNA (HIXON and MOUSTACCHI, 1978). In all cases the SOS repair system is primarily responsible for the production of mutations (GOODSON and ROWBURY, 1990).

To increase the frequency of mutation, the error-free mechanisms of photo-reactivation and excision repair must be prevented by carrying out all manipulations in the dark or under visible light > 600 nm and/or through the use of caffeine or similar inhibitors of excision repair (see Sect. 9).

8.1.4 Near Ultraviolet Radiation in the Presence of 8-Methoxypsoralen

Radiation at wavelengths of 320–400 nm has fewer lethal and mutagenic effects than far UV. But in the presence of a photo-sensitizing agent such as furocoumarin derivatives (BEN-HUR and SONG, 1984), increased mutation frequencies have been observed in a broad range of organisms, e.g., *Escherichia coli* (FUJITA and SUZUKI, 1978), *Streptomyces coelicolor* (TOWNSEND et al., 1971), *Saccharomyces cerevisiae* (BIANCHI et al., 1990), or mammalian cells (SAGE and BREDBERG, 1991). A very effective activator in this combination of near UV radiation and photo-sensitization is the bifunctional furocoumarin 8-methoxypsoralen (MOP) (Fig. 5). This so-called PUVA treatment is currently used in the photochemotherapy of certain skin diseases such as psoriasis or mycosis fungoides. 8-Methoxypsoralen intercalates between the base pairs of double-strand DNA; and after the absorption of near UV, an adduct is formed between MOP and a pyrimidine base. Absorption of a second photon causes the coupling of the pyrimidine–psoralen monoadduct with an additional pyrimidine (SMITH, 1988). MOP reacts mainly with thymine bases. Biadduct formation between complementary strands of nucleic acid results in crosslinks.

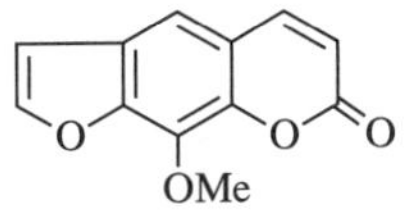

O O O OMe

Fig. 5. Structure of 8-methoxypsoralen.

In *E. coli*, near UV/MOP-induced DNA lesions are repaired by complex repair mechanisms; the exact process is not yet clear (HOLLAND et al., 1991). Both MOP induced monoadducts and crosslinks involve the *uvrABC* excision repair, the *recA* gene product, DNA polymerase I, helicase II, and DNA ligase (VAN HOUTEN et al., 1986). The UvrABC exonuclease as one of the involved enzymes excises the lesions; after enlargement of the incision gap by the 5′→3′ exonuclease activity of DNA polymerase I, the gap seems to be subject to the SOS-like long-patch excision repair. In recognition and incision of DNA a 55 kDa protein (*puvA* gene product) together with the regulatory gene *puvR* was involved (HOLLAND et al., 1991). Base pair substitutions and frameshift mutations occur as mutations.

8.1.5 Ionizing Radiation

Ionizing radiations include X-rays, γ-rays, and β-rays with low linear energy transfer values (LET), and radiations with high LET values, including α-particles and neutrons. Along the track of each high-energy ray, a train of ions is formed which can initiate a variety of chemical reactions, where high LET radiations follow single-hit kinetics and low LET radiations multi-hit kinetics (AUERBACH, 1976).

Though ionizing radiations are a widely used and studied agent (GLICKMAN et al., 1980; HUTCHINSON, 1985; KOZUBEK et al., 1989), the primary lesions of DNA responsible for the observed mutagenesis have not yet been determined. Gene mutations such as transitions, transversions, and frameshift mutations could be observed as fundamental effects. The majority of lesions found in DNA are single- and double-strand breaks, with lower frequencies cleavages of H-bonding between complementary base pairs, interstrand crosslinks or unidentified base damages (BERTRAM, 1988). Beside the direct radiation lesions, the available data strongly suggest that as an indirect effect of radiation the DNA is damaged by radiation-produced radicals, mainly $OH^{\bullet}$ (HUTCHINSON, 1985).

Most of single- and double-strand breaks and other DNA lesions are efficiently repaired. In human cells the repair process is finished after 1 to 2 hours.

In *E. coli*, single-strand breaks are repaired either rapidly by error-free nucleotide excision repair enzymes or slowly by inducing the error-prone SOS repair (BRIDGES and MOTTERSHEAD, 1972; KOZUBEK et al., 1990).

Analogous to the adaptive response in *E. coli,* treated with low levels of alkylating agents, an inducible *radioadaptive response* has been studied in mammalian cells (BOSI and OLIVIERI, 1989). Exposed to 0.02 Gy of X-rays in the G_1 phase, human lymphocytes became less susceptible to the induction of chromosome aberrations by subsequent exposure to 3 Gy of X-rays (WANG et al., 1991).

The molecular basis involved in repair of double-strand breaks, which are believed to cause the formation of chromosome-type aberrations, is not yet clear. In *E. coli*, it appears to be an inducible process, which requires the presence of a homologous DNA duplex, RecA protein and a *recN* gene product that is needed specifically for the repair of double-strand breaks (PICKSLEY et al., 1984).

UV radiation or chemical agents are normally preferable for mutagenesis in industrial strain development due to the fact that the chromosome mutations inducing double-strand breaks occur by ionizing radiation with a significantly higher probability than with all other mutagens.

8.2 Chemical Mutagenesis

A variety of chemicals are known which are mutagenic, and these may be classified into three groups according to their modes of action:

(1) Mutagens which affect non-replicating DNA.

(2) Base analogs, which are incorporated into replicating DNA due to their structural similarity with one of the naturally occurring bases.

(3) Frameshift mutagens, which enter into DNA during replication or repair and through this intercalation cause insertion or deletion of one or a few base pairs.

8.2.1 Mutagens Affecting Non-Replicating DNA

A number of chemicals are known which cause direct damage to non-replicating DNA by deamination or by formation of addition products via alkylating agents or hydroxylamine.

8.2.1.1 Nitrous Acid

HNO_2 oxidatively deaminates adenine, cytosine, and guanine to hypoxanthine, uracil and xanthine, respectively (SCHUSTER, 1960). Through the changed pairing properties of the deamination products (hypoxanthine pairs with cytosine, uracil with adenine) AT→GC and/or GC→AT bidirectional transitions result through mispairing in replication (GREEN and DRAKE, 1974). Fig. 6 shows the establishment of the mutation after two generations. Deamination of guanine results in xanthine which cannot pair with any base. Excision and recombination repair are involved in the elimination of deamination products (CLARKE, 1970). In addition to deaminations, HNO_2 induces crosslinks between the complementary strands (BECKER et al., 1964) which are assumed to result in deficiencies that occur relatively frequently besides point mutations.

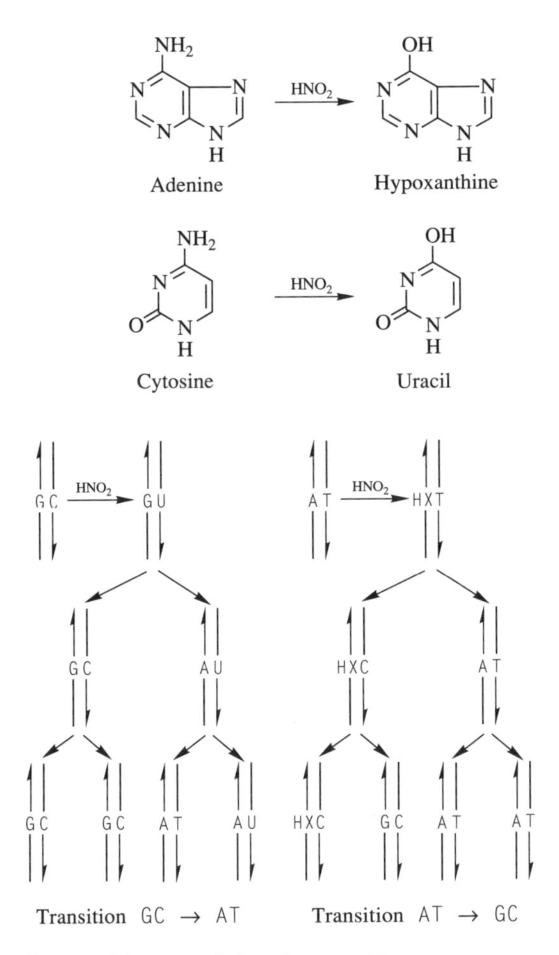

Fig. 6. Mutagenesis by nitrous acid.

8.2.1.2 Hydroxylamine

HA reacts with pyrimidines, but only the reaction with cytosine is mutagenic (SINGER and FRAENKEL-CONRAT, 1969), whereby N^4-hydroxycytosine and 5,6-dihydro-6-hydroxy-N^4-hydroxylaminocytosine are produced (for review see BUDOWSKI, 1976). In the DNA of phage T4, where cytosine is 5-substituted, only 5-hydroxymethyl-N^4-hydroxycytosine is formed (JANION and SHUGAR, 1965). The HA derivatives from cytosine exhibit tautomerization and then pair with adenine, so that through HA action GC→AT transitions are caused. Repair of HA lesions was studied in T4 and in λ bacteriophage by JANION (1982). Only nucleotide excision repair was found to be involved.

8.2.1.3 Alkylating Agents

Alkylating agents are another group of chemicals affecting non-replicating DNA. Except for ultraviolet radiation, alkylating agents are the most potent mutagenic systems for practical application. They comprise a variety of compounds of very different chemical structures acting as mutagens, carcinogens, and anticancer agents, whose common function is their ability to transfer small alkyl functionalities.

Most alkylating agents act as monofunctional agents which transfer only one alkyl group. Monofunctional alkylators are epoxides [e.g., diepoxybutane (DEB)], ethyleneimines [e.g., ethyleneimine (EI), VERSCHAEVE and KIRSCH-VOLDERS, 1990], the alkyl sulfates, consisting of diethyl-(DES) and dimethyl sulfate (DMS), the alkyl alkane-sulfonates [methyl- (MMS) and ethyl methanesulfonate (EMS), see EDER et al., 1989], the nitrosamides including methyl- (MNU) and ethylnitrosourea (ENU), N-methyl- (MNNG)-, ethyl- (ENNG)- and propyl-N′-nitro-N-nitrosoguanidine (PNNG), and the indirect acting dimethyl- (DMN) and diethyl nitrosamines (DEN) (JACKSON, 1964) which must undergo metabolic activation prior to exhibiting their effects.

Di- or polyfunctional agents can induce both intrastrand and interstrand DNA crosslinks. Included are sulfur and nitrogen mustards [e.g., di-(2-chloroethyl)sulfide, 'mustard gas'; di-(2-chloroethyl)amine, 'nitrogen mustard', (BAKER et al., 1984); ICR compounds due to their side chains (DE MARINI et al., 1984)], mitomycin C (MMC) (KUMARESAN and JAYARAMAN, 1990) which requires metabolic reduction to become biologically active. The alkylating agents have been reviewed by BROOKES (1990); the structures are given in Fig. 7.

Alkylating agents bind to a variety of sites on DNA following basically the rules of electrophilicity and nucleophilicity (for a review of chemical theories, see SWENSON, 1983). The transfer of alkyl groups can be accomplished

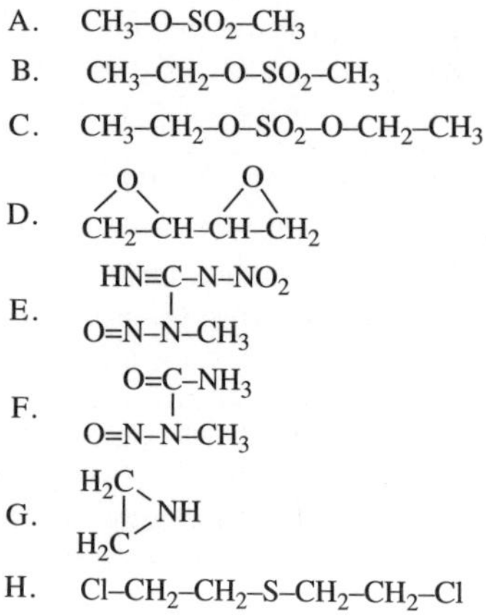

A. CH_3–O–SO_2–CH_3

B. CH_3–CH_2–O–SO_2–CH_3

C. CH_3–CH_2–O–SO_2–O–CH_2–CH_3

D. CH_2(O)–CH–CH–CH_2(O) (two epoxide rings)

E. HN=C–N–NO_2 / O=N–N–CH_3

F. O=C–NH_3 / O=N–N–CH_3

G. H_2C–NH–H_2C (ring)

H. Cl–CH_2–CH_2–S–CH_2–CH_2–Cl

Fig. 7. Structures of alkylating agents.
A. methyl methanesulfonate (MMS)
B. ethyl methanesulfonate (EMS)
C. diethylsulfate (DES)
D. diepoxybutane (DEB)
E. N-methyl-N′-nitro-N-nitrosoguanidine (MNNG)
F. N-methyl-N-nitrosourea
G. ethyleneimine
H. mustard gas.

either through a unimolecular S_N1 or a bimolecular S_N2 mechanism (ORLEMAN-GOLKAR et al., 1970; LAWLEY, 1984; SINGER, 1986). In principle, most of the nitrogen and the oxygen atoms in DNA are potential sites for attack by alkylating agents (BASU and ESSIGMANN, 1988, 1990).

A compilation of DNA adducts by several monofunctional alkylating agents is given in Tab. 3 (BERANEK, 1990). Main alkylation products are alkylpurines and, to a lesser extent, alkylpyrimidines. Besides, AP sites, single-strand breaks and phosphotriesters are formed. Breakage of the deoxyribose polyphosphate backbone of DNA takes place as a consequence of phosphate alkylation. However, the relative distribution of adducts in the DNA is strongly dependent on the nucleophilic selection of the alkylating agent used (SINGER and GRUNBERG, 1983) and on nearest neighbor effects (RICHARDSON and RICHARDSON, 1990).

In all cases 7-alkylguanine is the most common alkylation product. It has base pairing

Tab. 3. Alkylation Products of Double-Stranded DNA after *in vitro* Exposure to Monofunctional Alkylating Agents (Alkylation Patterns are Expressed as Percent of Total Alkylation) (SINGER and GRUNBERG, 1983; BERANEK, 1990)

Site of	Methylation Pattern			Ethylation Pattern		
Alkylation	MMS	MNU	MNNG	DES	EMS	ENU
Adenine						
N1-	2–4	1	1	2	2	0.2
N3-	10–11	8–9	12	10	4–5	3–6
N^6-	nd	nd	—	—	nd	nd
N7-	2	1–2	—	15	1–2	0.3–0.6
Cytosine						
O^2-	nd	0.1	—	nd	0.3	3
N3-	<1	<0,6	2	0.7	0.4–0.6	0.2–0.6
Guanine						
N1-	nd	nd	—	—	nd	nd
N3-	0.6	0.6–1.9	—	0.9	0.3–0.9	0.6–1.6
O^6-	0.3	6–8	7	0.2	2	8–10
N7-	81–83	65–70	67	67	58–65	11
Thymidine						
O^2-	nd	0.1–0.3	—	nd	nd	7–8
N3-	0.1	0.1–0.3	—	nd	nd	0.8
O^4-	nd	0.1–0.7	—	nd	nd	1–2.5
Phosphotriester	0.8	12–17	—	16	12–13	55–57

nd, alkylation adducts not detected or below limits of detection
–, data not reported

properties similar to guanine, and thus it is not mutagenic (GERCHMAN and LUDLUM, 1973; O'CONNOR et al., 1988). In some cases alkylation of the N7 may result in secondary lesions by labilizing the glycosylic bond resulting in AP sites, or by labilizing the C8-N9 bond of the imidazole ring generating a formamidopyrimidine (Fapy) residue. Some lesions, such as 3-methylpurines, are potentially lethal (LARSON et al., 1985). A small proportion of adducts is stable in DNA over long periods of time, as DEN ENGELSE et al. (1987) found with O^2-ethylthymidine in ENU treated tissues of rats. The most important premutational adducts are O^6-alkylguanine and, to a lesser extent, O^4-alkylthymine (COULONDRE and MILLER, 1977), and much evidence suggests that the adducts are responsible for most of the mutations in any prokaryotic and eukaryotic cells treated with monofunctional alkylating agents. The predominant mutations generated by S_N1 compounds are the GC→AT transitions (COULONDRE and MILLER, 1977) at guanines flanked (5′) by a purine residue (i.e., at 5′-R G-3′ sites), whereas the analogous S_N2 compounds lack this preference (ZIELENSKA et al., 1989). It was found by PERERA (1988) that, even at very low doses, the degree of initial alkylation *in vivo* is directly proportional to the administered dose. Guanine crosslinks have been found as alkylating product of the difunctional mustard gas in bacteriophages, bacteria, and mammalian cells.

Most mutations by alkylating agents are thought to arise by misreplication or misrepair of damaged DNA nucleotides, where factors such as adduct distribution or nearest neighbor effects contribute to the expression of a mutation.

It has been suggested that the occurrence of mutations in *E. coli* is dosage-dependent in relation to its repair systems: At a low level of alkylation of DNA the constitutive error-free systems perform the repair, and mutations seldom occur. At higher mutagenic doses, on the other hand, the performance of the constitutive repair system is not sufficient, and the adaptive repair enzymes are induced. At even higher doses, the enzymes involved in SOS repair are also induced. Between the error-free adaptive repair and the error-prone SOS system there is competition for the repair of DNA lesions. The frequency of mutation is critically dependent upon which of these repair systems is working (PEGG et al., 1985; LINDAHL et al., 1988; ROSSI et al., 1989).

The most commonly used alkylating agents are the alkyl alkane sulfonates MMS and EMS as well as the nitrosamides MNNG and NMG.

The EMS induced mutagenesis was reviewed by SEGA (1984). In bacteriophages, *Neurospora crassa,* and *Drosophila* methylation by MMS was five to ten times faster than ethylation by EMS, but ethylation was found to be more effective in causing mutations. EMS mutagenesis occurs in a wide variety of organisms, from viruses, bacteria and plants to mammals. Alkylation occurs via a mixed S_N1/S_N2 reaction mechanism thus alkylating significant levels of oxygen such as the O^6 of guanine in addition to the alkylation of nitrogen positions. The distribution of alkylating products is given in Tab. 3. In *Neurospora crassa* (MALLING and DE SERRES, 1968) 41% of the mutants involved a GC to AT transition, while 17% were AT to GC transitions. In addition, 9% of the mutants show small insertions or deletions. Errors may arise from mispairing of nucleotides to the template or by misincorporation of nucleotides during DNA synthesis (SAFFHILL et al., 1985).

One of the most powerful chemical mutagens in prokaryotic and eukaryotic cells is MNNG (DRAKE and BALTZ, 1976; GICHNER and VELEMINSKY, 1982), but it is hazardous in handling because of its carcinogenic effects. Reacting through a S_N1 mechanism (HORSFALL et al., 1990), MNNG treatment leads to 7-methylguanine, O^6-methylguanine, 3-methyladenine and phosphatetriesters as major classes of adducts although the last mentioned are not throught to lead to mutagenesis (SINGER, 1986). In *E. coli* and *Bacillus subtilis* 90% of the mutations induced are GC→AT transitions (COULONDRE and MILLER, 1977; HADDEN et al., 1983), to a small extent AT→GC transitions. Deletions and frameshift mutations are also found.

Besides the alkylation of non-replicating DNA, the main point of action of MNNG is the replication point of DNA, through a change in DNA polymerase during DNA replication. In this process, there is incorrect duplication in a short segment of the DNA until the defective polymerase is replaced by an intact molecule. This mode of action results in mutational hot spots (CERDÁ-OLMEDO and HANAWALT, 1968; CERDÁ-OLMEDO et al., 1968; DAWES and CARTER, 1974). When a mutation is induced in a specific locus, a large number of further mutations, so-called *comutations,* may be found in closely linked genes (GUEROLA et al., 1971). In *E. coli* 40% of comutations are concentrated in a region of about 50000 base pairs (about 1/60 of the genome). In *Streptomyces coelicolor* the comutation region is about twice as large.

This comutation effect and the so-called *sequential mutagenesis* in synchronized cultures by MNNG pulses, where a specific genome segment is replicated in the vast majority of the chromosomes at a specific time (CERDÁ-OLMEDO et al., 1968), have been used in several bacteria to induce mutations into certain sites of the genome, or to draw up genetic maps.

Under MNNG treatment a large proportion of mutants is found with a low killing rate: in *E. coli* 10% auxotrophs at a survival level of about 1%, in *Schizosaccharomyces pombe* 8% auxotrophs at a survival level of 20% (SINHA and CHATOO, 1975).

The premutational lesions, mainly the O^6-methylguanine and the phosphotriesters are subject to repair by specific alkyltransferases as part of the inducible adaptive response (KARRAN and LINDAHL, 1979). The 3-methyladenine is excised by a DNA glycosylase which generates an AP site (LINDAHL et al., 1988) that is then repaired by the excision repair (PEGG et al., 1985; SAFFHILL et al., 1985). In addition, evidence for the nucleotide excision repair (SAMSON et al., 1988), the mismatch correction of O^6-methylguanine and for the induction of SOS repair has also been demonstrated in *E. coli.*

8.2.2 Mutagens Affecting Replicating DNA

Only few mutagens act by being incorporated into replicating DNA or through intercalation during replication or repair.

8.2.2.1 Base Analogs

Base analogs are weak mutagens which are mainly used in phages and bacteria. Because of their structural similarities, base analogs such as 5-bromouracil (BU) (RYDBERG, 1977), 2-aminopurine (AP) (GOODMAN et al., 1977), or N^4-hydroxycytidine (SLEDZIEWSKA-GOJSKA and JANION, 1982) are incorporated into replicating DNA instead of the correct base.

The analogs tautomerize more frequently than the natural bases. The adenine analog BU, which is used in phages and bacteria pairs in the keto form with adenine, whereas BU in the enol form pairs with guanine. If the keto form of BU is incorporated, there is an AT→GC transition caused by tautomerization; if the incorporation takes place in the enol form, a GC→AT transition is caused (Fig. 8).

The adenine analog AP induces AT→GC transitions in prokaryotes. In Chinese hamster V79 cells AP was found to cause both gene mutations and sister-chromatid exchanges (SPEIT et al., 1990). AP, 2-amino-N^6-hydroxyadenine, 2-amino-N^6-methoxyadenine and 2-amino-N^6-methylhydroxyadenine, which are mutagenic in *E. coli*, were found to induce SOS response whereby mismatch repair is involved in the process of SOS induction (BEBENEK and JANION, 1985).

8.2.2.2 Frameshift Mutagens

Mutagens, which intercalate into the DNA molecule causing crosslinks, result in addition or deletion of a few nucleotides thus changing the reading frame (for reviews, see ROTH, 1974; DRAKE, 1989; FERGUSON and DENNY, 1990). Not all of the crosslinking agents cause frameshift mutations.

The most important class of frameshift mutagens are the acridines, such as acridine orange, proflavine, 9-aminoacridine, acriflavine, and the ICR compounds (for structures see Fig. 9).

Originally, the mutagenic power of acridines was detected in T2 and T4 phages (WACKERNAGEL and WINKLER, 1971). In the dark, acridines induce mainly frameshifts, in the presence of visible light transitions and trans-

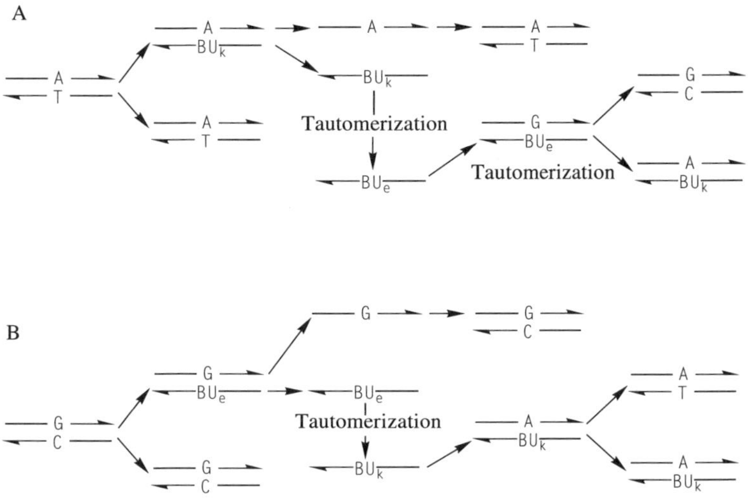

A

A T A BU_k A BU_k A T Tautomerization BU_e G BU_e Tautomerization G C A BU_k

B

G C G BU_e G C BU_e Tautomerization BU_k G G C A BU_k A T A BU_k

Fig. 8. Mutagenesis by the base analog 5-bromouracil (BU).
(A) BU is incorporated in its keto form (BU_k); tautomerization during replication causes an AT to GC transition.
(B) BU is incorporated in its enol form (BU_e); tautomerization during replication causes a GC to AT transition.

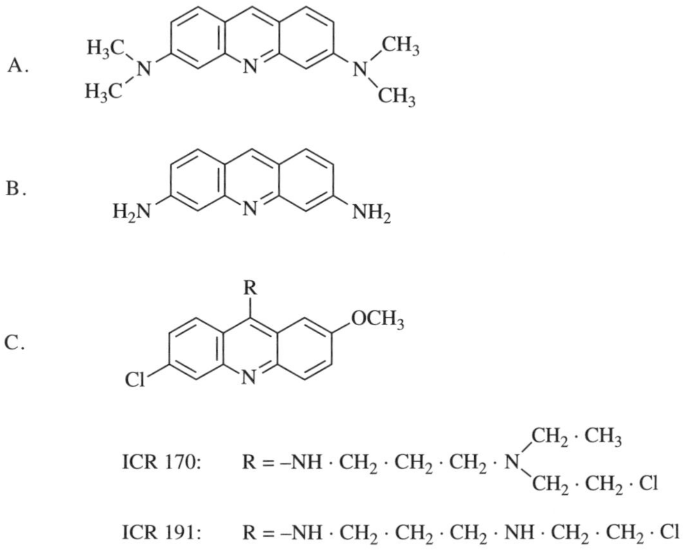

A. H_3C N H_3C N N CH_3 CH_3

B. H_2N N NH_2

C. R OCH_3 Cl N

Fig. 9. Structures of frameshift mutagens.
A. acridine orange
B. proflavine
C. ICR compounds ICR 170 and ICR 191.

ICR 170: R = –NH · CH_2 · CH_2 · CH_2 · N(CH_2 · CH_3)(CH_2 · CH_2 · Cl)

ICR 191: R = –NH · CH_2 · CH_2 · CH_2 · NH · CH_2 · CH_2 · Cl

versions. 9-Aminoacridine was shown to induce the loss of GC base pairs in the N-terminal region of the *lacI* gene of *E. coli* (Gordon et al., 1991). The ICR compounds ICR170 and ICR191, originally synthesized as antitumor agents by the Institute for Cancer Research, are powerful mutagens. Calos and Miller (1981) found by genetic and sequence analysis of ICR191 induced mutants in the *lacI* gene of *E. coli* that 97.9% of the mutants are

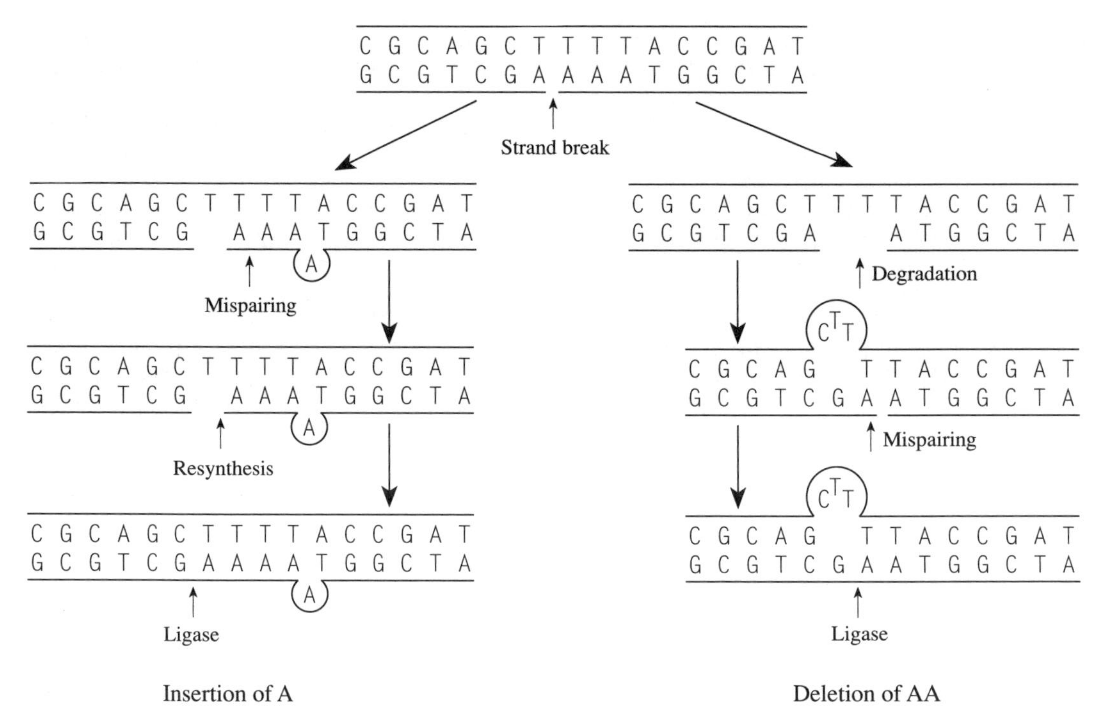

Fig. 10. Possible mechanism for frameshift mutagenesis by intercalating agents according to STREISINGER's model of strand slippage for the induction of insertions or deletions (STREISINGER et al., 1966).

frameshift mutations with addition or deletion of a single GC base pair from a $^{GGG}_{CCC}$ sequence. ICR170 has stronger mutagenic activities in eukaryotes such as *Drosophila* or *Neurospora* (AMES and WHITFIELD, 1966).

Homidium bromide (the former ethidium bromide) is a most effective mutagen for mitochondrial DNA producing respiratory deficient colonies (petite) in *Saccharomyces cerevisiae* (FUKUNAGA et al., 1984).

The photoreactive psoralens also act as crosslinking agents which induce frameshift mutations (see Sect. 8. 1.4).

DNA intercalators appear to interact preferentially with B-form DNA (YOON et al., 1988). With acridines, hot spots of frameshift mutations mainly occur in repetitive sequences (IMADA et al., 1970).

Different acridines have different sequence selectivities, e.g., mutagenesis by proflavine occurs predominantly in AT runs, by ICR compounds in GC regions of DNA (ROTH, 1974). Based on these data and on further observations in bacteriophage T4, STREISINGER et al. (1966) proposed their still accepted model of strand slippage for the induction of frameshift mutagenesis by acridines that is given in Fig. 10 in simplified form. *In vitro* and *in vivo* studies on various acridines suggest that frameshift mutations can occur in the absence of any enzyme of DNA repair (CONRAD and TOPAL, 1986). It must be suggested that there are other pathways of mutagenesis, where topoisomerase II is believed to be of importance.

9 Antimutagenesis

The mutagenicity of radiation or chemical agents may be modulated by specific compounds and treatments or by the genetic set-up of the organism. One well known example is caffeine which potentiates the mutagenic and lethal effects of many agents by inhibition of photo-reactivation and nucleotide excision repair (SELBY and SANCAR, 1990). The same is true for acriflavine, 8-methoxypsoralen, procaine, lidocaine (TODO and YONEI, 1983), or homidium, quinacrine and hycanthone, inhibitors of excision repair (CLARKE and SHANKEL, 1974). All these compounds exhibit the paradoxical property of also being able to act as frameshift mutagens.

Antimutagenesis is well documented by several examples (GEBHART, 1974; CLARKE and SHANKEL, 1974; HAHN, 1976). Mutagenic effects of radiation or of chemical mutagens are antagonized, and the frequency of spontaneous mutations is also reduced. Antimutagenesis may be physiologically induced by caffeine (now acting as antimutagen), spermine, quinacrine, acridines, Mn(II) or Co(II), gallic or tannic acids, heat shock, starvation for amino acids, thiamine, glucose or phosphate.

Antimutagenic effects may also be caused by antimutator alleles of genes involved in replication and repair. In the phage T4 genome several antimutator alleles of gene 43, the DNA polymerase gene, have been identified, which reduce mutation rates at many sites in the T4 genome, in some cases up to 100-fold (DRAKE and ALLEN, 1968).

Antimutagenesis has been documented in *Escherichia coli, Salmonella typhimurium, Saccharomyces cerevisiae,* and *Neurospora crassa.* In antimutator polymerases a higher ratio of exonuclease activity to polymerase activity was exhibited (SCHNAAR et al., 1973), but beyond that very little is known about the biochemical basis of antimutability.

10 Insertional Mutagenesis with Transposable Genetic Elements

Transposable genetic elements are classified according to their genetic organization and transposition mechanisms as IS elements, transposons, and transposing bacteriophages (SAUNDERS and SAUNDERS, 1987). They all are DNA segments with the capability to insert as discrete DNA sequences at various sites within a genome. They can be transposed from one position in a replicon to another site either on the same replicon (intramolecular transposition) or on another replicon present in the cell (intermolecular transposition). Integration and excision is independent of recA mediated homologous recombination. Transposable genetic elements have been found in a variety of organisms, including eubacteria (KLECKNER, 1981), streptomycetes (LYDIATE et al., 1986), archaebacteria (PFEIFER et al., 1984), *Drosophila* (CAMERON et al., 1979), and plants (DÖRING and STARLINGER, 1984).

Transposable genetic elements, when inserted at a given locus, destroy the function of the gene at the site of their integration and cause large-scale rearrangements of adjacent DNA sequences, including inversions, deletions, and duplications (NEVERS and SAEDLER, 1977). The transposable elements are used for a variety of *in vivo* and *in vitro* manipulations, and for analysis, such as isolation of mutants, localized mutagenesis, chromosomal mapping, isolation of deletions with one or both end points specified, selection and maintenance of chromosomal duplications, and isolation of new genes by transposon tagging (WIENAND and SAEDLER, 1988), where the gene to be isolated is marked by insertion of a known transposon and the resulting mutated gene is identified through its transposon-specific sequences as a probe.

IS elements are small insertion sequences of variable length (0.8-2.1 kb) which encode determinants involved in promoting transpositions. Besides insertion mutations, they cause chromosomal rearrangements such as deletions or inversions, whereby the IS element itself is not part of the deleted or inverted sequence.

The temperate *bacteriophage Mu* (37 kb) can generate random insertion mutations due to its ability to integrate at many different sites within the genomes of its hosts (such as *E. coli, Shigella dysenteriae, Citrobacter freudii*) (TOUSSAINT and RÉSIBOIS, 1983). By using broad host range plasmids Mu can also be introduced into various Gram negative bacteria.

Mu derivatives, such as the thermoinducible phage Mu*cts* with inactivated killing functions (BUKHARI, 1975), or the mini-Mus retain Mu integrative properties and can transpose to randomly distributed sites on the host genome upon incubation, but lack essential functions for the lytic development (FAELEN et al., 1979). After induction at 42 °C (CABEZON et al., 1975), imprecise excision of Mu DNA may occur leading to deletions of host DNA adjacent to the prophage.

Transposons comprise a central DNA segment flanked on either side by a copy of an IS. The IS elements encode the information for transposition, while the control DNA sequence encodes accessory determinants, e.g., antibiotic resistance, thus giving a selectable marker after insertion.

A comprehensive description of transposons, their application as mutagens, and the advantages of transposon mutagenesis as compared to classical methods can be found in Chapter 11 of this volume.

11 Site-Specific Mutagenesis

Physical and chemical mutagenesis as classical *in vivo* mutation techniques have a number of drawbacks. Mutagenesis is normally random with the exception of sequential mutagenesis or comutation techniques. The introduction of a desired mutation may be accompanied by undesirable mutations in adjacent genes or even unlinked genes depending on the mutagen. Location and nature of the induced mutation are unknown and can only be determined by time-consuming genetic analysis. The progress of genetic engineering technology allowed the development of *in vitro* mutagenesis techniques, which permit the direction of mutation to particular sequences or even individual nucleotides of a DNA molecule. A prerequisite for application of the site-specific mutagenesis in industrial strain improvement, however, is that the gene of interest has been cloned and at least partially characterized concerning its significance in the process to be improved.

Details about methodology and application of site-specific mutagenesis are given in Chapters 9 and 10 of this volume.

12 Industrial Application of Mutagenesis

With the exception of the food industry, only a few commercial fermentation processes use wild strains isolated directly from nature. For the production of enzymes, amino acids, antibiotics, or other metabolic products, the production strains must be specifically adapted to the fermentation process. Although a fermentation process may be optimized by changing the cultural conditions, growth behavior, regulation of biochemical pathways, and productivity are ultimately controlled by the genetic background of the organism. Thus, to improve the process, the organism's genome must be modified. This may be achieved in a strain development program by mutation and/or recombination combined with a suitable selection mechanism.

The objective of genetic strain improvement depends on the process. In general, the motivation for industrial strain development is economic, since the metabolite concentrations produced by wild strains are usually too low for economical processes. Through an extensive strain development program (which may require several years), yield increases up to 100 times or more can usually be attained. The success of these programs depends greatly on the substance to be improved. For instance, the yield products involving the activity of one or a few genes, such as enzymes, can be increased simply by raising the gene dose. However, with secondary metabolites, which are frequently the end products of complex, highly regulated biosynthetic pathways, a variety of changes in the genome may be necessary to permit the selection of high-yielding strains.

For a cost-effective process, strains with improved fermentation properties may also be needed. Depending on the system, it may be desirable to isolate strains with shorter fermentation times, without production of undesirable pigments, with reduced oxygen demand, with lower viscosity of the culture so that aeration is a lesser problem, with decreased foaming during fermentation, with the ability to metabolize inexpensive substrates, with tolerance to high concentrations of carbon or nitrogen sources, or with resistance to bacteriophages.

Wild strains frequently produce a mixture of chemically closely related substances. Mutants which synthesize one component as the main product are preferable, since they make possible a simplified process for product recovery. Changes in the genotype of microorganisms can lead to the biosynthesis of new metabolites. Thus, mutants that synthesize modified antibiotics may be selected.

The traditional method of empirical strain improvement by mutation and selection on the basis of direct titer measurement – the so-called *random screening* – has proven very effective in increasing production of classical antibiotics and other metabolites: The penicillin titer was increased from 20 U/mL (1943) to about 85000 U/mL (approx. 50 g/L; 1989); the titers of cephalosporin, streptomycin, erythromycin or oxytetracycline fermentations are now in the range of 20 g/L. Citric acid is produced with yields of more than 100 g/L; vitamin B_{12} titers are 6 g/L; the fermentation yields of L-glutamic acid are 100 g/L, for L-lysine 70 g/L; the flavor-enhancing nucleosides and nucleotides are produced in the range of 20–30 g/L (for details, see CRUEGER and CRUEGER, 1989).

Effective mutagenesis depends on the mutagenic treatment itself, the phenotypic expression of mutations, and on environmental factors prior to, during, and immediately after mutagenesis.

Mutagenesis is the source of genetic variation but no single mutagenic treatment will give all possible types of mutation. This phenomenon of mutagen specificity is not completely understood. The base composition of the target sequence in DNA, the DNA configuration, and the endogenous repair activities of an organism are influencing factors affecting not only its mutability by a given mutagen, but also the types of mutations induced. In most cases the exact type of molecular change required to improve a given strain is not known. It is thus advisable to use different types of mutagenic treatment or combinations of it to generate as wide a spectrum of mutant types as possible on which to carry out the subsequent selection (ROWLANDS, 1984). Another factor in the choice of a mutagen is safety. It must be remembered that all mutagens are potential carcinogens, and so effective precautions must be taken.

The conditions for mutagenesis must be optimized for each specific strain according to mutagen dose and environmental conditions. Mutagenic efficiency, that is the ratio of mutational to lethal events, depends on the interaction between the mutagen and the strain. Thus, a dose–response curve for survival and for mutagenesis should be constructed for each process (see Figs. 11 and 12), and the optimum dose of mutagen, in terms of mutants per survivors, should be determined. In the same strain, this optimum varies depending upon the type of the mutant required. This procedure is often very time-consuming and costly. Therefore, easily detectable changes such as mutations for resistance or reversion to auxotrophy are frequently used as a model to optimize conditions for mutagenesis. However, the result of these latter experiments need not have any bearing on optimal conditions for increased formation of a desired product. In addition to the strain-specific factors and the mutagen used, the type of mutation can be influenced by environmental factors such as temperature, the nature of the suspending medium during and after mutagenic treatment (CLARKE, 1975), or the addition of repair inhibitors such as caffeine or acriflavine (AUERBACH, 1976).

Another crucial point is the phenotypic expression of mutations. Many mutations which result in increased formation of metabolites are recessive. When a recessive mutation takes place in a uninuclear, haploid cell (e.g., bacteria, actinomycete spores and protoplasts, asexual conidia of fungi), a heteroduplex results from it and the mutant phenotype can only be expressed after a further growth step. This also applies to exponentially growing bacterial cells, which can contain 2–8 chromosomes. Pure mutant clones do not appear until several steps of replication have taken place. With the filamentous actinomycetes, special procedures for mutant expression must be used. In the course of strain development, actinomycetes can lose their sporulation ability. To obtain cells for plating, the heterokaryotic mycelium which results from mutagenesis is grown and then fragmented by ultrasonic

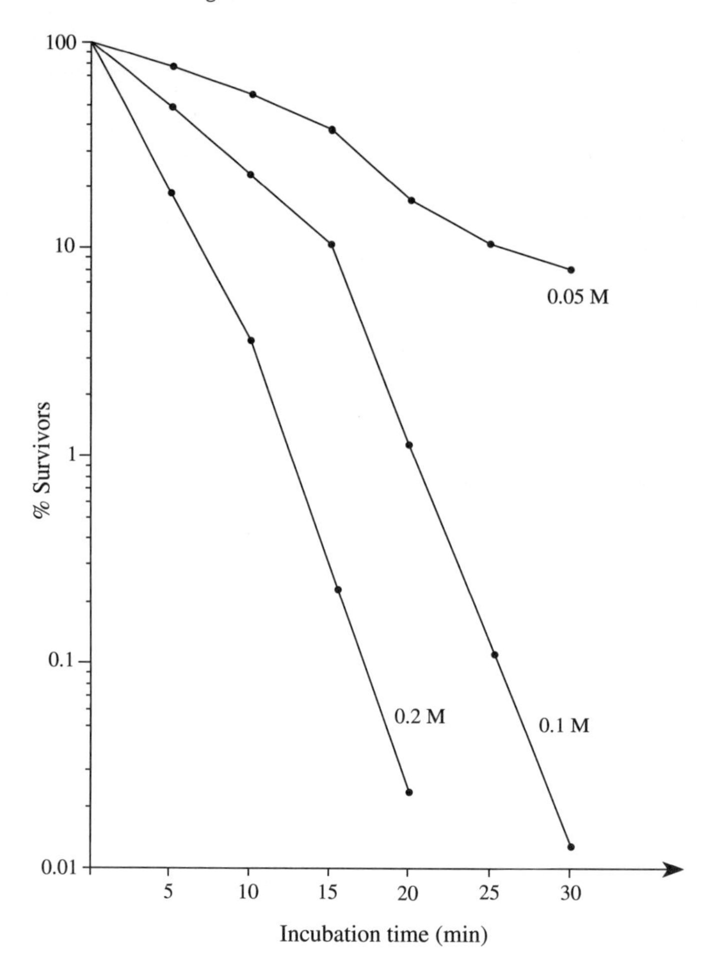

Fig. 11. Dose-response curve; survival of *Micromonospora inyoensis* after nitrous acid treatment.

treatment or shaking with glass beads. After filtration through paper, cotton, or an 8 μm membrane filter, mycelium fragments containing only one or a few nuclear bodies are used for plating. Homokaryotic material can be ultimately selected by repeating this segregation process. The preparation of protoplasts containing one or few nuclei is another way for attaining segregation.

In diploid or heterokaryotic eukaryotes, recessive mutations are allowed to undergo phenotypic expression after meiosis, haploidization, or mitotic recombination.

Delays in expression which are not directly the result of genetic effects are observed, such as mutations which cause changed ribosomes or mutations resulting in the loss of surface receptors (as in the development of bacteriophage resistance). In both cases, the wild-type structures must be diluted out during growth, before the mutation is recognizable phenotypically.

Besides an optimal mutagenesis, the method of selection is crucial for the effective screening of mutants. Usually a random selection of survivors from a mutagenized population can be examined for the desired characteristic in a fermentation process that closely mimics the large-scale process. This procedure is very costly, but it is often the only way to find mutants with increased productivity in industrial strains. With enzymes, activities can often be observed directly in colonies growing on plates by spraying with suitable reagents or by incor-

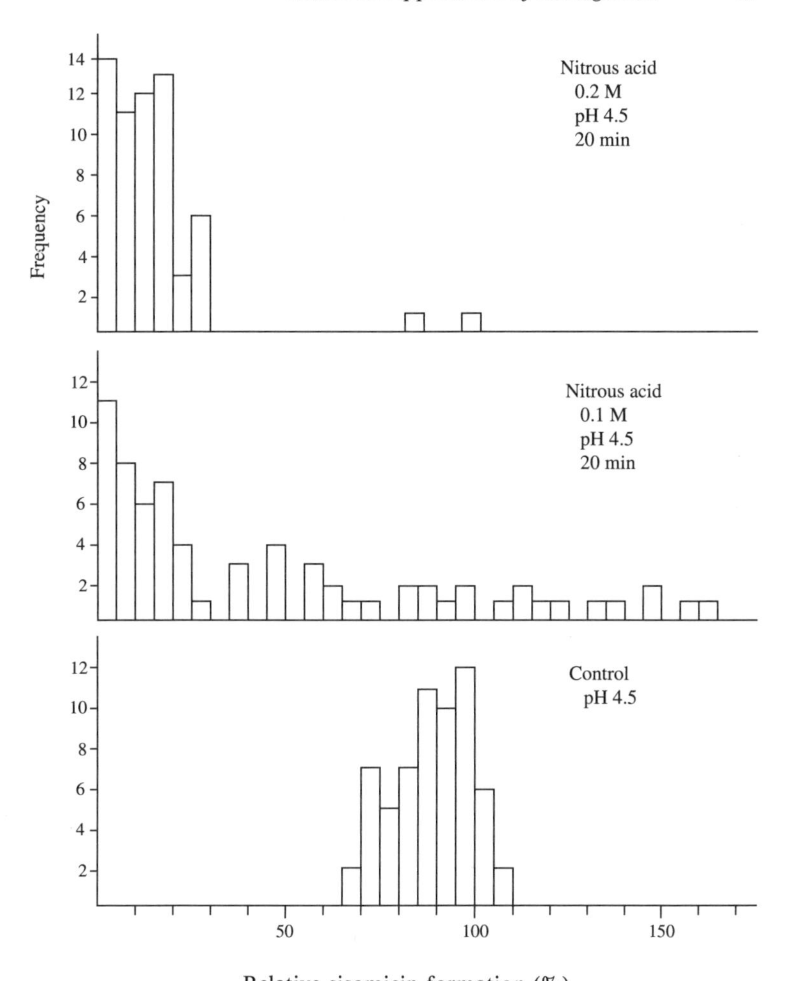

Fig. 12. Dose-response curve; sisomicin production by *Micromonospora inyoensis* in relation to mutagen treatment with nitrous acid (control strain = 100%).

porating indicator dyes into the culture medium. The best strains from such a mutation cycle are again mutated and selected, and increases of 5–20% are usually found. Mutants with high yield increases are much less frequent than those with only slight improvements. A gradual increase in yield is attained by continuing with this sequential mutation procedure. Due to the fact that unrecognized mutations can cause certain strains to show no increase in further mutation cycles, more than one strain (usually the 5–10 best mutants of one cycle) should be used as parent strains for the next step of mutation and selection.

Many factors determine the size of the mutational screening necessary to isolate improved strains: the frequency of mutation, the extent of yield increases, the amount of time required for a mutation selection cycle due to growth characteristics of the strain, and the accuracy of the screening assay. Thus the screening capacity can be a limiting factor in strain improvement. The use of automation in mutant screening helps to increase the number of isolates which can be processed per time unit, and statistically based screening designs maximize the probability of detecting improved strains.

Other more or less random screening approaches to increase the productivity of strains are the isolation of revertants from auxotrophs leading to deregulated strains or the isolation

Tab. 4. Increase of Glucose Isomerase Activity by Mutagenesis (CRUEGER and CRUEGER, 1984)

Strain	Mutagenic Agent	Increase of Productivity (%)
Streptomyces ATCC 21175	Ethyleneimine	62
Streptomyces olivaceus NRRL 3583	UV	16
S. olivochromogenes	UV	50
S. nigrificans	UV	198

of resistance mutants, where antibiotic resistance may be used to increase cell permeability or effectiveness of protein synthesis (CLARKE, 1976). Antimetabolite resistance can be used to select mutants which exhibit defective regulation (MALIK, 1979).

More rational programs are possible when data about biosynthesis and regulation are available. In the case of amino acid or nucleotide producing strains branched biosynthetic pathways often lead to different products. By using certain blocked mutants, only the desired product is synthesized (CLARKE, 1976).

Self-resistance to toxic end products, such as ethanol or antibiotics, or escape from end product feedback inhibition may also allow higher production yields by elimination of these characters.

Very few papers from industry with data on strain improvement and production yields are available to document the enormous progress made in the past by mutagenesis in industrial strain improvement. Examples of typical industrial processes are given for the improvement of an enzymatic process and of a primary and secondary metabolite.

For enzymes which are the product of only one or a few genes, the increase made by a single mutation step may be impressive, as given in Tab. 4 for several glucose isomerase producing strains (CRUEGER and CRUEGER, 1984).

Primary metabolites, such as amino acids, or nucleosides or nucleotides are produced via short unbranched or branched pathways which are efficiently regulated. Detailed knowledge about biosynthetic pathways, their regulation, and the fluxes through the pathways are available in most cases, allowing a rational mutation program (for details, see CRUEGER and CRUEGER, 1989). The flavor-enhancer 5'-IMP, which is produced in Japan with 2000 tons per year by a *Brevibacterium ammonia-*

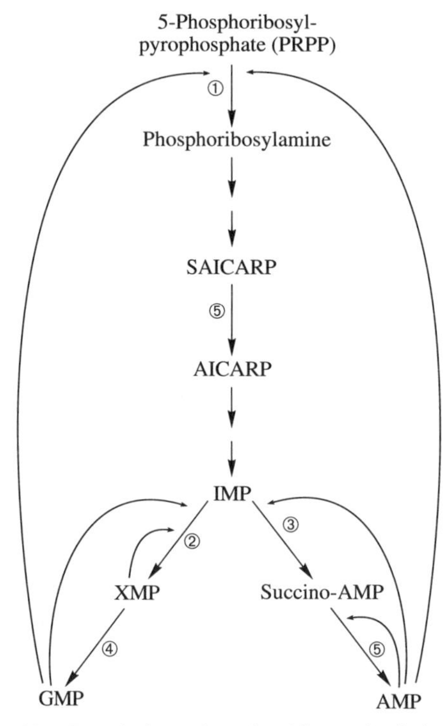

Fig. 13. Regulation of nucleotide synthesis in *Bacillus subtilis.* (1) PRPP amidotransferase; (2) IMP dehydrogenase; (3) adenylsuccinate synthetase; (4) GMP synthetase; (5) adenylsuccinate lyase. PRPP, 5-phospho-α-D-ribosylpyrophosphate; SAICARP, 1-(5'-phosphoribosyl)-4-(N-succino-carboxamide)-5-aminoimidazole; AICARP, 5-amino-1-(5'phosphoribosyl)-imidazole-4-carboxamide.

Tab. 5. Genealogy of 5′-IMP Producing Mutants of *Brevibacterium ammoniagenes* (KUNINAKA, 1986)

Strains	IMP (mg/mL)	HX[a] (mg/mL)	Total IMP[b] (mg/mL)
KY 3453 (wild type)			
↓ UV			
KY 13102 (ade^L)	1–2	8–10	9–12
↓ MNNG			
KY 13171 (ade^L, Mn^I)	7–8	4–6	11–13
↓ MNNG			
KY 13184 (ade^l, Mn^I, gua^-)	8–10	7–8	15–18
↓ MNNG			
Ky 13198 (ade^L, Mn^I, gua^-)	9–12	5–6	15–18
↓ MNNG			
KY 13361 (ade^L, Mn^I, gua^-)	12–16	2–3	15–19
↓ MNNG			
Ky 13363 (ade^L, Mn^I, gua^-)	18–20	trace	18–20
↓ MNNG			
KY 13369 (ade^L, Mn^I, gua^-)	20–27	trace	20–27

[a] HX, hypoxanthine calculated as IMP
[b] sum of IMP and HX
ade^L, adenine leaky mutant
Mn^I, Mn^{2+} insensitive mutant

genes strain, is a biosynthetic precursor of GMP and AMP (Fig. 13).

Starting with the wild strain KY 3454, the IMP productivity was increased stepwise through repeating mutational treatments from less than 1 mg/mL up to 20 - 27 mg/mL (FURUY and TOSHIBA, 1982; KUNINAKA, 1986). The genealogy of these mutants is given in Tab. 5. In the first step of strain development, the feedback inhibitory activity of the end product AMP against PRPP amido-transferase as the first enzyme of the pathway was eliminated by induction of an adenine leaky mutant. In the strains KY 3454 and KY 13102 which were sensitive to Mn^{2+}, the optimal manganese content of the medium was 0.01–0.02 mg/L; at higher levels (>0.025 mg/L) IMP production decreased dramatically. Under Mn^{2+} limitation, abnormal cells with reduced permeability are formed. They excrete hypoxanthine which is extracellularly phosphorylated to 5′-IMP. In addition, 5′-IMP is synthesized *de novo* in the subsequent fermentation and is excreted directly into the culture medium. By the following MNNG treatment, a permeability mutant could be isolated, whose IMP accumulation no longer was influenced by Mn^{2+} (1 mg/L). This membrane change resulted in a stimulation of IMP excretion from cells, which was a prerequisite for further improvement. The overall production of IMP by mutant KY 13171 has not yet been increased, but a change in IMP to hypoxanthine (HX) could be observed. In order to release feedback regulation operating in *de novo* IMP biosynthesis by GMP, a guanine requiring mutant, KY 13184, being devoid of IMP dehydrogenase, was isolated following MNNG treatment. All further mutants were obtained from KY 13184 through repeating mutation and selection cycles. First the IMP to HX ratio was shifted to IMP excretion without increasing the IMP *de novo* synthesis, and finally IMP synthesis was increased. After seven steps of mutation and selection, IMP production could be improved by a factor of 20.

The situation is much more complicated when improving secondary metabolites such as antibiotics, where the biosynthesis is encoded at least by five different classes of genes:

(1) structural genes which encode for enzymes involved in secondary metabolite biosynthesis

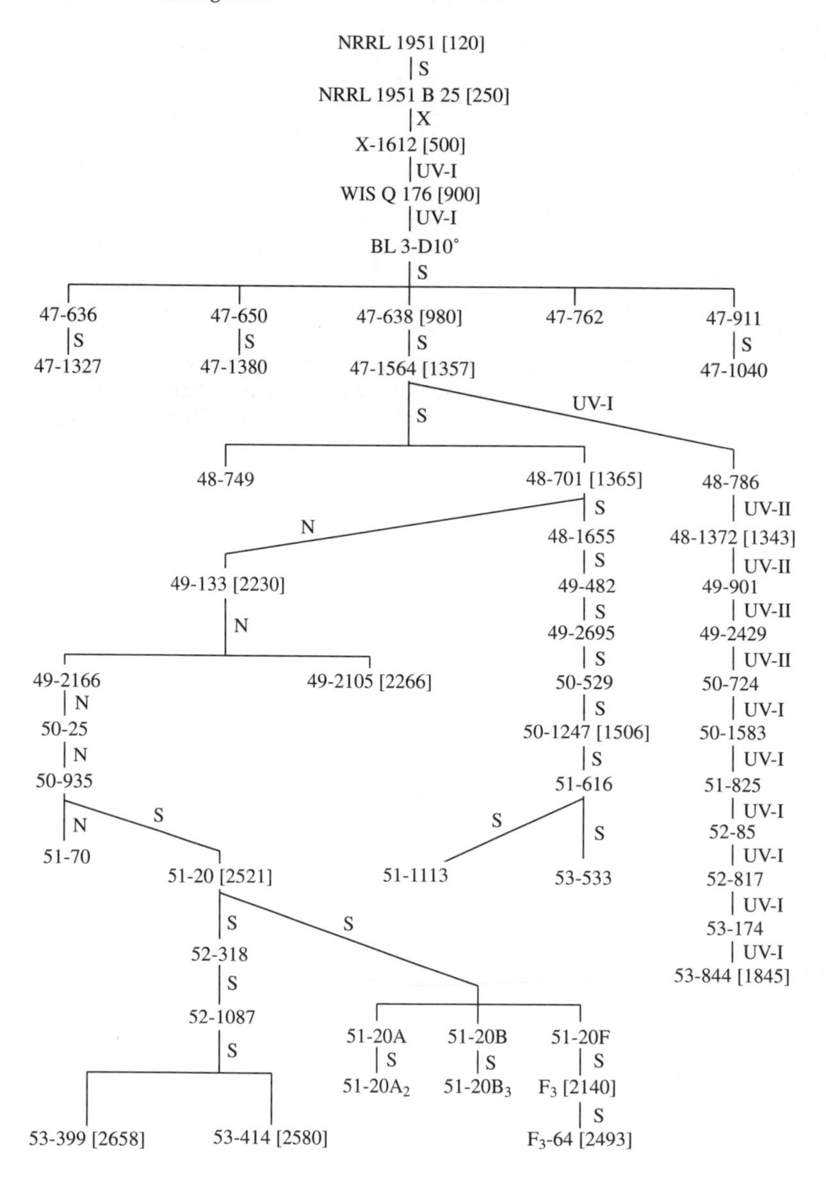

Fig. 14. Genealogy of the Wisconsin strain of *Penicillium chrysogenum*.
S, selection without mutagenesis; X, X-ray treatment; UV I, ultraviolet radiation at 275 nm; UV II, ultraviolet radiation at 252 nm; N, treatment with nitrogen mustard. Square brackets show yields in international units per milliliter. [a], pigment-free mutant (from BACKUS and STAUFFER, 1955).

(2) regulatory genes for secondary metabolism
(3) resistance genes, which keep antibiotic-producing strains immune to their own products
(4) permeability genes
(5) regulatory genes which control primary metabolism and thus indirectly affect the biosynthesis of secondary metabolites.

One published example of strain improvement is the genealogy of the famous Wisconsin strain of *Penicillium chrysogenum* (BACKUS and STAUFFER, 1955). As can be seen from Fig. 14 for a 20-fold increase of penicillin yields, a complex mutation and selection scheme was necessary including about 15 steps and several cell lines. Even with more knowledge about biosynthesis and regulation, the possibilities for more rational programs in antibiotic improvement remain restricted, unless key enzymes and bottleneck reactions of the process are elucidated.

During the last years, the same procedures used in microbial strain improvement have been used with plant cells. For example, in carrot suspension culture, resistances [to 5-methyltryptophan (WIDHOLM, 1977), or cycloheximide (MALIGA, 1980)] could be introduced by EMS and MNNG treatment. In cultures of *Dioscorea deltoidea* the steroid content was modified by NMU treatment (BUTENKO et al., 1987). Thus, it can be expected that plant cell cultures behave in this respect like microorganisms and are accessible to the successful methods of improvement used there.

13 Conclusions

Industrial strain improvement is a central part of the commercial development of fermentation processes. The current practice of strain improvement by mutation and selection in combination with breeding by sexual and parasexual recombination and protoplast fusion has been found to be an effective technique. In recent years new methodologies due to the development of genetic engineering techniques have become available for site-specific mutagenesis, amplification of desired genes, or analysis of gene functions, allowing rational screening programs. However, one difficulty in applying the techniques to improvement of existing processes is that the organisms of widest use in industry are in many cases unfortunately not the organisms for which the greatest amount of basic genetic information is available. There is an apparent gap between basic research and industrial application. Biosynthesis and regulation, along with the genetic fundamentals of industrially important microorganisms, must be understood before a rational program can be started. In the last ten years in several industrial companies steps to bridge the gap between basic knowledge and industrial application have been made, and with single-gene or single-operon determined products such as enzymes or amino acids a marked increase in the effectiveness of strain improvement by genetic engineering technology can be anticipated. Also in strain improvement programs for secondary metabolites the first efforts for using the new techniques look promising.

Research on the process has to be done before recombinant DNA techniques can be effectively used. Therefore, random screening is still a cost-effective procedure, and for reliable short-term strain development frequently may be the method of choice. As reviewed by ROWLANDS (1984) on the basis of current knowledge, mutagenesis can be optimized in terms of type of mutagen used on the nature of its initial action, and mutagen specifity effects can be taken into account. Mutant screening can be designed to allow maximum experience of the desirable mutant types. In addition, the screening capacities can be increased by automated procedures. Without doubt, the mutagenic techniques together with hybridization and recombinant DNA technology will in the future allow very effective improvements in industrial biotechnology.

14 References

AGER, D. D., HAYNES, R. H. (1990), Analysis of interactions between mutagens, I. Heat and ultraviolet light in *Saccharomyces cerevisiae, Mutat. Res.* **232**, 313–326.

AMES, B. N. (1983), Dietary carcinogens and anticarcinogens, *Science* **221**, 1256–1264.

AMES, B. N., WHITFIELD, H. J. (1966), Frameshift mutagenesis in *Salmonella, Cold Spring Harbor Symp. Quant. Biol.* **31**, 221–225.

AUERBACH, C. (1967), The chemical production of mutations, *Science* **158**, 1141–1147.

AUERBACH, C. (1976), *Mutation Research: Problems, Results and Perspectives,* London: Chapman & Hall.

AUERBACH, C., ROBSON, J. M. (1947), The production of mutations by chemical substances, *Proc. R. Soc. Edinburgh (B)* **62**, 271–283.

AVERY, O. T., MACLEOD, C. M., MCCARTY, M. (1944), Studies on the chemical nature of the substance inducing transformation of pneumococcal types, *J. Exp. Med.* **79**, 137–158.

BACKUS, M. P., STAUFFER, J. F. (1955), The production and selection of a family of strains in *Penicillium chrysogenum, Mycologia* **47**, 429–463.

BAILLY, V., VERLY, W. G. (1989), AP-endonucleases and AP lyases, *Nucleic Acids Res.* **17**, 3617–3618.

BAILLY, V., VERLY, W. G., O'CONNOR, T. R., LAVAL, J. (1989), Mechanism of DNA strand nicking at apurinic/apyrimidinic sites by *Escherichia coli* (formamidopyrimidine) DNA glycosylase, *Biochem. J.* **262**, 581–589.

BAKER, J. M., PARISH, J. H., CURTIS, J. P. E. (1984), DNA-DNA and DNA-protein crosslinking and repair in *Neurospora crassa* following exposure to nitrogen mustard, *Mutat. Res.* **132**, 171–179.

BASU, A. K., ESSIGMANN, J. M. (1988), Site-specifically modified oligodeoxynucleotides as probes for the structural and biological effects of DNA-damaging agents, *Chem. Res. Toxicol.* **1**, 1–18.

BASU, A. K., ESSIGMANN, J. M. (1990), Site-specifically alkylated oligodeoxynucleotides: Probes for mutagenesis, DNA repair and the structural effects of DNA damage, *Mutat. Res.* **233**, 189–201.

BEBENEK, K., JANION, C. (1985), Ability of base analogs to induce the SOS response: effect of a *dam* mutation and mismatch repair systems, *Mol. Gen. Genet.* **201**, 519–524.

BECKER, E. F., ZIMMERMANN, B. K., GEIDUSCHEK, E. P. (1964), Structure and function of cross-linking DNA, *J. Mol. Biol.* **8**, 377–391.

BELGUISE-VALLADIER, P., FUCHS, R. P. P. (1991), Strong sequence-dependent polymorphism in adduct-induced DNA structure: analysis of a single N-2-acetylaminofluorene residue bound within the *NarI* mutation hot spot, *Biochemistry* **30**, 10091–10100.

BEN-HUR, E., SONG, P. S. (1984), The photochemistry and photobiology of furocoumarins (psoralens), *Adv. Radiat. Biol.* **11**, 131–171.

BENZER, S. (1961), On the topography of the genetic fine structure, *Proc. Natl. Acad. Sci. USA* **47**, 403–416.

BERANEK, D. T. (1990), Distribution of methyl and ethyl adducts following alkylation with monofunctional alkylating agents, *Mutat. Res.* **231**, 11–30.

BERTRAM, H. (1988), DNA damage and mutagenesis of lambda phage induced by gamma-rays, *Mutagenesis* **3**, 29–33.

BIANCHI, L., BIANCHI, A., DALL'ACQUA, F., SANTAMARIA, L. (1990), Photobiologocal effects in *Saccharomyces cerevisiae* induced by the monofunctional furocoumarin 4,4′,6-trimethylangelicin (TMA) and the bifunctional furocoumarin 8-methoxypsoralen (8-MOP), *Mutat. Res.* **235**, 1–7.

BORGAONKAR, D. S. (1980), *Chromosomal Variation in Man: A Catalog of Chromosomal Variants and Anomalies,* New York: Alan R. Liss.

BOSI, A., OLIVIERI, G. (1989), Variability of the adaptive response to ionizing radiations in humans, *Mutat. Res.* **211**, 13–17.

BRIDGES, B. A., MOTTERSHEAD, R. P. (1972), Gamma ray mutagenesis in a strain of *Escherichia coli* deficient in DNA polymerase I, *Heredity* **29**, 203–211.

BRINK, R. A. (1973), Paramutation, *Annu. Rev. Genet.* **7**, 129–152.

BROOKES, P. (1990), The early history of the biological alkylating agents, 1918–1968, *Mutat. Res.* **233**, 3–14.

BRUTLAG, D., KORNBERG, A. (1972), Enzymatic synthesis of deoxyribonucleic acid. XXXVI. A proofreading function for the 3′→5′ exonuclease activity in deoxyribonucleic acid polymerase, *J. Biol. Chem.* **247**, 241–248.

BUDOWSKI, E. I. (1976), The mechanism action of hydroxylamine, *Progr. Nucleic. Acid Res. Mol. Biol.* **16**, 125–188.

BUKHARI, A. I. (1975), Reversal of mutator phage Mu integration, *J. Mol. Biol.* **96**, 87–99.

BUTENKO, R. G., SHAMINA, Z. B., NEGRUK, V. I., KARANOVA, S. L. (1987), Genetics of cultured plant cells producing commercially important products, in: *Genetics of Industrial Microorganisms, GIM 86* (ALACEVIC, M., HRANUELI, D., TOMAN, Z., Eds.), pp. 447–460, Zagreb: Organizational Committee of GIM 86.

CABEZON, T., FAELEN, M., DE WILDE, M., BOLLEN, A., THOMAS, R. (1975), Expression of ribosomal protein genes in *Escherichia coli, Mol. Gen. Genet.* **137**, 125–129.

CALCOTT, P. H., THOMAS, M. (1981), Sensitivity of DNA repair deficient mutants of *Escherichia coli* to freezing and thawing, *FEMS Microbiol. Lett.* **12**, 117–120.

CALOS, M. P., MILLER, J. H. (1981), Genetic and sequence analysis of frameshift mutations induced by ICR-191, *J. Mol. Biol.* **153**, 39–66.

CAMERON, J. R., LOH, E. Y., DAVIS, R. W. (1979), Evidence for transposition of dispersed repetitive DNA families in yeast, *Cell* **16**, 739–751.

CERDÁ-OLEMEDO, E., HANAWALT, P. C. (1968), The replication of the *Escherichia coli* chromosome studied by sequential nitrosoguanidine mutagenesis, *Cold Spring Harbor Symp. Quant. Biol.* **33**, 599–607.

CERDÁ-OLMEDO, E., HANAWALT, P. C., GUEROLA, N. (1968), Mutagenesis of the replication point by nitrosoguanidine: Map and pattern of replication of the *Escherichia coli* chromosome, *J. Mol. Biol.* **33**, 705–719.

CLARKE, C. H. (1970), Repair systems and nitrous acid mutagenesis in *E. coli* B/r, *Mutat. Res.* **9**, 359–368.

CLARKE, C. H. (1975), Mutagenesis and repair in microorganisms, *Sci. Prog.* (Oxford) **62**, 559–577.

CLARKE, C. H. (1976), Mutant isolation, in: *Second International Symposium of the Genetics of Industrial Microorganisms* (MACDONALD, K. D., Ed.), pp. 15–28, London: Academic Press.

CLARKE, C. H., JOHNSTON, A. W. B. (1976), Intragenic mutational spectra and hot spots, *Mutat. Res.* **36**, 147–164.

CLARKE, C. H., SHANKEL, D. M. (1974), Effects of ethidium, quinacrine and hycanthone on survival and mutagenesis of UV-irradiated Hcr^+ and Hcr^- strains of *E. coli* B/r, *Mutat. Res.* **26**, 473–481.

CONRAD, M., TOPAL, M. D. (1986), Induction of deletion and insertion mutations by 9-aminoacridine. An *in vitro* model, *J. Biol. Chem.* **261**, 16226–16232.

COOPER, P. K. (1982), Characterization of long patch excision repair of DNA in ultraviolet-irradiated *Escherichia coli:* an inducible function under Rec-Lex control, *Mol. Gen. Genet.* **185**, 189–197.

COULONDRE, C., MILLER, J. H. (1977), Genetic studies of the *lac* repressor. IV. Mutagenic specificity in the *lacI* gene of *Escherichia coli, J. Mol. Biol.* **117**, 577–606.

CRAWFORD, I. P., SIKES, S., BELSER, W. O., MARTINEZ, L. (1970), Mutants of *Escherichia coli* defective in the B protein of tryptophan synthetase. III. Intragenic clustering, *Genetics* **65**, 201–211.

CRUEGER, A., CRUEGER, W. (1984), Carbohydrates, in: *Biotechnology* (REHM, H.-J., REED, G., Eds.) Vol. 6a, pp. 421–457, Weinheim: Verlag Chemie.

CRUEGER, W., CRUEGER, A. (1989), *Biotechnology: A Textbook of Industrial Microbiology,* 2nd Ed., Sunderland: Sinauer.

DAWES, I. W., CARTER, L. A. (1974), Nitrosoguanidine mutagenesis during nuclear and mitochondrial gene replication, *Nature* **250**, 709–712.

DE MARINI, D. M., PHAM, H. N., KATZ, A. J., BROCKMAN, H. E. (1984), Relationships between structures and mutagenic potencies of 16 heterocyclic nitrogen mustards (ICR compounds) in *Salmonella typhimurium, Mutat. Res.* **136**, 185–199.

DEMEREC, M. (1937), Frequency of spontaneous mutations in certain stocks of *Drosophila melanogaster, Genetics* **22**, 469–478.

DEN ENGELSE, L. A., DE GRAAF, A., DE BRIJ, R. J., MENKVELD, G. J. (1987), O^2- and O^4-Ethylthymine and the ethyl phosphotriester dTp(Et)dT are highly persistent DNA modifications in slowly dividing tissues of the ethylnitrosourea treated rats, *Carcinogenesis* **2**, 751–757.

DHILLON, E. K. S., DHILLON, T. S. (1974), N-methyl-N′-nitro-N-nitrosoguanidine and hydroxylamine induced mutants in the rII region of the phage T4, *Mutat. Res.* **22**, 223–233.

DOETSCH, P. W., CUNNINGHAM, R. P. (1990), The enzymology of apurinic/apyrimidinic endonucleases, *Mutat. Res.* **236**, 173–201.

DÖRING, H. P., STARLINGER, P. (1984), Barbara McClintock's controlling elements: now at the DNA level, *Cell* **39**, 253–259.

DRAKE, J. W. (1969), Comparative rates of spontaneous mutation, *Nature* **221**, 1122.

DRAKE, J. W. (1970), *The Molecular Basis of Mutation,* San Francisco: Holden-Day.

DRAKE, J. W. (1989), Mechanism of mutagenesis, *Environ. Mol. Mutagen.* **14** (S16), 11–15.

DRAKE, J. W., ALLEN, E. F. (1968), Antimutagenic DNA polymerases of bacteriophage T4, *Cold Spring Harbor Symp. Quant. Biol.* **33**, 339–344.

DRAKE, J. W., BALTZ, R. H. (1976), The biochemistry of mutagenesis, *Annu. Rev. Biochem.* **45**, 11–37.

DULBECCO, R. (1949), Reactivation of ultraviolet-inactivated bacteriophage by visible light, *Nature* **163**, 949–950.

EDER, E., FAVRE, A., DEININGER, C., HAHN, H., KÜTT, W. (1989), Induction of SOS repair by monofunctional methanesulfonates in various *Escherichia coli* strains. Structure–activity relationships in comparison with mutagenicity in *Salmonella typhimurium, Mutagenesis* **4**, 179–186.

EKER, A. P. M., HESSELS, J. K. C., VAN DER VELDE, J. (1988), Photoreactivating enzyme from the green algae *Scenedesmus acutus.* Evidence for the presence of two different flavin chromophores, *Biochemistry* **27**, 1758–1765.

FAELEN, M., RÉSIBOIS, A., TOUSSAINT, A. (1979), Mini-Mu, an insertion element derived from temperente phage Mu-1, *Cold Spring Harbor Symp. Quant. Biol.* **43**, 1169–77.

FERGUSON, L. R., DENNY, W. A. (1990), Frameshift mutagenesis by acridines and other reversibly-binding DNA ligands, *Mutagenesis* **5**, 529–540.

FERSHT, A.R., KNILL-JONES, J. W., TSUI, W.-C. (1982), Kinetic basis of spontaneous misinsertion frequencies, proofreading specificities and cost of proofreading by DNA polymerases of *Escherichia coli, J. Mol. Biol.* **156,** 37–51.

FRAM, R. J., SULLIVAN, J., MARINUS, M. G. (1986), Mutagenesis and repair of DNA damage caused by nitrogen mustard, N,N′-bis(2-chloroethyl)-N-nitrosourea (BCNU), streptozotin, and mitomycin C in *E. coli, Mutat. Res.* **166,** 229–242.

FRANKLIN, W. A., LINDAHL, T. (1988), DNA deoxyribophosphodiesterase, *EMBO J.* **7,** 3617–3622.

FREESE, E. (1959), The difference between spontaneous and base analogue induced mutations of phage T4, *Proc. Natl. Acad. Sci. USA* **45,** 622–633.

FRIEDBERG, E. C. (1985), *DNA Repair,* New York: Freeman.

FUJITA, H., SUZUKI, K. (1978), Effect of near-UV light on *Escherichia coli* in the presence of 8-methoxypsoralen: Wavelength dependency of killing, induction of prophage, and mutation, *J. Bacteriol.* **135,** 345–362.

FUKUNAGA, M., MIZUGUCHI, Y., YIELDING, L. W., YIELDING, K. L. (1984), Petite induction in *Saccharomyces cerevisiae* by ethidium analogs. Action on mitochondrial genome, *Mutat. Res.* **127,** 15–21.

FURUY, A., TOSHIBA, S. (1982), Production of 5′-IMP by mutants of *Brevibacterium ammoniagenes,* in: *Genetics of Industrial Microorganisms, GIM 82* (IKEDA, Y., BEPPU, T., Eds.), pp. 259–262, Tokyo: Kodansha.

GEBHART, E. (1974), Antimutagens, data and problems, *Humangenetik* **24,** 1–32.

GERCHMAN, L. L., LUDLUM, D. B. (1973), The properties of O^6-methylguanine in templates for RNA polymerase, *Biochim. Biophys. Acta* **308,** 310–316.

GICHNER, T., VELEMINSKY, J. (1982), Genetic effects of N-methyl-N′-nitro-N-nitrosoguanidine and its homologs, *Mutat. Res.* **99,** 129–242.

GLASS, B., RITTERHOFF, R. K. (1956), Spontaneous mutation rates at specific loci in *Drosophila* males and females, *Science* **124,** 314–315.

GLICKMAN, B.W., RIETVELD,K., AARON, C. S. (1980), γ-Ray induced mutational spectrum in the *lacI* gene of *Escherichia coli;* Comparison of induced and spontaneous spectra at the molecular level, *Mutat. Res.* **69,** 1–12.

GLICKMAN, B. W., SCHAAPER, R. M., HASELTINE, W. A., DUNN, R. L., BRASH, D. E. (1986), The C-C (6-4) UV photoproduct is mutagenic in *Escherichia coli, Proc. Natl. Acad. Sci. USA* **83,** 6945–6949.

GOLDSTEIN, A., SMOOT, J. S. (1955), A strain of *Escherichia coli* with an unusually high rate of auxotrophic mutation, *J. Bacteriol.* **70,** 588–595.

GOODMAN, M. F., HOPKINS, R., GORE, W. C. (1977), 2-Aminopurine-induced mutagenesis in T4 bacteriophage: a model relating mutation frequency to 2-aminopurine incorporation in DNA, *Proc. Natl. Acad. Sci. USA* **74,** 4806–4810.

GOODSON, M., ROWBURY, R. J. (1990), Habituation to alkali and increased UV-resistance in DNA repair-proficient and -deficient strains of *Escherichia coli* grown at pH 9.0, *Lett. Appl. Microbiol.* **11,** 123–125.

GORDON, A. J. E., SCHY, W. E., GLICKMAN, B. W. (1990), Non-phenotypic selection of N-methyl-N′-nitro-N-nitrosoguanidine-directed mutation at a predicted hotspot site, *Mutat. Res.* **243,** 145–149.

GORDON, A. J. E., HALLIDAY, J. A., HORSFALL, M. J., GLICKMAN, B. W. (1991), Spontaneous and 9-aminoacridine-induced frameshift mutagenesis: Second-site frameshift mutation within the N-terminal region of the *lacI* gene of *Escherichia coli, Mol. Gen. Genet.* **227,** 160-164.

GREEN, M. M. (1976), Mutable and mutator loci, in: *The Genetics and Biology of Drosophila* (ASHBURNER, M., NOVITSKI, E., Eds.), pp. 929–946, London: Academic Press.

GREEN, R. R., DRAKE, J. W. (1974), Misrepair mutagenesis of bacteriophage T4, *Genetics* **78,** 81–89.

GRILLEY, M., HOLMES, J., YASHAR, B., MODRICH, P. (1990), Mechanisms of DNA-mismatch correction, *Mutat. Res.* **236,** 253–267.

GROSSMAN, L., YEUNG, A. T. (1990), The UvrABC endonuclease system of *E. coli* – A view from Baltimore, *Mutat. Res.* **236,** 213–221.

GUEROLA, N., INGRAHAM, J. L., CERDÁ-OLMEDO, E. (1971), Induction of closely linked multiple mutations by nitrosoguanidine, *Nature New Biol.* **230,** 122–125.

GUTZ, H. (1961), Distribution of X-ray and nitrous acid-induced mutations in the genetic fine structure of the *ad7* locus of *Schizosaccharomyces pombe, Nature* **191,** 1124–1125.

HADDEN, C. T., FOOTE, R. S., MITRA, S. (1983), Adaptive response of *Bacillus subtilis* to N-methyl-N′-nitro-N-nitrosoguanidine, *J. Bacteriol.* **153,** 756–762.

HAHN, F. E. (1976), Antimutagens and the prevention of chromosomal mutations to drug resistance, *Antibiot. Chemother.* **20,** 112–132.

HARMAN, D. (1981), The aging process, *Proc. Natl. Acad. Sci. USA* **78,** 7124–7128.

HASTINGS, P. J., QUATE, S.-K., VON BORSTEL, R. C. (1976), Spontaneous mutation by mutagenic repair of spontaneous lesions in DNA, *Nature* **264**, 719–722.

HIXON, S., MOUSTACCHI, E. (1978), The fate of yeast mitochondrial DNA after ultraviolet irradiation. I – Degradation during post-UV dark liquid holding in non-nutrient medium, *Biochem. Biophys. Res. Commun.* **84**, 288–296.

HOIJMAKERS, J. H. J., EKER, A. P. M., WOOD, R. D., ROBINS, P. (1990), Use of *in vivo* and *in vitro* assays for the characterization of mammalian excision repair and isolation of repair proteins, *Mutat. Res.* **236**, 223–238.

HOLLAND, J., HOLLAND, I. B., AHMAD, S. I. (1991), DNA damage by 8-methoxypsoralen plus near ultraviolet light (PUVA) and its repair in *Escherichia coli:* Genetic analysis, *Mutat. Res.* **254**, 289–298.

HORIUCHI, T., MAKI, H., SEKIGUCHI, M. (1989), Mutators and fidelity of DNA replication, *Bull. Inst. Pasteur* **87**, 309–336.

HORSFALL, M. J., GORDON, A. J. E., BURNS, P. A., ZIELENSKA, M., VAN DER VLIET, G. M. E., GLICKMAN, B. W. (1990), Mutational specificity of alkylating agents and the influence of DNA repair, *Environ. Mol. Mutagen.* **15**, 107–122.

HOWARD-FLANDERS, P. (1973), DNA repair and recombination, *Br. Med. Bull.* **29**, 226–235.

HUTCHINSON, F. (1985), Chemical changes induced in DNA by ionizing radiation, *Prog. Nucleic Acid Res. Mol. Biol.* **32**, 115–154.

IMADA, M., INOYE, M., EDA, M., TSUGITA, A. (1970), Frameshift mutation in the lysozyme gene of bacteriophage T4: Demonstration of the insertion of four bases and the preferential occurrence of base addition in acridine mutagenesis, *J. Mol. Biol.* **54**, 199–217.

ISHIKAWA, T. (1967), Mutagenic specificity at the *ad-8* locus in *Neurospora crassa, Jpn. J. Genet.* **42**, 43–50.

JACKSON, H. (1964), The effects of alkylating agents on fertility, *Br. Med. Bull.* **20**, 107–114.

JACOBSON, G. K. (1981), Mutations, in: *Biotechnology* (REHM, H.-J., REED, G., Eds.) Vol. 1, pp. 279–329, Weinheim: Verlag Chemie.

JANION, C. (1982), Effect of bacterial host repair systems on the viability of hydroxylamine and methyl methanesulfonate treated T4 and λ bacteriophages, *Mol. Gen. Genet.* **186**, 419–426.

JANION, C., SHUGAR, D. (1965), Reaction of hydroxylamine with 5-substituted cytosines, *Biochem. Biophys. Res. Commun.* **18**, 617–622.

JOHNSON, J. L., HAMM-ALVAREZ, S., PAYNE, G., SANCAR, G. B., RAJAGOPALAN, U. V., SANCAR, A. (1988), Identification of the second chromophore of *Escherichia coli* and yeast DNA photolyase as 5,10-methenyltetrahydrofolate, *Proc. Natl. Acad. Sci. USA* **85**, 2046–2050.

KARRAN, P., LINDAHL, T. (1979), Adaptive response to alkylating agents involves alteration *in situ* of O^6-methylguanine residues in DNA, *Nature* **280**, 76–77.

KARRAN, P., STEPHENSON, C., CAIRNS-SMITH, S., MACPHERSON, P. (1990), Regulation of O^6-methylguanine-DNA methyltransferase expression in the Burkitt's lymphoma cell line Raji, *Mutat. Res.* **233**, 23–30.

KATO, T. (1977), Effects of chloramphenicol and caffeine on postreplication repair in *uvrA*$^-$ *umuC*$^-$ and *uvrA*$^-$ *recoli* K-12, *Mol. Gen. Genet.* **156**, 115–120.

KELNER, A. (1949), Photoreactivation of ultraviolet-irradiated *Escherichia coli* with special reference to the dose-reduction principle and to ultraviolet-induced mutation, *J. Bacteriol.* **58**, 511–522.

KEOHAVONG, P., LIU, V. F., THILLY, W. G. (1991), Analysis of point mutations induced by ultraviolet light in human cells, *Mutat. Res.* **249**, 147–159.

KLEBANOFF, S. J. (1988), Phagocytic cells: Products of oxygen metabolism, in: *Inflammation: Basic Principles and Clinical Correlates* (GALLIN, J. I., GOLDSTEIN, I. M., SNYDERMAN, R., Eds.), New York: Raven.

KLECKNER, N. (1981), Transposable elements in prokaryotes, *Ann. Rev. Genet.* **15**, 341–404.

KOEHL, P., BURNOUF, D., FUCHS, R. P. P. (1990), Mutagenesis induced by a single acetylaminofluorene adduct within the NarI site is position dependent, *Environ, Sci. Res.* **40**, 105–112.

KOZUBEK, S., KRASAVIN, E. A., LIN, N., SOSKA, J., DRASIL, V., AMIRTAYEV, K. G., TOKAROVA, B., BONEV, M. (1989), Introduction of the SOS response in *Escherichia coli* by heavy ions, *Mutat. Res.* **215**, 49–53.

KOZUBEK, S., OGIEVETSKAYA, M. M., KRASAVIN, E. A., DRASIL, V., SOSKA, J. (1990), Investigation of the SOS response of *Escherichia coli* after γ-irradiation by means of the SOS chromotest, *Mutat. Res.* **230**, 1–7.

KUMARESAN, K.R., JAYARANAN, R. (1990), The *sir* locus of *Escherichia coli:* a gene involved in SOS-independent repair of mitomycinC-induced DNA damage, *Mutat. Res.* **235**, 85–92.

KUNINAKA, A. (1986), Nucleic acids, nucleotides and related compounds, in: *Biotechnology* (REHM, H.-J., REED, G., Eds.) Vol. 4, pp. 71–114, Weinheim: VCH.

LARSON, K., SHAM, J., SHENKAR, R., STRAUSS, B. (1985), Methylation-induced blocks to *in vitro* DNA replication, *Mutat. Res.* **150**, 77–84.

LAVAL, J., BOITEUX, S., O'CONNOR, T. R. (1990), Physiological properties and repair of apurinic/

apyrimidinic sites and imidazole ring-opened guanines in DNA, *Mutat. Res.* **233,** 73–79.

LAWLEY, P. D. (1984), Carcinogenesis by alkylating agents, in: *Chemical Carcinogenesis* (SEARLE, C. E., Ed.), 2nd Ed., *ACS Monograph* **182,** pp. 325–484, Washington, DC: American Chemical Society.

LIEB, M. (1987), Bacterial genes *mutL, mutS* and *dcm* participate in repair of mismatches at 5-methylcytosine sites, *J. Bacteriol.* **169,** 5241–5246.

LINDAHL, T. (1990), Repair of intrinsic DNA lesions, *Mutat. Res.* **238,** 305–311.

LINDAHL, T., SEDGWICK, B., SEKIGUCHI, M., NAKABEPPU, Y. (1988), Regulation and expression of the adaptive response to alkylating agents, *Annu. Rev. Biochem.* **57,** 133–157.

LITTLE, J. W., MOUNT, D. W. (1982), The SOS regulatory system of *Escherichia coli, Cell* **29,** 11–22.

LJUNGQUIST, S., ANDERSSON, A., LINDAHL, T. (1974), A mammalian endonuclease specific for apurinic sites in double-stranded deoxribonucleic acid, *J. Biol. Chem.* **249,** 1536–1540.

LOEB, L. A., CHENG, K. C. (1990), Errors in DNA synthesis: A source of spontaneous mutations, *Mutat. Res.* **238,** 297–304.

LOPRIENO, N. (1967), Intragenic mapping of chemically induced *ad-7* mutants of *Schizosaccharomyces pombe, J. Bacteriol.* **94,** 1162–1165.

LOVELESS, A. (1969), Possible relevance of O^6-alkylation of deoxyguanosine to the mutagenicity and carcinogenicity of nitrosamines and nitrosamides, *Nature* **223,** 206–207.

LYDIATE, D. J., IKEDA, H., HOPWOOD, D. A. (1986), A 2.6 kb DNA sequence of *Streptomyces coelicolor* A3 (2) which functions as a transposable element, *Mol. Gen. Genet.* **203,** 79–88.

MALIGA, P. (1980), Isolation, characterization, and utilization of mutant cell lines in higher plants, in: *International Review of Cytology* (VASIK, I. K. Ed.), Suppl. 11A, pp. 225–248, New York: Academic Press.

MALIK, V. S. (1979), Genetics of applied microbiology, *Adv. Genet.* **20,** 37–126.

MALLING, H. V., DE SERRES, F. J. (1968), Identification of genetic alterations induced by ethyl methanesulfonate in *Neurospora crassa, Mutat. Res.* **6,** 181–193.

MARGISON, G. P., COOPER, D. P., POTTER, P. M. (1990), The *E. coli ogt* gene, *Mutat. Res.* **233,** 15–21.

MCCLINTOCK, B. (1951), Chromosome organization and genetic expression, *Cold Spring Harb. Symp. Quant. Biol.* **16,** 13–43.

MISURACA, F., RAMPOLLA, D., GRIMAUDO, S. (1991), Identification and cloning of a *umu* locus in *Streptomyces coelicolor* A3 (2), *Mutat. Res.* **262,** 183–188.

MODRICH, P. (1989), Methyl-directed DNA mismatch correction, *J. Biol. Chem.* **264,** 6597–6600.

MULLER, H. J. (1927), Artificial transmutations of the gene, *Science* **66,** 84–87.

NEVERS, P., SAEDLER, H. (1977), Transposable genetic elements as agents of gene instability and chromosomal rearrangements, *Nature* **268,** 109–115.

O'CONNOR, T. R., LAVAL, J. (1989), Physical association of the 2,6-diamino-4-hydroxy-5-N-formamidopyrimidine-DNA glycosylase of *Escherichia coli* and an activity nicking DNA at apurinic/apyrimidinic sites, *Proc. Natl. Acad. Sci. USA* **86,** 5222–5228.

O'CONNOR, T. R., BOITEUX, S., LAVAL, J. (1988), Ring-opened 7-methylguanine residues in DNA are a block to *in vitro* DNA synthesis, *Nucleic Acids Res.* **16,** 5879–5894.

OSTERMAN-GOLKAR, S., EHRENBERG, L., WACHTMEISTER, C. A. (1970), Reaction kinetics of biological action in barley of monofunctional methanesulfonic esters, *Radiat. Bot.* **10,** 303–327.

PEGG, A. E., DOLAN, M. E., SCICCHITANO, D., MORIMOTO, K. (1985), Studies of the repair of O^6-alkylguanine and O^4-alkylthymine in DNA by alkyltransferase from mammalian cells and bacteria, *Environ. Health Perspect.* **62,** 109–114.

PERERA, F. P. (1988), The significance of DNA and protein adducts in human biomonitoring studies, *Mutat. Res.* **205,** 255–269.

PFEIFER, F., FRIEDMAN, J., BOYER, H. W., BETLACH, M. (1984), Characterisation of insertions affecting the expression of the bacterial opsin gene in *Halobacterium halobium, Nucleic Acids Res.* **12,** 2489–2497.

PICKSLEY, S. M., ATTFIELD, P. V., LLOYD, R. G. (1984), Repair of DNA double-strand breaks in *Escherichia coli* K12 requires a functional *recN* product, *Mol. Gen. Genet.* **195,** 267–274.

POTTER, P. M., WILKINSON, M. C., FITTON, J., CARR, F. J., BRENNAUD, J., COOPER, D. P., MARGISON, G. P. (1987), Characterization and nucleotide sequence of *ogt,* the O^6-alkylguanine-DNA alkyltransferase gene of *E. coli, Nucleic Acids Res.* **15,** 9177–9193.

RADMAN, M. (1975), SOS repair hypothesis: Phenomenology of an inducible DNA repair which is accompanied by mutagenesis, in: *Molecular Mechanisms for Repair of DNA* (HANAWALT, P. C., SETLOW, R. B., Eds.), pp. 355–367, New York: Plenum.

RADMAN, M., WAGNER, R. (1986), Mismatch repair in *Escherichia coli, Annu. Rev. Genet.* **20,** 523–538.

RAHN, R. O., PATRICK, M. H. (1976), Photochemistry of DNA; secondary structure, photosensibilisation, base substitution, and exogenous molecules, in: *Photochemistry and Photobiology of Nucleic Acids* (WANG, S. Y., Ed.), Vol. 2, pp. 97–145, New York: Academic Press.

RANT, C., SIMPSON, W. L. (1955), The effect of X-rays and ultraviolet light of different wavelengths on the production of cytochrome-deficient yeast, *Arch. Biochem. Biophys.* **57**, 218–228.

REBECK, G. W., SMITH, C. M., GOAD, D. L., SAMPSON, L. (1989), Characterization of the major DNA repair methyltransferase activity in unadapted *Escherichia coli* and identification of a similar activity in *Salmonella typhimurium, J. Bacteriol.* **171**, 4563–4568.

RHOADES, M. M. (1941), The genetic control of mutability in maize, *Cold Spring Harb. Symp. Quant. Biol.* **9**, 138–144.

RIAZUDDIN, S., LINDAHL, T. (1978), Properties of 3-methyladenine-DNA glycosylase from *Escherichia coli, Biochemistry* **17**, 2110–2118.

RICHARDSON, F. C., RICHARDSON, K. K. (1990), Sequence-dependent formation of alkyl DNA adducts: a review of methods, results, and biological correlates, *Mutat. Res.* **233**, 127–138.

ROBINS, P., CAIRNS, J. (1979), Quantitation of the adaptive response to alkylation agents, *Nature* **280**, 74–76.

ROSSI, S. C., CONRAD, M., VOIGT, J. M., TOPAL, M. D. (1989), Excision repair of O^6-methylguanine synthesized at the rat H-ras N-methyl-N-nitrosourea activation site and introduced into *Escherichia coli, Carcinogenesis* **10**, 373–377.

ROTH, J. (1974), Frameshift mutations, *Annu. Rev. Genet.* **8**, 319–346.

ROTHMAN, R. H., KATO, T., CLARK, A. J. (1975), The beginning of an investigation of the role of *recF* in the pathways of metabolism of ultraviolet-irradiated DNA in *Escherichia coli,* in: *Molecular Mechanisms for Repair* (HANAWALT, P. C., SETLOW, R. B., Eds.), pp. 283–291, New York: Plenum Publ. Corp.

ROWLANDS, R. T. (1984), Industrial strain improvement: mutagenesis and random screening procedures, *Enzyme Microb. Technol.* **6**, 3–10.

RUPP, W. D., WILDE, C. E., RENO, D. L., HOWARD-FLANDERS, P. (1971), Exchanges between DNA strands in ultraviolet-irradiated *Escherichia coli, J. Mol. Biol.* **61**, 25–44.

RYDBERG, B. (1977), Bromouracil mutagenesis in *Escherichia coli.* Evidence for involvement of mismatch repair, *Mol. Gen. Genet.* **152**, 19–28.

SAFFHILL, R., MARGISON, G. P., O'CONNOR, P. J. (1985), Mechanisms of carcinogenesis induced by alkylating agents, *Biochim. Biophys. Acta* **823**, 111–145.

SAGE, E., BREDBERG, A. (1991), Damage distribution and mutation spectrum: the case of 8-methoxypsoralen plus UVA in mammalian cells, *Mutat. Res.* **263**, 217–222.

SAKUMI, K., SEKIGUCHI, M. (1990), Structure and functions of DNA glycosylases, *Mutat. Res.* **236**, 161–172.

SAMSON, L., THOMALE, J., RAJEWSKI, M. (1988), Alternative pathways for the *in vivo* repair of O^6-alkylguanine and O^4-alkylthymine in *E. coli.* The adaptive response and nucleotide excision repair, *EMBO J.* **7**, 2261–2267.

SANCAR, G. B. (1990), DNA photolyases: Physical properties, action mechanism, and roles in dark repair, *Mutat. Res.* **236**, 147–160.

SANCAR, A., SANCAR, G. B. (1988), DNA repair enzymes, *Annu. Rev. Biochem.* **57**, 29–67.

SARGENTINI, N. J., SMITH, K. C. (1985), Spontaneous mutagenesis: The role of DNA repair, replication, and recombination, *Mutat. Res.* **154**, 1–27.

SAUNDERS, V. A., SAUNDERS, J. R. (1987), *Microbial Genetics Applied to Biotechnology,* pp. 3–13, London: Croom Helm.

SCHAAPER, R. M., DUNN, R. L. (1987), Spectra of spontaneous mutations in *Escherichia coli* strains defective in mismatch correction: The nature of *in vivo* DNA replication errors, *Proc. Natl. Acad. Sci. USA* **84**, 6220–6224.

SCHAAPER, R. M., RADMAN, M. (1989), The extreme mutator effect of *Escherichia coli mutD5* results from saturation of mismatch repair by excessive DNA replication errors, *EMBO J.* **8**, 3511–3516.

SCHAAPER, R. M., DUNN, R. L., GLICKMAN, B. W. (1987), Mechanisms of ultraviolet-induced mutation. Mutational spectra in the *Escherichia coli lacI* gene for a wild-type and an excision-repair-deficient strain, *J. Mol. Biol.* **198**, 187–202.

SCHLAGER, G., DICKIE, M. M. (1967), Spontaneous mutations and mutation rates in the house mouse, *Genetics* **57**, 319–330.

SCHNAAR, R. L., MUZYCZKA, N., BESSMAN, M. J. (1973), Utilization of aminopurine deoxynucleotide triphosphate by mutator, antimutator and wild-type DNA polymerases of bacteriophage T4, *Genetics* **73**, 137–140.

SCHUSTER, H. (1960), The reaction of nitrous acid with deoxyribonucleic acid, *Biochem. Biophys. Res. Commun.* **2**, 320–323.

SEARS, B. B., SOKALSKI, M. B. (1991), The *Oenothera* plastome mutator: effect of UV irradiation and nitroso-methyl urea on mutation frequencies, *Mol. Gen. Genet.* **229**, 245–252.

SEGA, G. A. (1984), A review of the genetic effects

of ethyl methanesulfonate, *Mutat. Res.* **134,** 113-142.

SEKIGUCHI, M., NAKABEPPU, Y. (1987), Adaptive response: induced synthesis of DNA repair enzymes by alkylating agents, *Trends Genet.* **3,** 51-54.

SELBY, C. P., SANCAR, A. (1990), Molecular mechanisms of DNA repair inhibition by caffeine, *Proc. Natl. Acad. Sci. USA* **87,** 3522–3525.

SETLOW, J. K., SETLOW, R. B. (1963), Nature of the photoreactivable ultraviolet lesion in deoxyribonucleic acid, *Nature* **197,** 560–562.

SHINAGAWA, H., IWASAKI, H., KATO, T., NAKATA, A. (1988), RecA protein-dependent cleavage of UmuD protein and SOS mutagenesis, *Proc. Natl. Acad. Sci. USA* **85,** 1806–1810.

SHWARTZ, H., SHAVITT, O., LIVNEH, Z. (1988), The role of exonucleolytic processing and polymerase-DNA association in bypass of lesions during replication *in vitro, J. Biol. Chem.* **263,** 18277–18285.

SIEGEL, E. C., BRYSON, V. (1964), Selection of resistant strains of *Escherichia coli* by antibiotics and antibacterial agents: role of normal and mutator strains, *Antimicrob. Agents Chemother.,* 629–634.

SIEGEL, E. C., BRYSON, V. (1967), Mutator gene of *Escherichia coli* B, *J. Bacteriol.* **94,** 38–47.

SINGER, B. (1986), O-Alkyl pyrimidines in mutagenesis and carcinogenesis: occurrence and significance, *Cancer Res.* **46,** 4879–4885.

SINGER, B., FRAENKEL-CONRAT, H. (1969), The role of conformation in chemical mutagensis, *Progr. Nucleic Acid Res. Mol. Biol.* **9,** 1–29.

SINGER, B., GRUNBERG, D. (1983), *Molecular Biology of Mutagens and Carcinogens,* New York: Plenum Press.

SINHA, U., CHATTOO, B. B. (1975), Biological effects of N-methyl-N′-nitro-N-nitrosoguanidine, *J. Sci. Ind. Res.* **34,** 499–505.

SKAAR, P. D. (1956), A binary mutability system in *Escherichia coli, Proc. Natl. Acad. Sci. USA* **42,** 245–249.

SLEDZIEWSKA-GOJSKA, E., JANION, C. (1982), Effect of proofreading and *dam*-instructed mismatch repair systems on N^4-hydroxycytidine-induced mutagenesis, *Mol. Gen. Genet.* **186,** 411-418.

SLITALY, S. N., LITTLE, J. W. (1987), Lysine-156 and serine-119 are needed for *lexA* repressor cleavage: a possible mechanism, *Proc. Natl. Acad. Sci. USA* **84,** 3987-3991.

SMITH, K. C. (1978), Multiple pathways of DNA repair in bacteria and their rates in mutagenesis, *Photochem. Photobiol.* **28,** 121–129.

SMITH, C. A. (1988), Repair of DNA containing furocoumarin adducts, in: *Psoralen DNA Photobiology* (GASPARRO, F. P., Ed.), Vol. II, pp. 87-116, Boca Raton: CRC Press.

SPEIT, G., GARCOV, S., HAUPTER, S., KÖBERLE, B. (1990), Genetic effects of 2-aminopurine in mammalian cells, *Mutagenesis* **5,** 185–190.

SPEYER, J. F. (1965), Mutagenic DNA polymerase, *Biochem. Biophys. Res. Commun.* **21,** 6–8.

STADLER, L. J. (1942), Some observations on gene variability and spontaneous mutation, *The Spragg Memorial Lectures,* Third Series, Michigan State College.

STREISINGER, G., OKADA, Y., EMRICH, J., NEWTON, J., TSUGITA, A., TERZHAGI, E., INOUYE, M. (1966), Frameshift mutations and the genetic code, *Cold Spring Harbor Symp. Quant. Biol.* **31,** 77–84.

SUSKIND, S. R., KUREK, L. I. (1959), On a mechanism of suppressor gene regulation of tryptophan synthetase activity in *Neurospora crassa, Proc. Natl. Acad. Sci. USA* **45,** 193–196.

SWENSON, D. H. (1983), Significance of electrophilic reactivity and especially DNA alkylation in carcinogenesis and mutagenesis, in: *Developments in the Science and Practice of Toxicology* (HAYES, A. W., SCHNELL, R. C., MIYA, T. S., Eds.), pp. 247–254, Amsterdam: Elsevier.

TAKANO, K., NAKAMURA, T., SEKIGUCHI, M. (1991), Roles of two types of O^6-methylguanine-DNA methyl-transferases in DNA repair, *Mutat. Res.* **254,** 37–44.

TANAKA, Y., YOH, M., TAKEDA, Y., MIWATANI, T. (1979), Induction of mutation in *Escherichia coli* by freeze-drying, *Appl. Environ. Microbiol.* **37,** 369–372.

TEO, F., SEDGWICK, B., KILPATRICK, M. W., MCCARTHY, T. V., LINDAHL, T. (1986), The intracellular signal for induction of resistance to alkylating agents in *E. coli, Cell* **45,** 315–334.

THOMAS, S. M., MACPHEE, D. G. (1985), Frameshift mutagenesis by 9-aminoacridine and ICR 191 in *Escherichia coli:* effect of *uvrB, recA* and *lexA* mutations and of plasmid pKM101, *Mutat. Res.* **151,** 49–56.

TODO, T., YONEI, S. (1988), Inhibitory effect of membrane-binding drugs on excision repair of DNA damage in UV-irradiated *Escherichia coli, Mutat. Res.* **112,** 97–107.

TOPAL, M. D., FRESCO, J. R. (1976), Complementary base pairing and the origin of substitution mutations, *Nature* **263,** 285–289.

TOUSSAINT, A., RÉSIBOIS, A. (1983), Phage Mu: transposition as a life-style, in: *Mobile Genetic Elements* (SHAOIRO, J. A., Ed.), pp. 105–158, New York: Academic Press.

TOWNSEND, M. E., WRIGHT, H. M., HOPWOOD, D. A. (1971), Efficient mutagenesis by near ultraviolet light in the presence of 8-methoxypsoralen in *Streptomyces, J. Appl. Bacteriol.* **34,** 799–801.

TREFFERS, H. P., SPINELLI, V., BELSER, N. O. (1954), A factor (or mutator gene) influencing mutation rates in *Escherichia coli, Proc. Natl. Acad. Sci. USA* **40,** 1064–1071.

VAN HOUTEN, B., GAMPER, H., HOLBROOK, S. R., HEARST, J. E., SANCAR, A. (1986), Action mechanism of UvrABC excision nuclease on a DNA substrate containing a psoralen crosslink at a defined position, *Proc. Natl. Acad. Sci. USA* **83,** 8077–8081.

VERLY, W. G., RASSART, E. (1975), Purification of *E. coli* endonuclease specific for apurinic sites in DNA, *J. Biol. Chem.* **250,** 8214–8219.

VERSCHAEVE, L., KIRSCH-VOLDERS, M. (1990), Mutagenicity of ethyleneimine, *Mutat. Res.* **238,** 39–55.

VOGEL, F., RATHENBERG, R. (1975), Spontaneous mutation in man, *Adv. Hum. Genet.* **5,** 223–318.

VOLKERT, M. R. (1988), Adaptive response in *Escherichia coli* to alkylation damage, *Environ. Mol. Mutagen.* **11,** 241–255.

VOLKERT, M. R., NGUYEN, D. C. (1984), Induction of specific *Escherichia coli* genes by sublethal treatments with alkylating agents, *Proc. Natl. Acad. Sci. USA* **81,** 4110–4114.

WACKERNAGEL, W., WINKLER, U. (1971), A mutation in *E. coli* enhancing the UV-mutability of phage λ but not of its infectious DNA in a spheroplast assay, *Mol. Gen. Genet.* **114,** 68–79.

WALKER, G. C. (1984), Mutagenesis and inducible responses to deoxyribonucleic acid damage in *Escherichia coli, Microbiol. Rev.* **48,** 60–93.

WALLACE, S. S. (1988), AP endonucleases and DNA glycosylases that recognize oxidative DNA damage, *Environ. Mol. Mutagen.* **12,** 431–477.

WANG, T. V., SMITH, K. C. (1981), Effect of *recB21, uvrD3, lexA101* and *recF143* mutations on ultraviolet radiation sensitivity and genetic recombination in *ΔuvrB* strains of *Escherichia coli, Mol. Gen. Genet.* **183,** 27–44.

WANG, T. V., SMITH, K. C. (1983), Mechanisms for the *recF*-dependent pathway of postreplication repair in UV-irradiated *Escherichia coli uvrB, J. Bacteriol.* **156,** 119–124.

WANG, T. V., SMITH, K. C. (1985), Mechanism of *sbcB*-suppression of the *recBC*-deficiency in postreplication repair in UV-irradiated *Escherichia coli* K12, *Mol. Gen. Genet.* **201,** 186–191.

WANG, Z.-Q., SAIGUSA, S., SASAKI, M. S. (1991), Adaptive response to chromosome damage in cultured human lymphocytes primed with low doses of X-rays, *Mutat. Res.* **246,** 179–186.

WEIGLE, J. J. (1953), Induction of mutation in a bacterial virus, *Proc. Natl. Acad. Sci. USA* **39,** 628–638.

WEISS, B., GROSSMAN, L. (1987), Photodiesterases involved in DNA repair, *Adv. Enzymol.* **60,** 1–34.

WHITTIER, R. F., CHASE, J. W. (1981), DNA repair in *E. coli* strains deficient in single-strand DNA binding protein, *Mol. Gen. Genet.* **183,** 341–347.

WIDHOLM, J. M. (1977), Isolation of biochemical mutants of cultured plant cells, in: *Molecular Genetic Modification of Eukaryotes* (RUBENSTEIN, I., PHILLIPS, R. L., GREE, C. E., DESNICK, R., Eds.), pp. 57–65, New York: Academic Press.

WIENAND, U., SAEDLER, H. (1988), Plant transposable elements: unique structures for gene tagging and gene cloning, in: *Plant DNA Infectious Agents* (HOHN, T., SCHELL, J., Eds.), pp. 205–228, Wien: Springer.

WILKINS, M. F. H., STOKES, A. R., WILSON, H. R. (1953), Molecular structure of deoxypentose nucleic acid, *Nature* **171,** 738–740.

WITKIN, E. M. (1976), Ultraviolet mutagenesis and inducible DNA repair in *Escherichia coli, Bacteriol. Rev.* **40,** 868–907.

WOJCIK, A., TUSCHL, H. (1990), Indications of an adaptive response in C57BL mice pre-exposed *in vivo* to low doses of ionizing radiation, *Mutat. Res.* **243,** 67–73.

WOLFF, S., WIENCKE, J. K. (1988), The induction of chromosome repair enzymes by 1cGy (1 rad) of X-rays to human lymphocytes, *Environ. Mol. Mutagen.* **2** (Suppl. 11), 114 (abstract).

YANOFSKY, C., COX, E. C., HORN, V. (1966), The unusual mutagenic specificity of an *E. coli* mutator gene, *Proc. Natl. Acad. Sci. USA* **55,** 274–281.

YOON, C., PRIVE, C. G., GOODSELL, D. S., DICKERSON, R. E. (1988), Binding of 9-aminoacridine to bulged base DNA oligomers from a frameshift hot spot, *Biochemistry* **27,** 8904–8914.

YOUNGS, D. A., SMITH, K. C. (1976), Genetic control of multiple pathways of post-replicational repair in *uvrB* strains of *Escherichia coli* K-12, *J. 2acteriol.* **125,** 102–110.

ZAMENHOF, P. J. (1966), A genetic locus responsible for generalized high mutability in *Escherichia coli, Proc. Natl. Acad. Sci. USA* **56,** 845–852.

ZELL, R., FRITZ, H.-J. (1987), DNA mismatch-repair in *Escherichia coli* counteracting the hydrolytic deamination of 5-methylcytosine residues, *EMBO J.* **6,** 1809–1815.

ZIELENSKA, M., HORSFALL, M. J., GLICKMAN, B. W. (1989), The dissimilar mutational consequences of S_N1 and S_N2 DNA alkylation pathways; clues from the mutational specificity of dimethylsulfate in the *lacI* gene of *Escherichia coli, Mutagenesis* **4,** 230–234.

2 Genetic Exchange Processes for Prokaryotes

BRUCE W. HOLLOWAY

Clayton, Victoria, Australia

1 Introduction

It is nearly fifty years since the first genetic exchange, transformation, was demonstrated by experimentation in a bacterium: the *Pneumococcus*, and shortly afterwards LEDERBERG and TATUM described the first conjugal genetic transfer in *Escherichia coli* K12. In 1952, ZINDER and LEDERBERG demonstrated the role of bacteriophage in genetic exchange in *Salmonella typhimurium*. Those seminal discoveries have been extended until now genetic exchange processes are known for a wide range of bacteria.

There have been three major impacts of this work. Firstly, the benefits of genetic analysis have been made available to many bacteria, which has been essential for the better understanding of the physiology, biochemistry, taxonomy, and ecology of microorganisms. The understanding of the various mechanisms of bacterial genetic exchange has enriched the genetic and microbiological literature, and this knowledge has provided concepts of vital importance to eukaryotic genetics.

Secondly, the ability to manipulate bacteria for commercial benefit by genetic exchange procedures has confirmed and extended the role of genetics in biotechnology. Finally, the intensive genetic analysis of certain bacteria, particularly *E. coli*, their phages, and their plasmids has provided the foundation for recombinant DNA technology which is applicable to all prokaryotes and eukaryotes.

Paradoxically, the very success of these techniques has diminished the importance of the classical genetic exchange procedures for the manipulation of bacteria to produce bacterial strains having desired properties. In summary, the combination of *in vitro* and *in vivo* techniques has enabled the manipulation of microorganisms used in biotechnology to create new products, increased production efficiency, and a new dimension for commercial and scientific interactions in biology.

A major outcome of the classical processes of gene exchange in bacteria has been that the data so obtained could be used for the construction of chromosome maps of microorganisms. As has been the case for the entire history of plant and animal breeding, increased complexity of the available genetic maps results in increased effectiveness of breeding programs. Linkage knowledge has also contributed to the better understanding of a range of biological phenomena, a classic case being the mechanisms of gene regulation in bacteria. Now, through recombinant DNA techniques, even more sophisticated and precise mapping can be achieved in which combined physical and genetic maps of microorganisms can be constructed, and this additional knowledge will flow on to the taxonomic, epidemiological, biotechnological, and environmental aspects of microbiology.

In the following account, emphasis will be placed on identifying significant developments in mechanisms of gene exchange in bacteria, its use with organisms valuable to biotechnology, the application of gene transfer processes to newly isolated organisms, and the way in which such processes are common to a variety of organisms. Comprehensive accounts of the best studied organisms are available in a variety of textbooks and standard works, and reference will be made to these. The starting point for any basic understanding of microbial genetics must be the classic volume of HAYES (1968). Other more recent general publications which provide extensive accounts of bacterial genetic exchange mechanisms include KING (1974), LOW and PORTER (1978), BALL (1984), SCAIFE et al. (1985), NEIDHARDT et al. (1987), BIRGE (1988), and DRLICA and RILEY (1990).

There are three major ways in which genetic material can be exchanged in bacteria: *conjugation* in which plasmids promote transfer of bacterial chromosome (Fig. 1); *transduction,* the bacteriophage-mediated transfer of chromosomal fragments (Fig. 2), and *transformation* in which fragments of bacterial genomic DNA are taken up by recipient bacteria (Fig. 3). In all cases, the acquired piece of chromosome is incorporated by recombination into the chromosome of the recipient organism and passed on to subsequent generations. Given the wide occurrence of these phenomena over many bacterial genera, variants on these major themes have been found, illustrating the diversity of origins of hereditary variation so familiar to geneticists.

The frequency of genetic transfer and the selection of recombinants are two important

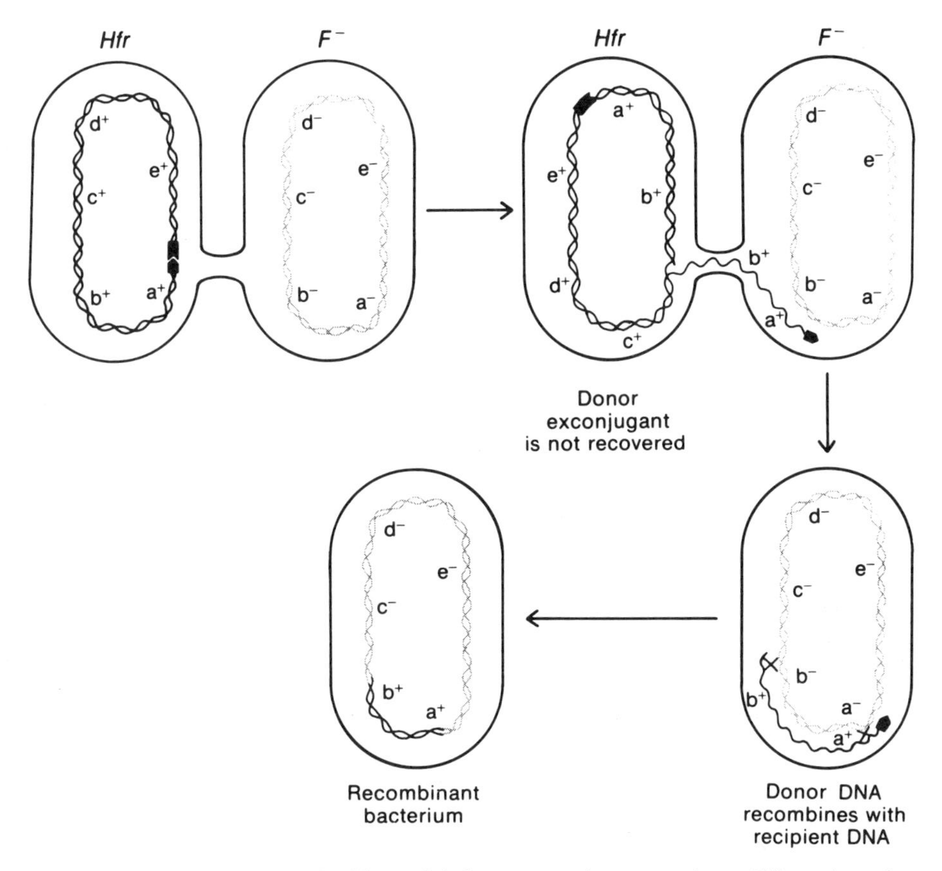

Fig. 1. Transfer in *Escherichia coli* K12 of bacterial chromosome between a donor Hfr strain and a recipient F^- parent. The F plasmid is integrated into the donor Hfr chromosome and part of it leads the chromosome into the recipient cell. Alleles of markers a, b, c, d, e are indicated by + or −, and the positions of crossovers to give the recombinant bacterium are indicated.
Reproduced by permission from *Principles of Genetics*, 2nd Ed. (1988) (FRISTROM, J. W., CLEGG, M. T., Eds.), p. 300. New York: W. H. Freeman and Company.

technical aspects of the development of gene transfer systems in bacteria. In general terms, the frequency of recombinant recovery needs to be at least 10^{-6}/donor cell or higher for effective genetic analysis. The most successful type of selective marker is the auxotroph, but this is not always available because some organisms do not grow on the defined media necessary for the isolation of auxotrophic mutants. Antibiotic resistance has been used for selection of recombinants, but expression difficulties are commonly encountered and the quantitation of recombinant frequency is difficult. The ability to use a variety of carbon or nitrogen substrates and the fermentation of carbohydrates are phenotypic characteristics which have provided a range of useful markers.

2 Gene Exchange in Prokaryotes by Conjugation

This is now known to be a common phenomenon of genetic exchange in bacteria. The

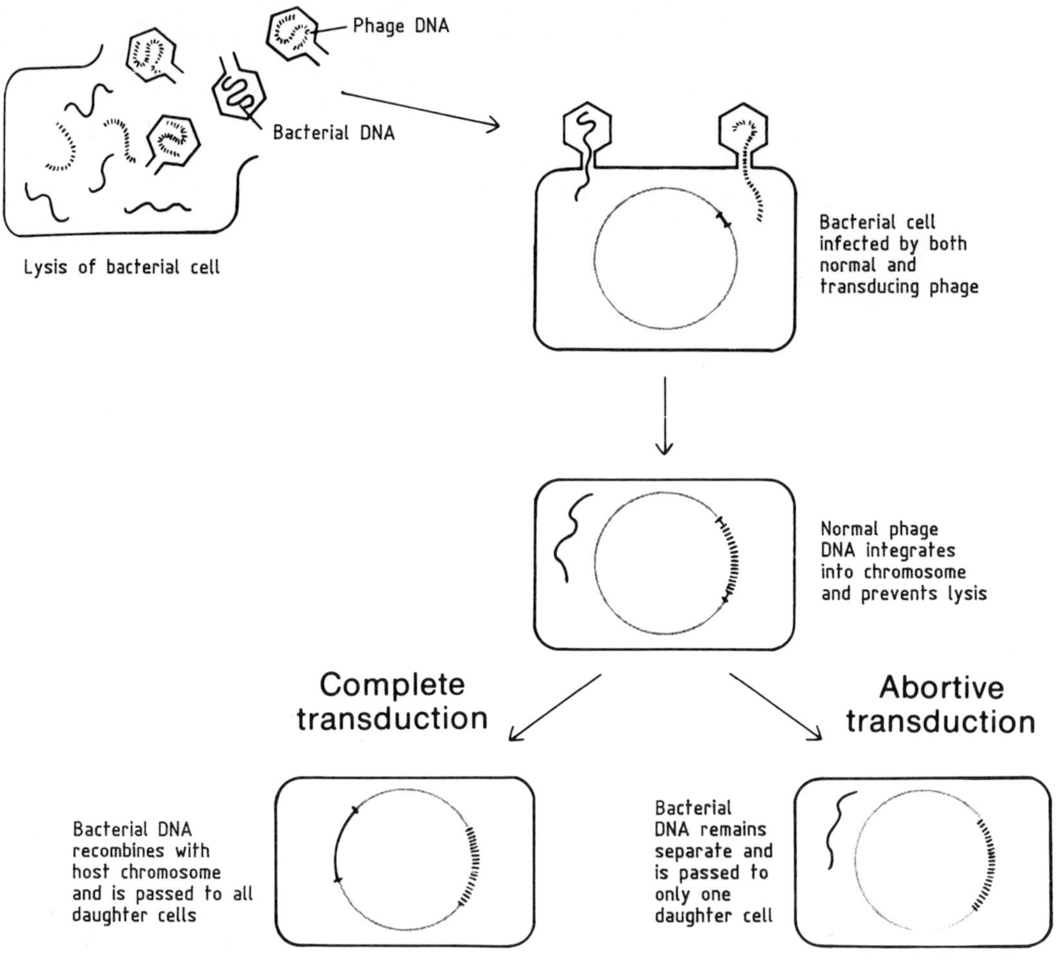

Fig. 2. General transduction in bacteria by a temperate phage. Transducing phage particles are produced following lysis of a donor bacterium. When such particles infect a recipient bacterium, integration of the chromosome fragment results in production of a recombinant cell.
Reproduced by permission from *Principles of Genetics*, 2nd Ed. (1988) (FRISTROM, J. W., CLEGG, M. T., Eds.), p. 291. New York: W. H. Freeman and Company.

first system to be described, the F plasmid (Fig. 4) promoted transfer of chromosome in *Escherichia coli* K12, remains the best characterized, but there are now many bacteria genera for which a variety of plasmids show the same basic features. The essential requirement is a plasmid which can code for the ability to conjugate with other bacterial cells, and in addition interact with the bacterial chromosome to result in mobilization of chromosome from the donor, plasmid carrying cell to a recipient cell which usually is free of that particular plasmid. Progeny of the recipient cell in which integration of the transferred donor chromosome fragment has taken place can then be analyzed for the acquisition of particular genetic markers.

The interaction between the plasmid and the donor cell chromosome is commonly through the formation of a plasmid–chromosome cointegrate. For example, the F plasmid integrates into the *E. coli* K12 chromosome by means of

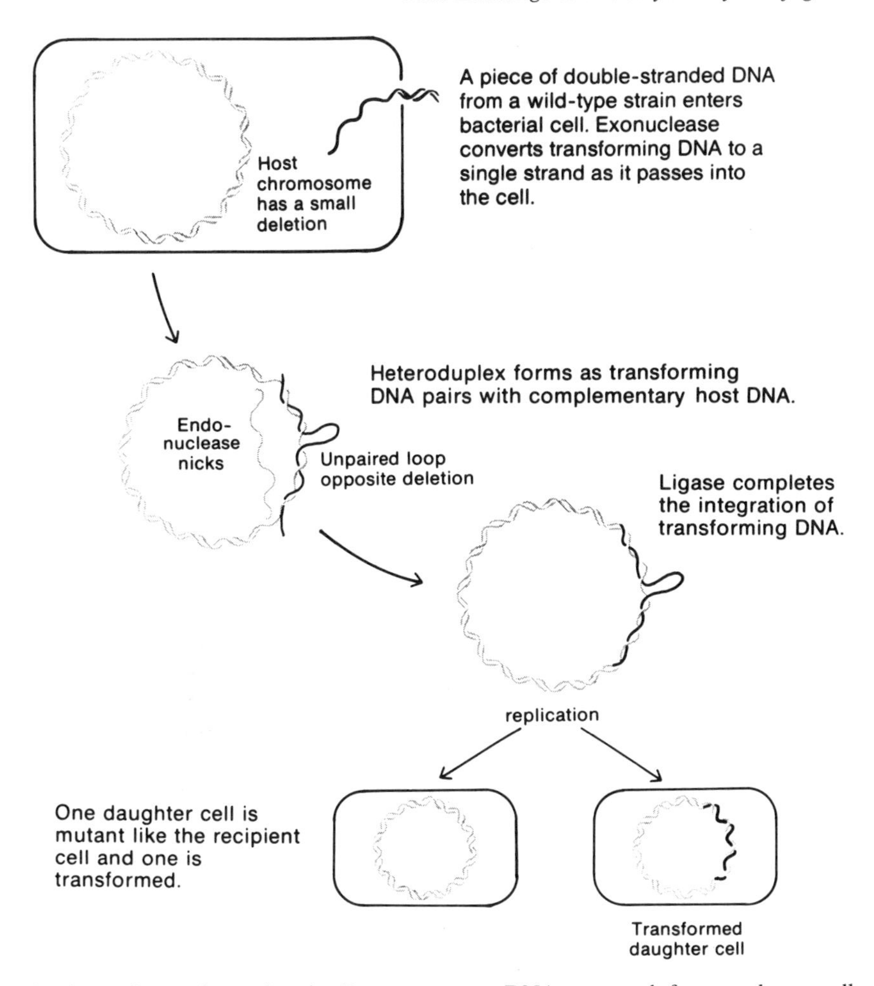

Fig. 3. Mechanism of transformation in *Pneumococcus*. DNA prepared from a donor cell enters the recipient bacterium and is integrated into the homologous region of the chromosome to result in formation of a recombinant bacterium.
Reproduced by permission from *Principles of Genetics*, 2nd Ed. (1988) (FRISTROM, J. W., CLEGG, M. T., Eds.), p. 288. New York: W. H. Freeman and Company.

insertion sequences (IS) which occur both on the F plasmid and the chromosome. The result is the formation of Hfr donors which transfer chromosomal genes at high frequencies (NEIDHARDT et al., 1987). Homology of resident IS sequences on plasmid and chromosome is not necessary. With R68.45, it appears that the insertion segment IS*21* present in some IncP-1 plasmids jumps from the plasmid to random chromosomal sites to enable plasmid–chromosome cointegrate formation leading to chromosome transfer from a large number of sites (WILLETTS et al., 1981).

There are two experimental issues involved in the study of conjugation in bacteria. The first is to understand the mechanism by which a plasmid can mobilize chromosome, and then how this system can be used to map the bacterial chromosome. The extensive studies on F and *E. coli* K12 have largely solved these issues (NEIDHARDT et al., 1987), and the *E. coli* map is now the most extensive of any bacterium

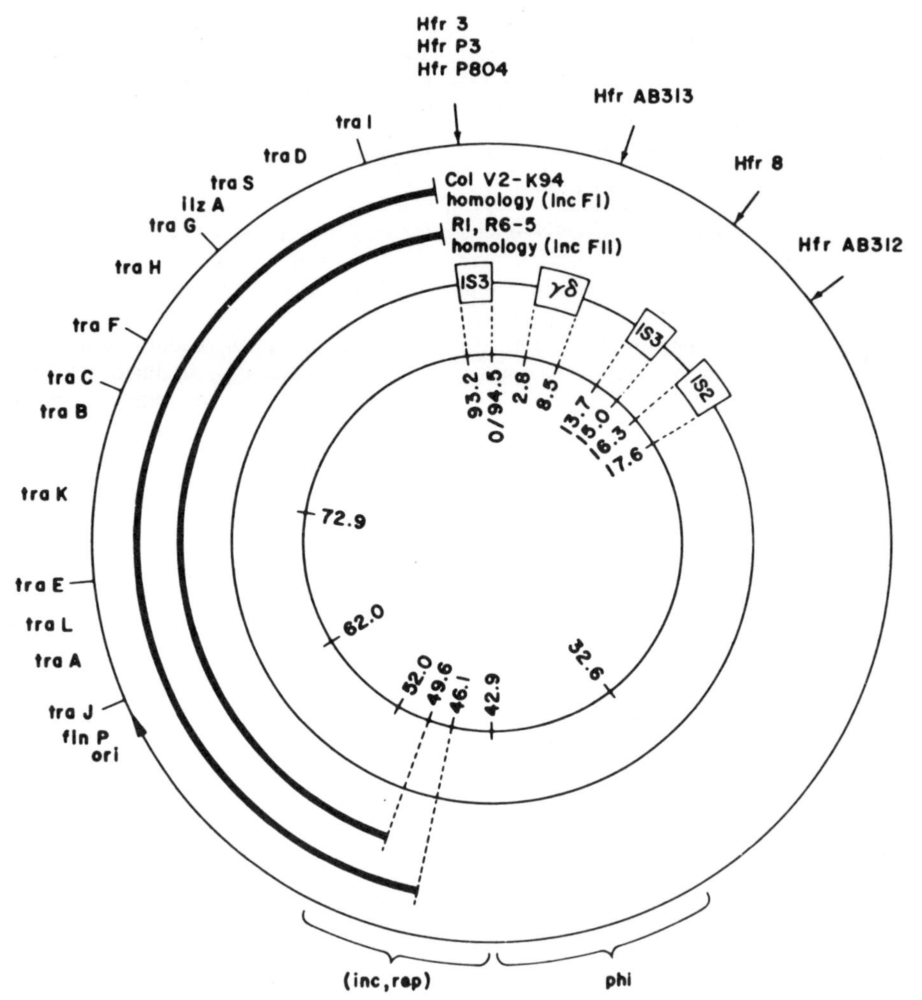

Fig. 4. F, the *Escherichia coli* sex factor. The inner circle gives certain physical coordinates in kilobase units (kb). The next circle indicates the locations of identified insertion elements. The two arcs show regions of extensive homology with colV2-K94, R1 and R6-5 plasmids. The outer circle indicates the location of some genetic loci involved in phage inhibition (*phi*), incompatibility (*inc*), replication (*rep*), transfer (*tra*), fertility inhibition (*fin*), and immunity to lethal zygosis (*ilz*). The origin of transfer replication (*ori*) is also indicated.
From SHAPIRO, J. A. (1977), *F, the E. coli Sex Factor*, p. 671, in *DNA Insertion Elements, Plasmids and Episomes*, p. 671 (BUKHARI, A. I., SHAPIRO, J. A., ADHYA, S. L., Eds.). Cold Spring Harbor, NY: Cold Spring Harbor Laboratory Press.

with over 1200 loci located (BACHMAN, 1990). This represents about 25% of the entire genome.

The second area of importance is the need to extend the benefits of conjugation analysis to other bacteria, these benefits manifesting themselves in terms of the ability to map genes and to construct recombinant strains with desired phenotypic properties, the latter feature being significant for the use of microorganisms in biotechnology. Bacteria of value to biotechnology are almost invariably not laboratory strains but have been selected as new isolates from nature because they display a metabolic activity of particular interest.

Despite the many and varied contributions of *E. coli* genetics and despite the extensive knowledge of the F plasmid (IPPEN-IHLER and MINKLEY, 1986), this F plasmid has been shown to mobilize chromosome effectively in only a few other bacteria. It has been used for the genetic analysis of some *Salmonella* species (SANDERSON and MACLACHLAN, 1987) and for various species of the phytopathogenic *Erwinia*. As a result, considerable effort has been put into finding conjugative plasmids which will mobilize chromosome in a variety of other organisms.

2.1 Conjugative Gene Exchange in Gram Negative Bacteria

Conjugative plasmids are frequently found in bacteria. They have been better characterized for Gram negative bacteria, but where Gram positive organisms have been studied, they seem to be just as common. In only relatively few instances, however, have conjugative plasmids been systematically examined for their ability to mobilize chromosome. For example, clinical strains of *Pseudomonas aeruginosa* commonly carry plasmids capable of mobilizing chromosome (DEAN et al., 1979), but their activity in this respect has not been tested for other species of *Pseudomonas*. It is known that the better characterized *Pseudomonas* sex factors, such as FP2, FP5, FP110, and FP39, only mobilize chromosome in *P. aeruginosa* (HOLLOWAY, 1986).

As early as 1962, it was shown that drug resistance plasmids could promote chromosome transfer in *E. coli* (SUGINO and HIROTA, 1962). The discovery by LOWBURY et al. (1969) of plasmids which could bridge wide generic boundaries in Gram negative bacteria was the first major step to extending the availability of conjugational plasmids. Subsequently, STANISICH and HOLLOWAY (1971) demonstrated that such broad host range plasmids could promote chromosome transfer. The plasmids isolated by LOWBURY and his colleagues were classified as IncPI, and it is these plasmids which have been effective in enabling genetic transfer in a wide range of Gram negative genera including *Acinetobacter, Agrobacterium, Azospirillum, Azotobacter, Erwinia, Escherichia, Klebsiella, Methylobacterium, Pseudomonas, Rhizobium, Rhodobacter*, and *Zymomonas* (HOLLOWAY, 1986). The properties of IncPI plasmids have been described in detail by THOMAS and SMITH (1987).

While native IncP1 plasmids can mobilize chromosome in a few bacteria with a recombinant frequency of recombination acceptable for genetic analysis, in most bacteria it has been necessary to select plasmid variants with enhanced frequencies of marker transfer. The most widely used variant is R68.45 (Fig. 5) (HAAS and HOLLOWAY, 1976, 1978). The ability of R68.45 to mobilize the chromosome of a range of bacteria (see Tab. 1) is due to the presence of a tandem copy of a 2.1 kb sequence (IS*21*) which has been found only in certain IncP1 plasmids, usually in only the one copy condition (WILLETTS et al., 1981). It has been demonstrated (SCHURTER and HOLLO-

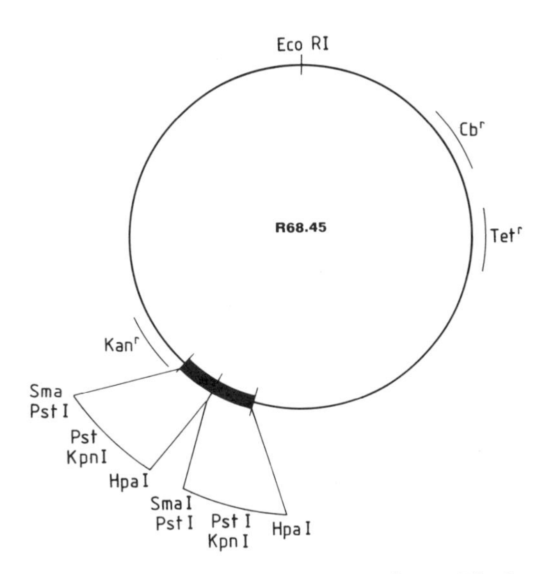

Fig. 5. R68.45, a wide host range sex factor. The location of the genes coding for antibiotic resistance are indicated (Cb^r), carbenicillin; Tet^r, tetracycline; Kan^r, kanamycin). The locations of some of the restriction enzyme sites are indicated. The tandem duplicate IS*21* regions are indicated near the Kan^r site.

Tab. 1. Bacterial Genera for which Genetic Transfer by Conjugation has been Demonstrated

Genus	Representative Plasmids	Reference[a]
Acinetobacter	RP4	HOLLOWAY, 1979; BALL, 1984
Acinetobacter		BAINBRIDGE and TYPAS, 1984
Agrobacterium	R68.45	BALL, 1984; HOLLOWAY, 1986
Alcaligenes		BAINBRIDGE and TYPAS, 1984
Azospirillum	R68.45	HOLLOWAY, 1986
Azotobacter	R68.45	KENNEDY and TOUKDARIAN, 1987
Citrobacter	F	HOLLOWAY, 1979
Enterococcus	pAD1	HOLLOWAY, 1979; BALL, 1984
Erwinia	F'Lac$^+$, R68.45	HOLLOWAY, 1979; BAINBRIDGE and TYPAS, 1984
Escherichia	F, ColV	NEIDHARDT et al., 1987
Klebsiella	R144 drd 3, R68.45	HOLLOWAY, 1979
Methylobacillus	RP4::ColE1	SEREBRIJSKI et al., 1989
Methylobacterium	R68.45	TATRA and GOODWIN, 1985
Proteus	R772	HOLLOWAY, 1979
Pseudomonas	FP2, R68.45, pMO514, pMO75	HOLLOWAY and MORGAN, 1986; O'HOY and KRISHNAPILLAI, 1987
Rhizobium	R68.45	HOLLOWAY, 1979; LONG, 1989
Rhodobacter	R68.45	HOLLOWAY, 1979; BALL, 1984
Salmonella	F	NEIDHARDT et al., 1987
Serratia	R471	HOLLOWAY, 1979; BAINBRIDGE and TYPAS, 1984
Staphylococcus	pCRG1690	BALL, 1984; STOUT and IANDOLO, 1990
Streptomyces	SCP1	HOPWOOD and CHATER, 1984
Vibrio		LOW and PORTER, 1978; GUIDOLIN and MANNING, 1987
Xanthobacter	not characterized	WILKE, 1980
Zymomonas	R68.45	HOLLOWAY, 1986

[a] To save space, the references provided are a guide to the literature of genetic exchange by conjugation for the relevant genus and a more extended bibliography is available using the references quoted. Not all plasmids promoting chromosomal transfer are listed. The generic names used are those current in 1991 and there are some differences from those used in the original publications.

WAY, 1987) that IS*21* has a promoter site at the left hand end. IS*21* transposes at high frequency from the tandem configuration, because a transposition gene on the left hand copy is transcribed from the copy of this promoter located in the right hand copy. IS*21* can promote chromosome transfer in a wide variety of bacteria suggesting that it has a very low specificity of insert sites for bacterial DNA.

Given the increasingly wide use of recombinant DNA techniques in identifying genes and gene products, do *in vivo* mechanisms for gene transfer, mapping, and the production of recombinants still have a place in the analysis of microorganisms and in biotechnology? For bacteria where both systems are available, compared to organisms for which only *in vitro* techniques are available, it is apparent that the knowledge from both systems is far superior to that where only *in vitro* data are available. For example, where a whole pathway needs to be manipulated, a knowledge of the linkage arrangements obtained by *in vivo* procedures will influence the cloning strategy adopted. Mapping data obtained by *in vivo* procedures for multigene functions may well reveal unexpected gene arrangements which reflect gene regulation patterns. An excellent example of the way in which *in vivo* and *in vitro* procedures for genetic transfer can complement each other and provide information of considerable practical significance for biotechnology is the genus *Streptomyces* (CHATER and HOPWOOD, 1983) (see below).

A comprehensive list of organisms for which conjugation has been demonstrated is provided in Tab. 1.

2.2 Conjugative Gene Exchange in Gram Positive Bacteria

For historical and other reasons, the knowledge of gene transfer by conjugation in Gram positive organisms has not progressed as quickly or attracted as many workers as has been the situation in Gram negative bacteria. Undoubtedly, additional facets of such conjugal transfer have been revealed by the elucidation of mechanisms of gene transfer in this group of organisms. Indeed one genus, *Streptomyces*, is now among the best known, with an extensive body of work on *in vivo* and *in vitro* aspects of genetic transfer, stimulated by the industrial importance of this organism. While conjugative plasmids have been commonly isolated in many Gram positive organisms, there is not the variety of such plasmids which have been described for the Gram negative organisms, particularly the human pathogenic species. One consequence of this is that there has been less reliance on *in vivo* genetic procedures for genome mapping in Gram positive organisms.

The genetics of *Streptomyces* is more complex than that of the simpler prokaryotes but conjugal gene transfer does take place by a mechanism similar to the prototype *E. coli*. It requires fertility factors which are plasmids, incomplete genomic segments being transferred. In *Streptomyces coelicolor* the plasmid SCP1 can be either autonomous (like F) or associated with the chromosome, a condition comparable to the Hfr stage. A second fertility factor, SCP2, is also present and can promote genetic exchange (HOPWOOD and CHATER, 1984).

Inheritance of chromosome markers is usually, but not always, bidirectional from a site on the chromosome which varies according to the plasmid. Variants of the native plasmid SCP2 have been used which promote chromosome transfer more efficiently. The availability of transduction (see below), protoplast fusion, liposome-protoplast fusion, and transformation of plasmids means that a variety of genetic exchange mechanisms have been used to construct a detailed chromosome map. The chromosome of *S. coelicolor* is circular but with an unusual gene distribution. There are two long, almost silent quadrants which intervene between the other quadrants which contain most of the genes so far identified (HOPWOOD and KIESER, 1990).

This range of genetic exchange mechanisms has provided the basis for extensive genetic and biochemical studies on the pathways of antibiotic production in this organism (HOPWOOD et al., 1990).

In *Enterococcus faecalis*, a variety of conjugative plasmids have been identified. One group, exemplified by pAD1, transfers at high frequency in broth matings, whereas others, for example, pAMβ1, require cell to cell contact for conjugation to occur (CLEWELL, 1981) (Fig. 6). pAD1 has been shown to transfer when induced by a small peptide pheromone called cAD1 which is released by recipient cells lacking the plasmid. The pheromone acts by enhancing cell to cell contact, through cell aggregation, production of specific surface proteins, and those functions necessary for physical transfer of DNA. The presence of pAD1 in

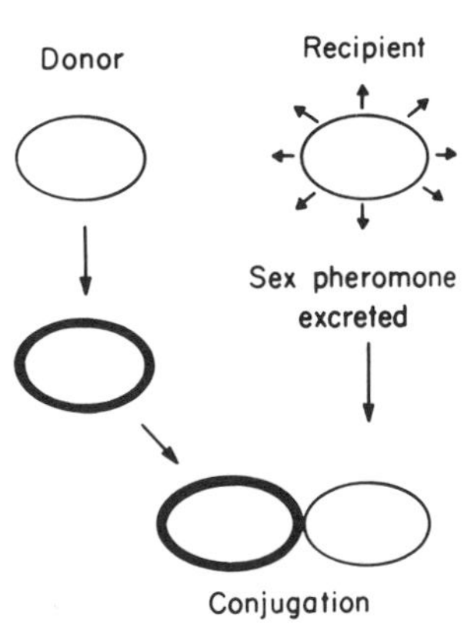

Fig. 6. Expression of sex pheromone by a recipient *Enterococcus faecalis* strain and the response by a donor containing a pheromone sensitive conjugative plasmid.
Reproduced by permission from CLEWELL, D. B. (1981), *Plasmids, Drug Resistance and Gene Transfer in the Genus Streptococcus, Microbiol Rev.* **45**, p. 415. Washington, DC: American Society for Microbiology.

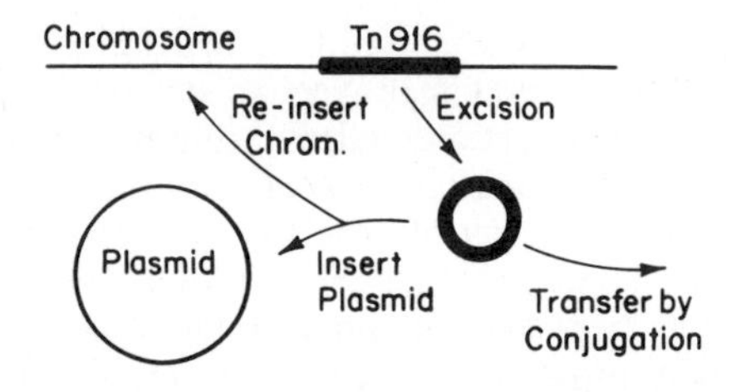

Fig. 7. Model mechanism by which Tn*916* acts as a transposon capable of conjugal transfer. Tn*916* excises from the donor chromosome and can either reinsert into the same chromosome, perhaps at a different location. Alternatively, it can insert into a resident plasmid or transfer to another cell by conjugation.
Reproduced by permission from CLEWELL, D. B. (1981), *Plasmids, Drug Resistance and Gene Transfer in the Genus Streptococcus, Microbiol. Rev.* **45**, p. 419. Washington, DC: American Society for Microbiology.

a cell prevents cAD1 synthesis, but does not affect production of pheromones specific for other plasmids (WEAVER and CLEWELL, 1990).

Conjugation in *E. faecalis* also enables the transfer of antibiotic resistance markers between different strains, although the mechanism is different from the R plasmids commonly found in Gram negative organisms. It has been shown (CLEWELL, 1981) that there is a transposon, Tn*916* (Fig. 7) on the chromosome of *E. faecalis* which can transpose to pAD1, which is then transferred by the mechanism described above, and in the recipient Tn*916* transposes to the chromosome, where it inserts at a range of sites.

The contribution of conjugation to the understanding of the genetic structure of Gram positive bacteria has not been as extensive as with Gram negative organisms.

The lactic acid bacteria are important for a variety of biotechnological uses, particularly in the food industry, and there has been considerable effort in establishing *in vivo* systems of genetic exchange (GASSON, 1990). Transduction has been demonstrated in *Lactococcus lactis* and *Streptococcus salivarius* ssp. *thermophilus*. This has been particularly useful for transduction of plasmids. Plasmids capable of conjugal transfer of lactose genes have been extensively studied, and while such genes are usually located on plasmids, there is some evidence for their chromosomal integration under some circumstances.

Insertion sequences have been identified in *L. lactis* plasmids, and there is evidence of recombination between a plasmid carrying an IS element and chromosomal DNA. Progeny that arise when the lactose plasmid pLP712 is transferred from *L. lactis* 712 to a lactose negative strain include some that have a much enhanced transfer ability for lactose fermentation. It has been shown that this is associated with a segment of chromosomal DNA which includes an insertion sequence ISSIS. High frequency transfer is associated with donor cell aggregation. Phenotypes other than lactose fermentation including bacteriocin production and bacteriophage resistance have been transferred by this system which has now been extended to other lactic streptococcal strains. The nature of the cell aggregation has been investigated in detail, and it has been shown that pheromones are not involved in contrast to the situation in *Enterococcus* (VAN DER LELIÈ et al., 1991).

Wide host range transmissible plasmids also occur in Gram positive bacteria. pAMβ1 can be transferred to *Bacillus subtilis, Lactococcus lactis, Enterococcus faecalis, Lactobacillus plantarum*, and *Pediococcus* sp. (GASSON, 1990).

Genetic analysis of clostridial species has been handicapped by the cultural difficulties inherent in growing anaerobes and the fact that most, for example, *Clostridium perfringens*, not an auxotroph, require at least eleven amino acids. Attempts to develop conjugation and transduction systems have not been successful (BRÉFORT et al., 1977). *C. perfringens* is the only species of that genus for which conjugative R plasmids have been found, but they have not been shown to promote chromosome transfer, and the future genetic analysis and manipulation of *Clostridium* strains will have to rely on recombinant DNA procedures (ROOD and COLE, 1991).

In *Staphylococcus aureus*, the increase in frequency of antibiotic resistant strains over the last twenty years and the importance of this for hospital associated infections has led to extensive studies on conjugative plasmids (LYON and SKURRAY, 1987). Transduction, transformation, conjugation and the so-called "phage mediated conjugation" have all been identified as mechanisms of genetic transfer in staphylococci. Most of this work has been aimed at characterizing antibiotic resistance determinants, and the chromosomal mapping aspects have been approached using transformation and protoplast fusion. However, a system of chromosomal gene transfer by conjugation has recently been described (STOUT and IANDOLO, 1990). A chromosomal copy of the transposon Tn*551*, a transposon specifying erythromycin resistance in *Staphylococcus aureus* having some homology to the well characterized Tn*3*, combined with a copy of Tn*551* co-resident on a gentamycin resistant conjugative plasmid, resulted in the mobilization of chromosomal genes during filter mating. Gene mobilization does not seem to be restricted to any specific region of the chromosome.

2.3 Chromosome Mobilization by Transposable Elements

Transposable elements which carry antibiotic resistance markers have permitted a wide range of genetic manipulations of bacterial genomes (KLECKNER et al., 1977; KLECKNER, 1981; BERG and HOWE, 1989). An artificial homology can be created between the conjugative plasmid and the chromosome by the presence of the same transposon in each component, and this has been done with such transposons as Tn*1*, Tn*3*, Tn*5*, Tn*10*, and Tn*501* in a range of bacterial genera including *Salmonella, Pseudomonas*, and *Vibrio* (HOLLOWAY, 1979; JOHNSON and ROMIG, 1979; CHUMLEY et al., 1979; KRISHNAPILLAI et al., 1981; DEAN and MORGAN, 1983). As described above, insertion sequences (IS) are now known to be essential for the mobilization of chromosome by plasmids such as F in *E. coli* and R68.45 in a variety of genera. Transposons carrying antibiotic resistance markers have proved valuable in a variety of bacterial genetic manipulations. They can be used as easily selectable markers at or near genes for which selection is not possible. They can provide sites for oriented chromosome transfer. This system has been particularly valuable for chromosome mapping in *Pseudomonas* (KRISHNAPILLAI et al., 1981; STROM et al., 1990).

Derivatives of the bacteriophage Mu have been of considerable value, particularly for *in vivo* cloning procedures (BERG et al., 1989; DÉNARIÉ et al., 1977). By isolating an RP4::Mu hybrid, it was possible to combine the high transposability of Mu with the wide host range of the IncP1 plasmid. This hybrid plasmid was shown to mobilize chromosome in *E. coli, Klebsiella pneumoniae*, and *Rhizobium* ssp. Another way to enhance the genetic transfer abilities of IncP1 plasmids is to include cloned chromosomal fragments into the plasmid to provide a region of homology (BARTH, 1979).

2.4 Transfer of Genetic Material by Prime Plasmids

Plasmids can act in two ways to transfer genetic material. They can either promote the transfer of the chromosome as described above, or they can acquire fragments of chromosome to form a hybrid plasmid which is transferred during conjugation at the frequency characteristic of the plasmid, usually orders of magnitude higher than that for chromosome transfer, resulting in high frequency transfer of some chromosomal markers.

The notion that recombination could take place between unrelated genomes was quite foreign to the genetic theory of the 1950s so that the data that led to the identification of the first F prime (or plasmid-bacterial chromosome hybrid) were at first difficult to interpret. However, it soon became clear that a fragment of bacterial chromosome could become part of the continuity of the plasmid genome, and a variety of F primes were characterized for *Escherichia coli* and *Salmonella typhimurium*. In effect these were the first *in vivo* cloning experiments, and they introduced to microbial genetics a highly efficient technique for transfer of selected segments of bacterial chromosome.

The use of R plasmids to mediate chromosome transfer has expanded the genera for which prime plasmids have become available. It has been found that R68.45 and similar de-

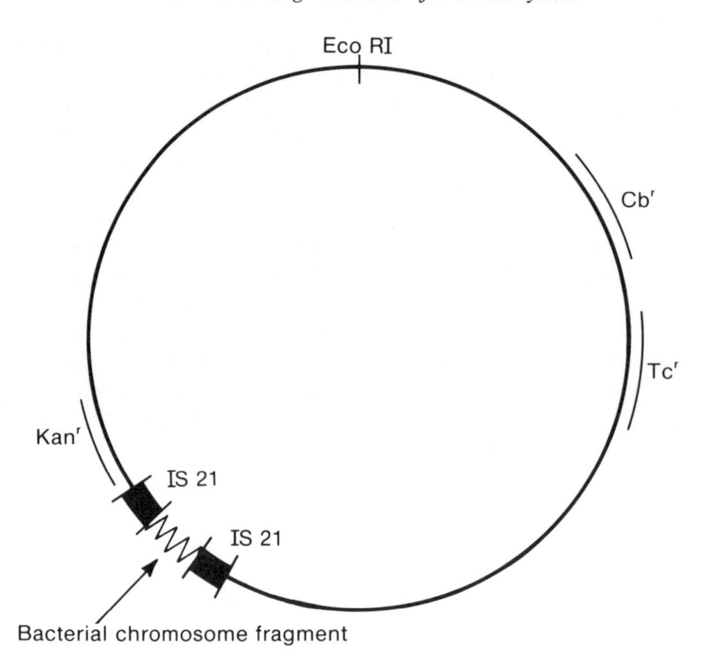

Fig. 8. The structure of an R' plasmid derived from R68.45. The positions of the drug resistance markers on the plasmid chromosome are indicated: Cb^r, carbenicillin resistance; Tc^r, tetracycline resistance; Kan^r, kanamycin resistance. A copy of IS*21* flanks the inserted fragment of the bacterial chromosome.

rivatives generate R prime plasmids in a variety of genera (Fig. 8). The wide host range nature of R68.45 has permitted development of an easy technique for the generation of R primes in many different genera.

The easiest procedure for isolating F primes in *E. coli* is to mate either an F+ of Hfr donor with a recombination deficient (Rec$^-$) donor. Thus any recombinants that arise for the chromosomal marker selected must result from the formation of primes, their transfer to the recipient, and their replication independent of the chromosome. All chromosomal markers transferred will not integrate and persist in subsequent generations because the Rec$^-$ phenotype will prevent this. Rec$^-$ forms of bacteria are difficult to isolate so by using R68.45 carried by the genus for which the prime plasmid is desired, then mating that to *Pseudomonas aeruginosa*, the recipient pseudomonad acts as a surrogate recombination deficient host, and R prime plasmids carrying desired segments of the chromosome of the donor organism can be isolated (HOLLOWAY, 1986). By this and other methods, R primes have been isolated for *Escherichia, Pseudomonas, Rhizobium, Klebsiella, Rhodobacter, Acinetobacter, Salmonella, Methylophilus*, and *Methylobacillus* (HOLLOWAY, 1986; TSYGANKOV et al., 1990).

R prime construction has a particular value in mapping of some organisms because they can carry large inserts, especially useful where genes affecting a particular function are not closely linked. This property has been found to be particularly useful in the genetic analysis of *Rhizobium* (NAYUDU and ROLFE, 1987).

2.5 Genetic Exchange between Unrelated Bacteria

Recent results have increased the possibility of transfer of genetic material between unrelated bacteria, particularly in nature. These findings have significant implications for those aspects of biotechnology which are concerned with the release of genetically engineered organisms into the environment. The properties of IncP-1 R prime plasmids described above demonstrate that chromosomal material can be readily transferred between different Gram negative genera, and in most cases the R prime plasmid is stable in the recipient host, provided it survives the entry process and potential attack by restriction enzymes (see below).

Recently TRIEU-CUOT et al. (1988) have demonstrated that genetic information can be transferred by conjugation from Gram positive to Gram negative bacteria. A shuttle vector with the *tra* functions of an enterococcal plasmid was transferred in conjugation from *Enterococcus faecalis* to *Escherichia coli*, albeit at a low frequency of 5×10^{-9} per donor colony. Using a similar plasmid vector, MAZODIER et al. (1989) demonstrated genetic transfer between *E. coli* and *Streptomyces* species. The conjugative enterococcal transposon Tn*916* has been found to transfer between a variety of Gram positive and Gram negative bacteria, and it is likely that this is by a conjugation mechanism (BERTRAM et al., 1991).

There are a number of reasons why intergeneric exchange of genetic material is such a rare event. It is clear from what is known of conjugation in *E. coli* K12 that many gene products contribute to the many interactions of the parent cells that lead to DNA transfer, and there is considerable specificity in the action of those gene products.

In addition, the entry of foreign DNA into a bacterium is limited by the activity of restriction endonucleases which have the potential to destroy the biological activity of incoming chromosomal or plasmid DNA. Endonucleases appear to be ubiquitous among bacteria, providing effective barriers against such acquisition of new genetic material. Endonucleases can even inhibit the exchange of DNA between species of the same genus. Experimentally there are ways of overcoming such endonuclease activities. The simplest is by heating the recipient cells at 50 °C for 30 min after which the mating is initiated. The endonuclease is inactivated by such treatment, enabling the donor DNA to enter and function (UETAKE et al., 1964). A further technique is to select mutations to *p*-fluorophenylalanine resistance, many of which are restriction deficient (ROLFE and HOLLOWAY, 1968).

A method which is limited to some strains of *Pseudomonas aeruginosa* is to grow the recipient organism at 43 °C, which leads to acquisition of a restriction deficient phenotype which is then retained for over 50 generations of growth at 37 °C. The explanation of this unusual phenomenon is unknown, but it is a very useful method by which *P. aeruginosa* can be made to acquire intact foreign genetic material at high efficiency (HOLLOWAY, 1965).

Granted that there is unencumbered entry of DNA into the bacterial cell, the stability of the resulting recombinant is dependent upon a variety of genetic and other factors. Recombination and integration into the recipient chromosome entirely depends on nucleotide homology, and without crossing over, integration, and retention of the incoming DNA will not occur. The presence of a separate replicon, an R prime plasmid for example, will enable stable preservation of the donor DNA in subsequent generations, provided there is replication of the plasmid, *pari passu* with the replication of the chromosome and with cell division. This type of stability is known to be under genetic control, and it has been shown that the *parB* locus of the *E. coli* plasmid R1 mediates effective plasmid stabilization and will stabilize the inheritance of a wide range of plasmids in a variety of Gram negative bacteria (GERDES, 1988).

3 Gene Exchange in Prokaryotes by Transduction

The transfer of bacterial chromosome fragments was first demonstrated by ZINDER and LEDERBERG (1952) in *Salmonella typhimurium* using the temperate phage now called P22. Phage preparations of P22 can transfer any marker on the chromosome hence the name generalized transduction for this type of genetic transfer. Subsequently a range of phages has been shown to display this phenomenon in a variety of bacterial genera including *Acetobacter, Acinetobacter, Agrobacterium, Alcaligenes, Bacillus, Citrobacter, Enterococcus, Erwinia, Escherichia, Flavobacterium, Methylosinus, Proteus, Pseudomonas, Rhizobium, Rhodobacter, Serratia, Staphylococcus, Streptococcus, Streptomyces*, and *Vibrio* (BALL, 1984; HAYES, 1968; PATTEE, 1990; LONG, 1984; GUIDOLIN and MANNING, 1987).

Generalized transduction is an excellent means of establishing linkage and gene order

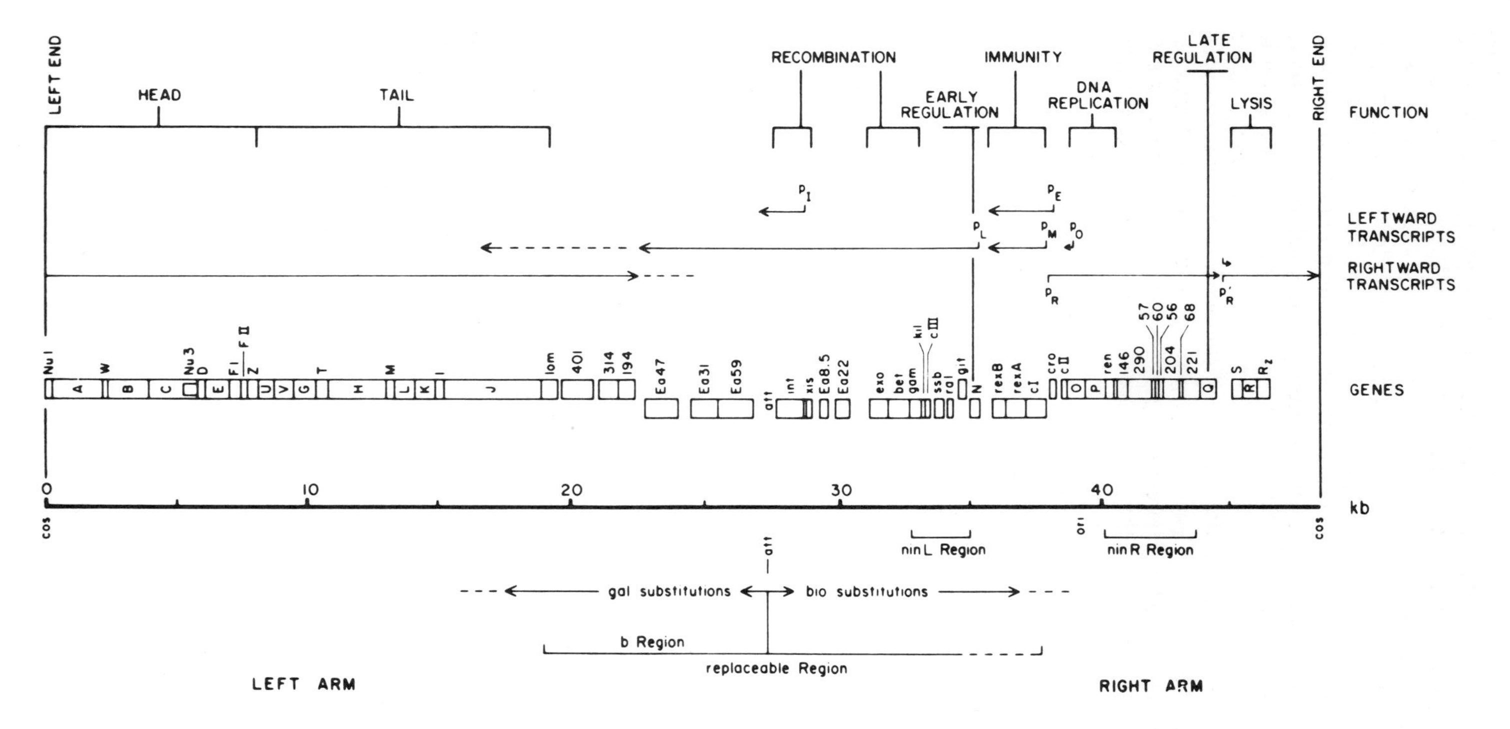

Fig. 9. Map of bacteriophage λ DNA. A scale drawing of the molecular map presented in Tab. 1 is presented above the kilobase scale, beginning and ending at the cohesive end site (*cos*). Gene clusters with related functions are indicated above the brackets, and the regulatory genes, *N* (involved in earlier regulation) and *Q* (involved in late regulation) are indicated by vertical lines. Known promoters are denoted by *p* with a subscript to indicate their unique points of origin: p_I, *int* protein promoter; p_E, establishment promoter for *c*I; p_M, maintenance promoter for *c*I; p_L, major leftward promoter; p_R, major rightward promoter; p_O, *oop* promoter; p'_R, late promoter. →, extent and direction of transcription; ----, readthrough. Map positions of genes as determined by analysis of open reading frames (ORFs) in the DNA sequence are indicated. Known genes (identified either genetically by mutation analysis or functionally by SDS-gel electrophoresis of protein product, or both) are indicated with letter names; whereas ORFs presumed from the sequence to code for protein products but with no previously known gene assignments, are given numbers corresponding to the coding capacity of the ORF. Major areas of substitution and deletion mutations are indicated below the scale: *att*, attachment site; *ori*, origin of replication.
Reproduced by permission from DANIELS, D. L., SCHROEDER, J. L., SZYBALSKI, W., SANGER, F. (1990), *A Molecular Map of Coliphage Lambda in Genetic Maps, Locus Maps of Complex Genomes*, 5th Ed., (O'BRIEN, S. J., Ed.). Cold Spring Harbor, NY: Cold Spring Harbor Laboratory Press.

of closely linked chromosomal genes. Its occurrence is closely associated to those bacteria for which bacteriophage can be readily found. There is no requirement for the transducing bacteriophage to be temperate, and transduction has been demonstrated with lytic phages, including the coliphage T1 (DREXLER, 1970), although experimental precautions have to be taken to minimize loss of recombinant transductants through phage lysis.

There is a second type of transduction: specialized transduction which is limited to a very few phages, the best studied being lambda in *Escherichia coli* (Fig. 9). Lysogens of lambda in *E. coli* may undergo abnormal excision of the prophage so that part of the chromosome adjacent to the prophage site is included in the phage genome, to the exclusion of some phage genes. Again, this is a form of *in vivo* cloning, and the appropriate segment of the chromosome, usually the galactose operon or a cluster of genes for biotin synthesis, are amplified together with the phage genome. On infection of a recipient bacterium by such a phage, transfer and integration of the donor chromosome fragment into the recipient chromosome occur at high frequency. By manipulating the site of integration of lambda into other regions of the *E. coli* chromosome, other genes may be transferred in this manner (NEIDHARDT et al., 1987).

Phages capable of specialized transduction are not common. Another example in *E. coli* is ϕ 80 which can transduce genes of the tryptophan operon in a manner similar to that found with lambda (NEIDHARDT et al., 1987), while D3 is a specialized transducing phage in *Pseudomonas aeruginosa* PAO (CAVENAGH and MILLER, 1986).

Transduction is ideal for linkage studies, and it has been used for construction of desired recombinants where a single gene change is required. In general, it has not been used extensively for construction of bacteria with desired phenotypes. However, the understanding gained by intensive study of the phage lambda was essential to permit the development of a range of cloning vehicles currently widely used for both prokaryotes and eukaryotes.

In agreement with the wide diversity of bacterial activity (summarized by HOPWOOD and CHATER, 1989), there are variants on the above three main themes of bacterial genetic exchange. *Rhodobacter capsulatus* displays the phenomenon of capsduction in which fragments of the chromosomal DNA are released in as yet not well defined packages. These packages, referred to as the gene transfer agent (GTA), contain uniform length fragments of bacterial chromosome, and these cannot replicate in the recipient unless integrated into the chromosome. Little is known about the molecular mechanism of this genetic exchange system which has not yet been identified in any other bacterium (SCOLNIK and MARRS, 1987).

A very restricted form of genetic exchange is found in those cases of lysogeny in which the bacterium acquires additional properties other than those associated with bacteriophage properties. This phenomenon, known as lysogenic conversion, has been reported for *Staphylococcus, Pseudomonas, Shigella, Corynebacterium*, and *Escherichia*. Usually the additional phenotypes acquired are related to virulence, such as changed surface antigens and toxin production. It is likely that examples of lysogenic conversion are more common than the published record suggests, but on the present evidence the correlation between acquisition of a prophage and increased virulence represents a very selective type of bacterial genetic exchange (HAYASHI et al., 1990; VERMA et al., 1991; COLEMAN et al., 1989).

The potential role of transduction as a natural mode of bacterial genetic exchange under a range of environmental conditions has been strengthened by a series of papers which have demonstrated that bacteriophages occur at much higher titers in natural waters than has been previously thought (BERGH et al., 1989; PROCTOR and FUHRMAN, 1990; KOKJOHN et al., 1991). These findings have considerable implications for the transfer of genes in nature, an area of significance for the release of genetically modified organisms for biotechnological purposes, particularly those having applications to agriculture.

4 Gene Exchange in Prokaryotes by Transformation

This was the earliest detected mechanism of genetic exchange in bacteria, first shown for the *Pneumococcus* by GRIFFITH in 1928, with the observation refined by the now classic experiments of AVERY, MACLEOD, and MCCARTY in 1946. Since then transformation has been demonstrated in a wide variety of genera including *Achromobacter, Acinetobacter, Agrobacterium, Azotobacter, Bacillus, Citrobacter, Haemophilus, Lactobacillus, Methylobacterium, Methylococcus, Micrococcus, Moraxella, Mycobacterium, Neisseria, Pseudomonas, Rhodobacter, Serratia, Streptococcus, Streptomyces*, and *Synechococcus* (BALL, 1984; STEWART and CARLSON, 1986).

With the increased importance of recombinant DNA technology, transformation has acquired additional importance as the technique for returning manipulated DNA back into a cell, so that it can be shown to exert a biological effect. Transformation now occupies a pivotal place in the strategy of reverse genetics. Any manipulation of an organism involving recombinant DNA procedures must involve return of the manipulated DNA to the genome of the organism. This invariably involves transformation. If a bacterium cannot be transformed, then in the absence of other gene transfer procedures, it cannot be manipulated by *in vivo* or *in vitro* procedures, and its potential for use in basic research and industry is reduced.

For all transformation systems the critical feature is the ability of the bacterial cells to take up DNA fragments, a condition known as competence. The mechanism of achieving a population of competent cells varies widely from organism to organism, and the percentage of competent cells in a population can be highly variable. Competence can be transient or long-lived. For *Bacillus* it is possible to store batches of competent cells by freezing them at −20 °C in an appropriate medium. At the other end of the spectrum, *Acinetobacter calcoaceticus* does not require any special cultural conditions for competence, and mere spreading of a mixture of an overnight culture and the DNA preparation onto an appropriate selective medium will result in high frequencies of transformation (NEIDLE and ORNSTON, 1986). STEWART and CARLSON (1986) have discussed the phenomena associated with the development of competence and have concluded that of all the mechanisms of bacterial genetic exchange, transformation is the least understood. Their review is the most comprehensive and most recent description of the problems associated with transformation. They make the important point that transformation, by the uptake of DNA chromosomal fragments, does take place in a variety of bacteria under conditions existing in native environments, there is in fact natural competence, and it occurs in both Gram positive and Gram negative organisms.

The conditions of competence required under laboratory conditions for linear DNA uptake, as is the case with preparations of bacterial chromosomal material, are usually quite different from those required for the uptake of circular DNA, as is the situation for plasmid DNA preparations. This latter situation is a common requirement for cloned DNA in plasmid vectors. However, these two situations are not necessarily simple. For example, in *Bacillus* monomeric plasmid molecules transform much less frequently than plasmid molecules which contain internal repeats (LOVETT, 1984).

New technology has come to the aid of transformation. The technique of electroporation in which high voltage, high current exponential pulses of controlled characteristics are passed through bacterial cells has been found to enhance entry of DNA into bacteria for which conditions of competence have not been established or at higher frequency for those in which transformation is an established technique. The process is technically simple, commercially produced equipment is available, it can be generally applied to many bacteria, and it is a procedure which is capable of being improved. Genera for which transporation has enabled effective entry of DNA include *Bacillus, Bordetella, Brevibacterium, Campylobacter, Clostridium, Corynebacterium,* Cyanobacteria, *Enterococcus, Erwinia, Escherichia,*

Haemophilus, Klebsiella, Listeria, Lactobacillus, Leuconostoc, Mycobacterium, Myxococcus, Pediococcus, Propionobacterium, Pseudomonas, Rhizobium, Salmonella, Staphylococcus, and *Streptococcus* (CHASSY et al., 1988; SHIGEKAWA and DOWER, 1988; MINTON et al., 1990).

5 Genetic Maps

One of the major contributions to our understanding of bacterial genomes made by the study of genetic exchanges has been the construction of chromosome maps.

Bacterial genomes vary in size from 585 kb for *Mycoplasma genitalium* to 9454 kb in *Myxococcus xanthus*. These data and the genome sizes for a variety of other bacteria have been summarized by KRAWIEC and RILEY (1990). The measurements have been made by pulsed field gel electrophoresis and as such include any plasmids present in the genome and are not strictly chromosome size measurements. For most free living Gram positive and negative bacteria, the genome size is in the range 2500–6000 kb. Circularity of the bacterial chromosome has been established by a variety of techniques for some twenty or so bacteria.

The first bacterial chromosome map was that constructed in *E. coli* K12 by conjugation using Hfr donors. The genetic demonstration of circularity preceded the physical demonstration of a circular chromosome by CAIRNS (1963). Until quite recently all bacterial chromosome maps have been constructed by gene exchange techniques, and the most recent compendium of genetic maps (O'BRIEN, 1990) lists the following bacteria: *Acinetobacter calcoaceticus, Bacillus subtilis, Caulobacter crescentus, Escherichia coli, Haemophilus influenzae, Neisseria gonorrhoeae, Proteus mirabilis, Proteus morganii, Pseudomonas aeruginosa, Pseudomonas putida, Rhizobium leguminosarum, Rhizobium meliloti, Salmonella typhimurium*, and *Staphylococcus aureus*. Other published bacterial maps include *Erwinia chrysanthemi* (HUGOUVIEUX-COTTE-PATTAT et al., 1989), a *Vibrio* species (ICHIGE et al., 1989) and *Azotobacter vinelandii* (BLANCO et al., 1990). The most comprehensive bacterial genetic map is that of *Escherichia coli* K12 and the latest map (BACHMAN, 1990) has 1403 loci placed, which may represent between one-third and one-half of the genes of this organism.

Conjugation has certain technical advantages for the construction of a circular chromosome map, but the same end can be achieved using other means of genetic transfer. For example, *Bacillus subtilis* 168 is the most extensively mapped Gram positive organism, with more than 700 loci placed in relation to each other. While transformation was initially used to map genes in this bacterium, generalized transduction using the phage PBS1 has proved to be more effective (PIGGOTT, 1990).

For another Gram positive organism, *Staphylococcus aureus*, a combination of techniques has been necessary to achieve a comprehensive map. Transformation was used to define the linkage arrangements within three groups of markers. Protoplast fusion then enabled these linkage groups to be combined into a single circular map (PATTEE, 1990).

The methylotrophs have been a particularly intransigent group of organisms in terms of genetic analysis. It is difficult to find markers, bacteriophages are practically unknown for the group, and conjugative plasmids are not effective for the species, which are either of industrial interest or which have been biochemically characterized and by virtue of their ability to use C-1 compounds are of particular interest (HOLLOWAY, 1984; LIDSTROM and STIRLING, 1990).

One solution to this problem has been to use IncP1 derivatives which are effective in forming prime plasmids in methylotrophs, then using these plasmids to complement known mutants of *Pseudomonas aeruginosa*. This method of mapping is known as complementation mapping, and *P. aeruginosa* is a particularly suitable host, because it expresses heterologous DNA with high efficiency. *P. aeruginosa* mutant recipients can be made restriction deficient by growth at 43 °C, enabling them to accept and express the incoming methylotrophic gene fragments (HOLLOWAY, 1965). By this method MOORE et al. (1983) using *Methylophilus methylotrophus* AS1 mapped 19 markers into four linkage groups, linkage being detected where a given prime plasmid complemented more than one marker of *P. aeruginosa*. This work was extended by using cosmids derived from an IncP1 plasmid and resulted in 28 markers being located for *M. methylotrophus* and 25 markers for *Methylophilus viscogenes*, in each case four linkage groups being involved (LYON et al., 1988). R primes have also been used for mapping in *Methylobacillus flagellatum* (TSYGANKOV et al., 1990). The use of such prime plasmid and cosmid complementation mapping techniques is an excellent example of the power of the wide host properties of the IncP1 plasmids in promoting genetic exchange within and between Gram negative bacteria.

The benefits of a chromosome map are numerous and varied. An understanding of genome arrangement has contributed to the definition of gene regulation, for example the lactose operon, the location of Hfr sites and hence insertion sequence locations in *E. coli*, and non-random patterns of gene linkages as discussed by KRAWIEC and RILEY (1990).

In any genome analysis of bacteria, it is important to differentiate between chromosomally located genes and plasmid-borne determinants. This is not necessarily a simple procedure. With conjugative plasmids, the correlation between transfer of the plasmid (as demonstrated by physical means) and acquisition of a phenotype by the recipient strain is compelling evidence. However, not all plasmids are conjugative, the larger ones (more than 100 kb) are sometimes difficult to demonstrate by physical procedures; and hence demonstrating that a gene is located on a plasmid may require a variety of techniques, most of which involve recombinant DNA procedures for conclusive evidence that a gene is chromosomal. Linkage to genes already located on the chromosome can be a somewhat circular argument, but there is doubt that a comprehensive linkage map is a valuable experimental tool. With limited data where different isolates of the same species have been used for mapping, there is no reason to believe that such different isolates will have major differences in gene arrangement. Hence providing an interstrain gene exchange system is available, a standard laboratory, genetically characterized strain can be used as a reference strain for demonstrating linkage, and hence chromosomal location of a gene of interest.

The second major technique for distinguishing chromosomal and plasmid genes is pulsed field gel electrophoresis. By a variety of techniques a circular physical map of restriction sites can be constructed for the chromosome of any bacterium. By using DNA probes either from cloned genes, or synthetic probes constructed by data available from amino acid sequence knowledge it is possible to locate any gene in relation to the physical map.

Despite the effectiveness of bacterial chromosome maps produced by gene exchange techniques, it is likely that most future maps of bacteria will be produced by physical techniques involving a combination of pulsed field gel electrophoresis, restriction endonuclease site mapping, insertion sequence location, restriction fragment length polymorphisms, genetic probes from cloned genes, polymerase chain reaction (PCR) and the use of computer banks of nucleotide and amino acid sequences which are now available internationally. The approaches involved are discussed by SMITH and CONDEMINE (1990) and KRAWIEC and RILEY (1990). Furthermore, these techniques are the best available for estimating both total genome size and total genome structures. A good example of such an analysis of genome structure was that carried out in *Rhodobacter sphaeroides* by SUWANTO and KAPLAN (1989a, b) which demonstrated that this organism has two distinct circular DNA structures each carrying essential genes as well as five plasmids.

6 Applications of Genetic Exchange in Bacteria to Biotechnology Processes: Production of *cis-cis*-Muconic Acid by *Pseudomonas putida*

A principal aim of biotechnology processes involving bacteria is the construction of a pro-

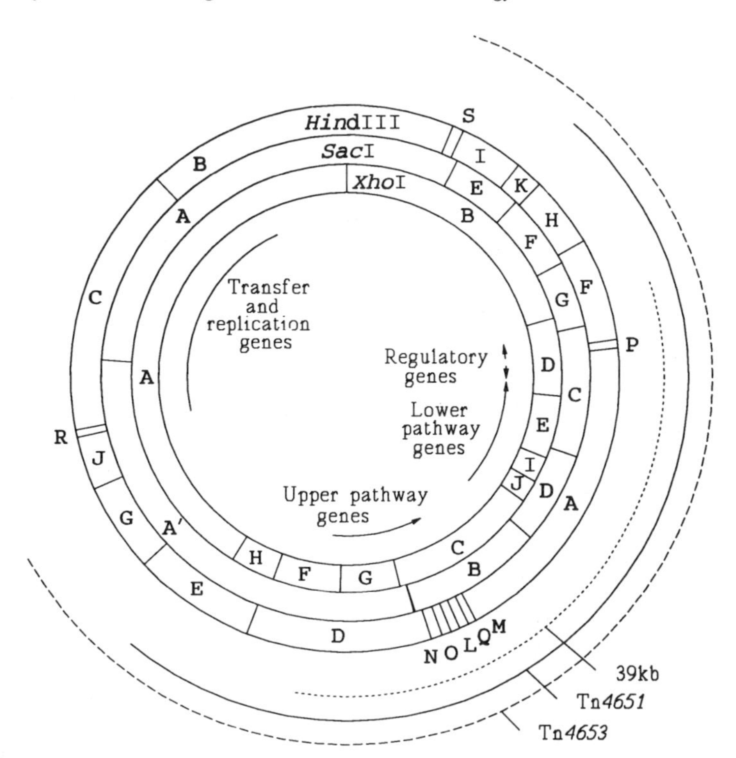

Fig. 10. Physical and genetic map of the TOL plasmid pWW0 showing the restriction sites for *Hin*dIII and *Xho*I (modified from FRANKLIN et al., 1981). The location of a *Spe*I site is indicated by an arrowhead and the extent of the transposons Tn*4651* (56 kb), Tn*4653* (70 kb), and a 39 kb excisable segment is indicated by arcs outside the map (TSUDA and IINO, 1987, 1988; SINCLAIR and HOLLOWAY, 1991).

duction strain with desired metabolic properties, which is stable, cost-efficient, and capable of being protected as intellectual property. Given that most bacteria which are used in this way are usually newly isolated from nature and genetically uncharacterized, genetic exchange processes are used for two purposes: the characterization of the genetic basis of the desired metabolic activity and the improvement of the efficiency of formation of the end product. An example will serve to demonstrate these uses of bacterial genetic manipulation.

Ever since the thalidomide disaster, the pharmaceutical industry and national drug regulatory agencies have been aware of the need for isomeric and chirally pure compounds in drugs intended for human and animal use. In addition, the presence of inactive or inhibitory isomers in agricultural products intended for crop application must result in unnecessary chemical pollution of the agricultural environment. Hence, more attention is being given to the availability of starting compounds or intermediates in multistep synthetic processes for pharmaceutical or agricultural chemical products which will enable an isomeric or chirally pure end product. Most products made by conventional chemical synthesis are impure in this respect, and while chemical procedures for separating isomers do exist, a more effective approach is the use of microorganisms to produce the desired pure compound. One such example is the use of a strain of *Pseudomonas putida* to produce *cis-cis*-muconic acid by microbiological oxidation of an inexpensive starting compound, toluene (MAXWELL, 1982, 1986; HSIEH et al., 1985; MAXWELL et al., 1987; SINCLAIR et al., 1986).

cis-cis-Muconic acid is an intermediate in the pathway of metabolism from toluene to β-ketoadipate, characterized principally in *Pseudomonas* but also found in some other Gram negative bacteria.

One of the best characterized plasmids in *P. putida* is the TOL plasmid pWW0 (FRANKLIN et al., 1981). The 117 kb of this plasmid is

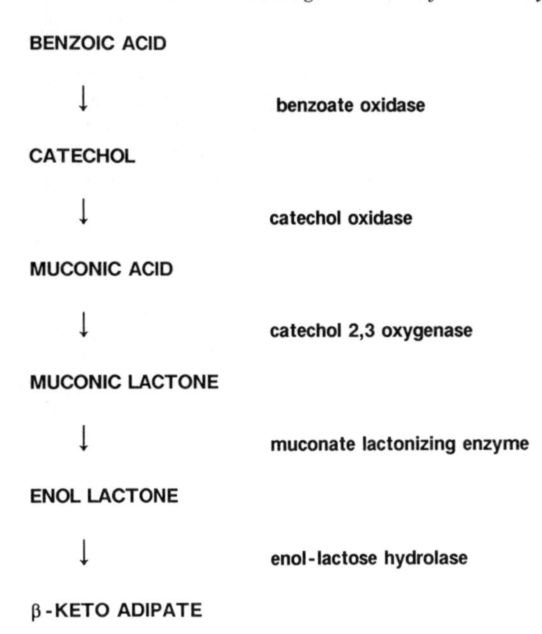

Fig. 11. The pathway of benzoic acid catabolism in *Pseudomonas putida* giving intermediates and enzymes involved (adapted from CLARKE and ORNSTON, 1975) and indicating the formation of *cis-cis*-muconic acid.

now known to comprise two transposons Tn*4651* (56 kb) and Tn*4653* (70 kb) as illustrated in Fig. 10.

This plasmid carries all genes necessary for the catabolism of toluene, *p*-xylene, and *m*-xylene. The chromosomal pathway for the metabolism of catechol is referred to as the *ortho*-pathway and the plasmid coded reactions, the *meta*-pathway. For the effective production of *cis-cis*-muconic acid two requirements were needed. The bacterium must metabolize products of the *meta*-pathway by means of the *ortho*-pathway and in addition allow the accumulation of muconic acid without further assimilation. The pathway from benzoate to *β*-ketoadipate is shown in Fig. 11.

A strain of *P. putida* was isolated from soil which could grow on toluene as sole carbon source. The presence of the *meta*-pathway was demonstrated, and by a series of mutations, the *meta*-pathway was blocked, then revertants were isolated which could grow on benzoate. Further mutation and selection enabled the isolation of a strain of *P. putida* which produced *cis-cis*-muconic acid at economically acceptable rates. Essential to this genetic manipulation was the demonstration that unlike most strains of *P. putida* which can grow on toluene, the Tn*4651* component of the pWW0 plasmid was present in the *P. putida* strain isolated and that this 56 kb segment was integrated into the bacterial chromosome, as it turned out at a site near the genes for benzoate metabolism.

The mapping of this 56 kb segment on the chromosome was done in the first instance by conjugation using Hfr strains of *P. putida* constructed using the conjugative plasmid pMO75 in which linkage of the TOL region was demonstrated to markers already located on the genetically characterized strain PPN (SINCLAIR et al., 1986; HOLLOWAY and MORGAN, 1986) (Fig. 12). Thus the utility of this conju-

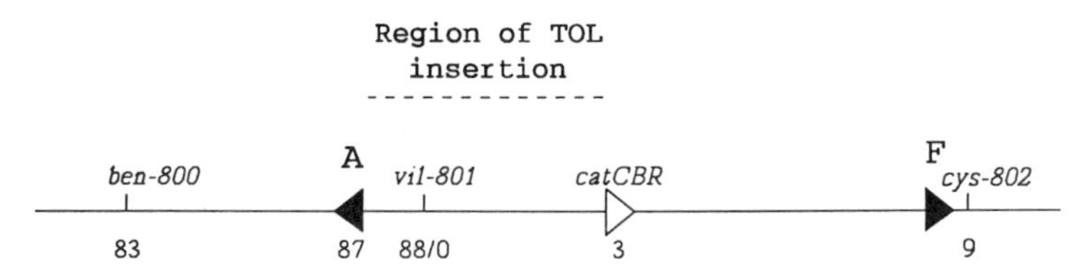

Fig. 12. Genetic map of *Pseudomonas putida* PPN in the 83 to 9 minute region of the chromosome. Solid arrowheads indicate the locations and directions of chromosome mobilization for the pMO22 Hfr origins A and F. The open arrowhead denotes the location and mobilization direction of the *cat*R::pMO75 Hfr origin. Numbers beneath the line show the map position in minutes on the 88 minute *P. putida* PPN genetic map. The dashed line above the map indicates the region of TOL insertion in *P. putida* MW1000. Abbreviations for genetic markers: *ben*, benzoate utilization; *cat*, catechol utilization; *cys*, cysteine requirement; *vil*, valine + isoleucine + leucine requirement (SINCLAIR et al., 1986; DEAN and MORGAN, 1983).

gative system was demonstrated by matings between the newly isolated strain and the characterized laboratory strain. This result stresses the importance of having gene exchange systems which are applicable to a variety of species and strains and the value of genetic mapping of a reference strain.

7 Discussion and Summary

Gene exchange in bacteria is a widespread phenomenon. There has been intensive study of only a relatively few bacteria, but the knowledge so gained has enabled sufficient genetic information to be acquired for a wide range of other bacteria. This information has been highly useful for a variety of practical purposes including nitrogen fixation, biodegradation, and pathogenicity of human pathogens.

Without this background of genetic information it is inconceivable that the development of cloning procedures and other methods for the recombinant DNA analysis could have developed so effectively or so rapidly. The success of these procedures is such that they are uniformly applicable to almost all bacteria, and most of the genetic information needed as a basis for the understanding of microbiological phenomena or the manipulation of bacteria for practical purposes will be obtained by these methods. Hence, while it is unlikely that microbial geneticists of the future will use the conventional gene exchange techniques for mapping or strain construction, these procedures have had a glorious history and the years following LEDERBERG and TATUM's discovery of conjugation and AVERY, McCarty, and McLEOD's demonstration of transformation will be seen as one of the golden eras of biological research.

This does not mean that *in vivo* conjugation or other classical procedures do not have a role to play in the manipulation of bacteria for practical purposes. Indeed, there are situations where manipulation of bacterial strains by recombinant DNA procedures could result in additional regulatory problems, and hence delays in production schedules for production of specialty chemicals by bacteria or the use of recombinant organisms for food processing. A particular example has been cited by GASSON (1990) who points out that bacteriophage resistant dairy starter cultures of *Lactococcus lactis* might be more readily adopted by the food industry if such strains were constructed by *in vivo* techniques. The production of specialty chemicals by bacteria is a further area where there may be economic inhibitors to the use of microbial strains modified by recombinant DNA techniques. The additional expense in meeting regulatory requirements for fermentation equipment, disposing of waste liquors and monitoring could inhibit the development of this potentially large market area. Given the increasing pressure to use chirally pure starting or intermediary materials in the synthesis of pharmaceutical products, microbial procedures have obvious advantages over traditional chemical synthetic methods. LIDSTROM and STIRLING (1990) have listed the potential uses of methylotrophs in the production of specialty chemicals, and *in vivo* techniques of genetic exchange are available for this group of organisms as described above.

Likewise, there is increased interest in the use of microorganisms for bioremediation, and it is unlikely that variants prepared by recombinant DNA procedures will be considered for use in the foreseeable future, given the public concerns on the release of genetically modified organisms. A case can be made that recombinants produced by *in vivo* procedures may be acceptable to regulatory authorities.

There is a need to develop recombinant organisms which can replace chemicals as pesticides in agricultural environments, and there is an understandable resistance to the release of such organisms lest they cause genetic disturbances in the natural flora and fauna. Perhaps as an initial stage, recombinants constructed by *in vivo* techniques could be used as marker organisms to determine the longevity of recombinant bacteria in a new environment. This has already been done with a lactose fermenting pseudomonad although in this case the organism was constructed by recombinant DNA procedures (BARRY, 1986). Field studies on plant–bacterial interactions using recombinant bacteria could probably proceed more quickly if more effort was put into *in vivo* recombinant procedures for constructing the microorganisms involved. The natural modifications of IncP-1 plasmids, for example R68.45 and similar derivatives and the use of R prime plasmids would satisfy most regulatory authorities as "natural" exchanges of genetic material.

In conclusion, it is worthwhile to remember that bacteria have been shown to exchange genetic material with eukaryotes. HEINEMANN and SPRAGUE (1989) demonstrated the conjugal transfer of DNA between *E. coli* and *Saccharomyces* and the interkingdom genetic exchange of *Agrobacterium* has been clearly documented (REAM, 1989). How many other such examples remain to be discovered?

Acknowledgements

Research work in the author's laboratory is supported by the Australian Research Council, the National Health and Medical Research Council and the Celgene Corporation. I wish to thank Dr V. Krishnapillai for his comments and criticisms of the manuscript, Martha Sinclair for providing Figures 10 and 12 and Mrs Jenny Elliston for preparing the manuscript.

8 References

BACHMAN, B. J. (1990), Linkage map of *Escherichia coli* K12, Edition 8, *Microbiol. Rev.* **54,** 130–197.

BAINBRIDGE, B. W., TYPAS, M. A. (1984), Miscellaneous bacteria, in: *Genetics and Breeding of Industrial Microorganisms*, pp. 94–114 (BALL, C. E., Ed.), Boca Raton, Florida: CRC Press.

BALL, C. (Ed.) (1984), *Genetics and Breeding of Industrial Microorganisms*. Boca Raton, Florida: CRC Press.

BARRY, G. F. (1986), Permanent insertion of foreign genes into the chromosomes of soil bacteria, *Bio/Tech* **4,** 446–449.

BARTH, P. T. (1979), Plasmid RP4, with *Escherichia coli* DNA inserted *in vitro*, mediates chromosomal transfer, *Plasmid* **2,** 130–136.

BERG, D. E., HOWE, M. M. (Eds.) (1989), *Mobile DNA*, Washington, DC: American Society for Microbiology.

BERG, C. M., BERG, D. E., GROISMAN, E. A. (1989), Transposable elements and the genetic engineering of bacteria, in: *Mobile DNA*, pp. 879–926 (BERG, D. E., HOWE, M. M., Eds.), Washington, DC: American Society for Microbiology.

BERGH, O., BORSHEIM, K. Y., BRATBAK, G., HELDAL, M. (1989), High abundance of viruses found in aquatic environments, *Nature* **340,** 467–468.

BERTRAM, J., STRÄTZ, M., DÜRRE, P. (1991), Natural transfer of conjugative transposon Tn*916* between gram positive and gram negative bacteria, *J. Bacteriol.* **173,** 443–448.

BIRGE, E. A. (1988), *Bacterial and Bacteriophage Genetics*, 2nd Ed. Berlin: Springer-Verlag.

BLANCO, G., RAMOS, F., MEDINA, J. R., TORTOLERO, M. (1990), Chromosomal linkage map of *Azotobacter vinelandii, Mol. Gen. Genet.* **224,** 241–247.

BRÉFORT, G., MAGOT, M., IONESCO, H., SEBALD, M. (1977), Characterization and transferability of *Clostridium perfringens* plasmids, *Plasmid* **1,** 52–66.

CAIRNS, J. (1963), The chromosome of *Escherichia coli, Cold Spring Harbor Symp. Quant. Biol.* **28,** 43–45.

CAVENAGH, M. M., MILLER, R. V. (1986), Specialized transduction of *Pseudomonas aeruginosa* PAO by bacteriophage D3, *J. Bacteriol.* **165,** 448–452.

CHASSY, B. M., MERCENIER, A., FLICKINGER, J. (1988), Transformation of bacteria by electroporation, *Trends Biotechnol.* **6,** 303–309.

CHATER, K. F., HOPWOOD, D. A. (1983), *Streptomyces* genetics, in: *The Biology of the Actinomycetes*, pp. 230–286 (GOODFELLOW, M., MORDARSKI, M., WILLIAMS, S. T., Eds.), Orlando, Florida: Academic Press.

CHUMLEY, F. G., HENZEL, R., ROTH, J. R. (1979), Hfr formation directed by Tn*10*, *Genetics* **91,** 639–655.

CLEWELL, D. B. (1981), Plasmids, drug resistance, and gene transfer in the genus *Streptococcus, Microbiol. Rev.* **45,** 409–436.

COLEMAN, D. C., SULLIVAN, D. J., RUSSELL, R. J., ARBUTHNOT, J. P., CAREY, B. F., POMEROY, H. M. (1989), *Staphylococcus aureus* bacteriophages mediating the simultaneous lysogenic conversion of β-lysine, staphylokinase and enterotoxin A: molecular mechanism of triple conversion, *J. Gen. Microbiol.* **135,** 1679–1697.

DEAN, H. F., MORGAN, A. F. (1983), Integration of R91-5::Tn*501* into the *Pseudomonas putida* PPN chromosome and genetic circularity of the chromosomal map, *J. Bacteriol.* **153,** 485–497.

DEAN, H. F., ROYLE, P., MORGAN, A. F. (1979), Detection of FP plasmids in hospital isolates of *Pseudomonas aeruginosa, J. Bacteriol.* **138,** 249–250.

DÉNARIÉ, J., ROSENBERG, C., BERGERON, B., BOUCHER, C., MICHEL, M., BARATE DE BERTALMIO, M. (1977), Potential of RP4::Mu plasmids for *in vivo* genetic engineering of gram negative bacteria, in: *DNA Insertion Elements, Plasmids and Episomes*, pp. 507–520 (BUKHARI, A. I., SHAPIRO, J. A., ADHYA, S. L., Eds.), Cold Spring Harbor: Cold Spring Harbor Laboratory Press.

DREXLER, H. (1970), Transduction by bacteriophage T1, *Proc. Natl. Acad. Sci. USA* **66,** 1083–1088.

DRLICA, K., RILEY, M. (Eds.) (1990), *The Bacterial Chromosome*, Washington, DC: American Society for Microbiology.

FRANKLIN, F. C. H., BAGDASARIAN, M., BAGDASARIAN, M. M., TIMMIS, K. N. (1981), Molecular and functional analysis of the TOL plasmid pWW0 from *Pseudomonas putida* and cloning of genes for the entire regulated aromatic ring *meta* cleavage pathway, *Proc. Natl. Acad. Sci. USA* **78,** 7458–7462.

GASSON, M. J. (1990), *In vivo* genetic systems in lactic bacteriology, *FEMS Microbiol. Rev.* **87,** 43–60.

GERDES, K. (1988), The *parB* (*hok/sok*) locus of plasmid R1: a general purpose plasmid stabilization system, *Bio/Tech* **6,** 1402–1405.

GUIDOLIN, A., MANNING, P. A. (1987), Genetics of *Vibrio cholerae* and its bacteriophages, *Microbiol. Rev.* **51,** 285–298.

HAAS, D., HOLLOWAY, B. W. (1976), R factor variants with enhanced sex factor activity in *Pseudomonas aeruginosa, Mol. Gen. Genet.* **144,** 243–251.

HAAS, D., HOLLOWAY, B. W. (1978), Chromosome mobilization by the R plasmid R68.45: A tool in *Pseudomonas* genetics, *Mol. Gen. Genet.* **158,** 229–237.

HAYASHI, T., BABA, T., MATSUMOTO, H., TERAWAKI, Y. (1990), Phage-conversion of cytotoxin production in *Pseudomonas aeruginosa, Mol. Microbiol.* **4,** 1703–1709.

HAYES, W. (1968), *The Genetics of Bacteria and their Viruses. Studies in Basic Genetics and Molecular Biology*, 2nd Ed. Oxford: Blackwell.

HEINEMANN, J. A., SPRAGUE, G. F. (1989), Bacterial conjugative plasmids mobilize DNA transfer between bacteria and yeast, *Nature* **340,** 205–209.

HOLLOWAY, B. W. (1965), Variations in restriction and modification of bacteriophage following increase of growth temperature of *Pseudomonas aeruginosa, Virology* **25,** 634–642.

HOLLOWAY, B. W. (1979), Plasmids that mobilize bacterial chromosome, *Plasmid* **2,** 1–19.

HOLLOWAY, B. W. (1984), Genetics of methylotrophs, in: *Methylotrophs: Microbiology, Biochemistry and Genetics*, pp. 87–106 (HOU, C. T., Ed.), Boca Raton: CRC Press.

HOLLOWAY, B. W. (1986), Chromosome mobilization and genomic organization in *Pseudomonas*, in: *The Biology of Pseudomonas. The Bacteria*, Vol. 10, pp. 230–286 (SOKATCH, J., Ed.), New York: Academic Press.

HOLLOWAY, B. W., MORGAN, A. F. (1986), Genome organization in *Pseudomonas, Annu. Rev. Microbiol.* **40,** 79–105.

HOPWOOD, D. A., CHATER, K. F. (1984), Streptomycetes, in: *Genetics and Breeding of Industrial Microorganisms*, pp. 7–42 (BALL, C., Ed.), Boca Raton: CRC Press.

HOPWOOD, D. A., CHATER, K. F. (1989), *Genetics of Bacterial Diversity*. London: Academic Press.

HOPWOOD, D. A., KIESER, T. (1990), The *Streptomyces* genome, in: *The Bacterial Chromosome*, pp. 147–162 (DRLICA, K., RILEY, M., Eds.), Washington, DC: American Society for Microbiology.

HOPWOOD, D. A., SHERMAN, D. H., KHOSLA, C., BIBB, M. J., SIMPSON, T. J., FERNANDEZ-MORENA, M. A., MARTINEZ, E., MALPARTIDA, F. (1990), "Hybrid" pathways for the production of secondary metabolites, in: *Proc. 6th Int. Symp. Genet. Indust. Microbiol.*, pp. 259–270 (HESLOT, H., DAVIES, J., FLORENT, J., BOBICHON, L., DURAND, G., PENASSE, L., Eds.), Paris: Société Française de Microbiologie.

HSIEK, J.-H., BARER, S. J., MAXWELL, P. C. (1985), Muconic acid productivity by a stabilized mutant microorganism population, *US Patent* 4,535,059.

HUGOURVIEUX-COTTE-PATTAT, N., REVERCHON, S., ROBERT-BAUDOUY, J. (1989), Expanded linkage map of *Erwinia chrysanthemi* strain 3937, *Mol. Microbiol.* **3,** 222–229.

ICHIGE, A., MATSUTANI, S., OISHI, K., MIZUSHIMA, S. (1989), Establishment of gene transfer systems for construction of the genetic map of a marine *Vibrio* strain, *J. Bacteriol.* **171,** 1825–1834.

IPPEN-IHLER, K. A., MINKLEY, E. G. (1986), The conjugation system of F, the fertility factor of *Escherichia coli, Annu. Rev. Genet.* **20,** 593–624.

JOHNSON, S. R., ROMIG, W. R. (1979), Transposon facilitated recombination in *Vibrio cholerae, Mol. Gen. Genet.* **170,** 93–101.

KENNEDY, C., TOUKDARIAN, A. (1987), Genetics of *Azotobacters:* applications to nitrogen fixation and related aspects of metabolism, *Annu. Rev. Microbiol.* **41,** 227–258.

KING, R. (Ed.) (1974), *Handbook of Genetics*, Vol. 1: *Bacteria, Bacteriophages, and Fungi*, New York: Plenum Press.

KLECKNER, N. (1981), Transposable elements in prokaryotes, *Annu. Rev. Genet.* **15,** 341–404.

KLECKNER, N., ROTH, J., BOTSTEIN, D. (1977), Genetic engineering *in vivo* using translocatable drug-resistance elements. New methods in bacterial genetics, *J. Mol. Biol.* **116,** 125–159.

KOKJOHN, T. A., SAYLER, G. S., MILLER, R. V. (1991), Attachment and replication of *Pseudomonas aeruginosa* bacteriophages under conditions simulating aquatic environments, *J. Gen. Microbiol.* **137,** 661–666.

KRAWIEC, S., RILEY, M. (1990), Organization of the bacterial chromosome, *Microbiol. Rev.* **54,** 502–539.

KRISHNAPILLAI, V., ROYLE, P., LEHRER, J. (1981), Insertions of the transposon Tn*1* into the *Pseudomonas aeruginosa* chromosome, *Genetics* **97,** 495–511.

LIDSTROM, M. E., STIRLING, D. I. (1990), Methylotrophs: genetics and commercial applications, *Annu. Rev. Microbiol.* **44,** 27–58.

LONG, S. R. (1984), Genetics of *Rhizobium* nodulation, in: *Plant Microbe Interactions*, pp. 265–306 (KOSUGE, T., NESTER, E., Eds.), New York: Macmillan.

LONG, S. R. (1989), *Rhizobium* genetics, *Annu. Rev. Genet.* **23,** 483–506.

LOVETT, P. S. (1984), *Bacillus*, in: *Genetics and Breeding of Industrial Microorganisms*, pp. 44–62 (BALL, C., Ed.), Boca Raton, Florida: CRC Press.

LOW, K. B., PORTER, D. D. (1978), Modes of gene transfer and recombination in bacteria, *Annu. Rev. Genet.* **12,** 249–287.

LOWBURY, E. J. L., KIDSON, A., LILLY, H. A., AYLIFFE, G. A., JONES, R. J. (1969), Sensitivity of *Pseudomonas aeruginosa* to antibiotics: Emergence of strains highly resistant to carbenicillin, *Lancet* **2,** 448–452.

LYON, B. R., SKURRAY, R. (1987), Antimicrobial resistance of *Staphylococcus aureus:* genetic basis, *Microbiol. Rev.* **51,** 88–134.

LYON, B. R., KEARNEY, P. P., SINCLAIR, M. I., HOLLOWAY, B. W. (1988), Comparative complementation mapping of *Methylophilus* spp. using cosmid gene libraries and prime plasmids, *J. Gen. Microbiol.* **134,** 123–132.

MAXWELL, P. C. (1982), Production of muconic acid, *US Patent* 4,355,107.

MAXWELL, P. C. (1986), Process for the production of muconic acid, *US Patent* 4,588,688.

MAXWELL, P. C., HSIEK, J.-H., FIESCHKO, J. C. (1987), Stabilization of a mutant microorganism population, *US Patent* 4,657,863.

MAZODIER, P., PETTER, R., THOMPSON, C. (1989), Intergeneric conjugation between *Escherichia coli* and *Streptomyces* species, *J. Bacteriol.* **171,** 3583–3585.

MINTON, N. P., BREHM, J. K., OULTRAM, J., SWINFIELD, T. J., SCHIMMING, S., WHELAN, S. E., THOMPSON, D. E. YOUNG, M. STAUDENBAUER, W. L. (1990), Development of genetic systems for *Clostridium acetobutylicum*, in: *Proc. 6th Int. Symp. Genet. Indust. Microbiol.*, pp. 759–770 (HESLOT, H., DAVIES, J., FLORENT, J., BOBICHON, L., DURAND, G., PENASSE, L., Eds.), Paris: Société Française de Microbiologie.

MOORE, A. T., NAYUDU, M., HOLLOWAY, B. W. (1983), Genetic mapping in *Methylophilus methylotrophus* AS1, *J. Gen. Microbiol.* **129,** 785–799.

NAYUDU, M., ROLFE, B. G. (1987), Analysis of R-primes demonstrates that genes for broad host nodulation of *Rhizobium* strain NGR-234 are dispersed on the Sym plasmid, *Mol. Gen. Genet.* **206,** 326–337.

NEIDHARDT, F. C., INGRAHAM, J. L., LOW, K. B., MAGASANIK, B., SCHAECHTER, M. (Eds.) (1987), *Escherichia coli and Salmonella typhimurium, Cellular and Molecular Biology*, Vol. 2, Washington, DC: American Society for Microbiology.

NEIDLE, E. L., ORNSTON, L. N. (1986), Cloning and expression of *Acinetobacter calcoaceticus* catechol 1,2-dioxygenase structural gene *catA* in *Escherichia coli, J. Bacteriol.* **168,** 815–820.

O'BRIEN, S. J. (Ed.) (1990), *Genetic Maps. Locus Maps of Complex Genomes*, 5th Ed., Cold Spring Harbor: Cold Spring Harbor Laboratory Press.

O'HOY, K., KRISHNAPILLAI, V. (1987), Recalibration of the *Pseudomonas aeruginosa* PAO chromosome map in time units using high-frequency-of-recombination donors, *Genetics* **115,** 611–618.

PATTEE, P. A. (1990), Genetic and physical mapping of the chromosome of *Staphylococcus aureus* NCTC83325, in: *The Bacterial Chromosome*, pp. 163–169 (DRLICA, K., RILEY, M., Eds.), Washington, DC: American Society for Microbiology.

PIGGOTT, P. J. (1990), Genetic map of *Bacillus subtilis* 168, in: *The Bacterial Chromosome*, pp. 107–145 (DRLICA, K., RILEY, M., Eds.), Washington, DC: American Society for Microbiology.

PROCTOR, L. M., FUHRMAN, J. A. (1990), Viral mortality of marine bacteria and cyanobacteria, *Nature* **343**, 60–62.

REAM, W. (1989), *Agrobacterium tumefaciens* and interkingdom genetic exchange, *Annu. Rev. Phytopathol.* **27**, 583–618.

ROLFE, B., HOLLOWAY, B. W. (1968), Genetic control of DNA specificity in *Pseudomonas aeruginosa, Genet. Res.* **12**, 99–102.

ROOD, J. I., COLE, S. T. (1991), Molecular genetics and pathogenesis of *Clostridium perfringens, Microbiol. Rev.*, **55**, 621–648.

SANDERSON, K. E., MACLACHLAN, P. R. (1987), F-mediated conjugation, F+ strains and Hfr strains of *Salmonella typhimurium* and *Salmonella abony* in *Escherichia coli* and *Salmonella typhimurium*, in: *Cellular and Molecular Biology*, Vol. 2 (NEIDHARDT, F. C., INGRAHAM, J. L., LOW, K. B., MAGASANIK, B., SCHAECHTER, M., Eds.), pp. 1138–1144, Washington, DC: American Society for Microbiology.

SCAIFE, J., LEACH, D., GALIZZI, A. (Eds.) (1985), *Genetics of Bacteria*, London: Academic Press.

SCHURTER, W., HOLLOWAY, B. W. (1986), Genetic analysis of promoters on the insertion sequence IS*21* of plasmid R68.45, *Plasmid* **15**, 8–18.

SCOLNIK, P. A., MARRS, B. L. (1987), Genetic research with photosynthetic bacteria, *Annu. Rev. Microbiol.* **41**, 701–726.

SEREBRIJSKI, I. G., KAZAKORA, S. M., TSYGANKOR, Y. D. (1989), Construction of Hfr donors of the obligate methanol-oxidizing bacterium *Methylobacillus flagellatum* KT, *FEMS Microbiol. Lett.* **59**, 203–206.

SHIGEKAWA, K., DOWER, W. J. (1988), Electroporation of eukaryotes and prokaryotes: A general approach to the introduction of macromolecules into cells, *BioTechniques* **6**, 742–751.

SINCLAIR, M. I., HOLLOWAY, B. W. (1991), Chromosomal insertion of TOL transposons in *Pseudomonas aeruginosa* PAO, *J. Gen. Microbiol.* **137**, 1111–1120.

SINCLAIR, M. I., MAXWELL, P. C., LYON, B. R., HOLLOWAY, B. W. (1986), Chromosomal location of TOL plasmid DNA in *Pseudomonas putida, J. Bacteriol.* **168**, 1302–1308.

SMITH, C. L., CONDEMINE, G. (1990), New approaches for physical mapping of small genomes, *J. Bacteriol.* **172**, 1167–1172.

STANISICH, V. A., HOLLOWAY, B. W. (1971), Chromosome transfer in *Pseudomonas aeruginosa* mediated by R factors, *Genet. Res.* **17**, 169–172.

STEWART, G. J., CARLSON, C. A. (1986), The biology of natural transformations, *Annu. Rev. Microbiol.* **40**, 211–235.

STOUT, V. G., IANDOLO, J. J. (1990), Chromosomal gene transfer during conjugation by *Staphylococcus aureus* is mediated by transposon-facilitated mobilization, *J. Bacteriol.* **172**, 6148–6150.

STROM, D., HIRST, R., PETERING, J., MORGAN, A. (1990), Isolation of high frequency of recombination donors from Tn*5* chromosomal mutants of *Pseudomonas* PPN and recalibration of the genetic map, *Genetics* **126**, 497–503.

SUGINO, Y., HIROTA, Y. (1962), Conjugal fertility associated with resistance factor R in *Escherichia coli, J. Bacteriol.* **84**, 902–910.

SUWANTO, A., KAPLAN, S. (1989a), Physical and genetic mapping of the *Rhodobacter sphaeroides* 2.4.1 genome: genome size, fragment identification and gene localization, *J. Bacteriol.* **171**, 5840–5849.

SUWANTO, A., KAPLAN, S. (1989b), Physical and genetic mapping of the *Rhodobacter sphaeroides* 2.4.1 genome: presence of two unique circular chromosomes, *J. Bacteriol.* **171**, 5850–5859.

TATRA, P. K., GOODWIN, P. M. (1985), Mapping of some genes involved in C-1 metabolism in the facultative methylotroph *Methylobacterium* sp. strain AM1 (*Pseudomonas* AM1), *Arch. Microbiol.* **143**, 169–177.

THOMAS, C. T., SMITH, C. A. (1987), Incompatibility group P plasmids: Genetics, evolution and use in genetic manipulation, *Annu. Rev. Microbiol.* **41**, 77–101.

TRIEU-CUOT, P., CARLIER, C., COURVALIN, P. (1988), Conjugative plasmid transfer from *Enterococcus faecalis* to *Escherichia coli, J. Bacteriol.* **170**, 4388–4391.

TSUDA, M., IINO, T. (1987), Genetic analysis of a transposon carrying toluene degrading gene on a TOL plasmid pWW0, *Mol. Gen. Genet.* **210**, 270–278.

TSUDA, M., IINO, T. (1988), Identification and characterization of Tn*4653*, a transposon covering the toluene transposon Tn*4651*, on TOL plasmid pWW0, *Mol. Gen. Genet.* **213**, 72–77.

TSYGANKOV, Y. D., KAZAKOVA, S. M., SEREBRIJSKI, I. G. (1990), Genetic mapping of the obligate methylotroph *Methylobacillus flagellatum:* Characteristics of prime plasmids and mapping of the chromosome in time-of-entry units, *J. Bacteriol.* **172**, 2742–2754.

UETAKE, H., TOYAMA, S., HAGIWARA, S. (1964), On the mechanism of host-induced modification. Multiplicity activation and thermolabile factor responsible for phage growth restriction, *Virology* **22**, 203–213.

VAN DER LELIÈ, D., CHAVARRI, F., VENEMA, G., GASSON, M. J. (1991), Identification of a new genetic determinant for cell aggregation association with lactose plasmid transfer in *Lactococcus lactis, Appl. Environ. Microbiol.* **57,** 201–206.

VERMA, N. K., BRANDT, J. M., VERMA, D. J., LINDBERG, A. A. (1991), Molecular characterization of the O-acetyl transferase gene of converting bacteriophage SF6 that adds group antigen 6 to *Shigella flexneri, Mol. Microbiol.* **5,** 71–75.

WEAVER, K. E., CLEWELL, D. B. (1990), Regulation of the PAD1 sex pheromone response in *Enterococcus faecalis:* Effects of host strain and *traA, traB*, and *C* region mutants on expression of an E region pheromone - inducible *lacZ* fusion, *J. Bacteriol.* **172,** 2633–2641.

WILKE, D. (1980), Conjugational gene transfer in *Xanthobacter autotrophicus* GZ29, *J. Gen. Microbiol.* **117,** 431–436, 1980.

WILLETTS, N. S., CROWTHER, C., HOLLOWAY, B. W. (1981), The insertion sequence IS*21* of R68.45 and the molecular basis for mobilization of the bacterial chromosome, *Plasmid* **6,** 30–51.

ZINDER, N., LEDERBERG, J. (1952), Genetic exchange in *Salmonella, J. Bacteriol.* **64,** 679–691.

3 Genetic Exchange Processes in Lower Eukaryotes

ULF STAHL

Berlin, Federal Republic of Germany

KARL ESSER

Bochum, Federal Republic of Germany

1 Introduction

Genetic exchange processes, in the broadest sense, are brought about by any combination of genetically different individuals leading to an offspring of genotypes different from that of either parent. This definition naturally implies a recombination of the genetic material concerned. Mutational events, which are naturally restricted to single individuals of cells, can be combined with different genetic backgrounds and will be distributed very effectively within a population, and the chance will be extended to give rise to selective advantages. This is true regardless of whether the mutation occurred spontaneously or was induced. Thus, it is evident that genetic exchange processes were, and still are, the important feature which permits the interaction of the three basic parameters of evolution: mutation, recombination, and selection.

Genetic exchange processes have been used since early times to breed crops or domesticated animals but it was done subconsciously without any understanding of the underlying mechanism and, even with wrong assumptions (Moses, Book 1 (Genesis) Chapter 30, Verse 32–42). Since the beginning of the present century, breeding has been carried out more consciously, at least with higher organisms. The few lower organisms, such as yeasts, which were previously of economic interest, were used in the traditional way by vegetative propagation and were not considered proper subjects for breeding. This situation changed abruptly when – mainly initiated by the discovery of penicillin – the industrial use of fungi and later on, of bacteria, created the concepts of biotechnology. At this point it was evident that for most microorganisms characteristics such as life cycles, mode of fertilization, or systems for a concerted breeding program (the combined use of mutation, recombination, and selection) were not known (ESSER, 1971). Furthermore, it became evident that most fungi of industrial interest do not even have a sexual life cycle (fungi imperfecti). Such fungi, and also bacteria, can only be hybridized by parasexual procedures.

Thus, in discussing genetic exchange in lower eukaryotes, we must distinguish between the following processes: sexual hybridization exclusively for lower eukaryotes possessing a sexual cycle; and parasexual hybridization for imperfect eukaryotes lacking a sexual cycle and certainly for tissue cultures of higher eukaryotes. Since most eukaryotic organisms used for biotechnological processes are fungi, we will restrict the following statements mainly to these organisms. The following literature reviews should be consulted for further information (ESSER and KUENEN, 1967; ESSER, 1971, 1981, 1986; VAN DEN ENDE, 1976; SMITH and BERRY, 1975; BURNETT, 1975; FINCHAM and DAY, 1979).

Since the practical application of any scientific knowledge is dependent on the level attained by basic research, the application of sexual and parasexual breeding in the past decade has been a stimulus to the field of biotechnology. In making use of appropriate genetic mechanisms and elaborated technology it was possible to improve yeasts and filamentous fungi of different genera. The disadvantage (at least partially) of sexual or parasexual recombination processes is that it is not possible to limit recombination to the desired genetic qualities. This is due to the fact that any possible recombination can take place whereby the genotypes are a random product. Undesired genetic qualities must, through continuous crossbreeding, be removed from the descendents, which can be relatively time-consuming. An additional problem is that sexual processes, such as genetic exchange and recombination, are only possible within a species. This species specificity can be overcome, at least partially, by parasexual processes.

A breakthrough in the efforts to transfer a specific gene or genetic characteristics from a donor to a recipient was achieved with the discovery of endogenous plasmids in yeast and filamentous fungi. Another fortunate discovery was that these plasmids may act interspecifically, and as a result the characterization of regulatory structures on a molecular level could be tackled. Moreover, the construction of vector/host systems for biotechnological problems could be initiated. In particular, the implementation of using fungi in genetic engineering succeeded in transferring specific genes to more manageable strains. In combination with formal genetics, concerted breeding programs became effectively possible (ESSER, 1989; ESSER and MOHR, 1990; MOHR and ESSER, 1992; MEINHARDT et al., 1990; ESSER et al., 1986; TIMBERLAKE, 1985; VAN DEN HONDEL, 1991).

2 Sexual Recombination

2.1 Principle

The succession of karyogamy and meiosis is the principle of sexual propagation, i.e., the fusion of two haploid nuclei which form a diploid nucleus and subsequent meiotic divisions. This permits a rearrangement and a reorganization of chromosomes (recombination). It follows from this that sexual processes are restricted to eukaryotes. For the concept of karyogamy, the mode of bringing the nuclei together is important, that is, whether there is plasmogamy between gametes, gametangia, somatic cells, or even between artificially produced protoplasts.

2.2 Life Cycles

The two landmarks of sexual propagation, karyogamy and meiosis, which characterize the alteration of the nuclear phase (diploid/haploid), are not subsequent events. They are preceded, or followed, by mitotic divisions which lead to the formation of somatic cells (generally in multicellular organisms). Different life cycles are established depending on the occurrence of mitosis in the haploid phase, in the diploid phase, or in both phases (Fig. 1a, b, c).

For some fungi, however, the onset of karyogamy is an important feature of their cycles in addition to karyogamy and meiosis. This is due to the fact that after the fusion of the mates karyogamy is postponed by a number of conjugated mitotic divisions of the gametic nuclei, thus establishing the so-called dikaryotic phase (Fig. 1d, e). That means that the nuclei of both sexual partners share a common cytoplasm but do not fuse. Although the sexual process has been initiated by plasmogamy, the resulting dikaryon remains haploid, albeit via complementation dominance, epistasy and other effects, typical of diploids, resimulated. Because of these peculiarities the following two cycles, which are often found in fungi, are treated separately from the others.

2.2.1 Haploid Life Cycle: Karyogamy – Meiosis – Mitosis

The somatic cells are haploid, and the only diploid stage is the zygote, since plasmogamy, karyogamy, and meiosis are subsequent events. This life cycle is found predominantly in the lower fungi, such as the Mucorales, and in many algae. Asexual propagation, if present, occurs by means of haploid spores.

2.2.2 Diploid Life Cycle: Karyogamy – Mitosis – Meiosis

The somatic cells of the organisms are diploid, only the gametes or their equivalents are haploid, since plasmogamy and karyogamy are

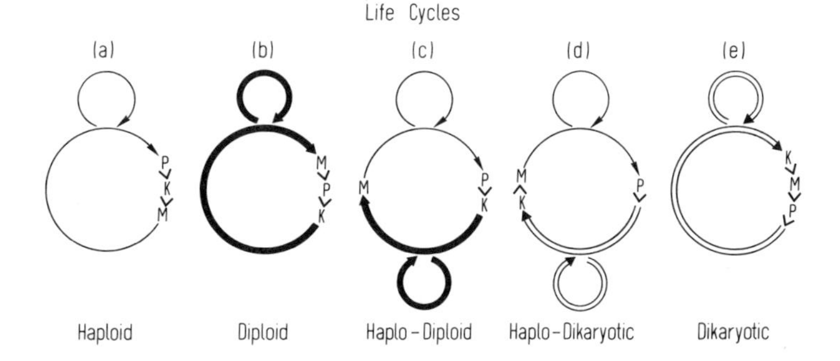

Fig. 1. Scheme of essential life cycles. – Thin line: haploid phase. Heavy line: diploid phase. Double line: dikaryotic phase. K: karyogamy. M: meiosis. P: plasmogamy. The smaller circles represent the possibility of asexual propagation.

followed by a series of mitotic divisions. This cycle is predominant in higher organisms, but it is also found in some fungi (e.g., *Saccharomyces cerevisiae* and water molds) and in algae. Asexual propagation, if present, occurs by means of diploid spores.

2.2.3 Haplo-Diploid Life Cycle: Karyogamy - Mitosis - Meiosis - Mitosis

The life cycle comprises two generations, i.e., haploid and diploid generations alternate. It occurs infrequently in fungi; for instance, in some Blastocladiales of the section *Eu-Allomyces* and in the alkane yeast, *Yarrowia lipolytica*, but more often in algae and predominantly in mosses, ferns, and higher plants. The two generations may be morphologically alike, as for cited examples of fungi or as in many algae. Or they may differ more or less in their habits, as in higher organisms. Asexual propagation by spores may occur in both nuclear phases.

2.2.4 Haplo-Dikaryotic Life Cycle: Karyogamy - Meiosis - Mitosis - Plasmogamy - Mitosis

In haploid organisms the onset of karyogamy is postponed by mitotic divisions after plasmogamy. This cycle is generally found in higher Ascomycetes. Asexual propagation is carried out mainly by haplospores.

2.2.5 Dikaryotic Life Cycle: Karyogamy - Meiosis - Plasmogamy - Mitosis

This cycle is widely encountered among higher Basidiomycetes. The whole somatic phase is dikaryotic, because plasmogamy occurs immediately after the formation of meiospores. In most of these fungi the regular distribution of the two parental nuclei within the dikaryon is brought about by a very special mode of cellular division called "clamp formation". Asexual propagation is carried out predominantly by dikaryotic spores but also by haploid spores.

We have confined ourselves to a general description of life cycles and abstained from the description of specific cycles because a description has been published elsewhere (ESSER and KUENEN, 1967; ESSER, 1986).

2.3 Breeding Systems

In the preceding outline of various life cycles the manifold restrictions interfering at various steps have not been considered. These may be described as a second set of parameters which are involved in the sexual process, the breeding systems. These systems regulate the outbreeding of a population by interfering with inbreeding via self-fertilization. In this manner they promote a continuous mixture of the genetic material, i.e., by hybridization. Numerous attempts have been made to describe the action and interaction of the various breeding systems mostly by using different and specific terms (KNIEP, 1928; HARTMANN, 1956; RAPER and RAPER, 1966; BURNETT, 1975; BERGFELD, 1977; BRIEGER, 1930; WHITEHOUSE, 1949).

Some time ago we tried to summarize the various breeding systems under a general point of view (ESSER, 1971), i.e., on the basis of monoecism and dioecism. However, we defined these terms, derived from botany, not by morphological but by physiological criteria. This was done with consideration of the capacity of an organism to contribute either both nuclei or only one to karyogamy. In the latter a partner is required to contribute the second nucleus. As can be seen from Fig. 2, a *monoecious organism* may act either as donor (male) or as acceptor (female) of a nucleus.

A *dioecious organism* has only one or the other capacity. We are dealing with morphological dioecism if this capability is correlated with sexual differentiation. Otherwise, we are dealing with physiological dioecism, if recognizable sexual differences are absent as, for instance, in some yeasts whose undifferentiated haploid cells fuse. Naturally, both modes of

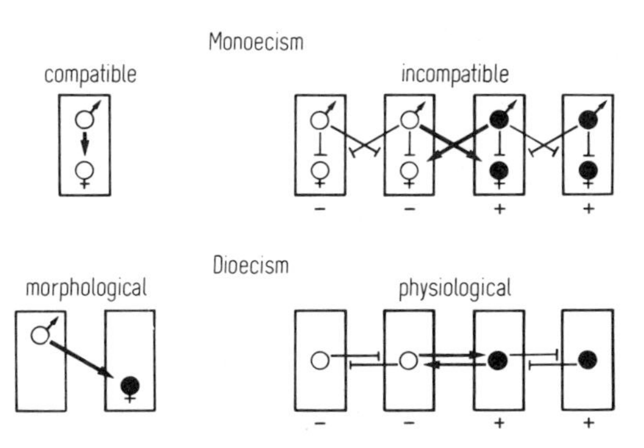

Fig. 2. Schematic representation of essential breeding systems. – Rectangles represent single individuals. Circles represent nuclei. Full or empty circles indicate the heterogeneity of nuclei. Sex symbols show the male or female determination of the nuclei. Arrows: karyogamy possible. Blocked arrows: no karyogamy. + and – indicate different mating types. For further explanation see text. (From ESSER, 1971).

dioecism are under genetic control by nuclear genes (see Fig. 2). The genetic diversity of physiological dioecism is mostly caused by a monoallelic (monogenetic) difference expressed as two mating types, called + and –, or a and α. Morphological dioecism may be under multigenetic control, or it may even be caused by sex chromosomes which are responsible for male and female individuals.

The distribution of breeding systems among microorganisms may be summarized as follows. A few cases of morphological dioecism are found in lower fungi. The same is true for physiological dioecism, with the exception of yeasts (Ascomycetes). But incompatibility determines the sexual processes in many higher Ascomycetes and Basidiomycetes (see list in ESSER and KUENEN, 1967).

We are well aware that each system of classification forced upon nature is incomplete and omits exceptions. But we think that the systems discussed above may, at least, give the non-biologists engaged in the field of biotechnology an overview of the many possibilities for control of recombination which nature has evolved in fungi.

During the last few years two typical physiological dioecious organisms, namely, *Saccharomyces cerevisiae* and *Schizosaccharomyces pombe,* have been elucidated (BEACH, 1983; NIELSEN and EGEL, 1989; NASMYTH, 1983; HERSKOWITZ, 1989; KLAR, 1989). Despite the evolutionary divergence between *S. cerevisiae* and *S. pombe* there are clear parallels in the molecular organization and function of the mating-type genes. In both yeasts both the plus and the minus cells possess a mating-type region which consists of three components: the so-called expression-cassette, containing the plus or the minus allele, and two storage loci, keeping in readiness a silent copy of both the plus and the minus allele. These two silent cassettes serve as a donor, as their information can be transposed to the expression cassette (normally determined as *mat*) during a switching of the mating type. In *S. pombe,* for instance, each of the three mating-type components consist of about 1100 base pairs (bp) and are separated by inserts of about 15 kb (KELLY et al., 1988). Depending on which of the two silent cassettes are inserted at the expression site, the cell is physiologically determined, plus or minus, since only the appropriate genetic information is transcribed. Although investigations have, until now, revealed some differences in the molecular organization of the mating-type alleles of the two yeasts (in *S. cerevisiae* the plus (a)-allele codes for a single polypeptide, the minus (α)-allele for two polypeptides; in *S. pombe* each of the two mating-type alleles for two polypeptides), it is evident that the gene products are regulatory proteins which activate, or repress, the transcription of numerous genes dispersed to different chromosomes.

As soon as haploid cells carrying different mating types and, therefore, excreting different sexual hormones (pheromones) undergo sexual fusion, the gene products of the plus/minus mating types, now in a common cytoplasm, regulate, as a complex, all the functions necessary for the sexual morphogenesis-like meiosis and sporulation.

Based on the recombinant organization of the mating-type region in *S. cerevisiae* and *S. pombe,* the mating type of the haploid cell can change itself from plus to minus (or from minus to plus). This so-called mating-type switching, whose frequency is strain-specific, is essentially a unidirectional gene conversion of non-homologous DNA. Both elicitation and resolution of the recombination are guided by flanking boxes of homologous DNA, present at all three cassette loci (KELLY et al., 1988).

It is evident from the foregoing that only dioecism favors hybridization, whereas monoecism restricts this intermixture of genetic material. In contrast to animals, morphological dioecism is rather rare in fungi and physiological dioecism is infrequent. Therefore, nature has evolved another breeding system which also causes a rather high degree of hybridization. This is homogenic incompatibility, which inhibits not only self-fertilization in monoecious organisms but also cross-fertilization between homogenic nuclei. Homogenic incompatibility, like dioecism, is controlled by genes; in the simplest case by a pair called + and −, or A and a. Unfortunately, the two types of hermaphrodite organisms are also called mating types, as in the breeding system of physiological dioecism. This may cause confusion, but a careful study of Fig. 2 should make the differences evident by showing that incompatibility overlaps monoecism.

Homogenic incompatibility may also be controlled by more than two loci, or even more complex genetic structures leading to 4 mating types. In this so-called tetrapolar mechanism (in contrast to the bipolar +/− mechanism) compatibility (= karyogamy) takes place only if the nuclei differ genetically at all the loci involved.

In comparison to physiological dioecious organisms, the molecular organization of the mating types of monoecious organisms seems to be different. As has been established mainly in the Ascomycetes *Neurospora crassa* (GLASS et al., 1988) and *Cochliobolus heterotrophus* (YODER et al., 1989), the two mating types (plus and minus) a and A, are not organized cassette-like, as in yeasts. Since the mating types consist of non-homologous DNA regions, one cannot use the term alleles to designate these very different loci which do not seem to have evolved from a single ancestral gene. Therefore, the term *idiomorph* has been proposed (METZENBERG, 1990).

Like in yeasts, each of the mating types, at least in *N. crassa* (STABEN and YANOFSKY, 1990) consists of the region of 3.2 kb (mating type a) and 5.3 kb (mating type A). Each of them codes for at least one or two polypeptides which are responsible for several functions, such as mating-type designations, heterokaryon incompatibility, and fruiting-body induction (GLASS and STABEN, 1990). The interaction and regulation of the mating-type products are presumed to be similar to these in yeasts.

The mating in Basidiomycetes is controlled by two pairs of mating-type factors, A and B, which may consist as in *Schizophyllum commune* (RAPER and RAPER, 1966) and *Coprinus cinereus* (CASSELTON et al., 1989) of two multiallelic loci each (α and β). In *S. commune* or *C. cinereus* the B-factors, in *Ustilago maydis,* the A-factor is responsible for cell fusion and nuclear migration. The A-factors (*S. commune, C. cinereus*) and the B-factor (*U. maydis*) seem to regulate the formation of clamp connections and hence the further sexual development and pathogenicity, respectively. In the corn pathogen *Ustilago maydis* the two mating-type factors A and B seem to resemble single loci (BANUETTE and HERSKOWITZ, 1989). As has long been known, according to formal genetic investigations, both mating-type factors of maters must be different to induce cell fusion and the transition of the homokaryons to a dikaryon. This mating process culminates eventually, depending on external factors, in fruiting, meiosis, and sporulation (ESSER, 1986).

In *Ustilago maydis*, where the molecular exploration of the mating types has made the greatest progress, the A-factor alleles (at present only two are known) consist of non-identical DNA (A_1 has a length of 9 kb and A_2 of 5 kb) and thus bear some resemblance to the a/A idiomorphs of *Neurospora crassa* (FROELIGER and KRONSTADT, 1990). It encodes, in addition, for a pheromone and a pheromone receptor (BÖLKER et al., 1992; GILLISSEN et al., 1992). The B-factor, in contrast, is a true multi-allelic locus (25 alleles known). The deduced polypeptides are very similar and contain a homeodomain, which suggests that the B-polypeptides are DNA-binding proteins (SCHULTZ, et al., 1990). A similar organization of the B-locus of *U. maydis* is found in the $A\alpha$-locus of *Schizophyllum commune* (NOVOTNY et al., 1991; ULLRICH et al., 1991) and the A-factor of *Coprinus cinereus* (KÜES and CASSELTON, 1992; KÜES et al., 1992). In this factor, however, several redundant gene pairs have been detected.

3 Restrictions

In the preceding sections we have seen that hybridization of the genetic material, even within a single species or race, does not occur arbitrarily. On the contrary, it is channeled into certain tracks allowing only specific cells or specific nuclei to fuse. But in addition to these restrictions involved in the life cycles, or caused by the breeding systems, there is another widespread restriction system which inhibits nuclear fusion even between male and female, or + and −, nuclei, and goes even further. If nuclei, compatible according to the previously described breeding systems within the frame of the appropriate life cycle, are brought together via plasmogamy, they cannot co-exist in this common physiological machinery. They disintegrate and the cell dies. This reaction is brought about by single genes. In the simplest case a single allelic difference is sufficient to cause this self-destruction of the cell. In order to distinguish this restriction mechanism from the sexual incompatibility described above, this phenomenon was called heterogenic or vegetative incompatibility (ESSER, 1971).

This restriction is a basic biological phenomenon which is widespread in nature in both plants and animals (ESSER and BLAICH, 1973). It occurs not only in the sexual phase, but also in the asexual phase as well as in prokaryotes. Naturally, according to the physiological diversities of the organisms, the molecular basis and the physiological mechanisms of heterogenic incompatibility are greatly varied: at the molecular level bacterial endonucleases restrict the compatibility of prokaryotic DNA (ARBER and LINN, 1969). Heterogenic incompatibility in fungi (sexual and asexual) occurs at the cellular level. In mammals the well-known tissue incompatibility is found after organ transplantation (BACH, 1976).

Heterogenic incompatibility is frequently found between strains of the same species

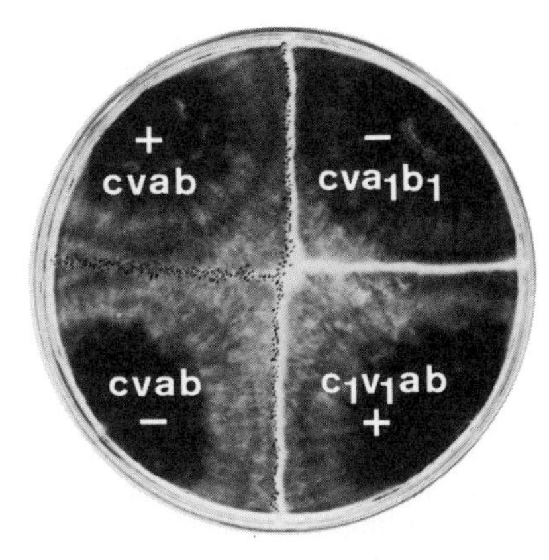

Fig. 3. Heterogenic incompatibility between four strains of the ascomycete *Podospora anserina*. - Left side of Petri dish: + and − strains of the same race are compatible and form fruiting bodies. The small black spots are the perithecia. Right side of Petri dish: + and − strains of different races do not form fruiting bodies due to heterogeneity of 4 loci. It can also be seen that digenic heterogeneity is not sufficient to suppress fruiting completely (compare left and right side). The white zones between some strains are the result of hyphal interaction leading to cellular death and show the effect of heterogenic incompatibility. (From ESSER, 1959).

from different geographical areas. An example of this phenomenon can be demonstrated in the ascomycetous-hyphal fungus *Podospora anserina* which was analyzed not only by formal genetic methods (ESSER, 1959), but also on a molecular level (ROSSIGNOL and PICARD, 1991). This type of restriction interferes in a most drastic manner with the exchange of genetic material. It decreases recombination, as may be seen from Fig. 3.

Since the molecular analysis of all the known incompatibility genes in *Podospora anserina* (at least nine) is at the very beginning, a general prediction concerning the organization and function cannot be made. As an example, the incompatibility genes, s and S, should be mentioned. These were recently determined as two-alleles of one gene (TURCQ et al., 1990). They both encode for 30 kDa polypeptides which differ from each other by 14 amino acids. As far as they are co-expressed after cell fusion, they obviously trigger the incompatibility reaction by increased proteolytic activity (TURCQ et al., 1991).

From a basic point of view heterogenic incompatibility initiates the speciation. But it is even more important to biotechnology, because one cannot expect that "fresh genetic information" taken from nature by screening techniques will be suitable for genetic exchange processes.

In conclusion it must be emphasized that any concerted breeding requires, in addition to the knowledge of the life cycle of the organism, a consideration of the potential impediments brought about by heterogenic incompatibility.

4 Practical Application of Meiotic Recombination

It is difficult to estimate to what extent genetic exchange processes in context with the sexual cycle have been used for strain improvement of lower eukaryotes: algae, fungi, and part of the mosses. Only fungi (yeasts and hyphal fungi) have, so far, biotechnological relevance. The most important handicap is the fact that only a few of these organisms have a sexual cycle. Therefore, from a practical point of view, a very limited number of fungi may be improved by sexual exchange processes. One may also suppose that companies which use fungi in their production process have more or less intensively tried to improve the performance of the strains by selection and mutation. The additional integration of the sexual recombination is, in most cases, only possible after some expensive research, which does not seem to justify the expected profit. But as may be seen from the few published examples listed in Tab. 1, strain improvement by sexual recombination can compensate for the time-consuming procedure. Especially in combination with mutation and selection, the yield of biomass or metabolites can be increased tremendously. In this connection, one should also remember that all the breeding successes of domesticated plants and animals are based on sexual recombination.

In contrast to higher eukaryotes, where the phenotypes and/or genotypes of the offspring of matings can only be evaluated by random analysis, only lower eukaryotes provide the possibility of tetrad analysis. With this evaluation scheme (Fig. 4) all four products of a single meiotic event can occasionally be ascertained. It permits direct conclusions about the number and mode or kind of the recombinational events which took place during meiosis. Depending on whether the products (sexual spores) of a single meiotic event are arranged randomly (unordered tetrad) as in Phycomycetes, Zygomycetes, in some Ascomycetes and in Basidiomycetes or as a linear row (ordered tetrad) as in several Ascomycetes, the informational content obtainable by a tetrad analysis differs considerably. As only ordered tetrads (as shown in Fig. 4) make it possible to discriminate between the first and second meiotic division, recombinational events leading to a distinct segregation can be assigned to one of the two divisions. Some additional predictions are possible, since the centromeres of the chromosomes can be used as reference points for the localization of genes under investigation. These are listed as examples (ESSER and KUENEN, 1967):

- Distance (in map or morgan units) between centromere and a distinct gene
- Localization of a gene on the left or right arm of a distinct chromosome
- Recognition of mutual influence of recombination events
- Elaboration of accurate chromosome maps.

In Tab. 1 lower organisms are listed which have been improved by sexual recombination. It does not contain perfect fungi which have only been the subject of fundamental research. Although the list is not complete (companies do not publish procedures leading to higher yields principally for protecting proprietary rights), two main fields of application of meiotic recombination can be ascertained: solid state fermentation of predominantly plant waste for biomass production (e.g., edible mushrooms, single-cell protein) and submerged cultures for the production of primary or secondary metabolites. Examples from both

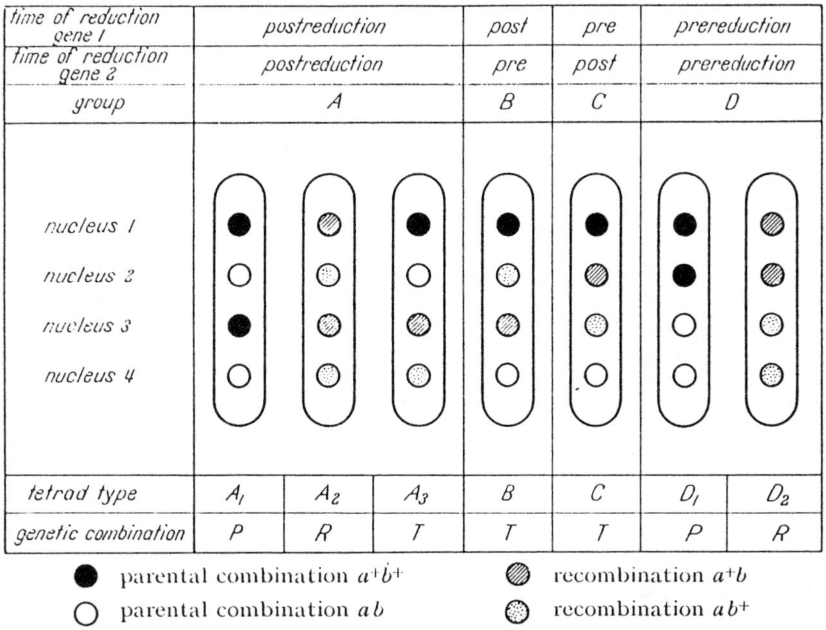

time of reduction gene 1	postreduction			post	pre	prereduction	
time of reduction gene 2	postreduction			pre	post	prereduction	
group	A			B	C	D	
nucleus 1							
nucleus 2							
nucleus 3							
nucleus 4							
tetrad type	A_1	A_2	A_3	B	C	D_1	D_2
genetic combination	P	R	T	T	T	P	R

parental combination a^+b^+

parental combination $a\,b$

recombination a^+b

recombination $a\,b^+$

Fig. 4. The seven types of ordered tetrads which are theoretically possible in the two-factor cross $a^+b^+ \times a\,b$. The nuclei, which arise through the two divisions of meiosis, are designated by circles. In eight-spored Ascomycetes each of the meiotic products divides by mitosis into two identical daughter nuclei. Explanation in the text.

Tab. 1. Survey of Biotechnologically Relevant Perfect Fungi which have been Improved by Sexual Hybridization

Organism	Goal	Reference
Agaricus bitorquis	Biomass	FRITSCHE, 1976
Yarrowia lipolytica	Single-cell protein	ESSER and STAHL, 1976
Agrocybe aegerita	Biomass, recycling or plant wastes	ESSER et al., 1974; MEINHARDT, 1980
Pleurotus ostreatus	Biomass	EGER et al., 1976
Saccharomyces cerevisiae	Biomass, ethanol	WINDISCH et al., 1976; RUSSEL et al., 1987
	Elimination of undesired wine-making properties	ESCHENBRUCH et al., 1982; THORNTON, 1982
	Ethanol tolerance	CHRISTENSEN, 1987; RUSSEL et al., 1987
Aspergillus nidulans	Penicillin	MERRICK, 1975a, b; MERRICK and CATEN, 1975
Claviceps purpurea	Ergot alkaloids	ESSER and TUDZYNSKI, 1978
Emericellopsis terricola	Penicillin	FANTANI, 1962
Pleurotus mutilis	Pleuromutilin	KNAUSEDER and BRANDL, 1976

groups have been chosen for further discussion.

5 Concerted Breeding

5.1 Concerted Breeding of Edible Mushrooms

Most of the reports dealing with the breeding of edible mushrooms contain data concerning the selection of appropriate wild isolates or the optimization of nutrients and/or environmental conditions. Only a few of the reports deal with the utilization of sexual recombination for strain improvement. This is not surprising, as for most of these fungi a knowledge of their cycles or of the genetic control of fruiting is still incomplete. This is also true for the interaction of genetic and environmental factors for the induction of fruit bodies.

Due to the early discovery by KNIEP (1928) and the detailed studies of RAPER and his collaborators (1966), the action of the mating-type factors responsible for the transition from the monokaryophase to the dikaryophase, and, therefore, a prerequisite for fruit body production, has been elucidated. However, this knowledge cannot be generalized, as, depending on the life cycle, the dikaryophase may already be established in germinating spores, as in *Agaricus bisporus* (RAPER and RAPER, 1972). In addition, the following steps depend on strain-specific morphogenetic genes: the genetic control of the differentiation of fruiting bodies and especially the conversion of the undifferentiated hyphal growth to a plektenchymatic fruiting body (STAHL and ESSER, 1976; MEINHARDT, 1980). As shown by the white rot fungus *Agrocybe aegerita,* which decays cellulose but mainly lignin, and can be grown on straw and wood wastes, subsequent action of at least three genes is required for the fruit body formation.

Once a dikaryon is established by the interaction of the incompatibility factors A and B, the plektenchymatic growth, leading via fruit initials (fi) to fruiting bodies (fb) (Fig. 5), can only be initiated if the corresponding genes are derepressed. Formal genetic analysis, not only for *A. aegerita* (MEINHARD and ESSER, 1981), but also for the wood rotting fungus, *Polyporus ciliatus* (STAHL and ESSER, 1976) showed that the threshold of undifferentiated hyphal growth to sexual differentiation (i.e., fruit body formation, karyogamy, meiosis, and spore development) is brought about by a suppressor gene, in its active (su^+) or inactive (su) form. As soon as it is "switched on" (e.g., by environmental factors), the whole sexual differentiation will be accomplished.

For practical purposes it is interesting to note that the dose of fi^+ and fb^+ genes (Fig. 6) is responsible for the length of time in fruiting as well as for the yield of biomass. Since similar control mechanisms for fruiting are also found in two other wood-decaying Basidiomycetes (*Polyporus ciliatus,* STAHL and ESSER, 1976; *Schizophyllum commune,* ESSER et al., 1979), the gene sequence for fruiting might also be present in other fungi. This should permit the improvement not only of biomass productivity of mushrooms but also the ability to decay plant wastes for feeding purposes.

Without being aware of the underlying mechanism, it was shown, especially for *Agaricus bitorquis,* that heterokaryons result-

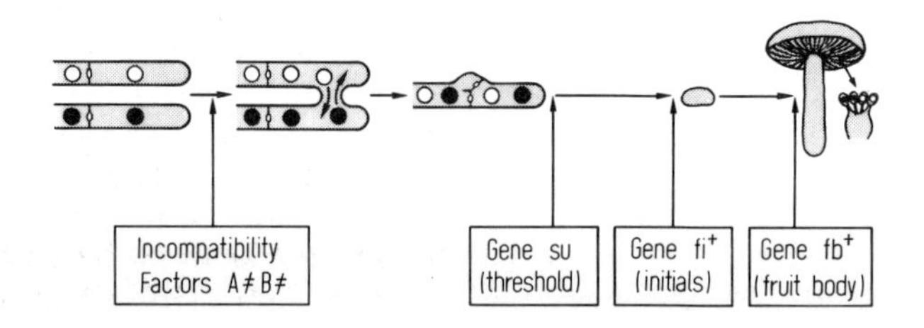

Fig. 5. Genetic control of fruiting in *Agrocybe aegerita.* – For further details see text.

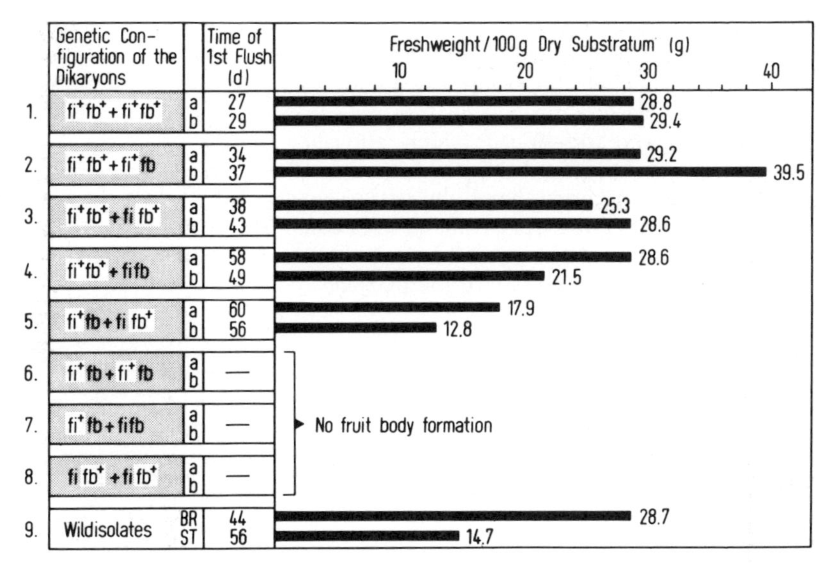

Fig. 6. Correlation between weight of fruit bodies and the dose of genes fi^+ and fb^+, a and b are parallel series of experiments grown on straw.

ing from the hybridization of two wild isolates, resulted in an increase of up to 50% in yield, compared to the parental strains (FRITSCHE, 1976). This so-called hybrid breeding, which is well known in the breeding of animals and higher plants, can be used to increase not only the biomass of mushrooms but also the yield of yeasts for single-cell proteins (ESSER and STAHL, 1976) or the production of metabolites in general (see below).

5.2 Concerted Breeding of Metabolite Producers

As it seemed to be less time-consuming to improve the yield of metabolites, for example, antibiotics, by selection and mutation, only a few attempts to increase the productivity of fungi by meiotic recombination have been published. Additionally, one has to have in mind that only a minority of relevant metabolite producers have a perfect life cycle.

An example is the successful application of sexual recombination to improve a metabolite by HOLT and McDONALD (1968a, b) and by MERRICK and CATEN (1975), who used *Aspergillus nidulans* as model organism for the production of penicillin. It was detected that most wild strains, isolated from nature, are separated by genetic barriers which only inhibit hyphal anastomosis (vegetative heterogenic incompatibility). Therefore, different strains represent, more or less, isolated gene pools (in analogy to *Podospora anserina,* see Fig. 3) which, certainly in nature, have a very restricted exchange capacity. Since the sexual phase is not affected by this incompatibility barrier, strains can be experimentally crossed. This has led to an increase of productivity of 60% after three subsequent intercrosses (MERRICK, 1975a, b).

In similar outbreeding experiments (see Tab. 1) the yield of other secondary metabolites (antibiotics, alkaloids) or primary metabolites (e.g., ethanol, CO_2) is increased. Even the transfer of maltotriose and dextrin from one species variant to another, or the elimination of undesired wine making properties is possible (ESCHENBRUCH et al., 1982; THORNTON, 1982; WINDISCH, 1978).

In conclusion: There have been some attempts to improve the productivity of biotechnologically important fungi by sexual recombination. However, during the last decade there have been hardly any reports dealing with this matter. It is difficult to ascertain whether this is due to difficulties in genetically handling these fungi or because during this time molecular methods were developed which allowed transfer of single genes between non-related organisms. On the other hand, by sexual recombination only members of one species can interact and genes of the parental strains will be transferred only randomly to the progeny.

6 Parasexual Recombination

6.1 Principle

All non-meiotic processes in vegetative cells leading to recombination are called parasexual. As a prerequisite in fungi, appropriate haploid nuclei must be brought via hyphal anastomosis or protoplast fusion into a common cytoplasm where they may fuse (karyogamy). In the resulting diploid (polyploid or aneuploid) nuclei, rearrangements of chromosomes (or parts of them) occur in the course of mitotic divisions, eventually followed by a stepwise (via aneuploid stages) haploidization. This type of parasexual recombination is termed mitotic recombination. It also takes place after cell fusion in cultures of higher plants and animals.

6.2 Mechanisms

Parasexual processes, i.e., in eukaryotes mitotic recombination, are not part of the ontogenesis of an organism. Therefore, specific "life cycles", as in sexual processes, do not exist. In nature, different nuclei are brought together in a common cytoplasm by hyphal (via anastomosis) or cell fusion. In as much as they can coexist, mitotic recombinations can occur, as a very rare event after karyogamy. Interestingly enough, recombinations during mitotic divisions were first detected in the fruit fly, *Drosophila* (STERN, 1936). Since its rediscovery by ROPER (1952) it has been applied to microorganisms and cell cultures. As primarily described in the haploid ascomycetous fungus *Aspergillus nidulans* (PONTECORVO, 1954) mitotic recombination consists of the following steps (see Fig. 7):

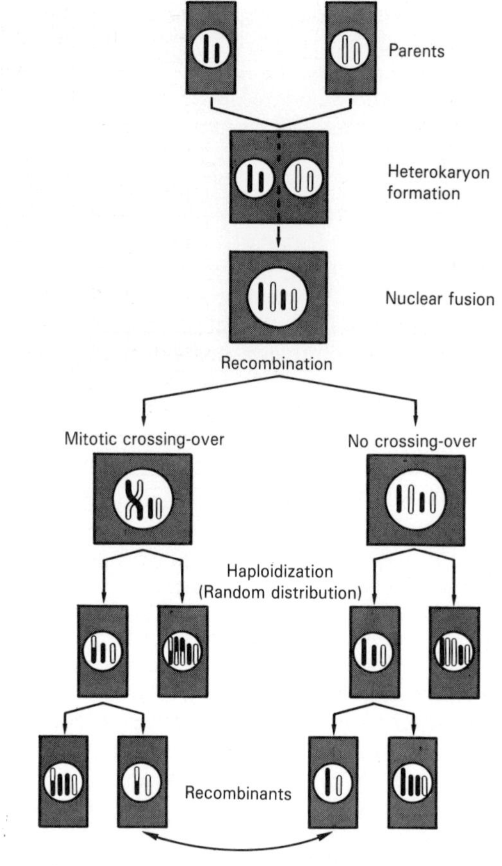

Fig. 7. Scheme of mitotic recombination.

- *Heterokaryon formation.* Two haploid (or diploid, polyploid, aneuploid) mycelia or cells fuse to form a heterokaryon.
- *Nuclear fusion* occurs on rare occasions in heterokaryons (at a frequency of approximately 10^{-6} per nuclear division) and results in the formation of a diploid (polyploid, aneuploid) heterozygous nucleus. Both the parental haploid and the fused diploid nuclei in the heterokaryon multiply by mitosis. Eventually, howev-

er, a visible segregation occurs so that haploid, heterokaryotic, as well as diploid heterozygous sectors of the mycelium can be distinguished.

- *Mitotic crossing-over* takes place in the diploid nuclei at the frequency of about 10^{-2} to 10^{-3} per nuclear division. It results in intrachromosomal exchanges in comparison to meiotic crossing-over.
- *Haploidization*. The still diploid nuclei are reduced to the haploid state through different stages of aneuploidy. The frequency is relatively constant at 10^{-3} per nuclear division. During these steps an additional recombination may take place (in comparison to interchromosomal meiotic exchanges), because the elimination of chromosomes occurs in a random fashion.

Mitotic recombination has been demonstrated in many other fungi (ESSER and KUENEN, 1967), in higher plants, and in animal tissue culture (HARRIS, 1970). There is no doubt that this type of recombination is a general phenomenon which can replace, at least in part, the meiotic recombination and may occur in nature, especially in imperfect fungi.

6.3 Restrictions

In many fungi the application of mitotic recombination is restricted by the lack of spontaneous hyphal or cell fusion. This cannot be completely overcome by the method of "forced heterokaryons", a fusion of complementary auxotrophs on a minimal medium.

In addition, restrictions occur in parasexual processes due to homogenic or heterogenic incompatibility (see above). However, not only these barriers but also transfer restrictions between different species, and even between genera may be overcome, in many cases, by specific laboratory procedures, since the main condition for parasexual recombination is the fusion of the participating hyphae or cells. Such procedures consist of the lytic removal of the cell wall under suitable osmotic conditions, which avoid the bursting of the cytoplasmic membrane. The protoplasts obtained may either fuse spontaneously (polyethyleneglycol) or by a very short pulse of high frequency current (electrofusion). Especially these methods have been successfully applied to fungi of different taxonomic relationships (for further information consult appropriate reviews or monographs).

Fig. 8 shows a survey of the application of the protoplast-induced fusion with yeasts of the *Candida* type as an example. The figure indicates that an appropriate selective system is essential. After fusion and regeneration the cells or hyphae with heterogenous nuclei, not the heterokaryotic fusion products, must be identified. In single-cell organisms, such as yeasts, this prerequisite for the subsequent identification of recombinants is easy to accomplish, since after some subsequent mitotic divisions one can be sure that the majority of the cells is mono-nucleate. In hyphal fungi the selection of regeneration products with heterogenous nuclei is more complicated. On the one hand, it is possible that the conidia of some species are mono-nuclear, at least at a very early stage of development. Hence after the spreading of a conidia suspension of a heterokaryotic and partly putative diploid mycelium on minimal medium, only conidia with diploid nuclei can germinate if "parental nuclei" (parental strains) with complementing auxotrophic defects were used. Another possibility for the differentiation of conidia with diploid or haploid nuclei is based on the different diameters: conidia with haploid nuclei show about two thirds of the diameter of diploids. It is possible to discriminate between the two types with an automatic photoscanner, a procedure primarily designed for the detection and discrimination of mutated and non-mutated cells due to visible characteristics.

The selection system used for recombinants (usually after the induction of haploidization) strongly depends on the abilities of the "parents" which should be combined in the siblings. Generally it can be said that due to the low probability of mitotic recombinational events (favorably about 10^{-5}), the selection system must be very efficient.

However, in this context it is not worthwhile discussing appropriate protocols of protoplast fusion for mitotic recombination, since the experimental conditions for protoplast production, protoplast fusion, protoplast re-

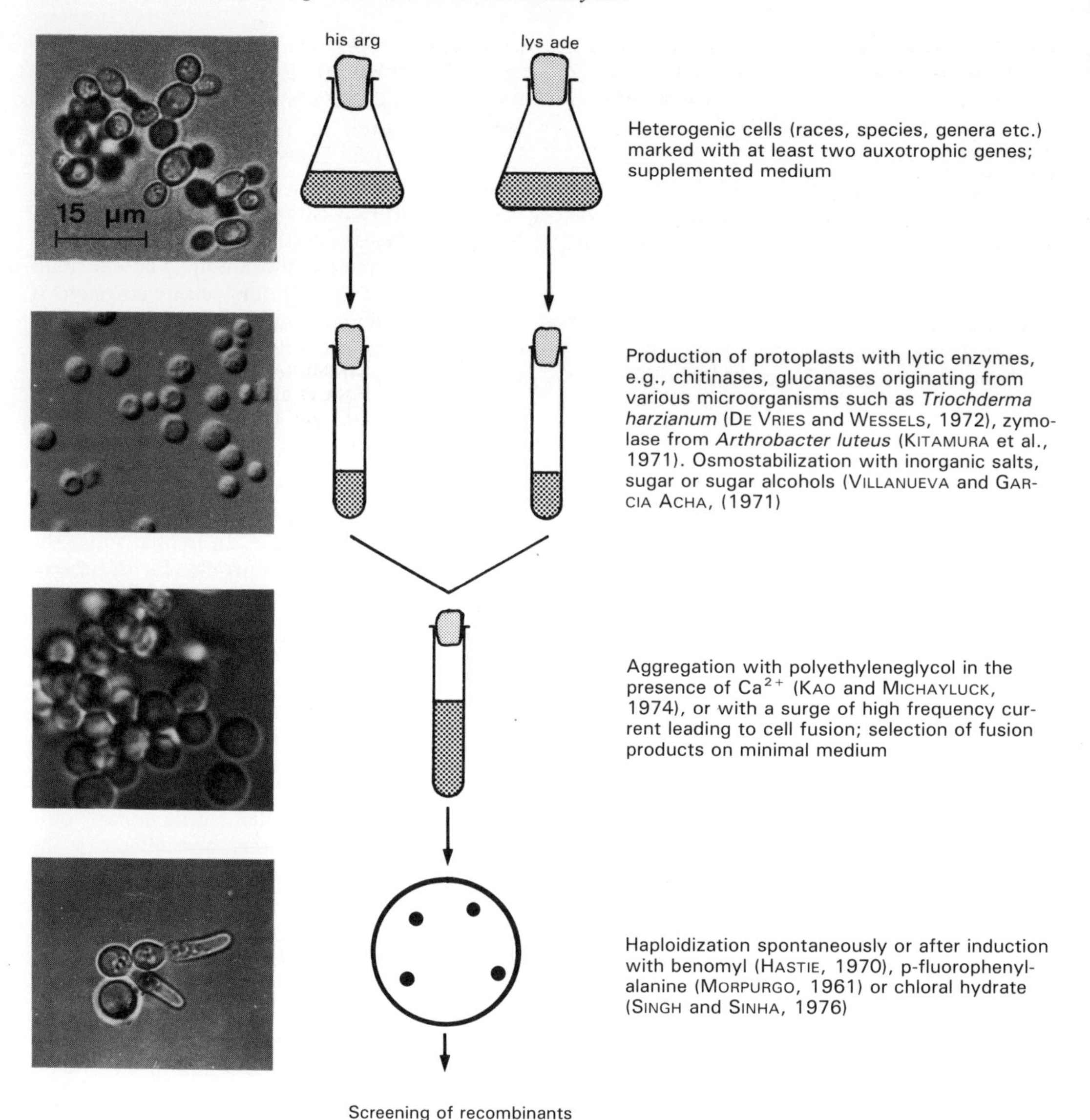

Fig. 8. Recombination of genetic material via protoplast fusion, exemplified in *Candida* yeast. (Adapted from KÜCK et al., 1980).

generation, and selection of the different stages have to be worked out on a strictly empirical basis, and strongly depend on the organisms to be combined. For detailed information and suggestions concerning the elaboration of a protocol for distinct organisms one of the already mentioned reviews or monographs should be consulted.

6.4 Industrial Applications

An attempt to review the use of mitotic recombination for biotechnological purposes requires the same reservations which have already been made in the discussion of meiotic recombination. Thus, in the following only

Tab. 2. Compilation of Biotechnologically Relevant Fungi Improved by Mitotic Recombination

Organism(s)	Fusion Method	Improved Character/Ability	Reference
intraspecific			
Cephalosporium acremonium	Spontaneous	Cephalosporin	HAMLYN and BALL, 1979
Aspergillus niger	Spontaneous	Citric acid	ILCZUK, 1971; BONATELLI et al., 1983; DAS and ROY, 1978; DAS, 1980; IKEDA, 1961; SEICHERTOVA and LEOPOLD, 1969
	Protoplast technique		MARTINKOVA et al., 1990; KIRIMURA et al., 1986; 1988a, b; FINA et al., 1986
	Protoplast t.	Glucoamylase	DAS and GHOSH, 1989
Penicillium chrysogenum	Spontaneous	Penicillin	ELANDER et al., 1973
Penicillium roquefortii	Protoplast t.	Alkaloids	SCRAPHIN et al., 1987
Saccharomyces cerevisiae	Protoplast t.	Ethanol yield	SEKI et al., 1983
	Protoplast t.	Low carbohydrate content in beer	JANDEROVA et al., 1990
	Protoplast t.	Transfer of VLPs (killer toxin secretion)	RÖCKEN, 1984a, b; VANDREJS et al., 1983
Saccharomyces diastaticus	Protoplast t.	Ethanol yield	SAKAI et al., 1986
interspecific			
Aspergillus oryzae × *A. sojae*	Protoplast t.	Soy sauce and miso	USHIJIMA et al., 1991
Saccharomyces uvarum × *S. diastaticus*	Protoplast t.	Low carbohydrate content in beer	STEWART et al., 1983; FREEMAN, 1981
Saccharomyces cerevisiae × *S. mollis*	Protoplast t.	Ethanol yield	LEGMANN and MARGALITH, 1986
intergeneric			
Saccharomyces cerevisiae × *Kluyveromyces lactis*	Protoplast t.	Transfer of killer-toxin complex	SUGISAKI et al., 1983
Saccharomyces cerevisiae × *Zygosaccharomyces fermentati*	Protoplast t.	Ethanol yield and cellobiose utilization	PINA et al., 1986
Saccharomyces cerevisiae × *Candida* sp.	Protoplast t.	Ethanol yield in combination with heat tolerance	MARTIN-RENDON et al., 1989

those data shown in Tab. 2 shall be discussed which seem to be transmittable to other organisms or characters.

Due to intraspecific recombination in filamentous fungi it was possible to increase not only the yield of primary metabolites (e.g., citric acid, glucoamylase), but also the yield of secondary metabolites (e.g., penicillin, cephalosporin, alkaloids). In some cases, it was not even necessary to complete the whole parasexual cycle, as strain improvement could already be found after heterokaryon formation or diploidization. In the citric acid production of *Aspergillus niger*, diploid strains showed an increased productivity of about 15% (KIRIMURA et al., 1988b; MARTINKOVA et al., 1990). After haploidization, some of the siblings had an even higher capacity (KIRIMURA et al., 1988a). Based on these experiments these authors recommend, especially for the change-over of *A. niger* strains from surface to submersed cultivation, the fusion and mitotic recombination of appropriate "parents". The alternative, the screening and subsequent mutational improvement of a submerse-capable wild strain, would be, in contrast, much more time-consuming and the success would be doubtful.

Intraspecific, but also interspecific recombination has been successfully used to improve brewing strains of the genus *Saccharomyces*. On the one hand, it was possible to combine the ability to produce glucoamylase from *S. diastaticus* with a brewing yeast, leading to a hybrid which is able to ferment dextrins giving low-calory light beer (FREEMAN, 1981). On the other hand, the phenolic flavor, which was typical for these hybrids and which originated from *S. diastaticus,* could be removed after repeated mitotic recombination with the brewing strain (STEWART et al., 1983; JANDEROVA et al., 1990). It should also be mentioned that it is possible to transfer the so-called killer character by protoplast fusion, which is usually determined by virus-like particles (VLP) from *Saccharomyces* or non-*Saccharomyces* yeasts to brewing or winery yeasts. By this procedure, which is indeed a parasexual process but does not comprise mitotic recombination, VLPs can be exclusively transferred. They enable the recipient strain to produce a toxin (killer protein) which inactivates *Saccharomyces* contaminants that do not carry VLPs (RÖCKEN, 1984a, b; TOKUNAGA et al., 1990; VANDREJS et al., 1983). Interestingly enough, even the transfer of VLPs from *Kluyveromyces lactis* to *S. cerevisiae* was possible (intergeneric transfer) leading to siblings which possess a broad contamination-protection against non-*Saccharomyces* yeasts.

As already mentioned, not only different species can be combined by mitotic recombination, but also different genera. Although only a few reports on successful improvements of intergeneric recombination are available, the examples available seem to have considerable biotechnological importance. As may be seen from Tab. 2, with intergeneric recombination the high ethanol yield of *S. cerevisiae* could be combined, on the one hand, with the ability to metabolize cellobiose (PINA et al., 1986) and, on the other hand, to increase its heat tolerance (MARTIN-RENDON et al., 1989). Using the hybrids able to convert cellobiose (hydrolysate of cellulose) to ethanol creates the possibility of recycling waste paper into a high-quality product at a low cost instead of producing a lower-grade quality of paper which is presently the case. Other very interesting organisms are the ethanol-producing recombinants with increased heat tolerance, as the higher the temperature of the fermenting broth, the lower the use of cooling water.

Since mitotic recombination does not depend on a sexual cycle, it seems, particularly with imperfect fungi, to be suitable for recombination processes. However, this method is seldom used (see Tab. 2). This is most likely due to the fact that there is very little possibility for recombination and moreover, is due to the laborious methods of selection. This is particularly so for many characteristics so that it is often less laborious to attain a strain improvement through mutation than by mitotic recombination.

7 Conclusion

It is obvious from the previous pages that considerable progress has been achieved in strain improvement of biotechnologically relevant fungi (in both sexual and parasexual cycles) by concerted breeding in the framework of classical genetics. Although the potential for both meiotic and mitotic recombination has not yet been completely exploited, it should be noted that in the last few years molecular methods for improving strains have been developed. As molecular methods are only suitable for improving directly coded products (e.g., enzymes), and less suitable for non-protein products (e.g., antibiotics, alkaloids, alcohol, acids, etc.), non-protein products will, in the future, be attained mainly by classical genetic methods. It will therefore depend on the organism (or on the improved quality) which method is chosen (or is more useful) and whether classical and/or molecular methods will be combined. In any case, the biology of the fungus concerned (life cycles, etc.) must be understood.

Thus, progress in strain improvement should in the future be made by a concerted application of classical and molecular genetics.

This is an updated, revised, and extended version of a paper published in the first edition of this series: Biotechnology Vol. 1, pp. 305–329, Verlag Chemie, Weinheim, 1981.

8 References

ARBER, W., LINN, S. (1969), DNA modification and restriction, *Annu. Rev. Biochem.* **38**, 467–500.

BACH, F. H. (1976), Genetics of transplantation: The major histocompatibility complex, *Annu. Rev. Genet.* **10**, 319–339.

BANUETTE, F., HERSKOWITZ, I. (1989), Different a alleles of *Ustilago maydis* are necessary for maintenance of filamentous growth but not for meiosis, *Proc. Natl. Acad. Sci. USA* **86**, 5878–5882.

BEACH, D. H. (1983), Cell type switching by DNA transposition in fission yeast, *Nature* **305**, 682–688.

BERGFELD, R. (1977), *Sexualität bei Pflanzen,* Stuttgart: Ulmer.

BÖLKER, M., URBAN, M., KAHMANN, R. (1992), The a mating type locus of *U. maydis* specifies cell signaling components, *Cell* **68**, 441–450.

BONATELLI, R., Jr., AZEVEDO, J. L., UMBUZEIRO VALENT, G. (1983), Parasexuality in a citric acid producing strain of *Aspergillus niger, Rev. Brasil. Genet.* **6**, 399–405.

BRIEGER, F. (1930), *Selbststerilität und Kreuzungssterilität im Pflanzenreich und Tierreich,* Berlin: Springer,

BURNETT, J. H. (1975), Mycogenetics, London - New York - Sydney - Toronto: Wiley.

CASSELTON, L. A., MUTASA, E. S., TYMON, A., MELLON, F. M., LITTLE, P. F. R., TAYLOR, S., BERNHAGEN, J., STRATMANN, R. (1989), The molecular analysis of basidiomycete mating type genes, in: *Proceedings of the EMBO-Alko Workshop on Molecular Biology of Filamentous Fungi,* Helsinki 1989 (NAVALAINEN, H., PENTTILÄ, M., Eds.), *Found. Biotech. Ind. Ferment. Res.* **6**, 139–148.

CHRISTENSEN, B. E. (1987), Cross-breeding of distillers' yeast by hybridization of spore derived clones, *Carlsberg Res. Commun.* **52**, 253–262.

DAS, A. (1980), Parasexual hybridization and citric acid production by *Aspergillus niger, Eur. J. Appl. Microbiol. Biotechnol.* **9**, 117–119.

DAS, A., GHOSH, A. (1989), Breeding by protoplast fusion for glucoamylase production, *Biotechnol. Lett.* **11**, 705–708.

DAS, A., ROY, P. (1978), Improved production of citric acid by a diploid strain of *Aspergillus niger, Can. J. Microbiol.* **24**, 622–625.

DE VRIES, O. M. H., WESSELS, J. G. H. (1972), Release of protoplasts from *Schizophyllum commune* by a lytic enzyme preparation from *Trichoderma viride, J. Gen. Microbiol.* **73**, 13–22.

EGER, G., EDEN, G., WISSIG, E. (1976), *Pleurotus ostreatus*-breeding potential of a new cultivated mushroom, *Theor. Appl. Genet.* **47**, 155–163.

ELANDER, R. P., ESPENSHADE, M. A., PATHAK, S. C., PAN, C. H. (1973), in: *Genetics of Industrial Microorganisms* (VANEK, Z., HOSTALEK, Z., CUDLIN, J., Eds.), pp. 239–253. Prague: Academia.

ESCHENBRUCH, R., CRESSWELL, K. J., FISCHER, B. M., THORNTON, R. J. (1982), Selective hybridization of pure culture wine yeasts. 1. Elimination of undesirable wine-making properties, *Eur. J. Appl. Microbiol. Biotechnol.* **14**, 155–158.

ESSER, K. (1959), Incompatibilitätsbeziehungen zwischen geographischen Rassen von *Podospora anserina:* III. Untersuchungen zur Genphysiologie der Barragebildung und Semiincompatibilität, *Z. Vererbungsl.* **90**, 29–52.

ESSER, K. (1962), Die Genetik der sexuellen Fortpflanzung bei den Pilzen, *Biol. Zentralbl.* **81**, 161–172.

ESSER, K. (1971), Breeding systems in fungi and their significance for genetic recombination, *Mol. Gen. Genet.* **110**, 86–100.

ESSER, K. (1981), *Impact of Basic Research on the Practical Application of Fungal Processes,* Vol. 22, pp. 19–37, Arlington, VA: Society for Industrial Microbiology.

ESSER, K. (1986), *Kryptogamen*, Berlin: Springer-Verlag.

ESSER, K. (1989), Anwendung von Methoden der klassischen und molekularen Genetik bei der Züchtung von Nutzpilzen: Fakten und Perspektiven. *Mushroom Science* XII, Part I. *Proc. 12th Int. Congr. Science and Cultivation of Edible Fungi*, Braunschweig, FRG, 1987.

ESSER, K., BLAICH, R. (1973), Heterogenic incompatibility in plants and animals, *Adv. Genet.* **17**, 107–152.

ESSER, K., KUENEN, R. (1967), *Genetik der Pilze. Genetics of Fungi,* translated by E. STEINER, Berlin - Heidelberg - New York: Springer.

ESSER, K., MOHR, G. (1990), Stammverbesserung von Hyphenpilzen durch die Gentechnik: Fakten und Perspektiven, *Bioengineering* **6**, 44–55.

ESSER, K., STAHL, U. (1976), Cytological and genetic studies of the life cycle of *Saccharomycopsis lipolytica, Mol. Gen. Genet.* **146**, 101–106.

ESSER, K., TUDZYNSKI, P. (1978), Genetics of the ergot fungus *Claviceps purpurea, Theor. Appl. Genet.* **53**, 145–149.

ESSER, K., SEMERDZIEVA, M., STAHL, U. (1974), Genetische Untersuchungen an dem Basidiomyceten *Agrocybe aegerita, Theor. Appl. Genet.* **45**, 77–85.

ESSER, K., SALEH, F., MEINHARDT, F. (1979), Genetics of fruit body production in higher basidiomycetes, *Curr. Genet.* **1**, 85–88.

FANTANI, A. A. (1962), Genetics and antibiotic production of *Emericellopsis* species, *Genetics* **47**, 161–177.

FINCHAM, J. R. S., DAY, P. R. (1979), *Fungal Genetics,* 4th Ed., Oxford – Edinburgh: Blackwell.

FREEMAN, R. F. (1981), Construction of brewing yeasts production of low carbohydrate beers, *Eur. Brew. Conv. Congr.,* 497 ff.

FRITSCHE, G. (1976), Welche Möglichkeiten eröffnet der viersporige Champignon *Agaricus bitorquis (Quél.) Sacc.* dem Züchter? *Theor. Appl. Genet.* **47**, 125–131.

FROELIGER, E. H., KRONSTAD, J. W. (1990), Mating and pathogenesis in *Ustilago maydis, Sem. Dev. Biol.* **1**, 185–193.

GILLISSEN, B., BERGEMANN, J., SANDMANN, C., SCHROER, B., BÖLKER, M., KAHMANN, R. (1992), A two-component regulatory system for self/nonself recognition in *Ustilago maydis, Cell* **68**, 647–658.

GLASS, N. L., STABEN, C. (1990), Genetic control of mating in *Neurospora crassa, Sem. Dev. Biol.* **1**, 177–184.

GLASS, N. L., VOLLMER, S. J., STABEN, C., GROTELUESCHEN, J., METZENBERG, R. L. (1988), DNAs of the mating type alleles of *Neurospora crassa* are highly dissimilar, *Science* **241**, 570–573.

HAMLYN, P. F., BALL, C. (1979), Recombination studies with *Cephalosporium acremonium,* in: *Genetics of Industrial Microorganisms* (SEBEK, O. K., LASKIN, A. I., Eds.), pp. 185–191, Washington, DC: American Society for Microbiology.

HARRIS, H. (1970), *Cell Fusion. The Dunham Lectures,* London – New York – Oxford: Oxford University Press.

HARTMANN, M. (1956), *Die Sexualität,* 2nd Ed., Stuttgart: Fischer.

HASTIE, A. C. (1970), Benlate-induced instability of *Aspergillus diploids, Nature* **226**, 771.

HERSKOWITZ, I. (1989), A regulatory hierarchy for cell specialization in yeast, *Nature* **342**, 749–757.

HOLT, G., MCDONALD, K. D. (1968a), Penicillin production and its mode of inheritance in *Aspergillus nidulans, Antonie van Leeuwenhoek* **34**, 409–416.

HOLT, G., MCDONALD, K. D. (1968b), Isolation of strains with increased penicillin yield after hybridization in *Aspergillus nidulans, Nature* **219**, 636–637.

IKEDA, Y. (1961), Potential application of parasexuality in breeding fungi, in: *Recent Advances in Botany,* pp. 383–386, Toronto: University of Toronto Press.

ILCZUK, Z. (1971), Genetik der Zitronensäure erzeugenden Stämme *Aspergillus niger.* III. Citronensäuresynthese durch erzwungene Heterokaryen zwischen auxotrophen Mutanten von *Aspergillus niger, Nahrung* **15**, 251–262.

JANDEROVA, B., CVRCKOVA, F., BENDOVA, O. (1990), Construction of the dextrin-degrading of brewing yeast by protoplast fusion, *J. Basic Microbiol.* **30**, 466–505.

KAO, K. N., MICHAYLUCK, M. R. (1974), A method for high frequency intergeneric fusion of plant protoplasts, *Planta* **115**, 355–367.

KELLY, M., BURKE, J., SMITH, M., KLAR, A., BEACH, D. (1988), Four mating-type genes control sexual differentiation in the fission yeasts, *EMBO J.* **7**, 1537–1547.

KIRIMURA, K., YAGUCHI, T., USAMI, S. (1986), Intraspecific protoplast fusion of citric acid producing strains of *Aspergillus niger, J. Ferment. Technol.* **64**, 473–479.

KIRIMURA, K., LEE, S. P., NAKAJIMA, I., KAWABE, S., USAMI, S. (1988a) Improvement in citric acid production by haploidization of *Aspergillus niger* diploid strains, *J. Ferment. Technol.* **66**, 375–382.

KIRIMURA, K., NAKAJIMA, I., LEE, S. P., KAWABE, S., USAMI, S. (1988b), Citric acid production by the diploid strains of *Aspergillus niger* obtained by protoplast fusion, *Appl. Microbiol. Biotechnol.* **27**, 504–506.

KITAMURA, K., KANEKO, T., YAMAMOTO, Y. (1971), Lysis of viable cells by enzymes of *Arthrobacter luteus, Arch. Biochem. Biophys.* **145**, 402–404.

KLAR, A. J. S. (1989), *Mobile DNA* (BERG, D. E., HOWE, M. M., Eds.), pp. 671–691, Washington, DC: American Society for Microbioloy.

KNAUSEDER, F., BRANDL, E. (1976), Pleuromutilins – fermentation, structure and biosynthesis, *J. Antibiot.* **29**, 125–131.

KNIEP, H. (1928), *Die Sexualität der niederen Pflanzen,* Jena: Fischer.

KÜCK, U., STAHL, U., LHERMITTE, A., ESSER, K. (1980), Isolation and characterization of mitochondrial DNA from the alkane yeast *Saccharomycopsis lipolytica, Curr. Genet.* **2**, 97–101.

KÜES, U., CASSELTON, L. A. (1992), Molecular and functional analysis of the a mating type genes of *Coprinus cinereus,* in: *Genetic Engineering, Principles and Methods* (SETLOW, J. K., Ed.), Vol. 14, pp. 14–18, New York: Plenum Press.

KÜES, U., RICHARDSON, W. V. J., TYMON, A. M., MUTASA, E. S., GÖTTGENS, B., GAUBATZ, S., GREGORIADES, A., CASSELTON, L. A. (1992), The combination of dissimilar alleles of the Aa and Ab gene complexes, whose proteins contain homeo domain motifs, determines sexual development in the mushroom *Coprinus cinereus, Genes Dev.* **6**, 568–577.

LEGMANN, R., MARGALITH, P. (1986), Ethanol formation by hybrid yeasts, *Appl. Microbiol. Biotechnol.* **23**, 198–202.

MARTIN-RENDON, E., JIMENEZ, J., BENITCZ, T. (1989), Ethanol inhibition of *Saccharomyces* and *Candida* enzymes. *Curr. Genet.* **15**, 7–16.

MARTINKOVA, L., MUSLIKOVA, M., UJCOVA, E., MACHEK, F., SEICHERT, L. (1990), Protoplast fusion in *Aspergillus niger* strains accumulating citric acid, *Folia Microbiol.* **35**, 143–148.

MEINHARDT, F. (1980), Untersuchungen zur Genetik des Fortpflanzungsverhaltens und der Fruchtkörper- und Antibiotikabildung des Basidiomyceten *Agrocybe aegerita, Bibl. Mycol.* **75**, Vaduz: J. Cramer.

MEINHARDT, F., ESSER, K. (1981), Genetic studies of the basidiomycete *Agrocybe aegerita* II. Genetic control of fruit body formation and its practical implications, *Theor. Appl. Genet.* **60**, 265–268.

MEINHARDT, F., KEMPKEN, F., KÄMPER, J., ESSER, K. (1990), Linear plasmids among eukaryotes: fundamentals and application, *Curr. Genet.* **17**, 89–95.

MERRICK, M. J. (1975a), Hybridization and selection for increased penicillin-titre in wild-type isolates of *Aspergillus nidulans, J. Gen. Microbiol.* **91**, 278–286.

MERRICK, M. J. (1975b), The inheritance of penicillin-titre in crosses between lines of *Aspergillus nidulans* for increased productivity, *J. Gen. Microbiol.* **91**, 287–294.

MERRICK, M. J., CATEN, C. E. (1975), The inheritance of penicillin-titre in wild-type isolates of *Aspergillus nidulans, J. Gen. Microbiol.* **86**, 283–293.

METZENBERG, R. L. (1990), The role of similarity and differences in fungal mating, *Genetics* **125**, 457–462.

MOHR, S., ESSER, K. (1992), Mobile genetic elements in eukaryotes: principles and applications in biotechnology, in: *Biotechnology Focus* (FINN, R. K., PRÄVE, P., SCHLINGMANN, M., CRUEGER, W., ESSER, K., THAUER, R., WAGNER, F., Eds.), Vol. 3, pp. 5–24, München: Hanser.

MORPURGO, G. (1961), *Aspergillus Newslett.* **2**, 10.

NASMYTH, K. (1983), Molecular analysis of a cell lineage, *Nature* **302**, 670–676.

NIELSEN, O., EGEL, R. (1989), Mapping of the double-strand breaks at the mating-type locus in fission yeast by genomic sequencing, *EMBO J.* **8**, 1537–1547.

NOVOTNY, C. P., STANKIS, M. M., SPECHT, C. A., YANG, H., GIASSON, L., ULLRICH, R. C. (1991), in: *More Manipulations in Fungi* (BENNETT, J. W., LASURE, L. L., Eds.), pp. 235–257, San Diego: Academic Press.

PINA, A., CALDERON, I. L., BENITCZ, T. (1986), Intergenetic hybrids of *Saccharomyces cerevisiae* and *Zygosaccharomyces fermentati* obtained by protoplast fusion, *Appl. Environ. Microbiol.* **51**, 995–1003.

PONTECORVO, G. (1954), The genetics of *Aspergillus nidulans, Adv. Genet.* **5**, 141–238.

RAPER, C. A., RAPER, J. R. (1966), Mutations modifying sexual morphogenesis in *Schizophyllum, Genetics* **54**, 1154–1168.

RAPER, J. R., RAPER, C. A. (1972), Life cycle and prospects for interstrain breeding in *Agaricus bisporus, Mushroom Sci.* **8**, 1–9.

RÖCKEN, W. (1984a), Crossing of brewers' yeasts by protoplast fusion, *Monatsschr. Brauwiss.* **37**, 76–82.

RÖCKEN, W. (1984b), Transfer of killer plasmids from a killer yeast to a bottom-fermentation brewers' yeast by protoplast fusion, *Monatsschr. Brauwiss.* **37**, 384–389.

ROPER, J. A. (1952), Production of heterozygous diploids in filamentous fungi, *Experientia* **8**, 14–15.

ROSSIGNOL, J.-L., PICARD, M. (1991), *Ascobolus immersus* and *Podospora anserina:* sex, recombination, silencing, and death, in: *More Manipulations in Fungi* (BENNETT, J. W., LASURE, L. L., Eds.), pp. 278–279, San Diego: Academic Press.

RUSSEL, J., BILINSKI, C. A., STEWART, G. G. (1987), Cross breeding of *Saccharomyces cerevisiae* and *Saccharomyces uvarum* (*carlsbergensis*) by mating of meiotic segregants: Isolation and characterisation of species hybrids, *Eur. Brew. Conv. Congr.,* Madrid, pp. 497 ff.

SAKAI, T., KOO, K., SAITOH, K., KATSURAGI, T. (1986), Use of the protoplast fusion for the development of rapid starch fermenting strains of *Saccharomyces diastaticus, Agric. Biol. Chem.* **50**, 297–306.

SCHULTZ, B., BANUETTE, F., DAHL, M., SCHLESINGER, R., SCHÄFER, W., MARTIN, T., HERSKOWITZ, I., KAHMANN, R. (1990), The b allele of *U. maydis,* whose combinations program pathogenic development, code for polypeptides containing a homeodomain-related motif, *Cell* **60,** 295–306.

SCRAPHIN, B., BOULET, A., SIMON, M., FAYE, C. (1987), Construction of a yeast strain devoid of mitochondrial introns and its use to screen nuclear genes involved in mitochondrial splicing, *Proc. Natl. Acad. Sci. USA* **84,** 6810–6814.

SEICHERTOVA, O., LEOPOLD, H. (1969), Die Aktivierung von Stämmen des *Aspergillus niger.* II. Die Benutzung der parasexuellen Hybridisation. *Zentralbl. Bakteriol. Abt. 2,* **123,** 564–570.

SEKI, T., MYOGA, S., LIMTONG, S., VEDONO, S., KUMNUANTA, J., TAGUCHI, J. (1983), Genetic construction of yeast strains for high ethanol production, *Biotechnol. Lett.* **5,** 351–356.

SINGH, M., SINHA, U. (1976), Chloral hydrate induced haploidization in *Aspergillus nidulans, Experientia* **32,** 1144–1145.

SMITH, J. E., BERRY, D. R. (Eds.) (1975), *The Filamentous Fungi,* Vol. 1: *Industrial Mycology,* London: Edward Arnold.

STABEN, C., YANOFSKY, C. (1990), The *Neurospora crassa* a mating-type region. *Proc. Natl. Acad. Sci. USA* **87,** 4917–4921.

STAHL, U., ESSER, K. (1976), Fruit body-production in higher basidiomycetes, *Mol. Gen. Genet.* **148,** 183–197.

STERN, C. (1936), Somatic crossing over and segregation in *Drosophila melanogaster, Genetics* **21,** 625–630.

STEWART, G. G., PANCHAL, C. J., RUSSEL, I. (1983), Current development in the genetic manipulation of brewing yeast strains – A review, *Eur. Brew. Conv. Congr.,* Helsinki, pp. 243–250.

SUGISAKI, Y., GUNGE, N., SAKAGUCHI, K., YAMASAKI, M., TAMURA, G. (1983), *Kluyveromyces lactis* killer toxin inhibits adenylate cyclase of sensitive yeast cells, *Nature* **304,** 464–466.

THORNTON, R. J. (1982), Selective hybridization of pure culture wine yeasts. II. Improvement of fermentation efficiency and inheritance of SO_2 tolerance, *Eur. J. Microbiol. Biotechnol.* **14,** 159–164.

TIMBERLAKE, W. E. (Ed.) (1985), *Molecular Genetics of Filamentous Fungi,* New York: Alan R. Liss, Inc.

TOKUNAGA, M., KAWAMURA, A., KITADA, K., HISHINUMA, F. (1990), Secretion of killer toxin encoded on the linear DNA plasmid pGKL 1 from *Saccharomyces cerevisiae, J. Biol. Chem.* **265,** 17274–17280.

TURCQ, B., DENAYROLLES, M., BÉUGUERET, J. (1990), Isolation of the two allelic incompatibility genes s and S of the fungus *P. anserina, Curr. Genet.* **17,** 297–304.

TURCQ, B., DELEU, C., DENAYROLLES, M., BÉGUERET, J. (1991), Two allelic genes responsible for vegetative incompatibility in the fungus *P. anserina* are not essential for cell viability, *Mol. Gen. Genet.* **228,** 265–269.

ULLRICH, R. C., SPECHT, C. A., STANKIS, M. M., YANG, H., GIASSON, L., NOVOTNY, C. P. (1991), Molecular biology of mating-type determination in *Schizophyllum commune,* in: *Genetic Engineering, Principles and Methods,* Vol. 13 (SETLOW, J. K., Ed.), pp. 279–306, New York: Plenum Press.

USHIJIMA, S., NAKADAI, T., UCHIDA, K. (1991), Interspecific electrofusion of protoplasts between *Aspergillus oryzae* and *Aspergillus sojae, Agric. Biol. Chem.* **55,** 129–136.

VAN DEN ENDE, H. (1976), *Sexual Interactions in Plants,* London – New York – San Francisco: Academic Press.

VAN DEN HONDEL, C. A. M. J. J., PUNT, P. J. (1991), Gene-transfer systems and vector development for filamentous fungi in: *Applied Molecular Genetics of Fungi* (PEBERDY, J. F., Ed.), BMS Symposium, Vol. 18, pp. 1–29, Cambridge: Cambridge University Press..

VANDREJS, V., PSENICKA, I., KUPCOVA, L., DOSTALOVA, R., JANDEROVA, B., BENDOVA, O. (1983), The use of killer factor in the selection of hybrid yeast strains, *Fol. Biol. Praha* **29,** 372–384.

VILLANUEVA, J. R., GARCIA ACHA, I. (1971), Production and the use of fungal protoplasts, in: *Methods in Microbiology* (BOOTH, C., Ed.), Vol. 4, pp. 665–718, New York: Academic Press.

WHITEHOUSE, H. L. K. (1949), Heterothallism and sex in fungi, *Biol. Rev.* **24,** 411–447.

WINDISCH, S. (1978), Developments in the breeding of yeasts for industrial use, *Lebensm. Wiss. Technol.* **11,** 338–340.

WINDISCH, S., KOWALSKY, S., ZANDER, I. (1976), Doughraising tests with hybrid yeasts, *J. Appl. Microbiol.* **3,** 213–221.

YODER, O. C., TURGEON, B. G., SCHAFER, W., CUIFETTI, L., BOHLMANN, H., VAN ETTEN, H. D. (1989), Molecular analysis of mating type and expression of a foreign pathogenicity gene in *Cochlio heterostrophus,* in: *Proc. EMBO-Alko Workshop Molecular Biology* of *Filamentous Fungi,* Helsinki 1989 (NEVALAINEN, H., PENNILÄ, M., Eds.), *Found. Biotech. Ind. Ferment. Res.* **6,** 189–196.

4 Cell Fusion

JOZEF ANNÉ

Leuven, Belgium

1 Introduction

Recombination and mutation are at the origin of genetic variability between living organisms. Until the advent of the more advanced genetic techniques recombination which is the result of exchange of hereditary traits between genetically different organisms was only possible in sexually reproducing species. Other means beyond sexual processes to isolate recombinants were first discovered for bacteria with the observation of DNA-mediated transformation and virus-mediated transduction. Several years later, the parasexual cycle was detected for filamentous fungi, and a method was developed to induce fusion of somatic cells, originally for mammalian cells, but later on procedures were introduced to fuse also cells of microbial and plant origin.

The basic event of induced cell fusion is the *in vitro* induction of membrane fusion. However, only naked membranes as they naturally occur in mammalian cells can be fused. In contrast, microbial organisms and plant cells are surrounded by a rigid cell wall, and their intact cells cannot be fused. Fusion experiments with these organisms became only possible after the development of a method by which the cell wall could be removed in order to produce viable wall-less cells, called protoplasts. As naked cells these protoplasts can be fused, and they have also the capacity to regenerate a new cell wall and to differentiate to a normal microbial cell or plant, when cultured under appropriate conditions.

A major advance of induced cell fusion is not only that sexual techniques of crossing can be bypassed allowing the hybridization of asexual species, but it permits interspecies and intergenera hybridization, because membranes of totally different species can be fused with subsequent eventual nuclear fusion and recombination phenomena. Such interspecies crosses can give rise to organisms with completely new properties which is of particular relevance for breeding purposes and in genetic mapping. Ample examples have already shown the usefulness of cell fusion in genetic mapping and strain improvement of bacteria and fungi of industrial interest, in somatic hybridization of plant and animal cells, for the production of hybridomas and as a means to gene transfer by the introduction of organelles, macromolecules such as DNA or RNA, nuclei or mitochondria.

This chapter reviews the different methods of membrane fusion and discusses the underlying physico-chemical events during the fusion process. Bacterial and fungal protoplast isolation and fusion will be described, and the possibilities of fusion with respect to genetic analysis and strain improvement will be surveyed and extensively illustrated with several examples. Plant protoplast fusion and mammalian cell fusion will only briefly be treated. For more information on this topic the reader is referred to other chapters in this volume.

2 Methods and Mechanisms of Membrane Fusion

Biological membranes consist of phospholipid bilayers 5 to 6 nm thick and containing different protein molecules (Fig. 1). Phospholipid bilayers have several important features including an enormous flexibility and the property of self-closing. In a phospholipid bilayer the molecules can freely move and bilayers to-

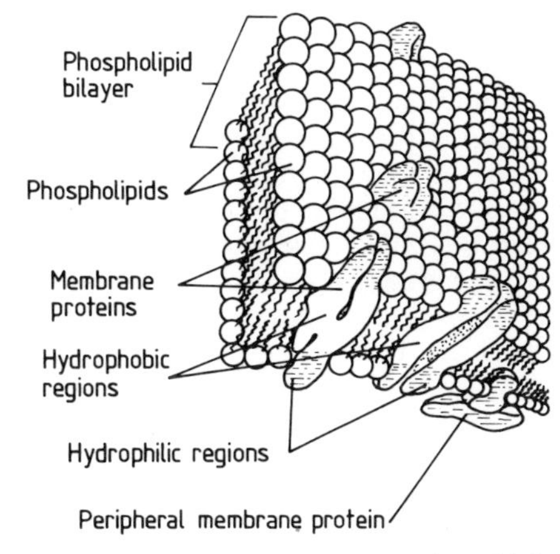

Fig. 1. Schematic representation of a biological membrane.

lerate all kinds of deformation without disrupting the bilayer structure. As a result of this flexibility, biological membranes can be fused either spontaneously or after induction by external means. Spontaneous fusion events outside biologically controlled processes have been mentioned for all types of cells, e.g., for murine neoplastic cells (HARRIS, 1970), fungal and plant protoplasts, but the frequency of fusion was very low in each case, even if fusion was forced by mechanical pressure (FERENCZY, 1981), and it remained restricted to cells of the same origin. In consequence, spontaneous fusion events were not really applicable to genetic studies. Therefore, other methods had to be devised to increase the fusion frequency.

2.1 Virus-Induced Cell Fusion

When mammalian cells are exposed to viruses, a cytopathic effect results in the formation of syncytia, a multinucleate type of cells in which the cytoplasm is not subdivided by membranes. These syncytia are usually produced by cell fusion. The first observations on syncytia formation in virus infection were reported as early as 1873 for smallpox, although the causative agent was not known at that time. Similar observations were made for varicella in 1889 and measles in 1910 (see HARRIS, 1970, for more details). Clear proof that viruses could give rise to syncytia formation was given by ENDERS and PEEBLES in 1954 who noticed that in tissue cell cultures syncytia are formed after infection with measles viruses.

An important discovery, and what can be considered as the onset of induced cell fusion, was made by OKADA et al. (1957). They first described that animal tumor cells in suspension can be rapidly fused under the influence of hemagglutinating virus of Japan (HVJ) or Sendai virus giving rise to the generation of multinucleate cells. Virus completely inactivated by UV retained its fusion capacity. HARRIS and WATKINS (1965) using UV inactivated Sendai virus detected the practical application of virus-induced cell fusion to produce heterokaryons between different animal cells, and they and others showed that even a wide species difference was no barrier to cell fusion. A number of enveloped DNA and RNA viruses have since been reported to produce syncytia (HOEKSTRA and KOK, 1989), but certain RNA viruses such as those that belong to the paramyxovirus group which include mumps, New Castle disease virus, and para-influenza viruses are better suited as fusion reagents. They stimulate membrane fusion during virus entry and not as a consequence of virus replication, allowing the use of inactivated virus for membrane fusion. Cell fusion by inactivated virus avoids complications such as cell death or an increased possibility of chromosome rearrangements resulting from the use of infectious virus. Instead of UV, alkylating agents including β-propiolactone can also be used for virus inactivation causing complete destruction of virus infectivity while leaving the cell fusion properties unaffected. In addition, chemically inactivated viruses are able to fuse cells in monolayers thus obviating the more laborious technique of working with cells in suspension. Induction of cell fusion by viruses has been used very often in genetic experiments on mammalian cells, but has now been replaced largely by other means of fusion.

2.2 Chemically Induced Cell Fusion

The successful applications of virus-induced fusion of mammalian cells to genetic studies and the promising perspectives of similar fusions of microbial and plant protoplasts have attracted several researchers. However, virus-induced fusions remained limited to mammalian cell types. They are not applicable, e.g., to microbial or plant protoplast fusions. Therefore, other means than virus-induced membrane fusions ought to be found to obtain membrane fusions.

POWER et al. (1970) devised first controlled conditions using high concentrations of $NaNO_3$ (0.25 M) to fuse plant protoplasts from different taxa in order to produce hybrid somatic cells. In spite of the confirmation of early results, the technique seemed not efficient, had uncertain reproducibility and an extremely low fusion frequency of less than 0.01% (LAZAR, 1983). It could also not be used for microbial protoplast fusions (ANNÉ, 1977). Better fusion results were obtained when the cells were treated with Ca^{2+} ions under alkaline conditions. Ca ions have been implicated as modulators in many biological fusion events, e.g., in secretion. Also Sendai-induced fusions require Ca^{2+}. The fusion capacity of Ca^{2+} solutions at high pH were detected for mammalian cells during experiments by TOISTER and LOYTER (1971). They investigated the biochemical reactions and the molecular changes taking place during the fusion events of avian erythrocytes induced by virus particles. Ca^{2+} under alkaline conditions led also to fusion of other cell types, including plant protoplasts, liver wort protoplasts, bacterial and fungal protoplasts.

Although Ca^{2+} at high pH was effective as a fusogen, the percentage of fused cells remained low. Much better results could be obtained with a new class of fusogens, independently detected by WALLIN et al. (1974) and KAO and MICHAYLUK (1974) who observed that the non-ionic water-soluble surfactant polyethylene glycol (PEG) could efficiently agglutinate plant protoplasts and that these protoplasts subsequently were fused at high frequency. It soon became clear that PEG-induced cell fusion was not cell-specific. In a short period of time it was shown that PEG was an efficient fusogen for any kind of cells (for a review, see FERENCZY, 1981) including also fungal protoplasts, bacterial protoplasts, and mammalian cells. It proved also to be suitable for the fusion of cells of different phylogenetic origin. Instead of PEG the non-ionic surfactant polyvinyl alcohol (PVA) with an average polymerization degree of 500–1500 can also be used for fusion with similar efficiency as PEG (NAGATA, 1978), but the latter compound has been much more generally applied as fusogen.

Optimal conditions for fusion have been determined for different systems. PEG of molecular weights ranging between 1500 and 6000 and at varying concentrations has been used. Optimum concentrations described varied between 25 and 50% (w/v) depending on the type of cell (HOPWOOD, 1981; ANNÉ, 1983). Fusion occurred almost immediately, and prolonged exposure of protoplasts to PEG seemed to reduce their survival. The addition of Ca^{2+} (mostly 0.01–0.1 M) was important to get a high fusion frequency.

Besides for somatic cell fusion, PEG treatment could be used as an efficient means to bring about the uptake of macromolecules such as DNA (HOPWOOD, 1981) and organelles like mitochondria or nuclei (FERENCZY, 1984; SIVAN et al., 1990) or even whole cells into larger cell types such as bacteria into fungal protoplasts (GUERRA-TSCHUSCHKE et al., 1991) or into plant protoplasts (HASEGAWA et al., 1983) or bacteria into mammalian cells. The latter method is currently applied efficiently to the direct gene transfer from bacteria to mammalian cells (SANDRI-GOLDIN et al., 1983; CAPORALE et al., 1990).

It has been observed that different batches of commercially available PEG preparations differed in their fusogenic activity (BALTZ and MATSUSHIMA, 1983). In addition, as a result of purification some batches of PEG almost completely lost their fusogenic activity. Therefore, it was suggested that PEG caused aggregation, not fusion, but some contaminating antioxidants like α-tocopherol or other phenolic compounds which are generally added to commercial-grade PEG were thought to be responsible for the fusogenic effect (HONDA et al., 1981). However, in the meantime it is proven that PEG itself is a fusogen, but the addition of lipid-soluble compounds including α-tocopherol, glyceryl monooleate, and retinol enhances PEG-induced cell fusion (SMITH et al., 1982). Variability in the behavior of different lots of PEG can therefore be partially explained by different amounts and types of contaminants in the various PEG preparations. It might also explain differences in optimal fusion conditions.

The importance of some lipids for inducing cell fusion, as just mentioned, has already been considered for a long time. Several lipids and phospholipids have been shown to induce cell fusion or to have at least a positive effect on the fusion process (LUCY et al., 1971). The most intensively studied lipids for fusion are lysolecithin and the above cited glyceryl monooleate (CROCE et al., 1971; CRAMP and LUCY, 1974) both of which have been used in mammalian cell fusion experiments. For plant protoplasts, a positively charged synthetic phospholipid has been mentioned for fusion (NAGATA et al., 1979). Also used for fusion in combination with PEG are liposomes, a sort of lipid crystal structures. They are made from natural or synthetic phospholipids arranged in bimolecular arrays. They can spontaneously fuse with biological membranes (MARTIN and MACDONALD, 1974; FELGNER et al., 1987; UCHIDA, 1988). It appeared that liposomes are also synergistic fusogens in combination with PEG, since in the presence of liposomes the requirement for high concentrations of the polymer for efficient protoplast fusion is lowered (MAKINS and HOLT, 1981). The main advantage of liposomes, however, is that biomolecules entrapped in the vesicles are protected from degradation by hydrolytic enzymes. Because of this advantage liposomes are increasingly used in transformation and transfection experiments with different cell types including mammalian cells (FELGNER et al., 1987), bacterial protoplasts (MAKINS and HOLT, 1981; CASO et al., 1987; BOIZET et al., 1988), yeast protoplasts (RUSSEL et al., 1983), and plant protoplasts (LURQUIN, 1979). Detailed description of the liposome mediated transformation and transfection is beyond the scope of this chapter.

Whereas contaminating lipids and phospholipids have a positive effect on PEG-induced fusion, other contaminants in PEG preparations have a negative effect on the fusogenic activity. These impurities consist of oxidative decomposition products such as aldehydes, ketones, and acids, which mainly cause cytotoxic effects reducing the viability of treated cells. The impurities are present in varying concentrations in different batches of PEG, and they are also influenced by storage and autoclaving. It has also been shown that heat treatment increases the carbonyl content of the solution enhancing cytotoxicity (CHAND et al., 1988). This effect has been demonstrated for different cell types. Autoclaved PEG solutions decreased yeast protoplast reversion and hybrid production (KAVANAGH et al., 1990; KOBORI et al., 1991), diminished fusion of plant protoplasts (CHAND et al., 1988), and gave reduced hybridoma yield (KADISH and WENC, 1983) as compared with membrane-sterilized PEG solutions. Purification of PEG solution by removing the carbonyl impurities promoted fusion and viability of plant protoplasts (KAO and SALEEM, 1986) and had a positive effect on PEG-mediated transformation in mammalian cells (KLEBE et al., 1984). For this reason membrane-sterilized PEG solutions are to be preferred.

2.3 Electrofusion

A totally new approach to *in vitro* cell fusion is the electrofusion method. This method is based on the combined action of dielectrophoresis and a transient change in membrane permeability obtained by electric pulses. The existence of a transient permeability change under the influence of high electric pulses had already been suggested by NEUMANN and ROSENHECK (1972). ZIMMERMANN and coworkers (for review, see ZIMMERMANN, 1986; ZIMMERMANN and URNOVITZ, 1987) investigated it more thoroughly, and they exploited this permeability change to obtain electro-induced fusions. These latter investigators suggested that fusion occurred as a consequence of what they called a reversible breakdown of the membrane caused by a structural alteration of the lipid bilayer in the membrane due to electric pulses. They came to these findings while measuring size distribution of *Escherichia coli* using a Coulter counter system. They observed a shift of the size distribution towards smaller volumes when a critical field strength was exceeded. They explained this shift by means of a reversible breakdown of the membrane. Irreversible breakdown leading to cell death had already earlier been mentioned for different cell types including bacteria, bacterial protoplasts, and yeast by applying a series of direct current (DC) pulses of increasing field strength and a duration of 2–20 μs. When the membrane is exposed to a DC pulse of high intensity, but of short duration (microseconds), a reversible breakdown can be achieved.

The breakdown voltage of most cell membranes is about 1 V, which corresponds to a field pulse in the kV/cm range. The voltage is strongly influenced by temperature. It decreases towards higher temperatures, and it also depends on membrane and system parameters. Membrane permeability is influenced by strength and duration of the electric pulse. Increased field strength and prolonged exposure times cause a considerable increase in membrane permeability that allows the uptake of large molecules. Too long pulse lengths at high field strength, however, give rise to an irreversible breakdown of the cell membrane.

Because of increased membrane permeability, cells subjected to an experimentally determined, correct high electric field pulse can efficiently be transformed or transfected. The first proof that this electrical method was effective for the introduction of DNA into cells was provided by AUER et al. (1976) demonstrating that DNA or RNA can be introduced into erythrocytes, after dielectric breakdown of the red blood cell membrane. Since these first experiments the electroporation method has been proven a valuable and even superior alternative to the chemical transfection or transformation methods for any type of cell; and microbial cells can be transformed without prior protoplasting (DELORME, 1989; CHAKRABORTY and KAPOOR, 1990).

On the other hand, when electropermeabilized cells are brought into close contact with each other, they can be fused. The aggregation required for fusion can be obtained by dielectrophoresis (Fig. 2). Dielectrophoresis comprises the migration of neutral particles such as cells or protoplasts in an alternating current (AC) electric field. In the presence of an electric field the cell or protoplast being a neutral but highly conductive particle becomes polarized and gives rise to dipole formation in which a transmembrane potential is created. The magnitude of this potential is proportional to the intensity of the external field and the diameter of the particle which has to be at least 0.3 μm (ZIMMERMANN, 1986). The positive charge of the dipole is directed towards the cathode, and the negative charge is nearest to the anode. In an AC field, the field strengths on both sites of the particles become unequal and the net force thus generated on the particle drives it towards the region of the higher field intensity, and this is independent of the arrangement of the polarity of the electrode. This means that in an AC field the polarized but neutral particle still moves towards the region of the highest field intensity. In addition, when the particles approach they attract each other as they are dipoles, and this leads to the formation of pearl chains (Fig. 2). The number of cells within a pearl chain depends upon the population density of the cells and the distance between the electrodes, but also on the pH of the medium (CHANG et al., 1989). When the cells in the pearl chain are subjected to a high-intensity DC pulse they fuse as a consequence of the reversible breakdown of the cell membranes.

Instead of aggregation by dielectrophoresis, electropermeabilized cells can also be fused by the addition of agglutinating agents such as PEG, PVA or spermine (CHAPEL et al., 1986), and even simple centrifugation is sufficient (TEISSIÉ and ROLS, 1986). Since agglutination may occur either before or after electroporation, it indicates that a long-lived fusogenic state of the membrane can be induced by high-field pulsation (MONTANÉ et al., 1990). The observation that this fusogenic state can last for several minutes is of particular interest from a practical point of view. Cells having large distinctive differences in their morphology or physiology and thus requiring their own specific field intensity for the induction of the fusogenic state can separately be electropermeabilized prior to fusion treatment.

Electric field induced permeabilization and fusion are now along with PEG treatments routinely used in cell biology. Electrofusion is a gentle procedure which gives under optimal conditions high fusion frequencies. Like PEG or PVA it is applicable to all types of naked cells. Due to the inherent variation of biological systems, optimal conditions have to be determined for each cell type. It is, therefore, impossible to standardize a specific method for electrofusion or electrotransfection. Compared to PEG or PVA it is claimed (ZIMMERMANN, 1986) that electrofusion has several advantages over the chemically induced fusion including the significantly higher fusion frequency – which is certainly the case for electrotransformation and transfection –, and it has a less harmful effect on the cell viability. As a consequence, electrofusion and electrotransformation are gradually more frequently applied. Several types of cell fusion apparatus can now commercially be obtained.

A number of examples of different cell types fused under the influence of high field electric pulses are listed in the following references: for *Streptomyces* and other Gram positive bacteria: OKAMURA et al. (1989); filamentous fungi: KÜNKEL et al. (1987), USHIJI-

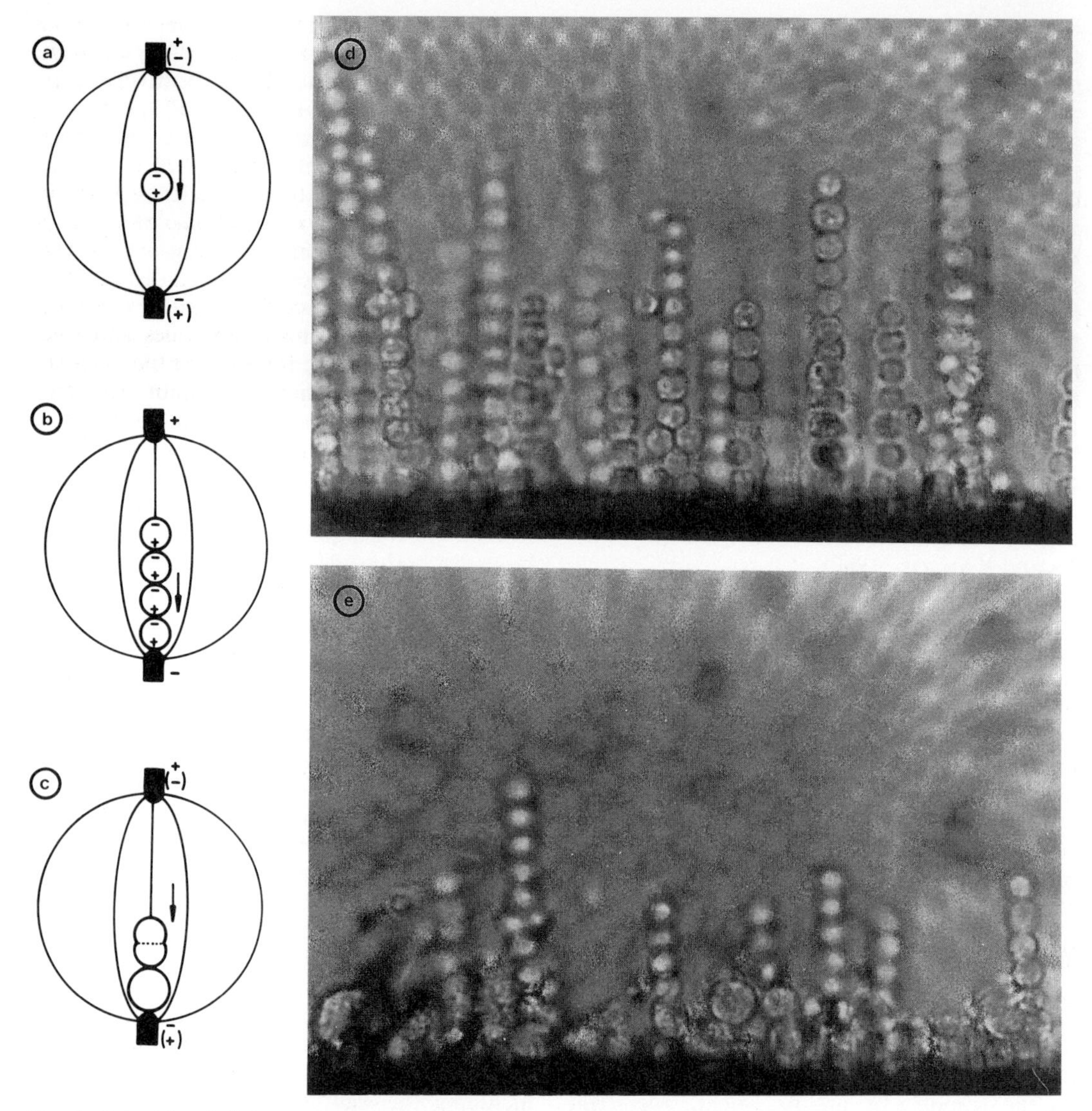

Fig. 2. Principle of dielectrophoresis and illustration of electrofusion.
(a) Cells appearing as induced dipoles in an inhomogeneous electric field migrate along the field lines as a result of a net force exerted on the particle. When the external electric voltage between the electrodes is reversed, the direction of the particle will not reverse.
(b) Polarized cells attract each other and form chains at the electrode in an inhomogeneous electric field.
(c) When the membrane is subsequently exposed to a direct current pulse of high intensity and short duration, membrane fusion can occur (ZIMMERMANN, 1986).
(d) Photograph of pearl-chain formation of fungal protoplasts.
(e) Photograph of pearl-chain containing fused protoplasts (photographs with courtesy from Y. TAMAI, Hokkaido University, Japan; TAMAI et al., 1989).

MA et al. (1991); yeast: HALFMANN et al. (1983), FÖRSTER and EMEIS (1986), VONDREJS et al. (1990); plant protoplasts: MONTANÉ et al. (1990), FINCH et al. (1990), KOOP and SCHWEIGER (1985), MORIKAWA et al. (1986); mammalian cells and the production of hybridomas: NEIL and URNOVITZ (1988), ONISHI et al. (1987), SCHMITT et al. (1989), FOUNG et al. (1990), HEWISH et al. (1989). On the other hand, when PEG induced fusions are carried out under optimal conditions, similar results are obtainable as with electrofusion (CHAND et al., 1988), and no special equipment is required.

A very promising method that is still under investigation is laser-induced cell fusion (WIEGAND et al., 1987). Laser beams with a very narrow diameter ranging between 0.3–0.5 μm can be produced thus allowing the individual fusion of selected pairs of target cells. In this manner, the time needed to select the cells after fusion can be reduced. Finally, a recent paper suggests an alternative approach to cell fusion, based on thermodynamic factors, i.e., thermo-osmotic forces and fluxes (ANTONOV, 1990). This method uses the induction of membrane defects by cooling the cell suspension to 0 °C followed by an aggregation forced either by dielectrophoresis or temperature-controlled centrifugation. The effectiveness of this method needs, however, still to be proven.

2.4 Molecular Mechanism of Membrane Fusion

Although membrane fusion is a very important event in living organisms, the underlying molecular mechanism of fusion is not yet completely understood. Our current knowledge of this complex and extremely rapidly occurring process is mainly based on the study of the interaction between pure lipid model membranes and on electron-microscopic investigations. However, lipid models are much simpler structures than biomembranes and electron-microscopic investigations are only possible after processing the biological samples. In consequence, the explanation of the mechanism remains a speculative model.

The prerequisite for fusion is that the membranes to be fused will be in such close contact that intramolecular interactions can take place between structures that are normally at an intermolecular distance. The barriers separating biomembranes include first of all a repulsive hydration force originating from bound water at the head group of the lipid molecules. When the distance between two apposing membranes is reduced to less than 2–3 nm, a powerful hydration repulsion occurs between hydrophilic surfaces (RAND, 1981). Other barriers include an exclusion volume of the plasma membrane (glyco)proteins and macromolecules and electrostatic repulsion forces (BLUMENTHAL, 1987). It is known that cell membranes carry net negative charges which for a considerable part originate from phosphate groups and ionized proteins. To obtain cell aggregation and subsequent fusion, barriers separating the membranes have to be removed or overcome. For example, an external pressure of 10^4–10^5 Pa must be applied to remove the water layer and establish a close contact between the membranes. It is obvious that the factors responsible for the induction of cell aggregation and subsequent membrane fusion may differ for the different classes of fusogens.

Paramyxoviruses, able to induce cell fusion as mentioned earlier, bind to the cell surface and then fuse with the cytoplasmic membrane under particular conditions (for review, see WHITE, 1990). Agglutination is realized by the spike glycoproteins dispersed on the outer site of the enveloped viruses causing cross-bridging of the adjacent cells by means of the virus particles. In Sendai viruses, two major glycoproteins, HN (hemagglutinin and neuramidase activity), and a fusion protein are present on the surface of the envelope. HN is responsible for attaching the virus to cell-surface sialic acid residues covalently linked to glycolipids and glycoproteins which act as primary cell-surface receptor sites (UCHIDA, 1988). The fusion protein is involved in the fusion reaction itself. A conformational change at the N-terminal sequence of the non-functional precursor fusion protein lies at its origin. As a consequence of this change a stretch of highly conserved and extremely hydrophobic amino acid residues is produced. It is suggested that the hydrophobic stretch of the fusion protein penetrates into the target membrane. The thus provided hydrophobic interaction between the viral membrane and the target membrane could induce membrane fusion. The mechanism of fusion is still unknown, but several morphological and biophysical studies suggest that as a consequence of the hydrophobic interaction water should be expelled causing a local dehydration at the interbilayer contact sites. This may induce a local transient disordering of the equilibrium bilayer configuration in the two approaching membranes (BURGER, 1991) allowing the interaction and intermixing of disturbed lipid molecules of these closely apposed membranes. Different models of intermediate fusion stages have been proposed (Fig. 3) including the existence of inverted hexagonals or H_{II} configurations (VERKLEIJ, 1984; SIEGEL, 1986).

Electron-microscopic investigations suggest that cell–cell fusion is achieved by the simultaneous fusion of a virus with two adjacent cells (KNUTTON, 1977). This conclusion was drawn from the fact that in suspensions treated with Sendai virus, virus particles with two adjacent cells were seen much more frequently. Finally, expansion of cells joined by small cytoplasmic connections to form spherical fused cells occurs by a process of permeable, osmotically induced cell swelling.

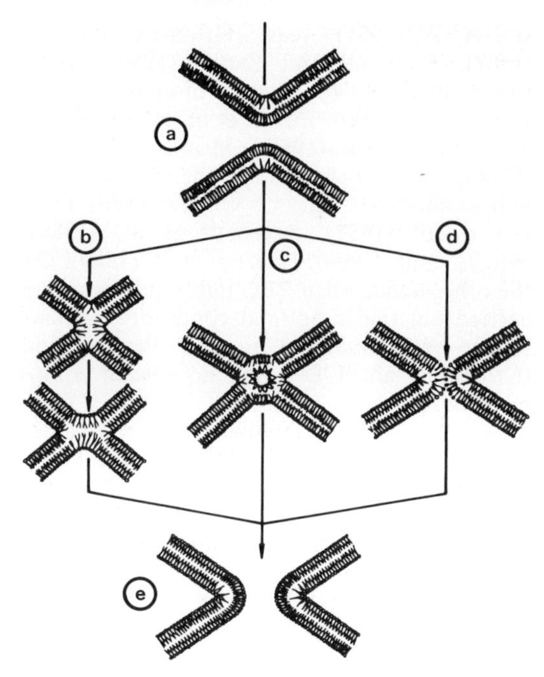

Fig. 3. Molecular models for membrane fusion (after BURGER, 1991). Following membrane apposition (a) three possible models are proposed: formation of monolayer stalks (b), inverted lipid micelles (c), or local point defects (d), which lead to membrane fusion (e).

Membrane apposition, the first requirement for membrane fusion is in virus-induced fusion brought about by the HN and fusion protein as explained above. In chemically induced fusions other factors have to be involved. Ca^{2+} are known to dehydrate the space between phospholipid bilayers, they cross-link membranes, and they have also the capacity to neutralize the negatively charged head groups of acidic phospholipids resulting in altered membrane fluidity and phase separation of the bilayer, factors believed to be important in membrane fusion. For PEG, neither aggregation nor fusion do arise from a direct interaction between PEG and the membranes, because PEG is excluded from the area of close contact of the apposed membranes as observed during electron-microscopic investigations (KNUTTON, 1979). It is believed, however, that the strong dehydrating effect of PEG is the driving force for membrane apposition and is at the origin of altered physico-chemical properties at the vicinity of the membrane surface (PRATSCH et al., 1989). Dehydration alone is not sufficient to explain the fusogenic properties of PEG, since the dehydrating agent Dextran does not act as a fusogen, indicating that other factors have to be involved as well. PEG decreases the surface potential of the membranes by several hundreds of millivolts (MAGGIO et al., 1976), and it may induce alterations in the orientation and hydration of the phospholipid head groups causing charge neutralization, segregation of the lipids and bilayer defects (BONI et al., 1984) probably inducing a type H_{II} non-bilayer structures between the two interacting apposed membranes. It is assumed that the combination of these effects makes PEG an efficient fusogen.

Also the fusion promoting activity of lysolecithin, monoglycerides and fatty acids is believed to be correlated to the possibility that these compounds promote the formation of H_{II} non-bilayer lipid structures (HOPE and CULLIS, 1981).

In electro-induced cell fusions the energy of the breakdown pulse provides another means of perturbation of the membrane phospholipids. An alteration in the organization of the polar heads may explain the membrane permeabilization. For electrofusion it has been suggested that there should be no stable intermediate structure. The real fusion intermediate should be a local disorder of lipid molecules that directly leads to membrane rupture after extensive thinning under the influence of the electric forces (LUCY and AHKONG, 1986).

Fusion is completed by the formation of a small aqueous pore connecting two originally separated aqueous compartiments. The extremely unstable configuration of the inverted micelle collapses probably by the system's tendency to reduce the curvature energy of the monolayer (LEIKIN et al., 1987). The pore is initially about 10–15 nm in diameter, but rapidly widens as swelling under the influence of the osmotic pressure proceeds (LUCY and AHKONG, 1986), and finally the content of the fused cells can be intermingled.

3 Bacterial Protoplast Fusion

3.1 Protoplast Formation and Regeneration

Shape and rigidity of the microbial cell is determined by the cell wall surrounding the cytoplasm with the cytoplasmic membrane. When deprived of its rigid cell-wall structure, the cell loses its characteristic shape, becomes osmotically sensitive and round, even if the cell wall is not completely removed. Cells with a completely removed wall are designated protoplasts. If some cell-wall remnants have remained, they are called spheroplasts, but for the sake of simplicity, in this chapter all osmotically labile cells will, in general, be referred to as protoplasts.

The first studies that demonstrated the possibility of isolating protoplasts from bacterial cells were by TOMCSIK and GUEX-HOLZER (1952) and by WEIBULL (1953) who worked with *Bacillus megaterium*, trying to prove that the substrate for lysozyme is the cell wall. WEIBULL (1953) showed in addition that it is possible to isolate a bacterial protoplast as an entity by dissolving the cell wall with lysozyme and stabilizing the protoplasts in 0.2 M sucrose or 7.5% polyethylene glycol. Stabilization of the protoplasts in hypertonic medium is necessary, because the intracellular osmotic pressure is very high ranging up to 3 MPa (30 atm). If there is no stabilizer in the medium, protoplasts take up too much water in order to equate chemical activities on both sides of the membrane and the protoplasts will swell and subsequently burst.

Since the first experiments on protoplast isolation, protoplasts have been produced from many different types of bacteria using different methods and conditions. In general, lysozyme, an enzyme that hydrolyzes the peptidoglycan layer of the cell wall, is used to convert the bacterial cell to a protoplast. Manifold examples can be found in the literature for different species (for review, see GUZE, 1968; HOPWOOD, 1981; HÜTTER and ECKHARDT, 1988).

Although walls of probably all bacteria contain peptidoglycan, lysozyme does not dissolve readily walls of all bacteria, not even of all Gram positive cells. In several cases, special treatment is required. Streptomycetes, Gram positive bacteria with a complex morphology of branched mycelia, are in most instances more readily protoplasted, when the cells are grown in a medium containing concentrations of glycine high enough to cause some cell growth retardation. The positive influence of glycine on protoplast formation is due to the fact that with an excess of glycine there is a competition with the normal murein amino acid component D-alanine, and terminal cross-linking reactions in cell-wall synthesis are inhibited (STROMINGER, 1968). As a consequence, damaged cell walls are produced that are more readily accessible for lysozyme.

The glycine concentration to obtain maximum protoplast yield in Streptomycetes ranges between 0.4 and 3.5% depending on the species and the growth conditions of the strain, and they have to be experimentally determined for each strain. In most instances S medium is used as growth medium (OKANISHI et al., 1974) and P medium containing 0.3 M sucrose as osmotic stabilizer in the presence of Ca^{2+} and Mg^{2+} (HOPWOOD et al., 1985a). For species related to *Streptomyces*, glycine is also added to the growth medium to improve protoplast formation, e.g., for *Micromonospora* (CASO et al., 1987; KIM et al., 1983), *Amycolatopsis orientalis* (MATSUSHIMA et al., 1987), *Actinoplanes brasiliensis* (PALLERONI, 1983), and also for other species such as *Clostridium acetobutolyticum* (ALLCOCK et al., 1982) and *Corynebacterium acetoacidophilum* (DEB et al., 1990).

In some cases, however, *Streptomyces* protoplasts could efficiently be produced without added glycine to the growth medium (BRADLEY, 1959; DOUGLAS et al., 1958; RODICIO et al., 1978; ANNÉ et al., 1990). Alternatively, it has been possible to produce protoplasts by growth in high concentrations of glycine without addition of lysozyme as reported for *Streptomyces pristinaespiralis* (HOPWOOD, 1981). A similar method of protoplast preparation has been successfully applied earlier to induce protoplasts from *Staphylococcus aureus* (STROMINGER, 1968) or certain Gram negative bacteria (MARTIN, 1983). Instead of glycine, antibiotics such as penicillin and cephalosporin, fosfomycin or D-cycloserine, all compounds that interfere with the murein cell-wall biosynthesis, can also induce protoplasts without the use of lysozyme, when added in sublethal concentrations during active growth. This approach to obtain protoplasts has often been successfully used for different Gram positive bacteria including *S. aureus* using penicillin (see GUZE, 1968) or fosfomycin (SCHMID, 1984), for *Brevibacterium flavum* (KANEKO and SAKACHUCHI, 1979), and also for Gram negative bacteria (see MARTIN, 1983).

In general, Gram negative bacteria are much more difficult to convert to protoplasts by lysozyme treatment than Gram positive bacteria. The reason is that the cell wall of Gram negative cells has a complex multi-component cell envelope, and the murein layer, the substrate for lysozyme, is protected by the complex outer membrane consisting of lipopolysaccharides, phospholipids, specific major proteins and the divalent Mg- and Ca-cations. Lysozyme can be applied to produce osmolabile cells of Gram negative bacteria – in this case spheroplasts are produced – only if it is used in combination with EDTA, which destabilizes the outer membrane allowing the penetration of lysozyme and the subsequent digestion of the murein layer.

Protoplasts normally do not divide, but under particular circumstances protoplasts from certain species are able to multiply as wall-defective bacteria. These have been called L-forms (L for Lister Institute, London, where they have been discovered). L-forms have mainly been studied in strains of clinical importance, but not in industrially important microorganisms. Only a few reports have described L-forms for Streptomycetes including *Streptomyces hygroscopicus, S. griseus, S. levoris* (see HÜTTER and ECKHARDT, 1988), but they have not been used in genetic studies.

Originally, protoplasts were used only for morphological investigations, not for genetic analysis. But after it became known that they could regenerate a new cell wall and revert to the normal cellular stage, they gained a lot of importance in pure and applied genetics through protoplast fusion and transformation experiments. Protoplast reversion demands particular circumstances. *Bacillus subtilis*, for which it was first discovered that bacterial protoplasts could revert to the bacillary cells, requires cultivation on solid medium in the presence of 25% gelatin, heat inactivated horse serum or casamino acids (LANDMAN and HALLE, 1963; GABOR and HOTCHKISS, 1979) or plasma expanders (AKAMATSU and SEKIGUCHI, 1981). On the other hand, for *Bacillus megaterium* or *B. licheniformis*, gelatin or casamino acids are not required (FODOR et al., 1975; FLEISCHER and VARY, 1985). Under optimal conditions, for *Bacillus* 80 to 100% protoplast reversion could occur. For many other species the maximum reversion frequencies that could be obtained so far, were much lower, e.g., less than 10% for Streptococci (GASSON, 1980) or *Nocardia mediterranei* (SCHUPP and DIVERS, 1986) and between 1 and 90% for Streptomycetes (HOPWOOD, 1981; MATSUSHIMA and BALTZ, 1986).

The basic conditions for *Streptomyces* protoplast regeneration were largely developed by OKANISHI et al. (1974) and adapted with minor modifications for different species. The available information suggests that each species and strain has its own optimal conditions for regeneration and that a multiplicity of parameters influences the results, including composition of the medium, the moisture content of the agar plates (about 20% loss of water), cell density at plating, the growth phase of the mycelium from which the protoplasts have been derived, i.e., mid to late exponential growth phase or early stationary phase; the temperature during growth prior to protoplasting, during embedding in the soft agar when applying the overlay method, and in the course of regeneration. For embedding, the lowest possible temperature should be chosen, and during regeneration lower temperatures are preferable.

3.2 Bacterial Protoplast Fusion

Bacteria are haploid microorganisms containing the entire genetic information on a single thread of double-stranded DNA, the bacterial chromosome, and, eventually, also on plasmid DNA. Transfer of genetic information between strains giving rise to recombinants can be obtained by different means. In transformation, genetic information is transferred using naked DNA, in transduction the DNA transfer is mediated with the aid of a bacteriophage, and in conjugation direct cellular contact is involved during which plasmid and/or chromosomal DNA is introduced into the mating cell. But such transfer systems are not known for all bacterial species and, in addition, recombination frequencies are very low and limited to rather closely linked genes, because the whole genome of a donor bacterium is seldom completely introduced into the recipient strain. In contrast, protoplast fusion can, in theory, be carried out with all species, if a suitable method for protoplast formation and reversion is available. Furthermore, in protoplast fusion not only closely linked genes are involved, but the whole genome can be introduced into the newly formed bacterial entity which consists of fused cells of genetically different strains belonging to the same or different species.

Fusion can be obtained either following PEG treatment or by electrofusion, as mentioned earlier. The principle of the fusion methodology consists of mixing the protoplasts of genetically different strains usually in a 1:1 ratio followed by the fusion treatment and cultivation of the fusion mixtures. For the selection of the fused protoplasts or of their recombinants, morphological, nutritional, or resistance markers have to be introduced into the strains. The recombinant progeny following protoplast fusion can be identified by the use of suitable selection media or by observation of changed morphology. Selection can be achieved in a direct or indirect manner. In the former method, the fused protoplasts are plated immediately on the selective media allowing only growth of nutritionally complementing strains, while in the indirect method the fused protoplasts are first allowed to revert on a rich non-selective medium prior to selection. The used selection method, however, influences the outcome of the results (HOPWOOD, 1981).

Instead of using mixtures of viable protoplasts, it has been shown that DNA of one partner killed by UV or heat treatment or killed by incubation with streptomycin can be rescued by fusing with viable protoplasts of the other partner, as demonstrated for *Bacillus, Micromonospora* and *Streptomyces* (MATSUSHIMA and BALTZ, 1986). This approach has not only the advantage that in some, although not in all, cases higher recombination frequencies can be obtained, but also that a prototrophic donor partner can be used. This is certainly of interest, if one intends to use protoplast fusion in strain improvement programs, since it is known that antibiotic production is lowered concomitantly with some auxotrophic mutations. In *Streptomyces* it has also been demonstrated that UV irradiation of both constituent parental protoplasts to a survival of 1–30% is a means to enrich for recombinants probably due to the fact that protoplasts with different, lethal UV damage can complement each other, and UV might also promote crossing-over events (HOPWOOD, 1981). Such an approach can, of course, not be applied in genetic mapping studies because of the risk of anomalous recombinations.

3.2.1 Protoplast Fusion and Genetic Mapping

A genetic map is a representation of relative distances separating non-allelic gene loci in a linkage structure. It can be constructed by the determination of the frequency of recombination that occurs between different genes, because recombination between two non-allelic gene loci is a function of the distance between these loci, i.e., the larger the distance, the more recombination events occur. In classical genetics, linkage maps are constructed by transformation, transduction, and conjugation experiments. This classical approach has the disadvantage that only relatively closely linked genes can be mapped at once. In contrast to the classical approach, protoplast fusion offers the possibility to produce recombinants over the whole length of the chromosome. Following protoplast fusion of genetically different bacterial strains, in a first instance a zygote, i.e., a heterozygous diploid, is produced. On further development, haploidization will occur giving rise to haploid recombinant progeny. In case multiple auxotrophic parents are used, it is expected that all possible combinations of parental markers could be recovered, allowing one to carry out an extensive genome mapping in a relatively short period of time. However, differences were observed for different species.

In *Bacillus* (see HOTCHKISS and GABOR, 1983), for which induced fusion of bacterial protoplasts was first reported, with *B. subtilis* and with *B. megaterium* ambiguous results were obtained with respect to recombination. Besides a small percentage (0.4–1.0%) of haploid recombinants, a large number of unstable complementing diploids and unstable non-complementing diploids – also named biparental diploids – were isolated following protoplast fusion of *Bacillus*. The phenotypically recombinant clones were either diploids containing silent genes until separation occurred, or complementing diploids in which both parental genes were phenotypically expressed. Unstable non-complementing diploids showed the phenotype of only one of the parent types in the supposed diploid, as if one chromosome remained unexpressed. Segregation analysis and DNA-transformation experiments have nevertheless shown that the non-complementary diploids carried the two parental chromosomes, but only one of the parental chromosomes was expressed. The molecular basis for the phenotypic suppression of one of the genotypes of the non-complementing diploid clones was demonstrated to be at the transcriptional level. A different DNA tertiary organization in one of the chromosomes was supposed to be at the origin of non-expression. Because of these problems, genetic mapping cannot be carried out by recombination analysis via protoplast fusion using the current methods. This kind of problems did occur not only in *B. subtilis*, but also in *B. megaterium* where similar non-complementing diploids have been isolated (FLEISCHER and VARY, 1985).

On the other hand, in *Bacillus stearothermophilus* (CHEN et al., 1986), in *Staphylococcus aureus*, and *Streptomyces* no real problems with genetic mapping have been experienced so far, and chromosomal mapping could be carried out with recombinants obtained following protoplast fusion. With *S. aureus*, the linkage relationships of markers based on protoplast fusion data were entirely consistent with the linkage relationships of markers previously defined by transformation (STAHL and PATTEE, 1983) and, in consequence, protoplast fusion can be used with these species for genetic mapping (TAM and PATTEE, 1986).

Conditions for efficient protoplast fusion and regeneration have been described for many different Streptomycetes (HÜTTER and ECKHARDT, 1988). The number of Streptomycetes involved in an extentive genetic analysis using protoplast fusion is, however, scarce and limited to *Streptomyces coelicolor* (HOPWOOD, 1981), *S. lividans* (HOPWOOD et al., 1983), *S. rimosus* (HRANUELI et al., 1983), and *S. clavuligerus* (ILLING et al., 1989). To estimate the results of genetic mapping obtained by protoplast fusion a comparison with the linkage map obtained by conjugation was made for all strains except *S. clavuligerus*. From the comparison it could be concluded that following protoplast fusion the proportion of multiple crossover classes among recombinants was higher by a factor of more than 10 and there was also a high proportion of progeny from crossovers widely separated on the chromosome. These differences with respect to the results obtained by conjugation are probably due to the formation of complete diploids following protoplast fusion, instead of merozygote formation by mating, and by the occurrence of eventual, multiple rounds of recombination during regeneration of the protoplasts prior to or at early branching. Because of the extra recombination associated with protoplast fusion, this approach can be most useful in the resolution of closely linked genes, but it might be less important for long-range mapping.

When applying protoplast fusion to genetic mapping one has also to take into account that protoplasting can cause genetic instability such as the loss of antibiotic production or resistance and the appearance of pleiotropic mutations (BALTZ and MATSUSHIMA, 1983). These genetic variations are probably associated with genome rearrangements involving large DNA deletions and/or amplifications.

The number of experiments involving protoplast fusion with Gram negative bacteria is limited to three species: *Escherichia coli, Providencia alcalifaciens* (HOPWOOD, 1981), and *Pseudomonas putida* (LEE et al., 1988). The reports only demonstrate the possibility that through protoplast fusion haploid, chromosomal recombinants can be obtained, but no genetic analysis has been carried out as yet.

3.2.2 Applications of Protoplast Fusion in Strain Improvement

Until the development of the novel techniques of protoplast fusion and plasmid mediated transformation, genetic recombination has not been explored widely as a method of strain improvement. Most breeding programs consisted mainly of mutation and screening for higher producing strains. The reason for the success of mutation is partially because microorganisms have a haploid life cycle and recessive mutation can readily be detected. Nevertheless, there are reasons to consider genetic recombination as a worthwhile alternative or a complementary strategy to strain improvement, not the least because highly mutated strains are less viable. Furthermore, high producer strains contain various mutations probably involving regulatory sequences. Hence, a further increase in yield will be more and more difficult. Using classical crossing techniques it has been shown that it is possible to obtain recombinants with production levels exceeding the parent strains. The advantage of the protoplast fusion technique is that the possibility to obtain recombinants is much increased, which allows testing of a large number of recombinants in a short period of time.

3.2.2.1 Intraspecies Protoplast Fusion in *Streptomyces*

Streptomycetes are industrially the most important group of bacteria. With more than 3000 species identified, they have been described as the greatest source of antibiotics with a wide variety of chemical structures. It is therefore not surprising that most investigations concerning the application of bacterial protoplast fusion involve species belonging to Streptomycetes. Both intra- and interspecies crosses have been used, but since most of this research has routinely been done in industrial laboratories, the number of reported results is not very high and for the greater part remains limited to optimizations of methods to obtain large numbers of recombinants. In this respect, UV or heat treatment or even the involvement of more than two different genotypes in the same mixture of protoplasts to be fused has often been mentioned to gain an efficient recovery of recombinants (BALTZ and MATSUSHIMA, 1983).

As mentioned before, following protoplast fusion multiple crossover events involving large regions of the genome occur thus giving rise to a high number of recombinants. This is an obvious advantage in strain improvement programs and is especially useful when combined with conventional mutation programs. When in a breeding program, e.g., divergent strains obtained after repeated mutagen treatment and selection are crossed, a range of new recombinants will arise with a redistribution of the mutated genomes. In this manner, a cross between a highly mutated, high-producing, slowly growing strain and a less mutated low-producing strain with excellent viability possibly may result in a fast-growing high producer. This has been demonstrated with *Nocardia lactamdurans*: two improved cephamycin C producing strains from an industrial strain development program were fused, and among the recombinants two cultures were isolated with 10–15% more antibiotic production than the best parent (WESSELING and LAGO, 1981). Similar results have been reported for strains of *Streptomyces griseus* subsp. *cryophileus* (KITANO et al., 1985) and for *S. rimosus* (VALLIN et al., 1986). In the latter case recombinant strains were isolated that produced 4–5 times higher amounts of oxytetracycline than the wild-type parent. Furthermore, by variations induced due to protoplasting the genetic variability may be increased as well.

3.2.2.2 Interspecies Protoplast Fusion in *Streptomyces*

New derivatives of secondary metabolites can be prepared in different ways including total synthesis or the semisynthetic preparation of new derivatives, the modification of the original structure through bioconversion and the genetic manipulation involving mutasynthesis (FLECK, 1979), gene transfer by recombinant DNA cloning techniques (FLOSS, 1987), and interspecies hybridization by mating, or more recently, by protoplast fusion. Using mating, only a few examples of true interspecies recombinants have been reported. Through protoplast fusion, in several instances the frequency of recombination could be increased (GODFREY et al., 1978) and also new antibiotic structures could be isolated (Tab. 1).

PEG-induced protoplast fusion resulted in a low number of primary fusants ($1 \cdot 10^{-4}$–$1 \cdot 10^{-6}$), whereas electrofused protoplasts showed a fusion frequency of more than 10.0% (OKAMURA et al., 1988), but in all cases the number of stable recombinants was very low. The low frequency of recovered primary fusion products may depend on the method used, but may also be due to the species used in the crossing. The different strains may contain different restriction–modification systems thus influencing the viability of the fusants. Some unstable recombinants, probably merozygotes, obtained after the interspecies crossing produced a new antibiotic, but on further selection most of them lost antibiotic production and only few stable hybrids producing new antibiotics remained. The low number of

Tab. 1. Interspecies Crosses with Streptomycetes Leading to New Antibiotics (for reference see OKAMURA et al., 1989)

Cross	Original Antibiotic	New Antibiotic
Streptomyces hygroscopicus × *S. violaceus*	Turimycin *Violamycin*	Iremycin
S. griseus × *S. tenjimariensis*	Streptomycin *Istamycin*	Indolimycin
S. fradiae × *S. narbonensis*	Mycaminose (tylosin precursor) *Narbonolide (narbomycin precursor)*	Neomycin
S. rimosus f. *paramomycinus* × *S. kanamyceticus*	Paramomycin *Kanamycin*	Neomycin
S. antibioticus × *S. fradiae*	Multhiomycin Neomycin	Unidentified new antibiotic

true recombinants probably indicates the low frequency of homology between the chromosomes of the species involved.

Nevertheless, the results, albeit limited in number, demonstrate the possibility of using interspecies crosses for the production of new antibiotic metabolites. These novel products are probably formed in "hybrid" biosynthetic pathways arising from the simultaneous action of enzymes encoded by the structural genes originating from the different Streptomycetes involved in the cross or as a result of the activation of latent or "silent" genetic information in the recipient strain (HOPWOOD et al., 1985b).

Alternative means to produce new hybrid antibiotics are the direct transfer of cloned antibiotic biosynthesis genes into another strain via transformation procedures – which also involves protoplasting (HOPWOOD et al., 1985b). This approach requires much preparative cloning work including the identification and isolation of the genes involved, but is, on the other hand, more directed.

3.2.2.3 Interspecies Protoplast Fusion with *Bacillus*

So far, only a few results have been reported for interspecies crosses with *Bacillus*. *B. thuringiensis israelensis* producing parasporal inclusions which upon ingestion kill susceptible insect larvae was crossed with *B. subtilis* (RUBINSTEIN and SANCHES-RIVAS, 1988). Selection was by complementation of auxotrophic markers and prototrophic fusants were obtained at high frequency (5–10%). A cross of *B. subtilis* with a mesophylic and cellulolytic *Cellulomonas* resulting in cellulolytic *B. subtilis* and fusants secreting a cell-bound aryl β-glucosidase of *Cellulomonas* was reported (GOKHALE and DEOBAGKAR, 1990). Another report (DEB et al., 1990) claimed the fusion of a xylan degrading *B. subtilis* with a lysine producing *Corynebacterium acetoacidophilum*. Different types of recombinants were discerned by biochemical identification, and the fusants expressed the parental characters although at varying levels. *B. thuringiensis* was fused with the Gram negative *Agrobacterium tumefaciens*, and fusion products occurring at a frequency of $0.2\text{–}5 \cdot 10^{-6}$ were selected by antibiotic resistance. Fusants showed mixed phenotypic and biochemical characters of both parents (PUNTAMBEKAR and RANJEKAR, 1989). In no case, however, were the genotypes of the fusants and their stability extensively analyzed. Therefore, care should be taken in the interpretation of the results. These limited results indicate, however, that it should be feasible to cross different species with *B. subtilis* or *B. thuringiensis* giving rise to viable fusants.

4 Fungal Protoplast Fusion

4.1 Protoplast Formation and Regeneration

The fungal cell wall has a complex structure composed mainly of polysaccharides including chitin, glucans with D-glucose units linked by $\beta_{1,3}$- and $\beta_{1,6}$-glucosidic bonds or consisting of $\alpha_{1,3}$-linked polysaccharides, or mannans as in yeast with *O*-glycosidic linked manno-oligosaccharides and also protein–polysaccharide complexes (Fig. 4). Among fungi there is a large heterogeneity in the cell-wall composition, which, in addition, depends on age. This means that for an efficient removal of the cell wall necessary to the formation of protoplasts for fungi, in contrast to bacteria, complex mixtures of cell-wall-degrading enzymes are required capable of degrading the different components of the cell wall (PEBERDY and FERENCZY, 1985). The first enzyme mixture described able to dissolve fungal (yeast) cell walls was the digestive juice of the snail *Helix pomatia* in 1914, though it was not until 1957 that its lytic properties – with bacterial protoplast formation with lysozyme as a model – were used to induce protoplast formation in yeast (EDDY and WILLIAMSON, 1957). Since then, protoplasts have been prepared from species representing all major taxonomic groups belonging to the different classes of Zygomycoti-

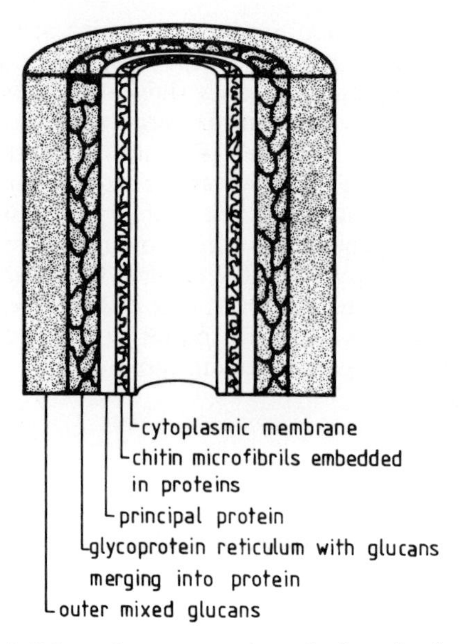

Fig. 4. Schematic representation of a fungal cell wall.

na, Ascomycotina, Basidiomycotina, and Deuteromycotina.

Because of the diversity of wall structure and arrangements within the fungi, a large variety of protoplast inducing systems have been described consisting of different crude enzyme preparations and stabilizers. Besides the snail gut juice which is commercially available as suc d'*Helix pomatia*, helicase, glusulase, or sulfatase and most often used for the preparation of yeast protoplasts, the cell-wall-lytic enzyme mixtures are crude extracellular preparations of different microorganisms including *Trichoderma harzianum* (Novozym – Novo Industri), *Arthrobacter luteus* (Zymolase – Kirin Brewery), *Irpex lacteus* (Driselase – Kyowa Hakko Kogyo Co.), *Penicillium funiculosum* (Cellulose CP – John & E. Sturge Ltd.), *Oxyporus* (cellulase – Merck), and several *Streptomyces* ssp. or even autolytic enzymes. Analysis of active enzymes in the cell-wall-lytic preparations revealed varying concentrations of chitinase, $\beta_{1,3}$- and $\beta_{1,6}$-glucanases and $\alpha_{1,3}$-glucanases as the major enzyme components. In some instances mixtures of the pure enzymes have been successfully applied. Most crude preparations are nowadays commercially available thus avoiding the need of preparing the mycolytic solutions in the laboratory as earlier required.

As a rule, protoplasts are prepared from mycelial cells or from unicellular yeast cells, but not from conidia, although a few reports have been published. Pretreatment of the organism is not required except for several yeast species for which preincubation with 2-mercaptoethanol or dithiothreitol prior to protoplasting could greatly enhance protoplast isolation. Inhibition of cell-wall biosynthesis with antibiotics, as used for bacterial protoplast formation, was not successful, except for some yeast species (Berliner and Reca, 1971), although chemicals specifically preventing or retarding the biosynthesis of the fungal cell wall, such as polyoxin or 2-deoxyglucose, are available. The former compound inhibits chitin and $\beta_{1,3}$-glucan synthesis, while 2-deoxyglucose retards $\alpha_{1,3}$-glucan synthesis. Protoplast formation can be monitored with the microscope. Filamentous fungi release their protoplasts as osmolabile spherical bodies through pores in the cell wall produced by the action of mycolytic enzymes (Fig. 5). The released protoplasts have a heterogeneous structure, because they originate from different parts of the mycelium. In contrast, in unicellular fungi, the protoplasts are, with a few exceptions of enucleate subprotoplasts, all alike containing the essential content of the cell.

Regeneration of protoplasts occurs rapidly within hours under suitable conditions, and the same general pattern is followed in all fungal species. This comprises the synthesis and release onto the cytoplasmic membrane surface of the chemical constituents of the cell wall. As a first result, there is an assembly of a microfiber network composed either of glucan and/or chitin followed by the deposit of an amorphous matrix in between the microfibers. After renewed wall formation, the cells restore the normal cell cycle, in the case of filamentous fungi by the outgrowth of hyphal branching (Fig. 5). Like bacterial protoplasts, fungal protoplasts do not divide, but if cell-wall formation is blocked, multinucleate protoplasts containing more than 15 nuclei can be observed. After withdrawal of the blocking agents, the multinucleate protoplasts regenerate giving rise to an apparently normal culture.

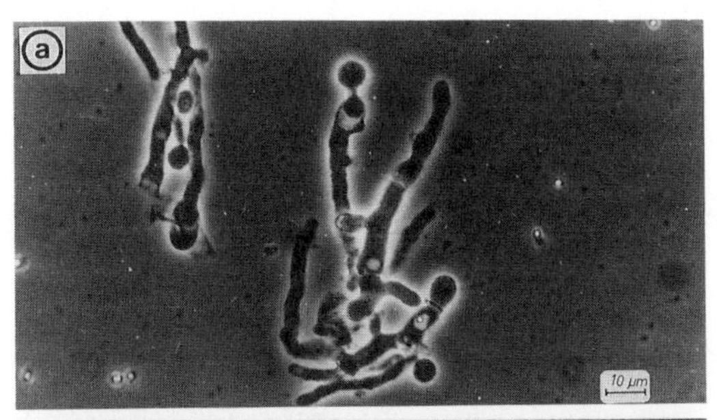

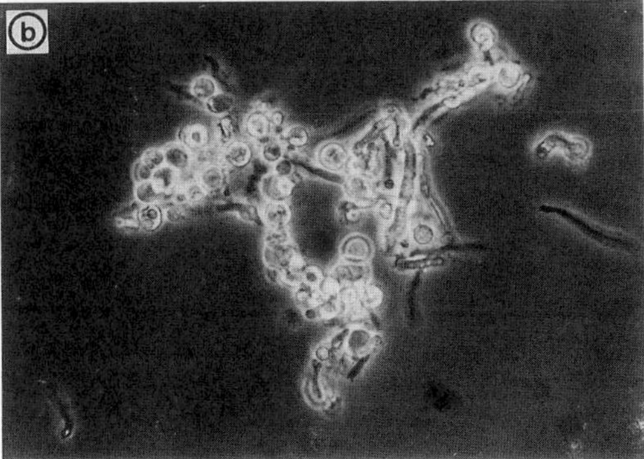

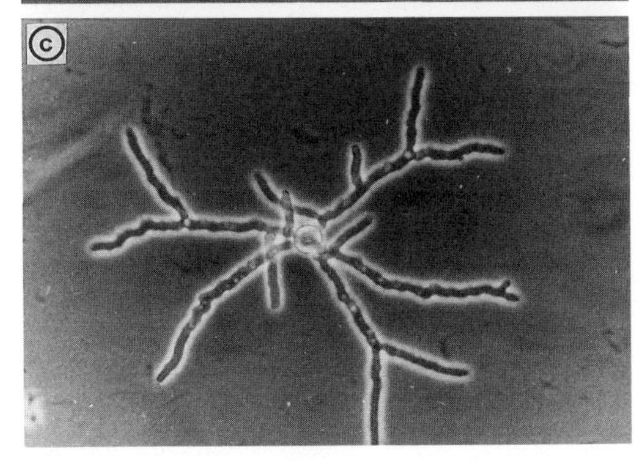

Fig. 5. Release and reversion of a fungal protoplast.
(a) Photomicrograph after 30 min incubation showing swollen hyphal tips and protoplast release.
(b) Completely digested mycelium after 3 h incubation in a protoplast inducing solution.
(c) Microcolony arisen from a regenerated protoplast after 12 h of regeneration (after ANNÉ et al., 1974, with permission).

Conditions for the regeneration of fungal protoplasts are not too critical, as long as the regeneration medium is made hypertonic. Spreading on the agar surface as well as embedding have been used. However, for many but not for all yeast species, embedding into gelatin or agar is required for regeneration of their protoplasts. For some species, such as *Schizosaccharomyces pombe, Saccharomycopsis lipolytica*, some *Candida* species, efficient regeneration can be achieved simply by spreading onto the surface of agar medium containing an osmotic stabilizer (MORGAN, 1983). The regeneration frequency of the protoplasts varied from less than 1% to more than 80% depending on the species, but also the osmotic stabilizer used can have a major influence on the regeneration frequency. In most instances, 0.7 M NaCl, 0.7 M KCl, 0.8–1.2 M sorbitol, or 0.6 M sucrose have been used, but other stabilizers were equally efficient in other cases.

4.2 Genetic Events at Fungal Protoplast Fusion

Fungi are eukaryotic microorganism usually having haploid nuclei in their normal life cycle, except some species, e.g., *Saccharomyces cerevisiae* that have both stable haploid and diploid stages. Apart from the Fungi Imperfecti (Deuteromycotina) and Mycelia Sterila, fungi reproduce sexually, and this sexuality is at the origin of genetic variability due to meiotic recombination. Besides the sexual cycle, some fungi have, in addition to or in place of a normal sexual reproduction, a parasexual cycle (Fig. 6) discovered by PONTECORVO and ROPER (1952). It comprises in a first step by hyphal anastomosis the formation of heterokaryons between genetically different strains. As a result mycelia are produced that have in the same cytoplasm two genetically different nuclei which upon fusion give rise to heterozygous diploids. The spontaneous fusion of nuclei is a rare event, but it can be increased by treatment of the heterokaryons with camphor or UV. The formed diploids are not fully stable, but they break down to a haploid progeny at a low frequency. This process of haploidization is initiated by defective mitotic divisions, termed non-disjunction. As a result of non-disjunction there is a progressive loss of chromosomes from the diploid nucleus via transient intermediate aneuploid stages to a haploid progeny (Fig. 6). In the heterozygous di-

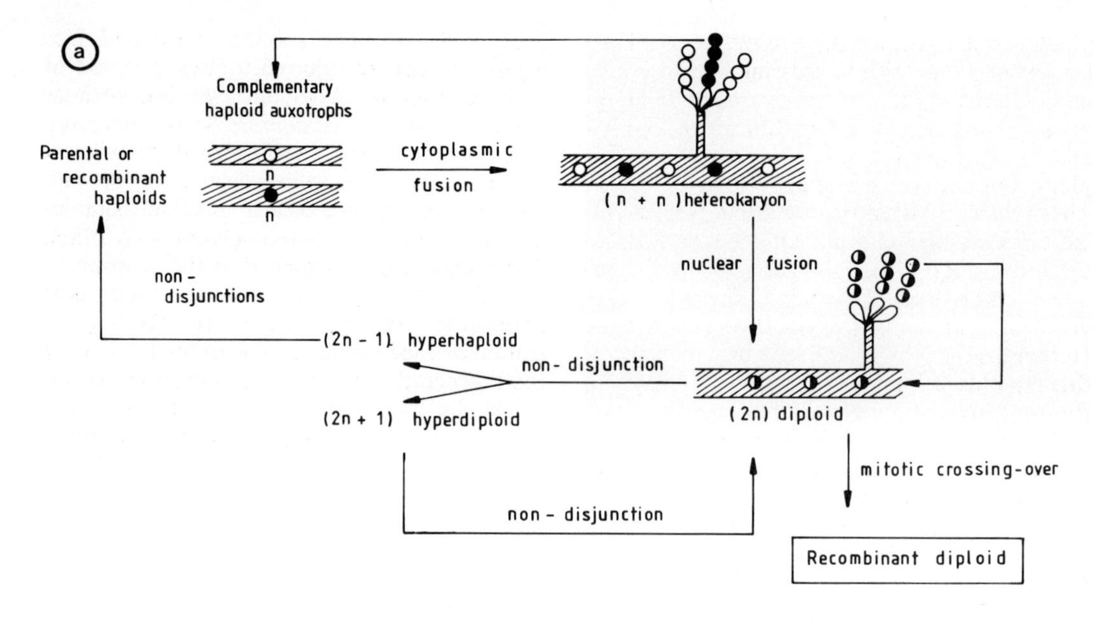

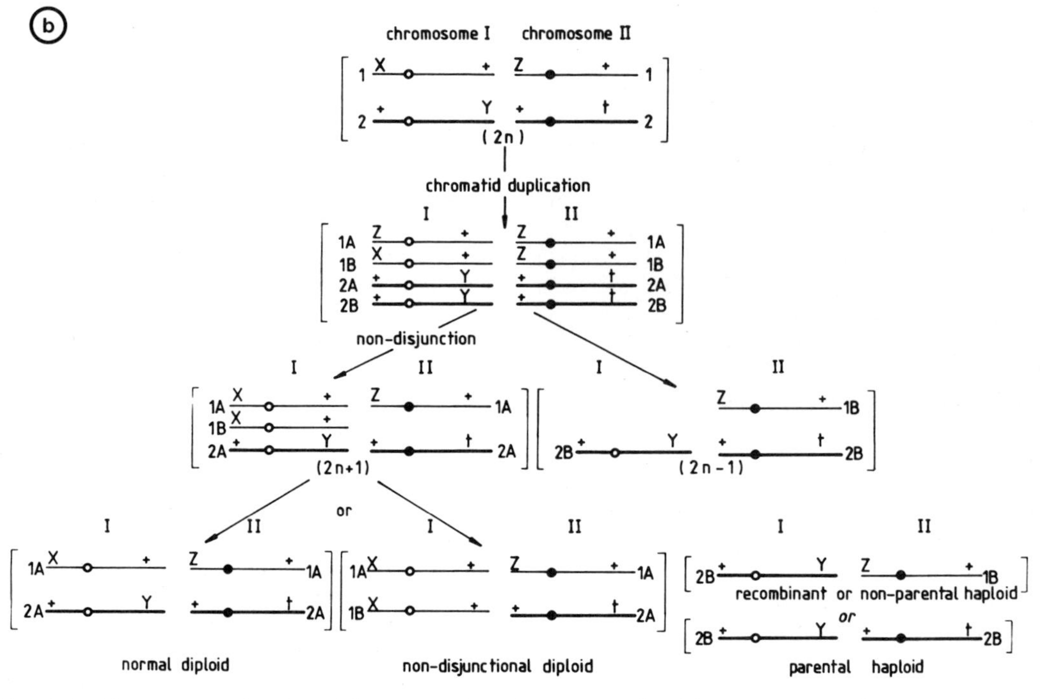

Fig. 6. (a) The parasexual cycle and (b) segregation of a somatic diploid by non-disjunction leading to normal or non-disjunctional diploids, recombinant or parental haploid (after ANNÉ et al., 1989), with permission).

ploid as well as during the consecutive haploidization process, mitotic crossing-over between homologous chromosomes can occur. Consequently, as a result of mitotic crossing-over and non-disjunctional breakdown of the diploid, haploid recombinants are produced. The frequency of mitotic crossing-over can be increased by the use of ionizing irradiation and by 5-fluorouracil. In this manner parental genome segregation is avoided, once a major problem in breeding between divergent strains. Haploidization can be induced by treatment of the diploids with the haploidizing agents *p*-fluorophenylalanine or benomyl.

For asexual species the parasexual cycle is the only way to obtain genetic recombination. With specific markers the different stages can easily be followed, e.g., when using complementary auxotrophs and spore color markers, sporulation in the heterokaryotic stage of filamentous fungi gives rise to conidia of the constituent partners, while diploid spores are prototrophic and show a wild-type spore color. In addition, diploid spores or cells have a larger size and on haploidization they produce different types of recombinants. Through recombination, redistribution of genes occurs which is certainly of interest in strain improvement programs, but it allows also to carry out genetic analysis and to locate genes on the different chromosomes.

The parasexual cycle is a natural but rare event and does not occur in all species. Protoplast fusion is a means to induce this cycle at high frequency and it can, in theory, be applied to all species. During the parasexual cycle induced by protoplast fusion, the same genetic processes occur as in the "natural" cycle and, in consequence, protoplast fusion can be suitable both in genetic mapping as well as in strain improvement programs.

4.3 Methodology and Selection Procedures

As mentioned above, fungal protoplasts can be fused after treatment with Ca^{2+} at high pH, following treatment with PEG or PVA and by electrofusion. In the vast majority of cases, PEG-induced protoplast fusion has been applied. Since the first reports on fungal protoplast fusion (ANNÉ and PEBERDY, 1975; FERENCZY et al., 1975), this method and its applications have been extensively reviewed (FERENCZY, 1981; ANNÉ, 1983; MATSUSHIMA and BALTZ, 1986; PEBERDY, 1989). Briefly, after protoplast preparation and their separation from mycelial remnants (when working with filamentous fungi) through a sintered glass filter No. 1, circa 10^6–10^7 protoplasts of genetically different strains are mixed (1:1) and centrifuged (e.g., 2 min in an Eppendorf centrifuge). The pelleted protoplasts are resuspended in 0.5 mL of a membrane sterilized solution of PEG (MW 4000 or 6000) containing 50 mM $CaCl_2$, 0.6 M sucrose and adjusted to pH 8.0. Subsequently, 0.1 mL suspensions and their serial dilutions are embedded in a soft selective hypertonic overlay. Alternatively, the protoplast suspension in PEG is diluted with a hypertonic solution, e.g., 0.7 M NaCl, washed, and subsequently embedded or plated onto a solidified hypertonic selection medium. For electrofusion, the fusion conditions differ depending on the type of fusion chambers and on the type of protoplasts used.

After fusion, as a first result, heterokaryons are produced. Selection of fused protoplasts is generally carried out by applying nutritionally complementing auxotrophs. This procedure is very accurate and useful for classical genetics, but for industrial strain improvement programs, the use of auxotrophs is, in general, unfavorable. The introduction of auxotrophic markers has often a noxious effect on the quality of the industrial strain and, therefore, unselected markers such as colony morphology or metabolite production are often used for industrial strains. Such screening, however, requires much more effort to select recombinant progeny.

Attempts at increasing fusion frequencies – which usually range between 0.03 and 3.0% of the number of cells surviving the fusion treatment – are rare. Similarly as described for bacterial protoplast fusion, heat killing of one of the fusion partners or UV treatment have been applied. Heat treatment gave rather confusing results or gave rise to colonies with altered morphology and an increased number of aneuploids (FERENCZY, 1984). UV irradiation of one parent gave a five- to tenfold increase of the fusion frequency (ANNÉ, 1983), but UV irradiation can cause additional, eventually unfavorable changes in the genomes. Therefore, drug resistance markers eventually located on the mitochondria as resistance to acriflavine, oligomycin, or fungicides might be more useful for screening fusion products and recombinants of industrially important strains. In addition, respiratory-deficient mutants, designated ϱ°, are very promising for facultative anaerobic yeast cells, for example, *Saccharomyces cerevisiae* and *Kluyveromyces lactis*. These mitochondrial mutants, also called "petite" cells, have mostly mitochondrially located mutations, but chromosomally located respiratory-deficient mutants also exist.

These respiratory-deficient mutants produce only microcolonies on a medium with a fermentable substrate as a consequence of the lack of several enzymatic activities, and they do not grow on non-fermentable substrates. The ϱ° mitochondrial mutants can be obtained by ethidium bromide treatment (10 μg/mL for 24 h in the dark). Ethidium bromide interferes with the replication of the mitochondrial chromosome resulting in partial or gross deletion of the mitochondrial DNA and, hence, the mitochondrial functions are abolished. Mitochondrial mutants, such as ϱ°, are especially interesting for crossing with industrial strains, because it avoids the need to use auxotrophic mutants of an industrial strain. Furthermore, many industrial yeast strains like brewing yeast are of polyploid or aneuploid nature from which it is difficult to obtain auxotrophs.

An interesting alternative to fusion of nucleated protoplasts is the introduction of isolated nuclei or mitochondria into a suitable protoplast. This transfer, sometimes also called transfusion, can be obtained, albeit at a low frequency, by PEG treatment (FERENCZY, 1984) or by electrofusion (SULO et al., 1989), and the transferred elements remain functionally active. The advantage of this transplantation technique is that prototrophic strains can be used as donor, and only the acceptor strain has to contain auxotrophic or other biochemical selection markers, except if enucleate protoplasts are used. In that case no auxotrophs are required. The "mini"-protoplasts can be produced from budding yeast following protoplast production from the buds present on exponentially growing budding yeast cells. The majority of these buds do not yet contain a nucleus, but only cytoplasmic elements including mitochondria. Anucleate protoplasts can accurately be obtained by linear gradient centrifugation (HRMOVA et al., 1984).

More efficient than nuclear transplantation is the combined use of miniprotoplasts and ϱ° mutant protoplasts. Their fusion provokes mixing of the nucleus of one strain with the cytoplasm of another strain. Mitochondria can be isolated from lysed protoplasts or from broken cells by fractional centrifugation. Fusion experiments between protoplasts and purified mitochondria or between enucleate protoplasts and ϱ° mutants is a way to introduce and investigate cytoplasmic inheritance of a character in a defined nuclear background.

As mentioned before, the selection of fused protoplasts occurs in a first instance by screening for heterokaryons and for the subsequent appearance of diploids and recombinants (see Sect. 4.2). When different species are crossed, it is of interest to identify the fate of the chromosomal DNA of the species involved, because in interspecies crosses anomalous segregation patterns resulting in loss or rearrangements of genetic information of one of the species are often observed. Recently, by means of pulsed field gel electrophoresis in combination with suitable chromosomal markers (WITTE et al., 1989; KOBORI et al., 1991), the individual chromosomes of the species could be visualized.

4.4 Fungal Protoplast Fusion and Strain Improvement

The industrial applications of fungi are considerable and very diverse. Different compounds are produced by these organisms: from low-molecular weight primary metabolites, such as ethanol or organic acids, to complex secondary metabolites including antibiotics and alkaloids. In addition, many enzymes and some vitamins are produced by fungi. In the food industry, fungi are indispensable for alcoholic fermentation, baking and cheese making, and several species are used for direct consumption. In spite of the long history of the application of these organisms, until recently strain improvement programs have been, as for bacteria, almost solely based on mutation and selection.

The basic knowledge of fungal genetics comes, with the exception of *Saccharomyces cerevisiae*, from non-industrial strains, e.g., *Aspergillus nidulans* and *Neurospora crassa*, and most industrially important species are genetically poorly characterized or not at all.

The major reason for the lack of interest in genetic breeding in strain improvement was in a first instance, apart from the obvious success of mutation and selection, the fact that most economically important species producing metabolites of high value have no sexual cycle. On the other hand, in sexually reproducing yeast many industrial strains show problems of homothallism, aneuploidy or triploidy, aberrations in mating behavior, poor sporulation and spore viability, which hinder the successful use of classical crossing for strain improvement.

An impetus to the use of genetic recombination in strain improvement was the discovery of the parasexual cycle in filamentous fungi. Discovered in *A. nidulans*, it was soon detected that the parasexual cycle could be induced also in industrially important fungi, e.g., *Penicillium* and *Aspergillus* spp. The principle of using the parasexual cycle in strain improvement is that two strains with particular interesting properties are combined to form a diploid or haploid recombinant with more desirable biochemical or morphological properties.

Although the application of the parasexual cycle for breeding is simple in theory, many obstacles remained including the difficulties to achieve the parasexual cycle in highly mutated, industrial strains. The latter obstacle could be removed by the development of the protoplast fusion technique. This method allows the induction of the parasexual cycle at high frequency not only for strains of the same species, but also for different species. Soon many reports on protoplast fusion appeared for a variety of fungal species because of the ease of manipulation, the early good prospects with respect to breeding, and the growing consciousness of the importance of genetic methods in breeding programs. Compared to bacterial protoplast fusion, a lot more reports appeared for fungi, and interspecies hybridization is more extensively investigated. Since several reviews have been published (POTRYKUS et al., 1983; ANNÉ, 1985; PEBERDY and FERENCZY, 1985; PEBERDY, 1989), the discussion will be limited to results with some industrially important species.

4.4.1 *Penicillium*

Industrially the most important *Penicillium* species is *P. chrysogenum*, because of the production of penicillin. Most commercial strains of *P. chrysogenum* are descendents of the NRRL 1951 strain, originally isolated from a moldy cantaloupe discarded by a small market in Peoria, Illinois. Through recurrent mutagenesis and selection of hundreds of thousands of strains, a series of highly yielding penicillin production strains has been developed from this former strain. As soon as the parasexual cycle was detected for *P. chrysogenum*, the idea of using recombination methods for strain improvement of this species was put forward. Several attempts have been made to use parasexuality for strain improvement. Heterozygous diploids synthesized between strains of relatively high penicillin yield gave, however, in general titers not greatly different from their parental strains. The reason is that genes concerned with increased penicillin production appeared in the main recessive, and independently induced mutations were often supposed to be allelic.

A few exceptions are known where heterozygous diploids have been shown to yield increased productivity (ROWLANDS, 1984). In an intensive large-scale industrial strain development program with *P. chrysogenum*, a stable heterozygous diploid strain that produced high concentration of phenoxymethyl penicillin was isolated and reported as having been used by Eli Lilly for commercial penicillin production. However, stable diploids are rather exceptional, both with regard to productivity and strain stability. On the other hand, diploids can be used as a source to obtain haploid recombinants with more attractive features than their progenitors. This approach could even be more successful than attempting to isolate higher yielding diploids, and examples are described in the literature, showing segregants with improved properties, such as the isolation of a spontaneous segregant that produced nearly 25 pct more penicillin than its diploid parent, which itself yielded better than its ancestor. In addition, from crosses carried out by protoplast fusion between a slowly growing *P. chrysogenum* strain producing high levels of penicillin V and no detectable amount of *p*-hydroxypenicillin V and a strain with faster growth but producing moderate levels of penicillin V and high yields of *p*-hydroxypenicillin V, recombinants could be recovered with the desired properties of fast growth and high levels of penicillin V, but low *p*-hydroxypenicillin V production (LOWE and ELANDER, 1983).

These examples clearly show the possibility of selecting improved recombinants after segregation of diploids obtained following protoplast fusion. However, in the isolation of recombinants difficulties were often encountered, not only to obtain stable recombinants, but segregants were mainly of one or another of the haploid parental type. This phenomenon known as parental genome segregation is probably due to differences in chromosomal morphology between the haploid parental strains as a result of translocation caused by multiple mutagen treatment. The occurrence of chromosomal rearrangements in production strains has in the meantime been physically demonstrated for *Cephalosporium acremonium* by gel electrophoretic separation of the chromosomes of strains from different lineage, obtained by recurrent mutagenesis and selection for improved cephalosporin C production (SMITH et al., 1991). Therefore, to obtain a maximum number of recombinants between production strains, the number of translocation for the strains involved should be low, indicating that "sister" crosses are more advisable.

The role of protoplast fusion remains *per se* limited to the efficient production of heterokaryons, and does not influence the subsequent genetic processes of diploidization and haploidization. The advantage of protoplast fusion is that heterokaryons can be produced at high frequency, which can reduce the introduction of selectable markers in the strains to be crossed to a minimum, allowing to use more easily parent strains with genetic markers that do not affect penicillin production, a method successfully applied at Gist Brocades (VEENSTRA et al., 1989).

Not only for strains used in the antibiotic industry, but also for other strains protoplast fusion experiments have been carried out for strain improvement. In *Penicillium caseicolum*, a species used in the dairy industry, protoplast fusion has been used to create strains with novel properties, which could be applied in the production of new dairy products. After fusion and selection, recombinant strains with changed morphological, lipolytical, and proteolytical properties could be isolated and an anti-mucor property was transferred to all fusants, probably because of its mitochondrial location (RAYMOND et al., 1986).

With respect to interspecies crosses in *Penicillium*, the possibility of obtaining new hybrids was explored mainly with respect to *P. chrysogenum*. Both closely and less closely related species (as classified by RAPER and THOM, 1949) were crossed (ANNÉ, 1985). The closely related species, i.e., *P. chrysogenum, P. notatum, P. cyaneofulvum*, and also the less closely related *P. citrinum* behaved in a cross similarly to intraspecies heterokaryons and diploids, except that the diploids did not show complementation for the spore color mutations. Mitotic segregation of the diploids resulted in recombinants that showed reassortment of the markers between haploid progeny, indicating a great deal of homology between the chromosomes of the species just mentioned. A similar behavior was shown by the progeny *P. caseicolum* × *P. album*.

It must be mentioned that recently *P. chrysogenum, P. notatum*, and *P. cyaneofulvum* on the one hand and *P. caseicolum* and *P. album* on the other are regarded as one species. Peculiar to these crosses – and unlike intraspecies crosses – was, however, the lack of spore color complementation, the high number of aneuploid-like segregants, and also the isolation of unusual, multiple auxotrophs that did not grow on selective media supplemented with all requirements of both parents. Such progeny might consist of non-complementary diploids or aneuploids as observed with *Bacillus* (see Sect. 3.2.1). The appearance of anomalous segregation patterns indicate that at least some genetic differences between the species should exist.

Tab. 2. Interspecies Crosses with *Penicillium chrysogenum* and Behavior of Fusion Progeny

Cross	Sporulation of Heterokaryons	Nuclear Fusion	Segregation of Hybrids
P. chrysogenum × *P. cyaneofulvum*	Both parents	+	Recombinants
P. chrysogenum × *P. citrinum*	Both parents	+	Recombinants
P. chrysogenum × *P. roqueforti*	*P. roqueforti*	+	*P. roqueforti*
P. chrysogenum × *P. stoloniferum*	*P. chrysogenum*	+	*P. chrysogenum*
P. chrysogenum × *P. patulum*	*P. chrysogenum*	+	*P. chrysogenum*
P. chrysogenum × *P. puberulum*	*P. chrysogenum*	+	*P. chrysogenum*
P. chrysogenum × *P. cyclopium*	No viable heterokaryons	−	−
P. chrysogenum × *P. lanosum*	No viable heterokaryons	−	−
P. chrysogenum × *P. nigricans*	No viable heterokaryons	−	−
P. chrysogenum × *P. baarnense*	Both parents	+	Both parents

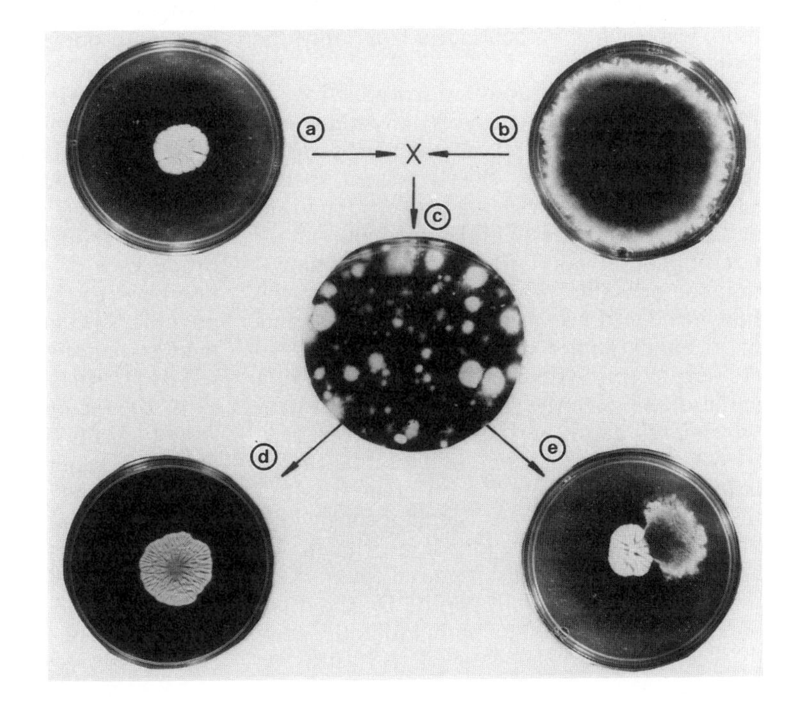

Fig. 7. Colony phenotypes of (a) *Penicillium chrysogenum*, (b) *P. roqueforti*, and of their fusion progeny, (c) fused regenerants on selection plate, (d) stable, and (e) unstable somatic hybrid segregating *P. roqueforti* on complete medium (after ANNÉ and PEBERDY, 1981, with permission).

Less related species showed a more extensive incompatibility as concluded from heterokaryon morphology. All viable heterokaryons obtained between less related species (Tab. 2) consisted of colonies with irregular morphology (Fig. 7), and they produced, in general, only one type of parental spores. Nuclear fusion also occurred, but it gave rise to different types of aneuploid – such as prototrophic hybrids that eventually broke down to one and only one of the complementing parents also observed in the sporulating heterokaryon. The genetic background of these hybrids is not known, but it might now be identified by electrophoretic techniques as pulsed field gel electrophoresis. It is furthermore speculated that the chromosomes of one species are preferentially lost, as it occurs in interspecies hybrids of higher organisms, both plants and animals.

Crosses carried out between more distant species did not give viable progeny. The lack of heterokaryon formation between crosses of non-related species is probably a reflection of profound physiological and biochemical differences between the species involved. To what extent mitochondria might be responsible for this incompatibility is not known, but a striking observation made for *Aspergillus* interspecies hybridization may point in that direction. It was noticed for *Aspergillus* that all species of the *A. nidulans* group which could be hybridized with each other had mitochondrial genomes which share a similar overall organization, though they might differ widely in size due to the presence or absence of inserts in the mitochondrial DNA (CROFT and DALES, 1983). Also somatic hybrids of animal cells are more stable when mitochondria from only one parent are present.

Investigations of antibiotic production of *P. chrysogenum* interspecies hybridization showed that diploid and recombinant progeny from crosses with the closely related species gave penicillin titers ranging between the lower and higher producers. On the one hand, an interesting observation was that complementation for penicillin production occurred in hybrids produced between penicillin-less mutants of *P. chrysogenum* and *P. patulum*, and in crosses *P. chrysogenum* × *P. puberulum* a variation between the relative amounts of pentyl-, heptyl- and benzylpenicillin was observed (ANNÉ, 1983). In addition, in hybrids from the latter cross genes from the different constitutive partners could be simultaneously expressed. Simultaneous expression of penicillin and griseofulvin was also reported for *P. chrysogenum* × *P. patulum* hybrids. On the other hand, no new metabolites with antibiotic activity were detected.

4.4.2 *Cephalosporium acremonium*

Cephalosporium acremonium (= *Acremonium chrysogenum*) is the only species commercially used for the production of the β-lactam antibiotic cephalosporin C. *C. acremonium* is an asexual species. In consequence, recombination can only be achieved by parasexual processes, but this parasexuality could hardly be induced by classical methods. Following protoplast fusion, heterokaryon induction and subsequent mitotic recombination were, however, easily obtained, and it allowed the construction of a preliminary genetic map of *C. acremonium* (HAMLYN et al., 1985). Genes which enhance antibiotic production could be localized and allied to specific linkage groups. In contrast to most other species, following heterokaryosis, nuclear fusion occurred immediately, and also the diploid stage was transient as chromosome segregation readily occurred resulting in a progeny of stable haploids and some unstable slowly growing aneuploids. The number of unstable heterozygotes increased, when using strains of more divergent lineage. Among the recombinants, strains with improved antibiotic titers and more interesting morphology could be detected.

The number of reports on interspecies crosses involving *Cephalosporium* is very limited. *C. acremonium* has been crossed with another *Cephalosporium* sp., with *Emericillopsis glabra* and *E. salmosynnemata*, closely related species having several physiological and biochemical properties in common including the production of cephalosporin C. Recombination between the different species seemed to be possible, since reassortment of the auxotrophic markers among the haploid progeny was observed (MINUTH and ESSER, 1983). Morphology of the recombinants, however, reflected always typically one of the parental types, probably as a result of selective chromosome loss of one of the parents as also observed with *Penicillium* interspecies hybrids. Antibiotic production was in every case lower than that of the parental strains. It is interesting that sporulation of *C. acremonium*-type recombinants could be improved by repeated interspecies crosses with a *Cephalosporium* strain which produced considerable numbers of conidia. Furthermore, repeated experiments to cross *C. acremonium* with *P. chrysogenum* in order to obtain modified β-lactam antibiotics have failed.

4.4.3 *Aspergillus*

Industrially very important among the Aspergilli is *A. niger*, mainly as source for the production of citric acid. Parasexual recombination in strain improvement programs to obtain strains with enhanced citric acid production and improved substrate-use efficiency has already been tried since the detection of the parasexual cycle (AZEVEDO and BONATELLI, 1982). This approach remained limited to closely related sister strains due to difficulties in obtaining heterokaryons by hyphal anastomosis. By protoplast fusion, however, heterokaryons could readily be obtained between *A. niger* strains of divergent origin. Following crossing, diploid strains with increased citric acid production in the solid phase compared to the parents could be obtained, and haploid recombinants from the higher producing diploids showed a further increase in citric acid production in both solid and submerged cultures (KIRIMURA et al., 1988; MARTINKOVA et al., 1990), indicating the usefulness of parasexual hybridization for strain improvement in this species. Similar results were obtained for the increase of α-amylase production in *A. niger*, for which recombinants could be isolated with improved enzymatic activity.

Much work concerning interspecies protoplast fusion has been carried out with the genetically well characterized *A. nidulans* (see PEBERDY, 1989). *A. nidulans* was crossed with several species classified in the *A. nidulans* group including *A. rugulosus, A. nidulans* var. *echinulatus, A. quadrilineatus*, and *A. violaceus*. All pairwise combinations between these five species were possible, except for the combinations of *A. quadrilineatus* with *A. violaceus* and with *A. nidulans* var. *echinulatus*, but the latter species could be crossed with *A. quadrilineatus*. Three other species of the *A. nidulans* group, i.e., *A. stellatus, A. unguis*, and *A. heterothallicus* did not produce viable interspecies heterokaryons. Crossing with *A. nidulans* containing well defined markers on each linkage group is helpful to examine the genetic relationship between the species and to investigate the fate of the genetic markers in the interspecies hybrids.

From the analysis of hybrid segregants it appeared that the species which could be crossed form a close association, but it suggested on the other hand differences in the linkage group. With respect to mitochondrial DNA similarities, as already mentioned, species which can be crossed have an overall mitochondrial DNA restriction map that is similar for the majority of restriction sites, although large differences in size may exist, and it was observed that there is a high probability that mitochondrial DNA recombination occurs spontaneously in interspecies diploids.

As in *Penicillium* interspecies hybrids, it was observed in *Aspergillus* that hybridization can have significance with regard to gene regulation or the modification of gene products, as detected by isozyme analysis and metabolite production.

A third group of Aspergilli that has been intensively investigated are the koji molds widely used throughout the orient in the food and spirit industry for fermentation of sake, miso, and soy sauce. Koji molds produce several hydrolytic enzymes such as proteases and glutaminase, and they are citric acid and acidic amylase hyperproducers. For strain-breeding purposes several of these strains including *A. oryzae* (also used in Western countries as a source for food-grade amylase), *A. soja, A. awamori* var. *kawachi, A. usamii* var. *shirousamii* have been protoplasted and fused either to form intraspecies diploids and recombinants or interspecies allodiploids and hybrids (USHIJIMA et al., 1990).

These species are all taxonomically related. Some of these species were also fused with *A. niger* (VIALTA and BONATELLI, 1990) for which interspecies hybrids could be obtained following mitotic crossing-over suggesting a closely phylogenetic relationship between these species. In several instances, diploid progeny, haploid recombinants, and hybrids with improved characteristics compared to the parental strains were obtained, and some of these approved strains were reported to be used in the fermentation industry. Intergeneric crosses were mentioned between *A. oryzae* and *Saccharomyces cerevisiae* (LEE et al., 1989), *A. oryzae* and *Monascus anka* (KIYOHARA et al., 1990), and *A. niger* and *Trichoderma viride* (KIRIMURA et al., 1990) to give rise to viable fusion progeny with intermediate properties of the constitutive parents. The isolation of these viable intergeneric hybrids is somewhat surprising, since attempts to fuse *Aspergillus* strains belonging to different *Aspergillus* species groups were in general not very successful.

4.4.4 *Trichoderma*

Trichoderma is well known as producer of cellulases and mycolytic enzymes. In addition, several species are being developed and utilized as biocontrol agents for plant pathogens in commercial agriculture. Because of these properties, different strains and species of *Trichoderma* are of industrial interest and are subjected to strain improvement by mutagenesis and to some extent by genetic recombination following protoplast fusion.

Among *T. reesei* and *T. viride* several isolates are high cellulase producers. Heterokaryosis by hyphal anastomosis is for most strains a rare event or not existing, but following protoplast fusion heterokaryosis could easily be achieved. For *T. reesei*, and in contrast to *T. viride*, it was reported that diploids could be obtained and some of this diploid progeny had improved hydrolytic activity (OGAWA et al., 1989).

As biocontrol agent against soil-born plant pathogenic fungi *T. harzianum* is very promising. The successful use of this species will be greatly enhanced if superior strains can be developed. Presently, several biotypes of *T. harzianum* are available, each of which has interesting properties. Combination of these desirable traits can give rise to superior strains (SIVAN and HARMAN, 1991).

Fused protoplasts of complementary auxotrophic strains could be selected as heterokaryons, but in intrastrain crosses at variance with a few reports no diploidization or recombination occurred. In interstrain crosses postfusion incompatibilities resulting from heterokaryosis were observed (PECCHIA and ANNÉ, 1989; STASZ and HARMAN, 1990). This incompatibility resulted in absence of growth or in slowly growing extremely imbalanced heterokaryotic progeny that on further cultivation gave rise to more rapidly growing sectors with a wide range of different morphological types. Incompatibility was suggested to have arisen from heterogeneity of some compatibility loci in the interstrain heterokaryons.

Despite the great variation in cultural and nutritional characteristics of the different types isolated from the unstable heterokaryotic colonies, there was no evidence that karyogamy had occurred as concluded from an extensive isozyme analysis (STASZ and HARMAN, 1990). Nevertheless, protoplast fusion between different strains of *T. harzianum* gave rise to phenotypically different variants, of which some strains had improved antagonistic properties to plant pathogenic fungi as compared to the parental strains involved in the fusion reaction. The genetic processes that caused these variants are, however, at present unknown. They probably differ from those occurring in the normal parasexual cycle, and nuclear-cytoplasmic interactions may be involved as well. Therefore, using the approach of transfer of isolated nuclei to obtain hybrids, as carried out by SIVAN et al. (1990), may be useful to understand the influence of the cytoplasmic elements on this variation.

Interspecies crosses in *Trichoderma* are also reported for *T. reesei, T. viride*, and *T. koningii*, which gave rise to different types (heterokaryons, allodiploids, or hybrids) of fusant progeny. No improved production of hydrolytic enzymes was obtained as mentioned by the only published report on this topic (TOYAMA et al., 1984). *T. harzianum*, on the other hand, was incompatible with the just mentioned species.

4.4.5 Edible Mushrooms

In Western countries the most important edible mushroom species are members of the genus *Agaricus*. Breeding by conventional genetic means in order to obtain, e.g., more virus-resistant types or variants with longer stalks has always been very difficult, because of the special character of sexual reproduction. Also mutagenesis is hardly used mainly due to problems in obtaining mutants from dikaryotic mycelium and the difficulties in isolating homokaryons. In Eastern countries other types of mushrooms are cultivated such as *Lentinus edodes, Flammulina velutipes, Pleurotus ostreatus* and other *Pleurotus* ssp., *Coprinus macrorhizus, Ganoderma* ssp., and *Lyophyllum ulmarium*. Breeding work with these species is often limited to collecting wild specimens of a superior quality or selection of a spontaneous mutant with improved properties for cultivation. In recent years, however, much attention has been paid to the genetical improvement of edible mushrooms using protoplast fusion. As for other fungi, protoplast fusion between basidiomycetes strains, for which natural mechanisms of mating do not operate, permits the generation of intraspecific heterokaryons. It also allows interspecies heterokaryosis by which new genetic traits can be introduced in a particular variant.

Several examples have been published (see PEBERDY, 1989) mainly from Japan and Korea showing the possibilities of intra- and interspecies fusion with particular interest in *Pleurotus ostreatus*. Not much genetic data are available, but attention has been paid to clamp connection and fruit body formation, which provides the crop for harvest. *Pleurotus ostreatus* has been crossed with several other compatible or incompatible mating types including *P. columbinus, P. sajor-caju, P. florida*, and *P. cornu-copiae*.

Interspecies crosses between compatible mating types resulted in fusion progeny of morphologically different variants, and clamp connection and fruit body formation was observed. Many variants showed higher fruit body formation. Interspecific fusion products between incompatible strains showed, however, no true clamp connection, and there was no fruit body formation, a phenomenon also observed when crossing *Coprinus macrorhizus* mutants of identical mating type (KIGUCHI and YANAGI, 1985). These results indicate that although fusion and the stability of the progeny of fused protoplasts are not affected by mating characters, clamp connections and fruit bodies were formed mainly on mycelium of compatible fusion products and not on the fusion progeny of incompatible species. Nevertheless, in some instances, e.g., in the fused progeny of the incompatible *P. ostreatus* and *P. eryngii* and of *P. florida* and *P. spodoleucus*, fruit body formation was observed, indicating the possibility of improving strains by fusing incompatible species (GO et al., 1989). However, which hybrids of incompatible strains will give rise to fruit body formation cannot be foreseen and can only be determined experimentally.

Concerning intergeneric crosses, a similar phenomenon as for other viable crosses in filamentous fungi occurred. In the hybrids one species was dominant over the other (YOO, 1989), even after transfer of isolated nuclei into the protoplasts (YOO et al., 1987) instead of carrying out fusions between whole protoplasts.

4.4.6 Yeasts

Yeasts are fungi that have in common a predominantly unicellular vegetative stage. Their life cycle consists of a haploid and diploid or dikaryotic phase of which one of both phases is prevailing depending on the species. Sexually reproducing yeasts have in general a bipolar mating system (α/a or $+/-$). Mating is initiated by the agglutination of the haploid cells of different mating types and following karyogamy gives rise to a zygote. From these diploid cells haploid ascospores (or sporidia in case of *Rhodosporidium*) are produced through meiosis during which recombination can occur. Because of the occurrence of meiotic recombination in sexually reproducing yeast cells, classical techniques of genetic breeding are potentially applicable in several industrially important yeasts. However, the majority of the industrial strains are polyploid or aneuploid, have aberrations in mating behavior, poor sporulation, or spore viability. Hence, conventional hybridization with these strains is difficult or impossible. Because of these hindrances, protoplast fusion has been intensively used as an alternative to sexual breeding. Many reports have already appeared on yeast protoplast fusion, either chemically induced or by electrofusion, and both intra- and interspecies fusions have been successfully carried out. Several extensive review articles have been published (SPENCER and SPENCER, 1983; MORGAN, 1983; PEBERDY and FERENCZY, 1985; INGOLIA and WOOD, 1986).

Intraspecies fusions have been mentioned mainly for *Saccharomyces cerevisiae*, but also for *Saccharomycopsis lipolytica, Saccharomyces diastaticus, Kluyveromyces lactis, Pichia guillermondii, Schizosaccharomyces pombe, Rhodosporidium toruloides, Candida utilis, Candida maltosa, Apiotrichum curvatum, Lodderomyces elongisporus, Hansenula wingei, Saccharomyces alleivius, Trichosporon adeninovorans*. Fusion occurred regardless of their mating type. Not only haploids of identical mating type could be fused but also haploids with diploids or diploids with diploids.

As a result of protoplast fusion heterokaryons were formed, or unstable diploids or stable diploids; but triploids and tetraploids, as a consequence of multiple fusion bodies, have also been mentioned. Except for *Saccharomycopsis lipolytica*, protoplast fusion between heterothallic strains of like mating-type resulted in the formation of sporulation-deficient hybrids, but when the diploids homozygous for the mating-type locus were crossed or fused with haploid or diploid cells or protoplasts of opposite mating type, they were able to develop spores. These observations indicate that mating-type alleles in most instances not only control the initial step of mating, i.e., cell recognition and agglutination, but also meiosis and ascospore formation.

Diploids can segregate mitotically giving rise to haploid recombinants, when treated with haploidizing agents such as *p*-fluorophenylalanine or methylbenzimidazol-2-yl carbamate. Spontaneous segregation to parental and recombinant types has also been observed following protoplast fusion, even in asexual species, but in the latter case karyogamy did not readily occur, in contrast to yeast from perfect genera for which the heterokaryotic stage is transient.

Following protoplast fusion mitotic or meiotic recombination between genetically different nuclei can occur, and therefore fusion offers the possibility of strain improvement also for those strains in which sexual conjugation and genetic recombination is not observed due to the lack of a mating system or as a consequence of polyploidy and aneuploidy or sterility. In addition, cytoplasmic fusion also takes place by protoplast fusion. It is known that some characteristics of industrial yeasts are controlled by genes located on the mitochondrial genomes, e.g., uptake of some sugars, starch utilization, flocculation, some flavor determinants and yeast killer factor. Not only can they contribute to strain improvement, but crosses involving ϱ° or petite mutants can help to elucidate or locate more precisely such functions on the mitochondria. In addition, for reasons mentioned earlier (Sect. 4.3) ϱ° mutants are often used in crosses aiming at strain improvement.

A number of experiments of which a few examples have been given show the applicability of intraspecies or closely related interspecies protoplast fusions for breeding. In this respect, it must be noted that many *Saccharomyces* spp. are so closely related that they have to be considered as the same species, the more so as these strains either crossed by rare mating or by protoplast fusion show no aberrant segregation pattern. *Saccharomyces diastaticus* has been widely used to improve brewery and distillery yeasts, because of its glucoamylase activity enabling the yeast to utilize dextrin and to ferment wort to produce low-carbohydrate beer (JANDEROVA et al., 1990). In this manner by fusing *S. uvarum* and *S. diastaticus* a brewery yeast strain was isolated which can utilize wort carbohydrates that normally are unfermentable. *S. diastaticus* has, however, the disadvantage that it carries a phenolic off-flavor (*pof*) gene which can cause phenolic off-flavors in beer making it unpalatable, but by using mutants pof or by backcrossing and selection for suitable recombinants this problem can be overcome. An interesting finding from protoplast fusion with brewery yeasts which usually sporulates poorly was also that the fusion products had a greatly increased degree of sporulation and showed normal asci formation. As a result, genetic analysis of the polyploid parents is rendered possible. Also wine yeasts have been improved using protoplast fusion with respect to fermentation performance and oenological properties (YOKOMORI et al., 1988).

Another type of strain improvement involves the oleaginous yeast *Apiotrichum curvatum*. Using protoplast fusion between a methionine auxotrophic mutant and an unsaturated fatty acid mutant of *A. curvatum*, hybrids were obtained with a changed fatty acid composition of their lipids thus producing a fatty acid approximating that of cacoa butter (VERWOERT et al., 1989).

Interspecies and intergenera hybridization is usually aimed at the combination of interesting properties of different species such as the capacity to grow on particular carbon sources for the production of ethanol, e.g., on cellulose hydrolysates such as cellobiose or on starch, under conditions of high glucose concentrations or with high acetic acid tolerance.

As previously demonstrated for filamentous fungi, the success of viable hybrid formation depends on the taxonomic relationship of the strains involved and may also be under the influence of mitochondrial compatibility as explained for *Aspergillus* (see Sect. 4.4.3). Therefore, the frequent use of petite mutants in the fusion of yeast protoplast experiments could explain the large number of different viable hybrids that have been isolated from interspecies protoplast fusion experiments. The hybrids obtained showed, however, different stabilities and segregation patterns. Several hybrids were unstable and spontaneously segregated into the original auxotrophic parent strains (*Saccharomyces cerevisiae* × *Kluyveromyces lactis; S. cerevisiae* × *Schwanniomyces castellii* or *Schwanniomyces allubius*); in other hybrids the fused hybrid nuclei underwent random loss of chromosomes to stabilize at various levels of aneuploidy prior to segregation. In these instances, usually only one of the parents was specifically recovered : *Schizosaccharomyces pombe* × *Schizosaccharomyces octoporus* segregating *S. octoporus; Saccharomycopsis lipolytica* × *Kluyveromyces lactis* segregating either *K. lactis* or *S. lipolytica; S. cerevisiae* × *Candida utilis* segregating *Candida utilis; Candida tropicalis* × *Candida boidinii*, basically with *C. boidinii* karyotypes; *S. cerevisiae* × *Kluyveromyces marxianus* with *Kluyveromyces* – like segregants; *S. cerevisiae* × *S. fermentati* (= *Torulospora delbrueckii*) showing two types of hybrids with the dominance of either of the parents. However, stable hybrid progeny has also been recovered exhibiting intermediate characteristics between the two parents and which could be of interest to industrial application. For example, *S. cerevisiae* × *Zygosaccharomyces fermentati* producing ethanol at high speed from cellobiose (PINA et al., 1986); *S. cerevisiae* × *Schwanniomyces castelli* showing excellent starch fermentation and ethanol production; *S. diastaticus* × *Saccharomyces rosei* fermenting starch more rapidly; *S. cerevisiae* × *S. mellis* or *S. rouxii* producing ethanol at high glucose concentration (LEGMANN and MARGALITH, 1983) and *S. cerevisiae* × *Candida holmii* exhibiting improved acetic acid tolerance (AARNIO and SUIHKO, 1991).

The genetic background of the hybrids is generally not known. Fusion products were usually characterized by their biochemical properties and extensive genetic analysis was not carried out. Furthermore, hybrids showed on anomalous segregation pattern with an apparent dominance of one of the partners. Because of this asymmetric segregation the genotypes could not be found out, but it was speculated that as for filamentous fungi one nucleus became dominant and that most of the chromosomes of the other parent were lost, although some genes could be retained. This assumption is now confirmed following chromosome separation using OFAGE techniques or pulsed field gel electrophoresis. Following OFAGE and with the aid of molecular probes it was demonstrated for the hybrids *K. marxianus* × *S. cerevisiae* that no intact *S. cerevisiae* chromosomes were present, but some *S. cerevisiae* sequences have been transmitted to the *K. marxianus* chromosomes (WITTE et al., 1989). Such transfer of genes can of course be of much value in strain breeding to obtain stable hybrids with improved properties.

5 Plant Protoplast Fusion

Protoplast fusion with plants is far more extensively investigated compared to microbial systems. As a result, there is an enormous amount of literature available concerning this topic. It is not the purpose of this chapter to give a complete overview of all published results. Therefore, the reader is referred to recently published review articles and books, e.g., PUITE et al., 1988; BAJAJ, 1989; GLIMELIUS et al., 1991, and to the next section. Selection procedures and the genetic processes occurring in plant somatic hybrids will only be briefly discussed, and, where applicable, they will be compared to those occurring in other systems. In addition, plant protoplast fusion will be evaluated in view of alternative genetic methods aimed at plant improvement.

5.1 Protoplast Formation and Regeneration

Cell walls of higher plant cells mainly consist of cellulose (β_{1-3}-glucan), hemicellulose, and pectin and, in consequence, they can be degraded by treatment of the plant tissue with mixtures of pectinases and cellulases. When digestion of plant tissue occurs in a suitable osmotic solution, such as 0.4 M mannitol or sorbitol, protoplasts can readily be isolated.

COCKING (1960) took the production of bacterial and fungal protoplasts by enzymatic removal of the cell wall as a model and demonstrated as the first that also from plant cells protoplasts could be obtained. Since that first report, plant protoplast technology has developed to a stage at which several practical applications are possible, e.g., the isolation of variants and mutants and the production of somatic hybrids (NEGRUTIU et al., 1984; SCHWEIGER et al., 1987; GLIMELIUS et al., 1991). Nowadays, many enzyme preparations with different degrees of purity and effectiveness are commercially available (see PUITE et al., 1988), and plant protoplasts can routinely be obtained in large quantities from almost every organ and tissue of a diversity of plant species (ROEST and GILISSEN, 1989).

Yield, viability, and quality of protoplasts depend on the enzyme preparations, but are also markedly influenced by the type of plant tissue, age, and environmental conditions of growth. Leaf mesophyll was the first and still is the most widely used source for protoplasts. Since the physiological state of the culture is critical, field-grown plants are not suitable. Therefore, plants are usually grown in controlled environment chambers and greenhouses, where light and temperature can be carefully controlled.

More recently, protoplasts have been produced from roots, cotyledons, and hypocotyls from young seedlings to reduce the need for labor-intensive greenhouse and growth-cabinet facilities, and somatic embryoids have also been used as a source of protoplasts (DAVEY, 1983) as well as meristem cells from shoot tips. Suspension cultures are another source routinely used by many workers with the best results obtained from cells taken at the early log phase. *In vitro* cultures, especially when maintained for extended periods of time, suffer, however, endopolyploidy, aneuploidy, and other forms of mitotic error (SHEPARD, 1981).

For some important species such as soybean (*Glycine max*), efficient protoplast liberation with commercial enzyme preparations has not been possible. For such species alternative tissue sources have sometimes been used as protoplast donors, e.g., immature pod tissue for soybeans, petioles of sweet potato (*Ipomea batatas*) leaves, or callus tissues.

Protoplasts offer an important biological entity for many physiological and biochemical studies, but their particular importance is their totipotency, i.e., their ability to regenerate to whole plant cells (Fig. 8), in many cases irrespective of the origin of the protoplasts. As a consequence of their totipotency, protoplasts are interesting tools for genetic manipulation, either in transformation or in protoplast fusion experiments. The regeneration process occurs in different stages. When protoplasts are placed in appropriate media, they may synthesize a cell wall and undergo division producing in a first instance cell colonies followed by embryogenic or non-embryogenic callus formation (Fig. 8). Some cells exhibit organogenesis in which shoot primordia and, subsequently, a shoot is formed. Following shoot formation and, eventually induced by an auxin hormone, a root may develop at the base of the shoot finally giving rise to a plant. In other species the cells grow to complete plants via embryogenesis.

Species and donor tissue used for protoplast isolation each have been found to affect greatly the subsequent culture and regeneration of the protoplasts. Therefore, the procedures for the regeneration from protoplasts to plants vary with the origin of the protoplasts. Conditions for efficient regeneration have to be established experimentally in each case. In recent years, considerable progress in plant protoplast regeneration has been made as a result of the enormous amount of work carried out in this field, which has led to different approaches. As reviewed by ROEST and GILISSEN (1989), more than 200 plant species belonging to 93 genera of 31 families have been regenerated into plants, or in some cases just

Fig. 8. Consecutive steps in the regeneration of plant protoplasts.
(a) Isolated protoplasts of *Zea mays* (bar: 20 μm); (b) cell colony formation of dividing protoplasts after 15 days incubation (bar: 18 μm); (c) embryonic maize protoplast-derived calli with the presence of somatic embryos (arrows) (bar: 1 mm); (d) plant regeneration from protoplasts isolated form embryonic cell suspensions; (e) maize platelets regenerated from protoplasts (photographs with courtesy from PRIOLI and SÖNDAHL, 1989, with permission).

into embryo-like structures, embryoids or shoots, and the number of plant species has still increased since.

The greatest success has been obtained with ornamental, drug, or crop species in the Solanaceae family, such as *Nicotiana, Solanum, Datura, Petunia*, and *Lycopersicon*. Plant regeneration of protoplasts from monocotyledons including cereal and grass species, and of some dicotyledonous species of the Papilionaceae, to which several important food crops belong, is a much more delicate and often unreliable process. A breakthrough came after VASIL and VASIL (1981) had demonstrated that protoplasts from embryogenic cell lines of the Gramineae species *Pennisetum americanum* could form cell colonies, somatic embryos, and plants. Since then, with this technique plants have been regenerated from protoplasts of other recalcitrant species such as rice, sugarcane, *Panicum, Lolium, Festuca, Dactylis*, or *Zea mays*. However, no successful regeneration has so far been realized for several other important cereal crops, which obviously poses a limitation on successful hybridization experiments with these species.

5.2 Somatic Hybridization with Plant Protoplasts

5.2.1 Selection Procedures

Fusion techniques for plant protoplasts are similar to those of other naked cell types. Equal numbers (10^6) of protoplasts of the two participating species suspended in a suitable medium are mixed and subsequently subjected to fusion treatment, either by chemical (PEG) or electric-field mediated fusion. Because plant protoplasts are more fragile than microbial protoplasts, they have to be treated more gently. Following fusion, the fusion mixture is plated onto a suitable selection medium for direct selection of the hybrids from the parental protoplasts or onto a medium allowing the propagation of fused and non-fused cells with hybrid selection in a later stage.

Selection is one of the major obstacles in the production of a fusion hybrid with desirable properties. Selection schemes for plants, reviewed by GLEDDIE et al. (1985) and LAZAR (1983), are much more diverse than those used in the microbial system. Most of the reported schemes are only applicable to a particular system, which precludes their widespread use in different fusion systems. They also depend on the availability of selectable markers for the particular species either acquired by mutation or based on existing natural differences. A series of biochemically selectable mutants have been produced via haploid or diploid protoplast isolation and culture (NEGRUTIU et al., 1984). The complementation of acquired metabolite deficiency such as nitrate reductase deficiency of different origin is one of the most frequently used selection techniques in higher plants, especially tobacco. Nitrate reductase deficient mutants do not grow with nitrate as the sole source of nitrogen. Complementation between other types of nutritional requirements such as between isoleucine and uracil requiring auxotrophic lines, eventually in combination with conditional lethal properties, e.g., temperature sensitivity, has also been successfully used for selection. Other selection schemes rely on the complementation of regenerative capability as a result of the somewhat variable, naturally existing differential responses of the parental species to the medium, the used culture conditions, and the need for phytohormones. Antimetabolite resistance (antibiotics, herbicides or other toxic compounds), either nuclear or organelle encoded, are also efficient.

Crosses between clones showing a different resistance give rise to double-resistant somatic hybrids which can easily be isolated. Chromosomally located resistance markers are, e.g., 5-methyltryptophan, *S*-(2-aminoethyl)-L-cysteine, methotrexate resistance, and kanamycin resistance. The latter selection marker can be introduced into several species by leaf disc transformation with *Agrobacterium tumefaciens* (ROGERS et al., 1986). Chloroplast or mitochondrially encoded resistance are, e.g., tentoxin, streptomycin, and oligomycin. It has to be mentioned, however, that the use of two different types of chloroplast-encoded resistant populations cannot result in a double-resistant cellular phenotype, since chloroplasts of either of the parental strains segregate randomly from the somatic hybrid. In consequence, the fusion progeny possesses only one type of the parental chloroplasts, and recombination between chloroplasts does not occur. This is in contrast to mitochondria and mitochondrially located resistance markers, because mitochondrial recombination frequently takes place.

Chlorophyll-deficient and albino lines can also be used as a selection criterion, because a restoration of photosynthetic properties can occur by complementation. Non-mutant wild-type protoplasts and their natural or induced fluorescence (from an incorporated fluorescent dye) are suitable, if a fluorescence-activated cell sorter (FACS) is at one's disposal.

Instead of using viable protoplasts for fusion, it is also possible to inactivate simultaneously the protoplasts of the two fusion partners using irreversible biochemical inhibitors. When protoplast suspensions of different cell lines are treated, each with an agent of different specificity (e.g., iodoacetate, rhodamine G-6) at lethal doses, fusion products can survive by complementation of the inhibited enzymes. With this approach, one could use cell lines of wild-type species to select hybrids (NEHLS, 1978), although this technique seemed to be successful only in case of unidirectional selection.

5.2.2 Gene Transfer via Somatic Hybridization

Following somatic fusion of plant protoplasts different types of progeny can arise. (1) True somatic hybrid plants containing the nuclei of both contributing species. Subsequent nuclear fusion, most frequently obtained by the parallel progression through mitosis of the heterospecific nuclei (HARMS, 1983), can give rise to allopolyploids, i.e., hybrids containing the sum of the complete chromosome set of the different parental species, eventually of different ploidy levels. In case the hybrid nuclei consist of the sum of the complete diploid chromosome complement of each parent, they are designated allotetraploid, and amphidiploids, when the hybrids contain the sum of the complete chromosome set, each eventually of different ploidy level.

(2) Cybrid plants containing the cytoplasms of both parental species, yet the nucleus of only one of the contributing species. As such, somatic hybridization differs from conventional sexual breeding not only by the fact that in this type of fusion sexually incompatible species can be crossed, but also that both parental cytoplasms with their organelles can be mixed. This is in contrast with sexual crosses, in which case there is usually unilateral exclusion of the male cytoplasm, because cytoplasmic organelles are maternally inherited in most crops. In cybrid plants, a widespread intermolecular recombination of mitochondrial genomes is possible. As mentioned before, chloroplasts do not hybridize and parental chloroplasts segregate randomly with the retention of one type of parental chloroplasts in one type of hybrid segregants.

Cybrid plants not only allow the particular study of the functionality of plant organelles in concert with a particular genome, but they also permit to alter traits, if they are chloroplast- or mitochondrially encoded. Some of them are of agronomic importance, e.g., because of their cytoplasmic male sterility (CMS), herbicide resistance, nectar production, or resistance to fungal toxins. In this manner, both CMS and fertile hybrids or cybrids could be produced following somatic cell hybridization between CMS lines and fertile lines in tobacco, petunia, rice, rapeseed, and carrot (TANNO-SUENAGA and IMAMURA, 1991; GLIMELIUS et al., 1991).

Numerous papers have been published on somatic hybridization between plants with the aim to obtain novel hybrid clones with improved properties. Fusions have been carried out between protoplasts of the same or closely related species, between species belonging to different genera, tribes, and families, and between clones with the same or different ploidy (HARMS, 1983; NEGRUTIU et al., 1989; PUITE et al., 1988). It soon appeared, however, that unless the partners are closely related, fusions between species gave rise to hybrid nuclei from which randomly chromosomes of one species were eliminated. In this manner, asymmetric hybrids were created that contained, in addition to a complete recipient genome, a varying number of chromosomes of the other partner. This was first observed for plant protoplasts with hybrids produced between *Glycine max* and *Nicotiana glauca* (KAO, 1977) and afterwards for many other interspecies crosses. Asymmetry of the chromosomal constitution has been mentioned for all intergeneric hybrids, and it may range from only a slight deviation from the chromosome complement of both constitutive parents to an almost undetectable genetic contribution of one parent. Moreover, it appears that in intergeneric protoplast fusion an extensive and progressing uniparental loss of chromosomes takes place before expression of morphogenesis is possible (HARMS, 1983). Furthermore, it appears that the tendency to eliminate chromosomes increases with the phylogenetic distances of the species fused. Loss of chromosomes is also common in mammalian cell hybrids (see Sect. 6 and also Chapter 5 of this volume) and, as earlier discussed, asymmetric hybrids are also frequently observed in interspecies fungal hybrids.

Chromosome constitution and elimination in plant hybrids can be detected in several ways: by the presence of marker chromosomes, by cytogenetic analysis if each parental species exhibits prominent differences in chromosome number and morphology, by isoenzyme analysis of, e.g., esterases, peroxidases, alcohol dehydrogenase, or acid phosphatase, and by restriction fragment length polymorphism which easily allows to establish the contribution of each parental genome to that of the hybrids, if suitable DNA probes are available. It also allows proof of the existence of recombination events in somatic hybrids.

Strong somatic incompatibility expressed in non-differentiating fusion products is supposed to be due to extensive genomic diversity. As a consequence and as already discussed above, fusions of protoplasts of phylogenetically remote species result in the unidirectional elimination of chromosomes of one species, giving rise to asymmetric hybrids. DUDITS and coworkers (1980) were the first to show that similarly as for mammalian hybrids (RODGERS, 1979) incompatibility responses could be bypassed, when the parental genome of one of the constitutive partners was significantly reduced. Nuclear-size reduction of donor protoplasts can be obtained by X- or γ-irradiation. As a result of irradiation the cell is inactivated due to aberration and fragmentation of the chromosomes, but the chromosome fragments can efficiently be transferred into the recipient chromosomes giving rise to highly asymmetric hybrids that contain only one or a few donor chromosomes. DUDITS et al. (1980) isolated *Daucus carota* + *Petroselinum hortense* somatic hybrids using X-irradiated *P. hortense* protoplasts as donor cells and chlorophyll deficiency of *D. carota* as a selection marker. Intergeneric hybrids between these species could not be obtained by symmetric fusion. Since the successful production of *D. carota* + *P. hortense* somatic hybrids using this asymmetric type of fusion, i.e., with X- or γ-irradiated protoplasts as donor cells, several other examples of successful hybridization using asymmetric fusions have been reported, e.g., *Nicotiana tabacum* + *Physalis minima, N. tabacum* + *Datura innoxia, N. glauca* + *N. langsdorffii, D. innoxia* + *P. minima, N. tabacum* + *Hordeum vulgaris, Solanum tuberosum-phureja* + *S. pinnatisectum, Hyoscyamus muticus* + *N. tabacum, N. tabacum* + *N. repanda* (for reviews, see NEGRUTIU et al., 1989; AGOUDGIL et al., 1990, BATES, 1990; YAMASHITA et al., 1989).

Currently, because of the success of asymmetric fusions and the disappointment of the production of symmetric hybrids between unrelated species, most fusions between less related species are carried out by asymmetric fusions. In an extensive study on intergeneric hybridization using γ-irradiated protoplasts of *Nicotiana plumbaginifolia* as donor, and non-irradiated protoplasts of several species as receptor it was shown, however, that irradiation damage alone seems not to be sufficient to remove all incompatibility barriers (NEGRUTIU et al., 1989), and the success of particular fusion combinations did not seem to depend solely on phylogenetic criteria. Furthermore, using asymmetric fusions, cybrids are more easily isolated with increased radiation doses.

In conclusion, examples of intra-, interspecies and intergeneric crosses which have led to fertile plants by somatic hybridization between plants which cannot be crossed sexually show that, in addition to the classical way of breeding and the newly developed tissue culture techniques, plant protoplast fusion can be valuable in crop improvement, when the right combination of plants is chosen in addition to an efficient selection system. The value of recombinant DNA techniques in plant breeding is discussed in Chapter 15.

6 Mammalian Cell Fusion

In contrast to microbial or plant cells, mammalian cells are not protected by a rigid cell wall, but the outer barrier is the readily fusable membrane double layer. Fusion of mammalian cells takes place in the course of a number of genetically controlled biological processes including polykaryon formation in bone and muscle. Mammalian cell fusion may also occur in a biological system by non-genetic processes, e.g., following infection of cultured cells with enveloped viruses such as influenza and Sendai virus. As mentioned before, Sendai virus has been used originally and during a relatively long period of time as a cell-fusion inducing agent, but this approach is now completely replaced by chemically (PEG) or electrically induced fusion. Mammalian cell fusion is carried out basically in a way similar to microbial of plant protoplast fusions, but with some proper modifications (HANSEN and STADLER, 1977; SCHMITT and ZIMMERMANN, 1989). Instead of fusing whole cells it is possible to fuse only the cytoplasmic part (cytoplast) of the cell and the nuclear part (karyoplast) of another cell similarly as discussed for fungal and plant cells. Fusions give rise to a reconstituted cell (cytoplast × karyoplast), a cybrid (cytoplast × whole cell), or a nuclear hybrid (karyoplast × whole cell). Various protocols of the production of enucleated cells and the isolation of nuclei have been described (SHAY, 1987).

6.1 Selection Systems of Fused Cells

The classical method of selecting fused cells is that of LITTLEFIELD (1964) who introduced the HAT (hypoxanthine-aminopterin-thymidine) selection system. This method relies on the functional presence of the enzymes thymidine kinase (TK) and hypoxanthine guanine phosphoribosyl transferase (HGPRT). These enzymes are required in the salvage pathways in which preformed nucleotides are recycled allowing the synthesis of DNA, when the main biosynthetic (*de novo*) pathway for purines and pyrimidines is blocked, e.g., under the influence of the folic acid antagonist aminopterin. Thus, if the cell is supplied with thymidine and hypoxanthine, DNA synthesis can still occur, provided the enzymes TK and HGPRT are present. Mutant cells deficient in TK or HGPRT will not grow, except in case they are fused with another cell which can complement the missing enzyme. In this manner, properly fused cells can be selected. In addition to TK^- and $HGPRT^-$ mutants, a number of other conditional mutants have been used, for example, drug resistance mutants, nutritional auxotrophs, temperature-sensitive mutants, but also irreversible biochemical inhibitors (see Sect. 5.2.1) have been used for selection (WRIGHT, 1978). The efficiency of fusion may be estimated as follows: (1) by fluorescence staining of the nuclei of each of the fusion partners with a different nuclear stain and determination of the number of double-stained cells in a fluorescence microscope or a two-color FACS or (2) by labeling the fusion partners with distinct surface markers which can be recognized after fusion by dye-labeled monoclonal antibodies.

6.2 Chromosome Behavior in Fused Cells

Fusions have been carried out mostly with rodent and human cell types, but also cells originating from other species, such as cattle, sheep, goat, monkey, chicken, have been involved in fusion experiments. When genetically different cell types (of the same or different species) are fused, heterokaryons are produced. It has been observed that in these heterokaryons DNA synthesis asynchronously occurs in most instances (HARRIS, 1970), and in culture these heterokaryons remain alive for several weeks, but they are not able to multiply indefinitely. Their continued reproduction depends upon the formation of daughter cells which contain a single nucleus. This monokaryotic stage from which the hybrid arises can be achieved by fusion of individual nuclei in a binucleate cell at mitosis. With polykaryotic heterokaryons irregular and abortive mitosis is more common resulting in abnormal nuclei and micronucleation, and the cell may not divide.

In intraspecific somatic hybrids, the karyotypic stability of the hybrid cell depends on the characteristics of the parental cells. For example, if a stable diploid cell is fused with another stable diploid, the hybrid will be a stable tetraploid, but in other instances there is a moderate degree of chromosome loss. In interspecific somatic hybrids, however, the chromosomes of one of the species are slowly and more or less randomly lost. Fusion products of closely related rodent species show a tendency for chromosome elimination that can affect the chromosomes of either parent. In contrast, human-rodent cell hybrids tend to lose human chromosomes preferentially and retain their rodent chromosome complement. Elimination of the chromosomes is truly random and occurs in an irregular but unidirectional manner. The mechanism by which the chromosomes of man disappear gradually on serial passage of the hybrid progeny is as yet unknown. Furthermore, it appears to be a general rule for primate–rodent crosses that these hybrids preferentially lose the primate chromosomes. Maintenance of hybrid cells on selective media under permissive or non-permissive conditions, however, influences the chromosome constitution of the hybrids. This means, e.g., that under selective pressure chromosome(s) containing the gene(s) complementing a defect necessary for growth will be uniformly retained in each hybrid clone. Conversely, it is also possible to promote the loss of specific human chromosomes, e.g., by imposing a non-permissive condition contingent on a particular human gene.

6.3 Applications of Mammalian Somatic Cell Hybrids and Heterokaryons

Somatic cell hybrids and heterokaryons obtained after fusion of mammalian cells can be used for several purposes, first of all for the exploitation of the spontaneous segregation of human chromosomes from human–rodent cell hybrids with the retention of limited and different sets of human chromosomes. These hybrids can be analyzed for the expression of a specific human gene product by analytical, biochemical, or immunochemical methods. Correlation of the expression or absence of expression of a particular gene with the presence or absence of a particular chromosome allows the assignation of this gene to its chromosome (for more details see Chapter 5). This method has been extensively used in human genetics for gene mapping to specific chromosomes.

By fusion, differentiated cell types can be combined, and the influence of one on the function of the other can be investigated. Hence, mammalian cell fusion experiments provide the possibility to investigate nucleocytoplasmic interactions and to study the control of gene expression in animal cells. In this way, it has been shown that following fusion between different cell types mammalian gene expression could be altered and, upon fusion, tissue-specific genes could be readily induced in numerous cell types that normally never express them (BLAU, 1989). In this respect the production of cybrids, reconstituted cells, and nuclear hybrids is of special importance. Hybrids obtained after cell fusion of whole cells often have, in addition to the nuclear genome of the two parental cells, the mitochondria of both cells. In contrast, nuclear hybrids have the nuclear genome of both cells, but mostly the mitochondrial genome of one cell, whereas cybrids have one nucleus in a mixed cytoplasm. This approach, therefore, allows more precisely to characterize whether the cell cytoplasm contains stable regulatory substances that might modulate nuclear gene expression or to investigate nuclear-nuclear interactions in the presence of one cytoplasm.

Hybridoma cells, originally devised by KÖHLER and MILSTEIN (1975), are another product of cell fusion. They are hybrid cells combining the property of antibody production of (non-dividing) spleen cells of immunized experimental animals, with the property of continuous growth of mouse cancer cells (myeloma cells). As a result, a cell line is produced that provides a continuous source of antibodies of predefined specificity, designated monoclonal antibodies (mAB), which can be used for the industrial production of mAB. Hybridoma technology is rapidly expanding: human–mouse, bovine–mouse, rabbit–mouse, sheep–mouse, and goat–mouse hybridomas have been described. Hybridoma technology has moved serology into a new era. Today hundreds of research and diagnostic procedures are based upon this technology, and mABs are being introduced as therapeutic agents. Because of the widespread use of this technology, several reviews have been written which describe monoclonal antibody production in great detail (KÖHLER, 1980; GODING, 1986; FRENCH et al., 1987; JAMES and BELL, 1987). It will also be considered in Volume 5 of this *Biotechnology* series.

7 General Conclusions

Cell fusion is a complex process involving destabilization and subsequent transient loss of the bilayer structure to a non-bilayer lipid structure of the interacting biomembranes. Induced cell fusion, first described for animal cells using enveloped viruses, has developed after introduction of chemically induced fusion and electrofusion, to a very efficient tool in pure and applied genetics for all sorts of living organisms.

As a result of membrane fusion an aqueous pore is formed between two adjacent cells, hence providing the opportunity of cytoplasmic mixing, nuclear fusion, and subsequent recombination or hybridization between the genomes of cells involved in the fusion process. As reviewed in this chapter, cell fusion for different organisms shows a great potential in different domains of biotechnology. Results with bacteria - although not extensively investigated - show the possibility of isolating strains with improved or modified properties, and protoplast fusion was proven to be useful in genetic analysis, but obscurities still remain such as the as yet unexplained phenomenon of non-parental diploids.

In fungi, intraspecies protoplast fusion is an efficient means to initiate the parasexual cycle, thus allowing mitotic recombination in all asexual species or in species which cannot be crossed sexually. Interspecies protoplast fusion provides the possibility of crossing related species and to create hybrids with novel properties. For less closely related species, however, incompatibility problems were experienced in most instances, and loss of chromosomes of one parental type was noticed, a phenomenon also occurring in other eukaryotic system including plants and animals, when cells of different species were crossed.

In plant interspecies protoplast fusion, it was shown that incompatibility could largely be overcome, when asymmetric fusions were used, an approach not thoroughly studied as yet with fungi. Because of the success of asymmetric fusions with plants, it should therefore be of interest to investigate to what extent asymmetric fusions could reduce or eliminate incompatibility barriers in fungal interspecies crosses. With plants, examples of plant modification following protoplast fusion indicate the possible value of protoplast fusion in crop improvement, and with animal cells, fusion has brought about an important contribution to immunology with the production of hybridoma cell lines and to human genetics by somatic cell genetics.

Notwithstanding these results, with the exception of mammalian cell fusion, the enthusiasm for cell fusion as a means for somatic hybridization, as seen in the first years after the development of the fusion technology, has diminished in several laboratories, and cell fusion is, to date, being used in many instances solely as a means of DNA uptake and not for somatic hybridization. Reasons for this, are in the first place the nowadays available efficient tools of recombinant DNA technology allowing the introduction of specific genes and regulatory sequences of various origins in different species. On the other hand, protoplast fusion lacks specificity and the outcome of the fusion process is a matter of luck as with classical breeding programs. Furthermore, the fact that interspecies protoplast fusion often showed disappointing and unpredictable results for interspecies hybridization due to incompatibility problems disencouraged several investigators.

Nevertheless, cell fusion will remain of value in breeding programs, especially of eukaryotic organisms, and in particular in cases in which desirable properties are encoded by genes which are not physically identified, or that are of polygenic nature, and for the transfer of traits encoded by cytoplasmically located elements. Finally, cell fusion can further be of interest in fundamental research, e.g., in the study of inheritance of cytoplasmic elements and of their interaction with the nuclear genome, and to identify factors involved in species incompatibility.

Acknowledgment
The aid of Dominique Brabants is gratefully acknowledged for typing the manuscript.

8 References

AARNIO, T. H., SUIHKO, M.-L. (1991), Electrofusion of an industrial baker's yeast strain with a sour dough yeast, *Appl. Biochem. Biotechnol.* **27**, 65–73.

AGOUDGIL, S., HINNISDAELS, S., MOURAS, A., NEGRUTIU, I., JACOBS, M. (1990), Metabolic complementation for a single gene function associated with partial and total loss of donor DNA in interspecific somatic hybrids, *Theor. Appl. Genet.* **80**, 337–342.

AKAMATSU, T., SEKIGUCHI, J. (1981), Studies on regeneration media of *Bacillus subtilis* protoplasts, *Agric. Biol. Chem.* **45**, 2887–2894.

ALLCOCK, E. R., REID, S. J., JONES, D. T., WOODS, D. R. (1982), *Clostridium acetobutylicum* protoplast formation and regeneration, *Appl. Environ. Microbiol.* **43**, 719–721.

ANNÉ, J. (1977), Somatic hybridization between *Penicillium* species after induced fusion of their protoplasts, *Agricultura* **25**, 1–117.

ANNÉ, J. (1983), Protoplasts of filamentous fungi in genetics and metabolite production, *Experientia Suppl.* **46**, 167–178.

ANNÉ, J. (1985), Taxonomic implications of hybridization of *Penicillium* protoplasts, in: *Advances in Penicillium and Aspergillus Systematics* (SAMSON, R. A., PITT, J. E., Eds.), pp. 337–350. New York: Plenum Press.

ANNÉ, J., PEBERDY, J. F. (1975), Conditions for induced fusion of fungal protoplasts in polyethylene glycol solutions, *Arch. Microbiol.* **105**, 201–205.

ANNÉ, J., PEBERDY, J. F. (1981), Characterisation of inter-species hybrids between *Penicillium chrysogenum* and *P. roqueforti* by iso-enzyme analysis, *Trans. Br. Mycol. Soc.* **77**, 401–408.

ANNÉ, J., EYSSEN, H., DE SOMER, P. (1974), Formation and regeneration of *Penicillium chrysogenum* protoplasts, *Arch. Microbiol.* **98**, 159–166.

ANNÉ, J., FÈVRE, M., SANGLIER, J. J. (1989), Améliorations génétiques des souches, in: *Biotechnologie des Antibiotiques* (LARPENT, J. P., SANGLIER, J. J., Eds.), pp. 220–287, Paris: Masson.

ANNÉ, J., VAN MELLAERT, L., EYSSEN, H. (1990), Optimum conditions for efficient transformation of *Streptomyces venezuelae* protoplasts, *Appl. Microbiol. Biotechnol.* **32**, 431–435.

ANTONOV, P. A. (1990), Thermofusion of cells, *Biochim. Biophys. Acta* **1051**, 279–281.

AUER, D., BRANDNER, G., BODEMER, W. (1976), Dielectric breakdown of the red blood cell membrane and uptake of SV40 DNA and mammalian cell RNA, *Naturwissenschaften* **63**, 391.

AZEVEDO, J. L., BONATELLI, JR. R. (1982), Genetics of the overproduction of organic acids, in: *Overproduction of Microbial Products* (KRUMPHANZL, V., SIKYTA, B., VANEK, Z., Eds.), pp. 439–450, London: Academic Press.

BAJAJ, Y. P. S. (1989), *Biotechnology in Agriculture and Forestry*, Vol. 8: *Plant Protoplasts and Genetic Engineering I.*, Berlin–Heidelberg: Springer-Verlag.

BALTZ, R. H., MATSUSHIMA, P. (1983), Advances in protoplast fusion and transformation in *Streptomyces, Experientia Suppl.* **46**, 143–148.

BATES, G. W. (1990), Asymmetric hybridization between *Nicotiana tabacum* and *N. repanda* by donor recipient protoplast fusion: transfer of TMV resistance, *Theor. Appl. Genet.* **80**, 481–487.

BERLINER, M. D., RECA, M. E. (1971), Studies on protoplast induction in the yeast phase of *Histoplasma capsulatum* by magnesium sulfate and 2-deoxy-D-glucose, *Mycologia* **6**, 1164–1172.

BIOZET, B., FLICKINGER, J. L., CHASSY, B. M. (1988), Transfection of *Lactobacillus bulgaricus* protoplasts by bacteriophage DNA, *Appl. Environ. Microbiol.* **54**, 3014–3018.

BLAU, H. M. (1989), How fixed is the differentiated state? *Trends Genet.* **5**, 268–272.

BLUMENTHAL, R. (1987), Membrane fusion, *Curr. Top. Membr. Transp.* **29**, 203–254.

BONI, L. T., HAH, J. V., HUI, S. W., MUKHERJEC, P., HO, J. T., JUNG, C. Y. (1984), Aggregation and fusion of unilamellar vesicles by polyethylene glycol, *Biochim. Biophys. Acta* **775**, 409–418.

BRADLEY, S. G. (1959), Protoplasts of *Streptomyces griseus* and *Nocardia paraguayensis, J. Bacteriol.* **77**, 115–116.

BURGER, K. (1991), The mechanism of influenza virus-membrane fusion, *Ph. D. Thesis*, University of Utrecht, The Netherlands.

CAPORALE, L. H., CHARTZAM, N., TOCCI, M., DE HAVEN, P. (1990), Protoplast fusion in microtiter plates for expression cloning in mammalian cells: demonstration of feasibility using membrane-bound alkaline phosphatase as a reporter enzyme, *Gene* **87**, 285–289.

CASO, J. L., HARDISSON, C., SUAREZ, J. E. (1987), Transfection in *Micromonospora* spp., *Appl. Environ. Microbiol.* **53**, 2544–2547.

CHAKRABORTY, B. N., KAPOOR, M. (1990), Transformation of filamentous fungi by electroporation, *Nucleic Acids Res.* **18**, 6737.

CHAND, P. K., DAVEY, M. R., POWER, J. B., COCKING, E. C. (1988), An improved procedure for protoplast fusion using polyethylene glycol. *J. Plant Physiol.* **133**, 480–485.

CHANG, D. C., HUNT, J. R., GAO, P-Q. (1989), Effects of pH on cell fusion induced by electric fields, *Cell Biophys.* **14**, 231–243.

CHAPEL, M., MONTANÉ, M.-H., RANTY, B., TEISSIÉ, J., ALIBERT, G. (1986), Viable somatic hybrids are obtained by direct current electrofusion of chemically aggregated plant protoplasts, *FEBS Lett.* **196**, 79–86.

CHEN, Z., WOJCIK, S. F., WELKER, N. E. (1986), Genetic analysis of *Bacillus stearothermophilus* by protoplast fusion, *J. Bacteriol.* **165**, 994–1001.

COCKING, E. C. (1960), A method for the isolation of plant protoplasts and vacuoles, *Nature* **187,** 962–963.

CRAMP, F. C., LUCY, J. A. (1974), Glyceryl monooleate as a fusogen for the formation of heterokaryons and interspecific hybrids, *Exp. Cell Res.* **87,** 107–110.

CROCE, C. M., SAWICKI, W., KRITCHEVSKY, D., KOPROWSKI, H. (1971), Induction of homokaryocyte, heterokaryocyte and hybrid formation by lysolecithin, *Exp. Cell Res.* **67,** 427–435.

CROFT, J. H., DALES, R. B. G. (1983), Interspecific somatic hybridization in *Aspergillus, Experientia Suppl.* **46,** 179–186.

DAVEY, M. R. (1983), Recent developments in the culture and regeneration of plant protoplasts, *Experientia Suppl.* **46,** 19–30.

DEB, J. K., MALIK, S., GHOSH, V. K., MATHAI, S., SETHI, R. (1990), Intergeneric protoplast fusion between xylanase producing *Bacillus subtilis* LYT and *Corynebacterium acetoacidophilum* ATCC 21476, *FEMS Microbiol. Lett.* **71,** 2887–2892.

DELORME, E. (1989), Transformation of *Saccharomyces cerevisiae* by electroporation, *Appl. Environ. Microbiol.* **55,** 2242–2246.

DOUGLAS, R. J., ROBINSON, J. B., CORKE, C. T. (1958), On the formation of protoplast-like structures from Streptomycetes, *Can. J. Microbiol.* **4,** 551–554.

DUDITS, D., FEJÉR, O., HADLACKZKY, G., KONCZ, C., LAZAR, G. B., HORVATH, G. (1980), Intergeneric gene transfer mediated by plant protoplast fusion, *Mol. Gen. Genet.* **179,** 283–288.

EDDY, A. A., WILLIAMSON, D. H. (1957), A method for isolating protoplasts from yeast, *Nature* **179,** 1252–1253.

ENDERS, J. F., PEEBLES, T. C. (1954), Propagation in tissue cultures of cytopathogenic agents from patients with measles, *Proc. Soc. Exp. Biol. Med.* **86,** 277–286.

FELGNER, P. L., GADEK, T. R., HOLM, M., ROMAN, R., CHAN, H. W., WENZ, M., NORTHROP, J. P., RINGOLD, G. M., DANIELSEN, M. (1987), Lipofection: a highly efficient, lipid-mediated DNA-transfection procedure, *Proc. Natl. Acad. Sci. USA* **84,** 7413–7417.

FERENCZY, L. (1981), Microbial protoplast fusion, in: *Genetics as a Tool in Microbiology* (GLOVER, S. W., HOPWOOD, D. A., Eds.), pp. 1–34. Cambridge: Cambridge University Press.

FERENCZY, L. (1984), Fungal protoplast fusion: basic and applied aspects, in: *Cell Fusion: Gene Transfer and Transformation* (BEER, JR., R. F., BASSETT, E. G., Eds.) pp. 145–169, New York: Raven Press.

FERENCZY, L., KEVEI, F., SZEGEDI, M. (1975), High-frequency fusion of fungal protoplasts, *Experientia* **31,** 1028–1029.

FINCH, R. P., HAMET, I. M., COCKING, E. C. (1990), Production of heterokaryons by the fusion of mesophyll protoplasts of *Porteresia coarctaca* and cell suspension-derived protoplasts of *Oryzae sativa*: a new approach to somatic hybridization in rice, *J. Plant Physiol.* **136,** 592–598.

FLECK, W. F. (1979), Genetic approaches to new Streptomycetes products, in: *Genetics of Industrial Microorganisms* (SEBEK, O. K., LASKIN, A. I., Eds.) pp. 117–122, Washington DC: American Society for Microbiology.

FLEISCHER, E. R., VARY, P. S. (1985), Genetic analysis of fusion recombinants and presence of noncomplementing diploids in *Bacillus megaterium, J. Gen. Microbiol.* **131,** 919–926.

FLOSS, H. G. (1987), Hybrid antibiotics – the contribution of the new gene combinations, TIBTECH **5,** 111–115.

FODOR, K., HADLACZKY, G., ALFÖDI, L. (1975), Reversion of *Bacillus megaterium* protoplasts to the bacillary form, *J. Bacteriol.* **121,** 390–391.

FÖRSTER, E., EMEIS, C. C. (1986), Enhanced frequency of karyogamy in electrofusion of yeast protoplasts by means of preceding G1 arrest, *FEMS Microbiol. Lett.* **34,** 69–72.

FOUNG, J., PERKINS, S., KAFADAR, K., GESSNER, P., ZIMMERMANN, U. (1990), Development of microfusion techniques to generate human hybridomas, *J. Immunol. Methods* **134,** 35–42.

FRENCH, D., KELLY, T., BUHL, S., SCHAFF, M. D. (1987), Somatic cell genetic analysis of myelomas and hybridomas, *Methods Enzymol.* **151,** 50–66.

GABOR, M. H., HOTCHKISS, R. D. (1979), Parameters governing bacterial regeneration and genetic recombination after fusion of *Bacillus subtilis* protoplasts, *J. Bacteriol.* **137,** 1346–1353.

GASSON, M. J. (1980), Production, regeneration and fusion of protoplasts in lactic streptococci, *FEMS Microbiol. Lett.* **9,** 99–102.

GLEDDIE, S., KELLER, W. A., SETTERFIELD, G. (1985). Production of new hybrid plants through protoplast fusion, in: *Biotechnology Handbook* (CHEREMISNOFF, O., OUELLETTE, R., Eds.), pp. 231–242, Lancester, PA: Technomic Publishing Co.

GLIMELIUS, K., FAHLESSON, J., LANDGREN, M., SJÖDIN, L., SUNDBERG, C. (1991), Gene transfer via somatic hybridization in plants, *TIBTECH* **9,** 24–30.

GO, S.-J., YOU, C.-H., SHIN, G.-C. (1989), Effects of incompatibility on protoplast fusion between intra- and interspecies in basidiomycete, *Pleurotus* spp., *Kor. J. Mycol.* **17,** 137–144.

GODFREY, O., FORD, L., HUBER, M. L. B. (1978), Interspecies matings of *Streptomyces fradiae* with *Streptomyces bikiniensis* mediated by conventional and protoplast fusion techniques, *Can. J. Microbiol.* **24,** 994–997.

GODING, J. W. (1986), *Monoclonal Antibodies, Principles and Practice*, 2nd Ed., New York: Academic Press.

GOKHALE, D. V., DEOBAGKAR D. N. (1990), Secretion of thermostable β-glucosidase by an intergeneric bacterial hybrid between *Cellulomonas* and *Bacillus subtilis, Appl. Biochem. Biotechnol.* **26,** 207–215.

GUERRA-TSCHUSCHKE, I., MARTIN, I., GONZALES, M. T. (1991), Polyethylene glycol-induced internationalization of bacteria into fungal protoplasts: electron microscopic study and optimization of experimental conditions, *Appl. Environ. Microbiol.* **57,** 1516–1522.

GUZE, L. B. (1968), Microbial protoplasts, spheroplasts and L-forms, Baltimore: Williams & Wilkins Company.

HALFMANN, H. J., EMEIS, C. C., ZIMMERMANN, U. (1983), Electrofusion and genetic analysis of fusion products of haploid and polyploid *Saccharomyces* yeast cells, *FEMS Microbiol. Lett.* **20,** 13–16.

HAMLYN, P. F., BIRKETT, J. A., PEREZ, G., PEBERDY, J. F. (1985), Protoplast fusion as a tool for genetic analysis in *Cephalosporium acremonium, J. Gen. Microbiol.* **131,** 2813–2823.

HANSEN, D., STADLER, J. (1977), Increased polyethylene glycol-mediated fusion competence in mitotic cells of a mouse lymphoid cell line, *Somatic Cell Genet.* **3,** 471–482.

HARMS, C. T. (1983), Somatic incompatibility in the development of higher plant somatic hybrids. *Qu. Rev. Biol.* **58,** 325–353.

HARRIS, H. (1970), *Cell Fusion*, Oxford: Clarendon Press.

HARRIS, H., WATKINS, J. F. (1965), Hybrid cells derived from mouse and man: artificial heterokaryons of mammalian cells from different species, *Nature* **205,** 640–646.

HASEGAWA, S., MATSUI, C., NAGATA, T., SYONO, K. (1983), Cytological study of the introduction of *Agrobacterium tumefaciens* spheroplasts into *Vinca rosea* protoplasts, *Can. J. Bot.* **61,** 1052–1057.

HEWISH, D. R., WERKMEISTER, J. A. (1989), The use of an electroporation apparatus for the production of murine hybridomas, *J. Immunol. Methods* **120,** 285–289.

HOEKSTRA, D., KOK, J. W. (1989), Entry mechanisms of enveloped viruses. Implications for fusion of intracellular membranes, *Biosci. Rep.* **9,** 273–305.

HONDA, K., MAEDA, Y., SASAKAWA, S., OHNO, H., TSUCHIDA, E. (1981), The components contained in polyethylene glycol of commercial grade (PEG-6000) as cell fusogen. *Biochem. Biophys. Res. Commun.* **101,** 165–171.

HOPE, M. J., CULLIS, P. R. (1981), The role of nonbilayer lipid structures in the fusion of human erythrocytes induced by lipid fusogens, *Biochim. Biophys. Acta* **640,** 82–90.

HOPWOOD, D. (1981), Genetic studies with bacterial protoplasts, *Annu. Rev. Microbiol.* **35,** 237–272.

HOPWOOD, D. A., KIESER, T., WRIGHT, H. M., BIBB, M. J. (1983), Plasmids, recombination and chromosome mapping in *Streptomyces lividans* 66, *J. Gen. Microbiol.* **129,** 2257–2269.

HOPWOOD, D. A., BIBB, M. J., CHATER, K. F., KIESER, T., BRUTON, C. J., KIESER, H. M., LYDIATE, D. J., SMITH, C. P., WARD, J. M., SCHREMPF, H. (1985a), *Genetic Manipulation of Streptomyces: A Laboratory Manual*, Norwich, UK: John Innes Foundation.

HOPWOOD, D. A., MALPARTIDA, F., KIESER, H. M., IKEDA, H., OMURA, S. (1985b), Cloning genes for antibiotic biosynthesis in *Streptomyces* spp.: production of a hybrid antibiotic, in: *Microbiology 1985* (LEIVE, L., Ed.), pp. 409–413. Washington, DC: American Society for Microbiology.

HOTCHKISS, R. D., GABOR, M. (1983), Chromosome interactions and expression in fused *Bacillus* protoplasts, *Experientia Suppl.* **46,** 149–154.

HRANUELI, D., PIGAC, J., SMOKVINA, T., ALACEVIC, M. (1983), Genetic interactions in *Streptomyces rimosus* mediated by conjugation and by protoplast fusion, *J. Gen. Microbiol.* **129,** 1415–1422.

HRMOVA, M., FARKAS, V., KOPECKA, M. (1984), Isolation of anucleated yeast protoplasts by means of density gradient centrifugation, *J. Microbiol. Methods* **2,** 257–263.

HÜTTER, R., ECKHARDT, T. (1988), Genetic manipulation, in: *Actinomycetes in Biotechnology* (GOODFELLOW, M., WILLIAMS, S. T., MORDANSKI, M., Eds.), pp. 90–184, London: Academic Press.

ILLING, G. T., NORMANSELL, I. D., PEBERDY, J. F. (1989), Genetic mapping in *Streptomyces clavuligerus* by protoplast fusion, *J. Gen. Microbiol.* **135,** 2299–2305.

INGOLIA, T. D., WOOD, J. S. (1986), Genetic manipulation of *Saccharomyces cerevisiae*, in: *Manual of Industrial Microbiology and Biotechnology* (DEMAIN, A. L., SOLOMON, N. A., Eds.), pp. 204–213, Washington, DC: American Society for Microbiology.

JAMES, K., BELL, G. T. (1987), Human monoclonal antibody production, *J. Immunol. Methods* **100,** 5–40.

JANDEROVA, B., CVRCKOVA, F., BENDOVA, O. (1990), Construction of the dextrin-degrading *pof* brewing yeast by protoplast fusion, *J. Basic Microbiol.* **30,** 499–505.

KADISH, J. L., WENC, K. M. (1983), Contamination of polyethylene glycol with aldehydes: implications for hybridoma fusion, *Hybridoma* **2,** 87–89.

KANEKO, H., SAKACHUCHI, K. (1979), Fusion of protoplasts and genetic recombination of *Brevibacterium flavum, Agric. Biol. Chem.* **43,** 1007–1013.

KAO, K. N. (1977), Chromosomal behaviour in somatic hybrids of soybean-*Nicotiana glauca, Mol. Gen. Genet.* **150,** 225–230.

KAO, K. N., MICHAYLUK, M. R. (1974), A method for high-frequency intergeneric fusion of plant protoplasts, *Planta* (Berlin) **115,** 355–367.

KAO, K. N., SALEEM, M. (1986), Improved fusion of mesophyll and cotyledon protoplasts with PEG and high pH-Ca^{2+} solution, *J. Plant Physiol.* **122,** 217–225.

KAVANAGH, K., GHANNOUM, M., MANSOUR, I., WHITTAKER, P. (1990), Autoclaved polyethylene glycol decreases yeast protoplast reversion and hybrid production, *Biotechnol. Tech.* **4,** 281–284.

KIGUCHI, T., YANAGI, S. O. (1985), Intraspecific heterokaryon and fruit body formation in *Coprinus macrorhizus* by protoplast fusion of auxotrophic mutants, *Appl. Microbiol. Biotechnol.* **22,** 121–127.

KIM, K. S., RYU, D. D. Y., LEE, S. Y. (1983), Application of protoplast fusion technique to genetic recombination of *Micromonospora rosaria, Enzyme Microb. Technol.* **5,** 273–280.

KIRIMURA, K., NAKAJIMA, I., LEE, S. P., KAWABE, S., USAMI, S. (1988), Citric acid production by the diploid strains of *Aspergillus niger* obtained by protoplast fusion, *Appl. Microbiol. Biotechnol.* **27,** 504–506.

KIRIMURA, K., ITOHIYA, Y., MATSUO, Y., ZHANG, M., USAMI, S. (1990), Production of cellulase and citric acid by the intergeneric fusants obtained via protoplast fusion between *Aspergillus niger* and *Trichoderma viride, Agric. Biol. Chem.* **54,** 1281–1283.

KITANO, K., NOZAKI, Y., IMADA, A. (1985), Strain improvement of a carbapenem antibiotic producer, *Streptomyces griseus* subsp. *cryophilus* C-19393, by protoplast fusion, *Agric. Biol. Chem.* **49,** 685–692.

KIYOHARA, H., WATANABE, T., IMAI, J., TAKIZAWA, N., HATTA, T., NAGAO, K., YAMAMOTO, A. (1990), Intergeneric hybridization between *Monascus anka* and *Aspergillus oryzae* by protoplast fusion, *Appl. Microbiol. Biotechnol.* **33,** 671–676.

KLEBE, R. J., HARRIS, J. V., HANSON, D. P., GAUNT, C. H. (1984), High-efficiency polyethylene glycol-mediated transformation of mammalian cells, *Somatic Cell Mol. Genet.* **10,** 496–502.

KNUTTON, S. (1977), Studies on membrane fusion. II. Fusion of human erythrocytes by Sendai virus, *J. Cell Sci.* **28,** 189–210.

KNUTTON, S. (1979), Studies on membrane fusion. III. Fusion of erythrocytes with polyethylene glycol, *J. Cell Sci.* **36,** 61–72.

KOBORI, H., TAKATA, Y., OSUMI, M. (1991), Interspecific protoplast fusion between *Candida tropicalis* and *Candida boidinii:* characterization of the fusants, *J. Ferment. Bioeng.* **72,** 439–444.

KÖHLER, G. (1980), *Hybridoma Techniques*, New York: Cold Spring Harbor Laboratories.

KÖHLER, G., MILSTEIN, C. (1975), Continuous cultures of fused cells secreting antibody of predefined specificity, *Nature* **256,** 495–497.

KOOP, H.-U., SCHWEIGER, H.-G. (1985), Regeneration of plants after electrofusion of selected pairs of protoplasts, *Eur. J. Cell Biol.* **39,** 46–49.

KÜNKEL, W., GROTH, I., JACOB, H.-E., RISCH, S., HARMISCH, M., MAY, R., BERG, H., KATENKAMP, U. (1987), Electrofusion of protoplasts of *Penicillium chrysogenum, Studia Biophys.* **119,** 35–36.

LANDMAN, O. E., HALLE, S. (1963), Enzymically and physically induced inheritance changes in *Bacillus subtilis, J. Mol. Biol.* **7,** 721–738.

LAZAR, G. B. (1983), Recent developments in plant protoplast fusion and selection technology, *Experientia Suppl.* **46,** 61–68.

LEE, J. S., LEE, M. R., LEE, Y. N. (1988), Spheroplast fusion of *Pseudomonas* spp. using plasmid as selective marker, *Kor. J. Microbiol.* **26,** 298–304.

LEE, J. S., LEE, S. Y., LEE, Y. N. (1989), Ethanol production from starch by protoplast fusion between *Aspergillus oryzae* and *Saccharomyces cerevisiae, Kor. J. Microbiol.* **27,** 221–224.

LEGMANN, R., MARGALITH, P. (1983), Interspecific protoplast fusion of *Saccharomyces cerevisiae* and *Saccharomyces mellis, Eur. J. Appl. Microbiol. Biotechnol.* **18,** 320–322.

LEIKIN, S. L., KOZLOV, M. M., CHERNOMODIK, L. V., MARKIN, V. S., CHIZMADZKEV, Y. A. (1987), Membrane fusion: overcoming of the hydration barrier and local restructuring, *J. Theor. Biol.* **129,** 411–425.

LITTLEFIELD, J. W. (1964), Selection of hybrids

from matings of fibroblasts *in vitro* and their presumed recombinants, *Science* **145**, 709–710.

LOWE, D. A., ELANDER, R. P. (1983), Contributions of mycology to the antibiotic industry, *Mycologia* **75**, 361–373.

LUCY, J. A., AHKONG, Q. F. (1986), An osmotic model for the fusion of biological membranes, *FEBS Lett.* **199**, 1–11.

LUCY, J. A., AHKONG, Q. F., CRAMP, F. C., FISHER, D., JOWELL, J. I. (1971), Cell fusion without viruses, *Biochem. J.* **124**, 469–478.

LURQUIN, P. F. (1979), Entrapment of plasmid DNA by liposomes and their interaction with plant protoplasts, *Nucleic Acids Res.* **6**, 3773–3784.

MAGGIO, B., AHKONG, Q. F., LUCY, J. A. (1976), Polyethylene glycol, surface potential and cell fusion, *Biochem. J.* **158**, 647–650.

MAKINS, J. F., HOLT, G. (1981), Liposome-mediated transformation of streptomycetes by chromosomal DNA, *Nature* **293**, 671–673.

MARTIN, F., MACDONALD, R. (1974), Liposomes can mimic virus membranes, *Nature* **252**, 161–163.

MARTIN, H. H. (1983), Protoplasts and spheroplasts of gram-negative bacteria with special emphasis on *Proteus mirabilis, Experientia Suppl.* **46**, 213–226.

MARTINKOVA, L., MUSILKOVA, M., UJCOVA, E., MACHEK, F., SEICHERT, L. (1990), Protoplast fusion in *Aspergillus niger* strains accumulating citric acid, *Folia Microbiol.* **35**, 143–148.

MATSUSHIMA, P., BALTZ, R. H. (1986), Protoplast fusion, in: *Manual of Industrial Microbiology and Biotechnology* (DEMAIN, A. L., SOLOMON, N. A., Eds.), pp. 170–183, Washington, DC: American Society for Microbiology.

MATSUSHIMA, P., MCHENNEY, M. A., BALTZ, R. H. (1987), Efficient transformation of *Amycolatopsis orientalis* (*Nocardia orientalis*) protoplasts by *Streptomyces* plasmids, *J. Bacteriol.* **169**, 2298–2300.

MINUTH, W., ESSER, K. (1983), Intraspecific, interspecific and intergeneric recombination in β-lactam producing fungi via protoplast fusion, *Eur. J. Appl. Microbiol. Biotechnol.* **18**, 38–46.

MONTANÉ, M.-H., DUPILLE, E., ALIBERT, G., TEISSIÉ, J. (1990), Induction of a long-lived fusogenic state in viable plant protoplasts permeabilized by electric fields, *Biochim. Biophys. Acta* **1024**, 203–207.

MORGAN, A. J. (1983), Yeast strain improvement by protoplast fusion and transformation, *Experientia Suppl.* **46**, 155–166.

MORIKAWA, H., SUGINO, K., HAYASHI, Y., TAKEDA, J., SENDA, M., HIRAI, A., YAMADA, Y. (1986), Interspecific plant hybridization by electrofusion in *Nicotiana, Bio/Technology* **4**, 57–60.

NAGATA, T. (1978), A novel cell-fusion method of protoplasts by polyvinyl alcohol, *Naturwissenschaften* **65**, 263.

NAGATA, T., EIBL, H., MELCHERS, G. (1979), Fusion of plant protoplasts induced by positively charged synthetic phospholipids, *Z. Naturforsch.* **34C**, 460–462.

NEGRUTIU, I., JACOBS, M., CABOCHE, M. (1984), Advances in somatic cell genetics of higher plants – the protoplast approach in basic studies on mutagenesis and isolation of biochemical mutants, *Theor. Appl. Genet.* **67**, 289–304.

NEGRUTIU, I., MOURAS, A., GLEBA, Y. Y., SIDOROV, V., HINNISDAELS, S., FAMELAER, Y., JACOBS, M. (1989), Symmetric versus asymmetric fusion combinations in higher plants, in: *Biotechnology in Agriculture and Forestry*, Vol. 8. *Plant Protoplasts and Genetic Engineering I* (BAJAJ, Y. P. S., Ed.), pp. 304–319, Berlin–Heidelberg: Springer-Verlag.

NEHLS, R. (1978), The use of metabolic inhibitors for the selection of fusion products of higher plant protoplasts, *Mol. Gen. Genet.* **166**, 117–118.

NEIL, G. A., URNOVITZ, H. B. (1988), Recent improvements in the production of antibody-secreting hybridoma cells. *TIBTECH.* **6**, 209–213.

NEUMANN, E., ROSENHECK, K. (1972), Permeability changes induced by electric impulses in vesicular membranes, *J. Membr. Biol.* **10**, 279–290.

OGAWA, K., OHARA, H., KOIDE, T., TOYAMA, N. (1989), Intraspecific hybridization of *Trichoderma reesei* by protoplast fusion, *J. Ferment. Bioeng.* **67**, 207–209.

OKADA, Y., SUZUKI, T., HOSAKA, Y. (1957), Interaction between influenza virus and Ehrlich's tumor cells. III. Fusion phenomenon of Ehrlich's tumor cells by the action of HVJ strain, *Med. J. Osaka Univ.* **7**, 709–717.

OKAMURA, T., NAGATA, S., MISONO, H., NAGASAKI, S. (1988), Interspecific electrofusion between protoplasts of *Streptomyces antibioticus* and *Streptomyces fradiae, Agric. Biol. Chem.* **52**, 1433–1438.

OKAMURA, T., NAGATA, S., MISONO, H., NAGASAKI, S. (1989), New antibiotic-producing *Streptomyces* TT-strain, generated by electrical fusion of protoplasts, *J. Ferment. Bioeng.* **67**, 221–225.

OKANISHI, M., SUZUKI, K., UMEZAWA, H. (1974), Formation and regeneration of Streptomycete protoplasts: cultural conditions and morphological study, *J. Gen. Microbiol.* **80**, 389–400.

ONISHI, K., CHIBA, J., GOTO, Y., TOKUNAGA, T. (1987), Improvement in the basic technology of electrofusion for generation of antibody-producing hybridomas, *J. Immunol. Methods* **100,** 181–189.

Palleroni, N. J. (1983), Genetic recombination in *Actinoplanes brasiliensis* by protoplast fusion, *Appl. Environ. Microbiol.* **45,** 1865–1869.

Peberdy, J. F. (1989), Fungi without coats-protoplasts as tools for mycological research, *Mycol. Res.* **93,** 1–20.

Peberdy, J. F., Ferenczy, L. (1985), *Fungal Protoplasts*, New York: Marcel Dekker Inc.

Pecchia, S., Anné, J. (1989), Fusion of protoplasts from antagonistic *Trichoderma harzianum* strains, *Acta Horticult.* **255,** 303–311.

Pina, A., Calderon, I. L., Benitez, T. (1986), Intergeneric hybrids of *Saccharomyces cerevisiae* and *Zygosaccharomyces fermentati* obtained by protoplast fusion, *Appl. Environ. Microbiol.* **51,** 995–1003.

Pontecorvo, G., Roper, J. A. (1952), Genetic analysis without sexual reproduction by means of polyploidy in *Aspergillus nidulans, J. Gen. Microbiol.* **6,** vii.

Potrykus, I., Harms, C. T., Hinnen, A., Hütter, R., King, P. J., Shilleto, R. D. (1983), Protoplasts 1983, *Experientia Suppl.* **46,** 1–269.

Power, J. B., Cummins, S. E., Cocking, E. C. (1970), Fusion of isolated plant protoplasts, *Nature* **225,** 1016–1018.

Pratsch, L., Herrmann, A., Schwede, I., Meyer, H. W. (1989), The influence of polyethylene glycol on the molecular dynamics within the glycocalyx, *Biochim. Biophys. Acta* **980,** 146–154.

Prioli, L. M., Söndahl, M. R. (1989), Plant regeneration and recovery of fertile plants from protoplasts of Maïze (*Zea Mays* L.), *Bio/Technology* **7,** 589–594.

Puite, K. J., Dons, J. J. M., Huizing, H. J., Kool, A. J., Koornneef, M., Krens, F. A. (1988), *Progress in Plant Protoplast Research*, Dordrecht: Kluwer Academic Publisher.

Puntambekar, U. S., Ranjekar, P. K. (1989), Intergeneric protoplast fusion between *Agrobacterium tumefaciens* and *Bacillus thuringiensis* subsp. *kurstaki, Biotechnol. Lett.* **11,** 717–722.

Rand, R. P. (1981), Interacting phospholipid bilayers: measured forces and induced structural changes, *Annu. Rev. Biophys. Bioeng.* **10,** 227–314.

Raper, K. B., Thom, C. (1949), *A Manual of the Penicillia*, Baltimore: Williams & Wilkins.

Raymond, P., Veau, P., Fèvre, M. (1986), Production by protoplast fusion of new strains of *Penicillium caseicolum* for use in the dairy industry, *Enzyme Microb. Technol.* **8,** 45–48.

Rodgers, A. (1979), Detection of small amounts of human DNA in human-rodent hybrids, *J. Cell Sci.* **38,** 391–403.

Rodicio, M.-L., Manzanal, M.-B., Hardisson, C. (1978), Protoplast-like structures formation from two species of enterobacteriaceae by fosfomycin treatment, *Arch. Microbiol.* **118,** 219–221.

Roest, S., Gilissen, L. J. W. (1989), Plant regeneration from protoplasts: a literature review, *Acta Bot. Neerl.* **38,** 1–23.

Rogers, S. G., Horsch, R. B., Fraley, R. T. (1986), Gene transfer in plants: production of transformed plants using Ti plasmid vectors, *Methods Enzymol.* **118,** 627–640.

Rowlands, R. T. (1984), Industrial strain improvement: rational screens and genetic recombination, *Enzyme Microb. Technol.* **6,** 290–300.

Rubinstein, C. P., Sanchez-Rivas, C. (1988), Production of protoplasts by autolytic induction in *Bacillus thuringiensis*: transformation and interspecific fusion, *FEMS Microbiol. Lett.* **52,** 67–72.

Russel, I., Jones, R. M., Weston, B. J., Stewart, G. G. (1983), Liposome-mediated DNA transfer in brewing and related yeast strains, *J. Inst. Brew.* **89,** 136.

Sandri-Goldin, R. M., Goldin, A. L., Levine, M., Glorioso, J. (1983), High-efficiency transfer of DNA into eukaryotic cells by protoplast fusion, *Methods Enzymol.* **101,** 402–411.

Schmid, E. N. (1984), Fosfomycin-induced protoplasts and L-forms of *Staphylococcus aureus, Chemotherapy* **30,** 35–39.

Schmitt, J. J., Zimmermann, U. (1989), Enhanced hybridoma production by electrofusion in strongly hypo-osmolar solutions, *Biochim. Biophys. Acta* **983,** 42–50.

Schmitt, J. J., Zimmermann, U., Gessner, P. (1989), Electrofusion of osmotically treated cells. High and reproducible yield of hybridoma cells, *Naturwissenschaften* **76,** 122–123.

Schupp, T., Divers, M. (1986), Protoplast preparation and regeneration in *Nocardia mediterranei, FEMS Microbiol. Lett.* **36,** 159–162.

Schweiger, H.-G., Dirk, J., Koop, H.-U., Kranz, E., Neuhaus, G., Spangenberg, G., Wolf, D. (1987), Individual selection, culture and manipulation of higher plant cells, *Theor. Appl. Genet.* **73,** 769–783.

Shay, J. W. (1987), Cell enucleation, cybrids, reconstituted cells, and nuclear hybrids, *Methods Enzymol.* **151,** 221–237.

Shepard, J. F. (1981), Protoplasts as source of disease resistance in plants, *Annu. Rev. Phytopathol.* **19,** 145–166.

Siegel, D. P. (1986), Inverted micellar intermediates and the transitions between lamellar, cubic, and inverted hexagonal lipid phases, *Biophys. J.* **49,** 1171–1183.

SIVAN, A., HARMAN, G. E. (1991), Improved rhizosphere competence in a protoplast fusion progeny of *Trichoderma harzianum, J. Gen. Microbiol.* **137,** 23–29.

SIVAN, A., HARMAN, G. E., STASZ, T. E. (1990), Transfer of isolated nuclei into protoplasts of *Trichoderma harzianum, Appl. Environ. Microbiol.* **56,** 2404–2409.

SMITH, A. W., COLLINS, K., RAMSDEN, M., FOX, H. M., PEBERDY, J. F. (1991), Chromosome rearrangements in improved cephalosporin C – producing strains of *Acremonium chrysogenum, Curr. Genet.* **19,** 235–237.

SMITH, C. L., AHKONG, Q. F., FISHER, D., LUCY, J. A. (1982), Is purified poly(ethylene glycol) able to induce cell fusion? *Biochim. Biophys. Acta* **692,** 109–114.

SPENCER, J. F. T., SPENCER, D. M. (1983), Genetic improvement of industrial yeasts, *Annu. Rev. Microbiol.* **37,** 121–142.

STAHL, M. L., PATTEE, P. A. (1983), Computer-assisted chromosome mapping by protoplast fusion in *Staphylococcus aureus, J. Bacteriol.* **154,** 395–405.

STASZ, T. E., HARMAN, G. E. (1990), Nonparental progeny resulting from protoplast fusion in *Trichoderma* in the absence of parasexuality, *Exp. Mycol.* **14,** 145–159.

STROMINGER, J. L. (1968), Enzymatic reactions in bacterial cell wall synthesis sensitive to penicillins and other antibacterial substances, in: *Microbial Protoplasts, Spheroplasts and L-Forms,* (GUZE, L. B., Ed.), pp. 55–61, Baltimore: Williams & Wilkins.

SULO, P., GRIAC, P., KLOBUCNIKOVA, V., KOVAC, L. (1989), A method for the efficient transfer of isolated mitochondria into yeast protoplasts, *Curr. Genet.* **15,** 1–6.

TAM, J. E., PATTEE, P. A. (1986), Characterization and genetic mapping of a mutation affecting apurinic endonuclease activity in *Staphylococcus aureus, J. Bacteriol.* **168,** 708–714.

TAMAI, A., MIORI, K., KAYAMA, T. (1989), Characterization of protoplasts from basidiospores of edible Basidiomycetes, *Res. Bull. Coll. Exp. Forests* **46,** 425–440.

TANNO-SUENAGA, L., IMAMURA, J. (1991), DNA hybridization analysis of mitochondrial genomes of carrot cybrids produced by donor–recipient protoplast fusion, *Plant Sci.* **73,** 79–86.

TEISSIÉ, J., ROLS, M. P. (1986), Fusion of mammalian cells in culture is obtained by creating the contact between cells after their electropermeabilization, *Biochem. Biophys. Res. Commun.* **140,** 258–266.

TOISTER, Z., LOYTER, A. (1971), Ca^{2+}-induced fusion of avian erythrocytes, *Biochim. Biophys. Acta* **241,** 719–724.

TOMCSIK, J., GUEX-HOLZER, S. (1952), Änderung der Struktur der Bakterienzelle im Verlauf der Lysozym-Einwirkung, *Schweiz. Z. Allgem. Pathol. Bakteriol.* **15,** 517–525.

TOYAMA, H., YOKOYAMA, T., SHINMYO, A., OKADA, H. (1984), Interspecific protoplast fusion of *Trichoderma, J. Biotechnol.* **1,** 25–35.

UCHIDA, T. (1988), Introduction of macromolecules into mammalian cells by cell fusion, *Exp. Cell Res.* **178,** 1–17.

USHIJIMA, S., NAKADAI, T., UCHIDA, K. (1990), Further evidence on the interspecific protoplast fusion between *Aspergillus oryzae* and *Aspergillus sojae* and subsequent haploidization, with special reference to their production of some hydrolyzing enzymes, *Agric. Biol. Chem.* **54,** 2393–2399.

USHIJIMA, S., NAKADAI, T., UCHIDA, K. (1991), Interspecific electrofusion of protoplasts between *Aspergillus oryzae* and *Aspergillus sojae, Agric. Biol. Chem.* **55,** 129–136.

VALLIN, C., RODRIGUEZ, A. R., ALONSO, E., BIRO, S. (1986), Increased oxytetracycline production in *Streptomyces rimosus* after protoplast fusion, *Biotechnol. Lett.* **8,** 343–344.

VASIL, V., VASIL, I. K. (1981), Somatic embryogenesis and plant regeneration from suspension cultures of pearl millet (*Pennisetum americanum*) *Ann. Bot.* **47,** 669–678.

VEENSTRA, A. E., VAN SOLINGEN, P., HUININGA-MUURLING, H., KOEKMAN, B. P., GROENEN, M. A. M., SMAAL, E. B., KATTEVILDER, A., ALVAREZ, E., BARREDO, J. L., MARTIN, J. F. (1989), Cloning of penicillin biosynthetic genes, in: *Genetics and Molecular Biology of Industrial Microorganisms* (HERSHBERGER, C. L., QUEENER, S. W., HEGEMAN, G., Eds.), pp. 262–269. Washington, DC: American Society for Microbiology.

VERKLEIJ, A. J. (1984), Lipidic intramembranous particles, *Biochim. Biophys. Acta* **779,** 43–63.

VERWOERT, I. I. G. S., YKEMA, A., VALKENBURG, J. A. C., VERBREE, E. C., NIJKAMP, H. J. J., SMIT, H. (1989), Modification of the fatty-acid composition in lipids of the oleoginous yeast *Apiotrichum curvatum* by intraspecific spheroplast fusion, *Appl. Microbiol. Biotechnol.* **32,** 327–333.

VIALTA, A., BONATELLI, JR., R. (1990), Parasexual analysis of *Aspergillus awamori* by using intraspecific diploids and interspecific hybrids with *A. niger, Rev. Bras. Genet.* **13,** 445–458.

VONDREJS, V., PAVLICEK, I., KOTHERA, M., PALKOVA, Z. (1990), Electrofusion of oriented *Schi-*

zosaccharomyces pombe cells through apical protoplast-protuberances, *Biochem. Biophys. Res. Commun.* **166,** 113–118.

WALLIN, A., GLIMELIUS, K., ERIKSSON, T. (1974), The induction and aggregation of *Daucus carota* protoplasts by polyethylene glycol, *Z. Pflanzenphysiol.* **74,** 64–80, 1974.

WEIBULL, C. (1953), The isolation of protoplasts from *Bacillus megaterium* by controlled treatment with lysozyme, *J. Bacteriol.* **66,** 688–695.

WESSELING, A. C., LAGO, B. D. (1981), Strain improvement by genetic recombination of cephamycin producers, *Nocardia lactamdurans* and *Streptomyces griseus, Dev. Ind. Microbiol.* **22,** 641–651.

WHITE, J. M. (1990), Viral and cellular membrane fusion proteins, *Annu. Rev. Physiol.* **52,** 675–697.

WIEGAND, R., WEBER, G., ZIMMERMANN, K., MONAJEMBASHI, S., WOLFRUM, J., GREULICH, K. O. (1987), Laser-induced fusion of mammalian cells and plant protoplasts, *J. Cell Sci.* **88,** 145–149.

WILEY, D. C., SKEHEL, J. J. (1987), The structure and function of hemagglutinin membrane glycoproteins of influenza virus, *Annu. Rev. Biochem.* **56,** 365–394.

WITTE, V., GROSSMANN, B., EMEIS, C. C. (1989), Molecular probes for the detection of *Kluyveromyces marxianus* chromosomal DNA in electrophoretic karyotypes of intergeneric protoplast fusion, *Arch. Microbiol.* **152,** 441–446.

WRIGHT, E. W. (1978), The isolation of heterokaryons and hybrids by a selective system using irreversible biochemical inhibitors, *Exp. Cell Res.* **112,** 395–407.

YAMASHITA, Y., TERADA, R., NISHIBAYASHI, S., SHIMAMOTO, K. (1989), Asymmetric somatic hybrids of *Brassica*: partial transfer of *B. campestris* genome into *B. oleracea* by cell fusion, *Theor. Appl. Genet.* **77,** 189–194.

YOKOMORI, Y., AKIYAMA, H., SHIMIZU, K. (1989), Breeding of wine yeast through protoplast fusion, *Yeast* **5,** S145–S150.

YOO, Y.-B. (1989), Fusion between protoplasts of *Ganoderma applanatum* and oidia of *Lyophyllum ulmarium, Kor. J. Mycol.* **17,** 197–201.

YOO, Y.-B., YOU, C.-H., SHIN, P.-G., PARK, Y.-H., CHANG, K.-Y. (1987), Transfer of isolated nuclei from *Pleurotus florida* into protoplasts of *Pleurotus ostreatus, Kor. J. Mycol.* **15,** 250–253.

ZIMMERMANN, U. (1986), Electrical breakdown, electropermeabilization and electrofusion, *Rev. Physiol. Biochem. Pharmacol.* **105,** 176–256.

ZIMMERMANN, U., URNOVITZ, B. H. (1987), Principles of electrofusion and electropermeabilization, Methods Enzymol. **151,** 194–221.

5 Gene Mapping in Animals and Plants

GÜNTER WRICKE

Hannover, Federal Republic of Germany

HERMANN GELDERMANN

Stuttgart, Federal Republic of Germany

W. EBERHARD WEBER

Hannover, Federal Republic of Germany

1 Introduction

Eukaryotes contain a large amount of DNA which is mainly organized in the chromosomes of the nucleus, to a small extent, in the mitochondria and, in plants, in the chloroplasts. The mammalian diploid cell DNA is about two meters long and is tightly packaged by a high level of folding into a nucleus which may be as small as 6×10^{-6} m. How this condensation is reached has not been fully explained or if there is a particular pattern which is general for all eukaryotic species. There are repeating subunits, termed nucleosomes, whose structure has been partly explained by electron microscopy investigations. Successive subunits are connected by variable lengths of DNA, called linker DNA.

A prerequisite for mapping genes on chromosomes has been – so far at least in the past – the recombination of genetic material during meiosis. This requires synapsis of homologous duplex DNAs. It involves a physical change of parts of the DNA which takes part during meiosis. A complex sequence of events must take place to guarantee an orderly exchange of material. Even with the application of recent molecular techniques the mechanism of meiosis is not fully explained. The different processes such as the formation of a synaptonemal complex, chiasma formation etc., are believed to be under strict genetic control. This permits an orderly conduct of meiosis and, in spite of a certain evironmental influence on the frequency of chiasma formation, is therefore the presupposition of a reliable estimate of the distance between genes. The closer together two genes are, the lower the recombination frequency between them. The recombination frequency can then be used to assign a linear order to the genes. The genetic distance between genes in map units is not strictly related to the physical distance since cross-over frequency can vary from one part of the chromosome to another while other regions are free or nearly free from the formation of chiasmata, as for example the constitutive heterochromatin. By *in situ* hybridization it was found that heterochromatin often contains highly repetitive DNA sequences.

The eukaryotic genome contains much more DNA in the nucleus than would be necessary to code for all proteins needed by the organism and much consists of repetitive sequences. The repetition frequency varies but, nevertheless, a certain classification can be made. Highly repetitive sequences mostly consist of relatively short sequences found in heterochromatic regions near the centromere and in the telomeres at the ends of the chromosomes. These segments are thought to stabilize special regions of the chromosome and to facilitate chromosomal rearrangement. Moderately repetitive DNA is distributed over the whole genome, located between regions of unique sequences. This part of repetitious DNA is divided into long interspersed repeat sequences, called LINEs, and short interspersed repeat sequences called SINEs.

With LINEs a genomic pattern of more than 6000 bp is assumed which occurs in variations and is distributed several thousand times throughout the genome. The members of each family are very similar but not identical. An example of this type in mammalia is the Alu family, consisting of about 300 bp in about half a million copies. The name is derived from the fact that the standard sequence is cleaved by the restriction endonuclease Alu I. The individual Alu sequences are dispersed widely over the genome.

Moderately and highly repetitive DNA which does not encode for proteins accounts in mammalia, for example, for about one third of the genome. Two thirds are unique sequences but there are long segments of non-coding single copy DNA between the genes.

Moreover, the genomic DNA for coding genes is often interrupted by introns which are excised from the messenger RNA in the so-called splicing process so that the ripe messenger RNA only consists of the series of exons.

All these non-coding regions can be up to 30 times longer than the coding segments. But all these peculiarities do not influence the process of genetic mapping which is concerned with identifying the position of genes and short segments of DNA and putting them into a linear order.

2 Markers for Gene Mapping

Mapping requires "landmarks" against which genes can be located. These points of reference are usually termed markers, and they are more or less measured by distinct phenotypes caused by chromosome structures, single genes or DNA sequences. Mapping of a gene by family studies requires to be marked via a mutation to an allele, using information from families to follow its segregation. In physical mapping, allelic variation is not necessary for mapping, but the association of a marker to a particular chromosome region should be definable.

2.1 Criteria for Markers

Basic criteria of markers which determine their application in genetic mapping using segregating populations are the number and frequencies of alleles per locus, expressed by the formula of polymorphism information content *(PIC)* (BOTSTEIN et al., 1980):

$$PIC = 1 - \sum_{i=1}^{k} p_i^2 - 2 \sum_{i=1}^{k-1} \sum_{j=i+1}^{k} p_i^2 p_j^2 \qquad (1)$$

with

k the number of alleles and

p_i, p_j the frequency of the ith and jth allele, respectively.

Further important criteria for markers are the direct identification of genotypes in the case of codominant or intermediate gene expression, and the number of loci included. Moreover, using earlier information, e.g., from comparative mapping, markers can be selected according to their distribution over the genome. Often markers are randomly selected but, in some experiments, markers at almost equal intervals could be considered. Several strategies have been reported for selection of specific markers linked to mapped loci of interest or belonging to a special gene family. In some cases, markers are available which are related to the physiology of a distinct trait.

2.2 Types of Markers

Different types of markers have been used for mapping. They include heterogeneity of diseases, chromosomes, antigens, proteins, enzymes, DNA etc. Although such distinctions are somewhat artificial, each type is characterized by special features.

Morphological Markers

Morphological markers include genes which act on traits of growth, body structure, color etc. For some plants such as maize, barley, tomatoes or peas it was possible to establish a reasonable number of morphological markers like leaf shape or type and color of flowers. However, allelic differences often exist only between breeding populations, whereas within a population of cultivated plants or farm animals the individuals have been selected for uniformity of special morphological values or qualities. Thus application of morphological traits in mapping is restricted to special cases.

The genomic location of mutations causing diseases is known by virtue of the wild-type genes. Thus genetic diseases are being localized by linkage of the gene(s) for the pathological phenotype to specific markers whose locations are known. A lot of diseases have been localized through the chromosomal assignment of the gene(s) for the deficiency. This makes it possible to compose the morbid anatomy of a genome according to the review for humans given by MCKUSICK (1986).

Markers of Chromosome Structure

Deletions or duplications alter the size of chromosome regions and sometimes become visible due to the high resolving power of modern cytogenetic techniques. There are many well documented situations of cytogenetic markers, however, they can be minute, sometimes barely visible. Again a cytogenetic aberration may affect gene loci, e.g., for proteins or enzymes, which may then be identified electrophoretically by their allozymes, gene dosage effects, or DNA sequences and used for mapping studies.

Immunological Markers

Immunological approaches identify particularly proteins. Variants of antigenic structures (alloantigens) were described in animals, e.g., blood groups, tissue cells, or soluble proteins. Family studies revealed several gene loci which have been used for mapping studies.

Enzyme and Protein Markers

Allelic variants of proteins or enzymes (allelic isoenzymes = allozymes) are important sources of biological variation. From the underlying structural genes the primary translation products are coded. Not every nucleotide change in the responsible DNA sequence results in alteration of the primary structure of a protein, and amino acid changes that do result cannot always be readily detected by biochemical techniques.

However, about one quarter of all amino acid substitutions result in a charge difference that can be detected electrophoretically (PAIGEN, 1971; HARRIS and HOPKINSON, 1974). Even though other methods will detect a higher percentage of structural gene changes, by far the most successful and widely applied method of detection has been electrophoresis followed by specific staining (the zymogram technique). For example, in the house mouse, the most thoroughly screened mammal for biochemical genetic variation, some 75% of the biochemically variant loci reported were detected by electrophoretic methods (CHAPMAN et al., 1979; PETERS, 1981). 2D-electrophoresis techniques split the cellular proteins into a great number of polypeptide spots (protein fingerprints), some of which could be assigned to one parental species and used as genetic markers.

Processing genes (PAIGEN, 1979) that code for the post-translational processing machinery in the phenotypic realization of a protein can be identified by the level and/or subcellular localization of enzyme activity as well as its electrophoretic mobility. In addition, variation in a processing gene may be expressed as differences of more than a single enzyme. For example, a gene affecting sialylation and consequently the electrophoretic mobility may influence a number of enzymes. As a general rule, alleles at processing genes influence the electrophoretic patterns of enzymes in a dominant-recessive manner as opposed to structural genes which typically express codominant allozyme patterns after electrophoresis.

Variants at regulatory loci are known and have contributed significantly to mammalian genetic maps, especially that of the mouse (PAIGEN, 1979). These variants affect the synthesis of specific proteins and are usually expressed as differences in enzyme activity. Moreover, temporal gene variation has been described which affects the relative tissue distribution of enzymes at different developmental stages.

Gene dosage effects are, at first sight, a very reasonable phenomenon: Two wild-type alleles give normal enzymatic activity, one deficient allele and one wild-type allele give half that value, and two deficient alleles give minimal activity. In some cases gene dosage effects can be demonstrated, but other examples have shown that the influence of gene regulation (compensation) can change the initially altered function.

DNA Markers

The conventional assessment of genes underlying genetic variants is deduced from a phenotype. A phenotype could be associated with the presence of a distinct allele only for a very limited number of traits, such as proteins, enzymes, or morphological traits. It is thus of central importance to use DNA sequences as probes for direct identification of genotypes. Since the DNA is present in all nucleated cells, the approach is no longer restricted to alterations of gene expression. DNA variants directly monitor alleles and are therefore more or less stable over generations as well as in cells of different tissues and stages of differentiation.

A great number of inherited differences in DNA sequence can be detected with a reasonable amount of experimental effort. The availability of DNA probes as markers dramatically increased the potential of both physical and genetic mapping (BOTSTEIN et al., 1980). With polymorphisms as markers, geneticists could study inheritance in existing pedigrees, since all individuals would be heterozygous at many loci. Extensive development of markers based on common variation in DNA sequence has produced a chromosomal overlay, within which to locate genes.

A large number of marker systems have been defined for several eukaryotic genomes. These systems reveal sequence variations in genomic DNA at the locus homologous to the probe.

The function of the DNA sequences used as probes for identification of markers may be known, or they may be arbitrary fragments. The probes can be sequences obtained by DNA synthesis, cDNA copied from isolated mRNA, or sequences cloned from gene libraries. DNA probes including long sets of selected sequences for the identification of multiple markers were described by BUFTON et al. (1986).

In most cases the identification of DNA variants starts with the generation of genomic DNA fragments by the action of restriction enzymes. These fragments can be detected, e.g., by agarose gel electrophoresis, Southern blotting and hybridization with a labeled DNA or RNA probe (SKOLNIK and WHITE, 1982). For the assay of DNA variants different approaches are available.

Restriction fragment length polymorphisms (RFLPs) reveal a main group of variants; variable restriction fragment lengths are produced by base alterations or by chromosomal rearrangements. A base-pair alteration in the DNA sequence can change the restriction site for an enzyme and thereby initiate restriction site variants (RSVs) detected by RFLPs (SKOLNIK and WHITE, 1982; JEFFREYS and FLAVELL, 1977). An alternative method for the determination of genetic variability at the DNA level is based on the use of allele-specific oligonucleotide probes. This method has been introduced for the prenatal diagnosis of genetic disease in man (CONNER et al., 1983) and has been proposed by BECKMANN (1988) for marker-based selection programs in plant and animal breeding.

Chromosomal rearrangements that change within the restriction fragments the number of elements of tandemly repeated DNA sequences, generate variable numbers of tandem repeats (VNTRs) in repetitive regions of DNA. In natural populations, large numbers of allelic variants at VNTR loci exist, and most individuals sampled will be heterozygous and therefore informative for linkage studies. VNTR regions are dispersed in all eukaryotic genomes and arise from unequal exchanges that alter the number of short tandem repeat elements in a subset (JEFFREYS et al., 1985, 1986). A hybridization probe containing the repeated core sequence (microsatellites with up to a 10 bp repeat, minisatellites with more than a 10 bp repeat) can detect many highly polymorphic loci simultaneously, and the resulting patterns are called DNA fingerprints or multilocus VNTRs. Mono-locus VNTRs can be detected by using primers for regions flanking a distinct VNTR site (NAKAMURA et al., 1987).

A further principal technique for the detection of DNA variants starts with the amplification of selected DNA intervals by using the polymerase chain reaction (PCR). Polymorphisms among the fragments generated include RSVs as well as VNTRs and are usually detected electrophoretically. WILLIAMS et al. (1990) proposed a method with random amplified polymorphic DNA (RAPD) using oligonucleotides of nine and ten bases as primers. Recently this method has been used to find additional markers in potatoes (KLEIN-LANKHORST et al., 1991), rye (PHILIPP, 1992), sugarbeets (UPHOFF and WRICKE, 1992), and *Brassica* (QUIROS et al., 1991).

Systematic screening has revealed that DNA variants are common in all eukaryotic genomes. Studies at the beta-globin locus and other loci have demonstrated that about one nucleotide site in 100 may be potentially polymorphic and about one nucleotide in 500 will actually differ between any two randomly chosen chromosomes.

2.3 Gene Clusters

Several gene families, gene clusters, or super genes have been described in animal and plant species. For example, the immunoglobulin genes consist of three clusters. Three genes are responsible for a light chain synthesis, two for the variable segments V and J, and one for the constant region which has two markers, kappa and lambda. The heavy chains are coded by four gene families of which three code for the variable region common to all immunoglobulins: the H family with a hundred genes, the D family with ten genes, and the J family with four genes. The constant region is coded by a 10-gene family. Further instances of gene clusters are the major histocompatibility complex (MHC) and the genes for coding hemoglobin chains, histones, interferons, or milk proteins.

In plants the high-molecular weight glutenins and also the gliadins are controlled by tightly linked gene clusters (GALILI and FELDMAN, 1984; ODENBACH and MAHGOUB, 1988).

These gene families have important evolutionary, functional, and ontogenic significance. Moreover, polymorphic alleles and numerous haplotypes can be used for exact mapping analysis.

3 Mapping by Recombination Rates

Gene mapping by recombination is possible with all types of markers. A prerequisite is allelic variation at the marker loci and Mendelian inheritance. Informative progenies must be in linkage disequilibrium. In this chapter mathematical methods to estimate linkage are discussed first. In the second part the creation of informative biological material is described.

3.1 Mathematical Methods

The recombination rate between marker loci is estimated in such a way that the expected frequencies of the marker classes fit the observed frequencies best. Usually maximum likelihood (ML) estimates are used (see, for example, MATHER, 1951; ALLARD, 1956; BAILEY, 1961). Given the marker classes, the observed frequency o_i and the expected frequency e_i for class i, the logarithm of the likelihood function, briefly the loglikelihood,

$$l = \sum_{i=1}^{k} o_i \ln(e_i) \tag{2}$$

has to be maximized. The e_i depends on the linkage parameters, that is the recombination rate c in the case of two linked loci. To find the maximum the derivation of Eq. (2) to c is set to zero:

$$\frac{\mathrm{d}l}{\mathrm{d}c} = \sum_{i=1}^{k} o_i \frac{\mathrm{d}\ln(e_i)}{\mathrm{d}c} = \sum_{i=1}^{k} \frac{o_i}{e_i} \frac{\mathrm{d}e_i}{\mathrm{d}c} = 0$$

The variance of c is obtained from

$$\frac{1}{\mathrm{var}(c)} = n \cdot \sum_{i=1}^{k} \frac{1}{e_i} \left(\frac{\mathrm{d}e_i}{\mathrm{d}c}\right)^2 \tag{3}$$

Linkage Between Two Marker Loci

Linkage can only be estimated from a population in linkage disequilibrium. Basic populations are the backcross and the F_2. In both cases the F_1 genotype must be heterozygous for both marker loci. In the case of backcrossing the parent used for the backcross must be homozygous at both loci for a codominant or recessive allele. Then four marker classes exist. In the case of the F_2 the number of marker classes depends on the type of inheritance and is 4 (both loci show dominance), 6 (one locus with codominance), or 9 (both loci with codominance). For the backcross and the F_2 with dominance explicit formulae can be given. To simplify notation, no difference is made between class and relative frequency observed in that class:

Backcross (AB/ab × aabb): $c = \mathrm{Aabb} + \mathrm{aaBb}$

F_2 (AB/ab selfed), dominance at both loci: The recombination rate c is estimated in three steps. First, an auxiliary variable x is calculated from the relative frequencies, then another auxiliary variable θ from x, and finally the recombination rate c from θ:

$$x = \mathrm{A.B.} - 2(\mathrm{A.bb} + \mathrm{aaB.}) - \mathrm{aabb}$$

$$\theta = (x \pm \sqrt{(x^2 + 8\mathrm{aabb})})/2$$

$$c = \begin{cases} \sqrt{\theta} \text{ for } \theta < 0.25 \text{ (repulsion)} \\ 1 - \sqrt{\theta} \text{ for } 0.25 < \theta \text{ (coupling)} \end{cases}$$

For the F_2 with dominance at both loci FISHER's product method can also be used with the same efficiency (BAILEY, 1961). For other F_2 situations Eq. (2) must be solved numerically. ALLARD (1956) prepared auxiliary tables, but generally computer programs such as LINKAGE-1 (SUITER et al., 1983), MAPMAKER (LANDER et al., 1987), or G-MENDEL (LIU and KNAPP, 1990) are used.

Biochemical and molecular markers show mostly codominance. But dominance also occurs, for example, with null alleles. The efficiency of different crossing types depends on the type of inheritance and the degree of linkage and is measured as the amount of information $I = 1/\text{var}(c)$ relative to the amount of information in case of a complete F_2 classification with 10 classes. The complete F_2 classification contains two classes of double heterozygous genotypes (coupling and repulsion type) which cannot be distinguished phenotypically, so that in a cross with two codominant markers, 9 classes can be observed. Fig. 1 shows the efficiency for the backcross and the three F_2 populations.

More Than Two Loci in One Linkage Group

In this case all possible pairs of loci have to be analyzed. As an example a linkage group ABC may be considered. Since crossover events between adjacent chromosome segments AB and BC may depend on interference, the recombination rate between A and C cannot be estimated from the recombination rates between A and B and between B and C

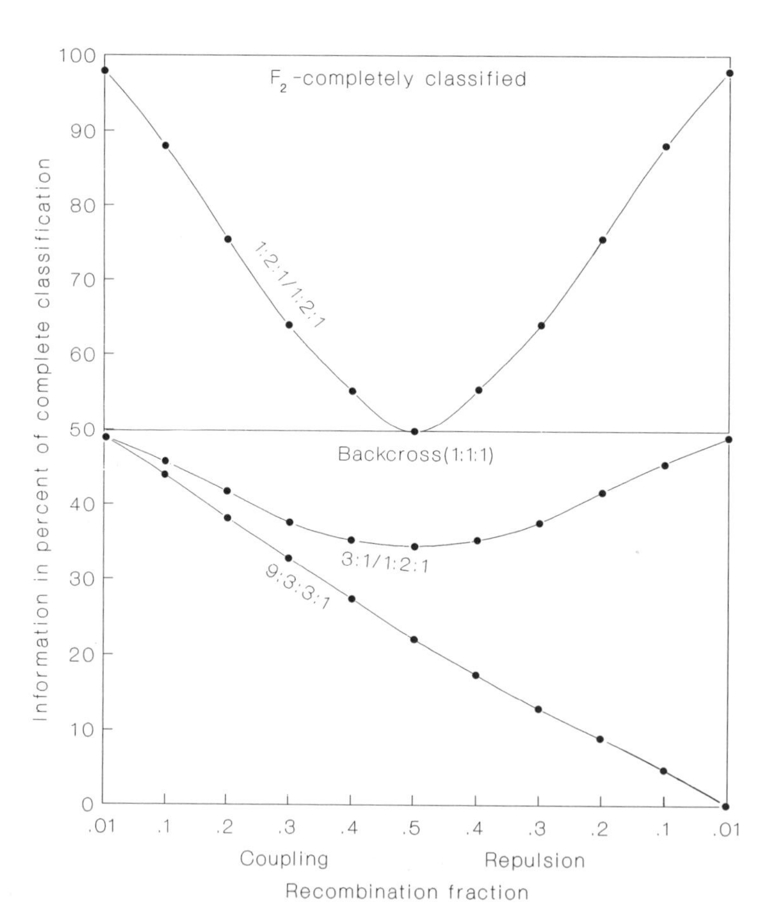

Fig. 1. Information content of the backcross and the three possible F_2 segregation types for two loci relative to the complete classification (from ALLARD, 1956, modified).

without an assumption of the degree of that interference.

For more than three loci higher-order interactions may exist, so that not only all pairwise recombination values, but also simultaneous recombination frequencies in linkage groups of 4 and more must be estimated (JONES, 1960; SCHNELL, 1963). The order of loci is indicated by the fact that the recombination value is smallest for neighboring loci. Parental types occur more frequently than exchange types. For a complete simultaneous solution, computer packages must be used (e.g., MAPMAKER by LANDER et al., 1987).

More than one crossover on a short chromosome segment is a rare event. Therefore, genetic distances based on recombination values can be added approximately if a dense map is available. In general, recombination values cannot be converted into map distances without further assumptions, since double crossovers have to be taken into accont. Without interference, HALDANE's mapping function

$$c=0.5(1-\exp(-2m)) \text{ or } m=-0.5\ln(1-2c)$$

with m = distance can be used (HALDANE, 1919).

Another formula often used in computer programs has been given by KOSAMBI (1944). He assumed that the coincidence, that is the ratio of actual double crossovers to the expected number of double crossovers without interference, itself depends linearly on the recombination rate. An overview on mapping functions has recently been given by CROW and DOVE (1990).

Disturbed Segregation Ratios for Single Loci

Selection against single alleles or allele combinations influence segregation ratios. This has no effect on the estimates of linkage in the case of codominance of at least one locus, as long as each locus is affected independently (WAGNER et al., 1992). Only the case of two dominant loci needs a modified estimating procedure since several genotypes are merged into one phenotypic marker class (HEUN and GREGORIUS, 1987).

In plants, segregation ratios of markers may be influenced by a linked incompatibility locus. Many types of incompatibility exist (DENETTANCOURT, 1977). In every case some zygotes will not be formed. For a gametophytic incompatibility system the male parent must be heterozygous for the marker and the incompatibility locus, so that the segregation ratio of the linked marker locus is distorted. Consider as an example gametophytic incompatibility at one locus S, linked with a marker locus A/a with a recombination rate c. Then in the cross $aaS_iS_j \times AS_i/aS_k$ the marker types Aa and aa occur in frequencies c and $(1-c)$, since only AS_k and aS_k gametes successfully pollinate the female parent. This is the way to localize the incompatibility locus. In rye this approach was used by WRICKE and WEHLING (1985) and GERTZ and WRICKE (1989).

Tests

The segregation data have to be tested against free recombination. The power of the test depends on the sample size. In the case of loose linkage, sample size must be large to detect linkage. The estimated recombination rate must be outside the confidence interval for free recombination ($c=0.5$). The 95% confidence interval is roughly $c \pm 2s$, s being the squareroot of var(c) from Eq. (3). The $2s$ values (twice the standard deviation for free recombination, $c=0.5$) are listed in Tab. 1.

Tab. 1. $2s$ Value for Free Recombination

Progenies	$2s$
Backcross (AB/ab × aabb)	$1/\sqrt{n}$
F_2 (AB/ab selfed), dominance at both loci	$1.5/\sqrt{n}$
F_2 (AB/ab selfed), codominance at one locus	$\sqrt{(1.5/n)}$
F_2 (AB/ab selfed), codominance at both loci	$1/\sqrt{n}$

The necessary size of the experiment to detect a recombination rate of c is then easily calculated. For example, for $c = 0.4$ the difference relative to free recombination is 0.1, and $2s$ must be less than 0.1. This yields for the backcross $n = 100$ observations and for the F_2 with two dominant loci $n = 225$ observations.

It is also possible to construct a confidence interval for the estimated c value, using $c \pm = 2s$. Since the confidence interval is not symmetric, a correction is necessary for small c values. Otherwise the upper limit is underestimated.

LOD Score

The LOD score z is defined as the $\log_{10}$ ratio of two probabilities for recombination rates, c and 0.5 (MORTON, 1955). The LOD score reaches its maximum for the ML estimate. The stronger the linkage, the higher the LOD score. The value depends on the sample size and is increased with n. Since the logarithm is taken, the z values of different samples with the same expected linkage can be added. An LOD score of 3 means that the probability of a linkage value c is $10^3 = 1000$-fold compared to the probability of free recombination. The necessary large LOD score is explained by the fact that only the two recombination rates c and 0.5 are considered.

For LOD scores, confidence intervals can be determined. If the LOD score z is calculated for different c values, z becomes a function of

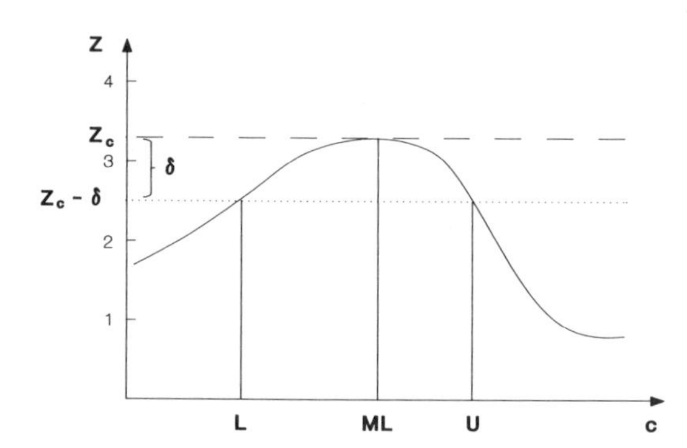

Fig. 2. Confidence interval with LOD score values. ML maximum likelihood estimate for the recombination rate, L and U, lower and upper limit, z, LOD score, $\delta = 0.834$, critical difference.

c reaching a maximum at the ML estimate c (Fig. 2). Let this value be z_c. Then, moving away from c in both directions, there will be lower (L) and upper (U) c values for which the LOD score will be $z_c - \delta$. With $\delta = 0.834$ an approximate 95% confidence interval can be constructed. Here the fact is used that z is approximatively distributed as $0.5 \log_{10} e\chi_1^2 = 0.217\chi_1^2$. The upper 5% value for χ_1^2 is 3.84 yielding $0.217 \cdot 3.84 = 0.834$.

Methods to Localize QTLs

Generally, several loci contribute to the value of a quantitative trait, and non-genetic effects modify the expression. Thus the genetic constitution of a single locus cannot be derived from the phenotypic value. However, markers can be used if specific chromosome segments linked to the marker have effects on the quantitative trait values. The quantitative trait can be analyzed for several markers separately. Such markers, in which different alleles are associated with different values for the trait, are possibly linked to one or several so-called QTLs.

For linkage between markers the backcross and F_2 are the most used generations, since the population must be in linkage disequilibrium. The F_1 genotype must be heterozygous for the markers and the QTL. The latter fact cannot be tested directly, but the probability can be increased if the F_1 genotype is obtained by crossing two homozygous genotypes differing in many marker loci. The genotypes should also differ greatly for the quantitative trait under investigation.

A QTL can be found using a single linked marker or flanking markers. Consider three loci AQB with A and B as markers, while Q stands for a QTL. Then the allele combination AB only remains together if there is a double crossover or no crossover. Consider the backcross $AQ_1B/aQ_2b \times aaQ_{22}bb$ with recombination rates c_1 between A and Q, c_2 between Q and B, and c betweeen A and B. The expected frequencies of the genotypes Q_{12} and Q_{22} within the different marker classes and without interference are given in Tab. 2.

The expected frequency of Q_{12} and Q_{22} genotypes is different for each marker class. The

Tab. 2. Frequencies of QTL Types within Classes of Flanking Markers A und B for a Backcross A Q_1 B/a Q_2 b × aa Q_{12} bb

Marker Class	Q_{12}	Q_{22}
AaBb	$(1-c_1)(1-c_2)/(1-c)$	$c_1c_2/(1-c)$
Aabb	$c_1(1-c_2)/c$	$c_2(1-c_1)/c$
aaBb	$c_2(1-c_1)/c$	$c_1(1-c_2)/c$
aabb	$c_1c_2/(1-c)$	$(1-c_1)(1-c_2)/(1-c)$

expected difference between AaBb and aabb is $(1-c_1\text{-}c_2)$ $(Q_{12}\text{-}Q_{22})/(1-c)$ and between Aabb and aaBb $(c_1\text{-}c_2)$ $(Q_{12}\text{-}Q_{22})/c$. Since c can be estimated from ordinary linkage analysis of markers, it is possible to estimate all three parameters, c_1, c_2, and $(Q_{12}\text{-}Q_{22})$ from the observed values of the quantitative trait for the four marker values. This advantage with flanking markers is lost, if only a single linked marker is available. Then only two marker classes exist. The corresponding frequencies are:

Marker Class	Q_{12}	Q_{22}
Aa	$(1-c)/2$	$c/2$
aa	$c/2$	$(1-c)/2$

The difference between Aa und aa is now $(1-2c)$ $(Q_{12}\text{-}Q_{22})/2$. Two parameters c and $(Q_{12}\text{-}Q_{22})$ must be estimated, but only one equation exists. Therefore, linkage and the difference for the trait are confounded. Nevertheless, even in this case at least the presence of a QTL can be checked, provided the difference is not zero.

Tests

While an exact identification of the phenotype of marker loci is usually possible, this is not so for the quantitative effect of the QTL. Even if a genotype is cloned, variation for the trait is still found resulting from non-genetic causes. The effects of other loci contributing to the trait under consideration are not the same for different genotypes within a marker class. They also contribute to the unexplained variation.

The variation s^2 within the marker class is a measure of the error, regardless of its origin, whether genetic or non-genetic. The difference between the means of the two marker classes in a backcross therefore has an error variance of $s_D{}^2 = 2s^2/n$ with n = number of observations within each class. The test statistic $t = D/s_D = D/\sqrt{(2s^2/n)}$ therefore can be increased by increasing the sample size. Since D and s^2 cannot be influenced, the necessary sample size necessary to detect the difference D at a significance level α is therefore $n > 2t_\alpha s^2/D^2$. For the F_2 with three marker classes the test statistic

$$F = \frac{\text{mean squares between classes}}{\text{mean squares within classes}}$$

from an analysis of variance with 2 and $3(n\text{-}1)$ d.f. is used. Here also unequal expected variances within marker classes can be taken into account (WELLER, 1986). Again, formulae exist to estimate the necessary sample size to detect a QTL. If progenies from single genotypes can be raised, as is the case for many plant species, more powerful experiments with plots and replications can be conducted so that the error variance can be drastically reduced.

Mixture Distributions and LOD Scores

Since each marker class contains more than one genotype, the frequency distribution for the trait within the marker class can be analyzed. It should have more than one mode since different QTL types are mixed, for example two types with backcrossing (see Tab. 2). The distribution therefore can be analyzed as a mixed distribution of overlapping unimodal distributions with different means. Statistical methods exist to estimate the means by ML (McLachlan and Basford, 1988). Then even with only one marker locus, both the recombination rate and the effect of a QTL can be estimated. Consider again the backcross with two overlapping distributions. In the case of free recombination, about one half of the observations must be classified into each distribution. But with close linkage one type occurs more often than the other. This allows one to estimate the recombination rate from the shape of the mixture distribution independently of the mean (see Tab. 2).

For each distance from the marker the likelihood of a QTL can be estimated. Since the likelihood can also be calculated under the assumption of no QTL, a LOD score can be given. The LOD score for the ML estimate can be used as test criterion for the existence of a QTL. This criterion leads to the same test as previously described.

With flanking markers A and B, a LOD score can be estimated for each position in the AB interval. Since the QTL is now linked with two markers, estimation is more effective. If the mode of the LOD score function exceeds a given value, the presence of a QTL is assumed. Also with this method, more than one QTL can be identified within the interval AB. The backcross is described by Lander and Botstein (1989) and the F_2 by Paterson et al., (1991).

Types of Progeny

For one marker locus the backcross yields 2 and the F_2 generation 3 classes. Single plants must be measured for the quantitative trait in both cases. But other types of progenies with 2 or 3 marker classes exist which allow one to test plots. Such types are clones and lines. Clones have the advantage that within one there is no genetic variation. This is also true for doubled haploid lines (DH) and recombinant inbred lines (RIL), which are described in more detail in Sect. 3.2. The efficacy of plot tests is thus connected with an increased effort in establishing the progenies.

The difference between the marker classes depends on the genetic value of the QTL and the types of progeny. In Tab. 3 the usual quantitative genetic model (Wricke and Weber 1986) is used. For 2 classes only one genetic parameter can be estimated so that for the backcross the dominance effect d cannot be separated from the additive effect a. This is only possible with 3 classes. Tab. 3 contains the F_2 and the F_3. Other selfing generations can also be used, but the F_2 is preferred to the F_3, the F_3 to the F_4, and so on. With derived lines plots can again be used, but an effective estimation of dominance is ruled out (Wricke and Weber, 1986). In Tab. 3, A is linked with Q_1, and the genotypic values for the QTL are a for Q_{11}, d for Q_{12} and $(-a)$ for Q_{22}.

Tab. 3. Differences between Marker Classes for Some Types of Progenies

a) 2 Classes

Type	Marker Classes	Differences
Backcross with aa	Aa aa	$(1-2c)\ (a-d)$
DH lines	AA aa	$2(1-2c)a$
RIL	AA aa	$2(1-2c)/(1+2c)a$

b) 3 Classes

Type	Differences	
	additive effect AA – aa	dominance effect Aa – (AA + aa)/2
F_2 generation	$2(1-2c)a$	$(1-2c)^2d$
F_2 lines in F_{2+t}	$2(1-2c)a$	$2^{-t}(1-2c)^2d$
F_3 generation	$3(3-2c)(1-2c)a/3$	$(3-10c+12c^2-8c^3\ (1-2c)d/3$

3.2 Biological Material

Gene mapping by recombination is only possible in segregating populations. In this section only "real" populations are considered, and mostly such populations would be families of related individuals. Two situations have to be considered: on the one hand, linkage analysis between two or more observable marker loci, and on the other, linkage analysis between marker loci and loci which do not directly cause observable gene effects.

Linkage Disequilibrium

As already stated, segregating populations must be in linkage disequilibrium. Therefore, populations with genotypic frequencies near linkage equilibrium (e.g., random mating groups) cannot be used for linkage studies. The degree of disequilibrium of a pair of loci must be known as well as the recombination value. Creation of a disequilibrium is possible by selfing, crossing, or doubling gametes of a heterozygous genotype. In all cases at least one parent must be double heterozygous for the pair of loci under study.

The probability of getting informative progenies is increased with the number of alleles existing at single loci and with intermediate frequencies, see Eq. (1), especially if a large number of loci pairs are informative and not a specific pair.

To simplify the situation, only a pair of loci is considered. In multipoint crosses the basic analysis is again the analysis of recombination rates between pairs of loci. In plants, most studies have been conducted on selfed progenies of double heterozygous genotypes AaBb or on backcross progenies of AaBb with aabb. In the case of dominance at both loci, the parent of the selfed progeny should be in coupling phase (AB/ab) to reduce the error of the estimated recombination rate, see Eq. (3). Often the phase is known in advance. The recurrent parent in the backcross must be double recessive (aabb). But most biochemical and molecular markers show codominance, and then the problem of choosing the right phase does not exist.

In plants, genotypes in advanced selfing generations can also be used. Recombinant inbred lines (RIL) are derived by the so-called single-seed descent method from an F_2 generation in self-fertilizing plants. By this method only one random seed per plant is propagated. After several generations the plants are nearly completely homozygous. Starting with the cross AABB × aabb, more parental AABB and aabb lines will be found than recombinant AAbb and aaBB lines. The expected frequencies are $0.5/(1+2c)$ for each of the two parental lines and $c/(1+2c)$ for each of the two recombinant lines (WRICKE and WEBER, 1986).

Doubled haploid lines (DH), derived from gametes of an AB/ab plant, are another type of progeny. The expected frequencies of the 4 homozygous types equal the gametic frequencies and are $(1-c)/2$ for each of the two parental and $c/2$ for each of the two recombinant lines. Since in recombination DH is restricted to one generation, fewer recombinant lines are expected than in RIL, and the recombination rate c is estimated with more precision.

In most animals, selfing is not possible but can be replaced by crossing. To achieve a good estimate of the recombination rate, both parents should be in coupling phase in the case of dominance at both loci. Parents in the opposite phase, that is AB/ab × Ab/aB, should be avoided as far as possible. Backcrosses can be made as for plants. Linkage can also be studied from half sib families, if the common parent is double heterozygous and the other parents represent a random sample of the population with known frequencies.

It is also possible to estimate the recombination rate, if two populations in linkage equilibrium with known, but different frequencies are crossed. The resulting hybrid population is not in linkage equilibrium (WRICKE and WEBER, 1986) so that after random mating a shift in gametic frequencies is expected, which can be used to estimate linkage. The estimate can be improved if the known gene frequencies in the two populations differ as much as possible. This leads to the conclusion that a cross of different inbred lines is the most efficient.

In some plant species, and in mice, near-isogenic lines (NIL) are available. They have been derived by backcrossing several generations with the same parent, retaining every time only genotypes with a target gene, for example, a gene for resistance, from the recurrent parent. These lines differ only in a small chromosome segment surrounding the target gene. The size of the segment depends on the number of backcross generations (HANSON, 1959; STAM and ZEVEN, 1981). NILs are also very useful for finding RFLP loci linked with the target gene. The use of NILs allows one to screen clones from a genomic library rapidly (YOUNG et al., 1988). Most genomic clones will not detect RFLPs, since the NILs differ only in a small segment. Those clones which are successful can be used for further studies to localize the RFLPs.

Linkage with Undefined Loci

As stated earlier, markers can be used to analyze linkage to loci with non-observable genetic effects. One important example in plants is self-incompatibility, that is when a plant cannot be successfully pollinated by its own pollen. Self-incompatibility is always connected with certain types of cross-incompatibility, causing distorted segregation ratios. Linked marker loci reflect that distortion and, therefore, allow the localization of the incompatibility loci.

Another case is an association between marker loci and quantitative trait values. Since quantitative traits are controlled by several loci (QTLs), the genotype of a specific locus cannot be derived from the observed phenotypic values. After gene mapping the linkage between a marker and a QTL can be used to estimate the genotypic value of the QTL for a known marker genotype.

To create the necessary linkage disequilibrium, the same techniques as before can be used. Since it is not known in advance which marker will be linked to the unknown locus, a large number of marker loci should be segregating. Sometimes preliminary map information can be used to select promising markers.

The efficiency of linkage detection is increased if two flanking marker loci are linked. If no recombination of the flanking markers takes place, it can be assumed that there was no crossover between any locus in the interval and these two markers, except in the rare case of a double crossover.

The use of genotypes segregating for several markers offers the possibility of detecting many loci for one or more quantitative traits simultaneously. For each segregating marker the genotypes of the progenies can be classified independently. But there is always the problem that single individuals must be analyzed. The quantitative traits can only be observed with errors caused by non-genetic effects and genetic effects of unlinked loci. To reduce this error, analysis of samples instead of single individuals is more effective.

In plants, multiplication is possible through cloning, selfing, or by the production of a half-sib family. In some species, such as barley, double haploids can be derived with reasonable effort from F_1 plants. Since they are homozygous, the production of a sample of identical genotypes can be created by selfing.

In animals, half-sib families must normally be used, but cloning techniques also exist. Early stages of embryos are divided for the production of twins or triplets. Another way is embryo cloning by the transfer of nuclei from the morula stage or stem cell cultures into enucleated fertilized egg cells.

An efficient method for detecting a QTL in the segment around the target gene is the use of near-isogenic lines (NILs). The localization is achieved by a comparison of the NILs for the quantitative trait. Masking genetic variance of unlinked loci is avoided for the most part, since the NILs differ only in a small segment.

4 Mapping by Physical Distance (Physical Mapping)

Physical maps specify the distances between landmarks along chromosomes by using, for example, chromosome structures and/or nucleotides as units. They are gaining in importance relative to genetic linkage maps, in general because physical maps describe the arrangement of genes at the fundamental level. As physical markers that can be followed gene-

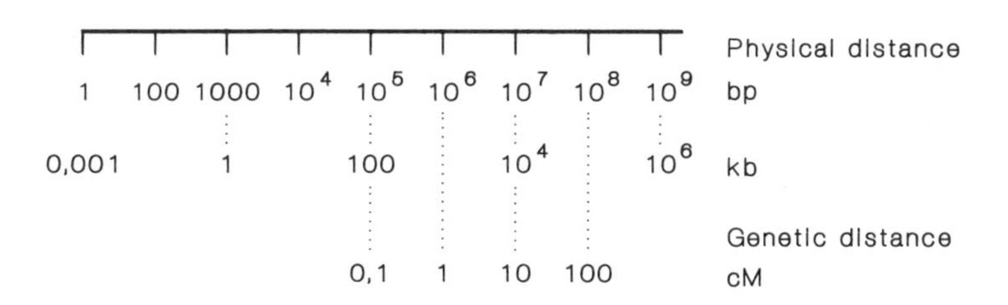

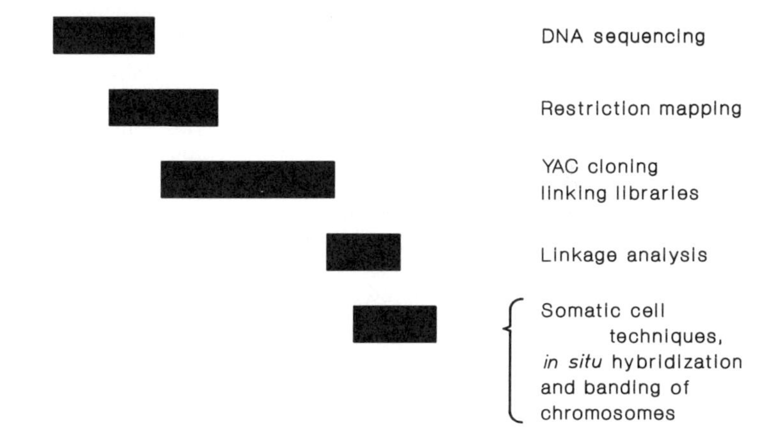

Fig. 3. Levels of analysis in mapping eukaryotic genomes. Approximate sizes of: a gene: 1000 to 10^6 bp, a chromosome band: 10^6 to 10^7 bp, a chromosome: 10^7 to 10^8 bp, a genome: $\geq 10^9$.

tically, DNA variants like RFLPs link the genetical (measured in centimorgans, cM) and physical maps (measured in base pairs, bp) at a large number of genomic sites.

The different types of physical mapping presuppose an inherent trade-off between the level of detail (resolution) in the map and the extent to which the map provides a convenient overview of the mapped markers. As shown in Fig. 3, at the "high end" of physical mapping, the banding patterns observed by light microscopy during cytological mapping allow an average chromosome to be subdivided into 10 to 20 regions, each of a length of 10 to 12 million nucleotides. The resolution of the "low-end" of the physical maps is obtained by restriction enzymes or DNA sequencing and describes distances of less than 10000 nucleotides. Thc 1000-fold gap in resolution can be bridged by cleaving the DNA into fragments of large size and then analyzing sequences that surround the cleaving sites.

Physical mapping is performed by using different methods, such as chromosome banding, somatic cell techniques, study of chromosomal aberrations, *in situ* hybridization, sorting and dissection of chromosomes, and DNA techniques.

4.1 Chromosome Banding

Banding identifies parts within chromosomes and was first used for the physical mapping of polytene chromosomes in *Drosophila* (BRIDGES, 1935). Later on, specific banding patterns were revealed also within unitene chromosomes. One technique was developed by CASPERSSON et al. (1969) with the alkylating fluorochrome quinacrine mustard. Another technique is the acetic-saline Giemsa technique (SUMNER, 1972). The different staining behavior of chromosomes after various treatments during culture and on slides creates banding structures on chromosomes which indicate different base configurations as well as histone and non-histone protein structures around the DNA.

International agreements on the general principles for chromosome banding exist for several species. A description of chromosome banding techniques possible for farm animals has been summarized by GUSTAVSSON (1990) and FRIES (1990). In humans, ZABEL et al. (1983) could assign series of genes to specific regions of chromosomes and showed up to 1000 Giemsa bands per haploid set of chromosomes. At this level of resolution one band represents on the average 3×10^3 kbp or about 3 cM.

Chromosome banding (C-banding) has provided information on the normal and abnormal localization of genes in eukaryotic chromosomes.

In plant breeding, chromosomes of one species have been used as sources of valuable disease resistance in another species. This widespread use has stimulated the analysis of the cytological structure of such chromosomes by C-banding techniques. An example is the use of chromosome 1R of rye (BAUM and APPLES, 1991) as a source of alien chromatin in bread wheat. This work has also stimulated the development of genetic maps incorporating protein and DNA markers.

Moreover, the chromosome banding has been used to analyze the localization of genes during the cell's life, especially that part of the interphase cycle during which gene expression and regulation takes place. For example, nucleolar genes coding for ribosomal RNA are usually associated with the secondary constriction of the chromosomes. These constrictions bear the nucleolus organizer regions (NORs) for which number and localization are specific. Progress in electron microscopy, autoradiography, and cytochemistry has made it possible to identify the NORs during nucleolus development and its association between structures and functions (BOUTEILLE and HERNANDEZ-VERDUN, 1979).

4.2 Somatic Cell Genetics in Mammalia

Somatic cell genetics is based on parasexual events that allow the fusion of somatic cells into hybrids, which can be considered as the transfer of the complete genome from a somatic donor to a recipient somatic cell. For such hybrid somatic cell techniques the use of mutant cell lines ensures the efficient selection of hybrid cells from parental cells (Fig. 4). For example, leucocytes have been fused with hypoxanthine phosphoribosyl transferase (HPRT)-deficient Chinese hamster cells in the presence of polyethylene glycol and subsequently grown in hypoxanthine-aminopterin-thymidine (HAT) medium. Complementing hybrid cells can then be cloned and grown to the quantities necessary for enzyme electrophoresis, DNA analysis, and karyotyping (GRZESCHEK, 1986; WOMACK and MOLL, 1986). Formation and isolation of somatic cell hybrids by selection systems and by cell cloning as well as techniques for biochemical and cytogenetic analysis of the fusion products have been greatly contributed by RUDDLE and his coworkers (BOONE and RUDDLE, 1969; RUDDLE, 1972, 1981).

As a general rule, in somatic cell genetics, chromosomes from a primary cell progenitor will be preferentially segregated when hybridized to a transformed cell line of another species. Thus, if mouse and hamster cells are combined with cells from another species, such as human or pig, there is progressive and preferential loss of non-mouse or non-hamster chromosomes (WEISS and GREEN, 1967). The following mutant cell lines were used (apart from hypoxanthine phosphoribosyl transferase (HPRT)-deficient Chinese hamster cells): thymidine kinase-deficient mouse LMTK cells, Chinese hamster lines auxotropic for phosphoribosylglycinamide synthetase (PRGS), and phosphoribosylaminoimidazole synthetase (PAIS). These rodent cell lines require different chromosomes for complementation and can therefore be used to retain selected hybrid cells.

One of the alternative fusion procedures which help to produce hybrid cells with few or single donor chromosomes is microcell fusion.

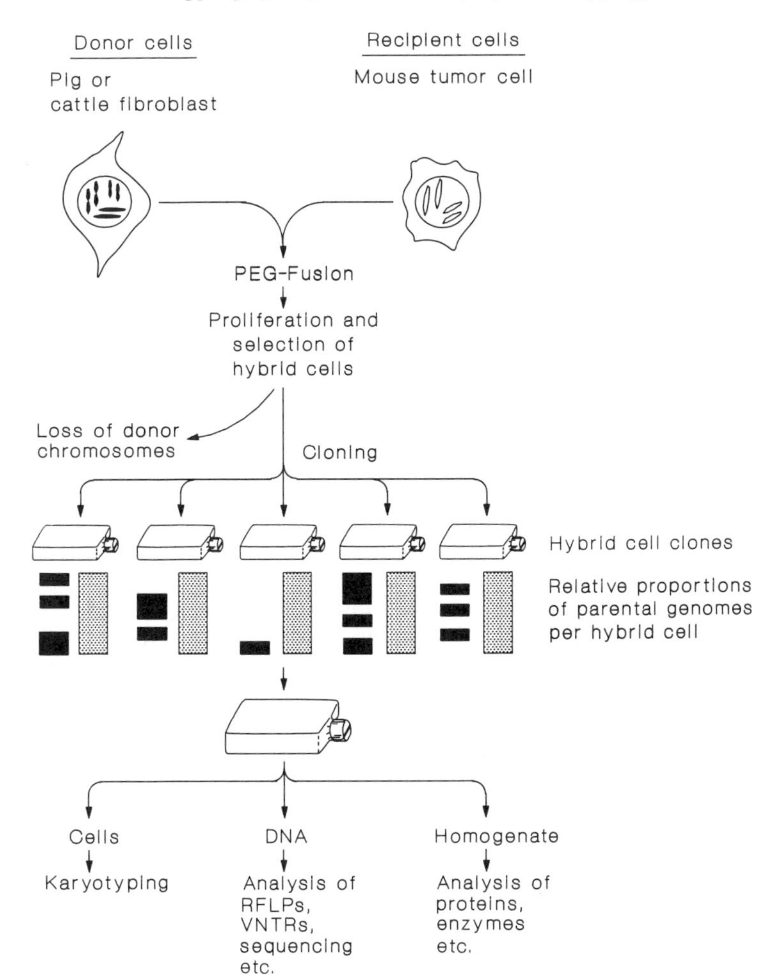

Fig. 4. Schematic diagram of construction and genetic analysis of hybrid cell clones from cattle donor and mouse recipient cells (from WOMACK, 1988, modified).

Metaphase nuclei are split into microcells which contain few or single chromosomes packed into a cell-membrane-derived envelope (EGE and RINGERTZ, 1974). After fusion with a recipient cell the production of hybrids with single donor chromosomes is obtained faster than with cell to cell fusion. Further improved techniques for producing special hybrid cells were developed by microcell-, chromosome-, and DNA-mediated gene transfer, mechanical dissection of chromosomes, and mechanical transfer of single chromosomes.

For several species panels of hybrid cell clones are generated, each retaining a partial complement of the genome of the segregating progenitor species. Two types of collection of panels can be designed: a single-chromosome mapping panel, in which the cells of each hybrid clone contain one unique donor chromosome, or a multiple-chromosome mapping panel containing clones with multiple donor chromosomes overlapping in such a way that a unique segregation pattern characterizes each chromosome. Panels containing single chromosomes can be generated by chance immediately after cell fusion or during a permanent process of chromosome segregation from established cell hybrids. Generally, a pragmatic approach has been adopted: a number of independent hybrid clones was generated and, after a period of growth and subcloning, analyzed for the presence of donor chromosomes by cytogenetic and biochemical techniques. Unfortunately, it has turned out that some donor chromosomes tend to be partially deleted or rearranged (KARMACK et al., 1984). Thus the retention of the desired chromosome intact in all cells of the clone has to be carefully controlled. The consequence of this limited stability of chromosome configuration is that so far no laboratory seems to have succeeded in producing a complete single-chromosome mapping panel.

For the karyotypic analysis of hybrid cells single chromosomes of the segregating progenitor species need to be distinguished from the hamster or mouse chromosomes. However, the accurate karyotypic definition of hybrid-cell panels is in some cases rather difficult, e.g., many of the smaller bovine chromosomes are not easy to distinguish from one another. Therefore, a special staining technique has been developed to facilitate distinction between the chromosomes of species. Furthermore, *in situ* hybridization with species-specific labeled DNA is used to mark the chromosomes in hybrid cells.

For the investigation of the chromosomal complement, the presence or absence of gene products or specific cellular phenotypes was confirmed. In most cases, protein or enzyme electrophoresis was used to determine the gene products. When a DNA probe was available, the presence of locus-specific DNA was directly determined by restriction fragment analysis (RUDDLE, 1981). The growing number of DNA probes gives a new dimension to somatic cell genetics, since any known DNA sequence can now be mapped in a panel of hybrid cells. For this purpose, hybrid cell DNA has been restricted so as to get fragments for electrophoresis. After hybridization with labeled DNA probes, the identified bands of genomic DNA allowed the determination of the presence or absence of chromosome-specific fragments. More recently, DNA sequences have been used as primers for an amplification of selected chromosome fragments. These fragments were compared for concordance with other genes and chromosomes analyzed in the same hybrid clones. Furthermore, gene technology permits engineering of DNA vector carrying prokaryotic (viral) or special eukaryotic genes which can serve as selectable markers in tissue culture cells (MULLIGAN and BERG, 1980; MILLER et al., 1983; GLASER and HOUSMAN, 1984). These DNA vectors integrate, probably at random, into chromosomes of various cell types and provide point of attachment for retaining their host chromosome or the genomic sequences neighboring the insertion site by selective media in the host cell.

Somatic cell methodology has developed during the last 20 years and is used in many fields of biological research. In genetics its major success has been in gene mapping. For example, in humans about 60% of the gene assignments achieved before 1983 were performed by somatic cell genetics. Gene mapping by somatic cell technique is based on clones of cell hybrids retaining the complete genome of one parent (recipient) and one or several chromosomes of the segregating genome (donor). The concurrent segregation of markers indicates that genes for the markers are located on the same chromosome (syntenic). The word synteny (RENWICK, 1971) expresses the fact that two or several genes are carried by the same chromosome. There are essentially two types of result: the establishment of syntenic groups and the assignment of genes or syntenic groups to specific chromosomes. Syntenic groups of genes were defined that were retained or segregated together. The assignment of syntenic groups to chromosomes can be made by karyotype analysis of the same panels of hybrid cells and the scoring of each clone for the presence or absence of each of the chromosomes. Concordance of retention of a gene or syntenic group with a particular chromosome is the basis for the assignment of that gene or group to the respective chromosome. Also, rearranged chromosomes in these hybrid cells are potentially useful in order to assign genes regionally relative to break points.

4.3 Aneuploids in Plants

Monosomics, nullisomics, and other aneuploids have been produced in plant species and are used for mapping studies. Many plant genes have been allocated using aneuploids.

For example, primary trisomics possess a chromosome additionally to the normal chromosome set, and the triple chromosomes allow dosage mapping studies. The staining intensity of protein or isoenzyme bands of electropherograms often reflects the actual dosage of the underlying gene(s). Thus, a diploid genotype homozygous for a given allele may display a band intensity of the corresponding gene product approximately double that of a heterozygote genotype which carries only a single copy of this allele. On the other hand, a trisomic line for the corresponding chromosome carries a triple dose of the locus in question. By comparing the dosage effects of several trisomics, it is possible to assign the locus to a certain chromosome.

Analysis and interpretation of results are simplified if the additional chromosome carries an allele which can be identified by the coded proteins. This is often possible when the chromosome is added to an alien genome. For several cultivated plant species full sets of addition lines exist in which all different chromosomes of one species are added to the genome of another species. For example, wheat addition lines contain the full set of 21 pairs of chromosomes of the hexaploid wheat and an added pair of intact alien chromosomes (Hart, 1979). Sugar beet (*Beta vulgaris*) is another example for cultivated plants (Lange et al., 1988; Reamon-Ramos and Wricke, 1992). By analyzing such lines, many biochemical and DNA markers can be associated with distinct chromosomes.

4.4 Use of Chromosomal Aberrations

Gene mapping by chromosome aberrations is basically the analysis of the combined segregation of a gene marker with a chromosome fragment (Renwick, 1971). Thus, individuals with deletions and duplications of small segments of chromosomes that survive as heterozygotes can be used directly for mapping purposes. In the case of individuals heterozygous for a deletion, the failure to complement phenotypically or the absence of assumed wild-type recombinants is utilized to localize the lesions. Likewise, the complementation of the homozygous mutant by a duplication indicates inclusion of the point mutant site within the duplicated segment. The combination of segregation data from chromosomal aberrations with data obtained by other mapping approaches (Ferguson-Smith and Aitken, 1982; Ott and DeMars, 1983) determines a region on a chromosome in which a given gene or a set of genes is localized and is called the smallest region of overlap (SRO). Potentially the method allows very fine mapping, and even the establishment of a gene order within a microscopically detectable chromosome band (Fig. 5).

Studies of individuals with unbalanced chromosome aberrations have greatly contributed to the human gene map, either by direct study of qualitative phenotypes or by gene dosage effects (Ferguson-Smith and Aitken 1982). Using the break points of chromosome rearrangements for the positioning of genes provides good linear maps of linked genes. However, these techniques are not an efficient method for assigning genes to specific chromosomal loci.

Regional chromosomal assignments have also been made in hybrid clones carrying broken or translocated chromosomes, and gene order can be established from pedigree analysis. Goss and Harris (1977) induced donor chromosome breakage by irradiation and subsequently transferred these cells to recipients by hybrid formation. In hybrid clones containing a single chromosome of the donor species, genes are present in the haploid state and single recessive mutations become detectable. The use of mutagenized subclones containing specific deletions in a chromosome makes possible the mapping of other genes carried on the chromosome under investigation.

4.5 *In situ* Hybridization of Chromosomes

In situ hybridization of suitable probes with fixed preparations of metaphase chromosomes is used to map single-copy DNA sequences in many species. For this purpose, Harper and Saunders (1981) improved a technique that was first developed by Gall and Pardue (1969). In principle, metaphase chromosome preparations on a microscope slide are hybrid-

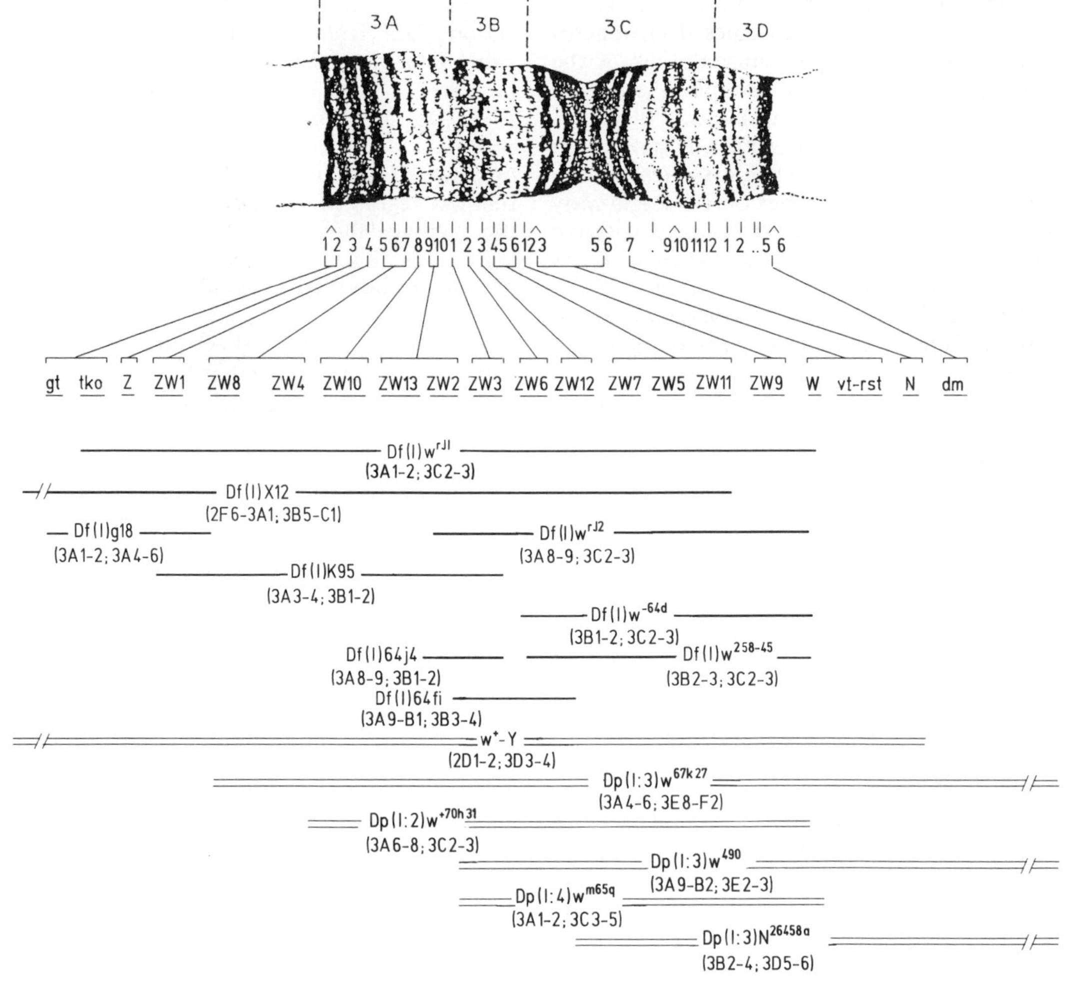

Fig. 5. Correlation of genetic and cytological maps for a region of the X chromosome of *Drosophila melanogaster,* as determined by deletion and duplication mapping (from B. H. JUDD, 1979, modified); gt, giant; tko, behavior mutant; W, white eye; Z, zeste eye; ZW 1–13, lethal mutants.

ized with highly labeled DNA probes, and the radioactive signal is detected by autoradiography. The probes are usually labeled with ^{3}H, ^{125}I, or ^{35}S, and the distribution of silver grains over a specific chromosome region is considered as a signal of hybridization. Procedures based on non-radioactive labeling were developed for a more precise localization of the probe signal on the chromosome and therefore for increased mapping resolution. For this purpose, LANDEGENT et al. (1985, 1987) demonstrated that the use of cosmid clones as probes for non-radioactive *in situ* hybridization resulted in an improved detection limit. CHERIF et al. (1989) have been able to locate, chromosomally, unique DNA sequences of less than 2 kb in length using biotinylated probes generated by oligolabeling and an improved immunofluorescence detection technique.

The method is limited to DNA sequences with a single or at least infrequent location in the genome. In most cases, coding gene sequences were used for *in situ* hybridization.

RAYBURN and GILL (1985) used a 120 bp

rye DNA probe by *in situ* hybridization to somatic metaphase chromosomes of common wheat. It was found to be a rapid, consistent, and reliable technique to detect repeated DNA sequences by *in situ* hybridization in wheat.

AMBROS et al. (1986) were the first to demonstrate the localization of a single-copy sequence in plant chromosomes by *in situ* hybridization. The same results were obtained in these experiments with either tritium-labeled or biotin-labeled probes.

An essential step in gene mapping by *in situ* hybridization is the unambiguous identification of chromosomes. *In situ* hybridization has therefore been combined with appropriate cytogenetic resolution and requires considerable cytogenetic experience. The resolution of mapping by *in situ* hybridization depends on the band resolution of the chromosomes and is in the range of 5–10 cM (RUDDLE, 1981).

In farm animal species, FRIES and coworkers used the *in situ* hybridization for gene mapping in cattle (FRIES et al., 1986, 1988; HEDIGER, 1988). Prior to hybridization quinacrine (Q)-banded chromosomes were photographed and then scored for silver grain distribution in the prephotographed spreads after autoradiography (Fig. 6).

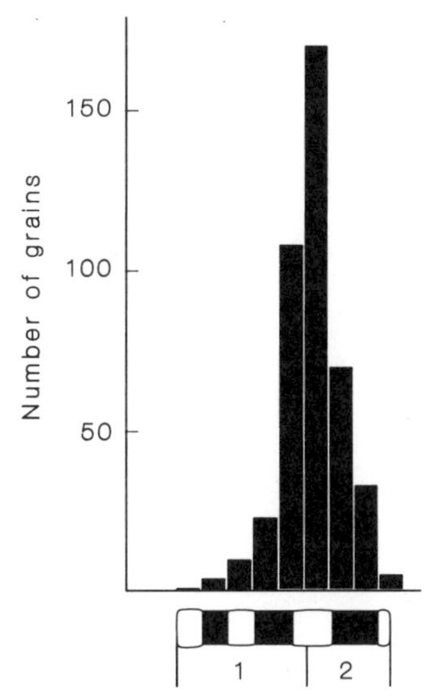

Fig. 6. Histogram illustrating the distribution of silver grains found over chromosome 20 in 63 horse metaphases after hybridization with a labeled DNA probe of porcine major histocompatibility (SLA) gene (from ANSARI et al., 1988, modified).

4.6 Sorting and Dissection of Chromosomes

The development of chromosome sorters offers a possible separation of single chromosomes. Individual chromosomes or chromosome segments can be isolated by microdissection methods via glass needles. Using an electronically controlled micromanipulator, LÜDECKE et al. (1989, 1990) dissected bands from normal 6-banded metaphase human chromosomes. Recently, HADANO et al. (1991) reported a microdissection method using a laser beam. Such microdissection methods will become more important in the future and will also include plant species.

Isolated metaphase chromosomes can be stained and sorted according to parameters onto nitrocellulose filters. A panel of filters (dot-blots) can then be used to map genes. In addition to confirming known localizations of genes, previously unassigned genes can be mapped (LEBO et al., 1984; GRUNEWALD et al., 1986). The resolving power of this direct method is limited to individual chromosomes or chromosome fragments. The equipment required for generating dot-blots is expensive and requires experienced handling. However, if dot-blots become available from central laboratories, they will be extremely useful for large scale mapping programs.

4.7 DNA Techniques

DNA techniques are of central importance for several approaches in physical gene mapping, e.g., the supply of DNA probes for somatic cell techniques and *in situ* hybridization. Moreover, mapping techniques can be applied directly at the DNA level.

The most widely used landmarks for gene mapping are cleavage sites of restriction enzymes, and the maps are calibrated by measuring the sizes of DNA fragments between identified cleavage sites (restriction mapping, Fig. 7a). A special type of physical map that pro-

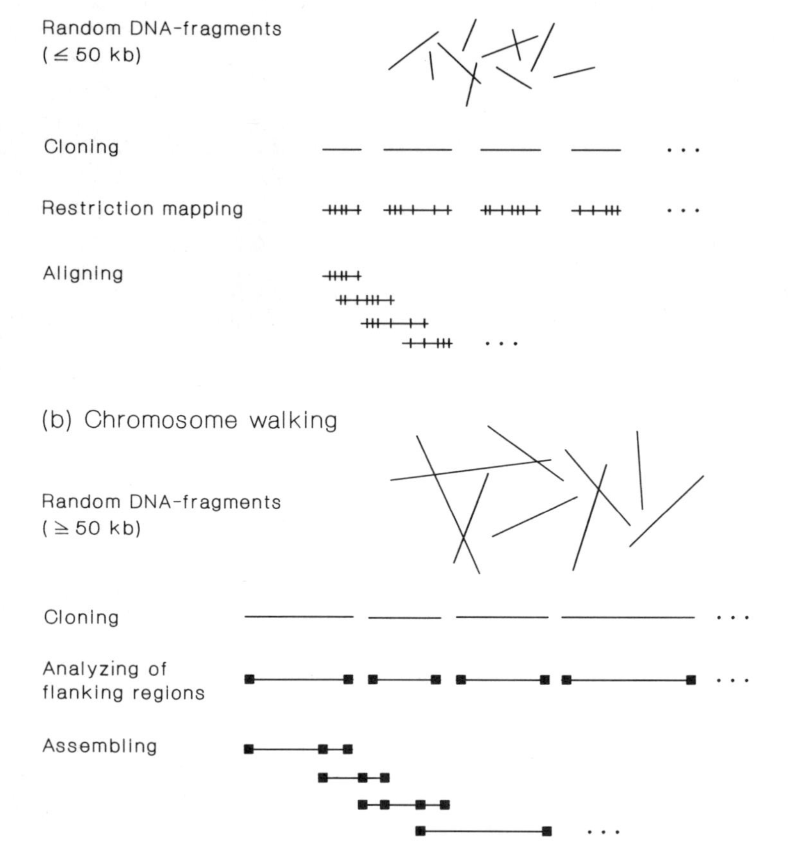

Fig. 7. DNA cloning techniques. A subset of fragments from each clone is analyzed. Patterns of restriction enzymes (a) or DNA sequences of flanking regions (b) can be used to assemble cloned fragments.

vides information on location of expressed genes is a complementary DNA (cDNA) map. The cDNAs can be obtained by using reverse transcription of mRNAs. cDNA also permits the localization of genes in other species and can be applied to genes that are expressed only in differentiated tissues and at particular stages of development and differentiation.

More extended physical maps are generated by pulsed-field electrophoresis of large DNA fragments to which multiple gene loci can be assigned. Pulsed-field gel electrophoresis allows one to separate fragments as large as 10 million nucleotides. A mammalian chromosome, on average, spans 10^5 kb of DNA, thus the order of 50 to 500 fragments per chromosome can result. Large fragments can be cloned in yeast artifical chromosomes (YACs, POUSTKA et al., 1984; COOKE, 1987). They are determined either by using distinct sets (obtained with two restriction enzymes) or by linking probes generated by cloning short DNA segments that surround each of the cleavage sites (POUSTKA and LEHRBACH, 1986). For example, THREADGILL and WOMACK (1990) have begun to apply this technique to multigene families in the cattle genome and have localized the entire casein gene family in a 400 kb segment on chromosome 6.

JUNG et al. (1990) have tried to use pulsed-field electrophoresis in mapping a chromosome fragment of sugar beet containing a gene for nematode resistance.

Methods are well established for isolation and cloning of a gene on the basis of its product. However, for many genes the products are unknown, and even for genes already identified by classical genetics, the mechanisms by which they act and their primary products are largely unknown. One attempt to clone genes without knowledge of their products is to screen transposon-mutagenized individuals and to use them for efficient mapping. The same technique can be used with T-DNA in plants. The other method for product-independent gene localization is offered by map-based cloning, often referred to as reverse genetics (ORKIN, 1986). This approach starts from already mapped RFLP markers and identifies the flanking regions by generating libraries with overlapping clones. A move or "walk" along chromosome (chromosome walking, Fig. 7b) from the known RFLP to the gene of interest has already been used to clone genes involved in hereditary diseases of humans (BENDER et al., 1979). For reasonable chromosomal distances large DNA segments were cloned in cosmid or yeast chromosome vectors as described above.

The ultimate physical map of a genome is the complete nucleotide sequence, now a realistic goal in eukaryotic genetics. By using automatic DNA sequencing about 5000 to 10000 nucleotides can be sequenced per day per person, and with multiplex DNA sequencing more than 100000 nucleotides are performed per day (HOOD et al., 1987; HEINRICH and DOMDEY, 1990). From this it follows that 100 persons need about one year to sequence the whole genome of a mammalian species, a task easily realized by cooperation of laboratories.

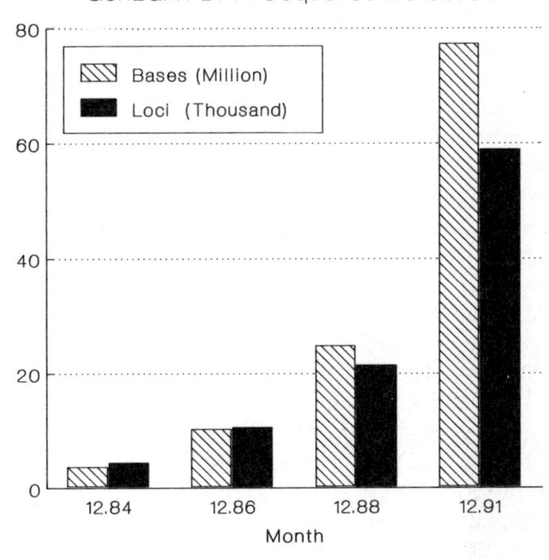

Fig. 8. Development of the GenBank nucleic acid sequence data base containing 24690876 bases from 21248 loci (Release 70.0, Dec. 1991). Collected and distributed by IntelliGenetics, Inc., in cooperation with the Los Alamos National Laboratory (GenBank 1992).

5 Maps in Animal and Plant Species

5.1 Data Bases of Genomic Structure

More than 30 million nucleotides from several eukaryotic organisms have been sequenced, and the number has doubled about every two years (Fig. 8). The increasing amount of genomic data and their complexity have resulted from special computer-based methods for data base development (SUHAI, 1990). Sophisticated data base structures available worldwide have been generated which reflect the complexity and the amount of genomic information entered. Three major organizations, the DNA Data Bank of Japan (DDBJ, Mishima), the EMBL Data Library (Heidelberg), and the GenBank (Los Alamos and Mountain View) independently collect and distribute DNA/RNA data. Each data base contains crossreferences to other relevant data bases, and since 1988 the three data collectors have merged their data bases. All of them use tables of sites and features to decribe the roles and locations of higher-order sequence domains and elements within the genome of an organism. A flexible framework is used to describe sequence regions which perform a biological function, affect or are the result of the expression of a biological function, interact with other molecules, affect replication or recombination, and have a secondary or tertiary structure etc.

For data on genome mapping the order of nucleotide blocks at the level of DNA sequences is one of the primary structural descriptions. However, each sequenced fragment is embedded in the hierarchy of "maps" of different order of resolution, and its position can be located relative to certain landmarks on the genome. The distances can be measured by different units: physical maps specify the distances between landmarks by counting the number of nucleotides, whereas genetic mapping data are described as recombination frequency per meiosis and scaled in centimorgans. At the highest level (lowest resolution) DNA loci are associated to chromosomes or represented relative to landmarks on the cytogenetical map, at the lowest level (highest resolution) the DNA sequence is located relative to restriction sites. Some physical markers, like RFLPs or altered chromosome structures, can be followed genetically; they thus link the genetical and physical maps at a potentially large number of genomic sites. This will facilitate finding the actual DNA sequence that corresponds to a gene, once such a gene is localized on the genetical linkage map and *vice versa.* Research on gene mapping is proceeding from lower to higher resolution, adding to the existing data more and more details of genomic structure and function. However, all stages of resolution of mapping presuppose an inherent interaction.

Implementation of specific data base procedures is available. Software is supplied to search the data bases, supports remote access via TELENET, and relevant subsets of the data base(s) can be captured, e.g., on personal computers.

The organization of genomic data bases reflects the fact that genome research has, until now, concentrated on the structure and function of single genes or small groups of genes. Thus, existing data banks are additive collections of information on many independent genes. However, molecular geneticists are now going to search for the logical structure of eukaryotic genomes and the interactions of different parts of a genome. The search of such highly branched structures needs new software tools like artificial intelligence programing techniques. For example, by analyzing complex genomes, data bases have to view simultaneously the data in different contexts, i.e., from structures (physical as well as genetic mapping data) and from a functional view (encoded gene products, their structures and interactions, etc.). Thus mathematical and physical methods of genomic mapping have to be combined with computational facilities in order to answer the new questions of theoretical molecular biology, and the limitations of existing genomic data bases will force intensive research.

5.2 Maps in Plants and Animals

Plants

Genetic maps are now available for several plant species including important crops. The number of analyzed markers is steadily increasing. The use of the large number of morphological markers in plants is limited. Deviating types often cannot be used in practice, if they are connected with undesired defects. Their mapping is more of theoretical interest.

Gene mapping can also include the C-banding pattern thus integrating genetic and physical maps. LINDE-LAURSEN (1979) combined banding patterns with biochemical markers in mapping studies in barley. CURTIS and LUKASZEWSKI (1991) analyzed the linkage relation between eleven C-bands on chromosome 1B in tetraploid wheat and storage protein genes Gli-B1 and Glu-B1. This method presupposes a clear banding pattern and suitable plant material which shows a sufficient C-banding polymorphism among the plants.

Tab. 4 summarizes the present stage of knowledge for marker loci in some cultivated plants. Data are taken from part 6 of *Genetic Maps* (O'BRIEN, 1990), if not otherwise stated.

For tomatoes TANKSLEY and MUTSCHLER, (1991) give a classical as well as a RFLP map. The estimated chromosome length, based on recombination values between adjacent loci, differs. But the two maps can be connected, since some isozyme markers are included in both. For maize the classical map, two RFLP maps, and a physical map exist (COE et al., 1990). The two RFLP maps share a few common loci. The classical map is related to the

Tab. 4. Present Stage of Known Markers in Some Cultivated Plants

Crop	*n*	Enzymes	RFLP Markers	Others
Maize	10	53	186[d]/263[e]	476
Wheat	21	43	179	160
Barley	7	47	155[f]	366
Rye	7	51	20[g]	117
Tomatoes	12	33	164	258
Soybeans	10	15	27[c]	198
Brassica campestris[a]	10	—	280	—
Sugar beet[b]	9	21	90	3
Garden peas	7	38	—	240

[a] SONG et al., 1991
[b] PILLEN et al., in press
[c] APUYA et al., 1988
[d] Brookhaven National Laboratories
[e] University of Missouri
[f] HEUN et al., 1991
[g] WRICKE, 1991

physical map as far as possible, and the classical map is also related to the RFLP maps. In soybeans few markers have been grouped so far.

Fig. 9 shows the map of chromosome 1 in maize, taken from COE et al. (1990), which is one of the best maps available in plants. Besides the traditional map (top: short arm; bottom: long arm) two RFLP maps developed by B. BURR at the Brookhaven National Laboratories (BNL) and D. A. HOISINGTON at the University of Missouri (UM) and a physical map exist. Some information of use in correlating the maps is available, especially for the long arm. On the right, marker loci on chromosome 1 are listed, loci which have not been fully located before. For some marker loci a region is identified, for others only the chromosome arm, and for some, only the chromosome itself.

Animals

The stages of gene mapping in some vertebrate species are summarized in Tab. 5; the most extensive data being available in humans. 4831 loci were analyzed from which 1743 loci are mapped or at least assigned to distinct chromosomes (MCKUSICK, 1990). 3641 of the loci have been analyzed by RFLPs for which 356 are already mapped (KIDD et al., 1991). More recently large gene mapping projects have been started for farm animal species providing quickly increasing information on localized genes. An example of mapping is given for the pig in Fig. 10.

5.3 Mitochondria and Chloroplasts

Mitochondria

Most mitochondrial proteins are under nuclear control. During evolution most of the genetic information supplied by the invading endobiontic predecessors of mitochondria has been transformed to the nucleus (KNOPP and BRENNICKE, 1991). Nevertheless, mitochondria, like chloroplasts in plants, also contain their own unique genetic system. In animals, mitochondrial genomes are remarkably uniform in size and structure. The circular molecules are about 15 to 17 kilobases in size. Most plant mitochondrial genomes display a complex multipartite structure. The circular master chromosome contains the entire sequence complexity. In higher plants the mitochondrial genomes are much larger in size than in animals and range from 200 kilobases to

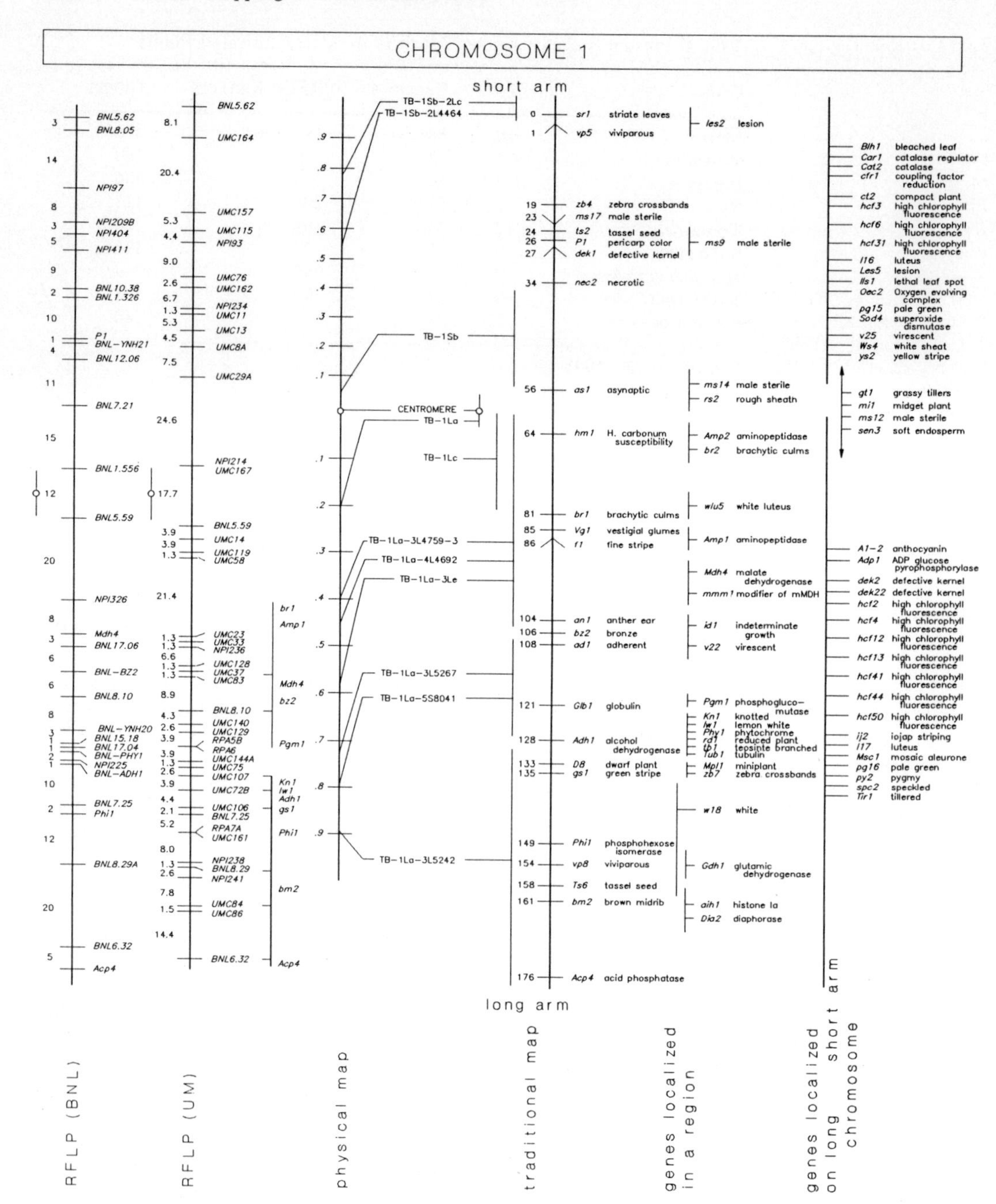

Fig. 9. Maps of chromosome 1 in maize (from Coe et al., 1990).

Tab. 5. Present Stage of Known Markers in Humans and Some Animals

Species	*n*	Assigned Markers	Linkage Groups	Enzymes	RFLP Markers
Human	23	1743[a]	24	—	356[b]
Mouse	20	1450[c]	21	—	585[c]
Pig	19	67[d]	9	25[d]	40[e]
Bovine	30	127[f]	13	8	73
Poultry	39	>200[g]	>8	—	few[h]

[a] McKusick (1989)
[b] Donis-Keller et al. (1987)
[c] Lalley et al. (1989)
[d] Echard et al. (1991)
[e] Archibald et al. (1991)
[f] Fries (1991)
[g] Bitgood and Somes (1990)
[h] Bulfield (1990)

over 2400 kilobases. In spite of this larger size, the number of genes encoding proteins is not much greater (Ecke et al., 1989).

Linkage and recombination is not restricted to nuclear chromosomal genes. Linkage groups are also found for genes in mitochondria like in chloroplasts. Therefore, mapping of chromosomes of the organelles is possible. Since restriction enzymes permit the cleavage of DNA molecules at specific sites, it is possible to identify the segments and place them in sequential order. By using different endonucleases reasonably large fragments can be obtained. Then by an "overlapping" method, the parts of an entire DNA molecule can be sequenced.

The essence of this analysis by double digestion is as follows: A certain segment of DNA is digested successively by two different enzymes and, in addition, it is simultaneously cleaved with these two enzymes. Thus each fragment of cleavage of enzyme A is digested with enzyme B and *vice versa*.

To take an example, we assume that a DNA molecule of some kilobase pairs is digested with two enzymes A and B (Fig. 11). After digestion with A we get a fragment of 2000 base pairs (bp) and after digestion with enzyme B a fragment of 2800 bp. When the segment with 2000 bp is subjected to enzyme B, it is cut into fragments of 1700 and 300. On the other hand, fragment 2800 is cut with enzyme A into segments of 1700 and 1100. Digestion by both enzymes yields segments 1700, 1100, and 300. We can conclude from this result that the single cut fragments overlap in the region of 1700 bp and construct the firtst part of a restriction map.

In a similar way the map can be continued at each end, and with such a procedure, piece by piece, the map can be extended. For several species such restriction maps already exist. Among the cultivated plants restriction maps for the complete mitochondrial genome exist already for several *Brassica* species, *Beta vulgaris,* and *Zea mays.* The genome sizes for these species are estimated by these methods to be about 200 to 570 kilobases (Lonsdale et al., 1988).

The mitochondrial genome of many mammals is well known today. Compared with that of plants, its organization is extremely compact. However, most of our current knowledge of mitochondrial genes stems from studies in *Saccharomyces cerevisiae,* which has a fivefold larger genome. This species has proved to be especially suitable for genetic analysis. Petite mutants that are defective in the ability to utilize oxygen in the metabolism of carbohydrates were studied some years ago (Ephrussi, 1953).

Base substitutions in mitochondria and in the nuclear genome supply information on the speed of evolution between the two genomes.

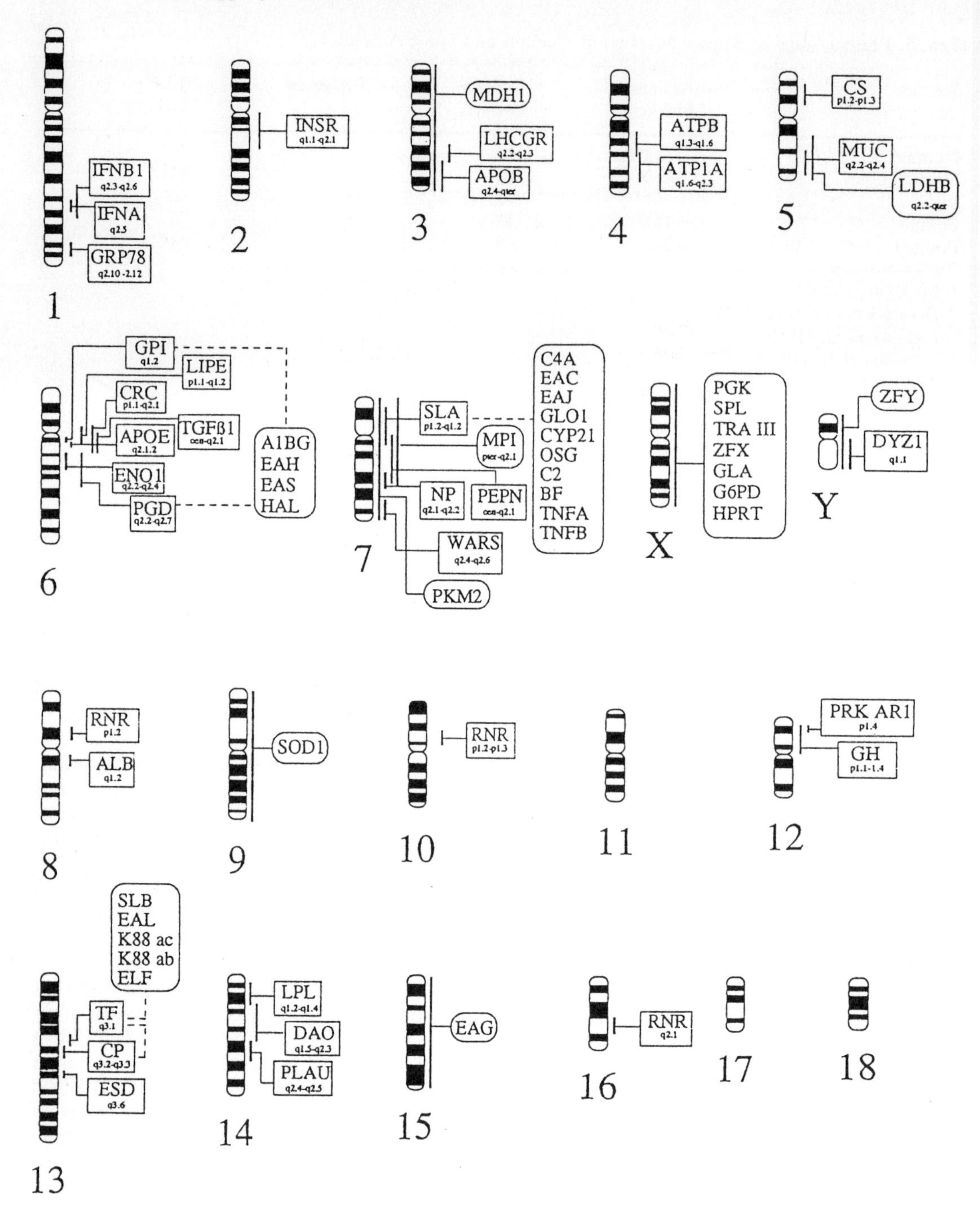

Fig. 10. Porcine gene map (from ECHARD et al., 1991).

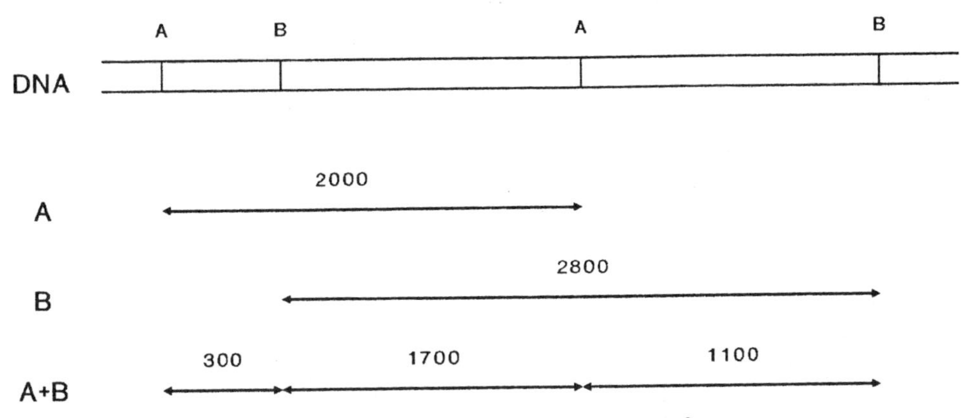

Fig. 11. Restriction mapping by double digestion of a certain DNA fragment.

Tab. 6. Restriction Enzyme Analysis of the Chloroplast Genome of *Vicia faba* with Five Enzymes (Ko, 1986)

Pst I	Xho I	Kpn I	Sal I	Sma I
35.0 (P1)	20.5 (X1)	24.5 (K1)	31.0 (Sla&b)	50.0 (Sml)
20.0 (P2)	19.5 (X2)	23.5 (K2)	16.5 (S2)	24.0 (Sm2)
15.5 (P3)	16.0 (X3)	20.0 (K3)	12.5 (S3a&b)	20.0 (Sm3)
14.5 (P4)	15.0 (X4)	12.5 (K4)	10.0 (S4)	16.0 (Sm4)
13.0 (P5)	14.0 (X5)	10.5 (K5)	8.5 (S5)	8.0 (Sm5)
10.0 (P6)	12.0 (X6)	8.1 (K6)	1.6 (S6)	4.4 (Sm6)
6.3 (P7)	10.5 (X7)	5.0 (K7a&b)		
5.2 (P8)	4.2 (X8)	4.3 (K8)		
1.8 (P9)	3.2 (X9)	4.1 (K9)		
1.1 (P10)	3.0 (X10a&b)	3.0 (K10)		
0.6 (P11)	1.3 (X11)	1.1 (K11)		
	0.6 (X12)	0.6 (K12)		
123.000	122.700	122.200	123.800	122.400

It has given interesting results. The mitochondrial genomes of higher plants evolve much more slowly than the nuclear genome, when the primary sequences are considered. This is not the case in mammals. Here the mitochondrial genomes evolved much faster in the past than the nuclear genomes which in turn seem to have the same rate of evolution as plant nuclear genomes. From considerable data WOLFE et al. (1987) calculated that the base substitution rates of plant and mammalian mitochondrial genomes differ by more than one order of magnitude.

In contrast, the mitochondrial genes that have been localized on physical maps show no conservation of their arrangement. Two *Brassica* species, *B. campestris* and *B. hirta,* are good examples (PALMER and HERBORN, 1987). The two mitochondrial genomes are similar in size and primary sequence. They differ, however, significantly in the arrangement of the genes. No less than ten independent inversions can be recognized between these two species.

Recombination seems to be a common process in mitochondria. In plants, cytoplasmic male sterility is of considerable practical importance for hybrid seed production, and in some cases it has been shown that male sterility seems to be the consequence of aberrant recombination events resulting in chimeric genes and novel polypeptides.

Chloroplasts

In addition to mitochondria in plants there is another main energy transducer, the chloroplast. As in mitochondria it retains control over synthesis of several of its proteins by its own unique DNA and RNA molecules and protein synthesizing machinery. Chloroplast DNA was first detected by SAGER and ISHIDA (1963) in *Chlamydomonas.* The chloroplast chromosome in most vascular plants ranges in size from 120 to 160 kilobase pairs (PALMER, 1985), and it consists of a single circular DNA molecule. The chloroplast genome is mainly maternally inherited. For several species a biparental inheritance is known, and in only a few cases has paternal inheritance been observed. Recently the nucleotide sequence of the entire circular DNA molecule of the liverwort *Marchantia* (OHYSAMA et al., 1986), tobacco (SHINOZAKI et al., 1986), and rice (HIRATSUKA et al., 1989) have been completely sequenced. For many other plant species, physical maps of important regions of the genome have been constructed, especially of the gene for the large subunit of ribulose biphosphate carboxylase. The number of genes has been estimated to be of the order of 150. The coding regions on plastid DNAs seem to be highly conserved (HENNIG and HERRMANN, 1986; HERRMANN, 1992). Gene probes from most species can therefore be used for cross-hybridization with the chloroplast DNA of other nonrelated species. By this method a simple mapping of interesting DNA regions is possible which in turn can be important for studies of gene expression. Restriction endonuclease pattern also reflects genome size. For the chloroplast genome of *Vicia faba* a restriction enzyme analysis has been made with five enzymes (KO, 1986). The results are given in Tab. 6.

Restriction fragment length polymorphisms of chloroplast DNA provide also a good way to measure evolutionary distances. Many recent publications have shown that chloroplast RFLP analysis can be used effectively for studying plant phylogenetics and systematics (HAGEMANN et al., 1989) and are therefore also very important for plant breeding research.

An important result of these studies was that the chloroplast genome evolves at a lower rate than the nuclear genome. This indicates a stringent selective pressure on some functional characters of the chloroplast genome. The remarkably constant length of chloroplast chromosomes from different higher plant species points in the same direction.

The complete genetic map of the *Nicotiana tabacum* chloroplast genome is given in Fig. 12 taken from WHITTIER and SUGIURA (1992).

5.4 Gene and Chromosome Homologies between Species

During evolution a diversity of organisms has been produced by genetic mechanisms in interaction with forces of the environment. A wide variety of phenotypes was created, and identical genes in the genetic backgrounds of different species can differ somewhat less in their phenotypic effects. Even phenotypic similarity would be no guarantee that the genes involved are homologous, which makes genetic comparison between species very difficult.

However, comparative considerations of the genetic material itself allow one to analyze diversity at the primary level of information. Differences in the organizational construction of genetic information have become accessible through different methods and techniques, especially after protein variants and DNA markers became standard tools for gene mapping. As the pace of gene mapping increased, comparative gene mapping gained in importance, and homologies were observed for chromosome banding patterns, for synteny and gene linkage maps as well as for gene structure and function.

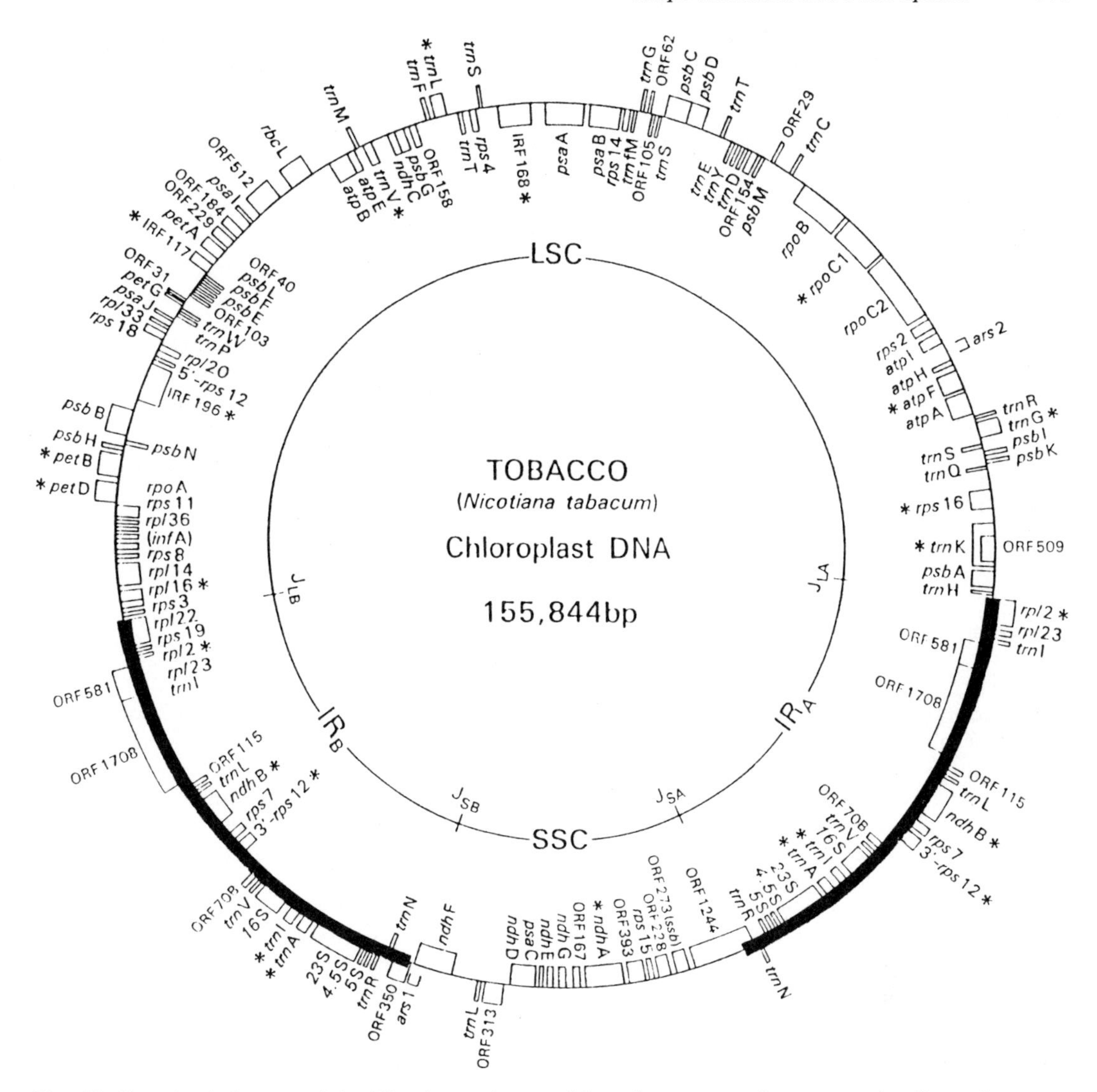

Fig. 12. Genetic circle map of the *Nicotiana tabacum* chloroplast genome drawn to scale. Genes shown on the outside of the circle are encoded on the A strand and transcribed counter-clockwise. Genes on the inside are encoded on the B strand and transcribed clockwise. Asteriks denote split genes. LSC, large single copy region; SSC, small single copy regions; IR, inverted repeat (from WHITTIER and SUGIURA, 1992).

5.4.1 Chromosome Homologies

For comparison of related species chromosome morphology was used primarily to indicate evolutionary developments. Similarities were found for size, for position of centromer (arm length ratio), for regions and number of nucleolar organizers, and for replication patterns (STRANZINGER, 1990). A more precise identification of single chromosomes became possible through different techniques that result in banding patterns (CASPERSSON et al., 1967, 1969).

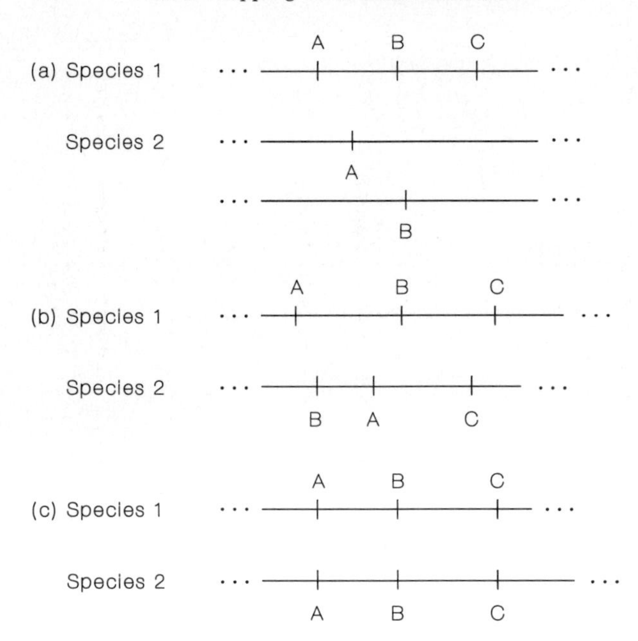

Fig. 13. Characterizing homology in chromosome segments (according to NADEAU, 1989).
(a) Homology segment: chromosome segment marked by at least one gene that has been mapped in two or more species.
(b) Synteny conservation: two or more pairs of homologous genes of any order on the same chromosome in two or more species.
(c) Linkage conservation: conservation not only of synteny but also of gene order.

An example of a comparison of chromosome morphology between species was demonstrated in Bovidae by WURSTER and BENIRSCHKE (1968). Only two autosomal chromosomes showed significant banding differences between cattle and sheep, the remaining chromosomes were very similar despite centric fusions in ovine chromosomes.

In contrast to the widely conserved linkage groups on the X chromosome, the banding patterns of X chromosomes of different mammalian species are not generally comparable. For example, between humans and pigs the C-banding shows a quite similar pattern, but among the related species of the Bovidae relatively large differences occur in the X chromosome banding (STRANZINGER, 1990).

5.4.2 Synteny and Gene Linkage Homologies

The recent introduction of biochemical and molecular methods for characterizing and mapping genes has dramatically increased the number of homologous genes mapped in more than one species. Analyses of recombination frequencies, somatic cell hybridization techniques, and the *in situ* hybridization of chromosomes were used to identify homologous areas of chromosomes. It has been of primary importance that for hybridization a great number of distinct DNA probes were used in different species and did show cross-hybridization of chromosomal material.

By applying the new techniques, large regions of chromosomes between species were found to be more or less homologous. Criteria for defining homologies between genes in different species are provided in the *Report of the Comparative Mapping Committee* (LALLEY and MCKUSICK, 1989). As illustrated in Fig. 13, the chromosomal distribution of homologous genes between different species can be characterized in three ways, depending on the number and order of genes in each segment.

Comparisons between syntenies and gene linkage groups of different species were made available in several reviews (O'BRIEN, 1990; WOMACK and MOLL, 1986; FRIES et al., 1989). For example, comparative mapping data are presently available for more than 25 species of mammals and the number of genes mapped in more than one species ranges from

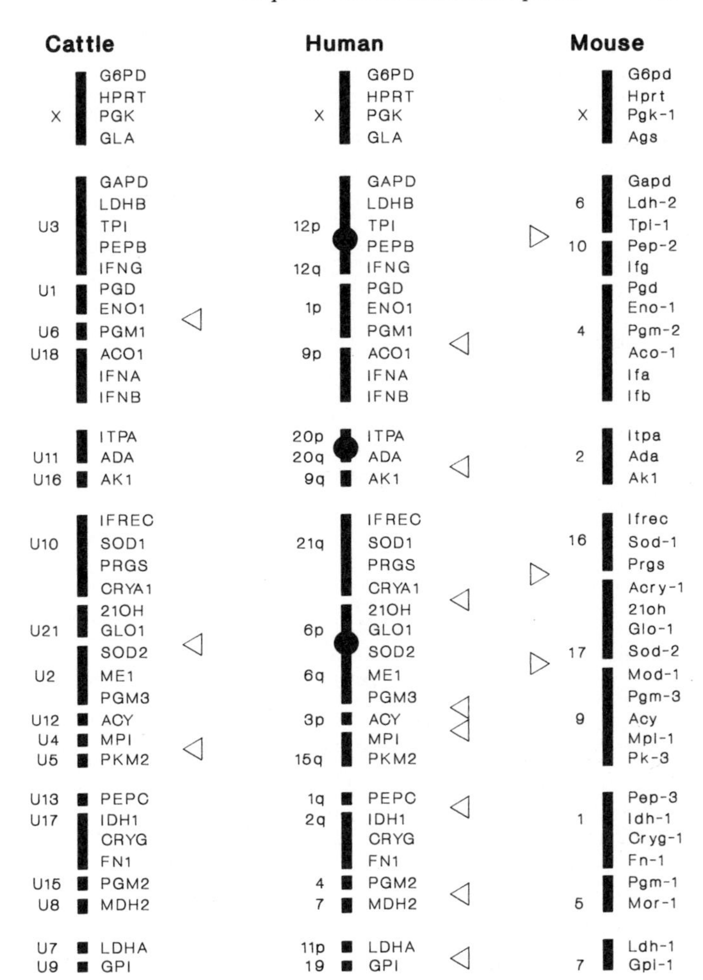

Fig. 14. Comparative map of genes that have been mapped in cattle, humans, and mice. The chromosomal location of the human homolog is given after each mouse gene. Only synteny and linkage assignments are included. Genes that have been assigned to a chromosome but not located within the linkage map are listed below each chromosome. Conserved linkages are highlighted (from NADEAU, 1989).

less than 20 to more than 250 for mouse and man.

Comparative mapping has proceeded particularly well in the comparison of human, cattle, and mouse gene maps (see Fig. 14). The human homologies of nearly half the length of the mouse recombination map are now known (NADEAU, 1989; SEARLE et al., 1989; LALLEY et al., 1989). Homologous autosomal genes are located on all autosomes. Each mouse chromosome has homologies with two to seven human chromosomes, and the known length of 26 conserved segments ranges from <1 cM to >30 cM with an average length of about 10 cM. Similarly, each human chromosome has homologies with up to six mouse chromosomes. So far, for instance, all known homologies of genes on human 17 are on mouse chromosome 11, but mouse 11 has homologies with five other human chromosomes (BUCHBERG et al., 1989). While the total number of genes mapped in mouse and man continues to increase exponentially, the synteny and linkage assignments approach saturation (Fig. 15). Extrapolation of the curve shown in Fig. 15b suggests that the number of conserved segments may be less than 80.

In plants, homology between potatoes and tomatoes has been analyzed by BANIERBALE et al. (1988). The comparative genetic map is shown in Fig. 16.

Another example of an indication of con-

a

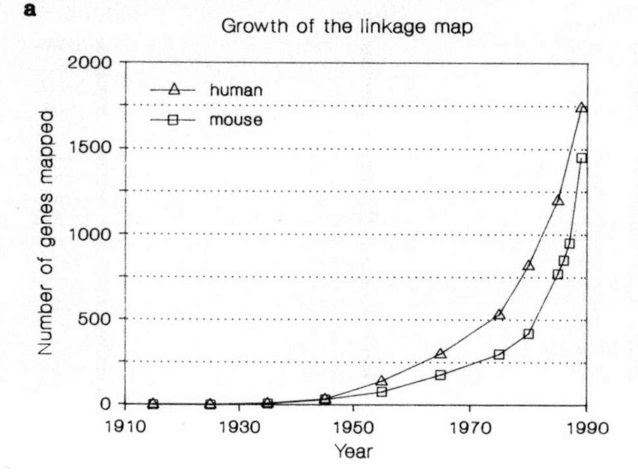

b

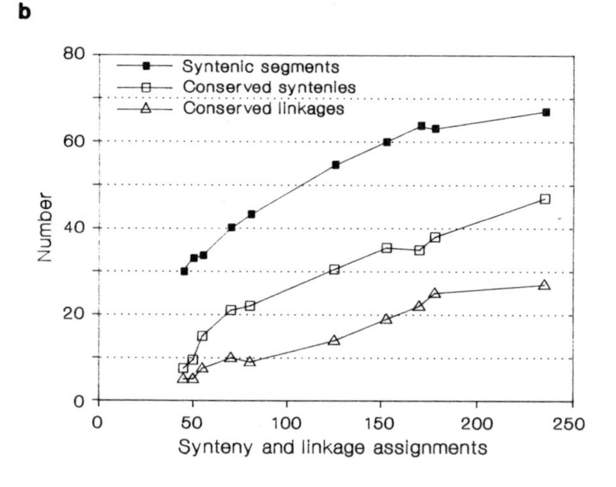

Fig. 15. Gene mapping data in mouse and human.
(a) Linkage mapping data (HGM 9, 1988).
(b) Number of conserved segments as a function of the number of genes in the comparative map (from NADEAU, 1989; DONIS-KELLER et al., 1987; O'BRIEN, 1990).

served linkage groups in plants was found in grasses. CORNISH et al. (1980) in *Lolium perenne* and FEARON et al. (1983) in *L. multiflorum* found linkage between one of the two loci which determine the genetic incompatibility system and a glucose phosphoisomerase locus (Gpi-2). In rye, *Secale cereale,* the S-locus for incompatibility was also located on the chromosome which bears Gpi-2 (WRICKE and WEHLING 1985). LEACH and HAYMANN (1987) thereupon studied the species *Alopecurus myosuroides, Phalaris coerulescens, Festuca pratensis, Holcus lanatus,* and also *Secale cereale* and found, in all cases, linkage with an incompatibility locus. The recombination values varied between 0.11 and 0.30.

The linkage, therefore, has been demonstrated for seven species of the Poaceae suggesting the conservation of a common chromosomal element in the evolution of this tribe. Especially for the subtribe Triticinae, many investigations on isozymes and seed storage proteins have shown that there is a highly conservative synteny of genes. (For literature see HART and GALE, 1990.)

For many mammalian species it has been found that the X chromosome carries the loci for distinct enzymes such as glucose-6-phosphate dehydrogenase (G-6-PD) or hypoxantine-guanine-phosphoribosyltransferase (HPRT). Between human and mouse at least 42 genes are known to be homologous and assigned to

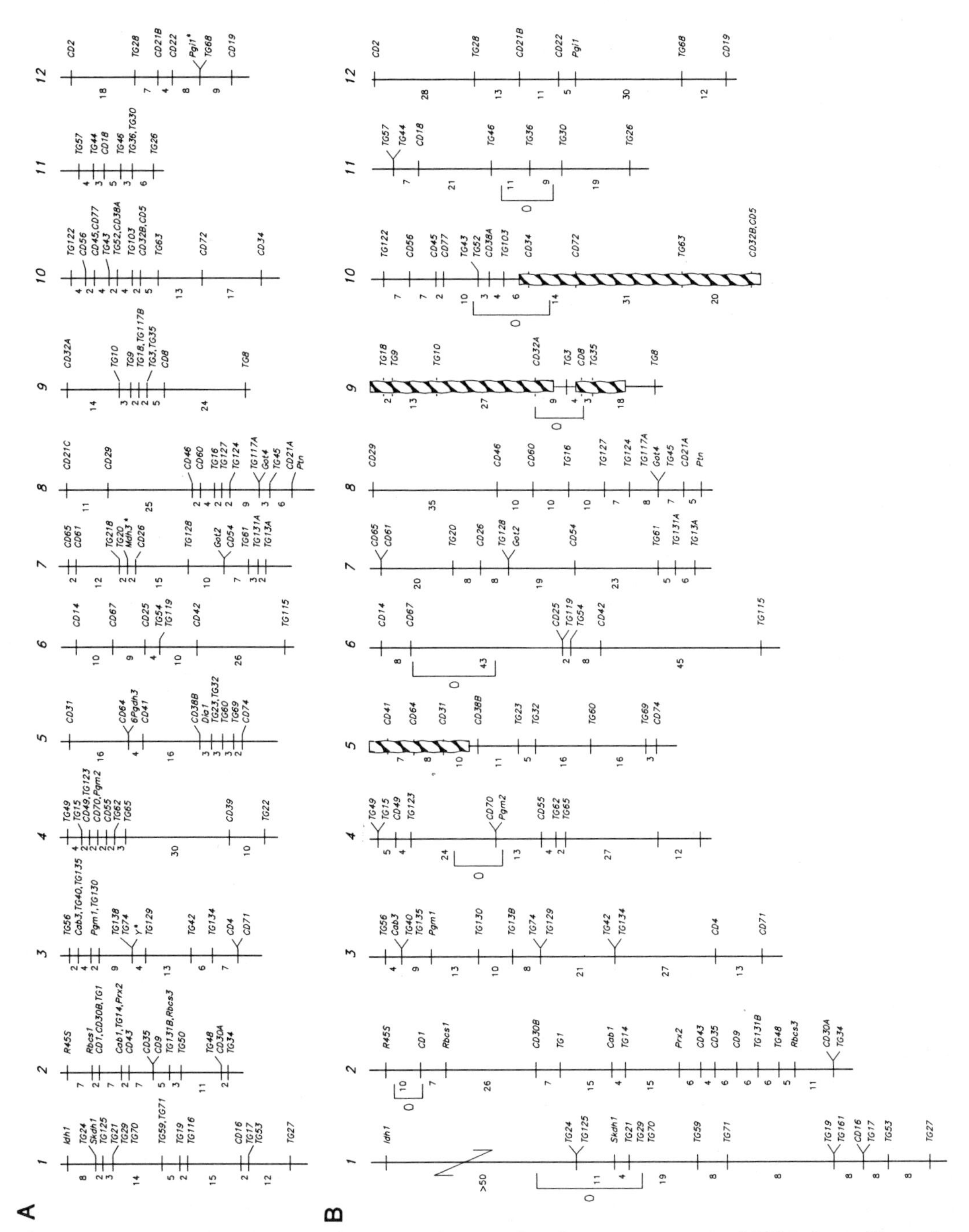

Fig. 16. Comparative genetic maps of potato and tomato based on a common set of DNA clones. Two point linkage distances between adjacent markers are based on KOSAMBI (1944) map units.
A, Potato: Asterisks indicate markers mapped by segregation of alleles from *Solanum phureja*.
B, Tomato: Hatched boxes represent intervals in which the order of markers is inverted between the two genomes. Brackets indicate approximate positions of centromeres.

the X chromosome (LALLEY et al., 1989), and for the Y chromosome 5 homologous genes are known between human and mouse. The conservation of genes on the X chromosome in mouse and man supports Ohno's law that chromosomal rearrangements involving the X chromosome and autosomes are strongly selected against (OHNO, 1967). Similarly, complete synteny conservation is found for all genes mapped on the Y chromosome.

Recombination distances between homologous genes within conserved segments can vary considerably between species (NADEAU, 1989). It remains to be determined whether differences in recombination frequency reflected intervening chromosome rearrangements or differences in the likelihood of recombination along segments of the same physical length.

5.4.3 Homologies of Gene Structures and Products

Studies of DNA sequences in several organisms have yielded similarities between genes and their products. LALLEY and MCKUSICK (1985) recommended the following criteria for identifying gene or gene product homologies between species:

- Similar nucleotide or amino acid sequence
- Similar immunological cross-reaction
- Formation of functional heteropolymeric molecules in interspecific somatic cell hybrids in cases of multimeric proteins
- Similar tissue distribution
- Similar developmental time of appearance
- Similar pleiotropic effects
- Identical subcellular locations
- Similar substrate specificity
- Similar response to specific inhibitors
- Cross-hybridization to the same DNA probe.

Eukaryotic genes are composed of a mosaic of several elements. Highly conserved gene elements were found in almost all studies. Some are present for functional reasons, e.g., the CAT and TATA box regions or the polyadenine tail encoding sequences. Others have a common evolutionary history, e.g., some exons which occur in several genes (e.g., MHC, some hormone genes, globin b genes).

On the level of gene products the homologies and differences are shown by electrophoresis, analysis of primary amino acid sequences, or effects of mutants on malformation syndromes. For example, over 80 malformation syndromes with mouse/man homology were reviewed by WINTER (1988). By identifying single gene defects causing similar developmental abnormalities in mouse and man, comparative gene mapping and molecular studies were performed. Examples include the testicular feminization syndrome, due to a mutation in the androgen receptor (LYON, 1988), and the hemoglobin mutants, in which distinct molecular changes produce comparable physiological effects (PETERS et al., 1985). An important application of comparative maps involves predicting the location of unknown loci in one species given their location in other species. An example is the retinoblastoma gene (RB 1), which is closely linked with the esterase D gene (ESD) on human chromosome 13q. The location of the murine homolog of ESD, Es 10, on chromosome 14 led to the prediction that the murine homolog of RB1 should also be on chromosome 14. Recent gene mapping studies confirmed this prediction (STONE et al., 1991). MCDONALD et al. (1988) mapped 12 murine homologs of markers for the Alzheimer's disease and Down syndrome loci on chromosome 21q22 in man. Seven of the murine homologs were located on chromosome 16, and there was no evidence that the gene order was disrupted. Four anomalous genes were located on mouse chromosome 10 and define a new conserved synteny.

These studies provide an excellent example of the way in which comparative mapping may help to elucidate the genetic basis of complex traits. In these cases, knowledge that mouse and/or human genes were appropriately located was important in finding the homology in other mammalian species, and genes of further homologous syndromes will no doubt be identified in this way in the future. Thus, it is very important to compare the map locations of genes responsible for phenotypic syndromes to identify those that may be true homologs and to predict linkage of those that have not been mapped in other species under investigation.

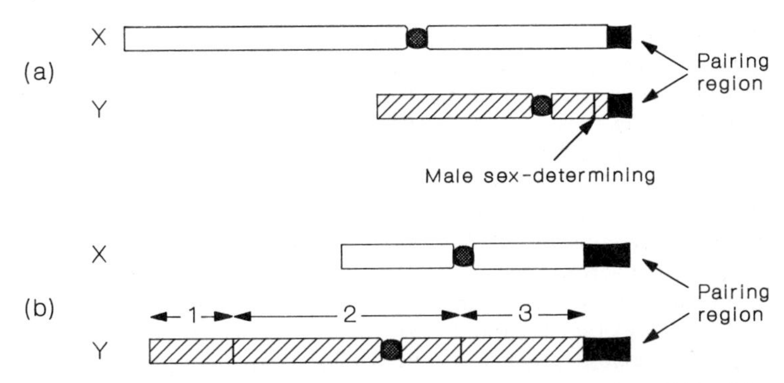

Fig. 17. Characteristic structure of sex chromosomes in eukaryotic species (CHARLESWORTH, 1991). The filled circles represent the centromeres. The clear regions of the X chromosome contain active genes that are absent from the hatched regions of the Y chromosome, with which it does not undergo recombinational exchange.
(a) Human X and Y chromosomes (ELLIS et al., 1989). The male-determining genes of the Y are located on the left of the pairing region.
(b) X and Y chromosomes of the red campion *Silene dioica* (WESTERGAARD, 1959). Region 1 carries gene(s) suppressing femaleness, region 2 promotes maleness, and region 3 promotes anther development. The clear region of the X promotes femaleness.

5.4.4 Evolutionary Aspects of Gene Arrangement Homologies

Gene and DNA sequence evolution seems to be influenced by their genomic location. Thus the chromosomal distribution of homologous genes may answer many important questions concerning genome organization and evolution (NADEAU, 1989):

- What are the relative contributions of chance and necessity in shaping genome organization and in determining the pattern of synteny and linkage conservation?
- Is conservation a consequence of the function of the gene within a given segment or of the distribution of sites at which chromosomes are most likely to rearrange?
- What is the relative importance of physiological factors, such as efficiency of DNA repair, and of population factors, such as generation time and population size, in determining the rate of disruption?

Recently, the mammalian genome has been described as a mosaic of long “isochores” (perhaps 200–1000 kb in length) of varying base composition (SHARP, 1991). Within each isochore, the G+C content is comparatively homogeneous. Isochores were found in warm-blooded vertebrates as well as in certain monocots, i.e., the grasses. They may be related to cytological banding patterns and seem to have been handled more or less as a unit during evolution.

The distribution of segment lengths is consistent with the argument that chromosomal rearrangements are randomly distributed within the genome (NADEAU, 1989). The lengths of conserved segments have also been used to estimate the number of linkage disruptions that have occurred since the lineages leading to the examined species diverged. This rate of disruption, about 1 per million years, is quite slow even in comparison with the fixation rate of about 1700 per million years for Robertsonian translocations (CAPANNA et al., 1976). For autosomes, the lengths of conserved segments suggest that about 138 (SD, 32) linkage disruptions have occurred since mouse and man diverged (NADEAU, 1989).

The evolution of chromosome architecture has been extensively studied for the sex chromosomes (CHARLESWORTH, 1991). Structurally distinct sex chromosomes are the most familiar mode of genetic sex determination and have evolved independently in many different taxa. Sexual reproduction with some degree of gamete dimorphism is nearly universal among eukaryotes, and the most familiar form of genetic sex determination involves structurally distinct sex chromosomes, called X (Z) and Y (W) (Fig. 17).

Characteristically, there is little or no recombinational exchange between X and Y chromosomes. For example, in mammals the order of genes on the X chromosome has been disrupted several times, even though synteny has been conserved. LYON (1988) suggested that as few as two inversions could account for these differences, and DAVISSON (1987) argued that at least one transposition in addition to the inversions may be involved. The absence of genetic exchange is associated with a lack of genetic homology between the relevant regions of the X and Y chromosomes. In mammals, the pairing region at one end of the Y contains loci homologous with loci on the X (Fig. 17), whereas the rest of the chromosome largely lacks functional loci apart from a small number required for male sex determination and fertility. Much of the Y seems to consist of highly repeated DNA sequences similar in nature to transposable elements and to satellite sequences of the type associated with centric heterochromatin. In many species, the repeated DNA sequences are variable in size, suggesting that these sequences are of no functional significance. Dosage compensation is a corollary of the lack of functional loci on the Y and the resulting hemizygosity of the X in the heterogametic sex (CHARLESWORTH, 1991).

There are numerous examples of intermediate levels of structural and genetic differentiation between X and Y chromosomes. There are, however, species which are perhaps at a very early stage of evolution of the dimorphism of the sex chromosomes; in plants *Asparagus officinalis* is an example. A clear 1:1 segregation of male and female plants is found, but in spite of intensive efforts no chromosomal differences could be found between X and Y. Finally by a trisomic analysis it was possible to find that chromosome 5 is the one which must bear the sex determining genes (LÖPTIEN, 1979). On the other hand, some plant species have well-differentiated sex chromosomes as, for example, *Melandrium album* (WESTERGAARD, 1959).

It is therefore probable that advanced systems originated from ones in which the X and Y were initially genetically largely homologous. The parallel evolution of advanced sex chromosome systems in different groups strongly suggests that relatively simple evolutionary forces have been involved. The classic explanation of the genetic inactivity of the Y chromosome (MULLER, 1918) was that the X and Y were originally homologous but lacked genetic exchange and remained permanently heterozygous. This was followed by a degeneration of the Y chromosome for dosage compensation and sex determination.

6 Applications and Prospects

The mapping of markers is not always necessary. TANKSLEY (1983) mentioned some examples in plants: outcrossing rate, seed control, control of cloning haploids, or population genetic studies.

Without outcrossing, no non-parental alleles will be found in the progeny of the plant other than rare mutation events. A single marker locus would be sufficient to check that hypothesis, if there are no common alleles for the parent and other plants. Otherwise certain types of outcrossing cannot be detected. The probability of not detecting outcrossing is reduced if several loci are observed. To estimate the outcrossing rate, the gene frequencies for the alleles in the population must be known.

The rate of pollen transferred into a population from the outside is of interest in commercial seed production. In this case, too, markers can be used to detect such transfers. ROEMER (1932) used the phenomenon of xenia formation to estimate the percentage of pollination with pollen from remote fields of rye with green kernels. The mother plants contained the recessive gene for yellow kernels. Thus cross-fertilization with pollen containing the gene for green kernels could be estimated.

In a crossing program markers can be used to identify successful crosses. In cases of dominance, the breeder uses the recessive parent as female and can identify hybrids by their dominant phenotype. In hybrid breeding, inbred lines must be propagated in isolation. Again markers are used to identify plants from unwanted outcrossing.

True haploids in anther culture programs can be identified, if the mother plant is heterozygous for a codominant marker locus, for example, showing two bands for an enzyme locus. Then the haploids and the spontaneous double haploid show only one band in contrast to regenerates from the diploid tissue of the mother plant.

In plant breeding, markers are used to identify strains and cultivars. This application is connected with the interest of the breeder to protect his rights. It requires a set of markers that will allow confirmation that an unknown population actually represents a given commercial variety.

Marker loci can also be used for population genetic studies, and the variability between and within populations can be estimated. A necessary condition for their use is that the marker locus is not fixed or nearly fixed. Usually markers are regarded as samples of the genome. With this approach well conserved and highly variable regions can be detected and used for studies of evolution.

In some cases mapping is needed. In certain wheat varieties a small piece of the rye genome has been introduced by chance. Markers can be used to localize the insertion place and to estimate the length of the introduced piece.

Gene tagging uses linkage of the marker loci to a target gene or a whole chromosome segment for indirect selection. To increase the probability that tagging is possible for an arbitrary target gene, the gene map should be dense. If the probability of no successful tagging due to recombination should not exceed 5%, the distance between a single marker and the target gene must be less than 5 cM (TANKSLEY, 1983). With flanking markers the undetected recombination only occurs in the case of a double crossover with the target locus located between the two exchanges. Therefore, a larger distance between the markers and the target gene can be accepted. Without interference the product of the two recombination values between the target gene and each of the marker loci must be less than the chosen risk. For a given risk of 5% the recombination rate between the marker loci must be less than 0.35. The maximum probability of a double crossover is given if the target gene is centered exactly between the markers, and the recombination rate between the target gene and each marker is $c = 0.22$.

Marker-based selection is also promising in selecting for the genetic background of the recurrent parent in a backcross program. Cycle length can be shortened from two to one generation for recessive target genes, if the markers show codominance (TANKSLEY, 1983).

Markers can also be used for indirect selection of quantitative traits. A large number of genotypes is screened for the marker, and the quantitative trait is only observed for selected genotypes. Sometimes measurement of quantitative traits is easier than the analysis of markers, and so preselection for the quantitative trait may be the better way. Only the extreme classes for the quantitative trait are investigated for the marker loci. It is expected that loci with major contribution to the quantitative effect segregate and may be linked to a marker. The number of genotypes for a marker analysis is drastically reduced, since no plant with an average expression for the quantitative trait is analyzed. LEBOWITZ et al. (1987) discussed the advantages of the two procedures.

The number of examples of practical application is still small. A very old one is the use of chlorophyll mutants in *Matthiola incana*. Only plants with double flowers are of interest for sale, but such plants are completely sterile. Therefore they must be produced by crossing heterozygous normal plants. Due to a linked lethal gene which only acts on the male side, the pollen parent only transmits the allele for double flower so that half of the plants in the progeny have the double flower character. No true breeding lines with double flowers exist. According to SAUNDERS (1915) in a book published in France in the middle of the 19th century by Chaté it was reported that the plants which would later have double flowers had been selected by the shape of the leaves and their pale-green color in comparison to the dark-green leaf rosette of the normal flowering ones. KAPPERT (1951a) found a monogenic inherited chlorophyll mutant in this species which was closely linked to the factor for double flowers. By discarding dark green seedlings which flower normally, selection of double flowering plants is possible at the seedling stage (KAPPERT, 1951b).

In tomatoes a target locus *Mi* for nematode resistance on chromosome 6, transferred from *Lycopersicon peruvianum,* is strongly linked to Asp-1 (MEDINA-FILHO, 1980). Two RFLP markers tagging the *Tm-2a* gene for resistance to tobacco mosaic virus were detected using NILs (YOUNG et al., 1988). In sugar beet nematode-resistant plants in progenies of addition lines can be selected by special GOT- and aconitase genotypes (JUNG et al., 1986). In Western white pines an RFLP marker is linked to blister rust allowing selection on young plants, while direct selection is only possible on three to five year old plants (cited by MELCHINGER, 1990). In rice two resistance genes *Pi-2(t)* and *Pi-4(t)* to the blast fungal pathogen *Pyricularia orycaea* have been tagged by RFLP markers using NILs. The distances from the marker loci were 2.8 and 15.3 cM (YU et al., 1991). A large increase in such lists is expected in the near future.

Some early results are available for marker-facilitated selection on quantitative traits. SAX (1923) found an association between a quantitatively inherited character, seed weight, and major genes for color in *Phaseolus vulgaris.* In crosses of pigmented beans with small white beans, the pigmented F2 segregates showed a greater mean seed weight than white segregates. In maize relationships have been found between biochemical markers and several agronomic traits (STUBER et al., 1987; ABLER et al., 1991). The comparison between conventional and marker-based selection in two F_2 generations of crosses between inbred lines gave a similar selection response (STUBER, 1989). In tomatoes several fruit traits were linked to RFLP and enzyme loci (TANKSLEY and HEWITT, 1988). The effects of QTL are environmentally dependent (PATERSON et al., 1991). In barley JENSEN (1989) found linkage between a gene controlling the length of rachilla hairs and a QTL for kernel weight. Linkage relationships between C-bands on chromosome 3 in barley and QTLs have been reported by KJAER et al. (1991).

7 References

ABLER, B. S. B., EDWARDS, M. D., STUBER, C. W. (1991), Isoenzymatic identification of quantitative trait loci in crosses of elite maize inbreds, *Crop Sci.* **31,** 267–274.

ALLARD, R. W. (1956), Formulas and tables to facility the calculation of recombination values in heredity, *Hilgardia* **24,** 235–278.

AMBROS, P. F., MATZKE, M. A., MATZKE, A. J. M. (1986), Detection of a 17 kb unique sequence (T-DNA) in plant chromosomes by *in situ* hybridization, *Chromosoma* **94,** 11–18.

ANSARI, H. A., HEDIGER, R., FRIES, R., STRANZINGER, G, (1988), Chromosomal localization of the major histocompatibility complex of the horse (ELA) by *in situ* hybridization, *Immunogenetics* **28,** 362–364.

APUYA, N. R., FRAZIER, B. L., KEIM, P., ROTH, E. J., LARK, K. G. (1988), Restriction fragment length polymorphisms as genetic markers in soybean, *Glycine max* (L.) merrill, *Theor. Appl. Genet.* **75,** 889–901.

ARCHIBALD, A., HALEY, C. S., ANDERSSON, L., BOSMA, A. A., DAVIES, W., FREDHOLM, M., GELDERMANN, H., GELLIN, J., GROENEN, M., GUSTAVSON, I., OLIVIER, L., TUCKER, E. M., VAN DE WEGHE, A. (1991), The Ec pig gene mapping project, *1st PiGMaP Project Meeting,* Edinburgh, 15 April, 1991.

BAILEY, N. T. J., (1961), *Introduction to the Mathematical Theory of Genetic Linkage,* Oxford: Clarendon Press.

BANIERBALE, M. W., PLAISTED, R. L., TANKSLEY, S. D. (1988), RFLP maps based on a common set of clones reveal modes of chromosomal evolution in potato and tomato, *Genetics* **120,** 1095–1103.

BAUM, M., APPLES, R. (1991), The cytogenetic and molecular architecture of chromosome 1R – one of the most widely utilized source of alien chromatin in wheat varieties, *Chromosoma* **101,** 1–10.

BECKMANN, J. S. (1988) Oligonucleotide polymorphisms: A new tool for genomic genetics, *Biotechnology* **6,** 1061–1064.

BENDER, W., SPIERER, P., HOGNESS, D. S. (1979), Gene isolation by chromosomal walking, *J. Supramol. Struct.* **10,** 32.

BITGOOD, J. J., SOMES, R. G. (1990), Linkage relationships and gene mapping, in: *Poultry Breeding and Genetics, Developments in Animal and Veterinary Sciences,* Vol. 22 (CRAWFORD, R. D., Ed.) pp. 469–498, Amsterdam-Oxford-New York-Tokyo: Elsevier.

BOONE, C. M., RUDDLE, F. H. (1969), Interspecific hybridization between human and mouse somatic cells: enzyme and linkage studies, *Biochem. Genet.* **3,** 119–136.

Botstein, D., White, R. L., Skolnik, M., Davis, R. W. (1980), Construction of a genetic linkage map using restriction fragment length polymorphisms, *Am. J. Hum. Genet.* **32,** 314–331.

Bouteille, M., Hernandez-Verdun, D. (1979), Localization of a gene: The nucleolar organizer, *Biomedicine* **30,** 282–287.

Bridges, C. B. (1935), Salivary chromosome maps, *J. Hered.* **26,** 60–64.

Buchberg, A. M., Brownwell, E., Nagata, S., Jenkins, N. A., Copeland, N. G. (1989), A comprehensive genetic map of murine chromosome 11 reveals extensive linkage conservation between mouse and human, *Genetics* **122,** 153–161.

Bufton, L., Mohandas, T. K., Magenis, R. E., Sheehy, R., Bestwick, R. K., Litt, M. (1986), A highly polymorphic locus on chromosome 16q revealed by a probe from a chromosomal cosmid library, *Hum. Genet.* **74,** 425–431.

Bulfield, G. (1990), Molecular genetics, in: *Poultry Breeding and Genetics, Developments in Animal and Veterinary Sciences,* Vol. 22, pp. 543–584, (Crawford, R. D., Ed.), Elsevier. Amsterdam-Oxford-New York-Tokyo.

Capanna, E., Gropp, H., Winking, H., Noack, G., Civitelli, M. V. (1976), Robertsonian metacentrics in the mouse, *Chromosoma* **58,** 341–353.

Caspersson, T., Farber, S., Foley, G. E., Kudynowski, J., Modest, E. J., Simonsson, E., Wagh, U., Zech, L. (1967), Chemical differentiation along metaphase chromosomes, *Exp. Cell. Res.* **49,** 219–222.

Caspersson, T., Zech, L., Modest, E. J., Foley, G. E., Wagh, U., Simonsson, E. (1969), Chemical differentiation with fluoroscent alkylating agents in *Vicia faba* metaphase chromosomes, *Exp. Cell. Res.* **58,** 128–140.

Chapman, V. M., Paigen, K., Siracusa, L., Womack, J. E. (1979), Biochemical variation: mouse, in: *Inbred and Genetically Defined Strains of Laboratory Animals, Part 1. Mouse and Rat* (Altman, P. L., Katz, D. D., Eds.), Bethesda, MD: Faseb.

Charlesworth, B. (1991), The evolution of sex chromosmes, *Science* **251,** 1030–1033.

Cherif, D., Bernard, O., Berger, R. (1989), Detection of single-copy genes by nonisotopic *in situ* hybridization on human chromosomes, *Hum. Genet.* **81,** 358–362.

Coe, E. H., Hoisington, D. A., Neuffer, M. G. (1990), Linkage of corn (maize) (*Zea mays* L.) 2N = 20), in: *Genetic Maps,* 5th Ed. (O'Brien, S. J., Ed.), Cold Spring Harbor, NY: Cold Spring Harbor Laboratory Press.

Conner. B. J., Reyer, A. A., Morin, C., Itakura, K., Teplite, R. L., Wallace, R. B. (1983), Detection of sickle-cell βs-globin allele by hybridization with synthetic oligonucleotide, *Proc. Natl. Acad. Sci. USA* **80,** 278–282.

Cooke, H. (1987), Cloning in yeast: An appropriate scale for mammalian genomes, *Trends Genet.* **3,** 173–174.

Cornish, M. A., Hayward, M. D., Laurence, M. J. (1980), Self-incompatibility in ryegrass. III. The joint regeneration of S and P61-2 in *Lolium perenne* L., *Heredity* **44,** 55–62.

Crow, J. F., Dove, E. F., (1990), Anecdotal, historical and critical commentaries on genetics; mapping functions, *Genetics* **125,** 669–671.

Curtis, C. A., Lukaszewski, A. J. (1991), Genetic linkage between C-bands and storage protein genes in chromosome 1B of tetraploid wheat, *Theor. Appl. Genet.* **81,** 245–252.

Davisson, M. T. (1987), X-linked genetic homologies between mouse and man, *Genomics* **1,** 213–227.

DeNettancourt, D. (1977), *Incompatibility in Angiosperms,* Berlin-Heidelberg-New York: Springer-Verlag.

Donis-Keller, H., Green, P., Helms, C., Cartinhour, S., Weiffenbach, B., Stephans, K., Keith, T. P., Bowden, D. W., Smith, D. R., Lander, E. S., Botstein, D., Akots, G., Rediker, K. S., Gravius, T., Brown, V. A., Rising, M. B., Parker, C., Powers, J. A., Watt, D. E., Kauffmann, E. R., Bricker, A., Phipps, P., Muller-Kahle, H., Fulton, T. R., Siu, Ng., Schumm, J. W., Braman, J. C., Knowlton, R. G., Barker, D. F., Crooks, S. M., Lincoln, S. E., Daly, M. J., Abrahamson, J. (1987), A genetic linkage map of the human genome, *Cell* **51,** 319–337.

Echard, G., Milan, D., Yerle, M., Lahbib-Mansais, Y. (1991), *The Gene Map of Pigs* (*Sus scrofa domestica* L.), Laboratoire de Genetique Cellulaire - INRA, France, Poly Copy.

Ecke, W., Pannenbecker, G., Wasmund, O., Michaelis, G. (1989), Extranuclear inheritance: Mitochondrial genetics, *Prog. Bot.* **50,** 198–206.

Ege, T., Ringertz, N. R. (1974), Preparation of microcells by enucleation of micronucleate cells, *Exp. Cell Res.* **87,** 378–382.

Ellis, N. A., Goodfellow, P. N., Goodfellow, P. J., Pym, B., Smith, M., Palmer, M., Frischauf, A. M. (1989). The pseudoautosomal boundary in man is defined by an Alu repeat sequence inserted on the Y chromosome, *Nature* **337,** 81–84.

EPHRUSSI, B. (1953), *Nucleo-cytoplasmic Relations in Microorganisms*, London: Oxford University Press.

FEARON, CH., HAYWARD, M. D., LAWRENCE, M. J. (1983), Self-incompatibility in ryegrass. V. Genetic control, linkage and seed-set in diploid *Lolium multiflorum* L., *Heredity* **50**, 35–45.

FERGUSON-SMITH, M. A., AITKEN, D. A.(1982), The contribution of chromosome aberrations to the precision of human gene mapping, *Cytogenet. Cell Genet.* **32**, 24–42.

FRIES, R. (1990), The gene for the β subunit of the FSH maps to bovine chromosome 15, *J. Hered.* **80**, 401–403.

FRIES, R. (1991), *The Bovine Gene Map,* Department of Animal Science, ETH, Zürich, Switzerland.

FRIES, R., HEDIGER, R., STRANZINGER, G. (1986), Tentative chromosomal localization of the bovine major histocompatibility complex by *in situ* hybridisation, *Anim. Genet.* **17**, 287–294.

FRIES, R., HEDIGER, R., STRANZINGER, G. (1988), The loci for parathyroid hormone and beta-globulin are closely linked and map to bovine chromosome 15 in cattle, *Genomics* **3**, 302–307.

FRIES, R., BECKMAN, J. S., GEORGES, M., SOLLER, M., WOMACK, J. E, (1989), The bovine gene map, *Anim. Genet.* **20**, 3–29.

GALILI, G., FELDMAN, M. (1984), Mapping of glutenin and gliadin genes located on chromosome 1B of common wheat, *Mol. Gen. Genet.* **193**, 293–298.

GALL, J., PARDUE, M. D. (1969), Formation and detection of RNA-DNA hybrid molecules in cytological preparations, *Proc. Natl. Acad. Sci. USA* **63**, 378–383.

GERTZ, A., WRICKE, G. (1989), Linkage between the incompatibility locus Z and a β-glucosidase locus in rye, *Plant Breed.* **120**, 255–259.

GLASER, T., HOUSMAN, D. (1984), Insertion of a selectable marker into various sites on human chromosome 11, *ICSU Rep.* **1**, 174–175.

GOSS, S. J., HARRIS, H. (1977), Gene transfer by means of cell fusions, I. and II., *J. Cell Sci.* **25**, 17–58.

GRUNEWALD, D., GEFFROTIN, C., CHARDON, P., FRELAT, G., VAIMAN, M. (1986), Swine chromosomes: Flow sorting and spot blot hybridization, *Cytometry* **7**, 582–588.

GRZESCHECK, K.-H. (1986), The role of somatic cell genetics in human gene mapping, *Experientia* **42**, 1128–1137.

GUSTAVSSON, I. (1990), Chromosomes of the pig, *Adv. Vet. Sci. Comp. Med.* **34**, 73–107.

HADANO, S., WATANABE, M., YOKOI, H., KOGI, M., KONDO, I., TUSUCHIYA, H., KANAZAWA, I., WAKASA, K., IKEDA, J. E. (1991), Laser microdissection and single unique primer PCR allow generation of regional chromosome DNA clones from a single human chromosome, *Genomics* **11**, 364–373.

HAGEMANN, R., HAGEMANN, M. M., METZLAFF, M. (1989), Extranuclear inheritance: Plastid genetics, *Prog. Bot.* **51**, 237–250.

HALDANE, J. B. S. (1919), The recombination of linkage values and the calculation of distance between the loci of linked factors, *J. Genet.* **8**, 299–309.

HANSON, W. D. (1959), Early generation analysis of length of heterozygous chromosome segments around a locus held heterozygous with backcrossing or selfing, *Genetics* **44**, 833–837.

HARPER, M. E., SAUNDERS, G. F. (1981), Localization of single copy DNA sequences on G-banded human chromosomes by *in situ* hybridization, *Chromosoma* **83**, 431–439.

HARRIS, H., HOPKINSON, D. A. (1974), *Handbook of Enzyme Electrophoresis in Human Genetics,* Amsterdam: North-Holland.

HART, G. E., (1979), Genetical and chromosomal relationships among the wheats and their relatives, *Stadler Genet. Symp.* **11**, 9–29.

HART, H. E., GALE M. G. (1990), *Biochemical/ Molecular Loci of Hexaploid Wheat in Genetic Maps*, Cold Spring Harbor, NY: Cold Spring Harbor Laboratory Press 6.28.

HEDIGER, R. (1988), Die *in situ* Hybridisierung zur Genkartierung beim Rind und Schaf, *Thesis* No. 8725, ETH, Zürich, Switzerland.

HEINRICH, P., DOMDEY, H. (1990), Techniques, strategies and stages of DNA sequencing, in: *Genome Analysis in Domestic Animals* (GELDERMANN, H., ELLENDORFF, F., Eds.), Weinheim-New York-Basel-Cambridge: VCH.

HENNIG, J., HERRMANN, R. (1986), Chloroplast ATP synthesis of spinach contains nine nonidentical subunit species, six of which are encoded by plastid chromosomes in two operons in a phylogenetically conserved arrangements, *Mol. Gen. Genet.* **203**, 117–128.

HERRMANN, R. G. (1992), Biogenesis of plastids in higher plants, in: *Cell Organelles* (HERRMANN, R., Ed.), Wien-New York: Springer-Verlag, in press.

HEUN, M., GREGORIUS, M. R. (1987), A theoretical model for estimating linkage in F_2-populations with distorted single gene segregation, *Biom. J.* **4**, 397–406.

HEUN, M., KENNEDY, A. E., ANDERSON, J. A., LAPITAN, N. L. V., SORRELIS, M. E., TANKSLEY, S. D. (1991), Construction of an RFLP map for barley (*Hordeum vulgare* L.), Paper No. 800 of the *Cornell Plant Breeding Series,* Ithaca, NY.

HGM 9 (Human Gene Mapping Workshop 9) (1988), Paris Conference 1987, *Cytogenet. Cell Genet.* **46,** 1-762.

HGM 10 (Human Gene Mapping Workshop 10) (1989) Paris Conference 1987, *Cytogenet. Cell Genet.* **51,** 1-1147.

HIRATSUKA, J., SHIMADA, H., WHITTIER, W., ISHIBASHI, R., SAKAMOTO, M., MORI, M., KONDO, C., HONJI, Y., SUN, C. R., MENG, B. Y., LI, Y. Q., KANNO, A., NISHIZAWA, Y., HIRAI, A., SHINOZAKI, K., SUGIURA, M. (1989), The complete sequence of the rice *(Oryza sativa)* chloroplast genome: Intermolecular recombination between distinct tRNA genes accounts for a major plastid DNA inversion during the evolution of the cereals, *Mol. Gen. Genet.* **217,** 185-194.

HOOD, L. E., HUNKAPILLER, M. W., SMITH, L. M. (1987), Automatic DNA sequencing and analysis of the human genome, *Genomics* **1,** 201-212.

JEFFREYS, A. J., FLAVELL, R. (1977), A physical map of the DNA regions flanking the rabbit β-globin gene, *Cell* **12,** 429-439.

JEFFREYS, A. J., WILSON, V., THEIN, S. L. (1985), Hypervariable minisatellite regions in human DNA, *Nature* **314,** 67-72.

JEFFREYS, A., J., WILSON, V., THEIN, S. L., WEATHERALL, D. J., PONDER, B. A. (1986), DNA "fingerprints" and segregation analysis of multiple markers in human pedigrees, *Am. J. Hum. Genet.* **39,** 11-24.

JENSEN, J. (1989), Estimation of recombination parameters between a quantitative trait locus (QTL) and two marker gene loci *Theor. Appl. Genet.* **78,** 613-618.

JONES, R. M. (1960), Linkage distributions and epistacy in quantitative inheritance, *Heredity* **15,** 153-159.

JUDD, B. H. (1979), Mapping the functional organization of eukaryotic chromosomes, *Cell Biol.* **2,** 223-265.

JUNG, C., WEHLING, P., LÖPTIEN, H. (1986), Electrophoretic investigations on nematode resistant sugar beets, *Plant Breed.* **97,** 39-45.

JUNG, C., KLEINE, M., FISCHER, F., HERRMANN, R. G. (1990), Analysis of DNA from a *Beta procumbens* chromosome fragment in sugar beet carrying a gene for nematode resistance, *Theor. Appl. Genet.* **79,** 663-672.

JUNG, C., KOCH, R., FISCHER, F., BRANDES, A., WRICKE, G., HERRMANN, R. G. (1991), DNA markers closely linked to nematode resistance genes in sugar beet (*Beta vulgaris* L.) using additions and translocations originating from wild beets of the *Procumbentes* section, *Mol. Gen. Genet.* **232,** 271-278.

KAPPERT, H. (1951a), Mendelismus und Blumenzüchtung, *Z. Pflanzenzüchtg.* **29,** 406-414.

KAPPERT, H. (1951b), Untersuchungen über den Mechanismus des Immerspaltens bei der Kulturlevkoje, *Der Züchter* **21,** 205-211.

KARMACK, M. E., BARKER, P. E., MILLER, R. C. L., RUDDLE, F. H. (1984), Somatic cell hybrid mapping panels, *Exp. Cell Res.* **152,** 1-14.

KIDD, K. K., TRACK, R. K., BOWCOCK, A. M., RICCINTI, F., HUTCHINGS, G., CHAN, H. S. (1991), Human DNA restriction fragment length polymorphisms (RFLPs), in: *Genetic Maps* (O'BRIEN, S. J., Ed.), Cold Spring Harbor, NY: Cold Spring Harbor Laboratory Press.

KJAER, B., HAAHR, V., JENSEN, B. (1991), Associations between 23 quantitative traits and 10 genetic markers in a barley cross, *Plant Breed.* **106,** 261-274.

KLEIN-LANKHORST, R., RIETVELD, P., MACHIELS, B., VERKERK, R., WEIDE, R., GEBHARDT, C., KOORNNEEF, M., ZABEL, P. (1991), RFLP markers linked to the root knot nematode resistance gene *Mi* in tomato, *Theor. Appl. Genet,* **81,** 661-667.

KNOPP, V., BRENNICKE, A. (1991), A mitochondria intron sequence in the 5'-flanking region of a plant nuclear lectine gene, *Curr. Genet.* **20,** 423-425.

KO, K. (1986), Molecular analysis of the chloroplast genomes in plants, in: *The Chondriom* (MANTEL, S. H., CHAPMAN, G. P., STREET, P. F., Eds.), pp- 117-141, Harlow, UK: Longman.

KOSAMBI, D. D. (1944), The estimation of map distances from recombination values, *Ann. Eugen.* **12,** 172-175.

LALLEY, P. A., MCKUSICK, V. A. (1985), Report of the Committee on Comparative Mapping, HGM 8, *Cytogenet. Cell Genet.* **40,** 536-566.

LALLEY, P. A., DAVISSON, M. T., GRAVES, J. A. M., O'BRIEN, S. J., WOMACK, J. E., RODERICK, T. H., CREAU-GOLDBERG, N., HILLYARD, A. L., DOOLITTLE, D. P., ROGERS, J. A. (1989), Report of the Committee on Comparative Mapping, *Cytogenet. Cell Genet.* **51,** 503-523.

LANDEGENT, J. E., JANSEN IN DE WAL, N., VAN OMMEN, J. B., BAAS, F., DE VIJLDER, J. J. M., VAN DUIJEN. P., VAN DER PLOEG, M., (1985), Chromosomal localization of unique gene by non-autoradiographic *in situ* hybridisation, *Nature* **317,** 175-177.

LANDEGENT, J. E., JANSEN IN DE WAL, N., DIRKS, R. W., BAAS, F., VAN DER PLOEG, M. (1987), Use of whole cosmid cloned genomic sequences for chromosomal localization by non-radioactive *in situ* hybridization, *Hum. Genet.* **77,** 366-370.

LANDER, E. S., BOTSTEIN, D. (1989), Mapping Mendelian factors underlying quantitative traits using RFLP linkage maps, *Genetics* **121,** 185–199.

LANDER, E. S., GREEN, P., ABRAHAMSON, J., BARLOW, A., DALY, M., LINCOLN, S., NEWBERG, L. (1987), Mapmaker: An interactive computer package for constructing primary linkage maps of experimental and natural populations, *Genomics,* **1,** 174–181.

LANGE, W., DE BOCK, TH. S. M., VAN GEYTAND, J. P. C., OLEO, M. (1988), Monosonic additions in beet (*Beta vulgaris*) carrying extra chromosomes of *B. procumbens.* II. Effects of the alien chromosome on *in vivo* and *in vitro* plant development, *Theor. Appl. Genet.* **76,** 656–664.

LEACH, C. R. (1988), Detection and estimation of linkage for a codominant structural gene locus linked to a gametophytic self-incompatibility locus, *Theor. Appl. Genet.* **75,** 882–888.

LEACH, C. R., HAYMANN, D. L. (1987), The incompatibility loci as indicators of conserved linkage groups in the Poaceae, *Heredity* **58,** 303–305.

LEBO, R. V., GORIN, F., FLETTERICK, R. J., KAO, T. T., CHEUNG, M. C., BRUCE, B. D., KAN, Y. W. (1984), High-resolution chromosome sorting and DNA spot-blot analysis assign McArdle's syndrome to chromosome 11, *Science* **225,** 57–59.

LEBOWITZ, R. J., SOLLER, M., BECKMANN, J. S. (1987), Trait-based analyses for the detection of linkage between marker loci and quantitative trait loci in crosses between inbred lines, *Theor. Appl. Genet.* **73,** 556–562.

LINDE-LAURSEN, I. (1979), Giemsa C-banding of barley chromosomes II. Banding patterns of trisomics and telotrisomics, *Hereditas* **89,** 37–41.

LITTLE, P. F. R. (1990), Gene mapping and the human genome mapping project, *Curr. Opin. Cell Biol.* **2,** 478–484.

LIU, B.-H., KNAPP, S. J. (1990), GMENDEL: a program for Mendelian segregation and linkage analysis of individual or multiple progeny populations using log-likelihood ratios, *J. Hered.* **81,** 407.

LONSDALE, D. M., BEARS, T., HODGE, T. P., MELVILLE, S. E., ROTTMANN, W. H. (1988), The plant mitochondrial genome: homologous recombination as a mechanism for generating heterogeneity, *Phil. Trans. R. Soc. London* **B 319,** 149–163.

LÖPTIEN, H. (1979), Identification of the sex chromosomes pai in asparagus (*Asparagus officinalis* L.), *Z. Pflanzenzüchtg.* **82,** 162–173.

LÜDECKE, H. J., SENGER, G., CLAUSSEN, U., HORSTHEMKE, B. (1989), Cloning defined regions of the human genome by microdissection of banded chromosomes and enzymatic amplification, *Nature* **338,** 348–350.

LÜDECKE, H. J., SENGER, G., CLAUSSEN, U., HORSTHEMKE, B. (1990), Construction and characterization of band-specific DNA libraries, *Hum. Genet.* **84** 512–516.

LYON, M.F. (1988), X-chromosome interaction and the location and expression of X-linked genes, *Am. J. Hum. Genet.* **42,** 8–16.

MATHER, K. (1951), *The Measurement of Linkage in Heredity,* 2nd Ed., London: Methuen & Co.

MCDONALD, G. P., PRICE, E. R., CHU, M. L., TIMPLE, R., ALLORE, R., MARKS, A., DUNN, R., COX, D. R. (1988), Assignment of four human chromosome 21 genes to mouse chromosome 10: Implications four mouse models of Down syndrome, *Am. J. Hum. Genet.* **43,** A151.

MCKUSICK, V. A. (1986), *Mendelian Inheritance in Man,* 7th Ed., Baltimore: Johns Hopkins University Press.

MCKUSICK, V. A. (1989), Mapping and sequencing the human genome, *N. Engl. J. Med.* **320,** 910–915.

MCKUSICK, V. A. (1990), The human gene map, in: *Genetic Maps.* (O'BRIEN, S. J., Ed.), 5th Ed., pp. 47–133, New York: Plain View.

MCLACHLAN, G. J., BASFORD, K. E, (1988), *Mixture Models: Inference and Applications to Clustering,* New York: Marcel Dekker.

MEDINA-FILHO, H. P. (1980), Linkage of *Aps-1, Mi* and other markers on chromosome 6, *Rep. Tomato Genet. Coop.* **30,** 26–28.

MELCHINGER, A. E. (1990), Use of molecular markers in breeding for oliogenic disease resistance, *Plant Breed.* **104,** 1–19.

MILLER, A. D., JOLLY, D. J., FRIEDMANN, T., VERMA, I. M. (1983), A transmissible retrovirus expressing human hypoxanthine phosphoribosyltransferase (HPRT): Gene transfer into cells obtained from humans deficient in HPRT, *Proc. Natl. Acad. Sci. USA* **80,** 4709–4713.

MORTON, N. E. (1955), Sequential tests for the detection of linkage, *Am. J. Hum. Genet.* **7,** 277–318.

MULLER, H. J. (1918), Genetic variability twin hybrids and constant hybrids in a case of balanced factors, *Genetics* **101,** 57–69.

MULLIGAN, R. C., BERG, P. (1980), Expression of a bacterial gene in mammalian cells, *Science* **209,** 1422–1427.

NADEAU, J. H. (1989), Maps of linkage and synteny homologies between mouse and man, *Trends Genet.* **5,** 82–86.

NAKAMURA, Y., LEPPERT, M., O'CONNELL, P., WOLFF, R., HOLM, T., CULVER, M., MARTIN, C., FUJIMOTO, E., HOFF, M., KUMLIN, E.,

WHITE, R. (1987), Variable number of tandem repeat (VNTR) markers for human gene maping, *Science* **235**, 1616–1622.

O'BRIEN, S. J. (1990),*Genetic Maps,* 5th Ed., Cold Spring Harbor, NY: Spring Harbor Laboratory Press.

ODENBACH, W., MAHGOUB, E. S. (1988), Relationship between HMW glutenin subunit composition and the sedimentation value in reciprocal sets of inbred backcross lines derived from two winter wheat crosses, *Proc. 7th Int. Wheat, Genet. Symp.,* Cambridge.

OHNO, S. (1967), *Sex Chromosomes and Sex-linked Genes,* Heidelberg-Berlin: Springer-Verlag.

OHYSAMA, K., FUKUTAWA, H., KOCHLI, T., SHIRAI, H., SAUN, T., SAUN, S., UMESONO, K., SHIKI, S., TAKEUCJI, M., CHANG, Z., AOTON, S., INOKUCKI, H., OZEKI, H. (1986), Chloroplast gene organisation deduced from complete sequence of liverwort *Marchantia polymorpha* chloroplast DNA, *Nature* **322,** 572–574.

ORKIN, S. H. (1986), Reverse genetics and human diseases, *Cell* **47,** 845–850.

OTT, H. T., DEMARS, R. (1983), The organization of human major histocompatibility class I genes as determinated with deletion mutants, in: *Recombinant DNA and Medical Genetics* (MESSER, A., Ed.), pp. 117–134, New York: Academic Press.

PAIGEN, K. (1971), The genetics of enzyme realization (a review), in: *Enzyme Synthesis and Degradation in Mammalian Systems,* (RECHCIGL, M., Ed.), pp. 1–47, Basel: Karger.

PAIGEN, K. (1979), Acid hydrolases as models of genetic control, *Annu. Rev. Genet.* **13,** 417–466.

PALMER, J. D. (1985), Comparative organization of chloroplast genomes, *Annu. Rev. Genet.* **19,** 325–354.

PALMER, J. D., HERBORN, L. A. (1987), Unicircular structure of the *Brassica hirta* mitochondrial genome, *Curr. Genet.* **11,** 565–570.

PATERSON, A. H., DAMON, S., HEWITT, J. D., ZAMIR, D., RABINWITCH, H. D., LINCOLN, S. E., LANDER, E. S., TANKSLEY, S. D. (1991), Mendelian factors underlying quantitative traits in tomato: comparison across species, generations and environments, *Genetics* **127,** 181–197.

PETERS, J. (1981), Enzyme and protein polymorphism, in: *Biology of the House Mouse.* (BERRY, R. J., Ed.), pp. 479–516, London: Academic Press.

PETERS, J., ANDREWS, S. J., LOUTIT, J. F., CLEGG, J. B. (1985), A mouse β-globin mutant that is an exact model of hemoglobin Rainier in man, *Genetics* **110,** 709–721.

PHILIPP, U. (1992), Erstellung einer genetischen Karte beim Roggen (*Secale cereale* L.) mit Hilfe molekularer Marker, *Dissertation,* Universität Hannover.

PILLEN, K., STEINRÜCKEN, G., WRICKE, G., HERMANN, R. G., JUNG, C. A linkage map of sugar beet (*Beta vulgaris* L.), *Theor. Appl. Genet.,* in press.

POUSTKA, A., LEHRBACH, H. (1986), Jumping libraries and linking libraries: The next generation of molecular tools in mammalian genetics, *Trends Genet.* **2,** 174–179.

POUSTKA, A., RACKWITZ, H. R., FRISCHAUF, A. M., LEHRBACH, H. (1984), Selective isolation of cosmid clones by homologous recombination in *Escherichia coli, Proc. Natl. Acad. Sci. USA* **81,** 4129–4133.

QUIROS, C. F., HU, J., THIS, P., CHEUVE, A. M., DELSENY, M. (1991), Development and chromosomal localization of genome-specific markers by polymerase chain reaction in *Brassica, Theor. Appl. Genet.* **82,** 627–632.

RAYBURN, A. L., GILL, B. S. (1985), Use of biotin-labelled probes to map specific DNA sequences on wheat chromosomes, *J. Hered.* **76,** 78–81.

REAMON-RAMOS, S. M., WRICKE, G. (1992), A full set of monosomic addition lines in *Beta vulgaris* from *Beta webbiana:* morphology and isozyme markers, *Theor. Appl. Genet.* **84,** 411–418.

RENWICK, J. H. (1971), Assignment and map-positioning of human loci using chromosomal variation, *Ann. Hum. Genet. Lond.* **35,** 79–97.

ROEMER, T. H. (1932), Über die Reichweite des Pollens bei Roggen, *Z. Pflanzenzüchtg.* **17,** 14–35.

RUDDLE, F. H. (1972), Linkage analysis using somatic cell hybrids, *Adv. Hum. Genet.* **3,** 173–235.

RUDDLE, F. H. (1981), A new era in mammalian gene mapping: somatic cell genetics and recombinant DNA methodologies, *Nature* **294,** 115–120.

SAGER, R., ISHIDA, M. R. (1963), Chloroplast DNA in *Chlamydomonas, Proc. Natl. Acad. Sci. USA* **50,** 725–730.

SAUNDERS, E. R. (1915), The double stock, its history and behaviour, *J. R. Hortic. Soc.* **40,** 450–472.

SAX, K. (1923), The association of size differences with seed-coat pattern and pigmentation in *Phaseolus vulgaris, Genetics* **8,** 552–560.

SCHNELL, F. W. (1963), The covariance between relatives in the presence of linkage, *Stat. Genet. Plant Breed., NAS-NRC* **982,** 468–483.

SEARLE, A. G., PETERS, J., LYON, M. F., HALL, J. G., EVANS, P. E., EDWARDS, J. H., BUCKLE, V. J. (1989), Chromosome maps of man and mouse, IV. *Ann. Hum. Genet.* **53,** 89–140.

SHARP, P. M. (1991), Genome organization and evolution, *TREE* **6,** 71–72.

SHINOZAKI, K., OHME, M., TANAKA, M., WAKASUGI, T., HAYASHIDA, N., MATSUBAYASHI, T., ZAITA, N., CHUNWONGSE, J., OBOKATA, J., YAMAGUCHI-SHINOZAKI, K., OHTO, C., TORAZAWA, K., MENG, B. Y., SUGITA, M., DENO, H., KAMOGASHIRA, T., YAMADA, K., KUSUDA, J., TAKAIWA, F., KATO, A., TOHDOH, N., SHIMADA, H., SUGIURA, M. (1986), The complete nucleotide sequence of the tobacco chloroplast genome: its gene organisation and expression, *EMBO J.* **5**, 2043–2049.

SKOLNIK, M. H., WHITE, R. (1982), Strategies for detection and characterization restriction fragment length polymorphisms (RFLPs), *Cytogenet. Cell Genet.* **32**, 58–67.

SONG, K. M., YSUZUKI, J., SLOCUM, M. K., WILLIAMS, P. H., OSBORN, T. C. (1991), A linkage map of *Brassica rapa* (syn. *campestris*) based on restriction fragment length polymorphism loci, *Theor. Appl. Genet.* **82**, 296–304.

STAM, P., ZEVEN, A. C. (1981), The theoretical proportion of the donor genome in near-isogenic lines of self-fertilizers bred by backcrossing, *Euphytica* **30**, 227–238.

STONE, D. M., JACKY, P. B., PRIEUR, D. J. (1991), The Giemsa banding pattern of canine chromosomes, using a cell synchronization technique, *Genome* **34**, 407–412.

STRANZINGER, G. (1990), Gene and chromosome homologies in different species, in: *Genome Analysis in Domestic Animals* (GELDERMANN, H., ELLENDORFF, F., Eds.), pp. 115–134, Weinheim-New York-Basel-Cambridge: VCH.

STUBER, C. W. (1989), Marker-based selection for quantitative traits, *Vortr. Pflanzenzüchtg.* **15**, 31–49.

STUBER, C. W., EDWARDS, M. D., WENDEL, J. F. (1987), Molecular marker-facilitated investigations of quantitative trait loci in maize. II. Factors influencing yield and its componenet traits, *Crop Sci.* **27**, 639–648.

SUHAI, S. (1990), Computer-aided analysis of biomolecular sequence and structure in genome research, in: *Genome Analysis in Domestic Animals,* (GELDERMANN, H., ELLENDORFF, F., Eds,), pp. 49–80, Weinheim-New York-Basel-Cambridge: VCH.

SUITER, K. A., WENDEL, J. F., CASE, J. S. (1983), Linkage-1: A PASCAL computer program for the detection and analysis of genetic linkage, *J. Hered.* **74**, 203–204.

SUMNER, A. T. (1972), A simple technique for demonstrating centromeric heterochromatin, *Exp. Cell Res.* **75**, 304–306.

TANKSLEY, S. D. (1983), Molecular markers in plant breeding, *Plant Mol. Biol. Rep.* **1**, 3–8.

TANKSLEY, S. D., HEWITT, J. (1988), Use of molecular markers in breeding for soluble solids content in tomato – a re-examination, *Theor. Appl. Genet.* **75**, 811–823.

TANKSLEY, S. D., MUTSCHLER, M. A. (1991), Linkage map of the tomato (*Lycopersicon esculentum*) (2N = 24). in: *Genetic Maps,* (O'BRIEN, S. J., Ed.), 5th Ed., Cold Spring Harbor, NY: Cold Spring Harbor Laboratory Press.

THREADGILL, D. W., WORMACK, J. E. (1990), Genomic analysis of the major bovine milk protein genes, *Nucleic Acids Res.* **18**, 6935–6942.

UPHOFF, H., WRICKE, G. (1992), Random amplified polymorphic DNA (RAPD) markers in sugar beet (*Beta vulgaris* L.): mapping the genes for nematode resistence and hypocotyl colour, *Plant Breed.* **109**, 168–171.

WAGNER, H., WEBER, W. E., WRICKE, G. (1992), Estimating linkage relationship of isozyme markers and morphological markers in sugar beet (*Beta vulgaris* L.) including families with distorded segregations, *Plant Breed.* **108**, 89–96.

WEISS, M. C., GREEN, H. (1967), Human-mouse hybrid cell lines containing partial complements of human chromosomes and functioning human genes, *Proc. Natl. Acad. Sci. USA* **58**, 1104–1111.

WELLER, J. I. (1986), Maximum likelihood techniques for the mapping and analysis of quantitative trait loci with the aid of genetic markers, *Biometrics* **42**, 627–640.

WESTERGAARD, M. (1959), The mechanism of sex determination in devecious flowering plants, *Adv. Genet.* **9**, 217–281.

WHITTIER, R. F., SUGIURA, M. (1992), Plastid chromosomes from vascular plant-genes, in: *Plant Gene Research,* Vol. 8, *Cell Organelles* (HERRMANN, R. G., Ed.), 164–182, Wien-New York: Springer.

WILLIAMS, J. G. K., KUBELIK, A. R., LIVAK, K. L., RAFALSKI, J. A., TINGEY, S. V. (1990), DNA polymorphisms amplified by arbitrary primers are useful as genetic markers. *Nucleic Acids Res.* **18**, 6531–6535.

WINTER, R. M. (1988), Malformation syndromes: a review of mouse/human homology, *J. Med. Genet.* **25**, 480–487.

WOLFE, K. H., LI, W. H., SHARP, P. M. (1987), Rates of nucleotide substitution vary greatly among plant mitochondrial, chloroplast, and nuclear DNAs, *Proc. Natl. Acad. Sci. USA* **84**, 9054–9058.

WOMACK, J. E. (1988), Molecular cytogenesis of cattle: A genomic approach to disease resistance and productively, *J. Dairy Sci.* **71**, 1116–1123.

WOMACK, J. E., MOLL, Y. D. (1986), Gene map of the cow: conservation of linkage with mouse and man, *J. Hered.* **77**, 2–7.

WRICKE, G. (1991), A molecular marker linkage map of rye for plant breeding, in: *Proc. EUCARPIA Meeting, Cereal Section,* Schwerin, pp. 72–78.

WRICKE, G., WEBER, W. E. (1986), *Quantitative Genetics and Selection in Plant Breeding,* Berlin: de Gruyter.

WRICKE, G., WEHLING, P. (1985), Linkage between an incompatibility locus and a peroxidase isozyme locus (Prx7) in rye, *Theor. Appl. Genet.* **71,** 289–291.

WURSTER, D. H., BENIRSCHKE, K. (1968), Chromosome studies in the superfamily Bovidae, *Chromosoma* **25,** 152–171.

YOUNG, N. D., ZAMIR, D., GANAL, M. W., TANKSLEY, S. D. (1988), Use of isogenic lines and simultaneous probing to identify DNA markers tightly linked to the Tm-2a gene in tomato, *Genetics* **120,** 579–585.

YU, Z., MACKILL, D. J., BONMAN, J. M., TANKSLEY, S. D. (1991), Tagging genes for blast resistance in rice via linkage to RFLP markers, *Theor. Appl. Genet.* **81,** 471–476.

ZABEL, B. U., NAYLOR, S. L., SAKAGUCHI, A. Y., BELL, G. I., SHOWS, T. B. (1983), High-resolution chromosomal localization of human genes for amylase, propiomelanocortin, somatostatin, and a DNA fragment (D3S1) by *in situ* hybridization, *Proc. Natl. Acad. Sci. USA* **80,** 6932–6936.

Note added in proof:

OLIVER, S. G. et al. (1992), The complete DNA sequence of yeast chromosome III, *Nature* **357,** 38–76.

This paper reports on the first complete sequence analyses of an entire chromosome and provides information which is important for our understanding of the genome of higher eukaryotes.

II. Molecular Genetics

6 Structure and Function of DNA

FRIEDRICH GÖTZ

Tübingen, Federal Republic of Germany

1 Composition and Primary Structure of Single-Stranded DNA

The fundamental investigations which led to the discovery of nucleic acids were performed by 1869 by FRIEDRICH MIESCHER in the laboratory of HOPPE-SEYLER in Tübingen, Germany. He isolated nuclei from pus cells obtained from surgical bandages and showed that the preparation was composed of a basic protein factor and an unusually high amount of a phosphorus-containing compound, the nucleic acid. He termed these constituents "nuclein" (MIRSKY, 1980). His studies were continued by ALTMAN 1889 who described a method for the preparation of protein-free nucleic acids from animal tissues and yeasts. Much time passed before the nature of the sugars in deoxy pentose and pentose nucleic acids was elucidated, which now form the basis of the names deoxyribonucleic acid (DNA) and ribonucleic acid (RNA). The proof, however, that DNA is the genetic material came out of studies in bacterial genetics many decades later. AVERY, MACLEOD, and MCCARTY (1944) demonstrated in their classical transformation experiment with *Pneumococcus* that DNA, a normal constitutent of all cells, is the chemical basis of heredity. In eukaryotic cells, such as animal or plant cells, DNA is organized in chromosomes surrounded by the nuclear membrane. On the other hand, prokaryotes such as bacteria, neither contain a nucleus nor is their DNA organized like a typical eukaryotic chromosome.

Long before DNA was established as the genetic material, MORGAN and his coworkers discovered the phenomenon of crossing-over. They located 30 distinct genes of the four *Drosophila* chromosomes on a genetic map and laid the groundwork for the theory of the linear order of gene loci (MORGAN, 1919).

DNA is a polynucleotide chain consisting of four deoxyribonucleotide repeating units. Each is composed of a cyclic furanoside-type sugar, the D-2′-deoxyribose, which is phosphorylated in the 5′-position and substituted at C1′ by a purine (adenine and guanine) or pyrimidine (thymine and cytosine) base. In oligo- or polynucleotides, the individual deoxyribonucleotides are linked via 3′,5′-phosphodiester bonds (Fig. 1). A proportion of the DNA bases may be specifically modified. In many plants, for instance, 5 to 10% of the cytosine residues are substituted by 5-methylcytosine; in animal DNA only 1–2% of the cytosine residues are methylated. In bacterial DNA, up to 1% of the cytosine and adenine residues are methylated (ARBER, 1974). In the DNA of the *Escherichia coli*-specific bacteriophages T2, T4, and T6, cytosine is completely replaced by 5-hydroxymethylcytosine, HMC (WYATT and COHEN, 1952); HMC can even be further modified by glucosylation (JESAITIS, 1956).

About 1950, E. CHARGAFF investigated the molar proportions of bases in DNAs of various organisms and discovered certain regularities in the DNA composition, the so-called Chargaff's rules: The sum of the purines is always equal to the sum of the pyrimidines; i.e., the sum of adenine (A) and cytosine (C) is equal to the sum of the keto bases guanine (G) and thymine (T) (CHARGAFF, 1951). The knowledge of the equivalence of A:T and of G:C was extremely important for the later established concept of the DNA double helix. Only in viruses (e.g., parvovirus) and bacteriophages (e.g., ϕX174 and M13) containing single-stranded DNA, are A and T or G and C not equimolar.

The physical properties of DNA are strongly influenced by the percentage of G + C content. G + C ratios as low as 20% and as high as 78% are known in lower eukaryotes and prokaryotes (ADAMS et al., 1986). Since the GC content of DNA is a characteristic marker of a genus or family, its determination has relevance in taxonomy (BALOWS et al., 1991). Depending on the level of development of an organism, the length of the total DNA varies from several micrometers in prokaryotes (e.g., *Mycoplasma*, 200 μm) to up to approximately 1 meter in human cells (ADAMS, et al., 1986).

Many questions with respect to the nature and function of DNA remain unanswered. For example: a) How must the long DNA fiber be structurally organized to fit in a small nucleus or bacterium? b) How is the DNA replicated and how is the nearly identical genetic information transferred to the progeny from gener-

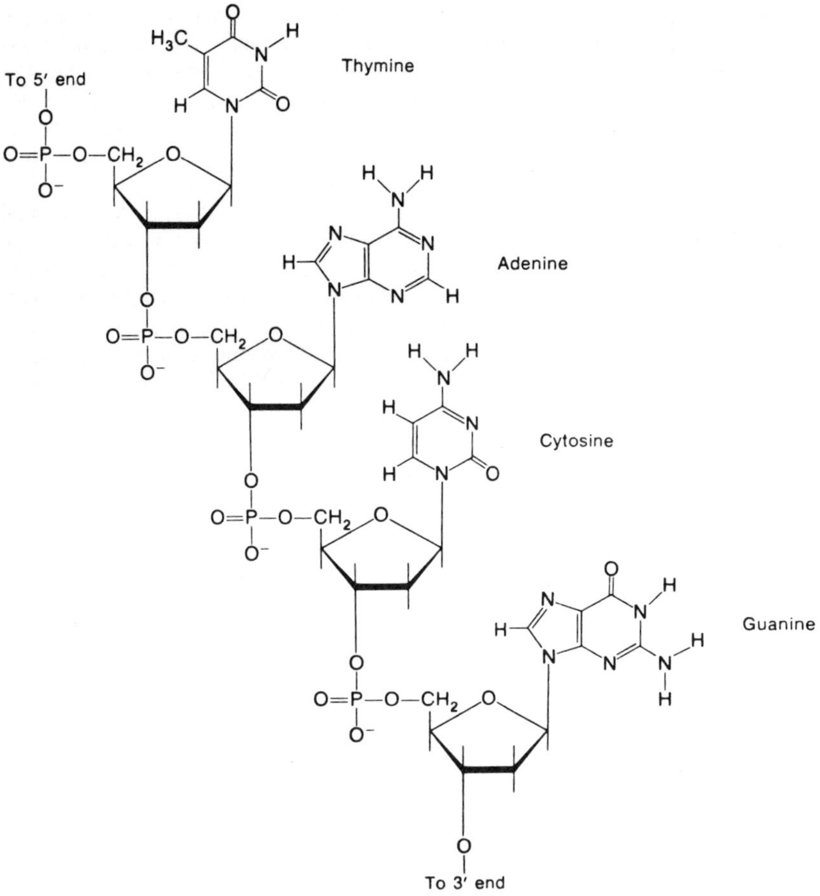

Fig. 1. A section of the primary structure of DNA, d(pTpApCpG).

ation to generation? c) How are genes expressed and how is gene expression regulated? The determination of the structure of DNA set the stage for the elucidation of some of the most important questions in biology.

I would like to recommend some excellent review articles and books on the biochemistry of nucleic acids (Adams et al., 1986), the three-dimensional structure of nucleotides, DNA and RNA (Dickerson, 1983; Felsenfeld, 1985; Saenger, 1984; Wells and Harvey, 1987), DNA replication (Kornberg, 1980), and unusual DNA structures (Wells, 1988).

2 X-Ray Crystal Structures of DNA

2.1 B-, A-, and Z-DNA

The most powerful method to determine the structure of nucleosides, nucleotides, or DNA fibers is X-ray crystallography. However, fiber diffraction analysis of quasi-crystalline polymers such as DNA and RNA gives only approximate overall structural information and cannot yield a satisfying structural model without additional data obtained by other methods. The ultimate goal of an X-ray crystal structure analysis is to be able to locate all atoms by analyzing the electron density. The first structure resolved with this method is the nucleoside cytidine (FURBERG, 1951). The first X-ray photographs of fibrous DNA already gave the impression that the bases were stacked upon one another (ASTBURY, 1947). WILKINS and colleagues showed that the DNA fiber is a helical molecule (WILKINS, 1963). By electro-titrimetric studies, GULLAND (1947) concluded that bases are linked by hydrogen bonding. On the basis of all this information, WATSON and CRICK (1953) proposed their double helix structure of DNA in which the two strands are held together by specific hydrogen bonds between the opposing base pairs A-T and G-C; with this model CHARGAFF's and GULLAND's data could be explained. Even at the time of WATSON and CRICK's discovery it was evident that DNA could exist in at least two forms: *B-DNA* is stable at a relative humidity of about 95% (the form WATSON and CRICK discovered) and is transformed to the *A-DNA* configuration when the humidity decreases. Below 75% humidity only the A-form is normally observed. A common feature of the A- and B-type DNA structures is that the double helix is right-handed. The DNA double helix contains a minor and a major groove. The floor of the major groove is paved with nitrogen and oxygen atoms that can make hydrogen bonds with certain residues of a DNA-binding protein and, therefore, can have a significant role in gene control. The minor groove is a poorer candidate for protein interaction. In B-DNA the base pairs are stacked nearly perpendicular to the helix axis which runs through the center of each base pair. In A-DNA the base pairs are tilted between 13 and 19 degrees from the perpendicular and are shifted toward the outside of the helix, whose axis lies in the major groove. B-DNA has an average of 10 bases per turn of the helix, while A-DNA has 11 bases. Since the B-DNA structure is dependent on the presence of high humidity, it was postulated that in most living cells the B-helix is predominant and plays the main role. These early studies were carried out with stretched natural DNA fibers.

Beginning in the late 1970s, when DNA synthesis made it possible to study single crystals of short molecules of any selected sequence, a number of studies showed that DNA can adopt various structures which are dependent on solvent, salts, temperature and DNA sequence. B-DNA, for example can be further subdivided in B-, B′-, C-, C′-, C″-, D-, E-, and T-DNA (LESLIE et al., 1980; ZIMMERMANN, 1982; SAENGER, 1984).

Compared with B-DNA, the D-DNA double helix is overwound and displays a very deep, narrow minor groove – a good cavity for trapping water and cations. D-DNA substitutes A-DNA only in AT-rich DNA sequences in synthetic DNAs; guanosine must be absent. In poly d(AT), the A-DNA conformation appears to represent only a metastable structured state (MAHENDRASINGAM et al., 1983). The only known natural DNA that can occur in D-DNA or T-DNA type conformation is phage T2 DNA (MOKUL'SKII et al., 1972). Phage T2 DNA is highly modified, containing 5-hydroxymethylcytosine instead of cytosine, and is in addition glycosylated to over 70%. The conformational parameters of this T-DNA are similar to those of D-DNA. The B-to-D helical transition occurs gradually with decreasing humidity.

It was a complete surprise, when in 1979 an entirely new type of DNA structure was reported which was neither A- nor B-DNA. The left-handed Z-DNA double helix is a repeating unit of purine-pyrimidine bases such as the hexamer d(CGCGCG) or its tetramer (WANG et al., 1979; DREW et al., 1980).

The existence of the various DNA structures shows that DNA is not a conformationally homogeneous molecule. Some of the main characteristic structural features of A-, B-, and Z-DNA will be briefly described. The DNA structures are shaped like a flexible ladder wrapped helically around a more or less central axis. The antiparallel, self-complementary strands of the double helix are held together by two and three hydrogen bonds between the bases A:T and G:C, respectively (Fig. 2).

A comparison of the structural properties of A-, B-, and Z-DNA as derived from single-crystal X-ray analysis is shown in Tab. 1. According to the computer-generated stereoscopic image (Fig. 3), the A-helix appears short and fat, the B-helix is slimmer and taller with a wide major groove and a narrow minor groove of comparable depth. The left-handed Z-helix

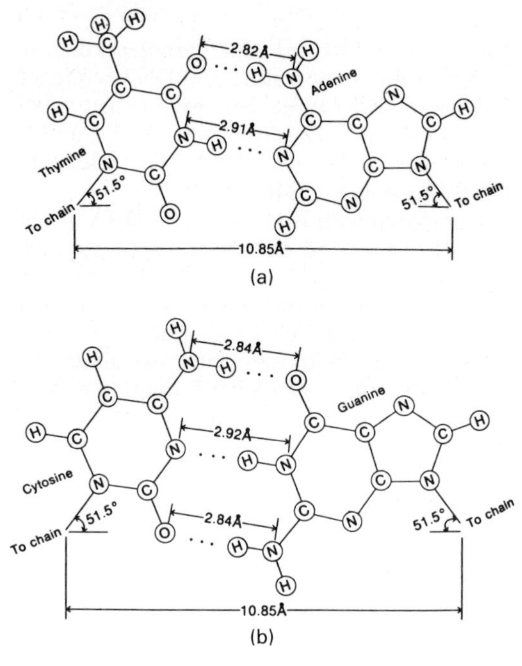

Fig. 2. Hydrogen bonds between the base pairs (a) thymine and adenine (T:A), and (b) cytosine and guanine (C:G).

is thin (the helix packing diameter is only 1.84 nm) and elongated (there are 12 bp within each helical turn of 4.56 nm), with a deep, narrow minor groove and a major groove that is pushed to the surface so that this groove is nearly eliminated. In B-DNA the helix axis runs through the base pairs, in A-DNA it is located in the major groove and in Z-DNA in the minor groove. A further characteristic of Z-DNA is that the glycosyl bond connecting deoxyribose with a pyrimidine is in the *anti* form whereas a purine is connected in the *syn* form (Fig. 4). In A- and B-DNA, the glycosyl bonds have only *anti* conformation. The alternating *anti-syn* configuration gives the backbone of Z-DNA its zigzag appearance, hence its designation as Z-DNA (DICKERSON, 1983). In Z-DNA, the mean value of the propeller twist, defined as the angle by which opposite base pairs are staggered, is only 4.4°.

2.2 DNA Bending

The principal crystalline form of DNA in X-ray diffraction studies is not only a result of its sequence. The double helical structures observed in natural or synthetic DNA are in-

Tab. 1. A Comparison of the Structural Properties of A-, B-, and Z-DNAs as Derived from Single-Crystal X-Ray Analysis

	Helix Type		
	A	B	Z
Overall proportions	Short and broad	Longer and thinner	Elongated and slim
Rise per base pair	0.23 nm	0.332 nm (0.019 nm)	0.38 nm
Helix-packing diameter	2.55 nm	2.37 nm	1.84 nm
Helix rotation sense	Right-handed	Right-handed	Left-handed
Base pairs per turn of helix	11	10	12
Pitch per turn of helix	2.46 nm	3.32 nm	0.456 nm
Mean propeller twist	+19°	−1.2° (4.1°)	−9°
Helix axis location	Major groove	Through base pairs	Minor groove
Major-groove proportions	Extremely narrow but very deep	Wide and of intermediate depth	Flattened out on helix surface
Minor-groove proportions	Very broad but shallow	Narrow and of intermediate depth	Extremely narrow but very deep
Glycosyl-bond conformation	*anti*	*anti*	*anti* at C, *syn* at G

Sources: DICKERSON et al. (1982); DICKERSON (1984).

fluenced by several other parameters such as humidity, solutes, counter ions (Na^+ or Li^+), or DNA modification. Some DNA sequences of selected synthetic DNA molecules and the resulting crystalline structures are summarized in Tab. 2.

The crystal structure analysis of the dodecamer d(CGCGAATTCGCG) revealed some interesting features. In this sequence, Z-DNA-compatible CGCG ends were coupled with a Z-DNA-incompatible AATT center, thus providing a test for the Z-forming power of the CGCG ends. The crystal structure of this molecule revealed not a Z- but rather a B-helix. With a closer inspection, however, there was a remarkable new finding that the helix is not straight but rather bent by 19°. It was assumed that this distortion was probably induced by crystal packing forces rather than being an intrinsic structural property of the complex (WING et al., 1980). However, it was later discovered that bending of DNA is very important for protein binding and gene regulation.

2.3 DNA–RNA Hybrids

DNA–RNA hybrids normally display an RNA-type double helix that is an A-helix structure. Duplexes between DNA and RNA are of biological importance, occurring when DNA sequences are transcribed into complementary RNAs. This process is catalyzed by RNA-polymerase, an enzyme that interacts with double-stranded helical DNA, separates the strands, and synthesizes RNA complementary to only one of the DNA strands, forming DNA–RNA hybrids. Such hybrids also occur when DNA is synthesized from RNA viruses by reverse transcriptase. DNA–RNA complexes (MILMAN et al., 1967) and some synthetic hybrids, such as poly (A) × poly d(T), poly d(I) × poly (C), and poly (I) × poly d(C), have been studied by X-ray fiber diffraction (ZIMMERMANN and PFEIFFER, 1981; O'BRIAN and MAC Ewan, 1970; CHAMBERLIN and PATTERSON, 1965). The former two adopt an A-RNA-type structure whereas the latter two more closely resemble A′-RNA. In general DNA–RNA hybrids are unable to exist as B-structures. There is, however, one exception. Under high humidity, the hybrid poly(A) × poly d(T) can transform from an A′-RNA-type into a double helix resembling B-form (ZIMMERMANN and PFEIFFER, 1981). Such hybrids are formed preferentially at the rho-independent transcription terminator regions. They represent a signal for bacterial RNA-polymerases to stop or disengage from the template.

2.4 B-DNA and B-to-A Transition

B-DNA, the stable DNA form under high humidity, is stabilized by water. The water molecules are in the vicinity of nearly every atom that can make hydrogen bonds with them, coating the helix with a layer of water one molecule deep. A second layer of water molecules runs down the minor groove as it spirals around the B-helix. DICKERSON (1983) postulated that the hydration in the minor groove spine is important for stabilizing the B-helix. When water molecules are removed by dehydration, or by introducing more NH_2 groups into the minor groove by increasing the GC content (ALDEN and KIM, 1979), the B-to-A transition is easier. Indeed it was shown by density gradient centrifugation studies and theoretical calculations that an A-T pair can bind one or two water molecules more than a G-C pair (TUNIS and HEARST, 1968; GOLDBLUM et al., 1978). All these findings may explain why the DNA conformation is largely dependent on its AT and GC content.

Surface accessibility calculations with a probe have shown that dominant base exposure occurs in the major groove of B-DNA and in the minor groove of A-DNA. These findings suggest that for protein–nucleic acid interactions with intact DNA double helices, specific contacts between base pairs and amino acid side chains occur preferentially in the major groove of B-DNA and in the minor groove of A-DNA. This is in agreement with data on protein B-DNA complexes with *E. coli* polymerase-*lac* promoter, Lac repressor-*lac* operator (ALDEN and KIM, 1979), lambda and *cro* repressors, and DNA–histone interactions in nucleosomes (KLUG et al., 1980; AZORIN et al., 1980).

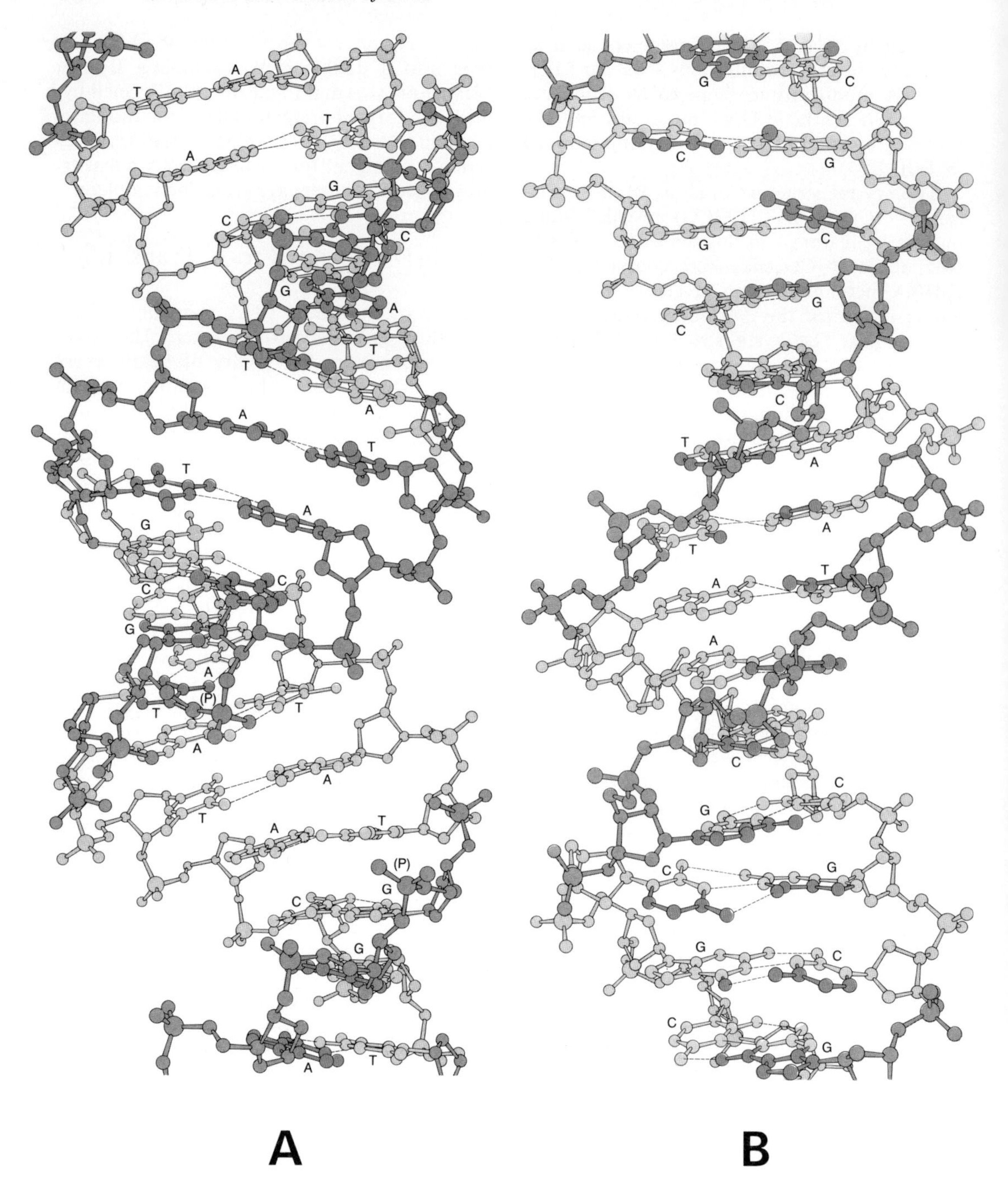

T
A
A
T
G
C
C
G
A
T
T
A
A
T
T
A
G
C
C
C
G
A
(P)
T
T
A
A
T
A
T
(P)
G
C
G
A
T
G
C
C
G
G
C
G
C
C
T
A
T
A
T
A
A
C
C
G
C
G
G
C
C
G
A
B

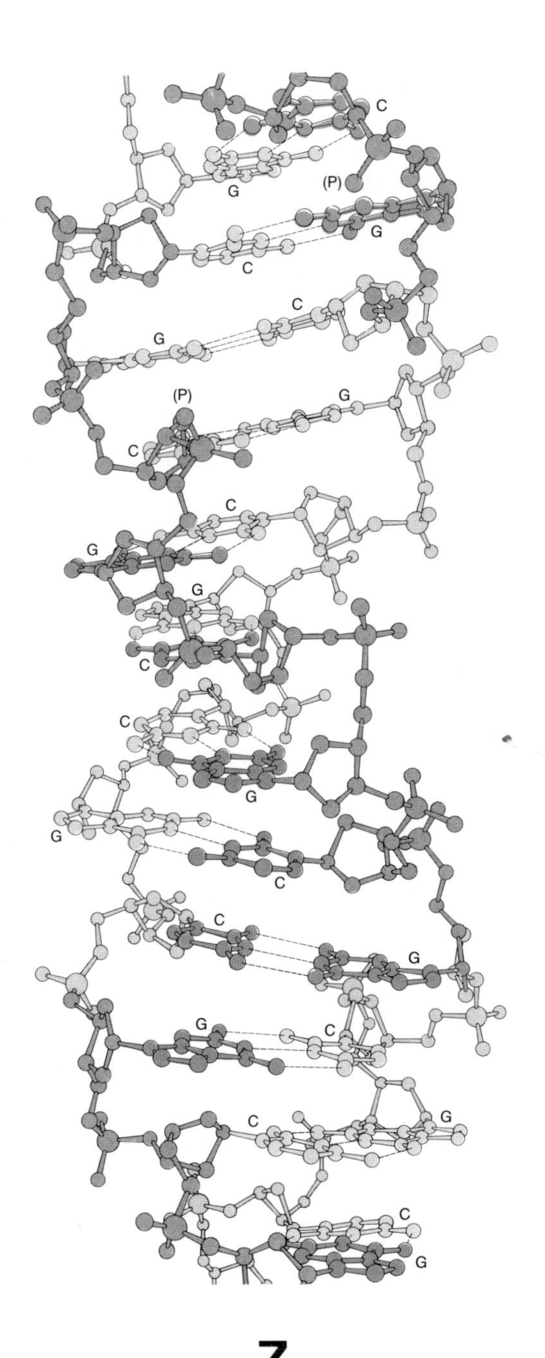

Fig. 3. Comparison of A-, B-, and Z-DNA helix in the perspective drawing. In A-DNA, the stereo pair was generated by extending the central six bases of d(GGTATACC); in B-DNA repetitive sequences of d(GCGAATTCG) and in Z-DNA repetitive sequences of d(GCGC) are shown (WANG et al., 1979). Hydrogen bonds between the bases are indicated by dotted lines. (Drawings are reproduced from SUZUKI et al., 1991).

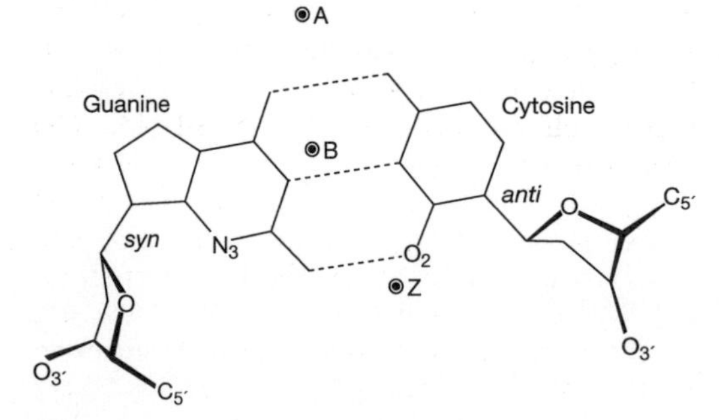

Fig. 4. The *syn-anti* configuration of the glycosyl bonds connecting guanine and cytosine in Z-DNA (according to DICKERSON, 1983).

2.5 Non-B-DNA in $(CA)_n$ Tracts

The X-ray structure of a non-palindromic DNA dodecamer containing the *Nar*I restriction site GGCGCC, a "hot spot" for AAF mutagenesis, has been described (TIMSIT et al., 1991). The duplex crystallizes in the B-form, but with major deviations from the canonical structure. The crystal packing of the B-DNA dodecamer d(ACCGGCGCCACA) is characterized by the reciprocal fit of double helices with specific base–backbone interactions in the major groove. The tilt of the bases leads to the disruption of the Watson–Crick pairing in the

Tab. 2. Influence of DNA Sequence on its Structure, as Determined by X-Ray Fiber Diffraction and Single-Crystal Studies

DNA Sequence[a]	DNA Conformation (% Relative Humidity)	Reference
Native DNA (Na^+)	A (75%)	FULLER et al., 1965
(Na^+)	B (92%)	LANGRIDGE et al., 1960
d(CCGG)	A	CONNER et al., 1982
d(GGCCGGCC)	A	WANG et al., 1982a
d(GGTATACC)	A	SHAKKED et al., 1983
d(GCCCGGGC)	A	HEINEMANN et al., 1988
d(GTGTACAC)	A	JAIN et al., 1987
d(GGATGGGAG)	A-like	MCCALL et al., 1986
RNA/DNA hybrid:		
r(GCC)d(TATACGC)	A-RNA	WANG et al., 1982b
d(CGCGAATTCGCG)[b]	B/bent	DICKERSON and DREW, 1981
d(GAATTC)	B	DICKERSON and DREW, 1981
d(CGCAAATTTGCC)	B	COLL et al., 1987
d(CGCAAAAAAGCG)	B	NELSON et al., 1987
d(ATAT)[c]	"Alternating" B	VISWAMITRA et al., 1982
d(AT)[d]	Hoogsteen/B	HOSUR et al., 1981
Poly d(AT) × Poly d(AT)	D (75%)	ARNOTT et al., 1974
	A(>98%)	DAVIES and BALDWIN, 1963
Phage T2-DNA	B (>95%)	MOKUL'SKII et al., 1972
	D (60%)	
d(CGCGCG)	Z	WANG et al., 1979
d(CGCG)	Z	DREW et al., 1980

[a] Only one strand is shown.
[b] Z-compatible CGCG ends are coupled with a Z-incompatible AATT center; the sequence contains an *Eco*RI site and is also self-complementary; the overall helix is not straight, but rather bent (curved).
[c] The sequence has a non-helical structure. The two base pairs at each end of the tetranucleotide form hydrogen bonds to two different adjacent tetranucleotide molecules, which yield short segments of right-handed antiparallel double helix.
[d] MAHENDRASINGAM et al. (1983) have also proposed a left-handed helix for poly d(AT).

major groove and to the formation of interactions with the 5′-neighbor of their complement. In the *c-Ha-ras* proto-oncogene, a similar structural feature is seen, and it is believed that this structure is sequence-dependent and favored by $(CA)_n$ tracts (TIMSIT et al., 1991), which are known to be involved in both recombination and transcription. It was suggested that the shifted structure is associated with the -CCACA- sequence of the dodecamer, leading to two phenomena: a) the strongly modified structure of the major groove can constitute a new recognition element for DNA-binding proteins, and b) the slippage of pairing induces a transient linear modification of the complementary code which could interfere with the processes associated with decoding of the genetic information, thus explaining their important biological role (RODGERS, 1983). It was suggested that the common unusual structural feature of CAC triplets can be used for gene regulation and recombination (CHEUNG et al., 1984). CAC sequences are required in a number of transcription regulation regions (COWIE and MEYERS, 1988). $(CA)_n$ tracts are involved in gene conversion, insertion, and enhancement of recombination. They may also be implicated in gene duplication mechanisms (SHEN et al., 1981).

The determination of crystalline DNA structures by X-ray diffraction raises the question as to whether these structures are relevant in nature where DNA molecules are in solution. X-ray analyses have already shown that the DNA structure is dependent on water content, counter ions, and salt concentration. It is, therefore, not surprising that a remarkable local helix variation in A- and B-DNA was observed, indicated by the large local deviations from the average values. DNA is not the rigid molecule as it first appeared.

3 DNA Structures in Solution

With the emergence of techniques such as circular dichroism (CD), Raman spectroscopy, and nuclear magnetic spin resonance (NMR), it was possible to study DNA structures in solution (NILSSON et al., 1986; PATEL et al., 1987). Raman spectroscopy, for example, allows determination of the sugar ring conformation characteristic of A- or B-DNA. CD spectroscopy can supply information on the extent of base–base interaction in the helical stack, overall stability of the helix, and can show gross differences in the helix structure. NMR-elucidated structures of DNA are not simple in refinement, since the differences in H–H distances in model A- and B-DNA helices are not of a great enough magnitude. A new method of directly observing native DNA structures with the scanning tunneling microscope (STM) has been reported (BEEBE et al., 1989). Uncoated double-stranded DNA dissolved in a salt solution is deposited on graphite and imaged in air with the STM. The resolution is such that the major and minor grooves can be distinguished.

3.1 B-to-Z Transition

It was a great surprise when POHL and JOVIN (1972) found for the first time that the poly d(GC) double helix in aqueous solution does not have a Z-DNA structure as expected from the X-ray diffraction studies, but has a B-DNA structure. However, when the salt concentration is increased to about 2.5 M (mol/L) NaCl, a conformational transition to Z-DNA occurs. This conformational transition between two different double helical forms is accompanied by changes in optical rotatory dispersion, circular dichroism, and ultraviolet absorption. Subsequently, other polymers with alternating purine-pyrimidine sequences, such as derivatives of poly d(GC) or poly d(AC), have been seen to change into the Z-form (BEHE and FELSENFELD, 1981; MÖLLER et al., 1984).

The transition from B- to Z-DNA, i.e., the conversion of right-handed to left-handed DNA, does not require strand separation, it very likely proceeds intrahelically. After an initial separation of base pairs, guanine flips over into *syn* conformation and the entire deoxycytidine rotates, thus retaining the *anti* orientation of the base. The B-Z transition, therefore, travels along the helix as a bubble (POHL and JOVIN, 1972; WANG et al., 1979).

Tab. 3. DNA Conformation in Solution (Circular Dichroism). Concentrations of Sodium Chloride and Ethanol at the Midpoint of the B–Z, B–A, or Z–A Transitions

Synthesized Oligonucleotide	B–Z Transition (NaCl, in mol L^{-1})	B–A Transition[a] (Ethanol, in %)
d(GTACGC)		80
d(GTACGCGC)	4.8	71
d(GTACGCGCGC)	4.2	70
d(GTACGCGCGCGC)	3.8	69
d(GTACGCGCGCGCGC)	3.5	65
d(GTACGCGCGCGCGCGC)	3.3	63; 75[b]
d(GTGCACGCGCGCGCGC)	3.8	68
d(ACGCGCGCGTGCGCGC)	4.0	72
d(CGCGCGTGCA)	4.8	71
Poly d(GC)	2.3	56; 70[b]

Source: LUTHMAN and BEHE, 1988

[a] A B–A transition occurs for polymers that have one ethanolic transition.

[b] For polymers that show two ethanolic transitions, the first is a B to Z and the second is a Z to A transition.

A number of reports indicate that regions of naturally occurring alternating purine-pyrimidine sequences are capable of forming Z-DNA under supercoiling tension or in the presence of specific binding proteins (KILPATRICK et al., 1984; AZORIN and RICH, 1985; HANIFORD and PULLEYBLANK, 1983). These regions contain A-T and G-C base pairs in various proportions and arrangements.

LUTHMAN and BEHE (1988) carried out a systematic study with eight synthesized double-stranded polydeoxynucleotides. The polymers have defined alternating purine-pyrimidine sequences with repeating units of 6–16 base pairs and contain 12.5–33% AT base pairs. A and T nucleotides are in nearest-neighbor positions in the series poly d-TA$(CG)_{2-7}$ (see Tab. 3). All the polymers, except poly d-TA$(CG)_2$, were shown by circular dichroism to undergo a B-Z transition at high NaCl concentrations. All polymers exhibit a B-A transition in the presence of ethanol. Poly $d(GC)_n$ and poly d-TA$(CG)_7$ undergo a B-Z-A transition in ethanol (Tab. 3). B-Z transitions were also studied with the oligonucleotides d(CGCGTACGCG) and d(TGCGCGCGCA) (CHEN, 1988). The former oligomer, in which alternating GC base pairs are interrupted by two TA base pairs, requires much more strenuous conditions to undergo the B-Z transition than does the latter, in which a long run of CG base pairs occurs uninterrupted. A duplex of the oligonucleotides d(GGGGGTTTTT) forms an A-B conformational junction in concentrated salt solution (WANG et al., 1989).

The B-Z transition is sequence- and salt concentration dependent. As a rule, one can say that the longer the runs of contiguous GC base pairs in the repeating unit of the polymer, the lower the NaCl concentration at which the B-Z transition may occur. Concentrated NaCl solution has both a dehydrating and an electrostatic screening effect on DNA molecules. Only a combination of both effects leads to the observed B-Z transition. Ethanol has only a dehydrating effect on the DNA, leading to a B-A transition, consistent with the structures of low- and high-humidity crystals of DNA fibers (see Tab. 1). Such a transition is also observed in circular dichroism studies. Only in the case of poly d(GC) and poly d-AT$(CG)_7$ are there two ethanolic transitions, the first a B-Z and the second, at higher ethanol concentrations, a Z-A transition. For example, poly d(GC) changes from the B- to the Z-form at 54% ethanol and changes to an A-conformation at concentrations above 70% (POHL, 1976).

3.2 Sequence, Probing, and Biological Function of Left-Handed Z-DNA

a) Z-DNA structure of alternating d(TG) and d(CA) sequences. Alternating purine-pyrimidine (poly d-TG and poly d-CA) sequences also adopt a left-handed structure, an important result since these sequences are widely distributed in eukaryotic DNA (HAMADA and KAKUNAGA, 1982). For example, the intervening sequences (IVS2) of three human fetal globin genes contain tracts of alternating dTG sequences approximately 40–60 base pairs in length which adopt left-handed Z-DNA helices under the influence of negative supercoiling. Since these simple sequences appear to be hotspots for recombination and gene conversion, unusual DNA conformations may participate in genetic expression. The conformation of B-Z junctions is not known. Because of the sensitivity of the junctions to single-strand-specific nucleases, it is assumed that the junctions possess elements of non-helical structure as found in random coil polynucleotides (KILPATRICK et al., 1984; SINGELTON et al., 1984).

It is interesting to note that several sequence studies have shown the presence of regions of alternating purine-pyrimidine residues in various DNAs (HAMADA and KAKUNAGA, 1982; TAVIANINI et al., 1984).

b) Left-handed DNA structure of consecutive d(AT) pairs. A series of synthetic oligodeoxynucleotides were cloned into the *Bam*HI site of pRW790, a small plasmid constructed for conformational studies. Supercoil relaxation studies using two-dimensional gel electrophoresis of topoisomers of each recombinant plasmid revealed the structures of Z-helices. From these results it was concluded that consecutive T-A base pairs, whether alternating (TATA) or contiguous (TTTT), can adopt a left-handed conformation (presumably Z) when flanked by reasonable short runs of alternating $(C\text{-}G)_n$ $(n = 3\text{–}5)$ (MCLEAN et al., 1986).

There is no doubt now that Z-DNA plays a much greater role in controlling biological functions than was thought when this was first proposed in 1970 (MITSUI et al., 1970) and later proven by X-ray diffraction studies (WANG et al., 1979; DREW et al., 1980). Because of the importance of Z-DNA, some further characteristics of this DNA structure are described. There are also excellent review articles on left-handed Z-DNA by RICH et al. (1984), LENG (1985), and WELLS (1988).

By means of physicochemical (RICH et al., 1984) and immunological techniques (REVET et al., 1984), it is firmly established that under physiological salt conditions, fragments $d(CG)_n$, $d(GT)_n$, and sequences of alternating purine and pyrimidine residues of negatively supercoiled DNAs with one non-alternating residue can be in the Z-conformation. The free energy of plasmid negative supercoiling appears to be one of the most important factors in inducing the B-Z transition. Forming Z-DNA is a means of reducing the number of superhelical turns (RICH et al., 1984).

c) Z-DNA and interaction with chemicals and drugs. In NaCl or $NaClO_4$, the equilibrium between the B- and Z-forms for poly d(GC) is independent of temperature, while in LiCl it is shifted towards the Z-form at high temperature. Opposite effects of temperature seem to occur with poly d(GC) in water/methanol or ethanol solution where an increase in temperature from 20 to 50 °C results in the complete transition from Z- to B-form; a reversible transition (KARAPETYAN et al., 1984). Chemical modifications of the bases play a role in the relative stabilities of B- and Z-DNA. Bromine atoms and the chemical carcinogen 3-N-acetylamino-4,6-dimethyldipyrido imidazole covalently bind to the C-8 of guanine residues stabilizing the Z-conformation. On the other hand, the carcinogen 4-nitroquinoline-1-oxide stabilizes the B-conformation (MÖLLER et al., 1984; HEBERT et al., 1984; BAILLEUL et al., 1984).

The anti-tumor drug *cis*-diamminedichloroplatinum(II) (*cis*Pt) binds to B- and Z-DNA. The drug forms a monodentate adduct with Z-DNA and a bidentate with B-DNA; the drug chelates two guanine residues. The monodentate leads to stabilization of Z-DNA, the bidentate induces large conformational changes (MALINGE and LENG, 1984; REVERT et al., 1984).

Z-DNA may support binding by both intercalation and bis-intercalation. Spectroscopic data indicate that the complex formed by ethidium bromide, a DNA intercalating drug, and Z-DNA resembles that with B-DNA, the binding constant being larger with B-DNA than with Z-DNA. There also appears to be no difference in the binding of the bifunctional intercalator bis(methidium) spermine to B- and Z-DNA. However, the chiral complex tris(4,7-diphenyl-1,10-phenanthroline)-ruthenium(II) (RuDIP) can be used as a probe to distinguish right- and left-handed DNA in solution. There are two enantiomers of RuDIP, both bind equally well to Z-DNA by intercalation, but only one enantiomer binds to B-DNA (BARTON et al., 1984).

DNA complexing drugs, such as the antibiotic daunomycin, complex to d(CGTACG) and influence the DNA structure (WANG et al., 1987). Proflavin, an intercalating drug, does not intercalate Z-DNA, but rather is sandwiched between adjacent stacked duplexes and binds on the outside in the deep (minor) groove of Z-DNA (WESTHOF et al., 1988).

The 1,10-phenanthroline copper ion complex and H_2O_2 acts as an artificial DNAse by cleaving B-DNA but not Z-DNA (POPE and SIGMAN, 1984). B- and Z-DNA are equally methylated on the N-7 of guanine residues by dimethyl sulfate. In alkaline conditions, the imidazole ring of methyl-7-guanosine residues opens up, yielding 2,6-diamino-4-oxo-5-methylformamidopyrimidine, a lesion which blocks DNA replication. This lesion can be removed enzymatically from modified poly d(GC) in the B-conformation but not in the Z-conformation by a DNA glycosylase (LAGRAVERE et al., 1984).

d) Antibodies against Z-DNA structure. Injection of Z-DNA into rabbits and mice induces the synthesis of antibodies specifically recognizing Z-DNA (HANAU et al., 1984; LEE et al., 1984; ZARLING et al., 1984). The presence of anti-Z-DNA antibodies in autoimmune and some other diseases has been reported (MADAIO et al., 1984). Anti-Z-DNA antibodies bind to chromosomes as shown by indirect immunofluorescence and peroxidase techniques. Questions were raised about the possible perturbations of the fixatives (acetic acid, alcohol, etc.) used for cytological preparations. Acetic acid may have dramatic effects at the molecular level, such as the removal of proteins which can change DNA accessibility, topological stress, or in facilitating the B-Z transition by protonation of the bases. Studies were performed on unfixed polytene chromosomes isolated from the salivary glands of *Chironomus thummi* larvae and from *Drosophila* (ROBERT-NICOUD et al., 1984; HILL and STOLLAR, 1983; HILL et al., 1984). At neutral pH and physiological ionic strength, there is only a background level binding of antibodies to Z-DNA. Exposure to acid pH leads to an abrupt increase in antibody binding. Nicking of acetic acid-treated chromosomes by DNAse I, topoisomerase or S1 nuclease greatly decreases Z-DNA immunoreactivity. The immunoreactivity also depends on the temperature and ionic strength.

e) Lack of restriction endonuclease and methylase activity on sites within or close to Z-DNA. The rate of B–Z transitions of poly d(GC) inserts in a plasmid seems to be more rapid than that of poly d(GC) induced by salt. Cleavage of these inserts by the restriction enzyme *Bss*HII (recognition site: GCGCGC) is strongly dependent upon the negative superhelical density, which promotes Z-conformation. When the poly d(GC) inserts are in Z-conformation, cleavage is greatly inhibited. When 50% of the inserts are in Z-conformation, as judged by an immunological assay with Z-DNA specific antibodies, *Bss*HII can still cleave to completion, which suggests a dynamic equilibrium between B- and Z-conformations.

The B–Z junction in supercoiled plasmids containing poly d(GC) inserts was also investigated with other restriction enzymes. The *Bam*HI recognition site (GGATCC) was placed various distances from the poly d(GC) insert in Z-conformation (AZORIN et al., 1984). Cleavage by *Bam*HI is slowed down when the number of base pairs is four or less between the insert and the recognition site. No inhibition occurs at a distance of eight base pairs.

Similar findings were obtained with two other enzymes, *Hha*I DNA methyltransferase and *Hha*I endonuclease, which are unable to methylate and cleave at GCGC sites when the recognition site in long GC-tracts (>30 bp) adopts Z-conformation (VARDIMON and RICH, 1984; ZACHARIAS et al., 1984). The inhibition of methylation is preferentially found in supercoiled plasmids with a supercoil-induced B-to-Z transition.

In summary, one can say that in a given plasmid the GCGC sites are poorly methylated or cleaved when they are in an insert in the Z-conformation as compared to the same site in B-conformation outside of the insert. However, a low but significant amount of methylation is observed under conditions in which the inserts are in the Z-conformation, explained by a dynamic B–Z equilibrium of the inserts. An interesting point is that since methylation of C-5 cytosine residues in poly d(GC) inserts stabilizes the Z-conformation, as the inserts of the B-conformation are methylated by methyltransferase, the B–Z transition of the inserts becomes easier and the enzyme is inhibited. This might be related to the regulatory effects of methylation.

f) Left-handed DNA in vivo. JAWORSKI et al. (1987) were the first to describe the existence of Z-DNA *in vivo*, showing that Z-DNA elicits a biological function in *Escherichia coli.* The *in vivo* assay employed was based on the earlier observations that a recognition site is not methylated or cleaved by its specific methylase or restriction enzyme, respectively, when the site is near or in a Z-helix. An *E. coli* clone with a plasmid containing the gene for a temperature-sensitive *Eco*RI methylase (MEcoRI) was cotransformed with various plasmids containing inserts which had varying capacities to form Z-DNA with a target *Eco*RI site in the center or at the ends of the inserts. Inhibition of methylation of the inserts was found for stable inserts with the longest left-handed (presumably Z) helices. *In vitro* methylation with purified MEcoRI agreed with the results obtained *in vivo.* Supercoiled-induced changes in the structure of the primary helix *in vitro* provided confirmation that Z-DNA is responsible for this behavior.

Other biological effects were also observed with Z-DNA. The stability of the inserts is influenced by their location in the plasmids. For example, insertion of a 56-bp insert, capable of forming a Z-helix, into the *Bam*HI site of pRW460 is not stable, i.e., deletions in the inserts are observed. When the same insert is cloned into the *Eco*RI site of pRW1560, the insert is quite stable. Deletions of (CG) inserts in the *Bam*HI site of pBR322 were reported earlier (MCLEAN et al., 1986) and could be a consequence of transcription, since the *Bam*HI site is in the tetracycline resistance gene. Alternatively, the effect may be due to an influence of neighboring sequences, as was shown for an AT-rich region on a local structural transition (SULLIVAN and LILLEY, 1986). It was also observed that GC tracts which are too long cause plasmid incompatibility.

g) Biological function of Z-DNA. It is now well-established that not only oligonucleotides in crystals, but also olignucleotides, double-stranded polynucleotides, and fragments of natural DNAs in solution can be in the Z-conformation.

It is tempting to speculate on the biological roles of Z-DNA. One has to consider short-range and long-range effects of the Z-DNA conformation. Short-range effects deal with the interactions between a fragment of DNA and ions, small molecules, proteins or enzymes, which can differ dramatically depending upon the conformation of the fragment. Long-range effects deal with a conformational change in a DNA fragment which is experienced by another fragment located far away along the DNA molecule.

Thus, Z-DNA could play a role in: (1) the phasing of nucleosomes; (2) DNA repair; (3) regulation of gene expression; (4) recombination, etc. In fact, the polymorphism of DNA is expected to affect all the cellular events involving DNA–DNA, DNA–RNA and DNA–protein interactions. Whether this polymorphism actually occurs *in vivo*, remains to be determined. Evidence for the biological function of Z-DNA follows.

It is well documented that DNA structures change upon formation of a complex with protein. This was found with binding of restriction enzymes to their respective DNA binding site. In the case of the *Eco*RI DNA binding site, the conformation of the oligonucleotide d(CGCGAATTCGCG) in solution and its crystalline complex with the restriction nuclease *Eco*RI were determined (MCCLARIN et al., 1986; THOMAS et al., 1989). Environmentally-induced conformational changes in B-type DNA occur.

The importance of DNA conformation (B-form versus Z-form) in the interaction with polypeptides, protein, and enzymes has already been reported (RICH et al., 1984). H-1 and H-3 histones bind more to Z-DNA than B-DNA, and the binding to Z-DNA is less sensitive to an increase in salt concentration (MURA and STOLLAR, 1984).

Poly d(CA) sequences are able to switch from B- to Z-conformation in physiological ionic conditions when cloned in supercoiled plasmids. These sequences are widely distributed in the eukaryotic genome (HAMADA and KAKUNAGA, 1982) and have been found in introns of human actin genes, human fetal gamma globin genes, and in the 5′-flanking region of the human parathyroid hormone gene. Transcription of the *Caenorhabditis elegans* $tRNA^{Pro}$ gene is inhibited when poly d(CA) sequences are placed in the flanking regions of the gene or between the split promoter (SANTORO et al., 1984).

Z-DNA binds tightly to *Ustilago* Rec1, a protein which promotes homologous pairing and strand exchange between DNAs. The binding reaction is strongly dependent on ATP, but complexes formed are rapidly dissociated by ADP. The parallel between the kinetics of Z-DNA binding and the synaptic pairing reaction leading to paranemic joint molecules suggests that formation of nascent heteroduplex structures in recombination is coupled with formation of left-handed Z-like DNA on the protein. Rec1 has a strong Z-DNA binding site that binds Z-DNA 75-times tighter than the B-form. It was proposed that Z-DNA is a key intermediate in the homologous pairing promoted by Rec1 protein (KMIEC et al., 1985).

pBR322 derivatives were also constructed containing tracts of d(CG) of varying length (KLYSIK et al., 1982). Sequences longer than approximately 50 bp suffer deletions; segments of 30 bp or less are in most cases stable. The d(CG) tracts seem to enhance a RecA-mediated recombination, when they are of suitable length and are cloned into certain sites of the plasmid. The d(CG)-containing plasmids are also less supercoiled (by 6–12 turns) relative to the control plasmid. It was later found that Z-DNA is preferentially bound by RecA in the presence of ATP; ADP inhibited binding (BLAHO and WELLS, 1987). RecA and Rec1 have a similar capacity for binding Z-DNA.

3.3 Does A-DNA Exist in Living Cells?

There is no doubt that sequence determines DNA structure. A circular dichroism study was conducted on the solution structure of several different oligonucleotides whose X-ray structures were solved. In solution the oligonucleotides can form structures that maintain geometrical elements which are typical of B- and A-DNA and their intermediate forms. The oligonucleotide d(GGATGGGAG) forms an A-DNA helix in the crystalline state (MCCALL et al., 1986). In aqueous solution, an A-DNA-like structure is only observed at temperatures below 10 °C. At temperatures between 10 and 25 °C, it assumes a structure intermediate between A- and B-DNA (GALAT, 1990).

The DNA in dormant spores of *Bacillus* species is associated with small, acid-soluble, double-stranded DNA binding proteins (SASP), whose amino acid sequence has been highly conserved in evolution. *In vitro* these proteins bind most strongly to DNA which readily adopts an A-like conformation (SETLOW, 1992). Associated with the conformational change in DNA is a change in its photochemistry, such that ultraviolet irradiation does not generate pyrimidine dimers, but rather a thyminyl–thymine adduct termed spore photoproduct (SP). In dormant spores of *Bacillus,* DNA has quite similar properties. Plasmid DNA has a much higher negative supercoiling, and the change in DNA conformation by SASP may prove advantageous to the spore, as its DNA is UV resistant. If A-like DNA can exist *in vivo*, why has it not been seen in other systems? The answer might be that A-like DNA results in a drastic reduction in RNA synthesis and, therefore, cell growth and in increased spontaneous mutation and often cell death. Conversion of a cell's genome to an A-like conformation in a dormant bacterial spore, in which gene expression as well as metabolic activity is absent, may not be risky. SETLOW (1992) stated that A-like DNA is a dormant adaptation.

4 Supercoiled DNA Structure and Function

VINOGRAD and coworkers found as early as 1965 that closed circular DNA molecules of polyoma virus are twisted and more compact than their nicked or linear counterparts. The twists and the compact DNA structure, which could be also visualized by electron microscopy (WEIL and VINOGRAD, 1963), were then referred to as supercoiled DNA (see review articles: DRLICA, 1984, 1990). Enzymes that regulate supercoiling were discovered by WANG (1971) and GELLERT et al. (1976). It is now clear that DNA supercoiling plays a central role in bacterial chromosome biology.

When circular DNA molecules are extracted from bacterial cells, they have fewer duplex turns relative to linear DNAs of the same length. This deficiency in duplex turns, also referred to as negative supercoiling, causes the molecule to writhe. Negative supercoiling is spontaneously relieved or relaxed by nicks or breaks in the DNA that allow strand rotation. Consequently, supercoiling is only found in circular DNA molecules in which strands cannot rotate.

Biological processes that separate strands relieve negative superhelical strain; these processes, such as DNA replication and transcription, tend to occur more readily in supercoiled than in relaxed DNA. Negative supercoiling also affects the three-dimensional configuration of DNA, facilitates loop formation and allows DNA to wrap around proteins (DRLICA, 1990). In a sense, negatively supercoiled DNA is energetically activated. Negative supercoiling not only compacts DNA but also enhances binding of intercalating agents such as ethidium bromide, and facilitates structures such as cruciforms and Z-DNA. The extent of supercoiling in extracted DNA reflects the intracellular enzymatic action of topoisomerases which alter the number of turns in DNA by breaking and rejoining, as shown in Fig. 5.

In only approximately 40% of a bacterial DNA is negative superhelical tension detectable; the residual portion of DNA is probably denatured by proteins and/or wrapped around proteins, such as the histone-like proteins HU (BROYLES and PETTIJOHN, 1986).

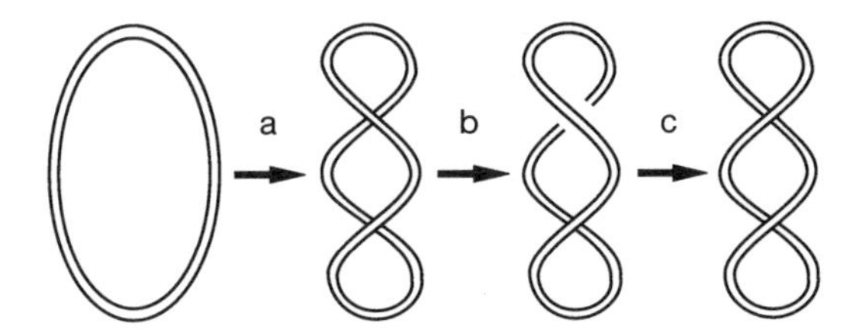

Fig. 5. Negative supercoils generated by gyrase. Both strands of the double helix are broken, a duplex helix is passed through the break and then the break is sealed (according to DRLICA,1990).

4.1 Supercoil Control by Gyrase and Topoisomerase I

There are two topoisomerases in bacteria which control DNA supercoiling (DRILCA, 1984). DNA gyrase (topoisomerase II) hydrolyzes ATP and introduces negative supercoils (COZZARELLI, 1980). In the absence of ATP, the enzyme removes supercoils. An inhibition of the gyrase activity by specific inhibitors (quinolones or coumarins) leads to a loss of negative supercoils in the DNA, and as a consequence, DNA replication, recombination, and transcription are affected. Gyrase is counterbalanced by topoisomerase I, which converts supercoiled DNA to the unstrained, energetically more favorable, relaxed state. Bacterial DNA is normally maintained in a negatively supercoiled state, regulated by the balance between the supercoiling activity of DNA gyrase and the opposing activity of topoisomerase I.

DNA gyrase is site-specific (LOCKSHON and MORRIS, 1985). Oxolinic acid or nalidixic acid form complexes with gyrase and DNA, and cause DNA cleavage in the presence of detergents such as sodium dodecylsulfate. Corresponding cleavage sites have been mapped, and it was estimated that the entire *E. coli* chromosome (4.4×16^6 base pairs) contains approximately 10000 gyrase binding sites (FRANCO and DRLICA, 1988). There are low-affinity and a small number of high-affinity sites. High-affinity binding sites on the chromosome are not clearly defined; chinolone cleaves the chromosome into about 50 pieces, and some of them may arise from cleavage at specific sites (DRLICA, 1990).

Such a strong DNA gyrase-binding site was discovered in bacteriophage Mu DNA (PATO et al., 1990). Mutants with single nucleotide changes have been isolated which are able to replicate in *gyrB* mutants of *E. coli*. These mutations strengthened the gyrase-binding site. Deletion of the site delayed onset of phage replication. Another example of a high-affinity binding site for gyrase is the 100 bp *par* sequence of plasmid pSC101 (WAHLE and KORNBERG, 1988). Within *par*, AT-rich sequences occur with a pronounced 10 bp periodicity that is shifted by 5 bp from a similar periodicity of GC-rich sequences. It was postulated that DNA gyrase contributes to the maintenance and/or segregation of bacterial replicons during cell division.

In *Escherichia coli* and *Salmonella typhimurium,* a family of extragenic palindromic (REP) sequences has been discovered in the chromosome. The REP family exists in hundreds of copies per genome, which are distributed throughout the chromosome and are always located outside structural genes. It has been estimated that these sequences comprise 0.5% of the chromosomal DNA. They may be involved in regulating intraoperonic gene expression; indeed, they are involved in protecting mRNA from degradation and in modulating translational initiation. However, both effects can be explained as a simple consequence of the palindromic secondary structures formed in mRNA. Since REP sequences are specifically bound by DNA gyrase, it was speculated that the REP sequences are involved in the higher-order structure of the bacterial chromosome (YANG and AMES, 1988). DNAse I footprinting revealed that gyrase protects 205 bp on a REP-containing DNA. The REP consensus sequence and the pBR322 strong gyrase cleavage sites have a high degree of homology.

4.2 Cellular Energetics and DNA Supercoiling

Since ATP is required for gyrase to introduce negative supercoils into DNA and since in the absence of ATP gyrase removes supercoils, it is not unexpected that the [ATP]:[ADP] ratio strongly influences the level of supercoiling (WESTERHOFF et al., 1988). Although topoisomerases tend to maintain supercoiling at a set level, this level varies in different growth environments (DRLICA, 1992). When *E. coli* cells are shifted from aerobic to anaerobic conditions, the ATP level drops and the linkage number of the pUC9 plasmid increases. Salt concentration also affects supercoiling. Addition of salt to culture media causes both supercoiling and an increase in the [ATP]:[ADP] ratio. Both later drop to a steady-state level that remains higher than that observed in the absence of salt. However, more evidence is necessary to convincingly show that changes in cellular energetics trigger DNA supercoiling.

4.3 Effect of Supercoiling on Transcription

Although the possibility of turning DNA during transcription was well-recognized, transcription was rarely viewed as a force that might actively supercoil DNA. Only special cases in which the polymerase is interacting with a DNA-bound regulatory protein, or when both the polymerase and the DNA are anchored on a cellular structure, have been considered in terms of DNA supercoiling driven by transcription. However, there are now several cases described in which transcription may lead to DNA supercoiling. The first observations were made with plasmid DNA isolated from *topA* mutants of *Escherichia coli* (PRUSS and DRLICA, 1986; LIU and WANG, 1987). Supercoiling of pBR322 is much more negative than that of the closely related plasmid pUC9, which is unable to transcribe the tetracycline resistance gene (*tet*). It was proposed that tracking of the transcription complex along DNA generates positive supercoils ahead of the complex and negative supercoils behind. Topoisomerase I would normally remove the negative supercoils and gyrase the positive ones. Thus, negative supercoils will accumulate in a *topA* mutant, and positive supercoils will accumulate when gyrase is inhibited (WU et al., 1988).

In pBR322 the ampicillin resistance gene (*amp*) and the *tet* gene are in opposite orientation. Only *tet* transcription affects supercoiling. There are several questions to answer:

(1) Is the entire *tet* gene or only its promoter necessary for the observed effect?
(2) Is the orientation and/or location of transcription important?
(3) Can the *tet* gene be substituted by a different gene?

Studies with pBR322 derivatives indicate that high negative supercoiling in *topA* mutants requires not only transcription, but also translation of the first 98 *tet*-codons; it is not required in the divergently transcribed *amp* gene. Since the N-terminal region of TetA is thought to insert into the inner membrane, it was postulated that supercoiling domains were created when DNA segments were anchored to a large cellular structure via coupled transcription, translation, and membrane insertion of a nascent protein (LODGE et al., 1989).

In another study, the *tet* gene of pBR322 was replaced by the chloramphenicol resistance gene, *cat,* which is transcribed from promoters whose activities could be controlled experimentally. Activation of transcription is accompanied by an increase in the plasmid's level of negative supercoiling. The number of added superhelical turns in the plasmid is proportional to the strength of the promoter (FIGUEROA and BOSSI, 1988).

Insertions of short, 12 bp long d(CG) tracts adopt Z-DNA *in vivo*, when they are placed upstream of the pBR322 *tet* promoter but not when they were cloned downstream of the *tet* gene (RAHMOUNI and WELLS, 1989). Interestingly, the ability of the (CG) inserts to undergo a B-to-Z transition *in vivo* is dependent on the position of the insertion site relative to the plasmid genes. Hence, if negative supercoiling is the predominant stabilizing factor for Z-DNA, these results reveal the existence of domains of negative supercoiling on a plasmid and their possible link with transcription. In this context, it has recently been discussed whether Z-DNA can be used as a probe for localized supercoiling *in vivo* (RAHMOUNI, 1992). DNA supercoiling and its role in prokaryotic transcription has been recently reviewed by PRUSS and DRLICA (1989).

Transcription can be a major contributor to the level of DNA supercoiling in bacteria and eukaryotes. The actual local supercoiling *in vivo* may be highly positive, negative or negligible, depending on the position of the promoter, the rate of transcription, and the efficiency of supercoil removal by topoisomerases. The waves of negative supercoiling generated behind the transcription machinery can be sufficient to transiently induce non-B DNA structural transitions.

4.4 Transcriptional Control by Supercoiling

a) Regulation of topoisomerase I and II gene expression. The relative amounts of topoisomerase I and II in cells are finely tuned to create just the right amount of negative supercoiling. Transcription of *gyrA/B* and *topA* genes themselves is responsive to supercoiling in such a way that changes in supercoiling level or activity are limited or corrected. Lowering negative supercoiling increases gyrase gene (*gyrA/B*) expression and decreases topoisomerase I gene (*topA*) expression (MENZEL and GELLERT, 1983; TSE-DINH, 1985). Increasing supercoiling raises levels of topoisomerase I expression (TSE-DINH and BERAN, 1988); regulation occurs at the transcriptional level. The *E. coli*-specific *topA* expression requires a negatively supercoiled DNA template for efficient transcription from the four *topA* promoters.

b) Regulation of the leucine operon. Another example of transcriptional control by supercoiling is the leucine operon (*leu*500) of *Salmonella typhimurium.* Transcription of this operon is activated by increased negative supercoiling. It is interesting to note that this regulation only functions when the promoter is in the chromosome but not in a plasmid, suggesting that the effective increase in supercoiling is influenced by flanking sequences of the chromosome (MARGOLIN et al., 1985; RICHARDSON et al., 1988). Indeed, there is a second promoter (P_X) in opposite orientation to P_{leu500}. The postulated model of "topological coupling" includes that transcription from P_X generates negative supercoiling in the vicinity of the *leu500* promoter. In *topA* mutants the negative supercoiling is not relaxed, and the *leu500* promoter can function. On a plasmid, even in the presence of the opposite *bla* promoter, the second promoter fails to generate a domain of supercoiling that is high enough to activate *leu500*, even in *topA* mutants, because of the alternative mechanism of relaxation. In the circular plasmid the domains of positive and negative supercoiling may be cancelled by rotation of the DNA on the opposite side of the circular molecule (LILLEY and HIGGINS, 1991).

The example provided by *leu500* suggests that promoters may influence one another via local effects on DNA supercoiling. One can envisage many situations in which control of one promoter may, in turn, control the activity of a second promoter. BECK and WARREN (1988) have shown that many promoters are organized in divergent pairs that are potentially subject to the kinds of effects discussed here. When RNA polymerase transcribes a gene, there is a competition between the introduction of negative supercoiling and the rate at which it can be relaxed. It is possible that strong promoters generate significant over-supercoiling even in the presence of topoisomerase I. Indeed, transcription-dependent supercoiling has been observed in plasmids containing strong promoters in the presence of normal levels of topoisomerase activity (FIGUEROA and BOSSI, 1988).

5 Influence of Environmental Changes on DNA Supercoiling

Environmental stresses such as osmolarity and anaerobicity can influence the supercoiling of cellular DNA.

5.1 Osmoregulation in Enteric Bacteria

HIGGINS et al. (1988) reported that plasmid supercoiling increases with increasing osmolarity of the growth medium. They proposed that increased supercoiling may be important in the osmotic induction of *proU*, a gene encoding a high-affinity transport system for the osmoprotective solute glycine betaine. Reducing gyrase activity by mutation or by the use of inhibitors reduces *proU* expression at both high and low osmolarity. In agreement with these results is a *topA* mutation that mimics an increase in osmolarity, facilitating *proU* expression even at low osmolarity. Apart from *topA*, they also identified a new gene, *osmZ*, which plays an important role in determining the *in vivo* level of DNA supercoiling (see details in Sect. 5.3). Mutations in *osmZ* are highly pleiotropic, increasing the frequency of site-specific recombination events (inversions that mediate fimbrial phase variation) and affecting the expression of a variety of chromosomal genes such as *ompF, ompC, fimA,* and the *bgl* operon. It was suggested that there is a class of "stress-regulated" genes that are regulated by a common mechanism in response to different environmental signals, which mediate changes in DNA supercoiling (BHRIAIN et al., 1989).

Escherichia coli strains harboring a deletion in *topA* are normally not viable in the absence of an additional compensating mutation which is thought to restore the otherwise over-supercoiled DNA to a normal level (PRUSS et al., 1982). Such compensation can be accomplished by mutation of the *gyrA/B* genes or the *toc* gene (topoisomerase one compensating) (RAJI et al., 1985). Recently, it was shown that the *toc* mutations involve the amplification of a region of chromosomal DNA which includes *tolC* (DORMAN et al., 1989). *tolC* mutants are highly pleiotropic and preferentially synthesize OmpC, even in media of low osmolarity (MISRA and REEVES, 1987).

5.2 Regulation of Anaerobic and Aerobic Growth

Using *Salmonella typhimurium* as a representative facultative anaerobe, strictly anaerobic and strictly aerobic mutants could be isolated (YAMAMOTO and DROFFNER, 1985). Strictly anaerobic mutants were found to carry a mutation in the DNA topoisomerase I gene (*topI*), while strictly aerobic mutants contained a defective DNA gyrase gene (*gyrA*). Each type of mutation appears to control expression of numerous operons and genes. Extracts of wild-type cells, cultured under vigorous aerobic conditions have topoisomerase I activity but no significant gyrase activity. In anaerobic wild-type cells, exactly opposite effects are observed. Aerobic cultures of wild-type and strictly aerobic mutants produce both superoxide dismutase and catalase, whereas anaerobic cultures of wild-type and strictly anaerobic mutants do not. It was concluded that topisomerase I activity, associated with relaxation of DNA, is necessary for expression of genes required for aerobic growth.

When facultative anaerobe enteric bacteria switch from aerobic to anaerobic growth, about 50 new proteins are synthesized, while the synthesis of other proteins is repressed (SMITH and NEIDHARDT, 1983a, b). For most anaerobically regulated genes, regulation is at the level of transcription. The expression of several induced genes requires Fnr, a positive regulatory protein; other genes require a functional *oxrC* which encodes phosphoglucose isomerase (JAMIESON and HIGGINS, 1986), implying that flux through the glycolytic pathway, or a glycolytic product, may play a role in the regulation of this class of genes.

There is another piece of evidence supporting the role of DNA supercoiling in the anaerobic response. TonB is a key protein in cellular physiology, being required for many energy-dependent outer membrane processes. Mutants lacking TonB are defective in all high iron transport systems and vitamin B12 uptake, and are resistant to many bacteriophages and colicins (SAUER et al., 1987). Expression of *tonB* is repressed anaerobically but is induced during aerobic growth, when iron is oxidized to an insoluble form and becomes limiting (HANTKE, 1981).

The mechanisms by which gene expression is regulated are still poorly understood. DORMAN and coworkers (1988) showed that the anaerobically regulated *tonB* promoter is very sensitive to changes in DNA supercoiling. TonB expression is reduced by anaerobiosis and increases both by novobiocin, an inhibitor of DNA gyrase, and by factors which increase DNA superhelicity. There is also excellent correlation between *tonB* expression from a plasmid and the level of supercoiling of that plasmid under a wide range of conditions. However, an induction of anaerobic gene expression is not always accompanied by local changes in DNA supercoiling. With certain photosynthetic genes or the cytochrome bc_1 genes of *Rhodobacter capsulatus,* no change in superhelicity is observed (COOK et al., 1989).

Changes in plasmid DNA supercoiling occur during several types of these nutrient upshifts and downshifts. The most dramatic change is seen in starved cells, in which two populations of differentially relaxed plasmids coexist.

5.3 Histone-Like Protein H1 and DNA Supercoiling

Changes in DNA supercoiling in response to environmental signals, such as osmolarity, temperature (GOLDSTEIN and DRLICA, 1984), or anaerobicity, appear to play an underlying role in the regulation of gene expression in bacteria. Extensive genetic analyses have implicated the *osmZ* gene in this regulatory process: *osmZ* mutations are highly pleiotropic and alter the topology of cellular DNA. Recently it was shown that the product of the *osmZ* gene is the histone-like protein H1 (HULTON et al., 1990).

The most abundant histone-like protein in *E. coli* is the highly basic protein HU, resembling eukaryotic histones. HU plays not only an important role in chromatin structure, but can compact DNA and constrain superhelical twists (BROYLES and PETTIJOHN, 1986).

The 15 kDa H1 ist also a highly abundant protein (20000 copies per cell) which is associated with the bacterial nucleoid (VARSHAVSKY et al., 1977; LAMMI et al., 1984). The amount of H1 is sufficient to bind about once every 400 bp of chromosomal DNA. H1 differs from HU and the other histone-like proteins in that it is neutral rather than basic. Although H1 binds relatively non-specifically to double-stranded DNA (FRIEDRICH et al., 1988), there are also cooperative binding sites present on a loose consensus sequence. Purified H1 compacts DNA and appears to affect transcription from a number of promoters.

Many of its properties, especially its abundance, have led to suggestions that H1 may be a major structural component of bacterial chromatin. It was also suggested that H1 might function as a general "silencer" of transcription (GORANNSON et al., 1990). The various phenotypes of *osmZ* mutants, however, suggest that H1 may play a more active role in regulation of chromatin structure, DNA topology, and gene expression. Furthermore, the findings suggest a mechanism by which environmental signals may alter DNA topology, providing a direct link between environmentally induced changes in DNA supercoiling and the regulation of gene expression (HULTON et al., 1990).

osmZ mutants exert pleiotropic effects (reviewed by HIGGINS et al., 1990): a) derepression of expression of the osmotically regulated, supercoiling-sensitive *proU* promoter; b) derepression of the *bgl* (β-glucoside) operon; c) mucoidy, probably owing to an overproduction of colanic acid which is normally synthesized during periods of osmotic stress to prevent dehydration (ANDERSON and ROGERS, 1963); and d) alteration in the ratio of the OmpC and OmpF porins.

5.4 Regulation of Virulence Gene Expression by Supercoiling

A number of environmental stresses such as osmotic shock or anaerobiosis have been shown to induce changes in DNA supercoiling that can directly affect the transcription of a specific subset of bacterial genes. At least some of the genes, those encoding the outer membrane porins and type 1 fimbriae (DORMAN and HIGGINS, 1987), play a role in bacterial virulence. Other virulence factors are known to be regulated by DNA supercoiling (reviewed by HIGGINS et al., 1990, and DORMAN, 1991).

The ability to enter intestinal epithelial cells is an essential virulence factor of salmonellae. For the invasion into tissue culture cells, a group of genes, *invA, B, C* and *D,* play a major role. Transcription of *invA* is enhanced eightfold when cells are grown in media with high osmolarity, a condition known to increase DNA superhelicity (GALAN and CURTISS, 1990). The osmoinducibility is independent of *ompR,* which controls the osmoinducibility of other genes. Gyrase inhibitors lead to a reduced expression of *invA*.

The *virR* gene of *Shigella flexneri,* implicated in the temperature regulation of plasmid-encoded virulence genes, is equivalent to the *osmZ* gene of *E. coli* (DORMAN et al., 1990).

6 DNA Bending

Until the late 1970s, it was assumed that any 200 bp segment of DNA could be approximated as a straight rod. It then became apparent that the DNA double helix may be curved or bent, either as a result of a specific intrinsic DNA sequence or because of protein binding.

6.1 Intrinsic Bending

Sequence-directed bent DNA structures have been found in prokaryotes and eukaryotes. Intrinsic bending of DNA can be conferred by particular oligonucleotide sequences, such as the conformationally rigid oligo (dA)-(dT) tracts (WU and CROTHERS, 1984; TRAVERS and KLUG, 1987; CALLADINE and DREW, 1986).

Bending was first detected in the kinetoplast DNA (kDNA) of parasitic hemoflagellate protozoa, the trypanosomes (MARINI et al., 1982). Subsequently, bent DNA structures (Fig. 6) were found in the origins of replication of lambda phage (ZAHN and BLATTNER, 1987), simian virus 40 (RYDER et al., 1986), adenovirus (ANDERSON, 1986), plasmid pSC101 (STENZEL et al., 1987), in the termini of SV40 replication and transcription (HSIEH and GRIFFITH, 1988), in yeast autonomously replicating sequence, ARS1 (SNYDER et al., 1986), and in the promoter sequences of several prokaryotic genes such as *hisR, rrnB,* and *ompF* (BOSSI and SMITH, 1984; GOURSE et al., 1986; MIZUNO, 1987; FIGUEROA et al., 1991). These examples imply that sequence-directed bends of the DNA helix may play critical roles

ARS1	
Domain B	cAAAtggtgtAAAAgactctaacAAAAtagcAAATTTcgtcAAAAAtgctaagAAAt
Domain C	aTTTTcttgtaTTTatcgtcTTTTcgctgtAAAAAc
HO ARS	aTTTTggatgAAAAAAAccaTTTTTagacTTTTc
Copy-I H4	aTTTcaacAAAAAgcgtacTTTacatataTTTattagacaagAAAAgcagattAAAt
Human ARS1	cTTTTacataTTTTgacccaTTTaatccccaTTTTg
Human ARS2	aTTTTgaactctaTTTcctagggAAAAAttgtacTTTTTC

Fig. 6. DNA sequence of bending elements in various autonomously replicating sequences (ARS). It is suggested that the bent structure *per se* is crucial for ARS function. (DNA sequence is from WILLIAMS et al., 1988.)

in providing local structural alterations that are required for DNA replication, transcription, or packaging.

Bent DNA segments were also found in the insertion sequence IS5, which is found in multiple copies on the *E. coli* chromosome. IS5 contains a sequence-directed bent DNA structure at its terminus, close to one of its 16 bp terminal repeats (MURAMATSU et al., 1988). The position of the bending region at the 3′-end of a large open reading frame suggests that it may be important either for transcription termination or more likely for transposition and integration. The bent DNA might facilitate the binding of the transposase and/or the integration of the host factor (IHF). The nucleotide sequence in the bending region revealed stretches of d(AT) approximately every 10.5 bp. Some bending sequences of various ARS elements have been listed by WILLIAMS et al. (1988). For example, the domain C of ARS1 has the following sequence:

d(aTTTTcttgta
TTTatcgtcTTTTcgctgtAAAAAc).

Important in the sequence is the periodicity of d(A/T) stretches. In other ARS elements, such as those in humans, similar sequences have been found.

The biological significance of bent DNA structures has been only gradually elucidated. However, the proximity of such structures to several origins of replication raises the possibility of a potential role of bent DNA structures in the initiation of replication. The presumption that these structures may serve as signals for the binding of proteins that play a role in initiation processes at the origin of replication, has been supported recently by the reports of KOEPSEL and KHAN (1987) on the recognition of the staphylococcal plasmid pT181 origin by the RepC protein and the recognition of the pSC101 origin by IHF-protein (STENZEL et al., 1987). RYDER et al. (1986) have described the role of bending in the binding of SV40 T-antigen to a high affinity site, which has been shown to facilitate DNA replication (DEB et al., 1986). Base substitution within the essential SV40 origin suggests a close correlation between A-tract bending and the initiation of DNA replication.

A study of DNA signals preceding strongly expressed proteins in methanogenic bacteria (archaebacteria) revealed a bent DNA sequence in front of the transcription start site (ALLMANSBERGER et al., 1988).

From the trypanosome *Crithidia fasciculata*, an endonuclease with nicking activity was isolated (LINIAL and SHLOMAI, 1988). This endonuclease recognizes preferentially the sequence-directed bends of various DNAs, such as the origins of replication of bacteriophage lambda and the simian virus 40 (SV40), as well as the bent DNA located within the autonomously replicating sequence (ARS1) of yeast.

6.2 DNA Bending as a Result of Protein Binding

DNA may become bent as a result of interaction with nucleosomes (MCGHEE and FELSENFELD, 1980), gyrasomes (KLEVAN and WANG, 1980), intersomes (BETTER et al., 1982), and Cro protein-DNA complexes (OHLENDORF et al., 1982).

Although many examples of protein-induced intrinsic DNA bending have been reported, only recently has clear evidence of the functional significance of the curvature as such emerged (Snyder et al., 1989; Goodman and Nash, 1989; Achtberger and McAllister, 1989; Travers, 1989). The integration of bacteriophage lambda into the chromosome is mediated by the intasome which consists of several molecules of DNA binding proteins, the phage-encoded integrase (Int) and the *E. coli* integration host factor (IHF) (Echols, 1986). When bound separately to their respective binding sites, IHF, but not Int, induces a substantial DNA bend, estimated to be at least 140° (Robertson and Nash, 1988; Thompson and Landy, 1988). A possible function of IHF would be to bend the DNA and thus define the three-dimensional path of the double helix in the intasome; IHF would act as an architectural element (Travers, 1989). The bivalent Int protein can simultaneously bind to two different segments of *attP,* the genetically defined attachment site on the phage DNA. These segments are separated by a DNA loop whose trajectory is defined by bound IHF (de Vargas et al., 1989). There is also evidence that IHF binding induces bending that would direct *attP* DNA to follow a left-handed superhelical path.

To test whether DNA bending *per se* is crucial for intasome function, Goodman and Nash (1989) adopted the strategy of "bend swapping". They replaced one IHF binding site in *attP* by the CRP binding site (CRP normally mediates cAMP-induced transcription and is also a DNA bending protein, but unrelated to IHF). In such constructs significant recombination is only observed in the presence of CRP, suggesting that only the bending of DNA is necessary for the assembly of a functional intasome. It was hypothesized that the sole *in vivo* role of IHF is to bend DNA at specific sites. Such a role would be fully consistent with its multifunctional effects on transcription, recombination, and DNA replication (Travers, 1989).

The effect of curved DNA extends to the activation of transcription initiation by *E. coli* RNA polymerase. In the complex formed between polymerase and promoter, the DNA is inferred to be wrapped around the protein for up to 60 bp upstream of the transcription start point. In many promoters the DNA upstream of the -35 region is often intrinsically bent. Achtberger and McAllister (1989) have shown that promoter activity is extremely sensitive to the direction of curvature in this region. When correctly phased bent DNA is placed 60–90 bp before the start point, beyond the previously observed limit of binding of a single polymerase molecule, transcriptional activity is conserved. This observation suggests either that the curved DNA serves simply to direct the polymerase to the promoter region, or that the curve allows the DNA to bend back to make additional contacts with the enzyme.

Bracco et al. (1989) demonstrated that the requirement for CRP for the *in vivo* activation of the *E. coli gal* promoter can be overcome by replacing the CRP-binding site by curved DNA. However, the extent of transcriptional activation is strongly dependent on the orientation of the inserted curve.

The principal conclusion from the studies on both the intasome and RNA polymerase is that the bending of DNA is an important determinant of both the function and the architecture of large nucleoprotein complexes. Both the direction and the magnitude of bending play a crucial role. On the other hand, the corresponding protein components may be sufficiently flexible to accommodate variations in the length of the DNA binding site as long as the DNA-topology is conserved (Travers, 1989, 1990).

6.3 Detection of DNA Bending

If $d(A)_n$ tracts are present in a DNA molecule, with n greater than 2, the helix axis is no longer straight. This effect becomes large when the $d(A)_n$ tracts are phased with the turn of the helix. The non-linearity of the DNA helix axis (curvature of the molecule) can be detected using several methods: circularization of DNA molecules (Ulanovsky and Trifonov, 1987), microscopy (Griffith et al., 1986), and gel migration anomaly (Marini et al., 1982; Wu and Crothers, 1984; Diekmann 1987, 1989). Results obtained from these measurements can be compared with DNA structures obtained from NMR (Sarma et al., 1988) and crystallographic analysis (Nelson et al., 1987; Coll et al., 1987).

DNA molecules having a curved helix axis migrate slower through the acrylamide gel pores than straight molecules, resulting in an anomalously slow electrophoretic mobility. Bent DNA segments tend to migrate more normally at high temperature. A quantitative analysis of the migration anomaly and the influence of temperature and salt has been presented by DIEKMANN (1989).

7 Cruciform Structure in Superhelical DNA

A palindromic or inverted repeat DNA sequence has the unique property that base pairing can potentially occur not only between the two complementary strands of DNA, but also within each single DNA strand. In the latter configuration, bases at the symmetric center of the palindrome are at the ends of "hairpin" or cruciform arms with respect to flanking, nonpalindromic sequences (Fig. 7). Cruciform DNA was first postulated by GIERER (1966). Since then, palindromes have been identified in DNA from many organisms and are often found at operator and transcription termination regions in bacterial DNA (ROSENBERG and COURT, 1979; HIDAKA et al. 1988), as well as in DNA replication origins of prokaryotes (MEIJER et al. 1979; ZYSKIND et al., 1983) and eukaryotes (CREWS et al., 1979). Considering the presence of palindromes at such important regulatory sites, it has been tempting to speculate about possible biological significance of cruciform structures.

The quantitative coupling between the free energy terms of cruciform formation and supercoiling was first applied to show that the binding of a *Lac* repressor to its operator, which contains an imperfect palindromic sequence, does not induce cruciform formation (WANG et al., 1974). Thermodynamic calculations indicate, however, that depending upon the length and perfection of a palindromic sequence, the cruciform structure may form at moderate negative superhelicity. Mechanical and statistical calculations also indicate that naturally occurring palindromes in various plasmids and phages can form cruciforms at negative superhelicities that are typical of supercoiled DNAs from natural sources (BENHAM, 1982).

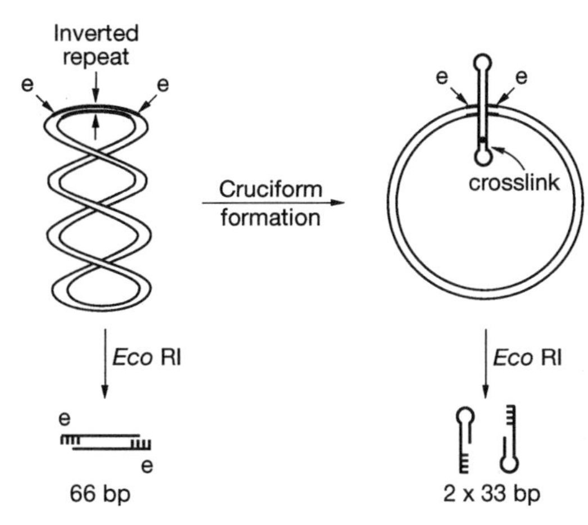

Fig. 7. DNA cruciform transition detected by a trimethylpsoralen crosslinking assay. The *lac* inverted repeat sequence of the 66-bp *lac* operator was cloned into the *Eco*RI site of plasmid pOEC12. Some of the thymidine residues are crosslinked by trimethylpsoralen. If the negative superhelical tension in the DNA was sufficiently high, the inverted repeat was in the cruciform, and crosslinks fix the structure. Conversely, if the torsinal tension was low the inverted repeat remained in the linear form and is fixed in that form. After excising the inverted repeat with *Eco*RI (at sites shown by e), the 66-bp fragment was produced from the linear form and two 33-bp fragments were released from the cruciform DNA (adapted from PETTIJOHN et al., 1988).

7.1 Formation and Probing of Cruciform DNA

Structural and physicochemical investigation of cruciforms in a chromosome or plasmid is difficult, since the feature of interest represents only a tiny part of the entire circular DNA molecule. Recently, however, a series of probing methods have become available to verify cruciform structures. A cruciform has two characteristic features, the single-stranded loop and the four-way junction. Most of the probes select the former as their target, being

Tab. 4. Enzyme and Chemical Probes Used to Study Cruciform Structures

Probe	Target	Result	Reference
S1 nuclease	Loop	Cleavage	LILLEY, 1980; PANAYOTATOS and WELLS, 1981
Micrococcal nuclease	Loop	Cleavage	DINGWALL et al., 1981
Bal31 nuclease	Loop	Cleavage	LILLEY and HALLAM, 1984
P1 nuclease	Loop	Cleavage	HANIFORD and PULLEYBLANK, 1985
Mung-bean nuclease	Loop	Cleavage	SHEFLIN and KOWALSKI, 1984
Bromoacetaldehyde	Loop (A, C)	Etheno adduct	LILLEY, 1983
Osmium tetroxide	Loop (T, C)	*cis*-Diester	LILLEY and PALECEK, 1984; BOUBLIKOVA and PALECEK, 1990
Bisulfite	Loop (C)	Deamination dU	GOUGH et al., 1986
Diethylpyrocarbonate	Loop (A)	Carboethoxylation	FURLONG and LILLEY, 1986; VOLOSHIN et al., 1989
T4 endonuclease VII	Junction	Cleavage	MIZUUCHI et al., 1982
T7 endonuclease I	Junction	Cleavage	DE MASSEY et al., 1987
Yeast resolvase	Junction	Cleavage	WEST and KORNER, 1985

either single-strand-specific enzymes such as S1 nuclease, or single-strand-selective chemicals such as bromoacetaldehyde. Probes of the junctions are fewer in number, but are structurally more discriminating (LILLEY et al., 1988). Some probes which have been used for varifying cruciform structures are listed in Tab. 4.

It has been observed that single-strand-specific nuclease S1 preferentially cleaves negatively supercoiled DNAs at sites that are near the dyads of palindromic sequences (LILLEY, 1980; PANAYOTATOS and WELLS, 1981; SINGLETON and WELLS, 1982). This preferential cleavage has generally been interpreted as a consequence of attack by the nuclease at the single-stranded hairpin loops of the cruciform structure. Another method to detect cruciforms is the crosslinking of double-helical DNA by psoralen (COLE, 1970). Crosslinking should fix cruciforms and thus provide a basis for their detection *in vivo*.

Recently, a *lac* repressor/T7 endonuclease hybrid protein was constructed (PANAYOTATOS and BACKMAN, 1989), a site-directed nuclease that requires two structural elements for activity: the *lac* operator and a target. Repressor-like binding directs the enzyme to the operator (which can be inserted in either orientation at desired sites of a plasmid) and nearby single-stranded DNA targets formed by cruciforms. This hybrid protein represents a powerful method for probing unusual structures in non-supercoil duplex DNA. (See Sect. 7.3 for more details on the use of this method).

Although both the physicochemical calculations and the nuclease digestion experiments argue in favor of the existence of cruciforms, there have been few direct experiments. In one elegant experiment, cruciform formation was directly observed by electron microscopy, when by joining the ends of a linear molecule, a giant palindromic DNA ring was formed which was negatively supercoiled by DNA gyrase (MIZUUCHI et al., 1982). Cruciform formation was also studied in a plasmid pBR322 derivative containing a 68 bp perfect palindromic sequence (COUREY and WANG, 1983). In relaxed DNA, the cruciform is unstable; in negatively supercoiled DNA, the cruciform DNA becomes the stable species. Nevertheless, even at native superhelical densities (typically around -0.06), cruciform formation is extremely slow unless the DNA is subjected to conditions that destabilize base pairing. It was, therefore, concluded that cruciform formation in a negatively supercoiled DNA may be kinetically forbidden under physiological conditions. Similar results were obtained when a perfect palindromic DNA sequence of 66 bp derived from the *lac* operon was cloned in pMB9 (SINDEN et al., 1983). Cruciform formation depends on the superhelical density of the plasmid DNA. Relaxed DNA contains no cruciforms. The cruciform structure rarely, if ever, exists *in vivo,* but after DNA isolation, more than 90% of the sequence is in cruciforms. These results suggest that plasmid DNA as organized *in vivo* either lacks sufficient torsinal tension to form this cruciform, or the palindrome is restrained in the linear form by other bound molecules.

Small cruciforms with seven and possibly fewer base pairs in the stem can apparently be extruded in the ϕX174 replicative form (MÜLLER and WILSON, 1987). However, these structures are very unstable and are lost by heat or by special conditions of the S1 reaction.

A 34 bp tract of the simple repeating dinucleotide $d(AT)_n$ cloned into a 2.4 kb plasmid undergoes a structural transition in response to negative superhelical coiling. The transition is characterized by several methods: a) two-dimensional gel electrophoresis, b) mapping of S1, P1, and T7 endonuclease 1 sensitive sites, and c) mapping of sites that are sensitive to modification by bromoacetaldehyde. The data indicated that the $d(AT)_n$ insert adopts a cruciform rather than a Z-conformation (HANIFORD and PULLEYBLANK, 1985). DNA sequences with $d(AT)_n$ tracts naturally occur relatively often in introns of eukaryotic genes, such as, for example, in the *Xenopus laevis* globin gene locus. This locus contains two separate $d(AT)_n$ tracts, $d(AT)_{37}$ and $d(AT)_{17}$, which are approximately 180 bp apart. Cloning of the gene locus in various rec^- *E. coli* strains is associated with a high frequency of deletion of these $d(AT)_n$ tracts and the intervening 180 bp. The deletion is independent of *rec*A, *rec*BC, *sbc*B and *rec*BC, *sbc*A, *rec*E strains (GREAVES and PATIENT, 1986). It was postulated that this instability is caused by the long inverted repeats and *in vivo* cruciform formation in *E. coli.*

Recently, the stability of palindromes flanked by 4 bp direct repeats was studied in *E. coli* (WESTON-HAFER and BERG, 1991). In the pBR322 *amp* gene, palindromic DNA sequences of varying length were inserted. With palindromes longer than 26 bp, the frequency of deletions is stimulated.

7.2 Cruciform DNA and Gene Regulation

Regulatory regions that control the activities of promoters in *cis* occur in both prokaryotic (MALAN et al., 1984) and eukaryotic genes. A central question is, therefore, how the information travels along a DNA molecule to result in the alteration of some physical property at a remote location. WELLS and coworkers (BURD et al., 1975) showed that the thermal melting of a DNA sequence could be strongly influenced by neighboring regions, an effect they termed "telestability". In these examples, the base composition of a block of sequence either potentiates or inhibits the ability of another part of the molecule to undergo structural transition. These effects are observed over long stretches of DNA. A structural transition that might be subject to similar contextual influence is the extrusion of cruciforms from inverted repeats in supercoiled DNA. Indeed, SULLIVAN and LILLEY (1986) discovered a striking dependence of cruciform formation in the ColE1 plasmid on a distant AT-rich sequence. They demonstrated that the kinetic character of cruciform extrusion is governed by the sequences that flank the inverted repeat. The base sequence of the inverted repeat itself is of minor importance. The unusual C-type kinetics, originally observed for the ColE1 cruciform, are conferred by the AT-rich sequence (more than 80% AT base pairs) that flanks the inverted repeat; the AT-rich sequence exerts its effect at either end. The flanking sequence allows cruciform extrusion to occur at relatively low temperature (28 °C) and low ionic strength. The observed effect operates in *cis,* is independent of polarity, and may be transmitted over a distance of at least 100 bp.

HORWITZ and LOEB (1988) constructed a promoter capable of adopting cruciform base pairing. The plasmid pX contains a 50 bp inverted AT-rich repeat, spanning from −23 to +27 with respect to the transcription start site, +1. Each repeated unit contains a −10 sequence and the most upstream of the units is spaced to align with a −35 sequence to complete the elements of a promoter. The control plasmid contains a 50-bp direct repeat composed of the same units as in pX. The influence of the cruciform was assessed by the promotion of *in vitro* transcription with *E. coli* RNA polymerase. Transcription in pX is repressed as the cruciform is extruded by increasing negative supercoiling. Transcription *in vivo* is induced as supercoiling is relaxed by growth conditions that inhibit DNA gyrase. A DNA conformational change is, therefore, capable of regulating the initiation of transcription.

7.3 *In vivo* Existence of Cruciform DNA

PANAYOTATOS and FONTAINE (1987) presented evidence for the existence of cruciform structures *in vivo*. They inserted the naturally occurring ColE1 palindrome, which forms a more stable cruciform *in vitro*, into a pBR322 derivative. On the same plasmid, the T7 endonuclease gene under control of the *lac*UV5 promoter is present. T7 endonuclease preferentially cleaves purified supercoiled pBR322 and ColE1 plasmids at the single-stranded regions exposed when palindromic sequences assume cruciform structures. Induction of the T7 endonuclease results not only in degradation of the genomic DNA, but also in intracellular nicking and linearization of the plasmid at the ColE1 palindrome site. These results indicated that cruciform structures exist intracellularly.

Cruciform formation in *E. coli* cells was also detected by an indirect approach which is based on the estimation of superhelicity of isolated DNAs as well as direct chemical probing of the intracellular DNA by chloracetaldehyde (VOGT et al., 1988; DAYN et al., 1991), using pUC19 containing $d(AT)_{32 \text{ and } 42}$ inserts. Cruciforms are formed after cells are exposed to various stresses, such as inhibition of protein synthesis, anaerobiosis, and osmotic shock. Since all these stimuli lead to an increase in plasmid superhelicity, it was suggested that it is the increase in superhelicity which entails the appearance of cruciform structure.

Inverted repeat sequences which are able to extrude to form cruciform junctions *in vitro* are found near replication origins in prokaryotes (STALKER et al., 1980) and eukaryotic viruses (HAY and DE PAMPLILIS, 1982). Cruciform extrusion might, therefore, serve as structural signal for initiation of replication at yeast and mammalian origins. In addition, inverted repeats have been implicated in transcription termination (ROSENBERG and COURT, 1979), and are present as intermediates during the replication of teleomere sequences (MOYER and GRAVES, 1981).

Several proteins are known that interact with X- and Y-junctions in DNA. Two bacteriophage enzymes (T4 endonuclease and T7 endonuclease I) cleave branched DNA structures *in vitro* to form linear duplex molecules containing ligatable nicks (DE MASSY et al., 1987). Mutant T4 phages, which are defective in the endonuclease VII gene, produce highly branched multimeric DNA that results in abortive infections. The endonuclease VII therefore appears to be required for the resolution of branches that are produced by the interlinked phage replication and recombination process.

Enzymes capable of resolving cruciform junctions *in vitro* have also been isolated from *Saccharomyces cerevisiae* (WEST et al., 1987). In gel electrophoretic binding assays it was found that human lymphoblast proteins bind cruciform structures in duplex DNA. The binding appears specific for DNA structure rather than sequence (ELBOROUGH and WEST, 1988).

8 Triplex DNA

8.1 Structure of Triplex DNA

Homopurine–homopyrimidine tracts are known to adopt unusual DNA structures (H-DNA) under superhelical stress. These are proposed to consist of an intramolecular triplex in which half the pyrimidine strand is unpaired and forms Hoogsteen base pairs with the remaining DNA duplex; the other half of the purine strand is left in an unstructured, single-stranded configuration (WELLS et al., 1988; MIRKIN et al., 1987; JOHNSTON 1988; HANVEY et al., 1988). For example, $d(A)_n$-$d(T)_n$ can form a triplex with another strand of $d(T)_n$

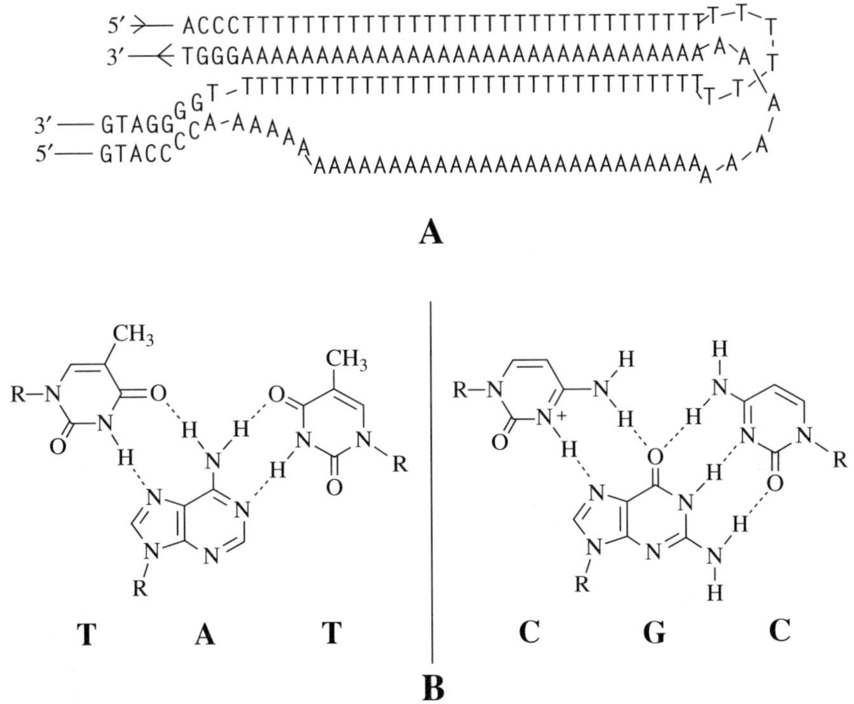

Fig. 8. (A) Schematic representation of a model for the formation of the intramolecular triple helix in the T_{69}-A_{69} insert in plasmid pUC19 (adapted from FOX, 1990). (B) Hoogsteen base pairing stabilizes the triplex.

to generate $d(A)_n$-$2d(T)_n$ (Fig. 8a). The alternating $d(AT)_n$ DNA polymer does not adopt a triplex structure. It is generally accepted that the third (polypyrimidine) strand is associated with the duplex via Hoogsteen (or reversed Hoogsteen) base pairs in the major groove (Fig. 8b).

Various probes have been used to detect the formation of these unusual DNA structures including single-strand-specific nucleases (S1 and P1), diethylpyrocarbonate (DEPC), and osmium tetroxide. These react with the long single-stranded loop of the purine strand and the short loop of the pyrimidine strand.

8.2 Occurrence of Triplexes

Triplexes have previously been detected in DNAs containing 25–100% (G+C) residues, and are generally formed at low pH, necessary for the formation of the C^+GC triplet (PULLEYBLANK et al., 1985; HANVEY et al., 1988).

For $d(AG)_n$ tracts, it was postulated that Hoogsteen G-C^+ base pairs alternate with Watson–Crick A-T base pairs. For short purine-pyrimidine sequences, a hinged DNA structure (H-DNA) with a single-stranded loop, and in long sequences a double-stranded loop, was proposed (CHRISTOPHE et al., 1985; LYAMICHEV et al., 1986; MIRKIN et al., 1987). A model for H-DNA formation, in which nucleation of the Hoogsteen base pairing near the middle of the repeats determines the ultimate structure, was proposed by HTUN and DAHLBERG (1989).

Poly d(G) prefers an A-like DNA conformation (MCCALL et al., 1985). However, oligo d(G) tracts are hypersensitive to S1 nuclease in supercoiled plasmids and in active chromatin, indicating that under torsional stress the sequence adopts an altered DNA conformation, very likely a triplex structure (KOHWI and KOHWI-SHIGEMATSU, 1988). In the presence of Mg^{2+}, a dG-dG-dC triplex is formed, whereas in the absence of Mg^{2+}, a dC^+-dG-dC triplex is formed. The relevance of such a triplex structure is supported by NMR studies of the homopurine $d(AG)_4$ and the homopyrimidine $d(TC)_4$ (RAJAGOPAL and FEIGON, 1989). Imino protons are directly observed from protonated cytosines in the triplex. The conformational behavior of $d(GA)_{22}$ sequences cloned in SV40 has been analyzed by BERNUES et al. (1989). This sequence can adopt different structural conformations, depending on the presence or absence of zinc ions or on pH: at neutral pH in the absence of Zn^{2+} the sequence exists predominantly as B-DNA; increasing proton concentration results in a B-H transition; increasing Zn^{2+} concentration results in the formation of *H-DNA. The patterns of chemical modification obtained at neutral pH in the presence of 4 mM $ZnCl_2$ are consistent with a purine-pyrimidine triplex.

Long stretches of homopurine, homopyrimidine commonly occur throughout many eukaryotic genomes, and several suggestions have been made concerning their biological relevance (MANOR et al., 1988).

Regarding the structure of runs of d(A) or d(T), much attention has previously been given to the occurrence of short runs of A which can cause DNA bending, as mentioned above. Less has been reported concerning longer runs of A or T which are also widespread. Although poly d(A) · 2 poly d(T) is known to form a stable triple helix (RILEY et al., 1966) and $d(A)_{10}$ and 2 $d(T)_{10}$ has been characterized by NMR studies (PILCH et al., 1990) no intramolecular triplexes consisting of only dA and dT bases have been detected. Indeed $d(A)_{20} \cdot d(T)_{20}$ does not adopt an unusual structure under superhelical stress (WELLS et al., 1988), although these sequences ought to be able to form intramolecular triplexes at physiological pHs since no extra protonation is required to form the additional A · T Hoogsteen base pair T · A · T triplet.

Recently plasmids containing long tracts of $d(A)_n \cdot d(T)_n$ have been prepared and their configurations examined in linear and supercoiled DNA (FOX, 1990). It was found that under superhelical stress and in the presence of magnesium, the sequence $T_{69} \cdot A_{69}$ adopts an intramolecular DNA triplex conformation at pH 8.0. Site-specific cleavage of the supercoiled plasmid by single-strand specific nucleases occurs within the A · T insert and the 5′-end of the purine strand is sensitive to reactions with diethylpyrocarbonate, while the central 5–6 bases of the pyrimidine strand are reactive with osmium tetroxide. In contrast, shorter inserts of $A_{33} \cdot T_{33}$ and $A_{23} \cdot T_{23}$ do not appear to form unusual structures. Since these results were obtained by *in vitro* studies, there remains the question as to whether triplex structures are of any biological relevance. No direct proof exists, however, many eukaryotic DNA sequences reveal long tracts of poly d(A) which are preferentially located in introns within longer repetitive stretches. Some of these sites are known to be hypersensitive to S1 nuclease (LA VOLPE et al., 1985).

8.3 Probes for the Triplex Structure

To test unusual DNA structures not only *in vitro* but also *in vivo*, a range of powerful new techniques has been developed to probe DNA structure in supercoiled plasmids. The new techniques comprise two-dimensional gel electrophoresis, restriction enzymes, single-strand-specific nucleases and chemical probes. Some of the specific reagents which have been used in this way are known from the Maxam and Gilbert DNA sequencing. The reagents bromoacetaldehyde (BAA), chloroacetaldehyde (CAA), diethylpyrocarbonate (DEPC), OsO_4, and dimethylsulfate (DMS) are selective toward different bases and are also sensitive to different conformational aberrations in DNA (JOHNSTON and RICH, 1985; WELLS et al., 1988).

8.4 G-DNA

Recently, evidence for a new DNA structure has been presented, the G-DNA (PANYUTIN et al., 1990). A plasmid was constructed which contains $d(G)n \cdot d(C)n$ inserts at the *Pst*I site of pUC18 with $n = 37$ and 27, respectively. After the short insert-containing restriction fragments are denatured, the $d(G)_n$- and $d(C)_n$-containing strands renature much more slowly than do those of arbitrary sequence. In addition, the $d(G)_n$ strand has an abnormally high mobility in PAGE. These data argue in favor of the existence of an intrastrand structure (G-structure) within $d(G)_n$-containing single-stranded fragments stabilized by formation of hydrogen bonds between guanines. The principal element of this G-DNA structure is, according to molecular modelling, a *quadruple helix* formed by pairwise antiparallel segments of the twice-folded $d(G)n$ stretch. Such compact structures may be formed from $d(G)_n$ stretches of telomeric sequences.

9 Conclusion

In the past, the structure of DNA was of minor interest to the biologist, because such studies have traditionally been the province of physical chemists. This is no longer the case since there is a growing awareness of the relevance of the various DNA structures in prokaryotic and eukaryotic cell biology and regulation. One of the most significant developments in our understanding of the biochemistry of DNA is our awareness that the double helix has considerable conformational flexibility. DNA is no longer looked upon as a static molecule, but rather a dynamic structure in which different conformations are in equilibrium with each other (RICH et al., 1984).

While B-DNA was already long ago regarded as the predominant naturally occurring DNA structure, one thought that the A-DNA configuration, which is only stable at low humidity, did not occur in living cells. However, there is evidence that in endospores of bacilli, DNA exists in an A-like structure. There is also no doubt that the Z-DNA structure occurs *in vivo*. Despite the fact that B- and Z-DNA structures are very different (B-DNA is right-handed and Z-DNA is left-handed), there is a reversible B-Z transition under certain circumstances.

In addition to the classical DNA structures, it was found that with certain DNA sequences the DNA structure is not straight, but bent. Both the Z-DNA structure and DNA bending are influenced by the degree of DNA supercoiling, a supramolecular DNA structure. Twenty years ago, WORCEL and BURGI (1972) discovered that the bacterial chromosome contains supercoiled DNA. Since then, supercoiling has become one of the best understood structural features of the bacterial chromosome. Supercoiling affects many activities of DNA, and it is now apparent that the chromosomal structure is altered in response to growth and environmental changes. Significant progress has been made in recent years concerning our understanding of DNA supercoiling and its implications in cellular processes (for reviews see COZZARELLI and WANG, 1990).

Other supramolecular structures include the cruciform and triplex structures. There is good evidence that all these DNA structures can be adopted *in vivo*. Because of the structural flexibility of DNA structures, the concept of "conformational microheterogeneity", i.e., neighboring segments with different secondary structures, has been developed (KLYSIK et al., 1981). This heterogeneity is presumed to play a role in key cellular processes, including replication, recombination, mutagenesis-carcinogenesis, repair, transcription, chromosomal organization, and virus packaging.

Some years ago it was found that certain regulatory proteins bind to specific DNA sites and can interact with a remote DNA site. Thus DNA can form loops of intervening DNA that affect the transcriptional status. Looped structures are clearly important in the effects of enhancer sequences found in eukaryotic organisms, and they also have been shown to influence transcriptional regulation in prokaryotes (see review by MATTHEWS, 1992).

10 References

ACHTBERGER, C. F., MCALLISTER, E. C. (1989), Rotational orientation of upstream curved DNA affects promoter function in *Bacillus subtilis*, *J. Biol. Chem.* **264**, 10451–10456.

ADAMS, R. L. P., KNOWLER, J. T., LEADER, D. P. (1986), *The Biochemistry of the Nucleic Acids*, London: Chapman and Hall.

ALDEN, C. J., KIM, S.-H. (1979), Solvent-accessible surfaces of nucleic acids, *J. Mol. Biol.* **132**, 411–434.

ALLMANSBERGER, R., KNAUB, S., KLEIN, A. (1988), Conserved elements in the transcription initation regions preceding highly expressed structural genes of methanogenic archaebacteria, *Nucleic Acids Res.* **16**, 7419–7436.

ANDERSON, J. N. (1986), Detection, sequence patterns and function of unusual DNA structures, *Nucleic Acids Res.* **14**, 8513–8533.

ANDERSON, E. S., ROGERS, A. H. (1963), Slime polysaccharides of the Enterobacteriaceae, *Nature* **198**, 714–715.

ARBER, W. (1974), DNA modification and restriction, *Prog. Nucleic Acid Res. Mol. Biol.* **14**, 1–37.

ARNOTT, S., CHANDRASEKARAN, R., HUKINS, D. W. L., SMITH, P. J. C., WATTS, L. (1974), Structural details of a double-helix observed for DNAs containing alternating purine-pyrimidine sequences, *J. Mol. Biol.* **88**, 523–533.

ASTBURY, W. T. (1947), X-ray studies of nucleic acids, *Symp. Soc. Exp. Biol. (Nucleic Res.)* **1**, 66–76.

AVERY, O. T., MCLEOD, C. M., MCCARTY, M. (1944), Studies on the chemical nature of the substance inducing transformation of pneumococcal types. Induction of transformation by a desoxyribonucleic acid fraction isolated from *Pneumococcus* type III, *J. Exp. Med.* **79**, 137–158.

AZORIN, F., RICH, A. (1985), Isolation of Z-DNA binding proteins from SV40 minichromosomes: Evidence for binding to the viral control region, *Cell* **41**, 365–374.

AZORIN, F., MARTINEZ, A. B., SUBIRANA, J. A. (1980), Organization of nucleosomes and spacer DNA in chromatin fibers, *Int. J. Biol. Macromol.* **2**, 81–92.

AZORIN, F., HAHN, R., RICH, A. (1984), Restriction endonucleases can be used to study B-Z junctions in supercoiled DNA, *Proc. Natl. Acad. Sci. USA* **81**, 5714–5718.

BAILLEUL, B., GALIEGUE-ZOUITINA, S., LOUCHEUX-LEFEBRE, M. H. (1984), Conformations of poly (dG-dC) × poly (dG-dC) modified by the O-acetyl derivative of the carcinogen 4-hydroxyaminoquinoline-1-oxide, *Nucleic Acids Res.* **12**, 7915–7927.

BALOWS, A., TRÜPER, H. G., DWORKIN, M. HARDER, W., SCHLEIFER, K. H. (Eds.) (1991), *The Procaryotes,* 2nd Ed., New York: Springer Verlag.

BARTON, J. K., BASILE, L. A., DANISHEFSKY, A., ALEXANDRESCU, A. (1984), Chiral probes for the handedness of DNA helices: Enantiomers of tris(4,7-diphenylphenanthroline)ruthenium(II), *Proc. Natl. Acad. Sci. USA* **81**, 1961–1965.

BECK, C. F., WARREN, R. A. J. (1988), Divergent promoters, a common from of gene organization, *Microbiol. Rev.* **52**, 318–326.

BEEBE, T. P., WILSON, T. E., OGLETREE, D. F., KATZ, J. E., BALHORN, R., SALMERON, M. B., SIEKHAUS, W. J. (1989), Direct observation of native DNA structures with the scanning tunneling microscope, *Science* **243**, 370–372.

BEHE, M., FELSENFELD, G. (1981), Effects of methylation on a synthetic polynucleotide: the B-Z transition in poly(dG-m^{5}dC) · poly(dG-m^{5}dC), *Proc. Natl. Acad. Sci. USA* **78**, 1619–1623.

BENHAM, C. J. (1982), Stable cruciform formation at inverted repeat sequences in supercoiled DNA, *Biopolymers* **21**, 679–696.

BERNUES, J., BELTRAN, R., CASANOVAS, J. M., AZORIN, F. (1989), Structural polymorphism of sequences: the secondary DNA structure adopted by a d(GA.CT)$_{22}$ sequence in the presence of zinc ions, *EMBO J.* **8**, 2087–2094.

BETTER, M., LU, C., WILLIAMS, R. C., ECHOLS, H. (1982), Site-specific DNA condensation and pairing mediated by the Int protein of bacteriophage lambda, *Proc. Natl. Acad. Sci. USA* **79**, 5837–5841.

BHRIAIN, N. N., DORMAN, C. J., HIGGINS, C. F. (1989), An overlap between osmotic and anerobic stress responses: a potential role for DNA supercoiling in the coordinate regulation of gene expression, *Mol. Microbiol.* **3**, 933–942.

BLAHO, J. A., WELLS, R. D. (1987), Left-handed Z-DNA binding by the RecA protein of *Escherichia coli, J. Biol. Chem.* **262**, 6082–6088.

BOSSI, L., SMITH, D. M. (1984), Conformational change in the DNA associated with an unusual promoter mutation in a tRNA operon of *Salmonella, Cell* **39**, 643–652.

BOUBLIKOVA, P., PALECEK, E. (1990), Osmium tetroxide, N,N,N′,N′-tetramethylethylenediamine: A new probe of DNA structure in the cell, *FEBS Lett.* **263**, 281–284.

BRACCO, L., KOTLARZ, D., KOLB, A., DIEKMANN, S., BUC, H. (1989), Synthetic curved DNA sequences can act as transcriptional activators in *Escherichia coli, EMBO J.* **8**, 4289–4296.

BROYLES, S. S., PETTIJOHN, D. E. (1986), Interaction of the *Escherichia coli* HU protein with DNA. Evidence for formation of nucleosome-like structures with altered DNA helical pitch, *J. Mol. Biol.* **187**, 47–60.

BURD, J. F., WARTELL, R. M., DODGSON, J. B., WELLS, R. D. (1975), Transmission of stability (telestability) in deoxyribonucleic acid, *J. Biol. Chem.* **250**, 5209–5113.

CALLADINE, C. R., DREW, H. R. (1986), Principles of sequence-dependent flexure of DNA, *J. Mol. Biol.* **192**, 907–918.

CHAMBERLIN, M. J., PATTERSON, D. L. (1965), Physical and chemical characterization of the ordered complexes formed between polyinosinic acid, polycytidylic acid and their deoxyribo-analogues, *J. Mol. Biol.* **12,** 410–428.

CHARGAFF, E. (1951), Structure and function of nucleic acids as cell constituents, *Fed. Proc.* **10,** 654–659.

CHEN, F.-M. (1988), Effects of A-T base pairs on the B-Z conformational transitions of DNA, *Nucleic Acids Res.* **16,** 2269–2281.

CHEUNG, S., ARNDT, K., LU, P. (1984), Correlation of *lac* operator DNA imino proton exchange kinetics with its function, *Proc. Natl. Acad. Sci. USA* **81,** 3665–3669.

CHRISTOPHE, D., CABRER, B., BACOLLA, A., TARGOVNIK, H., POHL, V., VASSART, G. (1985), An unusually long poly(purine)-poly(pyrimidine) sequence is located upstream from the human thyroglobulin gene, *Nucleic Acids Res.* **13,** 5127–5144.

COLE, R. S. (1970), Light-induced cross-linking of DNA in the presence of a furocoumarin (psoralen), studies with phage lambda, *E. coli,* and mouse leukemia cells, *Biochim. Biophys. Acta* **217,** 30–39.

COLL, M., FREDERICK, C. A., WANG, A. H. J., RICH, A. (1987), A bifurcated hydrogen-bonded conformation in d(AT) base pairs of the DNA dodecamer d(CGCAAATTTGCG) and its complex with distamycin, *Proc. Natl. Acad. Sci. USA* **84,** 8385–8389.

CONNER, B. N., TAKANO, T., TANAKA, S., ITAKURA, K., DICKERSON, R. E. (1982), The molecular structure of d(CpCpGpG), a fragment of right-handed double helical A-DNA, *Nature* **295,** 294–299.

COOK, D. N., ARMSTRONG, G. A., HEARST, J. E. (1989), Induction of anerobic gene expression in *Rhodobacter capsulatus* is not accompanied by a local change in chromosomal supercoiling as measured by a novel assay, *J. Bacteriol.* **171,** 4836–4843.

COUREY, A. J., WANG, J. C. (1983), Cruciform formation in a negatively supercoiled DNA may be kinetically forbidden under physiological conditions, *Cell* **33,** 817–829.

COWIE, A., MEYERS, R. (1988), DNA sequences involved in transcriptional regulation of the mouse β-globin promoter in murine erythroleukemia cells, *Mol. Cell. Biol.* **8,** 3122–3128.

COZZARELLI, N. R. (1980), DNA topoisomerases, *Cell* **22,** 327–328.

COZZARELLI, N. R., WANG, J. C. (1990), *DNA Topology and its Biological Effects,* Cold Spring Harbor, NY: Cold Spring Harbor Laboratory Press.

CREWS, S., OJALA, D., POSAKONY, J., NISHIGUCHI, J., ATTARDI, G. (1979), Nucleotide sequence of a region of human mitochondrial DNA containing the precisely identified origin of replication, *Nature* **277,** 192–198.

DAVIES, D. R., BALDWIN, R. L. (1963), X-ray studies of two synthetic copolymers, *J. Mol. Biol.* **6,** 251–255.

DAYN, A., MALKHOSYAN, S., DUZHY, D., LYAMICHEV, V., PANCHENKO, Y., MIRKIN, S. (1991), Formation of $(dA\text{-}dT)_n$ cruciforms in *Escherichia coli* cells under different environmental conditions, *J. Bacteriol.* **173,** 2658–2664.

DEB, S., DELUCIA, A. L., PARTIN, K., TEGTMEYER, P. (1986), Functional interactions of the Simian virus 40 core origin of replication with flanking regulatory sequences, *J. Virol.* **57,** 138–144.

DE MASSEY, B., WEISBERG, R. A., STUDIER, F. W. (1987), Gene 3 endonuclease of bacteriophage T7 resolves conformationally branched structures in double-stranded DNA, *J. Mol. Biol.* **193,** 359–376.

DE VARGAS, L. M., KIM, S., LANDY, A. (1989), DNA looping generated by DNA bending protein IHF and the two domains of lambda integrase, *Science* **244,** 1457–1461.

DICKERSON, R. E. (1983), The DNA helix and how it is read, *Sci. Am.* **249,** 87–102.

DICKERSON, R. E., DREW, H. R. (1981), Structure of a B-DNA dodecamer. II. Influence of base sequence on helix structure, *J. Mol. Biol.* **149,** 761–786.

DICKERSON, R. E., DREW, H. R., CONNER, B. N., WING, R. M., FRATINI, A. V., KOPKA, M. L. (1982), The anatomy of A-, B-, and Z-DNA, *Science* **216,** 475–485.

DIEKMANN, S. (1987), Temperature and salt dependence of the gel migration anomaly of curved DNA fragments, *Nucleic Acids Res.* **15,** 247–265.

DIEKMANN, S. (1989), The migration anomaly of DNA fragments in polyacrylamide gels allows the detection of small sequence-specific DNA structure variations, *Electrophoresis* **10,** 354–359.

DINGWALL, C., LOMONOSSOFF, G. P., LASKEY, R. A. (1981), High sequence specificity of micrococcal nuclease, *Nucleic Acids Res.* **9,** 2659–2673.

DORMAN, C. J. (1991), DNA supercoiling and environmental regulation of gene expression in pathogenic bacteria, *Infect. Immun.* **59,** 745–749.

DORMAN, C. J., HIGGINS, C. F. (1987), Fimbrial phase variation in *Escherichia coli:* dependence on integration host factor and homologies with other site-specific recombinases, *J. Bacteriol.* **169,** 3840–3843.

DORMAN, C. J., BARR, G. C., BHRIAIN, N. N., HIGGINS, C. F. (1988), DNA supercoiling and the anaerobic and growth phase regulation of *tonB* gene expression, *J. Bacteriol.* **170**, 2816–2826.

DORMAN, C. J., LYNCH, A. S., BHRIAIN, N. N., HIGGINS, C. F. (1989), DNA supercoiling in *Escherichia coli: topA* mutations can be suppressed by DNA amplifications involving the *tolC* locus, *Mol. Microbiol.* **3**, 531–540.

DORMAN, C. J., BHRIAIN, N. N., HIGGINS, C. F. (1990), DNA supercoiling and environmental regulation of virulence gene expression in *Shigella flexneri, Nature* **344**, 789–792.

DREW, H., TAKANO, T., TANAKA, S., ITAKURA, K., DICKERSON, R. E. (1980), High-salt d(CpGpCpG): A left-handed Z′ DNA double helix, *Nature* **286**, 567–573.

DRLICA, K. (1984), Biology of bacterial deoxyribonucleic acid topoisomerases, *Microbiol. Rev.* **48**, 273–289.

DRLICA, K. (1990), Bacterial topoisomerases and the control of DNA supercoiling, *Trends Genet.* **6**, 433–437.

DRLICA, K. (1992), Control of bacterial DNA supercoiling, *Mol. Microbiol.* **6**, 425–433.

ECHOLS, H. (1986), Multiple DNA–protein interactions governing high-precision DNA transactions, *Science* **233**, 1050–1056.

ELBOROUGH, K. M., WEST, S. C. (1988), Specific binding of cruciform DNA structures by a protein from human extracts, *Nucleic Acids Res.* **16**, 3603–3616.

FELSENFELD, G. (1985), DNA, *Sci. Am.* **253**, 58–67.

FIGUEROA, N., BOSSI. L. (1988), Transcription induces gyration of the DNA template in *Escherichia coli, Proc. Natl. Acad. Sci. USA* **85**, 9416–9420.

FIGUEROA, N., WILLS, N., BOSSI, L. (1991), Common sequence determinants of the response of a procaryotic promoter to DNA bending and supercoiling, *EMBO J.* **10**, 941–949.

FOX, K. R. (1990), Long $(dA)_n \cdot (dT)_n$ tracts can form intramolecular triplexes under superhelical stress, *Nucleic Acids Res.* **18**, 5387–5391.

FRANCO, R., DRLICA, K. (1988), DNA gyrase on the bacterial chromosome: Oxolinic acid-induced DNA cleavage in the *dnaA-gyrB* region, *J. Mol. Biol.* **201**, 229–233.

FRIEDRICH, K., GUALERZI, C. O., LAMMI, M., LOSSO, M. A., PON, C. L. (1988), Proteins from the procaryotic nucleoid: interaction of nucleic acids with the 15 kDa *Escherichia coli* histone-like protein H-NS, *FEBS Lett.* **229**, 197–202.

FULLER, W., WILKINS, M. H. F., WILSON, H. R., HAMILTON, L. D. (1965), The molecular configuration of deoxyribonucleic acid. IV. X-ray diffraction study of the A-form, *J. Mol. Biol.* **12**, 60–80.

FURBERG, S. (1951), The crystal structure of cytidine, *Acta Crystallogr.* **3**, 325–331.

FURLONG, J. C., LILLEY, D. M. J. (1986), Highly selective chemical modification of cruciform loops by diethyl pyrocarbonate. *Nucleic Acids Res.* **14**, 3995–4007.

GALAN, J. E., CURTISS III, R. (1990), Expression of *Salmonella typhimurium* genes required for invasion is regulated by changes in DNA supercoiling, *Infect. Immun.* **58**, 1879–1885.

GALAT, A. (1990), A note on sequence-dependence of DNA structure, *Eur. Biophys. J.* **17**, 331–342.

GELLERT, M., O'DEA, M. H., MIZUUCHI, K., NASH, H. (1976), DNA gyrase: an enzyme that introduces superhelical turns into DNA, *Proc. Natl. Acad. Sci. USA* **73**, 3872–3876.

GIERER, A. (1966), Model for DNA and protein interaction and the function of the operator, *Nature* **212**, 1480–1481.

GOLDBLUM, A., PERAHIA, D., PULLMAN, A. (1978), Hydration scheme of the complementary base-pairs of DNA, *FEBS Lett.* **91**, 213–215.

GOLDSTEIN, E., DRLICA, K. (1984), Regulation of bacterial DNA supercoiling: plasmid linking numbers vary with growth temperature, *Proc. Natl. Acad. Sci. USA* **81**, 4046–4050.

GOODMAN, S. D., NASH, H. A. (1989), Functional replacement of a protein-induced bend in a DNA recombination site, *Nature* **341**, 251–254.

GORANSSON, M., SONDEN, B., NILSSON, P., DAGBERG, B., FORSMAN, K., EMANUELSSON, K., UHLIN, B.-E. (1990), Transcriptional silencing and thermoregulation of gene expression in *Escherichia coli, Nature* **344**, 682–685.

GOUGH, G. W., SULLIVAN, K. M., LILLEY, D. M. J. (1986), The structure of cruciforms in supercoiled DNA: probing the single-stranded character of nucleotide bases with bisulphite, *EMBO J.* **5**, 191–196.

GOURSE, R., DO BOER, H. A., NOMURA, M. (1986), DNA determinants of rRNA synthesis in *Escherichia coli:* growth rate dependent regulation, feedback inhibition, upstream activation, antitermination, *Cell* **44**, 197–205.

GREAVES, D. R., PATIENT, R. K. (1986), *Rec*BC, *sbc*B independent, $(AT)_n$-mediated deletion of sequences flanking a *Xenopus laevis* β-globin gene on propagation in *E. coli, Nucleic Acids Res.* **14**, 4147–4158.

GRIFFITH, J., BLEYMAN, M., RAUCH, C. A., KITCHIN, P. A., ENGLUND, P. T. (1986), Visualization of the bent helix in kinetoplast DNA by electron microscopy, *Cell* **46**, 717–724.

GULLAND, J. M. (1947), The structure of nucleic acids, *Cold Spring Harbor Symp. Quant. Biol.* **12**, 95–103.

HAMADA, H., KAKUNAGA, T. (1982), Potential Z-DNA forming sequences are highly dispersed in the human genome, *Nature* **298**, 396–398.

HANAU, L. H., SANTELLA, R. M., GRUNBERGER, D., ERLANGER, B. F. (1984), An immunochemical examination of acetylaminofluorene-modified poly(dG-dC) × poly(dG-dC) in the Z-conformation, *J. Biol. Chem.* **259**, 173–178.

HANIFORD, D. B., PULLEYBLANK, D. E. (1983), Facile transition of poly [d(TG) · d(CA)] into a left-handed helix in physiological conditions, *Nature* **302**, 632–634.

HANIFORD, D. B., PULLEYBLANK, D. E. (1985), Transition of a cloned $d(AT)_n$-$d(AT)_n$ tract to a cruciform *in vivo, Nucleic Acids Res.* **13**, 4343–4363.

HANTKE, K. (1981), Regulation of ferric iron transport in *Escherichia coli* K12; isolation of a constitutive mutant, *Mol. Gen. Genet.* **182**, 288–292.

HANVEY, J. C., KLYSIK, J., WELLS, R. D. (1988), Influence of DNA sequence on the formation of non-B right-handed helices in oligopurine · oligopyrimidine inserts in plasmids, *J. Biol. Chem.* **263**, 7386–7396.

HAY, R. T., DE PAMPLILIS, M. L. (1982), Initiation of SV40 DNA replication *in vivo:* Location and structure of 5′ ends of DNA synthesized in the *ori* region, *Cell* **28**, 767–779.

HEBERT, E., LOUKAKOU, B., SAINT-RUF, G., LENG, M. (1984), Conformational changes induced in DNA by the *in vitro* reaction with the mutagenic amine: 3-N,N-acetoxyacetylamine-4,6-dimethyldipyrido(1,2-a:3′,2′-d)imidazole, *Nucleic Acids Res.* **12**, 8553–8556.

HEINEMANN, U., LAUBLE, H., FRANK, R., BLOCKER, H. (1988), Crystal structure analysis of an A-DNA fragment at 1.8 Å resolution: d(GCCCGGGC), *Nucleic Acid Res.* **15**, 9531–9550.

HIDAKA, M., AKIYAMA, M., HORIUCHI, T. (1988), A consensus sequence of three DNA replication terminus sites on the *E. coli* chromosome is highly homologous to the *terR* of the R6K plasmid, *Cell* **55**, 467–475.

HIGGINS, C. F., DORMAN, C. J., STIRLING, D. A., WADDELL, L., BOOTH, I. R., MAY, G., BREMER, E. (1988), A physiological role for DNA supercoiling in the osmotic regulation of gene expression in *Salmonella typhimurium* and *Escherichia coli, Cell* **52**, 569–584.

HIGGINS, C. F., HINTON, J. C. D., HULTON, C. S. J., OWEN-HUGHES, T., PAVITT, G. D., SEIRAFI, A. (1990), Protein H1: a role for chromatin structure in the regulation of bacterial gene expression and virulence, *Mol. Microbiol.* **4**, 2007–2012.

HILL, R. J., STOLLAR, D. B. (1983), Dependence of Z-DNA antibody binding to polytene chromosomes on acid fixation and DNA torsional strain, *Nature* **305**, 338–340.

HILL, R. J., WATT, F., STOLLAR, D. B. (1984), Z-DNA immunoreactivity of *Drosophila* polytene chromosomes, *Exp. Cell Res.* **153**, 469–482.

HORWITZ, M. S. Z., LOEB, L. A. (1988), An *E. coli* promoter that regulates transcription by DNA superhelix-induced cruciform extrusion, *Science* **241**, 703–705.

HOSUR, R. V., GOVIL, G., HOSUR, M. V., VISWAMITRA, M. A. (1981), Sequence effects in structures of the dinucleotides d-pApT and d-pTpA, *J. Mol. Struct.* **72**, 261–267.

HSIEH, C.-H., GRIFFITH, J. D. (1988), The terminus of SV40 DNA replication and transcription contains a sharp sequence-directed curve, *Cell* **52**, 535–544.

HTUN, H., DAHLBERG, J. E. (1989), Topology and formation of triple-stranded H-DNA, *Science* **243**, 1571–1576.

HULTON, C. S. J., SEIRAFI, A., HINTON, J. C. D., SIDEBOTHAM, J. M., WADDELL, L., PAVITT, G. D., OWEN-HUGHES, T., SPASSKY, A., BUC, H., HIGGINS, C. F. (1990), Histone-like protein H1 (H-NS), DNA supercoiling, and gene expression in bacteria, *Cell* **63**, 631–642.

JAIN, S., ZON, G., SUNDARALINGAM, M. (1987), The potentially Z-DNA forming sequence d(GTGTACAC) crystallizes as A-DNA, *J. Mol. Biol.* **197**, 141–145.

JAMIESON, D. J., HIGGINS, C. F. (1986), Two genetically distinct pathways for transcriptional regulation of anaerobic gene expression in *Salmonella typhimurium, J. Bacteriol.* **168**, 389–397.

JAWORSKI, A., HSIEH, W-T., BLAHO, J. A., LARSON, J. E., WELLS, R. D. (1987), Left-handed DNA *in vivo, Science* **238**, 773–777.

JESAITIS, M. (1956), Differences in the chemical composition of the phage nucleic acids, *Nature* **178**, 637–641.

JOHNSTON, B. H. (1988), The S1-sensitive form of $d(C\text{-}T)_n \cdot d(A\text{-}G)_n$: Chemical evidence for a three-stranded structure in plasmids, *Science* **241**, 1800–1804.

JOHNSTON, B. H., RICH, A. (1985), Chemical probes of DNA conformation: detection of Z-DNA at nuclotide resolution, *Cell* **42**, 713–724.

KARAPETYAN, A. T., MINYAT, E. E., IVANOV, V. I. (1984), Increase in temperature induces the Z to B transition of poly (d(G-C)) in water–ethanol solution, *FEBS Lett.* **173**, 243–246.

KILPATRICK, M. W., KLYSIK, J., SINGLETON, C. K., ZARLING, D. A., JOVIN, T. M., HANAN, L. H., ERLANGER, B. F., WELLS, R. D. (1984), Intervening sequences in human fetal globin genes adopt left-handed Z helices, *J. Biol. Chem.* **259**, 7268–7274.

KLEVAN, L., WANG, J. C. (1980), Deoxyribonucleic acid gyrase–deoxyribonucleic acid complex containing 140 base pairs of deoxyribonucleic acid and an $-_2\beta_2$ protein core, *Biochemistry* **19**, 5229–5234.

KLUG, A., RHODES, D., SMITH, J., FINCH, J. T., THOMAS, J. O. (1980), A low resolution structure for the histone core of the nucleosome, *Nature* **287**, 509–516.

KLYSIK, J., STIRDIVANT, S. M., LARSON, J. E., HART, P. A., WELLS, R. D. (1981), Left-handed DNA in restriction fragments and a recombinant plasmid, *Nature* **290**, 672–677.

KLYSIK, J., STIRDIVAN, S. M., WELLS, R. D. (1982), Left-handed DNA: Cloning, characterization, and instability of inserts containing different lengths of (dC-dG) in *Escherichia coli, J. Biol. Chem.* **257**, 10152–10158.

KMIEC, E. B., ANGELIDES, K. J., HOLLOMAN, W. K. (1985), Left-handed DNA and the synaptic pairing reaction promoted by *Ustilago* Rec1 protein, *Cell* **40**, 139–145.

KOEPSEL, R. R., KHAN, S. (1987), Static and initiator protein-enhanced bending of DNA at a replication origin, *Science* **233**, 1316–1318.

KOHWI, Y., KOHWI-SHIGEMATSU, T. (1988), Magnesium ion-dependent, novel triple-helix structure formed by homopurine-homopyrimidine sequences in supercoiled plasmid DNA, *Proc. Natl. Acad. Sci. USA* **85**, 3781–3785.

KORNBERG, A. (1980), *DNA Replication,* San Francisco: Freeman and Company.

LAGRAVERE, C., MALFOY, B., LENG, M., LAVAL, J. (1984), Ring-openend alkylated guanine is not repaired in Z-DNA, *Nature* **310**, 798–800.

LAMMI, M., PACI, M., PON, C. L., LOSSO, K. A., MIANO, A., PAWLIK, R. T., GIANFRANCESCHI, G. L., GUALERZI, C. O. (1984), Proteins from the procaryotic nucleoid: Biochemical and ^{1}H-NMR studies of three bacterial histone-like proteins, in: *Proteins Involved in DNA Replication* (HUBSCHER, J., SPADARI, S., Eds.), pp. 467–477, New York–London: Plenum Press.

LANGRIDGE, R., MARVIN, D. A., SEEDS, W. E., WILSON, H. R., HOOPER, C. W., WILKINS, M. H. F., HAMILTON, L. D. (1960), The molecular configuration of deoxyribonucleic acid. II. Molecular models and their Fourier transforms, *J. Mol. Biol.* **2**, 38–64.

LA VOLPE, A., SIMEONE, A., D'ESPOSITO, M., SCOTTO, L., FIDANZA, V., DE FALCO, A., BONCINELLI, E. (1985), Molecular analysis of the heterogeneity region of the human ribosomal spacer, *J. Mol. Biol.* **183**, 213–223.

LEE, J. S., WOODSWORTH, M. L., LATIMER, L. J. P. (1984), Functional groups on "Z" DNA recognized by monoclonal antibodies, *FEBS Lett.* **168**, 303–306.

LENG, M. (1985), Left-handed Z-DNA, *Biochim. Biophys. Acta* **825**, 339–344.

LESLIE, A. G. W., ARNOTT, S., CHANDRASEKARAN, R., RATLIFF, R. L. (1980), Polymorphism of DNA double helices, *J. Mol. Biol.* **143**, 49–72.

LILLEY, D. M. J. (1980), The inverted repeat as a recognizable structural feature in supercoiled DNA molecules, *Proc. Natl. Acad. Sci. USA* **77**, 6468–6472.

LILLEY, D. M. J. (1983), Structural perturbation in supercoiled DNA: Hypersensitivity to modification by a single-strand-selective chemical reagent conferred by inverted repeat sequences, *Nucleic Acids Res.* **11**, 3097–3112.

LILLEY, D. M. J., HALLAM, L. R. (1984), The thermodynamics of the ColE1 cruciform: comparisons between probing and topological experiments using single topoisomers, *J. Mol. Biol.* **180**, 179–200.

LILLEY, D. M. J., HIGGINS, C. F. (1991), Local DNA topology and gene expression: the case of the *leu-500* promoter, *Mol. Microbiol.* **5**, 779–783.

LILLEY, D. M. J., PALECEK, E. (1984), The supercoil-stabilised cruciform of ColE1 is hyper-reactive to osmium tetroxide, *EMBO J.* **3**, 1187–1192.

LILLEY, D. M. J., SULLIVAN, K. M., MURCHIE, A. I. H., FURLONG, J. C. (1988), Cruciform extrusion in supercoiled DNA – Mechanisms and contextual influence, in: *Unusual DNA Structures* (WELLS, R. D., HARVEY, S. C., Eds.), New York: Springer.

LINIAL, M., SHLOMAI, J. (1988), Bent DNA structures associated with several origins of replication are recognized by a unique enzyme from trypanosomatids, *Nucleic Acids Res.* **16**, 6477–6492.

LIU, L. F., WANG, J. C. (1987), Supercoiling of the DNA template during transcription, *Proc. Natl. Acad. Sci. USA* **84**, 7024–7027.

LOCKSHON, D., MORRIS, D. R. (1985), Sites of reaction of *Escherichia coli* DNA gyrase on pBR322 *in vivo* as revealed by oxolinic acid-induced plasmid linearization, *J. Mol. Biol.* **181**, 63–74.

LODGE, J. K., KAZIC, T., BERG, D. E. (1989), Formation of supercoiling domains in plasmid pBR322, *J. Bacteriol.* **171**, 2181–2187.

LUTHMAN, K., BEHE, M. J. (1988), Sequence dependence of DNA structure, *J. Biol. Chem.* **263**, 15535–15539.

LYAMICHEV, V. I., MIRKIN, S. M., FRANK-KAMENETSKII, M. D. (1986), Structures of homopurine-homopyrimidine tract in superhelical DNA, *J. Biomol. Struct. Dyn.* **3**, 667–669.

MADAIO, M. P., HODDER, S., SCHWARTZ, R. S., STOLLAR, B. D. (1984), Responsiveness of autoimmune and normal mice to nucleic acid antigens, *J. Immunol.* **132**, 872–876.

MAHENDRASINGAM, A., RHODES, N. J., GOODWIN, D. C., NAVE, C., PIGRAM, W. J., FULLER, W., BRAHMS, J., VERGNE, J. (1983), Conformational transitions in oriented fibers of the synthetic polynucleotide poly d(AT) × poly d(AT) double helix, *Nature* **301**, 535–537.

MALAN, T. P., KOLB, A., BUC, H., MCCLURE, W. R. (1984), Mechanism of CRP-cAMP activation of *lac* operator transcriptional initiation: activation of the P1 promoter, *J. Mol. Biol.* **180**, 881–909.

MALINGE, J-M., LENG, M. (1984), Reaction of *cis*-diamminedichloroplatinium II and DNA in B or Z conformation, *EMBO J.* **3**, 1273–1279.

MANOR, H., RAO, B. S., MARTIN, R. G. (1988), Abundance and degree of dispersion of genomic $d(GA)_n \cdot d(TC)_n$ sequences, *J. Mol. Evol.* **27**, 96–101.

MARAMUTSU, S., MASASHI, K., KOHARA, Y., MIZUNO, T. (1988), Insertion sequence IS5 contains a sharply curved DNA structure at its terminus, *Mol. Gen. Genet.* **214**, 433–438.

MARGOLIN, P., ZUMSTEIN, L., STERNGLANZ, R., WANG, J. C. (1985), The *Escherichia coli supX* locus is *topA*, the structural gene for DNA topoisomerase I, *Proc. Natl. Acad. Sci. USA* **82**, 5437–5441.

MARINI, J., LEVENE, S., CROTHERS, D. M., ENGLUND, P. T. (1982), Bent helical structures on kinetoplast DNA, *Proc. Natl. Acad. Sci. USA* **79**, 7664–7668.

MATHEWS, K. (1992), DNA looping, *Microbiol. Rev.* **56**, 123–136.

MCCALL, M., BROWN, T., KONNARD, O. (1985), The crystal structure of d(GGGGCCCC). A model for poly (dG) × poly (dC), *J. Mol. Biol.* **183**, 385–396.

MCCALL, M., BROWN, T., HUNTER W., KONNARD, O. (1986), The crystal structure of di(GGATGGGAG) forms an essential part of the binding site from transcription factor IIIA, *Nature* **322**, 661–664.

MCCLARIN, J. A., FREDERICK, C. A., WANG, B.-C., GREENE, P., BOYER, W. H., GRABLE, J., ROSENBERG, J. (1986), Structure of the DNA-*Eco*RI endonuclease recognition complex at 3 Å resolution. *Science* **234**, 1526–1541.

MCGHEE, J. D., FELSENFELD, G. (1980), Nucleosome structure, *Annu. Rev. Biochem.* **49**, 1115–1156.

MCLEAN, M. J., BLAHO, J. A., KILPATRICK, M. W., WELLS, R. D. (1986), Consecutive A-T pairs can adopt a left-handed DNA structure, *Proc. Natl. Acad. Sci. USA* **83**, 5884–5888.

MEIJER, M., BECK, E., HANSEN, F. G., BERGAMNS, H. E. N., MESSER, W., VON MEYENBURG, K., SCHALLER, H. (1979), Nucleotide sequence of the origin of replication of the *Escherichia coli* K12 chromosome, *Proc. Natl. Acad. Sci. USA* **76**, 580–584.

MENZEL, R., GELLERT, M. (1983), Regulation of the genes for *Escherichia coli* DNA gyrase: homeostatic control of DNA supercoiling, *Cell* **34**, 105–113.

MILMAN, G., LANGRIDGE, R., CHAMBERLIN, M. J. (1967), The structure of a DNA–RNA hybrid, *Proc. Natl. Acad. Sci. USA* **57**, 1804–1810.

MIRKIN, S. M., LYAMICHEV, V. I., DRUSHIYAK, K. N., DOBRYNIN, V. N., FILIPPOV, S. A., FRANK-KAMENETSKI, M. D. (1987), DNA H form requires a homopurine-homopyrimidine mirror repeat, *Nature* **330**, 495–497.

MIRSKY, A. E. (1980), The discovery of DNA, in: *Molecules to Living Cells,* San Francisco: Freeman and Company.

MISRA, R., REEVES, P. R. (1987), Role of *micF* in the *tolC* mediated regulation of OmpF, a major outer membrane protein of *Escherichia coli* K-12, *J. Bacteriol.* **169**, 4722–4730.

MITSUI, Y., LANGRIDGE, R., SHORTLE, B. E., CANTOR, C. R., GRANTI, R. C., KODAMA, M., WELLS, R. D. (1970), Physical and enzymatic studies on poly d(I-C) × poly d(I-C), an unusual double-helical DNA, *Nature* **228**, 1166–1169.

MIZUNO, T. (1987), Static bend of DNA helix at the activator recognition site of the *ompF* promoter in *Escherichia coli, Gene* **54**, 57–64.

MIZUUCHI, K., MIZUUCHI, M., GELLERT, M. (1982), Cruciform structures in palindromic DNA are favored by DNA supercoiling, *J. Mol. Biol.* **156**, 229–243.

MOKUL'SKII, M. A., KAPITANOVA, K. A., MOKUL'SKAYA, T. D. (1972), Secondary structure of phage T2 DNA, *Mol. Biol. (Moscow)* **6**, 714–731.

MÖLLER, A., NORDHEIM, A., KOZLOWSKI, S. A., PATEL, D. J., RICH, A. (1984), Bromination stabilizes poly(dG-dC) in the Z-DNA form under low-salt conditions, *Biochemistry* **23**, 54–62.

MORGAN, T. H. (1919), *The Physical Basis of Heredity,* Philadelphia: Lippincott.

MOYER, R. N., GRAVES, R. L. (1981), The mechanism of cytoplasmic orthopoxvirus DNA replication, *Cell* **27**, 391–401.

MÜLLER, U. R., WILSON, C. L. (1987), The effect of supercoil and termperature on the recognition of palindromic and non-palindromic regions in ϕX174 replicative form DNA by S1 and Bal31, *J. Biol. Chem.* **262,** 3730–3738.

MURA, C. V., STOLLAR, B. D. (1984), Interactions of H1 and H5 histones with polynucleotides of B- and Z-DNA conformations, *Biochemistry* **23,** 6147–6152.

NELSON, H. C. M., FINCH, J. T., LUISI, B. F., KLUG, A. (1987), The structure of an oligo(dA) × oligo(dT) tract and its biological implication, *Nature* **330,** 221–226.

NILSSON, L., CLORE, G. M., GRONENBORN, A. M. BRUNGER, A. T., KARPLUS, M. (1986), Structure refinement of oligonucleotides by molecular dynamics with nuclear Overhouser effect interproton distance restraints: application to 5′d(CGTACG)$_2$, *J. Mol. Biol.* **188,** 455–475.

O'BRIAN, E. J., MACEWAN, A. W. (1970), Molecular and crystal structure of polynucleotide complex: Polyinosinic acid plus polydeoxycytidylic acid, *J. Mol. Biol.* **48,** 243–261.

OHLENDORF, D. H., ANDERSON, W. F., FISHER, R. G., TAKEDA, Y., MATTHEWS, B. W. (1982), The molecular basis of DNA-protein recognition inferred form the structure of Cro repressor, *Nature* **298,** 718–723.

PANAYOTATOS, N., BACKMAN, S. (1989), II. A site-targeted recombinant nuclease probe of DNA structure, *J. Biol. Chem.* **264,** 15070–15073.

PANAYOTATOS, N., FONTAINE, A. (1987), A native cruciform DNA structure probed in bacteria by recombinant T7 endonuclease, *J. Biol. Chem.* **262,** 11364–11368.

PANAYOTATOS, N., WELLS, R. D. (1981), Cruciform structures in supercoiled DNA, *Nature* **289,** 466–470.

PANYUTIN, I. G., KOVALSKY, O. I., BUDOWSKY, E. I., DICKERSON, R. E., RIKHIREV, M. E. (1990), G-DNA: A twice-folded DNA structure adopted by single-stranded oligo d(G) and its implication for telomeres, *Proc. Natl. Acad. Sci. USA* **87,** 867–870.

PATEL, D. J., SHAPIRO, L., HARE, D. (1987), DNA and RNA: NMR studies of conformations and dynamics in solution, *Q. Rev. Biophys.* **20,** 35–112.

PATO, M. L., HOWE, M. M., HIGGINS, P. N. (1990), A DNA gyrase-binding site at the center of the bacteriophage Mu genome is required for efficient replicative transposition, *Proc. Natl. Acad. Sci. USA* **87,** 8716–8720.

PETTIJOHN, D. E., SINDEN, R. R., BROYLES, S. S. (1988), *Unusual DNA Structures* (WELLS, R. D., HARVEY, S. C., Eds.), pp. 103–113, New York: Springer Verlag.

PILCH, D. S., LEVENSON, C., SHAFER, R. H. (1990), Structural analysis of the (dA)$_{10}$·2(dT)$_{10}$ triple helix, *Proc. Natl. Acad. Sci. USA* **87,** 1942–1946.

POHL, F. M. (1976), Polymorphism of a synthetic DNA in solution, *Nature* **260,** 365–366.

POHL, F. M., JOVIN, T. M. (1972), Salt-induced co-operative conformational change of a synthetic DNA: Equilibrium and kinetic studies with poly (dG-C), *J. Mol. Biol.* **67,** 375–396.

POPE, L. E., SIGMAN, D. S. (1984), Secondary structure specificity of the nuclease activity of the 1,10-phenanthroline-copper complex, *Proc. Natl. Acad. Sci. USA* **81,** 3–7.

PRUSS, G. J., DRLICA, K. (1986), Topoisomerase I mutants: The gene on pBR322 that encodes resistance to tetracycline affects plasmid DNA supercoiling, *Proc. Natl. Acad. Sci. USA* **83,** 8952–8956.

PRUSS, G. J., DRLICA, K. (1989), DNA supercoiling and procaryotic transcription, *Cell* **56,** 521–523.

PRUSS, G. J., MANES, S. H., DRLICA, K. (1982), *Escherichia coli* DNA topoisomerase I mutants: increased supercoiling is corrected by mutations near gyrase genes, *Cell* **31,** 35–42.

PULLEYBLANK, D. E., HANIFORD, D. B., MORGAN, A. R. A. (1985), A structural basis for S1 nuclease sensitivity of double-stranded DNA, *Cell* **42,** 271–280.

RAHMOUNI, A. R. (1992), Z-DNA as a probe for localized supercoiling *in vivo, Mol. Microbiol.* **6,** 569–572.

RAHMOUNI, R. A., WELLS, R. D. (1989), Stabilization of Z DNA *in vivo* by localized supercoiling, *Science* **246,** 358–363.

RAJAGOPAL, P., FEIGON, J. (1989), Triple-strand formation in the homopurine:homopyrimidine DNA oligonucleotides d(G-A)$_4$ and d(T-C)$_4$ *Nature* **339,** 637–639.

RAJI, A., ZABEL, D. J., LAUFER, C. S., DEPEW, R. E. (1985), Genetic analysis of mutations that compensate for loss of *Escherichia coli* DNA topoisomerase I, *J. Bacteriol.* **162,** 1173–1179.

REVET, B., ZARLING, D. A., JOVIN, T. M., DELAIN, E. (1984), Different Z DNA forming sequences are revealed in ϕX174 RFI by high resolution darkfield immuno-electron microscopy, *EMBO J.* **3,** 3353–3358.

RICH, A., NORDHEIM, A., WANG, A. H.-J. (1984), The chemistry and biology of left-handed Z-DNA, *Annu. Rev. Biochem.* **53,** 791–846.

RICHARDSON, S. M. H., HIGGINS, C. F., LILLEY, D. M. J. (1988), DNA supercoiling and the *leu500* promoter mutation of *Salmonella typhimurium, EMBO J.* **7,** 1863–1869.

RILEY, M., MALING, B., CHAMBERLIN, M. J. (1966), Physical and chemical characterization of two- and three-stranded adenine-thymine and adenine-uracil homopolymer complexes, *J. Mol. Biol.* **20,** 359–389.

ROBERT-NICOUD, M., ARNDT-JOVIN, D. J., ZARLING, D. A., JOVIN, T. M. (1984), Immunological detection of left-handed Z-DNA in isolated polytene chromosomes. Effects of ionic strength, pH, temperature and topological stress, *EMBO J.* **3,** 721-731.

ROBERTSON, C. A., NASH, H. A. (1988), Bending of the bacteriophage lambda attachment site by *Escherichia coli* integration host factor, *J. Biol. Chem.* **263,** 3554–3557.

RODGERS, J. (1983), CACA sequences - the ends and the means? *Nature* **305,** 101–102.

ROSENBERG, M., COURT, D. (1979), Regulatory sequences involved in the promotion and termination of RNA transcription, *Annu. Rev. Genet.* **13,** 319–353.

RYDER, K., SILVER, S., DELUCIA, A. L., FANNING, E., TEGTMEYER, P. (1986), An altered DNA conformation in origin region I is determinant for the binding of SV40 large T antigene, *Cell* **44,** 719–725.

SAENGER, W. (1984), *Principles of Nucleic Acid Structure,* New York: Springer.

SANTORO, C., COSTANZO, F., CILIBERTO, G. (1984), Inhibition of eucaryotic tRNA transcription by potential Z-DNA sequences, *EMBO J.* **3,** 1553–1559.

SARMA, M. H., GUPTA, G., DARMA, R. H. (1988), Structure of a bent DNA; two-dimensional NMR studies on $d(GAAAATTTTC)_2$, *Biochemistry* **27,** 3423-3432.

SAUER, M., HANTKE, K., BRAUN, V. (1987), Ferric-coprogen receptor FhuE of *Escherichia coli:* processing and sequence common to all TonB-dependent outer membrane receptor proteins, *J. Bacteriol.* **169,** 2044–2049.

SETLOW, P. (1992), DNA in dormant spores of *Bacillus* species is in an A-like conformation, *Mol. Microbiol.* **6,** 563–567.

SHAKKED, Z., RABINOVICH, D., KENNARD, O., CRUSE, W. B. T., SALISBURY, S. A., VISWAMITRA, M. A. (1983), Sequence-dependent conformation of an A-DNA double helix, *J. Mol. Biol.* **166,** 183-201.

SHEFLIN, L. G., KOWALSKI, D. (1984), Mung bean nuclease cleavage of a dA + dT-rich sequence or an inverted repeat sequence in supercoiled PM2 DNA depends on ionic environment, *Nucleic Acids Res.* **12,** 7087–7104.

SHEN, S., SLIGHTOM, J., SMITHIES, O. (1981), A history of the human fetal globin gene duplication, *Cell* **26,** 191–203.

SINDEN, R. R., BROYLES, S. S., PETTIJOHN, D. E. (1983), Perfect palindromic *lac* operator DNA sequence exists as a stable cruciform structure in supercoiled DNA *in vitro* but not *in vivo, Proc. Natl. Acad. Sci. USA* **80,** 1797-1801.

SINGLETON, C. K., WELLS, R. D. (1982), Relationship between superhelical density and cruciform formation in plasmid pVH51, *J. Biol. Chem.* **257,** 6292–6295.

SINGLETON, C. K., KILPATRICK, M. W., WELLS, R. D. (1984), S1 nuclease recognises DNA conformational functions between left-handed helical $(dT\text{-}dG)_n \cdot (dC\text{-}dA)_n$ and contiguous right-handed sequences, *J. Biol. Chem.* **259,** 1963–1967.

SMITH, M. W., NEIDHARDT, F. C. (1983a), Proteins induced by anaerobiosis in *Escherichia coli, J. Bacteriol.* **154,** 336–343.

SMITH, M. W., NEIDHARDT, F. C. (1983b), Proteins induced by aerobiosis in *Escherichia coli, J. Bacteriol.* **154,** 344–350.

SNYDER, M., BUCHMAN, A. R., DAVIS, R. V. (1986), Bent DNA at a yeast autonomously replicating sequence, *Nature* **324,** 87–89.

SNYDER, U. K., THOMPSON, J. F., LANDY, A. (1989), Phasing of protein-induced DNA bends in a recombination complex, *Nature* **341,** 255-257.

STALKER, D. M., THOMAS, C. M., HELINSKI, D. R. (1981), Nucleotide sequence of the region of the origin of replication of the broad host range plasmid RK2, *Mol. Gen. Genet.* **181,** 8-12.

STENZEL, T. T., PATEL, P., BASTIA, D. (1987), The integration host factor of *Escherichia coli* binds to bent DNA at the origin of replication of the plasmid pSC101, *Cell* **49,** 709–717.

SULLIVAN, K. M., LILLEY, D. M. J. (1986), A dominant influence of flanking sequences on a local structural transition in DNA, *Cell* **47,** 817–827.

SUZUKI, D. T., GRIFFITHS, A. J. F., MILLER, J. H., LEWONTIN, R. C. (1991), *Genetik,* pp. 222–223, Weinheim–New York–Basel–Cambridge: VCH.

TAVIANINI, M. A., HAYES, T. E., MAGAZIN, M. D., MINTH, C. D., DIXON, J. E. (1984), Isolation, characterization, and DNA sequence of the rat somatostatin gene, *J. Biol. Chem.* **259,** 11798–11803.

THOMAS, G. A., KUBASEK, W. L., PETICOLAS, W. L., GREENE, P., GRABLE, J., ROSENBERG, J. M. (1989), Environmentally induced conformational changes in B-type DNA: comparison of the conformation of the oligonucleotide d(TCGCGAATTCGCG) in solution and its crystalline complex with the restriction nuclease *Eco*-RI, *Biochemistry* **28,** 2001–2009.

THOMPSON, J. F., LANDY, A. (1988), Empirical estimation of protein-induced DNA bending angles: applications to lambda site-specific recombination complexes, *Nucleic Acids Res.* **16,** 9687–9705.

TIMSIT, Y., VILBOIS, E., MORAS, D. (1991), Base-pairing shift in the major groove of $(CA)_n$ tracts by B-DNA crystal structures, *Nature* **354,** 167–170.

TRAVERS, A. (1989), Curves with a function, *Nature* **341,** 184–185.

TRAVERS, A. (1990), Why bend DNA, *Cell* **60,** 177–180.

TRAVERS, A., KLUG, A. (1987), Nucleoprotein complexes: DNA wrapping and writhing, *Nature* **327,** 280–281.

TSE-DINH, Y.-C. (1985), Regulation of the *Escherichia coli* DNA topoisomerase I gene by DNA supercoiling, *Nucleic Acids Res.* **13,** 4751–4763.

TSE-DINH, Y.-C., BERAN, R. K. (1988), Multiple promoters for transcription of the *Escherichia coli* DNA topoisomerase I gene and their regulation by DNA supercoiling, *J. Mol. Biol.* **202,** 735–742.

TUNIS, M. J. B., HEARST, J. E. (1968), On the hydration of DNA. II. Base composition dependence of the net hydration of DNA, *Biopolymers* **6,** 1245–1353.

ULANOVSKY, L. E., TRIFONOV, E. N. (1987), Estimation of wedge components in curved DNA, *Nature* **326,** 720–722.

VARDIMON, L., RICH, A. (1984), In Z-DNA the sequence GCGC is neither methylated by *Hha*I methyltransferase nor cleaved by *Hha*I restriction endonuclease, *Proc. Natl. Acad. Sci. USA* **81,** 3268–3272.

VARSHAVSKY, A. J., NEDOSPASOV, A., BAKAYEV, V. V., BAKAYEVA, T. G., GEORGIEV, G. (1977), Histone-like proteins in the purified *Escherichia coli* deoxyribonucleoprotein, *Nucleic Acids Res.* **4,** 2725–2745.

VINOGRAD, J., LEBOWITZ, J., RADLOFF, R., WATON, R., LAIPIS, P. (1965), The twisted circular form of polyoma viral DNA, *Proc. Natl. Acad. Sci. USA* **53,** 1104–1111.

VISWAMITRA, M. A., SHAKKED, Z., JONES, P. G., SHELDRICK, G. M., SALISBURY, S. A., KENNARD, O. (1982), Structure of the deoxytetranucleotide d-pApTpApT and a sequence-dependent model for poly (dA-dT), *Biopolymers* **21,** 513–533.

VOGT, N., MARROT, L., ROUSSEAU, N., MALFOY, B., LENG, M. (1988), Chloracetaldehyde reacts with Z-DNA, *J. Mol. Biol.* **201,** 773–776.

VOLOSHIN, O. N., SHLYAKHTENKO, L. S., LYUBCHENKO, Y. L. (1989), Localization of melted regions in supercoiled DNA, *FEBS Lett.* **243,** 377–380.

WAHLE, E., KORNBERG, A. (1988), The partition locus of plasmid pSC101 is a specific binding site for DNA gyrase, *EMBO J.* **7,** 1889–1895.

WANG, A. H.-J., QUIGLEY, G. J., KOLPACK, F. J., CRAWFORD, J. L., VON BOOM, J. H., VAN DER MARCEL, G., RICH, A. (1979), Molecular structure of a left-handed double helical DNA fragment at atomic resolution, *Nature* **282,** 680–686.

WANG, A. H. J., FUJI, S., VAN BOOM, J. H., RICH, A. (1982a), Molecular structure of the octamer d(GGCCGGCC): modified A-DNA, *Proc. Natl. Acad. Sci. USA* **79,** 3968–3972.

WANG, A. H. J., FUJI, S., VAN BOOM, J. H., VAN DER MARCEL, G. A., VAN BOECKEL, S. A. A., RICH, A. (1982b), Molecular structure of r(GCG)d(TATACGC): A DNA-RNA hybrid helix joined to double helical DNA, *Nature* **299,** 601–604.

WANG, A. H. J., UGHETTO, G., QUIGLEY, G. J., RICH, A. (1987), Interaction between an anthracycline antibiotic and DNA: molecular structure of daunomycin complexed to d(CpGpTpApCpG) at 1.2 Å resolution, *Biochemistry* **26,** 1152–1163.

WANG, J. C. (1971), Interaction between DNA and *Escherichia coli* protein, *J. Mol. Biol.* **55,** 523–533.

WANG, J. C., BARKLEY, M. D., BOURGEOIS, S. (1974), Measurements of unwinding of *lac* operator by repressor, *Nature* **251,** 247–249.

WANG, Y., THOMAS, G. A., PETICOLAS, W. L. (1989), A duplex of the oligonucleotides d(GGGGGTTTTT) and d(AAAAACCCCC) forms an A to B conformational junction in concentrated salt solutions, *J. Biomol. Struct. Dyn.* **6,** 1177–1187.

WATSON, J. D., CRICK, F. H. C. (1953), A structure for deoxyribose nucleic acid, *Nature* **171,** 737–738.

WEIL, R., VINOGRAD, J. (1963), The cyclic helix and cyclic coil forms of polyoma viral DNA, *Proc. Natl. Acad. Sci. USA* **50,** 730–738.

WELLS, R. D. (1988), Unusual DNA structures, *J. Biol. Chem.* **263,** 1095–1098.

WELLS, R. D., HARVEY, S. C. (Eds.) (1987), *Unusual DNA Structures,* New York: Springer.

WELLS, R. D., COLLIER, D. A., HANVEY, J. C., SHIMIZU, M., WOHLRAB, F. (1988), The chemistry and biology of unusual DNA structures adopted by oligopurine-oligopyrimidine sequences, *FASEB J.* **2,** 2939–2949.

WEST, S. C., KORNER, A. (1985), Cleavage of cruciform DNA structures by an activity from *Saccharomyces cerevisiae, Proc. Natl. Acad. Sci. USA* **82,** 6445–6449.

WEST, S. C., PARSONS, C. A., PICKSLEY, S. M. (1987), Purification and properties of a nuclease from *Saccharomyces cerevisiae* that cleaves DNA at cruciform junctions, *J. Biol Chem.* **262,** 12752–12758.

WESTERHOFF, H., O'DEA, M., MAXWELL, A., GELLERT, M. (1988), DNA supercoiling by DNA gyrase, *Cell Biophys.* **12,** 157–181.

WESTHOF, E., HOSUR, M. V., SUNDARALINGAM, M. (1988), Nonintercalative binding of proflavin to Z-DNA: Structure of a complex between d(5BrC-G-5BrC-G) and proflavin, *Biochemistry* **27,** 5742–5747.

WESTON-HAFER, K., BERG, D. E. (1991), Limits to the role of palindromy in deletion formation, *J. Bacteriol.* **173,** 315–318.

WILKINS, M. H. F. (1963), Molecular configuration of nucleic acids, *Science* **140,** 941–950.

WILLIAMS, J. S., ECKDAHL, T. T., ANDERSON. J. N. (1988), Bent DNA functions as a replication enhancer in *Saccharomyces cerevisiae, Mol. Cell. Biol.* **8,** 2763–2769.

WING, R., DREW, H., TAKANO, T., BROKA, C., TANAKA, S., ITAKURA, K., DICKERSON, R. E. (1980), Crystal structure analysis of a complete turn of DNA, *Nature* **287,** 755–758.

WORCEL, A., BURGI, E. (1972), On the structure of the folded chromosome of *Escherichia coli, J. Mol. Biol.* **71,** 127–147.

WU, H.-M., CROTHERS, D. M. (1984), The locus of sequence-directed and protein-induced DNA bending, *Nature* **308,** 509–513.

WU, H-Y., SHYY, S., WANG, J. C., LIU, L. F. (1988), Transcription generates positively and negatively supercoiled domains in the template, *Cell* **53,** 433–440.

WYATT, G. R., COHEN, S. S. (1952), A new pyrimidine base from bacteriophage nucleic acids, *Nature* **170,** 1072–1073.

YAMAMOTO, N., DROFFNER, M. L. (1985), Mechanisms determining aerobic or anaerobic growth in the facultative anaerobe *Salmonella typhimurium, Proc. Natl. Acad. Sci. USA* **82,** 2077–2081.

YANG, A., AMES, G. F.-L. (1988), DNA gyrase binds to the family of procaryotic repetitive extragenic palindromic sequences, *Proc. Natl. Acad. Sci. USA* **85,** 8850–8854.

ZACHARIAS, W., LARSON, J. E., KILPATRICK, M. W., WELLS, R. D. (1984), *Hha*I methylase and restriction endonuclease as probes for B to Z DNA conformational changes in d(GCGC) sequences, *Nucleic Acids Res.* **12,** 7677–7692.

ZAHN, K., BLATTNER, R. F. (1987), Direct evidence for DNA bending at the lambda replication origin, *Science* **236,** 416–422.

ZARLING, D., ARNDT-JOVIN, D. J., ROBERT-NICOUD, M., MCINTOSH, L. P., THOMAE, R., JOVIN, T. M. Immunoglobin recognition of synthetic and natural left-handed Z DNA conformations and sequences, *J. Mol. Biol.* **176,** 369–415.

ZIMMERMANN, S. B. (1982), The three-dimensional structure of DNA, *Annu. Rev. Biochem.* **51,** 395–427.

ZIMMERMANN, S. B., PFEIFFER, B. H. (1981), An RNA/DNA hybrid that can adopt two conformations: An X-ray diffraction study of poly (A) × poly d(T) in concentrated solution or in fibers, *Proc. Natl. Acad. Sci. USA* **78,** 78–82.

ZYSKIND, J. W., CLEARY, J. M., BRUSILOW, W. S. A., HARDING, N. E., SMITH, D. W. (1983), Chromosomal replication origin from the marine bacterium *Vibrio harveyi* functions in *Escherichia coli: ori*C consensus sequence, *Proc. Natl. Acad. Sci. USA* **80,** 1164–1168.

7 Principles of Gene Expression

Ralf Mattes

Stuttgart, Federal Republic of Germany

1 Overview

In most living organisms, the genetic information is usually stored in the cell's DNA. However, exceptions including some viruses, phages and viroids are known in which RNA stores the genetic information. RNA carries out the instructions encoded by the DNA and to this end, RNA copies of the coding information are made. The basic process of RNA synthesis is known as the transcription of genetic information and is catalyzed by specific enzymes called DNA-dependent RNA polymerases.

The synthesis of proteins occurs by a process called translation. This involves the concerted action of different components of the cellular machinery including the ribosomes, tRNA molecules charged with amino acids as well as various other factors.

The synthesis of nucleic acids and proteins begins at one end only and proceeds in one direction only (unidirectional). In the case of proteins, it proceeds from the amino terminus to the carboxyl terminus; in nucleic acids synthesis it occurs from the 5′ end to the 3′ end.

The initiation and correct termination of the transcriptional and translational processes are regulated by specialized molecules and signaling sequences.

The synthesis of RNA molecules and proteins, as prescribed by the information stored in the genetic material, is known as gene expression. The following sections describe the basic features and components involved in this process.

2 Transcription

Nucleic acids are linear polymers. The synthesis of a single-stranded RNA copy of one of the strands of the genetic material which is usually double-stranded DNA, is termed transcription. This process requires the substrate or template and DNA-dependent RNA polymerases which are modified by cofactors. Transcription is usually initiated at a nucleic acid sequence called promoter, while termination is mediated by other signal sequences and cellular factors.

2.1 DNA-Dependent RNA Polymerases

2.1.1 RNA Polymerase of *Escherichia coli*

In contrast to eukaryotes where three distinct RNA polymerase species (LEWIS and BURGESS, 1982) are known to exist, bacteria contain only one type of this enzyme (CHAMBERLIN, 1982) which is responsible for the selective synthesis of mRNA, proteins, and ribosomal RNA. The *Escherichia coli* enzyme is the most thoroughly characterized of prokaryotic RNA polymerases. Its structure, which is very similar in most bacterial species, is composed of four subunits ($2\alpha+\beta+\beta'$) which constitute the core enzyme. *In vitro*, this core enzyme is transcriptionally active on templates containing single-strand breaks. In order to initiate the transcription of native templates, the core enzyme must bind to a ligand called the sigma factor in order to form the biologically active holo-enzyme (see Sect. 2.1.2). Subsequently elongation of successfully initiated transcription is catalyzed by the core enzyme which has dissociated from the ligand (see Fig. 1 and Tab. 1).

Eukaryotic cells produce three different RNA polymerases for the selective synthesis of ribosomal RNA (polymerase I), mRNA (polymerase II), and small RNAs such as tRNA (polymerase III). These forms are distinguishable from each other by their different susceptibility to α-amanitin inhibition (II is inhibited) and their location within the nucleolus (I) or nucleoplasm (II and III).

The initiation of transcription by bacterial polymerase is inhibited by rifamycin which binds to the β-subunit of the holo-enzyme. Streptolydigin which binds to the β-subunit of the core enzyme inhibits the elongation of transcription.

Some bacteriophages, e.g., SP6, T3, T7 (BUTLER and CHAMBERLIN, 1982; MORRIS et al., 1986; DAVANLOO et al., 1984), encode their own RNA polymerases which consist of a

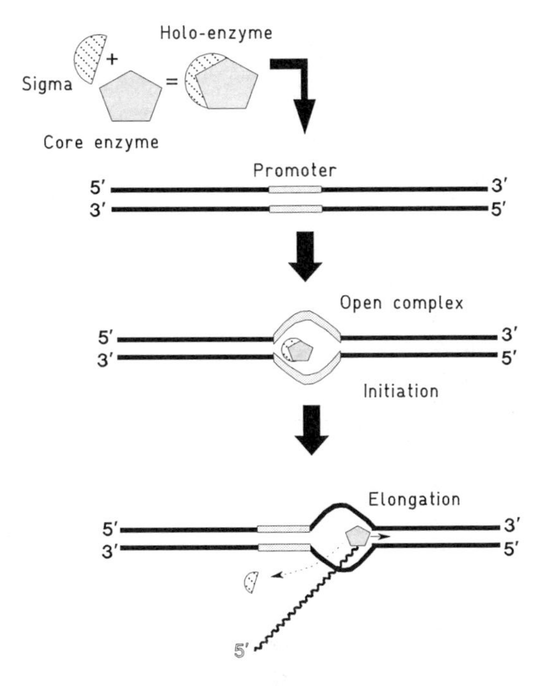

Fig. 1. Initiation of transcription.

single polypeptide chain and which are specific for the transcriptional initiation at promoters of phage genes.

2.1.2 Sigma Factors of Bacteria

The initiation of transcription in eubacteria is prompted by the ligand binding to the core enzyme to result in the holo-enzyme. The sigma factors (HELMANN and CHAMBERLIN, 1988) as ligands are required for the selective initiation of RNA polymerization at promoter sites. The affinity of the RNA polymerase can be enhanced a thousandfold. These sigma factors are characterized by their different molecular weights. Different classes of promoters are recognized selectively by RNA polymerase holo-enzymes containing different sigma factors (Tab. 2). Accordingly, the transcription of sets of genes can be coordinately initiated. The so-called house keeping genes of *E. coli*, which are active during normal growth, are transcriptionally initiated by the sigma factor 70. A different sigma factor initiates the transcription of 17 genes involved in the heatshock response, and when the cell is depleted of ammonium ions, yet another sigma factor activates a set of genes which encode enzymes able to use alternative nitrogen compounds (see Tab. 2). The various holo-enzymes are able to distinguish between the different consensus sequences of promoters. A far more complex pattern of sigma factor directed gene activation exists in the Gram positive bacterium *Bacillus subtilis* (LOSICK and PERO, 1981). In contrast to genes which are transcribed during the vegetative phase by the holo-enzyme containing sigma-43, at the end of the exponential growth phase during the sporulation of the cells, a concerted change of sigma factor species modulates the selective transcription of genes necessary for spore development. Again, a group of genes which is active at specific stages of sporulation, contains different promoters. Holo-enzymes modified by specific sigma factors, display selective affinity to particular promoters.

Similar modulation is effected by phage SPO1 after the infection of *B. subtilis*. The early genes of the phage DNA are transcribed by the host holo-enzyme with sigma-43, whereas the transcription of middle and late genes is selectively initiated by holo-enzymes containing phage-encoded factors gp28 or gp33 and gp34, respectively.

Tab. 1. Subunits of *Escherichia coli* RNA Polymerase Core Enzyme

Subunit	Gene	No. of Subunits per Core Enzyme	Molecular Weight (Daltons)	Proposed Function
α	*rpoA*	2	36512	Promoter binding
β	*rpoB*	1	150618	Ribonucleoside triphosphate binding
β'	*rpoC*	1	155613	DNA binding

Tab. 2. Bacterial Sigma Factors for RNA Polymerase

Sigma Type	Gene	Usage for	Consensus Sequence −35	−10
E. coli				
70	*rpoD*	House keeping functions	TTGACA	TATAAT
54	*rpoN*	Nitrogen-regulated genes	CTGGCA	TTGCA
32	*rpoH*	Heat-shock genes	CTTGAA	CCCCAT
gp55	T4 gene 55	Late genes of T4 phage	-	TATAAATA
B. subtilis				
43	*rpoD* (*sigA*)	House keeping functions	TTGACA	TATAAT
28	*sigD*	Flagellar and chemotactic genes	CTAAA	CCGATAT
29	*sigE*	Sporulation genes	TT-AAA	CATATT
30	*sigH*	Sporulation genes	?	?
32	*sigC*	unknown	AAATC	TA-TG-TT-TA
37	*sigB*	unknown	AGG-TT	GG-ATTG-T
gp28	SPO1 28	Middle genes of phage SPO 1	T-AGGAGA--A	TTT-TTT
gp33/34	SPO1 33,34	Late genes of phage SPO1	CGTTAGA	GATATT

Exceptions, however, are known as in the case of some RNA polymerases of bacteriophages (e.g., T7) which specifically initiate transcription with a single polypeptide enzyme.

In archeobacteria, a much more complex RNA polymerase core enzyme structure compared to that of *E. coli* has been described. A component seems to be required analogous to the sigma factors of eubacteria. A similar function is proposed for a specificity factor found in the mitochondria of yeast. For eukaryotes in general, the specificity of transcription appears to reside in many separate, promoter-specific binding proteins.

2.2 Sites of Initiation

2.2.1 Bacterial Promoters

Transcription within eubacteria is initiated upon the formation of a complex between the holo-enzyme of RNA polymerase and a signal sequence called the promoter. A comparison of a number of sigma-70 specific prokaryotic promoter regions reveals a consistent similarity in two regions of their DNA sequence (HAWLEY and MCCLURE, 1983). This indicates the ability of RNA polymerase to recognize promoter sequences specifically (MCCLURE, 1985). These promoter sequences are located 5′ of known transcriptional units.

The general structure of a promoter reveals that the nucleotide at which transcription starts (referred to as +1) is usually a purine. The first sequence of common nucleotides is located about 10 base pairs upstream of the RNA initiation site. This sequence is known as the −10 sequence, or the Pribnow box, named after its discoverer. It usually contains 6 base pairs which form the consensus sequence TATAAT. Approximately 15 to 20 nucleotides further upstream of the Pribnow box, the −35 element forming the TTGACA consensus sequence, occurs in most bacterial promoters. The nucleotide sequence between −35 and −10 does not seem to be critical, although the distance between these nucleotides is probably important for the proper functioning of the promoter. It is purported that the RNA polymerase first binds at the −35 sequence and that proper positioning at the −10 sequence follows. When bound to the promoter, the RNA polymerase holo-enzyme complex covers approximately 70 base pairs (between −50 and +20). An open promoter complex forms between residues −9 and +3, where the two strands of the DNA template are separate and transcription is initiated on the codogenic strand in a 5′→3′ direction.

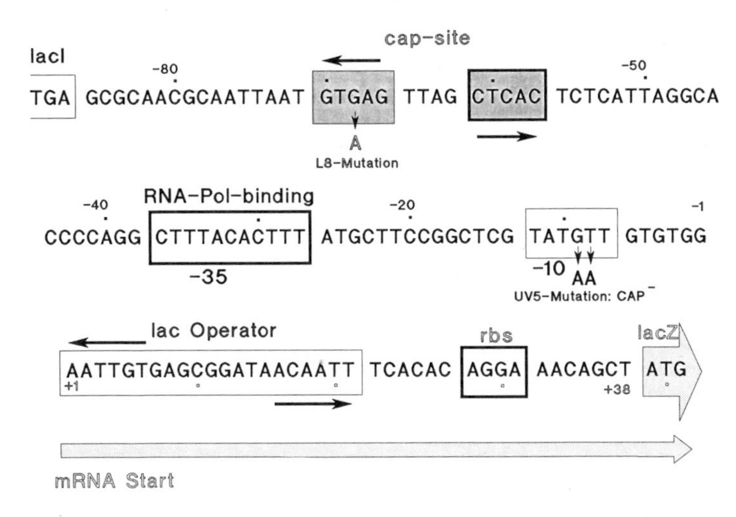

Fig. 2. Elements of the *Escherichia coli lac* promoter. The sequence shown begins at the stop codon of the *lacI* (repressor) gene. The interaction sites for the cAMP activator protein (CAP) and RNA polymerase (−35, −10) as well as for the binding of the repressor (operator) and ribosome (rbs) are boxed. Mutations known to modify the wild-type promoter are indicated by vertical arrows. Inverted repeat sequences are shown by horizontal arrows.

In some promoters, sequences upstream of the −35 and −10 signals are required to bind auxiliary protein factors which alter the DNA topology and modulate the efficiency of the promoter (see Sect. 8, e.g., upstream activating sequences). Because of the striking similarity between prokaryotic promoters that have been studied, elements of the well known *lac* promoter (Reznikoff and Abelson, 1980) of the *E. coli lac* operon (see Fig. 2) can serve as a model of the structure of such promoters.

2.3 Termination

In prokaryotes, successfully initiated transcription is continued by the core enzyme of RNA polymerase until it is stopped at signals called transcription terminators (Platt, 1986). Two different termination signals which have some common characteristics have been described. Both contain inverted repeat sequences which are thought to fold at the 3′ end of transcripts, into stem-loop structures of variable length (Fig. 3). The functioning of such terminators is known to be either dependent on or independent of rho. Rho is a RNA binding protein which, when bound to RNA, hydrolyzes ATP to ADP and inorganic phosphate. The pausing of the RNA polymerase at the stem-loop structure allows rho-dependent dissociation of template and RNA polymerase. Rho independent termination is effected by the pausing of RNA polymerase at stem-loop structures and dissociation of the complex.

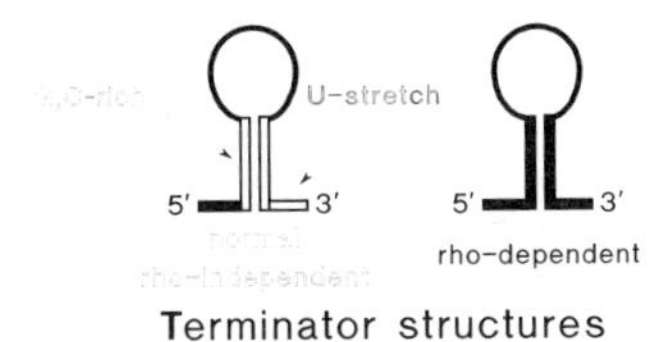

Fig. 3. Stem loop structures can act as terminators of transcription.

The termination of transcription can be regulated in a manner similar to that of the initiation reaction. So-called antiterminator proteins as well as other proteins exist which are able to inhibit the action of rho. Thus, mRNA formation can be modulated by the regulation of transcription termination.

3 Processing of RNA in Prokaryotes

The transcription products of prokaryotes usually serve without further processing as mRNA for translation. Three types of transcriptional units are distinguishable in bacteria (Fig. 4).

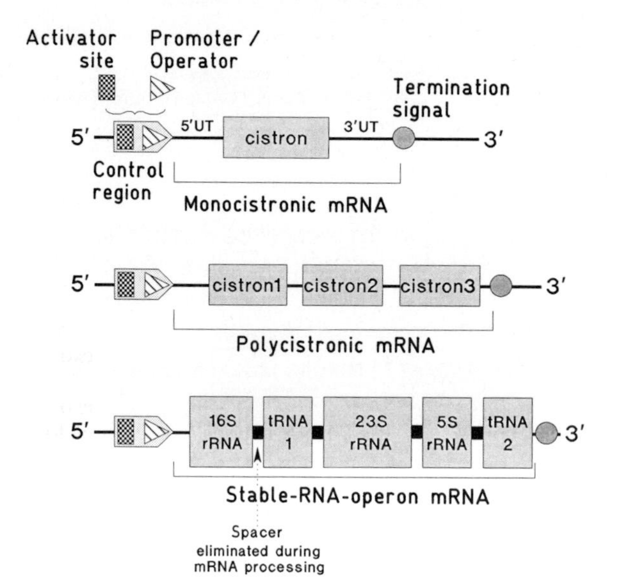

Fig. 4. Schemes of prokaryotic transcriptional units.

A characteristic feature of transcription in prokaryotes is the production of polycistronic mRNA, since most genes are grouped together under the control of a singular promoter. This holds true for genes encoding RNA components of the ribosomal subunits and for tRNAs. To obtain mature, functional forms of these molecules, the primary transcripts have to be cleaved at specific sites and modified at specific residues. These important processes are termed post-transcriptional modifications, or RNA processing.

3.1 Organization of Genes for Stable RNA

Seven operons (*rrnA-H*) or transcription units that encode rRNAs (rDNA) have been identified in the well studied *E. coli* genome (JINKS-ROBERTSON and NOMURA, 1987). The general structure of these *rrn* operons contains, in a 5′→3′ direction, the 16S, 23S, and 5S rDNA interspersed with tDNA sequences. All rDNA and tDNA segments within these operons are separated by spacer segments. Furthermore, most of the genes encoding the various tRNAs (tDNA) exist in numerous copies and are organized in operons (see Fig. 4). These gene clusters may contain either multiple copies of the same tDNAs, or arrays of different tDNAs interspersed with spacer elements. They are located ubiquitously in the genome of *E. coli*. A similar situation is found in other organisms and in the DNA of organelles of eukaryotes.

3.2 Processing of Stable RNA Gene Transcripts

The transcripts of these rDNA and tDNA operons require further processing to form the mature RNA products (GEGENHEIMER and APIRION, 1981). Three important RNases are known to participate in this process, namely RNase III, RNase P, and RNase D. The endonuclease RNase III initially cleaves the primary transcript into fragments that encode either a tRNA or one of the ribosomal RNA molecules. The recognition targets for the enzyme's activity are the stem-loop structures in the spacer regions which flank the RNA sequences. The transcripts of genes encoding tRNAs are matured by the action of RNase P which thereby creates the 5′ end of the tRNA (DEUTSCHER, 1984). An additional but as yet unidentified endonuclease cleaves the precur-

Tab. 3. Important Ribonucleases of *Escherichia coli*

Type	Gene	Substrate	Class
P	*rnpA + B*	5′ end of tRNA	endo
BN	?	3′ end of tRNA	exo
D	*rnd*	3′ end of tRNA	exo
T	?	CCA at 3′ end of tRNA	exo
III	*rnc*	tRNA, mRNA	endo
R	?	tRNA, mRNA	exo
E	*rne*	5S-rRNA	endo
I	*rna*	?	endo
II	*rnb*	?	exo
Polynucleotide-phosphorylase	*pnp*	?	exo
H	*rnh*	RNA-DNA hybrids	endo

sor at a hairpin-loop near the mature 3′ end. The exonuclease RNase D digests additional nucleotides, terminating at the 3′ CCA terminus. In other cases, an end is generated by the action of the exonuclease at which tRNA nucleotidyl transferase adds one or more of the invariant terminal nucleotides to complete the 3′ CCA sequence (Tab. 3).

Additional steps in the maturation process of tRNAs include modifications of the nucleotides and bases which are essential for tRNA physiological functions. These processes are accomplished at the precursor and processed stage by a battery of specific enzymes (BJÖRK et al., 1987).

3.3 Processing of mRNA

In contrast to the processing of the primary transcripts found in eukaryotes, similar post-transcriptional modifications in bacterial mRNA seem to be confined to the mRNA encoding stable RNA products. The identification of mRNA molecules having short half lives indicates that the degradation of mRNA has a profound influence on protein synthesis. This has been reported for some phage messengers. It is generally believed that the metabolic instability of mRNA molecules permits the rapid adaptation of the cell to environmental changes. However, no complete degradation pathway has yet been defined for any mRNA. In experiments in which mRNA was overproduced from cloned genes, it became apparent that constraints exist in the combination of the mRNA sequence and cellular factors which lead to a reduction of the numbers of mRNA molecules. Even in operons such as the *lac* operon of *E. coli* polycistronic mRNA molecules, processed to gene-size messengers with end points in the intercistronic region of *lacZ-lacY* and the untranslated region at the 3′ end of the operon, have been identified (MCCORMICK et al., 1991). Such processing may be related to the observation that genes of the *lac* operon are expressed at different levels despite their common transcriptional control; β-galactosidase is the dominant translational product with respect to the permease. The identification of so-called REP sequences (repetitive extragenic palindromic sequences) within the intergenic regions of various operons has been reported for *E. coli* and *Salmonella* (HIGGINS et al., 1988). These elements are characterized by a predilection for stem-loop structure formation. Such hairpin structures are thought to be terminator-like structures which inhibit 3′→5′ exonuclease degradation. The introduction of such REP sequences into genes, beyond the coding region, can enhance the stability of their mRNA.

In conclusion, reports are accumulating which suggest that post-transcriptional processing of mRNA occurs in bacteria. This processing seems to be implicated in the metabolic control of the cell.

4 Genetic Code

NIRENBERG and MATTHAEI elucidated the genetic code in 1961 by adding synthetic RNA polymers during *in vitro* translation of extracts of *Escherichia coli*. Triplets within the mRNA are decoded by adaptor molecules called tRNA which are charged specifically with amino acids. The triplet code within the mRNA is complemented by the anticodons within the tRNA molecules. Thus, triplets within the mRNA encode the amino acid sequence of the translated protein. The code itself is degenerate or redundant; all but two amino acids are encoded by more than one codon. The code is universal for all organisms analyzed, but differences in the sequence of the gene and its translated mRNA can occur, as observed in the genomes of eukaryotic organelles and some nuclear genes (see Sect. 4.5).

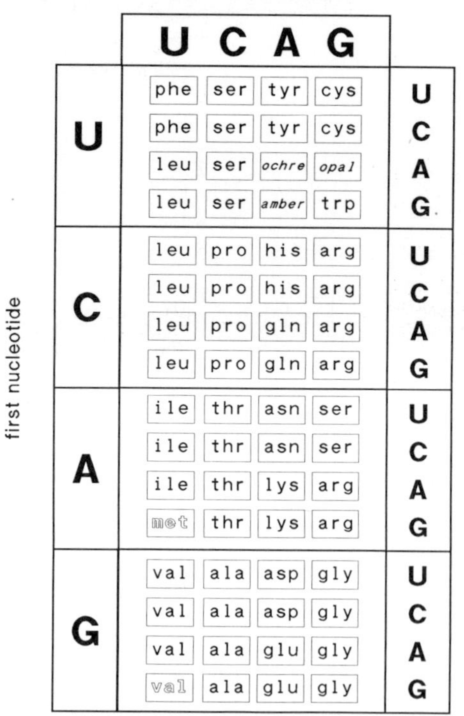

Fig. 5. The genetic code. Codons accepting fMet tRNA (AUG, GUG) are shown in open characters, stop codons in italics. Under certain circumstances, the opal codon UGA also accepts the tRNA for the 21st amino acid selenocysteine.

4.1 Degeneration

The only two amino acids that are encoded by a single codon, namely AUG and UGG, are methionine and tryptophan, respectively. All other 18 amino acids as well as the translational stop signal are encoded by 2 to 6 different codons (Fig. 5).

Thus, the determination of a peptide sequence from the gene sequence is unequivocal but the reverse is ambiguous. Most of the variations among codons encoding the same amino acid occur in the third position of the triplet. Mutations resulting in a nucleotide change in this position frequently do not alter the amino acid sequence and therefore, are designated as silent mutations.

4.2 Codon Usage

The redundancy of the genetic code is universal for most organisms. In several organisms studied, this redundancy is counteracted by the absence of genes for some of the tRNA molecules. In *E. coli*, only 45 tRNA species out of a possible 61 are encoded by 78 genes (KOMINE et al., 1990). Thus, some tRNAs are encoded by duplicated or multiple genes. The frequency of use of synonymous codons in *E. coli* genes approximately equals the number of tRNA gene species (see Fig. 6). This frequency of usage as well as the rate at which genes are expressed, varies considerably among different organisms (Fig. 7) (WADA et al., 1991). This is in moderate agreement with the amount of tRNA found for the different species in *E. coli*. The tRNA species for the two codons AGA and AGG, which encode arginine, are present at a very low level only in *E. coli* cells, whereas they are more abundant in *Saccharomyces cerevisiae*.

The tRNA genes are usually found clustered into cistrons in various organisms. In Gram positive organisms such as *Bacillus subtilis*, as many as 21 genes out of a possible 51, encoding only 31 different tRNAs, constitute a cistron. Specialized organisms such as *Mycoplas-*

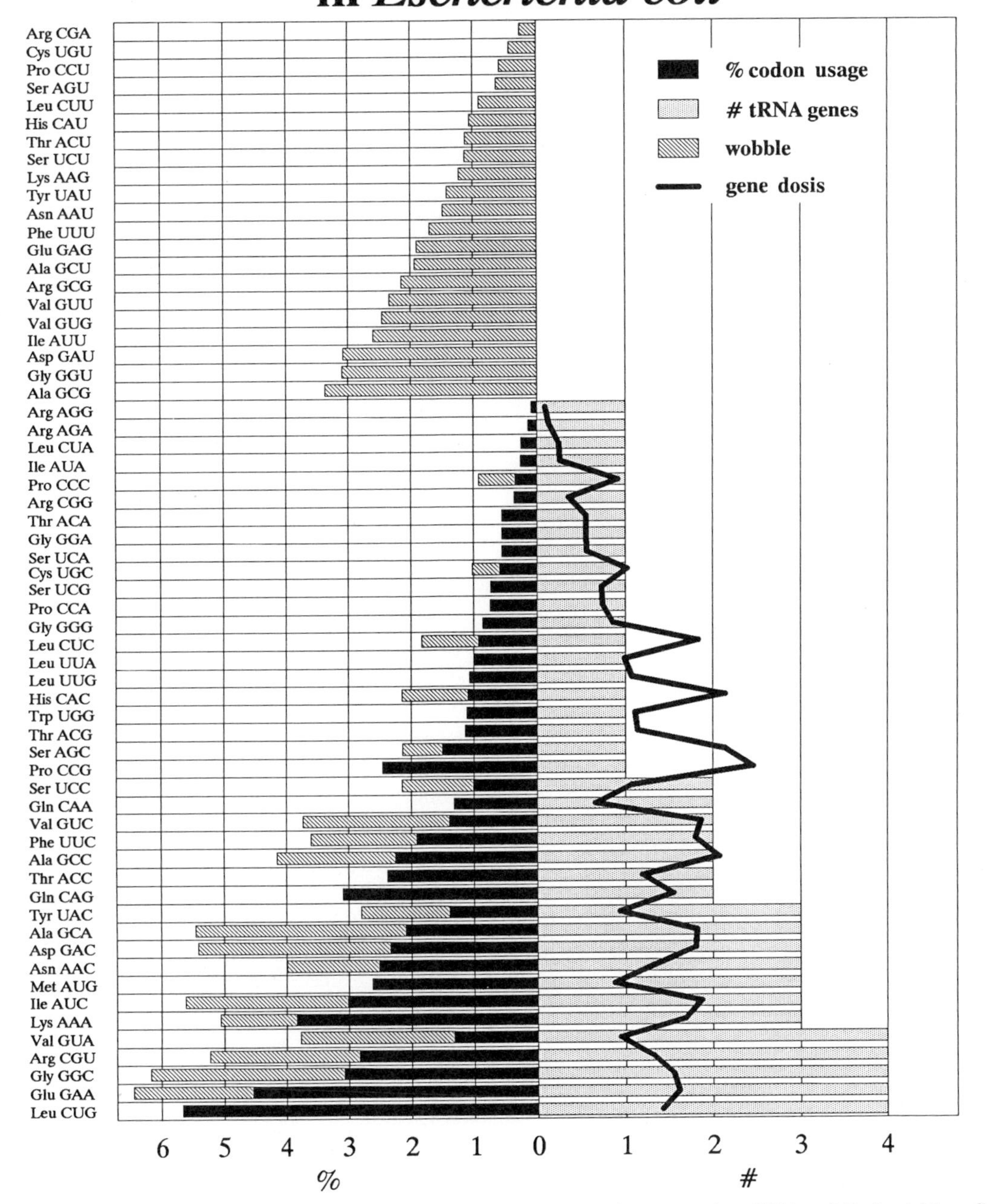

Fig. 6. Comparison of codon usage and number of genes known for respective tRNAs of *Escherichia coli*. The gene dosage approximately corresponds to the frequency of the codon usage. Codons without a known tRNA gene should be translated by cognate tRNAs due to the wobble hypothesis.

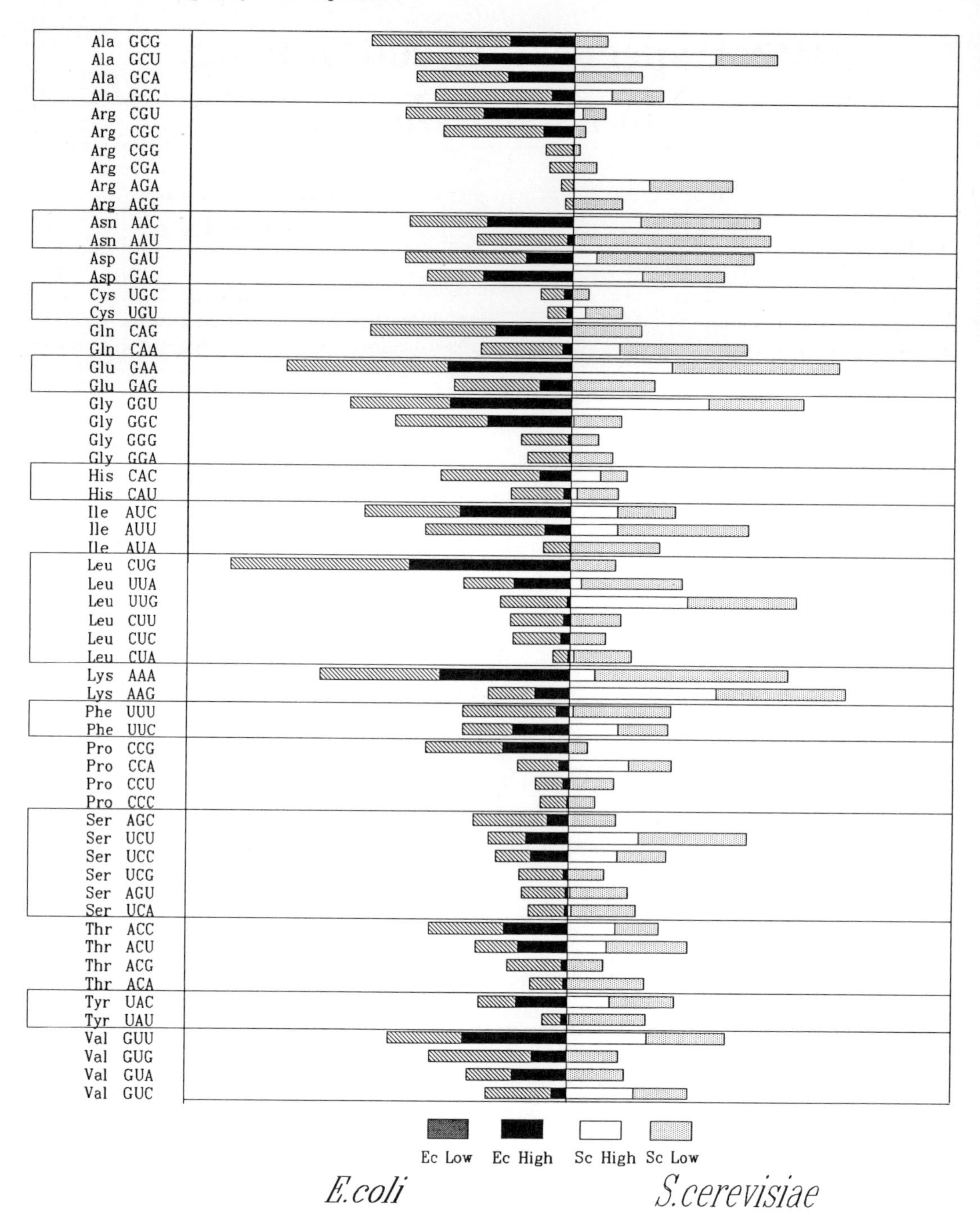

Fig. 7. Comparison of codon usage frequencies of *Escherichia coli* and *Saccharomyces cerevisiae*.

ma capricolum and wall-less eubacteria which are parasitic to tissues, contain only 30 tRNA genes encoding 29 tRNA species. Their codon usage consequently, is strongly biased towards A and T usage, and most of the four synonymous codons thus are read by a single anticodon or tRNA species (MUTO et al., 1990).

The solution to this paradox is believed to reside in the wobble hypothesis (CRICK, 1966) (see Sect. 5).

4.3 Exceptions from Universality: Suppression and RNA Editing

All prokaryotic organisms and most nuclear genes of eukaryotes appear to use the same genetic code. Exceptions are found in certain ciliate and mitochondrial genes, as well as in eubacteria. The universality concept of the genetic code depends on the codon–anticodon recognition process between the mRNA and relevant tRNA molecules. Two possible mechanisms of alteration are known to change this unambiguous relation between the codon–anticodon recognition and the incorporation of amino acids into the synthesized protein. First, mutations in the adaptor molecule tRNA can lead to altered anticodons or the binding of an amino acid other than that normally related to the code (anticodon) (PARKER, 1989). These phenomena, known as mutated suppressor tRNAs (Tab. 4), are well studied in bacteria and in mitochondria of some eukaryotes (BJÖRK et al., 1987).

A second class of DNA sequence alteration, which leads to changes in the genetic coding capacity of a gene, occurs at the level of altered mRNA sequences. A process called RNA editing has recently been elucidated in mitochondria of plants and trypanosomes (SIMPSON and SHAW, 1989). The mRNA of certain mitochondrial genes is post-transcriptionally modified by insertions and/or deletions of U residues (Fig. 8). Similar post-transcriptional modifications have been reported for other genes expressed in *Paramycovirus* infected cells and in mammalian tissues (CHEN et al., 1990).

In conclusion, several genes in various organisms do not display the amino acid sequence of their proteins which would be anticipated by the original concept of the universality of the genetic code. The intron sequences in eukaryotic genes, which are removed by the splicing process resulting in mature mRNA molecules, were the first observations in this context.

An additional exception to this previous concept is the translation of certain UGA stop codons found within functional mRNAs into selenocysteine, the so-called 21st amino acid. A specific tRNA molecule containing the anticodon UGA is charged with serine which is subsequently modified to selenocysteine (Fig. 9). The proper translation also requires the availability of an additional elongation factor

Tab. 4. Suppressors of Stop Codons in *Escherichia coli*

Gene	Old Name	Suppressed Stop Codon	Encoded Amino Acid	Alternative Gene Symbol
supB	Su_B	UAA, UAG	Gln	*glnU*
supC	Su-4	UAG, UAA	Tyr	*tyrT*
supD	Su-1	UAG	Ser	*serU*
supE	Su-2	UAG	Gln	*glnV*
supF	Su-3	UAG	Tyr	*tyrV*
supG	Su-5	UAA, UAG	Lys	?
supK	?	UGA	tRNA methylase	
supL	Su_β	UAA, UAG	Lys	*lysT*
supM	–	UAA, UAG	Tyr	*tyrU*
supP	Su-6	UAG	Leu	*leuX*
supU	Su-7	UAG	Trp	*trpT*

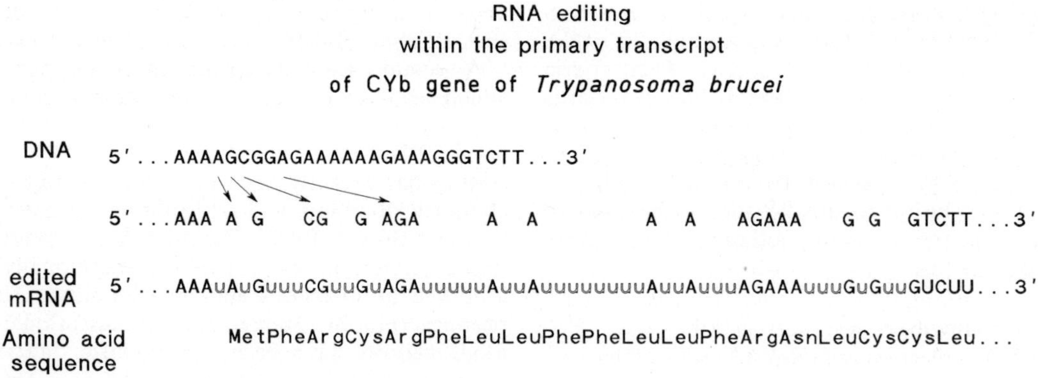

RNA editing
within the primary transcript
of CYb gene of Trypanosoma brucei
DNA 5'...AAAAGCGGAGAAAAAAGAAAGGGTCTT...3'
5'...AAA A G CG G AGA A A A A AGAAA G G GTCTT...3'
edited mRNA 5'...AAAuAuGuuuCGuuGuAGAuuuuuAuuAuuuuuuuuAuuAuuuAGAAAuuuGuGuuGUCUU...3'
Amino acid sequence MetPheArgCysArgPheLeuLeuPhePheLeuLeuPheArgAsnLeuCysCysLeu...

Fig. 8. Editing of mRNA:U residues are inserted at various positions.

Serine +
tRNA Sec UCA (selC)
Seryl–tRNA Sec UCA
Seleneocysteine Synthase (selA)
+R–Se⁻
Selenocysteyl–tRNA Sec UCA
SelB·GTP
SelB·GTP·Selenocysteyl–tRNA Sec UCA
Incorporation into protein
E.coli fdhF ..TGC GCT CGT GTC TGA CAC GGC CCA TCG.. Formate dehydrogenase
D.baculatus ..TCC TAC GAC CCG TGA CTG GGC TGT GCC.. Hydrogenase
Mouse ..GTC GCG TCT CTC TGA GGC ACC ACG ATC.. Glutathione peroxidase

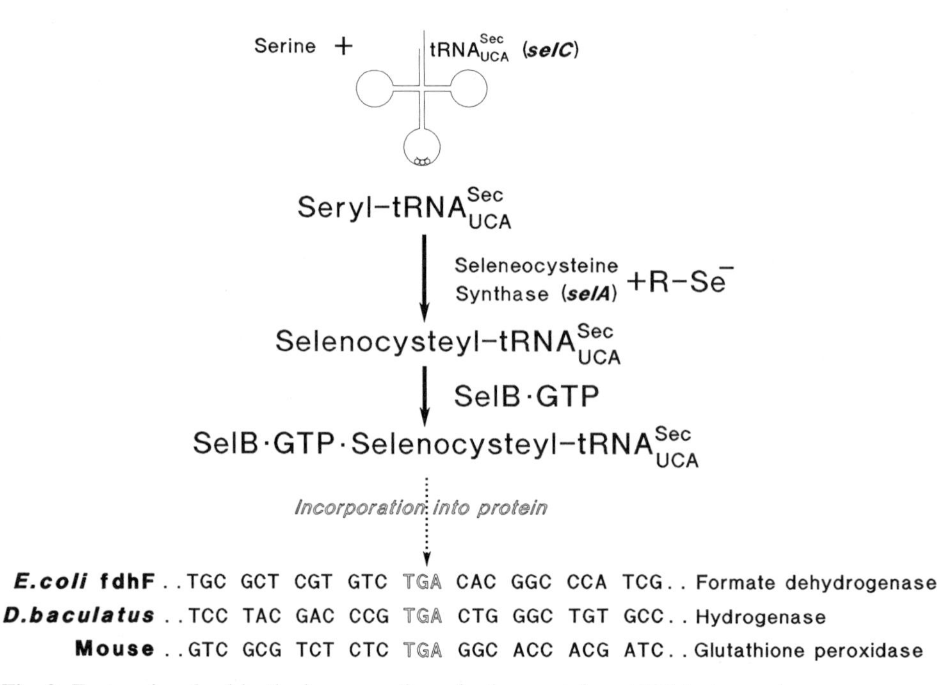

Fig. 9. Factors involved in the incorporation of selenocysteine at TGA stop codons.

(*selB*) in *E. coli*, instead of EF-Tu. However, the choice of certain UGA codons as opposed to the normal stop codons, seems to depend on the primary sequence of the neighboring triplets (Böck et al., 1991). It therefore becomes evident that the translational process of at least some mRNAs requires more than simply the codon–anticodon relationship. Similar observations have been made during the process of suppression using mutated so-called suppressor tRNAs (Eggertsson and Söll, 1988).

5 Translation

Translation is the process whereby the mRNA sequence is converted into an amino acid chain. This process requires a functional and hence, modified mRNA as template as well as several components such as proteins, protein RNA complexes, specialized tRNA adaptor molecules charged with amino acids, and additional factors. The principal components involved are termed ribosomes. The tRNA molecules are charged with amino acids by an ATP-dependent process which involves aminoacyl-tRNA synthetases and results in the formation of aminoacyl-tRNAs. The synthetase molecules involved are specific for the acceptance of the correct amino acid and its corresponding tRNA (SCHIMMEL, 1987).

5.1 Ribosomes of *Escherichia coli*

Ribosomes of prokaryotes and eukaryotes differ in some aspects and in molecular weight. Nevertheless, the *E. coli* ribosomes are regarded as being prototypic, since they are best understood (WITTMANN, 1983). They are composed of a small and a large subunit (see Fig. 10). The small subunit (30S) contains 21 different proteins (S1–S21) and a single rRNA molecule of 1542 nucleotides (16S). The large subunit (50S) is composed of 34 different proteins (L1–L34) and contains two rRNA molecules, the larger one is composed of 2904 nucleotides (23S) and the smaller species of 120 nucleotides (5S). The structure of eukaryotic ribosomes is not elucidated to such an extent, but their overall organization is comparable.

5.2 Initiation of Translation

The initiation of translation is well studied in *E. coli* and is discussed here with respect to the requirement of additional components. Besides a functional mRNA, additional initiation factors are needed. The ribosome is dissociated by the binding of the initiation factors 1 and 3 (IF-1 and IF-3) to the small subunit. This complex binds to an additional factor (IF-2), GTP and to the initiator tRNA (Fig. 11). The mRNA is attached to this complex at the region surrounding the AUG codon. The so-called initiator tRNA is a distinct species of Met-tRNA^{Met}.

This species of initiator tRNA differs in both pro- und eukaryotes from the Met-tRNA^{Met} which exclusively decodes internal methionine codons. In bacteria, this initiator Met-tRNA^{Met} contains a formylated amino group and is therefore designated Met-$\text{tRNA}_{\text{F}}^{\text{Met}}$. In eukaryotes, initiation is accomplished by a specific Met-$\text{tRNA}_{\text{I}}^{\text{Met}}$, which in contrast to that of bacteria lacks the formylation, but also serves uniquely in initiation. The special initiator function is evidently contained within the nucleotide sequence and/or the structure of the molecule.

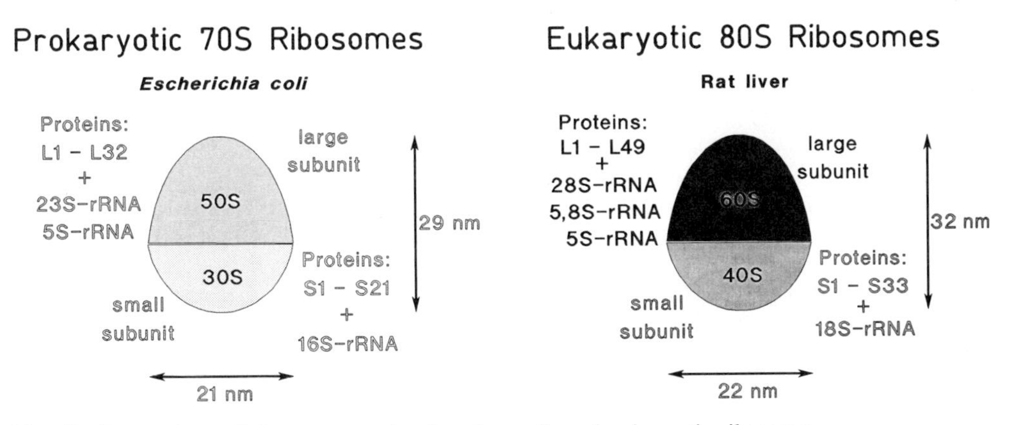

Fig. 10. Comparison of the components of prokaryotic and eukaryotic ribosomes.

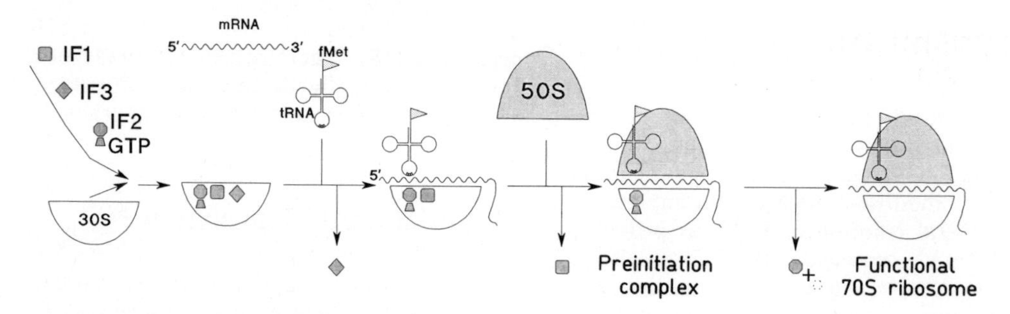

Fig. 11. Initiation of translation.

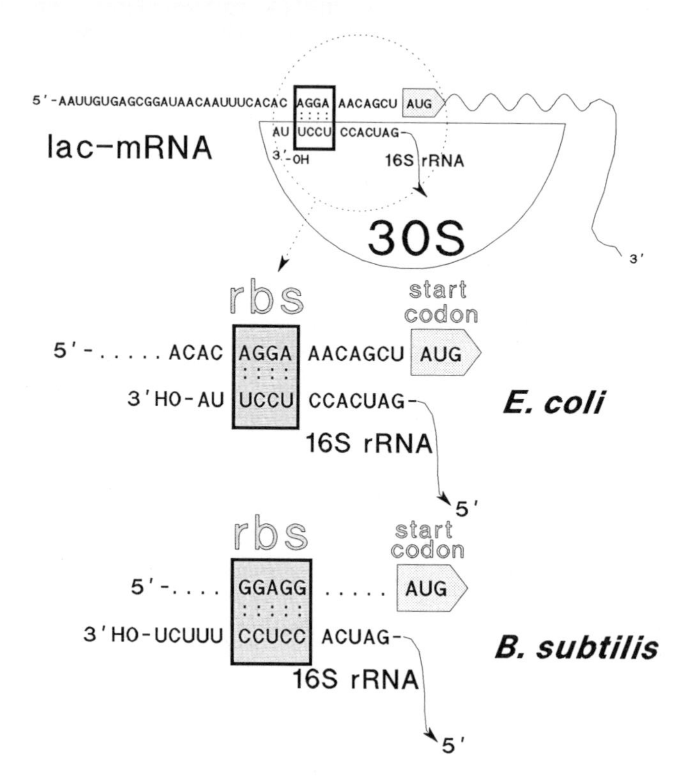

Fig. 12. Ribosomal binding site of mRNA and its homology to 16S rRNA.

The pre-initiation complex formed by these steps allows the formation of functional 70S ribosomes and the concomitant release of the initiation factors and hydrolysis of GTP to GDP.

How is the first AUG codon of the protein coding sequence recognized in prokaryotes? The nucleotide sequences upstream of the start codon are usually rich in purine residues and are complementary to the 3′ end of the 16S rRNA of the small ribosomal subunit. This sequence is termed the Shine–Dalgarno sequence or ribosomal binding site (rbs) (SHINE and DALGARNO, 1974). This sequence is also found upstream of every gene start codon of prokaryotic polycistronic mRNAs (Fig. 12).

The efficiency of the initiation of translation at a given coding sequence strongly depends on the distance of the rbs from the AUG codon and on the degree of its complementarity with the 16S rRNA. The latter is, in some degree, species-specific (see Fig. 12).

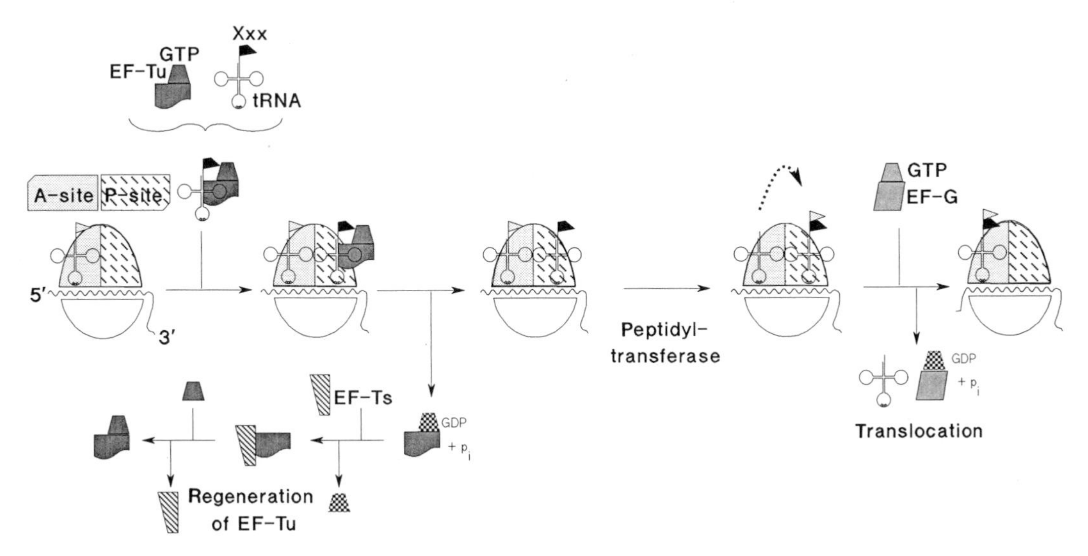

Fig. 13. Elongation of translation.

Thus comparisons of gene sequences reveal homologies in the rbs-AUG region and strengthen the identification of the rbs. Such comparisons have revealed that, in addition to the start codon AUG, GUG (usually encoding Val) also directs the incorporation of fMet-tRNA in about 10% of *E. coli* genes. Furthermore, even UUG (usually encoding Leu) is utilized, but less frequently. These findings are in agreement with reports that the interaction of the ribosomal small subunit with the mRNA during initiation seems to extend up to 12 nucleotides 3′ of the start codon. The concept of a translation initiation region (TIR) implies that more than merely the start codon and the rbs guide the translation initiation (SPRENGART et al., 1990).

5.3 Elongation and Termination

In addition to the start tRNA binding site known as the peptidyl- or P-site, the ribosome at the mRNA contains a second tRNA binding site called the aminoacyl- or A-site (JOHNSON et al., 1982). In order to occupy the A-site, aminoacyl-tRNAs have to be bound by the elongation factor EF-Tu and by GTP.

This ternary complex transfers the aminoacyl-tRNA corresponding to the next codon of the mRNA into the A-site by hydrolyzing GTP and releasing the EF-Tu-GDP complex. The peptidyl transferase activity (GARRETT and WOOLEY, 1982), located in the large subunit of the ribosome, catalyzes the transfer of the starting fMet group from its tRNA bound at the P-site to the amino group of the aminoacyl-tRNA in the A-site (Fig. 13).

The clearing of the tRNA at the P-site and the translocation of the peptidyl-tRNA from the A-site to the P-site are catalyzed by the elongation factor EF-G and require GTP hydrolysis. Thus, the next codon of the mRNA can receive the corresponding aminoacyl-tRNA. EF-Tu distinguishes between fMet-tRNA and Met-tRNA such that only Met-tRNA is inserted into the A-site.

The termination of the translational process occurs when one of the three termination codons (UAG, UAA, or UGA) enters the A-site of the ribosome. The lack of tRNAs to translate these codons together with specific release factors catalyzes the cleavage of the polypeptide chain from the tRNA and leads to release of both, as well as dissociation of the ribosome from the mRNA. Three release factors have been identified as RF-1, RF-2, and RF-3; RF-1 recognizes UAA or UAG, while RF-2 recognizes UAA or UGA. However, as mentioned in Sect. 4.3, the efficiency of the codon recognition may be influenced by the flanking sequences in the mRNA, and termination therefore is also affected.

5.4 Specific Features and Failures

Some additional features of the general process should be emphasized. In prokaryotes, translation is usually initiated concurrently with transcription, and multiple initiation is observed after the first ribosome has moved along the mRNA. Thus, translation occurs simultaneously at several ribosomes, and the complexes are called polysomes.

This ribosomal loading of the mRNA is seen as a factor which stabilizes the mRNA against nucleolytic attack, and which enhances the efficiency of protein synthesis.

The characteristic polycistronic nature of most prokaryotic mRNAs can also lead to independent translation initiations within the various start sites of the single cistrons. In addition, ribosomes which have reached a termination codon are dissociated by the initiation factor IF-3 and can be recycled for further translational initiation.

The fidelity of translation is foremost a function of the correct interaction between the tRNA molecule and specific aminoacyl-tRNA synthetase which links the correct amino acid to its corresponding tRNA. These enzymes are able to correct mistakes in the charging reaction. In addition, the A-site of the ribosome, in cooperation with EF-Tu, controls the fidelity of codon recognition. Certain chemicals such as aminoglycosides (e.g., streptomycin), bind to the S12 protein of the small subunit and alter the reading ability of C and U residues. This can lead to the incorrect incorporation of amino acids into the growing peptide chain. Furthermore, the availability of sufficient aminoacyl-tRNA molecules is required when a certain codon enters the A-site of the ribosome. The deficiency of appropriate adaptors at this time can decrease the translational velocity. This is called the pausing of ribosomes and occurs at codons which are recognized by rare tRNA species (BONEKAMP and JENSEN, 1988). This situation is reported to enhance the incorrect incorporation of amino acids into the growing peptide chain. In *E. coli*, codons AGA and AGG are read by tRNA species which occur infrequently, the so-called rare tRNAs. The occurrence of two such codons adjacent to each other can cause frame shifting of the ribosome (SPANJAARD and VAN DUIN, 1988). As mentioned in Sect. 4.2, these codons rarely occur in the coding sequences of *E. coli* genes, but are commonly found in other organisms. The introduction of foreign genes into host organisms by genetic exchange, viral infections or genetic engineering, can lead to problems in the translational processing of these foreign genes. Increasing the amount of the specific tRNA genes or their transcription in *E. coli*, encourages the proper translation (BRINKMANN et al., 1989). Similar results have been described for another tRNA in *Streptomyces* (LESKIW et al., 1991).

The unavailability of properly aminoacylated tRNAs at the ribosomal acceptor site during translation can lead to the so-called stringent response (GALLANT, 1979). The entry of uncharged tRNA at this site blocks further translocation of the ribosome. In consequence, the synthesis of rRNA and tRNA is depressed approximately tenfold. The stringent factor (coded by *relA*) is produced and converts GTP to pppGpp with the loss of ATP. The subsequent tranformation of pppGpp results in ppGpp which, as a second messenger, inhibits the initiation of transcription at promoters of *rrn* operons. In addition, the elongation of transcription is generally terminated.

This general effect of a modified nucleotide can be compared to the involvement of cAMP in transcriptional enhancement and supports the hypothesis of a second messenger.

6 Antibiotics Inhibitory to Transcription and Translation

The cellular components active in transcriptional and translational processes can be inhibited by a large number of natural or synthetic

Tab. 5. Inhibitors of Protein Synthesis

Antibiotic	Inhibitory Action in Prokaryotes	Inhibitory Action in Eukaryotes
Streptomycin Neomycin Kanamycin	Binds to protein S12 of 30S subunit Binding of initiator tRNA	Mitochondrial ribosomes
Chloramphenicol	Peptidyl transferase of 70S ribosome	Mitochondrial ribosomes
Cycloheximide	?	Peptidyl transferase
Tetracycline	Binding of aminoacylated tRNA	?
Puromycin	Premature chain termination (analog of aminoacyl tRNA)	dto.
Fusidic acid	Translocation, EF-G release	dto., eEF-2
Thiostreptone	EF-G, preventing translocation	?
Diphtheria toxin	–	Translocation by inactivating eIF-2

compounds. These have been used extensively as experimental tools in the dissection of the machinery involved. The substances are called antibiotics (FRANKLIN and SNOW, 1989). In general, the action of RNA polymerase is inhibited by the intercalating polypeptide antibiotic actinomycin D, but different enzymes are not affected to the same extent. Rifamycin or its synthetic analog rifampicin, affects prokaryotic RNA polymerases only by binding to the β-subunit of the holo-enzyme and thus, inhibiting initiation. The elongation is not affected by rifamycin, but by streptolydigin. The inhibitory effect of α-amanitin on eukaryotic RNA polymerases has already been mentioned (see Sect. 2.2).

The ribosomes of pro- and eukaryotes differ in their response to antibiotics, and some well understood examples are compiled in Tab. 5. Most agents, which are active on prokaryotic ribosomes, are less effective in their inhibition of eukaryotic ribosomes, but interact efficiently with ribosomes of the eukaryotic organelles.

7 Post-Translational Modification in Bacteria

In eukaryotes, a series of events have been described which lead to the modification of the primarily synthesized protein sequence (PFEFFER and ROTHMAN, 1987). The maturation of different peptides from a polypeptide precursor is common among various peptide hormones. Such processes are rare in bacteria, although a number of proteases have been described (MILLER, 1987). Two exceptions are generally known: the removal of the N-terminal formylated methionine and the cleavage of the signal peptide from proteins secreted into the periplasm. Two enzymes are involved specifically in these processes; the methionine aminopeptidase and the signal peptidase.

Methionine aminopeptidase (map) is able to remove the methionine from different polypeptides with variable efficiency, depending on the penultimate amino acid of the substrate. This has also been observed in a comparative analysis of the N-terminal sequences of cytosolic proteins (BEN-BASSAT et al., 1987). In prokaryotes, the N-formyl moiety of the terminal methionine is normally removed by deformylase prior to the action of the methionine aminopeptidase (ADAMS, 1968).

The removal of the signal peptide from proteins translocated across the inner membrane has been reported to depend on the signal peptidase I (*lepB*) in *E. coli*. The overall mechanism of protein translocation involves a series of specific host proteins and resembles, in its basic features, the better understood processes in eukaryotes. For a review of the current knowledge, see SAIER et al. (1989).

8 Regulatory Aspects in Prokaryotes

Most of the approximately 3000 genes in *Escherichia coli* are not expressed at the same time. Some are expressed at a constant level which is termed constitutive expression. In contrast, some genes, or related groups of genes, are regulated in a coordinated manner. In bacteria, genes that are regulated in this coordinated fashion, are often linked and transcribed from a single promoter into a single mRNA molecule. This type of messenger is referred to as a polycistronic transcript. The corresponding set of genes is called an operon. The basic mechanism of its regulation was proposed by JACOB and MONOD (1961).

8.1 Regulation of Transcription

The regulation of transcriptional levels can be controlled by regulatory proteins which inhibit or enhance the interaction of the promoter with RNA polymerase by binding to sites at or near the promoter. These regulatory proteins are therefore termed repressors or activators. The activity or binding properties of these proteins are influenced by molecules, usually metabolites, which serve as corepressors or coactivators.

8.1.1 Dual Control of the Lactose Operon

The regulation of transcriptional activity of the lactose operon occurs by two independent mechanisms: repression and activation (Fig. 14). The three genes constituting the *lac* operon, *lacZ* (β-galactosidase), *lacY* (permease), and *lacA* (transacetylase), are located behind the *lac* promoter (ZABIN and FOWLER, 1980). Upstream and downstream of this promoter, two sites known as the cyclic-AMP-binding protein or catabolite activator protein (CAP) sequence and the operator are regions of interaction with regulatory proteins. The repression of the transcriptional activity occurs at the operator where a so-called tetrameric *lac* repressor molecule is able to bind. The consequence of this reaction is the inhibition of transcription. The *lac* repressor protein is encoded by the *lacI* gene which is located upstream of the *lac* operon (MILLER, 1980). It is

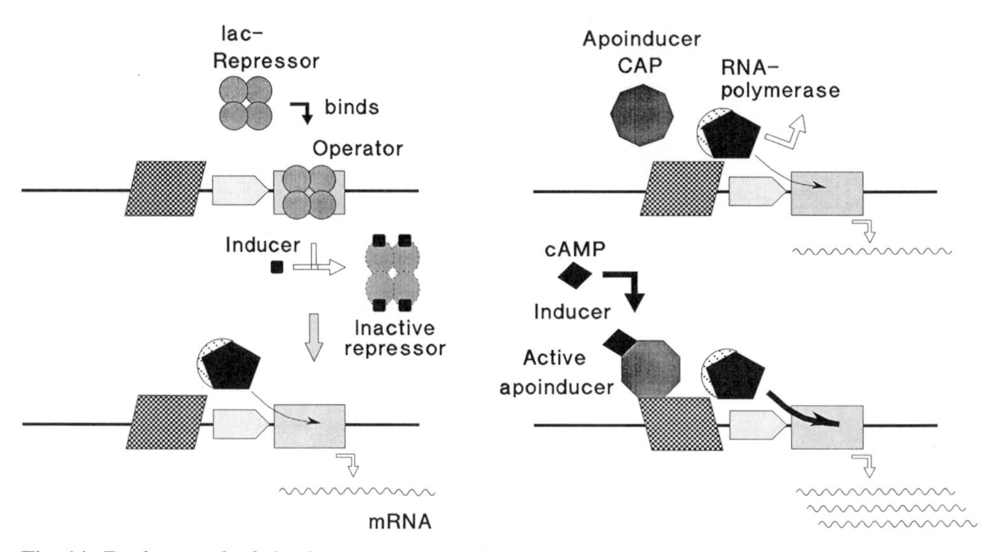

Fig. 14. Dual control of the *lac* operon. Negative control by *lac* repressor and positive control by cAMP activator protein (CAP). Both inducible systems are required for full transcriptional activity.

synthesized only in low amounts, usually about fifty molecules per cell. Its DNA binding capacity is abolished by the attachment of a variety of β-galactosidases, called inducers, to one of its identical subunits. This process, called induction, is believed to alter the conformation of the repressor and, as a result, the RNA-polymerase is able to initiate the transcription of the *lac* operon. Mutations which inactivate *lacI* therefore lead to the constitutive expression of the *lac* operon genes.

A second type of regulation exists, however, which occurs upstream of the promoter and which modulates the activity of the *lac* promoter independently, by activation (REZNIKOFF and ABELSON, 1980). Like in many other operons which encode sugar metabolizing enzymes, a second, positive regulatory signal is required for full activity of the *lac* promoter. The activator, called the catabolite activator protein (CAP) in complex with the coactivator cyclic AMP (cAMP), binds at the CAP site (DE CROMBRUGGHE et al., 1984) resulting in an increase of up to 50-fold of *lac* operon gene transcription. Two mechanisms discussed as being responsible for stimulating this regulation are an alteration of the DNA topology and the prevention of the polymerase from binding to a nearby weak promoter. As a result of this secondary regulation, the internal amount of cAMP, which is low if glucose is available to the cell, serves to prevent wasteful enzyme production.

8.1.2 Principles of Negative and Positive Regulation

Two independent regulatory systems are responsible for the dual control of the *lac* operon of *Escherichia coli*. A comparable regulatory system exists in other operons in this bacterium which encode enzymes for the degradation of various amino acids and nucleotides, and sugars such as galactose and arabinose.

The genetic analysis of mutants enables negative regulation to be distinguished from positive regulation (RAIBAUD and SCHWARTZ, 1984). Both systems can be induced or repressed (see Fig. 15). The transcription in an inducible, negatively controlled gene is abolished by a repressor, and this repressor can be inactivated by an inducer. An example of this type of regulation is the *lacI*-mediated control of the *lac* operon. A deletion or inactivating mutation of the repressor gene leads to recessive constitutive expression. Suppression of transcription in a repressible, negatively controlled gene requires the availability of a normally inactive repressor which is modified by a corepressor to become functional. The transcription is constitutive in the absence of this corepressor. In such a situation, an operon is termed derepressed which corresponds to the induced state of an inducible operon. Repressor gene deletions also result in recessive constitutive transcription which is typical of all negatively controlled genes. The regulation of genes for amino acid synthesizing enzymes is controlled in this repressible, negative manner. The repressor is active only in the presence of the respective amino acid. In contrast, positively controlled genes require a regulatory element as an activator of the initiation of transcription. Deletions of the gene lead to a recessive failure of transcription. The activator, called the apo-inducer, is primarily inactive and has to be modified by an inducer into the active form in order for transcriptions to occur. This type is termed inducible, positive control. The CAP-cAMP complex functions in this manner and constitutes the positive control mechanism of the *lac* and other operons. CAP activates transcription only in the presence of cAMP. In turn, repressible, positively regulated genes require a normally active apo-inducer which can be inactivated by a corepressor.

8.1.3 Attenuation

An interesting, additional control mechanism, which operates during initiated transcription and translation, is known to exist in the operons for, e.g., amino acid synthesizing enzymes as well as in some phage genes. Unique to this process is the transcription of a so-called leader mRNA which contains characteristic repeat structures. The model suggested by YANOFSKY (1988) implies that the translation of the leader mRNA is cotranscriptionally started. For example, in the *trp* operon of

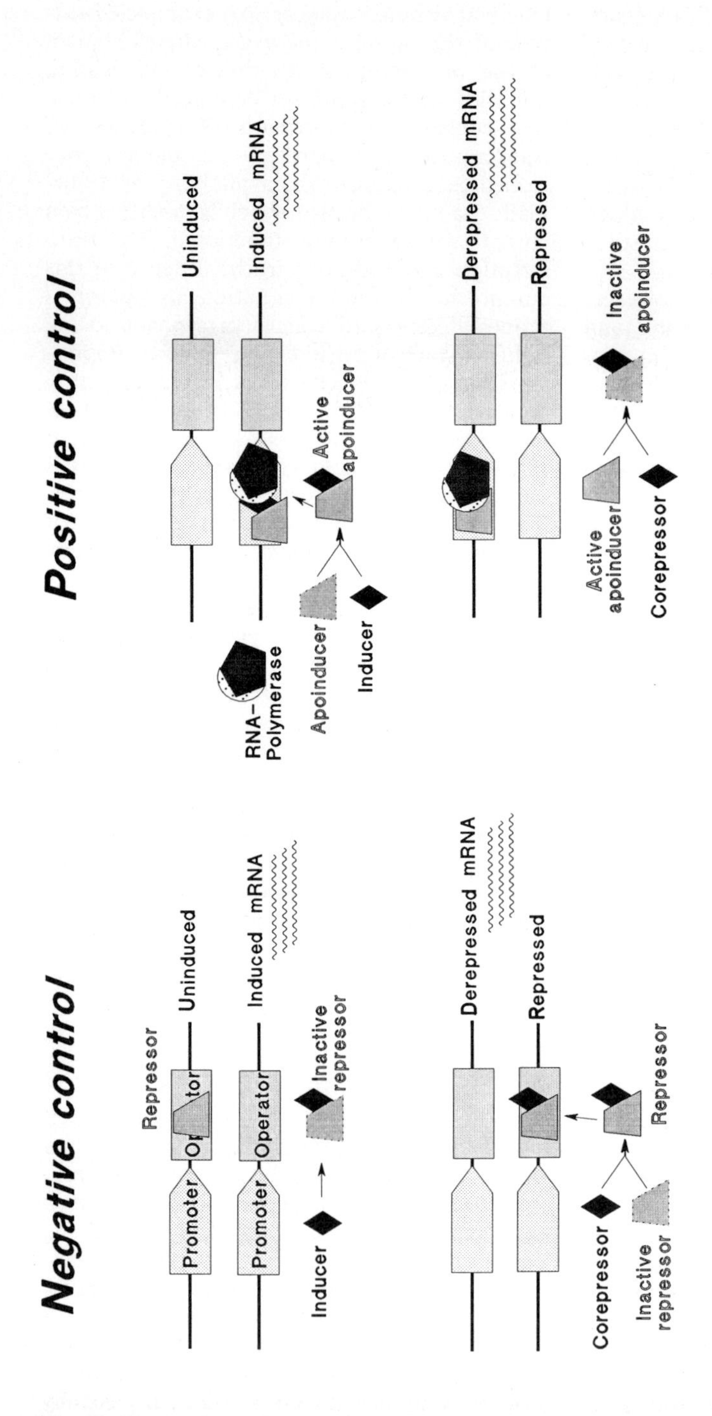

Fig. 15. Comparison of negative and positive control systems and their components.

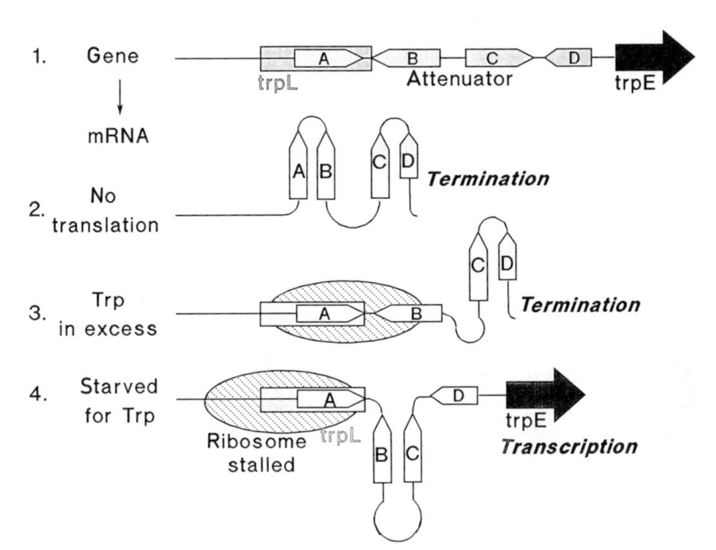

Fig. 16. Attenuation. Structural elements located behind the promoter and effects of translational activity on transcription elongation.

E. coli, the *trpL* element which consists of approximately 160 nucleotides situated between the promoter and the first structural gene *trpE*, contains the so-called attenuator sequence (Fig. 16). The sequence contains four regions (1+2, 3+4) which are able to form two hairpin structures which resemble those for rho-independent termination of transcription.

In addition to this, the leader sequence *trpL* codes for a 14 amino acid peptide which contains two adjacent tryptophan residues. Under conditions that enable the initiation of translation to occur, this part of the *trpL* mRNA is covered by ribosomes. If enough tryptophan is available to permit the production of charged $tRNA^{Trp}$, translation continues throughout the *trpL* open reading frame, although the second (3+4) hairpin structure leads to the termination of transcription further downstream. As a result, no tryptophan synthesizing enzyme is produced.

Whenever tryptophan becomes a limiting factor during transcription and translation, the amount of charged $tRNA^{Trp}$ will limit the translation of the *trpL* peptide. The ribosomes stall at the trp codons, and the pausing of translation allows a hairpin formation between the regions 2+3, to take place. This prevents the folding of the 3+4 terminator structure, and thus transcription can proceed beyond region 4. The overall effect is the expression of the genes encoding tryptophan synthesizing enzymes.

The occurrence of cognate amino acid codons found in the *E. coli trp* operon is seen in other amino acid biosynthetic operons. This type of regulation clearly shows the interdependence of transcription and translation and hence, allows the fine tuning of physiological responses.

The attenuation type control may also be regarded as an example of repressible, positive control, where the ribosome is the active apoinducer and the corresponding charged tRNA molecule acts as a corepressor.

8.1.4 Upstream Activating Sequences

Additional examples of positively regulated operons have been described recently, which show an activation similar to the CAP-cAMP complex of the *lac* promoter. Some promoters of stable RNA genes seem to require cellular factors such as FIS (factor of inversion stimulation) to enhance their transcriptional capacity (NILSSON et al., 1990). The corresponding interaction sites have been identified and are, to date, not known to depend on any additional inducer components. Some of these seem to affect the topology of the DNA, the so-called

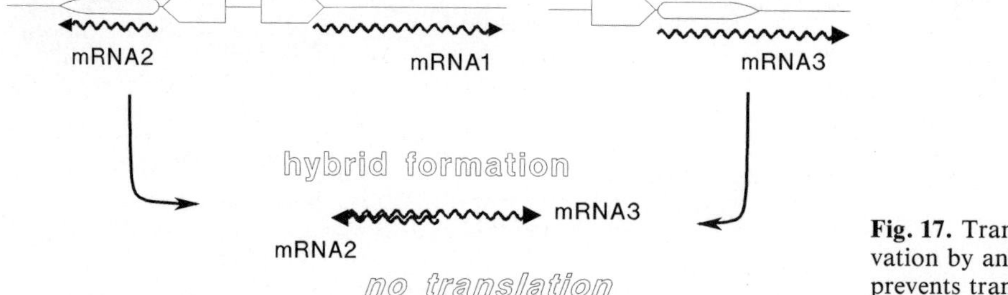

Fig. 17. Transcript inactivation by antisense RNA prevents translation.

bending, which may be accomplished by the binding of the protein factor to the DNA. Alternatively, the absence of such factor alters the conformation of the corresponding DNA sequence. As a result of a slight resemblance to eukaryotic enhancer sequences, they are termed upstream activating sequences (UAS). The activity or the amount of some of the cellular factors like FIS seem to depend on the growth stage or nutritional status of the cells.

8.2 Feedback Control

The transcription or translation of particular genes is regulated in a negative manner, by the respective gene product. Examples of transcriptional feedback control are found among regulatory genes of phages and resistance operons, e.g., genes for tetracyline resistance and mercury resistance. In these cases, operators which are able to bind the respective protein are located between promoter and coding sequence.

The feedback control of translation ensures the coordinate formation of ribosomal subunits (NOMURA et al., 1984). The genes for the proteins of the small and large subunits are intermingled in several operons with those for proteins which mediate the translational process (e.g., elongation factors). Ribosomal RNAs encoded by additional operons are also required for the subunit formation. In cases where an insufficiency of the appropriate amounts of rRNA molecules occurs, some of the ribosomal proteins bind to specific sequences of their mRNA, thereby preventing further translation.

Another type of feedback control ensures the proper relationship between the two outer membrane proteins of *E. coli* encoded by *ompC* and *ompF*. The positive regulator *ompR* activates transcription of *ompC* and a gene called *micF* (for mRNA interfering complementary RNA). This is commonly referred to as "antisense" RNA. The *ompF* gene appears to be constitutively transcribed. The production of *micF* mRNA, which is complementary to the *ompF* RNA, leads to the inhibition of *ompF* translation and in turn, the *ompC* protein becomes the dominant outer membrane constituent. This change from *ompF*- to *ompC*-type cells is mediated by *ompR* which responds to the osmotic conditions of the cellular environment. This type of control has been described for the modulation of genes in phages for primer formation in the replication of certain bacterial plasmids as well as for transposons. The concept of antisense RNA (EGUCHI et al., 1991) as a regulatory module has become a useful tool in pro- and eukaryotic genetic analyses (Fig. 17).

9 References

ADAMS, J. M. (1968), On the release of the formyl group from nascent proteins, *J. Mol. Biol.* **33,** 571–589.

BEN-BASSAT, A., BAUER, K., CHANG, S. Y., MYAMBO, K., BOOSMAN, A., CHANG, S. (1987), Processing of the initiation methionine from proteins: properties of the *Escherichia coli* methionine aminopeptidase and its gene structure, *J. Bacteriol.* **169,** 751–757.

BJÖRK, G. R., ERICSON, J. V., GUSTAFSON, C. E. D., HAGERVALL, T. G., JÖNSSON, Y. H., WIKSTRÖM, P. M. (1987), Transfer RNA modification, *Annu. Rev. Biochem.* **56,** 263–287.

BÖCK, A., FORCHHAMMER, K., HEIDER, J. LEINFELDER, W., SAWERS, G., VEPREK, B., ZINONI, F. (1991), Selenocysteine: the 21st amino acid, *Mol. Microbiol.* **5**, 515–520.

BONEKAMP, F., JENSEN, K. F. (1988), The AGG codon is translated slowly in *E. coli* even at very low expression levels, *Nucleic Acids Res.* **16**, 3013–3024.

BRINKMANN, U., MATTES, R. E., BUCKEL, P. (1989), High-level expression of recombinant genes in *Escherichia coli* is dependent on the availability of the *dnaY* gene product, *Gene* **85**, 109–114.

BUTLER, E. T., CHAMBERLIN, M. J. (1982), Bacteriophage SP6-specific RNA polymerase. I. Isolation and characterization of the enzyme, *J. Biol. Chem.* **257**, 5772–5778.

CHAMBERLIN, M. (1982), Bacterial DNA-dependent RNA polymerases, in: *The Enzymes*, Vol. 15 Part B, pp. 61–108 (BOYER, P., Ed.), New York: Academic Press.

CHEN, S., LI, X., LIAO, W. S. L., WU, J. H., CHAN, L. (1990), RNA editing of apolipoprotein B mRNA, *J. Biol. Chem.* **265**, 6811–6816.

CRICK, F. H. C. (1966), Codon-anticodon pairing: The wobble hypothesis, *J. Mol. Biol.* **19**, 548–555.

DAVANLOO, P., ROSENBERG, A. H., DUNN, J. J., STUDIER, F. W. (1984), Cloning and expression of the gene for bacteriophage T7 RNA polymerase, *Proc. Natl. Acad. Sci. USA* **81**, 2035–2039.

DE CROMBRUGGHE, B., BUSBY, S., BUC, H. (1984), Cyclic AMP receptor protein: Role in transcription activation, *Science* **224**, 831–838.

DEUTSCHER, M. (1984), Processing of tRNA in prokaryotes and eukaryotes, *Crit. Rev. Biochem.* **17**, 45–71.

EGGERTSSON, G., SÖLL, D. (1988), Transfer RNA-mediated suppression of termination codons in *Escherichia coli, Microbiol. Rev.* **52**, 354–374.

EGUCHI, Y., ITOH, T., TOMIZAWA, J. I. (1991), Antisense RNA, *Annu. Rev. Biochem.* **60**, 631–652.

FRANKLIN, T. J., SNOW, G. A. (1989), *Biochemistry of Antimicrobial Action*, London: Chapman and Hall.

GALLANT, J. A. (1979), Stringent control in *E. coli, Annu. Rev. Genet.* **13**, 393–415.

GARRETT, R. A., WOOLEY, P. (1982), Identifying the peptidyl transferase centre, *Trends Biochem. Sci.* **7**, 385–386.

GEGENHEIMER, P., APIRION, D. (1981), Processing of prokaryotic ribonucleic acid, *Microbiol. Rev.* **45**, 502–541.

HAWLEY, D. K., MCCLURE, W. R. (1983), Compilation and analysis of *Escherichia coli* promoter DNA sequences, *Nucleic Acids Res.* **11**, 2237–2255.

HELMANN, J. D., CHAMBERLIN, M. J. (1988), Structure and function of bacterial sigma factors, *Annu. Rev. Biochem.* **57**, 839–872.

HIGGINS, C. F., MCLAREN, R. S. NEWBURY, S. F. (1988), Repetitive extragenic palindromic sequences, mRNA stability and gene expression: evolution by gene conversion? a review, *Gene* **72**, 3–14.

JACOB, F., MONOD, J. (1961), Genetic regulatory mechanisms in the synthesis of proteins, *J. Mol. Biol.* **3**, 318–356.

JINKS-ROBERTSON, S., NOMURA, M. (1987), Ribosomes and tRNA, in: *Escherichia coli and Salmonella typhimurium*, Vol. 2, pp. 1358–1385 (NEIDHARDT, F. C., Ed.). Washington: American Society for Microbiology.

JOHNSON, A., ADKINS, H., MATTHEWS, E., CANTOR, C. (1982), Distance moved by transfer RNA during translocation from the A site to the P site on the ribosome, *J. Mol. Biol.* **156**, 113–140.

KOMINE, Y., ADACHI, T., INOKUCHI, H., OZEKI, H. (1990), Genomic organization and physical mapping of the transfer RNA genes in *Escherichia coli* K12, *J. Mol. Biol.* **212**, 579–598.

LESKIW, B. K., LAWLOR, E. J., FERNANDEZ-ABALOS, J. M., CHATER, K. F. (1991), TTA codons in some genes prevent their expression in a class of developmental, antibiotic-negative, *Streptomyces* mutants, *Proc. Natl. Acad. Sci. USA* **88**, 2461–2465.

LEWIS, M. K., BURGESS, R. R. (1982), Eukaryotic RNA polymerases, in: *The Enzymes*, Vol. 15, pp. 109–153 (BOYER, P., Ed.), New York: Academic Press.

LOSICK, R., PERO, J. (1981), Cascades of sigma factors, *Cell* **25**, 582–584.

MCCLURE, W. R. (1985), Mechanisms and control of transcription initiation in prokaryotes, *Annu. Rev. Biochem.* **54**, 171–204.

MCCORMICK, J. R., ZENGEL, J. M., LINDAHL, L. (1991), Intermediates in the degradation of mRNA from the lactose operon of *Escherichia coli, Nucleic Acids Res.* **19**, 2767–2776.

MILLER, C. G. (1987), Protein degradation and proteolytic modification, in: *Escherichia coli and Salmonella typhimurium*, Vol. 1, pp. 680–691 (NEIDHARDT, F. C., Ed.). Washington: American Society for Microbiology.

MILLER, J. H. (1980), The *lacI* gene: its role in lac operon control and its use as a genetic system, in: *The Operon* (MILLER, J. H., REZNIKOFF, W. S., Eds.), pp. 31–88, Cold Spring Harbor, NY: Cold Spring Harbor Laboratory Press.

MORRIS, C. E., KLEMENT, J. F., MCALLISTER, W. T. (1986), Cloning and expression of the bacteriophage T3 RNA polymerase gene, *Gene* **41**, 193–198.

MUTO, A., ANDACHI, Y., YUZAWA, H., YAMAO, F., OSAWA, S. (1990), The organization and evolution of transfer RNA genes in *Mycoplasma capricolum, Nucleic Acids Res.* **18**, 5037–5043.

NILSSON, L., VANET, A., VIJGENBOOM, E., BOSCH, L. (1990), The role of FIS in *trans* activation of stable RNA operons of *E. coli, EMBO J.* **9**, 727–734.

NIRENBERG, M. W., MATTHAEI, J. H. (1961), The dependence of cell-free protein synthesis in *E. coli* upon naturally occurring or synthetic polyribonucleotides, *Proc. Natl. Acad. Sci USA* **47**, 1588–1682.

NOMURA, M., GOURSE, R., BAUGHMAN, G. (1984), Regulation of the synthesis of ribosomes and ribosomal components, *Annu. Rev. Biochem.* **53**, 75–117.

PARKER, J. (1989), Errors and alternatives in reading the universal genetic code, *Microbiol. Rev.* **53**, 273–298.

PFEFFER, S. R., ROTHMAN, J. E. (1987), Biosynthetic protein transport and sorting by the endoplasmic reticulum and golgi, *Annu. Rev. Biochem.* **56**, 829–852.

PLATT, T. (1986), Transcription termination and the regulation of gene expression, *Annu. Rev. Biochem.* **55**, 339–372.

RAIBAUD, D., SCHWARTZ, M. (1984), Positive control of transcription initiation in bacteria, *Annu. Rev. Genet.* **18**, 173–206.

REZNIKOFF, W. S., ABELSON, J. N. (1980), The lac promoter, in: *The Operon* (MILLER, J. H., REZNIKOFF, W. S., Eds.), pp. 221–243, Cold Spring Harbor, NY: Cold Spring Harbor Laboratory Press.

SAIER, M. H., WERNER, P. K., MÜLLER, M. (1989), Insertion of proteins into bacterial membranes: mechanism, characteristics, and comparisons with the eukaryotic process, *Microbiol. Rev.* **53**, 333–366.

SCHIMMEL, P. (1987), Aminoacyl tRNA synthetases: general scheme of structure-function relationship in the polypeptides and recognition of transfer RNAs, *Annu. Rev. Biochem.* **56**, 125–158.

SHINE, J., DALGARNO, L. (1974), The 3′-terminal sequence of *E. coli* 16S rRNA: Complementarity to nonsense triplets and ribosome binding sites, *Proc. Natl. Acad. Sci. USA* **71**, 1342–1346.

SIMPSON, L., SHAW, J. (1989), RNA editing and the mitochondrial cryptogenes of kinetoplastid protozoa, *Cell* **57**, 355–366.

SPANJAARD, R. A., VAN DUIN, J. (1988), Translation of the sequence AGG-AGG yields 50% ribosomal frameshift, *Proc. Natl. Acad. Sci. USA* **85**, 7967–7971.

SPRENGART, M. L., FATSCHER, H. P., FUCHS, E. (1990), The initiation of translation in *E. coli:* apparent base pairing between 16S rRNA and downstream sequences of the mRNA, *Nucleic Acids Res.* **18**, 1719–1723.

WADA, K., WADA, Y., DOI, H., ISHIBASHI, F., GOJOBORI, T., IKEMURA, T. (1991), Codon usage tabulated from the GenBank genetic sequence data, *Nucleic Acids Res.* **19** *Suppl.*, 1981–1986.

WITTMANN, H. G. (1983), Architecture of prokaryotic ribosomes, *Annu. Rev. Biochem.* **52**, 35–66.

YANOFSKY, C. (1988), Transcription attenuation, *J. Biol. Chem.* **263**, 609–612.

ZABIN, I., FOWLER, A. V. (1980), β-Galactosidase, the lactose permease protein and thiogalactoside transacetylase, in: *The Operon* (MILLER, J. H., REZNIKOFF, W. S., Eds.), pp. 89–121, Cold Spring Harbor, NY: Cold Spring Harbor Laboratory Press.

8 DNA Sequencing

GUIDO VOLCKAERT
PETER VERHASSELT
MARLEEN VOET
JOHAN ROBBEN

Leuven, Belgium

1 Introduction

Genetic information is contained in DNA by a linear order of nucleotide bases. Unraveling this order is called *DNA sequencing* and is fundamentally a non-quantitative process. The order of nucleotides in a DNA molecule corresponds to the order of bases on the pentose-phosphate repeat unit or backbone of the polymer: therefore, in sequencing, 'nucleotide' and 'base' are often used interchangeably in a non-semantical manner.

DNAs contain thousands to millions of nucleotides, but a single sequencing experiment usually delivers the order of only a few hundreds of bases. Consequently, it is necessary to reduce the complexity of larger DNAs to much smaller fragments which are accessible for detailed analysis. In practical terms, this means that the target DNA should be enriched and produced in sufficient quantities for the particular sequencing technique. Before the discovery and isolation of restriction enzymes, this was an impossible task and DNA sequencing was a somewhat esoteric research field. The advent of recombinant DNA cloning technology enabled the nucleic acid biochemist to tackle this kind of problems efficiently and systematically. Two new sequencing methods emerged and quickly spread worldwide (SANGER et al., 1977a; MAXAM and GILBERT, 1977).

The four nucleotides in DNA (adenylate, cytidylate, guanylate, and thymidylate) are represented by the one-letter codes A, C, G, and T. For an arbitrary or unknown nucleotide the symbol N is used, and other standard codes were defined by the International Union of Biochemistry (CORNISH-BOWDEN, 1985).

It is of utmost importance to realize that the nucleotides by themselves are asymmetric chemical structures. This is indicated by the numbering of the C-atoms in the pentose ring. The positions nearest to the phosphodiester bond are 5′ and 3′, and are used to indicate the polarity of the DNA strands. To avoid confusion, a DNA sequence should always be written in the conventional 5′→3′ direction. (The nucleotide sequence A—C—T—G is chemically entirely different from G—T—C—A.) Most DNA molecules, however, exist as a complex of two chains (or strands) with opposite (i.e., anti-parallel) orientations, and the chains are intertwined into each other as a double-helix structure. To draw the double-stranded structure unequivocally, the 'upper' chain should be written in 5′→3′ orientation as top strand, the 'lower' chain in 3′→5′ orientation as bottom strand. The notion of 'upper' and 'lower' chain has neither physical nor biological meaning, since in a spatial model they switch from top to bottom strand every 10 base pairs.

The two chains are held together by hydrogen-bond interactions between the nucleic acid bases, the base pairs. Stable base pairing occurs only between A and T, and between G and C. Thus, the nucleotide sequence of one chain dictates the order of the nucleotides in the other, named complementary chain. As a consequence, knowing the sequence of one strand of a double-strand allows to deduce the sequence of the other strand and to define the complete primary structure of the entire DNA molecule. Conversely, experimental analysis of the second strand sequence can, and should, be used to verify the sequence of the first strand. The principle of complementarity is the basis of duplication of DNA *in vivo* and is fundamental to one of the most important current sequencing methods (Sect. 4.1).

2 Historical Account: Past, Present, and Future

2.1 The Pre-History of DNA Sequencing

The initial DNA sequencing techniques evolved from RNA sequencing methods. One reason for this was that there were no small DNAs available on which methods could be developed and evaluated. Early studies on RNA sequences were done on the tRNAs and 5S RNA, which contain 75–120 residues. The smallest pure DNAs, however, are bacteriophage (fd, ϕX174, etc.) and viral (SV40, polyoma, etc.) DNAs which contain about 5000 residues or base pairs.

Most of the work with RNA depended on the use of RNase T1 which splits the nucleotide chain specifically at G residues, and RNase A, which is specific for pyrimidine residues. DNases, on the contrary, either show little specificity, such as DNase I, or are extremely specific, such as the restriction enzymes. Later on, the latter would prove indispensable to dissect the large DNAs to fragments of manageable size for further analysis (DANNA and NATHANS, 1971).

Cleavage of an RNA with base-specific enzymes generates oligonucleotides. The main technical problem was the fractionation of the oligonucleotides or the complex mixtures of closely related oligomers that were produced on partial digestion of the RNA. Hence, much effort was devoted to the development of efficient techniques for fractionation. An impressive number of electrophoretic and chromatographic methods was generated, including two-dimensional separations known as fingerprints (SANGER et al., 1965; BROWNLEE and SANGER, 1969; VOLCKAERT et al., 1976). Because of the small amounts to be detected, very sensitive techniques for identification were required, and most of the analyses were done with RNA labeled *in vivo* to high specific activity with ^{32}P, one of the atoms in the repeating units of the polymer. RNase T1-generated oligonucleotides were usually small enough to be solved by further degradation with other nucleases. Ultimately, the relative order of the oligonucleotides in the RNA was determined by analysis of partial cleavage products to reconstitute the complete sequence. Taken together, this approach much resembles putting together the pieces of a puzzle.

SANGER and his colleagues (ADAMS et al., 1969) reported the first 'extensive' nucleotide sequence of part of a (relatively) large nucleic acid. The sequencing microtechniques with ^{32}P-labeled RNA were adapted to the problems raised by larger RNA molecules. Particular regions were isolated, e.g., by the ability of specific regions of the RNA to form a stable complex with another macromolecule (e.g., ribosomes) that renders this region resistant to digestion with nucleases while the uncovered nucleic acid is removed by hydrolysis. A more general strategy took advantage of the fact that RNA has a relatively high degree of secondary and tertiary structure and limited digestion with RNase T1 gives reproducibly fragments of up to several hundreds of nucleotides, which could be resolved by one- or two-dimensional polyacrylamide gel electrophoresis (DE WACHTER and FIERS, 1971). The complete MS2 phage genome (3569 nucleotides) has been sequenced by this approach (FIERS et al., 1976).

In the absence of DNases with similar specificity as RNase T1, an alternative was explored with quantitative depurination of the DNA by diphenylamine in formic acid (BURTON and PETERSON, 1960) followed by analysis of the pyrimidine tracts with the RNA fingerprinting methods (LING, 1972a, b). The nucleotide sequence of the unique, larger oligonucleotides isolated from such fingerprints could be completely solved with exonucleases, but the complexity of the fingerprints was far beyond the feasibility of complete sequencing of the genomes. Again, as in the RNA sequencing efforts, some specific regions could be approached by virtue of their interaction with specific proteins and the DNA segments isolated by digestion of the complex with nucleases to remove the unprotected part of the DNA. Thus, regions protected by RNA polymerase (promoter sites) (HEYDEN et al., 1972), by ribosomes (ROBERTSON et al., 1973), by phage λ repressor (MANIATIS and PTASHNE, 1973), and Lac repressor (GILBERT and MAXAM, 1973), were analyzed. Some regions were isolated by limited digestion of single-stranded DNA with an endonuclease and gel fractionation, a corollary that the DNA would have substantial secondary structure (ZIFF et al., 1973). Other targets for direct DNA sequencing were the termini of DNA fragments generated by restriction endonucleases (KELLY and SMITH, 1970; HEDGPETH et al., 1972; BIGGER et al., 1973) and of bacteriophage DNAs (WU and TAYLOR, 1971; ENGLUND, 1972; PADMANABHAN and WU, 1972). They were analyzed either by enzymatic labeling of the 5′ end, followed by limited nuclease digestion and analysis of the resulting oligonucleotides, or by DNA repair reactions with DNA polymerase I and α-^{32}P-labeled deoxynucleoside triphosphates as precursors. Repair reactions with ribo- and deoxyribonucleotides were subsequently used as a more general, but indirect approach to sequence determination of single-stranded DNA. In the presence of Mn^{2+} ions instead of Mg^{2+}, *Escherichia coli* DNA polymerase I incorporates ribonucleotide residues into the DNA chains (BERG et al., 1963). These residues are sensitive to cleavage with alkali or RNase, thus permitting use of the RNA fingerprinting techniques for subsequent isolation of the oligonucleotides (VAN DE SANDE et al., 1972; SANGER et al., 1973).

An even more indirect approach to DNA sequence determination was the use of *E. coli* RNA polymerase in copying reactions, so that the transcripts could be subject to RNA sequencing, at that time considered a technical advantage. Either cellular DNA or restriction fragments (e.g., from SV40 DNA, DANNA and NATHANS, 1971) have been used as templates. Depending on template, reaction conditions and the presence/absence of σ factor, either discrete fragments or a very heterogeneous population of molecules were obtained (LEBOWITZ et al., 1971; KLEPPE and KHORANA, 1972; GILBERT and MAXAM 1973; DAHR et al., 1974, MAROTTA et al., 1974; VOLCKAERT et al., 1977). Sometimes oligonucleotides were added to force initiation at specific places (GILBERT and MAXAM, 1973; VOLCKAERT et al., 1977).

A DNA sequence of 50 and 81 residues of bacteriophage f1 was obtained by SANGER and collaborators (SANGER et al., 1973, and SANGER et al., 1974, respectively) by the extension of a chemically synthesized octadeoxynucleotide primer on single-stranded template DNA with DNA polymerase I. Different sorts of repair reactions were used: (1) ribosubstitution repair mentioned above and (2) repair in the presence of Mg^{2+} with one or more of the four deoxynucleoside triphosphates absent. Partial fragments thus obtained were fractionated by two-dimensional cellulose acetate electrophoresis/DEAE cellulose thin-layer chromatography and individually analyzed. It was the first clear illustration of how repair reactions can initiate at a unique site by the addition of a designed synthetic primer. Further expansion of this concept is found in the 'plus and minus' sequencing method of SANGER and COULSON (1975). Enzymatic extensions starting from the primer (either synthetic or a strand from a restriction fragment) were manipulated so that four series of ^{32}P-labeled products were obtained, each containing intermediates specifically terminated at each of the four nucleotides. The samples were fractionated on a neutral polyacrylamide gel containing high concentrations of urea to keep the oligonucleotides single-stranded, which, under these conditions, migrate according to size and which were resolved over a range of more than 100 nucleotides (BARRELL et al., 1976). This brought together all elements for an entirely new approach for DNA sequencing.

2.2 'Modern' Sequencing Technology

The 'plus and minus' technique (SANGER and COULSON, 1975) may be considered as the incentive to the modern gel-based DNA sequencing technology. This technique (which itself never found wide application) not only was an inception to the now common dideoxynucleotide chain termination method of SANGER et al. (1977a), but also was the first report in which it was shown that single-stranded DNA fragments differing in size by only one nucleotide, can be resolved electrophoretically on denaturing polyacrylamide gels. Ever since, fractionation according to the relative molecular mass of the fragments has been a central principle of all widely applied new DNA sequencing techniques, including the automated methods.

A second principle at the heart of current DNA sequencing is the preparation of so-called 'nested sets', i.e., the collections of DNA fragments from which the DNA sequence can be derived (as explained further in Sect. 3.1). In the 'plus and minus' method, and in the more practical versions in which terminating analogs such as dideoxynucleoside triphosphates are used (SANGER et al., 1977a), the nested sets are prepared enzymatically by copying a DNA strand with a DNA polymerase (Fig. 1A). Alternatively, a DNA fragment can be treated with base-specific chemical reagents under conditions that generate partial degradation products forming a nested set (Fig. 1B). This concept was followed by MAXAM and GILBERT (1977) to develop their chem-

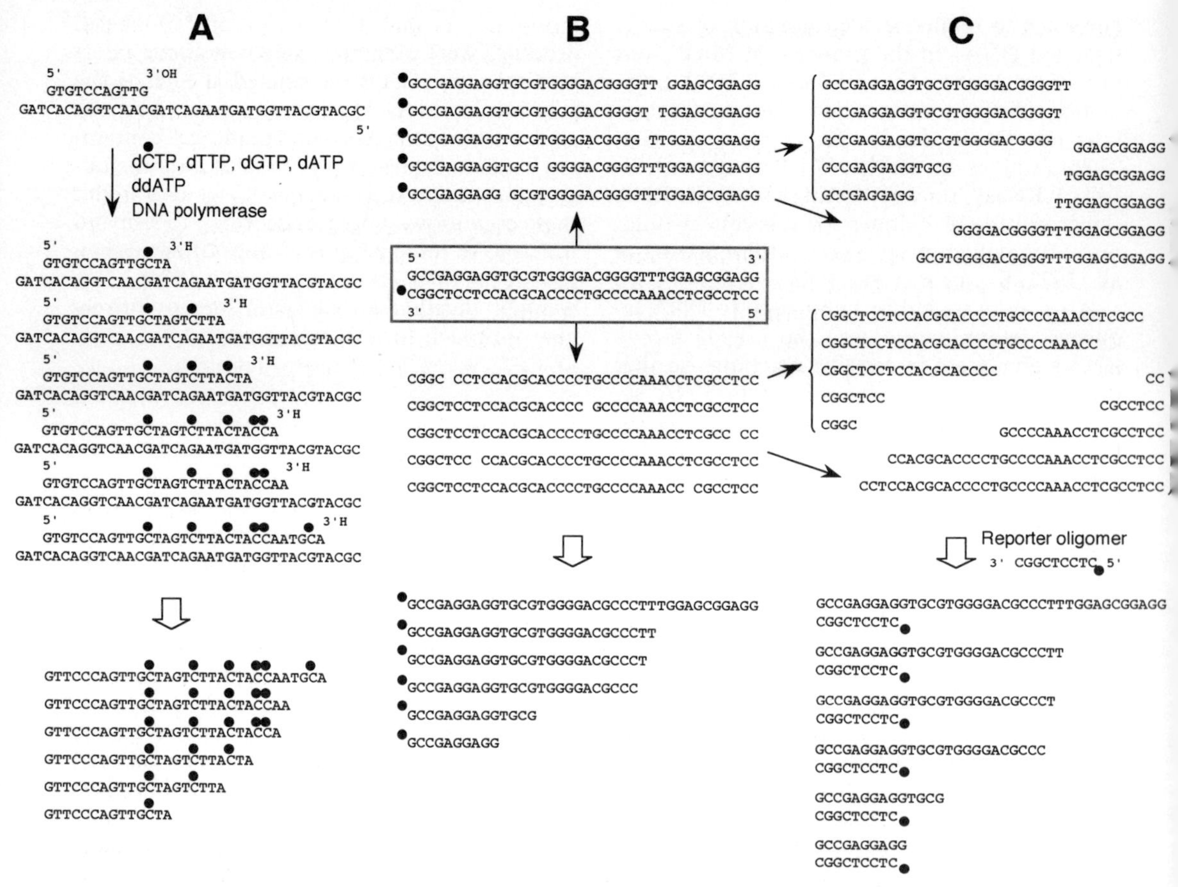

Fig. 1. Generation of nested sets in enzymatic (A) and chemical (B, C) DNA sequencing strategies. – Radioactive labels are represented by filled circles. The open arrows indicate the nested sets in order of length of the constituting fragments.
(A) Enzymatic generation of an A-specific nested set by extension of the primer (top strand) in the presence of dideoxy ATP.
(B) Sequencing with partial, base-specific, chemical degradations. Generation of a T-specific nested set is shown. The DNA fragment (boxed) is labeled at one end. Each strand is fragmented (black arrows), and the products form four nested sets (horizontal arrows from B to C). Only one of the collections is radioactively labeled.
(C) Detection by indirect labeling (Church and Gilbert, 1984). Degradation is done as in B, except that the DNA fragment is unlabeled. Each of the sets can be detected separately by hybridization to a suitable reporter oligonucleotide. In this example the labeled reporter 5′ CTCCTCGGC 3′ visualizes the uppermost set.

ical degradation sequencing method. These sequencing approaches have profoundly changed molecular biology during the past 15 years. The Maxam and Gilbert method originally received more attention and wider application as it was technically less exacting and seemed more reliable. Intense improvements and simplifications in the enzymatical technique, together with the efforts by commercial companies to supply appropriate high quality biochemicals and/or complete kits to the novices in the domain, have continuously stimulated the use of the enzymatical approach. Currently, one can estimate that more than 90% of all

DNA sequencing is done by the chain termination method.

2.3 New Trends in Development of Sequencing Techniques

The methods of MAXAM and GILBERT and of SANGER and COULSON have continuously undergone numerous incremental improvements in technology, but always relied on the nested set approach. They also, in particular the enzymatic method, were subject to automation (see Sect. 6). Many of those adaptations were the results of running sequencing projects, as witnessed by the ever growing nucleic acid data bases from EMBL (Heidelberg, Germany) and Genbank (Los Alamos, USA).

Some major extensions are genomic sequencing (CHURCH and GILBERT, 1984) and in particular the concept of multiplexing (CHURCH and KIEFFER-HIGGINS, 1988). These techniques, as well as the radical influence of the polymerase chain reaction (PCR) (SAIKI et al., 1985) are described in more detail in Sect. 4.4 of this chapter. Nevertheless, real major technological progress, increasing sequencing speed and volume substantially, should probably not be expected from these kinds of developments. Cost-effective sequencing of large genomes undoubtedly requires fundamentally newer techniques. Some of the ideas are described here briefly.

One particular idea is derived from flow cytometry. Since single cells can be detected after suitable fluorescent labeling, one might envisage that single molecules could be detectable if sufficiently strong fluorescent tags could be attached to every base in the DNA (four different tags corresponding to the four nucleotides). Several exonucleases release stepwise mononucleotides from one end of a single-stranded DNA molecule. In a system with an immobilized DNA molecule, those mononucleotides could be transported by a liquid stream towards a strong laser sensitive detector, which then would register and identify the nucleotides sequentially as they pass. This 'single-molecule-detection' approach for sequencing is being explored at the Los Alamos National Laboratory, USA (JETT et al., 1989). Different elements need to be worked out: fluorescent labeling of the bases, preparation of the DNA fragment to be sequenced, exonucleolytic release of labeled nucleotides, single base detection (DAVIS et al., 1991).

In another method (HYMAN, 1988), primer-template and DNA polymerase are immobilized, and stepwise synthesis is monitored by the release of inorganic pyrophosphate when the dNTP precursors are pumped through the column. A luminescent signal detects which nucleotide is added and quantifies the number of additions. The key problems with this method will probably be speed and synchronization of the reactions.

Direct visualization might be possible by scanning tunneling electron microscopy (ARSCOTT et al., 1989; BEEBE et al., 1989; DUNLAP and BUSTAMANTE, 1989; LINDSAY et al., 1989). DNA would be visualized and the images formed while scanning at several Ångström per second. This could give sequencing rates of kilobases per minute (LINDSAY and PHILIPP, 1991). Much investigation is required to solve technical problems such as sample preparation, location of samples, interpretation of data, etc.

Further analytical power may be found in mass spectrometry (ARLINGHAUS et al., 1991; see also Sect. 4.3.2.5). With fast atom bombardment mass spectrometry (FAB), for example, sequencing of short oligonucleotides has proven possible since some time (GROTJAHN et al., 1982).

Another approach is based on a mathematical analysis of the maximal complexity that random permutation of short sequences can generate. It proposes sequencing by DNA–DNA hybridization between oligonucleotide probes of known sequence and the target of unknown sequence (BAINS and SMITH, 1988; DRMANAC et al., 1989; KHRAPKO et al., 1989). The idea is that longer sequences can be obtained by the maximal and unique overlap of the constituent oligomers in a DNA, identified by hybridization, without knowledge of the frequency or the position of the oligomers. The procedures suggested by DRMANAC's and KHRAPKO's group (KHRAPKO et al., 1991; STREZOSKA et al., 1991) involve the use of a large battery of probes (from 65536 octamers up to more than 95000 undecamers), whereas

BAINS (1991) uses smaller targets and a limited number of probes (down to 256). DNA can be bound to a surface, with the probes in solution, or *vice versa*. Hybridization conditions need to be found that give clearcut 'yes/no' responses, with a 'no' as soon as a single mismatch is present. Obviously, these methods require substantial computer effort to assemble the final sequence from the hybridization data. Moreover, DNA synthesis and manipulation of large series of short oligomers is a tremendous task. Nevertheless, such technique could offer advantages in cost and speed.

3 Basic Concepts in Current DNA Sequencing

3.1 Nested Sets

The basis of all currently applied rapid sequencing techniques is the preparation of mixtures of derivatives of the DNA to be sequenced which are defined as 'nested set'. A nested set is a collection of single-stranded subfragments of a DNA molecule. The collection contains all possible chains starting at one end of the chain and extending to every position at which a particular nucleotide occurs. Hence, all subfragments have one terminus in common (called the 'fixed end'). Conversely, the second terminus ('variable end') is generated as a function of one of the nucleic acid bases (see Fig. 1). A T-specific nested set, for instance, consists of all single-stranded fragments that one can deduce from the same end in the DNA up to, or exactly past, each of the Ts that follow in the chain. Consequently, there are two kinds of nested sets, dependent on whether the presence or the absence of a given nucleotide at the variable end is the guiding principle. The former kind is used experimentally in enzymatic methods, whereas the latter is used in the chemical degradation technique, as described below. An optimal T-specific nested set is obtained when all T-positions in the fragment are equally represented as variable end in the reaction mixture. Fractionation of the mixture on a high-resolution, denaturing polyacrylamide gel will then produce a uniform band pattern, named *'ladder'* (Sect. 3.2 and Figs. 2 and 3), that corresponds to the relative positions of all Ts along the chain. If four nested sets, each specific for one of the four bases, are prepared and four corresponding ladders are run alongside each other on the gel, then the base sequence can be determined from the changing position of the band from one lane to the other. This is named 'reading' of the gel (in fact, of the DNA sequence) since one actually visually locates the consecutive bands on an autoradiographic picture of the gel.

How one generates nested sets depends on the particular sequencing technique. Nested sets can be obtained by (enzymatic) synthesis (Fig. 2 and Sect. 4.1: dideoxy chain termination method), or by partial chemical degradation of a DNA fragment (Fig. 3 and Sect. 4.2: chemical degradation method).

3.2 Polyacrylamide Gel Electrophoresis

DNA sequencing requires a separation method capable of distinguishing between long oligomers differing by only a single nucleotide in length. Actually, four samples, i.e., four nested sets with different specificity, must be separated simultaneously to permit accurate comparison between the ladders. Hence, a device permitting parallel separations is needed. Slab gels fulfill this requirement, and cross-linked polyacrylamide gels had been used extensively for fractionation of complex mixtures of RNA and DNA fragments in the early techniques for nucleic acid sequencing (DE WACHTER and FIERS, 1971). It was shown in the 'plus and minus' technique of SANGER and COULSON (1975) that in such semi-solid matrices single-stranded DNA fragments migrate according to molecular mass with single-nucleotide resolution up to relatively long chain lengths. To date, the same matrix is still used, although the general features of a sequencing gel have been dramatically improved to better master physico-chemical uniformity of the gel, migration of the fragments, and resolution between the bands.

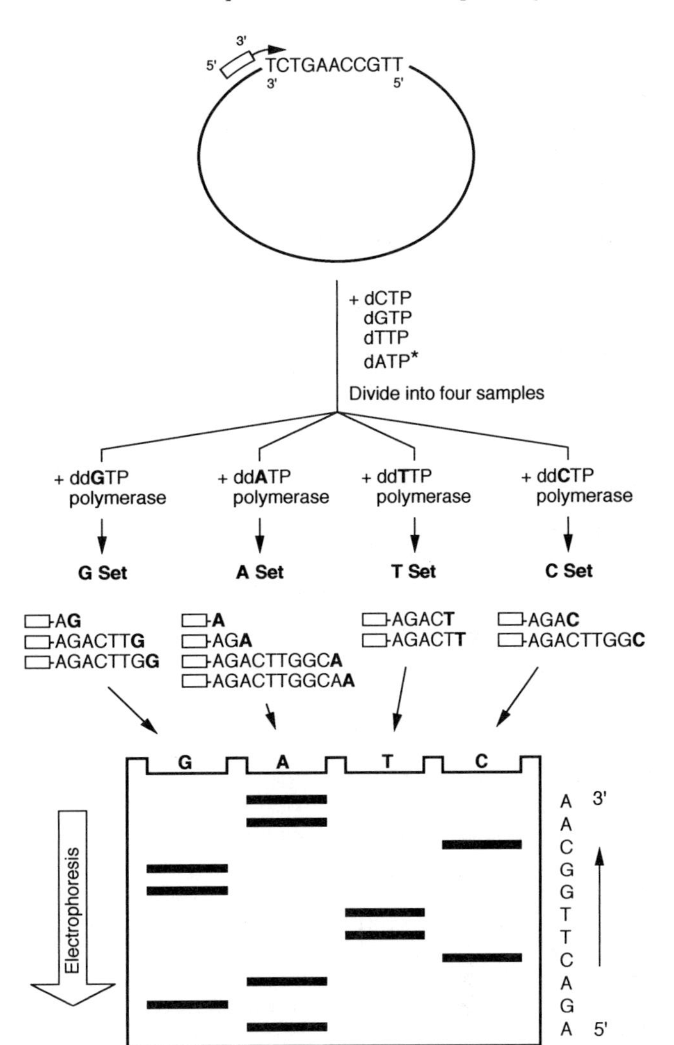

Fig. 2. Principle of dideoxy chain termination sequencing (SANGER et al., 1977).
A primer oligonucleotide, represented by an open bar, is annealed to a single-stranded template and extended (in the direction marked by the arrow-head) in the presence of dNTPs and ddNTPs. The latter are shown in boldface. One of the dNTPs is radioactively labeled in the α-phosphate moiety as represented by the asterisk (*).
The composition of the nested sets is shown. The products are fractionated in adjacent lanes on a polyacrylamide gel and the sequence deduced from the consecutive shift of the label from one lane to the other. The sequence is read from 5′ to 3′, from bottom to top, and shown at the right hand side. Note that the sequence read is complementary to the single-stranded template.

Nested sets contain single-stranded fragments which must be efficiently denatured to warrant migration of the fragments in true order of length. Urea is added to the gel for this purpose, and the Joule heating of the gel, caused by the high electrical field, also contributes to denaturation. Formamide can replace urea partly or completely to help resolving regions with strong intramolecular secondary structure (FRANK et al., 1981; FRANK and VOLCKAERT, unpublished results). It did, however, not find general application because it gives poorer resolution of the bands at longer distances in the gel.

Other variables pertinent to the resolution of the single-stranded DNA fragments are the percentage of acrylamide, the ratio to the cross-linker bis(acrylamide), the thickness of the gel, and the uniformity of heat distribution. Thin gels were introduced by SANGER and COULSON (1978). Thermostatically controlled gels were devised by GAROFF and ANSORGE (1981). Since migration of the fragments depends on the logarithm of their size, spacing between consecutive bands diminishes as the size increases and larger fragments are compressed. To improve spatial distribution, BIGGIN et al. (1983) developed buffer gradient

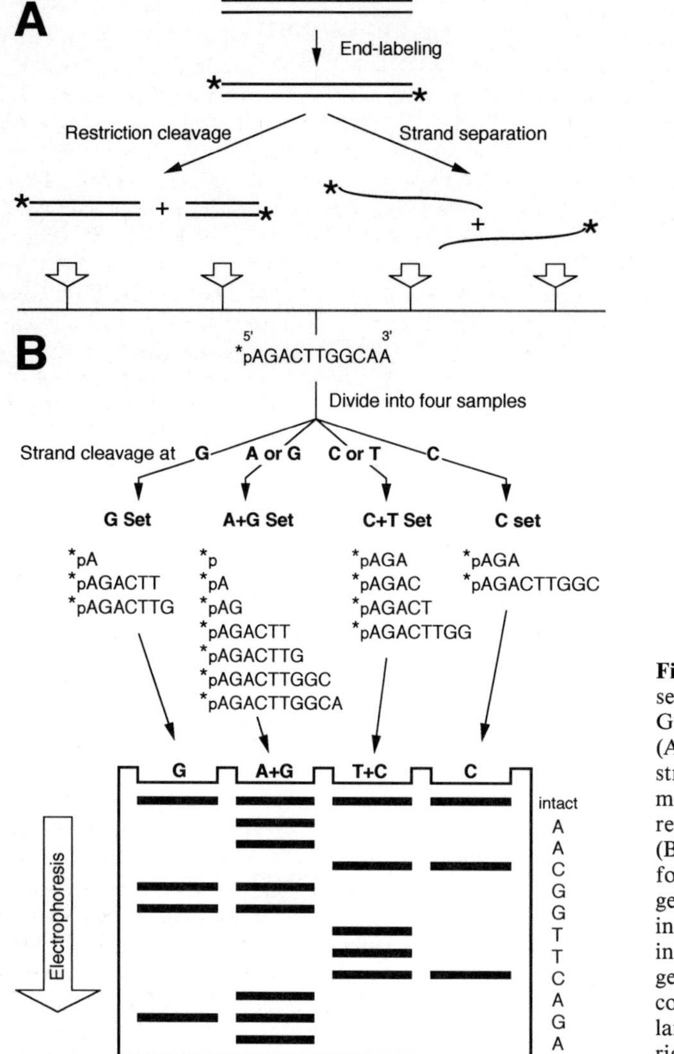

Fig. 3. Principle of chemical degradation sequencing according to MAXAM and GILBERT (1977, 1980).
(A) shows a strategy to prepare a single-stranded or double-stranded DNA fragment labeled at one end. The label is represented by an asterisk (*).
(B) Each of the fragments is treated with four base-specific chemical reagents to generate the corresponding nested sets as indicated. The products are fractionated in adjacent lanes on a polyacrylamide gel. The sequence is deduced from the consecutive shift of the label from one lane to the other(s) and shown at the right hand side.

gels in which longer fragments can move through a greater length of the gel thereby increasing the resolving power. From a typical sequencing gel of this kind the sequence of 250 to 400 nucleotides can be deduced. Other improvements came recently with chemically modified acrylamide monomers and other cross-linkers than N,N′-methylene-bis(acrylamide) (GELFI et al., 1990).

Detection of the nested sets has depended on radioactive labeling of the fragments for more than a decade. This was basically a matter of sensitivity since often less than 1 pmol of DNA template is processed. Moreover, some techniques generate several nested sets simultaneously (e.g., chemical degradation: see Sect. 4.2), and only one of those should be detected at a time. Radioisotopic labeling is now increasingly being replaced by fluorescent tagging and other techniques, which are described in later sections.

3.3 Impact of DNA Cloning on Sequencing Technology

One of the factors that delayed the development of techniques for DNA sequencing was the difficulty of obtaining pure DNA in sufficient amounts (Sect. 2.1). The discovery of restriction enzymes which cleave DNA at specific sequences allowed discrete samples to be isolated from larger DNAs. These fragments were used as primers in repair synthesis techniques with bacteriophage single-stranded DNA as template (e.g., the ϕX174 genome: SANGER et al., 1977b), or as targets to be sequenced (e.g., chemical sequencing of SV 40 DNA fragments: FIERS et al., 1978). Ordinarily, however, restriction fragments are present in complex mixtures and/or available in too limited quantities. DNA cloning provides a method for purification and *in vivo* amplification of those fragments. More recently, *in vitro* methods for amplification of specific DNA segments are replacing much of the cloning techniques to prepare DNA fragments for sequencing (see Sects. 3.4 and 4.4.3). The flow chart in Fig. 4 summarizes current routes to acquire sequence data.

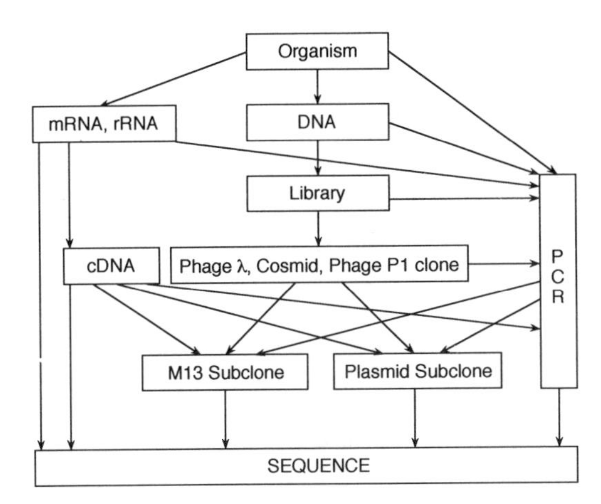

Fig. 4. Flow chart of major routes to acquisition of sequence data.
The arrows indicate phases which might be involved in isolation, cloning, subcloning and/or *in vitro* amplification (PCR) of target nucleic acid before sequences can be determined.

There are two levels at which DNA cloning has had a major impact on sequencing. First, a rather indirect effect, genome DNAs must be reduced to sizeable fragments for further manipulation. The sizes of those fragments may be several kb. The purpose of this step is to carve up a genome into large segments while keeping the region(s) of interest on a single or at most a few segments. At the same time, the region is transferred to *E. coli*, a suitable host for *in vivo* amplification of the DNA to large quantities. Vectors derived from phage lambda (BLATTNER et al., 1977; FRISCHAUF et al., 1983) and P1 phage (STERNBERG, 1990), and cosmid vectors (COLLINS and HOHN, 1978) permit large genomic segments to be cloned in *E. coli*. Transcribed sequences (cDNAs) are usually cloned in plasmid or phage vectors. A discussion of these methods is beyond the scope of this chapter, and details may be found in Chapter 10.

A second level of impact is the subcloning of those genomic or cDNA segments for subsequent sequencing, i.e., bringing each region of a few hundred bp with one of the ends within the resolution distance of a sequencing gel and thus within the reach of a sequence reading. A variety of strategies have been developed for this purpose. They can be divided into two main groups. One is called 'shotgun' cloning and consists of cleaving the DNA to small fragments and sequencing the fragments after cloning, regardless of their position and orientation in the original DNA. This method became particularly attractive when it was found that DNA can be fragmentated in a virtually random manner by sonication (DEININGER, 1983). A second group of strategies are procedures in which overlapping subclones are generated in a systematic and directional manner. The conceptual difference between the two groups of strategies is that in the former overlapping between gel readings is the result of, and hence realized after, sequencing, whereas in the latter overlapping is envisaged prior to sequencing so that fewer clones must be subjected to manipulation to nested sets. Some of the methods are compared and outlined in more detail in Sect. 5.

An early strategy devised for sequencing the ϕX174 genome used the single-stranded virion DNA as template and suitable restriction frag-

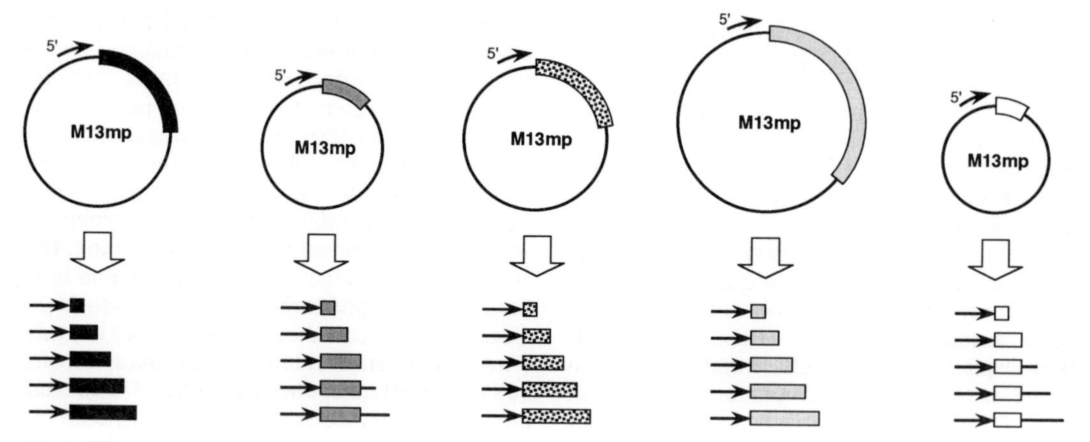

Fig. 5. The concept of subcloning in M13 vectors and sequencing with a 'universal' primer. Different DNA fragments (with different sizes) are cloned in the M13mp vector. The inserts are represented by different kinds of bars. The small black arrow is the (universal) primer sequence, while the large open arrows mark the generation of (one of) the nested sets obtained by enzymatic extension of the primer.

ments derived from the double-stranded replicative form as primers. This approach is obviously very limited since DNA molecules are usually double-stranded. A generally applicable extension of this strategy came with the development of (sub)cloning systems by which single-stranded DNA can be readily isolated in a similar fashion. Phage ϕX174 is difficult to turn into a cloning vector. However, another group of *E. coli* phages, filamentous phages (WEBSTER and CASHMAN, 1978), among which fd, f1, and M13 are prominent members, are better suited for this purpose since (1) they do not lyse the host, but continuously excrete viral particles, (2) their genome is single-stranded and has no strict size limit, (3) the genome is converted in the host to a double-stranded, plasmid-like, replicative form (RF) with which conventional cloning techniques can be applied to insert DNA fragments. It was shown by SCHREIER and CORTESE (1979) that single-stranded M13 DNA, isolated from virions, is suitable for dideoxy sequencing. By the same token, the use of a single primer for sequencing was demonstrated, regardless of the cloned insert (Fig. 5) (SANGER et al., 1980). MESSING and collaborators (MESSING et al., 1977, 1981; GRONENBORN and MESSING, 1978; MESSING and VIEIRA, 1982; YANISCH-PERRON et al., 1985) developed a series of vectors known as the M13mp series which became the major vehicles for subcloning for sequencing, in particular by shotgun approaches, and have had a tremendous impact on dideoxy chain termination sequencing. Other M13 derivatives were designed by, e.g., BARNES and BEVAN (1983) for sequencing by a systematic deletion approach.

M13mp vectors contain a portion of the *E. coli lacZ* gene including the operator, promoter, and 145 amino acid residues of the β-galactosidase (named the α-peptide). This part of the β-galactosidase is sufficient to complement a defective β-galactosidase encoded by the F sex factor in specialized, engineered host strains. A complementation assay is based on association of these polypeptides in cells infected by the M13 vector. The plaques display a blue perimeter on growth media containing isopropyl-β-D-thiogalactopyranoside (IPTG) as inducer, and 5-bromo-4-chloro-3-indolyl-β-galactoside (BCIG or X-gal) as substrate. Unique cloning sites are available in the *lac* region of the vectors in a strategic position where insertion of DNA fragments inactivates the α-peptide gene. By the lack of α-peptide, recombinant plaques remain colorless. Cells infected with individual recombinant phage are grown. Cells are discarded by centrifugation, and phage is collected from the supernatant. Pure single-stranded template DNA is then readily isolated by treatment of the phage with phenol to remove the protein coat.

A further extension of this technology for making single-stranded template DNA came with the development of plasmid derivatives, known as phasmids, fasmids, or phagemids. These were created by inserting the 'intergenic' region (where the *cis* functions for replication and packaging are located) of one of the single-stranded phages fd, f1, or M13 into multicopy plasmids such as the pUC class, so that they replicate in a single-stranded mode in *E. coli* host cells infected with a helper phage. The cells will not only secrete viable phage, but also particles containing one strand of the phasmid or recombinant derivative. These particles are a source of pure single-stranded DNA for sequencing.

For chemical sequencing discrete DNA fragments are needed, which can be labeled at one end. Since restriction cleavage is the major, if not sole source of suitable fragments, DNA cloning has been of overwhelming importance for the preparation of large amounts of DNA restriction fragments. Small plasmids replicating to high copy numbers are ideal vectors. In particular instances, specialized plasmids were designed which made the end-labeling strategy easier and applicable to large numbers of clones (FRISCHAUF et al., 1980; RÜTHER et al., 1981; VOLCKAERT et al., 1984a; and Sect. 4.2.2).

Another, more recent, sequencing technique which heavily relies on DNA cloning, is multiplex sequencing. It is described in Sect. 4.4.2.

3.4 Impact of the Polymerase Chain Reaction on DNA Sequencing

The polymerase chain reaction (PCR) is an enzymatic method for *in vitro* amplification of specific DNA fragments (SAIKI et al., 1985, 1988). This technique relies on the use of two oligonucleotides that are designed to hybridize with opposite DNA template strands and point to each other. Repeated cycles of denaturation of the DNA strands, annealing of the oligonucleotides primers, and DNA polymerase-catalyzed repair synthesis lead to an exponential increase of the target DNA.

PCR has altered many concepts and strategies in recombinant DNA technology. In the same way as restriction enzymes and DNA ligase were instrumental to the development of the first generation techniques for *in vitro* recombination of DNA fragments, oligonucleotide primers (replacing the restriction enzymes) and thermostable polymerase (replacing DNA ligase) are at the heart of a new era of DNA manipulation technology. Restriction recognition sites are substituted by the primer annealing sequences, and the multitude of specificities depends on chemical DNA synthesis. The importance of fast, reliable, cheap, automated DNA synthesis techniques and automated machines in this respect is undisputable. Although PCR is generally considered as a powerful analytical method, its real primary property is DNA synthesis, a property it shares with enzymatic DNA sequencing. The impact of PCR on DNA sequencing methods much resembles the previous impact of DNA cloning on those methods. Its influence is severalfold: (1) as an aid to cloning followed by sequencing of clone DNA (SCHARF et al., 1986); (2) as a substitute for cloning in template preparation (Sect. 4.4.3); (3) as direct sequencing approach, since Taq polymerase accepts dideoxynucleoside triphosphates as precursors (Sect. 4.4.3). Moreover, repair-termination protocols at high temperature give a distinct advantage to sequence G+C-rich DNA that often displays a high degree of intramolecular secondary structure.

The only requirement for PCR, and hence its limitation, is that the nucleotide sequences at both sides of the DNA to be sequenced are known, i.e., the priming regions (see Sect. 5.3 for procedures to circumvent this limitation). DNA fragments which previously had to be cloned in order to generate sufficient amounts of template, can now be prepared by PCR from very low, finite amounts of starting material. Some caution, however, is required with interpretation of sequencing data obtained by PCR. Some of the thermostable polymerases, including Taq polymerase, have no $3' \rightarrow 5'$ exonuclease (proofreading) activity and therefore are more prone to causing mutational errors than Klenow polymerase (SAIKI et al., 1988).

Taq polymerase has an observed error rate (mutations per nucleotide per cycle) of up to 2×10^{-4} (Tindall and Kunkel, 1988; Keohavong and Thilly, 1989). This will not disturbe sequencing patterns in those instances where PCR is used for the termination reactions starting from substantial amounts of template DNA, but may give rise to problems if template is amplified from extremely small amounts to detectable levels. Any synthesis error in the early cycles will co-amplify and may appear as a detectable signal on the sequencing gel. Alternatively, when this amplified DNA is purified by cloning prior to sequencing, errors may be fixed. Hence, clones from independent amplification experiments should be analyzed until consistently identical sequences are obtained.

General strategies for using PCR in sequencing are outlined in Sect. 4.4.3; approaching large targets by PCR in a systematic, directional manner is described in Sect. 5.3.

4 DNA Sequencing Techniques

4.1 Dideoxy Sequencing

In this method, developed by the pioneer in protein and RNA sequencing, Fred Sanger, DNA polymerase has a key role. The polymerization process of DNA replication can be simulated *in vitro* by the synthesis of DNA fragments. *In vivo* duplication of DNA, essential for survival of a cell, involves local melting of the two DNA strands and their enzymatic copying catalyzed by DNA polymerase. A multitude of DNA polymerases from different sources can be isolated in a biochemically pure and active form for *in vitro* use. These enzymes are usually multifunctional and have, besides polymerase, often exonuclease activities. The choice of DNA polymerase is an important aspect in current dideoxynucleotide sequencing.

In contrast to RNA polymerases, which initiate transcription of DNA at specific nucleotide sequences (*promoters*), no DNA polymerase is able to start copying *de novo*. The sole synthetic activity of DNA polymerase is the addition of a 5′-mononucleotide (i.e., a deoxynucleotide with the phosphate group attached to the 5′ position of the deoxyribose) to a recessed 3′-OH end of an incomplete DNA duplex structure. The complementary strand dictates which deoxynucleotide must be attached at each step of polymerization (Fig. 6). The mechanism of initiation of replication *in vivo* is rather complex. In dideoxy sequencing, DNA polymerization is initiated with an oligonucleotide, called the primer, that is complementary to the position where one wants to start the analysis, i.e., the fixed end of the nested set. In principle, a length of a few nucleotides is sufficient to permit initiation of DNA synthesis. However, a chain length of 15 or more nucleotides is usually required, and sufficient, to make the primer anneal to a single, selected site in the DNA. The limited availability of suitable primers was one of the factors initially delaying wide-spread application of the dideoxy-sequencing method. Small restriction fragments were used, but preparation of such fragments in suitable amounts was tedious and time-consuming. The development of fast and reliable DNA synthesis methods, followed by the development of automated machines (the so-called DNA synthesizers), that brought chemical DNA synthesis into laboratories lacking expertise in the specialized chemistry and not equipped accordingly, has reversed the situation completely. Oligonucleotide primers can now be readily made available in any designed sequence.

4.1.1 Biochemistry

The (four) deoxynucleoside triphosphates (dNTP) are the normal substrates for DNA synthesis. At each coupling step, pyrophosphate is released from the dNTP. The deoxynucleoside monophosphate is attached to the 3′-OH terminus and elongates the DNA chain by one unit (Fig. 6). At the same time, a new 3′-OH end becomes available as attachment point for the next nucleotide, since the dNTPs have a free 3′-OH. Thus, consecutive phosphodiester bonds are formed.

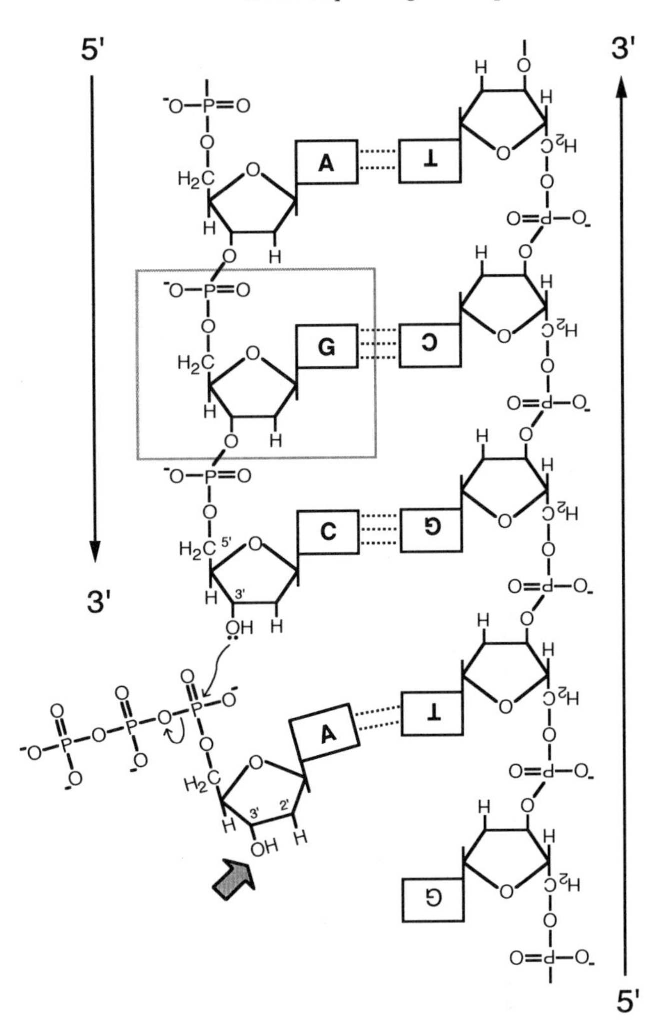

Fig. 6. Chemical structure in DNA synthesis.
The structure of a section of a partially double-stranded DNA is shown in the process of enzymatic extension. The nucleotide bases are represented by A, C, G, and T, and hydrogen bonds involved in base-pairing are indicated by dashed lines. The boxed region is a nucleotide, the basic monomer of DNA. Note the antiparallel character of the strands. An A (adenylate) residue is being added to the 3′ OH group. A is added since the base in the opposite strand is a T (thymidylate). Pyrophosphate is released by the reaction. The filled arrow marks to next 3′ OH group which is essential for further extension of the chain.

E. coli DNA polymerase I, Klenow polymerase, and many other DNA polymerases are able to use 2′–3′-dideoxynucleoside triphosphates (ddNTP) as substrate, albeit with a much lower efficiency. A dideoxynucleotide lacks the 3′-OH functional group, essential in esterification (Figs. 6 and 7). Consequently, its incorporation into a growing DNA chain prevents further elongation. In the method of SANGER et al. (1977a), four sequencing reactions are carried out in parallel, each with a dideoxy counterpart of a different dNTP. In a dideoxy-A reaction all four natural dNTPs and ddATP are present so that chain elongation is interrupted only now and then. The ratio ddATP to dATP determines the frequency of ddA incorporation, and hence, the average size of the synthesis product that always ends with an A (i.e., the variable end in the nested set). Thus, an A-specific nested set is prepared (Fig. 1A). With ddGTP, ddCTP, and ddTTP similar specific nested sets are prepared. Moreover, one of the four dNTPs is radioactively labeled in its α-phosphate moiety, either with a ^{32}P atom, as originally used by SANGER et al. (1977a), or with a ^{35}S atom by replacement of a P=O by P=S (i.e., an α-thiophosphate dNTP analog, Fig. 7) (BIGGIN et al., 1983). Consequently, the nested set products are radioactively labeled and readily detectable. The

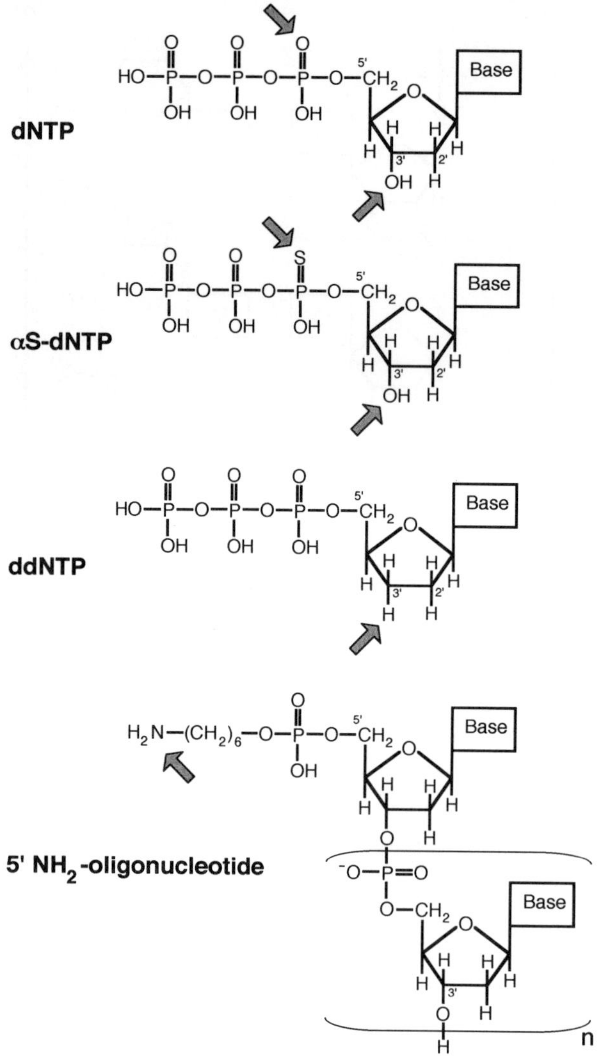

dNTP

αS-dNTP

ddNTP

5' NH_2-oligonucleotide

Fig. 7. Chemical structures of deoxynucleoside triphosphates and of a 5′-amino-oligonucleotide.
αS-dNTP and ddNTP are dNTP analogs. The relevant positions, i.e., the 3′ of the pentose sugar and the α-phosphate moiety are marked by filled arrows. In the bottom part, the structure of a 5′ amino-oligonucleotide is drawn.

reaction mixtures are fractionated as four adjacent 'termination ladders' on a denaturing polyacrylamide gel as described in Sects. 3.2 and 4.3 (Figs. 2 and 8).

A partially duplex DNA is required for template-dependent *in vitro* synthesis of DNA. When an oligonucleotide primer is annealed to single-stranded DNA, this condition is fulfilled. The 5′ end of the primer will be the fixed end of the nested set and should remain intact during the entire synthesis procedure. Therefore, DNA polymerases lacking 5′→3′ exonuclease activity are preferably used. The Klenow fragment of *E. coli* DNA polymerase I (called Klenow polymerase) (KLENOW and HENNINGSEN, 1970) has been the standard enzyme used in dideoxy chain termination sequencing for years. DNA polymerases with improved properties have been found later. Desirable properties include a low variability in termination efficiencies with ddNTPs and an ability to operate at higher temperatures to minimize the presence of polymerase-stalling secondary structures in the template. A first alternative is reverse transcriptase (ZAGURSKY et al., 1985), which can be used at slightly higher

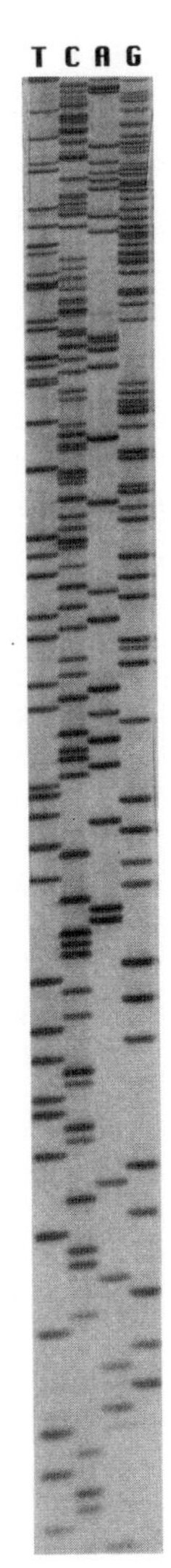

Fig. 8. Autoradiograph of a typical dideoxy chain termination sequencing gel. α-^{35}S-dCTP is used as labeled precursor. The order of the lanes is T, C, A, G. Hence, the sequence reads in 5′ to 3′ direction from bottom to top of the gel: ATCCTCTAGAGTC-GACCTGCAGTCCTTCCT etc. The top part of the gel picture is not shown.

temperatures. Bacteriophage T7 DNA polymerase is an other, extremely suitable alternative. The enzyme is highly processive, lacks 5′→3′ exonuclease activity, and gives a more uniform frequency of incorporation of dideoxynucleotides with concurrent chain termination. In addition, it seems less sensitive to secondary structure in the template (KRISTENSEN et al., 1988). In chemically modified versions of T7 polymerase, commercialized as Sequenase™, the very high 3′→5′ exonuclease activity is inactivated (TABOR and RICHARDSON, 1987a, b). The discovery and isolation of Taq polymerase and other thermostable DNA polymerases (Tab. 1 and references therein) have provided even more versatility in preparation of nested sets. On one hand, template is more efficiently denatured at the high reaction temperature, on the other hand, repeated rounds of denaturation and priming are feasible, known as cycle sequencing (Sect. 4.4.3.3), resulting in linear amplification of the extension–termination products. The gene for Taq polymerase has been cloned and expressed in *E. coli* (LAWYER et al., 1989). Undoubtedly, novel enzymes with improved characteristics

Tab. 1. List of Thermostable DNA Polymerases Used in PCR-Mediated Sequencing

Enzyme	Source	Reference	Supplier
AmpliTaq	*Thermus aquaticus*	INNIS et al., 1988	Perkin-Elmer Cetus
Taq	*Thermus aquaticus*	SHAFFER et al., 1990	Multiple suppliers
Vent and Vent (exo$^-$)	*Thermococcus litoralis*	MATTILA et al., 1991	New England Biolabs, Inc.
Pfu	*Pyrococcus furiosus*	not found	Stratagene Cloning Systems
Bst	*Bacillus stearothermophilus*	MCCLARY et al., 1991	Bio-Rad Laboratories
Tub	not specified	not found	Amersham International, plc

Several other polymerases, e.g., TetZ, Thermal SeqPol, Tth, etc., are supplied by different companies. The origin of the enzyme is not always clearly specified and literature data are rarely available.

will be developed by gene and protein engineering techniques in the near future.

A major improvement, alleviating the problem of sequence-dependent variations in termination frequencies, was the discovery by TABOR and RICHARDSON (1989) that manganese ions (Mn^{2+}) remove almost entirely the discrimination between dideoxynucleotides and deoxynucleotides by some DNA polymerases, with a consequent reduction in local peak intensity variations in fractionation ladders. This effect, unfortunately, is not observed with Taq polymerase.

Other than the standard dNTPs are sometimes used in the reactions to circumvent problems with keeping nested set products properly denatured in the sequencing gels, e.g., deoxyinosine triphosphate (dITP) (MILLS and KRAMER, 1979), 7-deaza dGTP (MIZUSAWA et al., 1986), and 7-deaza dATP (JENSEN et al., 1991).

In summary, three considerations are important in choosing a biochemistry for the preparation of a synthetic nested set: (1) Will the entire template efficiently denature and let the polymerase fill in the nucleotides? (2) Are dideoxy analogs incorporated with equal efficiency at every position? (3) Will all extension products in the nested set easily remain single-stranded or will some terminal regions tend to fold back as a loop structure (causing aberrant migration in a sequencing gel)?

4.1.2 Primers and Templates

Template DNA for dideoxy sequencing can be either single-stranded virion DNA or linearized double-stranded plasmid DNA converted to single strands with exonuclease III (SMITH, 1979; BARRELL et al., 1979). Small restriction fragments initially used as primers (ANDERSON et al., 1980) were soon superseded by chemically synthesized oligonucleotides (DUCKWORTH et al., 1981). The M13 (and phasmid) subcloning strategy provides a way to avoid the need for multiple primers, as outlined in Sect. 3.3 and Fig. 5. The M13 primer is designated 'universal primer' (SANGER et al., 1980), since it can be used for sequencing any clone, in any of the M13mp vectors. Several cloning sites, or series of cloning sites (arranged in a so-called polylinker), are available in M13mp2 to M13mp19. By the expansion of chemical DNA synthesis facilities, a variety of synthetic 'universal' primers homologous to different regions surrounding the polylinker region are now commercially available. Any other primer sequence may be synthesized, and it is possible to design, e.g., a new primer complementary to the template at the limit of a determined sequence to further extend the known sequence (STRAUSS et al., 1986). This is called 'primer walking sequencing'.

Synthetic primers, in contrast to restriction fragments, are usually complementary to only one strand of a DNA molecule. Hence, dideoxy sequencing with double-stranded DNA is feasible, in as far as the complementary strand does not prevent annealing of the primer to the template. Double-stranded DNA is denatured by, e.g., heating or alkaline treatment. This permits sequencing reactions to be performed on any double-stranded DNA, circular or linearized, but the best results are obtained with denatured, supercoiled DNA

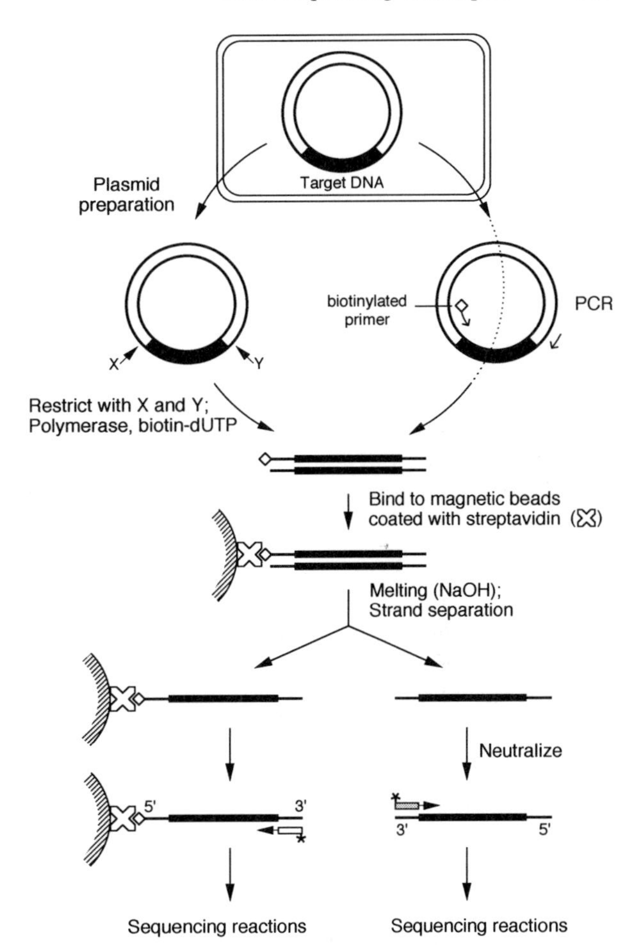

Fig. 9. Preparation of single-stranded DNA and solid-phase sequencing according to HULTMAN et al. (1989). A double-stranded DNA fragment is prepared by restriction digestion (left side) or PCR amplification (right side). A biotin group (shown as a diamond ◇) is incorporated at one of the ends by filling-in of an appropriate 5′ end with a biotin derivative of dUTP (left side) or by using one biotinylated primer in the PCR reaction (right side). Following binding to the magnetic beads (dashed area), separation of the strands by denaturation, and isolation of the beads, sequencing of the DNA is performed with primers annealing to the 3′ extremities of the single strands. Label or tag on the primers is marked by an asterisk (*). Note that the DNA strand at the left hand side remains bound to the beads during the sequencing reactions.

(CHEN and SEEBURG, 1985). Sequencing patterns obtained from M13 or phasmid single-stranded subclone DNA are usually of superior quality, when compared to those obtained from double-stranded systems, but the latter offer the advantage that no special subcloning or purification step is required and that the second strand is accessible for priming in the opposite orientation, with a 'reversed' primer.

A novel, general approach to the preparation of single-stranded DNA for sequencing was devised by HULTMAN et al. (1989). It is based on (1) the use of magnetic beads coated with streptavidin, and (2) the introduction of a biotin moiety in one of the DNA strands. The latter is done, either by filling in a 5′ protruding single-stranded end of a restriction fragment with biotin-dUTP and DNA polymerase, or by using one conventional and one biotinylated primer when amplifying a DNA fragment by PCR (Fig. 9 and Sect. 3.4). The DNA fragment is immobilized on the beads by virtue of the extremely stable interaction between streptavidin and biotin. The stability of the biotin–streptavidin complex allows to perform DNA manipulations, such as melting of the DNA strands with alkali or by heating, and elution with formamide, without destabilizing the complex. Thus, the DNA strands are separated by alkaline denaturation. The strand bearing the 5′ biotin group remains bound to the beads when these are collected and removed from the solution by a magnetic device. The other strand remains in solution and is sequenced after adjusting the pH. The former strand is sequenced while remaining attached to the beads. This is called 'solid phase DNA sequencing'.

4.2 Sequencing by Partial Chemical Degradation

MAXAM and GILBERT (1977) developed a different strategy for preparation of nested sets. The general principle of this approach is partial chemical degradation of a discrete, linear DNA molecule, usually a restriction fragment, and depends on reactions that specifically modify the different purine and pyrimidine bases. The extent of reaction is carefully limited to less than one base per molecule. Modified bases are destroyed under conditions that preserve the glycosidic bonds between unmodified bases and deoxyribose. Eventually, the unstable phosphodiester bonds at the sites of base destruction are hydrolyzed while leaving the other phosphodiesters intact.

A DNA strand, processed in this manner, is cleaved into two pieces. Similarly, a double-stranded DNA fragment is degraded to a mixture of subfragments, which can be seen as four collections, each retaining one of the four original termini (i.e., the two 5′ and two 3′ ends) as fixed end in one of four nested sets (Fig. 1B and C). To visualize one of these nested sets, the original fragment is radioactively labeled at the (5′ or 3′) terminus of choice (Sect. 4.2.2). High resolution polyacrylamide gel electrophoresis of the nested set is done as in dideoxy sequencing, but the chemical degradation ladder represents the positions of the nucleosides that were destroyed (hence, the absence of the target bases) (Figs. 3 and 10).

4.2.1 Chemistry of Base-Specific Degradation of DNA

A large number of chemical agents react with one or more of the nucleotide bases in DNA. These reactions result in modification of the base or change the ring structure, e.g., by opening or by increasing or decreasing the number of ring atoms. Many of those modifications influence the stability of the glycoside bond unfavorably. Hence, under suitable reaction conditions, usually heating the modified DNA in molar piperidine at 90 °C for 30 minutes, the modified base is released and the DNA strand cleaved without damaging any of the other phosphodiester bonds. The result of these manipulations thus is a base-specific fragmentation.

Sequencing is accomplished by dividing the target DNA into (at least) four subsamples and then treating the subsamples with a series of different base-specific chemical reagents that partially cleave the DNA. Conditions are adjusted for low levels of modification ('single hit per molecule conditions') and complete cleavage at the modified positions. Obviously only those reactions are suited that partially cleave the DNA in a strictly random fashion, so that each position within the respective required specificity is represented in the nested set, and result in uniform intensities in the separation pattern (see below). Although there are base-specific reactions available for all four bases, more typically single-base specific reactions are used in conjunction with purine- or pyrimidine-specific reactions (Fig. 3).

Ordinarily, a first sample is treated with dimethyl sulfate, which methylates a few percent of the guanines at position N^7. Subsequent treatment with piperidine leads to displacement of the ring-opened, methylated G and thereby cleaves the DNA at these sites. Although dimethyl sulfate also methylates adenine at position N^3, this modification does not lead to instability of the strand at this site under the conditions of piperidine treatment. A second sample is treated with formic acid, which protonates the N^7 of purines, followed by displacement of the affected purines and strand cleavage with piperidine. A third and fourth sample of the DNA is treated with hydrazine. Hydrazine destroys pyrimidine rings and leaves derivatives which are displaced by piperidine and lead to strand scission. In the third sample the reaction is conducted in high ionic conditions (1 mol/L NaCl), which suppresses reaction of hydrazine with thymine. In the fourth sample, in the absence of salts, both pyrimidines react so that a T+C specific nested set is obtained. (For further details on the nature of the chemical reactions, consult the review by MAXAM and GILBERT, 1980).

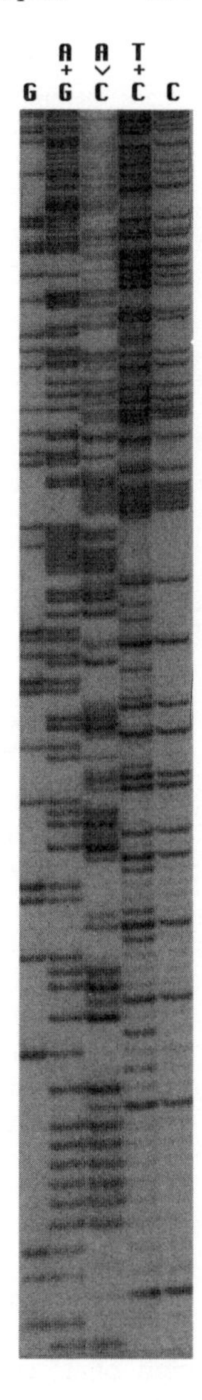

Fig. 10. Autoradiograph of a typical chemical degradation sequencing gel. DNA is labeled at the 3′ end with α-^{35}S-dCTP as shown in Fig. 11. The order of the lanes is G, A + G, A > C, T + C, C. Since label is introduced at the 3′ end, the ladder reads in 3′ to 5′ direction from bottom to top of the gel: AGCGGAAAAAACATGTA-CAAGTCTGGTCACAAGCCT etc. Only part of the gel picture is shown.

Among many other optional reactions (FRIEDMANN and BROWN, 1978; RUBIN and SCHMID, 1980; SAITO et al., 1984; AMBROSE and PLESS, 1985; DAVIES et al., 1990; ROSENTHAL et al., 1990a), two are quite frequently used: molar NaOH at 90 °C and $KMnO_4$. The former reacts primarily at adenine and to a lesser extent at cytosine residues producing a A>C-nested set, whereas the latter can be used as a thymine-specific agent, producing a T-nested set.

Preparation of nested sets by chemical degradation reactions involve several steps. To facilitate processing of the samples, solid phase procedures have been developed. Several supports were introduced with varying success. The one suggested by ROSENTHAL et al. (1986), an anion-exchange matrix (CCS) is commercially available and has also been used in automated systems (ROSENTHAL et al., 1990b).

4.2.2 End-Labeling of DNA

Chemical degradation of a DNA fragment does not generate unique nested sets (Fig. 3).

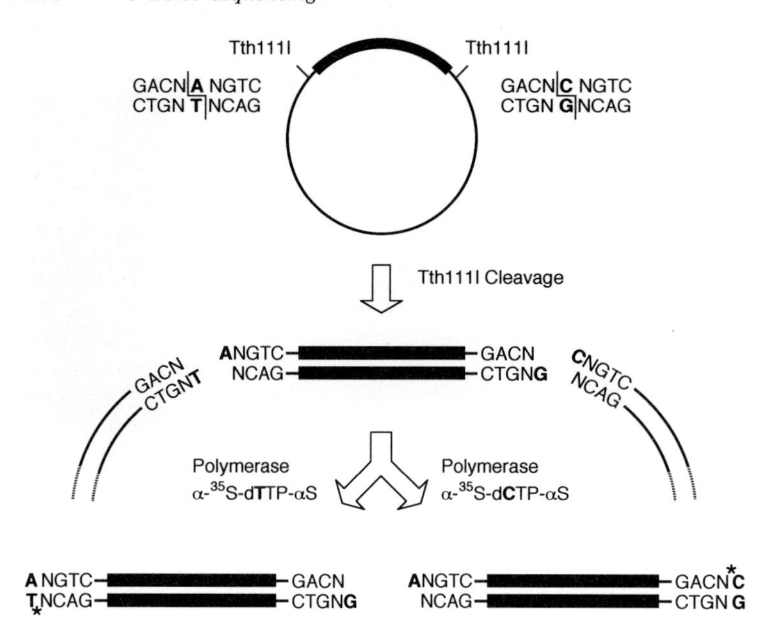

Fig. 11. Direct single-end labeling according to VOLCKAERT (1987). The cloning site of the vector is flanked by two *Tth*111I restriction sites that generate different single-base 5′-extensions upon cleavage. Inserts (filled bars) can thus be labeled at either end by choosing an appropriate α-^{35}S-labeled dNTP precursor for filling-in polymerization. The labeled DNA is directly amenable to subsequent chemical degradation reactions for sequencing. The radioactive label is marked by an asterisk (*).

To discriminate one of the collections against the other, MAXAM and GILBERT (1977, 1980) attached a specific radioactive label to one of the fixed ends (either a 5′ or 3′ end), prior to the base-specific chemical treatments. Those procedures are summarized in VOLCKAERT et al. (1984b). Labeling at 5′ ends is accomplished by phosphorylation with polynucleotide kinase and γ-^{32}P-ATP, after removing the unlabeled terminal phosphates with alkaline phosphatase. Labeling at the 3′ ends is feasible by filling in of recessed 3′ ends, or exchange of the ultimate nucleotide in blunt ends, with DNA polymerase and an appropriate α-^{32}P-labeled dNTP as precursor. Alternatively, any kind of 3′ end may be elongated by a single (labeled) residue with terminal transferase and α-^{32}P-cordicepin triphosphate.

The major drawback with those techniques is that either both 5′, or both 3′ ends become labeled so that an additional step is required to segregate the labels (Fig. 3). This is done by cleavage of the DNA fragment with a restriction enzyme, and fractionation and purification of the subfragments, or by separation of the DNA strands and isolation of the individual strands on a polyacrylamide gel. Those additional manipulations are one of the factors causing chemical degradation sequencing to be considered slow, tedious, and time-consuming.

Much improvement has been obtained by special (sub)cloning approaches (FRISCHAUF et al., 1980; RÜTHER et al., 1981; VOLCKAERT et al., 1984a; ECKERT, 1987; ARNOLD and PÜHLER, 1988). With the high copy number miniplasmids described by VOLCKAERT (1987), direct single-end labeling of subclone DNA is achieved (Fig. 11). A unique asymmetric restriction site is located in the vector, close to a polylinker where fragments are inserted. Cleavage of the plasmid DNA with the restriction enzyme generates a linear molecule with different ends, one of which is filled in by simple addition of an appropriate dNTP and polymerase. ^{35}S-nucleotide analogs are used for labeling and provide the same advantages as described for the use of this isotope in dideoxy sequencing (BIGGIN et al., 1983) (Sects. 4.1.1 and 4.3.1/2).

4.2.3 Limitations

In spite of the improvements by solid phase methods and subcloning approaches, chemical degradation sequencing remains slower and requires more manipulations than dideoxy sequencing with optimized protocols. The uniform representation of fragments in the chemical nested sets, and hence the reliability of sequence interpretation on sequencing gels was a long standing advantage, although similar results can now be obtained in dideoxy sequencing as well (Sect. 4.1.1).

Since essentially salt-free, dried samples are obtained for gel loading, strong secondary structure in the nested sets is more readily denatured. Moreover, artefacts due to polymerase stops in enzymatic copying are eliminated. Hence, some highly repetitive or G + C-rich sequences which fail to be unambiguously determined by dideoxy sequencing, often can be resolved by chemical sequencing.

The chemical degradation sequencing approach is, however, technically fundamentally different from the enzymatic approach. In the latter, a series of similar biochemical reactions is carried out with a single substrate as varying component. The former, however, requires several, entirely different chemical methods, which are less obvious to incorporate in facile, systematic, large-scale, routine procedures.

Nevertheless, chemical sequencing remains indispensable for direct 'genomic sequencing' (CHURCH and GILBERT, 1984; see Sect. 4.4.1) which is a method to analyze the structural characteristics of genome DNA which are lost by cloning, e.g., the methylation state of single bases in a given gene (NICK et al., 1986; SALUZ and JOST, 1986).

4.3 Fractionation of Nested Sets and Detection

Apart from a method to prepare nested sets, current DNA sequencing techniques require (1) a separation method capable of distinguishing between long oligomers differing by only a single nucleotide in length, and (2) a method of detection. In each case, performance criteria prevail, and both aspects of sequencing underwent dramatic changes in recent years. Nevertheless, polyacrylamide gel has not yet been superseded as matrix for separation of nested sets, but the slab gel configuration can be replaced by capillary gels when applying methods based on multi-wavelength fluorescent detection (Sect. 4.3.2.2) of the nested sets from a single lane (COHEN et al., 1990; DROSSMAN et al., 1990; LUCKEY et al., 1990; SWERDLOW and GESTELAND, 1990; SWERDLOW et al., 1990, 1991; KARGER et al., 1991).

Sequencing gels must be run at elevated temperature to warrant efficient melting of potential secondary structures in the DNA fragments. A temperature increase occurs by Joule heating from the electric field without special device. Uneven warming of the gel, however, may cause distortions, e.g., the so-called 'smiling' effect, if fragments migrate slower at the border sides than in the middle of the gel. Thermostatically controlled gels were devised by ANSORGE and collaborators (GAROFF and ANSORGE, 1981). Thin gels (SANGER and COULSON, 1978; ANSORGE and DE MAEYER, 1980; STEGEMANN et al., 1991) result in smaller temperature gradients across the gel thickness and can be run in substantially shorter running times with improved resolution. Extremely thin sequencing gels (as thin as 25 μm), in horizontal set-up withstand electric fields up to 250 volts per cm and permit a much faster throughput of readable bases (BRUMLEY and SMITH, 1991).

4.3.1 Post-Electrophoresis Visualization versus Real-Time Detection

Originally, nested sets were radiochemically labeled (Sect. 4.3.2.1), and gels were run during a fixed period of time to allow the smaller fragments to reach the bottom side of the gel. The electric power was then disconnected and the gel mold dismantled to allow autoradiography. This is still a common practice. The gel can be covered with a polyethylene foil and directly exposed to an X-ray film, but usually it is rinsed with acid to remove urea while fixing the DNA, and then dried (an essential step when the DNA is labeled with ^{35}S). Dry gels give sharper bands on an autoradiograph with increased sensitivity and resolution. To avoid distortion by any of the gel processing steps, the gel can be chemically bound to one of the glass plates during polymerization of the acrylamide (GAROFF and ANSORGE, 1981). Alternatively, it can be transferred to filter paper (BIGGIN et al., 1983) after electrophoresis and prior to rinsing and drying.

The spatial distribution of the fragments along the path of migration is non-linear, and discontinuous gels (with gradients of ionic strength or gel thickness) have been introduced that give a more linear relationship (BIGGIN et al., 1983; OLSSON et al., 1985), with a substantial increase of the readable distance from a single run.

Exposing the gel to an X-ray film produces an image of the fractionation pattern at the time that electrophoresis was stopped, and thus resembles a snap-shot (Figs. 8 and 10). Interruption of electrophoresis for autoradiography is irreversible, and to resolve as many bands as possible, the same sets may have to be loaded several times and run over longer periods of time. A snap-shot approach with post-electrophoresis detection is almost inevitable when radioactive labeling is used, due to the problems of on-line detection of very small amounts of radioactivity and the exquisite sensitivity of autoradiography, which may range from hours to days or even weeks. Conversely, some strategies based on non-radioactive detection (in particular the ones relying on fluorescent tagging of the nested sets: see next section), use real-time detection as sensitivity may be provided by electronic amplification of the signals. In real-time systems, separation and detection are integrated. Data are continuously acquired and recorded, and all products travel over the same distance, which results in longer stretches of sequences that are resolved. The major disadvantage is the need of specialized expensive instrumentation (Sect. 6).

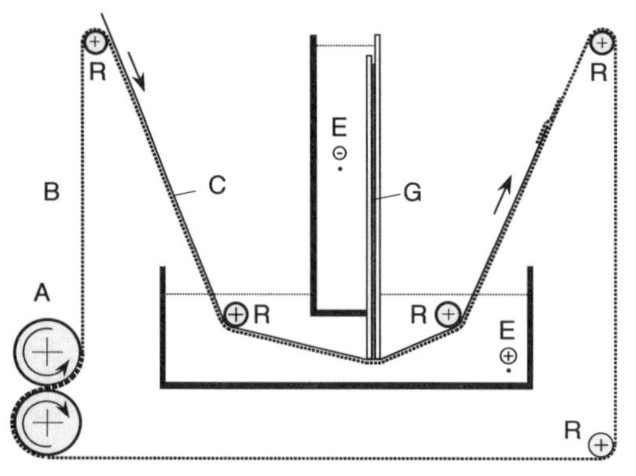

Fig. 12. Schematic representation of the direct transfer gel electrophoresis system developed by BECK and POHL (1984).
A, Motor-driven drum; R, rollers; B, supporting belt; C, moving cellulosenitrate membrane; E, electrophoresis buffer chambers; G, sequencing gel.

A special approach which is neither snap-shot nor real-time detection was devised by BECK and POHL (1984) (Fig. 12). A short sequencing gel (14 cm) is used and does not need to be dismantled for exposure. Instead, the fragments leave the matrix at the bottom of the gel and are continuously blotted and immobilized onto a cellulose nitrate membrane that moves across the bottom of the gel with a constant speed. Thus, the full separating power of the gel can be exploited. Eventually, the membrane is manually processed to visualize the bands, either by isotopic or non-isotopic methods (RICHTERICH et al., 1989).

4.3.2 Radioactive Labeling and Non-Radioactive Tagging

4.3.2.1 Radioactive Detection

Sensitivity was a paramount condition at the time when the sequencing techniques of MAXAM and GILBERT (1977) and SANGER and collaborators (SANGER and COULSON, 1975; SANGER et al., 1977a) were developed. Therefore, ^{32}P-labeling of the nested sets was used. Later, ^{35}S-labeled nucleotide analogs (Fig. 7) became available, and it was shown that ^{35}S could substitute ^{32}P in several applications. ^{35}S has a longer half-life than ^{32}P, and 10% of the energy. The latter results in sharper bands (less scattering) with better resolution, less damage from radiolysis, and considerably more safety for the experimenter. The use of radioisotope ^{33}P in dideoxy sequencing has recently been reported by ZAGURSKY et al. (1991). This isotope shows the same resolution and safety features as ^{35}S, but gives higher sensitivity.

End-labeled fragments are the usual substrates for chemical sequencing (Sect. 4.2.2). In dideoxy sequencing, labeling is ordinarily obtained by using an appropriate labeled dNTP and an unlabeled primer. Conversely, a pre-labeled primer can be used in combination with unlabeled nucleoside triphosphates. Most of the non-radioactive detection systems that

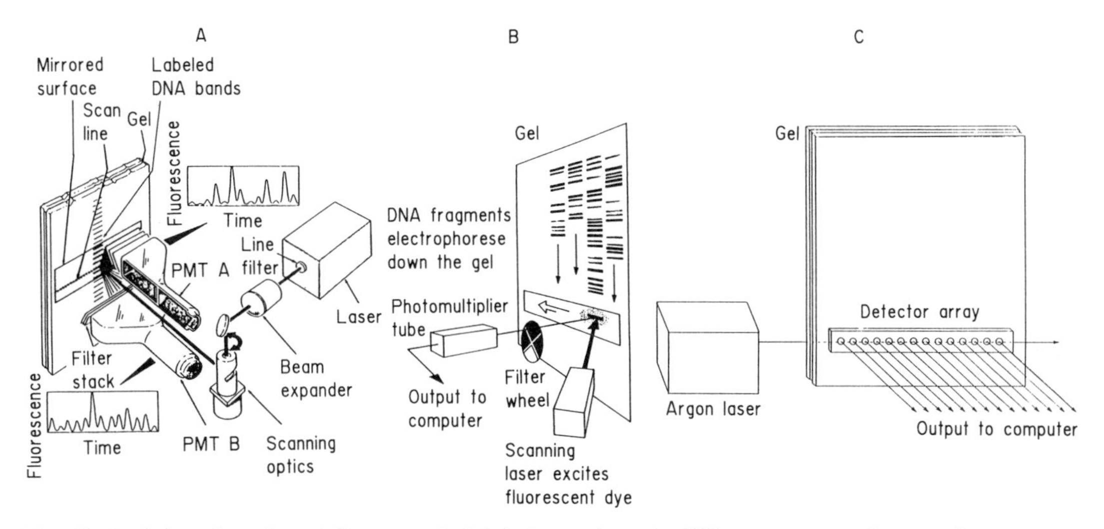

Fig. 13. Real-time detection of fluorescently labeled nested sets in different automated sequencing systems. Relative position of the separating gel, laser light source and capturing photomultiplier(s) in three automated DNA sequenators.
(A) Genesis2000, DuPont de Nemours & Company, Inc. (reprinted from PROBER et al., 1987, with permission; copyright 1987 by the AAAS).
(B) ABI 373A, Applied Biosystems Inc. (reprinted from CONNELL et al., 1987, with permission).
(C) A.L.F., Pharmacia-LKB AB (through the courtesy of M. VAN RANST and P. FITEN, Rega Institute, University of Leuven).

were developed during the past decade, adopted the latter approach.

4.3.2.2 Fluorescent Detection

Non-radioactive labeling is often called 'tagging'. Fluorescent tagging for sequencing can be realized by the attachment of a dye compound to a 5′-amino-oligonucleotide primer (i.e., an oligonucleotide with a 5′ terminal NH_2 instead of OH: Fig. 7) (SMITH et al., 1985) or to the bases of the dideoxynucleoside triphosphates (PROBER et al., 1987). For the use of these chromophoric substrates in dideoxy sequencing there exist three different methods:
(1) In the method of SMITH et al. (1986), four different dyes were chosen for the tagging of the primers: fluorescein, 4-chloro-7-nitrobenzo-2-oxa-1-diazole (NBD), Texas red, and tetramethylrhodamine, which are distinguishable by their emission spectra. Each dye primer corresponds to one base in four nesting reactions. The four reactions are mixed and run in a single lane, as the nested sets are distinguished and identified by different excitation and fluorescence emission wavelengths of the dyes (Fig. 13). An important prerequisite is that the tags should not have dramatically differential effects on electrophoretic mobility of the fragments.
(2) In a second approach, developed by ANSORGE et al. (1986), the potential problem of differential mobility of the dyes is avoided by using a single dye-labeled primer, with fluorescein as fluorophore. As a consequence, the setup requires four lanes in parallel as in a conventional strategy.
(3) The third approach, developed by PROBER et al. (1987), is based on the use of an unlabeled primer in combination with four different dye-labeled dideoxynucleoside triphosphates. In the original report, four succinylfluorescein dyes with closely spaced emission bands were used, but ddNTPs labeled with other fluorophores are now available (see Sect. 6). This approach has several distinct advantages: a single reaction produces four nested sets, any primer can be used, and 'false stops';

i.e., termination not caused by incorporation of a dideoxy analog, are not detected because the fragment remains unlabeled. A critical question, however, is whether a DNA polymerase will recognize with good efficiency and reliability the precursors which are modified in both the base and pentose moieties. This needs to be explored for each kind of ddNTP fluorophore.

In those three methods, the principal aim is to replace autoradiography and manual or semi-manual interpretation by real-time detection and computer interpretation. They are inherently dependent on sophisticated instrumentation, referred to as 'automated DNA sequencers' (Sect. 6).

4.3.2.3 Colorimetric Detection

Biotin-labeled dNTPs are bad substrates for incorporation by polymerase into nested sets and hence not suitable for their detection (BECK, 1990), but colorimetric detection for sequencing became feasible with the chemical synthesis of biotin-tagged primers (AGARWAL et al., 1986). As described in Sect. 4.1.2, biotin forms a stable complex with streptavidin. Biotin-tagged primer extension products are recognized by streptavidin, then localized by a sandwich complex with biotinylated alkaline phosphatase, and finally visualized by a colorimetric reaction with 5-bromo-4-chloro-3-indolyl phosphate (BCIP) and nitro blue tetrazolium (NBT) (BECK, 1987; RICHTERICH et al., 1989). In a similar manner, a hapten-tagged primer can be envisaged, in combination with specific anti-tag antibody detection. WAHLBERG et al. (1990) reported the use of the *E. coli* Lac operator sequence (21 bp) as tag. A fusion protein between *E. coli* Lac repressor and β-galactosidase selectively binds to the operator sequence in the DNA fragments in nested sets, and gives a color signal after addition of a chromogenic substrate for β-galactosidase.

4.3.2.4 Chemiluminescent Detection

Similar to colorimetric detection, a chemiluminescence reaction can be used for detection (BECK et al., 1989). Essentially as described above (Sect. 4.3.2.3), the biotin/streptavidin-phosphatase system is employed, but instead of the chromogenic dephosphorylation of BCIP, a chemiluminescent substrate is used. Dephosphorylation of 1,2-dioxetanes such as AMPPD (BECK et al., 1989; TIZARD et al., 1990) and CSPD (MARTIN et al., 1991) generates an unstable anion, which decomposes and emits light. The signals are recorded by exposure to either an X-ray or Polaroid film. A slightly different approach is used in multiplex sequencing (Sect. 4.4.2) and described in Sect. 4.3.3.

4.3.2.5 Stable Isotopes for Non-Radioactive Detection

Other strategies for non-radioactive labeling have recently been proposed and are based on the use of stable isotopes, which are detectable by resonance ionization spectroscopy (RIS), coupled with time-of-flight mass spectrometry (TOF-MS). The feasability of introducing ^{57}Fe as a primer tag was reported by JACOBSON et al. (1991); sequencing with ^{120}Sn-tagged primer was shown by SACHLEBEN et al. (1991). The use of a large number of stable isotopes could allow multispectral multiplexing (Sect. 4.4.2). Iron has 4 stable isotopes, tin has 10 stable isotopes. In contrast to the chemiluminescent strategies, detection is done directly from a dried gel, thus avoiding blotting or other transfer operations.

4.3.3 Direct Labeling versus Indirect Labeling by Hybridization

In all nested set techniques hitherto described in this chapter labeling is done by direct covalent linkage of the label or tag to the fragments. Such labeling is carried out either before or during the sequencing reactions.

Indirect end-labeling was introduced by CHURCH and GILBERT (1984) when developing the 'genomic sequencing' technique (Sect. 4.4.1). It is achieved by hybridization of an oligonucleotide to its complementary target se-

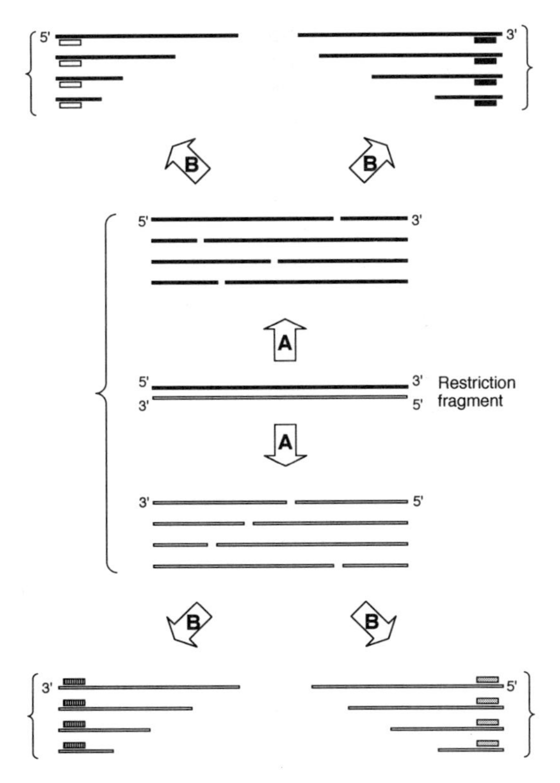

Fig. 14. Principle of the genomic sequencing technique of CHURCH and GILBERT (1984).
(A) DNA restriction fragments are subjected to the standard chemical degradation sequencing reactions, so that on average one break is introduced per molecule. Each strand generates a 5′ and a 3′ nested set.
(B) After gel fractionation, transfer to a membrane and immobilization of the fragments, each nested set can be detected by hybridization with a labeled or tagged oligonucleotide (short bars), complementary to a subterminal region of the corresponding extremity.

quence located near one of the sides of a DNA fragment that underwent chemical degradation (Fig. 14). Hybridization is carried out after fractionation of the nested set(s). Details are given in Sects. 4.4.1 and 4.4.2. Initially, hybridization required the use of ^{32}P end-labeled oligonucleotides, but now chemiluminescence is a more frequently performed and safer alternative. The chemiluminescent detection strategy outlined in Sect 4.3.2.4 is adapted, the difference being that the hybridizing oligonucleotide itself is labeled with biotin. The major advantage of indirect end-labeling is that the same sequencing gel can undergo several rounds of hybridization with different oligonucleotides, complementary to different targets. The selectivity of the oligonucleotide reporter molecules for the targets is of paramount importance.

4.4 Extensions of the Basic Sequencing Approaches

A somewhat esoteric system was developed by ECKSTEIN and collaborators (GISH and ECKSTEIN, 1988; NAKAMAYE et al., 1988) which involves enzymatic copying of DNA in the presence of one phosphorothioate deoxynucleoside triphosphate in each of four reactions. There is no chain termination but the phosphorothioate internucleotide bonds are subsequently randomly cleaved with alkylating agents to yield the nested sets.

Three other developments, implementing new ideas in the basic DNA sequencing methods, have had a much farther reaching impact.

4.4.1 Genomic Sequencing

'Genomic sequencing' is the name given to a method developed by CHURCH and GILBERT (1984) to analyze the DNA structure beyond the primary sequence of bases, e.g., the location of 5-methylcytosine and DNA-protein interactions. This information is normally lost by cloning in heterologous hosts and therefore can only be studied directly onto genomic DNA. (Other techniques adopted the same name for other reasons but do not have the same aim: see Sects. 4.4.3 and 5.)

The method of CHURCH and GILBERT (1984) is an extension of the MAXAM and GILBERT (1977) approach and is based on indirect labeling by hybridization (Sect 4.3.3). Total genomic DNA is digested completely with a restriction enzyme. All fragments are simultaneously subjected to partial chemical cleavage according to the Maxam and Gilbert procedures. Each of the (unlabeled) double-stranded

DNA fragments produces a mixture of four nested sets (Fig. 1C). At the end of the procedure, i.e., after fractionation of the nested sets on a sequencing gel, a single nested set is visualized by hybridization with a particular labeled or tagged synthetic oligonucleotide. This 'reporter molecule' is complementary to the region near one of the termini of the restriction fragment, which provides the fixed end of the nested set. The principle is shown in Fig. 1C and Fig. 14. Obviously, it is extremely important that the reporter oligonucleotide has a unique sequence within the entire genome. A particular prerequisite for hybridization to the high-resolution sequencing patterns obtained by electrophoresis is that the DNA subfragments are immobilized in a hybridizable form. Therefore, the fragments are transferred from the gel to a nylon membrane by electroblotting and are fixed to the membrane by a combination of UV irradiation and heating under vacuum. Limiting factors in the method of CHURCH and GILBERT are (1) sensitivity, since each hybridizing subfragment is present in only the femtomole range, and (2) the fact that partial nucleotide sequence information, i.e., the target sequence where the reporter must hybridize, must be available. On the other hand, since the reporter is not covalently attached to the fragments, it can be washed away after analysis, and the same gel blot can be re-used for hybridization with another oligonucleotide (see also Sect. 4.4.2).

A variant of the technique by which physical damage to the DNA introduced by UV or other footprinting techniques can be detected with a primer extension approach has been reported by BECKER et al. (1989). In two other variants chemical degradation is combined with *in vitro* amplification (Sect. 3.4). In 'ligation-mediated PCR' (Fig. 15) (MUELLER and WOLD, 1989; PFEIFER et al., 1989), genomic DNA is degraded with standard chemical degradation reactions. Subsequently, a complementary strand is synthesized from within the target region and an unphosphorylated oligonucleotide linker is ligated to this strand at the site(s) of chemical cleavage. (Note that the unphosphorylated linker is unable to ligate to the 3′ end of the filled in strand so that the site of chemical cleavage is specifically marked.) This results in a nested set flanked by two known sequences which is amplified by PCR and analyzed as in the Church and Gilbert method. Amplification increases sensitivity of detection. The other variant (SALUZ and JOST, 1989) uses linear PCR with a single primer in the target region, to amplify the chemical de-

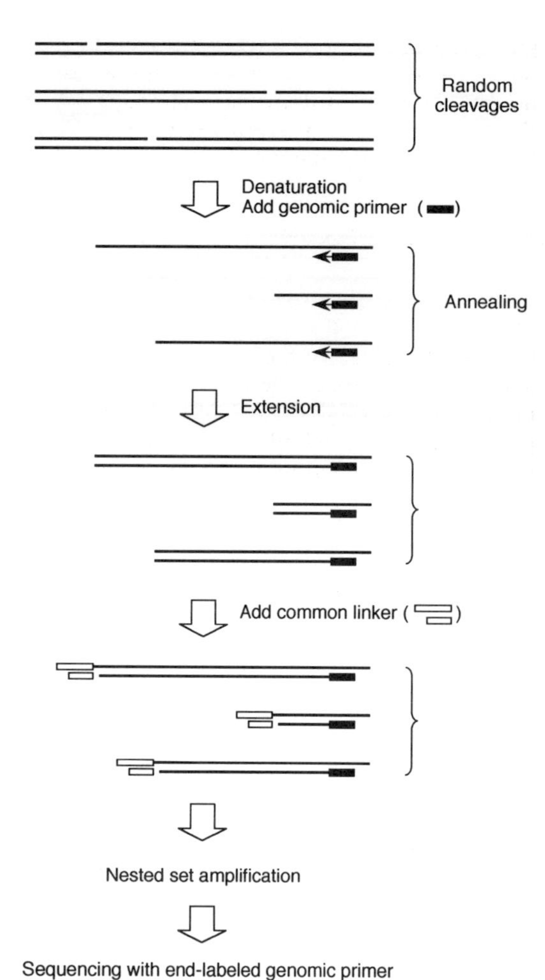

Fig. 15. Sequencing by ligation-mediated PCR. Chemical degradation cleavages are introduced in genomic DNA (see also Figs. 1 and 14). The unphosphorylated primer is extended up to the sites of chemical breakage. (Chemical degradation products do not act as primers since they bear a 3′-phosphate.) An unphosphorylated linker fragment ligates to the upper strand solely. Thus, the nested set products have known sequences at both sides for subsequent amplification by PCR, and sequencing with the original genomic primer.

gradation products in a straightforward fashion. This approach is much simpler as it does not require restriction cleavage of the genomic DNA, electroblotting, cross-linking to a membrane, and hybridization.

4.4.2 Multiplex Sequencing

Multiplex sequencing (CHURCH and KIEFFER-HIGGINS, 1988) is a further extension of the method of CHURCH and GILBERT (1984) and greatly increases the amount of data obtained from a single nested set producing experiment. Both methods share the following features: (1) numerous DNA fragments are treated simultaneously, (2) labeling of individual nested sets is done indirectly by hybridization, after gel fractionation of the nested sets.

On the other hand, they differ in purpose and strategy: (1) The key issue in the Church and Gilbert method is the analysis of structural modifications in genomic DNA, usually in a single target sequence which one wants to distinguish from all other fragments in a complex mixture; in multiplex sequencing, the key issue is sequence data acquisition and all fragments are targets, distinguished from each other by multiple consecutive hybridizations. (2) Genomic sequencing avoids cloning, to retain all the structural information in the DNA, whereas multiplex sequencing is heavily dependent on cloning to provide the target DNA with specific tags (see below and Fig. 16).

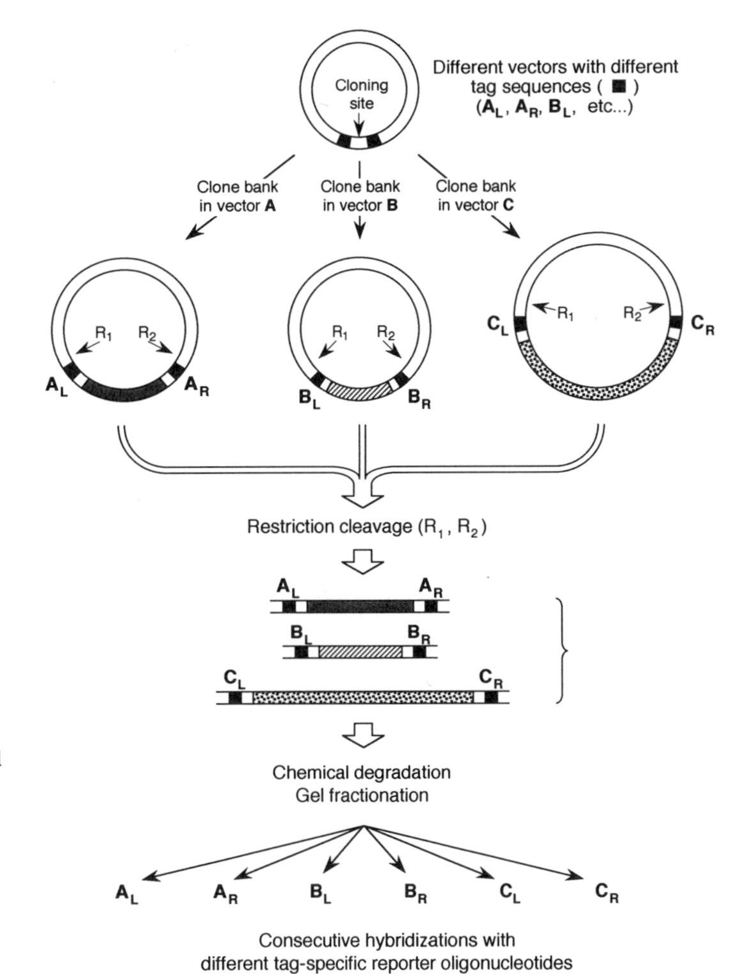

Fig. 16. Principle of multiplex sequencing (CHURCH and KIEFFER-HIGGINS, 1988).
DNA fragments cloned in different multiplex vectors are shown as shaded, hatched, and stippled bars, respectively. The inserts are flanked by unique reporter regions (A, B, C) at the left (L) and right (R) sides. DNAs, one from each bank, are mixed, cleaved with the restriction enzymes R1 and/or R2 to provide a fixed end in each nested set, subjected to chemical degradation sequencing reactions, and analyzed as shown in Fig. 14.

The term multiplexing is taken from electronics, describing the process by which many different electronic signals are mixed at one end of a circuit and decoded at the other end, after having traveled through a single channel. Translated to multiplex DNA sequencing, this means that a large number of fragments are mixed, chemically degraded to prepare an even larger number of nested sets in a single reaction, (note that each fragment falls into four nested sets as explained in Fig. 1C) and decoded after gel fractionation by virtue of a specific tag for each of the original fragments (Fig. 16).

Specific tags are provided by DNA cloning in a series of similar vectors. The vectors differ by the short stretches of sequence, acting as tags, flanking the cloning site at both sides. Thus, when a DNA insert is excised from the vector at unique restriction sites present at the distal sides of the short stretches, it carries a unique terminal sequence at both ends. This allows the mixture to be resolved following electrophoresis. As in the conventional chemical degradation method, four reactions with different base specificities are prepared and compared in four lanes. The superimposed sequence ladders are blotted from the gel and crosslinked to a nylon membrane, and detected one at a time by hybridization using tag-specific reporter oligonucleotides. ^{32}P-labeled oligonucleotides are now superseded by chemiluminescent methods described above (Sects. 4.3.2.4 and 4.3.3). In practice at least 40 rounds of hybridization are feasible with DNAs taken from 20 subclone libraries (i.e., one 'forward' and one 'backward' analysis with each DNA). This increases the throughput of sequencing extensively. Note that restriction cleavage of the cloned DNA is necessary to provide a fixed end to each nested set, the variable ends being produced by chemical degradation.

Extensions of the multiplex approach to dideoxy sequencing have been described by CHEE (1991). Several approaches are feasible (Fig. 17): (1) A set of different plasmid vectors is used as in the method of CHURCH and KIEFFER-HIGGINS (1988), with a universal sequencing primer separated from the target region by a vector-specific tagging sequence. Thus, in the dideoxy sequencing reactions the enzyme synthesizes through the tagging sequences so that a unique tag is included in every oligomer from a given nested set. A commercial version of this approach is available (Millipore Corp.). M13plex vectors (HELLER et al., 1991) combine this approach with subcloning in single-stranded phage vector (Sect. 3.3).

(2) A library of fragments cloned into any single vector is used with a set of tagged oligonucleotide primers. In tagged primers, a 5′ noncomplementary tail is attached to the vector-specific sequencing primer. Enzymatic reactions are done separately and pooled prior to gel fractionation.

(3) If four tagged primers are used, each base-specific reaction (A, C, G, and T, respectively) can be performed with a different primer, and the nested sets pooled for single-lane fractionation (as with multiple fluorescent tags, see Sect. 4.3.2.2) and chemiluminescent detection by four consecutive hybridizations.

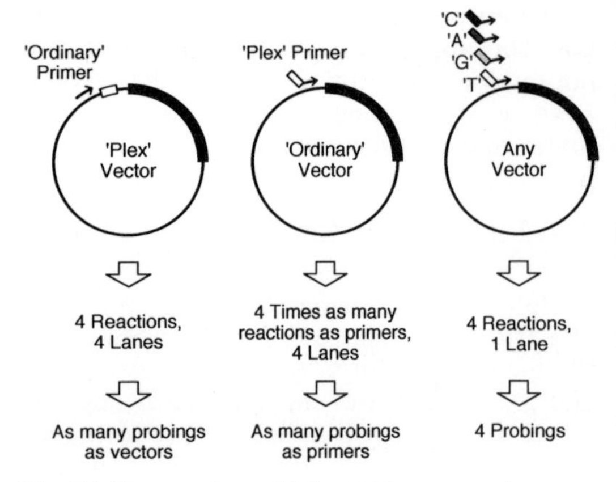

Fig. 17. Enzymatic multiplex DNA sequencing with dideoxy chain termination (CHEE, 1991).
From left to right: (1) a procedure using a series of multiplex vectors with different reporter sequences (open bar); the mixture of clones consists of one clone from each vector; (2) a procedure using one vector library of cloned fragments; each clone is analyzed with a primer with different reporter tails (open bar); (3) a procedure in which four primer tails (differently shaded bars) correspond to different reporter sequences; each primer is used for one of the four termination reactions; thus, each clone is run in a single lane.
The nested set products are analyzed by hybridization with the reporter oligonucleotides as in Fig. 14.

The latter strategy is similar to the proposals for multiplexing tagging with stable isotopes (Sect. 4.3.2.5).

4.4.3 PCR-Based Sequencing

PCR has two typical applications in DNA sequencing: (1) template preparation, and (2) amplification-assisted generation of dideoxy terminated nested sets. Several methods are applicable onto finite amounts of genome DNA and may be considered as direct genome sequencing strategies.

The purpose of using PCR for generating suitable templates for sequencing is to evade cloning of restriction fragments or products from PCR-mediated amplification. Numerous techniques have been described, either to generate single-stranded or double-stranded template DNA.

4.4.3.1 Generation of Single-Stranded DNA

Single-stranded DNA is produced by using unequal molar concentrations of the two amplification primers in the PCR, thus allowing the preferential accumulation of copies of one strand. A primer complementary to this strand, either the limiting primer from the asymmetric PCR reaction or a third (internal) primer is then used for the subsequent dideoxy sequencing reactions (WRISCHNIK et al., 1987; WONG et al., 1987; GYLLENSTEN and ERLICH, 1988; INNIS et al., 1988; SHYAMALA and AMES, 1989; ALLARD et al., 1991). Sequence analysis of DNA, asymmetrically amplified directly from a bacterial colony or plaque, has been demonstrated (WILSON et al., 1990a). Other methods for the production of single-stranded template DNA are 'blocking-primer PCR' (i.e., adding an excess of a third oligonucleotide, complementary to one of the primers, to eliminate one direction of synthesis: GYLLENSTEN, 1989), 'thermal asymmetric PCR' (based on a pronounced difference in melting temperature (T_m) between the primers and switching to high annealing temperature in the last 10 cycles of the amplification process:

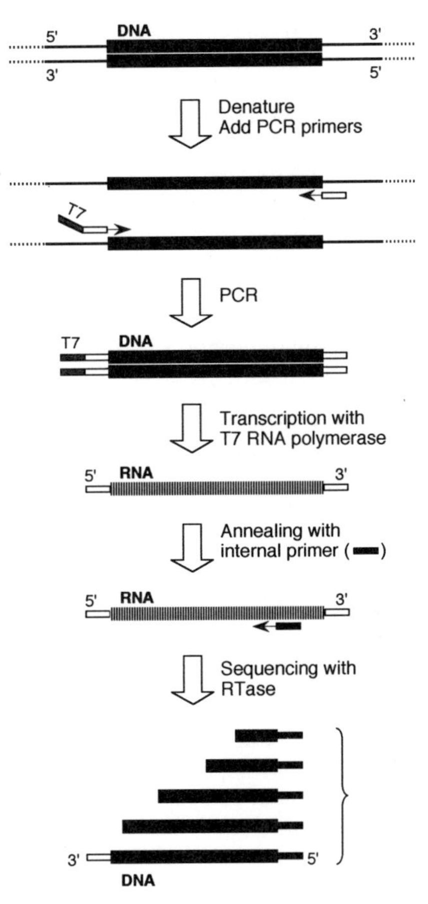

Fig. 18. Sequencing by GAWTS: genome amplification with transcript sequencing (STOFLET et al., 1988).
One of the PCR primers has a 5′ tail bearing the T7 RNA polymerase promoter sequence. The amplified product is transcribed with T7 RNA polymerase, and the RNA serves as template in subsequent dideoxy sequencing reactions with reverse transcriptase as polymerase and an internal primer.

MAZARS et al., 1991), or by procedures to remove one of the strands of a double-stranded PCR product (HIGUCHI and OCHMAN, 1989; GAL and HOHN, 1990). An interesting alternative to obtain single-stranded template is presented by GAWTS ('Genomic amplification with transcript sequencing': STOFLET et al., 1988) (Fig. 18). Genomic DNA is amplified with two specific primers, one of which contains the phage T7 promoter sequence at its 5′

end. After amplification, one DNA strand is transcribed manifold with T7 RNA polymerase and the RNA used as template for dideoxy sequencing with an end-labeled primer and reverse transcriptase as polymerase. It is worth noting that Taq polymerase also has reverse-transcriptase activity which can be exploited for direct amplification and subsequent sequencing of mRNA (SHAFFER et al., 1990).

4.4.3.2 Generation of Double-Stranded DNA

Double-stranded templates are generated by standard PCR protocols and may also be obtained directly from cells, colonies, or plaques (SCHOFIELD et al., 1989; GÜSSOW and CLACKSON, 1989; YANG et al., 1989; MERCIER et al., 1990; SARIS et al., 1990) and used for direct sequencing (NEWTON et al., 1988; SCHOFIELD et al., 1989; YANG et al., 1989). Many investigators, however, have experienced difficulties when trying to sequence double-stranded PCR fragments directly. These difficulties are caused by the presence of non-specific PCR products, the presence of residual oligonucleotides and dNTPs, and in particular by the ability of the two strands of the amplified product to rapidly reassociate after denaturation, thereby preventing efficient annealing of the sequencing primer or preventing the extension of the annealed sequencing primer. Procedures to circumvent these problems are compiled in Tab. 2.

Probably the most efficient way to generate pure template from any DNA is the solid-phase approach of HULTMAN et al. (1989) based on PCR with one biotin-labeled primer and one conventional primer, in combination with strand separation on magnetic beads that are coated with streptavidin (see Sect. 4.1.2 and Fig. 9 for details).

A further extension of this system is used in the authors' laboratory and is schematically represented in Fig. 19. DNA fragments are amplified, starting either directly with a colony or with plasmid DNA extracted from *E. coli*, and using tailor-made primers for PCR amplification in pGV451 (VOLCKAERT, 1987) and derivatives which are standard vectors for recombinant DNA manipulation in our laboratory. The primers are complementary to regions flanking multicloning sites in pGV451 and have a non-complementary 5′ tail (see below). There are two versions of each primer, namely one with and one without a biotin tag. This offers two combinations of one biotinylated and one conventional primer for PCR and the option to attach either one of the DNA strands to the magnetic beads according to the system of HULTMAN et al. (1989). Thus, either strand can be sequenced on solid phase *and* in solution. The primer-tail sequences correspond to two commercially available, fluorescently labeled primers, −21M13 and M13rev, respectively, of the ABI 373A sequen-

Tab. 2. Procedures for Purification of Double-Stranded PCR Products for Sequencing

Procedure	References
Purification of biotinylated PCR products on streptavidin-coated magnetic beads	HULTMAN et al., 1989
Purification of biotinylated PCR products on streptavidin agarose column	MITCHELL and MERRIL, 1989
Direct sequencing in low-melt agarose	KRETZ et al., 1989
Ultrafiltration (Centricon 30)	MELTZER et al., 1990; TAHARA et al., 1990
PEG precipitation	KUSUKAWA et al., 1990
Gel permeation (Sepharose CL-6B, Sephadex G50)	BECHHOFER, 1991; DUBOSE and HARTL, 1990; ZIMMERMAN and FUSCOE, 1991
Isolation from agarose gels by adsorption on glass milk	GIBBONS et al., 1991
Qiaex anion-exchange column chromatography	BOTH et al., 1991
HPLC on Gen-Pak FAX anion-exchange column	WARREN and DONIGER, 1991
HPLC on Zorbax GF-250 sizing column	KALNOSKI et al., 1991

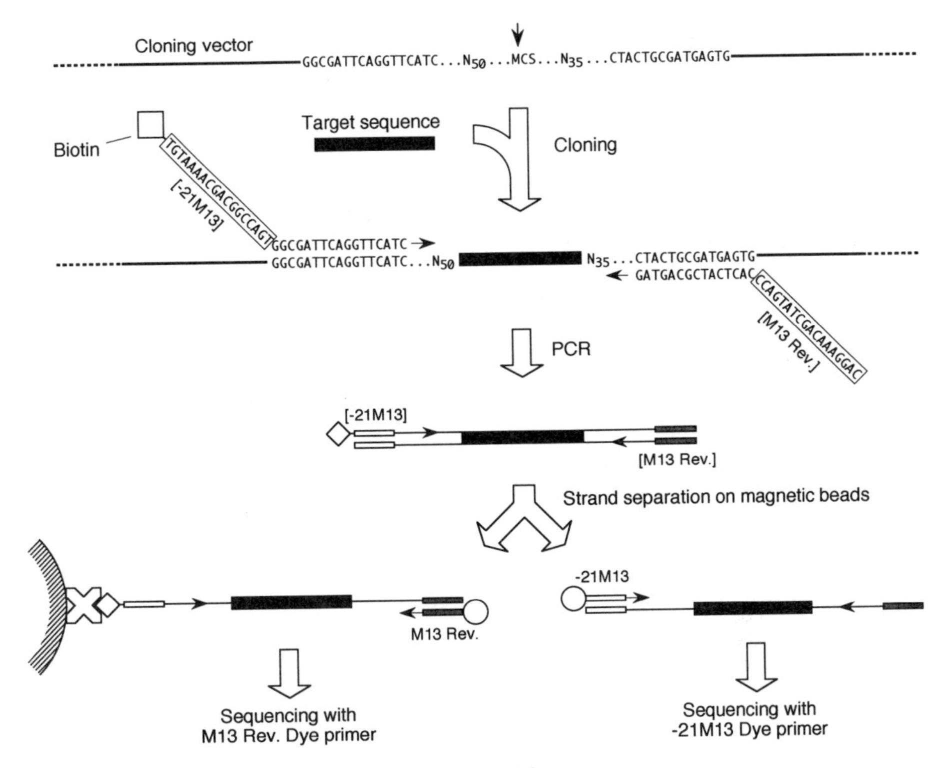

Fig. 19. Systematic semi-automated sequencing in pGV451 plasmids.
The target sequence (filled bar) is cloned in the multicloning region (MCS) of one of the pGV451-derived vectors (VOLCKAERT, 1987) and amplified by PCR with one biotinylated and one conventional primer. The primer tails provide the sequences of the sequencing primers (see below) for subsequent automated analysis of the nested sets in the ABI 373A system (Sect. 6.1).
After strand separation on magnetic beads (see Fig. 9 and text), the DNA is sequenced using fluorescently labeled sequencing primers −21M13 (open bar with arrow-head) and M13Rev (shaded bar with arrow-head). The fluorescent tag is represented by an open circle. Only one of the options is shown; in the second option the biotin tag is attached to the other PCR primer.
An example of such experiments is shown in Fig. 28 A.

ator (Applied Biosystems, Inc.) onto which the amplified DNA is ordinarily sequenced (Sect. 6.1). This approach combines maximal flexibility with minimal requirement of costly synthesis of modified primers and can be readily extended to other vector systems.

4.4.3.3 Generation of Dideoxy-Terminated Nested Sets by PCR

Amplification of template DNA with a single primer in the presence of dNTPs and an appropriate ddNTP leads to accumulation of a dideoxy-terminated nested set suitable for sequencing. This is named 'cycle sequencing'. The uses of radioactively labeled dNTP (LEE, 1991) and of radioactively end-labeled primer (SMITH et al., 1990a; MANAM and NICHOLS, 1991; RUANO and KIDD, 1991) have been described as approaches for detection of the nested sets. In addition, fluorescent labeling of nested sets by extension of dye-labeled primers, in conjunction with real-time detection by appropriate instrumentation (Sect. 6.1) is increasingly being applied (GONZALEZ-CADAVID et al., 1990; WILSON et al., 1990a, b; HULTMAN et al., 1991; OHNO et al., 1991).

The incorporation of dye-label by fluorescent ddNTP precursors is an alternative (Sect. 4.3.2.2) (SCHOFIELD et al., 1989; JONES et al., 1991; TRACY and MULCAHY, 1991).

5 Strategies for Approaching the Target Regions

A routine sequencing analysis experiment generally provides information on about 200 to 500 nucleotides. This may be sufficient in occasional control sequencing or in comparative studies of defined small areas in different genomes (e.g., in molecular systematics) or in diagnostic applications. More often, however, the sequences of larger regions must be determined. In the following sections (5.1 to 5.4), the problem of approaching sequence information in larger DNA segments is addressed and strategies are outlined.

5.1 Direct Sequencing versus Subcloning

An attempt to apply dideoxy chain termination sequencing directly to genomic DNA was successful with *Saccharomyces cerevisiae* (HUIBREGTSE and ENGELKE, 1986) but failed with larger genomes. In principle, the genomic sequencing method of CHURCH and GILBERT (1984) (Sect. 4.3.1) can be used more readily for direct sequencing, but this method is limited by the need for separate probes with known sequences and the need for appropriately located restriction sites.

Larger DNA segments cloned in plasmid, phage, or cosmid vectors in *Escherichia coli* represent a smaller complexity than entire genomes, and can be sequenced by several other direct sequencing methods, e.g., by oligomer walking (i.e., stepwise progressing by synthesis of a new primer complementary to the template at the limit of each determined sequence: STRAUSS et al., 1986), by using a library of primers (i.e., by fortuitously priming with a battery of shorter oligomers in a large series of analyses, speculating that a reasonably large fraction of the oligomers will have unique priming sites in the DNA clone and give readable ladders: STUDIER, 1989; SIEMIENIAK and SLIGHTOM, 1990), by the ExoMeth method (a multistep procedure in which one of the DNA strands is stepwise digested with exonuclease III, nested products are synthesized and labeled in the presence of 5-methyl-dCTP and exposed to frequently cutting enzymes that are inhibited by 5-methyl-cytidine; a selected zone of nested products moves progressively over a sequence of up to more than 5 kb: SORGE and BLINDERMAN, 1989). Multiplex sequencing of CHURCH and KIEFFER-HIGGINS (1988) (Sect. 4.4.2 and Fig. 16) combines cloning with direct genomic sequencing for larger projects, and has been combined with oligomer walking by OHARA et al. (1989) ('multiplex walking').

More often, however, strategies based on subcloning of the DNA segment are used (Sects. 5.2 and 5.4), or more recently and increasingly, methods based on amplification by PCR (Sects. 4.4.3 and 5.3) are chosen. Accuracy problems possibly associated with PCR sequencing have been stressed in Sect. 3.4.

5.2 Shotgun versus Deletion Strategies

In subcloning strategies large cloned DNA segments are fragmented so that relatively short stretches can be brought into juxtaposition with fixed primer binding sites for enzymatic sequencing, or with end-labeling sites for chemical sequencing.

Two distinct general concepts are: (1) shotgun cloning, and (2) engineering of progressively overlapping clones. It is impossible in the framework of this survey to cover the whole range of strategies which have been developed with the latter aim. By the same token, this may indicate that probably neither of those 'deletion' strategies combines all ideal properties with respect to versatility, reliability, ease of operation and speed. PCR- and transposon-based techniques are dealt with in Sects. 5.3 and 5.4, respectively.

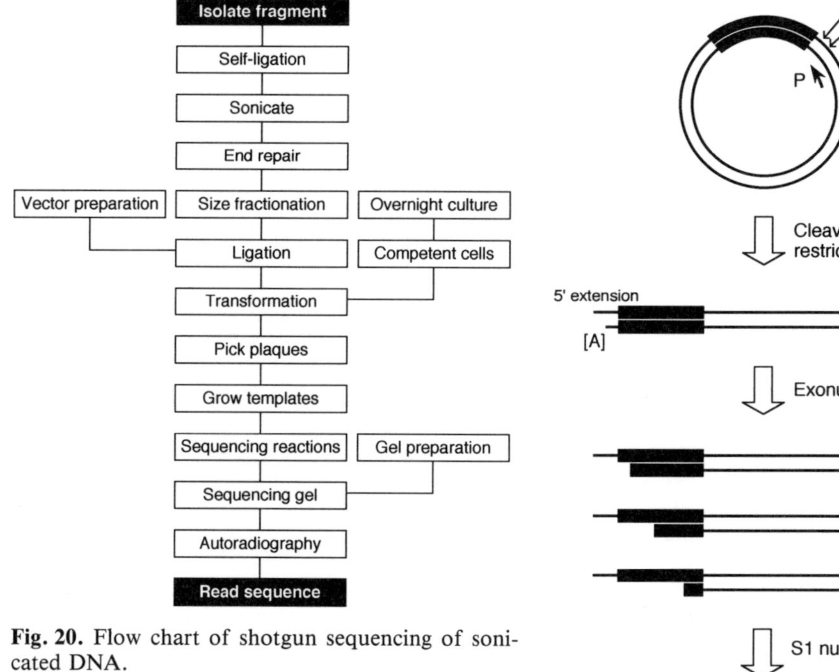

Fig. 20. Flow chart of shotgun sequencing of sonicated DNA.
All steps involved in preparation of sonicated DNA fragments, cloning of the fragments in an M13mp vector, isolation of single-stranded templates, and dideoxy sequencing are shown. Cloning produces a large set of subclones which are routinely analyzed and the gel readings are assembled to the final sequence with the help of computer software.
(Adapted from BANKIER et al., 1987)

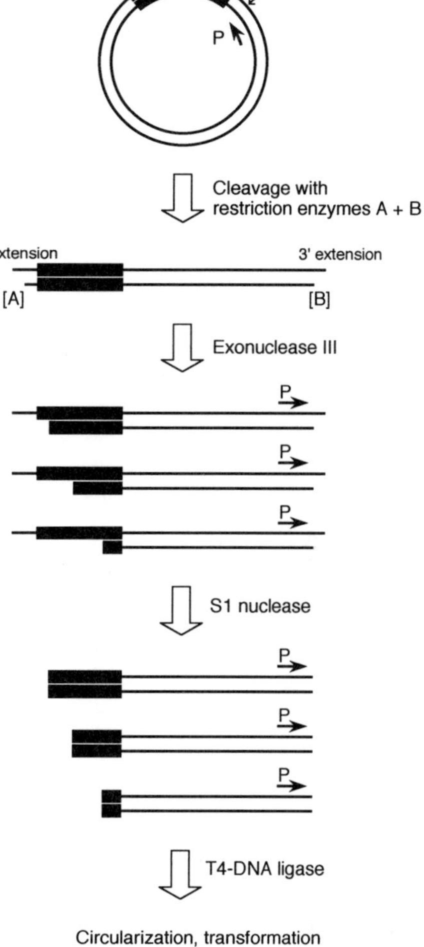

Fig. 21. Stepwise deletion of cloned target DNA with exonuclease III and nuclease S1 (HENIKOFF, 1984).
Target DNA is drawn as filled bars. The position of the sequencing primer annealing site (P) is indicated with an arrow. Restriction enzyme A generates a 5′ extension or blunt end; restriction enzyme B generates a four-base 3′ extension. Following enzymatic degradation as shown, circularization by ligation brings the primer sequence in front of different positions within the target DNA sequence.

Shotgun cloning is cloning of unfractionated fragmented DNA into an appropriate vector. Fragments are produced by restriction cleavage (or partial restriction digestion) or cleaved in a less sequence-dependent way by DNase I (ANDERSON, 1981) or sonication (DEININGER, 1983). The latter is by far the most versatile, efficient, and random procedure, and has been applied in major sequencing efforts (e.g., BAER et al., 1984; BANKIER et al., 1991), in particular in combination with subcloning in M13mp vectors (SANGER et al., 1980).

Shotgun cloning and sequencing is simple to execute (Fig. 20), and is very rapid for the acquisition of new primary sequencing data. Sequenced fragments are joined into contigs as soon as overlaps are found. (A contig is defined as a contiguous sequence assembled by ordering overlapping gel readings in a sequencing project: STADEN, 1986). Subsequent clones, however, may contain sequences al-

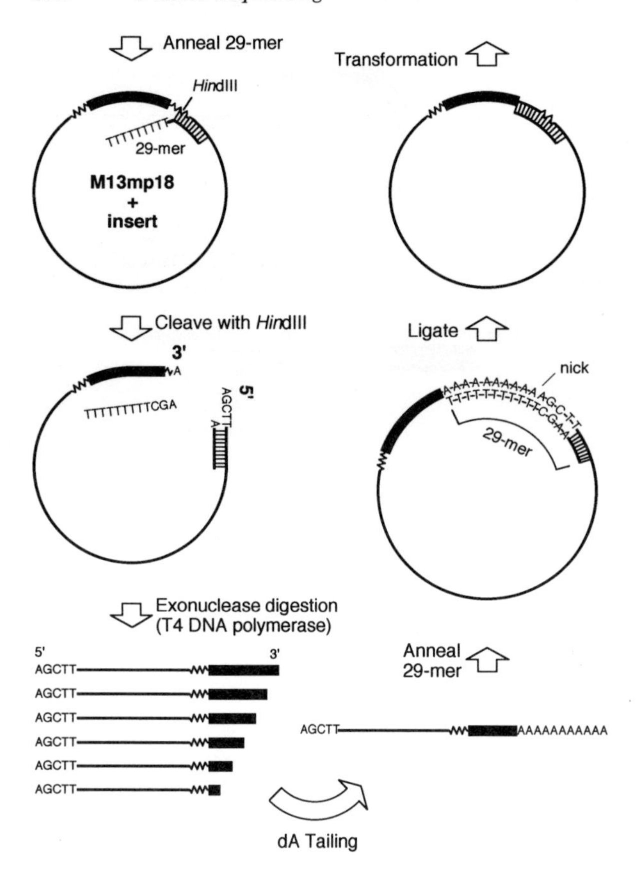

Fig. 22. Systematic deletion in single-stranded M13mp clones, according to DALE et al. (1985).
Target DNA is drawn as a filled bar. The single-stranded DNA is cleaved in a region where a double-stranded structure is created by hybridization of a synthetic 29-mer oligonucleotide that has an extra oligo$(dT)_n$ tail. This region contains one of the restriction sites at the 3′ side of the insert.
After shortening the linear strand by the exonucleolytic action of T4 DNA polymerase, an oligo$(dA)_n$ tail is attached to the 3′ ends with terminal deoxynucleotidyl transferase, and the DNAs are recircularized by joining the ends with the 29-mer oligonucleotide. The remaining nick is closed with T4 DNA ligase and the DNA used for transformation.

ready represented in the contigs, and the chance of finding further overlaps decreases as the project progresses. Typically, about 90% of the complete sequence is determined rapidly, but the remaining 10% can be difficult to isolate from the subclone bank. Each base in the fragment is determined on average six to eight times, but this redundancy may also be considered a built-in safeguard against sequencing errors. Nevertheless, supplemental cloning is required to close all gaps (i.e., regions of unknown sequence which join the contigs), e.g., by using restriction sites near the borders of the contigs. PCR (Sect. 4.4.3) may provide a faster way to solve this problem.

In spite of its limitations, it is noteworthy that the majority of large sequencing projects were primarily based on shotgun cloning. A full description of the experimental procedures involved and an assessment of the strategy can be found in BANKIER et al. (1987, 1991) and in DAVISON (1991).

The alternative to shotgun cloning when considering subcloning large DNA segments for sequencing, is the creation of systematic unidirectional deletions. Progressive shortening of DNA *in vitro* has been accomplished using enzymes such as DNase I (FRISCHAUF et al., 1980; HONG, 1982), Bal31 (PONCZ et al., 1982; BARNES and BEVAN, 1983), and exonuclease III in combination with S1 nuclease (HENIKOFF, 1984). The latter procedure is outlined in Fig. 21, as it is probably the more commonly applied deletion strategy. It requires the presence of two unique restriction cleavage sites near the DNA to be deleted: the site closer to this DNA must leave a 5′ protruding or blunt end, the other site must leave a 4-base 3′ protruding end. A 4-base 3′ protruding end resists the action of exonuclease

III. Hence in a carefully controlled reaction with exonuclease III, only the former end is progressively and quite uniformly shortened. The remaining single strand is easily removed with S1 nuclease, and the plasmid recircularized with ligase. (Note that S1 nuclease will also abolish the 3′ extension of the second site.) A shortened version of the method starting with single-stranded phagemid template and avoiding the restriction cleavages has been described by HENIKOFF (1990).

A unique strategy was developed by DALE et al. (1985), using a single-stranded M13 clone as initial template (Fig. 22). In a first step, the single-stranded DNA is linearized by hybridization of an oligonucleotide to the polylinker region flanking the target DNA and cleaved with an appropriate restriction enzyme. (This oligonucleotide also carries a 3′ short homopolymeric stretch required in a later step.) In the second step, the linear single strand is shortened to different extents by controlled exonucleolytic action of polymerase T4, and subsequently elongated with a short homopolymeric tail, complementary to the oligonucleotide tail mentioned in step 1. Finally, the oligonucleotide joins the ends of the single-stranded DNA and the remaining nick is sealed with T4 DNA ligase. The circular single-stranded (deletion) products are then used to transform *E. coli*.

5.3 Cloning versus *in vitro* Amplification

Before the invention of PCR techniques, template preparation for DNA sequencing heavily relied on *in vivo* amplification of cloned or subcloned DNA fragments (Sects. 3.3 and 5.2). With PCR, the intervention of DNA cloning can be minimized (Sects. 3.4 and 4.4.3). In its basic concept, however, *in vitro* DNA amplification by PCR is confined to fragments that are flanked by stretches of known sequence. Now, several techniques have been reported for amplification of the DNA segments that lie outside the boundaries of known sequences, extending the option to reduce the number of cloning steps, which are inherently slower and more tedious.

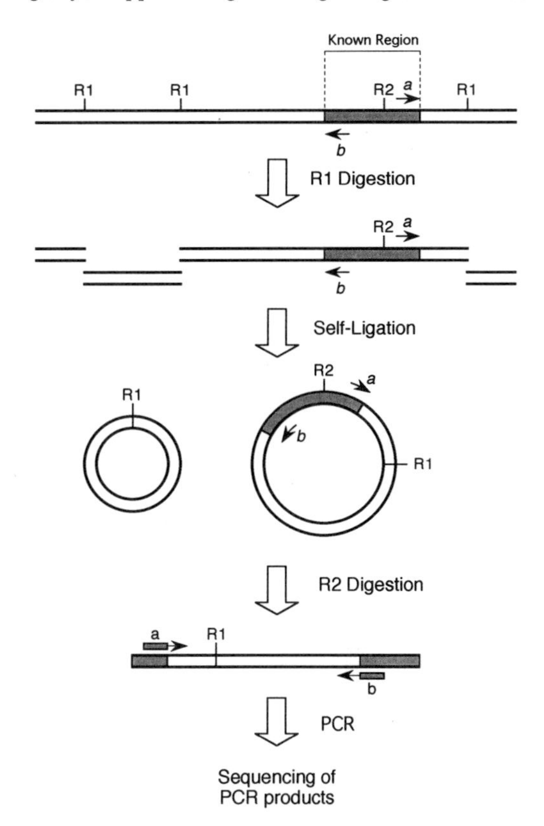

Fig. 23. Diagrammatic representation of inverse PCR (TRIGLIA et al., 1988).
Sequencing outside the boundaries of known sequence (shaded area) is enabled by restriction cleavage of genomic DNA and religation to adjust two PCR primer annealing sites (a and b) for amplification of unknown DNA. R1 and R2 are two restriction sites required in the unknown and known regions, respectively. Note that a complex mixture of fragments (and circles) is generated, but only one region is amplified by the use of appropriate specific primers.

In inverse PCR (TRIGLIA et al., 1988), genomic DNA is digested with a restriction enzyme that does not cut into the fragment of known sequence (Fig. 23). Restriction fragments are ligated under conditions that favor circularization and are re-opened by digestion with a restriction enzyme that cuts once into the known sequence. Hence, two subfragments of known sequence are now flanking the target region, and primers can be designed for amplification and sequencing. (Note that these

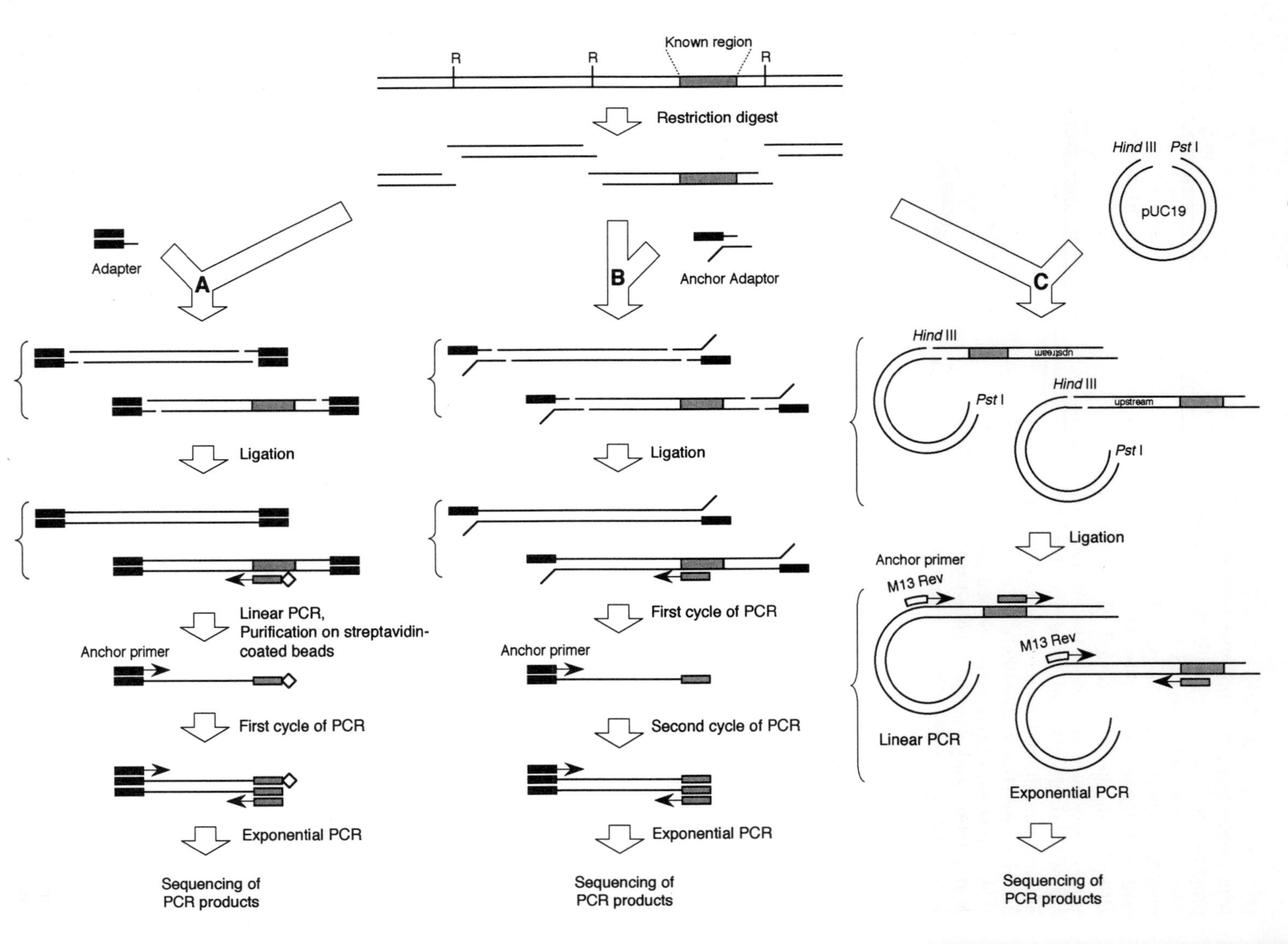
Known region
R
R
R
Restriction digest
Adapter
A
B
Anchor Adaptor
C
Hind III Pst I
pUC19
Hind III
upstream
Pst I
Hind III
upstream
Pst I
Ligation
Ligation
Ligation
Anchor primer
M13 Rev
M13 Rev
Linear PCR,
Purification on streptavidin-
coated beads
First cycle of PCR
Anchor primer
Anchor primer
Linear PCR
First cycle of PCR
Second cycle of PCR
Exponential PCR
Exponential PCR
Exponential PCR
Sequencing of
PCR products
Sequencing of
PCR products
Sequencing of
PCR products

primers point away from each other in the genomic DNA, but point towards each other after re-opening of the circularized fragment. Note also that the entire procedure is carried out in a complex mixture of fragments, but fragments other than the target do not contain appropriate annealing sites for both primers.)

Anchor PCR was originally conceived by LOH et al. (1989) for amplification of mRNA and has been modified for direct genomic DNA sequencing by numerous authors, e.g., KALMAN et al. (1990), ROBINSON et al. (1990), ROSENTHAL and JONES (1990), ROUX and DHANARAJAN (1990), COLLASIUS et al. (1991). Genomic DNA is digested with a restriction enzyme that does not cut the fragment of known sequence, and an extra DNA sequence is ligated to the fragment ends. This extra sequence, or anchor, is usually a synthetic fragment, which provides a second primer annealing site, with the first one located in the known sequence. The key problem in anchor PCR is that the anchor only ligates to all fragments at both sides which all become targets for subsequent PCR. How to limit amplification (and sequencing) to the sole fragment containing the first primer annealing site, is approached by the authors cited above in different ways. Three of those approaches are compared in Fig. 24.

The inverse PCR and anchor PCR techniques are limited by the need for suitable restriction endonuclease cleavage sites surrounding or cleaving the known sequence.

Conversely, in the 'targeted gene walking' approach of PARKER et al. (1991), genomic DNA is amplified in a series of PCR reactions with a 'target primer' (in a known sequence) and different non-specific primers of defined sequence. PCR products accumulate, and some of these apparently originate as a consequence of partial homology of the non-specific primers with genomic DNA that is located not too far from the target primer. These products are then sequenced using another (radioactively labeled) primer annealing at a short distance downstream of the target primer.

VERHASSELT et al. (1992) (Fig. 25) amplify DNA lying outside the boundaries of a known sequence using a specific primer homologous to this sequence and a semi-random primer. The latter consists of a 3′ terminal TCAG, preceded by a fully degenerated stretch of 13 nucleotides allowing annealing at multiple sites. An extra tail with a restriction site can be provided at the 5′ side of the primers to facilitate subsequent cloning of the amplified fragments. Multiple PCR products of different lengths are generated, having a fixed end (allocated by the specific primer) and a variable end. This array of overlapping clones can be used for systematic directional sequencing.

Fig. 24. Diagrammatic representation of anchored PCR.
Strategies for sequencing the upstream region of the known sequence (shaded area) are illustrated. R is an appropriate restriction site (which is *Hin*dIII in example C). The anchors are ligated to cleaved genomic DNA as shown.
(A) ROSENTHAL and JONES (1990): the target region is linearly amplified by PCR, using a single primer (shaded bar) annealing to the known DNA and carrying a biotin (shown as a diamond ◇). This strand is specifically extracted from the fragment mixture by attachment to magnetic beads coated with streptavidin (as in Fig. 9) and subsequently exponentially amplified with the anchor primer (filled bar) and the first primer.
(B) ROUX and DHANARAJAN (1990): An anchor adaptor is ligated to the target DNA fragment. This adaptor is a partial duplex structure with non-complementary tails at one side. Hence, generation of the second PCR primer annealing sequence (black bar) depends on the first cycle with the primer (shaded bar) complementary to the known DNA, and is limited to the target DNA flanking the known sequence.
(C) ROBINSON et al. (1990): Target DNA is ligated to one of the ends of double-digested pUC19. As the target can ligate in two orientations, a mixture is obtained, but one orientation will allow exponential amplification, whereas the other is amplified only linearly.

5.4 Transposon-Based Techniques

The use of a transposon to bring a priming site to the target, instead of bringing the target to the priming site by cloning in a vector is an attractive strategy to obviate *in vitro* reshuffling of DNA or costly synthesis of walking primers in sequencing projects. Obviously, the

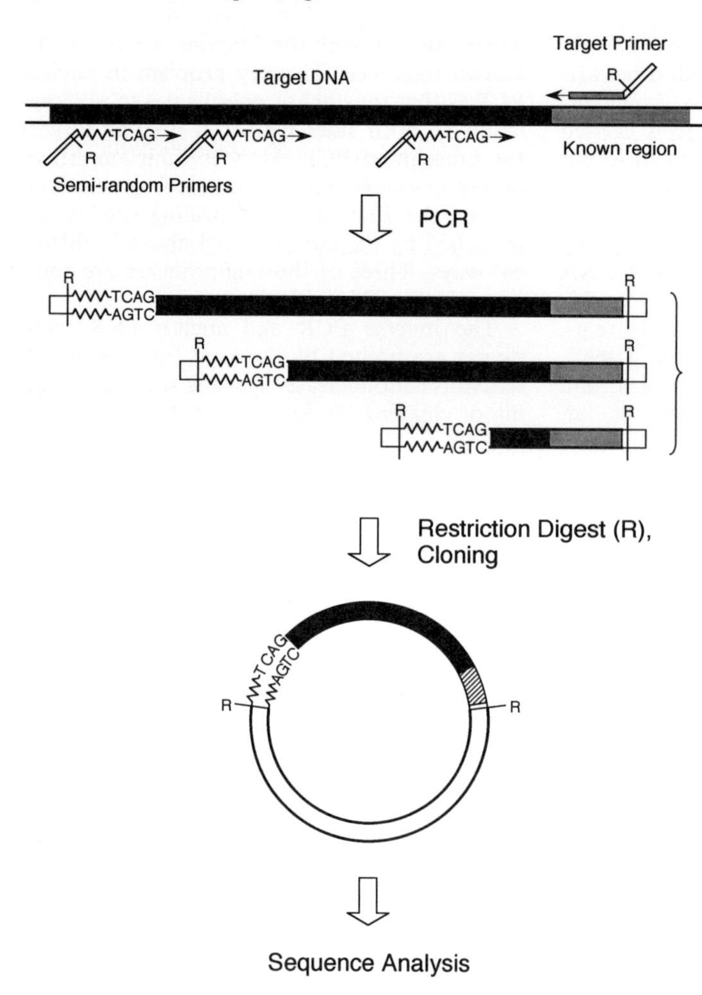

Fig. 25. Amplification of target DNA with semi-random primers (VERHASSELT et al., 1992).
The target primer (shaded bar) is complementary to the known DNA and points towards the region outside the boundaries of known sequence. Semi-random primers bind preferentially to sites assigned by their 3′ terminal TCAG, and annealing is stabilized by sequences from the random region (sawtooth lines). Obviously, the concentration of the oligonucleotides having sufficient complementarity with the template is much lower relative to the target primer concentration. Primers bear a tail sequence (open bars) to permit restriction cleavage (R) for subsequent cloning in an appropriate vector.

transposition events must be sufficiently random.

Although the underlying principle is straightforward and simple, some typical conditions must be met before a useful sequencing system is obtained: (1) Transposition is a very rare event, hence efficient genetical selection is required to isolate the mutants; this is merely a genetic problem, for which solutions have been engineered that will only be touched briefly here (for a more extensive discussion on this topic, see Chapter 2 and KLECKNER et al., 1991; GROISMAN, 1991). (2) The priming site for sequencing should be located near the border of the mobile element in order to immediately extend into the target DNA during the dideoxy termination reactions; however, since transposons are flanked by repeated sequences of often substantial length, it is not obvious that one can find unique sequences in those regions.

A truncated derivative of Tn*9* was engineered by AHMED (1985). It promotes deletions specifically by excision of DNA between the flanking IS1 at the left hand side of the element (IS1-L) and some distant site (Fig. 26). Target DNA is first cloned in the vector containing the transposon, and placed between the transposon and a gal^S gene (which signifies that the cells are killed by the presence of galactose). Unidirectional nested deletions are then obtained by selecting for Gal^R survivors

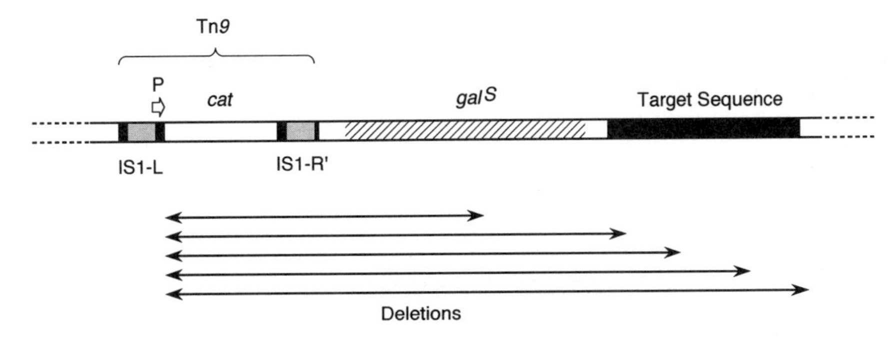

Fig. 26. Deletion by Tn*9* directed excision (AHMED, 1985).
The cloning vector used in this system contains a truncated derivative of Tn*9* which is separated from the cloned target DNA by a *gal*^{S} marker. Galactose resistant colonies appear by deletions extending from IS1-L and are shown by arrows. Plasmids harboring these deletions are sequenced directly using a primer (open arrow) derived from IS1-L.

that underwent deletions towards the target DNA (Fig. 26). To limit the number of sequencing runs, plasmid DNA with progressively overlapping deletions may be excised from an agarose gel. The terminal inverted repeats of IS1-L are only 23 bp long, and primer annealing sites for sequencing can be chosen from regions of nonhomology between those repeats. Sequencing was originally done on supercoiled DNA (CHEN and SEEBURG, 1985), and PENG and WU (1986) constructed a phasmid derivative (Sect. 3.3) to make isolation of single-stranded templates feasible.

LIU et al. (1987) developed a transposon-based sequencing strategy with γδ (Tn*1000*) as mobile element. Tn*1000* is flanked by terminal inverted repeats of 38 bp only, so that unique primer annealing sites may be derived from subterminal segments, in both directions. Transposon and target DNA are on two different plasmids, and since γδ transposes via a cointegrate mechanism, a selection strategy based on conjugal transfer of the cointegrate was worked out (STRATHMANN et al., 1991). Sites of insertion are readily mapped by PCR with one priming site in the transposon and one in the cloning site of the plasmid bearing the target DNA. The random character of insertion has been assessed by STRAUSBAUGH et al. (1990).

Although Tn*5* would be extremely well suited as mobile element (it makes simple insertions into dozens of sites per kb), it is not suitable because of its 1.5 kb inverted repeats. Transposable derivatives, however, have been engineered with just 19 bp from each Tn*5* or IS*50* end (CHOW and BERG, 1988; NAG et al., 1988). Tn*5supF* (PHADNIS et al., 1989) is a 264 bp mini-Tn*5* element with the *supF* gene as mobile marker. DNAs cloned in λ phage are the targets, and the *supF* gene is used to select transposition events, whereas unique sequences near each Tn*5supF* end serve as primer binding sites for sequencing. KRISHNAN et al. (1991) describe PCR methods for mapping of the insertions on phage plaques, and for using the PCR products as templates for sequencing.

6 Automation

DNA sequencing is a specialized, multistep procedure, prone of human error when handling large series of samples. Moreover, large, or even moderate sequencing projects increase the average number of individual analyses, thus expanding the routine character of the enterprise. Several elements of the sequencing process are amenable to automation: (1) the generation of a purified template, (2) the biochemical reactions that produce the nested sets, (3) the electrophoretic separation of the nested sets, (4) the deduction of a primary sequence from the electrophoretic data, and (5) the computational analysis and manual editing involved in assembling the final sequence. Automation of the more 'upstream' steps, i.e., cloning and/or deletion of DNA fragments, might be a more difficult task.

Initial efforts towards automation closely followed the basic manual sequencing procedures. Many reading systems were developed to extract the sequence data from autoradiograms, first with the aid of simple sonic digitizers, later on with film scanning devices (ELDER et al., 1986; WEST, 1988; ELDER, 1990; EBY, 1990). Considerable effort was invested to automate the manual chemical degradation and dideoxy chain termination procedures (reviewed in MARTIN and DAVIES, 1986). Thus, WADA and collaborators (WADA, 1984, 1987) constructed a microchemical robot that carried out all steps of the chemical degradation reactions in solution, including precipitation and centrifugation of the samples. Other groups attempted to shift gel fractionation of radioactively labeled nested sets from a snap-shot to real-time detection approach, sometimes with success (TONEGUZZO et al., 1988).

Nevertheless, substantial progress came with a move from radioactive to fluorescent detection (Sect. 4.3.2.2), which is far better suited to real-time detection. Concurrently, the fundamental concept of running four lanes per DNA fragment changed with the introduction of four distinguishable tags. The biochemistry of the enzymatic preparation of nested sets was adapted accordingly.

Due to massive investments by genome sequencing projects, this is an area of intense development and interest. We focus here on some instruments that are more generally accessible for the experimenter by virtue of their commercial availability. There is little doubt that more sophisticated methodologies and new instruments with enhanced features are yet to come.

6.1 Automation of Gel Separation and Data Collection

'Automated DNA sequencers' come with electrophoresis modules, detectors for fluorescent labels (Fig. 13), and base-calling software.

The 373A DNA Sequencing System (Applied Biosystems, Inc.), is based on the instrument developed at Caltech (Pasadena, USA) (SMITH et al., 1986). It is an electrophoretic system that automates the detection of fluorescently labeled nested sets generated during sequencing reactions and eventually deduces the sequence. It utilizes four differently colored dyes, each of the colors matching one of the four dideoxy termination reactions (as described by SMITH et al., 1986, and by PROBER et al., 1987, and outlined in Sect. 4.3.2.2). The dye is incorporated either at the 5′ or 3′ end of the nested set products by using dye-labeled primers or dye-labeled dideoxynucleotide triphosphates, respectively (Fig. 27). Thus, each base-specific termination reaction is identified by a different color dye. The dye-labeled ddNTPs are adapted to the laser detection system of the 373A machine and different from those described by PROBER et al. (1987). A maximum of 24 samples can be loaded, corresponding to a sequence analysis of 24 DNA fragments. As the dye-labeled nested set products migrate down the gel, they reach a fixed point where an argon-ion laser scans across the gel and excites the tags to detect the products in real-time mode. The emitted fluorescence passes sequentially through four filters, each passing light at the maximum emission wavelength for one of the dyes. The filtered light is directed to a photomultiplier tube, and the digitized output is transferred and stored in a computer. When electrophoresis is complete, the data are processed to determine the concentration of each of the four dyes at each point in time independently. A base-calling algorithm then corrects these data into sequence information. An example is shown in Fig. 28, and more examples with different sequencing strategies and run on either the ABI 373A or its predecessor 370A can be found in, e.g., GIBBS et al. (1989), MARDIS and ROE (1989), GONZALEZ-CADAVID et al. (1990), WILSON et al. (1990a), OHNO et al. (1991), OKUBO et al. (1991), TRACY and MULCAHY (1991). The analysis of nested sets prepared by chemical degradation of proper fluorescently labeled DNA fragments is possible on the ABI 373A (ROSENTHAL, personal communication), but not yet documented.

The principle of using four fluorescently la-

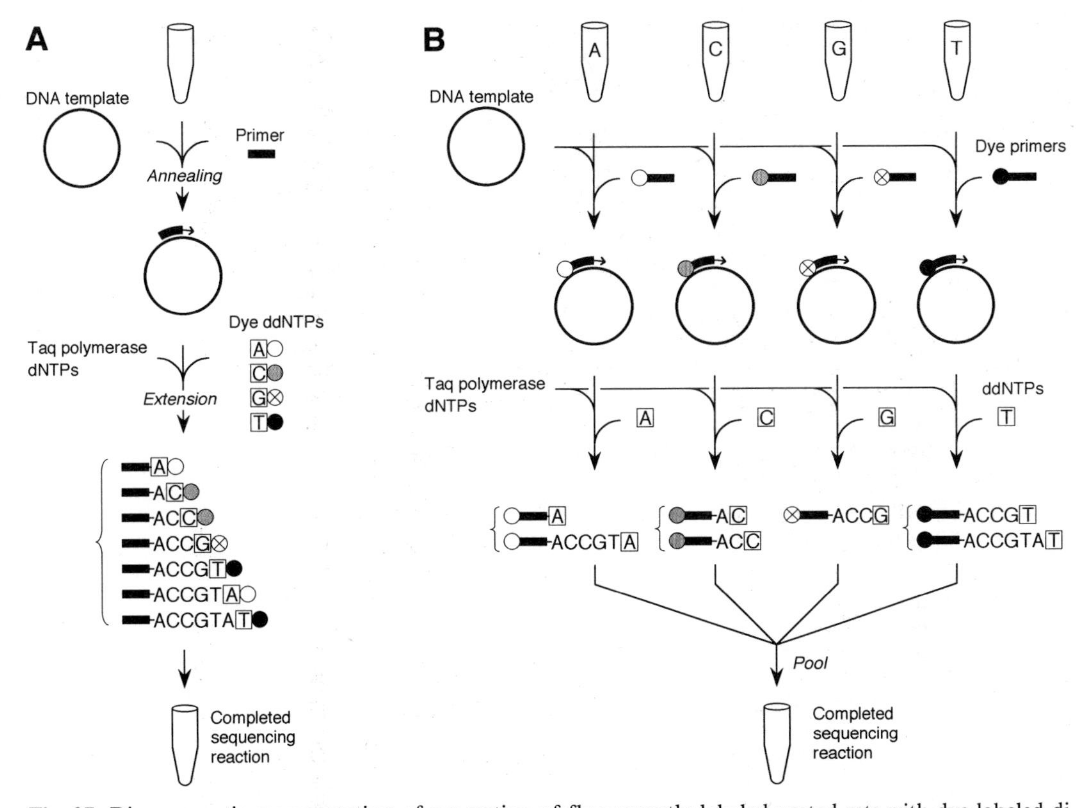

Fig. 27. Diagrammatic representation of generation of fluorescently labeled nested sets with dye-labeled dideoxynucleoside triphosphates (A) or dye-labeled primers (B).
Primer is a filled bar. Fluorescent tags are small circles, either open, filled, gray or crossed. In (A) a single reaction suffices; in (B) four separate reactions are required and combined before gel loading.

beled dideoxynucleotides had been previously invented by a group at DuPont de Nemours & Company, Inc. (Prober et al., 1987), which built a DNA sequencing instrument called Genesis2000, that apparently has been removed from the market. However, the instrument is still being used in sequencing projects (Schofield et al., 1989; Jones et al., 1991; Jensen et al., 1991). In this system, base tags with small differences in emission spectra are used (Sect. 4.3.2.2) and signals detected by two photomultipliers (two wavelengths) after excitation in the laser beam. The signal ratios are characteristic of the attached dye and are computed by an algorithm to assign the peaks and the bases.

Other instruments employ a single fluorophore. This approach offers simplicity to the biochemist in requiring only a single labeled primer, but has the disadvantage of requiring four lanes to run the nested sets separately, thus reducing the capacity of the gel. On the other hand, it is not necessary to compensate for differences in mobility (as observed with different dyes). The system developed at the European Molecular Biology Laboratory (EMBL) in Heidelberg (Ansorge et al., 1986, 1987) and its commercial counterpart (Automated Laser Fluorescent or A.L.F. DNA Sequencer, Pharmacia-LKB AB) can accommodate 40 lanes (corresponding to sequence analysis of 10 DNA fragments) and have a fixed laser beam traversing the entire gel width in a straight line, parallel to the plane of the glass plates. As the migrating nested set products intercept the beam, the fluorescent tag is excited and a series of fixed photodiodes, one for each of the 40 lanes, detect the emitted light. Raw

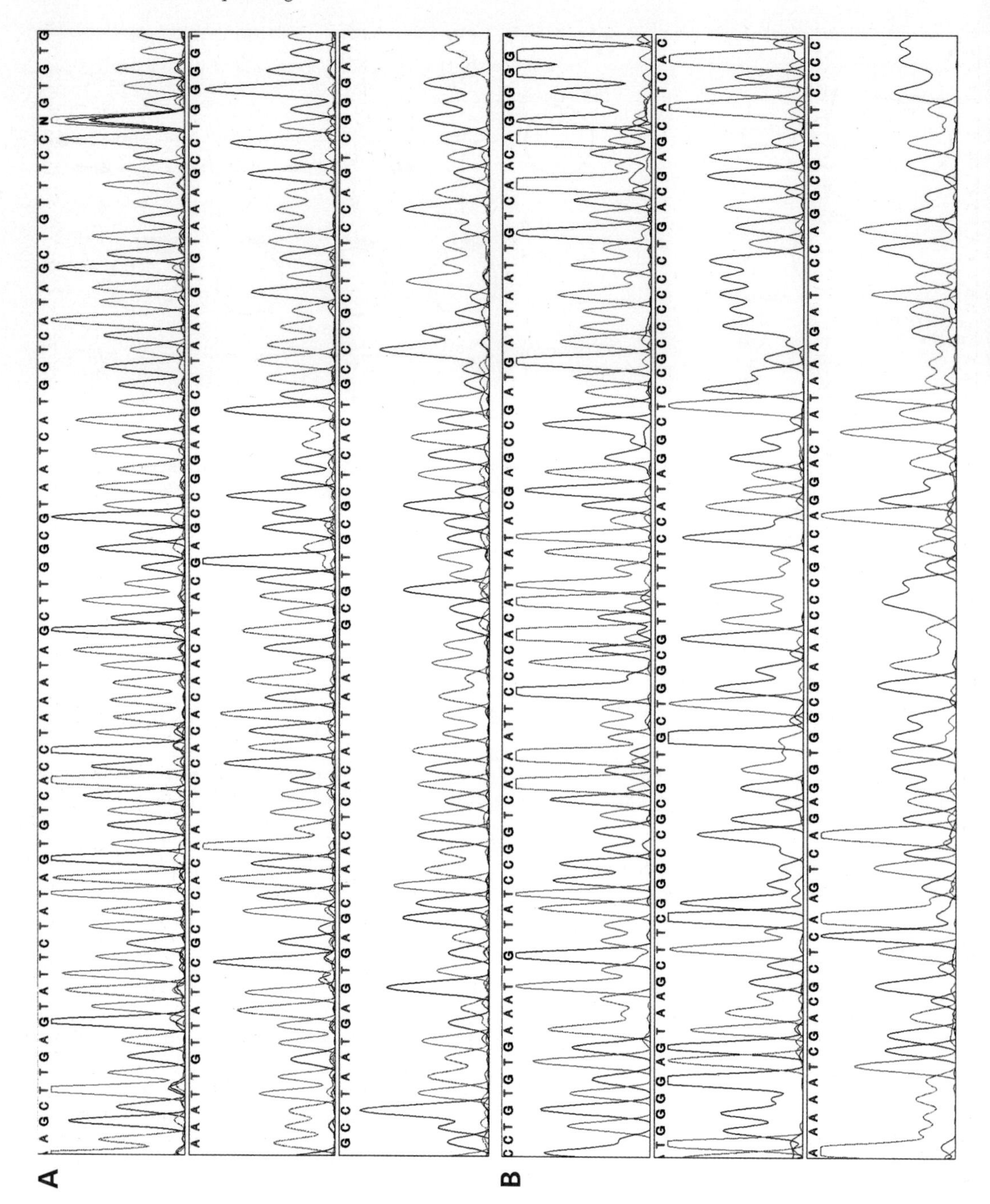
A
B

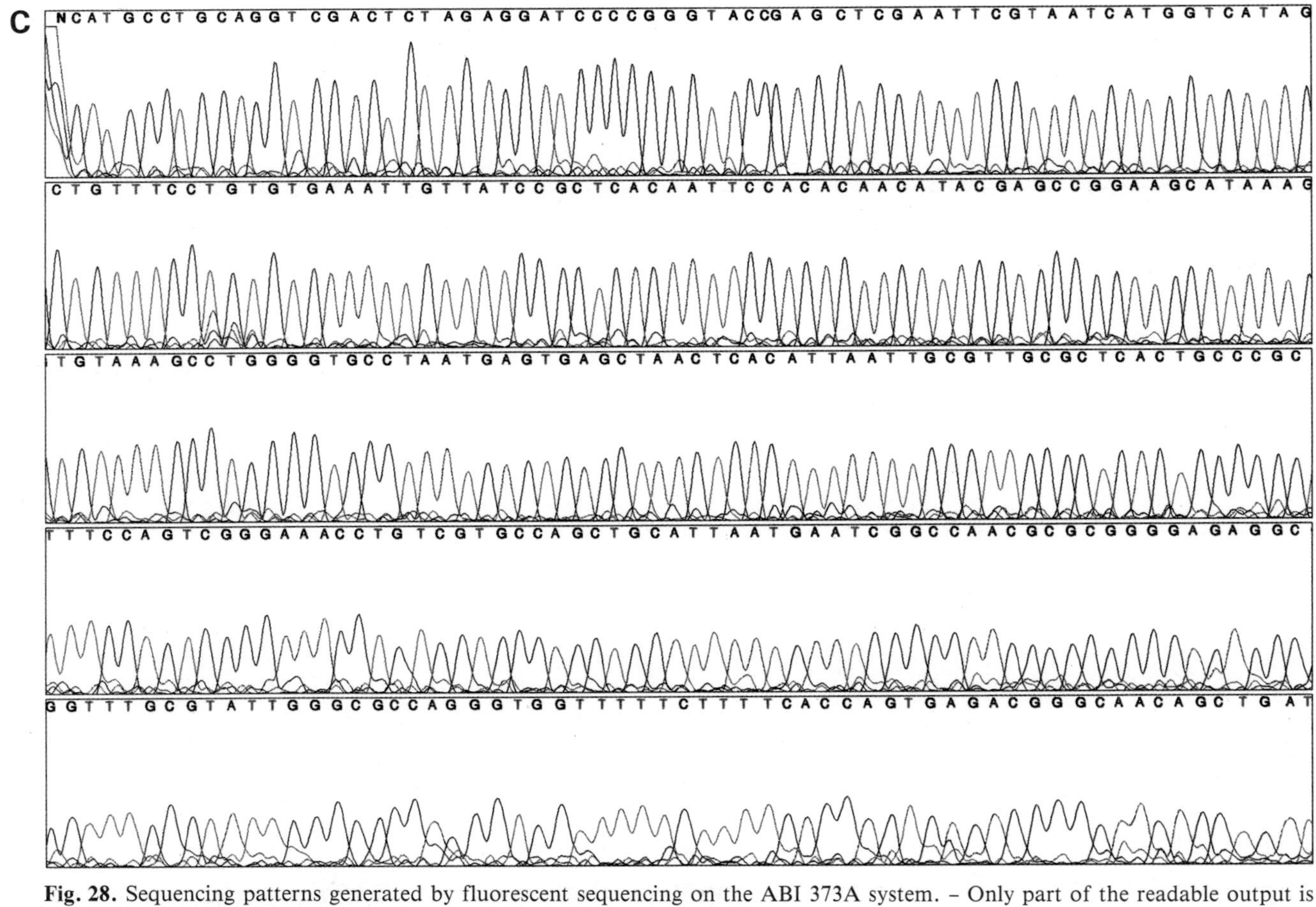

Fig. 28. Sequencing patterns generated by fluorescent sequencing on the ABI 373A system. – Only part of the readable output is shown. Colors of the peaks correspond to G: black, A: green, T: red, C: blue.
(A) Nested sets generated in a cycle procedure with Taq polymerase and fluorescently labeled primers. The strategy outlined in Fig. 19 was used.
(B) Nested sets generated in a cyclic procedure with Taq polymerase and fluorescently labeled dideoxynucleoside triphosphates. Double-stranded template was used with a 17-mer primer.
(C) Nested sets generated by sequencing PCR amplified DNA with fluorescently labeled primers and Sequenase® in the presence of Mn^{2+} (Tabor and Richardson, 1989).

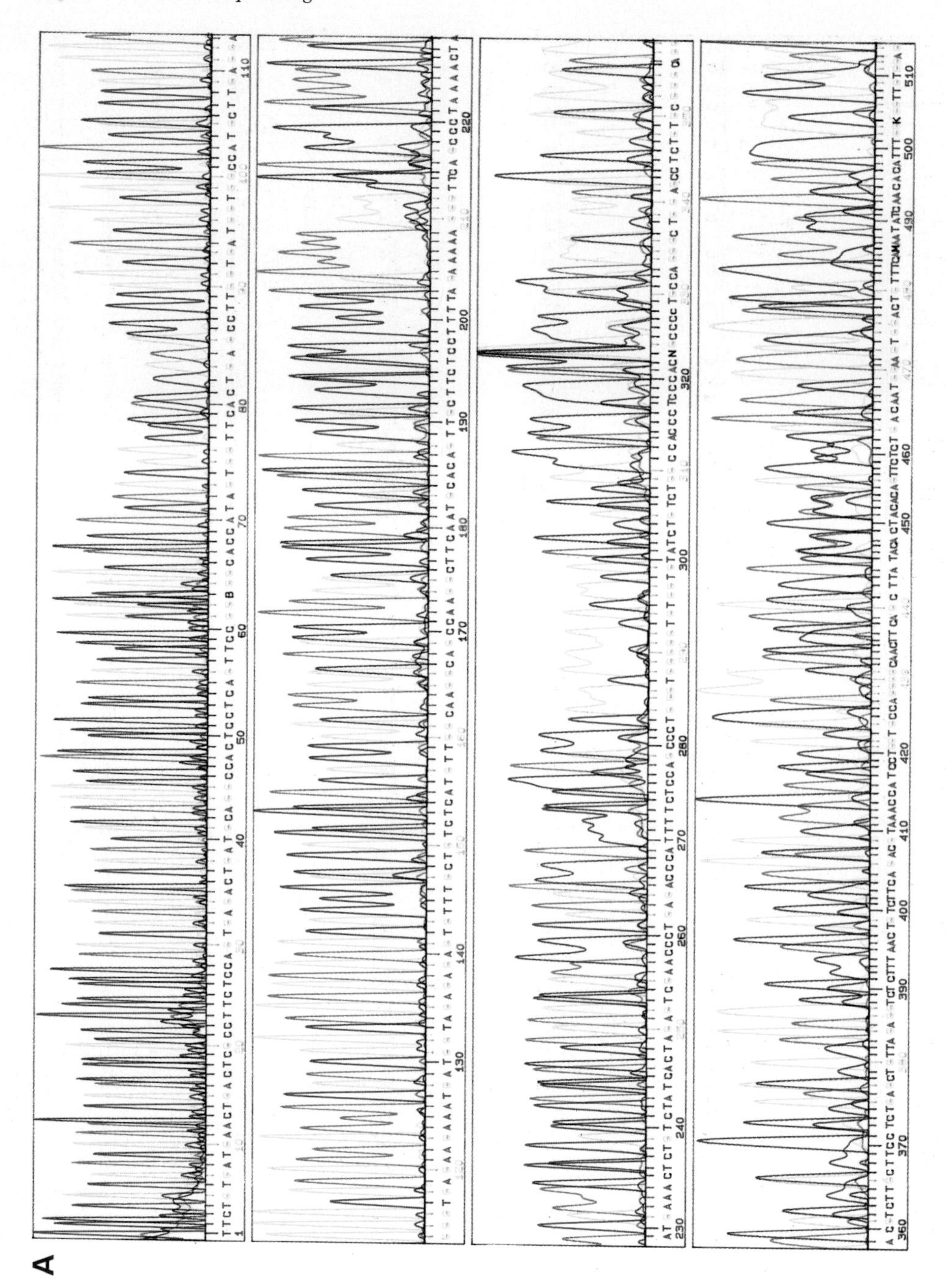
A

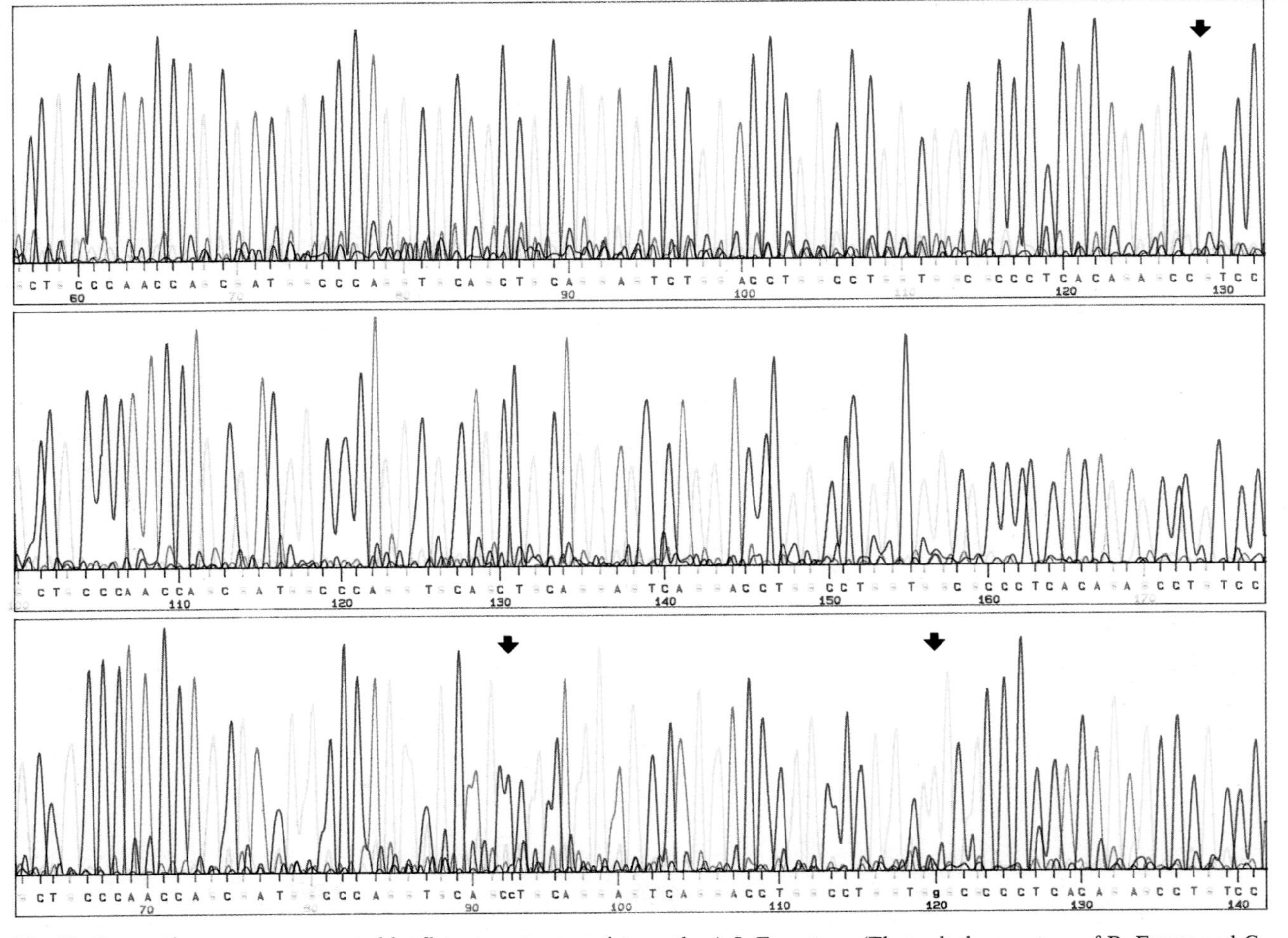

Fig. 29. Sequencing patterns generated by fluorescent sequencing on the A.L.F. system. (Through the courtesy of P. FITEN and G. OPDENAKKER, Rega Institute, University of Leuven). – Colors of the peak correspond to G: yellow, A: green, T: red, C: blue.
(A) Nested sets generated by using T7 DNA polymerase in the presence of Mn^{2+} ions. Reading distance exceeds 500 from the labeled primer.
(B) Solid phase automated sequencing of clones from a mutagenesis experiment. The same DNA segment in three clones is shown in three panels. The positions of the mutations are marked by arrows and correspond to a T deletion in the first panel, and a C and a G insertion in the bottom panel.

data are sent to a computer where they can be processed. Fluorescein-tagged primer is ordinarily used, although it was recently shown that fluorescein-12-dUTP can replace dTTP in sequencing reactions and provide internal tagging in combination with any primer complementary to the template (VOSS et al., 1991). An example of a DNA sequence analysis run on the A.L.F. machine is shown in Fig. 29. More examples are found in EDWARDS et al. (1990), VOSS et al. (1990), HULTMAN et al. (1991), KRISTENSEN et al. (1991), SYVÄNEN et al. (1991), STEGEMANN et al. (1991). Different procedures for fluorescent labeling of DNA followed by chemical degradation followed by analysis on the EMBL sequencing apparatus is described by VOSS et al. (1989) and ROSENTHAL et al. (1990b).

At the time of writing this chapter, a new commercial DNA sequencer, called BaseStation™ (Millipore Corp.) has been announced, using a single fluorescein-tagged primer and running 4 samples per DNA fragment. Instead of a laser, a CCD camera is used to visualize the bands, and band patterns and shapes are analyzed from these images.

Other automated systems for gel fractionation and data collection have been described, e.g., by KAMBARA et al. (1988, 1991), BRUMBAUGH et al. (1988), KOSTICHKA et al. (1992). References to automated systems using fractionation by capillary electrophoresis have been given in Sect. 4.3.

6.2 Robotization

Purification of DNA templates and enzymatic preparation of the nested sets are two tedious and repetitive steps in current dideoxy sequencing that have been automated by the use of robotic workstations. Automation of these steps might also minimize problems associated with human errors or variations in reaction conditions resulting from manipulation of large series of samples. Early realizations of automation of the dideoxy sequencing reactions have been reported by MARTIN et al. (1985) and FRANK et al. (1988). Most developments, however, have been achieved by making use of a commercially available robotic workstation: the Biomek®1000 (Beckman), a small bench-top robot pipettor originally designed for enzyme-linked immunoassays. It consists of a computer, an electronic interface unit, and an elevator/pod/tablet assembly and operates at ambient temperature. The computer controls all the functions of the instrument and software for programming of the operations is provided. Volumes down to 2 microliters can be accurately delivered with a single- or 8-channel pipetting device with disposable tips and using 96-well microtiter trays or microcentrifuge tubes as vessels. The system has been successfully used for semiautomated template purification (MARDIS and ROE, 1989; ZIMMERMAN et al., 1989; SMITH et al., 1990b), for performing the sequencing reactions (ZIMMERMAN et al., 1988; WILSON et al., 1988, 1990b; BANKIER and BARRELL, 1989; SCHOFIELD et al., 1989), and for an integrated automation of both steps (HULTMAN et al., 1991; SYVÄNEN et al., 1991). Template purification by ZIMMERMAN et al. (1989) is an automated version of a manual filtration method devised by KRISTENSEN et al. (1987).

The Catalyst™800 (Applied Biosystems, Inc.) is a multipurpose platform for micro-volume pipetting and thermal cycling (CATHCART, 1990). Currently, it performs optimized procedures for cycle sequencing (Sect. 4.1.1) using fluorescently labeled primers and terminators. Cycle sequencing is reminiscent of PCR, hence repeated heating and cooling is required, and the construction of the thermal cycler plate is designed to avoid uncontrolled sample evaporation. Liquid aspiration and delivery is syringe-based with a robotic arm and a single stainless steel probe that automatically calibrates itself with various modules on the work surface. Due to its recent introduction the performance of this system has not yet been documented in the scientific literature.

The purpose of these systems is to implement as much automation as possible in DNA sequencing. This might look unnecessary for the molecular biologist confronted with daily sequencing tasks of different kinds and size, or involved in rather modest sequencing projects. The feasibility of sequencing large genomes, however, is probably staked on further automation.

7 Storage, Retrieval, and Interpretation of Sequence Data

The scope of this chapter has been limited to laboratory techniques implicated in the actual manipulation of DNA in the process of sequence data acquisition. DNA segments can now readily be purified by cloning or amplified *in vitro*, and analyzed. This has resulted in a constantly growing amount of nucleotide sequence data. Yet, this information is useless until the significant features of the sequences can be deduced. Fortunately, highly developed computer methods are readily applicable to these problems and many different programs and software packages have been developed with different capabilities and performance. The capabilities include storing of primary data, searching for homologies and alignment of sequences, finding and joining overlapping fragments, analysis of secondary structure (e.g., RNA folding), search of typical patterns such as promoter sequences, finding open reading-frames and their decoding to the corresponding amino acid sequences, analysis of the translated sequences in terms of hydropathy, helicity, etc. Description at large of this kind of downstream analysis of DNA sequencing data should be consulted in the specialized literature, e.g., CABIOS (*Comput. Appl. Biosci.*). Some reviews and prospects are in LEWITTER and RINDONE (1987), BISHOP and RAWLINGS (1987), CANNON (1990), VON HEIJNE (1991), MURAL et al. (1992).

Nucleotide sequencing data are compiled and stored in a structured manner in several international databases, from where they can be retrieved (see FIELDS, 1992, and FUCHS et al., 1992). The best known and probably most up-to-date databases are the EMBL databank (Heidelberg, Germany), and Genbank (Los Alamos, USA). The databases are systematically updated and grow steadily as sequences are published or directly submitted to the central offices. Consequently, the accuracy of the data clearly depends on the individual researcher who submits the data. KRAWETZ (1989) estimated an error rate of 0.37 to 2.89 per 1000 in Genbank. This stresses the importance of sequencing both strands of the DNA and confirming every area with an extra experiment, considering the significant impact of inaccurate data on functional analysis of the sequences. One should also bear in mind that data in the databases are quite redundant in some research areas. Hence, general comparisons of, e.g., overall G + C content, overall codon usage, etc., between organisms, as well as individual data must be accessed with scrutiny.

The current initiatives towards unraveling entire genomes of organisms, such as *Escherichia coli, Bacillus subtilis, Caenorabditis elegans, Saccharomyces cerevisiae, Arabidopsis thaliana, Homo sapiens* will boost the growth of the current databases. They will undoubtedly soon outgrow their capacities. This is not only a problem of data storage, but also raises questions as to how sequences can be retrieved from enormous databases, or even more problematic, how sequence comparisons will be possible. Bioinformatics is requiring novel concepts in the near future.

8 References

ADAMS, J. M., JEPPESEN, P. G. SANGER, F., BARRELL. B. G. (1969), Nucleotide sequence from the coat protein cistron of R17 bacteriophage RNA, *Nature* **223**, 1009–1014.

AGARWAL, S., CHRISTODOULOU, C., GAIT, M. J. (1986), Efficient methods for attaching non-radioactive labels to the 5′ ends of synthetic oligodeoxyribonucleotides, *Nucleic Acids Res.* **14**, 6227–6245.

AHMED, A. (1985), A rapid procedure for DNA sequencing using transposon-promoted deletions in *Escherichia coli, Gene* **39**, 305–310.

ALLARD, M. W., ELLSWORTH, D. L., HONEYCUTT, R. L. (1991), The production of single-stranded DNA suitable for sequencing using the polymerase chain reaction, *BioTechniques* **10**, 24–26.

AMBROSE, B. J. B., PLESS, R. C. (1985), Analysis of DNA sequences using a single chemical cleavage procedure, *Biochemistry* **24**, 6194–6200.

ANDERSON, S. (1981), Shotgun DNA sequencing using cloned DNase I-generated fragments, *Nucleic Acids Res.* **9**, 3015–3027.

ANDERSON, S., GAIT, M. J., MAYOL, L., YOUNG, I. (1980), A short primer for sequencing DNA cloned in the single-stranded phage vector M13mp2, *Nucleic Acids Res.* **8,** 1731–1743.

ANSORGE, W., DE MAEYER, L. (1980), Thermally stabilized very thin (0.02–0.3 mm) polyacrylamide gels for electrophoresis, *J. Chromatogr.* **202,** 45–53.

ANSORGE, W., SPROAT, B. S., STEGEMANN, J. SCHWAGER, C. (1986), A non-radioactive automated method for DNA sequence determination, *J. Biochem. Biophys. Methods* **13,** 315–323.

ANSORGE, W., SPROAT, B. S., STEGEMANN, J., SCHWAGER, C., ZENKE, M. (1987), Automated DNA sequencing: ultrasensitive detection of fluorescent bands during electrophoresis, *Nucleic Acids Res.* **15,** 4593–4602.

ARLINGHAUS, H. F., THONNARD, N., SPAAR, M. T., SACHLEBEN, R. A., LARIMER, F. W., FOOTE, R. S., WOYCHIK, R. P., BROWN, G. M., SLOOP, F. V., JACOBSON, K. B. (1991), Potential application of sputter-initiated resonance ionization spectroscopy for DNA sequencing, *Anal. Chem.* **63,** 402–407.

ARNOLD, W., PÜHLER, A. (1988), A family of high-copy-number plasmid vectors with single end-label sites for rapid nucleotide sequencing, *Gene* **70,** 171–179.

ARSCOTT, P. G., LEE, G., BLOOMFIELD, V. A., EVANS, D. F. (1989), Scanning tunnelling microscopy of Z-DNA, *Nature* **339,** 484–486.

BAER, R., BANKIER, A. T., BIGGIN, M. D., DEININGER, P. L., FARRELL, P. J., GIBSON, T. J., HATFULL, G., HUDSON, G. S., SATCHWELL, S. C., SEGUIN, C., TUFFNELL, P. S., BARRELL, B. G. (1984), DNA sequence and expression of the B95-8 Epstein-Barr virus genome, *Nature* **310,** 207–211.

BAINS, W. (1991), Hybridization methods for DNA sequencing, *Genomics* **11,** 294–301.

BAINS, W., SMITH, G. C. (1988), A novel method for nucleic acid sequence determination, *J. Theor. Biol.* **135,** 303–307.

BANKIER, A. T., BARRELL, B. G. (1989), Sequencing single-stranded DNA using the chain-termination method, in: *Nucleic Acid Sequencing: A Practical Approach* (HOWE, C. J., WARD, S. E., Eds.), pp. 37–78, Oxford: IRL Press.

BANKIER, A. T., WESTON, K. M., BARRELL, B. G. (1987), Random cloning and sequencing by the M13/dideoxynucleotide chain termination method, *Methods Enzymol.* **155,** 51–93.

BANKIER, A. T., BECK, S., BOHNI, R., BROWN, C. M., CERNY, R., CHEE, M. S., HUTCHISON, C. A. III., KOUZARIDES, T., MARTIGNETTI, J. A., PREDDIE, E., SATCHWELL, S. C., TOMLINSON, P., WESTON, K. M., BARRELL, G. G. (1991), The DNA sequence of the human cytomegalovirus genome, *DNA Sequence 2,* 1–12.

BARNES, W. M., BEVAN, M. (1983), Kilo-sequencing: an ordered strategy for rapid DNA sequence data acquisition, *Nucleic Acids Res.* **11,** 349–368.

BARRELL, B. G., AIR, G. M., HUTCHISON, C. A. III. (1976), Overlapping genes in bacteriophage ϕX174, *Nature* **264,** 34–41.

BARRELL, B. G., BANKIER, A. T., DROUIN, J. (1979), A different genetic code in human mitochondria, *Nature* **282,** 189–194.

BECHHOFER, D. H. (1991), A method for sequencing polymerase chain reaction products can be used to sequence *Bacillus subtilis* "miniprep" plasmid DNA, *BioTechniques* **10,** 17–20.

BECK, S. (1987), Colorimetric-detected DNA sequencing, *Anal. Biochem.* **164,** 514–520.

BECK, S. (1990), Colorimetric-detected DNA sequencing, *Methods Enzymol.* **184,** 612–617.

BECK, S., POHL, F. M. (1984), DNA sequencing with direct blotting electrophoresis, *EMBO J.***3,** 2905–2909.

BECK, S., O'KEEFFE, T., COULL, J. M., KÖSTER, H. (1989), Chemiluminescent detection of DNA: Application for DNA sequencing and hybridization, *Nucleic Acids Res.* **17,** 5115–5123.

BECKER, M. M., WANG, Z., GROSSMANN, G., BECHERER, K. A. (1989), Genomic footprinting in mammalian cells with ultraviolet light, *Proc. Natl. Acad. Sci. USA* **86,** 5315–5319.

BEEBE, T. P., WILSON, T. E., OGLETREE, D. F., KATZ, J. E., BALHORN, R., SALMERON, M. B., SIEKHAUS, W. J. (1989), Direct observation of native DNA structures with the scanning tunneling microscope, *Science* **243,** 370–372.

BERG, P., FANCHER, H., CHAMBERLIN, M. (1963), in: *Symposium on Informational Macromolecules* (VOGEL, H. J., BRYSON, V., LAMPEN, J. O., Eds.), pp. 467–483, New York: Academic Press.

BIGGER, C. H., MURRAY, K., MURRAY, N. E. (1973), Recognition sequence of a restriction enzyme, *Nature* (*New Biol.*) **244,** 7–10.

BIGGIN, M. D., GIBSON, T. J., HONG, G. F. (1983), Buffer gradient gels and ^{35}S label as an aid to rapid DNA sequence determination, *Proc. Natl. Acad. Sci. USA* **80,** 3963–3965.

BISHOP, M. J., RAWLINGS, C. J. (1987), *Nucleic Acid and Protein Sequence Analysis, a Practical Approach*, Oxford: IRL Press.

BLATTNER, F. R., WILLIAMS, B. G., BLECHL, A. E., DENNISTON-THOMPSON, K., FABER, H. E., FURLONG, S. A., GRUNWALD, D. J., KIEFER, D. O., MOORE, D. D., SHELDON, E. L., SMITHIES, O. (1977), Charon phages: safer derivatives of bacteriophage λ for DNA cloning, *Science* **196,** 161–169.

BOTH B., KRUPP, G., STACKEBRANDT, E. (1991), Direct sequencing of double-stranded polymerase chain reaction-amplified 16S rDNA, *Anal. Biochem.* **199,** 216–218.

BROWNLEE, G. G., SANGER, F. (1969), Chromatography of ^{32}P-labelled oligonucleotides on thin-layers of DEAE-cellulose, *Eur. J. Biochem.* **11,** 395–399.

BRUMBAUGH, J. A., MIDDENDORF, L. R., GRONE, D. L., RUTH, J. L. (1988), Continuous, on-line DNA sequencing using oligodeoxynucleotide primers with multiple fluorophores, *Proc. Natl. Acad. Sci. USA* **85,** 5610–5614.

BRUMLEY JR., R. L., SMITH, L. M. (1991), Rapid DNA sequencing by horizontal ultrathin gel electrophoresis, *Nucleic Acids Res.* **19,** 4121–4126.

BURTON, K., PETERSON, G. B. (1960), The frequencies of certain sequences of nucleotides in deoxyribonucleic acid, *Biochem. J.* **75,** 17–27.

CANNON, G. (1990), Nucleic acid sequence analysis software for microcomputers, *Anal. Biochem.* **190,** 147–153.

CATHCART, R. (1990), Advances in automated DNA sequencing, *Nature* **347,** 310–310.

CHEE, M. (1991), Enzymatic multiplex DNA sequencing, *Nucleic Acids Res.* **19,** 3301–3305.

CHEN, E. Y., SEEBURG, P. H. (1985), Supercoil sequencing: a fast and simple method for sequencing plasmid DNA, *DNA* **4,** 165–170.

CHOW, W-Y., BERG, D. E. (1988), Tn*5tacl*, a derivative of transposon Tn*5* that generates conditional mutations, *Proc. Natl. Acad. Sci. USA* **85,** 6468–6472.

CHURCH, G. M., GILBERT, W. (1984), Genomic sequencing, *Proc. Natl. Acad. Sci. USA* **81,** 1991–1995.

CHURCH, G. M., KIEFFER-HIGGINS, S. (1988), Multiplex DNA sequencing, *Science* **240,** 185–188.

COHEN, A. S., NARAJIAN, D. R., KARGER, B. L. (1990), Separation and analysis of DNA sequence reaction products by capillary electrophoresis, *J. Chromatogr.* **516,** 49–60.

COLLASIUS, M., PUCHTA, H., SCHENKLER, S., VALET, G. (1991), Analysis of unknown DNA sequences by polymerase chain reaction (PCR) using a single specific primer and a standardized adaptor, *J. Virol. Methods* **32,** 115–119.

COLLINS, J., HOHN, B. (1978), Cosmids: a type of plasmid gene-cloning vector that is packageable *in vitro* in bacteriophage lambda heads, *Proc. Natl. Acad. Sci. USA* **75,** 4242–4246.

CONNELL, C., FUNG, S., HEINER, C., BRIDGHAM, J., CHAKERIAN, V., HERON, E., JONES, B., MENCHEN, S., MORDAN, W., RAFF, M., RECKNOR, M., SMITH, L., SPRINGER, J., WOO, S., HUNKAPILLAR, M. (1987), Automated DNA sequence analysis, *BioTechniques* **5,** 342–348.

CORNISH-BOWDEN, A. (1985), Nomenclature for incompletely specified bases in nucleic acid sequences: recommendations 1984, *Nucleic Acids Res.* **13,** 3021–3030.

DAHR, R., ZAIN, S., WEISSMAN, S. M., PAN, J., SUBRAMANIAN, K. (1974), Nucleotide sequences of RNA transcribed in infected cells and by *Escherichia coli* RNA polymerase from a segment of Simian virus 40 DNA, *Proc. Natl. Acad. Sci. USA* **71,** 371–375.

DALE, R. M. K., MCCLURE, B. A., HOUCHINS, J. P. (1985), A rapid single-stranded cloning strategy for producing a sequential series of overlapping clones for use in DNA sequencing: application to sequencing the corn mitochondrial 18S rDNA, *Plasmid* **13,** 31–40.

DANNA, K., NATHANS, D. (1971), Specific cleavage of Simian Virus 40 DNA by restriction endonuclease of *Hemophilus influenzae, Proc. Natl. Acad. Sci. USA* **68,** 2913–2917.

DAVIES, R. J. H., BOYD, D. R., KUMAR, S., SHARMA, N. D., STEVENSON, C. (1990), Preferential modification of guanine bases in DNA by dimethyldioxirane and its application to DNA sequencing, *Biochem. Biophys. Res. Commun.* **169,** 87–94.

DAVIS, L. M., FAIRFIELD, F. R., HARGER, C. A., JETT, J. H., KELLER, R. A., HAHN, J. H., KRAKOWSKI, L. A., MARRONE, B. L., MARTIN, J. C., NUTTER, H. L., RATLIFF, R. L., SHERA, E. B., SIMPSON, D. J., SOPER, S. A. (1991), Rapid DNA sequencing based upon single molecule detection, *Genet. Anal.* **8,** 1–7.

DAVISON, A. J. (1991), Experience in shotgun sequencing a 134 kilobase pair DNA molecule, *DNA Sequence* **1,** 389–394.

DEININGER, P. L. (1983), Random subcloning of sonicated DNA: application to shotgun DNA sequence analysis, *Anal. Biochem.* **129,** 216–223.

DE WACHTER, R., FIERS, W. (1971), Fractionation of RNA by electrophoresis on polyacrylamide gel slabs, *Methods Enzymol.* **21,** 167–178.

DRMANAC, R., LABAT, I., BRUKNER, I., CRKVENKAKOV, R. (1989), Sequencing of megabase plus DNA by hybridization: theory of the method, *Genomics* **4,** 114–128.

DROSSMAN, H., LUCKEY, J. A., KOSTICHKA, A. J., D'CUNHA, J., SMITH, L. M. (1990), High-speed separations of DNA sequencing reactions by capillary electrophoresis, *Anal. Chem.* **62,** 900–903.

DUBOSE, R. F., HARTL, D. L. (1990), Rapid purification of PCR products for DNA sequencing using sepharose CL-6B spin columns, *BioTechniques* **8**, 271–273.

Duckworth, M. L., Gait, M. J., Goelet, P., Hong, G. F., Sing, M., Titmas, R. C. (1981), Rapid synthesis of oligodeoxyribonucleotides. VI. Efficient mechanised synthesis of heptadecadeoxynucleotides by an improved solid phase phosphotriester route, *Nucleic Acids Res.* **9**, 1691–1706.

Dunlap, D. D., Bustamante, C. (1989), Images of single-stranded nucleic acids by scanning tunneling microscopy, *Nature* **342**, 204–206.

Eby, M. J. (1990), New sequence scanners come of age, *Bio/Technology* **8**, 1046–1049.

Eckert, R. L. (1987), New vectors for rapid sequencing of DNA fragments by chemical degradation, *Gene* **51**, 247–254.

Edwards, A., Voss, H., Rice, P., Civitello, A., Stegemann, J., Schwager, C., Zimmermann, J., Erfle, H., Caskey, C. T., Ansorge, W. (1990), Automated DNA sequencing of the human HPRT locus, *Genomics* **6**, 593–608.

Elder, J. K. (1990), Maximum entropy image reconstruction of DNA sequencing gel autoradiographs, *Electrophoresis* **11**, 440–444.

Elder, J. K., Green, D. K., Southern, E. M. (1986), Automatic reading of DNA sequencing gel autoradiographs using a large format digital scanner, *Nucleic Acids Res.* **14**, 417–424.

Englund, P. T. (1972), The 3′-terminal nucleotide sequences of T7 DNA, *J. Mol. Biol.* **66**, 209–224.

Fields, C. (1992), Data exchange and inter-database communication in genome projects, *Trends Biotechnol.* **10**, 58–61.

Fiers, W., Contreras, R., Duerinck, F., Haegeman, G., Iserentant, D., Merregaert, J., Min Jou, W., Molemans, F., Raeymakers, A., Van den Berghe, A., Volckaert, G., Ysebaert, M. (1976), Complete nucleotide sequence of bacteriophage MS2 RNA: primary and secondary structure of the replicase gene, *Nature* **260**, 500–507.

Fiers, W., Contreras, R., Haegeman, G., Rogiers, R., Van de Voorde, A., Van Heuverswyn, H., Van Herreweghe, J., Volckaert, G., Ysebaert, M. (1978), The complete nucleotide sequence of SV40 DNA, *Nature* **273**, 113–120.

Frank, R., Müller, D., Wolff, C. (1981), Identification and suppression of secondary structures formed from deoxy-oligonucleotides during electrophoresis in denaturing polyacrylamide-gels, *Nucleic Acids Res.* **9**, 4967–4979.

Frank, R., Bosserhoff, A., Boulin, C., Epstein, A., Gausepohl, H., Ashman, K. (1988), Automation of DNA sequencing reactions and related techniques: a workstation for micromanipulation of liquids, *Bio/Technology* **6**, 1211–1213.

Friedmann, T., Brown, D. M. (1978), Base-specific reactions useful for DNA sequencing: methylene blue-sensitized photooxidation of guanine and osmium tetraoxide modification of thymine, *Nucleic Acids Res.* **5**, 615–622.

Frischauf, A. M., Garoff, H., Lehrach, H. (1980), A subcloning strategy for DNA sequence analysis, *Nucleic Acids Res.* **8**, 5541–5549.

Frischauf, A. M., Lehrach, H., Poustka, A., Murray, N. (1983), λ Replacement vectors carrying polylinker sequences, *J. Mol. Biol.* **170**, 827–842.

Fuchs, R., Rice, P., Cameron, G. N. (1992), Molecular biological databases - present and future, *Trend Biotechnol.* **10**, 61–66.

Gal, S., Hohn, B. (1990), Direct sequencing of double-stranded DNA PCR products via removing of the complementary strand with single-stranded DNA of an M13 clone, *Nucleic Acids Res.* **18**, 1076–1076.

Garoff, H., Ansorge, W. (1981), Improvements of DNA sequencing gels, *Anal. Biochem.* **115**, 450–457.

Gelfi, C., Canali, A., Righetti, P. C., Vezzoni, P., Smith, C., Mellon, M., Jain, T., Shorr, R. (1990), DNA sequencing in HydroLink matrices: Extension of reading ability to >600 nucleotides, *Electrophoresis* **11**, 595–600.

Gibbons, I. R., Asai, D. J., Ching, N. S., Dolecki, G. J., Mocz, G., Phillipson, C. A., Ren, H., Tang, W. Y., Gibbons, B. H. (1991), A PCR procedure to determine the sequence of large polypeptides by rapid walking through a cDNA library, *Proc. Natl. Acad. Sci. USA* **88**, 8563–8567.

Gibbs, R. A., Nguyen, P.-N., McBride, L. J., Koepf, S. M., Caskey, C. T. (1989), Identification of mutations leading to the Lesch-Nyhan syndrome by automated direct DNA sequencing of *in vitro* amplified cDNA, *Proc. Natl. Acad. Sci. USA* **86**, 1919–1923.

Gilbert, W. Maxam, A. (1973), The nucleotide sequence of the Lac operator, *Proc. Natl. Acad. Sci. USA* **70**, 3581–3584.

Gish, G., Eckstein, F. (1988), DNA and RNA sequence determination based on phosphorothioate chemistry, *Science* **240**, 1520–1522.

Gonzalez-Cadavid, N., Gatti, R. A., Neuwirth, H. (1990), Automated direct sequencing of polymerase chain reaction-amplified fragments of the human Ha-ras gene, *Anal. Biochem.* **191**, 359–364.

GROISMAN, E. A. (1991), *In vivo* genetic engineering with bacteriophage Mu, *Methods Enzymol.* **204,** 180–212.

GRONENBORN, B., MESSING, J. (1978), Methylation of single-stranded DNA *in vitro* introduces new restriction endonuclease cleavage sites, *Nature* **272,** 375–377.

GROTJAHN, L., FRANK, R., BLÖCKER, H. (1982), Ultrafast sequencing of oligodeoxyribonucleotides by FAB-mass spectrometry, *Nucleic Acids Res.* **10,** 4671–4678.

GÜSSOW, D., CLACKSON, T. (1989), Direct clone characterization from plaques and colonies by the polymerase chain reaction, *Nucleic Acids Res.* **17,** 4000–4000.

GYLLENSTEN, U. B. (1989), PCR and DNA sequencing, *BioTechniques,* **7,** 700–709.

GYLLENSTEN, U. B., ERLICH, H. A. (1988), Generation of single-stranded DNA by the polymerase chain reaction and its application to direct sequencing of the HLA-DQA locus, *Proc. Natl. Acad. Sci. USA* **85,** 7652–7656.

HEDGPETH, J., GOODMAN, H. M., BOYER, H. W. (1972), DNA nucleotide sequence restricted by the RI endonuclease, *Proc. Natl. Acad. Sci. USA* **69,** 3448–3452.

HELLER, C., RADLEY, E., KHURSHID, F. A., BECK, S. (1991), M13plex vectors for multiplex DNA sequencing, *Gene* **103,** 131–132.

HENIKOFF, S. (1984), Unidirectional digestion with exonuclease III creates targeted breakpoints for DNA sequencing, *Gene* **28,** 351–359.

HENIKOFF, S. (1990), Ordered deletions for DNA sequencing and *in vitro* mutagenesis by polymerase extension and exonuclease III gapping of circular templates, *Nucleic Acids Res.* **18,** 2961–2966.

HEYDEN, B., NUSSLEIN, C., SCHALLER, H. (1972), Single RNA polymerase binding site isolated, *Nature* (*New Biol.*) **240,** 9–12.

HIGUCHI, R. G., OCHMAN, H. (1989), Production of single-stranded DNA templates by exonuclease digestion following the polymerase chain reaction, *Nucleic Acids Res.* **17,** 5865–5865.

HONG, G. F. (1982), A systematic DNA sequencing strategy, *J. Mol. Biol.* **158,** 539–549.

HUIBREGTSE, J. M., ENGELKE, D. R. (1986), Direct identification of small sequence changes in chromosomal DNA, *Gene* **44,** 151–158.

HULTMAN, T., STAHL, S., HORNES, E., UHLÉN, M. (1989), Direct solid phase sequencing of genomic and plasmid DNA using magnetic beads, *Nucleic Acids Res.* **17,** 4937–4946.

HULTMAN, T., BERGH, S., MOKS, T., UHLÉN, M. (1991), Bidirectional solid-phase sequencing of *in vitro*-amplified plasmid DNA, *BioTechniques* **10,** 84–93.

HYMAN, E. D. (1988), A new method of sequencing DNA, *Anal. Biochem.* **174,** 423–436.

INNIS, M. A., MYAMBO, K. B., GELFAND, D. H., BROW, M. A. D. (1988), DNA sequencing with *Thermus aquaticus* DNA polymerase and direct sequencing of polymerase chain reaction-amplified DNA, *Proc. Natl. Acad. Sci. USA* **85,** 9436–9440.

JACOBSON, K. B., ARLINGHAUS, H. F., SCHMITT, H. W., SACHLEBEN, R. A., BROWN, G. M., THONNARD, N., SLOOP, F. V., FOOTE, R. S., LARIMER, F. W., WOYCHIK, R. P., ENGLAND, M. W., BURCHETT, K. L., JACOBSON, D. A. (1991), An approach to the use of stable isotopes for DNA sequencing, *Genomics* **9,** 51–59.

JENSEN, M. A., ZAGURSKY, G. J., TRAINOR, G. L., COCUZZA, A. J., LEE, A., CHEN, E. Y. (1991), Improvements in the chain-termination method of DNA sequencing through the use of 7-deaza-2′-deoxyadenosine, *DNA Sequence* **1,** 233–239.

JETT, J. H., KELLER, R. A., MARTIN, J. C., MARRONE, B. L., MOYZIS, R. K., RATLIFF, R. L., SEITZINGER, N. K., SHERA, E. B., STEWART, C. C. (1989), High-speed DNA sequencing: An approach based upon fluorescence detection of single molecules, *J. Biomol. Struct. Dynam.* **7,** 301–309.

JONES, D. S. C., SCHOFIELD, J. P., VAUDIN, M. (1991), Fluorescent and radioactive solid phase dideoxy sequencing of PCR products in microtitre plates, *DNA Sequence* **1,** 279–283.

KALMAN, M., KALMAN, E. T., CASHEL, M. (1990), Polymerase chain reaction (PCR) amplification with a single specific primer, *Biochem. Biophys. Res. Commun.* **167,** 504–506.

KALNOSKI, M. H., MCCOY-HAMAN, M. F., HOLLIS, G. F. (1991), A rapid method for purifying PCR products for direct sequence analysis, *BioTechniques* **11,** 246–249.

KAMBARA, H., NISHIKAWA, T., KATAYAMA, Y., YAMAGUCHI, T. (1988), Optimization of parameters in a DNA sequenator using fluorescence detection, *Bio/Technology* **6,** 816–821.

KAMBARA, H., NAGAI, K., HAYASAKA, S. (1991), Real-time automated simultaneous double-stranded DNA sequencing using two-color fluorophore labeling, *Bio/Technology* **9,** 648–651.

KARGER, A. E. HARRIS, J. M., GESTELAND, R. F. (1991), Multiwavelength fluorescence detection for DNA sequencing using capillary electrophoresis, *Nucleic Acids Res.* **19,** 4955–4962.

KELLY, T. J., SMITH, H. O (1970), A restriction enzyme from *Hemophilus influenzae*. II. Base sequence of the recognition site, *J. Mol. Biol.* **51,** 393–409.

KEOHAVONG, P., THILLY, W. G. (1989), Fidelity of DNA polymerase in DNA amplification, *Proc. Natl. Acad. Sci. USA* **86,** 9253–9257.

KHRAPKO, K. R., LYSOV, Y. P., KHORLIN, A. A., SHICK, V. V., FLORENTIEV, V. L., MIRZABEKOV, A. D. (1989), An oligonucleotide hybridization approach to DNA sequencing, *FEBS Lett.* **256,** 118–122.

KHRAPKO, K. R., LYSOV, Y. P., KHORLIN, A. A., IVANOV, I. B., YERSHOV, G. M., VASILENKO, S. K., FLORENTIEV, V. L., MIRZABEKOV, A. D. (1991), A method for DNA sequencing by hybridization with oligonucleotide matrix, *DNA Sequence* **1,** 375–388.

KLECKNER, N., BENDER, J., GOTTESMAN, S. (1991), Uses of transposons with emphasis on Tn*10*, *Methods Enzymol.* **204,** 139–180.

KLENOW, H., HENNINGSEN, I. (1970), Selective elimination of the exonuclease activity of the deoxyribonucleic acid polymerase from *Escherichia coli* B by limited proteolysis, *Proc. Natl. Acad. Sci. USA* **65,** 168–175.

KLEPPE, R., KHORANA, H. G. (1972), Studies on polynucleotides. CXIX. Transcription of short double-stranded deoxyribonucleic acids of defined nucleotide sequences, *J. Biol. Chem.* **247,** 6149–6156.

KOSTICHKA, A. J., MARCHBANKS, M. L., BRUMLEY JR., R. L., DROSSMAN, H., SMITH, L. M. (1992), High speed automated DNA sequencing in ultrathin slab gels, *Bio/Technology* **10,** 78–81.

KRAWETZ, S. A. (1989), Sequence errors in GenBank: a means to determine the accuracy of DNA sequence interpretation, *Nucleic Acids Res.* **17,** 3951–3957.

KRETZ, K. A., CARSON, G. S., O'BRIEN, J. S. (1989), Direct sequencing from low-melt agarose with sequenase, *Nucleic Acids Res.* **17,** 5864–5864.

KRISHNAN, B. R., KERSULYTE, D., BRIKUN, I., BERG, C. M., BERG, D. E. (1991), Direct and crossover PCR amplification to facilitate Tn*5supF*-based sequencing of phage λ clones, *Nucleic Acids Res.* **19,** 6177–6182.

KRISTENSEN, T., VOSS, H., ANSORGE, W. (1987), A simple and rapid preparation of M13 sequencing templates for manual and automated dideoxy sequencing, *Nucleic Acids Res.* **15,** 5507–5516.

KRISTENSEN, T., VOSS, H., SCHWAGER, C., STEGEMANN, J., SPROAT, B., ANSORGE, W. (1988), T7 DNA polymerase in automated dideoxy sequencing, *Nucleic Acids Res.* **16,** 3487–3496.

KRISTENSEN, T., LARSEN, F., ENGEN, A., SOLHEIM, I., PRYDZ, H. (1991), Rapid and simple preparation of plasmids suitable for dideoxy DNA sequencing and other purposes, *DNA Sequence* **1,** 227–232.

KUSUKAWA, N., UEMORI, T., ASADA, K., KATO, I. (1990), Rapid and reliable protocol for direct sequencing of material amplified by the polymerase chain reaction, *BioTechniques* **9,** 66–72.

LAWYER, F. C., STOFFEL, S., SAIKI, R. K., MYAMBO, K. B., DRUMMOND, R., GELFAND, D. H. (1989), Isolation, characterization, and expression in *E. coli* of the DNA polymerase gene from *Thermus aquaticus, J. Biol. Chem.* **264,** 6427–6437.

LEBOWITZ, P., WEISSMAN, S. M., RADDING, C. M. (1971), Nucleotide sequence of a ribonucleic acid transcribed *in vitro* from λ phage deoxyribonucleic acid, *J. Biol. Chem.* **246,** 5120–5139.

LEE, J. S. (1991), Alternative dideoxy sequencing of double-stranded DNA by cyclic reactions using Taq polymerase, *DNA Cell Biol.* **10,** 67–73.

LEWITTER, F. I., RINDONE, W. P. (1987), Computer programs for analyzing DNA and protein sequences, *Methods Enzymol.* **155,** 582–593.

LINDSAY, S. M., PHILIPP, M. (1991), Can the scanning tunneling microscope sequence DNA?, *Genet. Anal.* **8,** 8–13.

LINDSAY, S. M., THUNDAT, T., NAGAHARA, L., KNIPPING, U., RILL, R. L. (1989), Images of the DNA double helix in water, *Science* **244,** 1063–1064.

LING, V. (1972a), Fractionation and sequences of the large pyrimidine oligonucleotides from bacteriophage fd DNA, *J. Mol. Biol.* **64,** 87–102.

LING, V. (1972b), Pyrimidine sequences from the DNA of bacteriophages fd, fl, and phiX174, *Proc. Natl. Acad. Sci. USA* **69,** 742–746.

LIU, L., WHALEN, W., DAS, A., BERG, C. M. (1987), Rapid sequencing of cloned DNA using a transposon for bidirectional priming: sequence of the *Escherichia coli* K-12 *outA* gene, *Nucleic Acids Res.* **15,** 9461–9469.

LOH, E. Y., ELLIOTT, J. F., CWIRLA, S., LANIER, L. L., DAVIS, M. M. (1989), Polymerase chain reaction with single-sided specificity: analysis of T cell receptor δ chain, *Science* **243,** 217–222.

LUCKEY, J. A., DROSSMAN, H., KOSTICHKA, A. J., MEAD, D. A., D'CUNHA, J., NORRIS, T. B., SMITH, L. M. (1990), High speed DNA sequencing by capillary electrophoresis, *Nucleic Acids Res.* **18,** 4417–4421.

MANAM, S., NICHOLS, W. W. (1991), Multiplex polymerase chain reaction amplification and direct sequencing of homologous sequences: point mutation analysis of the ras gene, *Anal. Biochem.* **199,** 106–111.

MANIATIS, T., PTASHNE, M. (1973), Multiple repressor binding at the operators in bacteriophage lambda, *Proc. Natl. Acad. Sci. USA* **70**, 1531–1535.

MARDIS, E. R., ROE, B. A. (1989), Automated methods for single-stranded DNA isolation and dideoxynucleotide DNA sequencing reactions on a robotic workstation, *BioTechniques* **7**, 840–850.

MAROTTA, C. A., LEBOWITZ, P., DHAR, R., ZAIN, S. B., WEISSMAN, S. M. (1974), Preparation of RNA transcripts of discrete segments of DNA, *Methods Enzymol.* **29**, 254–272.

MARTIN, W. J., DAVIES, R. W. (1986), Automated DNA sequencing: progress and prospects, *Bio/Technology* **4**, 890–895.

MARTIN, W. J., WARMINGTON, J. R., GALINSKI, B. R., GALLAGHER, M., DAVIES, R. W., BECK, M. S., OLIVER, S. G. (1985), Automation of DNA sequencing: a system to perform the Sanger dideoxy sequencing reactions, *Bio/Technology* **3**, 911–915.

MARTIN, C., BRESNICK, L., JUO, R.-R., VOYTA, J. C., BRONSTEIN, I. (1991), Improved chemiluminescent DNA sequencing, *BioTechniques* **11**, 110–113.

MATTILA, P., KORPELA, J., TENKÄNEN, T., PITKÄNEN, K. (1991), Fidelity of DNA synthesis by the *Thermococcus litoralis* DNA polymerase: an extremely heat-stable enzyme with proofreading activity, *Nucleic Acids Res.* **19**, 4967–4973.

MAXAM, A. M., GILBERT, W. (1977), A new method for sequencing DNA, *Proc. Natl. Acad. Sci. USA* **74**, 560–564.

MAXAM, A. M., GILBERT, W. (1980), Sequencing end-labeled DNA with base-specific chemical cleavages, *Methods Enzymol.* **65**, 499–560.

MAZARS, G.-R., MOYRET, C., JEANTEUR, P., THEILLET, C.-G. (1991), Direct sequencing by thermal asymmetric PCR, *Nucleic Acids Res.* **19**, 4783–4783.

MCCLARY, J., YE, S. Y., HONG, G. F., WITNEY, F. (1991), Sequencing with the large fragment of DNA polymerase I from *Bacillus stearothermophilus, DNA Sequence* **1**, 173–180.

MELTZER, S. J., MANE, S. M., WOOD, P. K., JOHNSON, L., NEEDLEMAN, S. W. (1990), Sequencing products of the polymerase chain reaction directly, without purification, *BioTechniques* **8**, 142–148.

MERCIER, B., GAUCHER, C., FEUGEAS, O., MAZURIER, C. (1990), Direct PCR from whole blood, without DNA extraction, *Nucleic Acids Res.* **18**, 5908–5908.

MESSING, J., VIEIRA, J. (1982), A new pair of M13 vectors for selecting either DNA strand of double-digest restriction fragments, *Gene* **19**, 269–276.

MESSING, J., GRONENBORN, B., MÜLLER-HILL, B., HOFSCHNEIDER, P. H. (1977), Filamentous coliphage M13 as a cloning vehicle. Insertion of a *Hind*II fragment of the lac regulatory region in M13 replicative form *in vitro, Proc. Natl. Acad. Sci. USA* **74**, 3642–3646.

MESSING, J., CREA, R., SEEBURG, P. H. (1981), A system for shotgun DNA sequencing, *Nucleic Acids Res.* **9**, 309–321.

MILLS, D. R., KRAMER, F. R. (1979), Structure-independent nucleotide sequence analysis, *Proc. Natl. Acad. Sci. USA* **76**, 2232–2235.

MITCHELL, L. G., MERRIL, C. R. (1989), Affinity generation of single-stranded DNA for dideoxy sequencing following the polymerase chain reaction, *Anal. Biochem.* **178**, 239–242.

MIZUSAWA, S., NISHIMURA, S., SEELA, F. (1986), Improvement of the dideoxy chain termination method of DNA sequencing by use of deoxy-7-deazaguanosine triphosphate in place of dGTP, *Nucleic Acids Res.* **14**, 1319–1324.

MUELLER, P. R., WOLD, B. (1989), *In vivo* footprinting of a muscle specific enhancer by ligation mediated PCR, *Science* **246**, 780–786.

MURAL, R. J., EINSTEIN, J. R., GUAN, X., MANN, R. C., UBERBACHER, E. C. (1992), An artificial intelligence approach to DNA sequence feature recognition, *Trends Biotechnol.* **10**, 66–69.

NAG, D. K., HUANG, H. V., BERG, D. E. (1988), Bidirectional chain-termination nucleotide sequencing: transposon Tn*5seq1* as a mobile source of primer sites, *Gene* **64**, 135–145.

NAKAMAYE, K. L., GISH, G., ECKSTEIN, F., VOSBERG, H.-P. (1988), Direct sequencing of polymerase chain reaction amplified DNA fragments through the incorporation of deoxynucleoside α-thiotriphosphates, *Nucleic Acids Res.* **16**, 9947–9960.

NEWTON, C. R., KALSHEKER, N., GRAHAM, A., POWELL, S., GAMMACK, A., RILEY, J., MARKHAM, A. F. (1988), Diagnosis of α1-antitrypsin deficiency by enzymatic amplification of human genomic DNA and direct sequencing of polymerase chain reaction products, *Nucleic Acids Res.* **16**, 8233–8243.

NICK, H., BOWEN, B., FERL, R. J., GILBERT, W. (1986), Detection of cytosine methylation in the maize alcohol dehydrogenase gene by genomic sequencing, *Nature* **319**, 243–246.

OHARA, O., DORIT, R. L., GILBERT, W. (1989), Direct genomic sequencing of bacterial DNA: the pyruvate kinase I gene of *Escherichia coli, Proc. Natl. Acad. Sci. USA* **86**, 6883–6887.

OHNO, K., TANAKA, M., INO, H., SUZUKI, H., TASHIRO, M., IBI, T., SAHASHI, K., TAKAHASHI, A., OZAWA, T. (1991), Direct DNA sequencing from colony: analysis of multiple deletions of mitochondrial genome, *Biochim. Biophys. Acta Gene Struct. Expression,* **1090,** 9–16.

OKUBO, K., HORI, N., MATOBA, R., NIIYAMA, T., MATSUBARA, K. (1991), A novel system for large-scale sequencing of cDNA by PCR amplification, *DNA Sequence* **2,** 137–144.

OLSSON, A., MOKS, T., UHLÉN, M., GAAL, F. B. (1985), Uniformly spaced banding pattern in DNA sequencing gels by use of field-strength gradient, *J. Biochem. Biophys. Methods* **10,** 83–90.

PADMANABHAN, R., WU, R. (1972), Nucleotide sequence analysis of DNA. IV. Complete nucleotide sequence of the left-hand cohesive and of coliphage 186 DNA, *J. Mol. Biol.* **65,** 447–467.

PARKER, J. D., RABINOVITCH, P. S., BURMER, G. C. (1991), Targeted gene walking polymerase chain reaction, *Nucleic Acids Res.* **19,** 3055–3060.

PENG, Z. G., WU, R. (1986), A simple and rapid nucleotide sequencing strategy and its application in analyzing a rice histone gene, *Gene* **45,** 247–252.

PFEIFER, G. P., STEIGERWALD, S. D., MUELLER, P. R., WOLD, B., RIGGS, A. D. (1989), Genomic sequencing and methylation analysis by ligation mediated PCR, *Science* **246,** 810–813.

PHADNIS, S. H., HUANG, H. V., BERG, D. E. (1989), Tn*5SupF*, a 264-base-pair transposon derived from Tn*5* for insertion mutagenesis and sequencing DNAs cloned in phage λ, *Proc. Natl. Acad. Sci. USA* **85,** 6468–6472.

PONCZ, M., SOLOWIEJOZYK, D., BALLENTINE, M., SCHWARTZ, F., SURREY, S. (1982), Nonrandom DNA sequence analysis in bacteriophage M13 by the dideoxy chain-termination method, *Proc. Natl. Acad. Sci. USA* **79,** 4298–4302.

PROBER, J. M., TRAINOR, G. L., DAM, R. J., HOBBS, F. W., ROBERTSON, C. W., ZAGURSKY, R. J., COCUZZA, A. J., JENSEN, M. A., BAUMEISTER, K. (1987), A system for rapid DNA sequencing with fluorescent chain-terminating dideoxynucleotides, *Science* **238,** 336–341.

RICHTERICH, P., HELLER, C., WURST, H., POHL, F. M. (1989), DNA sequencing with direct blotting electrophoresis and colorimetric detection, *BioTechniques* **7,** 52–59.

ROBERTSON, H. D., BARRELL, B. G., WEITH, H. L., DONELSON, J. E. (1973), Isolation and sequence analysis of a ribosome-protected fragment from bacteriophage ϕX174 DNA, *Nature (New Biol.)* **241,** 38–40.

ROBINSON, N. J., GUPTA, A., FORDHAM-SKELTON, A. P., CROY, R. R. D., WHITTON, B. A., HUCKLE, J. W. (1990), Prokaryotic metallothionein gene characterization and expression: chromosome crawling by ligation-mediated PCR, *Proc. R. Soc. Lond.* **B 242,** 241–247.

ROSENTHAL, A., JONES, D. S. C. (1990), Genomic walking and sequencing by oligo-cassette mediated polymerase chain reaction, *Nucleic Acids Res.* **18,** 3095–3096.

ROSENTHAL, A., JUNG, R., HUNGER, H.-D. (1986), Optimized conditions for solid-phase sequencing: simultaneous chemical cleavage of a series of long DNA fragments immobilized on CCS anion-exchange paper, *Gene* **41,** 1–9.

ROSENTHAL, A., SPROAT, B. S., BROWN, D. M. (1990a), A new guanine-specific reaction for chemical DNA sequencing using *m*-chloroperoxybenzoic acid, *Biochem. Biophys. Res. Commun.* **173,** 272–275.

ROSENTHAL, A., SPROAT, B. S., VOSS, H., STEGEMANN, J., SCHWAGER, C., ERFLE, H., ZIMMERMANN, J., COUTELLE, C., ANSORGE, W. (1990b), Automated sequencing of fluorescently labelled DNA by chemical sequencing, *DNA Sequence* **1,** 63–71.

ROUX, K. H., DHANARAJAN, P. (1990), A strategy for single-site PCR amplification of dsDNA: priming digested cloned or genomic DNA from an anchor-modified restriction site in a short internal sequence, *BioTechniques* **8,** 48–57.

RUANO, G., KIDD, K. K. (1991), Coupled amplification and sequencing of genomic DNA, *Proc. Natl. Acad. Sci. USA* **88,** 2815–2819.

RUBIN, C. M., SCHMID, C. W. (1980), Pyrimidine-specific chemical reactions useful for DNA sequencing, *Nucleic Acids Res.* **8,** 4613–4619.

RÜTHER, U., KOENEN, M., OTTO, K., MÜLLER-HILL, B. (1981), pUR222: a vector for cloning and rapid chemical sequencing of DNA, *Nucleic Acids Res.* **9,** 4087–4098.

SACHLEBEN, R. A., BROWN, G. M., SLOOP, F. V., ARLINGHAUS, H. F., ENGLAND, M. W., FOOTE, R. S., LARIMER, F. W., WOYCHIK, R. P., THONNARD, N., JACOBSON, K. B. (1991), Resonance ionization spectroscopy for multiplex sequencing of tin-labeled DNA, *Genet. Anal.* **8,** 167–170.

SAIKI, R. K., SCHARF, S., FALOONA, F., MULLIS, K. B., HORN, G. T., ERLICH, H. A., ARNHEIM, N. (1985), Enzymatic amplification of β-globin genomic sequences and restriction site analysis for diagnosis of sickle cell anemia, *Science* **230,** 1350–1354.

SAIKI, R. K., GELFLAND, D. H., STOFFEL, S., SCHARF, S. J., HIGUCHI, R., HORN, G. T., MULLIS, K. B., ERLICH, H. A. (1988), Primer-directed enzymatic amplification of DNA with a thermostable DNA polymerase, *Science* **239,** 487–491.

SAITO, I., SUGIYAMA, H., MATUURA, T., UEDA, K., KOMANO, T. (1984), A new procedure for determining thymine residues in DNA sequencing.

Photo-induced cleavage of DNA fragments in the presence of spermine, *Nucleic Acids Res.* **12**, 2879–2885.

SALUZ, H. P., JOST, J. P. (1986), Optimized genomic sequencing as a tool for the study of cytosine methylation in the regulatory region of the chicken vitellogenin II gene, *Gene* **42**, 151–157.

SALUZ, H. P., JOST, J.-P. (1989), A simple high-resolution procedure to study DNA methylation and *in vivo* DNA-protein interactions on a single-copy gene level in higher eukaryotes, *Proc. Natl. Acad. Sci. USA* **86**, 2602–2606.

SANGER, F., COULSON, A. R. (1975), A rapid method for determining sequences in DNA by primed synthesis with DNA polymerase, *J. Mol. Biol.* **94**, 444–448.

SANGER, F., COULSON, A. R. (1978), The use of thin acrylamide gels for DNA sequencing, *FEBS Lett.* **87**, 107–110.

SANGER, F., BROWNLEE, G. G., BARRELL, B. G. (1965), A two-dimensional fractionation procedure for radioactive nucleotides, *J. Mol. Biol.* **13**, 373–398.

SANGER, F., DONELSON, J. E., COULSON, A. R., KÖSSEL, H., FISCHER, D. (1973), Use of DNA polymerase I primed by a synthetic oligonucleotide to determine a nucleotide sequence in phage f1 DNA, *Proc. Natl. Acad. Sci. USA* **70**, 1209–1213.

SANGER, F., DONELSON, J. E., COULSON, A. R., KÖSSEL, H., FISCHER, D. (1974), Determination of a nucleotide sequence in bacteriophage f1 DNA by primed synthesis with DNA polymerase, *J. Mol. Biol.* **90**, 315–333.

SANGER, F., NICKLEN, S., COULSON, A. R. (1977a), DNA sequencing with chain-terminating inhibitors, *Proc. Natl. Acad. Sci. USA* **74**, 5463–5467.

SANGER, F., AIR, G. M., BARRELL, B. G., BROWN, N. L., COULSON, A. R., FIDDES, J. C., HUTCHISON, C. A., III, SLOCOMBE, P. M., SMITH, M. (1977b), Nucleotide sequence of bacteriophage ϕX174 DNA, *Nature* **265**, 687–695.

SANGER, F., COULSON, A. R., BARRELL, B. G., SMITH, A. J. H., ROE, B. A. (1980), Cloning in single-stranded bacteriophage as an aid to rapid DNA sequencing, *J. Mol. Biol.* **143**, 161–178.

SARIS, P. E. J., PAULIN, L. G., UHLÉN, M. (1990), Direct amplification of DNA from colonies of *Bacillus subtilis* and *Escherichia coli* by the polymerase chain reaction, *J. Microbiol. Methods* **11**, 121–126.

SCHARF, S. J., HORN, G. T., EHRLICH, H. A. (1986), Direct cloning and sequence analysis of enzymatically amplified genome sequences, *Science* **233**, 1076–1078.

SCHOFIELD, J. P., VAUDIN, M., KETTLE, S., JONES, D. S. C. (1989), A rapid semi-automated microtiter plate method for analysis and sequencing by PCR from bacterial stocks, *Nucleic Acids Res.* **17**, 9490–9490.

SCHREIER, P. H., CORTESE, R. (1979), A fast and simple method for sequencing DNA cloned in the single-stranded bacteriophage M13, *J. Mol. Biol.* **129**, 169–172.

SHAFFER, A. L., WOJNAR, W., NELSON, W. (1990), Amplification, detection, and automated sequencing of gibbon interleukin-2 mRNA by *Thermus aquaticus* DNA polymerase reverse transcription and polymerase chain reaction, *Anal. Biochem.* **190**, 292–296.

SHYAMALA, V., AMES, G. F. (1989), Amplification of bacterial genomic DNA by the polymerase chain reaction and direct sequencing after asymmetric amplification: application to the study of periplasmic permeases, *J. Bacteriol.* **171**, 1602–1608.

SIEMIENIAK, D. R., SLIGHTOM, J. L. (1990), A library of 3342 useful nonamer primers for genome sequencing, *Gene* **96**, 121–124.

SMITH, A. J. N. (1979), The use of exonuclease III for preparing single-stranded DNA for use as a template in the chain terminator sequencing method, *Nucleic Acids Res.* **6**, 831–848.

SMITH, L. M., FUNG, S., HUNKAPILLAR, M. W., HUNKAPILLAR, T. J., HOOD, L. E. (1985), The synthesis of oligonucleotides containing an aliphatic amine group at the 5′ terminus: synthesis of fluorescent DNA primers for use in DNA sequence analysis, *Nucleic Acids Res.* **13**, 2399–2412.

SMITH, L. M., SANDERS, J. Z., KAISER, R. J., HUGHES, P., DODD, C., CONNELL, C. R., HEINER, C., KENT, S. B. H., HOOD, L. E. (1986), Fluorescence detection in automated DNA sequence analysis. *Nature* **321**, 674–679.

SMITH, D. P., JOHNSTONE, E. M., LITTLE, S. P., HSIUNG, H. M. (1990a), Direct DNA sequencing of cDNA inserts from plaques using the linear polymerase chain reaction, *BioTechniques* **9**, 48–54.

SMITH, V., BROWN, C. M., BANKIER, A. T., BARRELL, B. G. (1990b), Semi-automated preparation of DNA templates for large scale sequencing projects, *DNA Sequence* **1**, 73–78.

SORGE, J. A., BLINDERMAN, L. A. (1989), Exo-Meth sequencing of DNA: eliminating the need for subcloning and oligonucleotide primers, *Proc. Natl. Acad. Sci. USA* **86**, 9208–9212.

STADEN, R. (1986), The current status and portability of our sequence handling software, *Nucleic Acids Res.* **14**, 217–231.

STEGEMANN, J., SCHWAGER, C., ERFLE, H., HEWITT, N., VOSS, H., ZIMMERMANN, J., AN-

SORGE, W. (1991), High speed on-line DNA sequencing on ultrathin slab gels, *Nucleic Acids Res.* **19,** 675–676.

STERNBERG, N. (1990), Bacteriophage P1 cloning system for the isolation, amplification and recovery of DNA fragments as large as 100 kilobase pairs, *Proc. Natl. Acad. Sci. USA* **87,** 103–107.

STOFLET, E. S., KOEBERL, D. D., SARKAR, G., SOMMER, S. S. (1988), Genomic amplification with transcript sequencing, *Science* **239,** 491–494.

STRATHMANN, M., HAMILTON, B. A., MAYEDA, C. A., SIMON, M. I., MEYEROWITZ, E. M., PALAZZOLO, M. J. (1991), Transposon-facilitated DNA sequencing, *Proc. Natl. Acad. Sci. USA* **88,** 1247–1250.

STRAUSBAUGH, L. D., BOURKE, M. T., SOMMER, M. T., COON, M. E., BERG, C. M. (1990), Probe mapping to facilitate transposon-based DNA sequencing, *Proc. Natl. Acad. Sci. USA* **87,** 6213–6217.

STRAUSS, E. C., KOBORI, J. A., SIU, G., HOOD, L. E. (1986), Specific-primer-directed DNA sequencing, *Anal. Biochem.* **154,** 353–360.

STREZOSKA, Z., PAUNESKU, T., RADOSAVLJEVIC, D., LABAT, I., DRMANAC, R., CRKVENJAKOV, R. (1991), DNA sequencing by hybridization: 100 bases read by a non-gel-based method, *Proc. Natl. Acad. Sci. USA* **88,** 10089–10093.

STUDIER, F. W. (1989), A strategy for high-volume sequencing of cosmid DNAs: random and directed priming with a library of oligonucleotides, *Proc. Natl. Acad. Sci. USA* **86,** 6917–6921.

SWERDLOW, H., GESTELAND, R. (1990), Capillary gel electrophoresis for rapid, high resolution DNA sequencing, *Nucleic Acids Res.* **18,** 1415–1419.

SWERDLOW, H., WU, S., HARKE, H., DOVICHI, N. J. (1990), Capillary gel electrophoresis for DNA sequencing. Laser-induced fluorescence detection with the sheath flow cuvette, *J. Chromatogr.* **516,** 61–67.

SWERDLOW, H., ZHONG ZHANG, J., YONG CHEN, D., HARKE, H. R., GREY, R., WU, S., DOVICHI, N. J., FULLER, C. (1991), Three DNA sequencing methods using capillary gel electrophoresis and laser-induced fluorescence, *Anal. Chem.* **63,** 2835–2841.

SYVÄNEN, A.-C., HULTMAN, T., AALTO-SETAELAE, K., SOEDERLUND, H., UHLÉN, M. (1991), Genetic analysis of the polymorphism of the human apolipoprotein E using automated solid-phase sequencing, *Genet. Anal.* **8,** 117–123.

TABOR, S., RICHARDSON, C. C. (1987a), DNA sequence analysis with a modified bacteriophage T7 DNA polymerase, *Proc. Natl. Acad. Sci. USA* **84,** 4767–4771.

TABOR, S., RICHARDSON, C. C. (1987b), Selective oxidation of the exonuclease domain of bacteriophage T7 DNA polymerase, *J. Biol. Chem.* **262,** 15330–15333.

TABOR, S., RICHARDSON, C. C. (1989), Effect of manganese ions on the incorporation of dideoxynucleotides by bacteriophage T7 DNA polymerase and *Escherichia coli* DNA polymerase I, *Proc. Natl. Acad. Sci. USA* **86,** 4076–4080.

TAHARA, T., KRAUS, J. P., ROSENBERG, L. E. (1990), Direct DNA sequencing of PCR amplified genomic DNA by the Maxam-Gilbert method, *BioTechniques* **8,** 366–368.

TINDALL, K. R., KUNKEL, T. A. (1988), Fidelity of DNA synthesis by the *Thermus aquaticus* DNA polymerase, *Biochemistry* **27,** 6008–6013.

TIZARD, R., CATE, R. L., RAMACHANDRAN, K. L., WYSK, M., VOYTA, J. C., MURPHY, O. J., BRONSTEIN, I. (1990), Imaging of DNA sequences with chemiluminescence, *Proc. Natl. Acad. Sci. USA* **87,** 4514–4518.

TONEGUZZO, F., GLYNN, S., LEVI, E., MJOLSNESS, S., HAYDAY, A. (1988), Use of a chemically modified T7 DNA polymerase for manual and automated sequencing of supercoiled DNA, *BioTechniques* **6,** 460–469.

TRACY, T. E., MULCAHY, L. S. (1991), A simple method for direct automated sequencing of PCR fragments, *BioTechniques* **11,** 68–75.

TRIGLIA, T., PETERSON, M. G., KEMP, D. J. (1988), A procedure for *in vitro* amplification of DNA segments that lie outside the boundaries of known sequences, *Nucleic Acids Res.* **16,** 8186–8186.

VAN DE SANDE, J. H., LOEWEN, P. C., KHORANA, H. G. (1972), Studies on polynucleotides. CXVIII. A further study of ribonucleotide incorporation into deoxyribonucleic acid chains by deoxyribonucleic acid polymerase I of *Escherichia coli, J. Biol. Chem.* **247,** 6140–6148.

VERHASSELT, P., VOET, M., VOLCKAERT, G. (1992), DNA sequencing by a subcloning-walking strategy using a specific and a semi-random primer in the polymerase chain reaction, *DNA Sequence* **2,** 281–287.

VOLCKAERT, G. (1987), A systematic approach to chemical DNA sequencing by subcloning in pGV451 and derived vectors, *Methods Enzymol.* **155,** 231–250.

VOLCKAERT, G., MIN JOU, W., FIERS, W. (1976), Analysis of ^{32}P-labeled bacteriophage MS2 RNA by a minifingerprinting procedure, *Anal. Biochem.* **72,** 433–446.

VOLCKAERT, G., CONTRERAS, R., SOEDA, E., VAN DE VOORDE, A., FIERS, W. (1977), Nucleotide sequence of Simian Virus 40 *Hind*H restriction fragment, *J. Mol. Biol.* **110,** 467–510.

VOLCKAERT, G., DE VLEESCHOUWER, E., BLÖCKER, H., FRANK, R. (1984a), A novel type of vectors for ultrarapid chemical degradation sequencing of DNA, *Genet. Anal.* **1**, 52–59.

VOLCKAERT, G., WINTER, G., GAILLARD, C. (1984b), DNA sequencing, in: *Advanced Molecular Genetics* (PÜHLER, A., TIMMIS, K. N., Eds.), pp. 249–280, Berlin: Springer-Verlag.

VON HEIJNE, G. (1991), Computer analysis of DNA and protein sequences, *Eur. J. Biochem.* **199**, 253–256.

VOSS, H., SCHWAGER, C., WIRKNER, U., SPROAT, B., ZIMMERMANN, J., ROSENTHAL, A., ERFLE, H., STEGEMANN, J., ANSORGE, W. (1989), Direct genomic fluorescent on-line sequencing and analysis using *in vitro* amplification of DNA, *Nucleic Acids Res.* **17**, 2517–2528.

VOSS, H., ZIMMERMANN, J., SCHWAGER, C., ERFLE, H., STEGEMANN, J., STUCKY, K., ANSORGE, W. (1990), Automated fluorescent sequencing of lambda DNA, *Nucleic Acids Res.* **18**, 5314–5314.

VOSS, H., SCHWAGER, C., WIRKNER, U., ZIMMERMANN, J., ERFLE, H., HEWITT, N. A., RUPP, T., STEGEMANN, J., ANSORGE, W. (1991), A new procedure for automated DNA sequencing with multiple internal labelling by fluorescent dUTP, *Methods Mol. Cell. Biol.* **3**.

WADA, A. (1984), Automatic DNA sequencing, *Nature* **307**, 193–193.

WADA, A. (1987), Automated high-speed DNA sequencing, *Nature* **325**, 771–772.

WAHLBERG, J., LUNDEBERG, J., HULTMAN, T., UHLÉN, M. (1990), General colorimetric method for DNA diagnostics allowing direct solid-phase genomic sequencing of the positive samples, *Proc. Natl. Acad. Sci. USA* **87**, 6569–6573.

WARREN, W., DONIGER, J. (1991), HPLC purification of polymerase chain reaction products for direct sequencing, *BioTechniques* **10**, 216–220.

WEBSTER, R. E., CASHMAN, J. S. (1978), Morphogenesis of the filamentous single-stranded DNA phages, in: *The Single-Stranded DNA Phages* (DENHARDT, D. T., DRESSLER, D., RAY, D. S., Eds.), pp. 557–569. Cold Spring Harbor, NY: Cold Spring Harbor Laboratory Press.

WEST, J (1988), Automated sequence reading and analysis, *Nucleic Acids Res.* **16**, 1847–1856.

WILSON, R. K., YUEN, A. S., CLARK, S. M., SPENCE, C., ARAKELIAN, P., HOOD, L. E. (1988), Automation of dideoxynucleotide DNA sequencing reactions using a robotic workstation, *BioTechniques* **6**, 776–787.

WILSON, R. K., CHEN, C., AVDALOVIC, N., BURNS, J., HOOD, L. (1990a), Development of an automated procedure for fluorescent DNA sequencing, *Genomics* **6**, 626–634.

WILSON, R. K., CHEN, C., HOOD, L. (1990b), Optimization of asymmetric PCR reaction for rapid fluorescent DNA sequencing, *BioTechniques* **8**, 184–189.

WONG, C., DOWLING, C. E., SAIKI, R. K., HIGUCHI, R. G., ERHLICH, H. A., KAZAZIAN, H. H. (1987), Characterization of β-thalassaemia mutations using direct sequencing of amplified single copy DNA, *Nature* **330**, 384–386.

WRISCHNIK, L. A., HIGUCHI, R. G., STONEKING, M., ERHLICH, H. A., ARNHEIM, N., WILSON, A. C. (1987), Length mutations in human mitochondrial DNA: direct sequencing of enzymatically amplified DNA, *Nucleic Acids Res.* **15**, 529–542.

WU, R., TAYLOR, E. (1971), Nucleotide sequence analysis of DNA. II. Complete nucleotide sequence of the cohesive ends of bacteriophage lambda DNA, *J. Mol. Biol.* **57**, 491–511.

YANG, J. L., MAHER, V. M., MCCORMICK, J. J. (1989), Amplification and direct nucleotide sequencing of cDNA from the lysate of low number of diploid human cells, *Gene* **83**, 347–354.

YANISCH-PERRON, C., VIEIRA, J., MESSING, J. (1985), Cloning vectors that yield high levels of single-stranded DNA for rapid DNA sequencing, *Gene* **27**, 183–191.

ZAGURSKY, R. J., BAUMEISTER, K., LOMAX, N., BERMAN, M. L. (1985), Rapid and easy sequencing of large linear double-stranded DNA and supercoiled plasmid, *Genet. Anal.* **2**, 89–94.

ZAGURSKY, R. J., CONWAY, P. S., KASHDAN, M. A. (1991), Use of ^{33}P for Sanger DNA sequencing, *BioTechniques* **11**, 36–38.

ZIFF, E. B., SEDAT, J. W., GALIBERT, F. (1973), Determination of the nucleotide sequence of a fragment of bacteriophage φX174 DNA, *Nature (New Biol.)* **241**, 34–37.

ZIMMERMAN, C. J., FUSCOE, J. C. (1991), Direct DNA sequencing of PCR products, *Environ. Mol. Mutagen.* **18**, 274–276.

ZIMMERMANN, J., VOSS, H., SCHWAGER, C., STEGEMANN, J., ANSORGE, W. (1988), Automated Sanger dideoxy sequencing reaction protocol, *FEBS Lett.* **233**, 432–436.

ZIMMERMANN, J., VOSS, H., KRISTENSEN, T., SCHWAGER, C., STEGEMANN, J., ERFLE, H., ANSORGE, W. (1989), Automated preparation and purification of M13 templates for DNA sequencing, *Methods Mol. Cell. Biol.* **1**, 29–34.

9 DNA Synthesis

Joachim W. Engels
Belinda Sprunkel
Eugen Uhlmann

Frankfurt am Main, Federal Republic of Germany

1 Introduction

The basic structure and chemical nature of DNA have been known since WATSON and CRICK'S fundamental paper (WATSON and CRICK, 1953). Starting from nucleotide chains built up of sugar, phosphate, and heterocycles (Fig. 1), the primary structure of the double helix can be deduced as two single strands running antiparallel to each other. An ordered structure results because of complementarity between the individual heterocycles and the hydrogen bonds that are therefore possible between them.

To describe such complex molecules in a short but comprehensible manner, the following nomenclature has been adopted. The heterocycles in the DNA, called nucleobases, are given the abbreviations A (adenine), C (cytosine), G (guanine), and T (thymine), while the fundamental sugar, 2′-deoxyribose, is indicated with d, and the carbons are numbered 1′ to 5′. Fig. 1 shows a section of a dodecamer whose X-ray structure (DICKERSON and DREW, 1981) forms the basis of our current understanding of B-DNA, one of several three-dimensional forms of DNA. In chemical DNA synthesis, the two complementary strands are constructed separately; i.e., only single-stranded oligonucleotides can be synthesized. These are then hybridized (formation of hydrogen bonds) and linked to larger DNA units by enzymatic coupling or, recently, chemical ligation (DOLINNAYA et al., 1991a, b).

It has become common practice to refer to all chemically synthesized single-stranded nucleic acid chains of defined length and sequence as oligonucleotides, even if they are longer than 100 residues in length (BLACKBURN and GAIT, 1990). The term polynucleotide refers to a single-stranded oligonucleotide of less defined length, usually obtained by polymerization reactions. Oligonucleotide synthesis where the individual repetitive unit consists of 2′-deoxyribonucleotides has gained the status of a reliable routine procedure. For ribonucleotides with repetitive units this is not yet fully the case. Routine methods are just emerging, the 2′-hydroxy protection being the critical point. Since there is a general underlying principle for oligonucleotide synthesis, Sect. 2.2 will concentrate on the overall strategy. This is followed by an overview of modified oligonucleotides (Sect. 2.3), which will play an important future role in fine tuning selective interactions with DNA or RNA.

DNA oligonucleotides and their analogs can be applied in biology and medicine (Sect. 3). Their use for diagnostic purposes will be the focus of Sects. 3.1 and 3.2. One of the most important recent techniques, the polymerase chain reaction, PCR, will be dealt with in more detail. The total synthesis of genes, by chemoenzymatic methods or lastly by pure chemical procedures, will be discussed in Sect. 3.3, followed by an outline of different strategies to arrive at double-stranded DNA. Besides the design and practical examples, the mutagenesis of synthetic genes will also be described (Sect. 3.4). More recently the selection principle has been applied to the question of identifying specific RNA and even DNA-single stranded sequences to interact with molecules of defined shape - a procedure which is termed applied molecular evolution (Sect. 3.5). In Sect. 3.6 the potential of oligonucleotides to specifically interact either with DNA or RNA will be covered. Here the possibility exists to build up a triple helix from the interaction of DNA or a double helix by binding to complementary RNA. These techniques represent powerful approaches to selectively interfere with either transcription or translation. Altogether we like to present some of the intriguing facets of the biopolymers named DNA and/or RNA.

2 Methods for the Synthesis of Oligonucleotides

2.1 Synthesis of Unmodified Oligonucleotides

Considering the chemical stability of nucleic acids, a range of precautions has to be considered. DNA is highly base-stable and fairly sta-

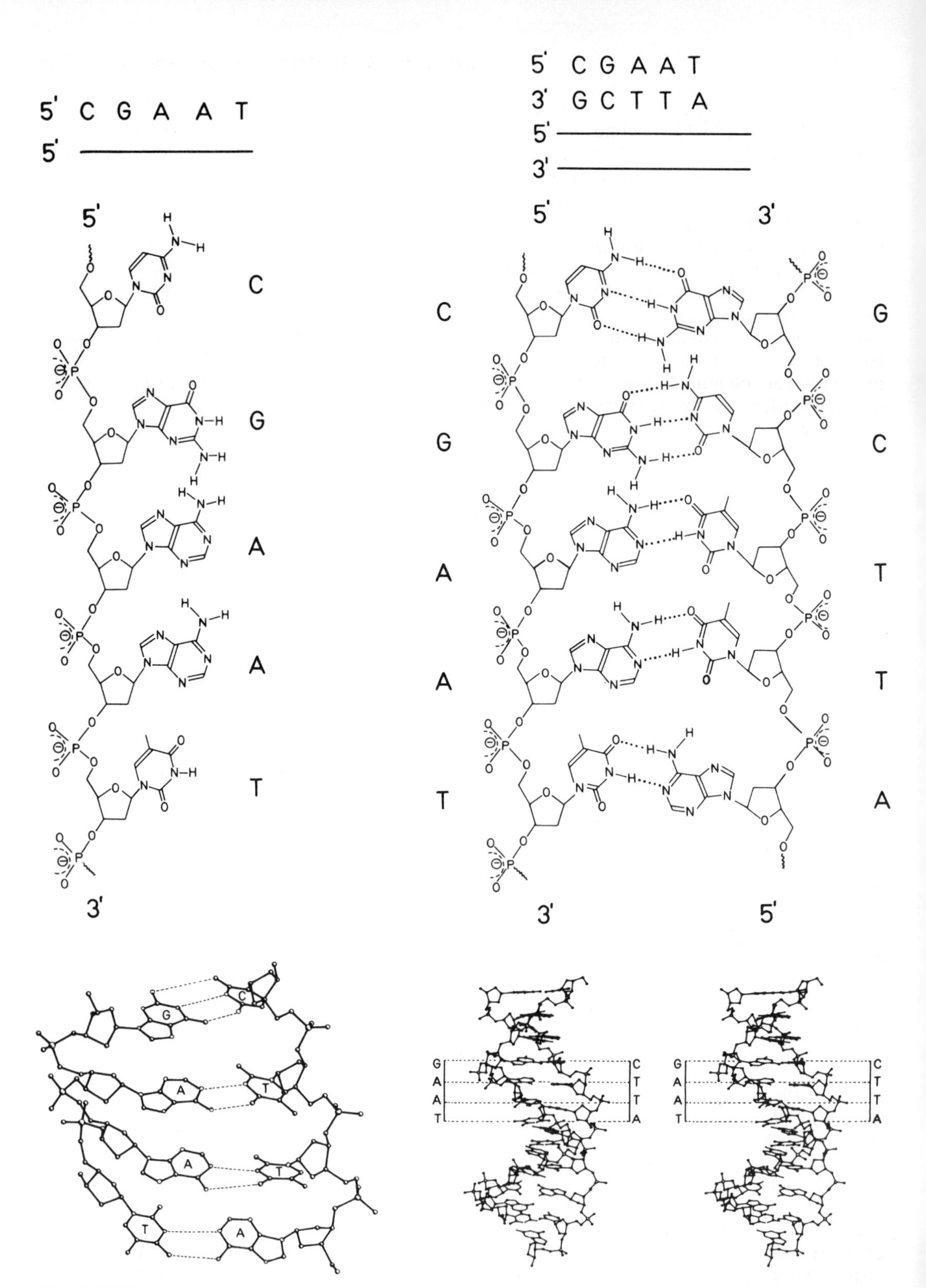

Fig. 1. DNA structure.

ble against acid. RNA is very labile to base and also labile under acidic conditions. Beside the backbone the sugar diol system can be cleaved oxidatively in the ribose system. This has been used to prepare derivatives post-synthetically. The glycosidic bond is prone to acid hydrolysis predominantly for purine nucleosides (depurination). For the heterocyclic moiety alkylation (most notably guanine), oxidation (guanine) and reduction (thymine, uracil) are potential side reactions. These side reactions certainly limit the choice of chemical reactions for oligonucleotide synthesis. Although a plethora of chemical approaches, protecting groups, and reagents have been tested for individual synthetic steps, the overall goal of a suitable method for all four nucleotides has been the focus of the synthetic efforts. Thus the general philosophy has been detailed optimization rather than completely new chemistry.

2.2 Strategies in Oligonucleotide Synthesis

Oligonucleotides are polymers built up by the polycondensation of nucleoside phosphates. In this process, synthesis is possible in two directions, by 3′–5′ and 5′–3′ phosphate coupling. Because of the higher reactivity of the primary hydroxyl 5′ group the 3′–5′ direction is used almost exclusively in synthesis, while nature uses the reverse direction, based on 5′-triphosphates.

Synthesis of a long oligomer or of a polymer can be carried out either by stepwise addition of monomers or by coupling of oligomeric building blocks. Since DNA consists of only four different nucleotides, 16 dimeric or 64 (4^3) trimeric oligonucleotide blocks can be prepared, and these can be used as building blocks in the synthesis of any desired DNA sequence. Both concepts have been realized. In choosing between them, two factors must be considered: the difficulty of synthesizing the oligonucleotide blocks and the yields obtainable by monomer addition. The diagram in Fig. 2 correlates the coupling yield with the length of the DNA and convincingly shows that for a yield of 99% per addition, which, astonishing-

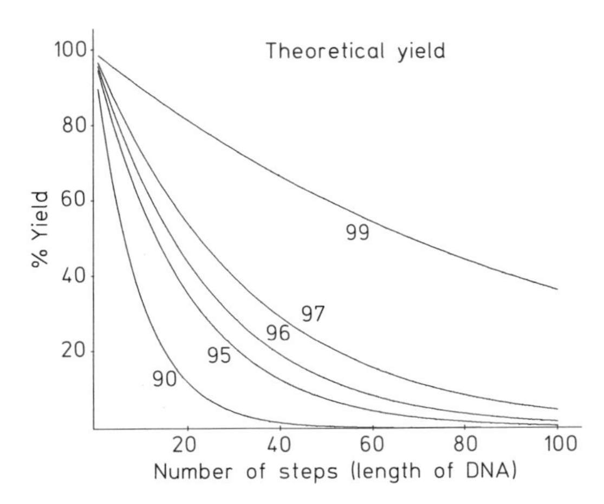

Fig. 2. Correlation of yield and chain length.

ly, can be achieved for all four nucleotides, pieces long enough to be used in gene synthesis can be synthesized in acceptable yields.

In practical solid-phase synthesis, the stepwise addition of monomeric building blocks has become standard. Nevertheless, the synthesis of dimers (Chow et al., 1981) and also of preformed trimers (Miyoshi et al., 1980) has been successfully performed.

Methods for synthesizing defined oligonucleotides are differentiated primarily by the way in which the phosphorus component is being introduced; one distinguishes the so-called triester (Fig. 3a), phosphite (Fig. 3b, X = Cl), phosphoramidite (Fig. 3b, $X = NiPr_2$), and H-phosphonate (Fig. 3c) methods. The corresponding protected nucleoside derivatives function as the starting compound. Although an enormous number of protecting groups are known (Greene and Wuts, 1991) and have also been tested in oligonucleotide synthesis (Amarnath and Broom, 1977; Reese, 1978; Narang, 1983; Itakura et al., 1984; Sonveaux, 1986; Hobbs, 1990; Beaucage and Iyer, 1992), only a few are routinely used, owing to commercialization of automated DNA synthesis.

Protecting groups are primarily those introduced by Khorana (1978), acyl groups on the exocyclic amino functions of adenine, guanine, and cytosine, and these acyl groups are of particular advantage in the synthesis of unmodified oligonucleotides. A comprehensive

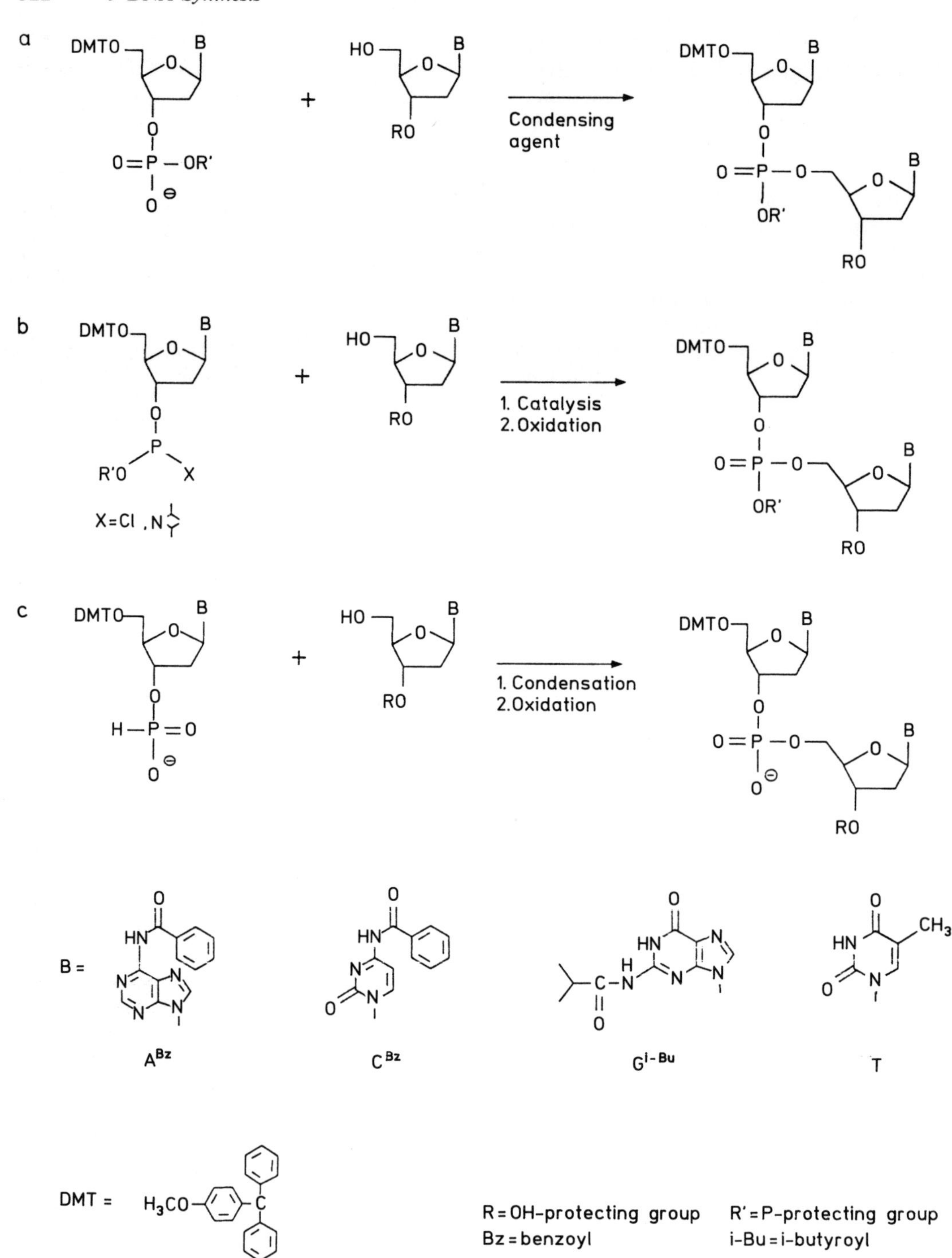

Fig. 3. Methods of oligonucleotide synthesis.

list of all the protecting groups used in oligonucleotide synthesis up to 1986 has been published by SONVEAUX (1986) and for the phosphoramidite chemistry by BEAUCAGE and IYER (1992).

For the synthesis of more labile modified oligonucleotides like phosphonates, more labile protecting groups have been recommended (FROEHLER and MATTEUCCI, 1983; SCHULHOF et al., 1987). Especially when synthetic methods using very reactive phosphorus derivatives are employed, it is important to consider undesired derivatization of the amide functions of guanine and thymine (REESE and UBASAWA, 1980; GAFFNEY et al., 1984). During the synthesis of longer guanosine stretches, a considerable amount of modifications took place which could be detected through cloning (YEUNG et al., 1988). Here, too, several protecting groups have been used to convert the amide to the iminoether (REESE and SKONE, 1984; HIMMELSBACH et al., 1984; JONES et al., 1981; KAMIMURA et al., 1983). For the phosphoramidite method, we have been able to show that optimization of the synthesis is preferable to excessive use of protecting groups (MAG and ENGELS, 1988). Also a change in the reagents during synthesis, N-methyl-imidazole for dimethylamino-pyridine, has been advocated (FARRANCE et al., 1989).

The corresponding protected nucleosides are then phosphorylated with suitable reagents, being either mono- or bifunctional. Due to the nature of the phosphorylating agent the resulting nucleoside phosphate is either in the tri- or pentavalent state. Based on the reactivity of the resulting phosphite or phosphate esters, protecting groups have to be introduced in order to block branching reactions. In the case of H-phosphonate chemistry this is not necessary, since due to the tautomerization of phosphoric acid–H-phosphonate, this phosphorus has already four bonds. In the following the individual methods of synthesizing oligonucleotides of defined sequence are outlined.

2.2.1 Diester Method

This method has been successfully used in the beginning of oligonucleotide chemistry but is not in use any more. A nucleoside monophosphate is reacted with a second nucleoside in the presence of an appropriate condensing agent. This can either be dicyclohexylcarbodiimide or a sterically hindered arylsulfonylchloride. The reaction mechanism is complex, the yields are not optimal, and the reactions are slow. Nevertheless, it is of historic importance, since the first successful gene synthesis is based on this approach (KHORANA, 1978).

2.2.2 Triester Method

For synthesis on a solid support by the Merrifield strategy, however, it has proved advantageous to synthesize stable monomeric nucleotide building blocks which can be stored ready to be used as needed in synthesis. A multitude of different functions have been used as protecting groups on phosphorus (SONVEAUX, 1986). Here, too, automation has brought simplification. A point worth mentioning is the chemical mechanism by which the phosphorus protecting group is removed. Reactions at phosphate phosphorus have been shown by WESTHEIMER (1968) to proceed by addition–elimination mechanisms, with the possibility of pseudorotation (GILLESPIE et al., 1971). To avoid potential problems, such as undesired hydrolysis, protecting groups are preferred in which attack of the reagent takes place directly at the protecting groups without affecting the phosphate center. Alkyl groups such as methyl and benzyl have been used here, as have been β-cyanoethyl, arylsulfonylethyl, arylethyl, and haloethyl groups, all of which can be removed by β-elimination. Isomerization of the phosphate bond is largely prevented, and strand breaks are avoided by this strategy. Based on these mechanistic considerations, optimized protecting groups have been introduced.

This synthetic method (Fig. 3a) was introduced by TODD (MICHELSON and TODD, 1955) and is based on suitable activation of a protected nucleoside phosphate diester with acid derivatives of the arylsulfonic acid azolide type. The activated phosphate diester then reacts with the unprotected 5′-OH of the nucleoside, giving the corresponding dinucleoside phosphate triester. A large number of such reagents have been used as their phosphoric and sulfonic acid derivatives (van Boom, 1977). Of the azolides, tetrazole and 3-nitrotriazole have proved particularly useful. The low initial reactivity of P(V) compounds can thus be increased to a sufficient level. Efimov (Efimov et al., 1985) successfully improved the phosphotriester synthesis of oligodeoxyribonucleotides by using oxygen nucleophiles. Pyridine-N-oxides proved to be advantageous, especially in an intramolecular fashion (Efimov et al., 1986). Decisive advantages of this method include the good stability of the starting diesters and a potential recovery of excess nucleotides (e.g., as barium salts, Gough et al., 1981). A disadvantage lies in the side reactions brought about by the condensing agent (e.g., sulfonation at the 5′-OH position). Recently an alternative strategy based on phosphotriester chemistry has been introduced. The method called HELP (high efficiency liquid phase) synthesis is based on standard coupling reactions where the starting nucleoside is fixed to a soluble polymer support, especially PEG (Bonora et al., 1990). By this method larger amounts of shorter oligonucleotides can be prepared more economically.

2.2.3 Phosphoramidite Method

Here, a suitably protected nucleoside phosphoramidite (Fig. 3b) is converted into a phosphite triester by reaction with a 5′-OH nucleoside in the presence of tetrazole. Subsequent oxidation leads to the corresponding triester (see Fig. 3a). The higher reactivity of P(III) compounds, most prominently dichloridites, has always made them attractive starting materials (Letsinger et al., 1975), but for routine oligonucleotide synthesis their poor handling and availability and especially the difficulty of storing the monomeric building blocks have hampered their application. With the introduction by Caruthers (Beaucage and Caruthers, 1981; Caruthers, 1991) of phosphoramidites, which can be activated by acid, the breakthrough was achieved. Although the initial choice fell on the dimethylamino groups as the amidite component and on the methoxy protecting group (Daub and van Tamelen, 1977), today, after further optimization of reaction conditions, the diisopropylamino compound (Adams et al., 1983), in combination with the β-cyanoethyl protecting group (Sinha et al., 1983, 1984), is regarded as the best compromise. The most important advantage of this method is the activation by the mild reagent tetrazole, which is practically free of side reactions. This method proved to be superior for automation, and most of the oligonucleotide synthesis has been accomplished by this method.

2.2.4 H-Phosphonate Method

Todd (Hall et al., 1957) initially recognized and used the synthetic potential of H-phosphonates. Here, a protected nucleoside H-phosphonate (Fig. 3c) is converted into a dinucleoside phosphonate diester by reaction with a 5′-OH nucleoside in the presence of a condensing agent, usually an acid chloride, be it pivaloyl chloride or adamantoyl chloride (Froehler et al., 1986; Garegg et al., 1986).

Subsequent oxidation leads to the corresponding triester. No intermediate protecting group is required on phosphorus, and the oxidation of the H-phosphonate is carried out only once at the end of the synthesis, not after each addition, as in the amidite method. On the other hand, side reactions (acylation) of the condensing agent (e.g., pivaloyl chloride) do occur, as in the triester method. Moreover, the yields in the H-phosphonate method do not match those of the phosphoramidite method, despite optimization (e.g., use of adamantoyl chloride as the condensing agent, Applied Biosystems, 1987).

H-phosphonate chemistry sometimes is the basis for synthesis of precious modified oligonucleotides, having the advantage of needing a smaller excess of incoming reagent.

2.3 Synthesis of Modified Oligonucleotides

A variety of modified oligonucleotides have found application as potential drugs (Uhlmann and Peyman, 1990). The basic principle underlying their use is their potential to

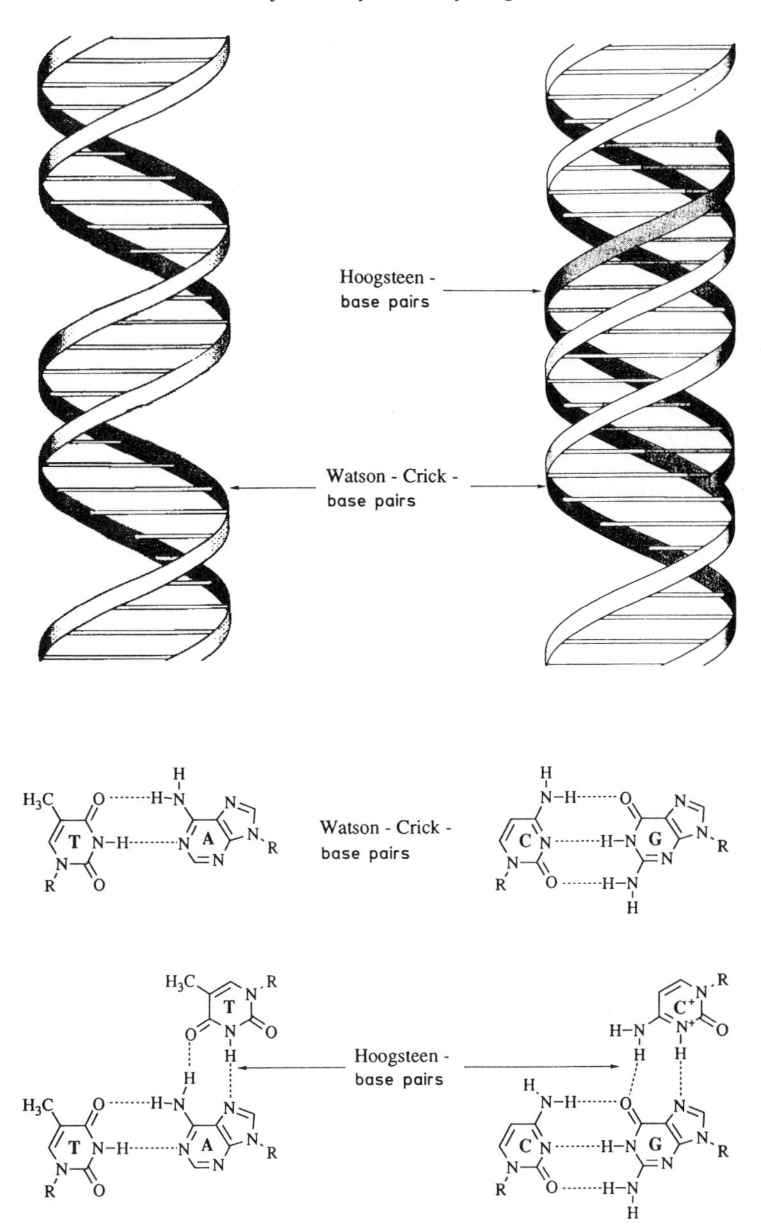

Fig. 4. Watson-Crick and Hoogsteen base pairs.

build up hydrogen bonds with DNA and RNA. This can be either accomplished by interacting with single- or double-stranded nucleic acids giving rise to a duplex or triplex. In the first case the known pattern of Watson–Crick hydrogen bonds yields a duplex. Utilizing the complementarity principle, a given mRNA (being the sense strand) can be embraced by an antisense strand to give a double-stranded helix. This so-called antisense concept blocks translation. On the other hand, a given DNA can be further recognized by hydrogen bonding, the Hoogsteen base pairs (Fig. 4) (ENGELS, 1991).

The possible array of donor and acceptor functions on the heterocyclic moieties allows for additional H-bonds. These triple helical arrangements (triplex) interfere with transcription (transcriptional arrest). For both approaches modified oligonucleotides have been tested. Most of the modifications utilized so far are either phosphorothioates or to a lesser extent dithioates or methylphosphonates. Recently, however, non-phosphate linked oligonucleotides have been advocated. Here a peptide backbone renders these complementary bases highly stable towards melting (NIELSON et al., 1991).

2.3.1 Phosphorothioates and Phosphorodithioates

Phosphorothioate oligodeoxynucleotides are a class of compounds in which one of the nonbridging phosphodiester oxygen atoms is replaced by a sulfur atom (STEC et al., 1984; STEIN et al., 1988a).

Like unmodified oligonucleotides, phosphorothioates can be synthesized by the aforementioned three basic methods: The phosphate triester using phosphorothioates, the H-phosphonate and the phosphoramidite approach. To construct oligonucleotides by the standard amidite method only the oxidation has to be changed using elementary sulfur S_8 in CS_2/pyridine instead of I_2/H_2O (STEC et al., 1984). But stepwise sulfurization is a slower reaction, and technical problems may arise. They can be circumvented by using new sulfurizing agents, 3H-1,2-benzodithiole-3-one-1,1-dioxide (IYER et al., 1990) or tetraethylthiuram disulfide (VU and HIRSCHBEIN, 1991) or by using the H-phosphonate approach following sulfur oxidation. The advantage of the latter results from the fact that oxidation with sulfur is necessary only once, after the synthesis is complete. The H-phosphonate approach albeit suffers from the inherent limitation that oligonucleotides with predetermined combinations of natural and phosphorothioate linkages cannot easily be prepared.

The introduction of a sulfur atom generates a chiral center at the phosphorus atom resulting in Rp and Sp diastereomers at each phosphodiester linkage. In normal solid phase synthesis one ends up with a complex mixture of diastereoisomers for oligonucleotides with phosphorothioate linkages.

Recently STEC (STEC et al., 1991) has found an elegant way to overcome this problem by introducing a stereo-controlled phosphorothioate synthesis utilizing base catalysis. Enantiomeric pure oxathiaphospholanes react in the presence of N-methylimidazole and a second 5'OH-nucleoside in a S_N2 fashion to control stereochemistry based on the configuration of the starting oxathiaphospholane.

One possibility to overcome the chirality problem is to substitute also the second phosphorus oxygen by sulfur to arrive at the so-called phosphorodithioates. They can be prepared from thiophosphoramidites (BJERGARDE and DAHL, 1991; BRILL, 1989; CARUTHERS, 1991), and synthesis was recently successful on a solid support using a dimer building block (BEATON, 1991). But these compounds are less reactive than usual phosphoramidites and in addition have the tendency to dismutate in the presence of the acidic catalyst used (tetrazole).

Deoxynucleoside-3'-phosphorothioamidates are prepared via a one-pot two-step procedure from a suitably protected deoxynucleoside bis(dimethylamino)phosphorous chloride and diisopropylethylamide (BJERGARDE and DAHL, 1991). The resulting deoxynucleoside phosphorothioamidite is used in solid-phase synthesis by addition of tetrazole and the 5' nucleoside. Oxidation with the help of sulfur in carbon disulfide/pyridine furnishes the dithioates. The resulting purity is less than for the monothioates, and the oligonucleotides show a melting temperature depression in the range of Tm 0.5–2 °C per phosphorodithioate linkage.

2.3.2 Methylphosphonates

Oligodeoxynucleoside methylphosphonates are attractive nucleotide analogs for use as antisense oligonucleotides (MILLER, 1991). In these compounds the negatively charged phosphate oxygen is replaced by a methyl group. So methylphosphonates are missing the negative charge on phosphorus which facilitates their passive diffusion into cells (MILLER et al., 1981; MARCUS-SECURA et al., 1987).

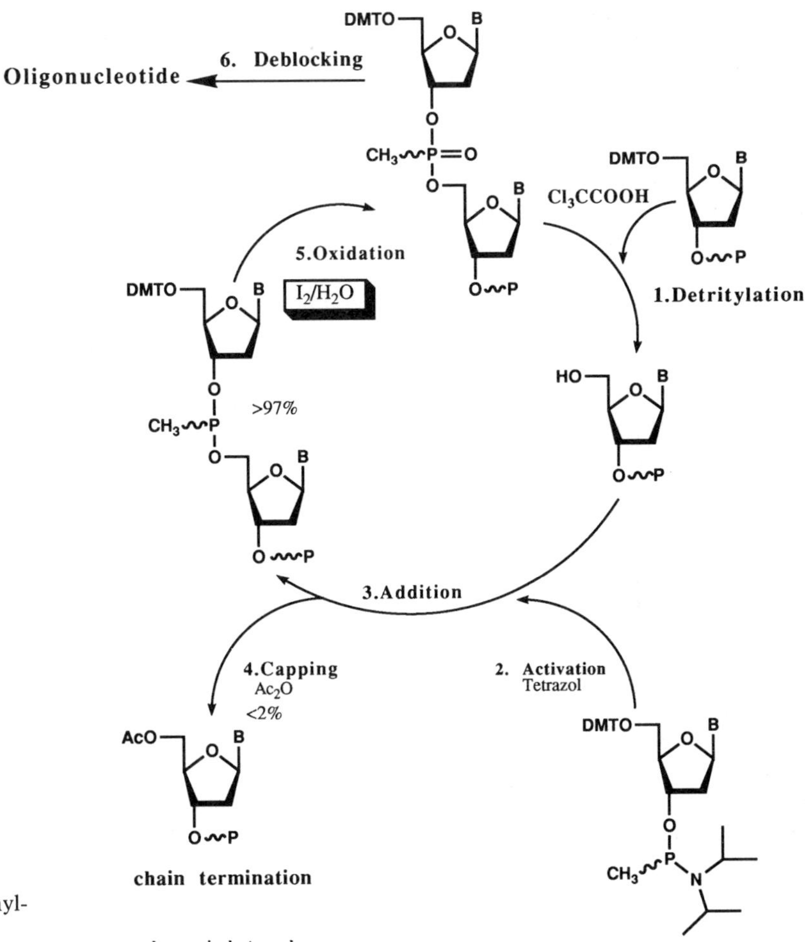

Fig. 5. Synthesis of methylphosphonates by the phosphoramidite method.

Based on electrostatic arguments, methylphosphonates of defined stereochemistry should hybridize stronger with complementary DNA and RNA (LIN et al., 1989). On the other hand, oligonucleotides with all phosphodiester bonds substituted by methylphosphonate linkages are poorly soluble in water. Here a balance has to be found between lipophilicity for better transport through membranes and water solubility. Moreover, these compounds are resistant to nuclease degradation (QUARTIN et al., 1989). Altogether they seem to be ideal candidates as potential oligonucleotide drugs, being stable and active as complementary binding agents.

However, like phosphorothioates, methylphosphonates bear a chiral center at the phosphorus. This means that an n-meric methylphosphonate is obtained as a mixture of 2^n diastereomers, which is an unsatisfactory result for many investigations, because all methylphosphonates or phosphorothioates synthesized on solid support are not clearly defined substances, but a mixture of many molecules with potentially quite different properties.

Methylphosphonates can be obtained either by the phosphotriester method (AGARWAL and RIFTINA, 1979) or by the phosphoramidite method using suitably protected 3′-O-(2′deoxynucleoside)-N,N-diisopropylamino-methylphosphoramidites to introduce methylphosphonate linkages into an oligonucleotide (LÖSCHNER and ENGELS, 1988).

The unambiguous assignment of the absolute configuration of the methyl group, be it Rp outside the groove or Sp pointing inside the groove, has been solved by nuclear Overhauser assignments (ROE spectroscopy) (LÖSCHNER and ENGELS, 1990). Fig. 5 shows the synthesis of methylphosphonates by the phosphoramidite method.

Since the internucleotide methylphosphonate bridge is more base labile than the natural internucleotide linkage, milder conditions are necessary for cleavage from the support and for deprotection. Thus ethylenediamine (EDA) or *tert*-butylamine (TBA) in ethanol can be used for this purpose.

2.3.3 Phosphate Triesters

Phosphate triesters like methylphosphonates lack the negative charge on the phosphorus. They are long known as products of alkylating agents, and some of them have already alkylating properties. This is especially the case with the methyl and ethyl triesters; the isopropyl triester is much more stable. The labile methylphosphotriesters were only successfully obtained by using a much more labile linker to the support (ALUL et al., 1991). The problem of chirality also occurs with the phosphate triesters.

Triesters can be synthesized using suitable phosphoramidites with methyl- or isopropyl groups instead of the protecting β-cyanoethyl group, since these compounds occur as intermediates in normal amidite chemistry.

2.3.4 Non-Oxygen Bridged Phosphate Analogs

An alternative possibility not to solve but to avoid the chirality problem is to replace the oxygen involved in the bridge on the phosphate group. Here isosteric NH-, S- and CH_2-bridges have been used. This leads to achiral and charged water-soluble compounds with different properties. Whereas uncharged methylphosphonate groups offer the advantage of enhanced membrane permeability and nuclease resistance, the reduced solubility of most non-charged oligonucleotide analogs in aqueous buffers limits their usefulness as antisense oligonucleotides.

It has previously been shown that oligomers bearing nucleoside units linked by 3′-O-P-N-5′ or 3′-N-P-O-5′ bonds are stable under neutral and alkaline conditions, but not under acidic conditions.

The bridged phosphoramidates and phosphorothioates can be synthesized using standard amidite chemistry. Here phosphoramidites are used which carry a trityl protected amino- or mercapto function at the 5′ end of the sugar moiety. For the bridged phosphoramidates the dimer building block synthesis has also been used (MAG and ENGELS, 1989). There is an initial formation of a phosphite imine, followed by a Michaelis-Arbusov-type transformation into the phosphoramidate catalyzed by LiCl.

A further modification comprises replacement of one of the oxygen atoms of the bridge by a methylene group. Synthesis can be achieved using modified phosphotriester chemistry and methylene bridges containing building blocks.

2.3.5 RNA Synthesis

The chemical difference between DNA and RNA lies mostly in the different sugars used, the latter being built up from ribose. Due to the extra hydroxyl group in position 2′ of the sugar, a much more difficult synthetic problem arises. The main difficulty being neighboring group participation and steric hindrance due to bulky protecting groups. A large variety of synthetic strategies has been put forward. Mainly due to the great success of the phosphoramidite method in DNA synthesis, it has also been implemented into the RNA field. Here the choice of the 2′-protection group has been the main focus. Whereas REESE (1989) has focused on different acid labile protecting groups, OGILVIE (WU et al., 1989) has gradually optimized the tertiary-butyl-dimethyl-silyl group. The latter building blocks have become commercially available, and oligoribonucleotides can be synthesized with yields better than 98% (LYTTLE et al., 1991). Although the procedure is similar to the solid-phase synthesis of oligodeoxynucleotides, some important modifications have to be considered. The activation by tetrazole probably due to steric reasons has to be prolonged, or even better, a stronger activation agent like *p*-nitrophenyl-tetrazole is advantageous. Deprotection of the silyl groups by fluoride anion imposes the problem of dealing later with a difficult to separate reagent. Thus careful work-up is a necessity. Furthermore, the inherent lability of RNA towards hydrolysis and its strong tendency towards secondary structure render the purification procedure more problematic.

2.4 Solid-Phase Synthesis

Nowadays oligonucleotide syntheses are routinely performed on a solid support, because oligonucleotide synthesis is a repetitive technique (GAIT, 1984). The polymer support has become an ideal combination of a protecting group and a handle to ease the automation of the individual reaction steps. The reaction cycle consists of repetitive steps which can be optimized by a computer. The underlying principle of solid-phase synthesis has been introduced by MERRIFIELD (1963) and LETSINGER (LETSINGER and KORNET, 1963) independently. As it is cited, the polymer used should be insoluble in the solvents used, it should be derivatized in order to allow coupling to the product to be synthesized, and its structure should not interfere with the chemistry to be performed. These prerequisites for an ideal polymer support have for a long time hampered the development of oligonucleotide synthesis. Due to the possibility of applying a large excess of reagent, chemical reactions are driven to completion. This can be easily achieved since a straightforward separation of reactant and reagent can simply be done by filtering off excess reagent. From the supports tested, such as polystyrene, cellulose, polyacrylamide, polytetrafluorethylene, and silica, a large amount of information has been gained. Today the most widely used supports are silica gel and especially defined glass beads, so-called controlled pore glass material (CPG) (see Fig. 6). This material has excellent mechanical and chemical properties and may be used with any solvent and coupling method (DAMHA et al., 1990). For short oligonucleotides and large-scale synthesis the polystyrene support may also be used (BARDELLA et al., 1990; MCCOLLUM and ANDRUS, 1991).

The synthesis via phosphoramidites according to CARUTHERS (1991) is currently the most efficient method for preparing oligonucleotides and will be discussed in detail (cf. Fig. 7). The coupling yield for unmodified oligodeoxyribonucleotides is greater than 99%.

Assembly of an oligonucleotide starts with preparing the appropriately protected nucleoside derivatives. Here the sugar as well as the heterocyclic base moiety have to be protected. The exocyclic amine functions of adenine and cyctosine can be protected by benzoylation. Transient protection of the sugar hydroxyls with trimethylsilyl groups followed by benzoylation is the most successful method in pre-

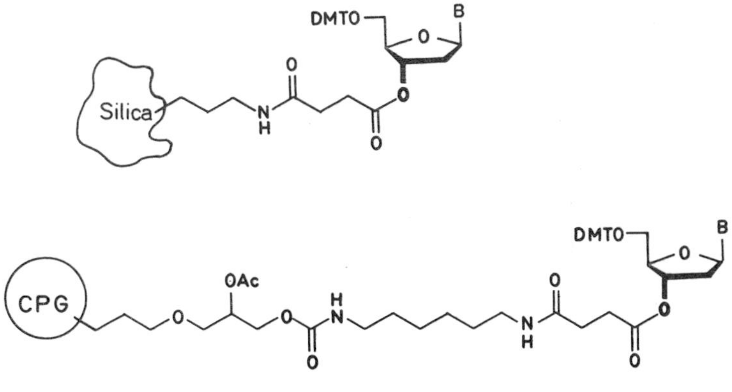

Fig. 6. Attachment of the starting nucleoside to the solid support.

paring protected nucleosides (TI et al., 1982). It is the method of choice today to prepare protected nucleosides in general. Thymine has been used unprotected in most cases. Guanine represents the most complicated case, since the exocyclic amino group as well as the the amide function are prone to side reactions. For amino protection, the isobutyryl group is widely used, because the benzoyl group proved to be too stable to be removed by mild ammonia treatment. Recently the phenoxyacetyl protecting groups have been recommended for amine functions being better removable by mild ammonia treatment than benzoyl or isobutyryl (SCHULHOF et al., 1987).

For protection of the 5′-hydroxyl group of the sugar moiety, the majority of synthetic strategies favor an acid labile protection. Here the 4,4′-dimethoxytrityl group (DMT) has been used with great advantage. Fairly mild acids such as dichloro- or trichloroacetic acid can be used for their cleavage. In addition, the trityl cation generated can be optically measured which introduces a method for detection of coupling yields.

After protecting base and sugar moiety, the nucleoside has to be phosphorylated for use in automated synthesis. This can be done by coupling a trivalent phosphitylating agent to the 3′-OH group.

The first nucleoside is linked via a spacer to the solid support by its 3′-hydroxyl function and a dicarboxylic acid (succinic acid), since in chemical oligonucleotide synthesis the chain grows in 3′–5′-direction. After removal of the 5′-OH protecting group of the immobilized nucleoside, the chain grows by nucleophilic attack to this 5′-hydroxyl to the incoming activated 3′-phosphoramidite function. In order to drive this reaction to completion, a larger excess has to be used. Ten equivalents are routinely used; for large scale five equivalents are sufficient. Following this reaction a so-called capping reaction has to be performed which guarantees the complete reaction of free 5′-OH groups in each cycle. Therefore, a highly reactive acylation or phosphorylation is needed. After capping of unreacted 5′-hydroxyls, the phosphorus is oxidized by the mild iodine reaction, and the cycle is to be re-entered according to the sequence to be synthesized. Historically, compared to the original cycle, the basic concept has been kept, but the details have been constantly optimized in order to guarantee almost quantitative yields.

The 5′-end of the oligonucleotide prepared can either remain or alternatively be directly phosphorylated chemically on the synthesizer (UHLMANN and ENGELS, 1986a, b). After completion of the synthesis on the solid support the oligonucleotides are freed from their protecting groups and cleaved from the support simply be treatment with concentrated aqueous ammonia. The synthetic oligonucleotides are then further purified by different methods.

2.5 Purification and Characterization

Usually the deprotected oligonucleotide is ready for further purification. Due to the great resolution of modern separation techniques, solid-phase synthesis has been successful. In the resulting mixture the desired oligonucleotide has to be separated from all possible failure sequences by either electrophoretic techniques or by high performance liquid chromatography (HPLC). The scale is usually in the μg up to mg quantities depending very much on the further use of the oligonucleotide. For the PCR the nmol scale is more than enough, as for antisense oligonucleotides a gram scale is desired.

For cloning and PCR purposes the purity of the oligomer obtained is generally sufficient for direct use, or only a fast cartouche method will be performed (MCBRIDE et al., 1988). Reversed phase chromatography is especially helpful for oligonucleotides which still retain the dimethoxytrityl group on the 5′-OH position.

Gel electrophoresis is the most powerful technique for resolving DNA and RNA. Longer oligonucleotides preferentially will be purified by this technique. The major limitation is the scale up (to a few mg). On the other hand, resolution up to single nucleotides is possible in the wanted range.

The handling of gels is well documented in the manuals of molecular biology (e.g., MANIATIS et al., 1989). Thick gels up to 5 mm have

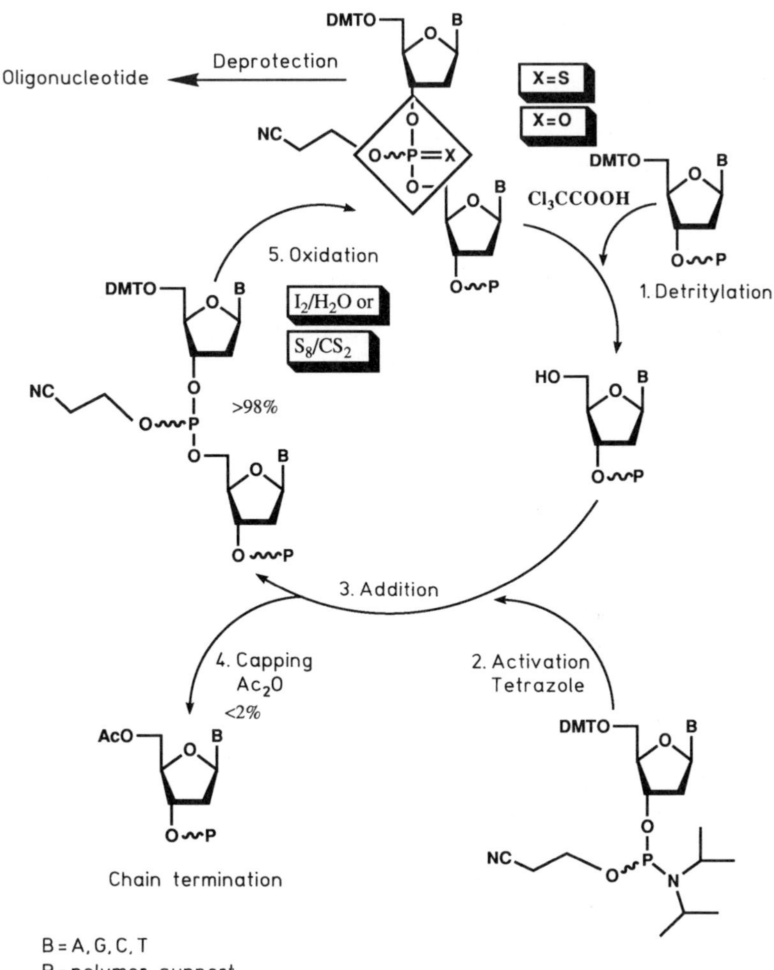

Fig. 7. The phosphoramidite cycle.

been routinely run in the author's laboratory in order to isolate mg-quantities even of modified oligonucleotides. HPLC is the method of choice for smaller oligonucleotides (up to 50mers) and larger quantities (mg-scale). Here a variety of different separation techniques has been tested. Reversed phase and anion exchange chromatography are the preferred methods.

For routine purification of a larger number of oligonucleotides the trityl on cartridge method has been developed (Lo et al., 1984; Horn and Urdea, 1988). Several purification cartridges of different companies are commercially available. This method has also been successfully automated (Ivanetich et al., 1991).

The analysis of oligonucleotides still poses a major problem to the chemist. Oligonucleotides are of polyanionic nature and only water-soluble compounds with difficult-to-define counterions and water content. Thus most of the conventional analytical techniques do not give correct answers. Furthermore, the scale of solid-phase synthesis for cloning purposes is in the range of 0.05 to 1.0 μmol giving rise to only several OD quantities of final oligonucleotide. As criteria for purity and shift mobility compared to size markers (Jing et al., 1986) gel electrophoresis and now more and more HPLC and capillary electrophoresis are used. Furthermore, UV spectra indicate the correct shape of a DNA curve, or sometimes only the quotient 260/280 nm is taken. As to the identity of the correct sequence, Maxam–Gilbert sequencing, especially on a solid phase, has been tested. More routinely a phosphodiesterase digest followed by HPLC analysis has proven to be helpful. This is especially true for modified oligonucleotides with base modifications. In the future nuclear magnetic resonance (NMR) (VAN DE VEN and HILBERS, 1988; HOSUR et al., 1988) and mass spectroscopy (MS) (MCCLOSKEY, 1991) will become the methods of choice for oligonucleotides being prepared on a larger scale, besides the X-ray analysis (KENNARD and HUNTER, 1989). For oligonucleotides new ionization methods are being developed, such as electrospray (WHITEHOUSE et al., 1985) and laser desorption (HILLENKAMP et al., 1991). These methods offer the potential to eventually be usable for DNA sequencing. NMR analysis of DNA and especially RNA is rather complex due to the narrow region of proton resonances of the sugar protons. Here the new 3D method in combination with heteronuclear-enriched DNA or RNA promises to be of future applicability (NIKONOWICZ and PARDI, 1992).

3 Application of Oligonucleotides in Biology and Medicine

3.1 Hybridization Probes

Nucleic acid hybridization assays have become a powerful analytical tool in biology research laboratories and are widely used in studies of gene structure and function. Over the last ten years, the use of DNA probes in molecular biology has increased tremendously. The term DNA probe is a functional description for synthetic DNA molecules which are used to identify the presence of specific target DNA within a given mixture. In the field of molecular cloning DNA probes are employed for recombinant library screening, *in situ* hybridization, DNA sequencing, and other methods (HAMES and HIGGINS, 1985).

Despite the quite different applications, the experimental procedures for most DNA assays are very similar. Under appropriate conditions the DNA probe will hybridize via hydrogen bonding according to Watson–Crick base pairing to the complementary target DNA (Fig. 4, Sect. 2.3), the more classical Southern blot (SOUTHERN, 1975). After the free unhybridized probe is removed, the DNA hybrid (target DNA with bound probe) can be visualized by a variety of direct and indirect detection methods using radioactive, fluorescent, chromogenic, or luminescent labels. The following section will show the possibilities which exist for labeling or visualizing oligonucleotides. In their review of 1984 on hybridization of nucleic acids immobilized on solid support, MEINKOTH and WAHL discussed radioactive labels in detail. A review only four years later listed a dozen direct and indirect labels for non-radioactive detection of probes (MATTHEWS and KRICKA, 1988).

3.1.1 Radioactive Labels

In radioactive labeling procedures nucleotides bearing one of a variety of possible isotopes ^{3}H, ^{32}P or ^{35}S are incorporated. In radioactive labels the radioactivity incorporated into the probe molecules allows direct detection by exposure to X-ray films. For that purpose the 3′ end of oligonucleotides may be labeled with terminal transferase by adding of one or more radioactive nucleotides (TU and COHEN, 1980). Alternatively the 5′ hydroxy group of an oligodeoxynucleotide may be enzymatically phosphorylated using γ-^{32}P-labeled or γ-^{35}S ATP and T4 polynucleotide kinase (HARRISON and ZIMMERMAN, 1986).

3.1.2 Direct and Indirect Non-Isotopic Labels of Oligonucleotides

In order to avoid radioactivity, a variety of indirect methods have been advocated mainly by making use of high affinity recognition methods such as biotin-(strept)avidin and digoxigenin and direct attachment of fluorophores.

In indirect labeling methods in the first step a component which itself does not generate a signal but can be specifically detected by a signal generating system is in correlation with the hybridization probe. Biotin labels are the most commonly studied indirect labels where the biotin incorporated directly into the probe molecules must be detected following binding of the biotin to a signal generating system. Biotin detection usually involves secondary labeling systems with avidin or streptavidin, which tightly bind to biotin, and a covalently linked reporter group, such as an enzyme, fluorescent moiety, or luminescent moiety. Oligonucleotides may be labeled at any position by reacting an aliphatic amino group such as allyl amino-dU, alkyne-containing alkylamino-dU (HARALAMBIDIS et al., 1987) with a biotinylating reagent, such as N-biotinyl-6-aminocaproic acid N-hydroxy-succinimide ester, as described by COOK et al. (1988). The introduction of an amino group at the 3′ end of an oligonucleotide which will be synthesized on a solid support needs a protected amino-containing moiety. This amino link is coupled to the CPG material prior to the addition of the first nucleotide (HARALAMBIDIS et al., 1990). For the introduction of an amino group at the 5′ end of an oligonucleotide aminoalkyl-phosphoramidites will be used (AGRAWAL et al., 1986; CHOLLET and KAWASHIMA, 1985). The biotin–streptavidin system has been applied in a variety of assays.

Enzyme labels which are directly attached to the probes include alkaline phosphatase and horseradish peroxidase (RENZ and KURZ, 1984), both with commercially available chromogenic substrates. Both methods are sensitive enough to substitute for radioactive detection in DNA sequencing (RICHTERICH et al., 1989). Horseradish peroxidase produces a color faster than alkaline phosphatase, but the sensitivity which can be achieved with alkaline phosphatase is greater (LEARY et al., 1983). A further possibility for indirect labeling are phosphorothioate linkages, which can be incorporated in oligonucleotides with phosphoramidite chemistry and subsequent S_8 oxidation. These linkages themselves are chemically reactive and can be used for post-hybridization fluorescent labeling of probes using the fluorophore monobrombimane (CONWAY et al., 1989; HODGES et al., 1989).

The ability to directly attach different fluorescent dyes to individual probe molecules allows the simultaneous detection of hybridization products involving more than one probe. N-hydroxy-succinimide esters of fluorescein and rhodamine derivatives are attached to aliphatic amine linker arms on synthetic oligonucleotides. A color complementation assay that allows rapid screening of specific genomic DNA sequences has been developed by CHEHAB and KAN (1989). It is based on the amplification of two or more DNA segments with fluorescent oligonucleotide primers such that the generation of a color, or combination of colors, can be visualized and used for diagnosis.

Automated DNA sequencing without blotting has been described using four different fluorophores per lane or one fluorophore for four lanes (PROBER et al., 1987; SMITH et al., 1986). Potentially, fluorescence energy transfer will be a practical basis for homogeneous hybridization assays. In one format, two probes bind to adjacent sequences on a target nucleic acid containing fluorescein and rhodamine. When both probes were bound, fluorescein emission was quenched and rhodamine emission was enhanced (CARDULLO et al., 1988).

The choice for labeling and for detection has expanded greatly over the past few years. Increasing competition between these systems should lead to an ever increasing substitution of non-radioactive versus radioactive methods (WETMUR, 1991). As generally with analytical methods the ease of handling paired with the proper detection specificity will dictate the method of choice.

3.2 Polymerase Chain Reaction (PCR)

Historical Background

The polymerase chain reaction is a recently developed procedure for the *in vitro* amplification of DNA sequences. In a very short time it has developed into an analytical tool which facilitates gene analyses and recombinant techniques, enables DNA detection from single cells (Li et al., 1988), and permits sequence determinations even from extinct species. Like the PCR process itself, the number of laboratories making use of this novel technique has been growing exponentially. Since the first description of the method more than 1200 publications involving PCR and its application have been reported (CAS online search June 1991).

The basic principle of enzymatic reaction was first described by Kleppe and Khorana in 1970 (Kleppe et al., 1971), but they had no possibility for carrying out PCR as an automated process as is done currently. The method was introduced in 1985 by Kerry B. Mullis at Cetus Corporation in Emeryville, California (Saiki et al., 1985). The invention grew from a theoretical scheme to perform limited dideoxynucleotide sequencing of unique human genes using synthetic oligonucleotides for the purpose of diagnosing common human disease mutations (Mullis, 1990; Gibbs, 1990). This method was originally used by a group at Cetus for the amplification of human β-globin DNA and for the prenatal diagnosis of sickle-cell anemia (Saiki et al., 1985, 1986; Embury et al., 1987). Initially, the PCR used the Klenow fragment of *Escherichia coli* DNA polymerase I to extend the annealed primers. This enzyme was inactivated by the high temperature required to separate the two DNA strands at the outset of each PCR cycle. Consequently, fresh enzyme had to be added during every cycle. The use of Taq DNA polymerase has transformed the PCR by allowing development of simple automated thermal cycling devices for carrying out the amplification reaction in a single tube containing the necessary reagents (Saiki et al., 1988a). Further advantages of these thermostable enzymes are the increasing specificity and yield of the amplification reaction. The reason for this is that the incorporation of Taq DNA polymerase into the PCR protocol allows the primers to anneal and extend at much higher temperatures than it was possible with the Klenow fragment, thus eliminating much of the non-specific amplification. Moreover, long PCR products can be amplified from genomic DNA, probably due to the reduction in the secondary structure of the template strands at elevated temperatures and for primer extension.

Very important for widespread application of PCR techniques was also the development of automated chemistry for the oligonucleotide synthesis (see Sect. 2.4). Now it is possible to purchase either an oligonucleotide synthesizer or the oligonucleotides themselves prepared by a commercial supplier.

3.2.1 General Methodology

3.2.1.1 PCR Amplification Scheme (see Fig. 8)

As indicated in the introduction, PCR involves repeated temperature cycling with three steps per cycle. The first step is denaturation at >90 °C in which the DNA to be amplified is separated into single strands. This is followed by lowering the temperature to allow annealing of primers to the different strands of the DNA.

One primer is complementary to the minus strand and the other one is complementary to the plus strand. In the last step the annealed strands will be extended with the Taq polymerase and deoxynucleotide triphosphates. This results in the synthesis of a plus and a minus strand containing the target DNA. These newly synthesized DNA strands are themselves templates for the PCR primers. Repeated cycles of denaturation, primer annealing and extension lead to exponential accumulation of the desired DNA fragment which is represented at a theoretical abundance of 2^n, where n is the number of PCR cycles performed.

Thus theoretically 20 cycles can generate a million copies of a single template. In practice, however, the gain per cycle depends on allele length with only very short amplification products approaching doubling per cycle. Fig. 9 shows the discrepancy between theoretical and actual yields of a PCR.

An important feature of this amplification scheme (Fig. 8) is that the majority of the amplification products that are present following many PCR cycles are double-stranded DNA fragments of discrete length. The strands that are synthesized as copies of the original tem-

PCR Amplification Scheme

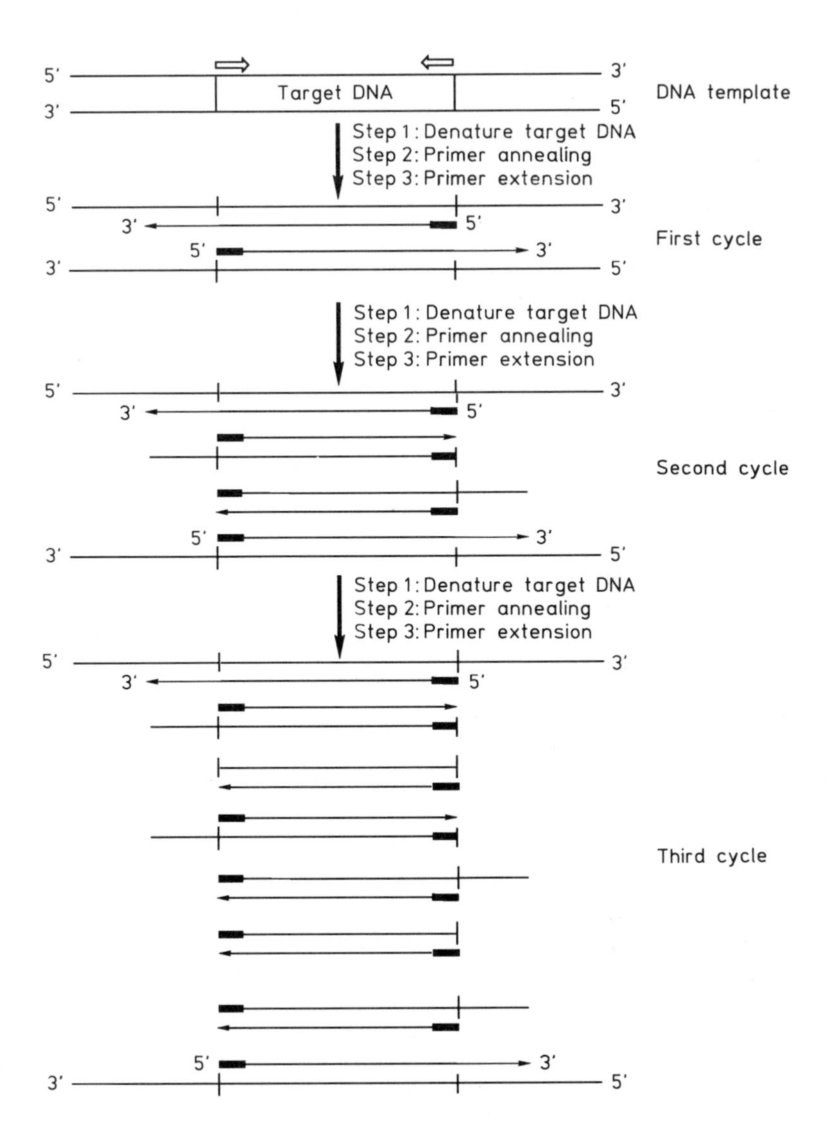

Fig. 8. Principles of the polymerase chain reaction amplification.

plate are bordered at the 5′ terminus by the oligonucleotide primer, while the 3′ terminus is determined by the position at which the DNA polymerase finishes its synthesis. In contrast, the products of polymerase extension resulting from priming of DNA strands that were produced during the PCR will have both their 5′ and 3′ termini defined by the position of the primers. After the third cycle the templates with defined ends begin to override the fragments without discrete length.

After 20–30 cycles the amplification reaction reaches a "plateau", which means that only little increase in DNA can be measured after 30 cycles of amplification reactions. The reason for this is the opposing competition of template–template hybrids and template primer hybrids. Another reason is the decreasing

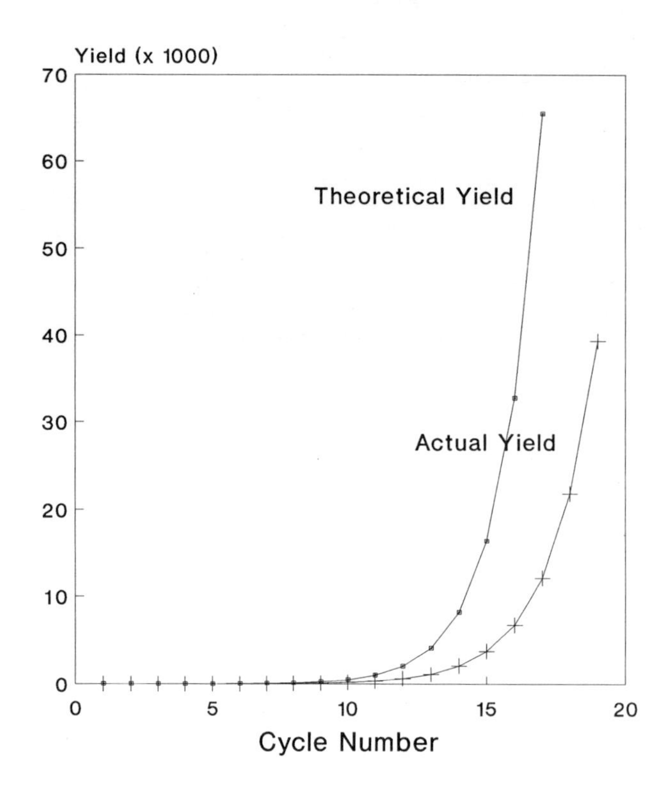

Fig. 9. Theoretical vs. actual yield of a PCR. Data describing the actual yield from HAFF and MEZEI (1989).

activity of the Taq polymerase in the reaction.

3.2.1.2 Ingredients for Performing PCR

The standard reaction involves the following components (the most often used concentrations are listed in Tab. 1): a pair of oligonucleotides complementary to the 3′ ends of the double-stranded DNA which will be amplified in the reaction, the template DNA, the buffer for the Taq polymerase, and the Taq polymerase itself. Tab. 1 summarizes the key chemical parameters for amplification.

A number of factors influence the specificity of Taq polymerase mediated amplification: the time of the primer extension step, the amount of enzyme used, concentration of Mg^{2+}, the nature of template DNA and primers, and the annealing temperature (MULLIS et al., 1986; SAIKI et al., 1988b; WU et al., 1991).

3.2.1.3 Influence of the Taq Polymerase

Thermus aquaticus YT1 was isolated from a hot spring in Yellowstone National Park (BROCK and FREEZE, 1969). Later on, the isolation of a thermostable polymerase from *T. aquaticus* YT1 revolutionized the PCR (SAIKI et al., 1988a). The 94 kDa enzyme has a specific activity of 200000 U/mg and a relatively high temperature optimum (T_{opt}) for DNA synthesis. Depending on the nature of the DNA template, GELFAND (1989) found an apparent T_{opt} of 75–80 °C with a K_{Cat} approaching 150 nucleotides per second per enzyme molecule. INNIS et al. (1988) reported progressive synthetic properties and an extension rate >60 nt/s at 70 °C with Taq DNA polymerase for a GC-rich 30mer primer and a significantly lower activity at 55 °C of 24 nt/s. This shows that activity decreases with the reduction of temperature. The marked attenuation of the Taq polymerase at lower temperatures can

Tab. 1. PRC Reactants

Component	Typical Size	Typical Concentration (μmol L^{-1})
Taq polymerase I	94 kDa	10^{-6} (2.5 U)
Mg^{2+}		1–10
dNTP		10^{-2}–1 (each)
Primer	15 –30 nt	10^{-4}–10^{-3} (each)
Target DNA		
Genome DNA	10^5–10^7 nt	10^{-17}–10^{-12} (1–10^5 copies/100 μL)
cDNA	10^2–10^4 nt	(1–10^5 copies/100 μL)
Intermediate strand	10^2–10^4 nt	10^{-16}–10^{-11} (final)
Short strand	10^2–10^3 nt	10^{-7}–10^{-4} (final)

nt, nucleotide(s)

have different reasons. The first one is the impaired ability of Taq polymerase to extend through regions of local intramolecular secondary structure on the template strand; the second one is a change in the ratio of the forward rate constant to the dissociation constant. Only little DNA synthesis occurs at very high temperatures (>90 °C). The amplification at these temperatures may be limited by the stability of the primer or primary strand and template strand duplex. The enzyme is very stable and is not denatured irreversibly by exposure to high temperature. For example, $T_{1/2}$ at 92.5 °C: 130 min; at 95 °C: 40 min, and at 97.5 °C: 5–6 min (GELFAND, 1989). Taq DNA polymerase has no 3′ to 5′ exonuclease activity, but a 5′ to 3′ exonuclease activity during polymerization. The misincorporation rate, less than 10^{-5} nucleotides per cycle, by Taq DNA polymerase during PCR was based on measuring the frequency of nucleotide substitutions in the sequence analysis of cloned PCR products (TINDALL and KUNKEL, 1988; GELFAND and WHITE, 1990; ECKERT and KUNKEL, 1991).

Other DNA polymerases, such as bacteriophage T4 DNA polymerase, appear to have a very low misincorporation rate in PCR (KEOHAVONG and THILLY, 1989). The heat lability of this enzyme, however, limits its utility. The thermostable VENT DNA polymerase, an enzyme recently isolated from *Thermococcus litoralis,* has the 3′ to 5′ exonuclease activity and may therefore have a lower misincorporation rate. Further studies will be required to determine if the capacity of this 3′ to 5′ exonuclease activity to degrade single-stranded molecules (e.g., oligonucleotide primers or PCR products prior to primer annealing) will pose problems for PCR amplification. In addition, DNA polymerases with 3′ to 5′ exonuclease activity probably cannot be used in sequence-specific priming reactions, because this activity removes the mismatched base at the 3′ end of the primer (ERLICH et al., 1991).

Genetic variants of the thermostable Taq DNA polymerase exhibit properties that can be useful for amplifying longer PCR products. For example, a mutant Taq DNA polymerase lacking a 5′ to 3′ exonuclease activity (ROSE, 1991) permits efficient amplification of long fragments. Many of the new thermostable polymerases have additional useful activities. The thermostable DNA polymerase from *Thermus thermophilus* (Tth) can reverse-transcribe RNA efficiently in the presence of $MnCl_2$ at high temperatures (ROSE, 1991). The DNA polymerase activity is enhanced by chelating Mn^{2+} and adding $MnCl_2$, allowing the cDNA synthesis and PCR amplification to be carried out in a single-enzyme, single-tube reaction.

3.2.1.4 The Nature of the DNA Template

A broad range of nucleic acid sources are suitable templates for PCR amplification. Purified DNAs from all parts of the evolutionary scale have been amplified, and many short protocols for the DNA purification have been reported (MERCIER et al., 1990; GÜSSOW and CLACKSON, 1989; WINBERG, 1991; KOGAN et al., 1987; MILLER et al., 1988; YAMADO et al., 1990). PCR is highly tolerant to impurities in the DNA sample being amplified. A biological sample normally should be deproteinized before introduction into the reaction mixture, if only to assure removal of proteases, nucleases, and phosphatases that might destroy reactants. In general the protocols for PCR amplification of crude DNA extracts have had the greatest success when the target fragments were relatively abundant.

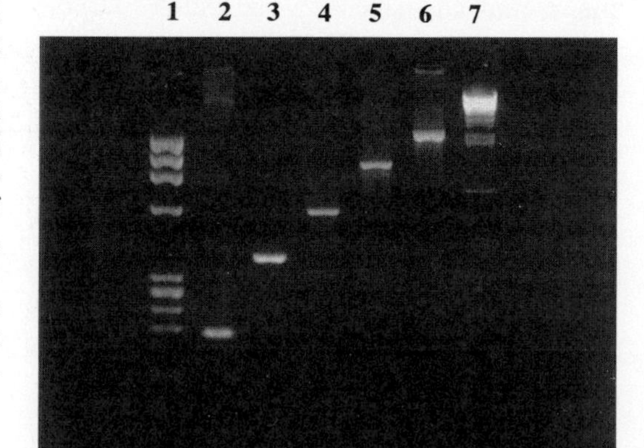

Fig. 10. Agarose gel electrophoresis of PCR amplified DNA.

3.2.1.5 Designing the Primers

Most primers will be between 20 and 30 bases in length. The minium length for amplification of allele-specific amplification in the human genome is given by the following mathematical rule:

$$1/(4 \times 10^9) = (1/4)^{16}.$$

The GC content of the primers should be similar to that of the fragment being amplified. Sequences with significant secondary structure should be avoided. For the optimal selection of oligonucleotide primers computer programs are very useful (LOWE et al., 1990). Sequences not complementary to the template can be added at the 5′ end of the primers. These exogenous sequences become incorporated into the double-stranded PCR product and provide a means of introducing restriction sites at the end of the amplified target sequence (SCHARF et al., 1986; STOFLET et al., 1988).

3.2.1.6 New Approaches to Improve the Specificity of PCR

One of these strategies is based on the observation that the Taq DNA polymerase retains considerable enzymatic activity at temperatures well below the optimum for DNA synthesis. Thus, in the initial heating step of the reaction, primers that anneal non-specifically to a partially single-stranded template region can be extended and stabilized before the reaction reaches 72 °C for extension of specifically annealed primers. Some of these non-specifically annealed and extended primers may be oriented with their 3′ hydroxyl directed toward each other, resulting in the exponential amplification of a non-target fragment. If the DNA polymerase is activated only after the reaction has reached temperatures >70 °C, non-target amplification can be minimized (ROSE, 1991). This can be accomplished by manual addition of an essential reagent (DNA polymerase, $MgCl_2$, primers) to the reaction tube at these temperatures, an approach called "hot start". Hot start not only improves specificity but minimizes the formation of the so-called "primer-dimer" (CLARK, 1988). Primer dimer is an amplification artifact often observed in the PCR product. It is a double-stranded fragment whose length is very close to the sum of the two primers and appears to occur when one primer is extended by the polymerase over the other primer.

"Nested" oligonucleotide PCR primers have been employed to improve the specificity of reactions that do otherwise not yield homogeneous products (GIBBS, 1990) (Fig. 11). This protocol uses a two-step reaction scheme beginning with amplification by the outer primer set, followed by initiation of a second reaction with the internally binding primers. The overall lowered complexity of the template for the second PCR ensures a more homogeneous final product. The applications for nested prim-

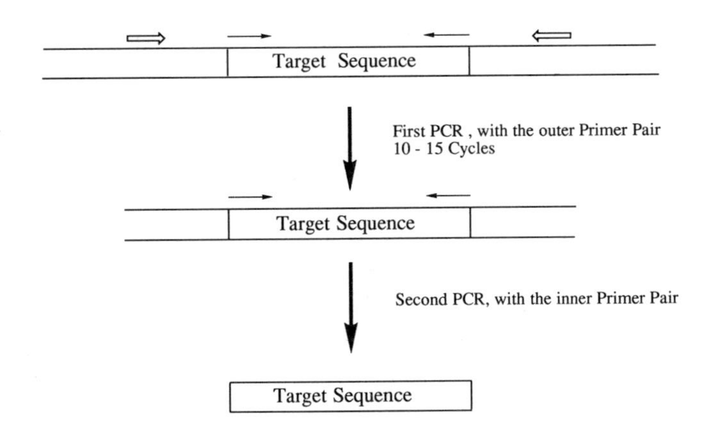

Fig. 11. The two-step reaction scheme for the nested priming reaction.

ing are amplifications of rare sequences (SCHOWALTER and SOMMER, 1989).

3.2.1.7 Qualitative Analysis of PCR Products

PCR product detection and identification can be accomplished by size-dependent or sequence-dependent means. For most applications ethidium bromide-stained gel electrophoresis is sufficiently sensitive to provide size-dependent detection (see Fig. 10) (MANIATIS et al., 1989). Acrylamide gels give greater resolution than agarose gels, and the sensitivity can be increased by silver staining. A possibility to increase the resolution of acrylamide is the use of a discontinuous buffer system as described by ALLEN et al. (1989). Other instrumentally oriented detection methods, such as HPLC and capillary electrophoresis, offer resolution approaching and sensitivity exceeding ethidium bromide-stained gels with obvious advantages for speed, automation, and quantitation. In addition, the desired PCR product can be purified for subsequent use (WARREN et al., 1991; KALNOSKI et al., 1991).

3.2.1.8 Contamination of PCR Reactions

Contamination of the amplification reaction with products of a previous PCR reaction (product carry-over), exogenous DNA, or other cellular material can create problems both in research and diagnostic application since the PCR can generate trillions of DNA copies from a template sequence (KWOK and HIGUCHI, 1989). The use of PCR for sensitive detection is complicated by the fact that the product of amplification serves as a substrate for the generation of more product. In general, attention to careful laboratory procedures – pre-aliquoting reagents, the use of dedicated pipettes, positive displacement pipettes, or tips with barriers preventing contamination of pipette barrels, and the physical separation of the reaction preparation from the area of reaction product analysis – minimize the risk of contamination. Multiple negative controls (non-template DNA added to the reaction) are necessary to monitor and reveal contamination. Several approaches to minimize the potential for PCR product carry-over have been developed, all based on interfering with the ability of the amplification products to serve as templates. One strategy utilizes the principle of restriction modification and excision repair systems of bacteria to pretreat PCR reactions and selectively destroy DNA synthesized in a previous PCR (ERLICH et al., 1991). In order to distinguish PCR products from sample template DNA, deoxyuridine triphosphate (dUTP) is substituted for deoxythymidine triphosphate (dTTP) in the PCR and is incorporated into the amplification products. The presence of this unconventional nucleotide allows the distinction of products of previous amplifications from the native DNA of the sample. The enzyme uracil N-glycosylase (UNG) present in the reaction premix, catalyzes the excision of uracil from any potential single- or double-stranded PCR carry-over DNA present in the reaction prior to the first PCR cycle.

Another possibility to avoid false positives in the PCR is the use of UV irradiation for the PCR reagents (OU et al., 1991). UV irradiation of DNA results in the formation of pyrimidine dimers and thus prevents them from being effective templates in subsequent PCR.

3.2.2 Applications of PCR

3.2.2.1 Diagnostic Applications of PCR

The initial diagnostic application of PCR was in the prenatal diagnosis of sickle cell anemia through the amplification of β-globin sequences. Hybridization of labeled oligonucleotide probes (SAIKI et al., 1985, 1986; EMBURY et al., 1987) or restriction-site analysis of the amplified products allowed the distinction of normal or mutant alleles. This genetic disease, which is caused by a nucleotide substitution in the β-globin gene, has served as the model system for developing a variety of simple diagnostic methods for detecting a known mutation. The amplification of a specific locus by PCR made the use of non-radioactive allele-specific oligonucleotide (ASO) probes in the dot blot hybridization test a rapid, general, and practical method for genetic typing (SAIKI et al., 1986, 1988b). This approach, which depends on the instability of binding of a probe mismatched with the target sequence (CONNER et al., 1983), has been applied not only to sickle cell anemia but to other genetic diseases as well as to the analysis of human leucocyte antigen (HLA) polymorphisms (ERLICH and BUGAWAN, 1989). The specificity of this method has been increased by the introduction of a thermostable DNA ligase isolated from *Ther-*

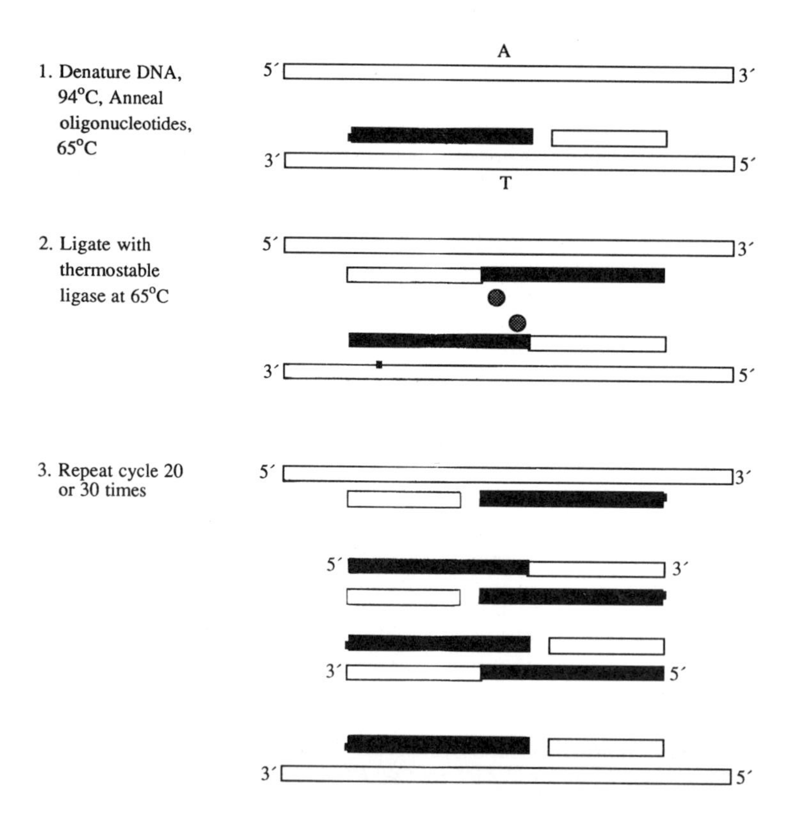

Fig. 12. The ligase chain reaction.

mus aquaticus (BARANY, 1991). The thermostable ligase can be used in conjunction with two pairs of complementary oligonucleotides to generate the target dependent on the exponential accumulation of the ligated oligonucleotides (the ligase chain reaction, see Fig. 12).

A further method of allelic discrimination that uses PCR is based on the effect of mismatches in the 3′ end of an oligonucleotide on the priming step (allele specific amplification). This approach requires that either two separate PCR amplifications be conducted (one with primers specific for the mutation and one with primers specific for the normal allele), or that the PCR products from the two alleles be distinguishable by length or by the fluorescence label on the two primers (CHEHAB and KAN, 1989). This method which depends on the specificity of priming in the first PCR cycle has recently been applied to sickle cell anemia (CHEBAB and KAN, 1989; WU et al., 1989).

3.2.2.2 Site-Directed Mutagenesis

Mutagenesis at the primer region in a PCR product has been described above. Here follows a general scheme for introducing a mutation in a PCR produced DNA fragment, anywhere along its chain. Two primary PCR reactions produce two overlapping DNA fragments, both bearing the same mutation introduced via primer mismatch, in the region of overlap. This mutation is not only restricted to base substitution. The overlap in sequence allows the two fragments to recombine in two possible ways after their mixture, denaturation, and renaturation. Only one of these combinations produces a structure with recessed 3′-OH ends that can be extended by a DNA polymerase to produce a complete duplex fragment. These extended segments can then serve as templates for the secondary re-amplification of the combined sequences using only the outermost two of the four primers employed to produce the primary fragments (HIGUCHI et al., 1988).

Alternatives to this method have been described by HEMSLEY et al. (1989) and PERRIN and GILLILAND (1990). For a detailed description of mutagenesis see Sect. 3.4.

3.2.2.3 PCR from mRNA

RNA can also be used as a template for PCR following reverse transcription (VERES et al., 1987; KAWASAKI, 1989). This is a useful procedure for the study of expressed gene sequences and retroviruses. In general the quality of the cDNA synthesis needs not to be as high as usually required for the construction of cDNA libraries, since incomplete cDNA strands will be lost during the PCR amplification. The polymerase can synthesize from both RNA and DNA templates, thus the initial reverse transcription step and the subsequent PCR amplification can be carried out by the same enzyme. The use of this thermostable polymerase should reduce the amount of secondary structure in the mRNA template, allowing more efficient synthesis of the cDNA strand (LONGO et al., 1990).

3.2.2.4 Generation of Single-Stranded DNA and its Application to Direct Sequencing of PCR Products

The principle of the ssDNA producing PCR reaction is shown in Fig. 13. The method is called asymmetric PCR. The two amplification primers are present in different molar amounts. During the first 10–15 cycles dsDNA will be produced exclusively. However, when the primer added in limiting amounts has been used up, an excess of ssDNA will be produced in each cycle. Theoretically, the amount of dsDNA should increase exponentially, whereas the production of ssDNA should only follow a linear growth (GYLLENSTEN and ERLICH, 1988). The ssDNA serves as a template in the following sequencing reaction as described by SANGER et al. (1977).

A promising method to generate ssDNA uses the streptavidin-biotin system (HULTMAN et al., 1989). A biotin residue can be introduced into a PCR product via the 5′ terminus of an oligonucleotide primer, and the strand is efficiently captured by using streptavidin-coated magnetic beads.

3.3 Total Synthesis of Genes and their Expression

3.3.1 Strategies for the Construction of Genes

Automated DNA synthesis results in single-stranded oligonucleotides which can be used to build up the functional double-stranded DNA (ENGELS and UHLMANN, 1989). There are two basic enzymatic reactions used to form new phosphodiester bonds, the DNA-ligase (see Sect. 3.2.2.2) and the DNA-polymerase reaction (see Sect. 3.2) (Fig. 14). T4-DNA ligase catalyzes the template-assisted formation of a phosphodiester linkage between adjacent 5′-phosphate and 3′-hydroxy groups under consumption of ATP. Whereas RNA-ligase is able to join single-stranded RNA molecules, T4-DNA ligase does not accept single-stranded DNA as substrate. The two oligonucleotides to be joined are brought into correct position by means of a complementary oligonucleotide. Basically, the ligation of blunt ends is also possible having the disadvantage of much lower efficiency and problems due to equivocal orientation of the ligated molecules. Standard oligonucleotide synthesis yields products with 5′-hydroxy groups which can be phosphorylated by polynucleotide kinase and ATP. We have developed a chemical phosphorylation method useful in automated DNA synthesis resulting in the desired oligonucleotide-5′-phosphates (UHLMANN and ENGELS, 1986a, b).

Asymmetric PCR
Generation of single-stranded DNA
Abundant Primer
Limiting Primer
Target Sequence
After Thermocycling
ssDNA
and
dsDNA

Fig. 13. A schematic drawing of the generation of single-stranded DNA with asymmetric PCR.

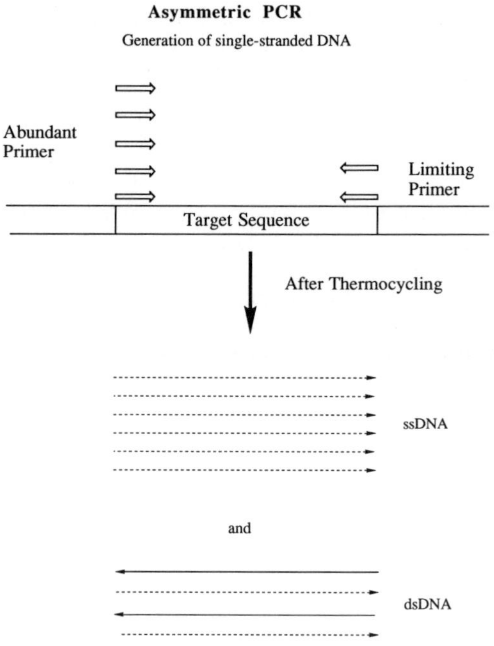

The second reaction type makes use of the repair activity of DNA polymerase I (Klenow fragment) which lacks the reverse 5′3′-exonu-

5′
3′
OH
P
P
HO
3′
5′
DNA ligase, ATP
5′
3′
3′
5′
DNA polymerase, dNTP
5′
HO
3′
5′

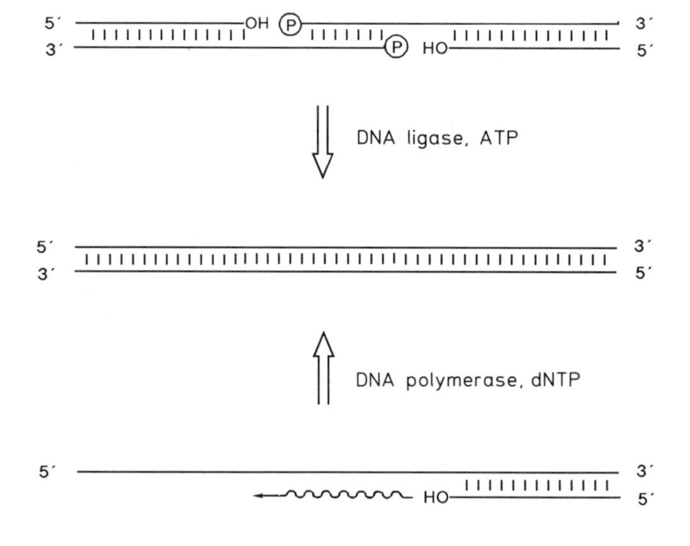

Fig. 14. The DNA ligase reaction and the DNA polymerase-catalyzed "fill-in" reaction.

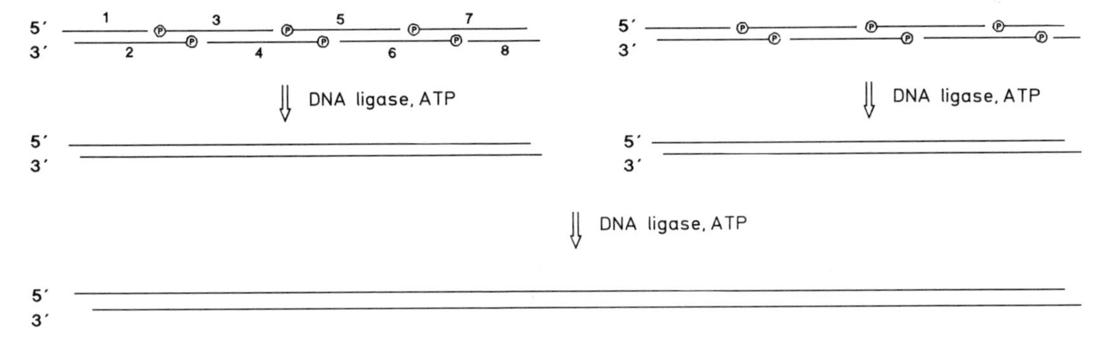

Fig. 15. Ligation strategy for gene synthesis according to KHORANA.

clease activity of the holoenzyme. After annealing of a short oligonucleotide to the 3′ end of a given template, the Klenow enzyme in the presence of the four deoxynucleoside-5′-triphosphates allows the synthesis of a full-length complementary copy to yield double-stranded DNA. Other polymerases like reverse transcriptase or Taq polymerase can also be used for this reaction.

Chemical ligation of oligonucleotides is an attractive alternative to the enzymatic reactions, because it should be cheaper and, most importantly, amenable to automation regarding solid-phase gene synthesis. Quite recently, the first non-enzymatic assembly of a biologically active gene has been reported (SHABAROVA et al., 1991) using cyanogen bromide as coupling reagent (DOLINNAYA et al., 1991a). Coupling yields seem to be somewhat higher with 1-ethyl-3(3′-dimethyl-aminopropyl)carbodiimide (DOLINNAYA et al., 1991a, b) than with cyanogen bromide. One major advantage of the chemical ligation over the enzymatic one is the high reaction rate of one to a few minutes compared to the DNA ligase catalyzed reaction which typically takes several hours.

The classical ligation strategy according to KHORANA and coworkers (KHORANA, 1968; AGARWAL et al., 1970) requires the synthesis of a set of overlapping oligonucleotides which after correct annealing are joined by DNA ligase to yield the gene or gene fragment (Fig. 15). The annealing of the oligonucleotides is achieved by first heating to 95 °C for five minutes and then slowly cooling the solution of the oligonucleotides. Usually, only the internal 5′ ends are phosphorylated in order to avoid concatenation of the formed gene fragment. Using this strategy, it was possible to join a set of 20 oligonucleotides in a single reaction to give a 280 bp structural gene (MULLENBACH et al., 1986). Larger genes can be divided into subfragments which are cloned individually ("subcloning"). A 1610 bp gene encoding tissue plasminogen activator has been synthesized totally from 101 oligonucleotides using the subcloning strategy (BELL et al., 1988). If longer oligonucleotides (>30 nucleotide units) are available, preformation of DNA duplices in pairs of two (oligo 1+2, oligo 3+4, etc.) is of advantage. The preformed duplices are then ligated sequentially or in a one-pot reaction via their sticky ends. Thus, wrong hybridization of oligonucleotides (e.g., oligo 1+oligo 5) can be avoided. Overhanging ends of 4 to 8 bases seem to be optimal following this ligation strategy.

As an alternative approach to ligation, KHORANA (1968) proposed the so-called "fill-in" method which has been elaborated later on by ROSSI et al. (1982) to construct double-stranded DNA. This method is based on the synthesis of relatively long oligonucleotides of more than 40 nucleotides in length which at their 3′ end share a short stretch of complementary bases (Fig. 16). After annealing of the two oligonucleotides via their 3′ ends, duplex DNA is obtained by means of the DNA polymerase (Klenow) reaction. Using this approach, approximately 40% of the chemical synthesis can be saved. There are some problems concerning this method. First, the fill-in reaction of the last nucleotide sometimes causes difficulties. Therefore, it is recom-

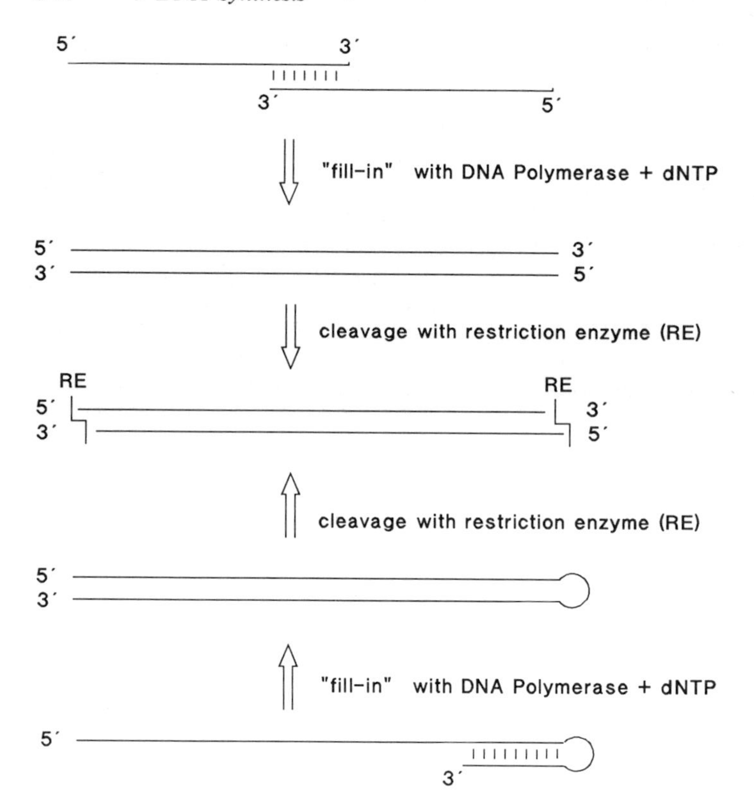

Fig. 16. Enzymatic "fill-in" reaction using two oligonucleotides with complementary 3′-ends or one selfpriming oligonucleotide.

mended to synthesize a few nucleotides more which are then removed by digestion with an appropriate restriction enzyme. At the same time, sticky ends are generated to allow for efficient further ligation.

By using longer sequences (>100 nucleotides) secondary structure problems are likely to get more pronounced which renders unequivocal annealing of the 3′ ends more difficult. Therefore, we have developed a variant method useful for long oligonucleotides which at their 3′ end have a short inverted repeat of their sequence. This leads to the formation of an unusable stable 3′-hairpin structure which serves as an efficient primer for the DNA-polymerase mediated synthesis of the second strand (UHLMANN and HEIN, 1987; UHLMANN, 1988). After the fill-in reaction of these so-called "selfpriming" or "autoprimer" oligonucleotides, overhanging ends are generated by cleavage with an appropriate restriction enzyme, thereby removing the intermediate loop structure. The advantage of this method is that no additional primer has to be synthesized and no extra annealing is necessary. In addition, the strand-separation activity of DNA polymerase which may remove the non-covalently annealed primer is circumvented. An obvious limitation of this strategy is the availability of long oligonucleotides to render the approach economic.

The direct cloning of a pair of annealed long synthetic oligonucleotides representing the coding and non-coding strands of a gene is a straightforward way to get access to small genes or gene fragments (HEIN et al., 1987). The size of these fragments is only limited by the length of the oligonucleotides available. Today the automated synthesis of oligonucleotides of up to about 200 nucleotides in length is feasible resulting in gene fragments encoding up to 65 amino acids. In the synthesis of a 1100 bp gene for bovine prochymosin this direct cloning method was combined with the sequential ligation of preformed DNA duplices (WOSNICK et al., 1987).

Regarding the design of synthetic genes it seems advisable to divide larger genes into fragments of 250 to 500 base pairs for cloning, since the observed mutation frequencies in the cloning of synthetic oligonucleotides are in the range of 0.05 to 0.32% per nucleotide. That means, the chance of finding a correct clone in the direct cloning of a 500 bp synthetic gene is 50% at best when based on an average mutation frequency of 0.15%.

3.3.1.1 Rapid Methods for Gene Synthesis

The so-called "shotgun" method relies on the chemical synthesis of a set of overlapping oligonucleotides which are ligated without prior purification into a suitable M13 vector and cloned in *Escherichia coli* (GRUNDSTRÖM et al., 1985). About one half of all analyzed clones harbored the desired sequence after shotgun cloning of a set of 15 oligonucleotides. It is obvious that by replacing two complementary oligonucleotides by a set of two "mutagenic" oligonucleotides a new mutant gene (see Sect. 3.4) can easily be produced in a separate shotgun reaction.

The strategy of cloning single-stranded oligonucleotides into a suitable plasmid vector will save about 50% of the chemical synthetic work. This can be achieved by cutting the plasmid with one enzyme that on cleavage creates a 5′ overhang and another enzyme giving a 3′ overhang. The single-stranded oligonucleotide can then be ligated into the double-cleaved vector in a patch-like manner (Fig. 17a) (CHAKHMAKHCHEVA et al., 1987; OLIPHANT et al., 1986). Although direct cloning of the resulting gapped duplex is possible, cloning efficiency is much higher by employing an *in vitro* fill-in reaction prior to the cloning step (DERBYSHIRE et al., 1986; MOUNTS et al., 1989). Ligation of long oligonucleotide sequences (CHEN et al., 1990) and their integration into plasmids assisted by short complementary oligonucleotides is another variation of this strategy being not limited to certain restriction enzyme pairs (ADAMS et al., 1988) (Fig. 17b).

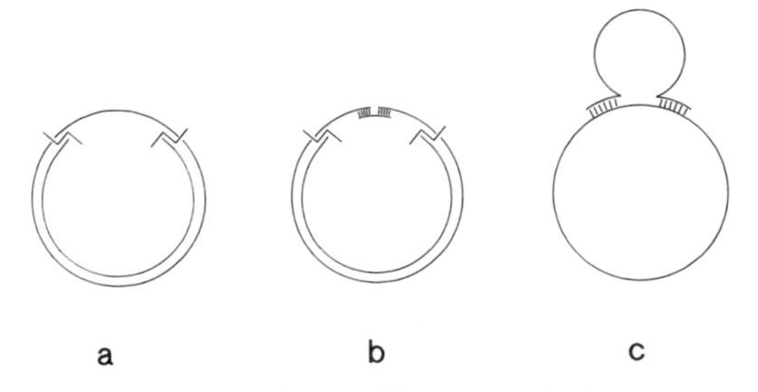

Fig. 17. Methods for rapid gene synthesis.
(a) Cloning of a single-stranded oligonucleotide
(b) Cloning of single-stranded oligonucleotides using an auxiliary oligonucleotide
(c) Insertional gene synthesis approach.

Insertional gene synthesis is a modification of M13 oligonucleotide-directed mutagenesis having the advantage that no restriction enzyme sites are required to build up and clone the gene (FRITZ, 1986). The method is based upon consecutive targeted insertions of long single-stranded oligonucleotides into a plasmid (Fig. 17c). By using just three long oligonucleotides (122, 123, and 133 bases) a 261 bp gene has been assembled in three successive cloning cycles (CICCARELLI et al., 1990). Interesting further developments of insertional gene synthesis make use of duplex vectors (MAZIN et al., 1990) or of the oligonucleotide-directed repair of a double-stranded DNA break (MANDECKI and BOLLING, 1988).

Rapid generation of DNA fragments by PCR amplification of crude synthetic oligonucleotides avoids the often cumbersome purification and ligation steps of traditional gene synthesis. The success of this method is based on the fact that failure sequences originating from chemical synthesis are doubled only once in the first PCR cycle, but are not amplified in the subsequent 25 to 30 PCR cycles. Using this selection/amplification procedure a 254 bp fragment could be synthesized from a crude 234mer and two appropriate primers (BARNETT and ERFLE, 1990). PCR-mediated gene synthesis has been described for construction of a gene encoding isozyme C of horseradish peroxidase (JAYARAMAN et al., 1991). This gene was obtained by ligating a total of 40 oligonucleotides in a single step followed by PCR amplification using the two outer oligonucleotides as PCR primers.

Similar to the solid-phase synthesis of oligonucleotides (see Sect. 2.4), total genes can be constructed by sequential ligation of gene fragments on supports (HOSTOMSKÝ and SMRT,

1987; HOSTOMSKÝ et al., 1987). If the linearized cloning vector is ligated in the last step to the ligated fragments on the support, then after cleavage from the matrix, only full-length ligated fragments having the correct ends will facilitate circularization of the gene/vector construct. A highly porous CPG 3000 support has been proposed which allows chemical and enzymatic gene synthesis (GRÖGER and SELIGER, 1988).

3.3.1.2 Design of Synthetic Genes

Since the genetic code is degenerated, that means, most of the amino acids are encoded by more than one nucleotide triplet, there is some flexibility in designing a gene for a specific protein. This fact, on one hand, allows the use of preferred codons for a specific host, and on the other hand, provides the flexibility to design restriction enzyme cleavage sites in regular distances within a given sequence. The preferred codon usage of *E. coli* and of other hosts is known, and there is evidence that codon usage determines translation rate in *E. coli* (SØRENSEN et al., 1989). For example, it has been shown that a synthetic gene with "optimal" codons expresses up to 16 times more interleukin-2 than a native cDNA sequence (WILLIAMS et al., 1988). In addition, mistranslation during over-expression of IGF-1 in *E. coli* using a synthetic gene that contains low frequency codons has been observed (SEETHARAM et al., 1988). Unique restriction enzyme cleavage sites provide a means for cloning and assembling DNA fragments, but also allow for subsequent alterations of the gene. The design of unique sites within a larger sequence is a relatively complex task and is only possible using appropriate computer programs.

3.3.2 Strategies for the Expression of Synthetic Genes

One of the decisive advantages of synthetic genes is that they can be designed to facilitate cloning and subsequent expression. Already in the design of the synthetic gene, provision can be made regarding the combination with desired promoters, operators, ribosome binding sites, signal sequences, fusion parts, and transcription terminators. Regarding the mode of expression there exist three basic routes: (1) direct expression, (2) expression as a fusion protein, (3) transport expression.

3.3.2.1 Direct Expression of Synthetic Genes

Direct expression of a gene under the control of a constitutive or inducible promoter is the most straightforward way, since it results directly in the desired protein having an additional methionine at the amino terminus. The methionine resulting from translation of the AUG start codon may be removed by cyanogen bromide or aminopeptidase cleavage. The direct expression mode can cause problems if the protein product is unstable inside the host or if it is toxic to the cells used. A "phage trojan horse" has been proposed for expression of lethal genes (HEITMAN et al., 1989). Duplication or multiplication of the synthetic gene within a plasmid has been used to enhance the expression yields. Two-cistron systems (SCHONER et al., 1987) or polycistronic tandem gene systems (LEE et al., 1984) in which two or n copies of the same structural genes are put behind each other, separated only by intercistronic regions, are easy to construct.

3.3.2.2 Fusion-Protein Expression of Synthetic Genes

The expression as a fusion protein is used to stabilize the expressed protein against degradation in the host organism. Thus, small polypeptides of less than 50 to 80 amino acids are often fused to larger proteins to achieve stabilization. Somatostatin, a tetradecapeptide could only be expressed as a fusion protein with β-galactosidase (ITAKURA et al., 1977). Thereby, the somatostatin was fused to about 1000 amino-terminal amino acids of β-galactosidase, linked by a methionine which allowed for cyanogen bromide cleavage of the expressed fusion protein to yield the tetradecapeptide.

To render the fusion expression more economic, considerably shorter fusion parts of β-galactosidase or other suitable proteins have been used. Thus, expression of bovine FGF was successful by fusing it to the first seven amino acids of β-galactosidase via an appropriate linker fragment (KNOERZER et al., 1989). Other fusion proteins contained TrpE (SMITH et al., 1982), interferon-γ (IVANOV et al., 1987), growth hormone (IKEHARA et al., 1986), or short homooligopeptide tails (SUNG et al., 1986) for stabilization. In some cases the stabilizing effect could be achieved by fusing multiple genes encoding the desired proteins after each other via a linker encoding a cleavable peptide, e.g., Asp-Pro, Asn-Gly or Ile-Gln-Gly-Arg or the like. Expression as a fusion protein often results in proteins in their biologically inactive form, so that cumbersome refolding experiments are necessary.

3.3.2.3 Transport Expression of Synthetic Genes

In the transport-expression mode the protein of interest is secreted into the periplasmic space of the host cell or even into the culture medium in a biologically active form. It therefore represents an attractive alternative to the above mentioned expression modes. In *E. coli*, transport expression can be effected, e.g., by the signal peptide of alkaline phosphatase (DODT et al., 1986; GRAY et al., 1985), and the protein is directed to the periplasmic space. We have used a synthetic DNA fragment encoding the signal peptide of alkaline phosphatase to achieve proinsulin secretion (UHLMANN et al., 1985). Surprisingly, transport expression in *E. coli* of pancreatic secretory trypsin inhibitor, when fused to ompA signal sequence, resulted in transport of the desired protein into the culture medium (MAYWALD et al., 1988). In the expression of IGF-1 using OmpF and LamB secretion leader sequences, only a small amount of IGF-1 was found in its native conformation in the medium, while the major portion of IGF-1 accumulated in the periplasmic space of *E. coli* (WONG et al., 1988). Interestingly, a synthetic gene, homologous to the one encoding the kil peptide, could be used to achieve transport expression of proteins into the culture medium of *E. coli* (BLANCHIN-ROLAND and MASSON, 1989). Secretion of proteins into the medium is also possible using other organisms like Staphylococci, *Bacillus subtilis, Streptomyces lividans* (KOLLER et al., 1989, 1991) or yeast (VLASUK et al., 1986).

3.3.3 Synthetic DNA Fragments Controlling Gene Expression

A decisive advantage of using synthetic genes for expression is that they can be designed to allow immediate expression after cloning without further alterations. Thereby, transcriptional or translational control units of chemical origin may be involved which are cosynthesized with the structural gene or are constructed as a building-block system. We have used a synthetic regulatory unit (ENGELS et al., 1985) built up from an idealized promoter, an operator, and a ribosome binding site for the expression of interferon-γ in *E. coli* (Fig. 18).

Transcription of genes can be controlled by synthetic promoters (ENGELS et al., 1985), operators (SADLER et al., 1983), and terminators (BRODIN et al., 1986). It is interesting that the idealized synthetic *lac* operator binds the *lac* repressor more strongly than the natural operator. Translational efficiency of the mRNA is mainly controlled by the ribosome binding site (RBS) and the secondary structure of the corresponding mRNA. In prokaryotes, initiation of translation is effected by the so-called Shine–Dalgarno (SD) sequence which involves four to eight nucleotides, preferably AAGGA. Synthetic RBSs have been used successfully for the expression of eukaryotic genes in bacteria. It should be pointed out that the region between the SD sequence and the AUG initiation codon is also of importance (DALBØGE et al., 1988) as are the bases upstream of the SD region (SCHAUDER and MCCARTHY, 1989). Translation termination is encoded by UAA, UAG, and UGA. Most synthetic genes contain two stop codons in tandem arrangement to avoid in any event overriding of the stop signal. Synthetic translation terminator fragments were proposed as a tool for dissecting translational direction of a gene (MARUYAMA et al., 1989).

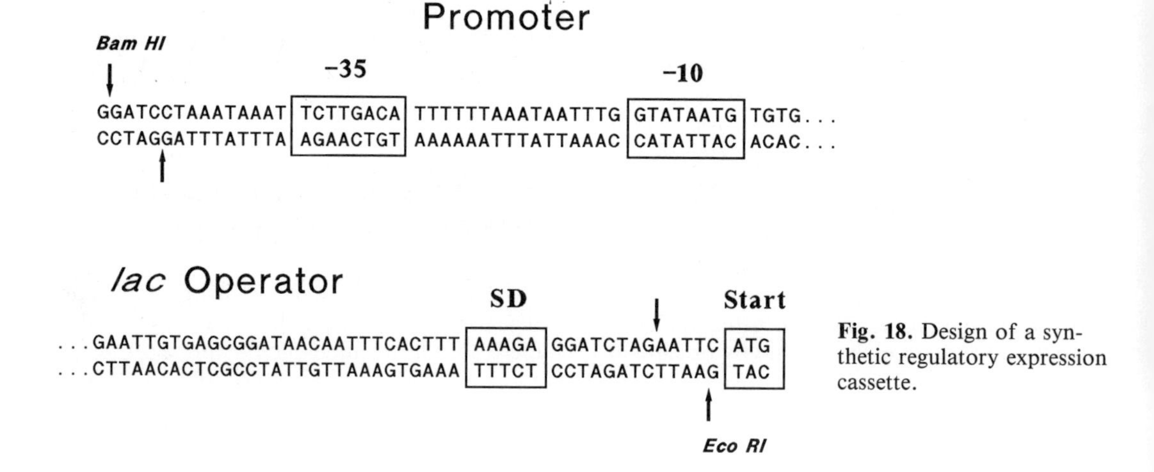

Fig. 18. Design of a synthetic regulatory expression cassette.

3.4 Mutagenesis of Synthetic Genes

Beside the above mentioned shotgun ligation (Sect. 3.3.1.1) for producing genes or mutant genes which is limited to synthetic genes, all methods known for mutagenesis of natural DNA can be used. There are two basic categories for *in vitro* mutagenesis: (1) the template-directed mutagenesis using a mutagenic primer which proceeds via a heteroduplex DNA intermediate and (2) cassette mutagenesis where a cassette within a target gene is replaced by a new cassette originating either from chemical synthesis or from mutagenic primer-directed PCR amplification.

3.4.1 Oligonucleotide-Directed Site-Specific Mutagenesis

Oligonucleotide-directed site-specific mutagenesis using single-stranded vectors like M13 has been frequently employed in the past for mutagenesis of artificial as well as natural DNA. It involves the use of a single-stranded DNA template, preferably M13, and a mutagenic oligonucleotide which is complementary to the template except for a mismatch which creates the mutation (see also "insertional gene synthesis", Sect. 3.3.1.1). After annealing of the mutagenic primer to the single-stranded DNA, the second strand is filled in by the Klenow enzyme. After circularization by means of DNA ligase a heteroduplex plasmid is generated which after recombination yields a mixture of both the mutagenic as well as the wild-type gene (ZOLLER and SMITH, 1982; SMITH, 1985). Due to the repair system of *Escherichia coli* mutagenesis efficiency in practice is much less than 50%. Several variations of this method were elaborated to enhance the efficiency, e.g., the gapped duplex method (KRAMER et al., 1984), the Kunkel method (KUNKEL, 1985), and the phosphorothioate method (TAYLOR et al., 1985) reaching mutation efficiencies of up to 95%.

3.4.2 PCR-Mediated Mutagenesis

PCR-mediated mutagenesis (see Sect. 3.2.2.3) is a logical further development of oligonucleotide-directed site-specific mutagenesis combined with primer-directed PCR-based fragment amplification. This procedure has the advantage that phage growth and purification as well as extensive screening steps are circumvented. In a general approach two fragments overlapping at the mutagenic site with their 5′ ends are produced in two primary PCR reactions (HIGUCHI et al., 1988). This overlap in sequence then allows amplification of the whole fragment in a further PCR reaction using two outer general primers (Fig. 19). This method relies on the presence of appropriate restriction enzyme cleavage sites flanking the

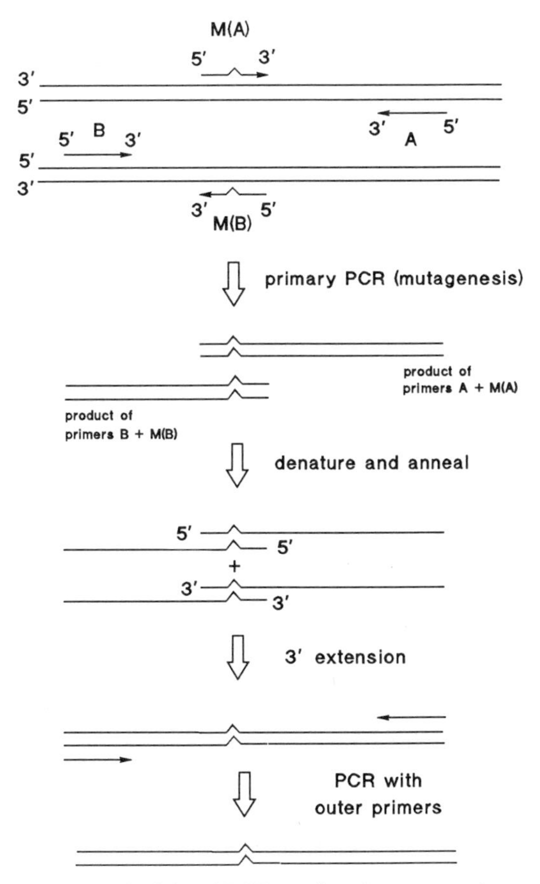

Fig. 19. Principle of PCR-mediated mutagenesis.

sequence to be mutated. A variation of this approach is the adaption of inverse PCR whereby a total plasmid is amplified and mutations can be introduced anywhere in the plasmid without requirement for appropriate restriction sites (HEMSLEY et al., 1989). Whereas the PCR-based procedures for mutagenesis mentioned so far make use of two specific primers to introduce a mutation, a general rapid mutagenesis procedure has been developed starting from a double-stranded DNA template and requiring just one specific primer in addition to a pair of universal primers (LANDT et al., 1990). The observed high mutagenesis efficiency using this protocol has been attributed to optimized dNTP concentrations and to purification of the resulting fragments.

3.4.3 Cassette Mutagenesis

Cassette mutagenesis is most convenient for the alteration of synthetic genes, since they can be designed specifically for this purpose. This method involves replacement of a DNA fragment flanked by two unique restriction enzyme cleavage sites by a corresponding fragment containing the desired mutation. In the design of a gene for growth hormone releasing hormone (GHRH) two unique restriction sites were placed next to codon 27 of methionine (ENGELS et al., 1987). This allowed for subsequent replacement by a synthetic DNA fragment containing the codon for leucine so that leu^{27}-GHRH could be liberated from a fusion protein with β-galactosidase by cyanogen bromide cleavage. If the mutagenic fragment is of chemical origin, this procedure is equivalent to a partial gene synthesis. However, the mutagenic fragment may as well result from PCR-mediated overlap extension (HIGUCHI et al., 1988) as described above. Cassette mutagenesis is furthermore very attractive for local random mutagenesis using ambiguously synthesized oligonucleotides (MATTEUCCI and HEYNEKER, 1983).

3.5 Applied Molecular Evolution

Mutational variation of genes and subsequent natural selection processes are the basic elements for evolutionary optimization. Random mutagenesis of genes or synthesis of random oligonucleotides, respectively, combined with a random screening step may be considered as the *in vitro* equivalent to the natural process. Since rational design of peptide drugs still poses a problem to the chemist, rational (not "irrational") random screening provides an attractive alternative. If we assume that for a special problem there exists a set of solutions from which we can deduce something like a "consensus shape" or "consensus sequence", then the task is to generate as many permutations of a chemical structure as possible, ideally an indefinite number, and subsequently screen the pool by means of an appropriate selection system for individual species which fit to a "selector" (Fig. 20). The initial step in-

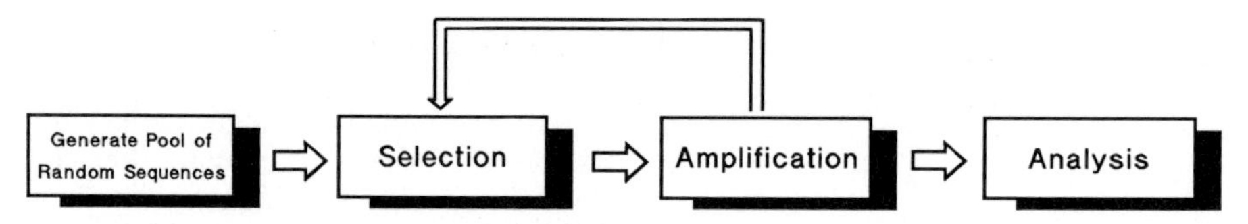

Fig. 20. Principle of applied molecular evolution.

volves generation of a library containing all random sequences ("sequence library"). The molecules possessing the desired properties are identified by repeated cycles of selection and amplification.

3.5.1 Random Oligonucleotides (Aptamers)

Synthesis of random oligonucleotide pools has been mentioned in Sect. 2.4. Already one decade ago, ambiguously synthesized oligonucleotides were used in a targeted random mutagenesis experiment to optimize the sequence immediately 5′ of the ATG start codon for the expression of the bovine growth hormone gene (MATTEUCCI and HEYNEKER, 1983). The selection methods used today are different from this early work in that repeated PCR amplification/selection steps are used instead of a single cloning/selection process. High affinity nucleic acid ligands for an enzyme were isolated by a procedure applying alternate cycles of ligand selection from a library of variant sequences and amplification of the bound species (TUERK and GOLD, 1990). In order to find novel RNA molecules which would bind to bacteriophage T4 DNA polymerase, a calculated number of 65536 (4^8) of an eight-nucleotide random sequence was synthesized. An eight-nucleotide loop sequence binds in nature to T4 DNA polymerase to control its activity. From this pool of DNA the corresponding RNA was prepared by *in vitro* transcription and screened with T4 DNA polymerase fixed to nitrocellulose for binding. The selected RNA molecules were amplified as double-stranded DNA which in turn was transcribed *in vitro*. Multiple rounds of selection and amplification resulted in accumulation of the best binding sequences. After systematic evolution of ligands by exponential enrichment (SELEX) TUERK and GOLD found two sequences, one of which was the known wild-type sequence, the other one was varied in four out of eight positions from the wild-type sequence. The binding constants of these two RNA molecules to T4 DNA polymerase were of similar order and could be used to prevent T4 infection when applied in excess. At the same time, SZOSTAK and ELLINGTON independently developed a similar method to the one described above for *in vitro* selection of RNA molecules that bind specific low molecular ligands (ELLINGTON and SZOSTAK, 1990). The idea behind this experiment was that some fraction of RNA molecules selected for binding to transition state analog affinity columns would probably catalyze a ribonucleolytic reaction and therefore may help to isolate novel ribozymes. Starting from DNA which comprised a T7 promoter 5′ to a 100 nucleotide random sequence, which was followed by a primer binding site to allow for a reverse transcription (Fig. 21), a random RNA pool was generated consisting of about 10^{13} different RNA molecules from a theoretical number of 10^{60} sequences. RNA molecules binding to dyes such as Cibacron Blue or Reactive Blue, which are related to NAD (nicotinamide adenine dinucleotide), were enriched by affinity purification over the corresponding dye columns. The specifically bound RNA was eluted, converted to cDNA, and amplified by PCR. After six cycles about 10^2 to 10^3 "aptamers" (from Latin "aptus", to fit) were found to bind to the affinity columns and were analyzed by DNA sequencing. That means, about one in 10^{10} random RNA molecules folds in such a way as to create a specific ligand binding site. Using a similar *in vitro* selection method a mutant form of an RNA enzyme that cleaves single-stranded RNA more efficiently than the wild-type enzyme has been identified (ROBERTSON and JOYCE, 1990).

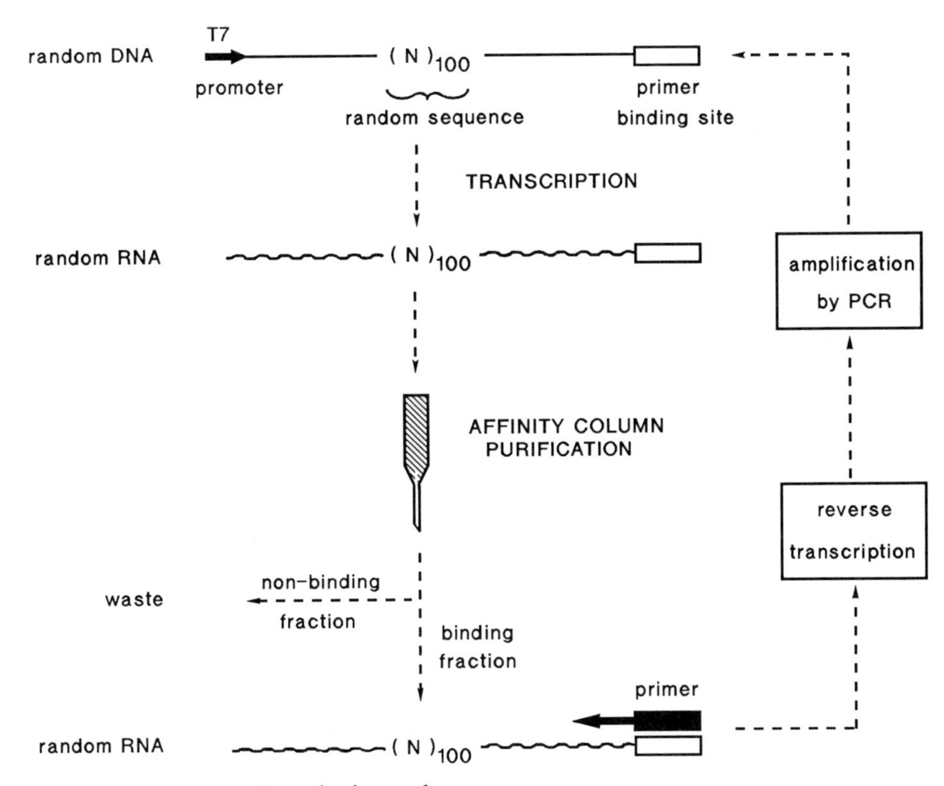

Fig. 21. Schematic diagram of the *in vitro* selection cycle.

Only recently, single-stranded aptamers with a highly conserved 14 to 17 base region were selected by *in vitro* evolution which bind and inhibit human thrombin (BOCK et al., 1992).

3.5.2 Random Cassette Expression (Peptide Library)

In analogy to the nucleotide sequence library described in the previous section, a "peptide library" or "epitope library" has been generated by inserting a random DNA sequence at a specific site into a phage DNA. The library consisted of a vast mixture of filamentous phage clones, each presenting one individual peptide fragment on the virion surface by expression of the corresponding DNA insertion (SCOTT and SMITH, 1990). Since the epitope library was constructed to code for a hexapeptide sequence, it was screened with two monoclonal antibodies which recognize a known hexapeptide epitope of myohemerythrin. Affinity purified phages were transfected into *E. coli* and propagated. After several cycles of amplification and selection, the amino acid sequences displayed on the phages were determined by DNA sequencing of the corresponding region of phage DNA. It is most interesting that this peptide library approach allows the identification of peptide ligands for such proteins for which no natural ligand is known. Using a 15 amino acid random sequence in a peptide library, a set of conserved peptide sequences was shown to bind streptavidin whose normal ligand is biotin (DEVLIN et al., 1990). This finding may have a potentially high impact on drug development, even for orphan receptors, whose ligands are not known.

3.6 Oligonucleotides as Inhibitors of Gene Expression

In recent years, three classes of potential therapeutic nucleic acids have been identified. The class of antisense oligonucleotides acts by employing Watson–Crick base pairing (see Sect. 2.3), whereas triplex forming oligonucleotides (TFOs) block gene expression by Hoogsteen base pairing (see Fig. 4). The action of sense oligonucleotides - aptamers are potential candidates of this class - is based on protein/nucleic acid interactions (Fig. 22). In order to allow *in vivo* application of either class of potential nucleic acid therapeutics, the compounds must fulfill the following requirements (STEIN and COHEN, 1988; ZON, 1988; MILLER and TS'O, 1988; HÉLÈNE and TOULMÉ, 1990; GOODCHILD, 1990; UHLMANN and PEYMAN, 1990):

(1) The oligonucleotide must be stable enough under *in vivo* conditions to survive in the serum and on its way to find its target.
(2) The oligonucleotide must be able to penetrate cell membranes to reach its site of action. In certain instances, this may not be required for sense oligonucleotides.
(3) The interaction between the oligonucleotide and its target must be sequence specific.
(4) The complex formed between the oligonucleotide and its target, be it via base pairing or protein/nucleic acid interaction, must be sufficiently stable under physiological conditions.

The first two requirements have to be met by all three classes of synthetic nucleic acids of potential therapeutic value. Therefore, stability and penetration will be a general subject to discuss, whereas target (DNA, RNA, and protein) recognition has to be dealt with in individual sections. The mechanism of action of the oligonucleotides is linked to the site of action as shown in Fig. 23. Transcriptional inhi-

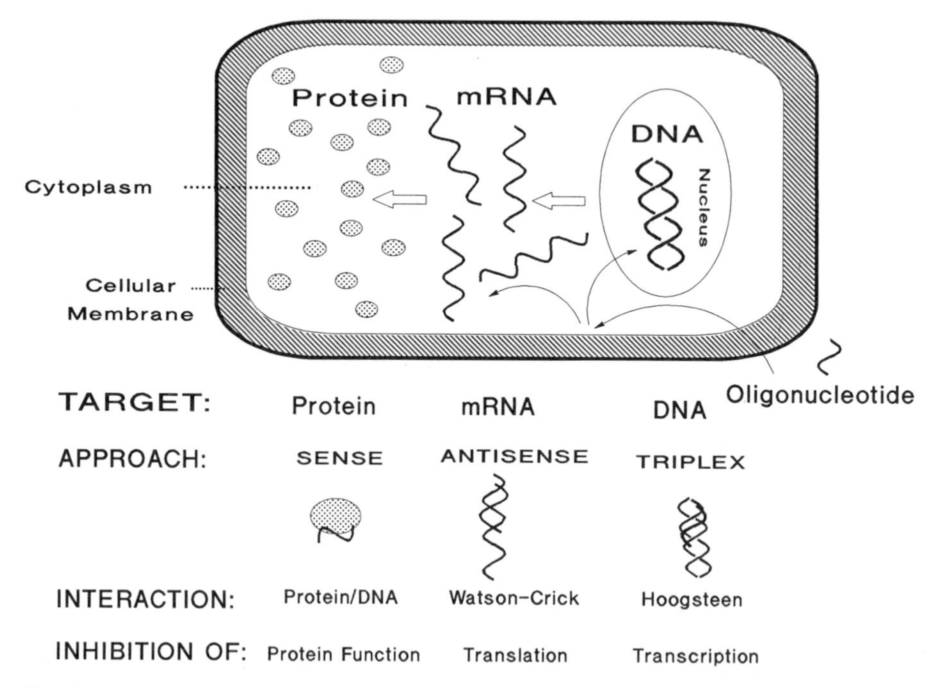

Fig. 22. Action of sense, antisense, and triplex-forming oligonucleotides.

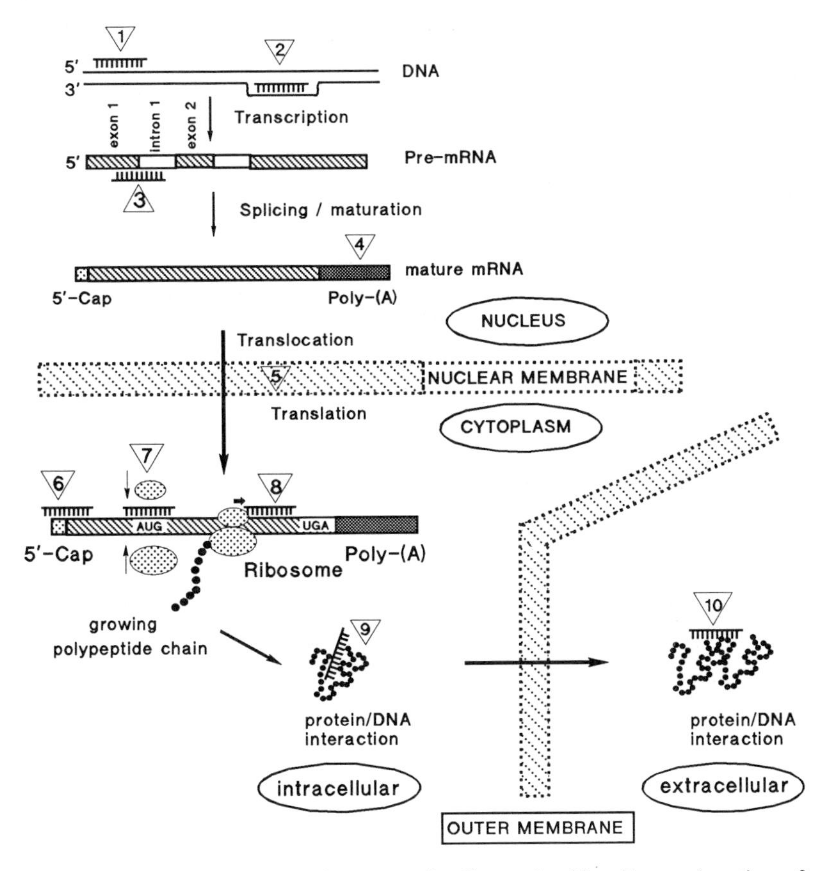

Fig. 23. Schematic view of potential intervention points of therapeutic oligonucleotides. For explanation of 1–10 see text.

bition on the DNA level is possible through triplex formation (1), especially in purine-rich promoter regions, or by Watson-Crick base-pairing through a strand-displacement reaction (2). Inhibition of pre-mRNA splicing (3) by binding of an oligonucleotide to an exon–intron splice junction as well as inhibition of mRNA maturation (4) are further possibilities to suppress gene expression. Binding of an oligonucleotide to the mature mRNA may interfere with the transport of the mRNA from the nucleus to the cytoplasm (5) where translation takes place. Competition of oligonucleotides for binding to mRNA with translational initiation factors (6), prevention of ribosome assembly (7), or hindrance of ribosome movement (8) are further possible sites of sequence-specific interventions. Finally, a sense oligonucleotide could bind to a growing polypeptide chain thereby interfering with proper folding of the protein or suppressing its intracellular function (e.g., transactivation) or transport out of the cell (9). Outside the cell, a sense oligonucleotide might directly block the function of a protein, for instance, by binding to the active site of an enzyme or to a receptor ligand binding site (10). Some of these strategic points of intervention were reviewed earlier by STEBBING (1984) considering single-stranded polynucleotides as antiviral agents.

From a general point of view one would conclude that inhibition at the DNA level is more effective than on the RNA or protein level, because from one or two copies of DNA several hundred or thousand species of mRNA are generated which in turn produce a high

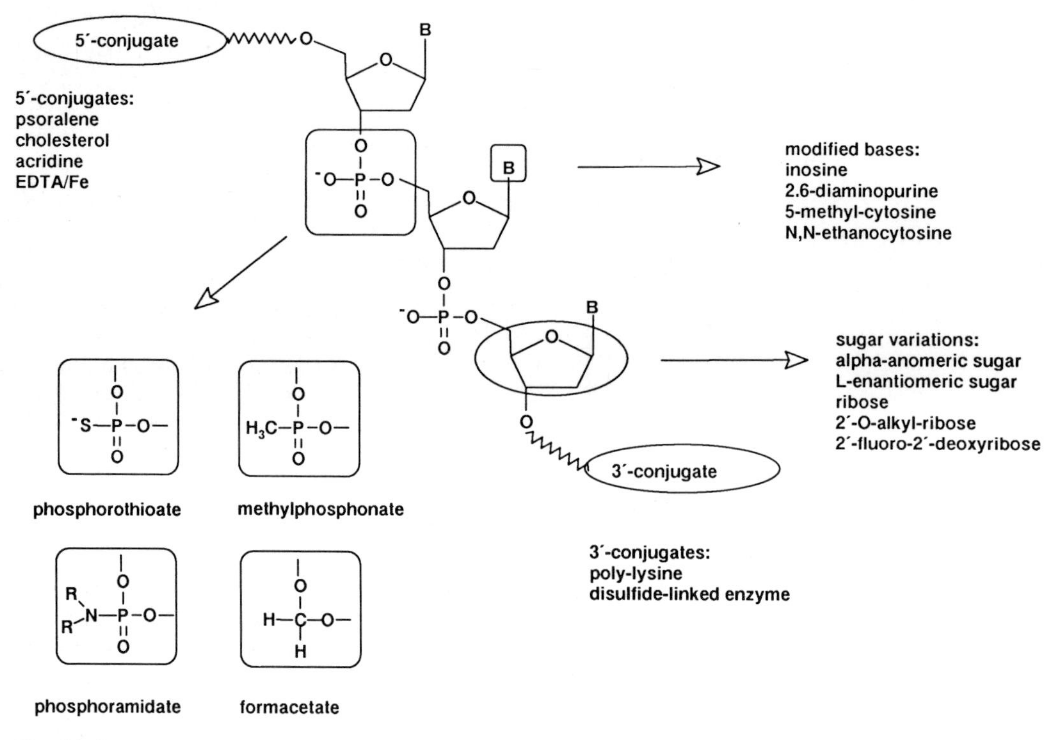

Fig. 24. Summary of different possibilities for modifying oligonucleotides.

number of protein molecules. However, there exist other mechanisms inside a cell which make things more complicated. The fact that RNase H, which occurs ubiquitously in plant and animal cells, cleaves the RNA strand of a duplex molecule of RNA and DNA renders the antisense oligonucleotide approach very attractive. In this case the target RNA is destroyed by nucleolytic cleavage.

3.6.1 Stability of Oligonucleotides to Nucleases

Normal oligonucleotides possessing the natural phosphodiester bond are rapidly degraded in serum by nucleases (WICKSTROM, 1986). It has been shown by others and us that a 3′-exonuclease activity is mainly responsible for the extracellular degradation. Since the 3′-exonuclease uses the 3′-hydroxy group of the 3′-terminal nucleotide unit for a nucleophilic attack on the neighboring phosphate center, modification of the 3′ end and/or alteration of the phosphodiester bond are possible ways to stabilize the oligonucleotides. Fig. 24 summarizes how oligonucleotides may be modified to change their properties. Most work has so far been done with methylphosphonate (see Sect. 2.3.2) and phosphorothioate (see Sect. 2.3.1) oligonucleotides which both are reasonably stable against nucleases. The non-ionic methylphosphonates were introduced by Ts'O and MILLER (MILLER et al., 1981), and the phosphorothioates whose resistance to nucleases was reported long ago by ECKSTEIN (ECKSTEIN 1983, 1985; DE CLERCQ et al., 1970) have been widely investigated by several groups (AGRAWAL et al., 1988; MATSUKURA et al., 1987, 1989; STEIN and COHEN, 1988; MAJUMDAR et al., 1989).

3.6.2 Oligonucleotide Uptake by Cells

Cellular uptake of oligonucleotides into the cytoplasm and nucleus is of great importance for the efficiency of antisense and triplex forming oligonucleotides to inhibit gene expression. Fortunately, it turned out that penetration of oligonucleotides is better than it had been expected for a polyanionic compound of this molecular weight (4000 to 8000 Da). In the micromolar range the intracellular concentration of oligonucleotides is a few percent of the concentration outside when investigated in cell culture (ZAMECNIK et al., 1986). The uptake of oligonucleotides is likely to be receptor-mediated (LOKE et al., 1989). But it is interesting that the oligonucleotides penetrate even to the cell nucleus when incubated in cell culture or microinjected into the cytoplasm (LEONETTI et al., 1991). Methylphosphonates seem to be taken up by passive diffusion (MILLER et al., 1981). In order to improve the cellular uptake of ionic oligonucleotides, they were modified by lipophilic groups (REGAN et al., 1990; SAISON-BEHMOARES et al., 1991), conjugated to cholesterol (DE SMIDT et al., 1991; STEIN et al., 1991b), poly-L-lysine (LEMAITRE et al., 1987; LEONETTI et al., 1988), or to specific polypeptide carriers (WAGNER et al., 1990). Packaging of oligonucleotides into liposomes which may be antibody-targeted (LEONETTI et al., 1990) represents another interesting possibility to improve cellular uptake.

3.6.3 Antisense Oligonucleotides

Antisense oligonucleotides recognize their target mRNA by Watson–Crick base pairing. The binding affinity of the oligonucleotide to its target can be characterized by the melting temperature T_M of the double-stranded nucleic acid, at which temperature 50% of the duplex is dissociated into single strands. In practice most of the known internucleotide phosphate modifications appear to have a negative effect on T_M as a result of sterical hindrance at the internucleotidic phosphorus. Although stereochemically pure methylphosphonates of the R configuration show increased T_M values compared to the phosphodiesters, the racemic mixture resulting from the ordinary synthesis shows an average depressed T_M value (LESNIKOWSKI et al., 1990). For the phosphorothioate oligonucleotides, binding affinity again is lowered compared to the phosphodiester oligonucleotides (STEIN et al., 1988a). The negative effect of phosphate modification can be compensated for by increasing the chain length. Binding affinity can also be enhanced by covalently linking intercalating agents to the oligonucleotide (ASSELINE et al., 1984). Oligonucleotides built up by α-anomeric nucleosides showed very high binding affinity to mRNA (MORVAN et al., 1986), but only poor inhibitory effect in a translation assay (GAGNOR et al., 1987). The explanation is that the duplex formed between an α-oligonucleotide and mRNA is not a substrate for RNase H. Phosphorothioate oligonucleotides, despite their lower binding affinity, show the most promising activity against HIV in cell culture at sub-micromolar concentrations. However, the antiviral effect is not sequence-specific, and even homopolymeric phosphorothioates are active against HIV in cell culture (MATSUKURA et al., 1987). Methylphosphonates are less active than phosphorothioates in cell culture experiments, although they are supposed to penetrate much better. Consistent with this finding is that the RNA strand is cleaved by RNase H if the complementary strand is a phosphorothioate oligodeoxynucleotide, but is not cleaved if it is a methylphosphonate oligodeoxynucleotide. It should be noted, however, that conjugation of methylphosphonates with photocross-linker units enhances their activity by a factor of at least 50 (TOULMÉ and HÉLÈNE, 1988).

Ribozymes are catalytic nucleic acids appearing in nature as part of RNA structures. The so-called "hammer-head" is presumably the simplest RNA autocleavage domain. In a systematic study HASELHOFF and GERLACH (1988) found that ribozymes can be directed against almost any RNA sequence by flanking the conserved catalytic domain by appropriate antisense sequences. At the cleavage site there is preference for a GUN triplet. A 13mer oligoribonucleotide is the smallest known ribozyme (JEFFRIES and SYMONS, 1989). In AIDS therapy plasmid-encoded ribozymes may be a good alternative to snythetic ribozymes (CHANG et al., 1990; SARVER et al., 1991).

Although at present no class of oligonucleotide derivatives fulfills all the requirements as stated in Sect. 3.6, there are many examples in the literature providing evidence that antisense oligonucleotides can be used effectively to inhibit gene expression in cell culture systems. For a review see UHLMANN and PEYMAN, 1990.

3.6.4 Triplex-Forming Oligonucleotides

Gene regulation and expression in cells is strongly controlled by many different proteins which activate, enhance, or repress transcription sequence-specifically at the DNA level. There are several non-nucleotide drugs which bind at least with some sequence specificity to DNA, e.g., to GC- or AT-rich regions. Most of these drugs, such as distamycin, netropsin, chromomycin, or Hoechst 33258, bind to the minor groove, whereas triplex-forming oligonucleotides (TFOs) bind via Hoogsteen base pairing (Fig. 4) to the major groove of DNA (NIELSEN, 1991). Triple-helix formation has been described for the first time by FELSENFELD et al. (1957) who found a 2:1 poly(U)·poly(A) RNA complex. At acidic pH poly(C) forms with poly(G) a triplex structure in which one cytosine is protonated. Both base pairing triplets T·AT and C^+·GC have been studied by two-dimensional NMR spectroscopy (DE LOS SANTOS et al., 1989). Triplex formation with the structural motif G·GC depends upon Mg^{2+}, but is relatively independent of pH (LIPSETT, 1963). As for T·AT and C^+·GC there exists a triplet of the "Hoogsteen" and of the "reverse Hoogsteen" type. HOGAN and coworkers showed that TFOs employing G·GC and T·AT motifs can bind to naturally occurring sites in duplex DNA at physiological pH (COONEY et al., 1988; DURLAND et al., 1991; POSTEL et al., 1991). By affinity cleaving experiments it has been found that a purine-rich TFO bound antiparallel to the Watson–Crick purine strand (DURLAND et al., 1991; BEAL and DERVAN, 1991). Further stabilization of triplex structures can be achieved via a G·TA motif, where guanine occurring within a pyrimidine TFO specifically recognizes a T·A base pair in DNA (GRIFFIN and DERVAN, 1989). In order to circumvent the acidic pH for triplex formation of C^+·GC which is not possible under *in vivo* conditions, several suggestions have been made. The base 5-methylcytosine has been used instead of cytosine which allowed triplex formation up to pH 7.4 (POVSIC and DERVAN, 1989). Introduction of pseudoisocytidine as its 2′-O-methyl-derivative into oligonucleotide analogs also permitted triplex formation at neutral pH (ONO et al., 1991). However, triple-helix formation is still limited to purine-rich tracts on DNA. HORNE and DERVAN (1990) reported on oligonucleotides containing a 3′3′-phosphodiester/dideoxy-D-ribose linkage which allowed binding to DNA by alternative-strand triple-helix formation. A similar 3′3′-switch TFO which allows "jumping" from one purine tract of one strand to another purine tract on the complementary strand was suggested by FROEHLER (MCCURDY et al., 1991).

Great efforts are made to use TFOs derivatized with azidoproflavine (LE DOAN et al., 1987; PRASEUTH et al., 1988), EDTA/Fe (MOSER and DERVAN, 1987), chloroethylamine (VLASSOV et al., 1988), or phenanthroline/Cu (FRANÇOIS et al., 1988) to modify or cleave DNA. One aspect is the construction of artificial restriction enzymes whose recognition site occurs very rarely in large DNA fragments. Another aspect is single-site cleavage by restriction enzymes of large DNA mediated by TFOs. Thus it has been demonstrated that a 340 kb yeast chromosome was cut just once by a reaction sequence of triplex formation, methylase protection, triple-helix disruption, and restriction enzyme cleavage (STROBEL and DERVAN, 1991).

Triple-helix formation is not restricted to double-stranded DNA. Single-stranded DNA targets can also form triplex structures where one part of the oligonucleotide binds via Watson–Crick hydrogen bonds, the other one after folding back via Hoogsteen hydrogen bonds (GIOVANNANGELI et al., 1991). Intramolecular single-strand DNA triplex formation has also been shown (HÄNER and DERVAN, 1990). Circular oligonucleotides with the potential of forming triplices with single-stranded DNA display very high binding affinities accompanied with increased selectivity (PRAKASH and KOOL, 1991; KOOL, 1991). In addition, cooperative binding of oligonucleotides to DNA by triplex formation was achieved by dimerization of the oligonucleotides by Watson–Crick base pairing to form a Y-shaped complex (DISTEFANO et al., 1991).

Unlike antisense sequences which have been extensively used to inhibit translation in cell culture assays, TFOs were investigated so far only in a limited number of experiments to inhibit gene expression. In *in vitro* experiments transcription of *c-myc* gene could be inhibited sequence-specifically using a 27mer oligonucleotide (COONEY et al., 1988). Triplex formation could be shown in the promoter regions of *c-myc*, epidermal growth factor receptor and insulin receptor genes (DURLAND et al., 1991). Thereby, TFOs bound in antiparallel to purine-rich regions by G·GC and T·AT Hoogsteen pairing with apparent dissociation constants in the 10^{-7} to 10^{-9} mol/L range at physiological pH. In CV-1 cell culture experiments inhibition of replication of SV40 could be demonstrated by triplex formation (BIRG et al., 1990). Triplex-mediated transcription arrest of *c-myc* in HeLa cells (POSTEL et al., 1991) as well as inhibition of IL-2 receptor-α transcription in lymphocytes (ORSON et al., 1991) further support the triple-helix concept for inhibiting gene expression.

3.6.5 Sense Oligonucleotides

Sense oligonucleotides are single- or double-stranded oligonucleotides that bind specifically to nucleic acid binding or processing proteins. Since their sequence is equivalent to the natural substrate or ligand, the term "sense" oligonucleotide is used. Single-stranded RNA molecules which mimic the region of viral RNA that interact directly with essential host or viral factors ("strategic sequences") were suggested as early as 1977 to inhibit virus replication (STEBBING et al., 1977). Those molecules are presumably the first representatives of single-stranded sense oligonucleotides. Double-stranded oligonucleotides were used to study regulatory sequences by competition in gel retardation and *in vitro* transcription assays (SINGH et al., 1986; HARRINGTON et al., 1988). In view of a potential therapeutic application, double-stranded oligonucleotides of varying length with different nuclear factor binding sites were effective in inhibiting *in vitro* transcription of adenovirus by more than 90% (WU et al., 1990). In the same experiment the phosphorothioate analogs inhibited transcription at one-tenth the concentration needed for the normal phosphodiesters. In a cell culture assay, phosphorothioate sense oligonucleotides specifically bound either octamer transcription factor or nuclear factor kappaB (BIELINSKA et al., 1990). Interestingly, modulation of gene expression was sequence-specific and the double-stranded phosphorothioates accumulated in cells more effectively than their phosphodiester parent molecules. In comparison, single-stranded antisense phosphorothioate oligonucleotides often show a lack in sequence specificity. They interact directly with proteins (STEIN et al., 1991a) and seem to penetrate less effectively than the normal phosphodiester oligonucleotides.

Using sense oligonucleotides as studying tools it could be suggested that Myb protein, a nuclear transcription factor encoded by the proto-oncogene *c-myb,* binds to and selectively activates LTR sequences of HIV (DASGUPTA et al., 1990). It ought to be mentioned that for single-stranded sense oligonucleotides the possibility that they act by hybridization to a complementary part of the RNA (antisense oligonucleotide mechanism) and not to RNA-binding proteins has to be excluded (CAMERON and JENNINGS, 1991). In antiviral assay systems the induction of the interferon cascade may also account for the biological effect (STEBBING, 1984; DE CLERCQ et al., 1970). The sense approach seems to be most attractive for the simultaneous suppression of different transcription factors which recognize a consensus sequence. In that case, different antisense oligonucleotides would be needed to block biosynthesis of all activators, whereas merely one sense oligonucleotide may trap all different activator proteins. By using the technique of applied molecular evolution, the optimal sense sequence could eventually be selected, provided that the different transcription factors are available for alternate affinity binding enrichment of the desired sense sequence.

4 References

ADAMS, S. P., KAVKA, K. S., WYKES, E. J., HOLDER, S. B., GALLUPPI, G. R. (1983), Hindered dialkylamino nucleoside phosphite reagents in the synthesis of two 51-mers, *J. Am. Chem. Soc.* **105**, 661–663.

ADAMS, S. E., JOHNSON, I. D., BRADDOCK, M., KINGSMAN, A. J., KINGSMAN, S., EDWARDS, R. M. (1988), Synthesis of a gene for the HIV transactivator protein TAT by a novel single-stranded approach involving *in vivo* gap repair, *Nucleic Acids Res.* **16**, 4287–4298.

AGARWAL, R. L., RIFTINA, F. (1979), Synthesis and enzymatic properties of deoxyribooligonucleotides containing methyl and phenylphosphonate linkages, *Nucleic Acids Res.* **6**, 3009.

AGARWAL, K. L., BÜCHI, H., CARUTHERS, M. H., GUPTA, N., KHORANA, H. G., KLEPPE, K., KUMAR, A., OHTSUKA, E., RAJBHANDARY, U. L., VAN DE SANDE, J. H., SGARAMELLA, V., WEBER, H., YAMADA, T. (1970), Total synthesis of the gene for an alanine transfer-ribonucleic acid from yeast, *Nature* **227**, 27–34.

AGRAWAL, S., CHRISTODOULOU, C., GAIT, M. J. (1986), Efficient methods for attaching non-radioactive labels to the 5′ ends of synthetic oligodeoxyribonucleotides, *Nucleic Acids Res.* **14**, 6227.

AGRAWAL, S., GOODCHILD, J., CIVEIRA, M., THORNTON, A. H., SARIN, P. S., ZAMECNIK, P. C. (1988), Oligodeoxynucleotide phosphoramidates and phosphorothioates as inhibitors of human immunodeficiency virus, *Proc. Natl. Acad. Sci. USA* **85**, 7079–7083.

ALLEN, R. C., GRAVES, G., BUDOWLE, B. (1989), Polymerase chain reaction amplification products separated on rehydratable polyacryamide gels and stained with silver, *Biotechniques* **7**, 736–744.

ALUL, R. H., SINGMAN, C. N., ZHANG, G., LETSINGER, R. L. (1991), Oxalyl-CPG: a labile support for synthesis of sensitive oligonucleotide derivatives, *Nucleic Acids Res.* **19**, 1527–1532.

AMARNATH, V., BROOM, A. D. (1977), Chemical synthesis of oligonucleotides, *Chem. Rev.* **77**, 183–217.

APPLIED BIOSYSTEMS (1987), *User Bull.* **44**, 1.

ASSELINE, U., DELAURUE, M., LANCELOT, G., TOULMÉ, F., THUONG, N. T., MONTENAY-GARESTIER, T., HÉLÈNE, C. (1984), Nucleic acid-binding molecules with high affinity and base sequence specificity: intercalating agents covalently linked to oligodeoxynucleotides, *Proc. Natl. Acad. Sci. USA* **81**, 3297–3301.

BARANY, F. (1991), Genetic disease detection and DNA amplification using cloned thermostable ligase, *Proc. Natl. Acad. Sci. USA* **88**, 189–193.

BARDELLA, F., GIRALT, E., PEDROSO, E. (1990), Polystyrene-supported synthesis by the phosphite triester approach: An alternative for the large scale synthesis of small oligodeoxyribonucleotides, *Tetrahedron Lett.* **31**, 6231–6234.

BARNETT, R. W., ERFLE, H. (1990), Rapid generation of DNA fragments by PCR amplification of crude, synthetic oligonucleotides, *Nucleic Acids Res.* **18**, 3094.

BEAL, P. A., DERVAN, P. (1991), Second structural motif for recognition of DNA by oligonucleotide-directed triple-helix formation, *Science* **251**, 1360–1363.

BEATON, G., BRILL, W. K.-D., GRANDAS, A., MA, Y.-X., NIELSON, J., YAU, E., CARUTHERS, M. H. (1991), Synthesis of oligonucleotide phosphorodithioates, *Tetrahedron* **47**, 2377–2388.

BEAUCAGE, S. L., CARUTHERS, M. H. (1981), Deoxynucleoside phosphoramidites – a new class of key intermediates for deoxypolynucleotide synthesis, *Tetrahedron Lett.* **22**, 1859–1862.

BEAUCAGE, S. L., IYER, R. P. (1992), Advances in the synthesis of oligonucleotides by the phosphoramidite approach, *Tetrahedron* **48**, 2223–2311.

BELL, L. D., SMITH, J. C., DERBYSHIRE, R., FINLAY, M., JOHNSON, I., GILBERT, R., SLOCOMBE, P., COOK, E., RICHARDS, H., CLISSOLD, P., MEREDITH, D., POWELL-JONES, C. H., DAWSON, K. M., CARTER, B. L., MCCULLAGH, K. G. (1988), Chemical synthesis, cloning and expression in mammalian cells of a gene coding for human tissue-type plasminogen activator, *Gene* **63**, 155–163.

BIELINSKA, A., SHIVDASANI, R. A., ZHANG, L., NABEL, L. (1990), Regulation of gene expression with double-stranded phosphorothioate oligonucleotides, *Science* **250**, 997–999.

BIRG, F., PRASEUTH, D., ZERIAL, A., THUONG, N. T., ASSELINE, U., LE DOAN, T., HÉLÈNE, C. (1990), Inhibition of Simian virus 40 DNA replication in CV-1 cells by an oligodeoxynucleotide covalently linked to an intercalating agent, *Nucleic Acids Res.* **18**, 2901–2908.

BJERGARDE, K., DAHL, O. (1991), Solid phase synthesis of oligodeoxyribonucleoside phosphorodithioates from thiophosphoramidites, *Nucleic Acids Res.* **19**, 5843–5850.

BLACKBURN, G. M., GAIT, M. J. (1990), *Nucleic Acids in Chemistry and Biology,* Oxford: IRL-Press.

BLANCHIN-ROLAND, S., MASSON, J.-M. (1989), Protein secretion controlled by a synthetic gene in *Escherichia coli, Protein Eng.* **2**, 473–480.

BOCK, L. C., GRIFFIN, L. C., LATHAM, J. A., VERMAAS, E. H., TOOLE, J. J. (1992), Selection of single-stranded DNA molecules that bind and inhibit human thrombin, *Nature* **355,** 564–566.

BONORA, G. M., SCREMIN, C. L., COLONNA, F. P., GARBESI, A. (1990), HELP (high efficiency liquid phase) new oligonucleotide synthesis on soluble polymeric support, *Nucleic Acids Res.* **18,** 3155–3159.

BRILL, W. K.-D., TANG, J.-Y., MA, Y.-X., CARUTHERS, M. H. (1989), Synthesis of oligodeoxynucleoside phosphorothioates via thioamidites, *J. Am. Chem. Soc.* **111,** 2321–2322.

BROCK, T. D., FREEZE, H. (1989), *Thermus aquaticus* gen. n. and sp. n., a nonsporulating extreme thermophile, *J. Bacteriol.* **98,** 289–297.

BRODIN, P., GRUNDSTRÖM, T., HOFMANN, T., DRAKENBERG, T., THULIN, E., FORSEN, S. (1986), Expression of bovine intestinal binding protein from a synthetic gene in *Escherichia coli* and characterization of the product, *Biochemistry* **25,** 5371–5377.

CAMERON, F. H., JENNINGS, P. A. (1991), Inhibition of gene expression by a short sense fragment, *Nucleic Acids Res.* **19,** 469–475.

CARDULLO, R. A., AGRAWAL, S., FLORES, C., ZAMECNIK, P. C., WOLF, D. E. (1988), Detection of nucleic acid hybridization by nonradioactive fluorescence resonance energy transfer, *Proc. Natl. Acad. Sci. USA* **85,** 8790.

CARUTHERS, M. H. (1991), Chemical synthesis of DNA and DNA analogues, *Acc. Chem. Res.* **24,** 278–284.

CHAKHMAKHCHEVA, O. G., EFIMOV, V. A., OVCHINNIKOV, Y. A. (1987), Versatile methodology for the construction of artificial genes, *Nucleosides Nucleotides* **6,** 321–324.

CHANG, P., CANTIN, E. M., ZAIA, J. A., LADNE, P. A., STEPHENS, D. A., SARVER, N., ROSSI, J. J. (1990), Ribozyme-mediated site-specific cleavage of the HIV-1 genome, *Clin. Biotechn.* **2,** 23–31.

CHEHAB, F. F., KAN, Y. W. (1990), Detection of sickle cell anaemia mutation by colour DNA amplification, *Lancet* **335,** 15–17.

CHEN, H.-B., WENG, J.-M., JIANG, K., BAO, J.-S. (1990), A new method for the synthesis of a structural gene, *Nucleic Acids Res.* **18,** 871–878.

CHOLLET, A., KAWASHIMA, E. H. (1985), Biotin-labeled synthetic oligodeoxyribonucleotides: chemical synthesis and use as hybridization probes, *Nucleic Acids Res.* **13,** 1529.

CHOW, F., KEMPE, T., PALM, G. (1981), Synthesis of oligodeoxyribonucleotides on silica gel support, *Nucleic Acids Res.* **9,** 2807–2817.

CICCARELLI, R. B., LOOMIS, L. A., MCCOON, P. E., HOLZSCHU, D. (1990), Insertional gene synthesis, a novel method of assembling consecutive DNA sequences within specific sites in plasmids. Construction of HIV-1 *tat* gene, *Nucleic Acids Res.* **18,** 1243–1248.

CLARK, J. M. (1988), Novel nontemplated nucleotide reactions catalyzed by prokaryotic and eukaryotic DNA polymerases, *Nucleic Acids Res.* **16,** 9677–9686.

CONNER, B. J., REYES, A. A., MORIN, C., ITAKURA, K., TEPLITZ, R. L., WALLACE, R. B. (1983), Detection of sickle cell beta S-globin allele by hybridization with synthetic oligonucleotides, *Proc. Natl. Acad. Sci. USA* **80,** 278.

CONWAY, N. E., FIDANZA, J., MCLAUGHLIN, L. W. (1989), The introduction of reporter groups at multiple and/or specific sites in DNA containing phosphorothioate diesters, *Nucleic Acids Res. Symp. Ser.* **43.**

COOK, A. F., VUOCOLO, E., BRAKEL, C. L. (1988), Synthesis and hybridization of a series of biotinylated oligonucleotides, *Nucleic Acids Res.* **16,** 4077.

COONEY, M., CZERNUSZEWICZ, G., POSTEL, E. H., FLINT, S. J., HOGAN, M. E. (1988), Site-specific oligonucleotide binding represses transcription of the human *c-myc* gene *in vitro, Science* **241,** 456–459.

DALBØGE, H., SØREN, C., JENSEN, E. B., CHRISTENSEN, T., DAHL, H.-H. (1988), Expression of recombinant growth hormone in *Escherichia coli:* Effect of the region between the Shine-Dalgarno sequence and the ATG initiation codon, *DNA Cell Biol.* **7,** 399–405.

DAMHA, M. J., GIANNARIS, P. A., ZABARYLO, S. V. (1990), An improved procedure for derivatization of controlled-pore glass beads for solid-phase oligonucleotide synthesis, *Nucleic Acids Res.* **18,** 3813–3821.

DASGUPTA, P., SAIKUMAR, P., REDDY, C. D., REDDY, E. P. (1990), Myb protein binds to human immunodeficiency virus 1 long terminal repeat (LTR) sequences and transactivates LTR-mediated transcription, *Proc. Natl. Acad. Sci. USA* **87,** 8090–8094.

DAUB, G. W., VAN TAMELEN, E. E. (1977), Synthesis of oligoribonucleotides based on the facile cleavage of methylphosphotriester intermediates, *J. Am. Chem. Soc.* **99,** 3526–3528.

DE CLERCQ, E., ECKSTEIN, F., STERNBACH, H., MERIGAN, T. C. (1970), The antiviral activity of thiophosphate-substituted polyribonucleotides *in vitro* and *in vivo, Virology* **42,** 421–428.

DE LOS SANTOS, C., ROSEN, M., PATEL, D. (1989), NMR Studies of DNA $(R+)_n \cdot (Y-)_n \cdot (Y+)_n$ triple helices in solution: Imino and amino proton markers of T·A·T and $C \cdot G \cdot C^+$ base-triple formation, *Biochemistry* **28,** 7282-7289.

DERBYSHIRE, K. M., SALVO, J. J., GRINGLEY, N. D. F. (1986), A simple and efficient procedure for saturation mutagenesis using mixed oligonucleotides, *Gene* **46,** 145–152.

DE SMIDT, P. C., LE DOAN, T., DE FALCO, S., VAN BERKEL, T. J. C. (1991), Association of antisense oligonucleotides with lipoproteins prolongs the plasma half-life and modifies the tissue distribution, *Nucleic Acids Res.* **19,** 4695–4700.

DEVLIN, J. J., PANGANIBAN, L. C., DEVLIN, P. E. (1990), Random peptide libraries: A source of specific protein binding molecules, *Science* **249,** 404–406.

DICKERSON, R. E., DREW, H. R. (1981), Structure of a B-DNA dodecamer, influence of base sequence on helix structure, *J. Mol. Biol.* **149,** 761.

DISTEFANO, M. D., SHIN, J. A., DERVAN, P. B. (1991), Cooperative binding of oligonucleotides to DNA by triple helix formation: Dimerization via Watson-Crick hydrogen bonds, *J. Am. Chem. Soc.* **113,** 5901–5902.

DODT, J., SCHMITZ, T., SCHÄFER, T., BERGMANN, C. (1986), Expression, secretion and processing of hirudin in *Escherichia coli* using alkaline phosphatase signal sequence, *FEBS Lett.* **202,** 373–377.

DOLINNAYA, N. G., SOKOLOVA, N. I., ASHIRBEKOVA, D. T., SHABAROVA, Z. A. (1991a), The use of BrCN for assembling modified DNA duplexes and DNA-RNA hybrids; comparison with water-soluble carbodiimide, *Nucleic Acids Res.* **19,** 3067–3072.

DOLINNAYA, N. G., TSYTOVICH, A. V., SERGEEV, V. N., ORETSKAYA, T. S., SHABAROVA, Z. A. (1991b), Structural and kinetic aspects of chemical reactions in DNA duplexes. Information on DNA local structure obtained from chemical ligation data, *Nucleic Acids Res.* **19,** 3077–3080.

DURLAND, R. H., KESSLER, D. J., GUNNELL, S., DUVIC, M., PETTITT, B. M., HOGAN, M. E. (1991), Binding of triple helix forming oligonucleotides to sites in gene promoters, *Biochemistry* **30,** 9246–9255.

ECKERT, K. A., KUNKEL, T. A. (1991), DNA polymerase fidelity and the polymerase chain reaction, *PCR Methods Appl.* **1,** 17–24.

ECKSTEIN, F. (1983), Phosphorothioatanaloga von Nucleotiden – Werkzeuge zur Untersuchung biochemischer Prozesse, *Angew. Chem.* **95,** 431–447.

ECKSTEIN, F. (1985), Nucleoside phosphorothioates, *Annu. Rev. Biochem.* **54,** 367–402.

EFIMOV, V. A., CHAKHMAKHCHEVA, O. G., OVCHINNIKOV, YU. A. (1985), Improved rapid phosphotriester synthesis of oligodeoxyribonucleotides using oxygen-nucleophilic catalysts, *Nucleic Acids Res.* **13,** 3651–3666.

EFIMOV, V. A., BURYAKOVA, A. A., DUBEY, I. Y., POLUSHIN, N. N., CHAKHMAKHCHEVA, O. G., OVCHINNIKOV, YU. A. (1986), Application of new catalytic phosphate protecting groups for the highly efficient phosphotriester oligonucleotide synthesis, *Nucleic Acids Res.* **14,** 6525–6540.

ELLINGTON, A. D., SZOSTAK, J. W. (1990), *In vitro* selection of RNA molecules that bind specific ligands, *Nature* **346,** 818–822.

EMBURY, S. H., SCHARF, J. S., SAIKI, R. K., GHOLSON, M. A., GOLBUS, M., ARNHEIM, N., ERLICH, H. A. (1987), Rapid prenatal diagnosis of sickle cell anemia by a new method of DNA analysis, *N. Engl. J. Med.* **316,** 656.

ENGELS, J. W. (1991), Antisense Oligonucleotide: Krankheit: Fehler in der Informationsübertragung, *Nachr. Chem. Tech.* **39,** 1250–1254.

ENGELS, J. W., UHLMANN, E. (1989), Gensynthese, *Angew. Chem.* **101,** 733–752; *Angew. Chem. Int. Ed. Engl.* **28,** 716–734.

ENGELS, J., LEINEWEBER, M., UHLMANN, E. (1985), Chemical and enzymatic synthesis of genes, in: *Organic Synthesis: An Interdisciplinary Challenge* (STREITH, J., PRINZBACH, H., SCHILL, G., Eds.), pp 205–214, Oxford: Blackwell Scientific Publications.

ENGELS, J. W., GLAUDER, J., MÜLLNER, H., TRIPIER, D., UHLMANN, E., WETEKAM, W. (1987), Enzymatic amidation of recombinant (leu^{27}) growth hormone releasing hormone-gly^{45}, *Protein Eng.* **1,** 195–199.

ERLICH, H. A., BUGAWAN, T. L. (1989), in: *PCR Technology* (ERLICH, H. A., Ed.), pp. 193–208, New York: Stockton Press.

ERLICH, H. A., GELFAND, D., SNINSKY, J. J. (1991), Recent advances in the polymerase chain reaction, *Science* **252,** 1643.

FARRANCE, I. K., SCOTT EADIE, J., IVARIE, R. (1989), Improved chemistry for oligodeoxyribonucleotide synthesis substantially improves restriction enzyme cleavage of a synthetic 35mer, *Nucleic Acids Res.* **17,** 1231–1245.

FELSENFELD, G., DAVIES, D. R., RICH, A. (1957), Formation of a three-stranded polynucleotide molecule, *J. Am. Chem. Soc.* **79,** 2023–2024.

FRANÇOIS, J.-C., SAISON-BEHMOARAS, T., CHASSIGNOL, M., THUONG, N. T., SUN, J.-S., HÉLÈNE, C. (1988), Periodic cleavage of Poly(dA) by oligothymidylates covalently linked to the 1,10-phenanthroline-copper complex, *Biochemistry* **27,** 2272–2276.

FRITZ, H. J. (1986), New developments in mutation construction and chemical gene synthesis, *Biol. Chem. Hoppe-Seyler* **367** Suppl. 86.

FROEHLER, B. C., MATTEUCCI, M. D. (1983), Dialkylformamidines: depurination resistant N^6-protecting group for deoxyadenosine, *Nucleic Acids Res.* **11,** 8031.

FROEHLER, B. C., NG, P. G., MATEUCCI, M. D. (1986), Synthesis of DNA via deoxynucleoside H-phosphonate intermediates, *Nucleic Acids Res.* **14,** 5399–5407.

GAFFNEY, B. L., MARKY, L. A., JONES, R. A. (1984), The influence of the 2-amino group on DNA conformation and stability II., *Tetrahedron* **40,** 3–13.

GAGNOR, C., BERTRAND, J. R., THENET, S., LEMAITRE, M., MORVAN, F., RAYNER, B., MALVY, C., LEBLEU, B., IMBACH, J.-L., PAOLETTI, C. (1988), DNA VI: comparative study of α- and β-anomeric oligodeoxyribonucleotides in hybridization to mRNA and in cell-free translation inhibition, *Nucleic Acids Res.* **15,** 10419–10436.

GAIT, M. J. (1984), *Oligonucleotide Synthesis, a Practical Approach,* Oxford: IRL Press.

GAREGG, P. J., LINDH, I., REGBERG, T., STAWINSKI, J., STROMBERG, R., HENRICHSON, C. (1986), Nucleoside H-phosphonates. III. Chemical synthesis of oligoribonucleotides by the hydrogenphosphonate, approach, *Tetrahedron Lett.* **27,** 4051–4054.

GELFAND, D. H. (1989), Taq DNA polymerase, Chapter 2 in: *PCR Technology* (ERLICH, H. A., Ed.) New York: Stockton Press.

GELFAND D. H., WHITE, T. J. (1990), A guide to methods and applications, in: *PCR Protocols* (INNIS, M. A., GELFAND, D. H., SNINSKY, J. J., WHITE, T. J., Eds.), p. 129, San Diego: Academic Press.

GIBBS, R. A. (1990), DNA amplification by the polymerase chain reaction, *Anal. Chem.* **62,** 1202–1214.

GILLESPIE, P., HOFFMANN, P., KLUSACEK, H., MARQUARDING, D., PFOHL, S., RAMIREZ, F., TSOLIS, E. A., UGI, I. (1971), Bewegliche Molekülgerüste - Pseudorotation und Turnstile - Rotation pentakoordinierter Phosphorverbindungen und verwandte Vorgänge, *Angew. Chem.* **83,** 691.

GIOVANNANGELI, C., MONTENAY-GARESTIER, T., ROUGÉE, CHASSIGNOL, M., THUONG, N., HÉLÈNE, C. (1991), Single-stranded DNA as a target for triple-helix formation, *J. Am. Chem. Soc.* **113,** 7775–7777.

GOODCHILD, J. (1990), Conjugates of oligonucleotides and modified oligonucleotides: A review of their synthesis and properties, *Bioconjugate Chem.* **1,** 165–197.

GOUGH, G. R., BRUNDEN, M. J., GILHAM, P. T. (1981), Recovery and recycling of synthetic units in the construction of oligodeoxyribonucleotides on solid supports, *Tetrahedron Lett.* **22,** 4177–4180.

GRAY, G. L., BALDRIDGE, J. S., MCKEOWN, K. S., HEYNEKER, H. L., CHANG, C. N. (1985), Periplasmic production of correctly processed human growth hormone in *Escherichia coli:* natural and bacterial signal sequences are interchangeable, *Gene* **39,** 247–254.

GREENE, T. W., WUTS, G. M. (1991), *Protective Groups in Organic Synthesis,* Second Ed., New York: J. Wiley & Sons.

GRIFFIN, L., DERVAN, P. B. (1989), Recognition of thymine·adenine base pairs by guanine in a pyrimidine triple helix motif, *Science* **245,** 967–971.

GRÖGER, G., SELIGER, H. (1988), A polymer support for chemical and enzymatic nucleic acid synthesis, *Nucleosides Nucleotides* **7,** 773–778.

GRUNDSTRÖM, T., ZENKE, W. M., WINTZERITH, M., MATTHES, H. W. D., STAUB, A., CHAMBON, P. (1985), Oligonucleotide-directed mutagenesis by microscale "shot-gun" gene synthesis, *Nucleic Acids Res.* **13,** 3305–3316.

GÜSSOW, D., CLACKSON, T. (1989), Direct clone characterization from plaques and colonies by the PCR, *Nucleic Acids Res.* **17,** 4000.

GYLLENSTEN, U. B., ERLICH, H. A. (1988), Generation of single stranded DNA by the polymerase chain reaction and its application to direct sequencing of the HLA-DQA locus, *Proc. Natl. Acad. Sci. USA* **85,** 7652–7656.

HAFF, L. A., MEZEI, L. M. (1989), *Measurement of PCR Amplification by Fluorescence, Amplifications - A Forum for PCR Users,* Issue **1,** 8.

HALL, R. H., TODD, A., WEBB, R. F. (1957), Nucleotides. Part XLI. Mixed anhydrides as intermediates in the synthesis of dinucleoside phosphates, *J. Chem. Soc.,* 3291–3296.

HAMES, B. D., HIGGINS, S. J. (Eds.) (1985), *Nucleic Acid Hybridization: A Practical Approach.* Oxford: IRL Press.

HÄNER, R., DERVAN, P. B. (1990), Single-strand DNA triple-helix formation, *Biochemistry* **29,** 9761–9765.

HARALAMBIDIS, J., CHAI, M., TREGEAR, G. W. (1987) Preparation of base-modified nucleosides suitible for non-radioactive label attachment and their incorporation into synthetic oligodeoxyribonucleotides, *Nucleic Acids Res.* **15,** 4857.

HARALAMBIDIS, J., ANGUS, K., POWNALL, S., DUNCAN, L., CHAI, M., TREGEAR, G. W. (1990), The preparation of polyamide-oligonucleotide probes containing multiple non-radioactive labels, *Nucleic Acids Res.* **18,** 501.

HARRINGTON, M. A., JONES, P. A., IMAGAWA, M., KARIN, M. (1988), Cytosine methylation does not affect binding of transcription factor Sp1, *Proc. Natl. Acad. Sci. USA* **85**, 2066–2070.

HARRISON B, ZIMMERMAN, S. B. (1986), T4 polynucleotide kinase: macromolecular crowding increases the efficiency of the reaction DNA termini, *Anal. Biochem.* **158**, 307.

HASELHOFF, J., GERLACH, W. L. (1988), Simple RNA enzymes with new and highly specific endoribonuclease activities, *Nature* **334**, 585–591.

HEIN, F., JANSEN, H. W., UHLMANN, E. (1987), Use of long oligonucleotides in gene synthesis: chemcial synthesis and cloning of a gene for salmon calcitonin, *Nucleosides Nucleotides* **6**, 489–490.

HEITMAN, J., FULFORD, W., MODEL, P. (1989), Phage Trojan horses: a conditional expression system for lethal genes, *Gene* **85**, 193–197.

HÉLÈNE, C., TOULMÉ, J.-J. (1990), Specific regulation of gene expression by antisense, sense and antigene nucleic acids, *Biochim. Biophys. Acta 1049,* 99–125.

HEMSLEY, A., ARNHEIM, N., TONEY, M. D., CORTOPASSI,G., GALES, D. J. (1989), A simple method for site directed mutagenesis using the polymerase chain reaction, *Nucleic Acids Res.* **17**, 6545.

HIGUCHI, R., KRUMMEL, B., SAIKI, R. (1988), A general method of *in vitro* preparation and specific mutagenesis of DNA fragments: study of protein and DNA interactions, *Nucleic Acids Res.* **16**, 7351–7367.

HILLENKAMP, F., KARAS, M., BEAVIS, R. C., CHAIT, B. T. (1991), Matrix-assisted laser desorption/ionization mass spectrometry of biopolymers, *Anal. Chem.* **63**, 1193–1203.

HIMMELSBACH, F., SCHULZ, B. S., TRICHTINGER, T., CHARUBALA, R., PFLEIDERER, W. (1984), The *p*-nitrophenylethyl (NPE) group. A versatile new blocking group for phosphate and aglyconeprotection in nucleosides and nucleotides, *Tetrahedron Lett.* **40**, 59–72.

HOBBS, J. B. (1990), Nucleotides and nucleic acids, *Organophosphorus Chem.* **21**, 201–321.

HODGES, R. R., CONWAY, N. E., MCLAUGHLIN, L. W. (1989), "Post-Assay" covalent labeling of phosphorothioate-containing nucleic acids with multiple fluorescent markers, *Biochemistry* **28**, 261–267.

HORN, T., URDEA, M. S. (1988), Solid support hydrolysis of apurinic sites in synthetic oligonucleotides for rapid and efficient purification on reverse-phase cartridges, *Nucleic Acids Res.* **16**, 11559–11571.

HORNE, D. A., DERVAN, P. B. (1990), Recognition of mixed-sequence duplex DNA by alternate-strand triple-helix formation, *J. Am. Chem. Soc.* **112**, 2435–2437.

HOSTOMSKÝ, Z., SMRT, J. (1987), Solid-phase assembly of DNA duplexes from synthetic oligonucleotides, *Nucleic Acids Res. Symp. Ser.* **18**, 241–248.

HOSTOMSKÝ, Z., SMRT, J., ARNOLD, L., TOCIK, Z., PACES, V. (1987), Solid-phase assembly of cow colostrum trypsin inhibitor gene, *Nucleic Acids Res.* **15**, 4849–4856.

HOSUR, R. V., GOVIL, G., MILES, H. T. (1988), Application of two-dimensional NMR spectroscopy in the determination of solution conformation of nucleic acids, *Magn. Reson. Chem. Biol.* **26**, 927–944.

HULTMAN, T., STAHL, S., HORNES, E., UHLEN, M. (1989), Direct solid phase sequencing of genomic and plasmid DNA using magnetic beads as solid support, *Nucleic Acids Res.* **17**, 4937–4946.

IKEHARA, M., OHTSUKA, E., TOKUNAGA, T., NISHIKAWA, S., UESUGI, S., TANAKA, T., AOYAMA, Y., KIKYODANI, S., FUJIMOTO, K., YANASE, K., FUCHIMURA, K., MORIOKA, H. (1986), Inquiries into the structure-function relationship of ribonuclease T1 using chemically synthesized coding sequences, *Proc. Natl. Acad. Sci. USA* **83**, 4695–4699.

INNIS, M. A., MYAMBO, K. B., GELFAND, D. H., BROW, M. D. (1988), DNA sequencing with *Thermus aquaticus* DNA polymerase and direct sequencing of polymerase chain reaction-amplified DNA, *Proc. Natl. Acad. Sci. USA* **85**, 9436–9440.

ITAKURA, K., HIROSE, T., CREA, R., RIGG, A. D., HEYNECKER, H., BOLIVAR, F., BOYER, H. (1977), Expression in *Escherichia coli* of a chemically synthesized gene for the hormone somatostatin, *Science* **198**, 1056–1063.

ITAKURA, K., ROSSI, J.J., WALLACE, R. B. (1984), Synthesis and use of synthetic oligonucleotides, *Annu. Rev. Biochem.* **53**, 323–356.

IVANETICH, K. M., AKIYAMA, J., SANTI, D. V., RESCHENBERG, M. (1991), Automated purification of synthetic oligonucleotides, *BioFeedback* **10**, 704–707.

IVANOV, I., GIGOVA, L., JAY, E. (1987), Chemical synthesis and expression in *Escherichia coli* of a human Val^8-calcitonin gene by fusion to a synthetic human interferon-gene, *FEBS Lett.* **210**, 56–60.

IYER, R. P., EGAN, W., REGAN, J. B., BEAUCAGE, S. L. (1990), 3H-1,2-benzodithiole-3-one 1,1-dioxide as an improved sulfurizing reagent in the solid phase synthesis of oligodeoxyribonucleoside phosphorothioates, *J. Am. Chem. Soc.* **112**, 1254–1255.

JAYARAMAN, K., FINGAR, S. A., SHAH, J., FYLES, J. (1991), Polymerase chain reaction-mediated gene synthesis: synthesis of a gene coding for isozyme c of horseradish peroxidase, *Proc. Natl. Acad. Sci. USA* **88,** 4084–4088.

JEFFRIES, A. C., SYMONS, R. H. (1989), A catalytic 13-mer ribozyme, *Nucleic Acids Res.* **17,** 1371–1377.

JING, G. Z., LUI, A. P., LEUNG, W. C. (1986), A method for the preparation of size marker for synthetic oligonucleotides, *Anal. Biochem.* **155,** 376.

JONES, S. S., REESE, C. B., SIBANDA, S., UBASAWA, A. (1981) The protection of uracil and guanine residues in oligonucleotide synthesis, *Tetrahedron Lett.* **22,** 4755–4758.

KALNOSKI, M. H., MCCOY-HAMAN, M. F., HOLLIS, G. F. (1991), A rapid method for purifying PCR products for direct sequence analysis, *BioTechniques* **11,** 246–249.

KAMIMURA, T., TSUCHIYA, M., KOURA, K., SEKINE, M., HATA, T. (1983), Diphenylcarbamoyl and propionyl groups: A new combination of protecting groups for the guanine residue, *Tetrahedron Lett.* **24,** 2775–2778.

KAWASAKI, E. (1989), in: *PCR Technology* (ERLICH, H. A., Ed.), p. 89, New York: Stockton Press.

KENNARD, O., HUNTER, W. N. (1989), Oligonucleotide structure: A decade of results from single crystal X-ray diffraction studies, *Q. Rev. Biophys.* **22,** 327–379.

KEOHAVONG, P., THILLY, W. G. (1989), Fidelity of DNA polymerases in DNA amplification, *Proc. Natl. Acad. Sci. USA* **86,** 9253–9257.

KHORANA, H. G. (1968), Nucleic acid synthesis, *Pure Appl. Chem.* **17,** 349–381.

KHORANA, H. G. (1978), Studies on nucleic acids: Total synthesis of a biologically functional gene, *Bioorg. Chem.* **1,** 351–393.

KLEPPE, K., OHTSUKA, F., KLEPPE, R., MOLINEUX, I., KHORANA, H. G. (1971) Studies on polynucleotides, *J. Mol. Biol.* **56,** 341–361.

KNOERZER, W., BINDER, H.-P., SCHNEIDER, K., GRUSS, P., MCCARTHY, J. E. G., RISAU, W. (1989), Expression of synthetic genes encoding bovine and human basic fibroblast growth factors (bFGFs) in *Escherichia coli, Gene* **75,** 21–30.

KOGAN, S. C., DOHERTY, M., GITSCHIER, J. (1987), An improved method for prenatal diagnosis of genetic diseases by analysis of amplified DNA sequences, *N. Engl. J. Med.* **317,** 985–990.

KOLLER, K.-P., RIEß, G., SAUBER, K., UHLMANN, E., WALLMEIER, H. (1989), Recombinant *Streptomyces lividans* secretes a fusion protein of Tendamistat and proinsulin, *Bio/Technology* **7,** 1055–1059.

KOLLER, K.-P., RIEß, G., SAUBER, K., VERTESY, L., UHLMANN, E., WALLMEIER, H. (1991), The Tendamistat expression-secretion system: synthesis of proinsulin fusion proteins with *Streptomyces lividans*, in: *Genetics and Product Formation in Streptomyces* (BAUMBERG, S., Ed.), pp. 227–233, New York: Plenum Press.

KOOL, E. T. (1991), Molecular recognition by circular oligonucleotides, increasing the selectivity of DNA binding, *J. Am. Chem. Soc.* **113,** 6265–6266.

KRAMER, W., DRUTSA, V., JANSEN, H.-W., KRAMER, B., PFLUGFELDER, M., FRITZ, H.-J. (1984), The gapped duplex DNA approach to oligonucleotide-directed mutation construction, *Nucleic Acids Res.* **12,** 9441–9456.

KUNKEL, T. A. (1985), Rapid and efficient site-specific mutagenesis without phenotypic selection, *Proc. Natl. Acad. Sci. USA* **82,** 488–492.

KWOK, S., HIGUCHI, R. (1989), Avoiding false positives with PCR, *Nature* **339,** 237.

LANDT, O., GRUNERT, H.-P., HAHN, U. (1990), A general method for rapid site-directed mutagenesis using the polymerase chain reaction, *Gene* **96,** 125–128.

LEARY, J. J., BRIGATI, D. J., WARD, D. C. (1983), Rapid and sensitive colorimetric method for visualizing biotin-labeled DNA probes hybridized to DNA or RNA immobilized on nitrocellulose: bioblots, *Proc. Natl. Acad. Sci. USA* **80,** 4045.

LE DOAN, T., PERROUAULT, L., PRASEUTH, D., HABHOUB, N., DECOUT, J.-L., THUONG, N. T., LHOMME, J., HÉLÈNE, C. (1987), Sequence-specific recognition, photocrosslinking and cleavage of the DNA double helix by an oligo-α-thymidylate covalently linked to an azidoproflavine derivative, *Nucleic Acids Res.* **15,** 7749–7760.

LEE, N., COZZITORTO, J., WAINWRIGHT, N., TESTA, D. (1984), Cloning with tandem gene systems for high level gene expression, *Nucleic Acids Res.* **12,** 6797–6812.

LEMAITRE, M., BAYARD, B., LEBLEU, B. (1987), Specific antiviral activity of poly(L-lysine)-conjugated oligodeoxyribonucleotide sequence complementary to vesicular stomatitis virus N protein mRNA initiation site, *Proc. Natl. Acad. Sci. USA* **84,** 648–652.

LEONETTI, J. P., RAYNER, B., LEMAITRE, M., GAGNOR, C., MILHAUD, P. G., IMBACH, J.-L., LEBLEU, B. (1988), Antiviral activity of conjugates between poly(L-lysine) and synthetic oligodeoxyribonucleotides, *Gene* **72,** 323–332.

LEONETTI, J.-P., MACHY, P., GEGLOS, G., LEBLEU, B., LESERMAN, L. (1990), Antibody-targeted liposomes containing oligodeoxyribonucleotides complementary to viral RNA selectively inhibit viral replication, *Proc. Natl. Acad. Sci. USA* **87**, 2448–2451.

LEONETTI, J. P., MECHTI, N., DEGOLS, G., GAGNOR, C., LEBLEU, B. (1991), Intracellular distribution of microinjected antisense oligonucleotides, *Proc. Natl. Acad. Sci. USA* **88**, 2702–2706.

LESNIKOWSKI, Z., JAWORSKA, M., STEC, W. J. (1990), Octa(thymidine methanephosphonates) of partially defined stereochemistry: synthesis and effect of chirality at phosphorus on binding to pentadecadeoxyriboadenylic acid, *Nucleic Acids Res.* **18**, 2109–2115.

LETSINGER, R. L., KORNET, M.J. (1963), Popcorn polymer as a support in multistep syntheses, *J. Am. Chem. Soc.* **85**, 3045–3046.

LETSINGER, R. L., FINNAN, J. L., HAEVNER, G. A., LUNSFORD, W. B. (1975), Phosphite coupling procedure for generating internucleotide links, *J. Am. Chem. Soc.* **97**, 3278–3279.

LI, H., GYLLENSTEN, U. B., CUI, X., SAIKI, R. K., EHRLICH, H. A., ARNHEIM, N. (1988), Amplification and analysis of DNA sequences in single human sperm and diploid cells, *Nature* **335**, 414–417.

LIN, S.-B., BLAKE, K. R., MILLER, P. S., TS'O, P. O. P. (1989), Use of EDTA derivatization to characterize interactions between oligodeoxyribonucleoside methylphosphonates and nucleic acids, *Biochemistry* **28**, 1054–1061.

LIPSETT, M. N. (1963), The interaction of poly c and guanine trinucleotide, *Biochem. Biophys. Res. Commun.* **11**, 224–228.

LO, K. M., JONES, S. S., HACKETT, N. R., KHORANA, H. G. (1984), Specific amino acid substitutions in bacterioopsin: Replacement of a restriction fragment in the structural gene by synthetic DNA fragments containing altered codons, *Proc. Natl. Acad. Sci. USA* **81**, 2285–2289.

LOKE, S. L., STEIN, C. A., ZHANG, X. H., MORI, K., NAKANISHI, M., SUBASNGHE, C., COHEN, J. S., NECKERS, L. M. (1989), Characterization of oligonucleotide transport into living cells, *Proc. Natl. Acad. Sci. USA* **86**, 3474–3478.

LONGO, M. C., BERNINGER, M. S., HARTLEY, J. L. (1990), Use of uracil DNA glycosylase to control carry-over contamination in polymerease chain reaction, *Gene* **93**, 125–128.

LÖSCHNER, T., ENGELS, J. W. (1988), Methylphosphoramidites: Preparation and application in oligonucleoside methylphosphonate synthesis, *Nucleosides Nucleotides* **7**, 729–732.

LÖSCHNER, T., ENGELS, J. W. (1990), Diastereomeric dinucleoside-methylphosphonates: Determination of configuration with the 2-D NMR ROESY technique, *Nucleic Acids Res.* **18**, 5083–5088.

LOWE, T., SHAREFKIN, J., YANG, S. Q., DIEFFENBACH, C. W. (1990), A computer program for selection of oligonucleotide primers for polymerase chain reactions, *Nucleic Acids Res.* **18**, 1757–1761.

LYTTLE, M. H., WRIGHT, P. B., SINHA, N. D., BAIN, J. D., CHAMBERLIN, A. R. (1991), New nucleoside phosphoramidites and coupling protocols for solid-phase RNA synthesis, *J. Org. Chem.* **56**, 4608–4615.

MAG, M., ENGELS, J. (1988), Synthesis and structure assignments of amide protected nucleosides and their use as phosphoramidites in deoxyoligonucleotide synthesis, *Nucleic Acids Res.* **16**, 3525–3543.

MAG, M., ENGELS, J. (1989), Synthesis and selective cleavage of oligodeoxyribonucleotides containing non-chiral internucleotide phosphoramidate linkages, *Nucleic Acids Res.* **17**, 5973–5988.

MAJUMDAR, C., STEIN, C. A., COHEN, J. S., BRODER, S., WILSON, S. H. (1989), Stepwise mechanism of HIV reverse transcriptase: primer function of phosphorothioate oligodeoxynucleotide, *Biochemistry* **28**, 1340–1346.

MANDECKI, W., BOLLING, T. J. (1988), *Fok* I method of gene synthesis, *Gene* **68**, 101–107.

MANIATIS, T., FRITSCH, E. F., SAMBROOK, J. J. (1989), *Molecular Cloning. A Laboratory Manual,* 2nd Ed., Cold Spring Harbor, NY: Cold Spring Harbor Laboratory Press.

MARCUS-SEKURA, C. J., WOERNER, A. M., SHINOZUKA, K., ZON, G., QUINNAN, G. V. JR. (1987), Comparative inhibition of chloramphenicol acetyltransferase gene expression by antisense oligonucleotide analogues having alkyl phosphotriester, methylphosphonate and phosphorothioate linkages, *Nucleic Acids Res.* **15**, 5749–5763.

MARUYAMA, I. N., HORIKOSHI, K., NAGASE, Y., SOMA, M., NOBUHARA, M., YASUDA, S., HIROTA, Y. (1989), A synthetic translation terminator gene: A tool for dissecting the translation direction of a gene, *Gene Anal. Tech.* **6**, 57–61.

MATSUKURA, M., SHINOZUKA, K., ZON, G., MITSUYA, H., REITZ, M., COHEN, J. S., BRODER, S. (1987), Phosphorothioate analogs of oligodesoxynucleotides: inhibitors of replication and cytopathic effects of human immunodeficiency virus, *Proc. Natl. Acad. Sci. USA* **84**, 7706–7710.

MATSUKURA, M., ZON, G., SHINOZUKA, K., ROBERT-GUROFF, M., SHIMADA, T., STEIN, C. A., MITSUYA, H., WONG-STAAL, F., COHEN, J. S., BRODER, S. (1989), Regulation of viral expression of human immunodeficiency virus *in vitro* by an antisense phosphorothiate oligodeoxynucleotide against rev (art/trs) in chronically infected cells, *Proc. Natl. Acad. Sci. USA* **86**, 4244–4248.

MATTEUCCI, M. D., HEYNEKER, H. L. (1983), Targeted random mutagenesis: the use of ambiguously synthesized oligonucleotides to mutagenize sequences immediately 5′ of an ATG initiation codon, *Nucleic Acids Res.* **11,** 3113–3132.

MATTHEWS, J. A., KRICKA, L. J. (1988), Analytical strategies for the use of DNA probes, *Anal. Biochem.* **169,** 1.

MAYWALD, F., BÖLDICKE, T., GROSS, G., FRANK, R., BLÖCKER, H. MEYERHANS, A., SCHWELLNUS, K. EBBERS, J., BRUNS, W., REINHARDT, W., SCHNABEL, E., SCHRÖDER, W., FRITZ, H., COLLINS, J. (1988), Human pancreatic secretory trypsin inhibitor (PSTI) produced in active form and secreted from *Escherichia coli, Gene* **68,** 357–369.

MAZIN, A. V., SPARBAEV, M. K., OVCHINNIKOVA, L. P., DIANOV, G. L., SALGANIK, R. I. (1990), Site-directed insertion of long single-stranded DNA fragments into plasmid DNA, *DNA Cell Biol.* **9,** 63–69.

MCBRIDE, L. J., MCCOLLUM, C., DAVIDSON, S., EFCAVITCH, J. W., ANDRUS, A., LOMBARDI, S. J. (1988), A new reliable cartridge for the rapid purification of synthetic DNA, *BioTechniques* **6,** 362–367.

MCCLOSKEY, J. A. (1991), Structural characterization of natural nucleosides by mass spectrometry, *Acc. Chem. Res.* **24,** 81–88.

MCCOLLUM, C., ANDRUS, A. (1991), An optimized polystyrene support for rapid efficient oligonucleotide synthesis, *Tetrahedron Lett.* **32,** 4069–4072.

MCCURDY, S., MOULDS, C., FROEHLER, B. (1991), Deoxyoligonucleotides with inverted polarity: synthesis and use in triple-helix formation, *Nucleosides Nucleotides* **10,** 287–290.

MEINKOTH, J., WAHL, G. (1984), Hybridization of nucleic acids immobilized on solid supports, *Anal. Biochem.* **138,** 267.

MERCIER, B., GAUCHER, C., FEUGEAS, O., MAZURIER, C. (1990), Direct PCR from whole blood, without DNA extraction, *Nucleic Acids Res.* **18,** 5908.

MERRIFIELD, R. B. (1963), Solid phase peptide synthesis. I. The synthesis of a tetrapeptide, *J. Am. Chem. Soc.* **85,** 2149–2154.

MICHELSON, A. M., TODD, A. R. (1955), Nucleotides part XXXII. Synthesis of a dithymidine dinucleotide containing a 3′:5′-internucleotidic linkage, *J. Chem. Soc.,* 2632–2638.

MILLER, P. S. (1991), Oligonucleoside methylphosphonates as antisense reagents, *Bio/Technology* **9,** 358–362.

MILLER, P. S., TS'O, P. O. P. (1988), Oligonucleotide inhibitors of gene expression in living cells; new opportunities in drug design, *Annu. Rep. Med. Chem.* **23,** 295–304.

MILLER, P. S., MCPARLAND, K.B., JAYARAMAN, K., TS'O, P. O. P. (1981), Biochemical and biological effects of nonionic nucleic acids, *Biochemistry* **20,** 1874–1880.

MILLER, S. A., DYKES, D. D., POLESKY, H. F. (1988), A simple salting out procedure for extracting DNA from human nucleated cells, *Nucleic Acids Res.* **16,** 1215.

MIYOSHI, K., MIYAKE, T., HOZUMI, T., ITAKURA, K. (1980), Solid-phase synthesis of polynucleotides. II. Synthesis of polythymidylic acids by the block coupling phosphotriester method, *Nucleic Acids Res.* **8,** 5473–5489.

MORVAN, F., RAYNER, B., IMBACH, J.-L., CHANG, D.-K., LOWN, J. E. (1986), DNA I. Synthesis, characterization by high-field ^{1}H-NMR, and base-pairing properties of the unnatural hexadeocyribonucleotide - [d(CpCpTpTpCpC)] with its complement β-[d(GpGpApApGpG)], *Nucleic Acids Res.* **14,** 5019–5035.

MOSER, H. E., DERVAN, P. B. (1987), Sequence-specific cleavage of double helical DNA by triple helix formation, *Science* **238,** 645–650.

MOUNTS, P., WU, T.-C., PEDEN, K. (1989), Method for cloning single-stranded oligonucleotides in a plasmid vector, *BioTechniques* **7,** 356–359.

MULLENBACH, G. T., TABRIZI, A., BLACHER, R. W., STEIMER, K. S. (1986), Chemical synthesis and expression in yeast of a gene encoding connective tissue activating peptide-III, *J. Biol. Chem.* **261,** 719–722.

MULLIS, K. B. (1990), Eine Nachtfahrt und die Polymerase-Kettenreaktion, *Spektrum Wiss.* **6,** 60–67.

MULLIS, K. B., FALOONA, F., SCHARF, S., SAIKI, R., HORN, G., ERLICH, H. (1986), Specific enzymatic of DNA *in vitro:* The polymerase reaction, *Cold Spring Harb. Symp. Quant. Biol.* **51,** 263–273.

NARANG, S. A. (1983), DNA synthesis, *Tetrahedron* **39,** 3–22.

NIELSEN, P. E. (1991), Sequence-selective DNA recognition by synthetic ligands, *Bioconjugate Chem.* **2,** 2–12.

NIELSEN, P. E., EGHOLM, M, BERG, R. H., BUCHARDT, O. (1991), Sequence-selective recognition of DNA by strand displacement with a thymine-substituted polyamide, *Science* **254,** 1497–1500.

NIKONOWICZ, E. P., PARDI, A. (1992), Three-dimensional heteronuclear NMR studies of RNA, *Nature* **355,** 184–185.

OLIPHANT, A. R., NUSSBAUM, A. L., STRUHL, K. (1986), Cloning of random sequence oligodeoxynucleotides, *Gene* **44,** 177–183.

ONO, A., TS'O, P. O. P., KAN, L.-S. (1991), Triplex formation of oligonucleotides containing 2′-O-methylpseudoisocytidine in substitution for 2′-deoxycytidine, *J. Am. Chem. Soc.* **113,** 4032–4033.

ORSON, F. M., THOMAS, D. W., MCSHAN, W. H., KESSLER, D. J., HOGAN, M. E. (1991), Oligonucleotide inhibition of IL-2 mRNA transcription by promoter region colinear triplex formation in lymphocytes, *Nucleic Acids Res.* **19,** 3435–3441.

OU, C. Y., MOORE, J. L., SCHOCHETMAN, G. (1991), Use of UV radiation to reduce false positivity in polymerase chain reaction, *BioTechniques* **10,** 442.

PERRIN, S., GILLILAND, G. (1990), Site-specific mutagenesis using asymmetric polymerase chain reaction and a single mutant primer, *Nucleic Acids Res.* **18,** 7433–7438.

POSTEL, E. H., FLINT, S. J., KESSLER, D. J., HOGAN, M. E. (1991), Evidence that a triplex-forming oligodeoxyribonucleotide binds to *c-myc* promoter in HeLa cells, thereby reducing *c-myc* mRNA levels, *Proc. Natl. Acad. Sci. USA* **88,** 8227–8231.

POVSIC, T. J., DERVAN, P. B. (1989), Triple helix formation by oligonucleotides on DNA extended to the physiological pH range, *J. Am. Chem. Soc.* **111,** 3059–3061.

PRAKASH, G., KOOL, E. T. (1991), Molecular recognition by circular oligonucleotides. Strong binding of single-stranded DNA and RNA, *J. Chem. Soc. Chem. Commun.,* 1161–1163.

PRASEUTH, D., PERROUAULT, L., LE DOAN, T., CHASSIGNOL, M., THUONG, N., HÉLÈNE, C. (1988), Sequence-specific binding and photocrosslinking of α and β oligodeoxynucleotides to the major groove of DNA via triple-helix formation, *Proc. Natl. Acad. Sci. USA* **85,** 1349–1353.

PROBER, J. M., TRAINOR, G. L., DAM, R. J., HOBBS, F. W., ROBERTSON, C. W., ZAGURSKY, R. J., COCUZZA, A. J., JENSEN, M. A., BAUMEISTER, K. (1987), A system for rapid DNA sequencing with fluorescent chain-terminating dideoxynucleotides, *Science* **238,** 336.

QUARTIN, R. S., BRAKEL, C. L., WETMUR, J. G. (1989), Number and distribution of methylphosphonate linkages in oligodeoxynucleotides affect exo- and endonuclease sensitivity and ability to form RNase H substrates, *Nucleic Acids Res.* **17,** 7253–7262.

REESE, C. B. (1978), The chemical synthesis of oligo- and polynucleotides by the phosphotriester approach, *Tetrahedron* **34,** 3143–3179.

REESE, C. B. (1989), The chemical synthesis of oligo- and polyribonucleotides, in: *Nucleic Acids and Molelcular Biology* (ECKSTEIN, F., LILLEY, D. M. J., Eds.). Vol. 3, pp. 164–181, Heidelberg: Springer-Verlag.

REESE, C. B., SKONE, P. A. (1984), The protection of thymine and guanine residues in oligodeoxyribonucleotide synthesis, *J. Chem. Soc. Perkin Trans.* **1,** 1263.

REESE, C. B., UBASAWA, A. (1980), Reaction between 1-arenesulphonyl-3-nitro-1,2,4-triazoles and nucleoside base residues. Elucidation of the nature of side-reactions during oligonucleotide synthesis, *Tetrahedron Lett.* **21,** 2265–2268.

REGAN, S., MARSTERS, J. C., BISCHOFBERGER, N. (1990), Synthesis, hybridization properties and antiviral activity of lipid-oligodeoxynucleotide conjugates, *Nucleic Acids Res.* **18,** 3777–3783.

RENZ, M., KURZ, C. (1984), A colorimetic method for DNA hybridization, *Nucleic Acids Res.* **12,** 3435.

RICHTERICH, P. (1989), Non-radioactive chemical sequencing of biotin labelled DNA, *Nucleic Acids Res.* **17,** 2181.

ROBERTSON, D. L., JOYCE, G. F. (1990), Selection *in vitro* of an RNA enzyme that specifically cleaves single-stranded DNA, *Nature* **344,** 467–468.

ROSE, S. (1991), *Workshop Perkin Elmer Cetus,* Heidelberg.

ROSSI, J. J., KIERZEK, R., HUANG, T., WALKER, P. A., ITAKURA, K. (1982), An alternative method for synthesis of double-stranded DNA segments, *J. Biol. Chem.* **257,** 9226–9229.

SADLER, J. R., SASMOR, H., BETZ, J. L. (1983), A perfectly symmetric *lac* operator binds the *lac* repressor very tightly, *Proc. Natl. Acad. Sci. USA* **890,** 6785–6789.

SAIKI, R. K., SCHARF, S., FALOONA, F., MULLIS, K. B., HORN, G. T., EHRLICH, H. A. (1985), Enzymatic amplification of β-globin genomic sequences and restriction site analysis for diagnosis of sickle cell anemia, *Science* **230,** 1350–1354.

SAIKI, R. K., BUGAWAN, T. L., HORN, G. T., MULLIS, K. B., ERLICH, H. A. (1986), Analysis of enzymatically amplified β-globin and HLA-DQα DNA with allele-specific oligonucleotide probes, *Nature* **324,** 163.

SAIKI, R. K., GELFAND, D. H., STOFFEL, S., SCHARF, S. J., HIGUCHI, R., HORN, G. T., MULLIS, K. B., ERLICH, H. A. (1988a), Primer directed enzymatic amplification of DNA with thermostable DNA polymerase, *Science* **239,** 487–491.

SAIKI, R. B., CHANG, C. A., LEVENSON, C. H., WARREN, T. C., BOEHM, C. D., KAZAZIAN, H. H., ERLICH, H. A. (1988b), Diagnosis of sickle cell anemia and β-thalassemia with enzymatically amplified DNA and nonradioactive allele-specific oligonucleotide probes, *N. Engl. J. Med.* **319,** 537.

Saison-Behmoares, T., Tocqué, B., Rey, I., Chassignol, M., Thuong, N. T., Hélène, C. (1991), Short modified antisense oligonucleotides directed against Ha-ras point mutation induce selective cleavage of the mRNA and inhibit t24 cells proliferation, *EMBO J.* **10,** 1111–1118.

Sanger, F., Nicklen, S., Coulson, A. R. (1977), DNA sequencing with chain-terminating inhibitors, *Proc. Natl. Acad. Sci. USA* **74,** 5463–5467.

Sarver, N., Cantin, E. M., Chang, P. S., Zaia, J. A., Ladne, P. A., Stephens, D. A., Rossi, J. J. (1991), Ribozymes as potential anti-HIV therapeutic agents, *Science* **247,** 1222–1225.

Scharf, S. J., Horn, G. T., Erlich, H. A. (1986), Direct cloning and sequence analysis of enzymatically amplified genomic sequences, *Science* **233,** 1076–1078.

Schauder, B., McCarthy, J. E. G. (1989), The role of bases upstream of the Shine-Dalgarno region and in the coding sequence in the control of gene expression in *Escherichia coli:* translation and stability of mRNA *in vivo, Gene* **78,** 59–72.

Schoner, B. E., Belagaje, R., Schoner, R. G. (1987), Expression of eukaryotic genes in *Escherichia coli* with a synthetic two-cistron system, *Methods Enzymol.* **153,** 401–417.

Schott, H. (1985), Nucleobases, nucleosides, nucleotides, in: *High Performance Liquid Chromatography in Biochemistry* (Henschen, A., Hupe, K. P., Lottspeich, F., Voelter, W., Eds.), p. 414, Weinheim: VCH Verlagsgesellschaft.

Schowalter, D. B., Sommer, S. S. (1989), The generation of radiolabeled DNA and RNA probes with polymerase chain reaction, *Anal. Biochem.* **177,** 90–94.

Schulhof, J. C., Molko, D., Teoule, R. (1987), The final deprotection step in oligonucleotide synthesis is reduced to a mild and rapid ammonia treatment by using labile base-protecting group, *Nucleic Acids Res.* **15,** 397–416.

Scott, J. K., Smith, G. (1990), Searching for peptide ligands with an epitope library, *Science* **249,** 386–390.

Seetharam, R., Heeren, R. A., Wong, E. Y., Bradford, S. R., Klein, B. K., Aykent, S., Kotts, C. E., Mathis, K. J., Bishop, B. F., Jennings, M. J., Smith, C. E., Siegel, N. R. (1988), Mistranslation in IGF-1 during overexpression of the protein in *Escherichia coli* using a synthetic gene containing low frequency codons, *Biochem. Biophys. Res. Commun.* **155,** 518–523.

Shabarova, Z. A., Merenkova, I. N., Oretskaya, T. S., Sokolova, N. I., Skripkin, E. A., Alexeyeva, E. V., Balakin, A. G., Bogdanov, A. A. (1991), Chemical ligation of DNA. The first non-enzymatic assembly of a biologically active gene, *Nucleic Acids Rs.* **19,** 4247–4251.

Singh, H., Sen, R., Baltimore, D., Sharp, P. A. (1986), A nuclear factor that binds to a conserved sequence motif in transcriptional control elements of immunoglobulin genes, *Nature* **319,** 154–158.

Sinha, N. D., Biernat, J., Köster, H. (1983), β-Cyanoethyl N,N-dialkylamino/N-morpholinomonochloro phosphoamidites, new phosphitylating agents facilitating ease of deprotection and work-up of synthesized oligo-nucleotides, *Tetrahedron Lett.* **24,** 5843–5846.

Sinha, N. D., Biernat, J., Köster, H. (1984), Polymer support oligonucleotide synthesis XVIII: Use of β-cyanoethyl-N,N-dialkylamino-/N-morpholino phosphoramidite of deoxynucleosides for the synthesis of DNA fragments simplifying deprotection and isolation of the final product, *Nucleic Acids Res.* **12,** 4539–4557.

Smith, M. (1985), *In vitro* mutagenesis, *Annu. Rev. Genet.* **19,** 423–462.

Smith, J., Cook, E., Fotheringham, I., Pheby, S., Derbyshire, R., Eaton, M. A. W., Doel, M., Lilley, D. M. J., Pardon, J. F., Patel, T., Lewis, H., Bell, L. D. (1982), Chemical synthesis and cloning of a gene for human β-urogastrone, *Nucleic Acids Res.* **10,** 4467–4482.

Smith, L. M., Sanders, J. Z., Kaiser, R. J., Hughes, P., Dodd, C., Connell, C. R., Heiner, C., Kent, S. B. H., Hood, L. E. (1986), Fluorescence detection in automated DNA sequence analysis, *Nature* **321,** 674.

Sonveaux, E. (1986), The organic chemistry underlying DNA synthesis, *Bioorg. Chem.* **14,** 274–325.

Sørensen, M. A., Kurland, C. G., Pedersen, S. (1989), Codon usage determines translation rate in *Escherichia coli, J. Mol. Biol.* **207,** 365–377.

Southern, E. M. (1975), Detection of specific sequences among DNA fragments separated by gel electrophoresis, *J. Mol. Biol.* **98,** 503.

Stebbing, N. (1984), Antiviral effects of single-stranded polynucleotides and their mode of action, *Int. Encycl. Pharmacol. Ther. Viral Chemother.* **1,** 437–477.

Stebbing, N., Grantham, C. A., Lindley, I. J. D., Eaton, M. A. W., Carey, N. H. (1977), *In vivo* antiviral activity of polynucleotide mimics of strategic regions in viral RNA, *Ann. N. Y. Acad. Sci.* **284,** 682–696.

Stec, W. J., Zon, G., Egan, W., Stec, B. (1984), Automated solid phase synthesis, separation and stereochemistry of phosphorothioate analogues of oligodeoxyribonucleotides, *J. Am. Chem. Soc.* **106**, 6077–6079.

STEC, W. J., GRAJKOWSKI, A., KOZIOLKIEWICZ, M., UZNANSKI, B. (1991), Novel route to oligodeoxyribonucleoside phosphorothioates. Stereocontrolled synthesis of P-chiral oligodeoxyribonucleoside phosphorothioates, *Nucleic Acids Res.* **19**, 5883–5888.

STEIN, C. A., COHEN, J. S. (1988), Oligodeoxynucleotides as inhibitors of gene expression: a review, *Cancer Res.* **48**, 2659–2668.

STEIN, C. A., SUBASINGHE, C., SHINOZUKA, K., COHEN, J. S. (1988a), Physicochemical properties of phosphorothioate oligodeoxynucleotides, *Nucleic Acids Res.* **16**, 3209–3221.

STEIN, C. A., MORI, K., LOKE, S. L., SUBASINGHE, C., SHINOZUKA, K., COHEN, J. S., NECKERS, L. M. (1988b), Phosphorothioate and normal oligodeoxyribonucleotides with 5′-linked acridine: charcterization and preliminary kinetics of uptake, *Gene* **72**, 333–341.

STEIN, C. A., NECKERS, L. M., NAIR, B. C., MUMBAUER, S., PAL, R. (1991a), Phosphorothioate oligodeoxycytidine interferes with binding of HIV-1 gp 120 to CD4, *J. Acquired Immune Deficiency Syndromes* **4**, 686–693.

STEIN, C. A., PAL, R., DE VICO, A. L., HOKE, G., MUMBAUER, S., KINSTLER, O., SARNGADHARAN, M. G., LETSINGER, R. L. (1991b), Mode of action of 5′-linked cholesteryl phosphorothioate oligonucleotides in inhibiting syncytia formation and infection by HIV-1 and HIV-2 *in vitro, Biochemistry* **30**, 2439–2444.

STOFLET, E. S., KOEBERL, D. D., SARKAR, G., SOMMER, S. S. (1988), Genomic amplification with transcript sequencing, *Science* **239**, 491–494.

STROBEL, S. A., DERVAN, P. B. (1991), Single-site enzymatic cleavage of yeast genomic DNA mediated by triple helix formation, *Nature* **350**, 172–174.

SUNG, W. L., YAO, F.-L., ZAHAB, D. M., NARANG, S. A. (1986), Short synthetic oligodeoxyribonucleotide leader sequences enhance accumulation of human proinsulin synthesized in *Escherichia coli, Proc. Natl. Acad. Sci. USA* **83**, 561–565.

TAYLOR, J. W., OTT, J., ECKSTEIN, F. (1985), The rapid generation of oligonucleotide-directed mutations at high frequency using phosphorothioate-modified DNA, *Nucleic Acids Res.* **13**, 8765–8785.

TI, G. S., GAFFNEY, B. L., JONES, R. A. (1982), Transient protection: Efficient one-flask syntheses of protected deoxynucleosides, *J. Am. Chem. Soc.* **104**, 1316–1319.

TINDALL, K. R., KUNKEL, T. A. (1988), Fidelity of DNA synthesis by the *Thermus aquaticus* DNA polymerase, *Biochemistry* **27**, 6008–6013.

TOULMÉ, J.-J., HÉLÈNE, C. (1988), Antimessenger oligodeoxyribonucleotides: an alternative to antisense RNA for artificial regulation of gene expression – a review, *Gene* **72**, 51–58.

TU, C.-P. D., COHEN, S. N. (1980), 3′-End labeling of DNA with [α-^{32}P]-cordycepin-5′-triphosphate, *Gene* **10**, 177.

TUERK, C., GOLD, L. (1990), Systematic evolution of ligands by exponential enrichment: RNA ligands to bacteriophage T4 DNA polymerase, *Science* **249**, 505–510.

UHLMANN, E. (1988), An alternative approach in gene synthesis: use of long self-priming oligodeoxynucleotides for the construction of double-stranded DNA, *Gene* **71**, 29–40.

UHLMANN, E., ENGELS, J. (1986a), Chemical 5′-phosphorylation of oligonucleotides valuable in automated DNA synthesis, *Tetrahedron Lett.* **27**, 1023–1026.

UHLMANN, E., ENGELS, J. (1986b), Automated 5′-phosphorylation of oligonucleotides, *Chem. Scri.* **26**, 217–219.

UHLMANN, E., HEIN, F. (1987), Chemoenzymatic synthesis of genes encoding medium-sized polypeptides by use of only one synthetic oligonucleotide, *Nucleic Acids Res. Symp. Ser.* **18**, 237–240.

UHLMANN, E., PEYMAN, A. (1990), Antisense oligonucleotides: A new therapeutic principle, *Chem. Rev.* **90**, 543–584.

UHLMANN, E., ENGELS, J., WETEKAM, W. (1985), Signal sequence: application of chemical DNA synthesis to studies of translocation of expressed proteins, *Nucleosides Nucleotides* **4**, 259–260.

VAN BOOM, J. H., BURGERS, P. M., VAN DER MAREL, G., VERDEGAAL, C. H., WILLE, G. (1977), Synthesis of oligonucleotides with sequences identical with or analogous to the 3′-end of 16S ribosomal RNA of *Escherichia coli:* preparation of A-C-C-U-C-C via the modified phosphotriester method, *Nucleic Acids Res.* **4**, 1047–1063.

VAN DE VEN, F. J. M., HILBERS, C. W. (1988), Nucleic acids and nuclear magnetic resonance, *Eur. J. Biochem.* **178**, 1–38.

VERES, G., GIBBS, R. A., SCHERER, S. E., CASKEY, C. T. (1987), The molecular basis of the sparse fur mouse mutation, *Science* **237**, 415–417.

VLASSOV, V. V., GAIDAMAKOV, S. A., ZARYTOVA, V. F., KNORRE, D. G., LEVINA, A. S., NIKONOVA, A. A., PODUST, L. M., FEDOROVA, O. S. (1988), Sequence-specific chemical modification of double-stranded DNA with alkylating oligodeoxyribonucleotide derivatives, *Gene* **72,** 313–322.

VLASUK, G. P., BENCEN, G. H., SCARBOROUGH, R. M., TSAI, P.-K., WHANG, J. L., MAACK, T., CAMARGO, M. J. F., KIRSHER, S. W., ABRAHAM, J. A. (1986), Expression and secretion of biologically active human atrial natriuretic peptide in *Saccharomyces cerevisiae, J. Biol. Chem.* **261,** 4789–4796.

VU, H., HIRSCHBEIN, B. L. (1991), Internucleotide phosphite sulfurization with tetraethylthiuram disulfide. Phosphorothioate oligonucleotide synthesis via phosphoramidite chemistry, *Tetrahedron Lett.* **32,** 3005–3008.

WAGNER, E., ZENKE, M., COTTON, M., BEUG, H., BIRNSTIEL, M. L. (1990), Transferrin-polycation conjugates as carriers for DNA uptake into cells, *Proc. Natl. Acad. Sci. USA* **87,** 3410–3414.

WARREN, W., WHEAT, T., KNUDSEN, P. (1991), Rapid analysis and quantitation of PCR products by high-performance liquid chromatography, *Biotechniques* **11,** 250–254.

WATSON, J. D., CRICK, F. H. (1953), Molecular structure of nucleic acids, *Nature* **171,** 737–738.

WESTHEIMER, F. H. (1968), Pseudo-rotation in the hydrolysis of phosphate esters, *Acc. Chem. Res.* **1,** 70–78.

WETMUR, J. G. (1991), DNA probes: Applications of the principles of nucleic acid hybridization, *Crit. Rev. Biochem. Mol. Biol.* **26,** 227–259.

WHITEHOUSE, C. M., DREYER, R. N., YAMASHITA, M., FENN, J. B. (1985), Electrospray interface for liquid chromatographs and mass spectrometers, *Anal. Chem.* **57,** 675–679.

WICKSTROM, E. (1986), Oligodeoxynucleotide stability in subcellular extracts and culture media, *J. Biochem. Biophys. Methods* **13,** 97–102.

WILLIAMS, D. P., REGIER, D., AKIYOSHI, D., GENBAUFFE, F., MURPHY, J. R. (1988), Design, synthesis and expression of human interleukin-2 gene incorporating the codon usage bias found in highly expressed *Escherichia coli* genes, *Nucleic Acids Res.* **16,** 10453–10467.

WINBERG, G. (1991), A rapid method for preparing DNA from blood, suited for PCR screening of transgenes in mice, *PCR Methods Appl.* **1,** 72–74.

WONG, E. Y., SEETHARAM, R., KOTTS, C. E., HEEREN, R. A., KLEIN, B. K., BRAFORD, S. R., MATHIS, K. J., BISHOP, B. F., SIEGEL, N. R., SMITH, C. E., TACON, W. C. (1988), Expression of secreted insulin-like growth factor-1 in *Escherichia coli, Gene* **68,** 193–203.

WOSNICK, M. A., BARNETT, R. W., VICENTINI, A. M., ERFLE, H., ELLIOTT, R., SUMNER-SMITH, M., DAVIES, R. W. (1987), Rapid construction of large synthetic genes: total chemical synthesis of two versions of the bovine prochymosin gene, *Gene* **60,** 115–127.

WU, T., OGILVIE, K. K., PON, R. T. (1989), Prevention of chain cleavage in the chemical synthesis of 2′-silylated oligoribonucleotides, *Nucleic Acids Res.* **17,** 3501–3517.

WU, H., HOLCENBERG, J. S., TOMICH, J., CHEN, J., JONES, P. A., HUANG, S.-H., CALAME, K. L. (1990), Inhibition of *in vitro* transcription by double-stranded oligodeoxynucleotides, *Gene* **89,** 203–209.

WU, D. Y., UGOZZOLI, L., PAL, B. K., QIAN, J., WALLACE, B. R. (1991), The effect of temperature and olgonucleotide primer length on the specificity and efficiency of amplification by the polymerase chain reaction, *DNA Cell Biol.* **10,** 233–238.

YAMADA, O., MATSUMOTO, T., NAKASHIMA, M., HAGARI, S. (1990), A new method for extracting DNA or RNA for polymerase chain reaction, *J. Virol. Methods* **27,** 203–210.

YEUNG, A. T., DINEHART, W. J., JONES, B. K. (1988), Modifications of guanine bases during oligonucleotide synthesis, *Nucleic Acids Res.* **16,** 4539–4554.

ZAMECNIK, P. C., GOODCHILD, J., TAGUCHI, Y., SARIN, P. S. (1986), Inhibition of replication and expression of human T-cell lymphotropic virus type III in cultured cells by exogenous synthetic oligonucleotides complementary to viral RNA, *Proc. Natl. Acad. Sci. USA* **83,** 4143–4146.

ZOLLER, M. J., SMITH, M. (1982), Oligonucleotide-directed mutagenesis using M13-derived vectors: an efficient and general procedure for the production of point mutations in any fragment of DNA, *Nucleic Acids Res.* **10,** 6497–6500.

ZON, G. (1988), Oligonucleotide analogues as potential chemotherapeutic agents, *Pharm. Res.* **5,** 539–549.

ZON, G., THOMPSON, J. A. (1986), A review of high-performance liquid chromatography in nucleic acids research. II. Isolation, purification, and analysis of oligonucleotides, *Bio-Chromatography* **1,** 22–32.

III. Genetic Engineering of Microorganisms

10 Principles of Genetic Engineering for *Escherichia coli*

HELMUT SCHWAB

Graz, Austria

1 Introduction

The first experiments in which DNA fragments were joined *in vitro* and the recombinant molecules reintroduced into living cells were performed nearly twenty years ago (JACKSON et al., 1972; COHEN et al., 1973). The basic information obtained in these early experiments, together with numerous new findings in all fields of bioscience as well as in chemical, physical, and computer sciences have led to the development of modern genetic engineering. This powerful methodology can be regarded as a set of biological, genetical, biochemical, chemical, and physical procedures that greatly facilitate the localization, isolation, characterization, modification, synthesis, and transfer of genetic material.

The application of genetic engineering as a research tool has changed nearly all areas of the biosciences and has dramatically accelerated the rate at which data can be obtained in these fields. This rapid accumulation of information in the biosciences has brought outstanding insights in many basic processes of living organisms. The methods of genetic engineering have also led to enormous developments in biotechnology. The application of genetically engineered organisms as production strains in industrial bioprocesses or as improved crops in agriculture is no longer a dream of the future but an integral part of present technologies.

Escherichia coli has played a central role in the development of genetic engineering techniques and applications. Although genetic engineering can now be applied to nearly all types of organisms, *E. coli* retains its central role. Nearly all strategies for applying genetic engineering methods include steps employing *E. coli* K12 strains as hosts for recombinant plasmids or phages. In general, all steps of cloning, characterization and modification of

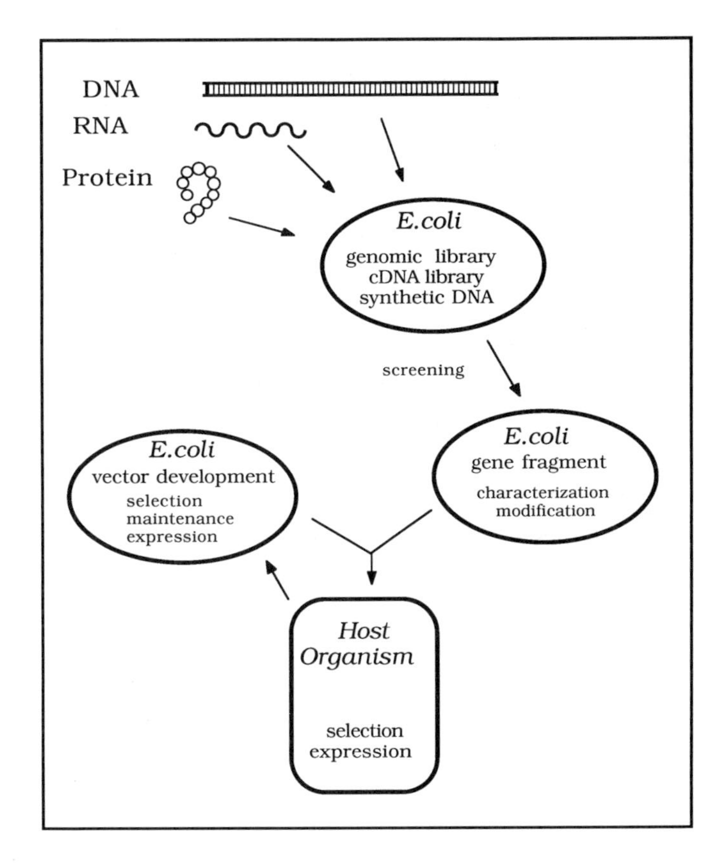

Fig. 1. Schematic representation of the role of *E. coli* in basic cloning strategies. Gene libraries constructed with *E. coli* host–vector systems or cloned synthetic DNA are the basis of most rDNA projects in biotechnology. *E. coli* host–vector systems are nearly exclusively used for all purposes of characterization and modification of DNA fragments. In addition, most of the work necessary for the development of suitable vector systems for any kind of organism and the work for the development of selection and expression systems for the desired target organisms is also performed with *E. coli* systems. Only the final constructs are then transferred into the desired hosts.

specific DNA fragments are performed with *E. coli* systems (Fig. 1). For genetic engineering applications with organisms other than *E. coli*, shuttle vector systems permitting the propagation of recombinant DNA in *E. coli* as well as in the specific host are usually used.

E. coli also plays an important role with respect to biotechnological applications. Although many prokaryotic and eukaryotic microorganisms have been developed as hosts for the expression of homologous and heterologous proteins within the last few years, *E. coli* K12 strains still represent potent hosts for the industrial production of various proteins.

2 Basic Cloning Techniques for *Escherichia coli*

Fundamental techniques in recombinant DNA technology are the isolation, modification, and joining of specific DNA molecules. The availability of proper vector systems as well as DNA transfer systems is also a prerequisite for genetic engineering experiments.

2.1 General Methods

2.1.1 Generation of DNA Fragments

In many cases high-molecular weight DNA is the starting material for the cloning of specific DNA fragments. A first step, therefore, is to generate DNA fragments within a suitable size range. Two principal mechanisms can be applied:

(a) mechanical breakdown by shear forces
(b) restriction endonuclease cleavage.

2.1.1.1 Mechanical Shearing

The generation of small DNA fragments for cloning by mechanical shearing can be achieved either by intensive mixing or by sonication. With both methods, the average fragment size obtained is determined by the intensity of the treatment. The ends of the resulting DNA fragments are not uniform. In addition to blunt ends, 5′- and 3′-overhanging single-stranded ends can occur. Thus, subsequent treatment of the fragments to generate uniform blunt ends (see Sect. 2.1.2) is necessary.

2.1.1.2 Restriction Endonuclease Cleavage

Restriction endonucleases cleave DNA at or in the vicinity of specific recognition sequences. Historically, the discovery of restriction endonucleases was of paramount importance to the development of recombinant DNA technology thereby making possible the enormous contribution of genetic engineering to biosciences and biotechnology.

Three types of restriction endonucleases are known. The type I endonucleases are complex enzymes having nuclease and methylase activity. These enzymes need ATP, S-adenosylmethionine and Mg^{2+} ions as cofactors. The cleavage takes place at a distance of about 1000 nucleotides from the recognition site. The simple type II endonucleases only need Mg^{2+} ions as a cofactor and cleave the DNA within their recognition sequences, typically a palindromic sequence of 4–8 bp. Nearly 1000 such enzymes are known and, depending on the enzyme, blunt ends, 3′-protruding ends (up to four bases), or 5′-protruding ends (up to six bases) can be formed. The type II restriction endonucleases are not associated with methylases; however, separate methylases that modify the same recognition sequences are part of type II restriction/modification systems and some have already been identified. Type III enzymes are complex enzymes which cleave the DNA at a specific distance (within approximately 25 bp) of the recognition sequence. Therefore, protruding ends generated by type III endonucleases are usually not compatible. Examples of restriction endonucleases are shown in Tab. 1. A compilation of known restriction endonucleases of all types as well as methylases has been given by Kessler and Manta (1990).

Tab. 1. Selected Examples of Restriction Endonucleases

Nuclease	Type	Recognition Sequence	Ends
*Dpn*I	II	G A*T C	blunt
*Scr*FI	II	C C*N G G	1 base 5′
*Taq*I	II	T*C G A	2 bases 5′
*Dde*I	II	C*T N A G	3 bases 5′
*Sau*3AI	II	*G A T C	4 bases 5′
*Eco*RI	II	G*A A T T C	4 bases 5′
*Not*I	II	G C*G G C C G C	4 bases 5′
*Mae*III	II	*G T N A C	5 bases 5′
*Hgi*CI	II	*G G Y R C C	6 bases 5′
*Hph*I	IIS	G G T T A (N)8 C C A A T (N)7	1 base 3′
*Sac*II	II	C C G C*G G	2 bases 3′
*Bgl*I	II	G C C N N N N*N G G C	3 bases 3′
*Sac*I	II	G A G C T*C	4 bases 3′
*Eco*AI	I	G A G N N N N N N N G T C A	—
*Eco*PI	III	A G A C C	—

*, Cleavage site
Data were taken from KESSLER and MANTA (1990).

For genetic engineering mainly the type II restriction endonucleases are of interest since these enzymes cut at defined positions and generate compatible ends. The reaction conditions are quite simple, the salt concentration can play an important role in activity and specificity of these enzymes. Improper salt conditions may result in "star" activities, i.e., the fidelity of sequence recognition is reduced, and the DNA is also cleaved at sequences which differ from the recognition sequence by one or more bp. A typical example is provided by the enzyme *Eco*RI which shows "star" activity at low salt concentrations.

The average number of cleavage sites for a specific enzyme within a given DNA molecule mainly depends on the type of the recognition sequence. For example, a tetranucleotide consisting of all four bases will be statistically present at every 256 (4^4) bp, a hexanucleotide containing all four bases will be found at every 4096 (4^6) bp. The frequency of the occurrence of a specific restriction site is also dependent on the general sequence structure of the substrate-DNA (non-statistical base distribution).

The generation of fragments by digestion of DNA with restriction endonucleases can be performed by different strategies:

- complete digestion of DNA by a single enzyme to generate fragments with identical termini
- complete digestion of DNA by a combination of two (or more) enzymes to generate fragments with different termini
- partial digestion of DNA by a frequently-cutting enzyme (usually having a tetranucleotide recognition sequence). By choosing appropriate enzyme concentrations this strategy allows the generation of overlapping DNA fragments of a defined size range.

2.1.2 Enzymatic Modification of DNA Molecules

In addition to restriction endonucleases, developed cloning strategies make use of a large variety of DNA-modifying enzymes which are now commercially available. These enzymes include:

- Specific methylases that are mainly used to methylate *in vitro* specific restriction sites to protect them from cleavage by the respective restriction endonuclease.

Tab. 2. DNA Modifying Enzymes and Their Action

Enzyme	Activities	Uses
DNA polymerase I (*E. coli*) holoenzyme	5′→3′-DNA polymerase 5′→3′-exonuclease 3′→5′-exonuclease	Nick translation, end labeling of DNA with 3′-protruding termini
Klenow fragment of *E. coli* DNA polymerase I	5′→3′-DNA polymerase 3′→5′-exonuclease	Fill-in of recessed 3′-termini (create blunt ends), end-labeling by filling recessed 3′-termini, synthesis of double-stranded DNA from single-stranded templates, sequencing
T4 DNA polymerase (bacteriophage T4)	5′→3′-DNA polymerase 3′→5′-exonuclease	All reactions as with Klenow fragment, end labeling of DNA with 3′-protruding termini, removal of protruding 3′-termini to create blunt ends
Reverse transcriptase	5′→3′-DNA polymerase RNA or DNA template	cDNA synthesis, sequencing
Terminal transferase	Addition of dNTP to 3′-hydroxyl termini of DNA molecules	Homopolymer tailing, labeling of 3′-termini
T4 DNA ligase	Creation of phosphodiester bonds between adjacent 3′-OH and 5′-PO_4 termini in DNA	Sticky end ligation, blunt end ligation
T4 polynucleotide kinase	Transfer of the γ-phosphate of ATP to a 5′-terminus of DNA or RNA	Phosphorylation of 5′-termini of DNA fragments, labeling of 5′-termini
Alkaline phosphatase	Hydrolysis of 5′-phosphates in DNA	Dephosphorylation of 5′-termini
DNAse I	Unspecific endonuclease	Introduction of random nicks (nick translation), DNAse footprinting
Lambda exonuclease	5′→3′-exonuclease	Generation of protruding 3′-termini in DNA
Nuclease *Bal* 31	3′→5′-exonuclease endonuclease	Shortening of DNA molecules in a controlled manner
Exonuclease III	3′→5′-exonuclease acts only on 3′-recessed termini or blunt ends	Generation of single-stranded regions in DNA, generation of nested sets of deletions in DNA in combination with nuclease S1 or mung-bean nuclease
Nuclease S1, mung-bean nuclease	Single-strand-specific nuclease (DNA and RNA)	Removal of single-stranded tails in DNA to create blunt ends

- Various DNA polymerases allowing the *in vitro* addition of nucleotides to DNA molecules or template-dependent synthesis of DNA strands.
- Various nucleases in addition to restriction endonucleases used for cleavage of DNA molecules and for removal of nucleotides from DNA molecules.
- DNA ligases to covalently link 3′-OH and 5′-PO_4 ends of DNA strands.
- Phosphatases and kinases to (de-)phosphorylate 5′-ends of DNA molecules.

Some important enzymes for the modification of DNA molecules and their properties are compiled in Tab. 2. Detailed information on DNA modifying enzymes is summarized in SAMBROOK et al. (1989) and AUSUBEL et al. (1987).

2.1.3 Joining of DNA Molecules

The process of joining DNA molecules *in vitro* usually consists of two steps. The first step can be regarded as the formation of a transient association between ends of two molecules of DNA. In a second step a stable hybrid molecule is generated by DNA ligase-mediated phosphodiester bond formation. The general strategy to obtain such a transient complex which is stable enough to allow efficient ligation is to generate compatible cohesive ends. It is also possible to join blunt-ended DNA molecules with T4 DNA ligase. Since it is easily possible to create blunt ends by various modification procedures (see Tab. 2), blunt-end ligation would be a universal method allowing the ligation of any DNA molecule. However, the efficiency of blunt-end ligation is rather low, and the reaction requires very high ligase concentrations and high concentrations of free DNA ends. This means that blunt-end ligation can be practically performed only with systems where at least one ligation partner has a rather low molecular weight and can be used at high molar concentrations. Blunt-end ligation is therefore a useful strategy for the ligation of short oligonucleotides to DNA molecules, or to ligate smaller DNA fragments into small plasmid vectors. The efficiency of blunt-end ligation can be increased, however, by adding condensing agents such as polyethylene glycol (PEG 8000) or hexamine cobalt chloride that cause the DNA molecules to aggregate (PHEIFFER and ZIMMERMAN, 1983; RUSCHE and HOWARD-FLANDERS, 1985).

For the efficient ligation of larger DNA molecules various strategies to generate compatible cohesive ends enabling the formation of transient intermediates prior to ligase treatment have been developed. Some of the most common strategies are described below.

2.1.3.1 Generation of Cohesive Ends by Restriction Endonucleases

Type II restriction endonucleases have been used to generate compatible cohesive ends on the DNA molecules to be ligated even in the first experiments that attempted to generate recombinant DNA molecules *in vitro*. In many cases it is not possible to cut both molecules to be ligated with the same restriction endonuclease in order to generate compatible sticky ends. In some cases combinations of enzymes that recognize different sites but generate identical single-stranded overhangs can be used. The central part of the recognition sequence of these enzymes has to be identical as illustrated in Fig. 2. This strategy is therefore limited to rather small numbers of enzyme combinations.

Another strategy is to partially fill in the recessed 3′-termini of DNA fragments with noncompatible 5′-protruding ends generated by different restriction endonucleases (HUNG and WENSINK, 1984). Using this method it is possible to generate compatible cohesive ends with a variety of combinations of restriction endonucleases which originally do not produce compatible termini. An example of this strategy is shown in Fig. 2. A cross index as published by KORCH (1987) allows one to rapidly find possible combinations.

2.1.3.2 Synthetic Linkers and Adapters

Linkers are small oligonucleotides, usually designed as palindromes to allow the formation of double-stranded DNA by self-annealing, that contain the sequence information for the cleavage sites of one or more restriction endonucleases. Such linkers can be used to add restriction sites to the ends of a blunt-ended DNA molecule. In this case phosphorylated linkers are used in high molar excess and ligated by blunt-end ligation with T4 DNA ligase to the target molecules. Many copies of the linker are ligated at both ends, and cohesive ends are generated by subsequent digestion with the respective restriction endonu-

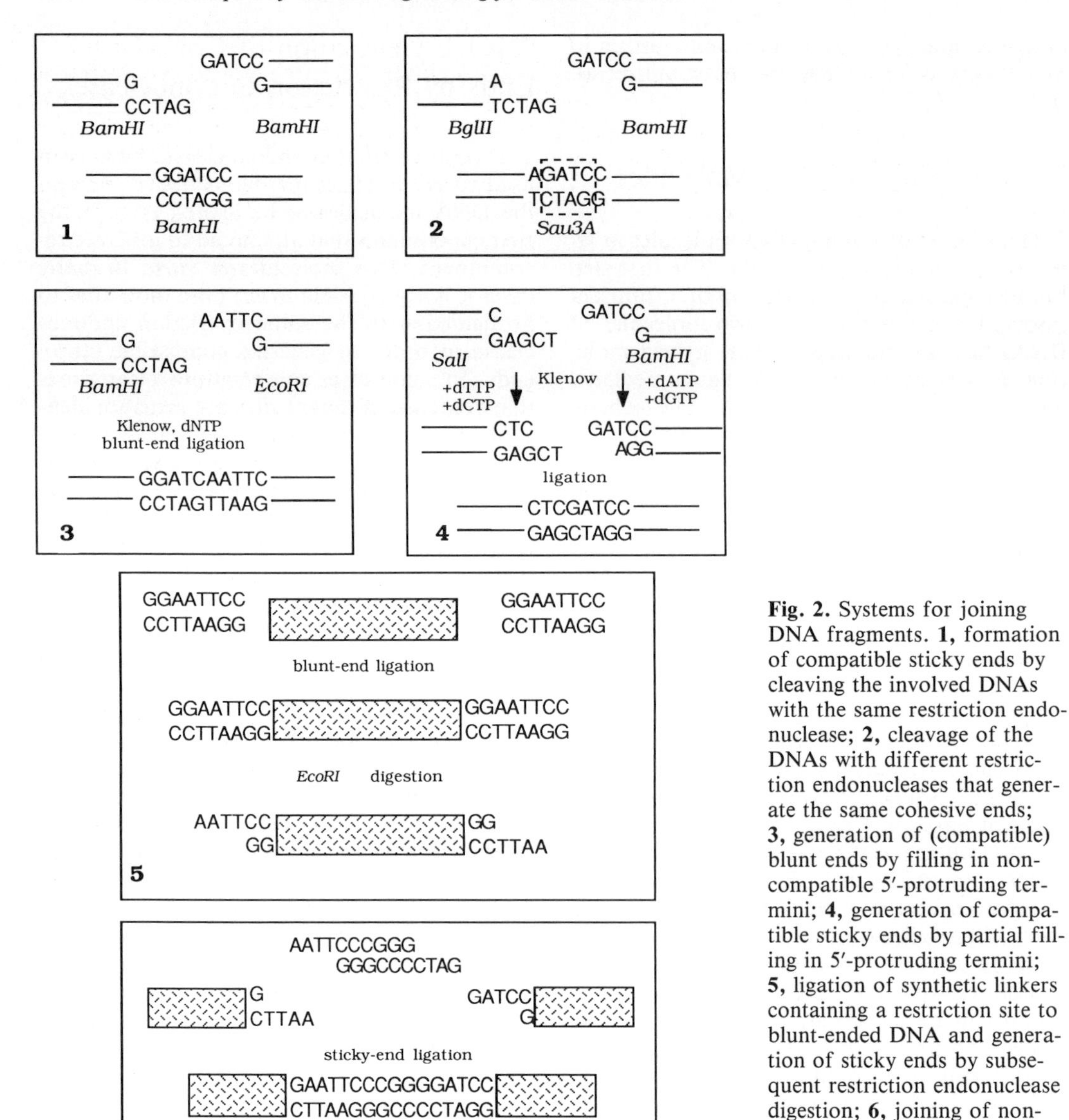

Fig. 2. Systems for joining DNA fragments. **1,** formation of compatible sticky ends by cleaving the involved DNAs with the same restriction endonuclease; **2,** cleavage of the DNAs with different restriction endonucleases that generate the same cohesive ends; **3,** generation of (compatible) blunt ends by filling in non-compatible 5′-protruding termini; **4,** generation of compatible sticky ends by partial filling in 5′-protruding termini; **5,** ligation of synthetic linkers containing a restriction site to blunt-ended DNA and generation of sticky ends by subsequent restriction endonuclease digestion; **6,** joining of non-compatible termini using synthetic adapters.

clease. This strategy can only be utilized if no site for this restriction endonuclease is present in the DNA fragment or when the DNA is protected by methylation with the respective methylase prior to the ligation of the linkers.

Another strategy is to use unphosphorylated linkers (Seth, 1984). In this case ligation is only possible at the phosphorylated 5′ end of the target DNA molecule. After a denaturation step and removal of the unligated oligonucleotides the DNA molecule contains a single strand of the linker at the 5′-ends. Molecules containing such single-stranded termini can be annealed and directly transformed into *E. coli* without ligation (see Sect. 2.1.3.3). Examples for the use of linkers are shown in Fig. 2.

Adapters are usually synthetic oligonucleotides that fit exactly to predetermined protruding ends of DNA molecules to be joined. In most cases these adapters are built up by two different oligonucleotides that are designed to contain a palindrome allowing efficient annealing of the two strands. The palindrome sequence usually contains a convenient restriction site (Fig. 2). Adapters represent a powerful tool and can be designed individually to realize a proper link between DNA fragments. Such adapters can be rather long oligonucleotides and may also include parts of genetic information replacing the original sequences.

2.1.3.3 Homopolymer Tailing

The enzyme terminal transferase is capable of adding dNTPs to the 3′-ends of a DNA molecule. Tailing of one DNA species with a specific nucleotide (e.g., dCTP) and the second DNA species with the complementary nucleotide (dGTP) results in single-stranded extensions of oligo-(dC) and oligo-(dG) at the 3′-ends of the DNA molecules. The tailing reaction can be performed in a controlled manner allowing the determination of the length of the extensions. For G/C tailing about 30 bp extensions are optimal, whereas for A/T tailing a length of about 100 bp is needed for the single-strand overhang. The joining of the tailed DNA molecules is performed by just annealing the complementary ends. The annealed molecules are directly transformed into *E. coli*, and the repair systems of the cells regenerate covalently closed DNA molecules.

2.1.4 DNA Transfer Systems for *E. coli*

An important step in all recombinant DNA experiments it to introduce the DNA molecules that have been manipulated *in vitro* into living cells of the desired host organism. For *E. coli* the classical method of transferring plasmid DNA into cells is transformation of cells made competent by treatment with calcium chloride. The transformation frequencies typically obtained with this simple technique, 10^6–10^7 per microgram of supercoiled plasmid DNA, are usually high enough for most applications. However, considerable differences in transformation efficiencies can be found with different strains of *E. coli* K12. Over the years the protocols for preparation of competent cells have been optimized. Rates of up to 10^9 transformants per microgram of supercoiled plasmid DNA are obtainable by exposing specifically improved strains to combinations of divalent cations and treating the cells with reducing agents, hexamine cobalt chloride and DMSO (HANAHAN, 1983).

Crucial points in the preparation of competent cells are the quality of chemicals and water, and the cleanliness of glassware and other labware used in the preparations. Several companies offer frozen competent cells that are usually of good and consistent quality but these are rather expensive. However, for experiments where high transformation rates are needed the use of such commercial preparations may be beneficial.

In recent years a new DNA-transfer technique based on the application of a high-voltage electric field to cells has been developed. This technique, called electroporation, was originally developed for eukaryotic cells but it has also become a valuable technique for the transformation of *E. coli* (DOWER et al., 1988). Transformation rates as high as 10^{10} transformants per microgram DNA have been obtained. The preparation of cell suspensions suitable for electroporation is considerably easier than the preparation of competent cells, and the cells can also be stored frozen for long periods of time. The electroporation protocol is very simple. The prepared cells and the DNA are mixed in a special electrode cuvette followed by the application of the electric field pulse.

An *in vitro* packaging system has been developed for the work with bacteriophage lambda-based vectors and cosmids. This system is based on two cell extracts of *E. coli* strains infected with special mutants of phage lambda. These lysates contain empty phage heads, phage tail components, and the proteins required for packaging of concatemeric DNA containing at least two *cos* sites. After packaging, the intact phage particles formed *in vitro* are used to infect cells of suitable *E. coli* strains susceptible to phage lambda infection. Since the latter step is highly efficient, nearly every phage particle can infect a cell giving rise to a recombinant clone. Although the packaging extracts can be prepared in any laboratory, most researchers prefer to use commercially available frozen packaging extracts which typically yield about 10^8 to 10^9 plaque forming particles per microgram of concatemerized phage DNA.

2.2 Vector Systems for *E. coli*

2.2.1 Plasmid Vectors

A small number of natural bacterial plasmids is the basis of now innumerable plasmid vectors developed for *E. coli*. These vectors are all characterized by three common features:

1. A replication system. This consists usually of an origin of replication including the necessary transcripts for priming. For some special replication systems the genetic information for transacting protein factors must also be present on the vector molecule. In general, the replication system determines the copy number of the plasmid vector. Low-copy plasmids usually exist in only a few copies per cell (<10). High-copy plasmids have a copy number of more than 20, and in special cases copy numbers of up to several hundred plasmid molecules per cell can be obtained.

2. Selectable marker(s). In most cases antibiotic resistance genes are used which allow dominant selection of cells carrying the vector molecule. The most commonly used markers are genes mediating resistance against the antibiotics ampicillin, chloramphenicol, kanamycin/neomycin, and tetracycline. For more specialized purposes other markers are occasionally also used.

3. Cloning site(s). These represent sequences on the plasmid vector which are located outside regions essential for plasmid maintenance or other required functions and contain suitable restriction endonuclease recognition sites for the insertion of foreign DNA.

2.2.1.1 General Purpose Cloning Vectors

The first plasmid vectors developed for *E. coli* were based on the pMB1 (or ColE1) replicon that is characterized by a moderate copy number. The most widely used vector of this family was plasmid pBR322. Many of the plasmid vectors in use today are descendants of these plasmids (BALBÁS et al., 1986). By removing the tetracycline resistance gene and regions involved in mobilization and copy number control, non-mobilizable vectors with a rather high copy number were developed. The most widely known series of such vectors is the pUC family. pUC plasmids contain a synthetic polylinker or multiple cloning site characterized by a tandem array of unique restriction sites that can be used singly or in combinations for cloning DNA fragments. A further advantage conferred by the use of multiple cloning sites is the option to subsequently excise cloned DNA fragments using any of the flanking sites. New restriction sites can therefore be available for subsequent cloning steps without further modification of the ends of the DNA fragment (e.g., the addition of linkers).

Vectors of the pUC family also carry a segment of the *E. coli lacZ* gene including the *lac* promoter and a small part of the coding sequence of β-galactosidase. The expressed N-terminal small polypeptide is capable of intraallelic (α-) complementation with an N-terminally truncated defective β-galactosidase polypeptide provided by a specific mutant host. The multiple cloning site is located within the coding region of the *lacZ* α-peptide. Cloning fragments into the multiple cloning site abolishes α-complementation. Since β-galactosidase activity can be easily monitored on X-Gal (5-bromo-4-chloro-3-indolyl-β-D-galactoside) plates, straight-forward identification of recombinant clones is possible. On X-Gal plates colonies of cells containing the empty vector are blue colored, whereas colonies of cells containing recombinant plasmids remain white.

The versatility of *E. coli* plasmid vectors was further enhanced by the introduction of promoters specific for bacteriophage RNA polymerases adjacent to the multiple cloning site of pUC related vectors. These promoters are de-

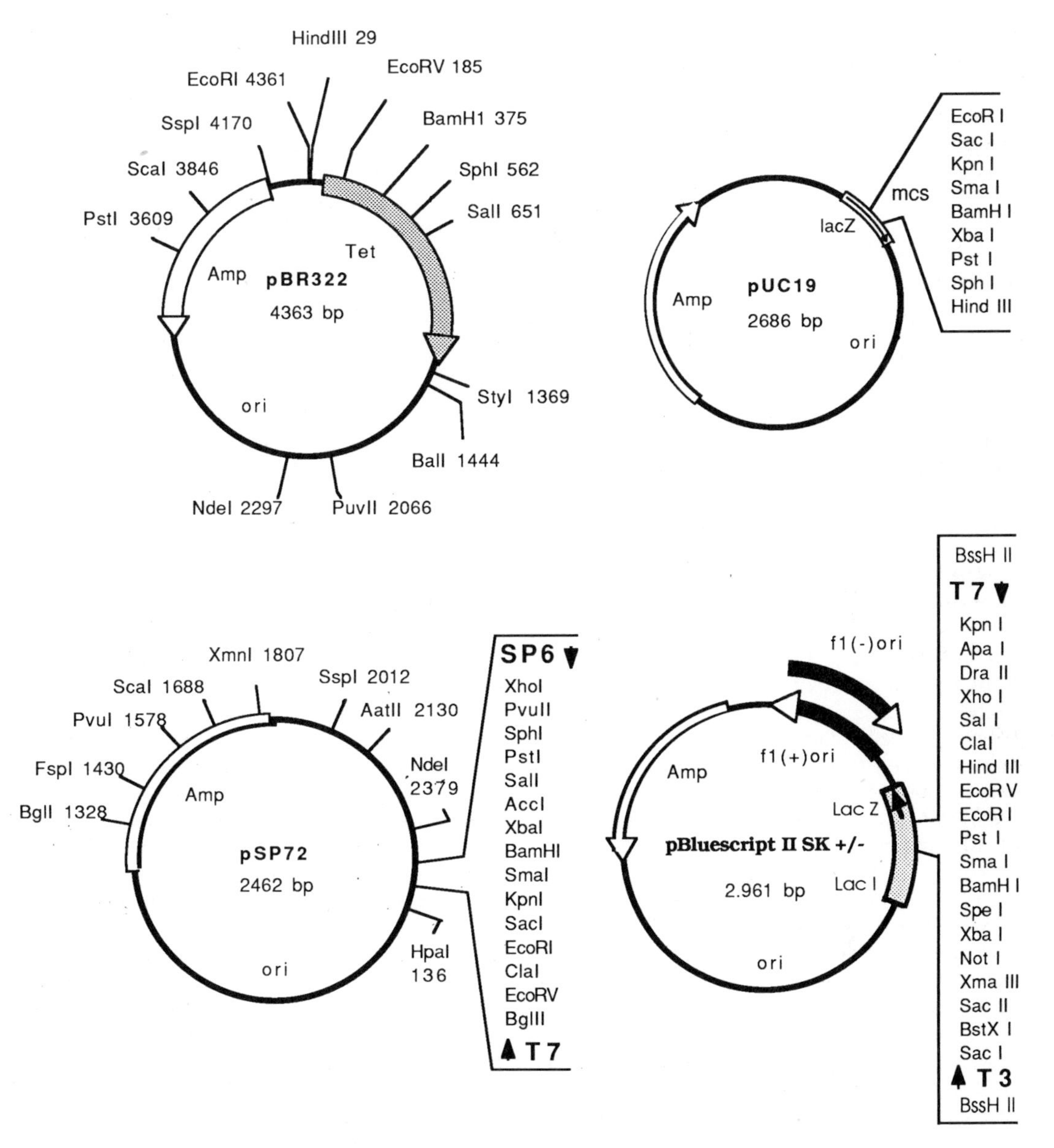

Fig. 3. Maps of general purpose plasmid vectors for *E. coli*. Amp, ampicillin resistance gene; Tet, tetracycline resistance gene; ori, origin of replication; f1(±), origin of single-strand replication derived from bacteriophage f1.

rived from bacteriophages T3, T7, or SP6. Usually a combination of two of these promoters is used which permits the specific *in vitro* transcription of both strands of a cloned DNA fragment.

Another improvement has been achieved by the introduction of a DNA fragment carrying the replication origin of a single-stranded bacteriophage such as M13 or f1. This allows the direct production of single-stranded DNA of one of the two strands of the recombinant plasmid thereby eliminating the need to subclone DNA fragments into phage M13 vectors (see below). Examples for each type of general purpose cloning vectors currently in use are shown in Fig. 3.

2.2.1.2 Cloning Vectors for Specific Purposes

A variety of vectors have been developed to solve specific problems. Although the vectors carrying phage promoters and single-strand origins were designed to fulfill particular requirements, these vectors are classified as general purpose vectors since they are widely used for all types of cloning work. The following types of special purpose vectors are important (vectors for expression of genes will be discussed in Sect. 6.2):

1. Probe vectors. These vectors are specifically designed for cloning DNA fragments coding for a certain gene function. Most commonly used vectors of this type are promoter probe vectors. In principle, a promoterless reporter gene is placed behind a suitable cloning site (frequently an mcs). Cloning of fragments into this site allows direct selection of promoter-active DNA fragments. Similar strategies have also been applied to the cloning of terminator sequences. Commonly used probe vectors are summarized in BALBAŚS et al. (1986) and BROSIUS and LUPSKI (1987).

2. Direct selection vectors. Various systems have been designed for the direct selection of recombinant clones. Such systems rely on the disruption of either genes being toxic under certain conditions or a negative regulator controlling the expression of an antibiotic resistance gene. The cloning sites are located within the toxic gene or the regulator gene, and by cloning fragments into these sites the gene functions are destroyed. This results in the ability of recombinants to grow under the conditions toxic for *E. coli* containing the intact toxic gene or, respectively, in the presence of the appropriate antibiotic. Although this strategy of direct selection for recombinant clones seems to represent a powerful tool, the disadvantages of such systems usually outweigh the advantages. The vectors usually are less well developed, and the use of specific hosts that is required very often results in low cloning efficiencies. Some vector systems of this type have been described by BALBÁS et al. (1986), BURNS and BEACHAM, (1984), and NILSSON et al. (1983).

3. "Reporter" vectors. Such vectors contain a reporter gene that is easy to monitor. Most commonly used reporter genes are the *E. coli lacZ* (β-galactosidase) gene, the *E. coli phoA* (alkaline phosphatase) gene and, as a recent development, the *lux* (luciferase) genes of bioluminescent bacteria (MEIGHEN, 1991). Useful cloning sites in front of such reporter genes allow the insertion of DNA fragments to be analyzed for expression activity. Vectors for the construction of both transcriptional and translational fusions have been developed. Translational fusions with reporter genes that are only expressed when transported into the periplasmic space (e.g., *phoA*) permit the detection of secretion-mediating signal sequences or even allow the analysis of membrane protein topology (EHRMANN et al., 1990).

4. Vectors for other specific purposes. In some cases it is desirable to use low-copy vectors, especially when a cloned DNA fragment encodes gene products harmful to the host cells. Low-copy vectors have been developed using mainly the replicons of plasmids F and R1 and the origin of replication of the *E. coli* chromosome. Recently, a low-copy vector based on the F-replication system was developed for the cloning of very large DNA fragments (more than 100 kb) (LEONARDO and SEDIVY, 1990). A further group of vectors useful for complementation studies are based on different compatible replication systems. Vectors such as pACYC177 and pACYC184 (CHANG and COHEN, 1978) based on the p15A replicon are, for example, compatible with plasmids based on the pMB1 (ColE1) replication system.

2.2.2 Bacteriophage Vectors

2.2.2.1 Vectors Based on Bacteriophage Lambda

The biology of bacteriophage lambda has been extensively studied. Based on this knowledge a variety of lambda vectors have been developed. These vectors are mainly used for the construction of gene libraries and offer several advantages for this purpose. The high efficiency of DNA transfer via *in vitro* packaging and phage infection represents a very important feature for the construction of gene libraries. Further advantages are provided by the possibility of using powerful direct selection systems for recombinant phages and by the possibility of efficient cloning of large DNA fragments. Selection systems for inserted DNA include

"size selection": lambda DNA cannot be packaged when the size is below 78% or above 105% of the length of wild-type lambda DNA. In addition, the use of *E. coli* hosts carrying the *pel* mutation severely inhibits lytic growth of phages of less than wild-type size,

"spi selection": the *red* and *gam* genes are present on the central stuffer fragment; removal allows growth on P2-lysogenic host strains, and

"*hfl*-selection": the cloning site is located within the *cI* gene. With an *hfl* host the vector cannot form plaques, only recombinant phages with the inactivated *cI* gene can form plaques.

Also the possibility of maintaining specific phage vectors in the lytic as well as in the lysogenic stage is useful for some applications. This is achieved by the use of a thermosensitive mutant allele of the *cI* repressor allowing lysogeny at low temperatures (30 °C) and induction of lytic growth by temperature shifts to 42 °C. Lambda vectors are also very useful with respect to screening by hybridization techniques. Plaques contain enough DNA of the respective sequences to be probed, and, therefore, no extra treatments to release DNA from clones are necessary.

Two different types of lambda-vectors can be distinguished:

- Replacement vectors are characterized by a central stuffer fragment flanked by regions containing proper restriction sites. The stuffer fragment contains regions of bacteriophage lambda that are not essential for lytic growth (about 1/3 of the lambda genome can be deleted). This central stuffer fragment is removed and can be replaced by foreign DNA allowing the insertion of fragments of up to 22 kb.
- Insertion vectors contain a single cloning site which usually includes several useful restriction sites for the insertion of DNA fragments. Smaller DNA fragments of up to 10 kb can be cloned with such vectors.

Prototypes of lambda replacement vectors are represented by EMBL3 (FRISCHAUF et al., 1983) and Charon4A (WILLIAMS and BLATTNER, 1979). These vectors have been widely used to construct genomic libraries, and EMBL3 is still in use. More recent developments in the construction of replacement vectors include the introduction of more restriction sites at the ends of the stuffer fragment including sites for rare cutting restriction enzymes such as *Sfi*I or *Not*I. This allows the excision of cloned fragments with a low probability of cutting the fragment at internal restriction sites. The further addition of bacteriophage promoters (T3, T7, SP6) adjacent to the cloning sites allows strand-specific transcription of the inserts directly from isolated DNA of recombinant lambda clones. Examples of currently used lambda replacement vectors are shown in Fig. 4.

Lambda gt10 (HUYNH et al., 1985) can be regarded as a prototype of insertion vectors. It contains a unique *Eco*RI site in the *cI* gene. In a derivative thereof, lambda gt11 (HUYNH et al., 1985), the *E. coli lac*Z gene has been introduced and a unique *Eco*RI site at the C-terminal end of the β-galactosidase coding region allows direct expression of translational fusions obtained by cloning cDNA inserts into this site. Insertion vectors, which have become more convenient to be used as additional restriction sites, including those for rarely cutting enzymes, are added. In a recent development, an entire plasmid vector containing a single-strand replication origin was inserted into phage lambda. These vectors (lambda ZAP) allow cloning of fragments into the polylinker of the inserted plasmid. Upon co-infection with M13 or f1 helper phages, the entire plasmid including the insert can be excised *in vivo*. This avoids any subcloning of fragments into plasmid vectors. Examples of lambda insertion vectors are shown in Fig. 4.

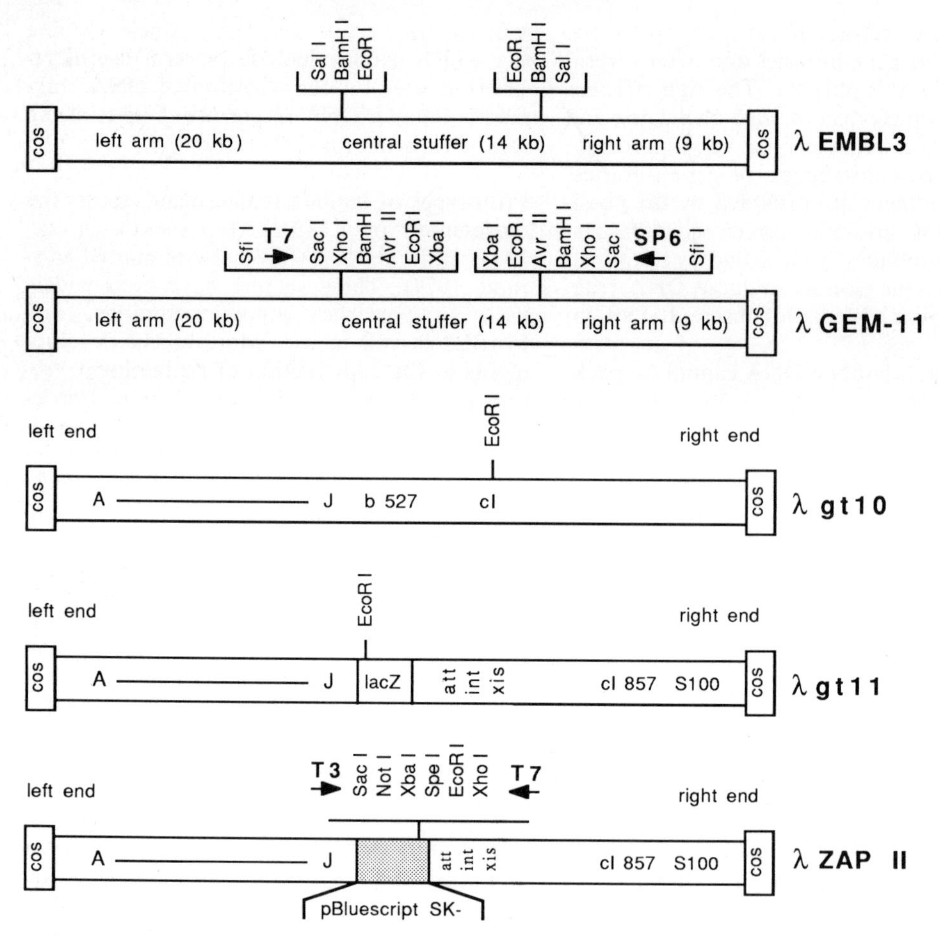

Fig. 4. Schematic maps of bacteriophage lambda vectors. For further explanation refer to the text. Lambda vectors GEM-11 and ZAP II are products of Promega and Stratagene, respectively.

2.2.2.2 Vectors Based on Single-Stranded Filamentous Bacteriophages

The development of vectors based on the filamentous phage M13 has greatly facilitated genetic engineering. M13 has a 6.4 kb genome which can exist in two forms. The replicative form consists of covalently closed circular double-stranded DNA and behaves like a plasmid. Its DNA can be isolated from infected cells by standard plasmid DNA isolation techniques and can be used in all cloning procedures in the same way as plasmid DNA. For packaging into phage particles, circular single-stranded DNA is produced in a special replication process from one specific strand. This DNA can be obtained by isolating DNA from phage particles after removal of host cells.

Prototypes of the vectors based on filamentous phages are the M13mp vectors (YANISH-PERRON et al., 1985). In principle, these vectors consist of the entire M13 genome into which a multicloning site and the *lacZ'* region as present in pUC plasmids (see Sect. 2.2.1.1) are inserted. Pairs of vectors with the polylinker inserted in opposite directions are available, allowing the insertion of DNA fragments in both orientations with respect to the single-strand replication origin and, therefore, production of single-stranded DNA of both strands of this fragment. Currently the M13mp vectors have been largely replaced by plasmids containing only the origin for single-strand replication (see Sect. 2.2.1.1). The latter support the production of single-stranded DNA in the same way as the M13 vectors, but are more convenient for cloning work.

2.2.3 Cosmid Vectors

Cosmids are hybrids between plasmid- and bacteriophage vectors and thus combine benefits of both systems. In principle, cosmid vectors are plasmids containing the genetic information necessary in *cis* for *in vitro* packaging of DNA into phage lambda heads. Only a short region of a few hundred base pairs of bacteriophage lambda including the *cos* site is needed. Since the plasmid part containing an origin of replication, a selectable marker and a proper cloning site, can be kept rather small, cosmids are capable of carrying large fragments of DNA (up to about 50 kb). Cosmids can be propagated in *E. coli* and handled *in vitro* as plasmids. For cloning large DNA fragments, the recombinant DNA can be packaged *in vitro* into phage lambda heads under the same conditions as needed for lambda vectors: a specific size range of the molecules (78–105% of wild-type lambda) and the presence of two copies of the *cos* site at the ends. Therefore, either concatemers consisting of vector-insert repetitions or constructions starting from cosmids having two copies of the *cos* site have to be used for packaging. Following packaging, the DNA can be transferred into *E. coli* by the efficient phage infection mechanism. After infection the DNA is circularized in the host cell, and the recombinant molecule represents a plasmid.

Two types of cosmid vectors have been developed:

(1) Vectors for propagation of large DNA fragments in *E. coli*. A prototype for such vectors is pJB8 (ISH-HOROWITZ and BURKE, 1981) which contains the pMB1 replicon, an ampicillin resistance gene, the *cos* site and a cloning site with several restriction sites. Further improvements of such vectors include the introduction of bacteriophage promoters adjacent to the cloning site allowing the *in vitro* generation of (labeled) transcripts of both ends of the insert, the incorporation of two *cos* sites (BATES, 1987), and the replacement of the plasmid origin of replication by a bacteriophage (lambda) origin of replication which is supposed to circumvent the stability problems (GIBSON et al., 1987). The general structure of a cosmid containing two *cos* sites and the principle of cloning with this vector are shown in Fig. 5.

(2) Vectors for shuttling cloned DNA fragments between *E. coli* and other prokaryotic or eukaryotic organisms. In principle, such cosmids contain, in addition to the elements needed for propagation in *E. coli*, selection markers and replicons for the target organisms. A large variety of these cosmids have been developed for various hosts. They are mainly used for complementation and rescue experiments.

2.3 Construction of Gene Libraries

Gene libraries are a vital resource in genetic engineering projects. Such libraries, once constructed, serve as a continuously available source for the isolation of specific genes or gene fragments. In principle, a gene library represents a pool of recombinant clones where the total genetic information of an organism, or parts thereof, is stored as small DNA fragments that are cloned into vector molecules. So far almost all libraries have been established using *E. coli* host-vector systems. However, new developments for cloning in yeast may play an important role in the future. The use of artificial chromosomes allows the cloning of fragments up to a size of several hundred base pairs. Initial attempts to construct libraries of higher eukaryotes in yeast have already been successful (BROWNSTEIN et al., 1989).

Generally there are two types of libraries, genomic DNA libraries and cDNA libraries. A general scheme of steps involved in the preparation of gene libraries is presented in Fig. 6. Since at the present time numerous libraries of

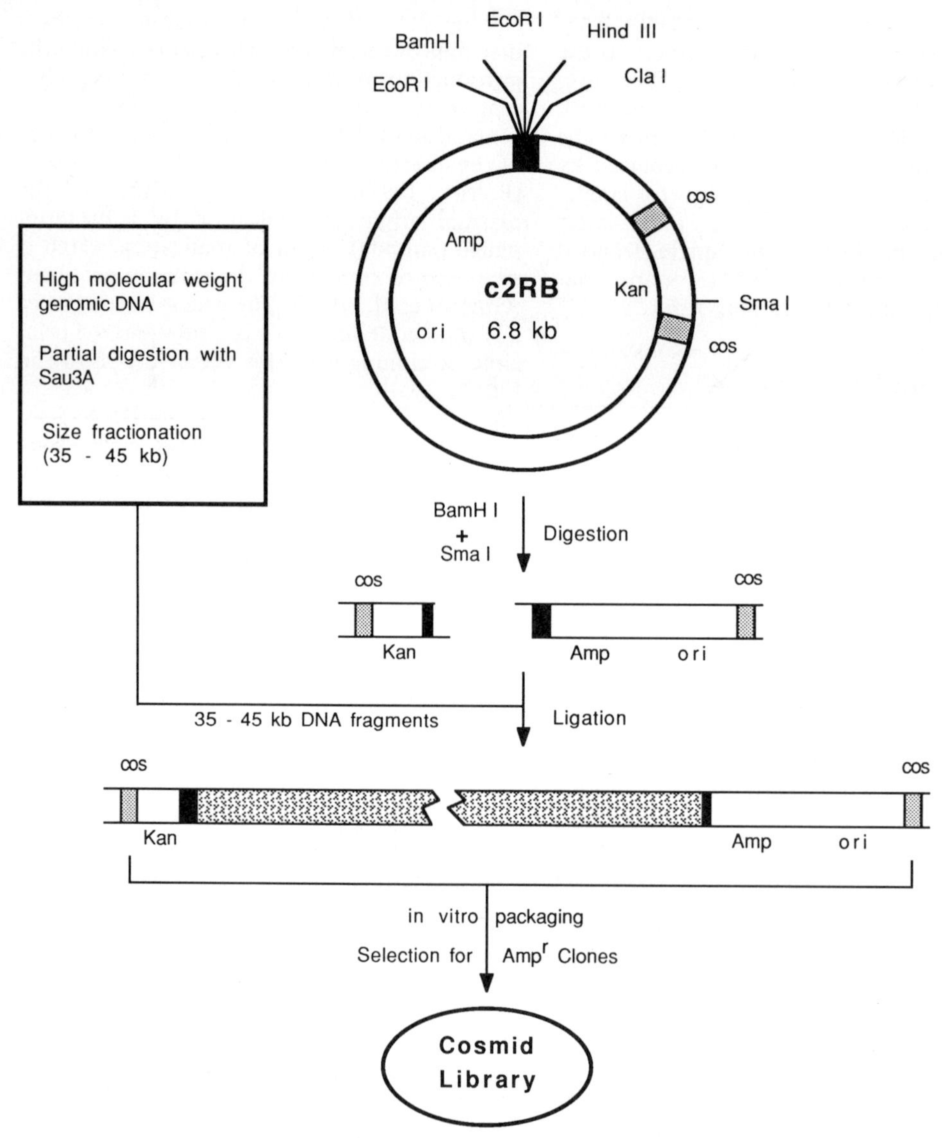

Fig. 5. Strategy for the construction of cosmid libraries using a double *cos*-site vector system according to BATES and SWIFT (1983). Amp, ampicillin resistance gene; Kan, kanamycin resistance gene; ori, origin of replication; cos, *cos* site of bacteriophage lambda.

a variety of organisms have been constructed, a desired library may be readily obtainable from a colleague. Furthermore, some companies already sell pre-made (mainly cDNA) libraries, and some companies also offer the construction of libraries for a fee.

2.3.1 Genomic Libraries

Genomic libraries are designed to contain the entire genome of an organism. In some cases libraries of specific parts of the genome

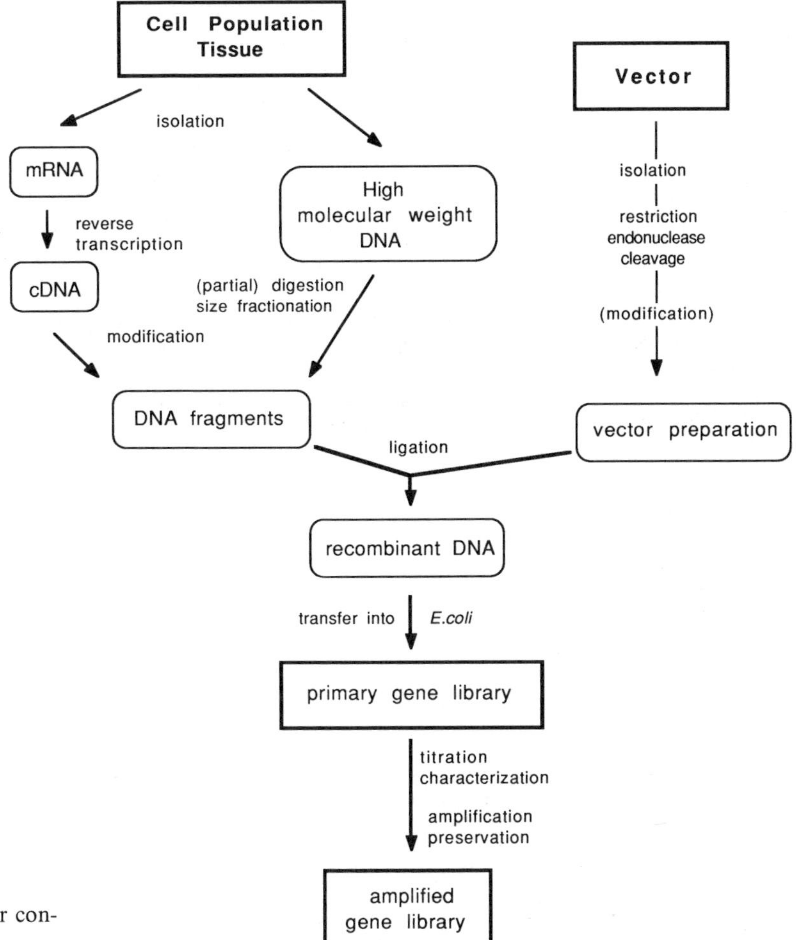

Fig. 6. General scheme for constructing gene libraries.

(e.g., specific chromosomes) or specific genomic fragments can be cloned to obtain a subgenomic library. To ensure that any specific gene can be isolated from a genomic library, overlapping fragments of isolated total DNA of an organism are cloned. The overlapping fragments are usually obtained by partial digestion with a frequently-cutting restriction endonuclease (e.g., *Sau*3A). The size of the fragments that are prepared from isolated high-molecular weight DNA by partial digestion and subsequent size fractionation, depends on the vector system used for the construction of the library.

The likelihood of finding a specific DNA fragment of interest (gene) in a genomic library can be estimated according to the following equation (CLARKE and CARBON, 1976).

$$N = \ln(1-P)/\ln(1-f)$$

N is the number of clones that must be screened to find a specific sequence with a probability of P. The factor f represents the fraction of the total genome present in the insert of a single recombinant of the library. f is therefore the quotient of the insert size divided by the size of the total genome. This equation is based on the assumption that the cloned DNA fragments randomly represent the sequences present in the genome. However, several facts have to be considered for practical work:

- The sites for a restriction enzyme are in reality not used with equal efficiency. Sequences immediately adjacent to restriction sites can influence the cleavage efficiency.
- Some fragments may contain sequences harmful to the vector or to the host cells and are therefore unclonable or underrepresented, especially after amplification.
- During isolation of genomic DNA, mechanical shearing introduces breaks into the DNA generating smaller fragments. If the size of these fragments comes close to the size of DNA fragments accepted by the vector, many fragments in the selected size range do not contain the restriction enzyme-generated compatible sticky ends at both termini and therefore cannot be cloned.

Bacteriophage lambda vectors are widely used for constructing genomic DNA libraries because they offer a high cloning efficiency. Moreover, the newer vectors are rather simple to use and provide various advantages (see Sect. 2.2.2). In addition, there are few problems with respect to genetic instability during propagation of lambda recombinants. Fragments of 15–20 kb can be cloned with these vectors. This size is sufficient for most purposes, including the cloning of various genes of higher eukaryotes.

However, some genes of higher eukaryotes may comprise rather long stretches of DNA reaching a size of more than 100 kb. Therefore, the possibility of cloning larger fragments is desirable. In that case, fewer independent clones would be needed for a representative library, and the probability of obtaining a more complex large gene within a single recombinant is much higher. Despite the advantages of efficiently cloning large DNA fragments (up to 45 kb), cosmid cloning (HOHN and COLLINS, 1988) represents a more difficult technique and introduces various problems. The most significant problems are instability of foreign DNA fragments upon propagation in *E. coli* hosts and a wide variation in the growth rate of clones containing recombinant cosmids resulting in over- or underrepresentation of certain sequences in amplified libraries. Cosmids are therefore used only when very large genes have to be cloned or when transfer of recombinants of the library to eukaryotic hosts (e.g., for complementation) is desired.

For the cloning of genes from prokaryotes, libraries constructed with plasmid vectors may be useful. Since the genome size of bacteria is rather small, and bacterial genes do not contain introns, the size range of fragments clonable with plasmid vectors (fragments of about 10 kb can be easily cloned when using carefully size-selected fragments) is sufficient to encode the largest bacterial genes and to construct representative libraries. Only a few thousand clones are needed to obtain a 99% probability of finding a specific gene in the library.

2.3.2 cDNA Libraries

In cDNA libraries double-stranded DNA copies of mRNA molecules are cloned. In contrast to genomic libraries, cDNA libraries contain only a specific selection of genes, depending on the source used for the isolation of RNA. There is also no statistical distribution of genes present in such libraries since mRNA species are present in specific cells in varying abundance. cDNA libraries reflect the specific expression patterns of cells. However, this can be an advantage since the abundance of a certain mRNA species can be specific to the tissue type, to the growth conditions used when culturing cells, or to the developmental stage of the cell. By isolating RNA from appropriately selected cell material it is possible to construct cDNA libraries that are enriched in copies of the gene of interest.

cDNA synthesis and cloning of cDNA still require sophisticated techniques. However, in recent years many commercial suppliers of molecular biology reagents have developed complete reagent kits that make cDNA synthesis and cloning more efficient and simple. Fig. 7 shows a schematic representation of presently used cDNA cloning strategies. The most critical points are first, the quality of the RNA preparation and second, a careful size selection of cDNA molecules after synthesis prior

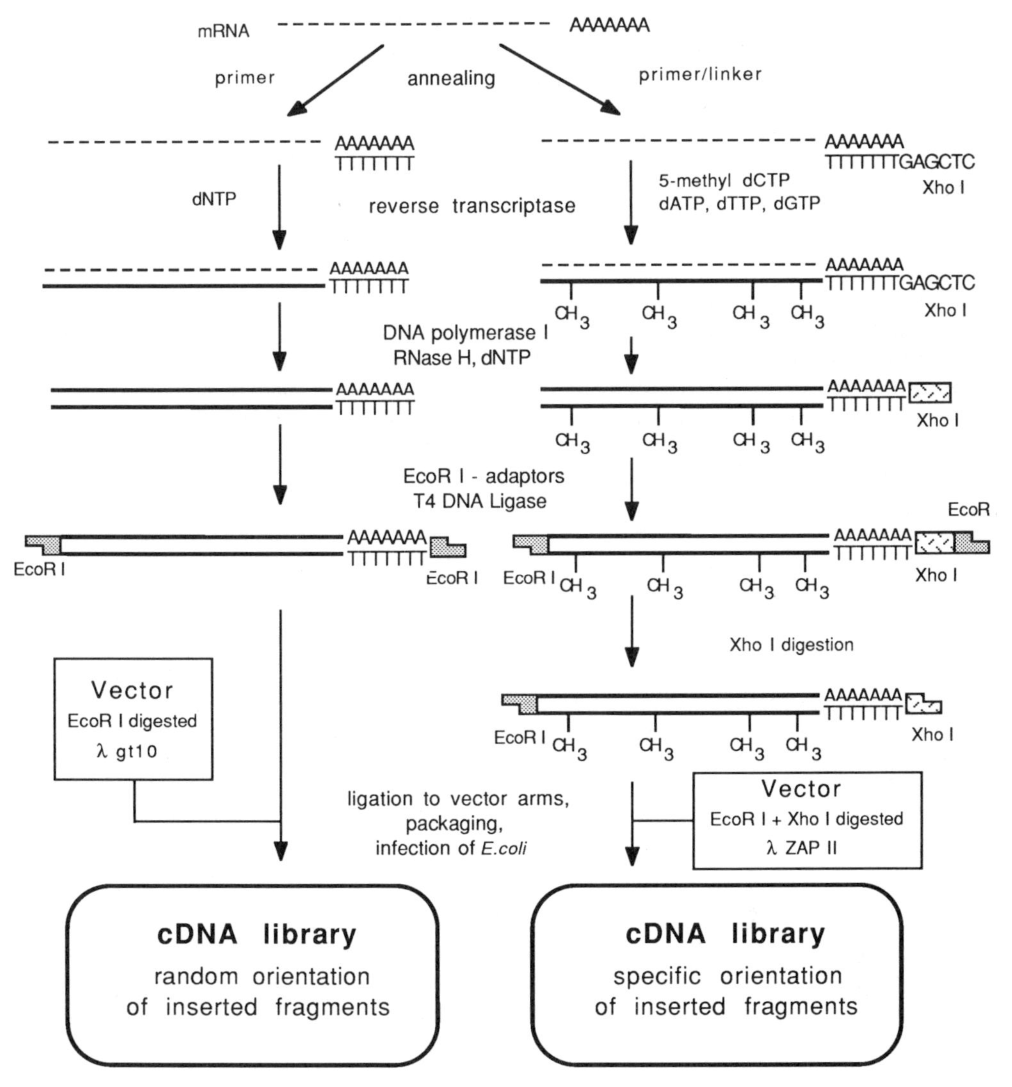

Fig. 7. Schematic representation of the construction of cDNA libraries with random (left side) or specific (right side) orientation of the inserted fragments.

to ligation to the vector. Efficient methods for the isolation of mRNA have already been developed for various types of cells and tissues (SAMBROOK et al., 1989; AUSUBEL et al., 1987; BERGER and KIMMEL, 1987). Attention must be directed to the elimination of nuclease activities. It has been shown that the use of total RNA instead of poly(A)$^+$ RNA purified by oligo(dT) cellulose chromatography is also reliable and does not result in a significantly higher proportion of rRNA-derived clones.

First strand cDNA synthesis is performed by AMV reverse transcriptase or MMLV reverse transcriptase. These enzymes are available as high-quality preparations that promote efficient synthesis. In our hands MMLV reverse transcriptase proved to be better suited for the efficient synthesis of full-length cDNAs. In the second strand reaction, nearly all protocols currently used follow the principle of replacement synthesis catalyzed by RNase H and *E. coli* DNA polymerase I (Okayama and Berg, 1982; Gubler and Hoffman, 1983). Cloning into a vector is usually performed with the aid of synthetic linkers and adapters. The use of specific primer adapters allows the orientation-specific directional cloning of cDNAs. *In vitro* methylation of cDNA or the use of methylated nucleotide analogs (e.g., 5-methyl dCTP) in cDNA synthesis allows the generation of cohesive ends for cloning by restriction endonuclease digestion without destroying the cDNA molecules at internal recognition sequences.

Since the synthesis of cDNA needs many steps and very often only a limited amount of RNA is available, a high efficiency in the cloning step should be achieved. The use of bacteriophage lambda insertion vectors is therefore recommended for cDNA cloning. In particular, the lambda ZAP vectors provide a very convenient system, since the cloned inserts can easily be obtained as plasmid clones by *in vivo* excision. Prior to cloning the cDNA into a vector, a size selection step by exclusion chromatography or agarose gel electrophoresis is essential for obtaining libraries containing clones with sufficiently large inserts.

2.4 Cloning of Chemically Synthesized Genes

The strategies and the methodology for chemical DNA synthesis are discussed in detail in Chapter 9 of this volume. Chemical DNA synthesis is mainly applied to the synthesis of short oligonucleotides used as primers, linkers, adapters, and probes for screening. However, the synthesis of entire genes based on known amino acid sequences may represent a useful strategy for the construction of clones meant to express a specific polypeptide. Using this procedure, genes can be specifically designed to accommodate the requirements for efficient expression specific to the host organism used for production (e.g., codon preference).

Several strategies for the synthesis of larger DNA fragments can be applied:

- Synthesis of overlapping oligonucleotides covering the entire sequence of interest on both strands. The gene is obtained by annealing and subsequently ligating these oligonucleotides. The resulting double-stranded DNA fragment can be cloned into a proper vector molecule. The cohesive ends for cloning can be included in the design of the oligonucleotides.
- As a variation of this strategy, the subfragments obtained by annealing respective pairs of oligonucleotides can be cloned separately into vectors. After amplification, these fragments are cut out and ligated to obtain the entire gene. Such a strategy is useful when the gene consists of several blocks that are to be altered independently in different constructs.
- Fill-in strategy. This strategy is based on synthetic oligonucleotides covering only parts of the sequence of both strands, but overlap for a short region at the ends. Following annealing of the oligonucleotides the remaining gaps are filled in enzymatically.

2.5 Subcloning and Modification of DNA Fragments

Subcloning of specific fragments from larger cloned DNA fragments (e.g., clones of a gene library) is frequently desired. In addition, to obtain fragments with particular features cloned DNA may require modification. General purpose plasmid vectors as described in Sect. 2.2.1.1 are usually used for these tasks. Specific fragments are typically obtained by restriction endonuclease cleavage of the original DNA fragment. The desired fragments can be isolated from agarose or in the case of very small fragments from polyacrylamide gels. In an alternative strategy, all fragments produced by restriction enzyme digestion are cloned into a vector. Transformants containing the cloned DNA fragment of interest are identified by restriction analysis of isolated plasmid DNA (mini-preparations). When two different restriction endonucleases are used to excise a fragment, directional cloning of that fragment is possible. In many cases it is necessary to modify the isolated fragments to generate suitable ends for further cloning steps. Various possibilities are described in Sect. 2.1.3.

Sometimes it is of interest to obtain a set of subfragments that are shortened progressively from one end. Such fragments can be obtained by exonuclease digestion under controlled conditions. Two systems can be applied: nuclease *Bal*31 or exonuclease III in combination with a single-strand-specific nuclease (see Tab. 2). The second strategy, developed basically by HENIKOFF (1984), has the advantage that 4 base 3′-protruding ends are protected from exonuclease III digestion. By cutting the recombinant plasmid containing the DNA fragment with two restriction enzymes, one generating a 3′-recessed end and the other generating a 3′-protruding end (such sites are usually available in the polylinker sequences) the direction of action of the exonuclease can be determined. Following degradation of the single-stranded region by S1 (or mung bean) nuclease the resulting blunt ends can either be directly religated, or a synthetic linker (unphosphorylated) is introduced at the deletion end point.

2.6 Polymerase Chain Reaction (PCR)

PCR is a rather simple but extremely efficient and fast method for *in vitro* enzymatic amplification of specific segments of DNA. Two oligonucleotide primers that flank the region of interest and that are complementary to opposite strands are used to direct polymerization of the desired DNA sequence. Extremely small quantities of template DNA, even just a few molecules in one reaction, can be amplified with this technique making PCR highly suited for the development of extremely sensitive methods for detecting specific DNA sequences. The principle of PCR is shown in Fig. 8. The PCR reaction was originally developed with the Klenow fragment of *E. coli*

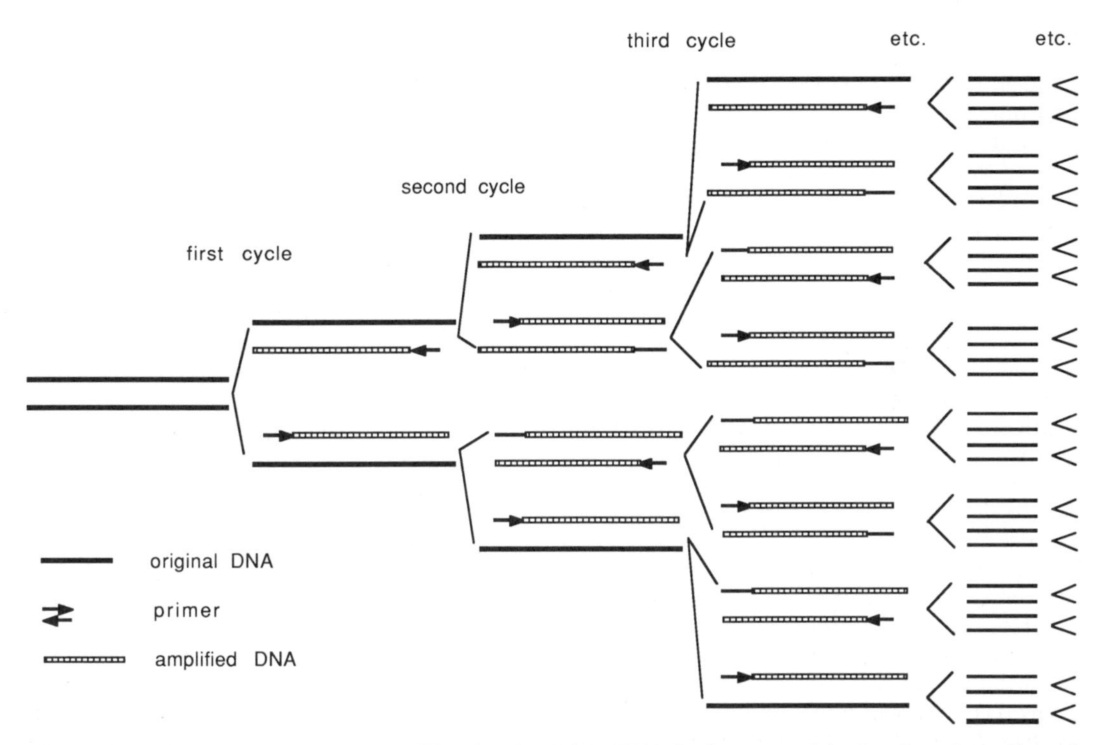

Fig. 8. General principle of DNA amplification by PCR. DNA is denaturated by heating. Annealing of specific primers and DNA synthesis are performed at lower temperatures. The use of a thermostable DNA polymerase allows the initiation of new rounds of replication by just running a temperature program which passes through the needed temperatures for denaturing, annealing, and synthesis. In each cycle a duplication of the sequences between the primer pair is achieved resulting in an exponential accumulation of the corresponding DNA fragments.

DNA polymerase I, however, fresh enzyme had to be added after each cycle at elevated temperature. A breakthrough was achieved by the introduction of a thermostable DNA polymerase from *Thermus aquaticus* (*Taq* DNA polymerase) (SAIKI et al., 1988). The thermostability of this enzyme makes an automated run of 20 to 30 cycles possible, resulting in many million-fold amplification of the specific DNA segment.

Various powerful applications for gene cloning exist for PCR (WHITE et al., 1989; ERLICH, 1989; INNIS et al., 1990). Some important methods are:

- Addition of useful ends to DNA fragments. This is achieved using specifically designed oligonucleotides which contain the desired end sequences.
- Alteraltion of specific bases in a sequence. Mismatches are incorporated into the sequence of the oligonucleotide primer.
- Amplification of cDNA. After the first strand cDNA has been synthesized, oligo(dG) is added at the 3′-ends of the cDNA. Oligo(dT) and oligo(dC) primers which contain an appropriate restriction site for cloning are then used in the amplification step.
- Production of single-stranded DNA (asymmetric PCR). One of the two primers is present at a limiting concentration resulting in the production of single-stranded DNA for sequencing or for analysis of DNA–protein interactions (footprinting).
- Cloning of unknown sequences adjacent to known sequences. Genomic DNA is cut with a restriction enzyme and the fragments circularized by ligation. Unknown sequences can be synthesized with primers that are complementary to the ends of the known sequence (OCHMAN et al., 1988).

A recent development takes advantage of the ability of the thermostable DNA polymerases from *Thermus aquaticus, Thermus flavus,* and *Thermococcus litoralis* to generate 3′-dA extensions. A specific vector which produces single dT overhangs upon cleavage with *Hph*I permits direct cloning of PCR-amplified fragments (MEAD et al., 1991).

For PCR applications where the amplified DNA is cloned and intended for further use it should be taken into account that the *in vitro* synthesis of DNA by *Taq* polymerase is characterized by a high rate of misincorporation (about 2×10^{-4}). In a 30-cycle reaction the overall error frequency is in the range of 0.25% (SAIKI et al., 1988). Any sequence obtained by PCR should therefore be confirmed by sequencing.

3 Identification of Specific Genes

The design and execution of screening procedures to identify specific DNA fragments containing the desired genetic information is one of the most crucial tasks of genetic engineering. Screening procedures may consume a large percentage of the manpower and financial resources available for a specific project. *E. coli* systems are employed for this task in most cases, although shuttling of clones to strains of the donor organism may be included in various strategies.

3.1 Selection for an Expressed Phenotype

These screening strategies are based on expression of gene functions encoded in the fragments of a gene library. These gene functions may either confer new capabilities to the host cells, or complement genetic defects in specific mutants used for detection.

3.1.1 Direct Selection in *E. coli* Hosts

Direct selection in *E. coli* represents the simplest method for the identification of specific clones. Strategies based on this approach are mainly used for the cloning of specific genes from other prokaryotic organisms. In this case, plasmid vectors that contain promoters adjacent to the cloning sites are usually used for the construction of the gene libraries. Cosmid libraries are also very useful since only a comparatively small number of clones need to be screened. In some cases libraries constructed with bacteriophage lambda vectors are also useful since genes under the control of an appropriate promoter included in the vector are also expressed from the lambda genome. Active gene products are therefore present in the plaques and can be detected by specific selection techniques. The clones of the gene library are simply plated onto selective media and, after incubation, positive colonies or plaques can be identified either by conferring the ability to grow or by mediating specific reactions on the screening plates. Such strategies have been successfully applied to cloning of various bacterial genes encoding enzymes of commerical interest (for some examples refer to DIDERICHSEN and CHRISTIANSEN, 1988; FRIEHS et al., 1986; REDDY et al., 1989; SPÖK et al., 1991).

A variety of genes have been isolated by complementation of specific mutants (e.g., auxotrophic mutants) of *E. coli*. This strategy has been successfully used to screen for various genes encoding enzymes of biosynthetic pathways, including numerous bacterial, and even a few fungal genes involved in amino acid biosynthesis (MARTIN, 1989).

3.1.2 Selection after Transfer to Specific Hosts

In many cases gene functions are not expressed or cannot be detected in *E. coli*. An alternative is provided by transferring clones of a library to other hosts that enable expression and detection of the encoded gene functions. The most commonly used strategy is to complement specific mutants of the organism from which the desired genes are to be cloned. An additional strategy would be the transfer of recombinant DNA from libraries to heterologous, but related host organisms that do not contain the specific genetic information, but that allow expression and detection of the respective gene functions. Both strategies have been widely used to clone a variety of genes of prokaryotic as well as of eukaryotic organisms. For example, such strategies were used to clone various genes involved in antibiotic biosynthesis from *Streptomyces* spp. (HUNTER and BAUMBERG, 1989). Cosmid libraries based on vectors that contain selection markers and systems for maintenance (e.g., origin of replication) in the selection host can be used (RAO et al., 1987). However, such strategies frequently suffer from poor rates of DNA transfer to the selection hosts particularly when certain mutant strains are used. Vectors based on mobilizable broad-host-range plasmids have proven to be very useful for applications with many Gram negative bacteria that cannot be efficiently transformed. Cosmid vectors based on broad-host-range plasmids have also been developed (FREY et al., 1983) for these purposes. For example, several genes involved in cobalamin (vitamin B 12) biosynthesis could be isolated by mobilization of recombinant clones to *Pseudomonas putida* and *Agrobacterium tumefaciens* mutants deficient in various steps of cobalamin biosynthesis (CAMERON et al., 1989).

3.1.3 Tagging

Tagging represents a technique where detectable DNA sequences (tag) are physically linked to sequences of interest. A library is then constructed from DNA isolated from the strain carrying the tagged gene. Since the tag sequences flank sequences belonging to the genes of interest these can then be identified by screening the library for clones containing the tag sequences. The screening is usually performed by efficient hybridization techniques (see below). An alternative method is to use tag sequences that contain functions necessary for *E. coli* vectors. In this case the flanking sequences can be recovered by simple transfer into *E. coli* of cleaved and religated DNA isolated from cells containing the tagged gene of interest. Any established recombinant clone should contain sequences of the desired gene (rescue techniques). Rescue strategies can be efficiently performed by using cosmid vectors. The isolated sequences usually do not contain a complete functional copy of the gene of interest (inactivated by tag sequences or truncated by rescuing), therefore, these can be used in a second procedure to screen a library constructed with DNA of the wild-type organism.

Tagging can be achieved by the following general approaches:

- Inactivation of a gene function by insertion of foreign sequences which simultaneously function as tag sequences (*in vivo* tagging).
- *In vitro* ligation of tag sequences to genomic DNA fragments (*in vitro* tagging) and transformation of a mutant defective for the desired gene function with the tagged DNA fragments. In phenotypically complemented cells the tag sequences are linked to sequences of the desired gene.

In Gram negative bacteria *in vivo* tagging can be efficiently performed by the use of transposons. A system based on the mobilization functions of a broad-host-range plasmid (SIMON et al., 1983) can be used in a wide variety of Gram negative bacteria. For example, the genes for the biosynthesis of a biodegradable polymer, poly-β-hydroxybutyrate (PHB), was isolated by this strategy (SCHUBERT et al., 1988).

In vitro tagging has been successfully used to clone various genes of higher eukaryotes (KINGSMAN and KINGSMAN, 1988). *In vivo* tagging with higher eukaryotes can be performed with retroviral vectors (GOFF, 1987). Transposable elements of lower eukaryotic organisms such as the *Ty* element of *Saccharomyces cerevisiae* or the P element of *Drosophila* have also been used for cloning genes of these organisms by tagging strategies.

Although tagging represents a very powerful strategy, it has to be taken into account that recombinants isolated by such a technique may represent sequences not related to the desired gene. With higher eukaryotes, usually many copies of a DNA fragment introduced into cells are randomly integrated into the genome accompanied by rearrangements. Therefore, the tag sequences can be connected to other sequences in the same cell. Problems also arise when strategies of insertional gene inactivation with transposons or retroviral vectors are used since multiple insertion events often occur with these elements. However, all clones containing tagged sequences taken together will represent a small library which is highly enriched for the gene of interest. With a few further steps of retransformation and reselection it is usually fairly straightforward to find the correct gene sequences.

3.2 Hybridization Techniques with Nucleic Acid Probes

Hybridization procedures using labeled nucleic acid probes are a very efficient means of screening libraries for specific recombinant clones sharing sequence homology to the probe sequences. In principle, DNA from plaques or colonies of clones of a library is transferred to solid supports (nitrocellulose or nylon membranes). After fixing the DNA to the membranes (baking at 80 °C for nitrocellulose membranes, UV crosslinking for nylon membranes) and blocking unspecific binding sites, the membranes are incubated with a labeled nucleic acid probe under conditions which permit only the specific binding of the probe to DNA sharing sequence homology. The stability of such hybrids depends mainly on the relatedness of the sequences of the nucleic acids, but also on the length and the type (DNA or RNA, G + C content) of the hybridizing sequences. After washing away unspecifically bound probe material, plaques or colonies containing the correct DNA sequences can be detected with high sensitivity by the label of the hybridized probe.

3.2.1 Preparation of Labeled Probes

3.2.1.1 Labeling and Detection Systems

The most common labeling technique for nucleic acids is based on the incorporation of radiolabeled nucleotides (usually with ^{32}P) which allows detection of very low quantities of labeled probes. Autoradiography is usually used to detect the presence of labeled probes. This detection method is quite flexible as the exposure time can be varied from a few minutes to several weeks. A second possibility is the labeling of nucleic acids with molecules which can be detected by immunological techniques that are also very sensitive.

Commonly used systems for such strategies are nucleotides labeled with biotin or with digoxigenin. Biotin is specifically recognized by avidin (WILCHEK and BAYER, 1988), which, in a second step, can be immunodetected by a specific antibody. Digoxigenin is directly recognized by a specific antibody. This antibody is coupled to an enzyme that indicates the location of the bound antibody by a specific reaction. Widely used systems employ alkaline phosphatase or horseradish peroxidase to catalyze the formation of colored substances from chromogenic substrates.

Another approach directly links an enzyme to the nucleic acid probe (e.g., by glutaraldehyde) thereby avoiding the use of a specific antibody. However, this strategy may strongly influence the binding of the probe to the target sequences. A disadvantage of systems based on color formation is inflexibility. Once the color-forming enzyme reaction has been performed, there exists no further possibility to vary the intensities of the signals.

Recent developments have increased the flexibility of enzyme-catalyzed detection systems. Enzyme reactions which result in the emission of light (chemiluminescence, BRONSTEIN and MCGRATH, 1989) are used which, as with autoradiography, allow (although within a limited range) different exposures from the same hybridized membrane (CARLSON et al., 1990; HANSEN and BRAMAN, 1991). As a general recommendation, radiolabeled probes should be used when a high sensitivity is needed or when the detection of differences in the intensity of the signals is important (e.g., differential hybridization, see below). The use of non-radioactive labeling systems is well suited to screens where probes that have long stretches of exact homology to the target sequences are used. Convenient reagent kits are supplied by several companies.

3.2.1.2 Labeling Reactions

A widely used technique for the preparation of radiolabeled or biotinylated nucleic acid probes is nick translation. Double-stranded DNA is simultaneously treated with DNase I and *E. coli* DNA polymerase I in the presence of nucleotide triphosphates, one of which is labeled. Starting from random nicks introduced by DNase I, the 5′→3′-exonuclease activity of DNA polymerase I removes nucleotides from the DNA while simultaneously the 5′→3′-polymerase activity extends the 3′-end of the nick and thereby incorporates the labeled nucleotide.

Another possibility is labeling by randomly-primed DNA synthesis using the Klenow fragment of DNA polymerase I and a mixture of short random oligonucleotides (e.g., hexamers) in the presence of a labeled nucleotide. The labeling efficiency is comparable to that obtained by nick translation.

Synthetic oligonucleotides to be used as probes are usually labeled at the 5′-end with polynucleotide kinase and (γ-^{32}P)ATP. One strategy designed to generate oligonucleotide probes with a high specific activity uses two complementary oligonucleotides which are phosphorylated at the 5′-ends. After annealing, the double strands are ligated to form concatemers which can subsequently be labeled by nick translation or random primed synthesis.

A very efficient technique to produce probes uniformly labeled to a high specific activity is to synthesize RNA *in vitro* in the presence of radiolabeled ribonucleotides using a bacteriophage RNA polymerase (T3, T7, SP6) from DNA sequences inserted downstream of the appropriate T3, T7, or SP6 promoter. By cleaving the DNA downstream of the sequences which are to be used as the probe, specific "run-off" transcripts of these sequences can be produced.

3.2.2 Screening Strategies

3.2.2.1 Screening for Sequences Homologous to Cloned DNA Fragments

The use of previously cloned DNA fragments as probes in hybridization screenings has several applications:

- Cloned cDNA can be used to screen for the respective genomic sequences in genomic libraries.
- By rescreening a cDNA library with an isolated cDNA clone, it is possible to obtain

larger (full-length) cDNA clones of the same gene.

- Using the border sequences of isolated DNA fragments, clones containing overlapping sequences adjacent to the previously cloned fragment can be isolated from genomic libraries (chromosome walking). This strategy is very useful for cloning genes which map in the vicinity of an already cloned gene. Furthermore, a set of overlapping clones spanning a larger region is very useful for a variety of purposes, namely cloning of the entire coding sequences of large genes or clusters of genes, sequencing of large genetic elements, and mapping. Chromosome walking can be performed very efficiently by using (cosmid) vectors which contain bacteriophage promoters (T3, T7, SP6) on each side of the cloning site. This permits the generation of labeled probes specific for a distinct end of the cloned fragment without the need of time-consuming subcloning of border sequences by direct *in vitro* transcription from these promoters (EVANS and WAHL, 1987).
- Cloned DNA fragments encoding a specific gene of one organism can be used to clone the corresponding genes from related organisms (heterologous hybridization). In this procedure the sequence homology between these genes should be sufficiently high such that the hybridization conditions can be stringent enough to allow specific detection of correct clones (WAHL et al. (1987).

3.2.2.2 Screening of Libraries with Synthetic Probes Based on Amino Acid Sequences

This widely used strategy is mainly applied in cases where amino acid sequence information for a protein is available. Oligonucleotides are then designed, based on the genetic code, synthesized, labeled, and used to screen an appropriate gene library. The main problems associated with this strategy relate to the degeneracy of the genetic code. One approach to obtaining a suitable probe preparation is the use of mixed short (20–30 bases) oligonucleotides containing all possible nucleotide sequences derived from the amino acid sequence. However, depending on the sequence structure, large numbers of different possibilities may have to be considered. This means that the portion of the correct oligonucleotide present in such preparations is rather low resulting in a low sensitivity. In addition, such preparations contain various sequences that while not highly related to the gene of interest may randomly show homology to other gene sequences and therefore can give rise to false signals. General considerations for designing such probes are: sequence complexity of mixed oligonucleotides, self-complementarity, G+C content, and length (WALLACE and MIYADA, 1987). The last point may have a great influence on the success of the screening. Short oligonucleotides (around 20 bases) have the disadvantage that with high-complexity libraries (higher eukaryotes) a good chance exists that some such sequences are randomly present. An alternative to the use of mixed short oligonucleotides is the use of single longer oligonucleotides (30–100 bases) that take into account codon usage preferences of the donor organism. Such "guess-mers" can be advantageous since the longer stretches of homology provided by the larger molecules result in more stable hybrids, and specific signals can be obtained even when a few bases do not perfectly match (WOOD, 1987).

3.2.2.3 Subtractive Hybridization

A special strategy based on hybridization with labeled cDNA populations facilitates the search for genes that are specifically expressed in a certain tissue, at a specific stage of development, or under certain environmental conditions. In protocols for this procedure a cDNA library is constructed from RNA isolated from cells where the genes of interest are expressed ("induced" stage). Replica filters of clones of this library are then screened by plaque (colony) hybridization with a labeled cDNA population obtained by reverse transcription (in the presence of a radiolabeled nucleotide) of RNA taken from cells that are not expected to express the genes of interest ("non-induced" stage). As a control, a replica filter of the same clones is also hybridized with a labeled cDNA population obtained from RNA of the "induced" stage. Clones which do not show a signal with cDNA of the "non-induced" stage, but show a signal with DNA of the "induced" stage should then represent genes specifically expressed in the "induced" stage (+/− screening). For the cloning of regulated genes of organisms with less complex genomes, such as lower eukaryotic microorganisms (fungi), genomic libraries instead of cDNA libraries can be used for screening.

The screening efficiency can be enhanced by the introduction of a further step where the cDNA population used as a probe for hybridization is specifically enriched for sequences only expressed in "induced" cells ("subtracted" probe). This can be achieved by solution hybridization of the labeled cDNA derived from RNA of "induced" cells with an excess of RNA of "non-induced" cells. cDNA molecules of sequences present in both populations form hybrids with the respective RNA molecules, and cDNA molecules of sequences only present in the population of the "induced" stage will remain single-stranded. The single-stranded material representing the "subtracted" probe can be separated from the double-stranded hybrids by chromatography on hydroxylapatite. More recent developments are based on the fact that biotinylated nucleic acids can be extracted by organic solvents. Hybridization of the labeled "induced" cDNA to "repressed" RNA biotinylated with photobiotin results in the formation of solvent-extractable hybrids (LOHMAN et al., 1988). A schematic presentation of a subtractive hybridization screening is shown in Fig. 9.

3.2.2.4 Hybrid Arrested/Selected Translation

Hybrid arrested/selected translation methodologies depend on the ability of encoded gene products of interest to be monitored in specific assays. In some cases a gene product can be monitored by *in vitro* translation of its RNA and subsequent detection of the polypeptide by SDS-PAGE and/or immunological techniques (RICCHIARDI et al., 1979). Another approach for detection may be based on *in vivo* activity of RNA transferred into tester cells (e.g., *Xenopus* oocytes).

"Hybrid arrested" translation is based on the ratio that a specific RNA, which represents the transcription product of a desired gene, is inactivated for translation by the addition of complementary DNA sequences due to the formation of RNA–DNA hybrids. DNA is isolated from clones of a cDNA library and hybridized in solution to RNA isolated from cells expressing the desired gene function. The resulting RNA is tested for the respective gene activity. No response should be found when DNA of the desired clone is added.

A similar strategy is "hybrid released translation". In this case DNA of cDNA clones is immobilized on a solid support (nitrocellulose or nylon membranes). The membrane is hybridized with RNA isolated from cells expressing the desired gene function. Only RNA molecules containing sequences complementary to the bound cDNA can hybridize and be retained by the membrane. The bound RNA is subsequently released from the membrane and subjected to the *in vitro* or *in vivo* test system. Both strategies (JAGUS, 1987) imply that each clone has to be tested individually, a process that is enormously laborious. Therefore, such strategies are rarely used, but have been successfully applied in cases where no other screening strategy was available (MARCH et al., 1985; LEMKE and AXEL, 1985).

3.3 Immunological Techniques

Screening systems based on specific antibodies generated to a protein of interest are quite similar to hybridization strategies. In this case the proteins generated in clones of an expression library are transferred and immobilized on a solid support (nitrocellulose membranes). The membranes are then incubated with a specific antibody which itself is usually recognized by a second antibody (reacting with species-specific determinants) being either radiolabeled or conjugated to an enzyme (e.g., alkaline phosphatase or horseradish peroxidase) catalyzing a detectable color reaction.

Screening techniques based on immunodetection of peptides expressed from recombinants in *E. coli* have been successfully used to isolate a variety of eukaryotic genes (ERLICH et al., 1978; YOUNG and DAVIES, 1983). A

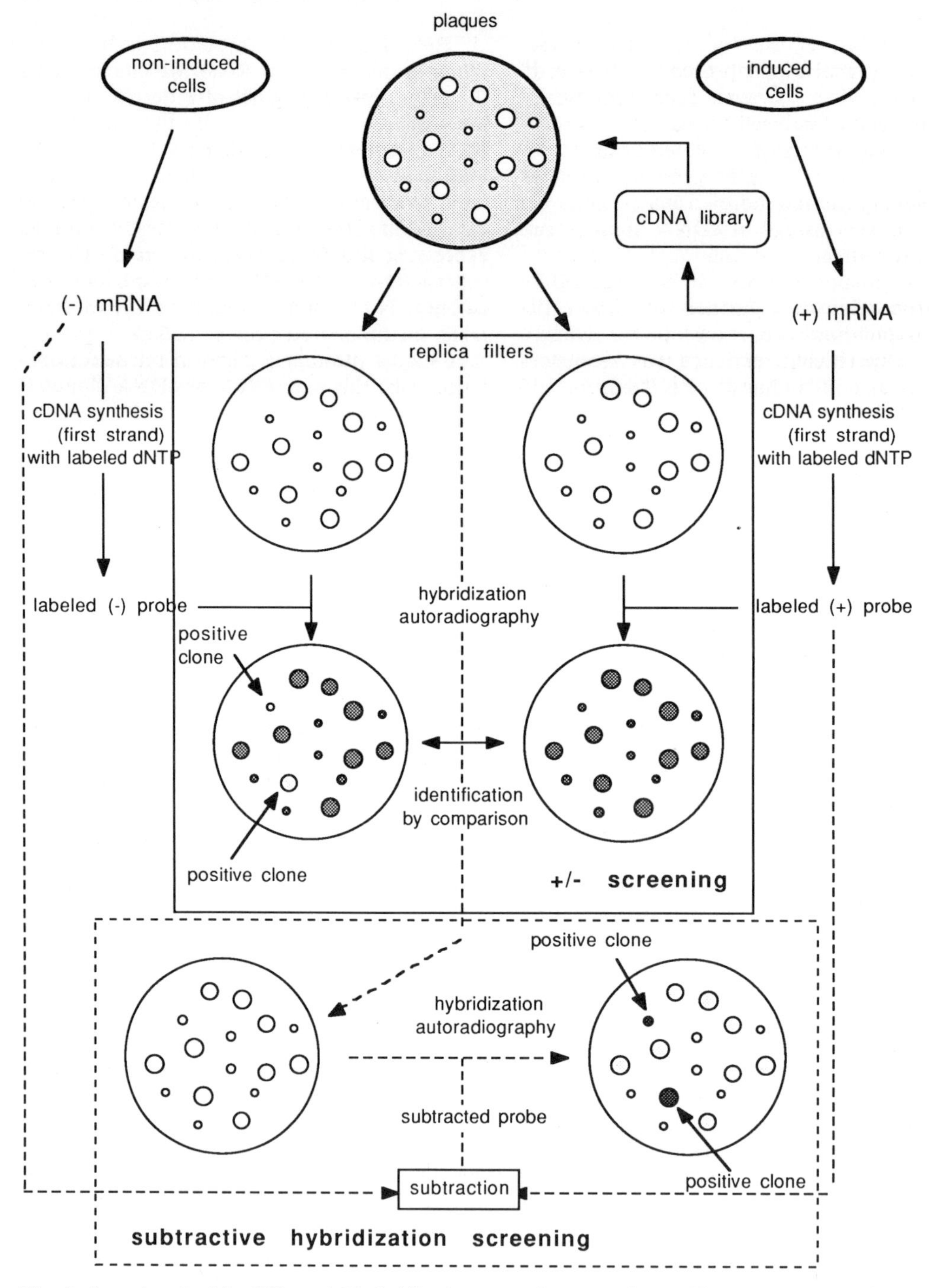

Fig. 9. Steps involved in differential hybridization screening procedures. The ± strategy needs comparison of signals obtained with replica filters hybridized with material derived from cells of the induced and the non-induced stage, respectively. The dotted lines refer to the strategy of using a subtracted probe which is highly enriched in sequences only expressed in the induced stage and, therefore, identification of clones is performed directly.

prerequisite for such strategies is expression of at least a portion of the protein in *E. coli*. A general strategy is to construct cDNA libraries in vectors containing an inducible *E. coli* promoter driving expression of cloned fragments. Furthermore, various vectors contain, in addition, the structural gene of a stable bacterial protein (e.g., β-galactosidase) with a proper cloning site at the C-terminal end. Cloning of cDNA fragments into this site can result in the formation of fusion proteins consisting of, e.g., β-galactosidase and part of the foreign protein. Due to incomplete transcription, preparations of cDNA fragments usually contain populations of molecules that have a variable amount of the sequence of interest. The presence of these fragments in addition to more rare fragments containing full-length coding sequence increases the probability that a fusion in the correct translational reading frame will occur.

Widely used vectors for the construction of expression libraries allowing the synthesis of fusion polypeptides are lambda gt11 (see Sect. 2.2.2.1) and some close relatives (gt18–gt21) (HUYNH et al., 1985; MIERENDORF et al., 1987). Although these lambda vectors can be maintained in the lysogenic state, screening is now mostly performed by transferring plaques obtained from lytic propagation. However, lambda vectors may have disadvantages in some cases, since only the proteins of the lysed host cells present in a plaque are available for immunodetection.

With plasmid vector-based libraries (standard plasmid expression vectors can be used) expression up to very high levels can be achieved resulting in an increased sensitivity of the immunological detection system (HELFMAN and HUGHES, 1987). Since cloning into plasmid vectors is less efficient and the release of proteins from cells is much more laborious, plasmid vectors are only used when immunodetection is not sensitive enough with plaques of recombinant lambda phages and when only small numbers of clones have to be screened (SAMBROOK et al., 1989).

4 Molecular Characterization of Cloned Genes

Various techniques have been developed for the molecular characterization of cloned genes. Such methods not only include the determination of the DNA structure, but give further information on encoded gene products, and on the molecular structure and the function of the cloned genes in the natural environment of the host organism.

4.1 Determination of the Molecular Structure of Cloned DNA Fragments

4.1.1 Restriction Analysis

Restriction analysis is a well established and rapid method for characterizing DNA fragments. Information on the location of restriction sites is very useful for the design of subcloning strategies and for experiments where specific regions of a DNA molecule are to be modified. About 150 different restriction enzymes are now commercially available. This assortment makes restriction analyses quite flexible. By digesting the DNA with combinations of restriction endonucleases it is possible to determine the relative location of restriction sites within DNA fragments. The fragment sizes of various restriction endonuclease digestions are determined by agarose (larger fragments) or by polyacrylamide (small fragments) gel electrophoresis, and restriction maps are established by carefully putting together the pieces of the "puzzle" which can be quite laborious.

SMITH and BIRNSTIEL (1976) have developed a strategy for restriction mapping based on labeling DNA fragments at one end and subsequent partial digestion. After electrophoresis and autoradiography the relative positions of the fragments can be easily determined. Recent developments of phage and cosmid cloning vectors make use of this strategy and allow rapid and easy establishment of restriction

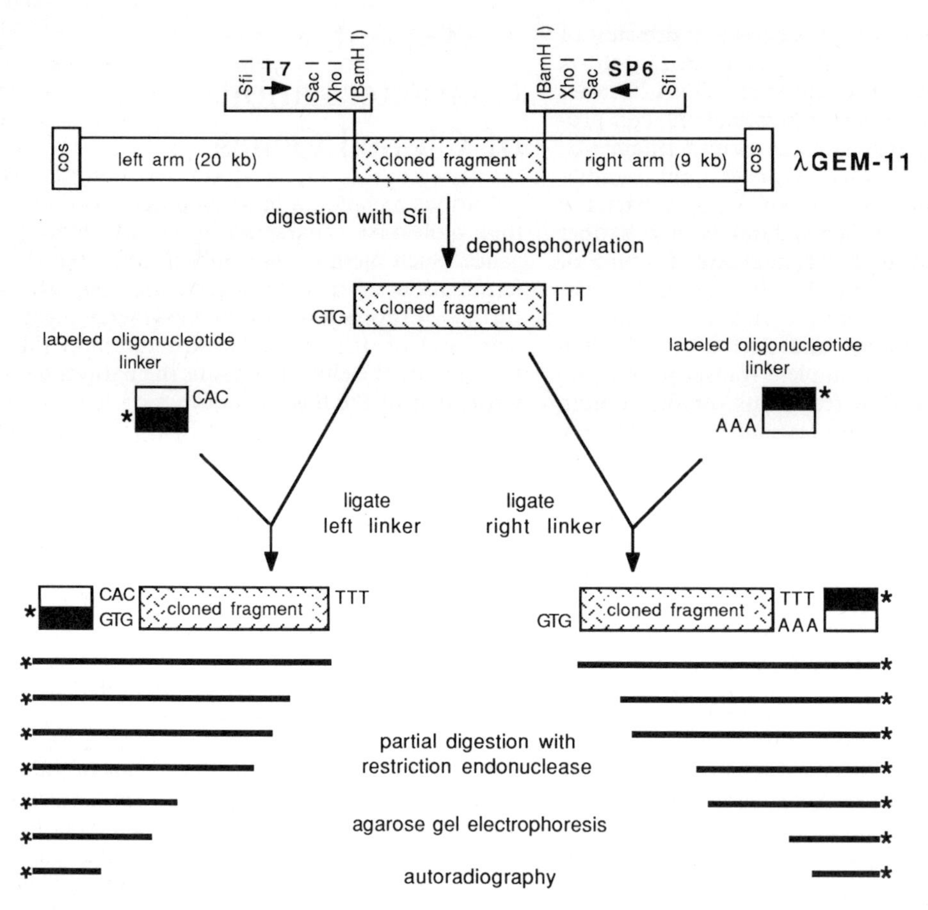

Fig. 10. Rapid mapping of restriction sites by partial digestion of end-labeled DNA fragments. In this example labeled synthetic oligonucleotide linkers specific for the 5′- and 3′-ends are ligated to the respective ends. Following partial digestion with a restriction endonuclease the sizes of the labeled fragments are determined by gel electrophoresis. The size of each labeled fragment is a direct measure for the distance of the restriction site from the labeled end. The use of the two end-specific linkers in parallel reactions permits rapid determination of the location of the sites from both ends.

maps (WAHL, 1989; TROUTMAN et al., 1991). The outline of this strategy is shown in Fig. 10.

4.1.2 DNA Sequencing

One can say that the development of efficient sequencing methods represents one of the central steps in the development of genetic engineering to a key technology of our century. Two different techniques were essentially simultaneously developed: the enzymatic chain termination method by SANGER et al. (1977) and the chemical method by MAXAM and GILBERT (1977). Many new developments (enzymes, vectors, electrophoresis equipment) have made sequencing a streamlined and efficient technique. Automated systems which are already in operation allow the determination of large numbers of bases per day. The availability of well designed reagent kits from various companies has made sequencing now also to a rather uncomplicated method. The strategies and methods as well as the new developments for DNA sequencing are described in detail in Chapter 8.

In parallel to the developments in sequencing methodology, the facilities and methods for computer-assisted handling and analysis of sequence data have emerged as powerful tools (VON HEIJNE, 1987). Sequence data libraries (Genbank, EMBL) provide online services making up-to-date sequence data immediately accessible.

4.2 Analysis of Gene Structure and Location in the Genome

Cloned DNA fragments can be used to obtain specific information about the molecular structure and chromosomal location of the corresponding genetic locus. Southern hybridization experiments (SOUTHERN, 1975; MEINKOTH and WAHL, 1984) with isolated chromosomal DNA of the donor organism using the cloned DNA fragments as probes allow the verification of the authenticy of the cloned fragment by comparing restriction fragments generated from chromosomal DNA and from the cloned fragment. The recently developed techniques for the separation of large DNA molecules by various types of pulsed field electrophoresis systems such as OFAGE (orthogonal field alternation gel electrophoresis, CARLE and OLSON, 1984), CHEF (contour clamped homogeneous electric field electrophoresis, CHU et al., 1986), TAFE (transverse alternating field gel electrophoresis, GARDINER et al., 1986) or RFE (rotating field gel electrophoresis, ZIEGLER et al., 1987) which has been further developed (CLARK et al., 1988) permit the separation of DNA molecules in the megabase range. Entire chromosomes from lower eukaryotes can be separated by these techniques (SMITH et al., 1987). After transferring the DNA of separated chromosomes to nitrocellulose or nylon membranes, the specific localization of the cloned gene can be determined by hybridizing the membranes with the labeled (see Sect. 3.2.1) cloned DNA fragment.

The use of rarely-cutting restriction endonucleases allows the generation of very large fragments from chromosomal DNA, including higher eukaryotes. Restriction maps of entire bacterial genomes or of large regions of chromosomes of higher eukaryotes can be established (SMITH et al., 1988; CONDEMINE and SMITH, 1990). Again, the location of a cloned gene on such specific restriction fragments can be determined by Southern hybridization using the cloned fragment as probe.

A further strategy permits the spatial localization of sequences within cytological preparations (*in situ* hybridization). Sites corresponding to cloned sequences can also be localized within larger chromosomes such as polytene chromosomes and human chromosomes (PARDUE, 1985).

The PCR technology has opened up a new field by making it possible to sequence DNA fragments amplified directly from genomic DNA (genomic sequencing) or from reverse transcribed cDNA starting from a point of known sequence. The sequence information of cloned DNA fragments provides the basis for the design of one primer directing DNA synthesis into the unknown region. After ligation of a standard primer to the extension product the reverse reaction can then also be performed (MIHOVILOVIC and LEE, 1989).

4.3 Analysis of Transcripts

Several features of transcribed mRNA molecules can be determined using cloned DNA fragments of the respective genes. One simple technique similar to Southern hybridization is the specific detection of transcripts by Northern hybridization (THOMAS, 1980). In addition to determining the size of the transcripts, information on the quantity of the specific mRNA can be obtained by probing parallel blots with sequences corresponding to genes of known expression levels. For detailed quantitative studies the dot blot technique is usually used (ANDERSON and YOUNG, 1985).

Similar to the localization of DNA sequences within chromosomes it is possible to determine the cellular localization of specific RNA molecules within cell populations or tissues by *in situ* hybridization of cytological preparations. For some examples see ZELLER et al. (1987), COSTA et al. (1988), or WEDDEN et

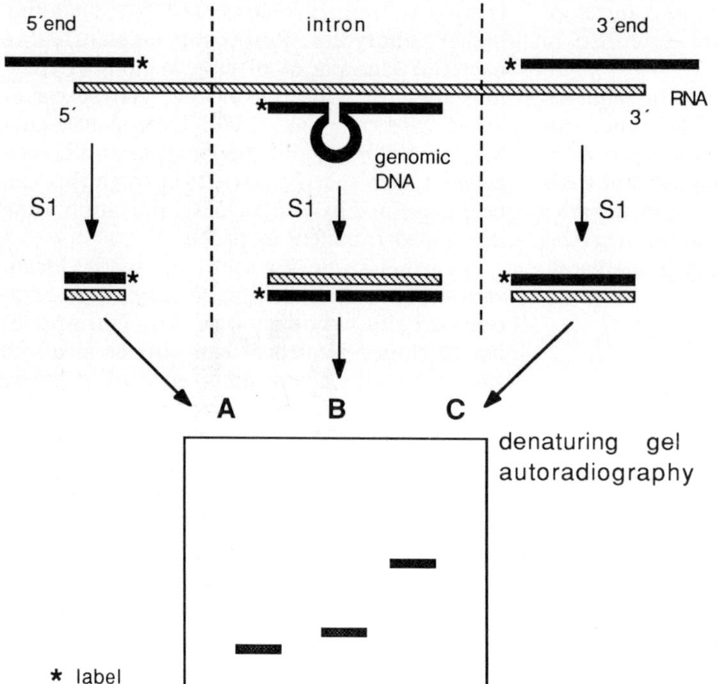

Fig. 11. Analysis of transcripts by S1 mapping. **A,** determination of the 5′-end; **B,** determination of intron location; **C,** determination of the 3′-end.

al. (1989). Biotinylated probes have also been successfully used for such studies (SINGER et al., 1986).

In addition, S1 mapping experiments can be carried out using the appropriate region of cloned DNA fragments (SHARP et al., 1980; WILLIAMS and MASON, 1985). In such experiments end-labeled DNA fragments are hybridized to RNA isolated from the donor organism. The hybridization mixture is then treated with S1 nuclease which digests all single-stranded regions of nucleic acids (DNA and RNA). Only the double-stranded structures formed between regions of complementarity contained in the labeled fragment and an mRNA molecule are protected. The size of the labeled DNA fragment that remains following nuclease treatment is determined by denaturing gel electrophoresis and subsequent autoradiography. By choosing appropriate regions of the cloned DNA fragment and the type of end labeling (5′-end or 3′-end), the transcription start site (5′-end) as well as the 3′-end can be determined. In addition it is possible to map the position of introns (SAMBROOK et al., 1989; AUSUBEL et al., 1987). By S1 mapping the locations of the 5′-ends or the borders of introns can only be determined within the range of the resolution of the gel used to fractionate the products of the reaction, typically within a few base pairs. The general principle of S1 mapping is shown in Fig. 11.

Using the sequence information of the cloned DNA fragment, proper oligonucleotide primers can be designed and used to exactly determine transcription start sites (5′-ends) by primer extension with reverse transcriptase. The end-labeled extension products are analyzed on a sequencing gel which permits the determination of 5′-ends at single base resolution (SAMBROOK et al., 1989; AUSUBEL et al., 1987). The principle and an example of the determination of transcription start sites by primer extension is shown in Fig. 12. Primer extension reactions can also be performed similarly to sequencing reactions using dideoxynucleotides for specific termination. This allows the direct determination of the nucleotide se-

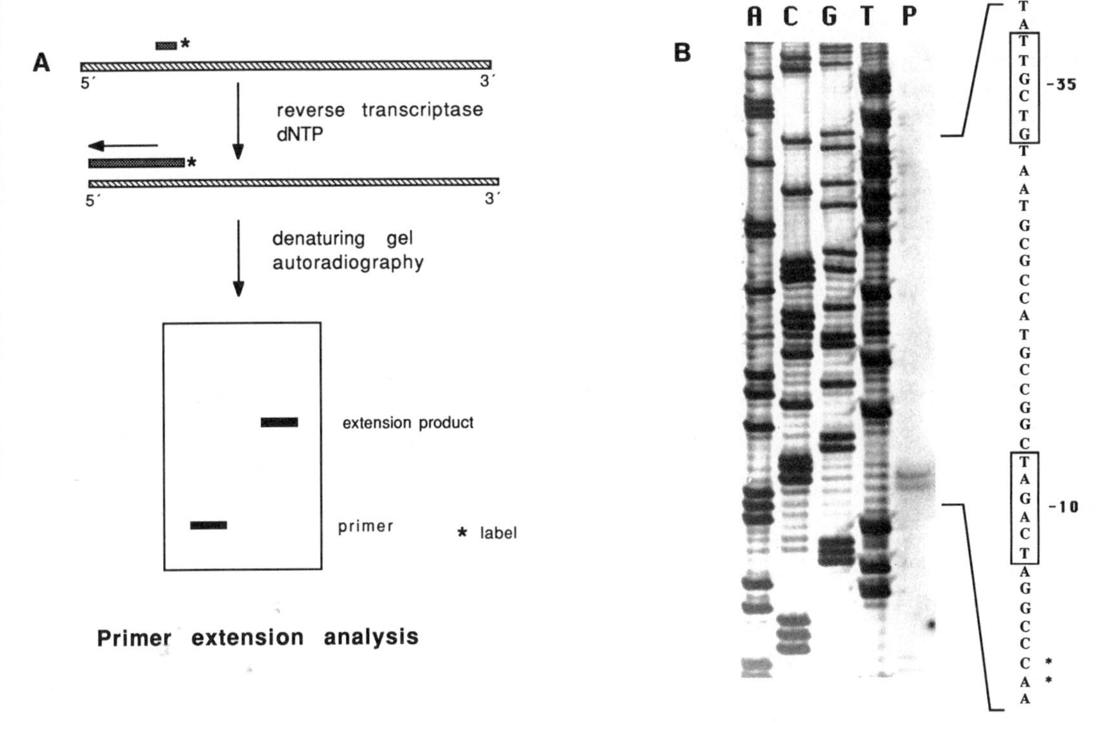

Fig. 12. Determination of transcription start sites by primer extension analysis. **A,** schematic representation of the general principle; **B,** example of the analysis of a bacterial promoter (EBERL and SCHWAB, unpublished data). In this case two start sites (*) are present. A, C, G, T, sequencing reactions (dideoxy method) with DNA of the corresponding region; P, primer extension reaction with isolated bacterial RNA. The DNA sequence of the coding strand is shown on the right side, the −10 and −35 regions of the promoter are enboxed.

quence in the surroundings of the start site (RNA sequencing).

4.4 Analysis of Encoded Gene Products

A variety of different methods have been developed to detect polypeptides encoded within a cloned DNA fragment. The main problem associated with these methods is to distinguish the specifically encoded polypeptides from the proteins of the system used for expression.

4.4.1 Minicell and Maxicell Systems

For polypeptides which can be expressed in *E. coli* (mainly bacterial genes) the "minicell" and "maxicell" *in vivo* expression systems provide good results. Both systems are based on specific mutants of *E. coli.*

The minicell system is based on a mutant that produces small anucleate cells (minicells). Plasmids present in the strain will segregate into these minicells. The minicells can be separated from normal cells by gradient centrifugation. Since minicells do not contain chromosomal DNA, plasmid-encoded polypeptides can be specifically labeled by incubation with (^{35}S)-methionine and analyzed by SDS-PAGE followed by autoradiography. The comparison of obtained bands with those of parallel experiments performed with the insert-free vector allows the identification of polypeptides specifically encoded on the cloned DNA fragment.

The maxicell system makes use of a UV sensitive mutant. Upon low-dose UV irradiation this strain extensively degrades its UV-damaged DNA. Since plasmids are very small compared to the bacterial chromosome, a large portion of the plasmid molecules will not be hit by UV quants. The undamaged plasmid DNA escapes degradation, whereas the chromosomal DNA is nearly completely degraded. Following UV irradiation, the cells can directly be incubated with radiolabeled methionine and the plasmid-encoded polypeptides detected as described for the minicell system. Both systems have been described in detail by STOKER et al. (1984).

4.4.2 Zubay System

Another system based on *E. coli* extracts permitting *in vitro* DNA-dependent coupled transcription–translation was developed by ZUBAY (PRATT, 1984). This cell-free system supports protein synthesis upon the addition of exogenous DNA. Although covalently closed circular DNA is a better substrate, linear DNA fragments containing just the cloned sequences to be analyzed can be used in this system. The Zubay system is commercially available minimizing the effort required to analyze a specific clone.

4.4.3 Specific Transcription from Phage Promoters

RNA prepared *in vitro* with bacteriophage RNA polymerases (T3, T7, SP6) from vectors containing the corresponding promoters as described in Sect. 2.2 can be used for the detection of encoded polypeptides by *in vitro* translation systems. This strategy also allows the detection of polypeptides of eukaryotic origin which cannot be directly expressed in *E. coli*. Capping can be performed by including an analog of the cap structure (e.g., GpppG or its methylated derivatives) in the synthesis reaction. Reticulocyte lysates or wheat germ extracts (CLEMENS, 1984) which are commercially available are usually used for *in vitro* translation of the synthesized RNAs.

A more recent development makes use of specific transcription from bacteriophage promoters *in vivo* with the aid of a cloned bacteriophage RNA polymerase gene expressed in the host cell from a regulated promoter. This strategy allows rapid and very efficient synthesis of plasmid-encoded polypeptides that can be expressed in *E. coli*. TABOR and RICHARDSON (1985) have developed such a system based on the T7 RNA polymerase and a T7 promoter based vector. The polymerase expression unit can either be supplied on a compatible second plasmid or on a bacteriophage M13 vector. With the latter system expression is induced by infection with the phage carrying the gene for T7 RNA polymerase under the control of a strong *E. coli* promoter.

5 Site-Specific Mutagenesis

Many methods have been developed to produce site-directed alterations in cloned DNA fragments. These methods are extremely useful for various aspects of analyzing gene functions as well as for the exploration of protein structure and function. In the field of biotechnological applications, the availability of site-specific mutagenesis methods paved the way for a large variety of powerful strategies for the construction of recombinant production strains. Finally, these techniques have allowed the development of protein engineering which is expected to revolutionize biotechnology within the next decade.

Besides some simple methods based on restriction endonucleases (deletions and insertions), the majority of the techniques now available have been developed on the basis of synthetic oligonucleotides. Three broad categories of site-directed mutagenesis experiments can be classified:

- Deletion or insertion of larger fragments at restriction sites,
- random changes within defined regions,
- defined alterations at specific locations.

5.1 Deletions and Insertions

Deletion or insertion of DNA sequences using particular restriction sites represent simple methods of altering DNA molecules at specific sites and are widely used. In addition, 5′- or 3′-protruding ends created upon cleavage with a given restriction endonuclease can also be blunted by filling in the ends with a DNA polymerase, or removed with nucleases (see Tab. 2). This results in the insertion or the deletion of a few bases within this restriction site. A further useful mutagenesis strategy is represented by the insertion of synthetic oligonucleotide linkers of defined sequence structure. The oligonucleotides may contain specific restriction sites for easy identification of mutated clones or for subcloning, or may contain specific signals for gene expression (e.g., stop codons in all three reading frames). These insertion/deletion strategies are limited in their application, however, since proper restriction sites have to be present at the points of interest and in many cases these are not available as required.

5.2 *In vitro* Mutagenesis Using Synthetic Oligonucleotides

The "classical" oligonucleotide-directed site-specific generation of mutations (SMITH, 1985) is based on the utilization of a mutagenizing oligonucleotide spanning the region of the desired mutation and containing the specific base alterations. These alterations may represent exchanges of one specific base or of a few distinct bases, as well as small deletions or insertions at defined positions. The oligonucleotide is then annealed to single-stranded DNA of the wild-type DNA fragment cloned into a vector allowing easy preparation of single-stranded DNA (M13 or phagemids, see Sect. 2.2). The annealed oligonucleotide serves as a primer for *in vitro* synthesis of the second strand, and the resulting double-stranded plasmid containing on one strand the wild-type sequence and on the other strand the mutated sequence (mismatch) is transformed into an *E. coli* host. The two strands are separated by replication and segregation upon cell division. The resulting population should contain the wild-type and the mutated plasmid in a 50:50 ratio. However, due to inefficient *in vitro* regeneration of the second (mutated) strand and due to active repair systems of the host cells recognizing the mismatches in the *in vitro* generated duplex molecule, the yield of clones containing the mutated copy can be very low. This can make the subsequent screening step, usually performed by colony hybridization with the mutagenizing oligonucleotide as the probe, quite laborious. Therefore, several strategies have been developed to overcome these difficulties and to obtain a high ratio of clones containing the mutated sequences (KRAMER and FRITZ, 1990).

One commonly used strategy (Fig. 13A) makes use of the fact that DNA isolated from a *dut*$^-$ *ung*$^-$ mutant contains uracil residues in place of thymine. When such DNA is used as template, the wild-type sequences are tagged with uracil, and this strand will be eliminated after transformation of the heteroduplex DNA into an *ung*$^+$ strain (KUNKEL, 1985; KUNKEL et al., 1987).

Another strategy (Fig. 13B) is based on the use of nucleotide thioanalogs (e.g., α-S-dCTP) in the *in vitro* extension reaction. Since some restriction endonucleases (e.g., *Nci*I) cannot cut at positions where a thioanalog is present, only the opposite template strand is cut. Exonuclease III can then be used to partially degrade the template strands starting from the generated nicks. This strand is subsequently resynthesized with DNA polymerase I resulting in a homoduplex molecule containing only mutant sequences on both strands. After transformation, a high yield of mutant clones can be obtained (TAYLOR et al., 1985).

A third strategy (Fig. 13C) allows a direct selection for the mutated strain. This strategy is based on two mutants of a vector each carrying an amber mutation in one of the two antibiotic resistance markers that are present. The target fragment to be mutagenized is cloned into one of these mutant vectors (1) which contains the mutation in one resistance marker (e.g., Cm). Single-stranded DNA is prepared from this construct. DNA of the second vector (2) containing the mutation in the other marker (Amp) is cut at the position where the target fragment was cloned into vector (1), denatured and annealed to the single-stranded DNA of the vector(1)-insert hybrid. This results in a duplex molecule that contains a region of single-stranded DNA correspond-

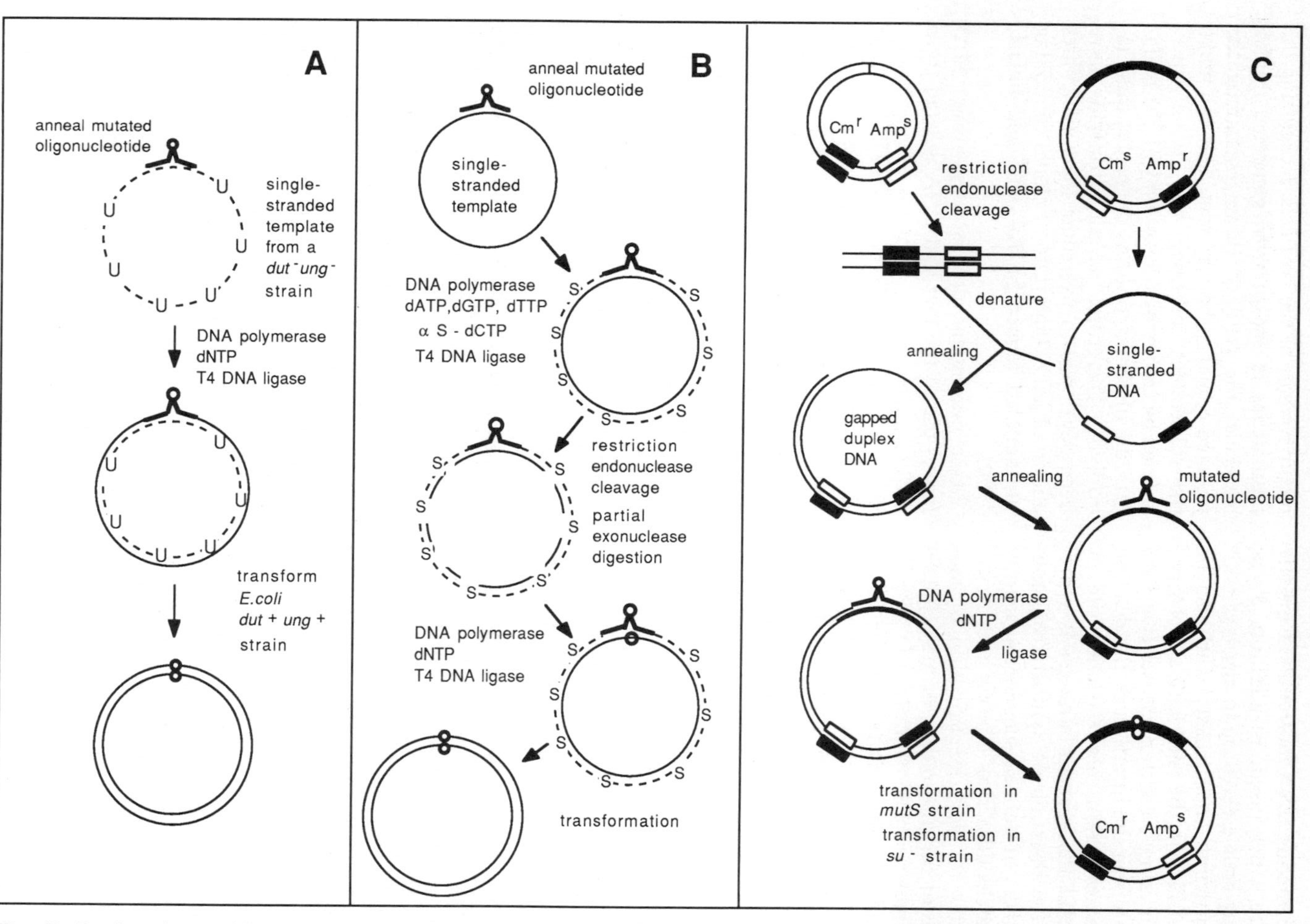

Fig. 13. *In vitro* site-specific mutagenesis using synthetic oligonucleotides. **A,** "Kunkel" method; **B,** strategy with thioanalogs; **C,** gapped duplex method with phenotypic selection. For further explanations refer to the text.

ing to the cloned target fragment (gapped duplex). The synthetic oligonucleotide containing the desired base changes is then annealed to the single-stranded region. The remaining single-stranded regions are then filled in, and the DNA is transformed into hosts deficient in mismatch repair (e.g., a *mutS* strain) and unable to suppress amber mutations. Selection for clones expressing the resistance marker mutated in the second plasmid (e.g., Cm) ensures that only molecules containing the mutated strands will be present in the obtained clones (STANSSENS et al., 1989).

This basic strategy of oligonucleotide-directed mutagenesis can also be used for the generation of random mutations at specific points within a defined region. A mixture of degenerate oligonucleotides can be applied instead of one specific oligonucleotide, and a series of clones containing alterations of all types specified by the degenerate oligonucleotide mixture can be obtained in one step (HÜBNER et al., 1988; HOROWITZ and DIMAIO, 1990).

5.3 Region-Specific Chemical Mutagenesis

Often in the functional analysis of a protein the introduction of random mutations within a specific DNA region is desired. One strategy makes use of the fact that sodium bisulfite can only mutagenize single-stranded DNA. The region of interest is cloned into a vector designed for generating single-stranded DNA. Bisulfite-treated single-stranded DNA is then reconstituted *in vitro* into double-stranded material with a DNA polymerase (see Tab. 2). A fragment comprising the region of interest is then recloned into the original construct replacing the non-mutagenized homolog (WEIHER and SCHALLER, 1982). Another possibility is to use "gapped duplex" molecules exposing the region of interest as single-stranded DNA (see Sect. 5.2). After mutagenizing the single-stranded gaps these are filled in with a DNA polymerase (EVERETT and CHAMBON, 1982).

A useful strategy based on chemical mutagenesis has been developed by MYERS et al. (1985). The target DNA is cloned into a single-strand vector (M13 or phagemid) and the single-stranded DNA is mutagenized *in vitro* with chemicals causing damage to specific bases (e.g., nitrous acid, formic acid, hydrazine) as desired. An oligonucleotide primer that spans an appropriate restriction site at the 3′-end of the desired region is then annealed and extended by reverse transcriptase. This enzyme incorporates random nucleotides at sites of damage with a high frequency. Following this reaction, the region of interest is excised by restriction endonucleases and is subsequently recloned replacing respective wild-type sequences.

5.4 Site-Directed Mutagenesis Using PCR

The polymerase chain reaction can also be used to efficiently and rapidly introduce any desired alteration into a cloned DNA fragment (KADOWAKI et al., 1989). Two oligonucleotides, one of which includes the desired mutations, are synthesized. Both oligonucleotides should include appropriate restriction sites at the ends. After annealing the oligonucleotides to the template to be mutagenized, the DNA fragment between the oligos is amplified by PCR (see Sect. 2.6). The amplified DNA is cut by the respective restriction endonucleases and recloned to replace the original fragment. In principle, all of the clones obtained should contain the desired altered sequences making this mutagenesis technique very efficient. However, because *Taq* polymerase exhibits a rather low fidelity *in vitro*, additional random mutations within the amplified region can arise. The amplified fragment must, therefore, be sequenced, a laborious prospect when the fragment is large. Recently, however, a thermostable polymerase containing a 3′-5′-exonuclease activity for proofreading, and presumably therefore demonstrating a higher fidelity, has become available through a commercial enzyme supplier. Another disadvantage may be the lack of convenient restriction sites flanking the region of interest. Several strategies, based on the elongation of fragments up to a convenient restriction site in a second PCR reaction, can be used to overcome this problem (LANDT et al., 1990; KAMMANN et al., 1989; HO et al., 1989).

6 Overexpression of Genes in *Escherichia coli*

The production of proteins is one of the main applications of genetic engineering in biotechnology. *E. coli* remains an important host system for the industrial production of proteins from cloned genes. Various efficient expression vector systems have been developed for *E. coli*, and a variety of *E. coli* mutants are available as host strains for different purposes. Furthermore, *E. coli* K12 strains can be easily grown to high cell densities in simple media (BAILEY et al., 1987; EPPSTEIN et al., 1989; RIESENBERG et al., 1991) and in addition, are classified as being of lowest risk (GRAS status). *E. coli* has been used not only successfully for the production of various heterologous proteins, including even functional fragments of human antibodies (SKERRA et al., 1991; PLÜCKTHUN, 1991), but also for the expression of genes encoding metabolic enzymes. Such recombinant strains can be used for producing metabolites or be used as biocatalysts for specific chemical reactions in organic synthesis (FAVRE-BULLE et al., 1991; SABATIÉ et al., 1991). One example of a specialized application is to use a recombinant *E. coli* strain that expresses a mammalian receptor protein as a tool for ligand screening (MARULLO et al., 1989).

In addition to the broad use of *E. coli* K12 hosts for production purposes, overexpression of heterologous proteins to very high levels with *E. coli* expression systems is an important tool for basic research that greatly facilitates efficient purification and analysis of such proteins. This is of particular importance in protein engineering projects and in studies of the functional elements of proteins by site-directed mutagenesis where large numbers of variants of a protein have to be produced and analyzed (WAGNER and BENKOVIC, 1990).

6.1 General Strategies for the Expression of Heterologous Genes in *E. coli*

The general principles of gene expression are described in detail in Chapter 7 of this volume. The requirements for efficient expression of heterologous genes in *E. coli* are:

1. *E. coli*-specific expression signals consisting of a strong (regulated) promoter for transcription initiation as well as a suitable Shine-Dalgarno sequence for translation initiation at the 5′-end of the structural gene. A potent transcription terminator at the 3′-end is usually needed to avoid interference with vector functions (e.g., plasmid replication).
2. A suitable basic vector system. The general properties (e.g., replication system, copy number, stability) of the vector system employed can play an important role in obtaining high yields of the desired gene product.
3. An appropriate host strain. The yields obtained with different host strains of *E. coli* K12 for the same expression plasmid can vary quite substantially (KAYTES et al., 1986).
4. A proper cultivation process. The fermentation conditions as well as the induction mode can be extremely important for the amount of recombinant protein present in the culture at the time of harvesting (KAPRÁLEK et al., 1991).

The assembly of the appropriate expression elements typically follows one of two alternative arrangements. In the first procedure the structural gene can be fused to *E. coli* expression signals in such a way that the intact native protein is directly expressed according to the sequence of the cloned structural gene (transcriptional fusions). In this case the polypeptide synthesized in *E. coli* starts with an N-terminal formylmethionine that is not naturally present in many proteins of interest. The formyl residue or the entire methionine may also be removed in a post-translational step, a process which is not controllable so far. Therefore, the final protein product may not be identical to the desired product and frequently will consist of a heterogeneous mixture of polypeptides being or not being processed at the N-terminus. In addition, when small polypeptides are expressed directly, they are often subject to rapid proteolysis resulting in very poor yields.

These problems can be overcome by constructing translational fusions, the second strategy. In this case the expression signals direct the synthesis of a bacterial (*E. coli*) gene (or parts thereof), and the sequences needed for the production of the desired polypeptide are fused in frame to this gene. Expression of this construct results in a fusion protein, and the desired protein can be obtained after proteolytic removal of the amino terminal fusion peptide. The strategy of constructing translational fusions can be applied in many variations including fusion to signal sequences for secretion, fusion to specific polypeptides that alter the overall physicochemical properties (e.g., solubility) of the protein, or allow easy purification by affinity chromatography (see below).

6.2 Expression Vector Systems

Over the years, a great variety of vector systems have been developed for the expression of heterologous genes in *E. coli*. Since the majority of technical developments occur in the industrial sector, many used vectors have not been described in the scientific literature. However, most of these vector systems are based on well characterized elements. A survey of various vectors has been given by RODRIGUEZ and DENHARDT (1988).

6.2.1 Basic Vector Features

In most cases plasmids derived from pBR322 or related plasmids, all based on the ColE1-type replication system, are used for the construction of expression vectors. These plasmids have moderate to high copy numbers (see Sect. 2.2.1) ensuring a high dosage of the genes to be expressed. Vectors based on such plasmids can be used when relatively high levels of the expressed gene product do not adversely affect the vitality of the host cells.

For the expression of polypeptides which are toxic to the host cells, vectors based on low copy plasmids (see Sect. 2.2.1) can be used. A special strategy is based on temperature-sensitive replication control mutants of plasmid R1. This plasmid normally replicates at a copy number of 1–2 copies per chromosome. The mutated plasmid can be amplified to extremely high copy numbers by elevation of temperature (UHLIN and NORDSTRÖM, 1978). Miniderivatives based on such mutants of R1 have been used to construct runaway-replication vectors (UHLIN et al., 1979) that are suitable for the expression of polypeptides toxic to *E. coli*.

6.2.2 Promoter Systems

Strong regulated promoters are usually used for *E. coli* expression vectors. The use of these promoters allows cultivation of the cells up to high densities under repressed conditions without concomitant synthesis of larger amounts of the recombinant protein. Production of the protein is turned on by derepressing the promoter at late stages of the culture when sufficient biomass is present. This strategy minimizes any adverse effects on cell growth caused by larger quantities of the protein to be produced. Frequently used systems are:

Temperature-regulated systems. Vectors are based on the bacteriophage lambda promoters p_L and p_R. Both promoters are regulated by the *c*I gene. A temperature-sensitive mutant allele (*c*I*ts*857), which represses transcription at low temperatures (around 30 °C) but is nonfunctional at elevated temperatures (around 40 °C), allows induction of gene expression by shifting the cultivation temperature. The repressor gene is usually provided by a specific host strain that carries an integrated defective bacteriophage lambda prophage containing this allele. A specific cassette containing the *c*I857 allele has been developed that facilitates cloning of the repressor gene directly into the expression plasmid (GEORGE et al., 1987). Since the repressor gene is then also present in multiple copies, cultivation of the cells at 37 °C is possible without a partial induction of expression. A general disadvantage of temperature induction is that the heat-shock system is also induced. Some of the heat-shock genes are potent proteases which can take part in rapid degradation of the product. This problem can be alleviated by using a wild-type *c*I allele and chemical induction by mitomycin C or nalidixic acid (SHATZMAN and ROSENBERG, 1987). However, this is only possible for laboratory-scale experiments.

Systems regulated by metabolic response. A variety of expression vectors that can be regulated by metabolic responses are based on the *lac* and *trp* promoter. The natural *lac* promoter is rather weak, and expression vectors usually contain the stronger *lac*UV5 mutant allele. This allele carries also a mutation in the CAP site and is, therefore, no more susceptible to carbon catabolite repression. The regulation of the *lac* promoter by the *lac* repressor, the product of the *lac*I gene, is not very stringent, even when the *lac*I^q allele (repressor overproduction) is used. Cloning of the repressor gene onto the multicopy expression plasmid improves the stringency (FÜRSTE et al., 1986). Induction is usually performed by the lactose analog isopropylthio-β-D-galactoside (IPTG). However, this is not very practical in large-scale production processes.

The *trp* promoter (YANSURA and HENNER, 1990) belongs to the strongest class of promoters known in *E. coli*. This promoter is also tightly regulated. Its basal expression level is rather low which eliminates problems due to undesired expression of the encoded polypeptides during the growth phase. Induction of the *trp* promoter is accomplished by starvation for tryptophan, a procedure that is well practicable in large-scale cultures of recombinant *E. coli*.

A hybrid promoter containing parts of the *lac* and the *trp* promoters is widely used for laboratory-scale overexpression of heterologous proteins. With respect to the transcription start site, this promoter (*tac* promoter) contains the sequences of the *trp* promoter upstream of -20 bp and the sequences of the *lac*UV5 promoter downstream of -20 bp. Thus, this hybrid promoter is regulated by the *lac* repressor but is much stronger than the *lac*UV5 promoter and even more efficient than the *trp* promoter in the absence of the *trp* repressor (DEBOER et al., 1983). The *tac* promoter is also functional in Gram negative bacteria other than *E. coli* (FÜRSTE et al., 1986).

Promoters inducible by phosphate starvation have also gained importance for use in expression vectors. Such systems are also of interest for large-scale processes, since the induction is easily accomplished by phosphate-limited growth conditions. The promoters of the *phoA* (KIKUCHI et al., 1981) and the *ugp* (SU et al., 1990) genes of *E. coli* have been particularly useful for vector constructions. The induction level with the *phoA* promoter was reported to be more than 1000-fold (WANNER, 1987).

Recently, an oxygen-responsive promoter was successfully used for oxygen-regulated expression of genes in *E. coli*. The promoter was obtained from the hemoglobin gene of *Vitreoscilla* and is induced at oxygen levels lower than 5% air saturation (KHOSLA et al., 1990).

Promoters driven by a specific RNA polymerase. As discussed in Sect. 4.4.3, promoters of genes from several bacteriophages are specifically recognized by the corresponding phage RNA polymerase. Such systems can be used for high-level selective expression of genes. The most common system is based on the promoter of the bacteriophage T7 Φ10 gene and the T7 RNA polymerase gene under the control of an inducible *E. coli* promoter (STUDIER and MOFFAT, 1986; STUDIER et al., 1990). Examples of expression vectors based on different promoter systems are shown in Fig. 14.

6.2.3 Elements for Efficient Translation

Expression in bacterial systems requires a ribosome binding site at the 5′-end of the mRNA. This element is determined by the Shine-Dalgarno (SD) sequence, a region of homology to the 3′-end of the 16S rRNA. The efficiency of translation strongly depends on the sequence structure of the SD element. At least four nucleotides of the general consensus AGGAGG should be present, and the optimum spacing between the SD element and the start codon is about nine nucleotides. The immediate surroundings of the SD sequence are also of great importance, and as a general rule the region between SD and the start codon should contain the nucleotides A or U. Optimal translation additionally requires that the region around the translation initiation site should not be able to form extensive secondary structures (STORMO, 1986). The initiation codon AUG mediates more frequent translation initiation than the two other possible codons, GUG or UUG. When the transcription start site is located far upstream of the translation initiation site, upstream out-of-frame transla-

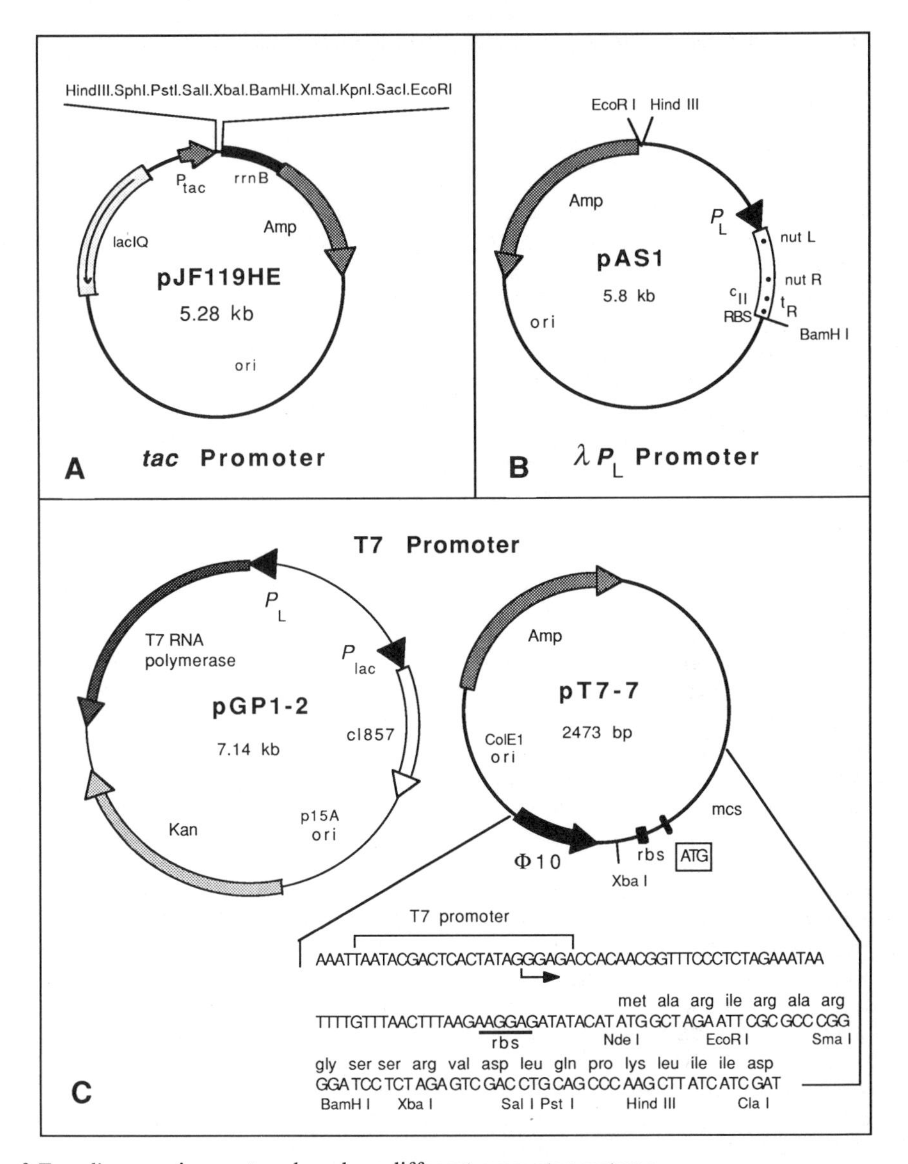

Fig. 14. Examples of *E. coli* expression vectors based on different promoter systems.
A: Vector based on the hybrid *tac* promoter. The *lacI^q* repressor gene is included in the expression plasmid to ensure tight regulation (FÜRSTE et al., 1986).
B: Vector based on the p_L promoter of bacteriophage lambda. A host containing the thermosensitive *c*I857 allele of the lambda repressor has to be used to allow induction by shifting to high temperatures. This vector (alternative name pSKF101) also contains the translational initiation site of the bacteriophage lambda *cII* gene including the start ATG. The *Bam*HI site (after blunting by S1 or mung-bean nuclease) can be used for cloning fragments immediately behind the start ATG. The *nut*L and *nut*R sites in combination with the t_{R1} provide additional regulation by the action of the N protein (ROSENBERG et al., 1983).
C: pT7-7 (TABOR, 1990) is a vector based on a T7 RNA polymerase specific promoter. This vector contains the promoter as well as the translation initiation region of the bacteriophage T7 Φ10 gene. Multiple restriction sites within the N-terminal end of the Φ10 gene permit the construction of translational fusions. The *Nde*I site which includes the start ATG can be used to insert fragments for direct expression. Induction of expression is performed by inducing the expression of the T7 RNA polymerase present on the compatible helper plasmid pGP1-2 (TABOR and RICHARDSON, 1985). The RNA polymerase gene is under the control of the p_L promoter. The bacteriophage lambda *c*I857 repressor allele under the control of the *lac* promoter permits induction of the system by shifting to high temperature.

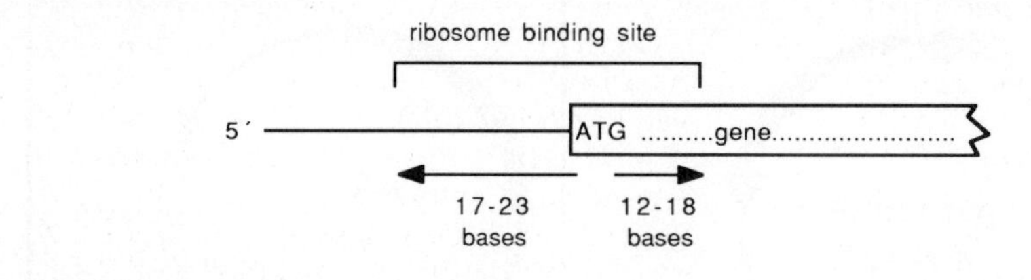

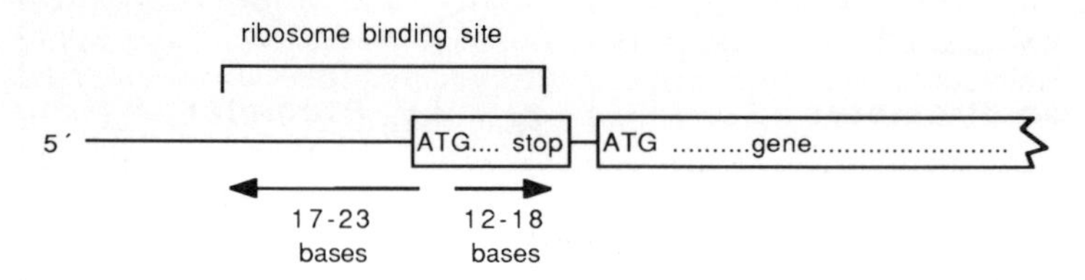

Fig. 15. Schematic representation of a "two-cistron" expression system. For further explanations refer to the text.

tion initiation can occur that interferes with translation from the desired start. It is therefore useful in such cases to include stop codons in all frames upstream of the SD sequence. GOLD and STORMO (1990) have described an optimized scheme for obtaining high-level translation based on the synthetic initiation region (stop codons in the upstream region are underlined):

5′ (UAANNUAANN)UAA-
GGAGGAAAAAAAAAUG-codons

It has also been suggested that the first bases within the 5′ coding region can dramatically influence the efficiency of translation (DEBOER and HUI, 1990). Since the sequence in this region cannot be altered too much without changing the amino acid sequence in the amino-terminal end of the synthesized polypeptide, this may be a serious problem. This potential problem can be alleviated by employing a two-cistron expression system. The first cistron contains sequence elements (codons for a few amino acids) that are optimized with respect to efficient translation initiation. The second cistron which encodes the desired polypeptide is translationally coupled to the first cistron (SCHONER et al., 1990). The principle of this strategy is outlined in Fig. 15.

6.3 Secretion Systems

Although *E. coli* is generally limited in its ability to secrete proteins into the culture medium, secretory systems for the export of synthesized recombinant proteins from *E. coli* are desirable for many reasons. These include simplified purification protocols (no cell disruption needed), the elimination of product degradation by intracellular proteases, the generation of polypeptides with correct amino-terminal amino acid sequences by processing during translocation, and a higher probability of correct folding and assembly of polypeptides after translocation into the periplasm. The cytoplasm is an unfavorable environment for disulfide bond formation, a fact that greatly facilitates inclusion body formation of overexpressed proteins accumulated in the cytoplasm (STADER and SILHAVY, 1990).

In *E. coli* secretion into the periplasm can be achieved in several ways. One possibility is the general secretory pathway (SCHATZ and BECKWITH, 1990) which, as in other organisms, requires an N-terminal signal peptide that is cleaved off during the translocation process. Fusion of the coding sequence of a protein to such a signal sequence usually results in the accumulation of the secreted protein within the periplasm. The most commonly used system is based on the signal sequence of *E. coli* alkaline phosphatase encoded by the *phoA* gene (OKA et al., 1985; MIYAKE et al., 1985; DODT et al., 1986). The use of signal sequences of outer membrane proteins such as the gene products of *ompA* (BECKER and HSIUNG, 1986; TAKAHARA et al., 1988) or ompF (NAGAHARI et al., 1986) can direct transport to the outer membrane followed by release into the medium.

More recently it was found that *E. coli* can secrete proteins directly into the culture medium by different pathways. A C-terminal signal directs the export of hemolysin across both bacterial membranes (KORONAKIS et al., 1989). Using this signal, it was also possible to direct secretion of heterologous proteins expressed in *E. coli* (MACKMAN et al., 1987). An additional system is based on staphylococcal protein A since it was recognized that two of the five IgG-binding domains are able to direct secretion of heterologous proteins fused to them. Secretion vectors were constructed based on modified synthetic domains that were already known to support the extracellular production of several polypeptides of pharmaceutical interest (JOSEPHSON and BISHOP, 1988).

It has also been recognized that several general limitations are responsible for poor secretion efficiency of *E. coli*, and tailoring of strains for this purpose would increase the yield of secreted protein (STADER and SILHAVY, 1990). One target is to increase the expression of components of the secretory pathways as these are believed to be available only in limited amounts. Another target is to overcome the barrier of the outer membrane. For example, "leaky" mutants have been isolated that are able to release proteins from the periplasmic space across the outer membrane into the medium.

Another possibility is to utilize a protein which was found to be essential for the release of bacteriocins such as colicin. Bacteriocin release protein (BRP) is a small polypeptide of 28 amino acids that causes a non-specific increase in the permeability of the inner and outer membranes (BATY et al., 1987). BRP was successfully used to increase the amount of human growth hormone, fused to the OmpA signal peptide, that is secreted (HSIUNG et al., 1989).

A related strategy for releasing cytoplasmic proteins is to utilize biological systems designed to disrupt bacterial cells. This approach can be regarded more as a biological alternative to mechanical methods of breaking up bacterial cells. Several enzymes are known to be involved in autolysis of *E. coli* cells. Three processes are known to trigger autolysis in *E. coli:* osmotic shock, the inhibition of peptidoglucan biosynthesis, and the activity of phage lysis genes (DABORA and COONEY, 1990). The use of bacteriophage lysis genes for this application seems especially promising. Such genes are put under the control of an inducible promoter and, following a period of cell growth, expression and accumulation of the gene products of interest, the lysis genes can be specifically turned on. One of the currently developed systems employs the E protein of bacteriophage ΦX174 (HENRICH and PLAPP, 1984). The lysing activity can also be increased by using hybrid proteins consisting of different bacteriophage lysis proteins (HARKNESS and LUBITZ, 1987). The main problem of such systems is to prevent any expression of the lysis protein during the cultivation of the cells, since even tightly regulated promoters mediate low levels of activity under non-induced conditions.

6.4 Fusion Systems for Efficient Downstream Processing

In addition to its role in the creation of multifunctional proteins (BÜLOW and MOSBACH, 1991), gene fusion is used primarily to facilitate purification of the protein of interest. The fusion polypeptide may have altered general properties that confer some advantage to the recovery of an active product. Alternatively, the portion fused to the protein of interest acts as a tag that can be exploited in the purification scheme. Examples of general fusion strategies facilitating product recovery (UHLÉN and MOKS, 1990) are shown in Fig. 16. Various options for site-specific cleavage of fusion polypeptides are now available and allow specific removal of the tag and the recovery of the desired protein in the correct primary structure. The most frequently used fusion and cleavage systems are summarized in Tab. 3.

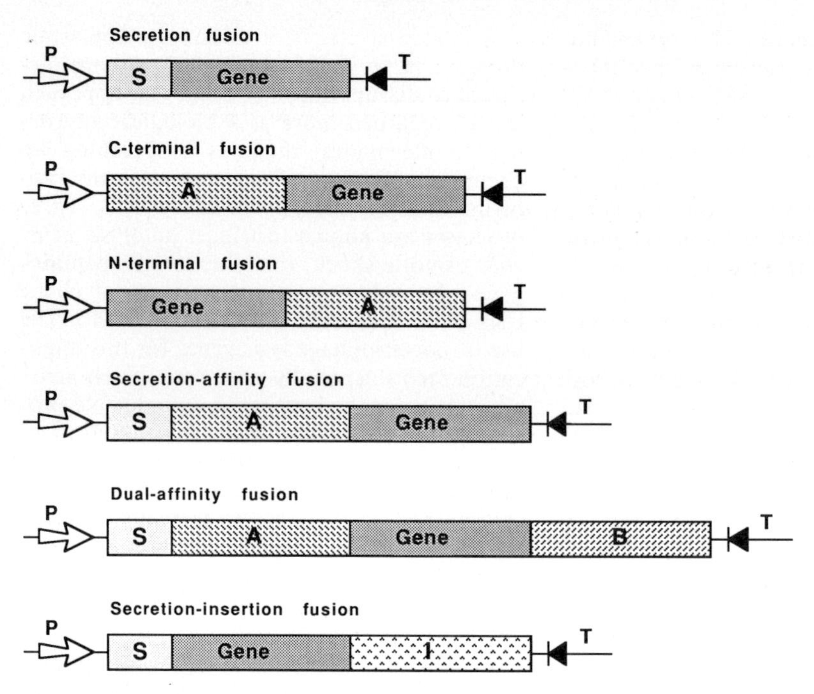

Fig. 16. Strategies for constructing translational fusions (after UHLÉN and MOKS, 1990). P, promoter; T, terminator; S, signal sequence; A, B, fusion peptides exploiting specific properties to the fusion protein (e.g., affinity tags); I, membrane integration domain. The use of membrane integration domains permits anchoring of proteins into one of the membrane systems or the cell wall which can be useful for the development of vaccines.

6.4.1 Fusion Proteins for Altered Solubility

Overexpression of heterologous polypeptides in *E. coli* can result in the formation of insoluble inclusion bodies. This can be a disadvantage or an advantage, depending on the nature of the polypeptide and its final use. The factors influencing the formation of inclusion bodies are not yet completely understood, but include the general solubility of the protein, environmental conditions, and the folding characteristics which, apart from other factors, generally depend on the presence and location of specific domains (MITRAKI and KING, 1989; SCHEIN, 1989).

Proteolysis of soluble (especially small) polypeptides has been frequently identified as a leading cause of low yields of recombinant proteins in *E. coli* systems (ENFORS et al., 1990). Proteolysis can be reduced by fusing the desired protein to a protecting polypeptide such as β-galactosidase. Another approach is to construct fusions where the protein of interest is fused to a polypeptide with high inclusion body-forming ability, since proteolysis can also be prevented by formation of insoluble inclusion bodies. One frequently used system is fusion to the *trpE* gene (YANSURA, 1990). A recent study worked out a set of correlations that can be used to predict the probability of inclusion body formation for a polypeptide based on its amino acid sequence (WILKINSON and HARRISON, 1991). On the other hand, a fusion partner with a high solubilizing ability can help in refolding of proteins following steps in the purification process where denaturing conditions have been applied. For example, two IgG binding domains (ZZ) of the *Staphylococcus* protein A were used to obtain high yields of correctly folded

Tab. 3. Methods for Site-Specific Cleavage of Proteins

Reagent/Method	Specificity	Remarks
Chemical Methods		
Cyanogen bromide	-X-Met*X-	acidic conditions
Formic acid	-X-Asp*Pro-X-	70%, heat
Hydroxylamine	-X-Asn*Gly-X-	pH 9, heat
2-(2-Nitrophenylsulphenyl)-3-methyl-3-bromoindolenine (BNPS-skatole)	-X-Trp*X-	50% acetic acid
Enzymatic Methods		
Exopeptidases		
Carboxypeptidase A	-X-X-Arg(Lys)*(X)n	all C-terminal amino acids except Arg or Lys
Carboxypeptidase B	-X-X*(Arg)n/(Lys)n	C-terminal Arg or Lys residues
Endopeptidases		
Collagenase	-X-Pro-X*Gly-Pro-X-	
Enterokinase	-X-Asp-Asp-Asp-Lys*X-	
Factor Xa	-X-Ile-Glu-Gly-Arg*X-	
(Ala^{64})-subtilisin	-Gly-Ala-His-Arg*-X-	
Thrombin	-X-Gly-Pro-Arg*X-	

Amino acids indicated are required for efficient cleavage. X, any amino acid; *, cleavage position.
Data were taken from BREWER and SASSENFELD, (1990); SASSENFELD (1990); UHLÉN and MOKS (1990).

human insulin-like growth factor I (IGF-I) protein (SAMUELSSON et al., 1991).

6.4.2 Tags for Purification of Proteins

In the first applications of using a specific tag for purification purposes, easily detectable marker enzymes were fused to the protein of interest. This permitted the fusion protein to be monitored during the various purification steps via the enzyme function. β-Galactosidase and chloramphenicol acetyltransferase have been successfully used for this approach (SASSENFELD, 1990). In recent years, various strategies for chromatographic purification of proteins based on fusion of a specific tag to a protein have been developed (BREWER and SASSENFELD, 1990; SASSENFELD, 1990; UHLÉN and MOKS, 1990; ENFORS et al. 1990).

These systems greatly facilitate the selective recovery of proteins from cell lysates or culture supernatants in a few steps. Affinity purification tags are based on the selective interaction of the tag with a specific ligand bound to a chromatography matrix. This interaction can be based on antigen-antibody interactions or on the binding ability of proteins to distinct molecules (e.g., enzyme substrates or analogs thereof). Another strategy is to fuse a stretch of several charged (e.g., arginine) or hydrophobic amino acids (e.g., phenylalanine) to the protein of interest such that ion exchange chromatography or hydrophobic chromatography, respectively, can be efficiently used in the purification scheme. An additional strategy is to use tags based on a run of amino acids, particularly histidine or cysteine, which specifically bind to metal ions. Commercially available resins with immobilized chelating groups, such as iminodiacetic acid (Chelating-Sepharose), can be first loaded with the metal of choice and then used to purify the fusion protein by metal affinity. A survey of the most common fusion systems used for purification is given in Tab. 4.

A new concept designed to avoid the copurification of degradation products is to fuse a different affinity handle to each end of a protein (dual-affinity fusion). Two successive affinity purification steps selecting respectively for the N-terminal and for the C-terminal tag

Tab. 4. Examples of Gene Fusion Systems Used for Protein Purification

Tag Peptide	Purification Principle/Matrix
β-Galactosidase	AC; APTG, TPEG-Sepharose
CAT	AC; *p*-aminochloramphenicol-Sepharose
GST	Glutathione agarose
PhoS	AC; Hydroxylapatite
Protein A	IAC; IgG-Sepharose
Z-peptide	IAC; IgG-Sepharose
Flag™-peptide	IAC; specific anti-Flag antibody
MBP	AC; crosslinked amylose, starch
Polyarginine	IEC; S-Sepharose
Polyphenylalanine	HC; Phenyl-Superose
Polycysteine	CB; Thiopropyl-Sepharose
Polyhistidine	MC; Ni(II), Cu(II), Co(II); Nitrolotriacetic-acid-Sepharose
His-Trp dipeptide	MC; Ni(II); Iminodiacetic-Sepharose

CAT, chloramphenicol acetyltransferase; GST, glutathione-S-transferase; PhoS, phosphate-binding protein; Z, IgG-binding fragment based on staphylococcal protein A; Flag, specifically designed synthetic peptide; MPB, maltose binding protein; APTG, *p*-aminophenyl-β-D-thiogalactoside; TPEG, *p*-aminophenyl-β-D-thiogalactosidase; AC, affinity chromatography; IAC, immunoaffinity chromatography; IEC, ion exchange chromatography; HC, hydrophobic interaction chromatography; CB, covalent binding; MC, metal chelate.
Data were taken from SASSENFELD (1990) and UHLÉN and MOKS (1990).

permit only intact fusion proteins to remain in the final eluate (UHLÉN and MOKS, 1990).

Another application of protein fusions is to influence the partitioning behavior of proteins in aqueous two-phase systems which are frequently used as a primary recovery step in purification schemes. For example, by fusing one or several copies of a specifically designed tetrapeptide (AlaTrpTrpPro) to a protein, it was possible to drastically increase the partition coefficient resulting in a greatly improved efficiency of the two-phase extraction process (KÖHLER et al., 1991). In this case the tag is used to direct the protein into a specific phase of the extraction system.

6.5 Genetic Stability

Overexpression of a specific gene generally stresses the host cell. Under these conditions, cells try to remove the source of the stress. With plasmid vector systems, this results in genetic instability problems mainly caused by enhanced plasmid loss. However, structural instability resulting in inactivation of the expression system by mutations, which in *E. coli* are frequently caused by insertion elements, can also be responsible for low levels of synthesis of the desired product (SCHWAB, 1988).

The presence of plasmid-free cells in the culture may drastically reduce the yield of the desired product. The application of a selective pressure, via the presence of an antibiotic resistance gene on the vector plasmid and addition of the respective antibiotic to the culture medium, remains a valuable strategy for laboratory experiments. However, for several reasons this is not practical in industrial production processes. Various systems have been developed to reduce the presence of plasmid-free cells (KUMAR et al., 1991). Molecular strategies include the introduction of stabilizing functions to the expression vectors to ensure stable maintenance during the segregation process. Such functions are present on various natural plasmids (SCHWAB, 1988). Some stabilizing elements, such as the *par* region of plasmid RP4 (SAURUGGER et al., 1986; GERLITZ et al., 1990), can be quite efficient, even under conditions of expression stress (CROUZET et al., in press; HAIGERMOSER et al., in press).

Another possibility is to make use of genes which confer post-segregational killing activity to cells having lost the plasmid (GERDES et al.,

1986). Several other maintenance systems have been developed that are based on providing in *trans* a gene encoding an essential function to a host cell that is defective in this function (DEGRYSE, 1991). Such systems can be quite efficient, as was reported for the use of a cloned *ssb* gene (PORTER et al., 1990), but the requirement for specific mutants as hosts reduces the flexibility of such systems.

Integration of the expression cassette into the chromosome is a further possibility. This usually results in high segregational stability, but only one copy of the gene to be expressed will be present. An efficient integration system for *E. coli* has been recently developed on the basis of the bacteriophage lambda integration system (DIEDERICH and MESSER, 1991).

Acknowledgement

I am grateful to Dr. Ellen Zechner for careful and critical reading of the manuscript, and for many excellent suggestions for correcting the manuscript. I thank also my wife Margit for helpful discussions and for help in correcting the manuscript.

7 References

ANDERSON, M. L. M., YOUNG, B. D. (1985), Quantitative filter hybridization, in: *Nucleic Acid Hybridization, a Practical Approach* (HAMES, B. D., HIGGINS, S. J., Eds.), pp. 73–111, Oxford: IRL Press.

AUSUBEL, F. M., BRENT, R., KINGSTON, R. E., MOORE, D. D., SEIDMAN, J. G., SMITH, J. A., STRUHL, K. (1987), *Current Protocols in Molecular Biology,* New York: John Wiley & Sons.

BAILEY, J. F., BLANKENSHIP, J., CONDRA, J. H., MAIGETTER, R. Z., ELLIS, R. W. (1987), High-cell-density fermentation studies of a recombinant *E. coli* that express atrial natriuretic factors, *J. Ind. Microbiol.* **2,** 47–52.

BALBÁS, P., SOBERÓN, X., MERINO E, ZURITA, M., LOMELI, H., VALLE, F., FLORES, N., BOLIVAR, F. (1986), Plasmid vector pBR322 and its special purpose derivatives - a review, *Gene* **50,** 3–40.

BATES, P. (1987), Double *cos* site vectors: simplified cosmid cloning, *Methods Enzymol.* **153,** 82–94.

BATES, P. F., SWIFT, R. A. (1983), Double *cos* site vectors: simplified cosmid cloning, *Gene* **26,** 137–146.

BATY, D., LLOUBÉS, R., GELI, V., LAZDUNSKI, C., HOWARD, S. P. (1987), Extracellular release of colicin is non-specific, *EMBO J.* **6,** 2463–2468.

BECKER, G. W., HSIUNG, H. M. (1986), Expression, secretion and folding of human growth hormone in *Escherichia coli.* Purification and characterization, *FEBS Lett.* **204,** 145–150.

BERGER, S. L., KIMMEL, A. R. (Eds.) (1987), Guide to molecular cloning techniques, *Methods Enzymol.* **152,** Orlando: Academic Press.

BREWER, S. J., SASSENFELD, H. M. (1990), Engineering proteins for purification, in: *Protein Purification Applications, A Practical Approach* (HARRIS, E. L. V., ANGAL, S., Eds.), pp. 91–111, Oxford: IRL Press.

BRONSTEIN, I., MCGRATH, P. (1989), Chemiluminiscence lights up, *Nature* **338,** 599–600.

BROSIUS, J., LUPSKI, J. R. (1987), Plasmids for the selection and analysis of prokaryotic promoters, *Methods Enzymol.* **153,** 54–68.

BROWNSTEIN, B. H., SILVERMAN, G. A., LITTLE, R. D., BURKE, D. T., KORSMEYER, S. J., SCHLESSINGER, D., OLSON, M. V. (1989), Isolation of single copy human genes from a library of yeast artificial chromosome clones, *Science* **244,** 1348–1351.

BÜLOW, L., MOSBACH, K. (1991), Multienzyme systems obtained by gene fusions, *Trends Biotechnol.* **9,** 226–231.

BURNS, D. M., BEACHAM, I. R. (1984), Positive selection vectors: a small plasmid vector useful for the direct selection of Sau3A-generated overlapping DNA fragments, *Gene* **27,** 323–325.

CAMERON, B., BRIGGS, K., PRIDMORE, S., BREFORT, G., CROUZET, J. (1989), Cloning and analysis of genes involved in coenzyme B12 biosynthesis in *Pseudomonas denitrificans, J. Bacteriol.* **171,** 547–557.

CARLE, G. F., OLSON, M. V. (1984), Separation of chromosomal DNA molecules from yeast by orthogonal field alternation gele electrophoresis, *Nucleic Acids Res.* **12,** 5647–5664.

CARLSON, D. D., SUPERKO, C., MACKEY, J., GASKILL, M. E., HANSEN, P. (1990), Chemiluminescent detection of nucleic acid hybridization, *Focus* **12,** 9–12.

CHANG, A. C. Y. C., COHEN, S. N. (1978), Construction and characterization of amplifiable multicopy DNA cloning vehicles derived from the P15A cryptic plasmid, *J. Bacteriol.* **134,** 1141–1156.

CHU, G., VOLLRATH, D., DAVIS, R. W. (1986), Separation of large DNA molecules by contour-clamped homogeneous electric fields, *Science* **234**, 1582–1585.

CLARK, S. M., LAI, E., BIRREN, W. B., HOOD, L. (1988), A novel instrument for separating large DNA molecules with pulsed homogeneous electric fields, *Science* **241**, 1203–1205.

CLARKE, L., CARBON, J. (1976), A colony bank containing synthetic ColE1 hybrids representative of the entire *E. coli* genome, *Cell* **9**, 91–99.

CLEMENS, M. J. (1984), Translation of eukaryotic messenger RNA in cell-free extracts, in: *Transcription and Translation – A Practical Approach* (HAMES, B. D., HIGGINS, S. J., Eds.), pp. 231–270, Oxford: IRL Press.

COHEN, S. N., CHANG, A. C. Y., BOYER, H., HELLING (1973), Construction of biologically functional plasmids *in vitro, Proc. Natl. Acad. Sci. USA* **70**, 3240–3244.

CONDEMIME, G., SMITH, C. L. (1990), Genetic mapping using large-DNA technology: alignment of SfiI and AvrII sites with the NotI genomic restriction map of *Escherichia coli* K-12, in: *The Bacterial Chromosome* (DRLICA, K., RILEY, M., Eds.), pp. 53–60, Washington, DC: American Society for Microbiology.

COSTA, M., WEIR, M., COULSON, A., SULSTON, J., KENYON, C. (1988), Posterior pattern formation in *C. elegans* involves position-specific expression of a gene containing a homeobox, *Cell* **55**, 747–756.

CROUZET, J., LÉVY-SCHIL, S., CAUCHOIS, L., CAMERON, B., Construction of a broad-host-range non-mobilizable stable vector carrying RP4 par-region, *Gene,* in press.

DABORA, R. L., COONEY, C. L. (1990), Intracellular lytic enzyme systems and their use for disruption of *Escherichia coli, Adv. Biochem. Eng. Biotechnol.* **43**, 13–30.

DEBOER, H. A., HUI, A. (1990), Sequences within ribosome binding site affecting messenger RNA translatability and methods to direct ribosomes to single messenger RNA species, *Methods Enzymol.* **185**, 103–114.

DEBOER, H. A., COMSTOCK, L. J., VASSER, M. (1983), The *tac* promoter: a functional hybrid derived from the *trp* and *lac* promoters, *Proc. Natl. Acad. Sci. USA* **80**, 21–25.

DEGRYSE, E. (1991), Stability of a host-vector system based on complementation of an essential gene in *Escherichia coli, J. Biotechnol.* **18**, 29–40.

DIDERICHSEN, B., CHRISTIANSEN, L. (1988), Cloning of maltogenic *alpha*-amylase from *Bacillus stearothermophilus, FEMS Microbiol. Lett.* **56**, 53–60.

DIEDERICH, L., MESSER, W. (1991), Ein Plasmidsystem zur Integration von DNA Fragmenten in die lambda Attachment Site attB des *E. coli*-Chromosoms, in: *Proceedings 15th Symposium on Plasmids and Gene Regulation,* Klein Aspach, Germany.

DODT, J. C., SCHMITZ, T., SCHAFER, T., BERGMANN, C. (1986), Expression, secretion and processing of hirudin in *E. coli* using the alkaline phosphatase signal sequence, *FEBS Lett.* **202**, 373–377

DOWER, W. J., MILLER, J. F. RAGSDALE, C.W. (1988), High efficiency transformation of *E. coli* by high voltage electroporation, *Nucleic Acids Res.* **16**, 6127–6145.

EHRMANN, M., BOYD, D., BECKWITH, J. (1990), Genetic analysis of membrane protein topology by a sandwich gene fusion approach, *Proc. Natl. Acad. Sci. USA* **87**, 7574–7578.

ENFORS, S.-O., HELLEBUST, H., KÖHLER, K., STRANDBERG, L., VEIDE, A. (1990), *Adv. Biochem. Eng. Biotechnol.* **43**, 31–42.

ERLICH, H. A. (Ed.) (1989), *PCR Technology: Principles and Applications for DNA Amplifications,* New York: Stockton Press.

ERLICH, H. A., COHEN, S. N., MCDEVITT, H. O. (1978), A sensitive radioimmunoassay for detecting products translated from cloned DNA fragments, *Cell* **13**, 681–689.

EPPSTEIN, L., SHEVITZ, J., YANG, X.-M., WEISS, S. (1989), Increased biomass production in a benchtop fermentor, *Bio/Technology* **7**, 1178–1181.

EVANS, G. A., WAHL, G. M. (1987), Cosmid vectors for genomic walking and rapid restriction mapping, *Methods Enzymol.* **152**, 604–610.

EVERETT, R. D., CHAMBON, P. (1982), A rapid and efficient method for region- and strand-specific mutagenesis of cloned DNA, *EMBO J.* **1**, 433–437.

FAVRE-BULLE, O., SCHOUTEN, T., KINGMA, J., WITHOLT, B. (1991), Bioconversion of *n*-octane to octanoic acid by a recombinant *Escherichia coli* cultured in a two-liquid-phase bioreactor, *Bio/Technology* **9**, 367–371.

FREY, J., BAGDASARIAN, M., FEISS, D., FRANKLIN, F. C. H., DESHUSSES, J. (1983), Stable cosmid vectors that enable introduction of cloned fragments into a wide range of Gram-negative bacteria, *Gene* **24**, 299–308.

FRIEHS, K., SCHÖRGENDORFER, K., SCHWAB, H., LAFFERTY, R. M. (1986), Cloning and phenotypic expression in *Escherichia coli* of a *Bacillus subtilis* gene fragment coding for sucrose hydrolysis, *J. Biotechnol.* **3**, 333–341.

FRISCHAUF, A.-M., LEHRACH, H., POLSTKA, A., MURRAY, N. M. (1983), Lambda replacement vectors carrying polylinker sequences, *J. Mol. Biol.* **170**, 827–842.

FÜRSTE, J. P., PANSEGRAU, W., FRANK, R., BLÖCKER, H., SCHOLZ, P., BAGDASARIAN, M., LANKA, E. (1986), Molecular cloning of the plasmid RP4 primase region in a multi-host-range tacP expression vector, *Gene* **43,** 119–131.

GARDINER, K., LAAS, W., PATTERSON, D. (1986), Fractionation of large mammalian DNA restriction fragments using vertical pulsed-field gradient electrophoresis, *Somatic Cell Mol. Genet.* **12,** 185–195.

GEORGE, H. J., WATSON, R. J., HARBRECHT, D. F., DELORBE, W. J. (1987), A bacteriophage λ cI857 cassette controls λ P_L expression vectors at physiologic temperatures, *Bio/Technology* **5,** 600–603.

GERDES, K., BECH, F. W., JORGENSEN, S. T., LOBNER-OLESSEN, A. (1986), Mechanism of post-segregational killing by the Hok gene product of the parB system of plasmid R1 and its homology with the RelF gene product of the *E. coli* relB operon, *EMBO J.* **5,** 2023–2029.

GERLITZ, M., HRABAK, O., SCHWAB, H. (1990), Partitioning of broad-host-range plasmid RP4 is a complex system involving site-specific recombination, *J. Bacteriol.* **172,** 6194–6203.

GIBSON, T. J., COULSON, A. R., SULSTON, J. E., MITTLE, P. F. R. (1987), Lorist 2, a cosmid with transcriptional terminators insulating vector genes from interference by promoters within the insert: Effect on DNA yield and cloned insert frequency, *Gene* **53,** 275–281.

GOLD, L., STORMO, G. T. (1990), High-level translation initiation, *Methods Enzymol.* **185,** 89–93.

GOFF, S. P. (1987), Gene isolation by retroviral tagging, *Methods Enzymol.* **152,** 469–481.

GUBLER, U., HOFFMAN, B. J. (1983), A simple and very effective method for generating cDNA libraries, *Gene* **25,** 263–269.

HAIGERMOSER, C., CHEN, G., GROHMANN, E., HRABAK, O., SCHWAB, H., Stability of r-microbes: Stabilization of plasmid vectors by the partitioning function of broad-host-range plasmid RP4, *J. Biotechnol.,* in press.

HANAHAN, D. (1983), Studies on transformation of *E. coli* with plasmids, *J. Mol. Biol.* **166,** 557–580.

HANSEN, C., BRAMAN, J. (1991), FLASH chemiluminescent detection with enzyme-conjugated oligonucleotide probes, *Strategies* **4,** 40–41.

HARKNESS, R. E., LUBITZ, W. (1987), Construction and properties of a chimeric bacteriophage lysis gene, *FEMS Microbiol. Lett.* **48,** 19–24.

HELFMAN, D. M., HUGHES, S. H. (1987), Use of antibodies to screen cDNA libraries prepared in plasmid vectors, *Methods Enzymol.* **152,** 451–457.

HENIKOFF, S. (1984), Unidirectional digestion with exonuclease III creates targeted breakpoints for DNA sequencing, *Gene* **28,** 351–359.

HENRICH, B., PLAPP, R. (1984), Use of a cloned bacteriophage gene to disrupt bacteria, *J. Biochem. Biophys. Methods* **10,** 25–34.

HO, S. N., HUNT, H. D., HORTON, R. M., PULLEN, J. K., PEASE, L. R. (1989), Site-directed mutagenesis by overlap extension using the polymerase chain reaction, *Gene* **77,** 51–59.

HOHN, B., COLLINS, J. (1988), Ten years of cosmids, *Trends Biotechnol.* **6,** 293–298.

HOROWITZ, B. H., DIMAIO, D. (1990), Saturation mutagenesis using mixed oligonucleotides and M13 templates containing uracil, *Methods Enzymol.* **185,** 599–611.

HSIUNG, H. M., CANTRELL, A., LUIRINK, J., OUDEGA, B., VEROS, A. J., BECKER, G. W. (1989), Use of bacteriocin release protein in *E. coli* for excretion of human growth hormone into the culture medium, *Bio/Technology* **7,** 267–271.

HÜBNER, P., IIDA, S., ARBER, W. (1988), Random mutagenesis using degenerate oligodeoxyribonucleotides, *Gene* **73,** 319–325.

HUNG, M.-C., WENSINK, P. C. (1984), Different restriction enzyme-generated sticky DNA ends can be joined *in vitro, Nucleic Acids Res.* **12,** 1863–1874.

HUNTER, I. S., BAUMBERG, S. (1989), Molecular genetics of antibiotic formation, in: *Microbial Products: New Approaches* (BAUMBERG, S., HUNTER, I., RHODES, M., Eds.), pp. 121–162, Cambridge: Cambridge University Press.

HUYNH, T. V., YOUNG, R. A., DAVIS, R. W. (1985), Constructing and screening of cDNA libraries in λgt10 and λgt11, in: *DNA Cloning: a Practical Approach,* Vol. 1 (GLOVER, D. M., Ed.), pp. 49–78. Oxford: IRL Press.

INNIS, M. A., GELFAND, D. H., SNINSKY, J. J., WHITE, T. J. (Eds.) (1990), *PCR Protocols: A Guide to Methods and Applications,* San Diego: Academic Press.

ISH-HOROWITZ, D., BURKE, J. F. (1981), Rapid and efficient cosmid cloning, *Nucleic Acids Res.* **9,** 2989–2998.

JACKSON, D. A., SYMONS, R. H., BERG, P. (1972), Biochemical method for inserting new genetic information into DNA of simian virus 40: circular SV40 DNA molecules containing lambda phage genes and the galactose operon of *E. coli, Proc. Natl. Acad. Sci. USA* **69,** 2904–2909.

JAGUS, R. (1987), Hybrid selection of mRNA and hybrid arrest of translation, *Methods Enzymol.* **152,** 567–572.

JOSEPHSON, S., BISHOP, R. (1988), Secretion of peptides from *E. coli*: a production system for the pharmaceutical industry, *Trends Biotechnol.* **6,** 218–223.

KADOWAKI, H., KADOWAKI, T., WONDISFORD, F. E., TAYLOE, S. I. (1989), Use of polymerase chain reaction catalyzed by Taq DNA polymerase for site-specific mutagenesis, *Gene* **76**, 161–166.

KAMMANN, M., LAUFS, J. SCHELL, J., GRONENBORN, B. (1989), Rapid insertional mutagenesis of DNA by polymerase chain reaction (PCR), *Nucleic Acids Res.* **17**, 5404.

KAPRÁLEK, F., JECMEN, P., SEDLÁCEK, J., FÁBRY, M., ZADRAZIL, S. (1991), Fermentation conditions for high-level expression of the *tac*-promoter-controlled calf chymosin cDNA in *Escherichia coli* HB101, *Biotechnol. Bioeng.* **37**, 71–79.

KAYTES, P. S., THERIAULT, N. Y., POORMAN, R. A., MURAKAMI, K., TOMICH, C. C. (1986), High level expression of human rennin in *Escherichia coli, J. Biotechnol.* **4**, 205–218.

KESSLER, C., MANTA, V. (1990), Specificity of restriction endonucleases and DNA modification methyltransferases - a review (edition 3), *Gene* **92**, 1–248.

KHOSLA, C., CURTIS, J. E., BYDALEK, P., SWARTZ, J. R., BAILEY, J. E. (1990), Expression of recombinant proteins in *Escherichia coli* using an oxygen-responsive promoter, *Bio/Technology* **8**, 554–558.

KIKUCHI, Y., YODA, K., MAMASAKI, M., TAMURA, G. (1981), The nucleotide sequence of the promoter and the amino-terminal region of alcaline phosphatase structural gene (phoA) of *Escherichia coli, Nucleic Acids Res.* **9**, 5671–5678.

KINGSMAN, S. M., KINGSMAN, A. J. (1988), *Genetic Engineering: An Introduction to Gene Analysis and Exploitation in Eukaryotes,* Oxford: Blackwell Scientific Publications.

KÖHLER, K., LJUNGQUIST, C., KONDO, A., VEIDE, A., NILSSON, B. (1991), Engineering proteins to enhance their partition coefficients in aqueous two-phase systems, *Bio/Technology* **9**, 642–646.

KORCH, C. (1987), Cross index for improving cloning selectivity by partially filling in 5′-extensions of DNA produced by type II restriction endonuclease, *Nucleic Acids Res.* **15**, 3199–3220.

KORONAKIS, V., KORONAKIS, E., HUGHES, C. (1989), Isolation and analysis of the C-terminal signal directing export of *Escherichia coli* hemolysin protein across both bacterial membranes, *EMBO J.* **8**, 595–605.

KRAMER, W., FRITZ, H.-J. (1990), Oligonucleotide-directed mutation construction, in: *Modern Methods in Protein- and Nucleic Acid Research* (TSCHESCHE, H., Ed.), pp. 19–36, Berlin: Walter de Gruyter.

KUMAR, P. K. R., MASCHKE, H.-E., FRIEHS, K., SCHÜGERL, K. (1991), Strategies for improving plasmid stability in genetically modified bacteria in bioreactors, *Trends Biotechnol.* **9**, 279–284.

KUNKEL, T. A. (1985), Rapid and efficient site-specific mutagenesis without phenotypic selection, *Proc. Natl. Acad. Sci. USA* **82**, 488–492.

KUNKEL, T. A., ROBERTS, J. D., ZAKOUR, R. A. (1987), Rapid and efficient site-specific mutagenesis without phenotypic selection, *Methods Enzymol.* **154**, 367–382.

LANDT, O., GRUNERT, H.-P., HAHN, U. (1990), A general method for site-specific mutagenesis using the polymerase chain reaction, *Gene* **96**, 125–128.

LEMKE, G., AXEL, R. (1985), Isolation and sequence of a cDNA encoding a major structural protein of peripheral myelin, *Cell* **40**, 501.

LEONARDO, E. D., SEDIVY, J. M. (1990), A new vector for cloning large eukaryotic DNA segments in *Escherichia coli, Bio/Technology* **8**, 841–844.

LOHMAN, K., AMASINO, R., SIVE, H. L., ST. JOHN, T. (1988), Differentially expressed genes by subtraction, *The Digest from Invitrogen* **1**, (3), 2.

MACKMAN, N., BAKER, K., GRAY, L., HAIGH, R., NICAUD, J.-M., HOLLAND, I. B. (1987), Release of a chimeric protein into the medium from *Escherichia coli* using the C-terminal secretion signal of haemolysin, *EMBO J.* **6**, 2835–2841.

MARCH, C. J., MOSLEY, B., LARSEN, A., CERRET, D. P., BRAEDT, G., PRICE, V., GILLIS, S., HENNEY, C. S., KRUNHEIM, S. P., GRABSTEIN, K., CANLON, P. J., HOPP, P., COSMAN, D. (1985), Cloning, sequence, and expression of two distinct interleukin I cDNAs, *Nature* **315**, 641.

MARTIN, J. F. (1989), Molecular genetics of amino acid-producing corynebacteria, in: *Microbial Products: New Approaches* (BAUMBERG, S., HUNTER, I., RHODES, M., Eds.), pp. 25–59, Cambridge: Cambridge University Press.

MARULLO, S., DELAVIER-KLUTCHKO, C., GUILLET, J.-G., CHARBIT, A., STROSBERG, A. D., EMORINE, L. J. (1989), Expression of human β-1 and β-2 adrenergic receptors in *E. coli* as a new tool for ligand screening, *Bio/Technology* **7**, 923–927.

MAXAM, A. M., GILBERT, W. (1977), A new method for sequencing DNA, *Proc. Natl. Acad. Sci. USA* **74**, 560–564.

MEAD, D. A., PEY, N. K., HERRNSTADT, C., MARCIL, R. A., SMITH, L. M. (1991), A universal method for the direct cloning of PCR amplified nucleic acid, *Bio/Technology* **9**, 657–663.

MEIGHEN, E. A. (1991), Molecular biology of bacterial bioluminiscence, *Microbiol. Rev.* **55**, 123–142.

MEINKOTH, J., WAHL, G. (1984), Hybridization of nucleic acids immobilized on solid supports, *Anal. Biochem.* **138,** 267–284.

MIERENDORF, R. C., PERCY, C., YOUNG, R. A. (1987), Gene isolation by screening λgt11 libraries with antibodies, *Methods Enzymol.,* 458–469.

MIHOVILOVIC, M., LEE, J. E. (1989), An efficient method for sequencing PCR amplified DNA, *BioTechniques* **7** (1), 14.

MITRAKI, A., KING, J. (1989), Protein folding intermediates and inclusion body formation, *Bio/ Technology* **7,** 690–697.

MIYAKE, T., OKA, T., NISHIZAWA, T., MISOKA, F., FUWA, T., YODA, K., YAMASAKI, M., TAMURA, M. (1985), Secretion of human interferon-α induced by using secretion vectors containing a promoter and signal sequence of alkaline phosphatase gene of *Escherichia coli, J. Biochem.* **97,** 1429–1436.

MYERS, R. M., LERMAN, L. S., MANIATIS, T. (1985), A general method for saturation mutagenesis of cloned DNA fragments, *Science* **229,** 242–246.

NAGAHARI, K., KANAYA, S., MUNAKATA, K., AOYAGI, Y., MIZUSHIMA, S. (1986), Secretion into the culture medium of a foreign gene product from *Escherichia coli*: use of the ompF gene for secretion of human β-endorphin, *EMBO J.* **4,** 3589-3592.

NILSSON, B., UHLEN, M., JOSEPHSON, S., GATENBECK, S., PHILIPSON, L. (1983), An improved positive selection vector constructed by oligonucleotide mediated mutagenesis, *Nucleic Acids Res.* **11,** 8019-8026.

OCHMAN, H., GERBER, A. S., HARTL, D. L. (1988), Genetic applications of an inverse polymerase chain reaction, *Genetics* **120,** 621–623.

OKA, T., BAKAMOTO, S., MIYASHI, K.-I., FUWA, T., YODA, K., YAMASAKI, M., TAMURA, G., MIYAKE, T. (1985), Synthesis and secretion of human epidermal growth factor by *Escherichia coli, Proc. Natl. Acad. Sci. USA* **82,** 7212–7216.

OKAYAMA, H., BERG, P. (1982), High-efficiency cloning of full-length cDNA, *Mol. Cell. Biol.* **2,** 161–170.

PARDUE, M. L. (1985), *In situ* hybridization, in: *Nucleic Acid Hybridization – A Practical Approach* (HAMES, B. D., HIGGINS, S. J., Eds.), pp. 179–202, Oxford: IRL Press.

PHEIFFER, B. H., ZIMMERMAN, S. B. (1983), Polymer-stimulated ligation: enhanced blunt- or cohesive-end ligation of DNA or deoxyoligonucleotides by T4 DNA ligase in polymer solutions, *Nucleic Acids Res.* **11,** 7853–7871.

PLÜCKTHUN, A. (1991), Antibody engineering: Advances from the use of *Escherichia coli* expression systems, *Bio/Technology* **9,** 545–551.

PORTER, R. D., BLACK, S., PANNURI, S., CARLSON, A. (1990), Use of the *Escherichia coli* ssb gene to prevent bioreactor takeover by plasmidless cells, *Bio/Technology* **8,** 47–51.

PRATT, J. M. (1984), Coupled transcription-translation in procaryotic cell-free systems, in: *Transcription and Translation – A Practical Approach* (HAMES, B. D., HIGGINS, S. J., Eds.), pp. 179–209, Oxford: IRL Press.

RAO, R. N., RICHARDSON, M. A. KUHSTOSS, S. (1987), Cosmid shuttle vectors for cloning and analysis of *Streptomyces* DNA, *Methods Enzymol.* **153,** 166–198.

REDDY, G. P., ALLON, R., MEVARECH, M., MENDELOVITZ, S., SATO, Y., GUTNICK, D. L. (1989), Cloning and expression in *Escherichia coli* of an esterase-coding gene from the oil-degrading bacterium *Acinetobacter calcoaceticus* RAG-1, *Gene* **76,** 145–152.

RICCHIARDI, R. P., MILLER, J. S., ROBERTS, B. E. (1979), Purification and mapping of specific mRNAs by hybridization selection and cell free translation, *Proc. Natl. Acad. Sci. USA* **76,** 4921.

RIESENBERG, D., SCHULZ, V., KNORRE, W. A., POHL, H.-D. KORZ, D., SANDERS, E. A., ROSS, A., DECKWER, W.-D. (1991), High cell density cultivation of *Escherichia coli* at controlled specific growth rate, *J. Biotechnol.* **20,** 17–28.

RODRIGUEZ, R. L., DENHARDT, D. T. (Eds.) (1988), *Vectors: A Survey of Molecular Cloning Vectors and Their Uses,* Stoneham, MA: Butterworth.

ROSENBERG, M., HO, Y. S., SHATZMAN, A. R. (1983), The use of pHC30 and its derivatives for controlled expression of genes, *Methods Enzymol.* **101,** 123–138.

RUSCHE, J. R., HOWARD-FLANDERS, P. (1985), Hexamine cobald chloride promotes intermolecular ligation of blunt end DNA fragments by T4 DNA ligase, *Nucleic Acids Res.* **13,** 1997–2008.

SABATIÉ, J., SPECK, D., REYMUND, J., HEBERT, C., CAUSSIN, L., WELTIN, D., GLOECKLER, R., O'REAGAN, M., BERNARD, S., LEDOUX, C., OHSAWA, I., KAMOGAWA, K., LEMOINE, Y., BROWN, S. W. (1991), Biotin formation by recombinant strains of *Escherichia coli:* influence of the host physiology, *J. Biotechnol.* **20,** 29–50.

SAIKI, R. K., GELFAND, D. H., STOFFEL, S., SCHARF, S. J., HIGUCHI, R., HORN, G. T., MULLIS, K. B., ERLICH, H. A. (1988), Primer-directed enzymatic amplification of DNA with a thermostable DNA polymerase, *Science* **239,** 487–488.

SAMBROOK, J., FRITSCH, E. F., MANIATIS, T. (1989), *Molecular Cloning – A Laboratory Man-*

ual, 2nd Ed., Cold Spring Harbor Laboratory Press.

SAMUELSSON, E., WADENSTEN, H., HARTMANS, M., MOKS, T., UHLÉN, M. (1991), Facilitated *in vitro* refolding of human recombinant insulin-like growth factor I using a solubilizing fusion partner, *Bio/Technology* **9**, 363–366.

SANGER, F., NICKLEN, S., COULSON, A. R. (1977), DNA sequencing with chain-terminating inhibitors, *Proc. Natl. Acad. Sci. USA* **74**, 5463–5467.

SASSENFELD, H. M. (1990), Engineering proteins for purification, *Trends Biotechnol.* **8**, 88–93.

SAURUGGER, P. N., HRABAK, O., SCHWAB, H., LAFFERTY, R. M. (1986), Mapping and cloning of the par-region of broad-host-range plasmid RP4, *J. Biotechnol.* **4**, 333–343.

SCHATZ, P. J., BECKWITH, J. (1990), Genetic analysis of protein export in *Escherichia coli, Annu. Rev. Genet.* **24**, 215–248.

SCHEIN, C. H. (1989), Production of soluble recombinant proteins in bacteria, *Bio/Technology* **7**, 1141–1147.

SCHONER, B. E., BELAGAJE, R. M., SCHONER, R. G. (1990), Enhanced translational efficiency with two-cistron expression system, *Methods Enzymol.* **185**, 94–103.

SCHUBERT, P., STEINBÜCHEL, A., SCHLEGEL, H. G. (1988), Cloning of the *Alcaligenes eutrophus* genes for synthesis of poly-β-hydroxybutyric acid (PHB) and synthesis of PHB in *Escherichia coli, J. Bacteriol.* **170**, 5837–5847.

SCHWAB, H. (1988), Strain improvement in industrial microorganisms by recombinant DNA techniques, *Adv. Biochem. Eng. Biotechnol.* **37**, 129–168.

SETH, A. (1984), A new method for linker ligation, *Gene Anal. Tech.* **1**, 99–103.

SHARP, P. A., BERK, A. J., BERGET, S. M. (1980), Transcription maps of adenovirus, *Methods Enzymol.* **65**, 750–768.

SHATZMAN, A. R., ROSENBERG, M. (1987), Expression, identification and characterization of recombinant gene products in *Escherichia coli, Methods Enzymol.* **152**, 661–673.

SIMON, R., PRIEFER, U., PÜHLER, A. (1983), A broad host range mobilization system for *in vivo* genetic engineering: transposon mutagenesis in Gram-negative bacteria, *Bio/Technology* **1**, 784–791.

SINGER, R. H., LAWRENCE, J. B., VILLNAVE, C. (1986), Optimization of *in situ* hybridization using isotopic and non-isotopic methods, *BioTechniques* **4**, 230–250.

SKERRA, A., PFITZINGER, I., PLÜCKTHUN, A. (1991), The functional expression of antibody F_v fragments in *Escherichia coli:* improved vectors and a generally applicable purification technique, *Bio/Technology* **9**, 273–278.

SMITH, M. (1985), *In vitro* mutagenesis, *Annu. Rev. Genet.* **19**, 423–462.

SMITH, C. L., MATSUMOTO, T., NIWA, O., FAN, J. B., YANAGIDA, M., CANTOR, C. R. (1987), An electrophoretic karyotype for *Schizosaccharomyces pombe* by pulsed field gel electrophoresis, *Nucleic Acids Res.* **15**, 4481–4490.

SMITH, C. L., KLCO, S. R., CANTOR, C. R. (1988), Pulsed-field gel electrophoresis and the technology of large DNA molecules, in: *Genome Analysis – A Practical Approach* (DAVIES, K. E., Ed.), pp. 41–72, Oxford: IRL Press.

SMITH, H. O., BIRNSTIEL, M. L. (1976), A simple method for restriction site mapping, *Nucleic Acids Res.* **3**, 2387–2398.

SOUTHERN, E. M. (1975), Detection of specific sequences among DNA fragments separated by gel electrophoresis, *J. Mol. Biol.* **98**, 503–517.

SPÖK, A., STUBENRAUCH, G., SCHÖRGENDORFER, K., SCHWAB, H. (1991), Molecular cloning and sequencing of a pectinesterase gene from *Pseudomonas solanacearum, J. Gen. Microbiol.* **137**, 131–140.

STADER, J. A., SILHAVY, T. J. (1990), Engineering *Escherichia coli* to secrete heterologous gene products, *Methods Enzymol.* **185**, 166–187.

STANSSENS, P., OPSOMER, C., MC. KEOWN, Y. M., KRAMER, W., ZABEAU, M., FRITZ, H.-J. (1989), Efficient oligonucleotide-directed construction of mutations in expression vectors by the gapped duplex DNA method using alternating selectable markers, *Nucleic Acids Res.* **17**, 4441–4454.

STOKER, N. G., PRATT, J. M., HOLLAND, B. (1984), *In vivo* gene expression systems in prokaryotes, in: *Transcription and Translation – A Practical Approach* (HAMES, B. D., HIGGINS, S. J., Eds.), pp. 153–177, Oxford: IRL Press.

STORMO, G. T. (1986), Translation initiation, in: *Maximizing Gene Expression* (REZNIKOFF, W., GOLD, L., Eds.), p. 195–224, Boston: Butterworths.

STUDIER, F. W., MOFFAT, B. A. (1986), Use of bacteriophage T7 RNA polymerase to direct selective high-level expression of cloned genes, *J. Mol. Biol.* **189**, 113–130.

STUDIER, W. F., ROSENBERG, A. H., DUNN, J. J., DUBENDORFF, J. W. (1990), Use of T7 RNA polymerase to direct expression of cloned genes, *Methods Enzymol.* **185**, 80–89.

SU, T.-Z., SCHWEIZER, H., OXENDER, D. L. (1990), A novel phosphate-regulated expression vector in *Escherichia coli, Gene* **90**, 129–133.

TABOR, S. (1990), Expression using the T7 RNA polymerase/promoter system, in: *Current Protocols in Molecular Biology, Suppl. 11* (AUSUBEL, F. M., BRENT, R., KINGSTON, R. E., MOORE, D. D., SEIDMAN, J. G., SMITH, J. A., STRUHL, K., Eds.), pp. 16.2.1-16.2.11, New York: John Wiley & Sons.

TABOR, S., RICHARDSON, C. C. (1985), A bacteriophage T7 RNA polymerase/promoter system for controlled exclusive expression of specific genes, *Proc. Natl. Acad. Sci. USA* **82,** 1074-1078.

TAKAHARA, M., SAGAI, H., INOUYE, S., INOUYE, M. (1988), Secretion of human superoxide dismutase in *Escherichia coli, Bio/Technology* **6,** 195-198.

TAYLOR, J. W., OTT., J., ECKSTEIN, F. (1985), The rapid generation of oligonucleotide-directed mutations at high frequency using phosphorothioate-modified DNA, *Nucleic Acids Res.* **13,** 8765-8785.

THOMAS, P. S. (1980), Hybridization of denatured RNA and small DNA fragments to nitrocellulose, *Proc. Natl. Acad. Sci. USA* **77,** 5201.

TROUTMAN, M., CONSIDINE, K., BRAMAN, J. (1991), FLASH nonradioactive gene mapping kit, *Strategies* **4,** 5.

UHLÉN, M., MOKS, T. (1990), Gene fusions for purpose of expression: an introduction, *Methods Enzymol.* **185,** 129-143.

UHLIN, B. E., NORDSTRÖM K. (1978), A runaway-replication mutant of plasmid R1drd-19: temperature-dependent loss of copy number control, *Mol. Gen. Genet.* **165,** 167-179.

UHLIN, B. E., MOLIN, S., GUSTAFSSON, P., NORDSTRÖM, K. (1979), Plasmids with temperature-dependent copy number for amplification of cloned genes and their products, *Gene* **6,** 91-106.

VON HEIJNE, G. (1987), *Sequence Analysis in Molecular Biology,* San Diego: Academic Press.

WAGNER, C. R., BENKOVIC, S. J. (1990), Site directed mutagenesis: a tool for enzyme mechanism dissection, *Trends Biotechnol.* **8,** 263-270.

WAHL, G. M. (1989), Rapid and definitive restriction mapping, *Strategies* **2,** 1-3.

WAHL, G. M., BERGER, S. L., KIMMEL, A. R. (1987), Molecular hybridization of immobilized nucleic acids: theoretical concepts and practical considerations, *Methods Enzymol.* **152,** 399-407.

WALLACE, B., MIYADA, C. G. (1987), Oligonucleotide probes for the screening of recombinant DNA libraries, *Methods Enzymol.* **152,** 432-442.

WANNER, B.L. (1987), Phosphate regulation of gene expression in *Escherichia coli,* in: *Escherichia coli and Salmonella typhimurium,* Vol. 2, *Cellular and Molecular Biology* (NEIDHARDT, F. C., INGRAHAM, J. L., LOW, K. B., MAGASANIK, B., SCHAECHTER, M., UMBARGER, H. E., Eds.), pp. 1326-1333, Washington, DC: American Society for Microbiology.

WEDDEN, S. E., PANG, K., EICHELE, G. (1989), Expression pattern of homeobox-containing genes during chick embryogenesis, *Dev. Biol.* **105,** 639-650.

WEIHER, H., SCHALLER, H. (1982), Segment-specific mutagenesis: extensive mutagenesis of a lac promoter/operator element, *Proc. Natl. Acad. Sci. USA* **79,** 1408-1412.

WHITE, T. J., ARNHEIM, N., ERLICH, H. A. (1989), The polymerase chain reaction, *Trends Genet.* **5,** 185-189.

WILCHEK, M., BAYER, E. A. (1988), The avidin-biotin complex in bioanalytical applications, *Anal. Biochem.* **171,** 1-32.

WILKINSON, D. L., HARRISON, R. G. (1991), Predicting the solubility of recombinant proteins in *Escherichia coli, Bio/Technology* **9,** 443-448.

WILLIAMS, B. G., BLATTNER, F. R. (1979), Construction and characterization of the hybrid bacteriophage lambda Charon vectors for DNA cloning, *J. Virol.* **29,** 555-575.

WILLIAMS, J. G., MASON, P. J. (1985), Hybridization in the analysis of RNA, in: *Nucleic Acid Hybridization - A Practical Approach* (HAMES, B. D., HIGGINS, S. J., Eds.), pp. 139-160, Oxford: IRL Press.

WOOD, W. I. (1987), Gene cloning based on long oligonucleotides, *Methods Enzymol.* **152,** 443-447.

YANISH-PERRON, C., VIEIRA, J., MESSING, J. (1985), Improved M13 phage cloning vectors and host strains: nucleotide sequences of the M13mp18 and pUC18 vectors, *Gene* **33,** 103-119.

YANSURA, D. G. (1990), Expression as trpE fusion, *Methods Enzymol.* **185,** 161-166.

YANSURA, D. G., HENNER, D. J. (1990), Use of *Escherichia coli trp* promoter for direct expression of proteins, *Methods Enzymol.* **185,** 54-60.

YOUNG, R. A., DAVIS, R. W. (1983), Efficient isolation of genes by using antibody probes, *Proc. Natl. Acad. Sci. USA* **80,** 1194-1198.

ZELLER, R., BLOCH, K. D., WILLIAM, B. S., ARCECI, R. J., SEIDMAN, C. E. (1987), Localized expression of the atrial natriuretic factor gene during cardiac embryogenesis, *Genes Dev.* **1,** 693-698.

ZIEGLER, A., GEIGER, K. H., RAGOUSSIS, J., ZALAY, G. (1987), A new electrophoresis apparatus for separating very large DNA molecules, *J. Clin. Chem. Biochem.* **25,** 578-579.

11 Genetic Engineering of Gram Negative Bacteria

URSULA B. PRIEFER

Bielefeld, Federal Republic of Germany

1 Introduction

Molecular genetics and molecular engineering of and in *Escherichia coli* has experienced a revolutionary change in the early 1970s with the development of techniques for cutting and joining DNA fragments *in vitro* and establishing recombinant DNA molecules in *E. coli* host cells by transformation. It was now possible to clone genes from any organism in *E. coli*, to mutate them *in vitro* in order to learn something about their function or to separate them from their original regulatory signals to get them expressed at higher levels or under different conditions.

Since *in vitro* genetic engineering depends on the availability of versatile and specialized cloning vectors and on an efficient transformation system, it was mainly restricted to work in *E. coli*, since these prerequisites were poorly developed in non-enteric Gram negative bacteria. However, the use of *E. coli* as host organism to study gene function and gene expression is not always ideal: many functions of Gram negative bacteria that are of medical, agricultural, and industrial interest, for example, symbiotic and pathogenic traits or complex metabolic pathways, are expressed only in the presence of a specialized genetic and environmental background. Appropriate conditions cannot be mimicked in *E. coli*. Thus, genetic engineering of Gram negative non-*E. coli* bacteria also depends on the possibility to reintroduce the genes under study into the host species from which they originated. So, *in vitro* genetics of Gram negative functions is usually a two-step procedure: cloning and manipulation of the DNA of interest is performed in *E. coli*, using the wide array of recombinant DNA technology available; subsequently, the recombinant molecule is re-introduced into species similar or identical to those whence they were derived. The introduced genetic material can be established in two ways: either extrachromosomally on replicative plasmids or by integration into the host's genome.

With the discovery of prokaryotic transposable elements in the late 1960s, an additional powerful technique for engineering Gram negative bacteria became available, i.e., the manipulation of genetic material *in vivo*, without the need of mapping or even cloning the gene of interest. Since transposons work in practically all genetic backgrounds, they are ideal tools for manipulating DNA not only in *E. coli*, but also directly in a wide range of Gram negative bacteria, provided, a system for introducing the transposon into the species under study is available.

So, to date, genetic engineering of Gram negative non-*E. coli* bacteria is based on

- *in vitro* techniques to clone and manipulate genes in *E. coli*
- vectors and methods to introduce genetic material into the organism of choice
- *in vivo* manipulation techniques using transposons.

The first topic is extensively surveyed in Chapter 10, the latter are the subject of this review. The interested reader is also referred to another review article (SIMON and PRIEFER, 1990) which covers vector systems and genetic manipulation techniques in Gram negative bacteria with special emphasis on nitrogen fixation research. There are also recent, specialized articles on genetic systems and techniques in Gram negative bacteria including *Haemophilus influenzae* (BARCAK et al., 1991), pathogenic *Neisseria* (SEIFERT and SO, 1991), *Caulobacter crescentus* (ELY, 1991), *Agrobacterium* (CANGELOSI et al., 1991), *Rhizobium meliloti* (GLAZEBROOK and WALKER, 1991), Rhodospirillaceae (DONOHUE and KAPLAN, 1991), *Pseudomonas* (ROTHMEL et al., 1991) and *Vibrio* (SILVERMAN et al., 1991). The scope of this chapter is, therefore, to give a general overview on techniques and tools principally available for Gram negative bacteria without referring to special aspects in certain species.

2 Introduction of Genetic Material into Gram Negative Bacteria: The Binary Broad Host Range Conjugation System

As outlined above, the introduction of genetic material into a bacterial host of choice is a crucial step in any genetic engineering experiment of Gram negative bacteria, independent of whether you want to establish a gene cloned and manipulated in *E. coli* or whether you want to deliver a transposon for *in vivo* manipulation.

Although transformation and transduction of genetic material has been described for some non-enteric Gram negative organisms (see Chapter 2 by B. W. HOLLOWAY), these systems are usually restricted to special strain lines and not generally applicable at high efficiencies. Fortunately, natural plasmids with a broad host range have been detected which can be transmitted via conjugation at high frequencies to a large variety of Gram negative species. Some promote their own transfer, i.e., members of the incompatibility classes P and W (self-transmissible plasmids), while others utilize the conjugative apparatus of other plasmids, such as IncQ derivatives (mobilizable plasmids). These plasmids have proved extraordinarily useful as a basis for the development of vector systems which allow efficient transfer of DNA across species barriers.

To limit the potential hazard of promiscuous and uncontrolled spread (and to keep the vectors as small as possible) it is advantageous to have the broad host range conjugative apparatus separated from the replication functions, i.e., the vectors should be preferably mobilizable but not self-transmissible. Most broad host range systems used to date, therefore, consist of two parts: the actual vector plasmid which is the carrier of the genetic material to be introduced and able to be efficiently mobilized, and a helper plasmid which provides the functions necessary for conjugal transfer (binary vector system).

2.1 Conjugative Helper Plasmids

The first conjugative helper plasmid originated from the study of the replication functions of the IncP group plasmid RP4, also known as RP1 and RK2 (BURKARDT et al., 1979). In this system, the RP4 transfer genes are cloned onto a ColE1 replicon to yield the Km^r plasmid pRK2013 (FIGURSKI and HELINSKI, 1979). This plasmid provides all functions necessary for RP4-mediated conjugative transfer and thus can mediate mobilization of any other replicon which carries the RP4 specific origin of transfer (*oriT*, *rlx*, *mob* site, GUINEY and HELINSKI, 1979). There are also derivatives of narrow host range *E. coli* vectors (e.g., pBR322, pACYC184, pACYC177) containing the promiscuous IncW or IncN type transfer genes (MORALES and SEQUEIRA, 1985; SELVERAJ and IYER, 1983). Usually, transfer of the genetic material is performed in a triparental mating, i.e., the donor harboring the mobilizable vector plasmid, carrying the DNA to be introduced, and the recipient strain are mated in the presence of an *E. coli* strain containing the conjugative helper plasmid. Naturally, all the helper plasmids mentioned above also promote their own transfer, but due to the *E. coli* specific replication systems, they cannot be stably maintained in most non-*E. coli* hosts.

Another IncP-1-based mobilization system has been constructed in our laboratory (SIMON et al., 1983a, b). The conjugative RP4 helper plasmid has been immobilized in the *E. coli* donor strain by its integration into the chromosome. To avoid problems with the antibiotic resistance markers (RP4 encodes resistance against ampicillin, kanamycin, and tetracycline, markers which are often used as selectable traits for cloning vectors), an RP4 derivative has been constructed and used for the chromosomal integration, in which all three resistance genes were inactivated. This system has the advantage that the helper plasmid is tracked in the chromosome of the "mobilizing" donor strains (known as *E. coli* strains SM10 and S17-1). Any mobilizable plasmid to be conjugally transferred can be introduced into this strain, e.g., by transformation and will be efficiently mobilized without concomitant self-transmission of the helper plasmid (compare also Fig. 8).

This mobilizing system has been originally developed for and applied to work in the plant symbiotic bacterium *Rhizobium*; but due to the extremely broad host range of plasmid RP4, it is applicable and used worldwide for practically all Gram negative species.

2.2 Mobilizable Vector Plasmids

The vector plasmids used to introduce and to establish genetic material into Gram negative bacteria should meet the same requirements as for *E. coli* specific cloning vectors, i.e., they should be of small size, contain single cleavage sites for commonly used restriction enzymes (preferably a multiple cloning site, mcs) and code for selectable traits that can be selected for in a variety of different organisms. In addition, for reasons outlined above, they should be mobilizable at high efficiencies into a broad range of host organisms. In general, the development of vectors for bacterial non-*E. coli* hosts has followed a course similar to that of *E. coli* vectors. General purpose vectors, regulatable expression vectors, and vectors for the identification and analysis of promoters are available for a wide range of genera.

In principle, onc can distinguish between two major types of vectors:

- vector plasmids which combine the broad host range mobilization ability with a broad host range replication system (replicative vectors; broad host range vectors) and
- vector plasmids which are able of being efficiently mobilized into a broad range of non-*E. coli* species, but which contain a narrow host range replication system, i.e., these vectors are unable to maintain themselves in non-*E. coli* species (integrative vectors; narrow host range vectors).

Within the last years, a huge number of such broadly applicable vector plasmids has been constructed. Any catalog would be arbitrary and incomplete, but it is probably helpful to mention at least some of them.

2.2.1 Replicative Vectors

Basic replicons for this type of vector plasmids may be plasmids that are indigenous in the host under study. However, such plasmids usually can replicate only in their original host species, and corresponding vector derivatives are therefore of limited general applicability. Broadly applicable replicative vectors (Tab. 1) are based on natural plasmids belonging to the incompatibility groups IncP, IncQ, and IncW, which display replication proficiency in a diversity of Gram negative prokaryotes. They therefore allow the introduction and extrachromosomal maintenance of any genetic material in practically all Gram negative host organisms.

One of the best known and most widely used IncP derived broad host range vectors is pRK290 (and its cosmid variants, DITTA et al., 1980; FRIEDMAN et al., 1982) which is derived from plasmid RK2. It still retains the RK2 encoded Tc resistance gene, the *oriT* site (*mob*) and those regions essential for broad host range replication. There is a number of pRK290 derivatives and modifications, which contain additional selection markers and a larger variety of cloning sites. Examples include pRK293, pRK310 (DITTA et al., 1985), and the series of pVK vectors (KNAUF and NESTER, 1982). All commonly used derivatives contain the RK2 *mob* site and are therefore efficiently mobilizable in the presence of the RP4 transfer system. Most of the IncP-derived vector plasmids have one major disadvantage, i.e., they are relatively large (around 20 kb) and therefore offer a low cloning capacity.

Because of their small size (8.7 kb) and their efficient mobilization by conjugative plasmids, IncQ plasmids such as RSF1010 (Fig. 1; BARTH and GRINTER, 1974; GUERRY et al., 1974) have attracted much interest as the basis for the development of replicative broad host range vectors. In addition, these plasmids naturally possess a high copy number (10–50 copies in *E. coli*, in contrast to RK2 and its derivatives with 5–8 copies per chromosome), which – in some cases – is advantageous. The

Tab. 1. Some General and Specific Purpose Broad Host Range Cloning Vectors

Vector[a]	Size (kb)	Selectable Markers[b]	Cloning Sites[c]	Remarks
RK2 (IncP-1) Derivatives				
pRK290 (1)	20	Tet	Bg, E	
pRK293 (2)	21.4	Kan, Tet	H, X (Kan) S (Tet)	
pRK310 (2)	20.4	Tet, LacZ	B, H, P (LacZ)	
pLAFR1 (3)	21.6	Tet	E	Cosmid vector
pVK100 (4)	23	Kan, Tet	H, X (Kan) S (Tet) E	Cosmid vector
RSF1010 (IncQ) Derivatives				
pKT248 (5)	12.4	Cam, Str	Sc (Str) S (Cam) Bs, E	
pKT231 (5)	13.0	Kan, Str	C, Sm, X, H (Kan) Sc (Str) B, Bg, E	
pMMB33 (6)	13.8	Kan	B, E	Cosmid vector
pKT240 (7)	12.9	Kan, Amp	C, H, Sm, X (Kan) B, Bs, E	Promoter probe vector, promoterless Str^r gene
pMMB24 (7)	12.7	Amp	B, H	p*tac* expression vector
pSUP104 (8)	9.5	Cam, Tet	B, S (Tet) E (Cam)	
pSUP106 (8)	9.9	Cam, Tet	B, S (Tet) E (Cam)	Cosmid vector
pML122 (9)	11.9	Gen	Bg (Gen) mcs	p*npt* expression vector
pML7 (9)	12	Gen	Bg (Gen) mcs	Promoter probe vector, promoterless *lacZ* gene
Sa (IncW) Derivatives				
pGV1106 (10)	8.7	Kan, Str	E (Str) Kp (Kan) B, Bg, P	Non-mobilizable
pSa151 (11)	13.1	Spc, Kan		
pUCD615 (12)	17.75	Kan, Amp	p (Amp), mcs	Promoter probe vector, promoterless *lux* operon

[a] References: (1) Ditta et al., 1980; (2) Ditta et al., 1985; (3) Friedman et al., 1982; (4) Knauf and Nester, 1982; (5) Bagdasarian et al., 1981; (6) Frey et al., 1983; (7) Bagdasarian et al., 1983; (8) Priefer et al., 1985; (9) Labes et al., 1990; (10) Leemans et al., 1982; (11) Tait et al., 1983; (12) Rogowsky et al., 1987

[b] Abbreviations: Tet, tetracycline; Kan, kanamycin; Cam, chloramphenicol; Str, streptomycin; Amp, ampicillin; Gen, gentamicin; Spc, spectinomycin

[c] Abbreviations: E, *Eco*RI; C, *Cla*I; H, *Hin*dIII; X, *Xho*I; S, *Sal*I; P, *Pst*I; Bg, *Bgl*II; Sc, *Sac*I; Bs, *Bst*EII; Sm, *Sma*I; Kp, *Kpn*I; mcs, multiple cloning site
Resistance markers in brackets refer to insertional inactivation.

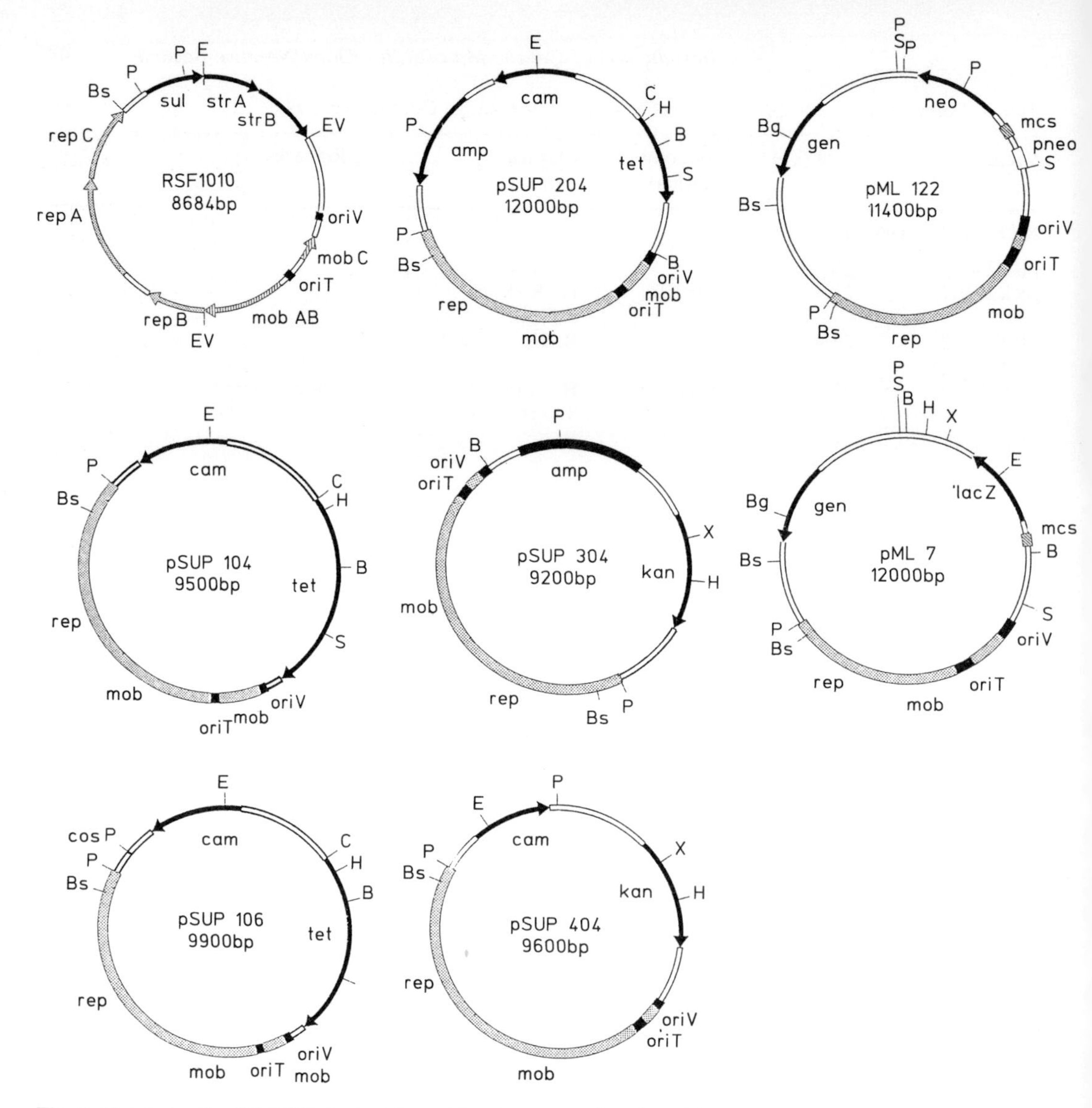

Fig. 1. Examples for RSF1010-derived broad host range vector plasmids.
The physical and genetic map of the IncQ plasmid RSF1010 is shown as well as some of the broad host range pSUP cloning vectors, including the cosmid derivative pSUP106, which were derived by cloning the replication and mobilization functions of RSF1010 into common *E. coli* cloning vectors. The part derived from RSF1010 is represented by the dotted fragment. Also illustrated are examples for pSUP104-based expression and promoter-probe vectors. Plasmid pML122 (expression vector) contains the promoter of the *npt* gene from Tn*5* (p*neo*) in front of a multiple cloning site (mcs); similar constructs are available which express genes cloned into mcs from the *lac* and the *tac* promoter, respectively (LABES et al., 1990). Plasmid pML7 represents one of the promoter probe vectors, derived from pSUP104. It contains a promoterless *lacZ* gene ('*lacZ*) downstream of a multiple cloning site (mcs).
Abbreviations:
Resistance markers: sul, sulfonamide; strA and strB, streptomycin; cam, chloramphenicol; kan, kanamycin; tet, tetracycline; amp, ampicillin; gen, gentamicin; 'neo, promoterless neomycin resistance gene; 'lacZ, promoterless β-galactosidase gene.
mob, genes involved in mobilization of RSF1010 and derivatives; *rep*, genes necessary for broad host range replication; *oriV*, origin of vegetative RSF1010 replication; *oriT*, origin of transfer replication; *cos*, cos site of phage lambda.
Restriction sites: B, *Bam*HI; Bs, *Bst*EII; Bg, *Bgl*II; C, *Cla*I; E, *Eco*RI; EV, *Eco*RV; H, *Hin*dIII; P, *Pst*I; S, *Sal*I; X, *Xho*I.

most commonly used RSF1010-derived constructs are the series of pKT vectors (BAGDASARIAN et al., 1981, 1982; BAGDASARIAN and TIMMIS, 1982; FREY et al., 1983) and the replicative pSUP vectors (Fig. 1; PRIEFER et al., 1985), which are both available as plasmid and cosmid vectors with different selectable markers and cloning sites. There are also derivatives which allow controlled expression of a cloned gene, e.g., through the regulatable *tac* and lambda pL promoters or through the *npt* promoter which is active in many Gram negative species (BAGDASARIAN et al., 1983; LEEMANS et al., 1987; LABES et al., 1990; Fig. 1). So-called promoter probe vectors, which contain an indicator gene (such as *lacZ*) devoid of its transcriptional and/or translational start signals, are especially suitable for the construction of transcriptional and translational gene fusions (NANO et al., 1985; LABES et al., 1990).

In general, the RSF1010-derived replicative vector plasmids are in the size range of 10–14 kb and thus have a larger insert capacity as compared to the IncP derivatives. Their high copy number facilitates DNA isolation and cloning procedures, but might - in some instances - cause problems and unexpected effects upon their introduction into the recipient strain of choice, especially if they carry cloned regulatory genes or regions. RSF1010 and its derivatives are naturally mobilizable by conjugative plasmids, including IncP plasmids, so that the IncP conjugative helper plasmids described above are equally well suited for efficient mobilization of the IncQ vector plasmids (GAUTIER and BONEWALD, 1980; WILLETS and CROWTHER, 1981; MEYER et al., 1982).

The IncW plasmid pSa has also served as a basis for the construction of broad host range vectors, yielding the vectors of the pGV family (LEEMANS et al., 1982) and the pSa vectors (TAIT et al., 1983). They have all lost their self-transferability, but some of them retained their ability to be mobilized by a separate helper plasmid containing the pSa transfer functions (see above).

2.2.2 Narrow Host Range Vectors

Vectors of this type are not able to replicate outside of *E. coli* and close relatives. Since they contain broad host range transfer or mobilization functions, they are well suited to deliver genetic material into any Gram negative cell, but maintenance is only achieved upon integration into the host's genome. Thus, they are especially useful as transposon carrier plasmids and also extremely helpful if stable integration of the introduced DNA via homologous recombination is possible.

In principle, the broad host range conjugative helper plasmids, such as pRK2013, described in the previous section, belong to this type of vector systems (Tab. 2), since they are transferable into, but cannot be maintained in many bacterial species due to their narrow host range replication apparatus (e.g., ColE1).

Loss or at least a severe reduction of its ability to become stably established in Gram negative bacteria was also observed for RP4 and other IncP-type plasmids, such as pPH1JI carrying the Mu bacteriophage genome (BOUCHER et al., 1977; VAN VLIET et al., 1978; HIRSCH and BERINGER, 1984). As a fact, this "suicide" effect, which was found more or less by chance, was the basis for the first transposon delivery system of broad applicability (DENARIE et al., 1977; BERINGER et al., 1978; HIRSCH and BERINGER, 1984).

Although these RP4 derivatives have been widely used to deliver transposons in a variety of Gram negative bacteria, they are relatively large and not suited for cloning, so that they cannot be used to introduce cloned and/or manipulated genetic material. An almost universal narrow host range system was developed by cloning the IncP-1-type specific recognition site for mobilization (*mob*-site, *oriT*) into commonly used *E. coli* vectors such as pACYC or pBR derivatives. This set of pSUP vectors (SIMON et al., 1983a, b; Fig. 2; Tab. 2) combines the advantages of the *E. coli* vectors (small size, single restriction sites, insertional inactivation) with the ability of being efficiently mobilized by the transfer functions of IncP plasmids (see above). Thus they can be used in *E. coli* like normal cloning vectors and can be introduced into all bacteria that can act as recipients for RP4; however, their replication host range is restricted to *E. coli* and close relatives. Consequently, they are extremely useful not only as transposon carrier plasmids for practically all Gram negative bacteria, but also to

Tab. 2. IncP-Type Transferable Narrow Host Range Vectors

Vector	Size (kb)	Replicon	Marker[a]	Cloning Sites[b]	Remarks[c]	References
pRK2013	48	ColE1	Kan	–	Tra$^+$	1
RP4::Mu	100	RP4	Amp, Kan, Tet	–	Tra$^+$, reduced replication due to Mu insertion	2, 3, 4
pSUP102	5.9	pACYC184	Cam, Tet	E (Cam), C, H, B, S (Tet)	Mob$^+$ (*oriT*$_{RP4}$)	5, 6
pSUP301	5.8	pACYC177	Amp, Kan	P (Amp) X, C, H (Kan)	Mob$^+$ (*oriT*$_{RP4}$)	5, 6
pSUP202	7.9	pBR325	Amp, Cam Tet	P (Amp) E (Cam) C, H, B, S (Tet)	Mob$^+$ (*oriT*$_{RP4}$)	5, 6
pSUP205	8.3	pBR325	Cam, Tet	E (Cam) C, H, B, S (Tet)	Mob$^+$ (*oriT*$_{RP4}$) cosmid vector	5, 6

[a] Kan, kanamycin resistance; Amp, ampicillin resistance; Tet, tetracycline resistance; Cam, chloramphenicol resistance

[b] E, *Eco*RI; C, *Cla*I; H, *Hin*dIII; B, *Bam*HI; S, *Sal*I; X, *Xho*I; markers in brackets refer to insertional inactivation

[c] Tra$^+$, self-transmissible; Mob$^+$, mobilizable in presence of RP4 transfer genes

References: 1 FIGURSKI and HELINSKI, 1979; 2 BOUCHER et al., 1977; 3 VAN VLIET et al., 1978; 4 DENARIE et al., 1977; 5 SIMON et al., 1983a; 6 SIMON et al., 1983b

deliver genetic material for genomic integration upon its cloning and engineering in *E. coli*.

3 Methods for Genomic Integration of Cloned DNA

The genetic material introduced can be maintained in the Gram negative host in two ways: either extrachromosomally on replicative plasmids or integrated into the host genome. For some purposes, integration into the genome by double crossing-over is indispensable, e.g., if the effect of a mutation on the host's phenotype is to be tested. But also in other cases, the integration of cloned DNA into the genome of the host organism is advantageous over its establishment on a replicative vector plasmid, especially with respect to stability or to undesired copy number effects. The following section summarizes the principles of genomic integration of cloned DNA by single and double crossing-over events.

3.1 Integration by Single Recombination Events

Stable integration into the genome is readily achieved by cloning the DNA fragment of interest into a narrow host range vector, which can be mobilized (e.g., one of the pSUP vectors described above), and by inserting the complete recombinant molecule into the host's genome via a single crossing-over between the cloned and the corresponding endogeneous DNA (Fig. 3). The recombination event can be selected for by the stable maintenance of a vector-encoded resistance marker. However, since integration depends on DNA homology, this method is restricted to the original host or to very close relatives. This limitation can be overcome by providing the vector with a se-

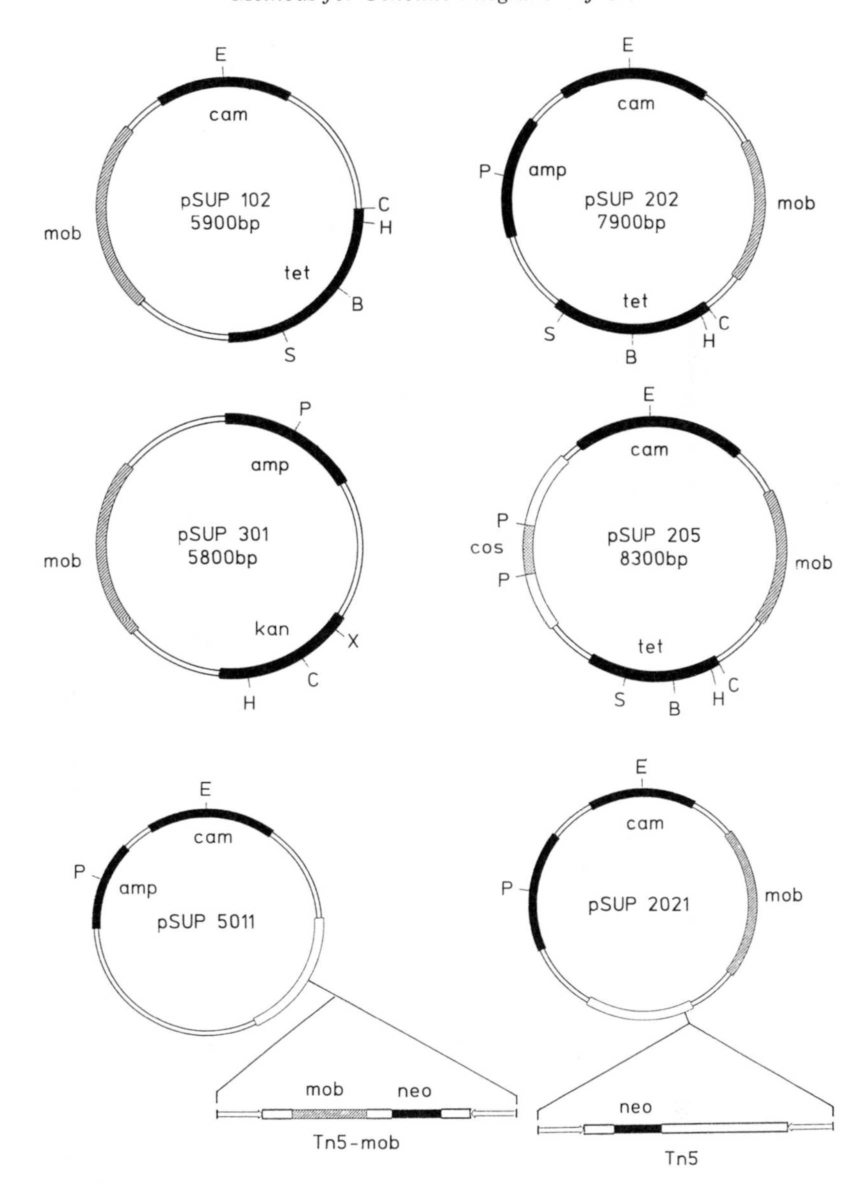

Fig. 2. Some of the mobilizable pSUP narrow host range cloning and transposon vectors.
The figure shows three of the original plasmid cloning vectors based on pACYC184 (pSUP102), pACYC177 (pSUP301), and pBR325 (pSUP202) as well as the cosmid derivative pSUP205. They all contain the *oriT* region (*mob*) of the IncP-1 plasmid RP4. Additionally, two transposon carrier vectors are shown; pSUP2021 contains a normal Tn*5* transposon inserted in the tetracycline resistance gene; pSUP5011 carries Tn*5*-*mob*. Abbreviations: amp, ampicillin; cam, chloramphenicol; kan, kanamycin; tet, tetracycline; neo, neomycin; *cos*, cos site of phage lambda; *mob, oriT* region of plasmid RP4.
Restriction sites: E, *Eco*RI; C, *Cla*I; H, *Hin*dIII; B, *Bam*HI; S, *Sal*I; X, *Xho*I.

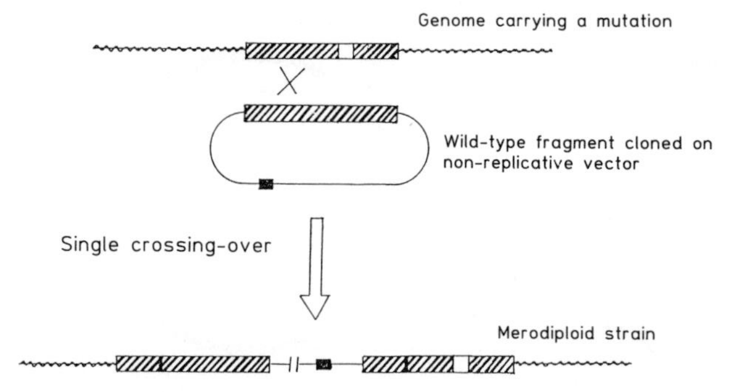

Fig. 3. Integration of cloned DNA into the genome.
A DNA fragment (in this example wild-type) is cloned into a narrow host range vector (e.g., one of the non-replicative pSUP vectors) and introduced via conjugation (or mobilization) into the recipient cell. Since the vector is not able to replicate in this host, selection for the maintenance of a vector-encoded resistance marker (indicated by a black box) results in a merodiploid strain in which the complete vector molecule, harboring the cloned DNA fragment, is integrated into the host's genome (in this case carrying a mutation in the corresponding DNA region) via a single recombination event.

quence which confers homology also to heterologous hosts. Such vectors have been constructed in our laboratory: the *recA* gene of the *Rhizobium meliloti* strain 2011 has been isolated and, provided with an internal multiple cloning site, inserted into a derivative of one of the narrow host range pSUP vectors described above (SELBITSCHKA et al., 1991, 1992). Thus, DNA originating from any organism can be cloned into the *R. meliloti recA* gene in *E. coli*, mobilized into *R. meliloti*, and stably integrated via the *recA*-mediated homology. Similarly, *recA* integration vectors for *R. leguminosarum* have been constructed and allow the integration of heterologous DNA into the *R. leguminosarum* genome.

Genomic integration via single crossing-over events gives rise to partial diploids carrying the cloned material (which might contain a wild-type gene, a mutated gene, or any other genetic construct, e.g., a gene or operon fusion) stably integrated additionally to the genomic copy (which again can be wild-type or mutated). Consequently, this strategy is especially helpful for the analysis of gene expression and regulation.

In addition, the construction of such merodiploid strains can be used as a first step in a so-called "poor man's cloning" experiment (*in-vivo* cloning, Fig. 4), for example, to isolate a transposon-marked mutant gene from the genome. The wild-type gene, cloned on a non-replicative, mobilizable vector, is introduced into the mutant strain and integrated into the genome via homologous, single crossing-over. In a second recombination event, the vector plasmid containing the mutated gene, can be excised, and, upon additionally introducing a conjugative helper plasmid, retransferred to *E. coli* for further analysis.

3.2 Gene Replacement by Double Recombination Events

The second type of genomic integration of DNA, which is extremely useful for the analysis of the function and genetic organization of a cloned DNA fragment, is based on a double crossing-over event and results in the exchange of the genomic DNA with the cloned DNA (Fig. 5). Usually, this technique is applied to replace wild-type DNA against the same, but altered DNA region: the cloned DNA fragment is manipulated in *E. coli* either *in vivo* by transposons or *in vitro* by inserting appropriate cartridges and transferred back into its original host where a double crossing-over between the genomic and the introduced DNA region leads to stable insertion of the engineered DNA fragment (gene replacement, marker exchange, homogenotization). This technique not only allows the analysis of gene

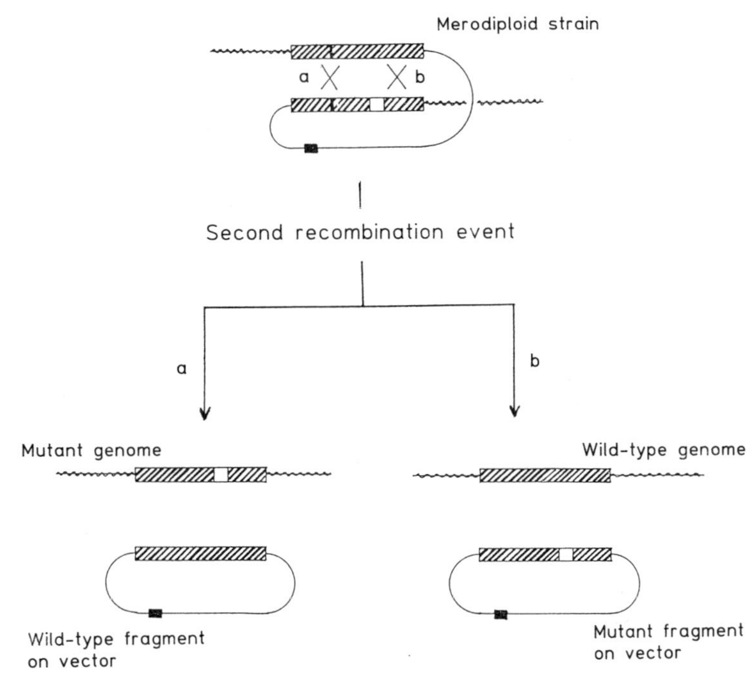

Fig. 4. *In vivo* cloning of a genomic DNA fragment.
The wild-type gene cloned on a non-replicative, mobilizable vector is introduced into the mutant strain via homologous single crossing-over (compare Fig. 3). Second recombination events lead to the excision of the vector molecule carrying either the wild-type DNA region (a) or the mutated DNA fragment (b). Upon introduction of a conjugative helper plasmid, the recombinant vector molecules can be transferred to *E. coli* for further investigations. If the genomic mutation is marked by an antibiotic resistance (e.g., by a transposon), both types of excised recombinant molecules can be easily distinguished by their resistance markers (in (a) only vector-encoded resistance, in (b) vector-encoded + transposon-encoded resistance).

function by the introduction of a mutation at a defined site in the host genome, but also the elucidation of the genetic organization by comparing the physical map with the phenotype resulting from the gene replacement.

The transfer of the manipulated DNA segment back into the original host may be achieved by broad host range (replicative) or narrow host range vectors. The first strategy, originally described by RUVKUN and colleagues for pRK290 (RUVKUN and AUSUBEL, 1981; RUVKUN et al., 1982), depends on the displacement of the replicative vector by introducing a plasmid of the same incompatibility group through a second mating: simultaneous selection for the incoming plasmid and maintenance of the mutated DNA fragment (resistance marker encoded by the transposon or cartridge) results in transconjugants in which the desired double crossing-over event has taken place (Fig. 5). The use of narrow host range vectors, preferably one of the mobilizable pSUP vectors, greatly facilitates the identification of homogenotization events (SIMON et al., 1986), since they are *per se* unable to replicate in the host organism and do not need to be displaced from the transconjugants (Fig. 5). After selection for the presence of the marker carried by the mutated DNA fragment, the clones only have to be screened for loss of the vector molecule (i.e., absence of a vector-encoded resistance marker) in order to distinguish between single crossing-over and the desired gene replacement by double crossing-over.

In any case, the marker exchange should be confirmed physically by appropriate hybridization experiments.

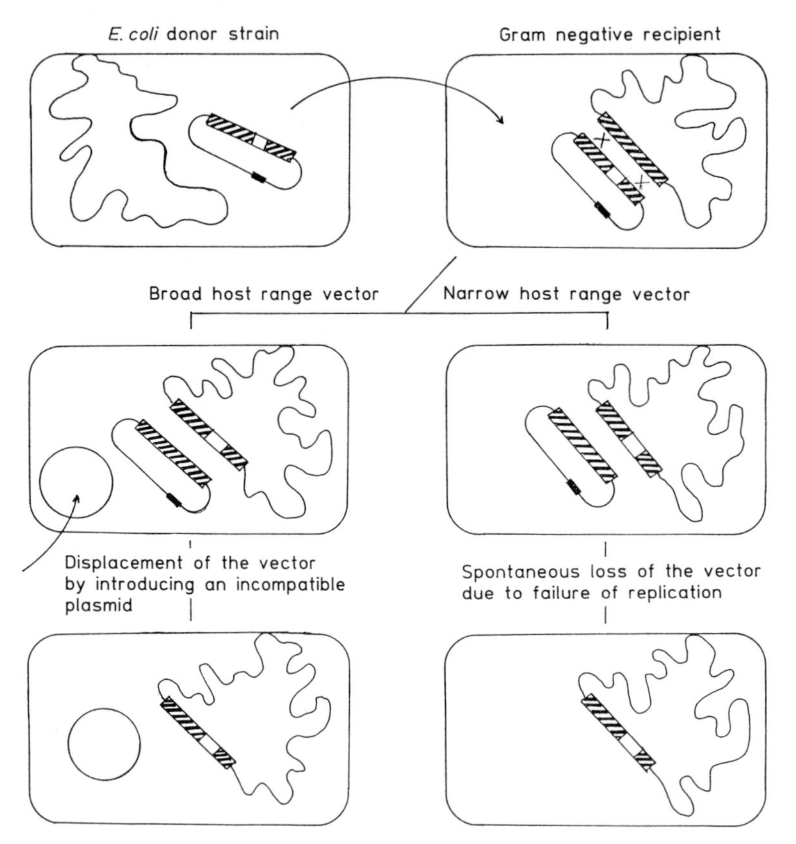

Fig. 5. Genomic integration of cloned DNA by double recombination events (gene replacement).
A cloned (here mutagenized) fragment is transferred from *E. coli* into the Gram negative recipient by conjugation. A double crossing-over event between the genomic and the introduced DNA regions leads to the replacement of the wild type (genomic) with the mutated (cloned) DNA fragment. In order to select these marker exchange events, the recombinant vector plasmid has to be eliminated from the recipient. If the vector molecule is able to replicate in the host (broad host range vector) it is usually displaced by introducing a second plasmid of the same incompatibility group and selecting for the maintenance of the incoming plasmid. If vectors with a narrow replication host range are used (e.g., non-replicative pSUP vectors), they are automatically segregated and lost in the host organism.

3.3 Introduction of Unmarked Mutations into the Genome

In many cases (for example, to construct strains with multiple mutations or for complementation analyses) it is advantageous to introduce an unmarked mutation into the genome of a given strain. For this purpose, a non-selectable mutation, such as a point mutation or a deletion, is generated in the DNA fragment, cloned in a mobilizable narrow host range vector, in *E. coli* and then introduced via bacterial mating and single crossing-over into the genome of the recipient strain. Integration of the recombinant plasmid can be identified by selection for a vector-specific antibiotic resistance (compare Fig. 3). A second recombination event (as outlined in Fig. 4) can lead to the excision of the vector molecule together with the wild-type sequence, leaving the modified sequence in the genome. Using narrow host range pSUP vectors, this method has been successfully applied, for example, in *Rhi-*

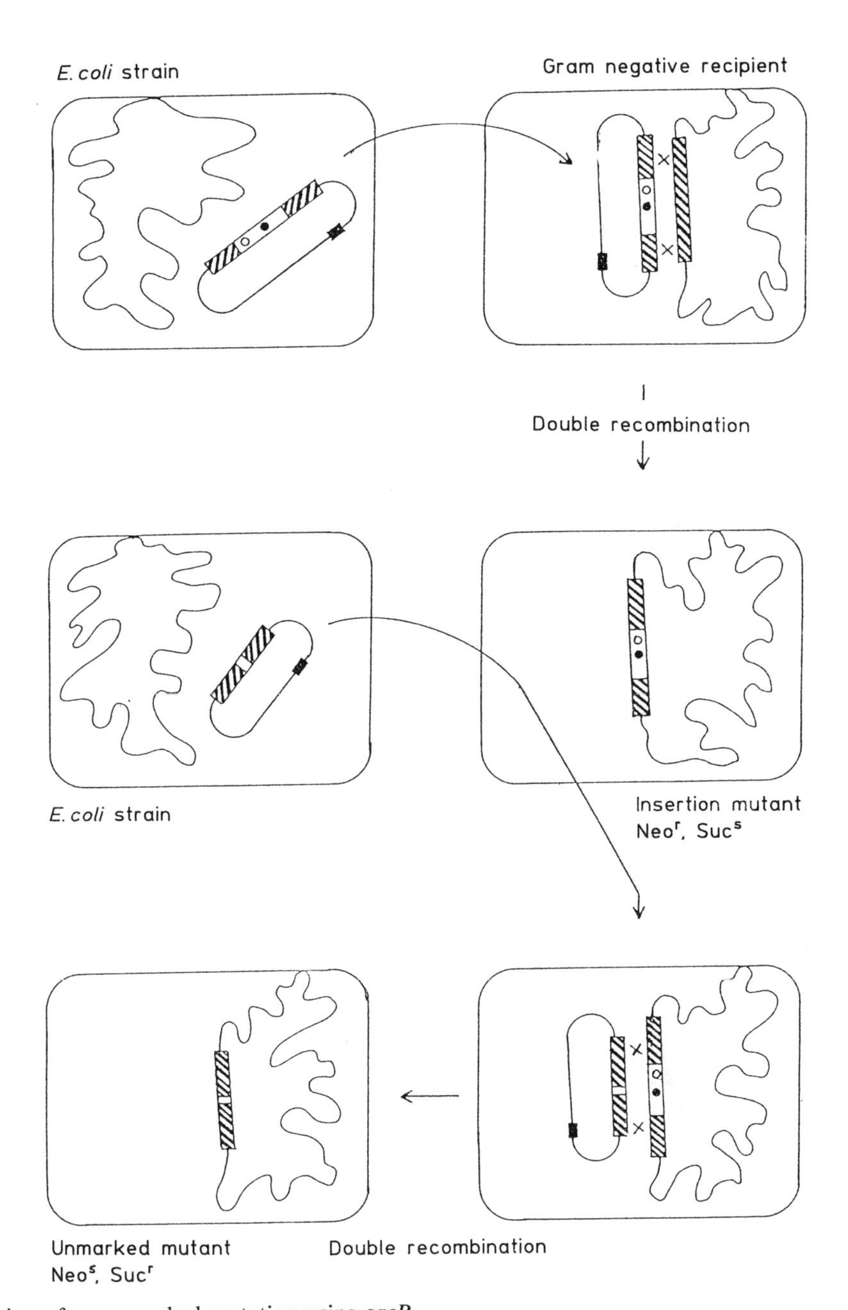

Fig. 6. Genomic integration of an unmarked mutation using *sacB*.
The fragment, cloned in a narrow host range vector, is mutagenized in *E. coli* by inserting the *nptI-sacB-sacR* cassette (symbolized by [○ ●]). The recombinant plasmid is introduced into the Gram negative recipient and double recombination events are isolated by selection for neomycin resistance (Neo^r; expressed from the *nptI* gene) and loss of the vector encoded resistance marker (marked by a black box). The insertion mutant, generated by this gene replacement is sucrose-sensitive (Suc^s) due to the presence of the *sacB* gene. In a second step, most of the *nptI-sacB-sacR* cartridge is removed from the cloned fragment, resulting in an unmarked point mutation (frame shift). The recombinant molecule is introduced from *E. coli* into the insertion mutant, where a second double recombination event can be selected on plates containing sucrose. Only cells which have lost the *nptI-sacB-sacR* cartridge by marker exchange are able to grow on high sucrose concentrations (Neo^s, Suc^r).

zobium (JAGADISH et al., 1985; NOTI et al., 1987); however, this method can be very time-consuming and tedious, since often many single colonies have to be tested to identify the desired second recombination event.

The procedure has been simplified by RIED and COLLMER (1987) by the development of a marker exchange–eviction method which is based on the exchange of the unmarked mutation with a DNA cassette containing a resistance marker (*nptI*, conferring neomycin resistance) and a dominant conditional lethal gene, the *sacB* gene of *Bacillus subtilis*. The gene *sacB* encodes the enzyme levansucrase, which accumulates in the periplasm of Gram negative bacteria and leads to lethal synthesis of the polymer levan in the presence of high sucrose concentrations (GAY et al., 1985). Therefore, Gram negative bacteria expressing this gene, cannot grow (or are at least drastically reduced in growth) on high-sucrose medium, and colonies which have lost the *sacB* gene can be positively selected on plates containing high concentrations of sucrose. The principle of this mutagenesis method is outlined in Fig. 6. In the first step, the DNA fragment, cloned into a narrow host range vector, is mutagenized by the introduction of the *nptI-sacB-sacR* cartridge (*sacR* refers to a *cis*-regulatory DNA region upstream of *sacB*; STEINMETZ et al., 1985), and the mutation is introduced into the host's genome by a double crossing-over event, which can be selected by the kanamycin/neomycin resistance provided by the DNA cassette. In the second step, the *nptI-sacB-sacR* cartridge is deleted from the cloned fragment using restriction enzymes which remove most of the cassette but leave a number of nucleotides behind, thus generating a frame-shift mutation at this site. The resulting deletion derivative (now carrying an unmarked mutation in the gene of interest) is introduced into the mutant strain, containing the *nptI-sacB-sacR* cassette in the genome. Transconjugants are then grown on sucrose medium to select for those clones which have lost the *nptI-sacB-sacR* cassette due to a second double recombination event. This approach has been successfully used, e.g., for the construction of unmarked mutants in *Erwinia chrysanthemi* (RIED and COLLMER, 1987, 1988) and, with a slightly altered cartridge, also for *Xanthomonas campestris* (R. SIMON, personal communication).

Recently, this method has been modified in the following way (KAMOUN et al., 1992): the unmarked mutation, e.g., a deletion, is constructed within the gene of interest in *E. coli* and cloned into a narrow host range vector that contains an antibiotic resistance gene and the levansucrase gene *sacB*. The recombinant plasmid is introduced into the desired organism (in this case, *Xanthomonas campestris*, KAMOUN et al., 1992) and integrated into the genome by single crossing-over (selection for the vector encoded resistance, cells become sucrose-sensitive). As described above (compare Fig. 4), a second recombination event can occur leading to the excision of the plasmid, which in this case can be positively selected on high sucrose concentrations. Sucrose-resistant clones (loss of *sacB*) finally have to be screened (phenotypically or by hybridization) to isolate those recombinants in which the wild-type gene was replaced by the unmarked mutation (excision of the plasmid together with the wild-type region, leaving the mutation in the genome).

4 *In vivo* Manipulation of Gram Negative Bacteria Using Transposons

Transposable elements are defined as genetic entities which are able to translocate from site to site within the genome. Based on physical and genetic criteria, they are grouped into insertion sequences (IS elements) and transposons. Also certain temperate phage genomes, such as Mu, can transpose and are often classified as transposable elements. It is not subject of this chapter to give a detailed description of transposable elements and their physical and genetic characteristics. The interested reader is referred to a series of reviews that have covered this field very thoroughly (BERG and HOWE, 1989; CALOS and MILLER, 1980; CAMPBELL et al., 1979; KLECKNER, 1981; SHAPIRO, 1983; STARLINGER, 1980).

It was not long after their discovery in the late 1960s that transposable elements, especially antibiotic resistance transposons, became useful and meanwhile indispensable tools for the genetic engineering of Gram negative bacteria (KLECKNER et al., 1977). Their first and most obvious application resulted from their mutagenic activity, i.e., transposons insert randomly in many genomic locations and so create a wide variety of mutants. Thus, they can be used as effective mutagens. The major advantage over classical mutagenic agents is that transposons carry easily selectable markers. A cell which is mutated by the insertion of a transposon is marked by the acquisition of the transposon-mediated trait (usually an antibiotic resistance). This has proved especially useful in isolating mutants in hard-to-select characters, e.g., nodulation in *Rhizobium* or pathogenicity in phytopathogenic bacteria. A further advantage of using transposons for mutagenesis is the fact that the mutation is physically and genetically marked by the insertion which facilitates mapping and cloning of the mutated gene.

4.1 Principles of Transposon Mutagenesis

Again, there is a number of reviews which summarize the techniques and general advantages of transposon-mediated mutagenesis of Gram negative bacteria (KLECKNER et al., 1977; BERG et al., 1989; HAAS, 1986; MILLS, 1985; SIMON, 1989).

In principle, there are two strategies of engineering Gram negative bacteria by transposons. Either the gene or DNA region of interest is already cloned in *E. coli*; in this case, the cloned DNA can be specifically manipulated by the *in vivo* methods available for *E. coli* and transferred back into its original host organism (on replicative plasmids or by integration into the genome) for further analysis (fragment-specific transposon mutagenesis). Alternatively, transposons can be used to randomly manipulate the genome of the Gram negative organism under study, thereby identifying and localizing genes of interest that have not been cloned previously (random or general transposon mutagenesis).

4.1.1 Fragment-Specific Transposon Mutagenesis

One way to isolate transposon insertions in the cloned sequence is to introduce the recombinant plasmid to be mutagenized into an *E. coli* strain that harbors a chromosomal transposon insertion (DITTA, 1986). Transposition events onto the plasmid occur spontaneously and can be identified by mating the plasmid out of the *E. coli* strain with selection for both, the vector- and transposon-mediated resistances. We reached better results with an alternative mutagenesis procedure which is outlined in Fig. 7. In this case, the transposon is provided on a lambda derivative, which is unable to be stably maintained in *E. coli* (DEBRUIJN and LUPSKI, 1984). The strain, harboring the vector plasmid with the cloned sequences to be mutagenized, is subjected to infection with lambda, carrying the transposon. Since the transposon carrier cannot be stably established, transposition must have occurred in cells which show the transposon-mediated antibiotic resistance. Transposition events onto the recombinant plasmid (compared to those into the chromosome) are then identified by a subsequent mating if the vector is mobilizable, or by transformation.

4.1.2 Random Transposon Mutagenesis

The rationale of a random transposon mutagenesis experiment is outlined in Fig. 8. Vectors suitable as transposon carriers are described in the first section. The classical transposon delivery vector is based on the RP4:Mu suicide plasmid (see above) loaded with transposon Tn*5* to yield plasmid pJB4JI (BERINGER et al., 1978). This system has been used successfully in a variety of diverse Gram negative organisms to isolate Tn*5*-induced mutants. However, it has been found that the suicide effect is not reliable in many cases, and also transposition of Mu sequences together with Tn*5* was found to occur (for more details see, e.g., SIMON, 1989; SIMON and PRIEFER, 1990; and references therein). Also pRK2013 and its derivatives have been adapted to carry various

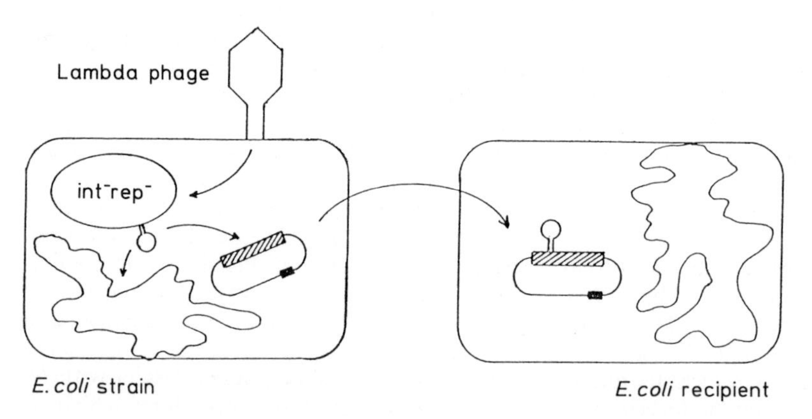

Fig. 7. Fragment-specific transposon mutagenesis.
The *E. coli* strain carrying the vector plasmid with the cloned sequence to be mutagenized is infected with a lambda phage derivative carrying the transposon inserted into its genome. Moreover, the phage genome is mutated (Δ b221, Oam, Pam) to prevent its stable establishment in a suppressor-negative *E. coli* strain (not able to integrate or replicate). Selection for the transposon-encoded resistance marker results in the isolation of cells in which transposition events (either into the chromosome or into the recombinant plasmid) have occurred. Transposon insertions in the recombinant vector molecule can be identified by a subsequent mating selecting transconjugants for acquisition of both, the vector- and transposon-encoded resistance marker. In case of IncP-type mobilizable vectors (such as the pSUP vectors), mutagenesis can be directly performed in the mobilizing donor strain S17-1 which facilitates the subsequent mobilization of transposon-carrying vector molecules (compare also Fig. 8).

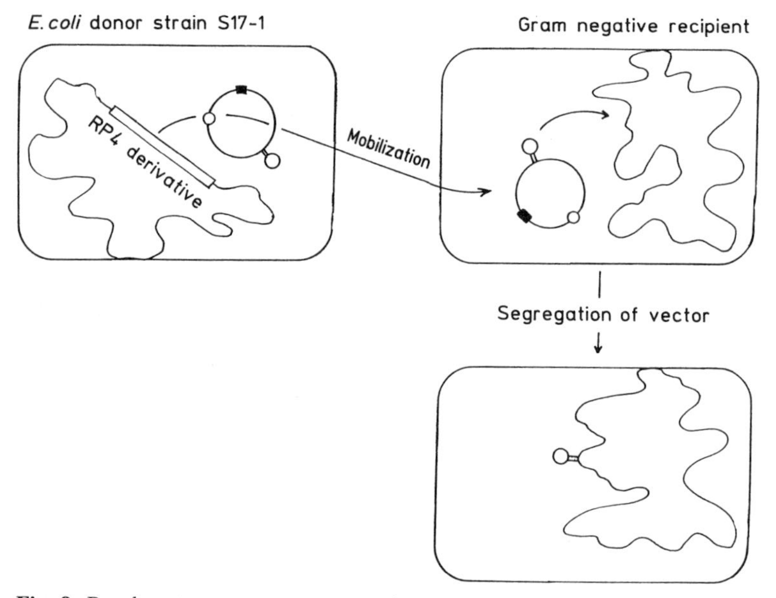

Fig. 8. Random transposon mutagenesis.
A narrow host range pSUP vector carrying the transposon is mobilized from the *E. coli* donor strain S17-1 (through transfer functions provided by the chromosomally integrated RP4 derivative acting in *trans* on the *oriT* (○) present on the pSUP vector molecule) into the Gram negative recipient to be mutagenized. Since the vector plasmid is not able to replicate outside *E. coli*, selection for the transposon-encoded antibiotic resistance results in transconjugants, in which the transposon has randomly inserted into the genome, while the transposon carrier is eliminated by segregation (loss of the vector-encoded resistance marker ■).

transposons (including Tn*7*, Tn*10*, Tn*5*, and Tn*501*) and have been used in a number of Gram negative bacteria (BOLTON et al., 1984; BULLERJAHN and BENZINGER, 1984; DEVOS et al., 1986; FORSTER et al., 1981; TURNER et al., 1984).

Naturally, the mobilizable narrow host range pSUP vectors are ideal transposon carrier vehicles, and derivatives harboring various transposons and transposon variants are available (SIMON et al., 1983a, 1986; SIMON, 1989). Some of them are shown in Fig. 2. They are being used for transposon mutagenesis in practically all Gram negative species.

4.2 Transposons Used in Genetic Engineering of Gram Negative Bacteria

Numerous transposons have been isolated from different sources of Gram negative bacteria and are being used for engineering Gram negative bacteria. They have been extensively reviewed by BERG et al. (1989). Examples (listed in Tab. 3) include Tn*1*/Tn*3*, which has been used, e.g., in strains of *Rhizobium* and *Pseudomonas* (CASADESUS et al., 1980; KRISHNAPILLAI et al., 1981), Tn*7* (used in *Rhizobium, Pseudomonas, Vibrio*, and *Xanthomonas*; BOLTON et al., 1984; CARUSO and SHAPIRO, 1982; THOMSON et al., 1981; TURNER et al., 1984), and Tn*10* (used mainly in *Shigella* and *Salmonella* strains; KLECKNER, 1981; BERG and BERG, 1987). There are also reporter gene transposons available, such as Tn*10-lac* (WAY et al., 1984) and Tn*3-HoHo1* (STACHEL et al., 1985) which carry a promoterless β-galactosidase gene (*lacZ*) as well as Mu-based *lac*-fusion elements (BAKER et al., 1983; CASADABAN and CHOU, 1984; CASTILHO et al., 1984; BREMER et al., 1985; GROISMAN and CASADABAN, 1986) which are often used in Gram negative species. Another very useful indicator gene transposon is Tn*4431* (SHAW and KADO, 1987; SHAW et al., 1988) which contains a promoterless *lux* operon allowing the study of gene expression using the bioluminescence phenotype, a trait which is especially helpful in the analysis of *in situ*, e.g., *in planta* expression of bacterial genes.

However, one of the earliest and still most often used transposons for manipulating non-*E. coli* Gram negative bacteria is Tn*5* and its derivatives (BERG and BERG, 1983; BERG, 1989; DEBRUIJN and LUPSKI, 1984; KLECKNER, 1981). Some of the features which made Tn*5* so popular can be listed as follows:

- high frequency of transposition in many Gram negative bacteria
- the usually low insertional specificity
- the low probability of genome rearrangements
- the normally strong polarity of the induced mutation
- the reliable marker gene (aminoglycoside 3′-phosphotransferase II) which confers resistance to kanamycin and neomycin which can be used in most Gram negative bacteria.

A further advantage of Tn*5* is the availability of a good genetic and physical map. The complete sequence of Tn*5* has been established (AUERSWALD et al., 1980; BECK et al., 1982; MAZODIER et al., 1983, 1985), and the presence and location of various restriction sites allowed the construction of Tn*5* derivatives without affecting its transposition properties. Again, only some examples can be mentioned (Tab. 4).

One group of *in vitro* constructed Tn*5* derivatives consists of transposons modified to contain different or additional selectable markers such as tetracycline, trimethoprim, or gentamicin resistance (JORGENSEN et al., 1979; SIMON et al., 1983a; RELLA et al., 1985; ZSEBO et al., 1984; HIRSCH et al., 1986; SASAKAWA and YOSHIKAWA, 1987). There are also Tn*5* derivatives carrying the *lacZ* (β-galactosidase) or *luxAB* (bacterial luciferase) genes as monitorable markers (DEVOS et al., 1986; BOIVIN et al., 1988). Another very helpful transposon variant is Tn*5-mob* (SIMON et al., 1983b; SIMON, 1984) or the more or less identical constructs Tn*5-oriT* (YAKOBSON and GUINEY, 1984) and Tn*5-A1* (SELVERAJ and IYER, 1983) which carry the IncP-specific recognition site for mobilization. Insertion of these transposons renders the target replicon (bacterial chromosome or plasmid) mobilizable by RP4 or other IncP-type transfer plasmids (see first section), thus facilitating the

Tab. 3. Examples for Transposons Used in Engineering Gram Negative Bacteria

Transposon	Marker	Length (kb)	Remarks
Tn*1*/Tn*3* (1, 2, 3)	Amp	5	Insertion preferentially into AT-rich regions
Tn*7* (4)	Tmp, Str, Spc	14	Hot spot in *E. coli* chromosome
Tn*10* (5)	Tet	9.3	Hot spots, stimulates deletions and inversions
Tn*501* (6, 7)	Mer	8.2	Transposition and resistance inducible by mercury
Tn*1721*/ Tn*1771* (8, 9)	Tet	11.2	Amplification of *tet* gene gives high levels of resistance
Tn*5* (10, 11)	Kan, Str	5.7	Cryptic Str resistance in *E. coli*, expressed in many other Gram negatives
Tn*10-lac* (12)	Tet, '*lac*	11	Transcriptional fusion to *lac* operon
Tn*3HoHo1* (13)	Amp, '*lac*	14	Transcriptional/translational fusion to *lac* operon
Tn*4431* (14)	Tet, '*lux*	15	Tn*1721* derivative, transcriptional fusion to *lux* operon

Abbreviations: Amp, ampicillin resistance; Tmp, trimethoprim resistance; Str, streptomycin resistance; Spc, spectinomycin resistance; Tet, tetracycline resistance; Mer, mercury resistance; Kan, kanamycin resistance; '*lac*, promoterless *lac* operon; '*lux*, promoterless *lux* operon from *Vibrio fischeri*

References: (1) HEDGES and JACOB, 1974
(2) RICHMOND and SYKES, 1972
(3) HEFFRON et al., 1975
(4) BARTH et al., 1976
(5) KLECKNER et al., 1975
(6) STANISICH et al., 1977
(7) BENNETT et al., 1978
(8) SCHMITT et al., 1979
(9) SCHÖFFL et al., 1981
(10) BERG et al., 1975
(11) BERG and BERG, 1983
(12) WAY et al., 1984
(13) STACHEL et al., 1985
(14) SHAW et al., 1988

mapping of chromosomal markers and the analysis of otherwise non-transmissible, cryptic plasmids. Another group of extremely useful Tn*5* derivatives are those carrying various promoterless indicator genes. Apart from generating mutations, these transposons additionally generate *in vivo* transcriptional or translational gene fusions. The need for constructing such fusions (either *in vitro* by cloning or *in vivo* by transposons) is obvious and is outlined in Chapter 4. There are Tn*5* derivatives carrying the most widely used indicator gene *lacZ* as well as variants harboring promoterless *gusA* (β-glucuronidase), *phoA* (alkaline phosphatase), or *luc* (firefly luciferase) genes (KROOS and KAISER, 1984; MANOIL and BECKWITH, 1985; SHARMA and SIGNER, 1990; SIMON et al., 1989; Tab. 4).

5 Specific Example: Identification and Characterization of LPS Biosynthesis Genes in *Rhizobium leguminosarum* bv. *viciae*

The following section will summarize as an example, how the above described methods and tools for genetic engineering of Gram negative bacteria were applied in our laboratory in order to clone and analyze genes in *Rhizobium leguminosarum* bv. *viciae* which are in-

Tab. 4. Tn*5* Derivatives Used in Genetic Engineering of Gram Negative Organisms

Derivative	Markers/ Characteristics	Comments
Tn*5* (1)	Neo, Kan, Str	*npt* gene gives resistance to Neo and Kan, *str* gene not expressed in *E. coli*
Tn*5*-*Tc* (2)	Neo, Kan, Tet	*tet* gene of RP4 cloned into *Bam*HI site
Tn*5*-*132* (3)	Tet	
Tn*5*-*TC1* (4)	Tet	
Tn*5*-*CM* (4)	Cam	Promoterless *cam* gene under the control of *npt* promoter
Tn*5*-*GM* (4)	Gen	
Tn*5*-*TP* (4)	Trp	
Tn*5*-*SM* (4)	Str	Promoterless *str* gene downstream of *npt* promoter
Tn*5*-*AP* (4)	Amp	
Tn*5*.*7* (5)	Trp, Str, Spc	
Tn*5*-*GmSpSm* (6)	Gen, Str, Spc	
Tn*5*-*233* (7)	Gen, Kan, Spc, Str	
Tn*5*-*751* (8)	Neo, Kan, Trp	
Tn*5*-*235* (7)	Neo, Kan, Lac	*lacZY* expressed constitutively from *npt* promoter
Tn*5*-*Lux* (9)	Neo, Kan, Lux	Constitutively expressed *luxAB* genes
Tn*5*-*oriT* (10)	Neo, Kan, Mob	Contains *oriT* of RP4
Tn*5*-*mob* (11)	Neo, Kan, Mob	Also called Tn*5*-*B10* (12) contains *oriT* of RP4
Tn*5*-*B11* (12)	Gen, Mob	Contains *oriT* of RP4
Tn*5*-*B12* (12)	Tet, Mob	Contains *oriT* of RP4
Tn*5*-*A1* (13)	Neo, Kan, Mob	Contains *oriT* of RP4
Tn*5*-*VB32* (14)	Tet, '*kan*	Transcriptional fusion to *npt* gene
Tn*5*-*lac* (15)	Neo, Kan, '*lac*	Transcriptional fusion to *lacZ* gene
Tn*5*-*B20* (12)	Neo, Kan, '*lac*	Transcriptional fusion to *lacZ* gene
Tn*5*-*B21* (12)	Tet, '*lac*	Transcriptional fusion to *lacZ* gene
Tn*5*-*phoA* (16)	Neo, Kan, '*phoA*	Translational fusion to *phoA* gene
Tn*5*-*gusA1* (17)	Neo, Kan, Tet, '*gusA*	Transcriptional fusion to *gusA* gene
Tn*5*-*gusA2* (17)	Neo, Kan, Tet, '*gusA*	Translational fusion to *gusA* gene
Tn*5*-*B40* (12)	Neo, Kan, '*luc*	Transcriptional fusion to *luc* gene
Tn*5*-*B41* (12)	Tet, '*luc*	Transcriptional fusion to *luc* gene

Abbreviations: Neo, neomycin; Kan, kanamycin; Tet, tetracycline; Trp, trimethoprim; Str, streptomycin; Spc, spectinomycin; Gen, gentamicin; Cam, chloramphenicol; Amp, ampicillin; *npt*, gene encoding neomycin phosphotransferase; *lacZ*, gene encoding β-galactosidase; *luxAB*, genes encoding the *Vibrio* luciferase; *luc*, gene encoding the firefly luciferase; *phoA*, gene encoding alkaline phosphatase; *gusA*, gene encoding β-glucuronidase

References:
(1) Berg and Berg, 1983
(2) Simon et al., 1983a
(3) Jorgensen et al., 1979
(4) Sasakawa and Yoshikawa, 1987
(5) Zsebo et al., 1984
(6) Hirsch et al., 1986
(7) DeVos et al., 1986
(8) Rella et al., 1985
(9) Boivin et al., 1988
(10) Yakobson and Guiney, 1984
(11) Simon, 1984
(12) Simon et al., 1989
(13) Selveraj and Iyer, 1983
(14) Bellofatto et al., 1984
(15) Kroos and Kaiser, 1984
(16) Manoil and Beckwith, 1985
(17) Sharma and Signer, 1990

volved in the synthesis of lipopolysaccharides (PRIEFER, 1989).

5.1 Isolation of LPS Mutants by Random Transposon Tn*5* Mutagenesis

To isolate mutants defective in the synthesis of a complete lipopolysaccharide molecule, the *Rhizobium leguminosarum* bv. *viciae* strain VF39 was subjected to a general Tn*5* mutagenesis using pSUP102 as transposon carrier. Plasmid pSUP1021 (= pSUP102::Tn*5*; SIMON et al., 1983a) was mobilized from the *E. coli* mobilizing donor S17-1 into *R. leguminosarum* VF39 by bacterial mating and Tn*5*-carrying transconjugants were selected on medium containing neomycin (compare Fig. 8). Neomycin-resistant transconjugants (= transposition of Tn*5* into the VF39 genome) were isolated at a frequency of 10^{-4} per recipient. This means that about 10^4 potential Tn*5*-induced mutants were generated in a single mutagenesis experiment. Mutants defective in LPS biosynthesis were selected from this pool visually by their colony morphology, since LPS mutants exhibit a rough colony surface as compared to the smooth, glossy wild-type colonies.

The fact that the mutants were physically marked by the insertion of Tn*5* was used to verify that they represented independent mutations. For this purpose, total DNA of the mutant strains was isolated, digested with *Eco*RI (Tn*5* does not contain a recognition site for *Eco*RI) and hybridized to a Tn*5* probe. For each mutant, only one fragment hybridized, indicating single Tn*5* insertions, and the hybridizing fragments differed from mutant to mutant, showing that independent Tn*5* insertions had occurred into different *Eco*RI fragments of the genome.

5.2 Complementation of the Tn*5*-Induced LPS Mutants

A gene bank of *R. leguminosarum* bv. *viciae* VF39, generated in the cosmid vector pSUP205 (SIMON et al., 1983a), was introduced into the individual mutants to isolate complementing cosmid clones. Since pSUP205 is not able to replicate in *Rhizobium*, cointegrates with RP4 were generated (SIMON et al., 1986) and conjugally transferred into the mutant strains. The principle of RP4-cointegrate formation has not been described above, but is explained in Fig. 9. The pSUP205 cosmid gene bank was introduced into an *E. coli* strain, carrying a tetracycline-sensitive derivative of RP4. Because of homologous DNA regions on RP4 and the cosmid vector pSUP205 (e.g., *oriT* region), single recombination events can occur leading to fused molecules. These cointegrates are now self-transferable and able to replicate in *Rhizobium*. Transconjugants harboring RP4/cosmid cointegrates were selected for tetracycline resistance, provided by the cosmid vector pSUP205 and tetracycline-resistant colonies were tested for complementation of the mutant phenotype. Complementing cosmids were retransferred to *E. coli* where spontaneous resolution of the cointegrates occurred. In this special case, the *E. coli* recipient used was a strain lysogenic for a temperature-inducible lambda phage (cIts), so that the complementing cosmids could be separated from RP4 and isolated by *in vivo* packaging into bacteriophage particles upon heat induction.

With this strategy, cosmids able to restore wild-type phenotype in the individual mutant strains, and therefore carrying the respective wild-type gene, were isolated. One of these cosmids, pRlCos4, is shown in Fig. 10.

5.3 *In vivo* Cloning of the Mutated Genes

With the strategy outlined above (Figs. 3 and 4), the Tn*5*-induced mutations were cloned from the VF39 genome. A complementing cosmid (e.g., pRlCos4, Fig. 10) was mobilized from *E. coli* S17-1 into the corresponding VF39 LPS mutant. Single crossing-over events were selected on media containing neomycin (selection for Tn*5*) and tetracycline (selection for the cosmid). Since pSUP205 cannot replicate in *Rhizobium*, this selection re-

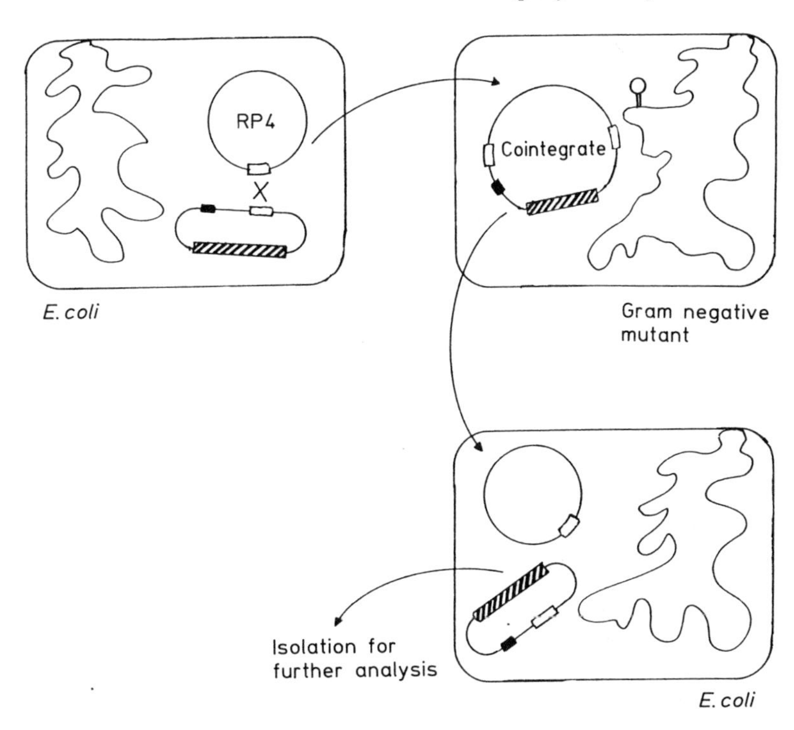

Fig. 9. Complementation of mutants by RP4 cointegrate formation.
If a narrow host range plasmid such as one of the pSUP vectors is to be used for complementation of a Gram negative mutant, replication in the non-*E. coli* host can be achieved by cointegrate formation with RP4. For this purpose, an RP4 derivative is introduced into the *E. coli* strain harboring the recombinant pSUP derivative. Since RP4 and pSUP share common DNA sequences (e.g., *oriT* region, □), single recombination events can occur leading to a fused molecule. This cointegrate, which is now able to replicate in the Gram negative host, can be selected for by the resistance marker provided by the pSUP vector (■). Plasmids of interest, e.g., complementing plasmids, can be retransferred by a normal mating to *E. coli*, where spontaneous resolution of the cointegrate can occur. The recombinant pSUP vector can be separated from RP4 and isolated for further analysis either by DNA isolation and transformation, or by *in vivo* packaging (in case of a cosmid derivative) using a lambda lysogenic *E. coli* recipient.

sults in the isolation of merodiploid strains in which the complete cosmid had integrated via a single homologous recombination event at the corresponding site in the VF39 genome (compare Fig. 3).

As outlined in Fig. 4, a second recombination event can spontaneously occur leading to the excision of either the wild-type cosmid or the cosmid carrying the genomic mutation (i.e., the Tn*5*-mutated gene). To isolate such cosmid molecules, a kanamycin/tetracycline-sensitive RP4 plasmid was conjugally introduced into selected merodiploid cells. Due to the RP4 specific *oriT* site on pSUP205, both cosmid types could be mobilized by the RP4 helper plasmid into an *E. coli* recipient strain; Tn*5*-carrying cosmid derivatives were recognized by their tetracycline/kanamycin resistance, whereas wild-type cosmids conferred only tetracycline resistance to the *E. coli* recipient.

Comparison of wild-type and Tn*5*-carrying cosmids by restriction analysis allowed the localization of the Tn*5* insertions (= mutated genes) on the physical map of the cosmid (Fig. 10).

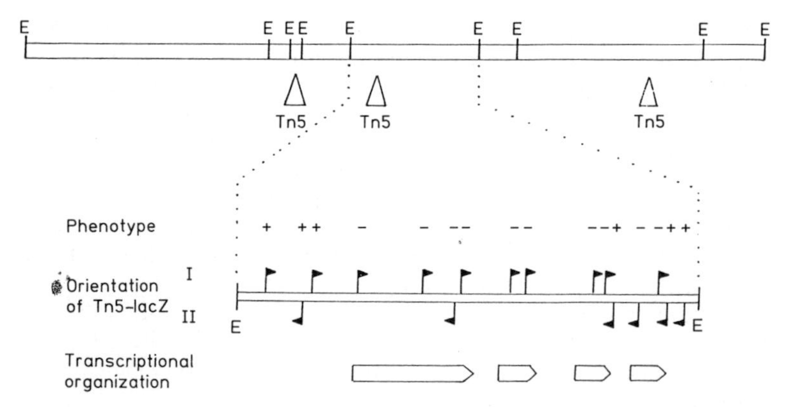

Fig. 10. Physical and genetic map of the *Rhizobium leguminosarum* DNA cloned in pRlCos4.
The upper part of the figure shows the restriction map for *Eco*RI (E) for the genomic insert in pRlCos4, identified by complementation of Tn*5*-induced LPS mutants of *Rhizobium leguminosarum* bv. *viciae*. The location of the Tn*5* insertions in three different mutants (indicated by triangles below the map) was achieved by *in vivo* cloning of the Tn*5* carrying cosmid derivates and comparison of the respective restriction maps. The analysis of one of the subcloned *Eco*RI restriction fragments by fragment-specific transposon mutagenesis is shown in more detail in the lower part. The orientation of Tn*5-lacZ* insertions is represented by flags. The phenotype of individual transposon insertions was analyzed upon their genomic integration by double crossing-over (+ no change in LPS production, − LPS mutant phenotype). The transcriptional organization of *lps* genes is indicated by open arrows and was determined by cross-complementation experiments and expression of β-galactosidase activity. For more details, see text.

5.4 Fragment-Specific Mutagenesis Using Tn*5*, Tn*5-lacZ* (Tn*5-B20*) and Tn*5-gusA1*

Individual *Eco*RI fragments of the complementing cosmid pRlCos4 were subcloned into vector pSUP205 and mutagenized with Tn*5*, Tn*5-lacZ* (Tn*5-B20*; SIMON et al., 1989) and Tn*5-gusA1* (SHARMA and SIGNER, 1990). For this purpose, the strategy outlined in Fig. 7 was used, i.e., *E. coli* strain S17-1, carrying the recombinant pSUP205 plasmids, was infected with the mutant lambda phage, carrying either Tn*5*, Tn*5-lacZ*, or Tn*5-gusA1*. Transposon insertions into the plasmid were selected by a mating (mobilization of pSUP205 through the transfer functions of the RP4 derivative integrated into the genome of S17-1) using another *E. coli* strain as recipient. Insertions into the cloned fragment (versus insertions into the vector part of the recombinant molecule) were identified and mapped by restriction enzyme analysis.

Individual transposon insertions were introduced into the genome of *R. leguminosarum* bv. *viciae* VF39, and their effect on LPS biosynthesis was analyzed. For the genomic integration of the mutated DNA fragments, the rationale described in Fig. 5 was used. The pSUP205 vectors, carrying individual transposon insertions in their cloned fragment, were mated from *E. coli* S17-1 into VF39 (wild type), where homologous recombination between the cloned and the genomic fragments may occur, leading to the exchange of the wild-type region with the region carrying the transposon insertion. Since the vector molecule pSUP205 cannot replicate in *Rhizobium*, it is spontaneously lost, and clones in which the marker exchange had occurred were easily identified by their neomycin resistance (Tn*5* marker) and tetracycline sensitivity (loss of pSUP205).

By these experiments, a number of transposon insertions were identified which, after marker exchange, resulted in a defective LPS synthesis. Thus, a number of regions were localized on cosmid pRlCos4, which are involved in LPS synthesis in *R. leguminosarum* bv. *viciae* (Fig. 10).

The cloned and genomic mutants generated by fragment-specific mutagenesis were also used to elucidate the transcriptional organization of the newly identified *lps* regions. The principle of the method applied is the following: a defined transposon insertion in the fragment (generated by fragment-specific mutagenesis) was cloned into the broad host range vector pSUP104 (PRIEFER et al., 1985) and introduced via mobilization into a specific genomic VF39 Tn*5* mutant (obtained by marker exchange). If introduction of the cloned mutation restored wild-type phenotype in the mutant strain, the two mutations can cross-complement each other and must therefore be located in two separate transcriptional units. If no complementation occurs, both mutations must be located in the same complementation group. By this cross-complementation assay, the various transposon insertions on the cloned fragments could be correlated with independent transcriptional units (Fig. 10).

5.5 Analysis of *lps* Gene Expression Using Reporter Gene Transposons

As indicated above, the fragment specific transposon mutagenesis was not only performed with Tn*5*, but also with the reporter gene transposons Tn*5-lacZ* and Tn*5-gusA1*. Use of these indicator transposons had the advantage that not only a mutation was generated by their insertion, but the mutated gene was simultaneously fused to a reporter gene. Expression of the reporter gene is only possible if the indicator transposon is inserted into a gene in the correct orientation, i.e., if the reporter gene is cotranscribed with the gene under study.

This fact was used to define the direction of transcription of the *lps* gene regions identified on pRlCos4. The orientation of the transposon insertions in the cloned fragment was determined by restriction analysis, and β-galactosidase or β-glucuronidase activity was monitored for each insertion. The measurements revealed that only insertions in orientation I (see Fig. 10) expressed the reporter gene, whereas insertions in the other orientation gave no enzyme activity. From this it was concluded that the *lps* genes on the cloned fragment were transcribed from left to right (Fig. 10).

The Tn*5-lacZ* and Tn*5-gusA1* fusions were also integrated by single crossing-over into the genome of VF39 (principle see Fig. 3), and host plants were inoculated with these merodiploid strains. Sections of the nodules formed were investigated and stained for β-galactosidase or β-glucuronidase activity. Using this approach, it was established that the *lps* genes under study were also expressed *in planta*, i.e., in the nodule.

6 Concluding Remarks

As pointed out at the beginning of this chapter, the aim of this review was to outline the principal methods and tools available for the engineering of Gram negative bacteria. Although the various techniques for engineering non-*E. coli* strains have been developed only within the last few years, an immense number of vectors and transposon derivatives has accumulated meanwhile which makes it impossible to compile a complete list. Moreover, there is no system and no vector, which is best suited for all purposes and for all organisms. For example, we and many other groups found Tn*5*, delivered on a narrow host range pSUP vector, to be an optimal agent for transposon mutagenesis of *Rhizobium* strains, but the same system gave severe problems in *Xanthomonas campestris* with respect to transposition frequency and undesired concomitant integration of the transposon carrier plasmid (SIMON, 1989). This reflects species- or even strain-dependent differences and emphasizes the fact that an optimal system for a specific species or strain cannot be predicted, it has to be developed and adapted individually in each case.

Acknowledgement

I wish to thank Judith Meyer for technical assistance in preparing this manuscript.

7 References

AUERSWALD, E. A., LUDWIG, G., SCHALLER, H. (1980), Structural analysis of Tn*5*, *Cold Spring Harbor Symp. Quant. Biol.* **45**, 107–113.

BAGDASARIAN, M., TIMMIS, K. N. (1982), Host vector systems for gene cloning in *Pseudomonas, Curr. Top. Microbiol. Immunol.* **96**, 47–67.

BAGDASARIAN, M., LURZ, R., RÜCKERT, B., FRANKLIN, F. C. H., BAGDASARIAN, M. M., FREY, J., TIMMIS, K. N. (1981), Specific purpose plasmid cloning vectors, II. Broad host range, high copy number, RSF1010-derived vectors and host-vector system for gene cloning in *Pseudomonas, Gene* **6**, 237–247.

BAGDASARIAN, M., BAGDASARIAN, M. M., LURZ, R., NORDHEIM, A., FREY, J., TIMMIS, K. N. (1982), Molecular and functional analysis of the broad host range plasmid RSF1010 and construction of vectors for gene cloning in gram-negative bacteria, in: *Drug Resistance in Bacteria*, pp. 183–197, Tokyo: Japan Scientific Societies Press.

BAGDASARIAN, M. M., AMMANN, E., LURZ, R., RÜCKERT, B., BAGDASARIAN, M. (1983), Activity of the hybrid *trp-lac* (*tac*) promoter of *Escherichia coli* in *Pseudomonas putida*. Construction of broad-host-range, controlled expression vectors, *Gene* **931**, 273–282.

BAKER, T. A., HOWE, M. M., GROSS, C. A. (1983), MudX, a derivative of Mud1 (*lac* Ap) which makes stable *lacZ* fusions at high temperature, *J. Bacteriol.* **156**, 970–974.

BARCAK, G. J., CHANDLER, M. S., REDFIELD, R. J., TOMB, J.-F. (1991), Genetic systems in *Haemophilus influenzae, Methods Enzymol.* **204**, 321–341.

BARTH, P. T., GRINTER, N. J. (1974), Comparison of the deoxyribonucleic acid molecular weights and homologies of plasmids conferring linked resistance to streptomycin and sulfonamides, *J. Bacteriol.* **120**, 618–630.

BARTH, P. T., DATTA, N., HEDGES, R. W., GRINTER, N. J. (1976), Transposition of a deoxyribonucleic acid sequence encoding trimethoprim and streptomycin resistances from R483 to other replicons, *J. Bacteriol.* **125**, 800–810.

BECK, E., LUDWIG, G., AUERSWALD, E. A., REISS, B., SCHALLER, H. (1982), Nucleotide sequence and exact location of the neomycin phosphotransferase gene from transposon Tn*5*, *Gene* **19**, 327–336.

BELLOFATTO, V., SHAPIRO, L., HODGSON, D. A. (1984), Generation of a Tn*5* promoter probe and its use in the study of gene expression in *Caulobacter crescentus, Proc. Natl. Acad. Sci. USA* **81**, 1035–1039.

BENNETT, P. M., GRINSTED, J., CHOI, C. L., RICHMOND, M. H. (1978), Characterization of Tn*501*, a transposon determining resistance to mercuric ions, *Mol. Gen. Genet.* **159**, 101–106.

BERG, D. E. (1989), Transposon Tn*5*, in: *Mobile DNA* (BERG, D. E., HOWE, M. M., Eds.), pp. 185–210, Washington, DC: American Society for Microbiology.

BERG, D. E., BERG, C. M. (1983), The prokaryotic transposable element Tn*5*, *Biotechnology* **1**, 417–435.

BERG, C. M., BERG, E. D. (1987), Uses of transposable elements and maps of known insertions, in: *Escherichia coli and Salmonella typhimurium: Cellular and Molecular Biology* (NEIDHARDT, F. C., INGRAHAM, J. L., LOW, K. B., MAGASANIK, B., SCHAECHTER, M., UMBARGER, H. E., Eds.), pp. 1071–1109, Washington, DC: American Society for Microbiology.

BERG, D. E., HOWE, M. M. (Eds.) (1989), *Mobile DNA*, Washington, DC: American Society for Microbiology.

BERG, D. E., DAVIES, J., ALLET, B., ROCHAIX, J.-D. (1975), Transposition of R factor genes to bacteriophage lambda, *Proc. Natl. Acad. Sci. USA* **72**, 3628–3632.

BERG, C. M., BERG, D. E., GROISMAN, E. A. (1989) Transposable elements and the genetic engineering of bacteria, in: *Mobile DNA* (BERG, D. E., HOWE, M. M., Eds.), pp. 879–925, Washington, DC: American Society for Microbiology.

BERINGER, J. E., BEYNON, J. L., BUCHANAN-WOLLASTON, A. V., JOHNSTON, A. W. B. (1978), Transfer of the drug-resistance transposon Tn*5* to *Rhizobium, Nature* **276**, 633–634.

BOIVIN, R., CHALIFOUR, F.-P., DION, P. (1988), Construction of a Tn*5* derivative encoding bioluminescence and its introduction in *Pseudomonas, Agrobacterium* and *Rhizobium, Mol. Gen. Genet.* **213**, 50–55.

BOLTON, E., GLYNN, P., O'GARA, F. (1984), Site specific transposition of Tn*7* into a *Rhizobium meliloti* megaplasmid, *Mol. Gen. Genet.* **193**, 153–157.

BOUCHER, C., BERGERON, B., BARATE DE BERTALMIO, M., DENARIE, J. (1977), Introduction of bacteriophage Mu into *Pseudomonas solanacearum* and *Rhizobium meliloti* using the R factor RP4, *J. Gen. Microbiol.* **98**, 253–263.

BREMER, E. T., SILHAVY, J., WEINSTOCK, G. M. (1985) Transposable lambda p*lac*-Mu bacteriophages for creating *lacZ* operon fusions and kanamycin resistance insertions in *Escherichia coli, J. Bacteriol.* **162**, 1092–1099.

BULLERJAHN, G. S., BENZINGER, R. H. (1984), Introduction of the mercury transposon Tn*501* into *Rhizobium japonicum* strains 31 and 110, *FEMS Microbiol. Lett.* **22,** 183–187.

BURKARDT, H.-J., RIESS, G., PÜHLER, A. (1979), Relationship of group P1 plasmids revealed by heteroduplex experiments: RP1, RP4, R68, and RK2 are identical, *J. Gen. Microbiol.* **114,** 341–348.

CALOS, M. P., MILLER, J. H. (1980), Transposable elements, *Cell* **20,** 579–595.

CAMPBELL, A., BERG, D. E., BOTSTEIN, D., LEDERBERG, E. M., NOVICK, R. P. (1979), Nomenclature of transposable elements in bacteria, *Gene* **5,** 197–206.

CANGELOSI, G. A., BEST, E. A., MARTINETTI, G., NESTER, E. W. (1991), Genetic analysis of *Agrobacterium, Methods Enzymol.* **204,** 384–397.

CARUSO, M., SHAPIRO, J. A. (1982), Interaction of Tn*7* and temperate phage F116L of *Pseudomonas aeruginosa, Mol. Gen. Genet.* **118,** 292–298.

CASADABAN, M. J., CHOU, J. (1984), *In vivo* formation of gene fusions encoding hybrid *β*-galactosidase proteins in one step with transposable Mu-*lac* transducing phage, *Proc. Natl. Acad. Sci. USA* **81,** 535–539.

CASADESUS, J., JANEZ, E., OLIVARES, J. (1980), Transposition of Tn*1* to the *Rhizobium meliloti* genome, *Mol. Gen. Genet.* **180,** 405–410.

CASTILHO, B. A., OLFSON, P., CASADABAN, M. J. (1984), Plasmid insertion mutagenesis and *lac* gene fusion with mini-Mu bacteriophage transposons, *J. Bacteriol.* **158,** 488–495.

DEBRUIJN, F. J., LUPSKI, J. R. (1984), The use of transposon Tn*5* mutagenesis in the rapid generation of correlated physical and genetic maps of DNA segments cloned into multicopy plasmids – A review, *Gene* **27,** 131–149.

DENARIE, J., ROSENBERG, C., BERGERON, B., BOUCHER, C., MICHEL, M., BORATE DE BERTALMIO, M. (1977), Potential of RP4::Mu plasmids for *in vivo* genetic engineering of Gram-negative bacteria, in: *DNA Insertion Elements, Plasmids and Episomes* (BUKHARI, A. I., SHAPIRO, J. A., ADHYA, S. L., Eds.), pp. 507–520, Cold Spring Harbor, NY: Cold Spring Harbor Laboratory Press.

DEVOS, G. F., WALKER, G. C., SIGNER, E. R. (1986), Genetic manipulations in *Rhizobium meliloti* utilizing two new transposon Tn*5* derivatives, *Mol. Gen. Genet.* **204,** 485–491.

DITTA, G. (1986), Tn*5* mapping of *Rhizobium* nitrogen fixation genes, *Methods Enzymol.* **118,** 519–528.

DITTA, G., STANFIELD, S., CORBIN, D., HELINSKI, D. R. (1980), Broad host range DNA cloning system for Gram-negative bacteria. Construction of a gene bank of *Rhizobium meliloti, Proc. Natl. Acad. Sci. USA* **77,** 7347–7351.

DITTA, G., SCHMIDHAUSER, T., YAKOBSON, E., LU, P., LIANG, Y.-W., FINLAY, D. R., GUINEY, D., HELINSKI, D. R. (1985), Plasmids related to the broad host range vector, pRK290, useful for gene cloning and for monitoring gene expression, *Plasmid* **13,** 149–153.

DONOHUE, T. J., KAPLAN, S. (1991), Genetic techniques in Rhodospirillaceae, *Methods Enzymol.* **204,** 459–484.

ELY, B. (1991), Genetics of *Caulobacter crescentus, Methods Enzymol.* **204,** 372–383.

FIGURSKI, D. H., HELINSKI, D. R. (1979), Replication of an origin-containing derivative of plasmid RK2 dependent on a plasmid function provided in trans, *Proc. Natl. Acad. Sci. USA* **76,** 1648–1652.

FORSTER, T. J., DAVIS, M. A., ROBERTS, D. E., TAKERHITA, K., KLECKNER, N. (1981), Genetic organisation of transposon Tn*10*, *Cell* **23,** 201–213.

FREY, J., BAGDASARIAN, M., FEISS, D., FRANKLIN, F. C. H., DESHUSSES, J. (1983), Stable cosmid vectors that enable the introduction of cloned fragments into a wide range of gram-negative bacteria, *Gene* **24,** 299–308.

FRIEDMAN, A. M., LONG, S. R., BROWN, S. E., BUIKEMA, W. J., AUSUBEL, F. (1982), Construction of a broad host range cosmid cloning vector and its use in the genetic analysis of *Rhizobium* mutants, *Gene* **18,** 289–296.

GAUTIER, F., BONEWALD, R. (1980), The use of plasmid R1162 and derivatives for gene cloning in the methanol-utilizing *Pseudomonas*, AM1, *Mol. Gen. Genet.* **178,** 375–380.

GAY, P., LE COQ, D., STEINMETZ, M., BERKELMAN, T., KADO, C. I. (1985), Positive selection procedure for entrapment of insertion sequence elements in gram-negative bacteria, *J. Bacteriol.* **164,** 918–921.

GLAZEBROOK, J., WALKER, G. C. (1991), Genetic techniques in *Rhizobium meliloti, Methods Enzymol.* **204,** 398–417.

GROISMAN, E. A., CASADABAN, M. J. (1986), Mini-Mu bacteriophage with plasmid replicons for *in vivo* cloning and *lac* gene fusing, *J. Bacteriol.* **168,** 357–364.

GUERRY, P., VANEMBDEN, J., FALKOW, S. (1974), Molecular nature of two non-conjugative plasmids carrying drug resistance genes, *J. Bacteriol.* **117,** 619–630.

GUINEY, D. G., HELINSKI, D. R. (1979), The DNA-protein relaxation complex of plasmid RK2: Location of the site-specific nick in the region of the proposed origin of transfer, *Mol. Gen. Genet.* **176,** 183–189.

HAAS, D. (1986), Jumping genes as tools in the analysis of bacterial metabolism, *Swiss Biotechnol.* **4**, 20–22.

HEDGES, R. W., JACOB, A. F. (1974), Transposition of ampicillin resistance from RP4 to other replicons, *Mol. Gen. Genet.* **132**, 31–40.

HEFFRON, F., RUBENS, C., FALKOW, S. (1975), Translocation of a plasmid DNA sequence which mediates ampicillin resistance: molecular nature and specificity of insertion, *Proc. Natl. Acad. Sci. USA* **72**, 3623–3627.

HIRSCH, P. R., BERINGER, J. E. (1984), A physical map of pPH1JI and pJB4JI, *Plasmid* **12**, 133–141.

HIRSCH, P. R., WANG, C. L., WOODWARD, M. J. (1986), Construction of a Tn*5* derivative determining resistance to gentamycin and spectinomycin using a fragment cloned from R1033, *Gene* **4**, 203–209.

JAGADISH, M. N., BOOKNER, S. D., SZALAY, A. A. (1985), A method for site-directed transplacement of *in vitro* altered DNA sequences in *Rhizobium, Mol. Gen. Genet.* **119**, 249–255.

JORGENSEN, R. A., ROTHSTEIN, S. J., REZNIKOFF, W. S. (1979), A restriction enzyme cleavage map of Tn*5* and localization of a region encoding neomycin resistance, *Mol. Gen. Genet.* **177**, 65–72.

KAMOUN, S., TOLA, E., KAMDAR, H., KADO, C. I. (1992), Rapid generation of directed and unmarked deletions in *Xanthomonas, Mol. Microbiol.* **6**, 809–816.

KLECKNER, N. (1981), Transposable elements in prokaryotes, *Annu. Rev. Genet.* **15**, 341–404.

KLECKNER, N., CHAN, R. K., TYE, B.-K., BOTSTEIN, D. (1975), Mutagenesis by insertion of a drug resistance element carrying an inverted repetition, *J. Mol. Biol.* **97**, 561–575.

KLECKNER, N., ROTH, J., BOTSTEIN, D. (1977), Genetic engineering *in vivo* using translocatable drug resistance elements: New methods in bacterial genetics, *J. Mol. Biol.* **116**, 125–159.

KNAUF, V. C., NESTER, E. W. (1982), Wide host range cloning vectors: a cosmid clone bank of *Agrobacterium* Ti plasmid, *Plasmid* **8**, 45–54.

KRISHNAPILLAI, V., ROYLE, P., LEHRER, J. (1981), Insertion of the transposon Tn*1* into the *Pseudomonas aeruginosa* chromosome, *Genetics* **97**, 295–311.

KROOS, L., KAISER, D. (1984), Construction of Tn*5* *lac*, a transposon that fuses *lacZ* expression to exogenous promoters, and its introduction into *Myxococcus xanthus, Proc. Natl. Acad. Sci. USA* **81**, 581–582.

LABES, M., PÜHLER, A., SIMON, R. (1990), A new family of RSF1010-derived expression and *lac*-fusion broad-host-range vectors for Gram-negative bacteria, *Gene* **89**, 37–46.

LEEMANS, J., LANGENAKENS, J., DEGREVE, H., DEBLAERE, R., VANMONTAGU, M., SCHELL, J. (1982), Broad-host-range cloning vectors derived from the W-plasmid Sa, *Gene* **19**, 361–364.

LEEMANS, R., REMAUT, E., FIERS, W. (1987), A broad-host-range expression vector based on the pL promoter of coliphage lambda: regulated synthesis of human interleukin 2 in *Erwinia* and *Serratia* species, *J. Bacteriol.* **169**, 1899–1904.

MANOIL, C., BECKWITH, J. (1985), Tn*phoA*: a transposon probe for protein export signals, *Proc. Natl. Acad. Sci. USA* **82**, 8192–8133.

MAZODIER, P., GIRAUD, E., GASSER, F. (1983), Genetic analysis of the streptomycin resistance encoded by Tn*5*, *Mol. Gen. Genet.* **192**, 155–162.

MAZODIER, P., COSSART, P., GIRAUD, E., GASSER, F. (1985), Completion of the nucleotide sequence of the central region of Tn*5* confirms the presence of three resistance genes, *Nucleic Acids Res.* **13**, 195–205.

MEYER, R., HINDS, M., BRASCH, M. (1982), Properties of R1162, a broad-host range, high-copy-number plasmid, *J. Bacteriol.* **150**, 552–562.

MILLS, D. (1985), Transposon mutagenesis and its potential for studying virulence genes in plant pathogens, *Annu. Rev. Phytopathol.* **23**, 297–320.

MORALES, V. M., SEQUEIRA, L (1985), Suicide vector for transposon mutagenesis in *Pseudomonas solanacearum, J. Bacteriol.* **163**, 1263–1264.

NANO, F. E., SHEPHERD, W. D., WATKINS, M. M., KUHL, S. A., KAPLAN, S. (1985), Broad-host-range plasmid vector for the *in vitro* construction of transcriptional/translational *lac* fusions, *Gene* **34**, 219–226.

NOTI, J. D., JAGADISH, M. N., SZALAY, A. A. (1987), Site-directed Tn*5* and transplacement mutagenesis: methods to identify symbiotic nitrogen fixation genes in slow-growing *Rhizobium. Methods Enzymol.* **154**, 197–217.

PRIEFER, U. B. (1989), Genes involved in lipopolysaccharide production and symbiosis are clustered on the chromosome of *Rhizobium leguminosarum* bv. *viciae, J. Bacteriol.* **171**, 6161–6168.

PRIEFER, U. B., SIMON, R., PÜHLER, A. (1985), Extension of the host range of *Escherichia coli* vectors by incorporation of RSF1010 replication and mobilization functions, *J. Bacteriol.* **163**, 324–330.

RELLA, M., MERCENIER, A., HAAS, D. (1985), Transposon insertion mutagenesis of *Pseudomonas aeruginosa* with a Tn*5* derivative: application to physical mapping of the *arc* gene cluster, *Gene* **33**, 293–303.

RICHMOND, M. H., SYKES, R. B. (1972), The chromosomal integration of a β-lactamase gene derived from the P-type R-factor RP1 in *E. coli, Genet. Res.* **20**, 231–237.

Ried, J. L., Collmer, A. (1987), An *nptI-sacB* cartridge for constructing directed, unmarked mutations in Gram-negative bacteria by marker exchange-eviction mutagenesis, *Gene* **57**, 239–246.

Ried, J. L., Collmer, A. (1988), Construction and characterization of an *Erwinia chrysanthemi* mutant with directed deletions in all of the pectate lyase structural genes, *Mol. Plant-Microbe Interact.* **1**, 32–38.

Rogowsky, P. M., Close, T. J., Chimera, J. A., Shaw, J. J., Kado, C. I. (1987), Regulation of *vir* genes of *Agrobacterium tumefaciens* by plasmid pTiC58, *J. Bacteriol.* **169**, 5101–5112.

Rothmel, R. K., Chakrabarty, A. M., Berry, A., Darzins, A. (1991), Genetic systems in *Pseudomonas, Methods Enzymol.* **204**, 485–514.

Ruvkun, G. B., Ausubel, F. M. (1981), A general method for site-directed mutagenesis in prokaryotes, *Nature* **289**, 85–88.

Ruvkun, G. B., Sundaresan, V., Ausubel, F. M. (1982), Directed transposon mutagenesis and complementation analysis of *Rhizobium meliloti* symbiotic nitrogen fixation genes, *Cell* **29**, 551–559.

Sasakawa, C., Yoshikawa, M. (1987), A series of Tn*5* variants with various drug-resistance markers and suicide vector for transposon mutagenesis, *Gene* **56**, 283–288.

Schmitt, R., Bernhard, E., Mattes, R. (1979), Characterization of Tn*1721* a new transposon containing tetracycline resistance genes capable of amplification, *Mol. Gen. Genet.* **172**, 53–65.

Schöffl, F., Arnold, W., Pühler, A., Altenbuchner, J., Schmitt, R. (1981), The tetracycline resistance transposons Tn*1721* and Tn*1771* have three 38-base-pair repeats and generate five-base-pair direct repeats, *Mol. Gen. Genet.* **181**, 87–94.

Seifert, H. S., So, M. (1991), Genetic systems in pathogenic Neisseriae, *Methods Enzymol.* **204**, 342–356.

Selbitschka, W., Arnold, W., Priefer, U. B., Rottschäfer, T., Schmidt, M., Simon, R., Pühler, A. (1991), Characterization of *recA* genes and *recA* mutants of *Rhizobium meliloti* and *Rhizobium leguminosarum* biovar. *viciae, Mol. Gen. Genet.* **229**, 86–95.

Selbitschka, W., Pühler, A., Simon, R. (1992), The construction of *recA*-deficient *Rhizobium meliloti* and *R. leguminosarum* strains marked with *gusA* of *luc* cassettes for the use in risk-assessment studies, *Mol. Ecol.* **1**, 9–19.

Selveraj, G., Iyer, V. N. (1983), Suicide plasmid vehicles for insertion mutagenesis in *Rhizobium meliloti* and related bacteria, *J. Bacteriol.* **156**, 1292–1300.

Shapiro, J. A. (Ed.) (1983), *Mobile Genetic Elements*, New York: Academic Press.

Sharma, S. B., Signer, E. R. (1990), Temporal and spatial regulation of the symbiotic genes of *R. meliloti* in planta revealed by transposon Tn*5*-*gusA, Genes Dev.* **4**, 344–356.

Shaw, J. J., Kado, C. I. (1987), Direct analysis of the invasiveness of *Xanthomonas campestris* mutants generated by Tn*4431*, a transposon containing a promoterless luciferase cassette for monitoring gene expression, in: *Molecular Genetics of Plant-Microbe Interactions* (Verma, D. P. S., Brisoon, N., Eds.), pp. 57–60, Dordrecht, The Netherlands: Martinus Nijhoff.

Shaw, J. J., Settles, L. G., Kado, C. I. (1988), Transposon Tn*4431* mutagenesis of *Xanthomonas campestris* pv. *campestris*: characterization of a nonpathogenic mutant and cloning of a locus for pathogenicity, *MPMI* **1**, 39–45.

Silverman, M., Showalter, R., McCarter, L. (1991), Genetic analysis in *Vibrio, Methods Enzymol.* **204**, 515–536.

Simon, R. (1984), High frequency mobilization of Gram-negative bacterial replicons by the *in vitro* constructed Tn*5-Mob* transposon, *Mol. Gen. Genet.* **196**, 413–420.

Simon, R. (1989), Transposon mutagenesis in non-enteric Gram-negative bacteria, in: *Promiscuous Plasmids of Gram-Negative Bacteria* (Thomas, C. M., Ed.), pp. 207–228, London: Academic Press, Inc.

Simon, R., Priefer, U. B. (1990), Vector technology of relevance to nitrogen fixation research, in: *Molecular Biology of Symbiotic Nitrogen Fixation* (Gresshoff, P. M., Ed.), pp. 13–49, Boca Raton, Florida: CRC Press.

Simon, R., Priefer, U., Pühler, A. (1983a), A broad host range mobilization system for *in vivo* genetic engineering: Transposon mutagenesis in Gram-negative bacteria, *Biotechnology* **1**, 784–791.

Simon, R., Priefer, U., Pühler, A. (1983b), Vector plasmids for *in vivo* and *in vitro* manipulations of Gram-negative bacteria, in: *Molecular Genetics of the Bacteria-Plant Interaction* (Pühler, A., Ed.), pp. 98–106, Heidelberg: Springer-Verlag.

Simon, R., O'Connell, M., Labes, M., Pühler, A. (1986), Plasmid vectors for the genetic analysis and manipulation of Rhizobia and other gram-negative bacteria, *Methods Enzymol.* **118**, 640–659. Academic Press.

Simon, R., Quandt, J., Klipp, W. (1989), New derivatives of transposon Tn*5* suitable for mobilization of replicons, generation of operon fusions and induction of genes in Gram-negative bacteria, *Gene* **80,** 161–169.

STACHEL, S. E., AN, G., FLORES, C., NESTER, E. W. (1985), A Tn*3 lacZ* transposon for the random generation of *β*-galactosidase gene fusions: application to the analysis of gene expression in *Agrobacterium, EMBO J.* **4,** 891–898.

STANISICH, V. A., BENNETT, P. M., RICHMOND, M. H. (1977), Characterization of a translocation unit encoding resistance to mercuric ions that occurs on a nonconjugative plasmid in *Pseudomonas aeruginosa, J. Bacteriol.* **129,** 1227–1233.

STARLINGER, P. (1980), IS elements and transposons, *Plasmid* **3,** 242–259.

STEINMETZ, M., LE COQ, D., AYMERICH, S., GONZY-TREBOUL, G., GAY, P. (1985), The DNA sequence of the gene for the secreted *Bacillus subtilis* enzyme levansucrase and its genetic control sites, *Mol. Gen. Genet.* **200,** 220–228.

TAIT, R. C., CLOSE, T. J., LUNDQUIST, R. C., HAGIYA, M., RODRIGUEZ, R. L., KADO, C. I. (1983), Construction and characterization of a versatile broad host range DNA cloning system for gram-negative bacteria, *Biotechnology* **1,** 269–275.

THOMSON, J. A., HENDSON, M., MAGNES, R. M. (1981), Mutagenesis by insertion of drug resistance transposon Tn*7* into *Vibrio* species, *J. Bacteriol.* **148,** 374–378.

TURNER, P., BARBER, C., DANIELS, M. (1984), Behaviour of the transposons Tn*5* and Tn*7* in *Xanthomonas campestris* pv. *campestris, Mol. Gen. Genet.* **195,** 101–107.

VAN VLIET, F., SILVA, B., VAN MONTAGU, M., SCHELL, J. (1978), Transfer of RP4::Mu plasmids to *Agrobacterium tumefaciens, Plasmid* **1,** 446–455.

WAY, J. C., DAVIS, M. A., MORISATO, D., ROBERTS, D. E., KLECKNER, N. (1984), New Tn*10* derivatives for transposon mutagenesis and for construction of *lacZ* operon fusions by transposition, *Gene* **32,** 369–379.

WILLETS, N., CROWTHER, C. (1981), Mobilization of the non-conjugative IncQ plasmid RSF1010, *Genet. Res.* **37,** 311–316.

YAKOBSON, E. A., GUINEY, D. G. (1984), Conjugal transfer of bacterial chromosomes mediated by the RK2 plasmid transfer origin cloned into transposon Tn*5*, *J. Bacteriol.* **160,** 451–453.

ZSEBO, K. M., WU, F., HEARST, J. E. (1984), Tn*5.7* construction and physical mapping of pRPS404 containing photosynthetic genes from *Rhodopseudomonas capsulata, Plasmid* **121,** 192–184.

12 Genetic Engineering of Gram Positive Bacteria

WOLFGANG WOHLLEBEN
GÜNTHER MUTH
Saarbrücken, Federal Republic of Germany

JÖRN KALINOWSKI
Bielefeld, Federal Republic of Germany

1 Introduction

Genetic engineering techniques were developed for *Escherichia coli* by analyzing the genetics and biochemistry of this model organism. Consequently, gene expression systems, particularly for the rapid expression of recombinant gene products, originate from those developed for *E. coli* (STADER and SILHAVY, 1990). It soon became apparent, however, that the *E. coli* system had many disadvantages with respect to secretion, low yields, and the incorrect processing and folding of protein products. In recent years, other bacteria, fungi and higher eukaryotes have been adapted for use in expression systems, each of which has its particular characteristics (see Chapters 10–11 and 13–18 of this volume).

In this chapter, we summarize the properties and features of Gram positive bacteria with respect to the cloning, manipulation and expression of homologous and heterologous genes. The emphasis is placed on bacilli, streptomycetes and corynebacteria which are the organisms of choice with respect to the genetic engineering *and* biotechnological application of Gram positive bacteria. The principles of genetic engineering applicable to other Gram positive bacteria such as staphylococci, streptococci, clostridia, lactococci and lactobacilli, are very similar and are summarized at the end of this chapter.

2 Genetic Engineering of Bacilli

2.1 Biology of Bacilli

Bacilli form a group of Gram positive, aerobic soil bacteria which are characterized by their ability to respond to conditions of nutrient limitations by forming endospores. These endospores – so called because they are produced within the parent vegetative cells – show a complex morphological ultrastructure. Most striking, however, is their remarkable resistance to heat, desiccation, radiation and other adverse environmental conditions.

Bacillus is the best studied Gram positive microorganism with respect to fundamental research and industrial application. Several species are currently used for the commercial production of toxins, antibiotics, and different enzymes used in food processing:

B. thuringiensis is characterized by its ability to produce crystalline inclusions during sporulation. These inclusions consist of proteins which exhibit a highly specific insecticidal activity (HÖFTE and WHITELY, 1989).

B. brevis produce important peptide antibiotics such as gramicidin and tyrocidine.

A great variety of food processing enzymes, mainly proteases, glucanases, and amylases, is produced by *B. subtilis, B. licheniformis, B. megaterium,* and *B. amyloliquefaciens. Bacillus* is interesting from the academic point of view, because of its complex regulation systems which control sporulation, competence, motility, antibiotic formation and the production of degradative enzymes (LOSICK et al., 1986).

The ability to secrete extracellular enzymes directly into the medium made *B. subtilis* one of the most studied systems with respect to the expression of heterologous genes (PALVA et al., 1983; CHANG, 1987; HENNER, 1990). The advantages of using the *Bacillus* system include:

(1) the non-pathogenic nature of bacilli (DE BOER and DIDERICHSEN, 1991),
(2) the absence of endotoxin or other toxicogenic products,
(3) a well-established fermentation technology for industrial and food products,
(4) well-characterized organisms in terms of genetics, physiology, and recombinant DNA technology.

To date, two major limitations have hindered the wide-spread application of *B. subtilis* in the production of heterologous products: the instability of the cloning vectors (structural and segregational instability) and the instability of the products (degradation of proteins by extracellular proteases). These difficulties, however, seem to have been overcome, since recent publications describe the development of stable cloning vectors. To solve the problem of protease degradation, two promising approaches have been reported:

(1) the generation of protease-deficient strains by gene replacement techniques,
(2) the use of *B. brevis,* a strain naturally devoid of high proteolytic activity, as an alternative expression system.

2.2 Methods for the Genetic Manipulation of Bacilli

Besides *Escherichia coli, Bacillus subtilis* is the best genetically characterized bacterium. In 1989 more than 700 gene loci had been mapped on the circular chromosome of *B. subtilis* (PIGGOT, 1989). Recently, this genetic linkage map was correlated with a physical map which was drawn up from the results of the pulsed-field gel electrophoresis of *Sfi*I and *Not*I fragments (AMJAD et al., 1990; ITAYA and TANAKA, 1991). The exact chromosome size was determined to be 4165 kb (ITAYA and TANAKA, 1991). The correlation of the physical map and the genetic linkage map is shown in Fig. 1.

Regulation of gene expression of *B. subtilis* was found to be more complex than that of *E. coli.* The transcription of particular operons in *Bacillus* is controlled by global, pleiotropic regulatory systems. One example is the *spo0A* gene product which belongs to the family of phosphorylation-activated proteins of two-component systems (TRACH et al., 1990). Mutations in *spo0A* impede the sporulation process during the early stage and concomitantly repress the expression of antibiotic biosynthetic, protease and competence genes (LOSICK et al., 1986). The production of degradative enzymes is under the control of another two-component system. The *DegS* gene product is the protein kinase, whereas *DegU* represents the effector molecule which is phosphorylated by *DegS* (STEINMETZ and AYMERICH, 1990).

Additionally, the gene expression is regulated by multiple forms of RNA polymerase containing different σ-factors (Tab. 1). The primary form found in vegetative cells contains σ^A (formerly σ^{43}) which is encoded by the *rpoD* gene. Each σ-factor confers the specificity to the RNA polymerase for a characteristic class of promoters (BENSON et al., 1990; ERRINGTON and ILLING, 1992; DEBARBOUILLE et al., 1991), allowing the exact timing and spatial localization (endospore formation) of transcription (LOSICK et al., 1986).

2.2.1 Cloning Vectors Derived from Staphylococcal Drug-Resistance Plasmids

A major breakthrough in the development of vector systems for *Bacillus subtilis* was the observation that small drug-resistance plasmids from *Staphylococcus aureus* were able to replicate in *B. subtilis.* Furthermore, the antibiotic resistance genes encoded by the *S. aureus* plasmids were also expressed in *B. subtilis.* Such plasmids, e.g., pT181 (IORDANESCU et al., 1978; KHAN and NOVICK, 1983), pC194 (IORDANESCU, 1976), pUB110 (LACEY and CHAPRA, 1974; MCKENZIE et al., 1986), pE194 (HORINOUCHI and WEISBLUM, 1982), have been manipulated to produce a range of cloning vectors suitable for particular applications. Plasmids pUB110 and pT181 are presented in Fig. 2.

The small staphylococcal plasmids transpired to form a family of highly interrelated replicative elements (NOVICK, 1989). A characteristic common to all these plasmids is their replication mode. In contrast to plasmids isolated from Gram negative bacteria, these plasmids were shown to replicate by the rolling circle mechanism (RCR) via single-stranded DNA (ss-plasmids) intermediates (TE RIELE et al., 1986; GRUSS and EHRLICH, 1989). In general, ss-plasmids contain five characteristic elements (NOVICK, 1990):

(1) the *rep* gene, a site-specific nuclease for the initiation of the leading strand synthesis,
(2) a leading strand origin (*ori*),
(3) the *pre* gene, a plasmid recombination function,
(4) the resistance determinant,
(5) *palA* or minus-origin (*mo*), the lagging strand origin for the initiation of lagging strand synthesis by host-encoded RNA polymerase.

In general, this orientation-specific minus-origin is functional only in the original host

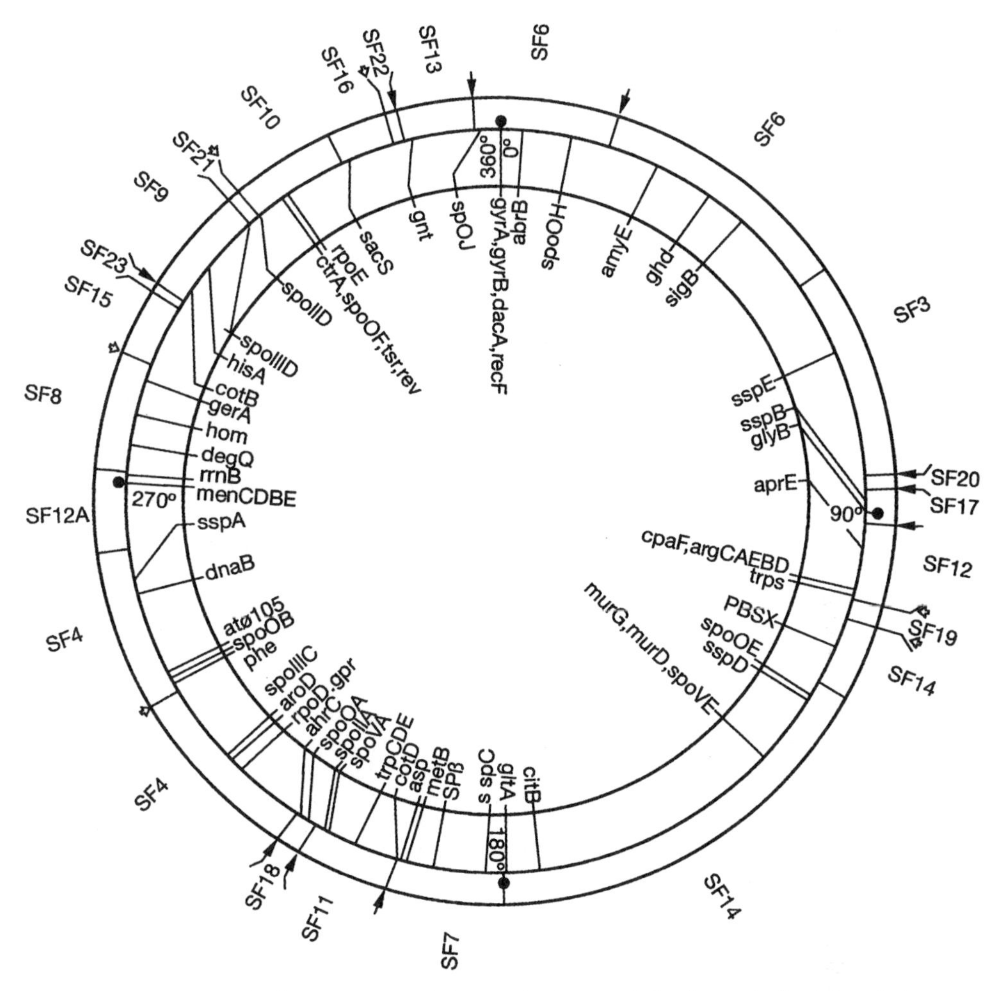

Fig. 1. Compilation of the genetic linkage map and the physical map of *Bacillus subtilis*. A restriction map of the 23 *Sfi*I restriction fragments of the *B. subtilis* chromosome is correlated with the genetic linkage map. The positions of selected loci are given on the inner circle. The outer double circle shows the location of the 23 *Sfi*I fragments. The origin of genome replication was taken as point zero. (The map is taken from AMJAD et al., 1990).

strain. If the minus-origin becomes deleted or is inactive, single-stranded plasmids accumulate in large amounts (GRUSS and EHRLICH, 1989). The accumulation of ss-DNA was suggested to be a major factor determining both the segregational (GRUSS and EHRLICH, 1989) and structural plasmid instability in *B. subtilis* (JANNIERE and EHRLICH, 1987). Since the production of ss-DNA and the RCR mechanism also stimulate homologous (NIAUDET et al., 1984) and illegitimate (JANNIERE et al., 1987) recombination, deletion formation can occur between short direct repeats of only 9 to 20 bp in length (BRON et al., 1991). The frequency of deletion formation was found to be directly proportional to the amounts of ss-DNA present (BRON et al., 1991). A model proposed by EHRLICH (1989) explains the structural instability by template-switching errors during replication. In this model, the replication machinery is assumed to "slip" from one repeat to another, resulting in the deletion

Tab. 1. σ-Factors of *Bacillus subtilis*

σ-Species	Gene	Function
σ^A (σ^{43})	*rpoD*	Major vegetative form
σ^B (σ^{37})	*sigB*	Minor vegetative form
σ^C (σ^{32})	*sigC*	Minor vegetative form
σ^H (σ^{30})	*spoOH*	Competence, onset of sporulation
σ^D (σ^{28})	*sigD*	Chemotaxis, flagellar operons
σ^E (σ^{29})	*spoIIGB*	Sporulation-specific
σ^F	*spoIIAC*	Sporulation, early genes
σ^G	*spoIIIG*	Sporulation, prespore-specific
σ^K	*spoIVCB/ spoIIIC*	Sporulation, mother cell-specific
σ^L	*sigL*	Nitrogen assimilation, levanase operon

Additional σ-factors are encoded by certain virulent *B. subtilis* phages.

of the intervening sequence in the plasmid progeny (EHRLICH, 1989). Therefore, the cloning efficiency in *B. subtilis* using vectors which accumulate ss-DNA is usually low, and small DNA inserts (<1 kb) are generally obtained.

However, recent publications described stably replicating vector systems for *B. subtilis.* Although plasmid pTA1060 (BRON et al., 1987) replicates by the RCR mechanism, it was found to be stable. When pTA1060 carries the minus-origin which is efficiently recognized in *B. subtilis,* no ss-DNA is accumulated (BRON et al., 1991), and even large DNA inserts could be cloned without affecting the stability (HAIMA et al., 1990). Based on plasmid pTA1060, cloning vectors were developed that included the *E. coli* β-galactosidase LacZα complementation system and that permitted clones to be selected directly in *B. subtilis* (HAIMA et al., 1990).

A completely different type of plasmid comprise the large, broad-host-range plasmids, such as pAMβ1 (CLEWELL et al., 1974) which was originally isolated from *Streptococcus faecalis* and plasmid pIP404 (GARNIER and COLE, 1988) from *Clostridium perfringens.* Plasmid pAMβ1 is a conjugative plasmid 26.5 kb in size, which mediates resistance to the macrolide, lincosamide, and streptogramin B group (MLS) of antibiotics. Plasmid pIP404 is a 10.2 kb bacteriocinogenic plasmid which can be mobilized by certain R-factors (BREFORT et al., 1977). In contrast to the small staphylococcal plasmids, these broad-host-range plasmids were found to be segregationally and structurally stable. The analysis of their replication mechanism revealed unidirectional theta-replication in *B. subtilis* (BRUAND et al., 1991; SWINFIELD et al., 1990; GARNIER and COLE, 1988), suggesting a correlation between the replication mode and the plasmid stability. In comparision with the ss-plasmids, the frequency of deletions occurring between short repeats were found to be 100-fold lower in plasmid pAMβ1 (BRON et al., 1991; JANNIERE et al., 1990). Cloning vectors derived from plasmid pAMβ1 were constructed that replicated stably and were able to carry large DNA inserts (JANNIERE et al., 1990).

2.2.2 DNA Transfer by Transformation and Transduction

The most important method of genetic exchange in *Bacillus subtilis* is the uptake of DNA utilizing natural competence (SPIZIZEN, 1958). Competence, a physiologically and genetically determined property, is usually expressed post-exponentially and associated with the onset of sporulation. Competent cells are able to bind, process, and internalize exogenous DNA efficiently (DUBNAU, 1991).

High-molecular-weight DNA is bound to receptors at the cell surface (not sequence-specific) and is rapidly fragmented by double-strand cleavage (ARWERT and VENEMA, 1973). A sin-

A

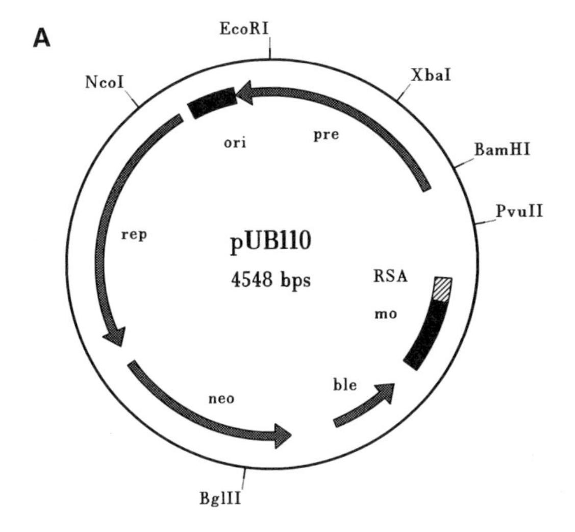

B

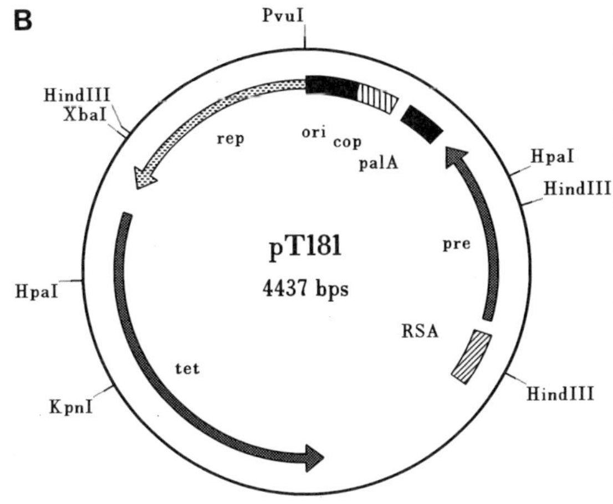

Fig. 2. Structure of the *Staphylococcus aureus* drug-resistance plasmids pUB110 (A) and pT181 (B). The staphylococcal plasmids pUB110 and pT181 are able to replicate in *B. subtilis* by the rolling circle (RCR) mechanism (Novick, 1989). Plasmid pUB110 has a copy number of 50 and mediates neomycin and bleomycin resistance, while plasmid pT181 encodes tetracycline resistance.
Abbreviations: rep, initiator protein determinant; pre, plasmid recombination function; neo, neomycin phosphotransferase gene; ble, bleomycin resistance determinant; tet, tetracycline resistance determinant; ori, origin of replication; mo, lagging strand (minus) origin; palA, lagging strand conversion signal; cop, replication control system; RSA, recombination site.

gle DNA strand is transported into the cytoplasm, whereas the second strand is degraded by a membrane-localized nuclease (Dubnau and Cirigliano, 1972). The incoming single strand is rescued by homologous recombination with chromosomal fragments. The transformation of competent *B. subtilis* cells with plasmid DNA is generally inefficient, since only plasmid concatemers yield transformants (Canosi et al., 1978). Only multimeric forms possess homology which allow the occurrence of recircularization by homologous recombination. Monomeric plasmids lack regions of homology and, consequently, cannot become established in the cell (Canosi et al., 1978). Monomers can be transformed into competent *B. subtilis* cells by "plasmid rescue", using either plasmids with internal direct repeats (Michel et al., 1982) or using a host strain already carrying a homologous plasmid (Contente and Dubnau, 1979). Since transformants obtained in this way contain both the recombinant and the resident plasmid, they have to be screened further.

These difficulties can be circumvented by the PEG-induced protoplast transformation method developed by Chang and Cohen (1979). *B. subtilis* cells are converted into protoplasts by the action of lysozyme, DNA is efficiently introduced in the presence of polyethyleneglycol (PEG) and the protoplasts are regenerated on complex media. Since plasmid DNA which is circular and double-stranded is employed, the transformation frequency is very high. As many as 10^7 transformants per µg DNA can be obtained. More recently, electroporation has been described for bacilli as another simple and highly efficient transformation method (Luchansky et al., 1988).

In contrast to transformation where only relatively small DNA fragments (<40 kb) are assimilated, large DNA molecules can be introduced by transduction. The *Bacillus* phage PBS1 was found to be able to transduce DNA fragments of 150–200 kb in size (Hoch, 1991). This transduction system replaced the method of conjugation to establish the chromosomal map of *B. subtilis*.

Conjugation systems in *Bacillus* are of minor importance. The *Streptococcus faecalis* conjugative transposon Tn*916* was transferred to *B. subtilis* and *B. thuringiensis* and was shown to be capable of mobilizing plasmids pUB110 and pC194 (NAGLICH and ANDREWS, 1988). Furthermore, plasmid DNA was transferred at low frequency from *E. coli* to *Bacillus* by interspecific mating (TRIEU-CUOT et al., 1987).

2.2.3 Transposon Mutagenesis and DNA Amplification in Bacilli

Most of the genetic systems described for *Escherichia coli* have been modified and adapted for the application in *B. subtilis*. This includes cloning and gene expression systems as well as those for integrating genes into the chromosome (CHANG, 1987; HAIMA et al., 1990; HENNER, 1990).

Methods of transposon mutagenesis have been established in *B. subtilis* to generate mutants efficiently. Several transposons derived from Gram positive bacteria were shown to be functional in *B. subtilis* (COURVALIN and CARLIER 1987; YOUNGMAN et al., 1983). Additionally, derivatives of the *E. coli* transposon Tn*10* were successfully used in *B. subtilis* when the transposase gene was fused to the appropriate expression signals (PETIT et al., 1990b).

Furthermore, a technique is available that allows the inducible amplification of chromosomal DNA fragments. A recent publication by PETIT et al. (1990b) describes the amplification and overexpression of the *Clostridium thermocellum* endoglucanase A gene in *B. subtilis*. To achieve this, the temperature-sensitive plasmid pE194 was integrated adjacent to the endoglucanase A gene which was located between two direct repeats. When the integrated pE194 plasmid was allowed to replicate, the endoglucanase gene and one repeat were amplified. Amplification levels of up to 250 copies were obtained, resulting in the overproduction of endoglucanase A (PETIT et al., 1990b).

2.3 Gene Expression Systems Using Inducible Transcriptional Signals

To date, a great variety of eukaryotic and prokaryotic genes have been expressed in *Bacillus subtilis* (RAPOPORT and KLIER, 1990). Various signal sequences from naturally secreted *Bacillus* and *Staphylococcus* proteins have been used (Tab. 2) to obtain secretion into the culture supernatant.

For the most part, constitutive promoters were used in these studies. However, since foreign proteins may be toxic to *B. subtilis* cells, particularly when overexpressed, regulatable promoter systems are of special interest. Therefore, well-established, inducible expression systems of *E. coli* were modified and transferred to *B. subtilis*. In addition, endogenous bacillar systems were developed to allow inducible gene expression to occur.

2.3.1 Expression Systems Derived from *Escherichia coli*

(1) YANSURA and HENNER (1984) transferred the well-characterized Lac-system to *B. subtilis*. The expression plasmid contained the *lacI* operator region fused to a *B. subtilis* promoter. The *lac* repressor gene located on a second plasmid, was also transcribed by a *B. subtilis* promoter. When located together within one host cell, the two plasmids effected an IPTG-inducible gene expression. Repression levels of approximately 100 were obtained. This system was optimized by using stronger promoters such as the bacteriophage T5 promoter PN25 (PESCHKE et al., 1985) and by integrating the *lac* repressor gene into the chromosome of *B. subtilis* (LE GRICE, 1990). Using the Lac-system, several eukaryotic proteins (mouse dihydrofolate reductase, HIV-1 reverse transcriptase, human TPA) have been expressed at high levels (LE GRICE et al., 1987; WANG et al., 1989).

(2) BREITLING et al. (1990) showed that the temperature-inducible repressor system of bacteriophage λ also functioned in *B. subtilis*. They demonstrated by S1 mapping, that the P_R promoter of phage λ was active in *B. subtilis*. The temperature-sensitive *CI857* repressor

Tab. 2. Heterologous Products Expressed and Secreted in *Bacillus*

Leader	Product	Amount (mg/mL)	Reference
Neutral protease	Human growth hormone	40	NAKAYAMA et al., 1988
Staphylokinase (sak42D)	α-Amylase	15	BREITLING et al., 1990
	Human interferon α1	15	BREITLING et al., 1989
Middle wall protein (MWP)	Human epidermal growth factor	240	YAMAGATA et al., 1989
	Swine pepsinogen	11	TAKAO et al., 1989
Levansucrase (*sacB*)	Human interferon	0.5	SCHEIN et al., 1986
	Endoglucanase (*Clostridium thermocellum*)	10	PETIT et al., 1990b
α-Amylase	Diphtheria toxin tox228	4	HEMILÄ et al., 1989
	Pertussis toxin	0.5	SARIS et al., 1990
	subunits	100	
	α-Galactosidase (*Cyamopsis tetragonoloba*)	10	OVERBEEKE et al., 1990
	Pectin methylesterase	500	HEIKINHEIMO et al., 1991
	Pneumolysin	10	TAIRA et al., 1989
Nuclease (*S. aureus*)	Staphylococcal nuclease	50	KOVACEVIC et al., 1985
Subtilisin	Protein A	3000	VASANTHA and THOMPSON, 1986
	Natriuretic factor	0.5	WANG et al., 1988
	Bovine pancreatic ribonuclease A	1	VASANTHA and FILIPULA, 1989

was fused to the staphylokinase (*sak42D*) transcriptional and translational signals to obtain expression of the repressor. The amylase (*amy*) gene expression could be induced by a factor of 1400 by shifting the incubation temperature from 37 to 42 °C.

(3) The *tet* regulatory element of the *E. coli* transposon Tn*10* was also used to construct inducible expression vectors. The *tet* operator was fused with the strong *xyl* promoter derived from the *B. subtilis* xylose operon (GÄRTNER et al., 1988). In the presence of the *tetR* gene, the hybrid *tet-xyl* promoter could be induced 100-fold by sublethal doses of tetracycline. The application of this system has permitted high-level expression of the *B. megaterium* glucose dehydrogenase and human single-chain urokinase-like plasminogen activator (GEISSENDÖRFER and HILLEN, 1990).

2.3.2 Endogenous Bacillar Expression Systems

(1) A strategy similar to the λ/CI-repressor system (BREITLING et al., 1990) was employed by OSBURNE et al. (1985). They developed a heat-inducible expression system by using the early promoter–operator sequence of the *B. subtilis* temperent phage Φ105 and the temperature-sensitive repressor from the mutant Φ105cts23.

(2) Levansucrase, encoded by the *sacB* gene is synthesized and secreted in large quantities during the logarithmic growth phase of *B. subtilis*. The transcription of *sacB* is regulated by the linked *sacR* locus which consists of a constitutive promoter, followed by a palindromic sequence which acts as a transcriptional terminator (STEINMETZ

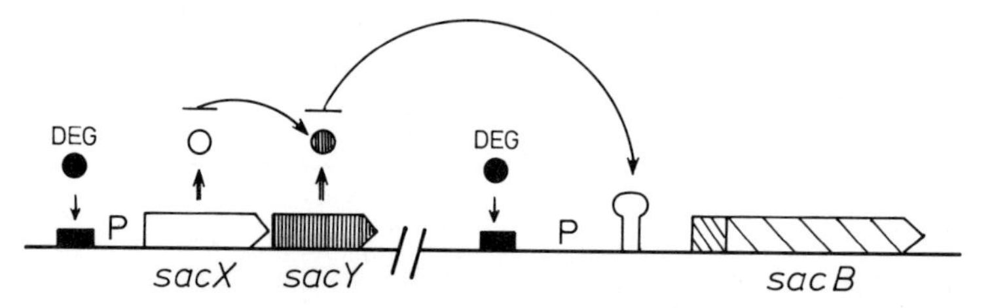

Fig. 3. Regulation of the *sacB* gene expression in *Bacillus subtilis.* The *sacX* and *sacY* genes constitute an operon. In the absence of sucrose, *sacX* is phosphorylated and inhibits *sacY* gene expression. In the presence of sucrose, the active *sacY* gene product acts as an antiterminator allowing transcription of the *sacB* gene. In addition, the transcription of both *sacB* and *sacX/Y* is controlled by the Deg regulators.

et al., 1985). In the presence of sucrose the positive regulatory *sacY* gene is expressed. *SacY*, acting as an anti-terminator allows the transcription of *sacB* to occur. The expression of the *sacY* gene itself is controlled by the regulatory genes *degS/degU* and *degQ*. Fig. 3 summarizes the control of the *sacB* expression in *B. subtilis.* Several expression and secretion vectors based on the *sacB* gene have been developed (EDELMAN et al., 1988; WONG, 1989; ZUKOWSKI et al., 1988). These vectors consist of the *sacB* promoter, the *sacR* regulatory region, and the signal sequence of the *sacB* gene, followed by engineered restriction sites for the convenient in-frame-fusion of foreign DNA fragments. When employing the levansucrase system, the expression levels can be further enhanced by using as host strains, *B. subtilis* carrying up-mutations in the regulatory *degS, degU,* or *degQ* genes (RAPOPORT and KLIER, 1990). Alternatively, the *degQ* gene can be overexpressed on a plasmid (WU et al., 1991).

(3) The xylose isomerase gene of *Bacillus megaterium,* a producer of glucose dehydrogenase is under transcriptional control of the strong, tightly regulated promoter P_A. The regulation is mediated by the repressor *xylR* which binds to the promoter in the absence of xylose. The genes encoding the glucose dehydrogenase (*gdhA*) from *B. megaterium, β*-galactosidase (*lacZ*) from *E. coli,* mutarotase (*mro*) from *Acetobacter calcoaceticus* and the human single-chain, urokinase-like plasminogen activator (rscupa) were fused to the P_A promoter in *B. megaterium.* The addition of xylose resulted in a 130-fold to 350-fold induction. Enzymatically active enzymes accumulated in the cytoplasm in amounts of up to 20 and 30% of the soluble protein (RYGUS and HILLEN, 1991).

2.4 Construction of Host Strains Deficient in Extracellular Protease Production for the Synthesis of Heterologous Products

As mentioned above, the main limitation of the *Bacillus* system for the expression of heterologous proteins is the instability of the products due to proteolytic degradation. *Bacillus subtilis* produces two major and at least five minor extracellular proteases (SLOMA et al., 1991). The two major extracellular proteolytic enzymes, alkaline (subtilisin) and neutral protease A, account for more than 90% of the proteolytic activity (KAWAMURA and DOI, 1984). To circumvent the problem of protease degradation, various research groups (KAWAMURA and DOI, 1984; YANG et al., 1984; STAHL and FERRARI, 1984; WU et al., 1991; SLOMA et al., 1991) isolated the relevant protease gene (Tab. 3), inactivated the cloned gene, and used the mutated gene to replace the corresponding chromosomal gene. Using this strategy, *B. subtilis* strains deficient in six ex-

Tab. 3. Extracellular *Bacillus subtilis* Proteases which Have Been Cloned and Inactivated

Protease	Gene	Reference
Neutral protease	*nprA*	YANG et al., 1984
Alkaline protease	*apr*	STAHL and FERRARI, 1984
Extracellular protease	*epr*	SLOMA et al., 1988
Metalloprotease	*mpr*	SLOMA et al., 1990a
Bacillopeptidase F	*bpr*	SLOMA et al., 1990b WU et al., 1990
Neutral protease B	*nprB*	TRAN et al., 1991
Extracellular serine protease	*vpr*	SLOMA et al., 1991

tracellular proteases were constructed (WU et al., 1991; SLOMA et al., 1991). The resulting strains possessed only about 0.3% extracellular proteolytic activity, showed normal growth, and were not impaired in their ability to secrete proteins. However, SLOMA et al. (1991) demonstrated that the elimination of the minor extracellular protease Vpr resulted in the induction of a further, until then unknown, protease gene. This approach should allow the elimination of the residual proteolytic activity of *B. subtilis*.

A completely different approach to circumvent the problem arising from protease activity is the use of a host which is naturally devoid of proteases. *Bacillus brevis* strains that secrete large amounts of protein, but possess only low proteolytic activity (0.02–1.6% of the *B. subtilis* activity) were isolated (MIYASHIRO et al., 1980). *B. brevis* strains 47 and HPD31 accumulate 12–25 g/L of extracellular proteins. These proteins consist mainly of the two major cell wall proteins OWP (outer wall protein) and MWP (middle wall protein). The cell wall proteins are synthesized not only during the logarithmic phase, but also during the stationary phase. The nucleotide sequence analysis of the cloned genes (ADACHI et al., 1989) showed that the OWP and MWP genes are organized in one operon. Several putative promoter regions were identified on a 300 bp fragment by S1 mapping. The P2 promoter responsible for the constitutive expression in *B. brevis* was found to be ineffective in *B. subtilis*. The OWP and MWP proteins are synthesized as preproteins which contain a leader peptide of 23 and 24 amino acids in length, respectively. The 5′ region of the MWP gene together with either an endogenous *B. brevis* plasmid (pWT481) or the *Staphylococcus aureus* plasmid pUB110 were used in the construction of expression–secretion vectors. Several mammalian and bacterial genes encoding secretory proteins were fused to the 5′ sequence of the MWP gene and expressed in a biologically active form. Although the expression levels differed for the different proteins, the proteins remained stable even after prolonged incubation. Using these vectors together with the described *B. brevis* host strains, 240 mg/L of biologically active human epidermal growth factor was produced in the culture supernatant (YAMAGATA et al., 1989).

3 Genetic Engineering of Streptomycetes

3.1 Biology of *Streptomyces*

Bacteria belonging to the group of Actinomycetales are well known for their ability to synthesize various compounds which are produced on an industrial scale. The best examined genus of this group is *Streptomyces*. Beside useful therapeutical compounds such as antimicrobial agents and other pharmacologically active molecules, herbicides and a variety of extracellular enzymes (e.g., amylases, cellulases, proteases, and nucleases) are produced by these filamentous soil bacteria (DESH-

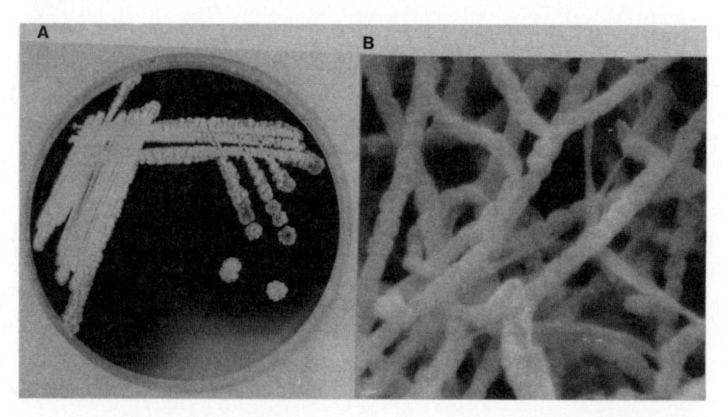

Fig. 4. Photograph of *Streptomyces* colonies on solid medium (A) and electron micrograph of a *Streptomyces* mycelium (B). During growth on solid medium, most *Streptomyces* strains excrete various secondary metabolites. (A) An agar plate containing *S. coelicolor* "Müller" colonies, grown for 6 days, was photographed. *S. coelicolor* "Müller" produces a heptaene antibiotic, lysozyme, various proteases, and a diffusible pigment. (B) The scanning electron micrograph shows the presence of aerial hyphae during the process of septation (courtesy of Hoechst AG, Frankfurt).

PANDE et al., 1988). Furthermore, these bacteria are used in bioconversion reactions. Because of their ability to synthesize antibiotics, streptomycetes have been employed for industrial fermentation since the early fifties. Consequently, much knowledge, information, and experimental experience have accumulated with respect to these non-pathogenic bacteria. To date, the endogenous synthetic properties have predominantly been exploited. In many laboratories, however, research is in progress developing the use of streptomycetes for the synthesis of heterologous proteins by exploiting their ability to excrete products into the culture medium.

The genus *Streptomyces* includes a large number of different species which were classified and grouped using physiological, morphological, and genetic criteria (WILLIAMS et al., 1983; LANGHAM et al., 1989; KÄMPFER et al., 1991).

All are capable of cellular differentiation. On solid media, the life cycle begins with the germination of a spore which gives rise to a multicellular substrate mycelium of branching hyphae. Next, an aerial mycelium grows on the substrate mycelium, followed by the fragmentation of the aerial hyphae into chains of spores (Fig. 4), each containing one chromosome (CHATER et al., 1982; CHATER, 1989). The synthesis of a multiplicity of compounds which are of interest to man may be correlated – at least in some cases – with this differentiation process and its multilevel regulation (CHATER, 1989). The extracellular enzymes, however, may play a role in the dissolution of solid food sources.

It is likely that the potential to produce secondary metabolites is even greater than is known at present, since the *Streptomyces coelicolor* genome is about 8000 kb in size which is twice that of *E. coli* (KIESER et al., 1992).

The differentiation and antibiotic production have been studied most extensively in *S. coelicolor* A3(2). Various reviews are available of *S. coelicolor* genetics (HOPWOOD, 1967; HOPWOOD et al., 1973), and the first comprehensive review of cloning in streptomycetes was published ten years ago (CHATER et al., 1982). Subsequently, a widely used laboratory manual on genetic manipulation techniques (HOPWOOD et al., 1985a) and excellent reviews of gene cloning and vector systems (e.g., HOPWOOD et al., 1987; RAO et al., 1987; KIESER and HOPWOOD, 1991) have become available. Within the scope of this chapter it is not possible to give a detailed summary of all these data. After describing the commonly used system and referring to relevant publications on gene cloning, new DNA techniques which are based on the application of temperature-sensitive vectors will be presented. In addition to

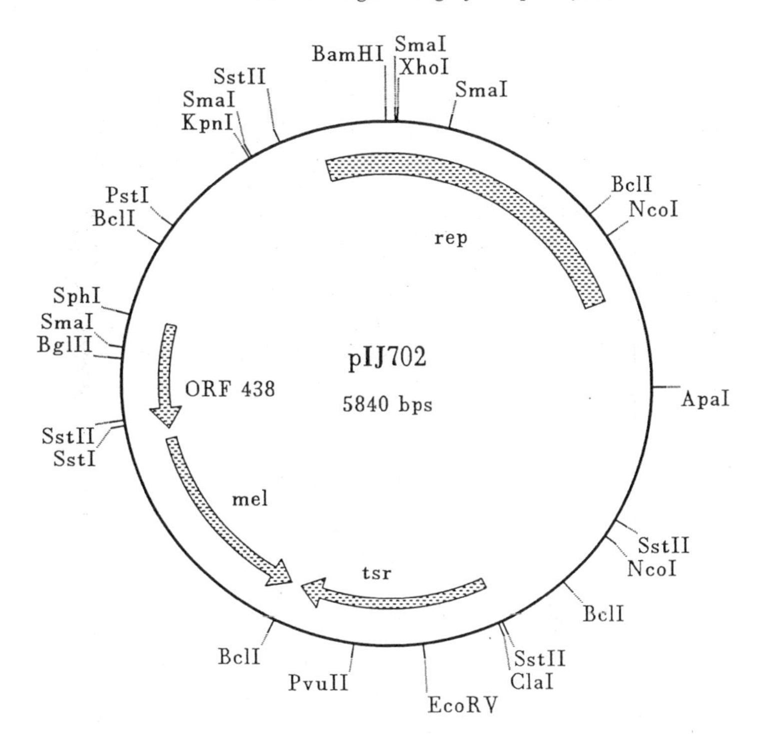

Fig. 5. Restriction map of *Streptomyces* vector pIJ702. The replication region (rep) of *S. lividans* plasmid pIJ101 (KIESER et al., 1982) was fused with the thiostrepton resistance gene (*tsr*) and the *mel* operon (consisting of the two genes *mel* and *ORF438*) to result in pIJ702 (KATZ et al., 1983). The relevant restriction sites are given; *Bgl*II and *Sph*I are commonly used since successful cloning into these sites can be detected by insertional inactivation (modified from KATZ et al., 1983; BERNAN et al., 1985; HOPWOOD et al., 1985a).

this, important aspects of gene expression and some examples of the application of genetic engineering techniques will be given.

3.2 Methods for the Genetic Manipulation of *Streptomyces*

3.2.1 *Streptomyces* Plasmid and Phage Cloning Vectors

Plasmids and phages have been used for the construction of vectors. Endogenous plasmids are present in many *Streptomyces* strains (HOPWOOD et al., 1986b), and phages can easily be isolated from soil samples, by standard procedures (HOPWOOD et al., 1985a). Since the first report of SCHREMPF et al., (1975), various high-copy-number and low-copy-number plasmids, as well as very large and very small plasmids have been characterized (for a review see HOPWOOD et al., 1986b). The first vectors were constructed by combining plasmid pIJ101 (KIESER et al., 1982), or SLP1.2 (BIBB et al., 1981), or SCP2* (BIBB et al., 1977) with antibiotic resistance genes isolated from antibiotic-producing streptomycetes (BIBB et al., 1980; THOMPSON et al., 1980, 1982a, b). Thereby the genes for neomycin phosphotransferase from *S. fradiae* (*aphI*), viomycin phosphotransferase from *S. vinaceus* (*vph*), and ribosomal 23SrRNA methylase determining resistance to thiostrepton from *S. azureus* (*tsr*) proved particularly useful (THOMPSON et al., 1982b). Subsequently, other replicons and new marker genes, in particular the tyrosinase (*mel*) gene (KATZ et al., 1983), were included in the construction of more sophisticated vectors. In addition to the *Streptomyces* marker genes, several *E. coli* genes such as the neomycin (*aphII*) and the bleomycin (*ble*) resistance gene from Tn*5* and the gentamicin resistance gene (*aacC1*) from Tn*1696* were used to establish vector systems (HOPWOOD et al., 1985a; WOHLLEBEN et al., 1989).

Such vectors (HOPWOOD et al., 1987) have proven useful in the isolation of many *Streptomyces* genes. Most genes of streptomycetes were cloned into vector pIJ702 (Fig. 5) which can be selected on thiostrepton-containing medium. Successful cloning events which result in the insertional inactivation of the *mel*-operon, give rise to white colonies which are easily distinguishable from the brown colonies harboring the original vector (KATZ et al., 1983).

Cloning vectors developed for use in other groups of bacteria have not been used successfully in *Streptomyces* species, since only one plasmid not originating from an actinomycete is known to replicate in streptomycetes. This is the broad-host-range plasmid RSF1010 (SCHOLZ et al., 1989; GORMLEY and DAVIES, 1991).

Among the phages, the temperate actinophage ΦC31 (LOMOVSKAYA et al., 1980) has been used extensively as a cloning vector (CHATER, 1986; HOPWOOD et al., 1987; KIESER and HOPWOOD, 1991). ΦC31 has some characteristics in common with *E. coli* phage λ: both possess cohesive ends, an *attP*-site for integration into the host chromosome (KUHSTOSS and RAO, 1991a), and a repressor gene (*c*) for the control of lysogeny. Deletion derivatives of these phages have been widely used for cloning foreign DNA of up to approximately 9 kb in size (HARRIS et al., 1983). Additionally, phage variants which lack the *attP*-site can be employed for mutational cloning experiments (CHATER and BRUTON, 1983). Recently, ΦC31 vectors containing the promoterless catechol dioxygenase gene (*xylE*) as a reporter cassette were constructed and employed to monitor the transcription of actinorhodin biosynthesis in *S. coelicolor* A3(2) (BRUTON et al., 1991).

Until now, cosmid systems for *Streptomyces* have not been available. However, hybrid plasmids were constructed consisting of a *Streptomyces* replicon and the cohesive ends (*cos*-site) of either actinophage R4 (MORINO et al., 1985) or ΦC31 (KOBLER et al., 1991) or a region of FP43 including the origin of headful packaging (*pac*-site) (HAHN et al., 1991c).

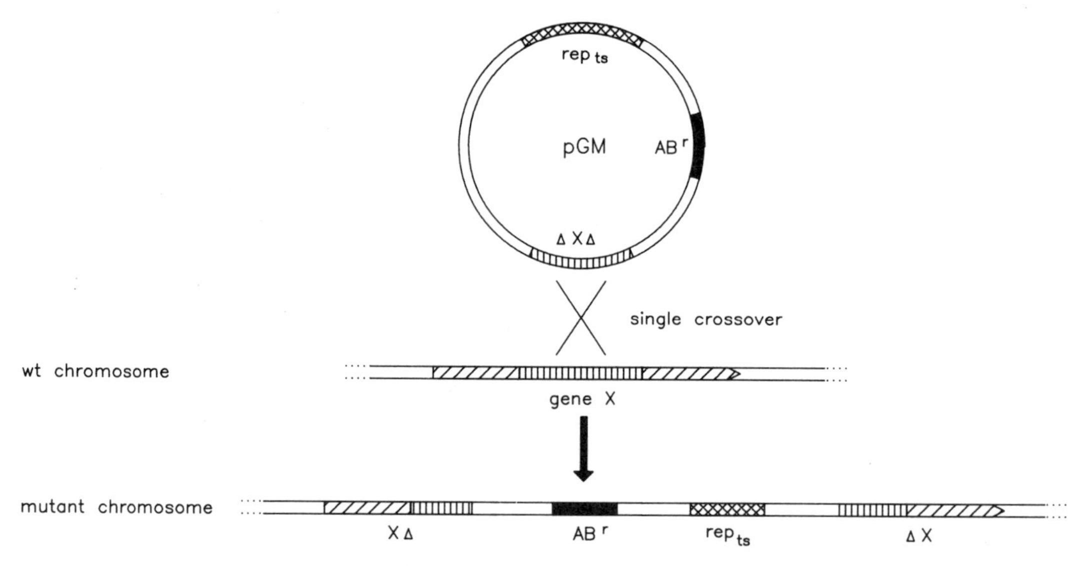

Fig. 6. Scheme of gene disruption. In order to disrupt a gene X, an internal fragment of the gene lacking its transcriptional start and translational stop signals (ΔXΔ) is cloned into a pGM vector which contains the replication region, rep_{ts}, of the inherently temperature-sensitive pSG5 (MUTH et al., 1989). The plasmid can integrate via homologous recombination into the chromosomal gene and thus disrupt it. Disruption mutants can be isolated at non-permissive temperatures on selection medium (AB). Instead of the functional gene X, the mutant contains two incomplete copies of the gene which either lack its 3′ region (XΔ) or its 5′ region (ΔX).
Symbols: AB^r, antibiotic resistance gene; rep_{ts}, replication region of plasmid pSG5; ▥ internal region of gene X; ▨ border region of gene X.

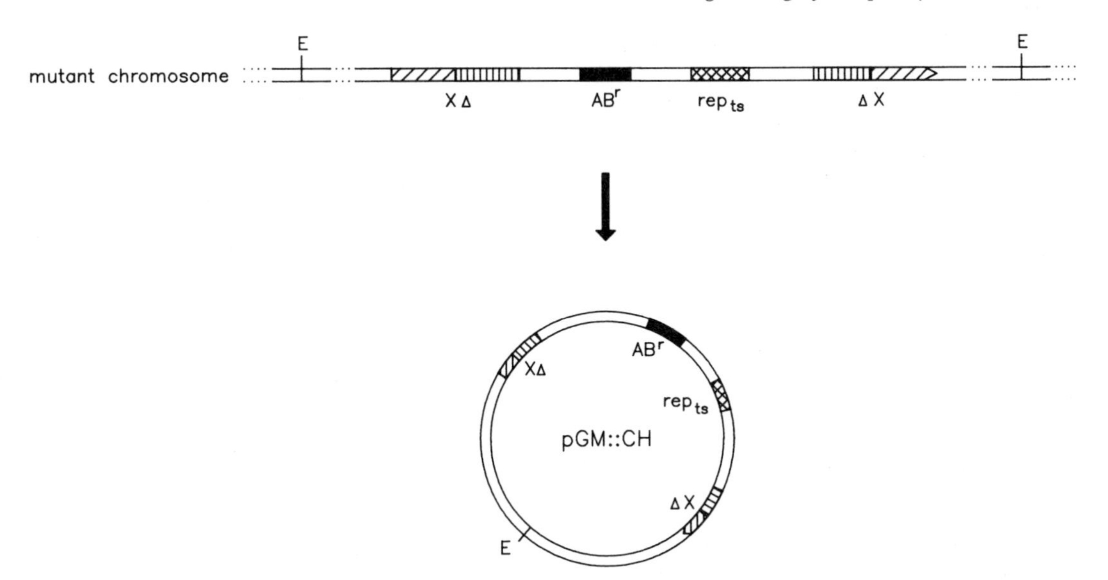

Fig. 7. Scheme of mutational cloning. Gene disruption mutants (Fig. 6) can be used to identify DNA fragments adjacent to the mutated gene. Following the isolation of genomic DNA from the mutant, this DNA is restricted with an enzyme (E) which does not cleave the plasmid originally used. The religation, transformation into a suitable recipient, and the selection for the vector marker, AB^r, results in a hybrid plasmid, pGM::CH, which carries DNA fragments flanking the gene X. Symbols: see Fig. 6.

These hybrid plasmids were efficiently packaged into R4, ΦC31, or FP43 phage particles *in vivo,* but no *in vitro* packaging system has yet been developed. However, these plasmids can be used for the phage-mediated transduction of genes, thus overcoming the difficulties associated with the transformation of some strains (MORINO et al., 1986, 1988; MCHENNEY and BALTZ 1988, 1991a; KOBLER et al., 1991).

The advantageous characteristics of plasmid cloning vectors and those of the *attP*-deleted phage vectors are combined in the pGM vectors. These are multi-copy plasmids present in approximately 55 copies per chromosome (LABES et al., 1990). The pGM vectors were derived from the inherently temperature-sensitive *Streptomyces ghanaensis* plasmid pSG5, by cloning suitable marker genes into the 1.6 kb minimal replication region (MUTH et al., 1988, 1989). The vectors obtained were stable at permissive temperatures, even over a long period without selection pressure (ROTH et al., 1991). At temperatures above 34 °C, the vectors are lost rapidly (MUTH et al., 1989; ROTH et al., 1991). However, if a region of homology with the chromosome is present, the plasmids integrate via homologous recombination.

The selection for the vector marker(s) at non-permissive temperatures results in bacteria carrying chromosomally integrated plasmids. Such integration leads to the disruption of a gene if a fragment not containing transcriptional start and translational stop signals, has been cloned (Fig. 6). Insertion mutants can be used to re-isolate the integrated plasmid together with adjacent fragments (mutational cloning/plasmid rescue) as shown in Fig. 7. Following the isolation of total DNA, the restriction with an enzyme which does not cleave the integrated plasmid, religation and transformation into a suitable recipient (preferentially *S. lividans*), plasmids are obtained which harbor pieces of chromosomal DNA.

In this way, antibiotic biosynthetic genes can be cloned as was demonstrated for the PTT-biosynthetic cluster. Following the isolation and characterization of the PTT-resistance gene (*pat*) (STRAUCH et al., 1988; WOHL-

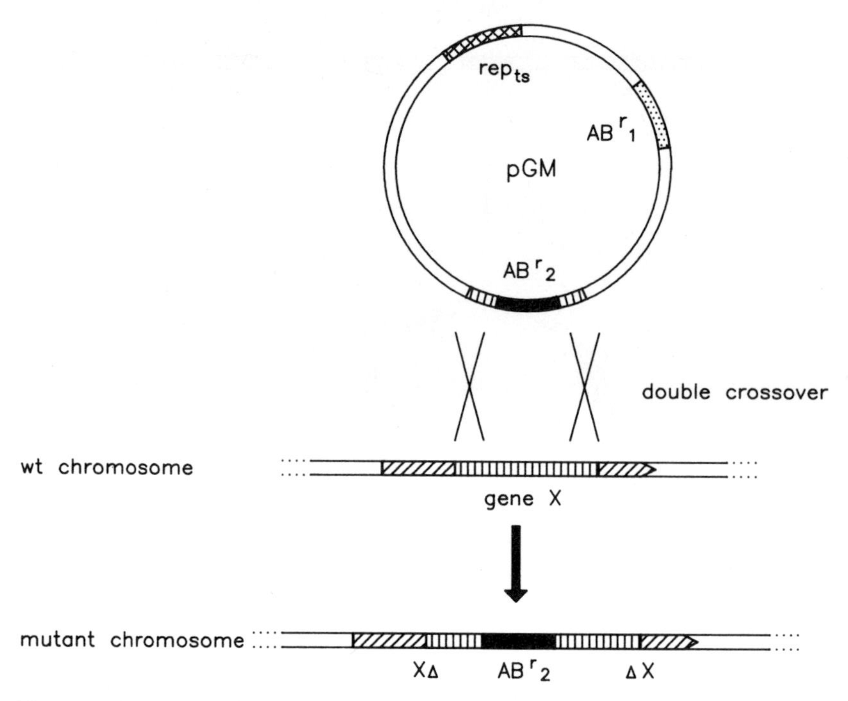

Fig. 8. Scheme of gene replacement. Gene replacement can be carried out by inserting a cassette, AB^r_2, into a cloned gene. The cloned gene may be complete or only an internal fragment as shown in the figure. After double recombination occurs in the regions flanking the cassette, the chromosomal marker is exchanged for that on the vector. Following a temperature shift, the plasmid is eliminated, and mutants are selectable by their resistance pattern: resistant to AB_2, sensitive to AB_1.

LEBEN et al., 1988), an internal fragment was cloned into pGM160 in order to disrupt the *pat* gene after the plasmid had integrated via homologous recombination (MUTH et al., 1989). The plasmid with adjacent fragments was recovered in *S. lividans* after plasmid rescuing as described above. Thus, biosynthetic genes could be identified and characterized (ALIJAH et al., 1991; WOHLLEBEN et al., 1992a). This strategy could easily be adopted for other biosynthetic clusters, since it is known that antibiotic biosynthetic genes including the resistance gene itself, are physically linked (MARTÍN and LIRAS, 1989).

Since gene disruption mutants of certain strains display instabilities at permissive temperature, it is desirable to insert a cassette or to replace an internal fragment of a gene by a cassette. After double recombination has occurred, such replacement events (Fig. 8) can be found with high frequency by selecting for the presence of the cassette marker and for the loss of the vector marker. The thiostrepton resistance gene *tsr* (THOMPSON et al., 1980) has proved to be the most suitable marker for streptomycetes and is normally used as the cassette marker. As a consequence, vectors containing an alternative marker for primary selection are needed for such experiments. The vector pGM11 (Fig. 9) which carries the *aphII* gene from Tn*5* (AUERSWALD et al., 1982) is an example of a vector well suited for gene replacement experiments, as well as for conventional cloning and for promoter probing. The temperature-sensitive plasmids can also serve as integration vectors and transposon delivery vehicles.

Various transposons have been described recently and have shown to be helpful tools in *Streptomyces* genetics. Since no naturally occurring transposable elements containing selectable markers have been described to date,

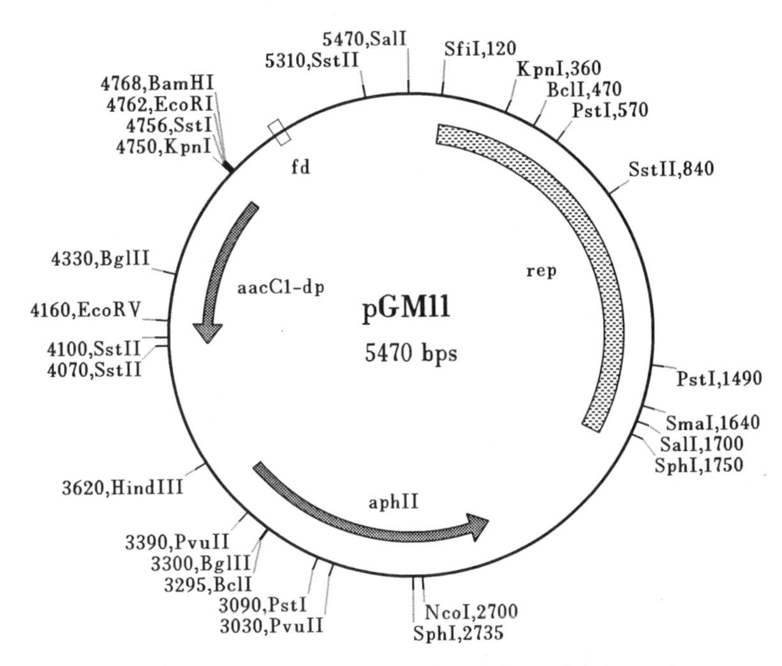

Fig. 9. Restriction map of the *Streptomyces* vector pGM11. pGM11 consists of the replication region, rep_{ts}, of the temperature-sensitive plasmid pSG5 (MUTH et al., 1988), the neomycin resistance gene, *aphII,* of Tn*5* (AUERSWALD et al., 1982), the promoterless gentamicin resistance gene, *aacC1-dp*, of Tn*1696* (WOHLLEBEN et al., 1989; LABES and WOHLLEBEN, in preparation) and the terminator (fd) of phage fd. Relevant restriction sites and their coordinates are given. pGM11 is especially suitable for promoter-probing after fragments have been cloned into the *Bam*HI, *Eco*RI, and *Sst*I sites, and for gene replacement using a thiostrepton-resistance cassette.

recombinant transposons, such as Tn*5096* and Tn*4560,* were constructed, Tn*5096* (SOLENBERG and BALTZ, 1991) is a derivative of IS*493* (SOLENBERG and BURGETT, 1989) of *S. lividans* which contains an apramycin resistance gene, and Tn*4560* (CHUNG, 1987) is a derivative of Tn*4556,* a class II transposable element from *S. fradiae* (CHUNG, 1987; OLSON and CHUNG, 1988; SIEMIENIAK et al., 1990) which contains a viomycin resistance gene. Since these two transposons were shown to transpose into different sites in the genomes of several *Streptomyces* sp., they were used for classical transposon mutagenesis (HAHN et al., 1991b; SOLENBEREG and BALTZ, 1991; SCHAUER et al., 1991). An elegant method for identifying independent transposition events was described by SOLENBERG and BALTZ (1991): 20–100 colonies harboring Tn*5096* on a temperature-sensitive pGM-vector were incubated for 2 days at 29 °C followed by an incubation at 39 °C for approximately 2 weeks. Only cells containing a copy of Tn*5096* on the chromosome continued to grow and formed clearly visible sectors. Cells from these sectors were easily analyzed. Recently, promoter-probe transposons were described which contained at one end of the transposon a promoterless reporter cassette which is able to create gene fusions. Tn*5099* (HAHN et al., 1991a), carrying the *xylE* gene, and Tn*5351* and Tn*5353* (SOHASKEY et al., 1992) harboring the *Vibrio harveyi* luciferase (*lux*) gene are available to monitor the transcriptional regulation in *Streptomyces.*

The transposons have been transferred to the cells by employing non-replicative plasmids (see below), the pGM-vectors and temperature-sensitive variants of pIJ702 (KATZ et al., 1983) such as pMT660 (BIRCH and CULLUM, 1985), and the transducible pMT660-derivative pRHB106 (MCHENNEY and BALTZ, 1991b).

Due to various restriction phenomena present in several *Streptomyces* strains, plasmid DNA isolated from heterologous bacteria is sometimes difficult to introduce into *Streptomyces* cells (transformation techniques see below). To overcome these problems, restriction-deficient *S. fradiae* mutants were generated (MATSUSHIMA et al., 1987), and DNA isolated from a modification-deficient *E. coli* was used for the successful transformation of *S. avermitilis* (MACNEIL, 1988). Another alternative is the use of phage origins for single-stranded (ss) DNA replication. The transformation and subsequent integration of ss-DNA in streptomycetes has been shown to be up to 100 times more efficient than that using double-stranded DNA (HILLEMANN et al., 1991).

Integration vectors are not only of interest with respect to transposon delivery and the generation of mutations, but they can also effect the stable chromosomal integration of genes.

A family of plasmids able to integrate at a specific site into the host chromosome was described, two plasmids of which proved to be valuable vectors: SLP1 from *S. coelicolor* A3(2) (BIBB et al., 1981) and pSAM2 from *S. ambofaciens* (PERNODET et al., 1984). In both cases, the integration proceeded via recombination between an *attP*-site on the plasmid and an *attB*-site on the chromosome. The *attB*-sites are located in a $tRNA_{Tyr}$ gene in the case of SLP1 (OMER and COHEN, 1984), and a $tRNA_{Pro}$ gene in the case of pSAM2 (MAZODIER et al., 1990), and function not only in the original *Streptomyces* host (MAZODIER et al., 1990), but also in Mycobacteria (MARTIN et al., 1991). Integration vectors (KUHSTOSS et al., 1989) as well as mobilizable *E. coli-Streptomyces* vectors have been constructed from pSAM2 (SMOKVINA et al., 1990).

Special integration vectors which can be used for the amplification of cloned genes were developed by ALTENBUCHNER and CULLUM (1987). They utilized the ability of spontaneously chloramphenicol-sensitive *S. lividans* mutants to give rise to arginine auxotrophs with a frequency of about 25% (ALTENBUCHNER and CULLUM, 1984). This mutation is accompanied by the approximately 100-fold amplification of a 5.7 kb chromosomal DNA element (ALTENBUCHNER and CULLUM, 1985). ALTENBUCHNER and CULLUM combined this amplifiable *S. lividans* sequence with selectable *Streptomyces* genes and an *E. coli* replicon and transformed the resulting plasmid into *S. lividans*. The selection for the *Streptomyces* marker resulted in colonies which carried the plasmid integrated into the amplifiable unit. The subsequent selection for chloramphenicol sensitivity and arginine auxotrophy allowed the identification of strains with co-amplified cloned genes.

3.2.2 DNA Introduction by PEG-Mediated Transformation, Transduction, and Conjugal Transfer

In theory, conjugation, transformation, and transduction can be used to introduce DNA into *Streptomyces*. Transformation methods exist to transfer DNA into *Streptomyces*. The standard technique is polyethyleneglycol-mediated transformation of protoplasts (BIBB et al., 1978). Employing *S. lividans*, approximately 10^5 transformants can be obtained when 10 ng plasmid DNA are transferred. Similar values are obtained for other strains when the DNA originates from that particular strain. The transformation efficiency of foreign DNA decreases drastically, probably due to restriction phenomena present in the recipient cell. Modifications of the standard transformation protocol (HOPWOOD et al., 1985a), such as heat attenuation (ENGEL, 1987) and embedding the regenerating protoplasts (SHIRAHAMA et al., 1981), have been proposed, but have to be individually optimized for each particular strain.

Electroporation, a method which is quite efficient for some other actinomycetes such as *Amycolatopsis* (LAL et al., 1991), can be applied to streptomycetes, but results in transformation rates approximately 10^3 times lower than PEG-mediated protoplast transformation (MACNEIL, 1987).

A further option is the use of a conjugation system. A prerequisite for conjugation is the insertion into the plasmid, of the *mob*-site of a broad-host-range plasmid of Gram negative bacteria such as RP4 (THOMAS et al., 1979) or RSF1010 (BUCHANAN-WOLLASTON et al., 1987). The plasmid can then be mobilized from *E. coli* S17-1 which carries a chromosomally integrated RP4-derivative (SIMON et al., 1983) to streptomycetes. Thus, transfer rates of about 10^{-5} transconjugants per recipient (MAZODIER et al., 1989; WOHLLEBEN and PIELSTICKER, 1989) can be obtained. If shuttle plasmids are used, the DNA can be established extrachromosomally. However, if the transferred plasmid does not contain a *Streptomyces* replicon, integration is possible via the cloned fragments. Thus disruption or replacement events can occur (see above) which may be utilized for the generation of mutants. In the case of the PTT biosynthesis in *S. viridochromogenes* Tü494, defined mutants were obtained after the insertion of mobilizable, suicide vectors (WOHLLEBEN et al., 1992a). Natural transduction does not play a major role in *Streptomyces* genetics. To date, only an *S. venezuelae* phage (STUTTARD, 1979) has been used to analyze the chloramphenicol biosynthetic genes (VATS et al., 1987). However, plasmid transduction utilizing *cos*- or *pac*-sites from actinophages (see above) is a promising alternative to transformation (MCHENNEY and BALTZ, 1988; HAHN et al., 1991b).

Whereas in *E. coli* and many other Gram negative cloning hosts, recombination-deficient mutants are commonly used, such recipients are not available for streptomycetes. The cloning and sequencing of an *recA* homolog in *Mycobacterium tuberculosis* was recently reported (NAIR and STEYN, 1991; DAVIS et al., 1991), but no hybridization to *S. coelicolor* A3(2) could be detected (NAIR and STEYN, 1991). At present, only *S. lividans* mutants containing a defect in their intraplasmid recombination have been characterized (CHEN et al., 1987; KIESER et al., 1989). However, *Streptomyces recA* mutants should become available in the near future, since a *recA* internal fragment was identified very recently in various *Streptomyces* strains and was cloned from *S. cattleya* (WOHLLEBEN et al., 1992b).

An additional problem associated with streptomycetes arises from the well-known genetic instability (for reviews see BIRCH et al., 1990; LEBLOND et al., 1990; SCHREMPF, 1991). The large-scale gene amplifications and deletions, as well as structural plasmid instabilities are possibly caused by the participation of a RecA-like protein (YOUNG and CULLUM, 1987).

3.2.3 Gene Expression in *Streptomyces*

3.2.3.1 Gene Regulation in *Streptomyces*

Gene expression in streptomycetes is a complex process. Due to the morphological differentiation of these bacteria, a cascade of regulatory mechanisms exists (HOPWOOD, 1988; CHATER, 1990). The synthesis of antibiotics and all other secondary metabolites is included in this regulation and, additionally, is regulated in an individual manner particular to a certain biosynthetic cluster. Gene regulation cannot be dealt with in detail in the scope of this chapter, only the most important principles can be described.

In *Streptomyces coelicolor* and probably also in other strains, a $tRNA_{Leu}$-like gene (*bldA*) is responsible for one level of control (LESKIW et al., 1991a). This tRNA recognizes the UUA codon which is present in only 13 of 100 sequenced *Streptomyces* genes (LESKIW et al., 1991b). These 13 genes are mainly involved in the regulation of antibiotic biosynthesis or in antibiotic resistance and antibiotic export. However, even in *S. coelicolor* regulatory and resistance genes are known to exist which do not possess TTA codons. [It should be mentioned that additional *bld*-mutants exist which are also involved in the morphological and physiological differentiation (CHAMPNESS, 1988; CHATER, 1989)].

The next level of control of secondary metabolism seems to be determined by the *afsB* and *afsC* genes which, when mutated, result in defects in antibiotic production and synthesis of factor A (*afsA*) in *S. coelicolor* (HORINOUCHI et al., 1983). The gene product deduced from the DNA sequence of the complementing gene *afsR* (formerly named *afsB*) (HORINOUCHI et al., 1986; STEIN and COHEN, 1989) revealed a similarity to DNA-binding proteins which contain a helix-turn-helix motif and A- and B-type ATP-binding consensus sequences (HORINOUCHI et al., 1990). The overproduction of the *S. coelicolor* antibiotics actinorhodin and undecylprodigiosin can be obtained after the introduction of the cloned *afsR* gene into *S. coelicolor* (HORINOUCHI and BEPPU, 1984; HORINOUCHI et al., 1989a). The regulatory activity of the AfsR-protein is probably modulated by phosphorylation in *S. coelicolor* and *S. lividans,* although not in *S. griseus* (HONG et al., 1991).

Whereas the A-factor itself seems not to play a role in antibiotic production in *S. coelicolor,* it is involved in the regulation of antibiotic production in other strains, the best studied example being *S. griseus.* The A-factor probably exerts its effect by direct interaction with an A-factor-binding protein (MIYAKE et al., 1989, 1990) and indirect interaction with a region upstream of the transcriptional start point of *strR* which codes for a putative regulatory protein (VUJAKLIJA et al., 1991).

The regulatory genes of some antibiotic biosynthetic pathways are known, for example, those controlling bialaphos/PTT (ANZAI et al., 1987), actinorhodin (HOPWOOD et al., 1986a), methylenomycin (CHATER and BRUTON, 1985), undecylprodigiosin (NARVA and FEITELSON, 1990), streptomycin (DISTLER et al., 1990) and daunorubicin synthesis (STUTZMAN-ENGWALL et al., 1992).

The disruption of the regulatory region of the methylenomycin cluster leads to an increment in production, thus suggesting a negative regulation. A pathway-specific regulation has been demonstrated to be present in other cases. The presence of additional cloned copies of the regulatory genes *brpA, actII-orf4, redD-orf1, strR,* and *dnrR*$_2$ leads to an overproduction of the corresponding antibiotic. Since some of these regulators seem to function in heterologous strains (MUTH et al., 1986; STROHL et al., 1991), a promising strategy to identify activators would be to screen gene banks in *Streptomyces lividans* and to identify the actinorhodin production by its color.

Additional regulation may occur at the level of transcription, mediated by different RNA polymerase holoenzymes. Since the detection of at least two different RNA polymerases in *S. coelicolor* (WESTPHELING et al., 1985), evidence for at least seven different σ-factors has been provided (e.g., BUTTNER, 1989; TANAKA et al., 1991). These enzymes may differentially recognize different promoter sequences of *Streptomyces* genes (for a review on *Streptomyces* promoters, see STROHL, 1992). For example, at least four transcription start points were mapped, and three different RNA polymerase holoenzymes were identified in the case of an agarase gene (*dagA*) (BUTTNER et al., 1987, 1988). Similar features can be found for other genes of which the neomycin resistance gene (*aphI*) of *S. fradiae* (JANSSEN et al., 1989) and the galactose operon (*gal*) gene of *S. lividans* (FORNWALD et al., 1987) are the best studied examples.

The situation is even more complex, because evidence exists for the presence of an additional control mechanism via ppGpp and GTP (stringent response) in several species (OCHI 1987, 1990; STRAUCH et al., 1991), although not in *S. clavuligerus* (BASCARAN et al., 1991).

An additional problem arises from the fact that translation in streptomycetes apparently does not always require a conserved ribosome-binding site (JANSSEN et al., 1989). Genes are known to exist, in which the translational start codon represents the very first nucleotide of the mRNA, such as the *aphI* gene of *S. fradiae* (JANSSEN et al., 1989) and the *afsA* gene of *S. griseus* (HORINOUCHI et al., 1989b).

Finally, it should be mentioned that secondary metabolite biosynthesis is normally regulated negatively at the level of transcription by easily assimilable phosphate, carbon and nitrogen (LIRAS et al., 1990). However, many enzymes involved in primary metabolism are stimulated by phosphate, thus providing greater amounts of precursors required for antibiotic biosynthesis. It remains to be ascertained whether, for example, phosphate-regulated promoters involved in antibiotic biosynthesis can be manipulated to overcome phosphate repression. Such a promoter was recently identified in the candicidin biosynthetic cluster of *S. griseus* (ASTURIAS et al., 1990).

3.2.3.2 Expression Systems in *Streptomyces*

The above data illustrate the difficulties involved in attempting to engineer the expression of genes which form part of biosynthetic clusters, and which are apparently involved in a complex regulatory network. The expression of individual genes, however, is less complicated and promises to be easier to manipulate.

Firstly, constitutive promoters may be used. The promoters P1 and P2 of the *aphI* gene (JANSSEN et al., 1989) and the *melC* promoter of the tyrosinase gene (KATZ et al., 1983; BERNAN et al., 1985) are often utilized because they are easily available in the commonly used *Streptomyces* vectors.

Secondly, the natural promoter of the neomycin resistance gene (*aphII*) of Tn5 (AUERSWALD et al., 1982) which is recognized by the *E. coli* and *Streptomyces* transcription systems, allows increased expression in many cases (VIGAL et al., 1991). In order to further increase the *aphII*-expression level, a synthetic promoter sequence was designed (DENIS and BRZEZINSKI, 1991) which can be used for the expression of desired genes.

Thirdly, a promoter-up mutation of the erythromycin resistance gene (*ermE*) from *Saccharopolyspora erythraea* (BIBB et al., 1985), the *erm** promoter, has been characterized (BIBB and JANSSEN, 1986).

Fourthly, phage promoters, which proved to be suitable for expression systems in other bacteria, can be employed. Such promoters were recently characterized (LABES and WOHLLEBEN, in preparation).

Finally, an inducible promoter (*tipA*) is available which delivers a high level of expression after induction with thiostrepton (MURAKAMI et al., 1989) and which is available in different expression vectors (SMOKVINA et al., 1990; KUHSTOSS and RAO, 1991b).

Promoter strength remains to be investigated systematically, but each promoter has to be optimized individually, as its activity seems to depend on the gene to be expressed. For example, in the case of an α-amylase inhibitor gene, the expression by different promoters varied by a factor of 50; whereby the *erm** promoter delivered the best results (SCHMITT-JOHN and ENGELS, 1992).

3.3 Exploitation of Endogenous Biosynthetic Abilities of *Streptomyces*

Overproducing *Streptomyces* strains are obtained for use in industry by conventional strain improvement programs and not by genetic engineering. These include mutagenesis and protoplast fusion. Attempts to increase yield by cloning of activator genes (see above) were only successful on a laboratory scale, no positive effect in industrial high-producers has been reported. For a non-random manipulation, it is necessary to understand the complex regulation and the involvement of primary metabolites in antibiotic production. Most of the mutations leading to high-producers probably affect the intracellular supply of precursors and co-factors, the response to available nutrients, and the export of the antibiotic, but do not affect the specific biosynthetic genes (HOPWOOD, 1989; CHATER, 1990). Genetic engineering techniques may be helpful in the near future in two other fields of research:

- the detection of silent biosynthetic gene(s) and
- the generation of new antibiotically active compounds (hybrid antibiotics).

Silent antibiotic biosynthetic genes were found serendipitously by JONES and HOPWOOD (1984). They detected in *S. lividans* the phenoxazinone synthesis gene which is part of the actinomycin biosynthetic cluster in *S. antibioticus*. During the analysis of σ-factor genes, a gene (*bar*) mediating resistance to the antibiotic PTT (= bialaphos) was identified in *S. coelicolor*. This strain does not produce PTT (BUTTNER, 1989; BEDFORD et al., 1991).

A novel, more systematic strategy was followed by MALPARTIDA et al. (1987). They identified producers of polyketide antibiotics by Southern hybridization experiments. Polyketides include many different antibiotics such as tetracyclines, anthracyclines, macrolides, polyethers, and ansamycins which all share a common pattern of biosynthesis (HOPWOOD and SHERMAN, 1990). After cloning the whole actinorhodin biosynthetic pathway from *S. coelicolor* A3(2) (MALPARTIDA and HOPWOOD, 1984), the 22 kb gene cluster was analyzed in detail (MALPARTIDA and HOPWOOD, 1986). It was shown that the polyketide synthase (PKS) genes of various producers were similar (SHERMAN et al., 1989; HOPWOOD and SHERMAN, 1990). By applying *actI* and *actIII* DNA probes in hybridization experiments, homologous DNA fragments were detected not only in known polyketide producers, but also in a chloramphenicol-producing *S. venezuelae* and in actinomycin-producing *S. parvulus* and *S. antibioticus* (MALPARTIDA et al., 1987). It is likely that this screening procedure used to detect new antibiotic producers could be extended to other chemical classes of antibiotics.

Chemically modified antibiotics, mainly derivatives of the β-lactams penicillin and cephalosporin have been produced in industry on a large scale. Genetic engineering techniques are helpful in generating novel structures. By combining the biosynthetic genes of different members of the same chemical class, such novel structures have already been obtained (Fig. 10). HOPWOOD et al. (1985b) reported the production of mederrhodin A and B and dihydrogranatirhodin after transfer of actinorhodin biosynthetic genes into producers of medermycin and granaticin. Subsequently, the pathways of the structurally related carbomycin and spiramycin were combined by cloning, resulting in the synthesis of isovaleryl-spiramycin (EPP et al., 1989).

In both examples, however, only slight variations were introduced into the parental structures. Enzymes catalyzing the final steps of the biosynthesis are apparently sufficiently unspecific to accept various structurally related precursors.

Even if generating novel antibiotic structures requires other strategies (HOPWOOD, 1989), it is, nevertheless, worthwile to follow this strategy to design new antibiotics (HUTCHINSON et al., 1989; STROHL et al., 1989).

It could be speculated that the site-directed mutagenesis of key enzymes of the polyketide, β-lactam, and polypeptide-antibiotic syntheses would result in changes in the specificity of such enzymes which would lead to new antibiotics. However, further basic research is required to identify the structure and function of the enzymes, and the domains of the relevant biosynthetic gene.

3.4 Use of *Streptomyces* as a Host for the Excretion of Heterologous Products

The ability of streptomycetes to excrete degrading enzymes has not yet been exploited on a large scale, although a large variety of interesting enzymes is known. During the last years, the cloning of many such enzymes was described, and the secretion signals were characterized (listed in ENGELS and KOLLER, 1991, or MANSOURI and PIEPERSBERG, 1991). In some cases, the cloning of the genes on multicopy vectors led to overproduction, in one case, even after transfer of the gene into an industrial high-producer (BRÄU et al., 1991). But the application of this method in large-scale production has not yet been reported.

At present, interest in this field is primarily focussed on the problem of constructing an expression–secretion system for heterologous (eukaryotic) proteins in streptomycetes.

The secretion of bovine growth hormone (GRAY et al., 1984), interleukin-2 (MUÑOZ et al., 1985; LICHENSTEIN et al., 1988), interferon-α2 (PULIDO et al., 1986), interferon-α1 (NOACK et al., 1988), and a tumor necrosis factor (TNF) (CHANG and CHANG, 1988) has been reported. The protein was secreted either as a fusion with the leader sequence of either staphylokinase (interferon-α1), a β-galactosidase (interleukin-2), a tyrosinase-ORF (TNF), or without any cloned secretion signals.

More systematic studies were performed employing a subtilisin inhibitor (SSI) gene (OBATA et al., 1989) which was used for the expression of the *Enterococcus faecalis* sex pheromone cAD1 (TAGUCHI et al., 1989), and more recently (TAGUCHI et al., 1992), for the production of the antibacterial peptide apidaecin 1b (AP1). In the case of cAD1, 28 mg of the SSI-cAD1 fusion protein were purified from one liter of culture medium. After cyanogen bromide cleavage and HPLC purification, 336 μg per liter could be recovered in

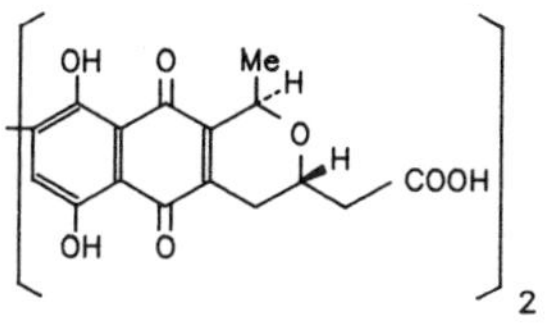

Actinorhodin

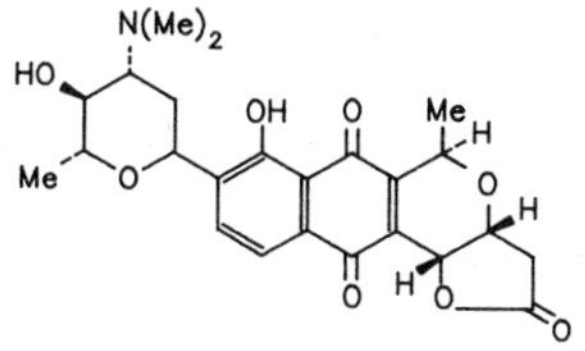

Medermycin

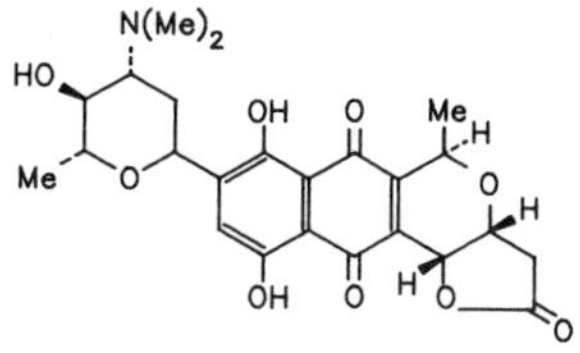

Mederrhodin A

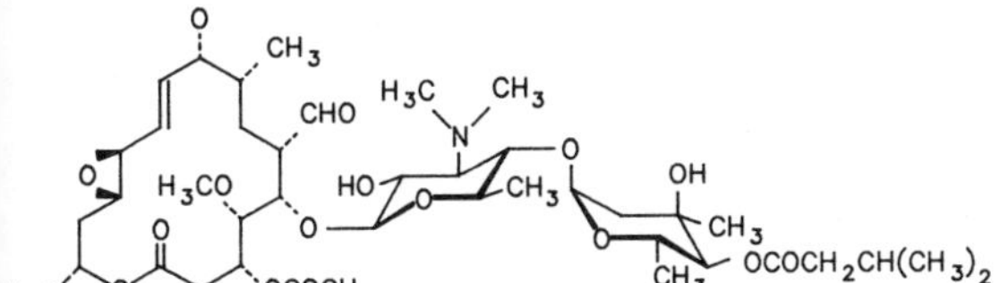

Carbomycin

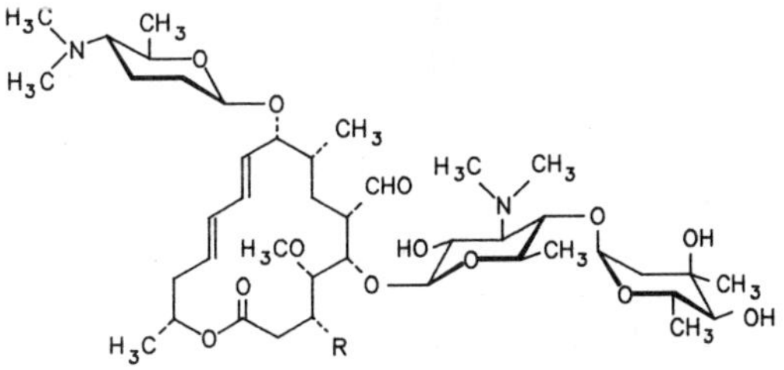

Spiramycin

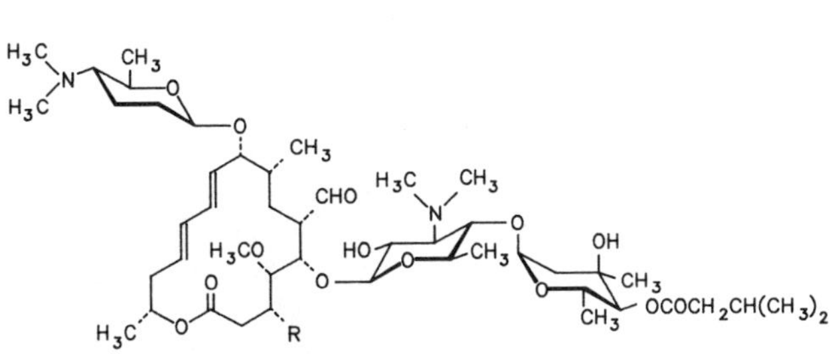

Isovaleryl Spiramycin

Fig. 10. Structures of the two "hybrid" antibiotics mederrhodin A and isovaleryl-spiramycin and of the "basic" antibiotics actinorhodin, medermycin, carbomycin, and spiramycin. The structures were taken from HOPWOOD et al. (1985b) and EPP et al. (1989).

comparison to 600 ng in the original host *E. faecalis*. AP1 was produced extracellularly as a fusion protein of SSI and AP1, since the genes were joined by a 12 bp nucleotide sequence which corresponded to the amino acid sequence specifying the cleavage by blood coagulation factor Xa. The inhibitory activity of AP1 was detected before and after the cleavage of the fusion protein by factor Xa. The production of the SSI–AP1 fusion protein was estimated to be greater than 200 mg/L of the liquid culture medium.

The majority of the published results concern an α-amylase inhibitor (tendamistat) and its leader sequence. The tendamistat gene that originates from *Streptomyces tendae* was cloned in *S. lividans* (KOLLER and RIESS, 1989). After the proinsulin gene of the monkey *Macaca fascicularis* was fused downstream of the tendamistat gene and separated by a linker sequence, a fusion protein was secreted (KOLLER et al., 1989a) which still exhibited α-amylase inhibiting activity. Since the α-amylase inhibitory activity can easily be detected and roughly quantified by the appearance of blue halos in a plate assay, this system allows the identification of secreting *S. lividans* cells which carry tendamistat fusions. Biologically active des-Thr(B30) insulin was obtained after tryptic digestion of the tendamistat–proinsulin fusion protein. However, the formation of the disulfide linkages was found to be incorrect. The correct folding could only be obtained after modifying the C-peptide (KOLLER et al., 1989b). Thus a yield of 30–40 mg insulin per liter was obtained without strain improvement or fermentation optimization. A further yield increase was obtained by using a shortened α-amylase inhibitor gene which was detectable only by immuno-blotting (VERTESY et al., 1991).

Fusions utilizing the tendamistat leader sequence were only used to secrete hirudin (BENDER et al., 1990a) and interleukin-2 (BENDER et al., 1990b). However, in these and further cases (VIGAL et al., 1991), the signal structure seems to require optimization for efficient processing and translocation to result in high yields of active protein.

Another promising approach is the stimulation of extracellular enzyme production by the cloning of activator genes as shown for the *saf* gene of *S. griseus* cloned in *S. lividans* (DAZA et al., 1990).

Unfortunately, the progress made in this field has not been fully published, since research takes place primarily in industrial laboratories. However, reports such as the expression of soluble CD-4 (sCD-4) which serves as the receptor of HIV-I infection of T-cells, show the attractiveness of the *Streptomyces* secretion systems (BRAWNER et al., 1990a). The promoter and signal sequence of the protease inhibitor LTI of *S. longisporus* direct the expression and secretion of sCD-4 in *S. lividans*. *Streptomyces*-produced sCD-4 displaying the correctly folded disulfide bridges is secreted in amounts of more than 200 mg/L. It is biologically active as shown by immunoprecipitation, and prevents HIV from binding to the T-cells (BRAWNER et al., 1990b).

4 Genetic Engineering of Corynebacteria

4.1 Biology of Corynebacteria

Corynebacteria are taxonomically defined by *Bergey's Manual of Systematic Bacteriology* Volume 2 (1986) as Gram positive, irregular, non sporulating rods (Fig. 11). These fac-

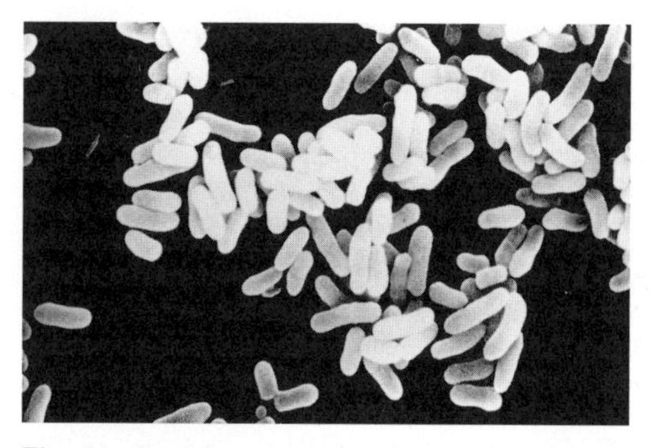

Fig. 11. Scanning electron micrograph of *Corynebacterium glutamicum* ATCC 13032 cells (courtesy of D. KAPP, Bielefeld).

ultatively anaerobic or strictly aerobic bacteria possess short-chain corynomycolic acids containing 22 to 36 carbon atoms anchored in their cell walls. The genus *Corynebacterium* contains animal pathogens like *Corynebacterium diphtheriae,* the etiologic agent of diphtheria, but most species of corynebacteria are nonpathogenic inhabitants of soil and water. The group formerly comprising the plant-pathogenic corynebacteria has been renamed *Clavibacter* (DAVIS, 1986).

Data from DNA reassociation studies (STACKEBRANDT and WOESE, 1981) and 16s-RNA sequencing (STACKEBRANDT and SCHLEIFER, 1984) are not in accordance with each other with respect to the positioning of the genus *Corynebacterium* in the taxonomic tree. Studies of the chemotaxonomy of the various members of corynebacteria and the related taxa, coryneform bacteria (SEILER, 1983), offer a more systematic view. SEILER divides the coryneform bacteria into classes A to F, where E and F include corynebacteria. Apart from *Corynebacterium,* class E also contains strains of the genera *Arthrobacter, Brevibacterium,* and *Microbacterium,* and group F also includes *Cellulomonas, Nocardia, Oerskovia,* and *Micrococcus.* These results correlate closely with recent hybridization studies conducted on a limited number of these bacterial species (LIEBL et al., 1991) and with studies of the host range of plasmids (SCHÄFER et al., 1990).

Members of the corynebacteria offer a broad spectrum of biotransformations and are used in a variety of biotechnological processes (Tab. 4). Industrially, the most important feature of corynebacteria is the fermentative production of amino acids. Strains used in these processes mainly include members of the groups E1 and E2 of SEILER (1983), the so-called *Corynebacterium glutamicum-Brevibacterium ammoniagenes* taxon. Since most genetic engineering is conducted on organisms of this subgroup of corynebacteria, this chapter will focus on *C. glutamicum, B. lactofermentum,* and closely related bacteria.

Although the amino acid production of *C. glutamicum* has been known since the 1950s, the first reports on the genetic engineering of *C. glutamicum* and its close relative *B. lactofermentum,* only appeared in 1984 (MIWA et al., 1984; SANTAMARIA et al., 1984).

Since then, the repertoire of genetic engineering methods as well as the understanding of the biochemical pathways and genetic systems in these organisms have increased significantly and will be presented in the following.

4.2 Methods for the Genetic Engineering of Corynebacteria

Essential prerequisites for the manipulation of bacteria of any kind are the availability of vectors derived from plasmids or bacteriophages (Sect. 4.2.1) and efficient DNA transfer systems (Sect. 4.2.2), These permit cloning, expression, disruption as well as replacement

Tab. 4. Examples of Biotechnological Processes Performed by Corynebacteria

Process	Organism	Reference
Production of amino acids	*C. glutamicum*[a]	YAMADA et al., 1972
Production of nucleotides	*B. ammoniagenes*[b]	OGATA et al., 1976
Production of antibiotics	*Corynebacterium* sp.	SUZUKI et al., 1972
Producton of surfactants	*C. hydrocarboclastum*	ZAJIC et al., 1977
Vitamin C precursors	*Corynebacterium* sp.	ANDERSON et al., 1985
Cheese ripening	*B. linens*	LEE et al., 1985
Bioconversion of steroids	*C. simplex*	CONSTANTINIDES, 1980
Terpenoid oxidation	*Corynebacterium* sp.	YAMADA et al., 1985
Degradation of carbohydrates	*Corynebacterium* sp.	CARDINI and JURTSHUK, 1970

[a] *C., Corynebacterium*
[b] *B., Brevibacterium*

Tab. 5. Endogenous Corynebacterial Plasmids Used in Vector Constructions

Name	Source	Size (kb)	Marker	Reference
pHM1519	*C. glutamicum*	3.055	cryptic	MIWA et al., 1984
pBL1	*B. lactofermentum*	4.45	cryptic	SANTAMARIA et al., 1984
pCC1	*C. callunae*	4.2	cryptic	SANDOVAL et al., 1984
pCL1	*C. lilium*	4.1	cryptic	CHAN KWO CHION et al., 1991
pBY503	*B. stationis*	15	cryptic	SATOH et al., 1990
pAG3	*C. melassecola*	4.5	cryptic	TAKEDA et al., 1990
pXZ10145	*C. glutamicum*	5.3	Cm	ZHEN et al., 1987
pGA1	*C. glutamicum*	4.9	cryptic	SONNEN et al., 1991
pNG2	*C. diphtheriae*	14.4	Em	SERWOLD-DAVIS et al., 1987
pCG2	*C. glutamicum*	6.8	cryptic	OZAKI et al., 1984
pAG1	*C. melassecola*	20.4	Tc	TAKEDA et al., 1990
pTP10	*C. xerosis*	50	Tc, Cm, Em, Km	KONO et al., 1983
pCG4	*C. glutamicum*	26	Spc, Str	KATSUMATA et al., 1984

Abbreviations: Cm, chloramphenicol resistance; Em, erythromycin resistance; Km, kanamycin resistance; Spc, spectinomycin resistance; Str, streptomycin resistance; Tc, tetracycline resistance; *C., Corynebacterium*; *B., Brevibacterium*

of genes, and transposon mutagenesis. By using these methods, the metabolic processes that are of biotechnological relevance can be optimized *in vitro* and *in vivo*. As an example, the genetic analysis and manipulation of the lysine biosynthetic pathway will be described (Sect. 4.2.3).

4.2.1 Vectors for Corynebacteria

Various attemps have been made to isolate endogenous plasmids from coryneform bacteria (Tab. 5) and to fuse these plasmids with commonly used *E. coli* cloning vectors in order to construct plasmid-based vector systems for corynebacteria. Small, cryptic plasmids displaying a high copy-number were used in these experiments, and some larger plasmids carrying antibiotic resistance genes were isolated. In several cases, identical or very similar plasmids were found independently, in different isolates of coryneform bacteria and consequently, were named differently. To avoid confusion, SONNEN et al. (1991) grouped plasmids of three different families and renamed them according to their first description in the scientific literature.

Of the plasmids included in Tab. 5, the pBL1 and pHM1519 families are the most widely distributed among laboratories involved in the genetic engineering of corynebacteria (Fig. 12). Both replicons were used to build bifunctional vectors which replicated in *E. coli* and corynebacteria as well as vectors which were only functional in the latter. Plasmid vectors based on these replicons are found in 5 to 30 copies per corynebacterial cell.

Although the host range of all replicons listed in Tab. 5 remains to be determined conclusively, the published data indicate the same host range for pHM1519 and pBL1 which includes at least the members of the *Corynebacterium glutamicum-Brevibacterium ammoniagenes* taxon.

The host range differs only in two cases observed so far. pBL1-derived vectors replicate in *B. linens* (SANDOVAL et al., 1985) whereas pHM1519-based plasmids are unable to replicate in this organism (SCHÄFER et al., 1990). The opposite occurs in *Brevibacterium* sp. R312 (CHAN KWO CHION et al., 1991), where only pHM1519 derivatives of the two plasmid families replicate.

Of all corynebacterial plasmids tested to date, pNG2 and its relatives have the widest host range which includes *C. glutamicum* and *E. coli* (SERWOLD-DAVIES et al., 1987, 1990). A broad-host-range derivative of pHM1519 was mentioned by ARCHER et al. (1990), but a detailed description of the mutation leading to

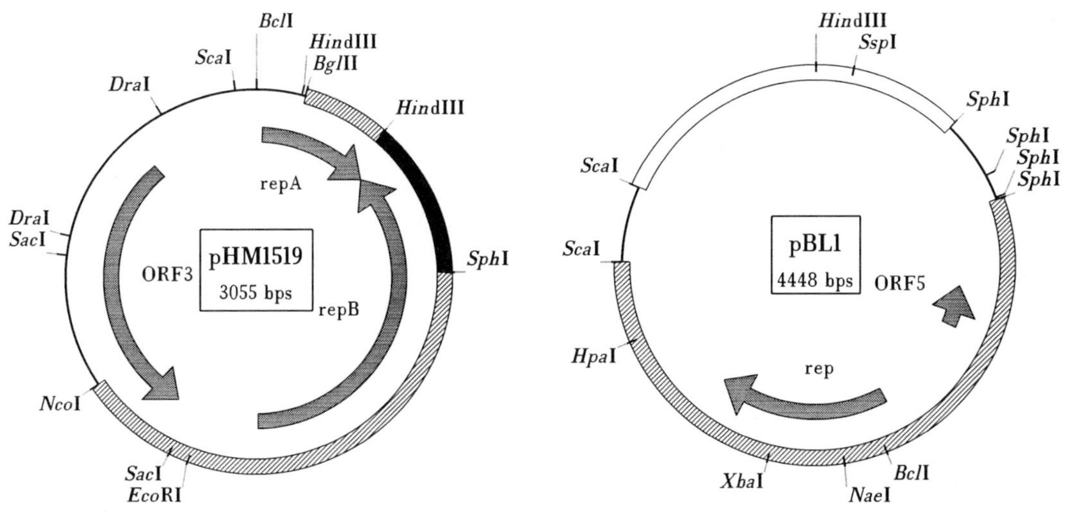

Fig. 12. Restriction maps of the coryneform plasmids pHM1519 and pBL1. The structure of pHM1519 was adapted from ARCHER et al. (1990) and TRAUTWETTER and BLANCO (1991) and that of pBL1 was deduced from the published nucleotide sequences (FILPULA et al., 1986; YAMAGUCHI et al., 1986). Important restriction enzyme sites and large open reading frames deduced from the nucleotide sequences are shown. The purported minimal replicon is shown as a hatched bar and the *ori*V region determined for pHM1519 in black. The open bar indicates a region which is of unknown relevance to plasmid maintenance.

the extended host range has not yet been published.

Compatibility tests indicate that at least pHM1519, pBL1, pCC1, and pGA1 can coexist stably within one cell (SONNEN et al., 1991).

The majority of the plasmid vectors derived from these replicons are segregationally stable and are retained in >90% of the cells, after 25 generations without selective pressure. When cultured under antibiotic selection, no plasmid loss is usually detected. Plasmids based on corynebacterial replicons do not, in general, display structural instability in *C. glutamicum,* although the occurrence of deletions (SANTAMARIA et al., 1985) and a considerably high segregational instability of shuttle vectors were reported in *B. lactofermentum* (SMITH et al., 1986).

The complete nucleotide sequences of two members of the pBL1 family is known (YAMAGUCHI et al., 1986; FILPULA et al., 1986), and the structure deduced from nucleotide sequencing of the pHM1519 replicon was also presented (ARCHER et al., 1990). The DNA sequence of the plasmid, however, remains to be published. In addition, the sequences of the minimal replicon of pNG2 from *C. diphtheriae* (SERWOLD-DAVIS et al., 1990) and of the stability region of plasmid pBY503 from *B. flavum* (KURUSU et al., 1991) are available.

For vectors based on bacteriophages, only report exists of the incorporation of a corynephage-derived *cos*-fragment into a corynebacterial plasmid vector (MIWA et al., 1985).

Only a few endogenous antibiotic resistance genes, which can be used as marker genes, have been found in amino acid-producing corynebacteria: resistance to streptomycin (KATSUMATA et al., 1984), trimethoprim (TAKAGI et al. 1986), chloramphenicol (ZHEN et al., 1987), and tetracycline (TAKEDA et al., 1990). However, antibiotic resistance markers from other organisms were also found to be expressed well in *C. glutamicum, B. lactofermentum,* and its relatives. These include resistance to erythromycin from *C. diphtheriae* (SERWOLD-DAVIS et al., 1990), resistance to erythromycin, tetracycline, kanamycin, and chloramphenicol from *C. xerosis* (KASSING et al., in preparation), resistance to kanamycin and chloramphenicol from *B. subtilis* (YOSHI-

HAMA et al., 1985), and resistance to chloramphenicol and hygromycin from streptomycetes (MARTÍN, 1989). In addition, the kanamycin and bleomycin resistance genes from transposon Tn*5* (SANTAMARIA et al., 1984; GUERRERO et al., 1992), the kanamycin resistance gene from Tn*903* (MENKEL et al., 1989), and the chloramphenicol resistance gene from Tn*9* (YEH et al., 1986) also proved suitable for selection purposes in these bacteria.

Using these genes, a variety of cloning vectors were constructed, some of which have specific applications such as promoter probing and integration into the chromosome. These vectors constitute the most recent developments in the field and will be described below.

During the last five years, numerous transcription-signal-probing vectors were constructed for amino acid-producing corynebacteria (MORINAGA et al., 1987; BARAK et al., 1990; CARDENAS et al., 1991; EIKMANNS et al., 1991; BARDONNET and BLANCO, 1991). In order to monitor transcription, antibiotic resistance genes for chloramphenicol and kanamycin, as well as the *E. coli* genes *lacZ* encoding β-galactosidase and *gusA* (formerly *uidA*) encoding β-glucuronidase, were used in these vectors. Promoters and terminators of *B. lactofermentum* (MARTÍN et al., 1990; BARDONNET and BLANCO, 1991) and *C. glutamicum* (EIKMANNS et al., 1991) were characterized by using these systems. In addition, known promoters of *E. coli* were evaluated and the functioning of P_{tac} and P_{lacUV5}, P_{trp}, and $\lambda P_R P_L$, and their corresponding regulators LacI, TrpR, and CI was demonstrated in *B. lactofermentum* (TSUCHIYA and MORINAGA, 1988).

The integration vectors constitute a special case of vectors, since they do not necessarily require a functional replication region in corynebacteria. These vectors are useful for gene disruption and gene replacement as well as for transposon delivery experiments (see also Sect. 3.2.1). Basically, two strategies were followed to establish such vector systems. The first approach was to construct vectors showing a temperature-sensitive replication. By using hydroxylamine mutagenesis, variants of a pHM1519-based corynebacterial vector were obtained that were unable to replicate at growth temperatures of 38 °C. By selecting for the antibiotic resistance marker carried by the plasmid, integration into the chromosome was achieved when a homologous DNA region was present on the plasmid (A. SCHWARZER, unpublished).

Such a vector system has two disadvantages in corynebacteria, one of which is the relatively laborious procedure of obtaining integrants containing no free plasmid DNA, since the corynebacterial vector is present in multiple copies and needs several generations of cell growth to be cured. The second drawback is that, at permissive temperatures, the recombination of the homologous sequences flanking the integrated plasmid can result in its excision from the genome.

To employ the second strategy which involves the so-called suicide vectors lacking a functional corynebacterial replicon, efficient methods are required to introduce foreign DNA into the strain to be manipulated. Both conjugation and electroporation, using vectors capable of replicating in *E. coli* and corynebacteria yield high numbers of transconjugants (see Sect. 4.2.2). Consequently, these methods proved useful for introducing mobilizable *E. coli* vectors containing a region of homology, thereby leading to gene disruption and gene replacement (SCHWARZER and PÜHLER, 1991). By combining the techniques of gene disruption and replacement, a defined deletion mutant was constructed which lacked all vector sequences and antibiotic resistance markers which resulted in a genetically stable strain. Such a strain could also be considered safe with respect to environmental release and horizontal gene transfer. The introduction of suicide vectors is possibly hampered by restriction systems that are not inactivated by a heat treatment (Sect. 4.2.2) of the recipient cells, as described by SCHÄFER et al. (1990). *Corynebacterium melassecola,* for example, is a highly restricting strain which is very difficult to transform, even using replicating vectors. REYES et al. (1991) reported a system of initially electroporating *Brevibacterium lactofermentum,* using a shuttle vector. This strain is characterized by a weak restriction system. Plasmid DNA isolated from these transformants was then digested by an appropriate restriction enzyme, to release a replicon-less fragment. This fragment was religated and subsequently, electroporated into the highly-restricting host *C. melassecola,* yielding true integrants.

4.2.2 DNA Transfer into Corynebacteria by Transformation, Electroporation, and Conjugation

Different systems for introducing DNA into corynebacteria have been established. These include transformation of spheroplasts and protoplasts, transfection of protoplasts, transduction, electroporation, conjugation, and fusion of protoplasts.

The first methods developed for corynebacteria were based on the standard techniques applied to other Gram positive bacteria which used the chaotropic agent polyethyleneglycol (KATSUMATA et al., 1984; SANTAMARIA et al., 1985). In these procedures, penicillin G was added to a bacterial culture which was at the logarithmic phase of growth, and the cells were incubated for several hours. Thereafter, the cells were harvested and treated with lysozyme to form protoplasts which were transformed in the presence of polyethyleneglycol (PEG) and Ca^{2+} ions. The frequency of transformation was 10^5 to 10^6 transformants per microgram of DNA. YOSHIHAMA et al. (1985) and BEST and BRITZ (1986) used a different strategy to render the cells more sensitive to lysozyme. Growth in the presence of glycine and subsequent lysozyme treatment, yielded osmotically sensitive, but rod-shaped cells, indicating that some cell-wall material was retained. These cells were termed spheroplasts and were transformed at a frequency up to 10^4 transformants per microgram of DNA. The spheroplasts formed in this way were converted to protoplasts by the action of the lytic enzyme achromopeptidase (THIERBACH et al., 1988).

In order to obtain a high frequency of transformation, the DNA to be transformed must be correctly methylated to avoid restriction by the host. *C. glutamicum* ATCC 13032 was shown to have an efficient restriction system by comparison with *B. lactofermentum* ATCC 21798. Consequently, the yield of transformation was at least 3 orders of magnitude lower when *E. coli* DNA was used to transform *C. glutamicum* as opposed to *B. lactofermentum* (THIERBACH et al., 1988).

Similarly, transfection was performed using protoplasts of *B. lactofermentum* (SANCHEZ et al., 1986), and *Corynebacterium lilium* (YEH et al., 1986), a strain apparently identical to *C. glutamicum.* The frequency of DNA transfer was fairly high (between 10^4 and 10^5 per μg DNA).

Nonetheless, only one report exists of the establishment of a functional vector system based on phages (MIWA et al., 1985). These authors constructed a cosmid vector using a DNA fragment containing the cohesive ends of a *Brevibacterium* phage. *Escherichia coli-Corynebacterium* shuttle vectors carrying the *cos* fragment were transduced into *B. lactofermentum* and *C. glutamicum* by the aid of the intact phage. Although this system was described as being suitable for cloning genes in *C. glutamicum* and *B. lactofermentum,* to our knowledge there is no report on the application of this system.

Polyethyleneglycol proved useful not only in the transformation, but also in the fusion of protoplasts of amino acid-producing corynebacteria. KARASAWA et al. (1986) described a PEG-induced fusion of protoplasts of a strain of *B. lactofermentum* which produced L-lysine and L-threonine, with a lysine auxothroph. By selecting for the resistance to two different amino acid analogs, one of which was carried by each of the parent strains, fusants were found which displayed lysine auxotrophy and produced threonine. Due to its lysine requirement, the fusant strain produced significantly more threonine than its parent.

In another experiment, a high lysine production was combined with a high glucose consumption rate.

Protoplast fusion can also be induced by electric shock, as described by ROLS et al., (1987). Both electric shock and PEG-induced fusion methods yield a maximum frequency of 10^{-3} to 10^{-4} fusants per viable cell.

Besides the systems described above, mainly two methods exist at present for the easy, highly efficient and time-saving transfer of plasmids to corynebacteria.

Of these methods, electroporation is the fastest and most widely used. Since a vast number of reports on electroporation of corynebacteria exist, only some interesting details of this method will be discussed. In theory, electroporation can be performed using different devices, but the equipment of choice produces an exponentially decaying pulse of duration of approximately 5 milliseconds and a field strength of up to 12.5 kV per cm.

In general, cells harvested during early to middle log phase are used in concentrations of 10^9 to 10^{10} cells per mL in these electroporation experiments. Different bacterial strains were found to behave slightly differently in respect to field strength. DUNICAN and SHIVNAN (1989) found 12.5 kV cm^{-1} to be optimal for electroporating *C. glutamicum,* whereas BONNASSIE et al. (1990) obtained a maximum transfer rate to *B. lactofermentum* at 5.7 kV cm^{-1}. Although the results obtained by different groups are not strictly comparable, some mutant strains seem to be far more readily transformed than their parent strains. These strains were originally isolated as rifampicin-resistant or lysozyme-sensitive mutants and were more efficiently transformed. Whether this has something to do with DNA restriction is unclear at the moment; some research groups report significant differences between electrotransfer of homologous and heterologous DNA (HAYNES and BRITZ, 1990) whereas others observed no difference (BONNASSIE et al., 1990). Since some strains generally yield low electrotransformation frequencies, different pretreatment methods were introduced to increase the yield of transformed cells. These include repeated freezing and thawing of the cells and the formation of spheroplasts before electroporation (WOLF et al., 1989), the addition of penicillin G (SATOH et al., 1990) or ampicillin (BONNASSIE et al., 1990), or glycine and isonicotinic acid or tween 80 (HAYNES and BRITZ, 1989) to the culture medium. These treatments apparently render the bacterial cell wall more permeable and allow exogenous DNA and cell membrane to form contact more readily. The maximum yield obtained by this method was 4×10^7 transformants for *C. glutamicum* ASO19 and 4.7×10^7 for *B. lactofermentum* (HAYNES and BRITZ, 1989).

This DNA transfer frequency is comparable to that obtained using a different method, namely conjugal transfer of mobilizable plasmids from Gram negative *E. coli* to Gram positive corynebacteria (SCHÄFER et al., 1990). This system depends on the mobilization functions of the Gram negative broad-host-range plasmid RP4 to transfer plasmids containing the RP4 *oriT* (origin of transfer), to different corynebacteria by conjugation. The conjugation between *E. coli* and corynebacteria is performed using nitrocellulose filters to which the bacteria can attach. This system is applicable without modification, to a wide range of different coryneform strains (SCHÄFER et al., 1990). When wild-type corynebacterial strains were conjugated a relatively low transfer frequency of up to 10^{-6} transconjugants per donor cell was obtained. In contrast, a highly efficient plasmid transfer displaying a frequency of up to several times 10^{-2} was obtained when restriction-deficient mutants were used.

The recipient cells can be rendered more fertile by applying a heat treatment. The heat treatmant of *C. glutamicum* ATCC 13032 consists of an incubation of approximately 9 minutes at 48 to 49°C. Thereafter, non-growing cultures remain in a competent state for several days. When transferred to fresh medium and incubated at optimal growth temperature and aeration, the cells lose their high fertility within a few hours. Since this method also dramatically increases the transfer of heterologous DNA to *C. glutamicum* by transformation of spheroplasts (THIERBACH et al., 1988), it is speculated that this heat treatment inactivates the restriction system of corynebacteria. This phenomenon seems to contradict the mechanism of conjugation postulated for Gram negative bacteria, where single-stranded DNA, which is supposed to be resistant to endonucleases, is thought to be transferred during the conjugation process.

4.2.3 Gene Expression in Corynebacteria

As in other bacteria, the expression of homologous and heterologous genes in corynebacteria requires a detailed knowledge of the structure and function of translational and transcriptional signals in these organisms. To date, no comprehensive investigation of such signals has been published, but there is no indication that translational start signals from other bacteria including *E. coli* should not function properly in corynebacteria. The situation is more complicated with respect to transcriptional signals. By using promoter-probing vectors (see above), TSUCHIYA and MORINAGA (1988) tested highly efficient *E. coli* promoters, such as P_{trp}, P_{lacUV5} and P_{tac}, and $\lambda P_R P_L$ in corynebacteria. The efficiency of these promoters, as well as their regulatory properties determined by the corresponding repressors Trp, LacI, and CI, were found to be roughly comparable to those in *E. coli.* This permitted the construction of inducible, high-expression vectors using the *lacI/P*$_{tac}$ system (EIKMANNS et al., 1991; LIEBL et al., 1992). Despite the fact that strong promoters from *E. coli* efficiently initiate transcription in corynebacteria, the information available on endogenous promoters in corynebacteria is limited. Only a few transcriptional start points have been mapped to date (reviewed by FOLLETTIE et al., 1990; MARTÍN, 1989), and two classes of promoters have been defined by the latter author. Promoters of the first class were recognized in corynebacteria and *E. coli* and were similar to typical σ70 promoters of *E. coli.* The second class contained promoters which were specific for corynebacteria and which lacked such sequence similarities.

The transcriptional termination signals of *E. coli* seem to be hardly recognized by corynebacteria. Neither the strong terminator from phage fd (CARDENAS et al., 1991) nor the Ω terminator cartridge (BARDONNET and BLANCO, 1991) seem to terminate transcription efficiently in *B. lactofermentum.* To terminate transcription, an endogenous terminator from the tryptophan operon of *B. lactofermentum* was used (CARDENAS et al., 1991).

The expression of secreted enzymes constitutes a special case of gene expression. However, protein secretion in corynebacteria is currently not well understood. This is surprising, since corynebacteria such as *C. glutamicum* have a great potential for use as industrial protein producers due to the nearly complete lack of secreted enzymes, the lack of secreted proteases, and the permissiveness with respect to heterologous transcriptional and translational signals.

The first investigations of protein secretion in *C. glutamicum* were undertaken by LIEBL and SINSKEY (1988) who cloned a gene encoding an extracellular DNase from *C. glutamicum.* They showed that the secretion of heterologous proteins, including a lipase from *Staphylococcus hyicus,* a heat-labile nuclease and protein A from *S. aureus,* was possible by this organism (LIEBL et al., 1989, 1992). In the case of the nuclease from *S. aureus,* the processing by the signal peptidase was found to be identical to that of *S. aureus* and *Bacillus subtilis,* and a second, post-secretory processing occurred in all three organisms (LIEBL et al., 1992). In addition, an amylase from *Bacillus amyloliquefaciens* (SMITH et al., 1986) and a cellulase from *Cellulomonas fimi* (PARADIS et al., 1987) were shown to be secreted by *Brevibacterium lactofermentum,* indicating the feasibility of employing these organisms for the production of heterologous proteins.

4.3 Engineering L-Lysine Production in *Corynebacterium glutamicum*

Economically, lysine is one of the most important amino acids produced by fermentation and is used mainly as an additive in feed production. In *C. glutamicum* as in other bacteria including *Escherichia coli,* the amino acids lysine, threonine, methionine, isoleucine, and diaminopimelate derive part or all of their carbon atoms from aspartate (Fig. 13). Lysine biosynthesis is influenced mainly by the aspartokinase and the aspartate semialdehyde dehydrogenase, the only two enzymes which are active in the common pathway of aspartate-derived amino acids, and those of the lysine-specific branch. Since the intermediate of lysine synthesis, *meso*-diaminopimelate, is an essential building block of the corynebacterial cell wall, enzyme(s) for the incorporation of *meso*-DAP into peptidoglycan could also influence the diaminopimelate and lysine pools. Finally, an essential factor in the efficient production of lysine is its excretion, a fact which renders corynebacteria particularly suitable for industrial use. An additional advantage of corynebacteria with regard to the industrial amino acid production, is the apparent lack of degradation of many amino acids. For example, no lysine decarboxylase as is present in *E. coli,*

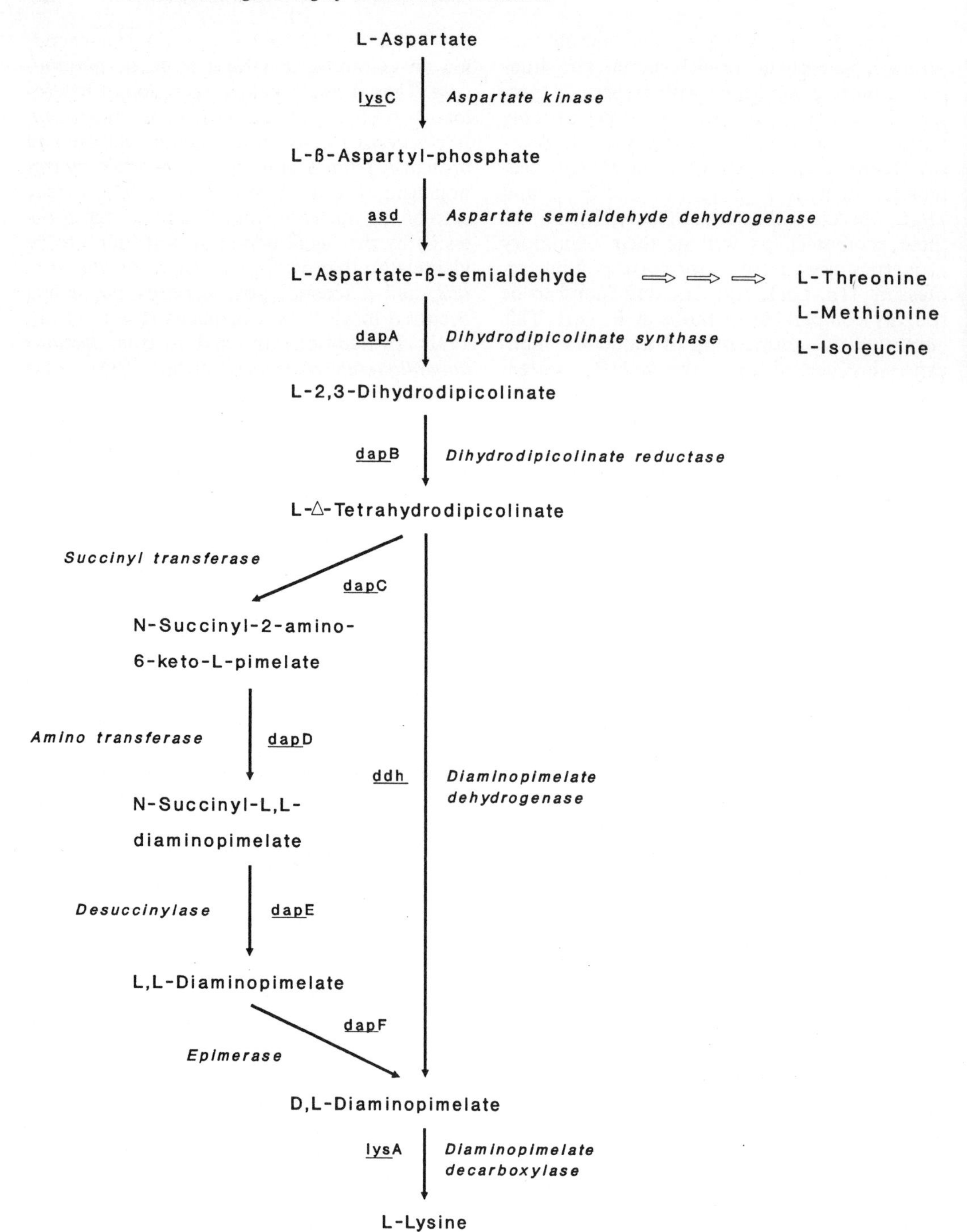

Fig. 13. The pathway of lysine biosynthesis in *Corynebacterium glutamicum*. The names of the biosynthetic enzymes as well as their corresponding genes are given.

nor any other lysine-degrading activity was found in corynebacteria. A third point that greatly facilitates the construction of mutants for use as amino acid producers is the simple regulation of at least the lysine pathway. A concerted feedback inhibition by lysine and threonine is exerted on the first enzyme, aspartokinase (TOSAKA and TAKINAMI, 1978). In addition, the first enzyme of the threonine branch, homoserine dehydrogenase, is inhibited by threonine, and its synthesis is repressed by methionine (FOLLETTIE et al., 1990). The absence of isoenzymes in *C. glutamicum* is also an unusual phenomenon, since, for example, both *B. subtilis* (GRAVES and SWITZER, 1990) and *E. coli* were shown to possess three isoenzymes of aspartokinase, two of which were fused to homoserine dehydrogenase isoenzymes in the latter organism (COHEN and SAINT-GIRONS, 1987).

In the light of these findings, a lysine overproduction can be obtained by preventing the concerted feedback inhibition of aspartokinase. Thereafter, individual enzyme levels can be raised by the amplification or overexpression of genes that constitute metabolic bottlenecks in the pathway.

In general, there are two different methods to impede the feedback inhibition of aspartokinase. One method is to select for homoserine auxotrophs. Such an auxotroph can be cultured in the presence of a controlled amount of homoserine which maintains the internal threonine concentration low enough to prevent feedback inhibition. Thus, by preventing both the inhibition of aspartokinase and the removal of aspartate semialdehyde from the pathway, a significant lysine production could be initiated. The second method involves the use of amino acid analogs to select mutant strains having an aspartokinase which is insensitive to feedback inhibition (SHIIO, 1982). The lysine analog most widely used is S-(2-aminoethyl)-L-cysteine (AEC), which together with threonine is lethal for the *C. glutamicum* wild type (TOSAKA and TAKINAMI, 1978). AEC-resistant mutant strains often show a deregulated aspartokinase enzyme and a considerably high lysine production of up to 30 g/L.

To clone genes involved in the biosynthesis of lysine, three strategies were followed. The first method involved the complementation of *E. coli* auxotrophs which enable the following genes to be cloned: those for dihydrodipicolinate synthase and dihydrodipicolinate reductase (*dapA, dapB*; YEH et al., 1988; CREMER et al., 1990), *meso*-diaminopimelate dehydrogenase (*ddh;* ISHINO et al., 1988; YEH et al., 1988), and diaminopimelate decarboxylase (*lysA*; YEH et al., 1988). Although the *ddh* gene specifies an enzymatic activity not present in *E. coli, dapD* mutants of *E. coli* were successfully complemented by the addition of *ddh* which provided an alternative pathway for *meso*-DAP biosynthesis.

The second strategy was to clone the AEC resistance genes from mutants into the *C. glutamicum* wild type. By using this approach, the genes for an AEC resistant aspartokinase (*lysC*^{FBR}) and the gene for aspartate semialdehyde dehydrogenase (*asd*) which was located immediately downstream of the *lysC* gene, were cloned (KALINOWSKI et al., 1990). To isolate the aspartokinase gene from the *C. glutamicum* wild type, this *asd* gene was employed to complement an *E. coli asd* mutant strain, thereby confirming the presence of the whole *lysC-asd* gene cluster (KALINOWSKI et al., 1991).

By a third strategy, a gene for lysine transport was cloned (SEEP-FELDHAUS et al., 1991). This strategy used the fact that lysine and AEC are transported into the cell by the same transport system. In this way, mutants defective in lysine/AEC uptake were AEC resistant, but could be converted to become AEC sensitive by the cloning of an intact lysine uptake gene from the wild type. Unfortunately, it transpired that the cloned gene was responsible for lysine uptake only (SEEP-FELDHAUS et al., 1992) and did not participate in the specific lysine excretion mechanism (BRÖER and KRÄMER, 1991).

Of all lysine biosynthetic genes cloned to date, only *lysC* and *asd* which are transcribed convergently and *dapA* and *dapB* which are transcribed divergently, were found to be located adjacent to each other on the *C. glutamicum* chromosome.

The structure of the lysine biosynthetic pathway was further analyzed using gene disruption and gene replacement techniques (SCHWARZER and PÜHLER, 1991). The genes *dapA, ddh,* and *lysA* were subjected to gene disruption using internal DNA fragments. As expected, *lysA* integrants were lysine auxotrophic, whereas no *dapA* integrants appeared, probably because those mutants were not efficiently supplemented by exogenous diaminopimelate (SCHWARZER, unpublished). Unexpectedly, *ddh* mutants constructed by the gene disruption technique showed no auxotrophy which indicated the presence of a fully functional alternative pathway for *meso*-DAP synthesis (SCHRUMPF et al., 1991). The same authors showed that the enzyme activities of the *E. coli* pathway for *meso*-DAP formation were measurable, and that both pathways contributed to lysine synthesis in high-producing strains of *C. glutamicum*.

All of the biosynthetic genes cloned were examined with respect to their effect on lysine production (CREMER et al., 1991). Only the genes *lysC*^{FBR} encoding feedback-resistant aspartokinase and *dapA* encoding dihydrodipicolinate synthase had a positive effect on lysine production when amplified in the *C. glutamicum* wild-type strain. This effect was additive when both genes were combined on one plasmid.

However, the genetic engineering of lysine production in amino acid-producing corynebacteria is still in its infancy. A plethora of genes, not directly involved in lysine biosynthesis, were shown to or purported to influence efficient overproduction. These include citrate synthase, phosphoenolpyruvate carboxylase and genes forming the precursors and energy supply of lysine synthesis (TOSAKA et al., 1985) as well as global cellular regulatory systems such as the stringent response (DEBABOV, 1990).

The important task for the near future is to use the arsenal of methods developed in order to identify, clone and genetically engineer the above genes.

5 Genetic Engineering of Miscellaneous Gram Positive Bacteria

Besides bacilli, streptomycetes, and corynebacteria described above, other Gram positive bacteria are also of biotechnological significance. Many of these, such as staphylococci, streptococci, and clostridia, comprise both pathogenic and non-pathogenic, commercially useful species. For biotechnological purposes, non-pathogenic strains are normally applied that are often derived from strains which have already been employed for years in food technology.

The genus *Staphylococcus* consists of approximately 20 distinct species (NOVICK, 1990). Most of the molecular and genetic studies have been conducted on *S. aureus* and, in particular, strain NCTC8325. Detailed information has been compiled, including a chromosomal map containing some hundred loci (NOVICK, 1990).

Whereas *S. aureus* strains are typical pathogens which produce toxins, hemolysins, protein A and coagulases, *S. carnosus* is a non-pathogenic organism which is used as starter culture in food processing. The following criteria make the latter a promising species for biotechnological application (GÖTZ, 1990):

- *S. carnosus* can be cultivated easily on simple media.
- High transformation rates (10^6 transformants per µg DNA) can be obtained by protoplast transformation (GÖTZ et al., 1983; GÖTZ and SCHUHMACHER, 1987).
- Plasmid vectors such as pCT20, pCA43, and pCA44 were constructed from the staphylococcal plasmids pC194, pT181, and the arsenic-resistance plasmid pSX267, respectively (KELLER et al., 1983; KREUTZ and GÖTZ, 1984).

Thus, it was possible to clone and express various genes from *Staphylococcus, Bacillus,* and *Clostridium*. A lipase gene from *S. hyicus* was analyzed in detail, and a 40-fold overproduction could be reached in *S. carnosus* (GÖTZ, 1990; GÖTZ et al., 1985).

Many serious illnesses are caused by streptococci. To understand the mechanisms of pathogenicity, much effort has been directed towards developing gene cloning systems for these strains, which could also be applied to non-pathogenic strains of streptococci.

Streptococci can undergo transduction, transformation, and conjugation; whereby transformation is the most practical technique

(CAPARON and SCOTT, 1990). *Streptococcus pneumococcus* and a few other species are naturally competent (AVERY et al., 1944). Protoplast transformation and electroporation are the standard techniques for DNA introduction (CAPARON and SCOTT, 1990).

Although streptococci are of minor importance to biotechnology, the conjugative transposons detected in these bacteria (for a review see CLEWELL and GAWRON-BURKE, 1986) are generally important to the genetic manipulation in many Gram positives.

The best characterized of the non-plasmid, conjugative systems is Tn*916* (FRANKE and CLEWELL, 1981) which was discovered in *Streptococcus faecalis* (now renamed as *Enterococcus faecalis*). This transposon was transferred from its location in the donor to a new location in the recipient. The transfer required close contact between the donor and recipient, e.g., by filter mating, and the Tn*916*-encoded resistance to tetracycline was used as the selection marker. The host range was not limited to streptococci, but also included bacilli, clostridia, and others. The transposition of Tn*916* and related elements proceeds by excision of a free, non-replicative, covalently closed, circular intermediate which is the substrate for integration (SCOTT et al., 1988). Conjugative transposition requires the presence of a functional integrase in both donor and recipient which need not be from the same species or even from the same genus (STORRS et al., 1991). Recombinant derivatives of Tn*916* and other conjugative transposons are available, including shuttle transposons which can be selected on various antibiotic-containing media and used in a variety of Gram positive bacteria (CAPARON and SCOTT, 1990).

Renewed interest in the fermentation of clostridia has been generated due to the use of renewable resources (JONES and WOODS, 1986). Although the manipulation under strictly anaerobic conditions is time-consuming and difficult, conjugal transfer and transposon mutagenesis systems (BERTRAM and DÜRRE, 1989; LIN and JOHNSON, 1991) have been developed as well as other genetic methods such as the electroporation using staphylococcal plasmids (YOUNG et al., 1989).

Recently, the first report of the cloning of homologous fermentative genes from the acetone-formation and the butyrate-formation pathway of *Clostridium acetobutylicum* was published (MERMELSTEIN et al., 1992). A new *Bacillus subtilis/Clostridium acetobutylicum* shuttle vector was used, which permitted the analysis of the complex anaerobic metabolism.

Lactobacilli and lactococci are widely used in food fermentation. Their main contribution to the fermentation process is acidification which prevents the growth of food-poisoning bacteria. Acidification is also a prerequisite for the coagulation of milk in the production of cheese and yoghurt and the coagulation of soluble meat proteins which is necessary to form a matrix during sausage fermentation (SMID et al., 1991; KNAUF et al., 1992).

In addition to conventional genetic engineering techniques (for reviews see DE VOS, 1987; LUCHANSKY et al., 1989), new systems have been developed during the last two years which include composite transposons (ROMERO and KLAENHAMMER, 1991) and endogenous conjugative transposons for transposon mutagenesis (RAUCH and DE VOS, 1992), integration vectors (LEENHOUTS et al., 1991; PETZEL and MCKAY, 1992; POLZIN and MCKAY, 1992), plasmid transduction (RAYA and KLAENHAMMER, 1992), and chromosomal mapping (LE BOURGEOIS et al., 1992). These genetic engineering techniques, however, are mainly used in the analysis and cloning of endogenous lactococcal genes (SHEARMAN et al., 1992) rather than for the expression of heterologous genes.

6 Concluding Remarks

Many Gram positive bacteria are now available for the application of genetic engineering techniques. This has permitted the analysis of many endogeneous properties such as pathogenicity and the synthesis of excreted products. However, most of these studies are still in their infancy.

The development of molecular genetic methods opens promising prospects for using Gram positive bacteria to obtain secreted heterologous proteins. In this respect, the Gram positive bacteria seem to be particularly well suited by comparison with other systems discussed in this volume.

Acknowledgement

We wish to thank K. Krey for preparing the illustrations. The authors' research work was funded by the Bundesministerium für Forschung und Technologie (Grants 0318787A, 0319374A and 038409) and the Deutsche Forschungsgemeinschaft (Wo 485/1-1).

7 References

ADACHI, T., YAMAGATA, H., TSUGAGOSHI, N., UDAKA, S. (1989), Multiple and tandemly arranged promoters of the cell wall gene operon in *Bacillus brevis, J. Bacteriol.* **171**, 1010–1016.

ALIJAH, R., DORENDORF, J., TALAY, S., PÜHLER, A., WOHLLEBEN, W. (1991), Genetic analysis of the phosphinothricin-tripeptide biosynthetic pathway of *Streptomyces viridochromogenes* Tü494, *Appl. Microbiol. Biotechnol.* **34**, 749–755.

ALTENBUCHNER, J., CULLUM, J. (1984), DNA amplification and an unstable arginine gene in *Streptomyces lividans* 66, *Mol. Gen. Genet.* **195**, 134–138.

ALTENBUCHNER, J., CULLUM, J. (1985), Structure of an amplifiable DNA sequence in *Streptomyces lividans* 66, *Mol. Gen. Genet.* **201**, 192–197.

ALTENBUCHNER, J., CULLUM, J. (1987), Amplification of cloned genes in *Streptomyces, Bio/Technology* **5**, 1328–1329.

AMJAD, M., CASTRO, J. M., SANDOVAL, H., WU, J. J., YANG, M., HENNER, D. J., PIGGOT, P. J. (1990), An SfiI restriction map of the *Bacillus subtilis* 168 genome, *Gene* **101**, 15–21.

ANDERSON, S., MARKS, C. B., LAZARUS, R., MILLER, J., STAFFORD, K., SEYMOUR, S., LIGHT, D., RASTETTER, W., ESTELL, D. (1985), Production of 2-keto-L-gluconate, an intermediate in L-ascorbate synthesis by a genetically modified *Erwinia herbicola, Science* **230**, 144–149.

ANZAI, H., MURAKAMI, T., IMAI, S., SATOH, A., NAGAOKA, K., THOMPSON, C. J. (1987), Transcriptional regulation of bialaphos biosynthesis in *Streptomyces hygroscopicus, J. Bacteriol.* **169**, 3482–3488.

ARCHER, J. A. C., FOLLETTIE, M. T., SINSKEY, A. J. (1990), Biology of *Corynebacterium glutamicum,* a molecular approach, in: *Proc. 6th Int. Symp. Genetics of Industrial Microorganisms* (HESLOT, H., DAVIES, J., FLORENT, J., BOBICHON, L., DURANT, G., PENASSE, L., Eds.), pp. 27–33, Strasbourg: Société Française de Microbiologie.

ARWERT, F., VENEMA, G. (1973), Transformation in *Bacillus subtilis:* Fate of newly introduced transforming DNA, *Mol. Gen. Genet.* **123**, 185–198.

ASTURIAS, J. A., LIRAS, P., MARTÍN, J. F. (1990), Phosphate control of *pabS* gene transcription during candicidin biosynthesis, *Gene* **93**, 79–84.

AUERSWALD, E. A., LUDWIG, G. R., SCHALLER, H. (1982), Structural analysis of Tn*5, Cold Spring Harbor Symp. Quant. Biol.* **45**, 107–113.

AVERY, O. T., MACLEOD, C. M., MACCARTY, M. (1944), Studies on the chemical nature of the substance inducing transformation of pneumococcal types, *J. Exp. Med.* **79**, 137–158.

BARAK, I., KOPTIDES, M., JUCOVIC, M., SISOVA, M., TIMKO, J. (1990), Construction of a promoter-probe shuttle vector for *Escherichia coli* and brevibacteria, *Gene* **95**, 133–135.

BARDONNET, N., BLANCO, C. (1991), Improved vectors for transcriptional signal screening in corynebacteria, *FEMS Microbiol. Lett.* **84**, 97–102.

BASCARAN, V., SANCHES, L., HARDISSON, C., BRANA, A. F. (1991), Stringent response and initiation of secondary metabolism in *Streptomyces clavuligerus, J. Gen. Microbiol.* **137**, 1625–1634.

BEDFORD, D. J., LEWIS, C. G., BUTTNER, M. J. (1991), Characterization of a gene conferring bialaphos resistance in *Streptomyces coelicolor* A3(2), *Gene* **104**, 39–45.

BENDER, E., VOGEL, R., KOLLER, K.-P., ENGELS, J. (1990a), Synthesis and secretion of hirudin by *Streptomyces lividans, Appl. Microbiol. Biotechnol.* **34**, 203–207.

BENDER, E., KOLLER, K.-P., ENGELS, J. (1990b), Secretory synthesis of human interleukin-2 by *Streptomyces lividans, Gene* **86**, 227–232.

BENSON, A.-K., STEVENSON, A., HALDENWANG, W. G. (1990), Mutation in *Bacillus subtilis* which influence the activity of a promoter recognized by a minor form of RNA polymerase ($E\text{–}\sigma^B$), in: *Genetics and Biotechnology of Bacillus* (ZUKOWSKI, M. M., GANESAN, A. T., HOCH, J. A., Eds.), Vol. 3, pp. 13–22, San Diego: Academic Press.

BERNAN, V., FILPULA, D., HERBER, W., BIBB, M., KATZ, E. (1985), The nucleotide sequence of the tyrosinase gene from *Streptomyces antibioticus* and characterization of the gene production, *Gene* **37**, 101–110.

BERTRAM, J., DÜRRE, P. (1989), Conjugal transfer and expression of streptococcal transposons in *Clostridium acetobutylicum, Arch. Microbiol.* **151,** 551-557.

BEST, G. R., BRITZ, M. L. (1986), Facilitated protoplasting in certain auxotrophic mutants of *Corynebacterium glutamicum, Appl. Microbiol. Biotechnol.* **23,** 288-293.

BIBB, M. J., JANSSEN, G. R. (1986), Unusual features of transcription and translation of antibiotic resistance genes in antibiotic-producing *Streptomyces,* in: *Fifth Int. Symp. Genetics of Industrial Microorganisms,* (ALACEVIC, M., HRANUELI, D., TOMAN, Z., Eds.), pp. 309-318, Karlovac: Ognjen Prica Printing Works.

BIBB, M. J., FREEMAN, R. F., HOPWOOD, D. A. (1977), Physical and genetical characterization of a second sex factor, SCP2, for *Streptomyces coelicolor* A3(2), *Mol. Gen. Genet.* **154,** 155-166.

BIBB, M. J., WARD, M. J., HOPWOOD, D. A. (1978), Transformation of plasmid DNA into *Streptomyces* at high frequency, *Nature,* **274,** 398-40.

BIBB, M. J., SCHOTTEL, J. L., COHEN, S. N. (1980), A DNA cloning system for interspecies gene transfer in antibiotic-producing *Streptomyces, Nature* **284,** 526-531.

BIBB, M. J., WARD, J. M., KIESER, T., COHEN, S. N., HOPWOOD, D. A. (1981), Excision of chromosomal DNA sequences from *Streptomyces coelicolor* forms a novel family of plasmids detectable in *Streptomyces lividans, Mol. Gen. Genet.* **184,** 230-240.

BIBB, M. J., JANSSEN, G. R., WARD, J. M. (1985), Cloning and analysis of the promoter region of the erythromycin-resistance gene (ermE) of *Streptomyces erythraeus, Gene* **38,** E357-E368.

BIRCH, A., CULLUM, J. (1985), Temperature-sensitive mutants of the *Streptomyces* plasmid pIJ702, *J. Gen. Microbiol.* **131,** 1299-1303.

BIRCH, A., HÄUSLER, A., HÜTTER, R. (1990), Genome rearrangement and genetic instability in *Streptomyces* spp., *J. Bacteriol.* **172,** 4138-4142.

BONNASSIE, S., BURINI, J.-F., OREGLIA, J., TRAUTWETTER, A., PATTE, J.-C., SICARD, A. M. (1990), Transfer of plasmid DNA to *Brevibacterium lactofermentum* by electrotransformation, *J. Gen. Microbiol.* **136,** 2107-2112.

BRÄU, B., HILGENFELD, R., SCHLINGMANN, M., MARQUARDT, R., BIRR, E., WOHLLEBEN, W., AUFDERHEIDE, K., PÜHLER, A. (1991), Increased yield of a lysozyme after self-cloning of the gene in *Streptomyces coelicolor* "Müller", *Appl. Microbiol. Biotechnol.* **34,** 481-487.

BRAWNER, M., FORNWALD, J., ROSENBERG, M., POSTE, G., WESTPHELING, J. (1990a), Heterologous gene expression in *Streptomyces,* in: *Proc. 6th Int. Symp. Genetics of Industrial Microorganisms* (HESLOT, H., DAVIS, J., FLORENT, J., BOBICHON, L., DURANT, G., PENASSE, L., Eds.), pp. 379-392, Strasbourg: Société Française de Microbiologie.

BRAWNER, M., TAYLOR, D., FORNWALD, J. (1990b), Expression of the soluble CD-4 receptor in *Streptomyces, J. Cell. Biochem.* **14A,** 103.

BREFORT, G., MAGOT, M., IONECO, H., SEBALD, M. (1977), Characterization and transferability of *Clostridium perfringens* plasmids, *Plasmid* **1,** 52-66.

BREITLING, R., GERLACH, D., HARTMANN, M., BEHNKE, D. (1989), Secretory expression in *Escherichia coli* and *Bacillus subtilis* of human interferon α genes directed by staphylokinase signals, *Mol. Gen. Genet.* **217,** 384-391.

BREITLING, R., SOROKIN, A. V., BEHNKE, D. (1990), Temperature-inducible gene expression in *Bacillus subtilis,* mediated by the CI857 encoded repressor of bacteriophage lambda, *Gene* **93,** 35-40.

BRÖER, S., KRÄMER, R. (1991), Lysine excretion by *Corynebacterium glutamicum,* 1. Identification of a specific secretion carrier system, *Eur. J. Biochemistry* **202,** 131-135.

BRON, S., BOSMA, P., VAN BELKUM, M., LUXEN, E. (1987), Stability function in the *Bacillus subtilis* plasmid pTA1060, *Plasmid* **18,** 8-15.

BRON, S., HOLSAPPEL, S., VENEMA, G., PEETERS, B. P. H. (1991), Plasmid deletion formation between short direct repeats in *Bacillus subtilis* is stimulated by single-stranded rolling-circle replication intermediates, *Mol. Gen. Genet.* **226,** 88-96.

BRUAND, C., EHRLICH, S. D., JANNIERE, L. (1991), Unidirectional theta replication of the structurally stable *Enterococcus faecalis* plasmid pAMβ1, *EMBO J.* **10,** 2171-2177.

BRUTON, C. J., GUTHRIE, E. P., CHATER, K. F. (1991), Phage vectors that allow monitoring of transcription of secondary metabolism genes in *Streptomyces, Bio/Technology* **9,** 652-656.

BUCHANAN-WOLLASTON, V., PASSIATORE, J. E., CANNON, F. (1987), The mob and oriT mobilization functions of a bacterial plasmid promote its transfer to plants, *Nature* **328,** 172-175.

BUTTNER, M. J. (1989), RNA polymerase heterogeneity in *Streptomyces coelicolor* A3(2), *Mol. Microbiol.* **3,** 1653-1659.

BUTTNER, M. J., FEARNLEY, I. M., BIBB, M. J. (1987), The agarase gene (*dagA*) of *Streptomyces coelicolor* A3(2): nucleotide sequence and transcriptional analysis, *Mol. Gen. Genet.* **209,** 101–109.

BUTTNER, M. J., SMITH, A.- M., BIBB, M. J. (1988), At least three different RNA polymerase holoenzymes direct transcription of the agarase gene (*dagA*) of *Streptomyces coelicolor* A3(2), *Cell* **52,** 599–607.

CANOSI, U., MORELLI, G., TRAUTNER, T. A. (1978), The relationship between molecular structure and transformation efficiency of some *Staphylococcus aureus* plasmids isolated from *Bacillus, Mol. Gen. Genet.* **166,** 259–267.

CAPARON, M. G., SCOTT, J. R. (1990), Genetic manipulation of pathogenic streptococci, *Methods Enzymol.* **204,** 556–586.

CARDENAS, R. F., MARTÍN, J. F., GIL, J. A. (1991), Construction and characterization of promoter-probe vectors for Corynebacteria using the kanamycin-resistance reporter gene, *Gene* **98,** 117–121.

CARDINI, G., JURTSHUK, P. (1970), The enzymatic hydroxylation of *n*-octane by *Corynebacterium* sp. strain 7EC1C, *J. Biol. Chem.* **245,** 2789–2796.

CHAMPNESS, W. C. (1988), New loci required for *Streptomyces coelicolor* morphological and physiological differentiation, *J. Bacteriol.* **170,** 1168–1174.

CHANG, S. (1987), Engineering for protein secretion in Gram-positive bacteria, *Methods Enzymol.* **153,** 507–516.

CHANG, S.-Y., CHANG, S. (1988), Secretion of heterologous proteins in *S. lividans,* in: *Biology of Actinomycetes '88,* (OKAMI, Y., BEPPU, T., OGAWARA, H., Eds.) pp. 103–107, Tokyo: Japan Scientific Societies Press.

CHANG, S., COHEN, S. N. (1979), High frequency transformation of *Bacillus subtilis* protoplasts by plasmid DNA, *Mol. Gen. Genet.* **168,** 111–115.

CHAN KWO CHION, C. K. N., DURAN, R., ARNAUD, A., GALZY, P. (1991), Cloning vectors and antibiotic-resistance markers for *Brevibacterium* sp. R312, *Gene* **105,** 119–124.

CHATER, K. F. (1986), *Streptomyces* phages and their application to *Streptomyces* genetics, in: *The Bacteria,* Vol. IX (QUEENER, S. W., DAY, L. E., Eds.), pp. 119–158, Orlando: Academic Press.

CHATER, K. F. (1989), Multilevel regulation of *Streptomyces* differentiation, *Trends Genet.* **5,** 372–373.

CHATER, K. F. (1990), The improving prospects for yield increase by genetic engineering in antibiotic-producing Streptomycetes, *Bio/Technology* **8,** 115–121.

CHATER, K. F., BRUTON, C. J. (1983), Mutational cloning in *Streptomyces* and the isolation of antibiotic production genes, *Gene* **26,** 67–78.

CHATER, K. F., BRUTON, C. J. (1985), Resistance, regulatory and production genes for the antibiotic methylenomycin are clustered, *EMBO J.* **4,** 1893–1897.

CHATER, K. F., HOPWOOD, D. A., KIESER, T., THOMPSON, C. J. (1982), Gene cloning in *Streptomyces, Curr. Top. Microbiol. Immunol.* **96,** 69–95.

CHEN, C. W., TSAI, J. F.-Y., CHUANG, S.-E. (1987), Intraplasmid recombination in *Streptomyces lividans* 66, *Mol. Gen. Genet.* **209,** 154–158.

CHUNG, S. T. (1987), Tn4556, a 6,8-kilobase-pair transposable element of *Streptomyces fradiae, J. Bacteriol.* **169,** 4436–4441.

CLEWELL, D. B., GAWRON-BURKE, C. (1986), Conjugative transposons and the dissemination of antibiotic resistance in streptococci, *Annu. Rev. Microbiol.* **40,** 635–659.

CLEWELL, D. B., YAGI, Y., DUNNY, G. M., SCHULTZ, S. K. (1974), Characterization of three plasmid desoxyribonucleic acids molecules in a strain of *Streptococcus faecalis,* identification of a plasmid determining erythromycin resistance, *J. Bacteriol.* **117,** 283–289.

COHEN, G. N., SAINT-GIRONS, I. (1987), Biosynthesis of threonine, lysine, and methionine, in: *Escherichia coli* and *Salmonella typhimurium* (NEIDHARDT, F. C., Ed.), Vol. 1, pp. 429–444, Washington: American Society for Microbiology.

CONSTANTINIDES, S. (1980), Steroid transformations at high substrate concentrations using immobilized *Corynebacterium simplex* cells, *Biotechnol. Bioeng.* **22,** 119–136.

CONTENTE, S., DUBNAU, D. (1979), Marker rescue transformation by linear plasmid DNA in *Bacillus subtilis, Plasmid* **2,** 555–571.

COURVALIN, P., CARLIER, C. (1987), *Tn1545,* a conjugative shuttle transposon, *Mol. Gen. Genet.* **206,** 259–264.

CREMER, J., EGGELING, L., SAHM, H. (1990), Cloning of the *dapA dapB* cluster of the lysine-secreting bacterium *Corynebacterium glutamicum, Mol. Gen. Genet.* **220,** 478–480.

CREMER, J., EGGELING, L., SAHM, H. (1991), Control of the lysine biosynthesis sequence in *Corynebacterium glutamicum* as analyzed by overexpression of the individual corresponding genes, *Appl. Environ. Microbiol.* **57,** 1746–1752.

DAVIS, E. O., SEDGWICK, S. G., COLSTON, M. J. (1991), Novel structure of the *recA* locus of *Mycobacterium tuberculosis* implies processing of the gene product, *J. Bacteriol.* **173,** 5653–5662.

DAVIS, M. J. (1986), Taxonomy of plant pathogenic corynebacteria, *Annu. Rev. Phytopathol.* **24,** 115–140.

DAZA, A., GIL, J. A., VIGAL, T., MARTÍN, J. F.(1990), Cloning and characterization of a gene of *Streptomyces griseus* that increases production of extracellular enzymes in several species of *Streptomyces, Mol. Gen. Genet.* **222,** 384–392.

DEBABOV, V. G. (1990), A genetic and molecular-physiological study of glutamate-producing corynebacteria, in: *Proc. 6th Int. Symp. Genetics of Industrial Microorganisms* (HESLOT, H., DAVIES, J., FLORENT, J., BOBICHON, L., DURANT, G., PENASSE, L., Eds.), pp. 353–362, Strasbourg: Société Française de Microbiologie.

DEBARBOUILLE, M., MARTIN-VERSTRAETE, I., KUNST, F., RAPOPORT, G. (1991), The *Bacillus subtilis sigL* gene encodes an equivalent of σ^{54} from Gram-negative bacteria, *PNAS* **88,** 9092–9096.

DE BOER, A. S., DIDERICHSEN, B. (1991), On the safety of *Bacillus subtilis* and *B. amyloliquefaciens,* a review, *Appl. Microbiol. Biotechnol.* **36,** 1–4.

DENIS, F., BRZEZINSKI, R. (1991), An improved aminoglycoside resistance gene cassette for use in Gram-negative bacteria and *Streptomyces, FEMS Microbiol. Lett.* **81,** 261–264.

DESHPANDE, B. S., AMBEDKAR, S. S., SHEWALE, J. G. (1988), Biologically active secondary metabolites from *Streptomyces, Enzyme Microb. Technol.* **10,** 455–473.

DE VOS, W. M. (1987), Gene cloning and expression in lactic strepotcocci, *FEMS Microbiol. Rev.* **46,** 281–295.

DISTLER, J., MANSOURI, K., MAYER, G., PIEPERSBERG, W. (1990), Regulation of biosynthesis of streptomycin, in: *Proc. 6th Int. Symp. Genetics of Industrial Microorganisms* (HESLOT, H., DAVIES, J., FLORENT, J., BOBICHON, L., DURANT, G., PENASSE, L., Eds.), pp. 379–392, Strasbourg: Société Française de Microbiologie.

DUBNAU, D. (1991), Genetic competence in *Bacillus subtilis, Microbiol. Rev.* **55,** 395–424.

DUBNAU, D., CIRIGLIANO, C. (1972), Fate of transforming DNA following uptake by competent *Bacillus subtilis.* III. Formation and properties of products isolated from transformed cells which are derived entirely from donor DNA, *J. Mol. Biol.* **64,** 9–29.

DUNICAN, L. K., SHIVNAN, E. (1990), High frequency transformation of whole cells of amino acid producing coryneform bacteria using high voltage electroporation, *Bio/Technology* **7,** 1067–1070.

EDELMAN, A., JOLIFF, G., KLIER, A., RAPOPORT, G. (1988), A system for the inducible secretion of proteins from *Bacillus subtilis* during logarithmic growth, *FEMS Microbiol. Lett.* **52,** 117–120.

EHRLICH, S. D. (1989), Illegitimate recombination in bacteria, in: *Mobile DNA* (BERG, D., HOWE, M., Eds.), pp. 797–829, Washington: American Society for Microbiology.

EIKMANNS, B. J., KLEINERTZ, E., LIEBL, W., SAHM, H. (1991), A familiy of *Corynebacterium glutamicum/Escherichia coli* shuttle vectors for cloning, controlled gene expression, and promotor probing, *Gene* **102,** 93–98.

ENGEL, P. (1987), Plasmid transformation of *Streptomyces tendae* after heat attenuation of restriction, *Appl. Environ. Microbiol.* **53,** 1–3.

ENGELS, J. W., KOLLER, K.-P. (1991), Gene expression and secretion of eucaryotic foreign proteins in *Streptomyces,* in: *Transgenesis, Application of Gene Transfer,* (MURRAY, J. A. H., Ed.), pp. 33–53, Buckingham: University Press.

EPP, J. K., HUBER, M. L. B., TURNER, J. R., GOODSON, T., SCHONER, B. E. (1989), Production of a hybrid macrolide antibiotic in *Streptomyces ambofaciens* and *Streptomyces lividans* by introduction of a cloned carbomycin biosynthetic gene from *Streptomyces thermotolerans, Gene* **85,** 293–301.

ERRINGTON, J., ILLING, N. (1992), Establishment of cell-specific transcription during sporulation in *Bacillus subtilis, Mol. Microbiol.* **6,** 689–695.

FILIPULA, D., ALLY, A. H., NAGLE, J. (1986), Complete nucleotide sequence of a native plasmid of *Brevibacterium lactofermentum, Nucleic Acids Res.* **14,** 514.

FOLLETTIE, M. T., PEOPLES, O. P., ARCHER, J. A. C., SINSKEY, A. J. (1990), Metabolic engineering of corynebacteria, in: *Proc. 6th Int. Symp. Genetics of Industrial Microorganisms* (HESLOT, H., DAVIES, J., FLORENT, J., BOBICHON, L., DURANT, G., PENASSE, L., Eds.), pp. 315–325, Strasbourg: Société Française de Microbiologie.

FORNWALD, J. A., SCHMIDT, F. J., ADAMS, C. W., ROSENBERG, M., BRAWNER, M. E, (1987), Two promoters, one inducible and one constitutive, control transcription of the *Streptomyces lividans* galactose operon, *PNAS* **84,** 2130–2134.

FRANKE, A. E., CLEWELL, D. B. (1981), Evidence for a chromosome-borne resistance transposon (Tn*916*) in *Streptococcus faecalis* that is capable of "conjugal" transfer in the absence of a conjugative plasmid, *J. Bacteriol.* **145,** 494–502.

GARNIER, T., COLE, S. T. (1988), Identification and molecular genetic analysis of replication functions of the bacteriocinogenic plasmid pIP404 from *Clostridium perfringens, Plasmid* **19,** 151–160.

GÄRTNER, D., GEISSENDÖRFER, M., HILLEN, W. (1988), Expression of the *Bacillus subtilis xyl* operon is repressed at the level of transcription and is induced by xylose, *J. Bacteriol.* **170,** 3102–3109.

GEISSENDÖRFER, M., HILLEN, W. (1990), Regulated expression of heterologous genes in *Bacillus subtilis* using the Tn*10* encoded *tet* regulatory elements, *Appl. Microbiol. Biotechnol.* **33,** 657–663.

GORMLEY, E. P., DAVIES, J. (1991), Transfer of plasmid RSF1010 by conjugation from *Escherichia coli* to *Streptomyces lividans* and *Mycobacterium smegmatis, J. Bacteriol.* **173,** 6705–6708.

GÖTZ, F. (1990), *Staphylococcus carnosus:* a new host organism for gene cloning and protein production, *J. Appl. Bacteriol. Symp. Suppl.,* 49S–53S.

GÖTZ, F., SCHUHMACHER, B. (1987), Improvements of protoplast transformation in *Staphylococcus carnosus, FEMS Microbiol. Lett.* **40,** 285–289.

GÖTZ, F., KREUTZ, B., SCHLEIFER, K. H. (1983), Protoplast transformation of *Staphylococcus carnosus* by plasmid DNA, *Mol. Gen. Genet.* **189,** 340–342.

GÖTZ, F., POPP, F., KORN, E., SCHLEIFER, K. H. (1985), Complete nucleotide sequence of the lipase gene from *Staphylococcus hyicus,* cloned in *Staphylococcus carnosus, Nucleic Acids Res.* **13,** 5895–5906.

GRAVES, L. M., SWITZER, R. L. (1990), Aspartokinase III, a new isozyme in *Bacillus subtilis* 168, *J. Bacteriol.* **172,** 218–223.

GRAY, G., SELZER, G., BUELL, G., SHAW, P., ESCANAZ, S., HOFER, S., VOEGELI, P., THOMPSON, C. J. (1984), Synthesis of bovine growth hormone by *Streptomyces lividans, Gene* **32,** 21–30.

GRUSS, A., EHRLICH, S. D. (1989), The family of highly interrelated single-stranded deoxyribonucleic acid plasmids, *Microbiol. Rev.* **53,** 231–241.

GUERRERO, C., MATEOS, L. M., MALUMBRES, M., MARTÍN, J. F. (1992), The bleomycin resistance gene of transposon Tn*5* is an excellent marker for transformation of corynebacteria, *Appl. Microbiol. Biotechnol.* **36,** 759–762.

HAHN, D. R., SOLENBERG, P. J., BALTZ, R. H.(1991a), Tn*5099,* a *xylE* promoter probe transposon for *Streptomyces* spp., *J. Bacteriol.* **173,** 5573–5577.

HAHN, D. R., SOLENBERG, P. J. MCHENNEY, M. A., BALTZ, R. H. (1991b), Transposition and transduction of plasmid DNA in *Streptomyces* spp., *J. Ind. Microbiol.* **7,** 229–234.

HAHN, D. R., MCHENNEY, M. A., BALTZ, R. H. (1991c), Properties of the streptomycete temperate bacteriophage FP43, *J. Bacteriol.* **173,** 3770–3775.

HAIMA, P., BRON, S., VENEMA, G. (1987), The effect of restriction on shotgun cloning and plasmid stability in *Bacillus subtilis* Marburg, *Mol. Gen. Genet.* **209,** 335–342.

HAIMA, P., VAN SINDEREN, D., BRON, S., VENEMA, G. (1990), An improved β-galactosidase α-complementation system for molecular cloning in *Bacillus subtilis, Gene* **93,** 41–47.

HARRIS, J. E., CHATER, K. F., BRUTON, C. J., PIRET, J. M. (1983), The restriction mapping of c gene deletions in *Streptomyces* bacteriophage ϕC31 and their use in cloning vector development, *Gene* **22,** 167–174.

HAYNES, J. A., BRITZ, M. L. (1989), Electrotransformation of *Brevibacterium lactofermentum* and *Corynebacterium glutamicum:* growth in tween 80 increases transformation frequencies, *FEMS Microbiol. Lett.* **61,** 329–334.

HAYNES, J. A., BRITZ, M. L. (1990), The effect of growth conditions of *Corynebacterium glutamicum* on the transformation frequency obtained by electroporation, *J. Gen. Microbiol.* **136,** 255–263.

HEIKINHEIMO, R., HEMILÄ, H., PAKKANEN, R., PALVA, I. (1991), Production of methylesterase from *Erwinia chrysanthemi* B374 in *Bacillus subtilis, Appl. Microbiol. Biotechnol.* **35,** 51–55.

HEMILÄ, H., GLODE, L. M., PALVA, I. (1989), Production of diphtheria toxin CRM228 in *B. subtilis, FEMS Microbiol. Lett.* **65,** 193–198.

HENNER, D. J. (1990), Inducible expression of regulatory genes in *Bacillus subtilis, Methods Enzymol.* **185,** 223–228.

HILLEMANN, D., PÜHLER, A. W., WOHLLEBEN, W. (1991), Gene disruption and gene replacement in *Streptomyces* via single stranded DNA transformation of integration vectors, *Nucleic Acids Res.* **19,** 727–731.

HOCH, J. A. (1991), Genetic analysis in *Bacillus subtilis, Methods Enzymol.* **204,** 305–321.

HÖFTE, H., WHITELY, H. R. (1989), Insecticidal crystal proteins of *Bacillus thuringiensis, Microbiol. Rev.* **53,** 242–255.

HONG, S.-W., KITO, M., BEPPU, T., HORINOUCHI, S. (1991), Phosphorylation of the AfsR product, a global regulatory protein for secondary-metabolite formation in *Streptomyces coelicolor* A3(2), *J. Bacteriol.* **173** 2311–2318.

HOPWOOD, D. A. (1967), Genetic analysis and genome structure in *Streptomyces coelicolor, Bacteriol. Rev.* **31,** 373–402.

HOPWOOD, D. A. (1988), Towards an understanding of gene switching in *Streptomyces,* the basis of sporulation and antibiotic production, *Proc. R. Soc. London* **B235,** 121–138.

HOPWOOD, D. A. (1989), Antibiotics: opportunities for genetic manipulation, *Phil. Trans. R. Soc. London* **324,** 549–562.

HOPWOOD, D. A., SHERMAN, D. H. (1990), Molecular genetics of polyketides and its comparison to fatty acid biosynthesis, *Annu. Rev. Genet.* **24,** 37–66.

HOPWOOD, D. A., CHATER, K. F., DOWDING, J. E., VIVIAN, A. (1973), Advances in *Streptomyces coelicolor* genetics, *Bacteriol. Rev.* **37,** 371–405.

HOPWOOD, D. A., BIBB, M. J., CHATER, K. F., KIESER, T., BRUTON, C. J., KIESER, H. M., LYDIATE, D. J., SMITH, C. P., WARD, J. M., SCHREMPF, H. (1985a), Genetic *Manipulation of Streptomyces: A Laboratory Manual,* Norwich: John Innes Foundation.

HOPWOOD, D. A., MALPARTIDA, F., KIESER, H. M., IKEDA, H., DUNCAN, J., FUJII, I., RUDD, B. A. M., FLOSS, H. G., OMURA, S. (1985b), Production of "hybrid" antibiotics by genetic engineering, *Nature* **314,** 642–644.

HOPWOOD, D. A., BIBB, M. J., CHATER, K. F., JANSSEN, G. R., MALPARTIDA, F., SMITH, C. P. (1986a), Regulation of gene expression in antibiotic-producing *Streptomyces,* in: *Regulation of Gene Expression 25 Years on* (BOOTH, I. R., HIGGINS, C. F., Eds.), pp.251–276, Cambridge: Cambridge University Press.

HOPWOOD, D. A., KIESER, T., LYDIATE, D. J., BIBB, M. J. (1986b), *Streptomyces* plasmids: Their biology and use as cloning vectors, in: *The Bacteria* (QUEENER, S. W., DAY, L. E., Eds.), Vol. IX, pp. 159–230, Orlando: Academic Press.

HOPWOOD, D. A., BIBB. M. J., CHATER, K. F., KIESER, T. (1987), Plasmid and phage vectors for gene cloning and analysis in *Streptomyces, Methods Enzymol.* **153,** 116–165.

HORINOUCHI, S., BEPPU, T. (1984), Production in large quantities of actinorhodin and undecylprodigiosin induced by *afsB* in *Streptomyces lividans, Agric. Biol. Chem.* **48,** 2131–2133.

HORINOUCHI, S., WEISBLUM, B. (1982), Nucleotide sequence and functional map of pE194, a plasmid that specifies inducible resistance to macrolide, lincosamide, and streptogramin type B antibiotics, *J. Bacteriol.* **150,** 804–814.

HORINOUCHI, S., HARA, O., BEPPU, T. (1983), Cloning of a pleiotropic gene that positively controls biosynthesis of A-factor, actinorhodin, and prodigiosin in *Streptomyces coelicolor* A3(2) and *Streptomyces lividans, J. Bacteriol.* **155,** 1238–1248.

HORINOUCHI, S., SUZUKI, H., BEPPU, T. (1986), Nucleotide sequence of afsB, a pleiotropic gene involved in secondary metabolism in *Streptomyces coelicolor* A3(2) and *"Streptomyces lividans", J. Bacteriol.* **168,** 257–269.

HORINOUCHI, S., MALPARTIDA, F., HOPWOOD, D. A.,BEPPU, T. (1989a), afsB stimulates transcription of the actinorhodin biosynthetic pathway in *Streptomyces coelicolor* A3(2) and *Streptomyces lividans, Mol. Gen. Genet.* **215,** 355–357.

HORINOUCHI, S., SUZUKI, H., NISHIYAMA, M., BEPPU, T. (1989b), Nucleotide sequence and transcriptional analysis of the *Streptomyces griseus* gene (*afsA*), responsible for A-factor biosynthesis, *J. Bacteriol.* **171,** 1206–1210.

HORINOUCHI, S., KITO, M., NISHIYAMA, M., FURUYA, K., HONG, S.-K., MIYAKE, K., BEPPU, T. (1990), Primary structure of AfsR, a global regulatory protein for secondary metabolite formation in *Streptomyces coelicolor* A3(2), *Gene* **95,** 49–56.

HUTCHINSON, C. R. (1988), Prospects for the discovery of new (hybrid) antibiotics by genetic engineering of antibiotic-producing bacteria, *Med. Res. Rev.* **8,** 557–567.

HUTCHINSON, C. R., BORELL, C. W., OTTEN, S. L., STUTZMAN-ENGWALL, K. J., WANG, Y.-G. (1989), Drug discovery and development through the genetic engineering of antibiotic-producing microorganisms, *J. Med. Chem.* **32,** 929–937.

IORDANESCU, S. (1976), Three distinct plasmids originating in the same *Staphylococcus aureus* strain, *Arch. Roum. Pathol. Exp. Microbiol.* **35,** 111–118.

IORDANESCU, S., SURDEANU, M., LATTA, P. D., NOVICK, R. (1978), Incompatibility and molecular relationship between small staphylococcal plasmids carrying the same resistance marker, *Plasmid* **1,** 468–479.

ISHINO, S., MIZUKAMI, T., YAMAGUCHI, K., KATSUMATA, R., ARAKI, K. (1987), Nucleotide sequence of the meso-diaminopimelate D-dehydrogenase gene from *Corynebacterium glutamicum, Nucleic Acids Res.* **15,** 3917.

ITAYA, M., TANAKA, T. (1991), Complete physical map of the *Bacillus subtilis* 168 chromosome constructed by a gene-directed mutagenesis method, *J. Mol. Biol.* **220,** 631–648.

JANNIERE, L., EHRLICH, S. D. (1987), Recombination between short repeat sequences is more frequent in plasmids than in the chromosome of *Bacillus subtilis, Mol. Gen. Genet.* **210,** 116–121.

JANNIERE, L., BRUAND, C., EHRLICH, S. D. (1990), Structurally stable *Bacillus subtilis* cloning vectors, *Gene* **87,** 53–61.

JANSSEN, G. R., WARD, J. M., BIBB, M. J. (1989), Unusual transcriptional and translational features in the aminoglycoside phosphotransferase gene (*aph*) from *Streptomyces fradiae, Genes Dev.* **3,** 415–429.

JONES, D. T., WOODS, D. R. (1986), Acetone-butanol fermentation revisited, *Microbiol. Rev.* **50,** 484-524.

JONES, G. H., HOPWOOD, D. A. (1984), Molecular cloning and expression of the phenoxazinone synthase gene from *Streptomyces antibioticus, J. Biol. Chem.* **259,** 14151-14157.

KALINOWSKI, J., BACHMANN, B., THIERBACH, G., PÜHLER, A. (1990), Aspartokinase genes *lysCα* and *lysCβ* overlap and are adjacent to the aspartate β-semialdehyde dehydrogenase gene *asd* in *Corynebacterium glutamicum, Mol. Gen. Genet.* **224,** 317-324.

KALINOWSKI, J., CREMER, J., BACHMANN, B., EGGELING, L., SAHM, H., PÜHLER, A. (1991), Genetic and biochemical analysis of the aspartokinase from *Corynebacterium glutamicum, Mol. Microbiol.* **5,** 1197-1204.

KÄMPFER, P., KROPPENSTEDT, R. M., DOTT, W. (1991), A numerical classification of the genera *Streptomyces* and *Streptoverticillium* using miniaturized physiological tests, *J. Gen. Microbiol.* **137,** 1831-1891.

KARASAWA, M., TOSAKA, O., IKEDA, S., YOSHII, H. (1986), Application of protoplast fusion to the development of L-threonine and L-lysine producers, *Agric. Biol. Chem.* **50,** 339-346.

KATSUMATA, R., OZAKI, A., OKA, T., FURUYA, A. (1984), Protoplast transformation of glutamate-producing bacteria with plasmid DNA, *J. Bacteriol.* **159,** 306-311.

KATZ, E., THOMPSON, C. J., HOPWOOD, D. A. (1983), Cloning and expression of the tyrosinase gene from *Streptomyces antibioticus* in *Streptomyces lividans, J. Gen. Microbiol.* **129,** 2703-2714.

KAWAMURA, F., DOI, R. H. (1984), Construction of a *Bacillus subtilis* double mutant deficient in extracellular alkaline and neutral protease, *J. Bacteriol.* **160,** 442-444.

KELLER, G., SCHLEIFER, K. H., GÖTZ, F. (1983), Construction and characterization of plasmid vectors for cloning in *Staphylococcus aureus* and *Staphylococcus carnosus, Plasmid* **10,** 270-278.

KHAN, S. A., NOVICK, R. P. (1983), Complete nucleotide sequence of pT181, a tetracycline-resistance plasmid from *Staphylococcus aureus, Plasmid* **10,** 251-259.

KIESER, T., HOPWOOD, D, A. (1991), Genetic manipulation of *Streptomyces:* integration vectors and gene replacement, *Methods Enzymol.* **204,** 430-458.

KIESER, T., HOPWOOD, D. A., WRIGHT, H. M., THOMPSON, C. J. (1982), pIJ101, a multi-copy broad host-range *Streptomyces* plasmid: functional analysis and development of DNA cloning vectors, *Mol. Gen. Genet.* **185,** 223-238.

KIESER, H. M., HENDERSON, D. J., CHEN, C. W., HOPWOOD, D. A. (1989), A mutation of *Streptomyces lividans* which prevents intraplasmid recombination has no effect on chromosomal recombination, *Mol. Gen. Genet.* **220,** 60-64.

KIESER, H. M., KIESER, T., HOPWOOD, D. A. (1992), A combined genetic and physical map of the *Streptomyces coelicolor* A3(2) chromosome, *J. Bacteriol.* **174,** 5496-5507.

KNAUF, H. J., VOGEL, R. F., HAMMES, W. P. (1992), Cloning, sequence, and phenotypic expression of *katA,* which encodes the catalase of *Lactobacillus sake* LTH677, *Appl. Environ. Microbiol.* **58,** 832-839.

KOBLER, L., SCHWERTFIRM, G., SCHMIEGER, H., BOLOTIN, A., SLADKOVA, I. (1991), Construction and transduction of a shuttle vector bearing the *cos* site of *Streptomyces* phage C31 and determination of its cohesive ends, *FEMS Microbiol. Lett.* **18,** 347-354.

KOLLER, K.-P., RIESS, G. (1989), Heterologous expression of the alpha-amylase inhibitor gene cloned from an amplified genomic sequence of *Streptomyces tendae, J. Bacteriol.* **171,** 4953-4957.

KOLLER, K.-P., RIESS G., SAUBER, K., UHLMANN, E., WALLMEIER, H. (1989a), Recombinant *Streptomyces lividans* secretes a fusion protein of tendamistat and proinsulin, *Bio/Technology* **7,** 1055-1059.

KOLLER, K.-P., RIESS, G., SAUBER, K., VERTESY, L., UHLMANN, E., WALLMEIER, H. (1989b), The tendamistat expression-secretion system: Synthesis of proinsulin fusion proteins with *Streptomyces lividans,* in: *Genetics and Product Formation in Streptomyces* (BAUMBERG, S., KRÜGEL, H., NOACK, D., Eds.), pp. 227-233, London: Plenum Press.

KONO, M., SASATSU, M., AOKI, T. (1983), R plasmids in *Corynebacterium xerosis* strains, *Antimicrob. Agents Chemother.* **23,** 506-508.

KOVACEVIC, S., VEAL, L. E., HSIUNG, H. M., MILLER, J. R. (1985), Secretion of staphylococcal nuclease by *Bacillus subtilis, J. Bacteriol.* **162,** 521-528.

KREUTZ, B., GÖTZ, F. (1984), Construction of *Staphylococcus* plasmid vector pCA43 conferring resistance to chloramphenicol, arsenate, arsenite and antimony, *Gene* **32,** 301-304.

KUHSTOSS, S., RAO. R. N. (1991a), Analysis of the integration function of the streptomycete bacteriophage C31, *J. Mol. Biol.* **222,** 897-908.

KUHSTOSS, S., RAO, R. N. (1991b), A thiostrepton-inducible expression vector for use in *Streptomyces* spp., *Gene* **103,** 97-99.

KUHSTOSS, S., RICHARDSON, M. A., NAGARAJA RAO, R. (1989), Plasmid cloning vectors that integrate site-specifically in *Streptomyces* spp. *Gene* **97**, 143–146.

KURUSU, Y., SATOH, Y., INUIE, M., KOHAMA, K., KOBAYASHI, M., TERESAWA, M., YUKAWA, H. (1991), Identification of plasmid partition function in coryneform bacteria, *Appl. Environ. Microbiol.* **57**, 759-764.

LABES, G., SIMON, R., WOHLLEBEN, W. (1990), A rapid methold for the analysis of plasmid content and copy number in various Streptomycetes grown on agar plates, *Nucleic Acids Res.* **18**, 2197.

LACEY, R., CHAPRA, I. (1974), Genetic studies of a multiresistant strain of *Staphylococcus aureus, J. Med. Microbiol.* **7**, 285-297.

LAL, R., LAL, S., GRUND, E., EICHENLAUB, R. (1991), Construction of a hybrid plasmid capable of replication in *Amycolatopsis mediterranei, Appl. Environ. Microbiol.* **57**, 665-671.

LANGHAM, C. D., WILLIAMS, S. T., SNEATH, P. H. A., MORTIMER, A. M. (1989), New probability matrices for identification of *Streptomyces, J. Gen, Microbiol.* **135**, 121-133.

LEBLOND, P., DEMUYTER, P., SIMONET, J. M., DECARIS, B. (1990), Genetic instability and hypervariability in *Streptomyces ambofaciens:* towards an understanding of a mechanism of genome plasticity, *Mol. Microbiol.* **4**, 707-714.

LE BOURGEOIS, P., LAUTIER, M., MATA, M., RITZENTHALER, P. (1992), New tools for the physical and genetic mapping of *Lactococcus* strains, *Gene* **111**, 109-114.

LEE, C. W., LUCAS, S., DESOMAZEAUD, M. J. (1985), Phenylalanine and tyrosine catabolism in some cheese coryneform bacteria, *FEMS Microbiol. Lett.* **26**, 201-205.

LEENHOUTS, K. J., KOK, J., VENEMA, G. (1991), Replacement recombination in *Lactococcus lactis, J. Bacteriol* **173**, 4794-4798.

LE GRICE, S. F. J. (1990), Regulated promoter for high level expression of heterologous genes in *Bacillus subtilis, Methods Enzymol.* **185**, 201-214.

LE GRICE, S. F. J., BEUCK, V., MOUS, J. (1987), Expression of biologically active human T-cell lymphotropic virus type III reverse transcriptase in *Bacillus subtilis, Gene* **55**, 95-103.

LESKIW, B. K., LAWLOR, E. J., FERNANDES-ABALOS, J. M., CHATER, K. F. (1991 a), TTA codons in some genes prevent their expression in a class of developmental, antibiotic-negative, *Streptomyces mutants. PNAS* **88**, 2461-2465.

LESKIW, B. K., BIBB, M. J., CHATER, K. F. (1991 b), The use of a rare codon specifically during development? *Mol. Microbiol.* **5**, 2861-2867.

LICHENSTEIN, H., BRAWNER, M. E., MILES, L. M., MEYERS, C. A., YOUNG, P. R., SIMON, P. L., ECKHARDT, T. (1988), Secretion of the interleukin-1β and *Escherichia coli* galactokinase by *Streptomyces lividans, J. Bacteriol.* **170**, 3924-3929.

LIEBL, W., SINSKEY, A. J. (1988), Molecular cloning and nucleotide sequence of a gene involved in the production of extracellular DNAase by *Corynebacterium glutamicum,* in: *Genetics and Biotechnology of Bacilli,* (GANESAN, T., HOCH, J. A., Eds.), Vol. 2, pp. 383-388, San Diego: Academic Press.

LIEBL, W., SCHLEIFER, K.-H., SINSKEY, A. J. (1989), Secretion of heterologous proteins by *Corynebacterium glutamicum,* in: *Genetic Transformation and Expression* (BUTLER, L. O., HARWOOD, C., MOSELEY, B. E. B., Eds.), pp. 553-559, Andover: Intercept Ltd.

LIEBL, W., EHRMANN, M., LUDWIG, W., SCHLEIFER, K.-H. (1991), Transfer of *Brevibacterium divaricatum* DSM 20297T, *"Brevibacterium flavum"* DSM20411, *"Brevibacterium lactofermentum"* DSM20412 and DSM1412, and *Corynebacterium lilium* DSM20137 to *Corynebacterium glutamicum* and their distinction by rRNA gene restriction patterns, *Int. J. Syst. Bacteriol.* **41**, 255-260.

LIEBL, W., SINSKEY, A. J., SCHLEIFER, K.-H. (1992), Expression, secretion, and processing of staphylococcal nuclease by *Corynebacterium glutamicum, J. Bacteriol.* **174**, 1854-1861.

LIN, W.-J., JOHNSON, E. A. (1991), Transposon Tn*916* mutagenesis in *Clostridium botulinum, Appl. Environ. Mcrobiol.* **57**, 2946-2950.

LIRAS, P., ASTURIAS, J. A., MARTÍN, J. F. (1990), Phosphate control sequences involved in transcriptional regulation of antibiotic biosynthesis, *TIBTECH* **6**, 184-189.

LOMOVSKAYA, N. D., CHATER, K. F., MKRTUMIAN, N. M. (1980), Genetics and molecular biology of *Streptomyces* bacteriophages, *Microbiol. Rev.* **44**, 206-229.

LOSICK, R., YOUNGMAN, P., PIGGOT, P. (1986), Genetics of endospore formation in *Bacillus subtilis, Annu. Rev. Genet,* **20**, 625-669.

LUCHANSKY, J. B., MURIANA, P. M., KLAENHAMMER, T. R. (1988), Application of electroporation for transfer of plasmid DNA to *Lactobacillus, Lactococcus, Leuconostoc, Listeria, Pediococcus, Bacillus, Staphylococcus, Enterococcus* and *Propionibacterium, Mol. Microbiol.* **2**, 637-646.

LUCHANSKY, J. B., KLEEMANN, E. G., RAYA, R. R., KLAENHAMMER, T. R. (1989), Genetic transfer systems for delivery of plasmid DNA to *Lactobacillus acidophilus* ADH: conjugation, electroporation, and transduction., *J. Dairy Sci.* **72**, 1408-1417.

MACNEIL, D. J. (1987), Introduction of plasmid DNA into *Streptomyces lividans* by electroporation, *FEMS Microbiol. Lett.* **42**, 239–244.

MACNEIL, D. J. (1988), Characterization of a unique methyl-specific restriction system in *Streptomyces avermitilis, J. Bacteriol.* **170**, 5607–5612.

MALPARTIDA, F., HOPWOOD, D. A. (1984), Molecular cloning of the whole biosynthetic pathway of a *Streptomyces* antibiotic and its expression in a heterologous host, *Nature* **309**, 462–464.

MALPARTIDA, F., HOPWOOD, D. A. (1986), Physical and genetic characterization of the gene cluster for the antibiotic actinorhodin in *Streptomyces coelicolor* A3(2), *Mol. Gen. Genet.* **205**, 66–73.

MALPARTIDA, F., HALLAM, S. E., KIESER, H. M., MOTAMEDI, H., HUTCHINSON, C. R., BUTLER, M. J., SUGDEN, D. A., WARREN, M., MCKILLOP, C., BAILEY, C. R., HUMPHREYS, G. O., HOPWOOD, D. A. (1987), Homology between *Streptomyces* genes coding for synthesis of different polyketides used to clone antibiotic biosynthetic genes, *Nature* **325**, 818–821.

MANSOURI, K., PIEPERSBERG, W. (1991), Genetics of streptomycin production in *Streptomyces griseus:* nucleotide sequence of five genes. *strFGHIK,* including a phosphatase gene, *Mol. Gen. Genet.* **228**, 459–469.

MARTIN, C., MAZODIER, P., MEDIOLA, M. V., GICQUEL, B., SMOKVINA, T., THOMPSON, C. J., DAVIES, J. (1991), Site specific integration of the *Streptomyces* plasmid pSAM2 in *Mycobacterium smegmatis, Mol. Microbiol.* **5**, 2499–2502.

MARTÍN, J. F. (1989), Molecular genetics of amino acid-producing Corynebacteria, in: *Society for General Microbiology Symposium* (BAUMBERG, S., HUNTER, I., RHODES, M., Eds.),Vol. 44, pp. 25–59, Cambridge: University Press.

MARTÍN, J. F., LIRAS, P. (1989), Organization and expression of genes involved in the biosynthesis of antibiotics and other secondary metabolites, *Annu. Rev. Microbiol.* **43**, 173–206.

MARTÍN, J. F., CADENAS, R. F., MALUMBRES, M., MATEOS, L. M., GUERRERO, C., GIL, J. A. (1990), Construction and utilization of promoter-probe and expression vectors in corynebacteria. Characterization of corynebacterial promoters, in: *Proc. 6th Int. Symp. Genetics of Industrial Microorganisms* (HESLOT, H., DAVIES, J., FLORENT, J., BOBICHON, L., DURANT, G., PENASSE, L., Eds.), pp. 283–292, Strasbourg: Société Française de Microbiologie.

MATSUSHIMA, P., COX, K. L., BALTZ, R. H. (1987), Highly transformable mutants of *Streptomyces fradiae* defective in several restriction systems, *Mol. Gen. Genet.* **206**, 393–400.

MAZODIER, P., PETTER, R., THOMPSON, C. (1989), Intergeneric conjugation between *Escherichia coli* and *Streptomyces* species, *J. Bacteriol.* **171**, 3583–3585.

MAZODIER, P., THOMPSON, C., BOCCARD, F. (1990), The chromosomal integration site of the *Streptomyces* element pSAM2 overlaps a putative tRNA gene conserved among actinomycetes, *Mol. Gen. Genet.* **222**, 431–434.

MCHENNEY, M. A., BALTZ, R. H. (1988), Transduction of plasmid DNA in *Streptomyces* spp. and related genera by bacteriophage FP43, *J. Bacteriol.* **170**, 2276–2282.

MCHENNEY, M. A., BALTZ, R. H. (1991a), Transduction of plasmid DNA containing the *ermE* gene and expression of erythromycin resistance in Streptomycetes, *J. Antibiot.* **44**, 1267–1269.

MCHENNEY, M. A., BALTZ, R. H. (1991b), Transposition of Tn*5096* from a temperature-sensitive transducible plasmid in *Streptomyces* spp., *J. Bacteriol.* **173**, 5578–5581.

MCKENZIE, T., HOSHING, T., TANAKA, T., SUEOKA, N. (1986), The nucleotide sequence of pUB110, some salient features in relation to replication and its regulation, *Plasmid* **15**, 93–103.

MENKEL, E., THIERBACH, G., EGGELING, L., SAHM, H. (1989), Influnece of increased aspartate availability on lysine formation by a recombinant strain of *Corynebacterium glutamicum* and utilization of fumarate, *Appl. Environ. Microbiol.* **55**, 684–688.

MERMELSTEIN, L. D., WELKER, N. E., BENNETT, G. N., PAPOUTSAKIS, E. T. (1992), Expression of cloned homologous fermentative genes in *Clostridium acetobutylicum* ATCC824, *Bio/Technology* **10**, 190–195.

MICHEL, B., NIAUDET, B., EHRLICH, S. D. (1982), Intramolecular recombination during plasmid transformation of *Bacillus subtilis* competent cells, *EMBO J.* **1**, 1565–1571.

MIWA, K., MATSUI, H., TERABE, M., MAKAMORI, S., SANO, K., MOMOSE, H. (1984), Cryptic plasmids in glutamic acid-producing bacteria, *Agric. Biol. Chem.* **48**, 2901–2903.

MIWA, K., MATSUI, K., TERABE, M., ITO, K., ISHIDA, M., TAKAGI, H., NAKAMORI, S., SANO, K. (1985), Construction of novel shuttle vectors and a cosmid vector for the glutamic acid-producing bacteria *Brevibacterium lactofermentum* and *Corynebacterium glutamicum, Gene* **39**, 281–286.

MIYAKE, K., HORINOUCHI, S., YOSHIDA, M., CHIBA, N., MORI, K., NOGAWA, N., MORIKAWA, N., BEPPU, T. (1989), Detection and properties of A-factor-binding protein from *Streptomyces griseus, J. Bacteriol.* **171**, 4298–4302.

Miyake, K., Kuzuyama, T., Horinouchi, S., Beppu, T. (1990), The A-factor-binding protein of *Streptomyces griseus* negatively controls streptomycin production and sporulation, *J. Bacteriol.* **172**, 3003–3008.

Miyashiro, S., Enei, H., Hirose, Y., Udaka, S. (1980), Effect of glycine and L-isoleucine on protein production by *Bacillus brevis* no. 47, *Agric. Biol. Chem.* **44**, 105–112.

Morinaga, Y., Tsuchiya, M., Miwa, K., Sano, K. (1987), Expression of *Escherichia coli* promoters in *Brevibacterium lactofermentum* using the shuttle vector pEB003, *J. Biotechnol.* **5**, 305–312.

Morino, T., Takahashi, H., Saito, H. (1985), Construction and characterization of a cosmid of *Streptomyces lividans, Mol. Gen. Genet.* **198**, 228–233.

Morino, T., Takagi, K., Nakamura, T., Takita, T., Saito, H., Takahashi, H. (1986), Studies of cosmid transduction in *Streptomyces lividans* and *Streptomyces parvulus, Agric. Biol. Chem.***50**, 2493–2497.

Morino, T., Takagi, K.-I., Nakamura, T., Takita, T., Saito, H., Takahashi, H. (1988), Interspecific transfer and expression of melanine gene(s) on cosmids in *Streptomyces* strains, *Appl. Microbiol. Biotechnol.* **27**, 517–520.

Muñoz, A., Pérez-Aranda, A., Barbero, J. L. (1985), Cloning and expression of human interleukin-2 in *Streptomyces lividans* using the *Escherichia coli* consensus promoter, *Biochem. Biophys. Res. Commun.* **133**, 511–519.

Murakami, T. Holt, T. G., Thompson, C. J. (1989), Thiostrepton-induced gene expression in *Streptomyces lividans, J. Bacteriol.* **171**, 1459–1466.

Muth, G., Wohlleben, W., Pühler, A., Wöhner, G., Marquardt, R. (1986), Farbmarker für Klonierungen in *S. lividans, Eur. Patent Application* 0257416.

Muth, G., Wohlleben, W., Pühler, A. (1988), The minimal replicon of the *Streptomyces ghanaensis* plasmid pSG5 identified by subcloning and Tn*5* mutagenesis, *Mol. Gen. Genet.* **211**, 424–429.

Muth, G., Nußbaumer, B., Wohlleben, W., Pühler, A. (1989), A vector system with temperature-sensitive replication for gene disruption and mutational cloning in Streptomycetes, *Mol. Gen. Genet.* **219**, 341–348.

Naglich, J. G., Andrews, R. E. (1988), In 916 – dependent conjugal transfer of pC194 and pUB110 from *Bacillus subtilis* into *Bacillus thuringiensis* subsp. *israelensis, Plasmid* **20**, 113–126.

Nair, S., Steyn, L. M. (1991), Cloning and expression in *Escherichia coli* of a *recA* homologue from *Mycobacterium tuberculosis, J. Gen. Microbiol.* **137**, 2409–2414.

Nakayama, A., Ando, K., Kawamura, K., Mita, I., Fukazawa, K., Hori, M., Honjo, M., Furutani, Y. (1988), Efficient secretion of the authentic mature human growth hormon by *Bacillus subtilis, J. Biotechnol.* **8**, 123–134.

Narva, K. E., Feitelson, J. S. (1990), Nucleotide sequence and transcriptional analysis of the *redD* locus of *Streptomyces coelicolor* A3(2), *J. Bacteriol.* **172**, 326–333.

Niaudet, B., Janniere, L., Ehrlich, S. D. (1984), Recombination between repeated DNA sequences occurs more often in plasmids than in the chromosome of *Bacillus subtilis, Mol. Gen. Genet.* **197**, 46–54.

Noack, D., Geuther, R., Tonew, M., Breitling, R., Behnke, D. (1988), Expression and secretion of interferon-α1 by *Streptomyces lividans:* use of the staphylokinase signals and amplification of a *neo* gene, *Gene* **68**, 53–62.

Novick, R. P. (1989), Staphylococcal plasmids and their replication, *Annu. Rev. Microbiol.* **43**, 537–565.

Novick, R. P. (1990), Genetic systems in staphylococci, *Methods Enzymol.* **204**, 587–636.

Obata, S., Taguchi, S., Kumagai, I., Miura, K. (1989), Molecular cloning and nucleotide sequence determination of gene encoding *Streptomyces* subtilisin inhibitor (SSI), *J. Biochem.* **105**, 367–371.

Ochi, K. (1987), Metabolic initiation of differentiation and secondary metabolism by *Streptomyces griseus:* significance of the stringent response (ppGpp), and GTP content in relation to A factor, *J. Bacteriol.* **169**, 3608–3616.

Ochi, K. (1990), *Streptomyces* relC mutants with an altered ribosomal protein ST-L11 and genetic analysis of a *Streptomyces griseus relC* mutant, *J. Bacteriol.* **172**, 4008–4016.

Ogata, K., Kinoshita, S., Tsunoda, T., Aida, K. (1976), *Microbial Production of Nucleic Acid Related Substances,* New York: Wiley.

Olson, E. R., Chung, S.-T. (1988), Transposon Tn*4556* of *Streptomyces fradiae:* nucleotide sequence of the ends and the target sites, *J. Bacteriol.* **170**, 1955–1957.

Omer, C. A., Cohen, S. N. (1984), Plasmid formation in *Streptomyces:* excision and integration of the SLP1 replicon at a specific chromosomal site, *Mol. Gen. Genet.* **196**, 429–438.

Osburne, M. S., Craig, R. J., Rothstein, D. M. (1985), Thermoinducible transcription system for *Bacillus subtilis* that utilizes control elements from temperature phage ϕ105, *J. Bacteriol.* **163,** 1101–1108.

OVERBEEKE, N., TERMORSHUIZEN G. H. M., GUISEPPIN, M. L. F., UNDERWOOD, D. R., VERRIPS, C. T. (1990), Secretion of the α-galactosidase from *Cyamopsis tetragonoloba* (guar) by *Bacillus subtilis, Appl. Environ. Microbiol.* **56,** 1429–1434.

OZAKI, A., KATSUMATA, R., OKA, T., FURUYA, A. (1984), Functional expression of the genes of *Escherichia coli* in Gram-positive *Corynebacterium glutamicum, Mol. Gen. Genet.* **196,** 175–178.

PALVA, I., LEHTOVAARA, P., KAARIAINEN, L., SIBAKOV, M., CONTELL, K., SCHEN, C. H., KASHIWAGI, K., WEISSMANN, C. (1983), Secretion of interferon by *Bacillus subtilis, Gene* **22,** 229–235.

PARADIS, F. W., WARREN, R. A. J., KILBURN, D. G., MILLER, JR., R. C. (1987), The expression of *Cellulomonas fimi* cellulase gene in *Brevibacterium lactofermentum, Gene* **61,** 199–206.

PERNODET, J.-L., SIMONET, J.-M., GUERINEAU, M. (1984), Plasmids in different strains of *Streptomyces ambofaciens:* free and integrated form of plasmid pSAM2, *Mol. Gen. Genet.* **198,** 35–41.

PESCHKE, U., BEUCK, V., BUJARD, H., GENTZ, R., LE GRICE, S. F. J. (1985), Efficient utilization of *Escherichia coli* transcriptional signals in *Bacillus subtilis, J. Mol. Biol.* **186,** 547–555.

PETIT, M.-A., BRUAND, C., JANNIERE, L., EHRLICH, S. D. (1990a), Tn*10*-derived transposons active in *Bacillus subtilis, J. Bacteriol.* **172,** 6736–6740.

PETIT, M.-A., JOLIFF, G., MESAS, J. M., KLIER, A., RAPOPORT, G., EHRLICH, S. D. (1990b), Hypersecretion of a cellulase from *Clostridium thermocellum* in *Bacillus subtilis* by induction of chromosomal DNA amplification, *Bio/Technology* **8,** 559–562.

PETZEL, J. P., MCKAY, L. L. (1992), Molecular characterization of the integration of the lactose plasmid from *Lactococcus lactis* subsp. *cremoris* SK11 into the chromosome of *L. Lactis* subsp. *lactis, Appl. Environ. Microbiol.* **58,** 125–131.

PIGGOT, P. J. (1989), Genetic map of *Bacillus subtilis* 168, in: *The Bacterial Chromosome* (DRLICA, K., RILEY, M., Eds.), pp. 107–145, Washington: American Society for Microbiology.

POLZIN, K. M., MCKAY, L. L. (1992), Development of a lactococcal integration vector by using IS*981* and a temperature-sensitive lactococcal replication region, *Appl. Environ. Microbiol.* **58,** 476–484.

PULIDO, D., VARA, A., JIMÉNEZ, A. (1986), Cloning and expression in biologically active form of the gene for human interferon α2 in *Streptomyces lividans, Gene* **45,** 167–174.

RAO, R. N., RICHARDSON, M. A., KUHSTOSS, S. (1987), Cosmid shuttle vectors for gene cloning and analysis of *Streptomyces* DNA, *Methods Enzymol.* **153,** 166–198.

RAPOPORT, G., KLIER, A. (1990), Gene expression using *Bacillus, Curr. Opin. Biotechnol.* **1,** 21–27.

RAUCH, P. J. G., DE VOS, W. M. (1992), Characterization of the novel nisin-sucrose conjugative transposon Tn*5276* and its insertion in *Lactococcus lactis, J. Bacteriol.* **174,** 1280–1287.

RAYA, R. R., KLAENHAMMER, T. R., (1992), High-frequency plasmid transduction by *Lactobacillus gasseri* bacteriophage ϕadh, *Appl. Environ. Microbiol.* **58,** 187–193.

REYES, O., GUYONVARCH, A., BONAMY, C., SALTI, V., DAVID, F., LEBLON, G. (1991), 'Integron'-bearing vectors: a method suitable for stable chromosomal integration in highly restrictive corynebacteria, *Gene* **107,** 61–68.

ROLS, M. P., BANDIERA, P., LANEELLE, G., TEISSIE, J. (1987), Obtaining of viable hybrids between Corynebacteria by electrofusion, *Stud. Biophys.* **119,** 37–40.

ROMERO, D. A., KLAENHAMMER, T. R. (1991), Construction of an IS*946*-based composite transposon in *Lactococcus lactis* subsp. *lactis, J. Bacteriol.* **173,** 7599–7606.

ROTH, M., MÜLLER, G., NEIGENFIND, M., HOFFMEIER, C., GEUTHER, R. (1991), Partitioning of plasmids in *Streptomyces:* Segregation in continuous culture of a vector with temperature-sensitive replication, in: *Genetics and Product Formation in Streptomyces,* (BAUMBERG, S., KRÜGEL, H., NOACK, D., Eds.), pp. 305–313, London: Plenum Press.

RYGUS, T., HILLEN, W. (1991), Inducible high-level expression of heterologous genes in *Bacillus megaterium* using the regulatory elements of the xylose-utilization operon, *Appl. Microbiol. Biotechnol.* **35,** 549–599.

SANCHEZ, F., PENALVA, M. A., PATINO, C., RUBIO, V. (1986), An efficient method for the introduction of viral DNA into *Brevibacterium lactofermentum* protoplasts, *J. Gen. Microbiol.* **132,** 1767–1770.

SANDOVAL, H., AGUILAR, A., PANIAGUA, C., MARTÍN. J. F. (1984), Isolation and physical characterization of plasmid pCC1 from *Corynebacterium callunae* and construction of hybrid derivatives, *Appl. Microbiol. Biotechnol.* **19,** 409–413.

SANDOVAL, H., DEL REAL, G., MATEOS, L. M., AGUILAR, A., MARTÍN, J. F. (1985), Screening of plasmids in non-pathogenic corynebacteria, *FEMS Microbiol. Lett.* **27,** 93–98.

SANTAMARIA, R., GIL, J. A., MESAS, J. M., MARTÍN, J. F. (1984), Characterization of an endogenous plasmid and development of cloning vectors and a transformation system in *Brevibacterium lactofermentum, J. Gen. Microbiol.* **130,** 2237–2246.

SANTAMARIA, R. I., GIL, J. A., MARTÍN, J. F. (1985), High-frequency transformation of *Brevibacterium lactofermentum* protoplasts by plasmid DNA, *J. Bacteriol.* **162,** 463–467.

SARIS, P., TAIRA, S., AIRAKSINEN, U., PALVA, A., SARVAS, M., PALVA, I., RUNEBERG-NYMAN, K. (1990), Expression and secretion of pertussis toxin subunits in *Bacillus subtilis, FEMS Microbiol. Lett.* **68,** 143–148.

SATOH, Y., HATAKEYAMA, K., KOHAMA, K., KURUSU, Y., YUKAWA, H. (1990), Electrotransformation of intact cells of *Brevibacterium flavum* MJ-223, *J. Ind. Microbiol.* **5,** 159–166.

SCHAFER, A., KALINOWSKI, J., SIMON, R., SEEP-FELDHAUS, A.-H., PÜHLER, A. (1990), High-frequency conjugal plasmid transfer from Gram-negative *Escherichia coli* to various Gram-positive coryneform bacteria, *J. Bacteriol.* **172,** 1663–1666.

SCHAUER, A. T., NELSON, A. D., DANIEL, J. B. (1991), Tn4563 transposition in *Streptomyces coelicolor* and its application to isolation of new morphological mutants, *J. Bacteriol.* **173,** 5060–5067.

SCHEIN, C. H., KASHIWAGI, K., FUJISAWA, A., WEISSMANN, C. (1986), Secretion of mature IFN-α2 and accumulation of uncleaved precursor by *Bacillus subtilis* transformed with a hybrid (α-amylase signal sequence-IFN-α2 gene, *Bio/Technology* **4,** 719–725.

SCHMITT-JOHN, T., ENGELS, J. W. (1992), Promoter constructions for efficient secretion expression in *Streptomyces lividans, Appl. Microbiol. Biotechnol.* **36,** 493–498.

SCHOLZ, P., HARING, V., WITTMANN-LIEBOLD, B., ASHMAN, K., BAGDASARIAN, M., SCHERZINGER, E. (1989), Complete nucleotide sequence and gene organization of broad-host-range plasmid RSF1010, *Gene* **75,** 271–288.

SCHREMPF, H. (1991), Genetic instability in *Streptomyces* in: *Genetics and Product Formation in Streptomyces,* (BAUMBERG, S., KRÜGEL, H., NOACK, D. Eds.), pp. 245–252, London: Plenum Press.

SCHREMPF, H., BUJARD, H., HOPWOOD, D. A., GOEBEL, W. (1975), Isolation of covalently closed circular deoxyribonucleic acid from *Streptomyces coelicolor, J. Bacteriol.* **121,** 416–421.

SCHRUMPF, B., SCHWARZER, A., KALINOWSKI, J., PÜHLER, A., EGGELING, L., SAHM, H. (1991), A functionally split pathway for lysine synthesis in *Corynebacterium glutamicum, J. Bacteriol.* **173,** 4510–4516.

SCHWARZER, A., PÜHLER, A. (1991), Manipulation of *Corynebacterium glutamicum* by gene disruption and replacement, *Bio/Technology* **90,** 84–87.

SCOTT, J. R., KIRCHMAN, P. A., CAPARON, M. G. (1988), An intermediate in transposition of the conjugative transposon Tn*916*, *PNAS* **85,** 4809–4813.

SEEP-FELDHAUS, A.-H., KALINOWSKI, J., PÜHLER, A. (1991), Molecular analysis of the *Corynebacterium glutamicum lysI* gene involved in lysine uptake, *Mol. Microbiol.* **5,** 2995–3005.

SEILER, H. (1983), Identification key for coryneform bacteria derived by numerical taxonomic studies, *J. Gen. Microbiol.* **129,** 1433–1471.

SERWOLD-DAVIS, T. M., GROMAN, N., RABIN, M. (1987), Transformation of *Corynebacterium diphtheriae, Corynebacterium ulcerans, Corynebacterium glutamicum,* and *Escherichia coli* with the *C. diphtheriae* plasmid pNG2, *PNAS* **84,** 4464–4968.

SERWOLD-DAVIS, T. M., GROMAN, N., KAO, C. C. (1990), Localization of an origin of replication in *Corynebacterium diphtheriae* broad host range plasmid pNG2 that also functions in *Escherichia coli, FEMS Microbiol. Lett.* **66,** 119–124.

SHEARMAN, C. A., JUDY, K., GASSON, M. J. (1992), Autolytic *Lactococcus lactis* expressing a lactococcal bacteriophage lysine gene, *Bio/Technology* **10,** 196–199.

SHERMAN, D. H., MALPARTIDA, F., BIBB, M. J., KIESER, H. M., BIBB, M. J., HOPWOOD, D. A. (1989), Structure and deduced function of the granaticin-producing polyketide synthase gene cluster of *Streptomyces violaceoruber* Tü22, *EMBO J.* **8,** 2717–2725.

SHIIO, I. (1982), Metabolic regulation and over-production of amino acids, in: *Overproduction of Metabolic Products* (KRUMPHANZL, V., SIKYTA, B., VANEK, Z., Eds.), pp. 463–472, London: Academic Press.

SHIRAHAMA, T., FURUMAI, T., OKANISHI, M. (1981), A modified regeneration method for Streptomycete protoplasts, *Agric. Biol. Chem.* **45,** 1271–1273.

SIEMIENIAK, D. R., SLIGTHOM, J. L., CHUNG, S.-T. (1990), Nucleotide sequence of *Streptomyces fradiae* transposable element Tn*4556*: a class-II transposon related to Tn*3*, *Gene* **86,** 1–9.

SIMON, R., PRIEFER, U., PÜHLER, A. (1983), A broad host range mobilization system for *in vivo* genetic engineering: transposon mutagenesis in

Gram negative bacteria, *Bio/Technology* **1**, 784–791.

SLOMA, A., ALLY, A., ALLY, D., PERO, J. (1988),Gene encoding a minor extracellular protease in *Bacillus subtilis, J. Bacteriol.* **170**, 5557–5563.

SLOMA, A., RUDOLPH, C. F., SULLIVAN, B. J., THERIAULT, K. A., ALLY, D., PERO, J. (1990a), Gene encoding a novel extracellular metalloprotease in *Bacillus subtilis, J. Bacteriol.* **172**, 1024–1029.

SLOMA, A., RUFO, G. A., RUDOLPH, C. F., SULLIVAN, B. J., THERIAULT, K. A., PERO, J. (1990b), Bacillopeptidase F of *Bacillus subtilis.* purification of the protein and cloning of the gene, *J. Bacteriol.* **172**, 1470–1477.

SLOMA, A., RUFO, G. A., THERIAULT, K. A., DWYNER, M., WILSON, S. W., PERO, J. (1991), Cloning and characterization of the gene for an additional extracellular serine protease of *Bacillus subtilis, J. Bacteriol.* **173**, 6889–6895.

SMID, E. J., POOLMAN, B., KONINGS, W. N. (1991), Casein utilization by lactococci, *Appl. Environ. Microbiol.* **57**, 2447–2452.

SMITH, M. D., FLICKINGER, J. L., LINEBERGER, D. W., SCHMIDT, B. (1986), Protoplast transformation in coryneform bacteria and introduction of an alpha-amylase gene from *Bacillus amyloliquefaciens* into *Brevibacterium lactofermentum, Appl. Environ. Microbiol.* **51**, 634–639.

SMOKVINA, T., MAZODIER, PHILIPPE, BOCCARD, F., THOMPSON, C. J., GUERINEAU, M. (1990), Construction of a series of pSAM2-based integrative vectors for use in actinomycetes, *Gene* **94**, 53–59.

SOHASKEY, C. D., IM, H., SCHAUER, A. T. (1992), Construction and application of plasmid- and transposon-based promoter-probe vectors for *Streptomyces* spp. that employ a *Vibrio harveyi* luciferase reporter cassette, *J. Bacteriol.* **174**, 367–376.

SOLENBERG, P. J., BALTZ, R. H. (1991), Transposition of Tn*5096* and other IS*493* derivates in *Streptomyces griseofuscus, J. Bacteriol.* **173**, 1096–1104.

SOLENBERG, P. J., BURGETT, S. G. (1989), Method for selection of transposable DNA and characterization of a new insertion sequence, IS*493*, from *Streptomyces lividans, J. Bacteriol.* **171**, 4807–4813.

SONNEN, H., THIERBACH, G., KAUTZ, S., KALINOWSKI, J., SCHNEIDER, J., PÜHLER, A., KUTZNER, H. J. (1991), Characterization of pGA1, a new plasmid from *Corynebacterium glutamicum* LP-6, *Gene* **107**, 69–74.

SPIZIZEN, J. (1958), Transformation of biochemically deficient strains of *Bacillus subtilis* by deoxyribonucleate, *PNAS* **44**, 1072–1078.

STACKEBRANDT, E., SCHLEIFER, K.-H. (1984), Molecular systematics of actinomycetes and related organisms, in: *Biological, Biochemical, and Biomedical Aspects of Actinomycetes* (BOJALIL, L. F., Ed.), pp. 485–504, London: Academic Press.

STACKEBRANDT, S. E., WOESE, C. R. (1981), Towards a phylogeny of the actinomycetes and related organisms, *Curr. Microbiol.* **5**, 197–202.

STADER, J. A., SILHAVY, T. J. (1990), Engineering *Escherichia coli* to secrete heterologous gene products, *Methods Enzymol.* **185**, 166–187.

STAHL. M. L., FERRARI, E. (1984), Replacement of the *Bacillus subtilis* subtilisin structural gene with an *in vitro* derived deletion mutation, *J. Bacteriol.* **158**, 411–418.

STEIN, D., COHEN, S. N. (1989), A cloned regulatory gene of *Streptomyces lividans* can suppress the pigment deficiency phenotype of different developmental mutants, *J. Bacteriol.* **171**, 2258–2261.

STEINMETZ, M., AYMERICH, S. (1990), The *Bacillus subtilis* sac-deg constellation, how and why?, in: *Genetics and Biotechnology of Bacilli* (ZUKOWSKI, M. M., GANESAN, A. T., HOCH, J. A., Eds.), Vol. 3, pp. 303–311, San Diego: Academic Press.

STEINMETZ, M., LE COQ, D., AYMERICH, S., GONZY-TREBOUL, G., GAY, P. (1985), The DNA sequence of the gene for the secreted *Bacillus subtilis* enzyme levansucrase and its genetic control sites, *Mol. Gen. Genet.* **200**, 220–228.

STORRS, M. J., POYART-SALMERON, C., TRIEU-CUOT, P., COURVALIN, P. (1991), Conjugative transposition of Tn*916* requires the excisive and integrative activities of the transposon-encoded integrase, *J. Bacteriol.***173**, 4347–4352.

STRAUCH, E., WOHLLEBEN, W., PÜHLER, A. (1988), Cloning of a phosphinothricin *N*-acetyltransferase gene from *Streptomyces viridochromogenes* Tü494 and its expression in *Streptomyces lividans* and *Escherichia Coli, Gene* **63**, 65–74.

STRAUCH, E., TAKANO, E., BAYLIS, H. A., BIBB, M. J. (1991), The stringent response in *Streptomyces coelicolor* A3(2), *Mol. Microbiol.* **5**, 289–298.

STROHL, W. R. (1992), Compilation and analysis of DNA sequences associated with apparent Streptomycete promoters, *Nucleic Acids Res.* **20**, 961–974.

STROHL, W. R., BARTEL, P. L., CONNORS, N. C., ZHU, C.-B., DOSCH, D. C., BEALE, J. M., JR, FLOSS, H. G., STUTZMAN-ENGWALL, K., OTTEN, S. L., HUTCHINSON, C. R. (1989), Biosynthesis of natural and hybrid polyketides by anthracyclin-producing Streptomycetes, in: *Genetics and Molecular Biology of Industrial Microorganims* (HERSHBERGER, C. L., QUEENER, S. W., HEGEMAN, G., Eds.), pp. 68–84, Washington: American Society for Microbiology.

STROHL, W. R., BARTEL, P. L., LI, Y., CONNORS, N. C., WOODMAN, R. H. (1991), Expression of polyketide biosynthesis and regulatory genes in heterologous Streptomycetes, *J. Ind. Microbiol.* **7**, 163–174.

STUTTARD, C. (1979), Transduction of auxotrophic markers in a chloramphenicol-producing strain of *Streptomyces, J. Gen. Microbiol.* **110**, 479–482.

STUTZMAN-ENGWALL, K. J., OTTEN, S. L., HUTCHINSON, C. R. (1992), Regulation of secondary metabolism in *Streptomyces* spp. and overproduction of daunorubicin in *Streptomyces peucetius, J. Bacteriol.* **174**, 144–154.

SUZUKI, T., HONDA, H., KATSUMATA, R. (1972), Production of antibacterial compounds analogous to chloramphenicol by *n*-paraffin-grown bacteria, *Agric. Biol. Chem.* **36**, 2223–2228.

SWINFIELD, T.-J., OULTRAM, J. D., THOMPSON, E. T., BREHM, J. K., MINTON, N. P. (1990), Physical characterisation of the replication region of the *Streptococcus faecalis* plasmid pAMβ1, *Gene* **87**, 79–90.

TAGUCHI, S., KUMAGAI, I., NAKAYAMA, J., SUZUKI, A., MIURA, K. (1989), Efficient extracellular expression of a foreign protein in *Streptomyces* using secretory protease inhibitor (SSI) gene fusions, *Bio/Technology* **7**, 1063–1066.

TAGUCHI, S., MAENO, M., MOMOSE, H. (1992), Extracellular production system of heterologous peptide driven by a secretory protease inhibitor of *Streptomyces, Appl. Microbiol. Biotechnol.* **36**, 749–753.

TAIRA, S., JALONEN, E., PATON, J. C., SARVAS, M., RUNEBERG-NYMAN, K. (1989), Production of pneumolysin, a pneumococcal toxin, in *Bacillus subtilis, Gene* **77**, 211–218.

TAKAGI, H., MORINAGA, Y., MIWA, K., NAKAMORI, S., SANO, K. (1986), Versatile cloning vectors constructed with genes indigenous to a glutamic acid-producer, *Brevibacterium lactofermentum, Agric. Biol. Chem.* **50**, 2597–2603.

TAKAO, M., MORIOKA, T., YAMAGATA, H., TSUKAGOSHI, N., UDAKA, S. (1989), Production of swine pepsinogen by protein-producing *Bacillus brevis* carrying swine pepsinogen cDNA, *Appl. Micobiol. Biotechnol.* **30**, 75–80.

TAKEDA, Y., FUJII, M., NAKAJYOH, Y., NISHIMURA, T., ISSHIKI, S. (1990), Isolation of a tetracycline resistance plasmid from a glutamate-producing *Corynebacterium, Corynebacterium melassecola, J. Ferment. Bioeng.* **70**, 177–179.

TANAKA, K., SHIINA, T., TAKAHASHI, H.(1991), Nucleotide sequence of genes *hrdA, hrdC* and *hrdD* from *Streptomyces coelicolor* A3(2) having similarity to *rpoD* genes, *Mol. Gen. Genet.* **229**, 334–340.

TE RIELE, H., MICHEL, B., EHRLICH, S. D. (1986), Are single-stranded circles intermediates in plasmid DNA replication? *EMBO J.* **5**, 631–637.

THIERBACH, G., SCHWARZER, A., PÜHLER, A. (1988), Transformation of spheroplasts and protoplasts of *Corynebacterium glutamicum, Appl. Microbiol. Biotechnol.* **29**, 356–362.

THOMAS, C. M., STALKER, D., GUINEY, D., HELINSKI, D. R. (1979), Essential regions for the replication and conjugal transfer of the broad host range plasmid RK2, in: *Plasmids of Medical, Environmental and Commercial Importance* (TIMMIS, K. N., PÜHLER, A., Eds.), pp. 375–385, Amsterdam: Elsevier/North-Holland Biomedical Press.

THOMPSON, C. J., WARD, J. M., HOPWOOD, D. A. (1980), DNA cloning in *Streptomyces:* Resistance genes from antibiotic-producing species, *Nature* **286**, 525–527.

THOMPSON, C. J., WARD, J. M., HOPWOOD, D. A. (1982a), Cloning of antibiotic resistance and nutritional genes in *Streptomyces, J. Bacteriol.* **151**, 668–677.

THOMPSON, C. J., KIESER, T., WARD, J. M., HOPWOOD, D. A. (1982b), Physical analysis of antibiotic-resistance genes from *Streptomyces* and their use in vector construction, *Gene* **20**, 51–62.

TOSAKA, O., TAKINAMI, K. (1978), Pathway and regulation of lysine biosynthesis in *Brevibacterium lactofermentum, Agric. Biol. Chem.* **42**, 95–100.

TOSAKA, O., YOSHIHARA, Y., IKEDA, S., TAKINAMI, K. (1985), Production of L-lysine by fluoropyruvate-sensitive mutants of *Brevibacterium lactofermentum, Agric. Biol. Chem.* **49**, 1305–1312.

TRACH, K., BURBULYS, D., SPIEGELMANN, G., PEREGO, M., VAN HOY, B., STRAUCH, M., DAY, J., HOCH, J. A. (1990), Phosphorylation of the Spo0A protein: a cumulative environsensory activation mechanism, in: *Genetics and Biotechnology of Bacilli* (ZUKOWSKI, M. M., GANESAN, A. T., HOCH, J. A., Eds.), Vol. 3, pp. 357–365, San Diego: Academic Press.

TRAN, L., WU, X.-C., WONG, S. L. (1991), Cloning and expression of a novel protease gene encoding an extracellular neutral protease from *Bacillus subtilis, J. Bacteriol.* **173**, 6364–6372.

TRAUTWETTER, A., BLANCO, C. (1991), Structural organization of the *Corynebacterium glutamicum* plasmid pCG100, *J. Gen. Microbiol.* **137**, 2093–2101.

TRIEU-CUOT, P., CARLIER, C., MARTIN, P., COURVALIN, P. (1987), Plasmid transfer by conjugation from *Escherichia coli* to Gram-positive bacteria, *FEMS Microbiol. Lett.* **48**, 289–294.

TSUCHIYA, M., MORINAGA, Y. (1988), Genetic control systems of *Escherichia coli* can confer inducible expression of cloned genes in coryneform bacteria, *Bio/Technology* **6**, 428–430.

VASANTHA, N., FILIPULA, D (1989), Expression of bovine pancreatic ribonuclease A coded by a synthetic gene in *Bacillus subtilis, Gene* **76**, 53–60.

VASANTHA, N., THOMPSON, L. D. (1986), Secretion of a heterologous protein from *Bacillus subtilis* with the aid of protease signal sequences, *J. Bacteriol.* **165**, 837–842.

VATS, S., STUTTARD, C., VINING, L. C. (1987), Transductional analysis of chloramphenicol biosynthesis genes in *Strepomyces venezuelae, J. Bacteriol.* **169**, 3809–3813.

VERTESY, L., TRIPIER, D., KOLLER, K.-P., RIESS, G. (1991), Disulphide bridge formation of proinsulin fusion proteins during secretion in *Streptomyces, Hoppe-Seyler, Z. Physiol. Chem.* **327**, 187–192.

VIGAL, T., GIL, J. A., DAZA, A., GARCIA-GONZALEZ, M. D., VILLADAS, P., MARTÍN, J. F. (1991), Effects of replacement of promoters and modification of the leader peptide region of the *amy* gene of *Streptomyces griseus* on the synthesis and secretion of α-amylase by *Streptomyces lividans, Mol. Gen. Genet.* **231**, 88–96.

VUJAKLIJA, D., UEDA, K., HONG, S.-K., BEPPU, T., HORINOUCHI, S. (1991), Identification of an A-factor-dependent promoter in the streptomycin biosynthetic gene cluster of *Streptomyces griseus, Mol. Gen. Genet.* **229**, 119–128.

WANG, L.-F., WONG, S.-L., LEE, S.-G., KALYAN, N. K., HUNG, P. P., HILLIKER, S., DOI, R. H. (1988), Expression and secretion of human atrial natriuretic α-factor in *Bacillus subtilis* using the subtilisin signal peptide, *Gene* **69**, 39–47.

WANG, L.-F., HUM, W. T., KALYAN, N. K., LEE, S. G., HUNG, P. P, DOI, R. H. (1989), Synthesis and refolding of human tissue-type plasminogen activator in *Bacillus subtilis, Gene* **84**, 127–133.

WESTPHELING, J., RANES, M., LOSICK, R. (1985), RNA polymerase heterogeneity in *Streptomyces coelicolor, Nature* **313**, 22–27.

WILLIAMS, S. T., GOODFELLOW, M., ALDERSON, G., WELLINGTON, E. M. H., SNEATH, P. H. A., SAKIN, M. J. (1983), Numerical classification of *Streptomyces* and related genera, *J. Gen. Microbiol.* **129**, 1743–1813.

WOHLLEBEN, W., PIELSTICKER, A. (1989), Investigation of plasmid transfer between *Escherichia coli* and *Streptomyces lividans,* in: *Dechema Biotechnology Conferences* (BEHRENS, D., DRIESEL, A. J., Eds.), Vol. 3, pp. 301–305, Weinheim: VCH Verlagsgesellschaft.

WOHLLEBEN, W., ARNOLD, W., BROER, I., HILLEMANN, D., STRAUCH, E., PÜHLER, A. (1988), Nucleotide sequence of the phosphinothricin *N*-acetyltransferase gene from *Streptomyces viridochromogenes* Tü494 and its expression in *Nicotiana tabacum, Gene* **70**, 25–37.

WOHLLEBEN, W., ARNOLD, W., BISSONNETTE, L., PELLETIER, A., TANGUAY, A., ROY, P. H., GAMBOA, G. C., BARRY, G. F., AUBERT, E., DAVIES, J., KAGAN, S. A. (1989), On the evolution of Tn*21*-like multiresistance transposons: Sequence analysis of the gene (*aacC1*), for gentamicin acetyltransferase-3-I (AAC(3)-I), another member of the Tn*21*-based expression cassette, *Mol. Gen. Genet* **217**, 202–208.

WOHLLEBEN, W., ALIJAH, R., DORENDORF, J., HILLEMANN, D., NUßBAUMER, B., PELZER, S. (1992a), Identification and charcaterization of phosphinothricin-tripeptide biosynthetic genes in *Streptomyces viridochromogenes, Gene* **115**, 127–132.

WOHLLEBEN, W., HARTMANN, V., HILLEMANN, D., KREY, K., MUTH, G., NUSSBAUMER, B., PELZER, S. (1992b), Transfer and establishment of DNA in *Streptomyces,* in: *Proc. 11th Eur. Meeting on Genetic Transformation,* Budapest: Intercept.

WOLF, H., PÜHLER, A., NEUMANN, E. (1989), Electrotransformation of intact and osmotically sensitive cells of *Corynebacterium glutamicum, Appl. Microbiol. Biotechnol.* **30**, 283–289.

WONG, S. L. (1989), Development of an inducible and enhancible expression system in *Bacillus subtilis, Gene* **83**, 212–223.

WONG, S. L., KAWAMURA, F., DOI, R. H. (1986), Use of the *Bacillus subtilis* subtilisin signal peptide for efficient secretion of β-lactamase during growth, *J. Bacteriol.* **168**, 1005–1009.

WU, X.-C., NATHOO, S., PANG, A. S. H., CARNE, T., WONG, S. L. (1990), Cloning, genetic organization, and characterization of a structural gene encoding bacillopeptidase F from *Bacillus subtilis, J. Biol. Chem.* **265**, 6845–6850.

WU, X.-C., LEE, W., TRAN, L., WONG, S. L. (1991), Engineering a *Bacillus subtilis* expression–secretion system with a strain deficient in six extracellular proteases, *J. Bacteriol.* **173**, 4952–4958.

YAMADA, K., KINOSHITA, S., TSUNODA, T., AIDA, K. (1972), *The Microbial Production of Amino Acids,* New York: Wiley.

YAMADA, Y., WON SEO, C. OKADA, H. (1985), Oxidation of acyclic terpenoids by *Corynebacterium* sp., *Appl. Environ. Microbiol.* **49**, 960–963.

YAMAGATA, H., ADACHI, T., TSUBOI, A., TAKAO,

M., SASAKI, T., TSUKAGOSHI, N., UDAKA, S. (1987), Cloning and characterization of the 5' region of the cell wall protein gene operon in *Bacillus brevis* 47, *J. Bacteriol.* **169**, 1239–1245.

YAMAGATA, H., NAKAHAMA, K., SUZUKI, Y., KAKINUMA, A., TSUKAGOSHI, N., UDAKA, S. (1989), Use of *Bacillus brevis* for efficient synthesis and secretion of human epidermal growth factor, *PNAS* **86**, 3589–3593.

YAMAGUCHI, R., TERABE, M., MIWA, K., TSUCHIYA, M., TAKAGI, H., MORINAGA, Y., NAKAMORI, S., SANO, K., MOMOSE, H., YAMAZAKI, A. (1986), Determination of the complete nucleotide sequence of *Brevibacterium lactofermentum* plasmid pAM330 and analysis of its genetic information, *Agric. Biol. Chem. Tokyo* **50**, 2771–2778.

YANG, M. Y., FERRARI, E., HENNER, D. J. (1984), Cloning of the neutral protease gene of *Bacillus subtilis* and the use of the cloned gene to create an *in vivo*-derived deletion mutation, *J. Bacteriol.* **160**, 15–21.

YANSURA, D. G., HENNER, D. J. (1984), Use of *Escherichia coli lac* repressor and operator to control gene expression in *Bacillus subtilis, PNAS* **81**, 439–443.

YEH, P., OREGLIA, J., PREVOTS, F., SICARD, A. M. (1986), A shuttle vector system for *Brevibacterium lactofermentum, Gene* **47**, 301–306.

YEH, P., SICARD, A. M., SINSKEY, A J. (1988), General organization of the genes specifically involved in the diaminopimelate-lysine biosynthetic pathway of *Corynebacterium glutamicum, Mol. Gen. Genet.* **212**, 105–111.

YOSHIHAMA, M., HIGASHIRO, K., RAO, E. A., AKEDO, M., SHANABRUCH, W. G., FOLLETTIE, M. T., WALKER, G. C., SINSKEY, A. J. (1985), Cloning vector system for *Corynebacterium glutamicum, J. Bacteriol.* **162**, 591–597.

YOUNG, M., CULLUM, J. (1987), A plausible mechanism for large-scale chromosomal DNA amplification in Streptomycetes, *FEBS Lett.* **212**, 10–14.

YOUNG, M., MINTON, N. P., STAUDENBAUER, W. L. (1989), Recent advances in the genetics of clostridia, *FEMS Microbiol. Rev.* **63**, 301–325.

YOUNGMAN, P. J., PERKINS, J. B., LOSICK, R. (1983), Genetic transposition and insertional mutagenesis in *Bacillus subtilis* with *Streptococcus faecalis* transposon Tn*917*, *PNAS* **80**, 2305–2309.

ZAJIC, J. E., GUIGNARD, H., GERSON, D. F. (1977), Emulsifying and surface active agents from *Corynebacterium hydrocarboclastum, Biotechnol. Bioeng.* **19**, 1295–1301.

ZHEN, Z.-X., MA, C.-P., YAN, W.-Y., HE, P.-F., MAO, Y.-X., SUN, W., LEI, Z.-Z., ZHU, P., WH, J.-F. (1987), Restriction map of plasmid pXZ10145 of *Corynebacterium glutamicum* and construction of an integrated plasmid, *Chin. J. Biotechnol.* **3**, 183–188.

ZUKOWSKI, M., MILLER, L., COGSWELL, P., CHEN, K. (1988), Inducible expression system based on sucrose metabolism genes of *Bacillus subtilis,* in: *Genetics and Biotechnology of Bacilli* (GANESAN, A. T., HOCH, J. A., Eds.), Vol. 2, pp. 17–22, San Diego: Academic Press.

13 Genetic Engineering of Yeast

PETER E. SUDBERY

Sheffield, United Kingdom

1 Introduction

In historical terms *Saccharomyces cerevisiae* has long been a subject of both biochemical and genetical research. The first protocol for molecular transformation was described in 1978 (HINNEN et al., 1978), and a wide variety of vectors, promoters and a capability for precise engineering of chromosomal sequences were developed shortly afterwards. This greatly facilitated the investigation of fundamental aspects of cell and molecular biology. Consequently the yeast has become an important eukaryotic model system in such fields as cell division, gene regulation, and protein sorting.

As a eukaryote, *S. cerevisiae* has several advantages compared to *Escherichia coli* in producing recombinant proteins. It has a secretory pathway which is very similar to that of higher organisms. Protein folding and many post-translational modifications are successfully carried out including formation of disulfide bridges and glycosylation of secreted proteins. Although in the latter process the nature of the sugar residues may be different from the native product. It also has a long history of use in biotechnology such as beer and wine fermentation and as a leavening agent to produce bread and associated products. This has provided a technological base for large-scale fermentations and confidence that no harmful pyrogens or toxic substances will contaminate the product. It thus has *GRAS* status. In comparison to mammalian tissue culture systems it is cheaper, more convenient, and gives higher product yields. Nevertheless, many large proteins or proteins subject to complex post-translational modifications can only successfully be produced in mammalian tissue culture.

A compilation of reviews of the molecular and cell biology of yeasts has been edited by WALTON and YARRANTON (1989a) and laboratory protocols and their background by GUTHRIE and FINK (1991).

Other yeasts have been used for recombinant protein production. Methylotrophic yeasts which exploit the powerful and regulatable methanol oxidase promoter have been particularly successful (SUDBERY and GLEESON, 1989).

2 Molecular Techniques for Gene Manipulation in *Saccharomyces cerevisiae*

2.1 Molecular Transformation

The first technique to be described for molecular transformation of yeast cells involves the production of protoplasts with enzymes such as β-glucuronidase or zymolyase in an osmoticum such as sorbitol (HINNEN et al. 1978; BEGGS, 1978). The protoplasts are then resuspended in an osmoticum containing $CaCl_2$ and mixed with the transforming DNA. Polyethylene glycol (PEG) is added and the cells plated out under selective conditions embedded in a top agar overlay to facilitate protoplast regeneration. This is still a very commonly used technique and is possibly the most efficient in terms of transformation frequency per microgram of transforming DNA. It suffers from three disadvantages:

(1) PEG causes cell fusion and so diploids and polyploids may result;
(2) the transformant colonies are embedded in the agar overlay which makes subsequent screening difficult;
(3) it is both laborious and time-consuming.

Transformation without cell fusion can be achieved by treating whole cells with lithium salts followed by PEG (ITO et al., 1983) or by use of electroporation (BECKER and GUARANTE, 1991). Both of these procedures are more convenient than the protoplast method, and both methods have the advantage that the colonies grow on the surface of the agar plate used for selection of transformants. If a large number of transformants is not required, a rapid and convenient method has been described (CONSTANZO and FOX, 1988) which uses cells directly from an agar plate. Competence is produced by agitating cells with glass beads.

The actual method of transformation is poorly understood with all these methods and has not been investigated. Not all strains are equally susceptible to transformation, and the transformability of a strain may vary from method to method. Therefore, it is important to choose a strain with good transformability at the start of a research program.

2.2 Selection Markers

The most commonly used markers for the selection of transformants are based upon auxotrophic mutations in the recipient strain, the vector carrying the wild-type gene. Thus genes such as *LEU*2, *URA*3, *TRP*1, *HIS*3, etc. are often used for vector construction. Care should be exercised when using the *LEU*2 gene, as two different versions have been independently cloned. One of them lacks the full promoter sequence and only complements effectively in high copy number. This version is sometimes referred to as *LEU*2d and is present in pJDB219 (BEGGS, 1978) and pJDB207 (BEGGS, 1981) but not YEp13 (BROACH et al., 1979). It is sometimes used to force the high copy number of an expressing plasmid.

The use of such genes requires a suitably marked recipient strain; and since the mutations are recessive, this means that the recipient is haploid or a specially constructed homozygous diploid. Industrial strains are often polyploid and have poorly defined genetic systems. A suitably marked recipient may therefore be unobtainable, and dominant selectable markers must be employed. Those available are: resistance to G418 conferred by the kanamycin resistance gene of Tn903 which encodes neomycin phosphotransferase II (WEBSTER and DICKSON, 1983); copper resistance conferred by copperthionein encoded by the *Saccharomyces cerevisiae CUP*^{R} genes (BUTT et al., 1984), chloramphenicol acetyltransferase (JIMINEZ and DAVIES, 1980), and hygromycin B phosphotransferase (GRITZ and DAVIES, 1983).

A useful class of markers are those which can be both selected for and selected against (counterselectable). The most commonly used marker of this type is *URA*3 which encodes orotidine-5′-phosphate decarboxylase in the uracil biosynthetic pathway. This enzyme also converts 5-fluoroorotic acid to 5-fluororacil which is toxic. URA3^{+} cells are therefore killed by 5-FOA, but ura3^{-} cells are not. Loss of a functioning gene can be selected for by growth on 5-FOA. This technique is used in gene disruption and replacement techniques as described below. The *URA*3 gene can of course be used in forward selection in transformation of *ura3*$^{-}$ cells. 5-FOA resistance is thus a useful way of introducing a marker into a strain so that it may be used as a transformation recipient.

Other counterselectable markers that may be used are: *LYS*2, *CAN*1, and *CYH*2. *LYS*2 is counterselectable by growth on α-aminoadipate as a sole nitrogen source (CHATOO et al., 1979). *CAN*1 encodes arginine permease and is counterselectable by resistance to the arginine analog canavine (HOFFMAN, 1985). Sensitivity is dominant to resistance, and so loss of *CAN*1 is selectable. Selection for *CAN*1 is also possible but requires a *can*1 (Arg^{-}) recipient. *CYH*2 encodes the ribosomal protein L29 (KAUFER et al., 1983); when this is mutated in *cyh*2 cells they are resistant to cyclohexamide. Sensitivity is dominant, so loss of *CYH*2 can be selected for. There is no positive selection for *CYH*2, so it must be introduced on a plasmid which contains another selectable marker.

2.3 Vectors

Nearly all yeast vectors are shuttle vectors which can be propagated in both *E. coli* and *S. cerevisiae.* This allows cloning steps to be carried out using *E. coli* as the host, the final construct then introduced into *S. cerevisiae.* For a review including plasmid maps of commonly used yeast vectors see ROSE and BROACH (1991).

2.3.1 Integrating Vectors

*Y*east *I*ntegrating *p*lasmids (YIp) consist of a selectable gene cloned into a general-purpose *E. coli* plasmid such as pBR322 of pUC18 (STRUHL et al., 1979). YIp plasmids lack an origin of replication, prototrophs can therefore only arise by chromosomal integration or by replacement or conversion of the auxotrophic allele (HINNEN et al., 1978). Integration

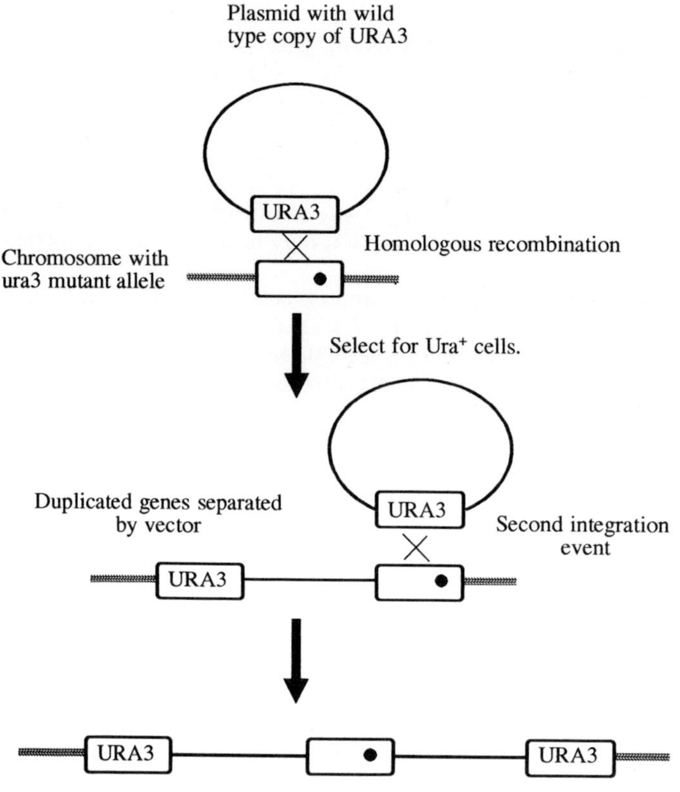

Fig. 1. Chromosomal integration by single or multiple homologous recombination events.

usually involves a single reciprocal recombination event with the homologous chromosomal sequence giving rise to direct repeats of the sequence, one which will be defective due to the original mutation (Fig. 1). Occasionally there may be two or more rounds of recombination giving rise to multiple tandem repeats. Gene conversion or a double cross-event results in a single functioning copy of the gene. Recombination can occasionally occur at a different locus possibly due to the presence of repeated sequences such as Ty elements or δ sequences. It is known that part of a Ty element is located on the chromosomal fragment carrying the *LEU2* gene in YEp13 (ROTHSTEIN, 1991). This may account for the apparently non-homologous recombination detected by HINNEN et al. (1978) in one class of transformants arising from a *LEU2* based integrating vector.

Transformation frequencies using YIp vectors are low (1–10/μg DNA). ORR-WEAVER et al. (1981) showed that they could be greatly increased (10- to 1000-fold) by the introduction of a double-stranded break in a region of homology with a chromosomal sequence (Fig. 2). Again multiple rounds of recombination may occur giving rise to an array of tandem repeats. If there are two yeast sequences on the plasmid, recombination is targeted to the chromosomal homolog of the sequences in which the break occurred (see below).

Transformants which arise from either intact circular plasmids or from linearized plasmids are stable, but recombination can occur between the direct repeats generated by the integration event resulting in the production of a "pop-out" and loss of the integrated vector (see Fig. 7). The two copies of the gene may be distinguishable from each other because of a mutation on the chromosomal copy or because of *in vitro* mutagenesis of the plasmid copy. If this is the case, the nature of the copy left be-

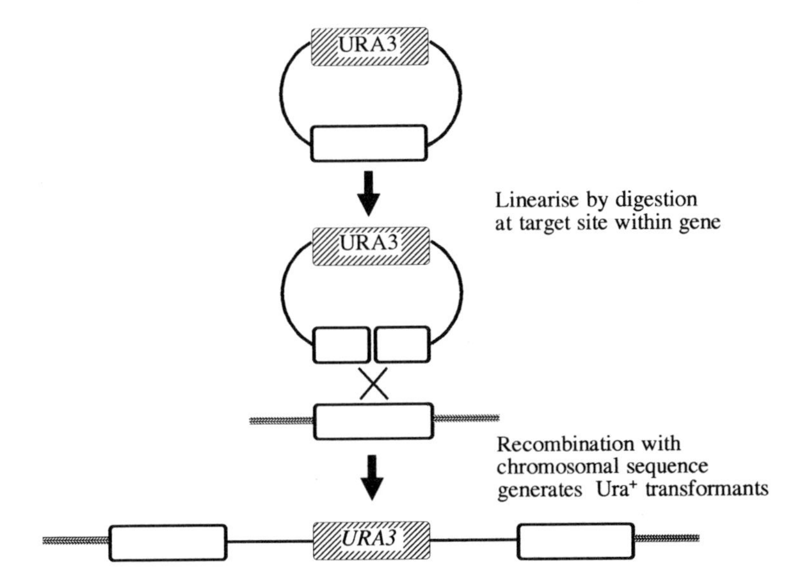

Fig. 2. Targeted integration of one of several yeast sequences on a plasmid by the introduction of a break within one of them.

hind depends on where the recombination event between the tandem repeats occurred. Fig. 7 shows how this affects the fate of a mutated gene originally introduced on a plasmid.

2.3.2 Yeast Episomal Plasmids

Saccharomyces cerevisiae is unusual in having an autonomously replicating nuclear plasmid known as the 2 μm circle (after its contour length). For a review see MEACOCK et al. (1989). Strains lacking the plasmid (cir^- strains) are phenotypically indistinguishable from those that harbor it (cir^+), it may, therefore, be an example of selfish DNA. It is 6318 bp in length and has a copy number of between 50 and 100 copies per cell representing 1–2% of total DNA. Its sequence reveals four open reading frames (ORFs A–D) that code for functions necessary for plasmid stability and recombination, a *cis*-acting region, *STB,* that is necessary for stability, an origin of replication and two perfect 600-bp direct repeats. Site-specific recombination between these repeats is catalyzed by the 2 μm FLP protein, the product of ORF A, and results in the interconversion of two plasmid forms known as the A and B forms. 2 μm replication is limited to one round per cycle, yet a single plasmid molecule introduced into a cell will quickly amplify up to the normal copy number. This copy number control is thought to result from the interconversion of the two forms during the course of DNA replication which re-orients the replication forks so that instead of converging, they follow each other – resulting in oligomeric structures arising without re-initiation.

The 2 μm plasmid can be used for the construction of independently replicating vectors. Collectively these are known as Yeast Episomal plasmids (YEp). pJDB219 and pJDB248 were the first such plasmids to be constructed (BEGGS, 1978). They contained the whole 2 μm plasmid combined with pBR322 for replication in *E. coli* plus the *LEU*2 gene for selection. They transform with high frequency 10^4 to 10^5 transformants/μg DNA and are relatively stable, plasmidless cells arise at a rate of about ~1% per generation. They will transform cir^- cells, since they contain all the functions needed for plasmid maintenance. They may have a very high copy number (~400 copies/cell) possibly due to the *LEU*2d selection marker which as discussed above only complements in high copy number. They are less stable in cir^+ cells due to competition with the endogenous plasmid.

Smaller vectors containing just the origin and *STB* locus have been constructed; commonly used examples of these are pJDB207

(BEGGS, 1981) and YEp13 (BROACH et al., 1979). These plasmids also transform with high efficiencies and are relatively stable. In contrast to pJDB219 and pJDB248, they will only transform cir$^+$ cells, since they lack functions which must be supplied in *trans*. Copy number is variable but is probably in the region of 50–100 copies per cell.

2.3.3 Disintegration Plasmids

Recombination can occur between the endogenous 2 μm circle and 2 μm sequences in YEp vectors. Selection for the presence of the original plasmid may result in rearrangements where the selectable marker is retained but the heterologous gene has been lost. Even plasmids containing the whole 2 μm circle are not completely stable. This is thought to be due to the unavoidable disruption of 2 μm sequences in vector construction. Furthermore, bacterial sequences including antibiotic resistance genes are present on all standard vectors. Large cultures containing such vectors may thus constitute an environmental hazard and pose an additional problem in the purification of a recombinant protein.

These problems led to the construction by CHINNERY and HINCHCLIFFE (1989) of so-called "disintegration vectors" based on the 2 μm circle. Sequences necessary for propagation in bacterial cells, are bounded by targets for the 2 μm *FLP* gene product. Upon transformation of yeast cells this results in the excision and consequent loss of these bacterial sequences. The expression cassette is cloned into the *REP*1 locus, a region of the 2 μm circle. Its disruption was found to have the least effect on stability. Such constructs are thus completely stable, lack undesirable bacterial sequences, and are present in high copy number to maximize expression of a heterologous gene.

2.3.4 Yeast Replicating Plasmids

Shot-gun cloning of chromosomal DNA into YIp vectors and selection for increased transformation frequency led to the recovery of *A*utonomously *R*eplicating *S*equences (*ARS*) which drive plasmid replication in the yeast cell (STRUHL et al., 1979; STINCHCOMBE et al., 1979). Such plasmids are termed *Y*east *R*eplicating *p*lasmids (YRp). They are very unstable with a loss rate much higher than YEp plasmids. Copy number is somewhat lower than that of YEp plasmids, and transformation frequency is high.

2.3.5 Centromeric Vectors

YRp plasmids may be stabilized by the addition of a centromere (CLARKE and CARBON, 1980). Such plasmids are termed *Y*east *C*entromeric *p*lasmids (YCp). They are very stable with a loss rate very much lower than 1% per generation. Copy number is low (1–5/cell) which makes them very useful for complementation studies where it is necessary to maintain gene copy number close to the normal level. During meiosis YCp plasmids generally segregate in a Mendelian fashion.

2.3.6 Yeast Artificial Chromosomes

Cloning of yeast telomeres together with the availability of selectable markers, origins of replication, and centromeres allows the construction of linear plasmids with all the elements of a normal chromosome (MURRAY and SZOSTAK, 1983). These are known as *Y*east *A*rtificial *C*hromosomes (YAC). Such constructs are only stable provided that they are at least 20 kb in size and stability increases further as the size approaches that of natural chromosomes (100–1000 kb). Large YAC constructs are as stable as natural chromosomes. The requirement for such large amounts of DNA makes them powerful vectors for the construction of libraries from complex genomes such as the human one (BURKE et al., 1987). Average insert sizes of over 500 kb have been reported. In addition, sequences which are unclonable in *E. coli* cosmid and lambda vectors are successfully cloned in YAC vectors.

YAC vectors are normally propagated in bacteria as circular plasmids (Fig. 3). Restriction enzyme target sites are arranged to produce two arms upon digestion, each of which

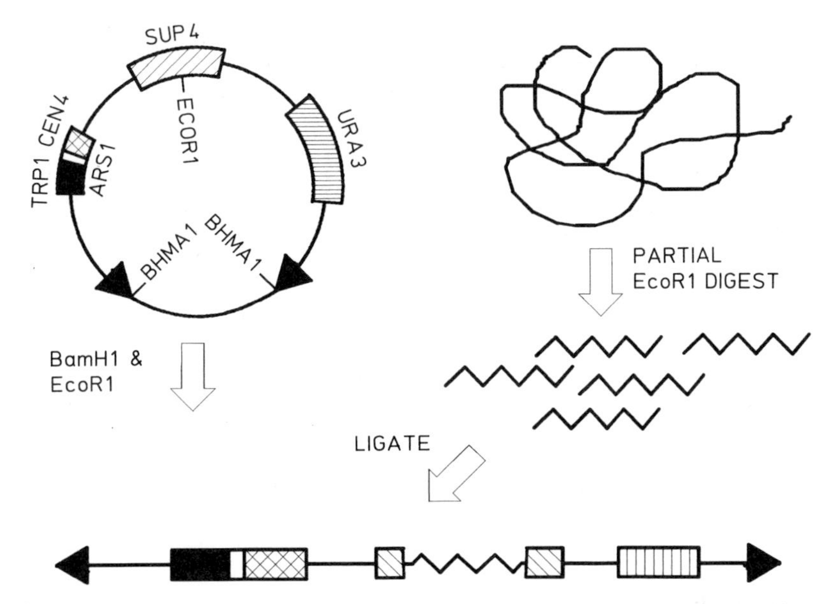

Fig. 3. Use of a YAC vector (top left) to clone large fragments of donor DNA (top right). The resulting linear plasmid (bottom) is transformed directly into yeast.

contains a different selectable marker and terminates at one end in a telomere, the other in a blunt end. In addition, one of the arms contains an *ARS* element. The two arms are purified away from a linking fragment and ligated with donor DNA fragmented so as to leave blunt ends. The ligation mixture is used to transform yeast cells, and the selection conditions are such as to require the presence of both arms, the insert interrupts a third selectable marker such as *SUP*4, which allows nonrecombinant structures to be recognized.

2.4 Chromosome Engineering

Free ends of DNA molecules in the yeast cell are highly recombinogenic with homologous chromosomal sequences. This has been exploited to provide a variety of techniques which allow specific changes to be engineered into chromosomes. These techniques have been reviewed in detail by ROTHSTEIN (1991). With all the techniques described below, it is essential to examine by hybridization the chromosome structure of the transformants to verify that the expected events have in fact occurred. It is very common to find a proportion of transformants showing unexpected structures due to tandem integration, integration at unexpected sites or to other, undefined reasons.

2.4.1 Targeted Integration and Allele Rescue

As described above (Sect. 2.3.1), digestion at a unique restriction site within a gene carried by the plasmid stimulates integration at the corresponding chromosomal locus (ORR-WEAVER et al., 1981). The recombination leads to gene duplication as if the plasmid had been circular and a single cross-over had occurred within the homologous sequence (Fig. 2). As well as a single copy of the linear plasmid integrating, multiple tandem copies may also occur (Fig. 1). Prototrophic transformants may also arise by replacement or conversion of the auxotrophic allele. The frequency of this latter event depends on a number of variables such as the length of the targeted sequence and the length of homology in the selectable marker.

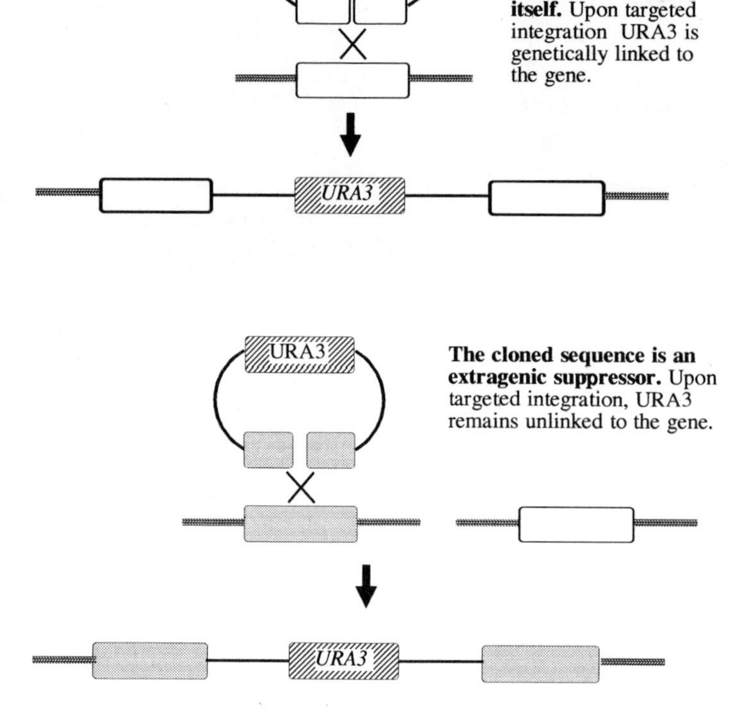

Fig. 4. The use of targeted integration to determine whether a gene itself or an extragenic suppressor has been cloned. The gene subject to the original mutation is shown by an open box. A possible extragenic suppressor is shown by the shaded box.

Targeted integration is a means of determining whether a sequence which has been cloned by complementation of a mutant allele is the gene itself or an extragenic suppressor (Fig. 4). As well as the complementing sequence, the plasmid generally has a selection marker such as *URA*3 or *LEU*2, and the recipient will carry the appropriate auxotrophic allele. The plasmid is linearized within the complementing sequence, and prototrophic transformants are selected. As integration has been targeted to the chromosomal copy of the complementing sequence, the selection marker is now closely linked genetically to it. If the complementing sequence is indeed the wild-type copy of the gene, then the selection marker will now be closely linked to it. If, on the other hand, the complementing sequence is an (unlinked) extragenic suppressor, then the selection marker will not be linked to the original mutant allele. The two possibilities can be distinguished from each other by appropriate genetic crosses.

If, instead of a break, a gap is produced, integration still occurs *without any deletion* in either of the duplicated copies (Fig. 5). If the transforming plasmid contains an *ARS* element, it is possible to recover intact circular plasmids in which the gap has been repaired using chromosomal sequence as a template. This may be used to determine the position of a mutant allele within a cloned sequence. If a gapped plasmid can complement a mutation in the recipient, then the sequence provided by repair cannot contain the mutation (Fig. 5). If complementation does not occur, then it may be inferred that the gap spans the site of the mutation. Furthermore, since the repaired plasmid now carries the mutant allele, it can be rescued from the yeast cell, and the nature of the mutation can be determined.

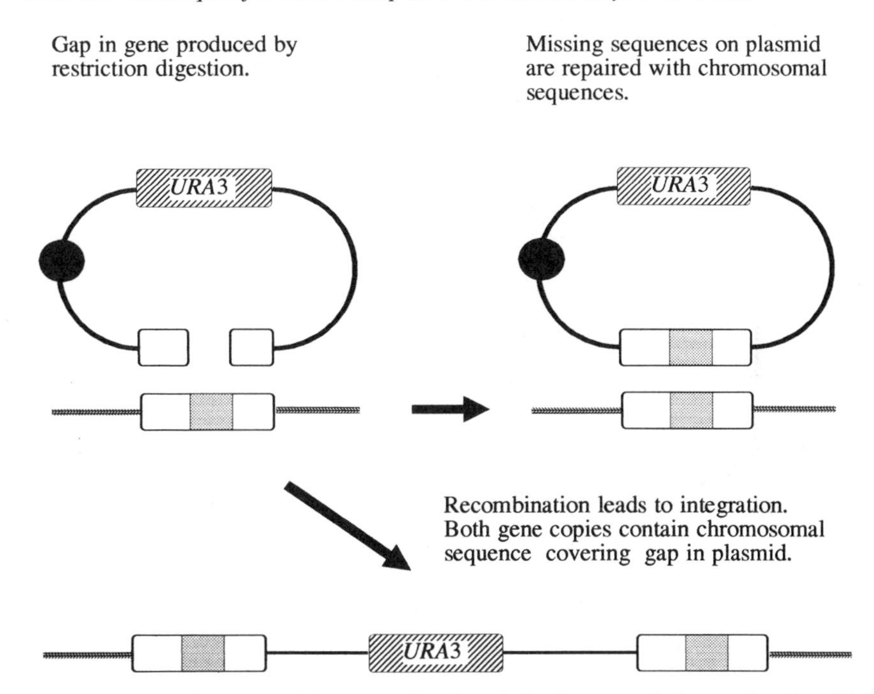

Fig. 5. Allele rescue by a gapped plasmid. If the original mutation lies within the area missing on the plasmid (light shading), then the repaired plasmid will carry the mutant sequence but will fail to complement the mutation. The solid circle represents a replication origin.

2.4.2 Gene Disruption

Once a cloned copy of a gene is available, the chromosomal copy may be disrupted or deleted in a single step. There are two methods by which this can be achieved. The first method known as the one-step gene disruption (ROTHSTEIN, 1991) is illustrated in Fig. 6. A selectable marker is engineered into the plasmid copy of the gene. This may be done in such a way as to lead to a simple insertion of the marker or to an insertion/deletion with some or all of the gene being deleted as well as the insertion of the markers. If the intention is to create a null-allele, insertion/deletion is preferable to simple insertion as the latter may result in a gene product with at least some residual activity. The plasmid is then digested to produce a linear sequence which consists of the selectable marker flanked on either side by DNA from the target sequence. The fragment is used to transform a suitably marked host.

Transformants can only arise by integration of the fragment, and this involves a double cross-over with each end of the linear fragment recombining with DNA on each side of the target. The result is the replacement of the chromosomal copy with the cloned sequence containing the insertion or insertion/deletion (Fig. 6). It is normal to carry out the disruption in a diploid so that if the disruption is lethal viability is maintained by the second copy of the gene. After sporulation the effect of a lethal disruption will be a 2:0 segregation of viability in the tetrads, none of the surviving spores carrying the selection marker. If all the spores are viable, those carrying the disruption will be those with the marker. The effect on the phenotype can then be examined.

Genes may also be inactivated by a two-step method (SCHERER and DAVIS, 1979). This involves generating a mutated version of the gene on a plasmid containing a counterselectable marker such as *URA3* (SCHERER and DAVIS, 1979). The plasmid is targeted to the chro-

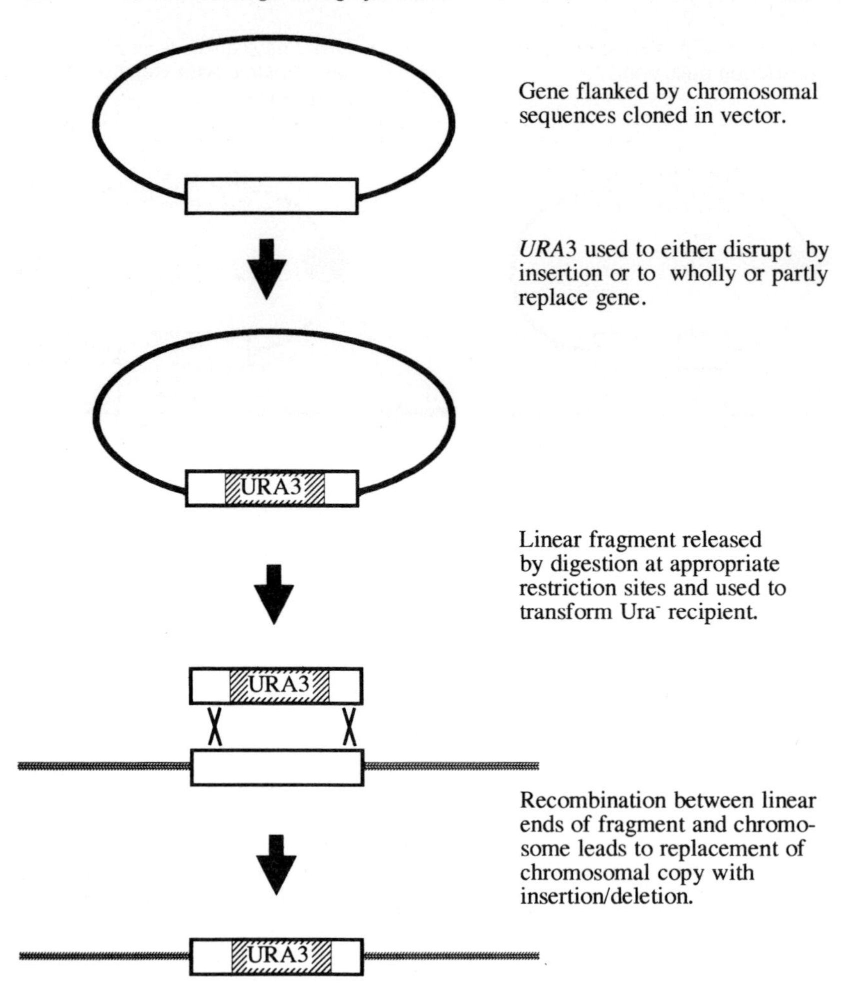

Fig. 6. Gene disruption or deletion by the replacement of a chromosomal sequence with a linear fragment.

mosomal copy of the gene to produce a gene duplication on either side of the *URA*3 sequence carried by the plasmid. Selection for Ura^- cells using 5-FOA identifies cells in which the *URA*3 gene has been deleted through homologous recombination between the duplicated copies of the gene (Fig. 7). This results either in the loss of the original gene to leave only the mutated version or the loss of the mutated version to leave the original. The method relies upon the ability to distinguish between these two outcomes either by hybridization or through the phenotypic effects of the mutation.

Gene disruption is used to determine whether a functioning copy of a cloned gene is essential for viability. This may be important where a gene is recognized through a mutant allele where it is not clear whether the phenotype results from a loss or an alteration of gene function. Alternatively, a sequence may have been cloned through some form of reverse genetics, for example, when a cDNA library has been screened with an antibody or with an oligonu-

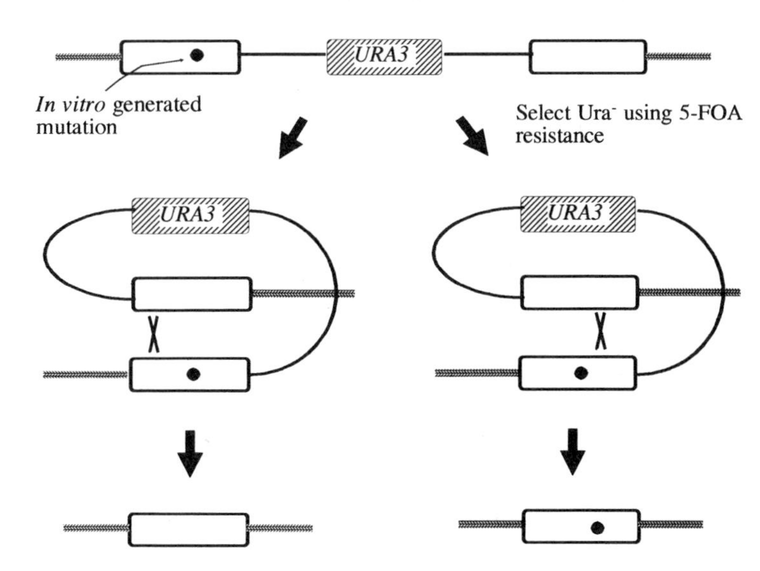

Fig. 7. Gene disruption or replacement of an integrated sequence by homologous recombination between tandemly duplicated sequences. A mutation introduced *in vitro* has been targeted to the homologous chromosomal sequence. The position of the recombination event determines whether the original or mutated allele remains on the chromosome.

cleotide probe derived from the amino acid sequence of a purified protein. In this case, there is no formal indication as to whether the gene is essential, whatever the apparent importance of the protein to the cell.

Functional redundancy may be revealed when deletion of an apparently essential function does not affect viability. There are well documented examples where deletion of one sequence does not affect viability, but deletion of two or more redundant copies is lethal. Functional redundancy does not necessarily involve duplication of identical or even similar sequences. So its existence may not be apparent from hybridization experiments where the sequence has been used to probe genomic DNA. The phenotype of the null allele can also give important information on the biological role of the gene.

An example of the application of these techniques is the analysis of the *CLN*1, 2, and 3 genes encoding the G1 cyclins which regulate the passage of cells through G1 (reviewed by REED, 1991). *CLN*1 and *CLN*2 were recognized as clones which will when in high copy number suppress a *cdc*28 mutation (HADWIGER et al., 1989). Gene targeting revealed that they were not CDC28 sequences but extragenic suppressors. They are recognizably homologous to each other. *CLN*3 was first recognized through a dominant mutation, *whi*1, leading to small cell size (SUDBERY et al., 1980). In none of these cases was it initially known what the phenotype of a null-allele would be. Gene disruption studies revealed that only one functioning copy of any gene in the set is required for viability. This indicated functional redundancy among the genes, even though the sequence of *CLN*3 was only weakly homologous to the other two genes (NASH et al., 1988).

2.4.3 Plasmid Shuffling, Gene Replacement, and Oligonucleotide Mutagenesis

Following the disruption of a non-essential gene, its biological function may be investigated by introducing versions mutated *in vitro* on a plasmid. Plasmid shuffling allows this to be achieved with an essential gene (SIKORSKI

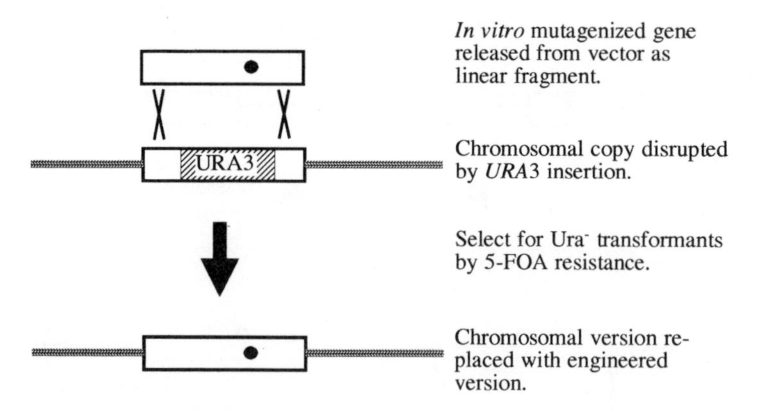

Fig. 8. Introduction of an *in vitro* generated mutation by the replacement of a disrupted locus with a linear fragment carrying the mutation.

and BOEKE, 1991). The gene disruption is engineered in a diploid, and a plasmid carrying the wild-type gene is introduced on a plasmid containing a counterselectable marker such as *URA*3. The diploid is sporulated, and progeny containing the disruption are identified. They are viable because they will be covered by the wild-type gene on the plasmid. A second plasmid containing the mutated version of the gene is now introduced and counterselection applied against the first plasmid causing its loss. In effect, the plasmid with the wild copy and the plasmid bearing the mutation are shuffled.

It may be inappropriate to locate the mutated version on a plasmid, since there may be effects of copy number, or the normal chromosomal environment may be important for normal gene function. In this case, the chromosomal copy of the gene may be replaced by the mutated version. This can be done either with the two-step method of SCHERER and DAVIS (1979) or a technique based on the one-step gene disruption procedure described above. Both methods result in a stable copy of the mutated sequence, because only a single copy remains on the chromosome and thus no further recombination can take place.

The method based on gene disruption uses a counterselectable marker such as *URA*3 to generate a disruption of the gene to be mutated. This is then transformed with a linear fragment containing the mutation, and Ura⁻ transformants are selected using 5-FOA (Fig. 8). These arise through a double recombination event in which the resident disrupted version is replaced with the fragment bearing the mutation. This method relies on the viability of disrupted strains. It may be used where gene redundancy permits this or the gene product is not essential for viability.

If the sequence carrying the mutation is a truncated form of the gene, then the integration event will result in the wild-type locus carrying the truncation and thus ensuring its inactivation (Fig. 9). It results in transplacement of the wild-type gene with the mutated form in a single step (SHORTLE et al., 1984). It suffers from the problem that further recombination between the remaining duplicated sequences can lead to the regeneration of the wild-type sequence. A further drawback is that sequences missing in the truncation are not available for mutagenesis.

In certain situations it is possible to mutagenize a gene directly with oligonucleotides. The protocol requires the availability of a mutant where it is possible to select for at least partly functional revertants. The sequence of the gene and the change caused by the mutation must also be known. The oligonucleotide is designed to repair the original mutation and to introduce further changes nearby. This method has been used to study amino-terminal processing and to introduce mutations into an evolutionary invariant region of iso-1-cytochrome *c* (MOERSCHELL et al., 1991). It could also be used to generate null-alleles in any counterselectable marker. In principle, mutations could be introduced into any gene, even where there is no selection for functional revertants. This could be done by using co-transformation with a separate selectable marker to

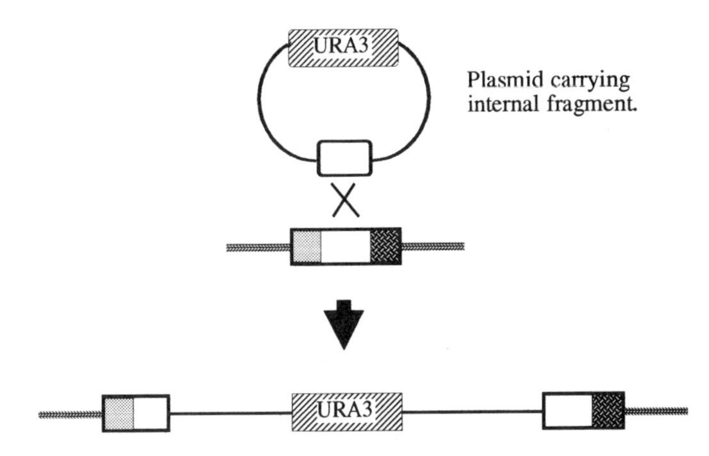

Fig. 9. Gene disruption by homologous recombination with an internal fragment.

introduce the oligonucleotide into a cell and mutations in the gene screened for using colony hybridization.

2.5 Expression of Heterologous Protein

For reviews of heterologous protein expression in yeast see KINGSMAN et al. (1985, 1987), KING et al. (1989), HRISCH et al. (1989), SCHNEIDER and GUARANTE (1991), MOIR and DAVIDOW (1991), and ROMANOS et al. (1992).

2.5.1 Expression Cassettes

Heterologous genes are normally expressed by the construction of a cassette in which the gene is placed between a promoter and a transcriptional terminator.

Yeast promoters resemble those from higher organisms in that that they are controlled by trans-activating or trans-repressing factors binding to upstream elements called UAS (upstream activating sequences) or URS (upstream repressing sequences), respectively. The structure of yeast promoters has been reviewed by MELLOR (1989). Yeast promoters that have been used for the expression of heterologous genes fall into three classes. Firstly, those derived from genes which are normally expressed at high levels but show limited regulation. Secondly, promoters with tight regulation but which usually show lower levels of expression than the first class. Thirdly, hybrid promoters which combine UAS elements from powerful promoters with regulatory elements.

Commonly used promoters of the first class are those derived from the *PGK, ADH*1, *PYK, GPD, and GUT*2 promoters. These all code for enzymes involved in glycolysis which are present in high levels during glucose fermentation. The products of each of these genes accounts for between 1–5% of total cell protein under these conditions. These promoters do show some regulation – levels are elevated 10- to 20-fold on glucose compared to a non-fermentable carbon source such as lactate (TUITE et al., 1982). Nevertheless, this does not approach the induction ratio found with promoters discussed below which are normally used when regulated expression is important. Expression of heterologous proteins is of course dependent on the particular protein being expressed. However, levels rarely approach that possible with the homologous gene of the promoter. For example, a multi-copy plasmid carrying the *PGK* gene and promoter results in 50% of total soluble protein being phosphoglycerokinase. If the same promoter and plasmid are used to express interferon-β, the heterologous protein represents less than 5% of total soluble protein (MELLOR et al., 1985). The reasons for this are not completely understood. It results from effects on both transcript level and efficiency of translation. The former effect may be partly explained by a down-stream activating sequence (DAS) in the *PGK* coding region (MELLOR et al., 1987).

A regulated promoter is necessary for the expression of a protein which is toxic or causes a significant slowing of growth. Commonly used regulatable promoters are those from the *GAL*1 and *PHO*5 genes. The *GAL*1 gene which encodes galactokinase is transcribed from a divergent promoter along with the *GAL*10 gene (for a review see SCHNEIDER and GUARANTE, 1991). This promoter is regulated by the *GAL*4 and *GAL*80 proteins. It is induced over 1000-fold from barely detectable levels to 0.8% total cell protein by a switch of the carbon source from glucose to galactose. The control mechanism involves both repression by glucose and induction by galactose. Rapid induction of the promoter requires growth on a non-repressing carbon source such as lactate so that the only requirement for expression is the addition of an inducer. The *GAL*1 promoter is rarely used industrially for the large-scale production of proteins because of the difficulty in switching the carbon source, the expense of galactose as a carbon source, and the foaming caused when a medium containing galactose is aerated to the high levels necessary for high cell density.

*PHO*5 encodes acid phosphatase induced by low levels of inorganic phosphate in the medium (see review by SCHNEIDER and GUARANTE, 1991). Like the *GAL*1 promoter, the induction ratio is high. Phosphate depletion necessary for induction is inconvenient on a large scale. However, a temperature-sensitive mutation in the *pho*80 repressor of the *PHO*5 gene allows control through a temperature shift (KRAMER et al., 1984).

The ideal type of promoter for industrial expression of proteins would be one that was turned on by glucose exhaustion at the end of batch growth in a fermenter. The promoter for the *PRB*1 gene shows such control (MOEHLE et al., 1987) and has been successfully used for the high-level expression of human serum albumin (SLEEP et al., 1990). Ironically, a mutation that led to increased levels of expression abolished this regulation (SLEEP et al., 1991).

Hybrid promoters which combine the high expression levels of one promoter with the regulatory elements from another have been created. One of these is based on the *PGK* promoter into which a MATα-2 operator has been incorporated and which is therefore repressed by the MATα-2 protein (WALTON and YARRANTON, 1989b). Conditionality is provided by a *sir*3^{ts} mutation in the recipient strain. This results in a temperature-dependent switch in mating type and therefore Maα-2 expression. At 23 °C the strain is functionally a-mating type. At 34 °C mating type switching occurs, the strain becomes functionally α-mating type, MATα-2 is produced, and the hybrid promoter is repressed. A temperature down-shift of a culture growing at 34 °C therefore results in induction of the promoter.

Conditional promoters have also been produced by incorporating mammalian steroid response elements into a powerful promoter. A second plasmid carries the steroid receptor that interacts with the response elements and causes its activation upon addition of the steroid to the medium. An example of this approach is the incorporation of estrogen response elements into the *PGK* promoter. A second plasmid carries a gene for an estrogen receptor. A β-galactosidase reporter gene cloned downstream of the hybrid promoter is induced by the addition of estrogen hormone to the culture medium (METZGER et al., 1988).

As well as mRNA abundance, the translation initiation environment should be considered carefully (reviewed by BROWN, 1989). Efficient translation requires an A at position -3 relative to the AUG start codon, but there is no invariant consensus sequence similar to the ribosome binding sequence in bacteria (HAMILTON et al., 1987). An empirical approach may be necessary to obtain the maximum possible product yield. One approach is to exactly reproduce the sequence 5′ to the AUG codon of the gene normally expressed by the promoter. Alternatively, the native initiation environment of the gene to be expressed may be used as it is found that this is often very effective even from mammalian genes. Lastly, the gene may be fused 3′ to an efficiently expressed gene thus relying on a proven initiation environment. This is what is done anyway when a yeast secretion leader sequence is used to direct secretion of the gene product (see below).

For internal expression a 3′ fusion to the ubiquitin coding sequence has been shown to be very effective. Some sequences clearly reduce translational efficiency. For example, deletion of 10 contiguous G residues sharply increased expression of hepatitis B core antigen (HAMILTON et al., 1987). Thus, placing a coding sequence directly into a vector polylinker region may not result in efficient expression.

Transcriptional terminators are necessary for maximum expression (MELLOR et al., 1983). The exact nature of the sequences responsible for transcriptional termination are still not fully defined, although certain consensus sequences have been suggested (BENNETZEN and HALL, 1982; ZARET and SHERMAN, 1984). It is normal to use a 200–300 bp fragment from the sequence immediately 3′ of the coding region of a highly expressed gene such as *PGK* or *ADH*1.

The expression cassettes are normally placed on multicopy plasmids to ensure the maximum copy number. Vectors based on the 2 μm circle are the most common, since these are much more stable than *ARS*-based vectors. Very large-scale industrial production places a high premium on stability. In this situation disintegration vectors are the best choice. Alternatively, stability may be achieved by integration into a chromosomal location, which may also increase expression for other reasons. Stable vectors allow yeast cells to be grown in rich non-defined medium without selection. This is desirable for two reasons: firstly, higher biomass and therefore high product yields can be obtained on rich as opposed to minimal-defined medium. Secondly, large-scale fermentations are more expensive with defined, synthetic media.

2.5.2 Cell Targeting and Secretion

The processes involved in protein targeting and secretion within the yeast cell have been the subject of extensive genetic and molecular analysis and in outline at least they are understood (DESHAIES et al., 1989). The secretion pathway for invertase was elaborated in a classic series of genetic studies (NOVICK et al., 1981; SCHEKMAN, 1985) and has served as the paradigm for the yeast secretory pathway. It is similar to that which occurs in cells from higher organisms. Proteins which are normally secreted from such cells can also be secreted when expressed in yeast. Protein folding, processing, glycosylation, and disulfide bond formation will occur. In N-linked glycosylation, the correct asparagine residues are modified, but an important difference is that yeast glycosylation is of the high mannose type and proteins which normally have a complex outer layer will not have an authentic pattern when expressed in yeast.

The native peptide signal sequence of the heterologous protein is often very effective in directing secretion and producing a protein in the culture supernatant which is authentically processed. Examples of this are interferons-α1 and α2 (HITZMAN et al., 1983), wheat α-amylase (ROTHSTEIN et al., 1984), and human serum albumin (SLEEP et al., 1990). Alternatively, the heterologous sequence may be fused inframe to a yeast secretion leader sequence. The most commonly used is taken from the mating pheromone prepro-α-factor. α-Factor is a 13-residue peptide secreted from α-mating type cells, which binds to receptor protein on the surface of a-mating type cells. Binding to this receptor induces cellular changes including cell cycle arrest, which are a necessary preliminary to cell fusion. It is produced as a 165 amino acid prepro-protein containing four copies of the mature α-factor sequence separated by spacer regions (KURJAN and HERSKOWITZ, 1982). After translocation across the membrane of the endoplasmic reticulum and removal of the signal sequence, the pro-protein is processed in the Golgi apparatus by three proteases – a carboxypeptidase, the *KEX*2 protease, and the *STE*13 dipeptidyl aminopeptidase (JULIUS et al., 1983, 1984). Overexpression of a heterologous protein often results in the accumulation of proteins with glu-ala residues at the amino terminal indicating that the *STE*13 dipeptidyl aminopeptidase is rate-limiting to processing. This occurs even when α-factor itself is overexpressed (KING et al., 1989). It is normal, therefore, to design a leader sequence in which the fusion of the leader sequence to the coding sequence of the mature heterologous protein occurs after the lys-arg dipeptide removing the need for the action of the *STE*13 dipeptidyl aminopeptidase.

2.5.3 Oversecreting Mutants

Once expression of a secreted product has been obtained, yields may be increased further by the isolation of mutants which increase product yield. This is usually done with some form of colony screen of the amount of secreted product after mutagenesis. For example, SMITH et al. (1985) isolated strains secreting elevated levels of calf prochymosin by screening mutagenized colonies of an expressing strain for elevated clot formation on milk plates. Three genes were identified in this way, one of them, called *PMR*1, was cloned and is thought to act during secretion in the ER to Golgi transition or in the early stages of processing in the Golgi (MOIR and DAVIDOW, 1991). Mutations in *PMR*1 not only elevate levels of chymosin secretion but also several other heterologous proteins indicating that this may be of general utility.

SLEEP et al. (1991) used antibodies to human serum albumin (HSA) to produce visible halos of immunoprecipitated material around colonies. They used this in successive rounds of mutagenesis and colony screening for elevated levels of HSA production. In this case at least some of the mutations were not thought to act during secretion. One of them was shown to abolish the regulation of the *PRB*1 promoter used to program HSA expression. Again, strains isolated by increased production of one product also showed elevated production of another, in this case α1-antitrypsin.

3 Other Yeasts

Besides *Saccharomyces cerevisiae* transformation systems have been described for a total of 11 other yeast species (reviewed by SUDBERY, 1992). The methodology heavily depends on that developed in *S. cerevisiae*. The introduction of DNA requires both a selectable marker and a suitable recipient strain. The most common form of marker is one derived from a biosynthetic pathway equivalent to the *S. cerevisiae LEU*2, *URA*3, *TRP*1, *HIS*4, etc. The sequence may be derived from the homologous system, but often it is found that the sequence derived from *S. cerevisiae* is effective. The homologous gene may be cloned through complementation of the corresponding mutation in *E. coli* or by probing a gene library with the *S. cerevisiae* sequence. The recipient is marked with the appropriate mutation which must be sufficiently stable to prevent revertants occurring at a similar or higher rate than transformants. The mutant strain is usually identified by enzymic assay of mutants with the appropriate phenotype.

Maintenance of the introduced sequence demands either autonomous replication or chromosomal integration. Autonomous replication is usually based on ARS-like sequences isolated in the same way as they were originally isolated in *S. cerevisiae*. Although sequences with ARS activity in *S. cerevisiae* can be isolated from the genome of a second yeast, these do not usually act in the same manner when returned to that yeast. Active *ARS* sequences must be selected for in the same yeast in which they will be used. However, heterologous sequences may have elements which fortuitously promote replication. The origin of replication of the 2 μm plasmid of *S. cerevisiae* does not function in other yeasts, indeed, its presence has been reported to be inhibitory to transformation in *Kluyveromyces fragilis* (DAS et al., 1984).

Integration can be brought about by transformation with vectors lacking an origin of replication. In many yeasts targeted integration or gene disruption as described above can also be used. The stringency of the recombination is variable, so that in some cases integration may take place at non-homologous sites, or more complex events may occur.

The most important of the other yeasts in terms of contribution to fundamental knowledge has been *Schizosaccharomyces pombe*. Studies on the cell cycle and the role of the p34 kinase product of the *CDC*2 gene have a major influence on the whole field of cell proliferation research in eukaryotes (reviewed by NURSE, 1990). Laboratory methods for this yeast have been reviewed recently (MORENO et al., 1991; NASIM et al., 1989). Because of the relatively limited use of *S. pombe* in biotechnology (apart from its role in the production of pombe beer in West Africa), it will not be reviewed here further.

3.1 Methylotrophic Yeasts

3.1.1 Introduction

Apart from *S. cerevisiae,* perhaps the most successful yeasts for heterologous protein expression are the methylotrophic yeast species *Hansenula polymorpha* and *Pichia pastoris.* In both cases the powerful and regulatable promoter for the gene known as methanol oxidase (MOX) in *H. polymorpha* or alcohol oxidase (AOX) in *P. pastoris* is used. For reviews see GLEESON and SUDBERY (1988), SUDBERY and GLEESON (1989), HAGENSON (1991).

Alcohol (methanol) oxidase catalyzes the oxidation of methanol to hydrogen peroxide and formaldehyde - the first step in methanol utilization. During growth on methanol the enzyme accounts for some 30–40% of total cell protein, sequestered in peroxisomes which can occupy up to 80% of the cell volume. During growth on glucose the enzyme activity is reduced to very low levels. The regulation of expression is brought about by a combination of glucose or ethanol repression and methanol induction. In *H. polymorpha* this binary control means that under derepressing conditions such as glucose limitation or growth on suboptimal carbon sources such as glycerol, significant levels of AOX activity are observed. In *Pichia pastoris* there is an absolute requirement for methanol in order for any significant expression to be seen.

3.1.2 Techniques, Selection Markers, and Vectors for Molecular Transformation of Methylotrophic Yeasts

Techniques for molecular transformation for *Pichia pastoris* were developed by CREGG et al. (1985). DNA was introduced by the protoplast method. The selectable marker used was the sequence homologous to the *S. cerevisiae HIS4* gene complementing a *P. pastoris* mutant defective in histidinol dehydrogenase. Either the homologous *HIS4* sequence from *P. pastoris* or the heterologous sequence from *S. cerevisiae* would function. Sequences called PARS acting in the same manner as the ARS sequences of *S. cerevisiae* were isolated. These resulted in a high frequency of transformation ($>10^5$ transformants per μg DNA) and in unstable autonomous replication of the plasmid for over 50 generations. Fragments carrying the *S. cerevisiae LEU2* or *HIS4* genes also replicated autonomously showing that they contain ARS activity in *P. pastoris* (but not in *S. cerevisiae*).

Integration was effectively achieved by transforming with linear molecules produced by the digestion of plasmids carrying the *P. pastoris HIS4* gene with *StuI* for which the *HIS4* contains a unique site. Alternatively, linear fragments carrying the *HIS4* gene between the 5′ and 3′ ends of the AOX locus result in the replacement of the resident AOX locus (CREGG et al., 1987). In the latter case the AOX coding sequence is replaced with a heterologous gene, so that the AOX function is lost in the replacement event. Since *P. pastoris* contains a second AOX locus, this does not result in a loss of methanol growth capacity, but it does slow down the growth rate. Multiple integration events leading to increased gene dosage elevate product yield without reducing stability.

Transformation systems for *Hansenula polymorpha* have been developed by several groups (ROGGENKAMP et al., 1986; GLEESON et al., 1986; TIKHOMIROVA et al., 1986). ROGGENKAMP et al. (1986) made use of the *S. cerevisiae URA3* sequence to complement the equivalent mutation in *H. polymorpha.* Replication was from an ARS-like sequence called HARS isolated from the *H. polymorpha* genome. GLEESON et al. (1986) used the *S. cerevisiae LEU2* sequence present on YEp13 (BROACH et al., 1979). This plasmid complements a Leu$^-$ *Hansenula polymorpha* mutant defective in *β*-isopropylmalate dehydrogenase, the enzyme coded for by *LEU2* in *S. cerevisiae* and the plasmid replicated autonomously. TIKHOMIROVA et al. (1986) also used a system based on the *S. cerevisiae LEU2* gene and fragments from *Candida utilis* mitochondrial DNA to program replication; this resulted in the *in vivo* concatamerization of the plasmids. An optimized protocol for *H. polymorpha* transformation has been reported by BERARDI and THOMAS (1991) which the reader is advised to use.

Besides auxotrophic markers, dominant selectable markers have also been used for transformation. TIKHOMIROVA et al. (1988) used the copper resistance *CUP*1^r gene to select copper resistance, and we have used resistance to the aminoglycoside G418, coded for by the *NPRT II* gene of Tn5 (VEALE, 1989; SUDBERY et al., 1988). In the case of copper resistance, TIKHOMIROVA et al. (1988) showed that although it was an inefficient way of selecting transformants, it was possible to increase copy number by growing transformed strains on successively increasing concentrations of copper. We inserted the *NPRT II* gene between the AOX promoter and terminator and transformed the construct on a linear fragment with the free ends homologous to the two flanking regions of the AOX locus - effectively the one-step gene disruption procedure described above in *S. cerevisiae*. The transformants had a disrupted AOX gene and were thus unable to grow on methanol. They were resistant to very high concentrations of G418 (>20 mg mL^{-1}). This procedure allows stable integration of a heterologous gene into the genome.

Vectors relying on *ARS* sequences are unstable. Integration can be selected for by a regime of alternating growth under selective and non-selective conditions. This may result from homologous recombination (FELLINGER et al., 1991) or non-homologous recombination (JANOWICZ et al., 1991). In the latter case multiple tandem repeats were generated which greatly aided the expression of a heterologous gene (HBsAg). Vectors which simply consist of the *S. cerevisiae LEU2* gene cloned into a general-purpose *E. coli* vectors, such as pUC18, are very effective in allowing transformation. We have found (SUDBERY, unpublished observations) that such plasmids do replicate initially, but quickly integrate when grown under appropriate conditions (described above). We find them the most convenient vector base for introducing genes into *H. polymorpha*.

3.1.3 Expression of Heterologous Genes

A number of heterologous proteins have been expressed in both *Pichia pastoris* (HAGENSON, 1991) and *H. polymorpha* (JANOWICZ et al., 1991; FELLINGER et al., 1991) using the AOX promoter and terminator. Generally high cell densities (>100 g dry wt/L) are produced by growth on glycerol (i.e., in the absence of expression) followed by a production phase induced by the addition of methanol. Product yields have been high, at least as good or better than the comparable product in *S. cerevisiae*. In the case of hepatitis B surface antigen (HBsAg) the yield is 50- to 100-fold higher, and the product naturally aggregates into 22 nm virus-like particles necessary for immunogenicity (CREGG et al., 1987). Expression can be internal or secreted via the *S. cerevisiae* invertase or MFα leader sequences. Glycosylation occurs during secretion. In a study of *S. cerevisiae* invertase expressed in *P. pastoris*, GRINNA and TSCHOPP (1989) showed that glycosylation occurred at 9 out of the possible 14 asparagine N-linked glycosylation sites as occurs in the native protein in *S. cerevisiae*. In contrast to *S. cerevisiae*, however, the product from *P. pastoris* contains much shorter and more uniformly sized chains which, it is claimed, more closely resemble the pattern found in animal cells.

4 References

BECKER, D. M., GUARANTE, L. (1991), High-efficiency transformation of yeast cells by electroporation, in: *Guide to Yeast Genetics and Molecular Biology* (GUTHRIE, C., FINK, G. R., Eds.), *Methods Enzymol.* **194**, 183–187.

BEGGS, J. D. (1978), Transformation of yeast by a replicating hybrid plasmid, *Nature* **275**, 104–109.

BEGGS, J. D. (1981), in: *Molecular Genetics in Yeast* (VON WETTSTEIN, D., Ed.), Vol. 16, p. 383, *Alfred Benzon Symposium*, Copenhagen.

BENNETZEN, J. D., HALL, B. D. (1982), The primary structure of the *Saccharomyces cerevisiae* gene for alcohol dehydrogenase 1, *J. Biol. Chem.* **257**, 3018–3025.

BERARDI, E., THOMAS, D. Y. (1991), An effective transformation method for *Hansenula polymorpha, Gene* **27**, 23–33.

BOEKE, J. D., LACROUTE, F., FINK, J. R. (1984), A positive selection method for mutants lacking orotidine-5′-phosphate dehydrogenase activity in yeast: 5′-fluoro-orotic acid resistance, *Mol. Gen. Genet.* **197**, 345–347.

BROACH, J. R., STRATHERN, J. N., HICKS, J. B. (1979), Transformation in yeast: development of a hybrid cloning vector and isolation of the CAN1 gene, *Gene* **8**, 121–133.

BROWN, A. J. P. (1989), Messenger RNA translation and degradation in *Saccharomyces cerevisiae,* in: *Molecular and Cell Biology of Yeasts* (WALTON, E. F., YARRANTON, G. T., Eds.), pp. 70–98, Glasgow–London: Blackie/New York: Van Nostrand Reinhold.

BURKE, D. T., CARLE, G. F., OLSON, M. V. (1987), Cloning of large segments of exogenous DNA into yeast by means of artificial chromosomes, *Science* **236**, 806–812.

BUTT, T. R., STERNBERG, E., HERD, J., CROOKE, S. T. (1984), Cloning and expression of a yeast copper metallothionein gene, *Gene* **27**, 23–33.

CHATOO, B. B., SHERMAN, F. F., AZUBALIS, D. A., FJELLSTEDT, T. A., MEHNERT, D., OGUR, M. (1979), Selection of *lys2* mutants of the yeast *Saccharomyces cerevisiae* by the utilisation of α-amino adopate, *Genetics* **93**, 51–65.

CHINNERY, S. A., HINCHCLIFFE, E. (1989), A novel class of vector for yeast transformation, *Curr. Genet.* **16**, 21–25.

CLARKE, L., CARBON, J. (1980), Isolation of a yeast centromere and construction of functional small circular chromosomes, *Nature* **287**, 504–509.

CONSTANZO, M. C., FOX, T. D. (1988), Transformation of yeast by agitation with glass beads, *Genetics* **120**, 667–670.

CREGG, J. M., BARRINGER, K. J., HESSLER, A. Y., MADDEN, K. R. (1985), *Pichia pastoris* as a host for transformations, *Mol. Cell. Biol.* **5**, 3376–3385.

CREGG, J. M., TSCHOPP, J. F., STILLMAN, C., SIEGEL, R., AKONG, M., CRAIG, W. S., BUCKHOLTZ, R. G., MADDEN, K. R., KELLARIS, P. A., DAVIS, G. R., SMILEY, B. L., CRUZE, J., TERRAGROSSA, R., VELICELEBI, G., THILL, G. (1987), High-level expression and efficient assembly of hepatitis B surface antigen in the methylotrophic yeast *Pichia pastoris, Bio/Technology* **5**, 479–485.

DAS, S., ELLERMAN, E., HOLLENBERG, C. P. (1984), Transformation of *Kluyveromyces lactis, J. Bacteriol.* **158**, 1165–1167.

DESHAIES, R. J., KEPES, F., BOHNI, P. C. (1989), Genetic dissection of the early protein secretion in yeast, *Trends Genet.* **5**, 87–93.

FELLINGER, A. J., VEALE, R. A., SUDBERY, P. E., BOM, I. J., VERBAKEL, J. M. A., VERRIPPS, C. T. (1991), Expression of the α-galactosidase of *Cyamposis tetragonobla* (Guar) in *Hansenula polymorpha, Yeast* **7**, 463–474.

GLEESON, M. A. G., SUDBERY, P. E. (1988), Genetic analysis in the methylotrophic yeast *Hansenula polymorpha, Yeast* **4**, 293–303.

GLEESON, M. A. G., ORTORI, S. O., SUDBERY, P. E. (1986), Transformation of the methylotrophic yeast *Hansenula polymorpha, J. Gen. Microbiol.* **132**, 3459–3465.

GRINNA, L. S., TSCHOPP, J. F. (1989), The distribution and general features of linked oligosaccharides from the methylotrophic yeast *Pichia pastoris, Yeast* **5**, 106–115.

GRITZ, L., DAVIES, J. (1983), Plasmid encoded hygromycin B resistance: the sequence of hygromycin B phosphotransferase and its expression in *Escherichia coli* and *Saccharomyces cerevisiae, Gene* **25**, 179–188.

GUTHRIE, C., FINK, G. R. (Eds.) (1991), *Guide to Yeast Genetics and Molecular Biology, Methods Enzymol.* **194**.

HADWIGER, J. A., WITTENBERG, C., DE BARROS LOPES, M. A., RICHARDSON, H. E., REED, S. I. (1989), A family of cyclin homologs that control G1 phase in yeast, *Proc. Natl. Acad. Sci. USA* **86**, 6255–6259.

HAGENSON, M. J. S. (1991), Production of recombinant proteins from the methylotrophic yeast *Pichia pastoris,* in: *Purification and Analysis of Recombinant Proteins* (SEETHARAM, R., SHARMA, S. K., Eds.), New York: Marcel Dekker Inc.

HAMILTON, R., WANATEBE, C. K., DE BOER, A. (1987), Compilation and comparison of the sequence context around the AUG start codons of *Saccharomyces cerevisiae, Nucleic Acids Res.* **15**, 3581–3593.

HINNEN, A., HICKS, J. B., FINK, G. R. (1978), Transformation of yeast, *Proc. Natl. Acad. Sci. USA* **75**, 1929–1933.

HITZMAN, R. A., LEUNG, D. W., PERRY, L. J., KOHR, W. J., LEVINE, H. L., GOEDDEL, D. R. (1983), Secretion of human interferon by yeast, *Science* **219**, 260–265.

HOFFMAN, W. (1985), Molecular characterisation of the *CAN1* locus of *Saccharomyces cerevisiae, J. Biol. Chem.* **260**, 11831–11837.

HRISCH, H. H., RENDUELES, P. S., WOLF, D. H. (1989), Yeast (*Saccharomyces cerevisiae*) proteinases: structure, characteristics and function, in: *Molecular and Cell Biology of Yeasts* (WALTON,

E. F., Yarranton, G. T., Eds.), pp. 134–187, Glasgow–London: Blackie/New York: Van Nostrand Reinhold.

Ito, H., Fukada, Y., Murata, K., Kimura, A. (1983), Transformation of intact yeast cells treated with alkali cations, *J. Bacteriol.* **153**, 163–168.

Janowicz, Z. A., Melber, K., Merkelbach, E., Jacobs, E., Harford, N., Comberbach, M., Hollenberg, C. P. (1991), Simultaneous expression of the S and L surface antigens of hepatitis B, and formation of mixed particles in the methylotrophic yeast *Hansenula polymorpha, Yeast* **7**, 431–445.

Jiminez, A., Davies, J. (1980), Expression of a transposable genetic element in *Saccharomyces, Nature* **287**, 869–871.

Julius, D., Blair, L., Brake, A., Sprague, G., Thorner, J. (1983), Yeast α-factor is processed from a larger precursor polypeptide: the essential role of a membrane bound dipeptidyl amino peptidase, *Cell* **32**, 838–852.

Julius, D., Schekman, R., Thorner, J. (1984), Glycosylation and processing of prepro-α-factor through the yeast secretory pathway, *Cell* **36**, 309–318.

Kaufer, A. F., Fried, H. M., Schindinger, W. F., Jasin, M., Warner, J. R. (1983), Cyclohexamide resistance in yeast: the gene and its proteins, *Nucleic Acids Res.* **11**, 3123.

King, E. F., Walton, E. F., Yarranton, G. T. (1989), The production of proteins and peptides from yeast, in: *Molecular and Cell Biology of Yeasts* (Walton, E. F., Yarranton, G. T., Eds.), pp. 107–133, Glasgow–London: Blackie/New York: Van Nostrand Reinhold.

Kingsman, S. M., Kingsman, A. J., Dobson, M. J., Mellor, J., Roberts, N. A. (1985), Heterologous gene expression in *Saccharomyces cerevisiae,* in: *Biotechnology and Genetic Engineering Reviews* (Russell, G. E., Ed.), pp. 377–416, Newcastle upon Tyne: Intercept.

Kingsman, S. M., Kingsman, A. J., Mellor, J. (1987), The production of mammalian proteins in *Saccharomyces cerevisiae, Trends Biotechnol.* **5**, 53–57.

Kramer, R. A., DeChiara, T. M., Schaber, M. D., Hilliker, S. (1984), Regulated expression of a human interferon gene in yeast: regulation by phosphate concentration or temperature, *Proc. Natl. Acad. Sci. USA* **81**, 367–370.

Kurjan, J., Herskowitz, I. (1982), Structure of yeast pheromone gene (MFα): a putative α-factor precursor contains four copies of tandem copies of mature α-factor, *Cell* **30**, 933–943.

Meacock, P. A., Breiden, K. W., Cashmore, A. M. (1989), The two micron circle: model replicon and yeast vector, in: *Molecular and Cell Biology of Yeasts* (Walton, E. F., Yarranton, G. T., Eds.), pp. 330–359, Glasgow–London: Blackie/New York: Van Nostrand Reinhold.

Mellor, J. (1989), The activation and repression of transcription by the promoters of *Saccharomyces cerevisiae,* in: *Molecular and Cell Biology of Yeasts* (Walton, E. F., Yarranton, G. T., Eds.), pp. 1–37, Glasgow–London: Blackie/New York: Van Nostrand Reinhold.

Mellor, J., Dobson, M. J., Roberts, N. A., Tuite, M. F., Emtage, J. S., White, S., Lowe, P. A., Patel, T., Kingsman, A. J., Kingsman, S. M. (1983), Efficient synthesis of enzymatically active calf chymosin in *Saccharomyces cerevisiae, Gene* **24**, 1–14.

Mellor, J., Dobson, M. J., Roberts, N. A., Kingsman, A. J., Kingsman, S. M. (1985), Factors affecting heterologous gene expression in *Saccharomyces cerevisiae, Gene* **33**, 215–226.

Mellor, J., Dobson, M. J., Kingsman, A. J., Kingsman, S. M. (1987), A transcriptional activator is located in the coding region of the yeast PGK gene, *Nucleic Acids Res.* **15**, 6243.

Metzger, D., White, J. H., Chambon, P. (1988), The oestrogen receptor functions in yeast, *Nature* **334**, 31–36.

Moehle, C. M., Aynardi, M. W., Kolodny, M. R., Park, F. J., Jones, E. W. (1987), Protease B of *Saccharomyces cerevisiae*: Isolation and regulation of the *PRB*1 structural gene, *Genetics* **115**, 255–263.

Moerschell, R. P., Das, G., Sherman, F. (1991), Transformation of yeast directly with synthetic oligonucleotides, in: *Guide to Yeast Genetics and Molecular Biology* (Guthrie, C., Fink, G. R., Eds.), *Methods Enzymol.* **194**, 362–372.

Moir, D. T., Davidow, L. S. (1991), Production of proteins by secretion in *Saccharomyces cerevisiae,* in: *Guide to Yeast Genetics and Molecular Biology* (Guthrie, C., Fink, G. R., Eds.), *Methods Enzymol.* **194**, 491–507.

Moreno, S., Klar, A., Nurse, P. (1991), Molecular genetic analysis of fission yeast *Schizosaccharomyces pombe,* in: *Guide to Yeast Genetics and Molecular Biology* (Guthrie, C., Fink, G. R., Eds.), *Methods Enzymol.* **194**, 795–823.

Murray, A., Szostak, J. W. (1983), Construction of artificial chromosomes in yeast, *Nature* **305**, 185–192.

Nash, R., Tokiwa, G., Anand, S., Futcher, A. B. (1988), The *WHI*1$^+$ gene of *S. cerevisiae* tethers cell division to cell size and is a cyclin homolog, *EMBO J.* **7**, 4335–4346.

Nasim, A., Young, P., Johnson, B. F. (1989),

The Molecular Biology of Fission Yeast, New York: Academic Press.

NOVICK, P., FERRO, S., SCHEKMAN, R. (1981), Order of events in the yeast secretory pathway, *Cell* **25,** 461.

NURSE, P. (1990), Universal control mechanism regulating the onset of mitosis, *Nature* **344,** 503–508.

ORR-WEAVER, T. L., SZOSTAK, J. W., ROTHSTEIN, R. J. (1981), Yeast transformation a model system for the study of transformation, *Proc. Natl. Acad. Sci. USA* **78,** 6354-6358.

REED, S. I. (1991), G1-specific cyclins: In search of an S-phase specific promoting factor, *Trends Genet.* **7,** 95–99.

ROGGENKAMP, R., HANSEN, H., ECKART, M., JANOWITZ, Z., HOLLENBERG, C. P. (1986), Transformation of the methylotrophic yeast *Hansenula polymorpha* by replication and integration vectors, *Mol. Gen. Genet.* **202,** 302–308.

ROMANOS, M. A., SCORER, C. A., CLARE, J. J. (1992), Foreign gene expression in yeast: a review, *Yeast* **8,** 423–488.

ROSE, M. D., BROACH, J. R. (1991), Cloning genes by complementation in yeast, in: *Guide to Yeast Genetics and Molecular Biology* (GUTHRIE, C., FINK, G. R., Eds.), *Methods Enzymol.* **194,** 195–239.

ROTHSTEIN, R. (1991), Targeting, disruption, replacement, and allele rescue: integrative transformation in yeast, in: *Guide of Yeast Genetics and Molecular Biology* (GUTHRIE, C., FINK, G. R., Eds.), *Methods Enyzmol.* **194,** 281–301.

ROTHSTEIN, S. J., LAZARUS, C. M., SMITH, W. E., BAULCOMBE, D. C., GATENBY, A. A. (1984), Secretion of a yeast α-amylase expressed in yeast, *Nature* **308,** 662–665.

SCHEKMAN, R. (1985), Protein localisation and membrane traffic in yeast, *Annu. Rev. Cell. Biol.* **1,** 115–143.

SCHERER, S., DAVIS, R. W. (1979), Replacement of chromosomal elements with altered DNA sequences constructed *in vitro, Proc. Natl. Acad. Sci. USA* **76,** 4951–4955.

SCHNEIDER, J. C., GUARANTE, L. (1991), Vectors for expression of cloned genes in yeast: regulation, overproduction and underproduction, in: *Guide to Yeast Genetics and Molecular Biology* (GUTHRIE, C., FINK, G. R., Eds.), *Methods Enzymol.* **194,** 373–388.

SHORTLE, D., NOVICK, P., BOSTEIN, D. (1984), Construction and genetic characterisation of temperature sensitive mutant alleles of the yeast actin gene, *Proc. Natl. Acad. Sci. USA* **81,** 4889–4893.

SIKORSKI, R. S., BOEKE, J. D. (1991), *In vitro* mutagenesis: from cloned gene to mutant yeast, in: *Guide to Yeast Genetics and Molecular Biology* (GUTHRIE, C., FINK, G. R., Eds.), *Methods Enzymol.* **194,** 302-317.

SLEEP, D., BELFIELD, G. P., GOODEY, A. R. (1990), The secretion of human serum albumin from the yeast *Saccharomyces cerevisiae* using 5 different leader sequences, *Bio/Technology* **8,** 42–46.

SLEEP, D., BELFIELD, G. P., BALLANCE, D. J., JONES, S. S., EVANS, L. R., MOIR, P. D., GOODEY, A. R. (1991), *Saccharomyces cerevisiae* strains that overexpress heterologous proteins, *Bio/Technology* **9,** 183–187.

SMITH, R. A., DUNCAN, M. J., MOIR, D. T. (1985), Heterologous protein secretion from yeast, *Science* **229,** 1219–1244.

STINCHCOMBE, D. T., STRUHL, K., DAVIES, R. W. (1979), Isolation and characterisation of a yeast chromosomal replicator, *Nature* **282,** 39–43.

STRUHL, K., STINCHCOMBE, D. T., SCHERER, S., DAVIES, R. W. (1979), High frequency transformation of yeast: autonomous replication of hybrid DNA molecules, *Proc. Natl. Acad. Sci. USA* **76,** 1035-1039.

SUDBERY, P. E. (1992), The Genetics of the lesser known industrial yeasts, in: *The Yeasts,* Vol. 6 (ROSE, A. H., HARRISON, J. S., WHEALS, A. W., Eds.), Academic Press, in press.

SUDBERY, P. E., CARTER, B. L. A., GOODEY, A. R. (1980), Genes that control cell proliferation in *Saccharomyces cerevisiae, Nature* **228,** 401-404.

SUDBERY, P. E., GLEESON, M. A., VEALE, R. A., LEDEBOER, A. M., ZOETMULDER, M. C. M. (1988), *Hansenula polymorpha* as a novel expression system for the expression of heterologous proteins, *Trans. Biochem. Soc.* **16,** 1081.

SUDBERY, P. E., GLEESON, M. A. G. (1989), Genetic manipulation of methylotrophic yeasts, in: *Molecular and Cell Biology of Yeasts* (WALTON, E. F., YARRANTON, G. T., Eds.), pp. 304-329, Glasgow–London: Blackie/New York: Van Nostrand Reinhold.

TIKHOMIROVA, L. P., IKONOMOVA, R. N., KUTZETSOVA, E. N. (1986), Evidence for autonomous replication and stabilisation of recombinant plasmids in transformants of the yeast *Hansenula polymorpha, Curr. Genet.* **10,** 741–747.

TIKHOMIROVA, L. P., IKONOMOVA, R. N., KUTZETSOVA, E. N., FODOR, I. I., BYSTRYKH, L. V., AMINOVA, L. R., TROTSENKO, Y. A. (1988), Transformation of methylotrophic yeast *Hansenula polymorpha:* Cloning and expression of genes, *J. Basic Microbiol.* **28,** 343–351.

TUITE, M. F., DOBSON, M. J., ROBERTS, N. A., KING, R. M., BURKE, D. C., KINGSMAN, S. M., KINGSMAN, S. J. (1982), Regulated high efficiency expression of human interferon-alpha in *Saccharomyces cerevisiae, EMBO J.* **1**, 603–608.

VEALE, R. A. (1989), Heterologous gene expression in *Hansenula polymorpha, Ph.D Thesis,* University of Sheffield, Sheffield, UK.

WALTON, E. F., YARRANTON, G. T. (Eds.) (1989a), *Molecular and Cell Biology of Yeasts,* Glasgow–London: Blackie/New York: Van Nostrand Reinhold.

WALTON, E. F., YARRANTON, G. T. (1989b), Negative regulation of mating type, in: *Molecular and Cell Biology of Yeasts* (WALTON, E. F., YARRANTON, G. T., Eds.), pp. 43–69, Glasgow–London: Blackie/New York: Van Nostrand Reinhold.

WEBSTER, T. D., DICKSON, R. C. (1983), Direct selection of *Saccharomyces cerevisiae* resistant to the antibiotic G418 following transformation with a DNA carrying the kanamycin-resistance gene of Tn903, *Gene* **26**, 243–252.

ZARET, K. S., SHERMAN, F. (1982), DNA sequence required for efficient transcription termination in yeast, *Cell* **28**, 563–573.

14 Genetic Engineering of Filamentous Fungi

GEOFFREY TURNER

Sheffield, United Kingdom

1 Introduction

Filamentous fungi have been of considerable importance in biotechnology for many years. They have been used for food production and as food for many centuries, as directly edible mushrooms and in the production of oriental fermented food and drink. More recently, they have been used for the production of primary and secondary metabolites and enzymes. Their direct consumption for food as mushrooms (mostly Basidiomycetes) is still the most economically important use, while production of beta-lactam antibiotics is the most significant fermentation use.

Microfungi, particularly the Euascomycetes *Aspergillus nidulans* and *Neurospora crassa,* have been used widely for fundamental studies on genetics, and detailed evidence for the 1 gene-1 enzyme hypothesis came from studies on *Neurospora crassa* (BEADLE, 1946). *Ascobolus* and *Sordaria* have also provided models for meiotic recombination (Chapter 3). Both fundamental and applied aspects of fungal biology have been the driving forces behind the development of fungal vectors and genetic manipulation techniques. While approaches found successful for *Saccharomyces cerevisiae* provided the initial guidelines for these developments, it was often found that filamentous Ascomycetes did not behave exactly the same as the yeasts (Hemiascomycetes), and alternative methods were devised. One of the first problems encountered in early attempts to devise transformation systems was that yeast genes did not seem to function in filamentous fungi, mostly as a result of promoter differences, e.g. (ULLRICH et al., 1985).

Transformation of *N. crassa* and *A. nidulans* was rapidly followed by other fungi, including those of commercial importance for biotechnology and as plant pathogens. By far the majority of these are either Ascomycetes or Deuteromycetes (no observed sexual stage) which are closely related to the Ascomycetes. One reason for the rapid progress with these species was the observation that a gene from one Ascomycete species usually expresses satisfactorily in another Ascomycete or Deuteromycete species. However, greater expression barriers seem to occur between the different fungal classes, and transformation systems for Basidiomycetes, which include edible mushrooms, have generally necessitated prior isolation of suitable Basidiomycete genes. At the time of writing, no transformation system has yet been reported for the edible species *Agaricus bisporus,* though molecular genetic studies have begun (HINTZ et al., 1989). For a detailed review of fungal transformation, see (FINCHAM, 1989).

2 Basic Techniques

2.1 Methods for Introducing DNA into Fungi

Transformation procedures are largely based on those devised for yeast, that is, release of protoplasts from mycelium using lytic enzymes such as Novozym 234, stabilization of resulting protoplasts in osmotic buffer, incubation with DNA, Ca^{2+} ions and polyethylene glycol, and regeneration on osmotically buffered solid medium. While this method is still the usual approach, more convenient alternatives have been tried, including electroporation (GOLDMAN et al., 1990) and treatment of whole mycelium with lithium acetate (DHAWALE et al., 1984). Electroporation has not been successful for filamentous fungi without prior removal of the cell wall, which to some extent defeats the objective of convenience. It has also been possible to introduce DNA into whole mycelium using the biolistic or shotgun approach, in which DNA-coated tungsten balls are shot into the mycelium (ARMALEO et al., 1990). However, this method is not convenient for most laboratories.

Quality of protoplasts is rather hard to define except in terms of "transformability", but does seem to be variable and important. Therefore, many attempts at improving transformation efficiency have centered on DNA entry. These have included variations in polyethylene glycol concentration, molecular weight, and time of exposure, addition of nuclease inhibitors to protoplasts, heat treatment of protoplasts, addition of protease inhibitors to the lytic enzyme. Sometimes transformation problems arise in a previously successful system for no obvious reason, leading investigators to check water quality, glassware cleaning, lytic enzyme batch and so on, until frequencies improve. This reflects the large amount of uncertainty about the factors involved in achieving optimum transformation efficiency.

2.2 Transformation Vectors

2.2.1 Selectable Markers

2.2.1.1 Transformation of a Mutant Recipient

Transformation of a nutritional mutant recipient with a prototrophic (wild-type) gene is the common method in fundamental genetic research of fungi such as *Aspergillus nidulans* and *Neurospora crassa*. This is because a wide range of mutants were already available and can be easily introduced into different backgrounds via a sexual cross. Examples of such selection systems are given in Tab. 1. Some of these systems are particularly useful because two-way selection is possible, that is, the marker (wild-type allele) can be selected *for* during transformation, and *against* for gene replacement or excision.

This ability to select for the mutant allele further opened the possibility of convenient mutant isolation in fungal species where suitable recipient strains were not available. Isolation of auxotrophic mutants is a time-consuming process, and positive selection for a mutant strain is a great advantage. All three such selection systems, fluoroorotate, chlorate, and selenate are now finding wide application.

2.2.1.2 Transformation of Wild-Type Recipient Strains

A wide variety of nutritional mutants is not generally available for important commercial species such as *Aspergillus niger* and *Penicillium chrysogenum*. Indeed, it is usually regarded as undesirable to introduce such defects into production strains. In such cases, an alternative type of selection system is desirable. The *amdS* gene of *A. nidulans* encodes acetamidase, which hydrolyzes acetamide to ammonium and acetate, providing both a nitrogen and a carbon source. This gene provided an early example of transformation by mutant complementation in *A. nidulans* (TILBURN et al., 1983). However, many fungi grow very poorly on acetamide as nitrogen source, and insertion of the *A. nidulans amdS* gene into wild-type strains results in greatly improved growth on acetamide medium. This was first observed with *A. niger* (KELLY and HYNES, 1985), but has since been used with other fungi (Tab. 2).

Drug resistance markers provide the most common way of transforming a wild-type recipient strain. A list of such markers is provided in Tab. 2. There are two main types of selection marker: (A) drug-resistant alleles of fungal genes and (B) bacterial resistance genes fused to fungal promoters. In most cases, lower transformation frequencies are observed with drug selection systems, though the benlate selection system of *N. crassa* is very efficient.

2.2.2 Integration and Replication

2.2.2.1 Integration

Transforming DNA is usually integrated into the chromosomes by both homologous (type I) and non-homologous (type II) recombination during vegetative growth. Homologous integration of circular and linear DNA resembles that seen in yeast (see Chapter 13, Fig. 1), except that linearization of circular DNA by cutting in a homologous sequence does not result in a dramatic increase in transformation frequency (see Chapter 13, Fig. 2). The relative frequencies with which homologous and non-homologous (ectopic) integration events are observed depends on the species and the length of the homologous region. In *Aspergillus nidulans* and *Penicillium chrysogenum*, over half of the integration events are homologous (BULL et al., 1988; YELTON et al., 1984), while in *Neurospora crassa*, such

Tab. 1. Examples of Nutritional Selectable Markers[a]

Gene	Function	Species of Origin	Other Species Transformed
acuA^{+} [b]	Acetyl CoA synthetase	*Ustilago maydis*	
ade-2^{+}	unknown	*Schizophyllum commune*	*Phanerochaete chrysosporium*
amdS^{+} [b]	Acetamidase	*Aspergillus nidulans*	*A. niger*[1] *Penicillium chrysogenum*[2] *Trichoderma reesei*[3]
argB^{+}	L-Ornithine carbamoyl transferase	*A. nidulans*	*A. oryzae*[4] *A. niger* *T. reesei*[3]
leu^{+}	unknown	*Mucor circinelloides*	
met-2^{+}	Homoserine-O-transacetylase	*Ascobolus immersus*	
niaD^{+}	Nitrate[b c] reductase	*A. nidulans* *A. niger* *A. oryzae* *P. chrysogenum*	
pyr-3^{+}	Dihydroorotase	*U. maydis*	
pyr-4^{+} [b c]	Orotidine-5′-phosphate	*Neurospora crassa*	*A. nidulans*
pyrG^{+}	decarboxylase	*A. nidulans* *A. oryzae* *A. niger*	
qa-2^{+}	Catabolic dehydroquinase	*N. crassa*	
sC^{+} [b]	ATP sulfurylase	*A. nidulans*	*A. niger*[5]
trpC^{+} *trp1*$^{+}$	Tryptophan[d] synthesis	*A. nidulans* *A. niger* *N. crassa* *P. chrysogenum* *S. commune* *Coprinus cinereus*	
ura-5^{+} [b]	Orotidylate pyrophosphorylase	*Podospora anserina*	

[a] adapted from VAN DEN HONDEL and PUNT (1991)
[b] 2-way selectable marker
[c] heterologous transformation used widely
[d] multifunctional enzymes encoded

[1] KELLY and HYNES (1985), [2] BERI and TURNER (1987), [3] PENTTILÄ et al. (1991), [4] CHRISTENSEN et al. (1988), [5] BUXTON et al. (1989)

events can be less than 5% (DHAWALE and MARZLUF, 1985). When homologous recombination events occur, they often result in gene conversion so that the recipient mutant allele is repaired to wild-type without integration of the whole of the transforming vector (type III). Homologous integration is particularly important for a number of manipulation techniques such as gene disruption and deletion (Sect. 2.4.1).

It is possible to reverse the integration event and select for excision of the plasmid if a two-way selection marker is used in the vector. For instance, when oligomycin-resistant transfor-

Tab. 2. Inhibitor Resistant Markers[a]

A. Fungal Genes from Resistant Mutants

Gene	Function	Origin	Inhibitor
*oliC*r	ATP synthase subunit 9	*Aspergillus nidulans* *A. niger* *Penicillium chrysogenum*	Oligomycin
tub-2	β-Tubulin	*Neurospora crassa* *Septoria nodorum*	Benomyl

B. Bacterial Genes with Fungal Promoter

Gene	Function	Origin	Inhibitor	Species Transformed
bar	Phosphinothricin acetylase	*Streptomyces hygroscopicus*	Bialophos	*N. crassa*
*bs*r	Deaminase	*Bacillus cereus*	Blasticydin S	*Rhizopus niveus*[1]
ble	Phleomycin binding protein	*Escherichia coli* *Streptomyces hindustanus*		*Aspergillus* spp. *P. chrysogenum*
hph	Hygromycin B phosphotransferase	*E. coli*		*Aspergillus* spp. *P. chrysogenum* Plant pathogens
sulI	Dihydropteroate synthetase	*E. coli*		*P. chrysogenum*

[a] adapted from VAN DEN HONDEL and PUNT (1991)
[1] YANAI et al. (1991)

mants of *A. nidulans* are plated on media containing triethyltin (WARD et al., 1986), or *pyr4* *k*(equivalent to URA3 of yeast) transformants are plated on media containing fluoroorotate, sectors of improved growth are observed. They result from recombination events at the integration site, some of which lead to plasmid excision. This can provide a basis for gene replacement strategies (see Chapter 13, Fig. 7).

Multiple integration events are quite common in transformation of filamentous fungi and can be a way of increasing gene dose in commercial applications (GWYNNE et al., 1987). Strains carrying multiple integrated copies are generally stable, unless there is significant selection pressure against them.

Although increased gene dose is often seen in such strains, there is no simple correlation with copy number, and other factors, such as the position of integration, appear to play a role in the level of expression achieved (CHRISTENSEN et al., 1988). Furthermore, expression of multiple copies of a gene may be limited by concentration of a regulatory protein or transcription factor (GWYNNE et al., 1987).

2.2.2.2 Replicating Vectors

Since integrative vectors often give low frequencies (typical transformation frequencies being from <1 to 100 transformants/μg transforming DNA/10^7 protoplasts), by analogy with yeasts, replicating vectors might be expected to result in higher transformation frequencies and more convenient cloning vectors. Unfortunately, natural plasmids which can be used as a basis for replicating vectors are extremely rare in filamentous fungi. Some species of *Neurospora* contain mitochondrial plasmids of unknown function (CHAN et al., 1991; SCHULTE and LAMBOWITZ, 1991), but

Tab. 3. Examples of Autonomous Replication of Vectors

Species	Replicating Element
Ustilago maydis	ARS
Podospora anserina	ARS TEL
Aspergillus nidulans	ARS[a]
Phanerochaete chrysosporium	plasmid
Mucor circillenoides	ARS
Absidia glauca	ARS
Phycomyces blakesleeanus	ARS

ARS, autonomously replicating sequence from chromosomal DNA, fortuitous or added; TEL, telomeric sequences from *Tetrahymena thermophila*

[a] AMA1 sequence seems to function in some other fungal species

From VAN DEN HONDEL and PUNT (1991) and references therein

these have not proven useful in the construction of autonomously replicating vectors.

A plasmid of *Phanerochaete chrysosporium,* pME, has recently been detected, and part of it has been incorporated into a stable, replicating vector with a kanamycin resistance marker which is maintained extrachromosomally in the absence of selection (RANDALL et al., 1991). The native circular plasmid exists at low copy number, and was relatively difficult to detect.

In the absence of natural plasmids, and by analogy with yeasts, it has been possible to construct replicating vectors using chromosomal sequences carrying putative origins of replication, autonomously replicating sequences (ARS) (Tab. 3). However, this has not proven to be a generally applicable technique and may sometimes require that the vector also carries a sequence preventing stable integration into the fungal genome. A replicating vector of *Podospora anserina* was constructed by adding telomeric sequences from the *Tetrahymena thermophila* ribosomal DNA (a self-replicating element in the protozoon) to a *ura5* selectable marker gene (PERROT et al., 1987). The plasmid becomes linearized in *Podospora* following transformation, and replicates at low copy number without integration. To date, it has not found any practical use for gene cloning.

More recently, a sequence isolated from *A. nidulans,* AMA1, confers on vectors which carry it the ability to replicate autonomously (GEMS et al., 1991), and leads to an increase in transformation frequency of 100-fold or more compared to integrating vectors. Although present in 10–30 copies per nucleus, such replicating vectors are unstable, and are rapidly lost without selection, resembling the behavior of ARS vectors of yeast. The AMA1 sequence contains inverted repeats separated by a unique sequence. Interestingly, the AMA1 sequence increases transformation efficiency and leads to unstable replication in other filamentous fungi, including *A. niger* and *A. oryzae.* Since chromosomal origins of replication might be expected to occur at relatively frequent intervals along the chromosomes, it is not clear why such behavior is so rare, and why in particular AMA1 behaves in this way, unless the inverted repeat it carries leads to lethality or extremely high instability on chromosomal integration.

An ARS-based vector which replicates autonomously has also been reported for the corn smut pathogen *Ustilago maydis,* a Hemibasidiomycete (TSUKUDA et al., 1988). A partially-repeated chromosomal sequence *ans1* isolated from *A. nidulans* as an ARS in yeast gave enhancement of transformation frequency in *A. nidulans* when incorporated into certain vectors, but transformation was still by integration (BALLANCE and TURNER, 1985). Although functional ARSs are rare in Ascomycetes, Basidiomycetes, and Deuteromycetes, they are common in Phycomycetes. Indeed, stable integration is difficult to achieve with this class, and unstable replication is normal (Tab. 3).

2.3 Vectors for Gene Isolation

Gene isolation by complementation of a mutant strain has been possible in a number of species (TURNER, 1991). This requires either an adequate frequency of fungal transformation (>100/μg plasmid DNA), or the use of cosmid vectors, so that fewer clones need to be screened for the desired gene. DNA prepared from plasmid or cosmid DNA libraries constructed in *Escherichia coli* is used to transform the fungal mutant, with selection for a wild-type transformant. Since most transformation systems to date have been integrative, different approaches have been taken to recover the transforming plasmid and its insert:

(1) Fungal DNA is isolated from the transformant, partially digested, religated, and used to transform *E. coli* to ampicillin resistance, in order to recover the original transforming plasmid, sometimes with host DNA flanking the site of integration (JOHNSTONE et al., 1985; BALLANCE and TURNER, 1986). It is also possible to rescue integrated plasmids at low frequency by transformation of *E. coli* using uncut fungal DNA.
(2) Integrated cosmids can be repackaged by mixing undigested fungal DNA with lambda packaging mix, and transfecting *E. coli* cells (YELTON et al., 1985). Rescue of both plasmid (1) and cosmid DNA molecules from uncut fungal DNA probably results from low frequency excision of the integrated DNA from the chromosome.
(3) The original bacterial cosmid library can be kept as separate colonies frozen in microtiter dishes. Cosmid DNA is then isolated from pools of individual clones, and used to transform the fungus. Any pool which gives positive results can then be subdivided into smaller groups, each of which are retested for fungal transformation. Eventually, a single positive clone is identified. This approach, known as "sib selection", avoids having to rescue a plasmid from the transformed fungus, and has been used extensively with *Neurospora crassa* (AKINS and LAMBOWITZ, 1985).

To facilitate gene isolation, and build an ordered library and a database for cloned genes, cosmid libraries of *A. nidulans* and *N. crassa* are now available from the Fungal Genetics Stock Center (Department of Microbiology, University of Kansas Medical Center, Kansas City, Kansas 66103, USA).

2.4 Gene Inactivation and Replacement

2.4.1 Gene Disruption and Deletion

Defined mutations can be generated by the use of transformation with an appropriate integrating vector, as in yeast, but this does depend on the ease with which homologous recombination can be achieved, and is therefore species-dependent. The best examples to date are with *Aspergillus nidulans.* Disruption makes use of a vector consisting of a selectable marker and an internal fragment of the gene to be disrupted (see Chapter 13, Fig. 9). This technique was used to help identify the gene encoding the first step in penicillin biosynthesis in *A. nidulans,* ACV synthetase, where disruption of the gene led to loss of penicillin production (SMITH et al., 1990). This mutant is being used to study regulation of the biosynthetic pathway. Prior to this, the only Npe (no penicillin) mutations obtained within the biosynthetic gene cluster, using conventional mutant screening, were large deletions or rearrangements affecting more than one gene (MAKINS et al., 1983; MACCABE et al., 1990). A similar approach to gene identification has been used in *Cephalosporium acremonium,* though simple, homologous recombination appears to be more difficult to achieve in this species (HOSKINS et al., 1990).

Disruption may not always be an ideal way to generate a mutation. If the gene of interest is small, then an internal fragment may be too small to obtain a satisfactory frequency of homologous recombination. Further, disruption results in the generation of a truncated protein, which may still have some function. An alternative approach is deletion by use of a linear fragment consisting of homologous ends and carrying an internal selectable marker (see Chapter 13, Fig. 6). The required fragment is assembled in a suitable *E. coli* plasmid, and excised prior to transformation. For example, a cluster of genes expressed during conidiation in *A. nidulans, spoC,* was identified via stage-specific mRNA. In order to attempt to identify their function by finding a mutant phenotype, the whole region (38 kb) was deleted by this "transplacement" method (ARAMAYO et al.,

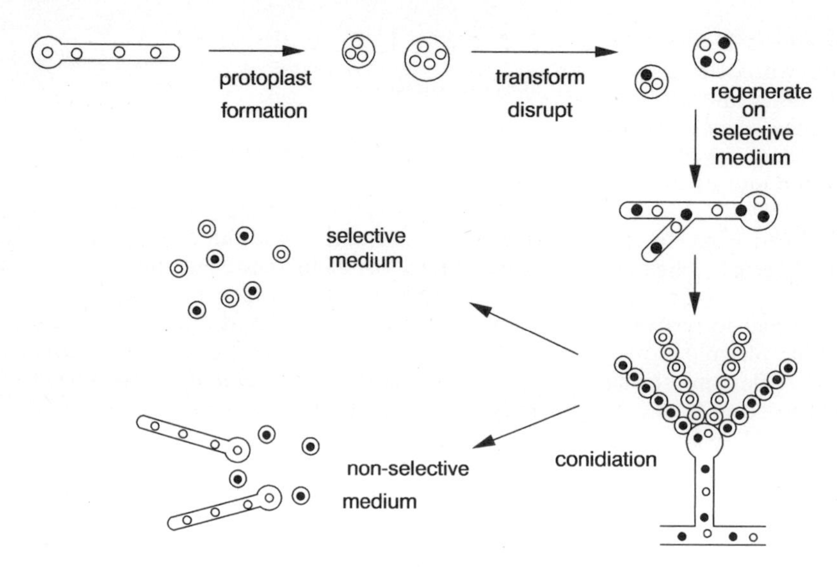

Fig. 1. Heterokaryon disruption of an essential gene in *Aspergillus nidulans.* The transforming DNA carries a selectable marker permitting only transformed recipients (filled nuclei) to grow on selective medium, and an internal fragment of the target gene. Homologous interaction at the target gene also inactivates the gene of interest (see Chapter 13, Fig. 9). Initial transformants are heterokaryotic because of the multinucleate protoplasts. If the disrupted gene is essential, then neither transformed (filled nuclei) nor recipient (open nuclei) nucleotype can grow on selective medium, but complementation can occur in a heterokaryon. Conidia are uninucleate, and neither type can germinate on selective medium. On non-selective medium, only the recipient nucleotype can grow.

1989). In this case, no marked effect on sporulation was observed. This approach has also been of value commercially. Production of bovine chymosin by an engineered strain of *Aspergillus awamori* was improved by removing the major protease activity, aspergillopepsin A, from the host. This was also achieved by deletion using a linear fragment (BERKA et al., 1990).

It is possible that disruption of a gene of unknown functional may be lethal. Where this possibility exists, a heterokaryon can be generated by transformation, in which a nucleus carrying the lethal mutation is carried in a heterokaryon containing also a viable nucleotype (Fig. 1). This technique, called heterokaryon disruption (OSMANI et al., 1988; OAKLEY et al., 1990) is most valuable in a species such as *Aspergillus nidulans,* where conidia are uninucleate. On sporulation of the heterokaryon, mixed conidial types are obtained, both untransformed, non-disrupted and transformed, disrupted conidia. The heterokaryotic state needs to be maintained by selection for each nucleotype, and this can be obtained by the use of suitable nutritional requirements. Individual conidia can be observed under the microscope to assess viability and behavior. For example, a *pyrG* (uridine auxotroph) mutant of *A. nidulans* was transformed with a vector carrying an internal fragment of the gamma-tubulin gene *mipA* and the *N. crassa* $pyr4^+$ gene as a selectable marker. Transformants were subcultured on selective medium, and $mipA^-$ $pyr4^+$ conidia examined for growth on selective medium. Although germination occurred, nuclear division did not, and growth soon ceased, demonstrating the lethal nature of the gamma-tubulin deletion.

2.4.2 Repeat-Induced Point Mutation (RIP)

Gene targeting by homologous recombination is more difficult in *Neurospora crassa,*

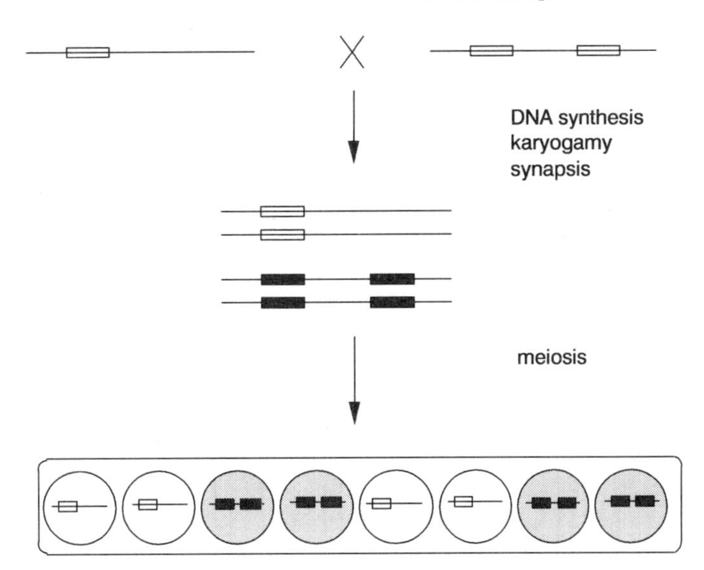

Fig. 2. Repeat-induced point mutation in *Neurospora crassa*. A sequence duplication (open bars) has been introduced into one parent by transformation. A single chromosome is depicted. During the cross, the duplicated sequences are altered by mutation (filled bars). If the mutation arose before DNA replication, then 4 spores in the ascus carry the same mutation (SELKER, 1990). Mutational inactivation can also occur when duplicated sequences are on different chromosomes.

and macroconidia are multinucleate, so that techniques developed for *A. nidulans* are not appropriate. However, an unusual gene inactivation phenomenon occurs during a sexual cross, and can be usefully exploited (Fig. 2). When transformed strains of *N. crassa* carrying two or more copies of the introduced gene go through the sexual cycle, the duplicated genes are often methylated and mutated by a process originally called rearrangement induced premeiotically, and later repeat-induced point mutation (RIP) (SELKER, 1990). This provides a convenient way of inactivating a particular gene.

A similar phenomenon is observed in *Ascobolus immersus,* but the methylation is not followed by mutational inactivation, and on subsequent subculture, gene function is restored (GOYON and FAUGERON, 1989). This phenomenon is thus rather limited, and it is the techniques devised for *A. nidulans* which are more applicable to commercial strains of fungi, most of which are Deuteromycetes lacking a sexual cycle.

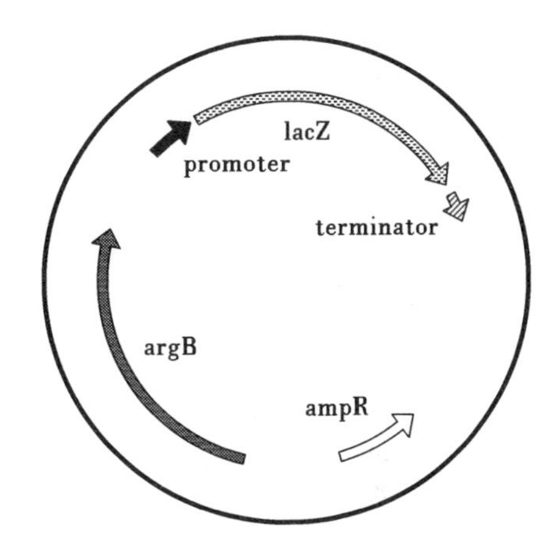

Fig. 3. A reporter gene vector for studying promoter behavior in *Aspergillus nidulans. lacZ, ampR,* β-galactosidase and ampicillin resistance genes of *E. coli*; *argB,* selectable marker for transformation of arginine auxotrophic recipient of *A. nidulans*; single copy integration at the host *argB* locus is easily obtained. The 3′ non-coding region of a fungal gene is included ("terminator") added to terminate transcription of the *lacZ* gene.

2.5 Analysis of Gene Expression

A series of useful reporter gene vectors have been constructed for *Aspergillus nidulans* using *E. coli lacZ* (β-galactosidase) (Fig. 3) and *uidA* (β-glucuronidase). They carry *argB* as a selectable marker for fungal transformation, and sites for insertion of fungal gene 5′ regions upstream of the reporter genes. They have been used in the analysis of fungal pro-

Tab. 4. Strong Promoters Used in Expression Systems

Gene and Organism of Origin		Inducer	Reference
Aspergillus niger var. *awamori*	Glucoamylase *glaA*	Starch	1
	ATP synthetase subunit 9	C	2
A. oryzae	α-Amylase	Starch	3
Trichoderma reesei	*Cellobiohydrolase I* *cbhI*	Cellulose	4
A. nidulans	Ethanol dehydrogenase *alcA*	Ethanol; threonine	5
	Glyceraldehyde phosphate dehydrogenase *gpdA*	C	6
	Triosephosphate isomerase *tpiA*	C	7

C, constitutive expression
1, Cullen et al. (1987); 2, Ward (1990); 3, Christensen et al. (1988); 4, Harkki et al. (1989); 5, Gwynne et al. (1987); 6, Punt et al. (1990a); 7, Upshall et al. (1987)

moter/regulatory sequences (Punt et al., 1990) and assessment of promoter strength in *A. nidulans,* where it is possible to insert the vector readily as a single copy at the *argB* locus by homologous integration. A disadvantage of the *lacZ* reporter gene is that under some growth conditions, native fungal β-galactosidase activity is expressed, interfering with assay of the reporter gene activity. While this is not a major problem in *A. nidulans,* other fungi possess considerable β-galactosidase activity, and this system cannot be used. β-Glucuronidase is absent from many more fungi, and avoids this problem (Punt et al., 1991), but the indicator compound, X-GLU, for detection of enzyme activity in solid media is much more expensive than X-GAL.

2.6 Promoters for Expression Systems

Applications of molecular techniques to fungal biotechnology usually require a strong promoter, and sometimes a regulatory sequence to control gene expression. Examples are given in Tab. 4. Since molecular techniques are now applied to a wide range of fungi, the question of promoter host range arises. While promoters of Ascomycetes and Deuteromycetes can be used widely within these classes, expression in a different host is not necessarily optimal. However, even near-normal gene regulation is often observed when genes are transferred between species (Connerton et al., 1990).

3 Genetic Manipulation of Fungi for Industry

3.1 Antibiotics

The commercial β-lactam antibiotic producers *Penicillium chrysogenum* and *Cephalosporium acremonium* have been the main targets for this work. Transformation has been used extensively in identification of cloned genes of the antibiotic biosynthetic pathways, and in attempts to improve antibiotic yield by insertion of additional copies of such genes. Since production strains generated over many years of conventional mutagenesis and screening often have multiple gene copies and high levels of transcription (SMITH et al., 1989), reports of success with such simple approaches have been rare. The best reported example is the yield improvement obtained in *C. acremonium* following insertion of additional copies of the gene encoding the expandase/hydroxylase enzyme, identified as a bottleneck in the pathway by observation of intermediate accumulation (SKATRUD et al., 1989). It should be noted that a dominant selectable marker was used, to avoid having to introduce undesirable mutations into the recipient strain prior to transformation.

Variation exists in the β-lactam antibiotic biosynthetic pathways in microorganisms, and some attempts have also been made at the construction of hybrid pathways by interspecies transfer of genes for particular steps using available transformation techniques and cloned genes (GUTIERREZ et al., 1991; CANTWELL et al., 1990).

3.2 Protein Production

3.2.1 Fungal Enzymes

A wide range of fungal enzyme products (e.g., amylases, cellulases) are used in the food industry (*Aspergillus niger, A. oryzae*) and the paper pulp industry (*Trichoderma reesei, Phanerochaete chrysosporium*). Most of these are secreted enzymes of relatively low value, which are made in large amounts from strains already improved by conventional procedures. Nevertheless, many of the genes encoding these enzymes have been isolated, and attempts have been made to improve yields by transformation (PENTTILÄ et al., 1991). For this purpose, a simple vector needs only the desired gene and a selectable marker. Construction of a commercial strain by molecular techniques may require several genetic manipulation steps (e.g., to insert a gene, to delete a protease). Hence, it is important to have a range of transformation systems available.

Some of the genes encoding useful enzymes, such as the α-amylase (Taka-amylase) of *A. oryzae* and the glucoamylase of *A. niger/A. awamori,* already possess some of the strongest fungal promoters known, particularly when they are derived from commercially-improved strains, and these promoters have been exploited for construction of hybrid expression systems. Fungal lipases such as those from *Humicola* and *Rhizopus* species are now used in detergents. Their production has been enhanced by expression in *A. oryzae,* using the α-amylase promoter of this organism (HUGE-JENSEN et al., 1989).

3.2.2 Mammalian Proteins

3.2.2.1 Chymosin

Bovine chymosin or rennin is used in the cheese industry, and because of its relatively high value and supply problems, it has been a target for microbial expression for some time. As a result of difficulties encountered in earlier attempts at expression in bacterial and yeast systems, *A. awamori* has been developed as a host (BERKA et al., 1991; WARD, 1990), using a wide range of genetic manipulation techniques. The preprochymosin gene is expressed using transcription signals derived from the host glucoamylase gene. Further improvements in secretion were obtained by fusing the whole of the glucoamylase gene to the prochymosin sequence (WARD et al., 1990). Secretion can be obtained using either the glucoamylase signal sequence or the natural prochymosin presequence. Although many studies with fungi have shown a rough correlation between gene copy number and expression level, this is by no means a direct relationship, and for commercial purposes, the empirical approach of screening all transformants for optimum expression, irrespective of copy number, has been adopted. The position and mode of integration in the genome clearly affects expression level.

In addition to the deletion of a protease gene as described earlier, further improvements in yield were achieved by standard mutagenesis and screening (LAMSA and BLOEBAUM, 1990). Thus, a combination of optimum vector design and use, followed by traditional methods of strain improvement, is likely to be important in fungal biotechnology. Since the traditional approach is time-consuming and the results are not necessarily reproducible, it is an advantage if an improved secreting strain can be "reused" for another protein product by excising the vector sequences and retaining the improved strain background. It is here that a two-way selectable marker is especially valuable.

3.2.2.2 Therapeutics

A number of pilot studies have been carried out to examine the potential of fungal systems to express mammalian proteins of therapeutic importance (GWYNNE et al., 1987; UPSHALL et al., 1987). Initial yields were low, as with early chymosin experiments (CULLEN et al., 1987), and none of them have been taken forward to commercial production to date. The vectors and methods used are based on the principles already outlined for chymosin. One additional modification was the use of a glucoamylase-interleukin-6 fusion gene with a KEX2-like (protease) cleavage site between the two sequences, leading to improved yields of secreted IL6 protein (CONTRERAS et al., 1990).

4 Future Prospects

The techniques described above are applicable to a wide range of species, making many fungal biotechnology processes accessible to genetic manipulation, and introducing the possibility of using fungi as hosts for expression of non-fungal proteins. Stable replicating vectors would be an additional useful tool, but remain elusive. It will be interesting to see whether the *Phanerochaete* plasmid can function in other species. Construction of a stable "minichromosome" might be an interesting future development, but would probably require the isolation of a fungal centromeric sequence. No such sequence has yet been isolated from a filamentous fungus, despite some unpublished attempts. The very large centromeric regions (>100 kb) of *Schizosaccharomyces pombe* were eventually isolated in functional form only by use of *Saccharomyces cerevisiae* YAC vectors (HAHRENBERGER et al., 1989). Transposons would also be useful additions to fungal molecular techniques, for gene tagging and isolation, and evidence for such elements exists in some species, including *Neurospora* (KINSEY, 1990), but no practical systems have yet emerged.

5 References

AKINS, R. A., LAMBOWITZ, A. M. (1985), General method for cloning *Neurospora crassa* nuclear genes by complementation of mutants, *Mol. Cell. Biol.* **5**, 2272–2278.

ARAMAYO, R., ADAMS, T. H., TIMBERLAKE, W. E. (1989), A large cluster of highly expressed genes is dispensable for growth and development in *Aspergillus nidulans, Genetics* **122**, 65–71.

ARMALEO, D., YE, G.-N., KLEIN, T. M., SHARK, K. B., SANFORD, J. C., JOHNSTON, S. A. (1990), Biolistic nuclear transformation of *Saccharomyces cerevisiae* and other fungi, *Curr. Genet.* **17**, 97–103.

BALLANCE, D. J., TURNER, G. (1985), Development of a high frequency transforming vector for *Aspergillus nidulans, Gene* **36**, 321–331.

BALLANCE, D. J., TURNER, G. (1986), Gene cloning in *Aspergillus nidulans:* isolation of the isocitrate lyase gene *(acuD), Mol. Gen. Genet.* **202**, 271–275.

BEADLE, G. W. (1946), Genes and the chemistry of the organism, *Am. Sci.* **34**, 31–53.

BERI, R., TURNER, G. (1987), Transformation of *Penicillium chrysogenum* with the *Aspergillus nidulans amdS* gene as a dominant selective marker, *Curr. Genet.* **11**, 639–641.

BERKA, R. M., WARD, M., WILSON, L. J., HAYENGA, K. J., KODAMA, K. H., CARLOMAGNA, L. P., THOMPSON, S. A. (1990), Molecular cloning and deletion of the gene encoding aspergillopepsin A from *Aspergillus awamori, Gene* **86**, 153–162.

BERKA, R. M., KODAMA, K. H., REY, M. W., WILSON, L. J., WARD, M. (1991), The development of *Aspergillus niger* var. *awamori* as a host for the expression and secretion of heterologous gene products, *Biochem. Soc. Trans.* **19**, 681–685.

BULL, J. H., SMITH, D. J., TURNER, G. (1988), Transformation of *Penicillium chrysogenum* with a dominant selectable marker, *Curr. Genet.* **13**, 377–382.

BUXTON, F. P., GWYNNE, D. I., DAVIES, R. W. (1989), Cloning of a new bidirectionally selectable marker for *Aspergillus* strains, *Gene* **84**, 329–334.

CANTWELL, C. A., BECKMANN, R. J., DOTZLAF, J. E., FISHER, D. L., SKATRUD, P. L., YEH, W. K., QUEENER, S. W. (1990), Cloning and expression of a hybrid *Streptomyces clavuligerus cefE* gene in *Penicillium chrysogenum, Curr. Genet.* **17**, 213–221.

CHAN, B. S. S., COURT, D. A., VIERULA, P. J., BERTRAND, H. (1991), The kalilo linear senescence-inducing plasmid of *Neurospora* is an invertron and encodes DNA and RNA-polymerases, *Curr. Genet.* **20**, 225–237.

CHRISTENSEN, T., WOELDIKE, H., BOEL, E., MORTENSEN, S. B., HJORTSHOEJ, L. T., HANSEN, M. T. (1988), High level expression of recombinant genes in *Aspergillus oryzae, Bio/Technology* **6**, 1419–1422.

CONNERTON, I. F., FINCHAM, J. R. S., SANDEMAN, R. A., HYNES, M. J. (1990), Comparison and cross-species expression of the acetyl-CoA synthetase genes of the ascomycete fungi, *Aspergillus nidulans* and *Neurospora crassa, Mol. Microbiol.* **4**, 451–460.

CONTRERAS, R., CARREZ, D., KINGHORN, J. R., VAN DEN HONDEL, C. A. M. J. J., FIERS, W. (1990), Efficient kex2-like processing of a glucoamylase-interleukin-6 fusion protein by *Aspergillus nidulans* and secretion of mature interleukin-6, *Bio/Technology* **9**, 378–381.

CULLEN, D., GRAY, G. L., WILSON, L. J., HAYENGA, K. J., LAMSA, M. H., REY, M. W., NORTON, S., BERKA, R. M. (1987), Controlled expression and secretion of bovine chymosin in *Aspergillus nidulans, Bio/Technology* **5**, 369–376.

DHAWALE, S. S., MARZLUF, G. A. (1985), Transformation of *Neurospora crassa* with circular and linear DNA and analysis of the fate of the transforming DNA, *Curr. Genet.* **10**, 205–212.

DHAWALE, S. S., PAIETTA, J. V., MARZLUF, G. A. (1984), A new rapid and efficient transformation procedure for *Neurospora, Curr. Genet.* **8**, 77–79.

FINCHAM, J. R. S. (1989), Transformation in fungi, *Microbiol. Rev.* **53**, 148–170.

GEMS, D., JOHNSTONE, I. L., CLUTTERBUCK, A. J. (1981), An autonomously replicating plasmid transforms *Aspergillus nidulans* at high frequency, *Gene* **98**, 61–67.

GOLDMAN, G. H., VAN MONTAGU, M., HERRERA-ESTRELLA, A. (1990), Transformation of *Trichoderma harzianum* by high-voltage electric pulse, *Curr. Genet.* **17**, 169–174.

GOYON, C., FAUGERON, G. (1989), Targeted transformation of *Ascobolus immersus* and *de novo* methylation of the resulting duplicated DNA sequences, *Mol. Cell. Biol.* **9**, 2818–2827.

GUTIERREZ, S., DIEZ, B., ALVAREZ, E., BARREDO, J. L., MARTIN, J. F. (1991), Expression of the *penDE* gene of *Penicillium chrysogenum* encoding isopenicillin N acyltransferase in *Cephalosporium acremonium* - production of benzylpenicillin, *Mol. Gen. Genet.* **255**, 56–64.

GWYNNE, D. I., BUXTON, F. P., WILLIAMS, S. A., GARVEN, S., DAVIES, R. W. (1987), Genetically engineered secretion of active human interferon and a bacterial endoglucanase from *Aspergillus nidulans, Bio/Technology* **5**, 713–719.

HAHRENBERGER, K. M., BAUM, M. P., POLIZZI, C. M., CARBON, J., CLARKE, L. (1989), Construction of functional artificial minichromosomes in the fission yeast *Schizosaccharomyces pombe, Proc. Natl. Acad. Sci. USA* **86**, 577–581.

HARKKI, A., UUSITALO, J., BAILEY, J., PENTTILÄ, M., KNOWLES, J. (1989), A novel fungal expression system: secretion of active calf chymosin from the filamentous fungus *Trichoderma reesei, Bio/Technology* **7**, 596–603.

HINTZ, W. E. A., ANDERSON, J. B., HORGEN, P. A. (1989), Relatedness of three species of *Agaricus* inferred from restriction fragment length polymorphism analysis of the ribosomal DNA repeat and mitochondrial DNA, *Genome* **32**, 173–178.

HOSKINS, J. A., O'CALLAGHAN, N., QUEENER, S. W., CANTWELL, C. A., WOOD, J. S., CHEN, V. J., SKATRUD, P. L. (1990), Gene disruption of the pcbAB gene encoding ACV synthetase in *Cephalosporium acremonium, Curr. Genet.* **18**, 523–530.

HUGE-JENSEN, B., ANDREASEN, F., CHRISTENSEN, T., CHRISTENSEN, M., THIM, L., BOEL, E. (1989), *Rhizomucor miehei* triglyceride lipase is processed and secreted from transformed *Aspergillus oryzae, Lipids* **24**, 781–785.

JOHNSTONE, I. L., HUGHES, S. G., CLUTTERBUCK, A. J. (1985), Cloning an *Aspergillus nidulans* developmental gene by transformation, *EMBO J.* **4**, 1307–1311.

KELLY, J. M., HYNES, M. J. (1985), Transformation of *Aspergillus niger* by the *amdS* gene of *Aspergillus nidulans, EMBO J.* **4**, 475–479.

KINSEY, J. A. (1990), *Tad,* a LINE-like transposable element of *Neurospora,* can transpose between nuclei in heterokaryons, *Genetics* **126**, 317–323.

LAMSA, M., BLOEBAUM, P. (1990), Mutation and screening to increase chymosin yield in a genetically-engineered strain of *Aspergillus awamori, J. Ind. Microbiol.* **5**, 229–238.

MacCabe, A. P., Riach, M. B. R., Unkles, S. E., Kinghorn, J. R. (1990), The *Aspergillus nidulans npeA* locus consists of three contiguous genes required for penicillin biosynthesis, *EMBO J.* **9**, 279–287.

Makins, J. F., Holt, G., Macdonald, K. D. (1983), The genetic location of three mutations impairing penicillin production in *Aspergillus nidulans, J. Gen. Microbiol.* **129**, 3027–3033.

Oakley, B. R., Oakley, C. E., Yoon, Y., Jung, M. K. (1990), Gamma-Tubulin is a component of the spindle pole body that is essential for microtubule function in *Aspergillus nidulans, Cell* **61**, 1289–1301.

Osmani, S. A., Engle, D. B., Doonan, J. H., Morris, N. R. (1988), Spindle formation and chromatin condensation in cells blocked at interphase by mutation of a negative cell cycle control gene, *Cell* **52**, 241–251.

Penttilä, M., Teeri, T. T., Nevalainen, H., Knowles, J. K. C. (1991), The molecular biology of *Trichoderma reesei* and its application to biotechnology, in: *Applied Molecular Genetics of Fungi* (Peberdy, J. F., Caten, C. E., Ogden, J. E., Bennett, J. W., Eds.), pp. 85–102, Cambridge: Cambridge University Press.

Perrot, M., Barreau, C., Begueret, J. (1987), Nonintegrative transformation in the filamentous fungus *Podospora anserina:* stabilization of a linear vector by the chromosomal ends of *Tetrahymena thermophila, Mol. Cell. Biol.* **7**, 1725–1730.

Punt, P. J., Dingemanse, M. A., Kuyvenhoven, A., Soede, R. D. M., Pouwels, P. H., van den Hondel, C. A. M. J. J. (1990), Functional elements in the promoter region of the *Aspergillus nidulans gpdA* gene encoding glyceraldehyde-3-phosphate dehydrogenase, *Gene* **93**, 101–109.

Punt, P. J., Greaves, P. A., Kuyvenhoven, A., van Deutekom, J. C. T., Kinghorn, J. R., Pouwels, P. H., van den Hondel, C. A. M. J. J. (1991), A twin-reporter vector for simultaneous analysis of expression signals of divergently transcribed, contiguous genes in filamentous fungi, *Gene* **104**, 119–122.

Randall, T., Reddy, C. A., Boominathan, K. (1991), A novel extrachromosomally maintained transformation vector for the lignin-degrading basidiomycete *Phanerochaete chrysosporium, J. Bacteriol.* **173**, 776–782.

Schulte, U., Lambowitz, A. M. (1991), The labelle mitochondrial plasmid of *Neurospora intermedia* encodes a novel DNA polymerase that may be derived from a reverse transcriptase, *Mol. Cell. Biol.* **11**, 1696–1706.

Selker, E. U. (1990), Premeiotic instability of repeated sequences in *Neurospora crassa, Annu. Rev. Genet.* **24**, 579–614.

Skatrud, P. L., Tietz, A. J., Ingolia, T. D., Cantwell, C. A., Fisher, D. L., Chapman, J. L., Queener, S. W. (1989), Use of recombinant DNA to improve production of cephalosporin C by *Cephalosporium acremonium, Bio/Technology* **7**, 477–485.

Smith, D. J., Bull, J. H., Edwards, J., Turner, G. (1989), Amplification of the isopenicillin N synthetase gene in a strain of *Penicillium chrysogenum* producing high levels of penicillin, *Mol. Gen. Genet.* **216**, 492–497.

Smith, D. J., Burnham, M. K. R., Bull, J. H., Hodgson, J. E., Ward, J. M., Browne, P., Brown, J., Barton, B., Earl, A. J., Turner, G. (1990), β-Lactam antibiotic biosynthetic genes have been conserved in clusters in prokaryotes and eukaryotes, *EMBO J.* **9**, 741–747.

Tilburn, J., Scazzocchio, C., Taylor, G. G., Zabicky-Zissman, J. H., Lockington, R. A., Davies, R. W. (1983), Transformation by integration in *Aspergillus nidulans, Gene* **26**, 205–221.

Tsukuda, T., Carleton, S., Fotheringham, S., Holloman, W. K. (1988), Isolation and characterisation of an autonomously replicating sequence from *Ustilago maydis, Mol. Cell. Biol.* **8**, 3703–3709.

Turner, G. (1991), Strategies for cloning genes from filamentous fungi, in: *Applied Molecular Genetics of Fungi* (Peberdy, J. F., Caten, C. E., Ogden, J. E., Bennett, J. W., Eds.), pp. 29–43, Cambridge: Cambridge University Press.

Ullrich, R. C., Novotny, C. P., Specht, C. A., Froeliger, E. H., Munoz-Rivas, A. M. (1985), Transforming Basidiomycetes, in: *Molecular Genetics of Filamentous Fungi* (*UCLA Symposia on Molecular and Cellular Biology,* New Series, Vol. 34) (Timberlake, W. E., Ed.), pp. 39–57, New York: Alan R. Liss, Inc.

Upshall, A., Kumar, A. A., Bailey, M. C., Parker, M. D., Favreau, M. A., Lewison, K. P., Joseph, M. L., Maroganore, J. M., McKnight, G. L. (1987), Secretion of active human tissue plasminogen activator from the filamentous fungus *Aspergillus nidulans, Bio/Technology* **5**, 1301–1304.

van den Hondel, C. A. M. J. J., Punt, P. J. (1991), Gene transfer systems and vector development for filamentous fungi, in: *Applied Molecular Genetics of Fungi* (Peberdy, J. F., Caten, C. E., Ogden, J. E., Bennett, J. W., Eds.), pp. 1–28, Cambridge: Cambridge University Press.

Ward, M. (1990), Chymosin production in *Aspergillus,* in: *Molecular Industrial Mycology: Sys-*

tems and Applications in Filamentous Fungi (LEONG, S. A., BERKA, R., Eds.), pp. 83–105, New York: Marcel Dekker.

WARD, M., WILKINSON, B., TURNER, G. (1986), Transformation of *Aspergillus nidulans* with a cloned, oligomycin-resistant ATP synthase subunit 9 gene, *Mol. Gen. Genet.* **202,** 265–270.

WARD, M., WILSON, L. J., KODAMA, K. H., REY, M. W., BERKA, R. M. (1990), Improved production of chymosin in *Aspergillus* by expression as a glucoamylase-chymosin fusion, *Bio/Technology* **8,** 435–440.

YANAI, K., HORIUCHI, H., TAKAGI, M., YANO, K. (1991), Transformation of *Rhizopus niveus* using a bacterial blasticidin-S resistance gene as a dominant selectable marker, *Curr. Genet.* **19,** 221–226.

YELTON, M. M., HAMER, J. E., TIMBERLAKE, W. E. (1984), Transformation of *Aspergillus nidulans* by using a *trpC* plasmid, *Proc. Natl. Acad. Sci. USA* **81,** 1470–1474.

YELTON, M. M., TIMBERLAKE, W. E., VAN DEN HONDEL, C. A. M. J. J. (1985), A cosmid for selecting genes by complementation in *Aspergillus nidulans*: selection for the developmentally-regulated *yA* locus, *Proc. Natl. Acad. Sci. USA* **82,** 834–838.

IV. Genetic Engineering of Plants

15 Genetic Engineering of Plant Cells

GÜNTER KAHL
KURT WEISING

Frankfurt am Main, Federal Republic of Germany

1 General

Plant genetic engineering started with the discovery that the plant pathogen *Agrobacterium tumefaciens* transfers a set of genes from one of its plasmids, later on coined Ti-plasmid, into infected plants, where they are expressed and responsible for a tumorous phenotype. This discovery catalyzed plant genetic research to dimensions never witnessed in the past. About five years later the first transgenic plant was created, exploiting the relatively well-explored Ti-plasmid vector system. After another five years or so, the first transgenic plant expressing an agronomically desirable trait has been introduced. And to-day, after a truly exciting and rewarding period of plant molecular biology, we are facing an unprecedented and almost frightening wealth of literature, documenting an ever-increasing community of plant researchers and an ever-increasing pile of data. Such tremendous progress is hard to compress into one single chapter, where each section deserves a chapter in its own. Nevertheless, we have striven to cover nearly all aspects of plant genetic engineering, which has remarkably broadened our understanding of how genes work, and additionally introduced plants with desirable traits spanning from disease resistance to the production of pharmaceutically important proteins. We dedicate this overview to all colleagues who have contributed in one way or another.

2 The Vectors for Genetic Engineering of Plant Cells

2.1 Natural Genetic Engineering of Plant Cells: *Agrobacterium tumefaciens* and *A. rhizogenes*

2.1.1 Introduction

Virulent strains of the Gram negative soil bacteria *Agrobacterium tumefaciens* and *Agrobacterium rhizogenes* incite neoplastic growth on many dicots and gymnosperms (De Cleene and De Ley, 1976, 1981, Porter, 1991) but also a few monocots (Hooykaas-Van Slogteren et al., 1984; De Cleene, 1985; Schäfer et al., 1987; Bytebier et al., 1987; Dommisse et al., 1990; Raineri et al., 1990). Both bacterial species carry characteristic large plasmids. While the Ti (tumor-inducing) plasmid of *A. tumefaciens* is responsible for the development of so-called crown gall tumors on appropriate host plants (Fig. 1), *A. rhizogenes* strains carrying Ri (root-inducing) plasmids are the causative agents for a disease called hairy root. Both diseases are the consequence of successful genetic engineering of the plant by *Agrobacterium:* the bacteria attach to cell walls of wound-exposed and wound-activated plant cells, excise a defined part of their Ti (or Ri) plasmid and transfer this so-called

Fig. 1. Typical appearance of crown gall tumors on a leaf of *Bryophyllum daigremontianum* (left) and a stem of *Kalanchoe prolifera* (right). The stem tumor was incited by *Agrobacterium tumefaciens,* strain C58 (nopaline-type Ti plasmid), and grows into shoot- or leaf-like structures (so-called teratomata).

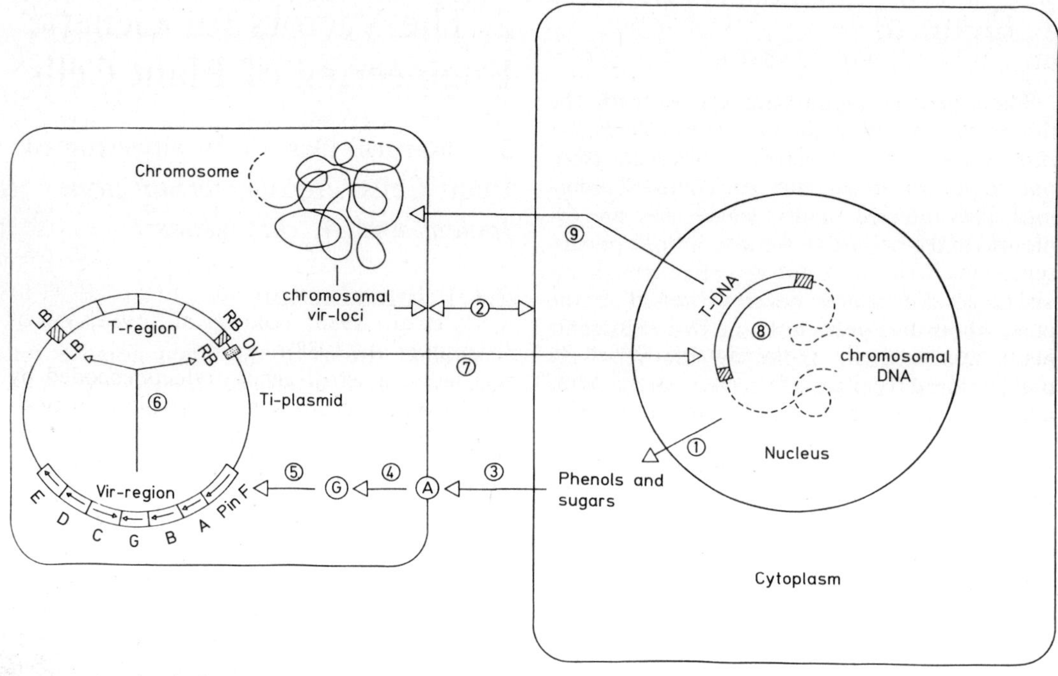

Fig. 2. A simplified scheme of *Agrobacterium*-mediated gene transfer into wounded plant cells that synthesize phenolic compounds as part of their defense and wound-healing strategies. These phenols serve as attractants for the ubiquitously occurring soil bacterium *Agrobacterium tumefaciens* (1) that approaches the wound site via positive chemotaxis. The bacterium attaches to certain compounds of the wound-exposed plant cell wall (2), a process dependent on chromosomal virulence genes of the bacterium. The phenols are recognized by a bacterial membrane protein (A, 3) encoded by gene A of the virulence (vir) region of the Ti plasmid. Recognition involves autophosphorylation and phosphorylation of a second vir protein (G, 4) that acts as a DNA-binding protein (5) and activates the vir region. The product of virD is a site-specific endonuclease that nicks the bottom strand of the T-region (6) at the left and right border sequences (LB, RB). The excised T-strand is packaged into virE proteins and piloted by virD proteins into the plant cell nucleus (7). The process is enhanced by a sequence close to the right border (overdrive, OV), which is characteristic for octopine strains of *Agrobacterium*. The T-strand is covalently integrated into the host plant genome (8) and expressed. The constitutive transcription of T-DNA genes 1, 2, and 4 leads to the accumulation of auxins and cytokinins that incite permanent proliferative growth (tumor formation). The product of the ocs and nos genes of the T-DNA, so-called opine synthases, catalyze the formation of opines in the host cells. These compounds are secreted (9) and serve as carbon, nitrogen, and energy source for the parasitic bacterium.

T-region into the nucleus of the host plant, where it becomes stably integrated (transferred DNA or T-DNA). The expression of T-DNA genes in the host cell's genome leads to the transformed phenotype.

Since the *Agrobacterium* system is still the method of choice for stable transformation of susceptible plant species, we will briefly outline the mechanisms of T-DNA excision, transfer, integration, and expression. The whole process is shown schematically in Fig. 2. For more detailed information the reader should consult recent reviews of the topic (ZAMBRYSKI, 1988; ZAMBRYSKI et al., 1989; HOOYKAAS, 1989; GELVIN, 1990; HOWARD and CITOVSKY, 1990; KADO, 1991).

2.1.2 The Virulence (vir) Region and T-DNA Mobilization

Three genetic components are required for the successful engineering of plant cells by virulent *Agrobacterium* strains. The first component is a set of so-called virulence (vir) genes located on the Ti (or Ri) plasmid. The vir regulon consists of about 25 genes organized in seven transcriptional units: virA, virB, virC, virD, virE, virG (STACHEL and NESTER, 1986) and virF (MELCHERS et al., 1990). These vir genes are coding for functions involved in plant cell recognition, bacterial attachment, and the excision, transfer, and probably the integration of T-DNA into the target genome. While some vir gene regions are essential for tumorigenesis (virA, virB, virG and virD), others affect the efficiency of transfer and hence the host range (virC and virE). Some genes of the latter category are observed in some types of Ti and Ri plasmids, but not in others (e.g., virF).

The regulation of vir genes is complex. In the absence of any induction, virA and virG are constitutively expressed at a low level. All vir genes are, however, inducible by a variety of plant phenolic compounds (see Tab. 1; STACHEL et al., 1985; 1986a; BOLTON et al., 1986; SPENCER and TOWERS, 1988; MELCHERS et al., 1989a; MESSENS et al., 1990; HESS et al., 1991), and by a group of plant monosaccharides that act synergistically (ANKERBAUER and NESTER, 1990; CANGELOSI et al., 1990; SHIMODA et al., 1990). Vir gene induction also depends on low temperature (below 28 °C) and acidic pH (ALT-MÖRBE et al. 1989), and is further enhanced by a variety of other factors such as certain opines (VELUTHAMBI et al., 1989), phosphate starvation (WINANS et al., 1988; WINANS, 1990), and flavonoid compounds (ZERBACK et al., 1989). In addition, virC and virD are negatively regulated by the chromosomal ros locus (CLOSE et al., 1987; TAIT and KADO, 1988; AOYAMA et al., 1991).

The second component necessary for T-DNA transfer, also located on the Ti (Ri) plasmid, is the T-DNA itself. Although it carries all genes (oncogenicity genes, onc genes) necessary for the transformation of plant cells with resulting crown gall or hairy root phenotype, respectively, none of these genes is needed for T-DNA excision and transfer. In essence, the only *cis*-requirement for T-DNA-transfer is the presence of two conserved, direct repeats of about 25 bp at both ends of the T-region, the so-called T-DNA borders (WANG et al., 1984; PERALTA and REAM, 1985; SLIGHTOM et al., 1986). Except for nopaline-type plasmids, however, an additional element called "overdrive" is located outside the right border and functions as a transformation enhancer (PERALTA et al., 1986; TORO et al., 1988; VELUTHAMBI et al., 1988).

Finally, a set of gene products encoded by the bacterial chromosome is involved in positive chemotactic movement of the bacterial cell towards plant wound sites, bacterial attachment to wound-exposed plant cell walls (DOUGLAS et al., 1985), the induction of vir genes by monosaccharides (CANGELOSI et al., 1990), and the regulation of virC and virD (TAIT and KADO, 1988). The mechanism of transfer seems to be essentially identical for both Ti and Ri plasmid-derived T-DNAs. *In vivo,* successful infection requires wounding of the plant tissue, which induces a series of metabolic pathways and the subsequent accumulation of metabolites necessary for the wound-healing process (KAHL, 1982).

Some of these metabolites represent effective chemoattractants for agrobacteria (ASHBY et al., 1987; 1988; SHAW, 1991), and can be broadly categorized into two classes. Whereas an assortment of monosaccharides (SHIMODA et al., 1990) is probably responsible for the long-distance attraction of virulent and non-virulent strains of *Agrobacterium* to the rhizosphere of wounded plants (SHAW, 1991), phenolic precursors and intermediates of lignin biosynthesis (e.g., acetosyringone) play a more specific role. At low concentrations ($<10^{-7}$ M for acetosysringone), these phenolics act as specific chemoattractants for virulent *Agrobacterium* strains only. VirA and virG gene products mediate this kind of chemotaxis (SHAW et al., 1988). At higher concentrations, however, the same phenolics act as specific inducers of the vir gene regulon via a two-component (virA and virG) signal transduction mechanism that is similar to other bacterial regulatory systems (STACHEL and ZAMBRYSKI, 1986a, b). The first step in this process is

Tab. 1. Chemical Compounds Identified as *vir* Gene Inducers

Compound	Structure of Some Prominent Inducers	Reference
Aromatic Phenolics		
Acetosyringone	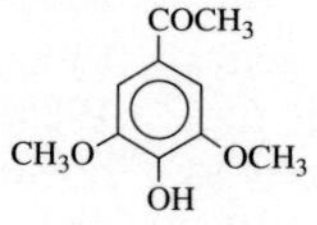 Acetosyringone	STACHEL et al., 1985 SPENCER and TOWERS, 1988 MELCHERS et al., 1989a
α-Hydroxy-acetosyringone		STACHEL et al., 1985
Syringic acid	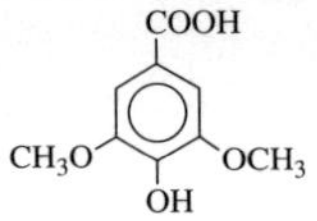 Syringic acid	STACHEL et al., 1985 SPENCER and TOWERS, 1988 MELCHERS et al., 1989a
Syringaldehyde	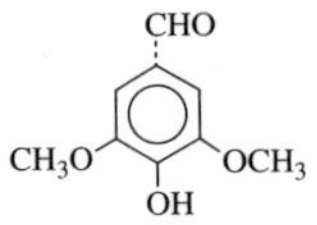 Syringaldehyde	STACHEL et al., 1985 SPENCER and TOWERS, 1988 MELCHERS et al., 1989a
Catechol		BOLTON et al., 1986
Protocatechuic acid		BOLTON et al., 1986
β-Resorcylic acid		BOLTON et al., 1986
Vanillin	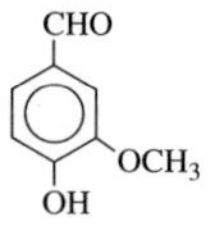 Vanillin	BOLTON et al., 1986 SPENCER and TOWERS, 1988 MELCHERS et al., 1989a
Acetovanillone	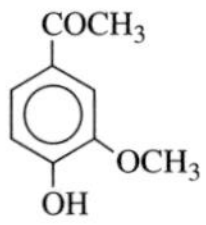 Acetovanillone	SPENCER and TOWERS, 1988 MELCHERS et al., 1989a
Vanillalacetone		SPENCER and TOWERS, 1988
Homovanillic acid		MELCHERS et al., 1989a
Vanillyl alcohol		MELCHERS et al., 1989a
Ferulic acid		ASHBY et al., 1988 SPENCER and TOWERS, 1988 MELCHERS et al., 1989a
Ethylferulate	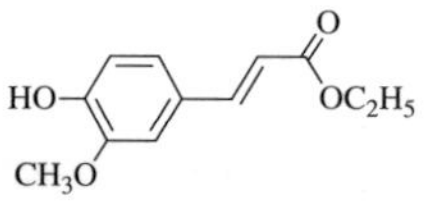 Ethylferulate	MESSENS et al., 1990

Tab. 1. Continued

Compound	Structure of Some Prominent Inducers	Reference
Coniferyl alcohol		SPENCER and TOWERS, 1988 MELCHERS et al., 1989a
Dehydroconiferyl alcohol		HESS et al., 1991
Sinapinic acid	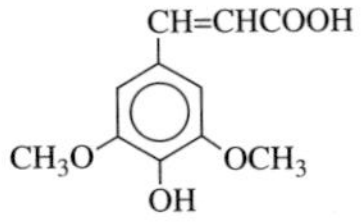 CH=CHCOOH; CH_3O; OCH_3; OH Sinapinic acid	STACHEL et al., 1985 SPENCER and TOWERS, 1988 MELCHERS et al., 1989a
Sinapyl alcohol		SPENCER and TOWERS, 1988
2′,4′,4-Trihydroxy-3-methoxy-chalcone		SPENCER and TOWERS, 1988
3,5-Dimethoxy-4-hydroxy-benzene		MELCHERS et al., 1989a
3,4,5-Trihydroxybenzoate (gallic acid)		BOLTON et al., 1986
Pyrogallic acid		BOLTON et al., 1986
p-Hydroxybenzoic acid		BOLTON et al., 1986
3-Ethoxy-4-hydroxy-benzaldehyde		MELCHERS et al., 1989a
3,4-Dihydroxy-benzaldehyde		MELCHERS et al., 1989a
3,4,5-Trimethoxy-benzaldehyde		MELCHERS et al., 1989a
4-Hydroxy-3-methyl-acetophenone		MELCHERS et al., 1989a
Guaiacol		MELCHERS et al., 1989a
Monosaccharides		
D-Glucose	OH OH OH OH HO OH Inositol	SHIMODA et al., 1990
D-Glucuronic acid		CANGELOSI et al., 1990
2-Desoxy-D-glucose		ANKENBAUER and NESTER, 1990
6-Desoxy-D-glucose		
D-Mannose		
D-Xylose		
D-Galactose		
D-Galacturonic acid		
L-Arabinose		
D-Fucose		
Inositol		
Cellobiose		
D-Idose		
D-Talose		
D-Lyxose		
Opines		
Octopine	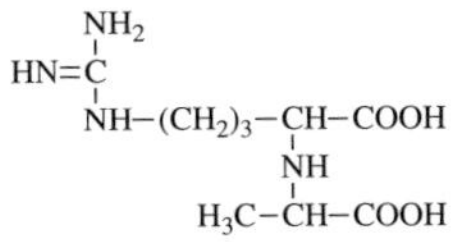 NH_2; HN=C; NH–$(CH_2)_3$–CH–COOH; NH; H_3C–CH–COOH Octopine	VELUTHAMBI et al., 1989
Nopaline		
Leucinopine		
Succinamopine		

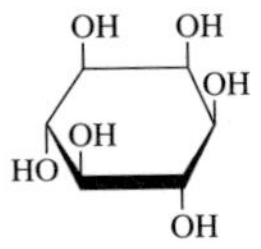

Tab. 1. Continued

Compound	Structure of Some Prominent Inducers	Reference
Flavonoids		
Kaempferol-3-glucosyl-galactoside Quercetin-3-glucosyl-galactoside Rutin Myricetin-3-galactoside Narcissin Apigenein-7-glucoside	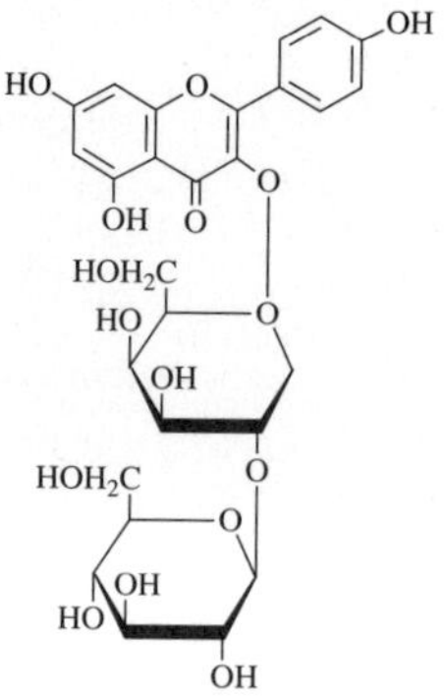 Kaempferol-3-glucosyl-galactoside	ZERBACK et al., 1989

the recognition of phenolic signal molecules by the virA gene product, which is a transmembrane sensor protein (LEROUX et al., 1987; MELCHERS et al., 1988; 1989b; WINANS et al., 1989) capable of autophosphorylation (HUANG et al., 1990b; JIN et al., 1990a). A mechanism for the recognition of phenolic signal molecules has recently been proposed that involves protonation of a basic amino acid residue on the receptor site (HESS et al., 1991). Binding of an appropriate signal molecule then induces a conformational change of the VirA protein and its subsequent autophosphorylation at a histidine residue in the cytoplasmic C-terminal domain (JIN et al., 1990a; HUANG et al., 1990b). In its autophosphorylated form the VirA protein becomes competent to phosphorylate the virG product (JIN et al., 1990b) which is a sequence-specific DNA-binding protein recognizing the conserved "vir box" region located upstream of all vir operons (JIN et al., 1990b, c; PAZOUR and DAS, 1990; WINANS, 1990). While VirG phosphorylation is not necessary for binding of the protein to the vir box (ROITSCH et al., 1990), this modification seems to be an important step in the vir activation cascade (possibly by increasing the VirG affinity for the binding site) that finally leads to the coordinated transcription and translation of all virA/virG-inducible operons. As soon as the Vir proteins are synthesized, T-DNA excision starts with the concerted action of VirD1 and VirD2 encoded by the virD locus. These proteins possess a site-specific endonuclease activity which specifically cuts within the conserved 25 bp T-DNA borders, the origin of transfer (oriT) (ALT-MÖRBE et al., 1986; YANOFSKY et al., 1986; JAYASWAL et al., 1987; WANG et al., 1987; VELUTHAMBI et al., 1987; WANG et al., 1990). VirD1 also has topoisomerase activity that may assist the endonuclease function in this excision process (GHAI and DAS, 1989). Excision by virD products seems to be a rate-limiting step, since virD overexpression resulted in a faster formation and an increase in the number of tumors in some plant species (WANG et al., 1990). Cutting results in three recognizable T-DNA intermediates: single-stranded (ALBRIGHT et al., 1987; STACHEL et al., 1986b; 1987; VELUTHAMBI et al., 1988; DE VOS and ZAMBRYSKI, 1989) and double-stranded linear T-DNA molecules (VELUTHAMBI et al., 1987; STECK et al., 1989), and double-stranded T-DNA circles (KOUKOLIKOVA-NICOLA et al., 1985; MACHIDA et al., 1986).

Single-stranded T-DNA molecules ("T-strands") are thought to be generated unidirectionally by strand displacement synthesis, starting at the right border in a process similar to bacterial conjugation (STACHEL and ZAMBRYSKI 1986b; ALBRIGHT et al., 1987; STACHEL et al., 1986b; 1987). *Double-stranded T-DNA* may either result from excision or from semiconservative replication (KADO, 1991). Covalently closed *T-DNA circles* were first detected by lambda *in vitro* packaging of bacterial DNA after infecting tobacco protoplasts with *A. tumefaciens* (KOUKOLIKOVA-NICOLA et al., 1985). These circles probably originate from site-specific reciprocal recombination between both 25 bp borders (MACHIDA et al., 1986; TIMMERMAN et al., 1988). Although it is not yet clear which intermediate is actually transported into the plant cell, some evidence favors the T-strand model. In a strategy called agroinfection (GRIMSLEY et al., 1986a), BAKKEREN et al. (1989) replaced the entire T-DNA, except for the 25 bp borders, by the cauliflower mosaic virus genome. Rescue of virus from infected plants allowed the analysis of individual amplified T-DNA molecules. The border arrangement of these T-DNAs suggested that the transfer involves linear, not circular T-DNA molecules. The characterization of the virE product as a single-stranded DNA-binding protein that cooperatively associates with the T-strand (GIETL et al., 1987; CHRISTIE et al., 1988; CITOVSKY et al., 1988; 1989; DAS, 1988; SEN et al., 1989) further suggested that single-stranded T-DNA is the transfer intermediate. The high efficiency of protoplast transformation with single-stranded plasmids (RODENBURG et al., 1989) also supports the view that single-stranded DNA is an effective transforming agent.

2.1.3 T-DNA Transfer

Whatever its secondary structure, the T-DNA is probably transferred as a DNA-protein complex. Among the proteins that seem to be cotransferred are the single-strand specific virE gene product (see above), and the virD2 protein that is covalently attached to the 5′ end of T-strands (HERRERA-ESTRELLA et al., 1988; WARD and BARNES, 1988; YOUNG and NESTER, 1988; HOWARD et al., 1989) and protects it from exonucleolytic degradation (DÜRRENBERGER et al., 1989). There is mounting evidence from several groups that in addition to its capping properties the virD2 protein serves as a "pilot" protein guiding the T-DNA to the plant cell nucleus. First, the VirD2 protein carries a nuclear translocation signal in its amino acid sequence (WANG et al., 1990) and is targeted to the nuclei of transgenic tobacco that harbors the virD2 gene under the control of a CaMV 35S promoter (KADO, 1991). Second, the VirD2 nuclear translocation signal is able to target a β-galactosidase fusion protein to the nucleus of transgenic plants (HERRERA-ESTRELLA et al., 1990). Once within the nucleus, the endonuclease activity of VirD2 might even help to integrate the T-DNA intermediate into the plant genome.

Our knowledge of the functions of virB, virC, and virF genes is still fragmentary. Since the majority of virB-encoded proteins is probably membrane-bound (THOMPSON et al., 1988; WARD et al., 1990), they are thought to be involved in the trans-membrane passage of the T-DNA-complex, possibly by providing a transfer channel (KULDAU et al., 1990). VirD4 protein, also located in the inner membrane (OKAMOTO et al., 1991), might assist in this process. VirB proteins might also provide the activation energy for T-DNA transport across membranes: VirB11 was reported to exhibit ATPase activity and autophosphorylating properties (CHRISTIE et al., 1989). Since virC is needed for the transformation of some plant species but not for others, it is thought to play an ancillary role in tumorigenesis (YANOFSKY and NESTER, 1986). Two lines of evidence suggest that VirC is somehow involved in the T-DNA excision process. First, the VirC protein binds to the overdrive region located close to the right T-DNA border (TORO et al., 1988; 1989). Second, VirC enhances VirD-mediated T-strand production in *Escherichia coli* when VirD1 and VirD2 are limiting (DE VOS and ZAMBRYSKI, 1989).

Most recently, the octopine-strain-specific virF region was also shown to influence host range. In its presence, T-DNA transfer to maize cells, but not to *Nicotiana glauca* cells, was effectively inhibited (JARCHOW et al., 1991). VirF mutants that were able to infect maize were shown to be extracellularly complementable by intact virF to restore the inhibitory phenotype. In the light of these results, VirF seems to interact with the T-DNA transfer complex, thereby facilitating its uptake by certain hosts while rendering it more difficult for others. According to this model, VirF might play a central role in discriminating between host and non-host plants (JARCHOW et al., 1991).

2.1.4 T-DNA Integration and Expression

Once transferred to the plant nucleus, the Ti- or Ri-plasmid-derived T-DNA(s) is (are) covalently integrated into the plant genome in one or several copies, either as a contiguous piece of DNA, or divided into two or more parts (reviewed by HOOYKAAS, 1989; REAM, 1989). Early experiments showed the integration sites to be randomly distributed throughout the plant genome (URSIC et al., 1983; CHYI et al., 1986; WALLROTH et al., 1986). Multiple insertions as well as aberrant integration patterns occurred, including tandemerization in direct (ZAMBRYSKI et al., 1980) and inverted repeats (JORGENSEN et al., 1987; MARKS and FELDMANN et al., 1989), rearrangements and truncation of T-DNA (PEERBOLTE et al., 1986; VAN LIJSEBETTENS et al., 1986) as well as of plant target DNA sequences (GHEYSEN et al., 1987; JOUANIN et al., 1989). After its integration, T-DNA adopts eukaryotic features of chromatin organization in terms of nucleosomal organization and DNase I-(hyper)sensitivity (SCHÄFER et al., 1984; KAHL et al., 1987).

Most recently, evidence has been presented that the insertion site is not actually random. T-DNA tagging experiments (KONCZ et al., 1989; HERMAN et al., 1990) as well as investigations of the chromatin structure of T-DNA in transgenic plants (WEISING et al., 1990) rather suggested a preferential integration into actively transcribed regions of the genome. Moreover, homologous recombination with short plant sequence motifs was shown to be involved in the integration process (MATSUMOTO et al., 1990; MAYERHOFER et al., 1991). It was suggested that the open chromatin structure of highly expressed sections of the target genome might facilitate recombination and therefore serve as a preferential target for integration of the T-DNA intermediate (HERMAN et al., 1990). The whole process is shown schematically in Fig. 2.

Since vir genes derived from Ri and Ti plasmids share extensive homology (HIRAYAMA et al., 1988) and are able to complement each other (HOOYKAAS-VAN SLOGTEREN et al., 1984), the mechanisms of T-DNA mobilization and integration into the host genome are probably conserved. However, the physiological basis of the transformed phenotype conferred by Ri and Ti plasmids, respectively, differs considerably. In any case, successful transformation relies on the transcription of the transferred genes. Analysis of open reading frames, transcript mapping, and insertion and deletion analysis revealed a set of seven to about thirteen T-DNA genes depending on the type of Ti or Ri plasmid (see, e.g., WILLMITZER et al., 1982; 1983; WHITE et al., 1985; SLIGHTOM et al., 1986). In crown-gall cells, T-DNA genes 1, 2, and 4 code for enzymes catalyzing auxin and cytokinin formation (INZÉ et al., 1984; AKIYOSHI et al., 1984; BUCHMANN et al., 1985; THOMASHOW et al., 1986). Their cooperative transcription leads to high phytohormone levels within transformed cells triggering high mitotic activity and tumor formation. The auxin and cytokinin effects on the presumptive tumor cell are further modulated by T-DNA genes 5 (KONCZ and SCHELL, 1986; KÖRBER et al., 1991) and 6b (SPANIER et al., 1989; TINLAND et al., 1989; 1990). The product of gene 5 was recently found to influence the plant-cell response to auxin by catalyzing the autoregulated synthesis of an auxin antagonist, indole-3-lactate, from tryptophan (KÖRBER et al., 1991). This mechanism might provide an upper limit of auxin levels and thus prevent toxic effects of an auxin overdose. The mechanism by which gene 6b exerts its influence is still unknown.

Two independent mechanisms that influence the active pool of phytohormones seem to operate in the generation of the hairy root phenotype. One mechanism is based on auxin overproduction: some regions of agropine-type Ri T-DNAs share homology with Ti-T-DNA, which includes the "auxin" genes 1 and 2 capable of root induction (VILAINE and

CASSE-DELBART, 1987; CAMILLERI and JOUANIN, 1991). These genes, however, play only a minor rolc in hairy root transformation. The major effect is exerted by another set of genes present in all Ri-T-DNAs, the so-called rol (root loci) genes rolA, rolB, rolC, and rolD (WHITE et al., 1985). The analysis of the regulated and synergistic function of individual rol genes in transgenic plants provided considerable evidence that rolA, B, and C expression is the main molecular basis for the aberrant phenotype resulting from hairy root transformation (VILAINE and CASSE-DELBART, 1987; SPENA et al., 1987; SINKAR et al., 1988a; CAPONE et al., 1989a, b; SCHMÜLLING et al., 1988; 1989).

The products of these genes were found to increase the competence of transformed tissues to respond to auxins (CARDARELLI et al., 1987; SHEN et al., 1988; SPANO et al., 1988), thereby enhancing the root-inducing effect of endogenous as well as of T-DNA-derived genes involved in auxin production. The increase in auxin sensitivity is probably mediated by the release of active auxins from inactive glucosylated storage forms by a specific indole-β-glucosidase encoded by the rolB locus (ESTRUCH et al., 1991b). The phytohormone balance of Ri-T-DNA-transformed cells is further perturbed by active cytokinins that are released by a rolC-encoded cytokinin-β-glucosidase in a similar mechanism (ESTRUCH et al., 1991a). Thus two of the Ri-T-DNA-specific oncogenes interfere with normal plant development by releasing phytohormones from their glucoside conjugates rather than by catalyzing their synthesis. In this context, it is interesting to note that rolB, and to a lesser extent also rolC, genes are themselves inducible by auxin (MAUREL et al., 1990; CAPONE et al., 1989b; 1991).

2.1.5 Opines and the Genetic Colonization Concept

In both kinds of neoplasia, specific genes of the T-DNA are constitutively expressed that encode enzymes involved in the synthesis of so-called opines, unusual amino acid derivatives or sugars, according to which the bacteria are classified. These opines, the most common being octopine, nopaline, mannopine, and agropine, accumulate in transformed cells, but cannot be metabolized by them. Instead, they are secreted (MESSENS et al., 1985). Secretion usually occurs at the transition zone from root to stem ("crown") in nature, and therefore the opines are released into the soil where only *Agrobacterium* can take them up and use them as nitrogen, carbon, and energy source. Catabolism of these compounds proceeds through enzymes encoded by the Ti plasmid. The presence of such genes selectively favors *Agrobacterium* which therefore has conquered an ecological niche. The cooperation of opine-synthesizing and opine-catabolizing genes is part of a parasitism which starts with the transfer of the T-region genes into the recipient plant cell ("genetic colonization").

Once established, the transformed state is usually stable. However, spontaneous reversion to a normal phenotype was observed in both crown gall (AMASINO et al., 1984) and hairy root (SINKAR et al., 1988b). Molecular analysis revealed that the inactivation of T-DNA genes by cytosine methylation (AMASINO et al., 1984; SINKAR et al., 1988b) or T-DNA deletion (HÄNISCH TEN CATE et al., 1990) is responsible for the reverted phenotype. The existence in plants of mechanisms to silence foreign gene expression (as, e.g., methylation, see Sect. 5.4) will be of tremendous importance for the application of plant genetic engineering (e.g., in plant breeding).

2.2 Design and Construction of T-DNA-Derived Vectors for Plant Cell Transformation

2.2.1 Early Approaches

By exploiting the *Agrobacterium*/Ti plasmid system, it became possible to transfer foreign DNA sequences into plant cells as early as 1980 (HERNALSTEENS et al., 1980). However, the early cointegrate vectors, based on modified Ti plasmids that still carried wild-type T-DNA genes, incited tumors and did not allow regeneration of transgenic plants. A considerable step forward was the development of so-called *disarmed vectors* by deleting the oncogenic regions of the T-DNA that interfere with normal plant development (ZAMBRYSKI et al., 1983). Since the border sequences are the only *cis*-requirements for T-DNA transfer, T-DNA genes could be totally replaced by "passenger" DNA. The creation of disarmed vectors led, however, to selection problems. While plant cells transformed by cointegrate vectors carrying full-length T-DNA could be simply selected by their T-DNA-encoded ability to grow on media without phytohormones, the absence of T-DNA genes challenged the development of new selection markers.

Bacterial antibiotic resistance genes that offered themselves as putative markers were silent in plant cells if controlled by their own promoter. Therefore, chimeric genes were constructed that combined plant-, plant virus-, or T-DNA-derived promoter and polyadenylation signals with the coding region of a bacterial antibiotic resistance marker. Expression of these chimeric genes in plant cells allowed selection of transformants on media containing the appropriate antibiotic (BEVAN et al., 1983; HERRERA-ESTELLA et al., 1983a, b; FRALEY et al., 1983), and their regeneration to transgenic plants (DE BLOCK et al., 1984; HORSCH et al., 1984).

2.2.2 Selectable Markers and Reporter Genes

A variety of *selectable markers* have now been introduced to plant genetic engineering (Tab. 2). Among these, the Tn5-derived neomycin phosphotransferase (npt II) gene is still one of the markers of choice. In many plant species, npt II gene expression provides high levels of resistance to kanamycin, neomycin, and the synthetic antibiotic G-418, thus allowing unambigous selection of transformants (Fig. 3). However, alternative markers were also developed for at least two reasons. First, there is no single drug-resistance gene that will function uniformly in all plant species. Second, additional markers are needed for repeated transformations of individual plants. Therefore, attention focused on the phosphinothricin acetyltransferase genes from *Streptomyces hygroscopicus* (THOMPSON et al., 1987; DE BLOCK et al., 1987) and *S. viridochromogenes* (WOHLLEBEN et al., 1988), conferring resistance to the herbicides bialaphos and phosphinothricin, and on the hygromycin phosphotransferase (hpt) gene from *E. coli* introduced by WALDRON et al. (1985) and VAN DEN ELZEN et al. (1985b). The hpt marker was successfully used for the transformation of several plant species which are not efficiently selectable on kanamycin (e.g., *Arabidopsis thaliana,* LLOYD et al., 1986).

Another category of marker genes is referred to as *reporter genes* (Tab. 2). Their expression in transgenic plants is scorable by sensitive assays (see, e.g., the comparative study by TÖPFER et al., 1988). Reporter genes proved to be valuable tools in the evaluation of qualitative and quantitative aspects of foreign gene expression. In particular, coding regions of reporter genes were combined with various promoters in order to study the role of *cis*-acting sequences in the regulation of gene activity. Most commonly used is the β-glucuronidase (gus) gene from *E. coli* (JEFFERSON et al., 1987; JEFFERSON, 1989). Its main advantages are the easy performance of a transient expression assay, the availability of a large assortment of suitable colorigenic and fluorigenic substrates, the low endogenous gus activity in most plant species, and the availability of histochemical techniques that allow analysis of cell-specific reporter gene activity in transgenic plants. A recently developed microassay allows the quantitative determination of GUS activities in individual cells, e.g., in PEG-transformed or microinjected tobacco protoplasts (SPÖRLEIN et al., 1991a).

Other commonly used reporters are the Tn7-derived cat gene (GORMAN et al., 1982), and the luciferase genes from the bacterium *Vibrio harveyi* (KONCZ et al., 1987a; 1990; OLSSON et al., 1988) or from the firefly *Photinus pyralis* (OW et al., 1986; BARNES, 1990). A particular advantage of luciferase assays is their high sensitivity, ease and versatility, and the opportunity to visualize gene expression directly in transgenic tissues (reviewed by KONCZ et al., 1990). A new reporter gene for transgenic plants encodes a calcium-sensitive luminescent protein from the marine coelenterate *Aequorea victoria.* If supplied with the hydrophobic luminophore coelenterazine, apoaequorin reports on transient changes of cyto-

Tab. 2. Scorable Reporter and Selectable Marker Genes for Plant Transformation

Gene	Enzyme	Useful as Reporter	Marker	Resistance against	References
npt II	Neomycin phosphotransferase	+	+ +	Neomycin, kanamycin, G-418	BEVAN et al. (1983)
npt I	Neomycin phosphotransferase	+	+	Neomycin, kanamycin, G-418	FRALEY et al. (1983)
cat	Chloramphenicol acetyltransferase	+ +	(+)	Chloramphenicol	HERRERA-ESTRELLA et al. (1983a, b)
hpt	Hygromycin phosphotransferase	—	+ +	Hygromycin	WALDRON et al. (1985) VAN DEN ELZEN et al. (1985b)
pat	Phosphinothricin acetyltransferase	—	+	Phosphinothricin	DE BLOCK et al. (1987)
gus	Glucuronidase	+ +	—	—	JEFFERSON et al. (1987)
luxA/B	Bacterial luciferase	+ +	—	—	KONCZ et al. (1987)
luc	Firefly luciferase	+ +	—	—	OW et al. (1986)
ocs	Octopine synthase	+	(+)	Aminoethylcysteine	DAHL and TEMPÉ (1983)
nos	Nopaline synthase	+	—	—	ZAMBRYSKI et al. (1983)
dhfr	Dihydrofolate reductase	—	+	Methotrexate	EICHHOLTZ et al. (1987)
lacZ	β-Galactosidase	+	—	—	HELMER et al. (1984) MATSUMOTO et al. (1988)
bsr	Blasticidin S desaminase	—	+	Blasticidin	KAMAKURA et al. (1990)
AAC (3)	Gentamycin acetyltransferase	—	+	Gentamycin	HAYFORD et al. (1988) CARRER et al. (1991)
SPT	Streptomycin phosphotransferase	—	+	Streptomycin	JONES et al. (1987)
ble		—	+	Phleomycin, bleomycin	HILLE et al. (1986), PEREZ et al. (1989)
sul	Dihydropteroate synthase	—	+	Sulfonamide	GUERINEAU et al. (1990)

plasmic calcium levels in transgenic *Nicotiana plumbaginifolia* plants (KNIGHT et al., 1991).

2.2.3 Binary Vectors

The availability of scorable and selectable marker genes provided the basis for routine plant transformation using disarmed Ti plasmid vectors and opened the way to further refinements, e.g., the so-called split-end-vector (SEV) system that carried the T-DNA border sequences on separate plasmids (FRALEY et al., 1985), and a variety of improved cointegrate vectors (DEBLAERE et al., 1985; MATZKE and MATZKE, 1986). The next major step in vector improvement, however, was the design of *binary vectors*. This concept was based on the observation that Ti-plasmid-en-

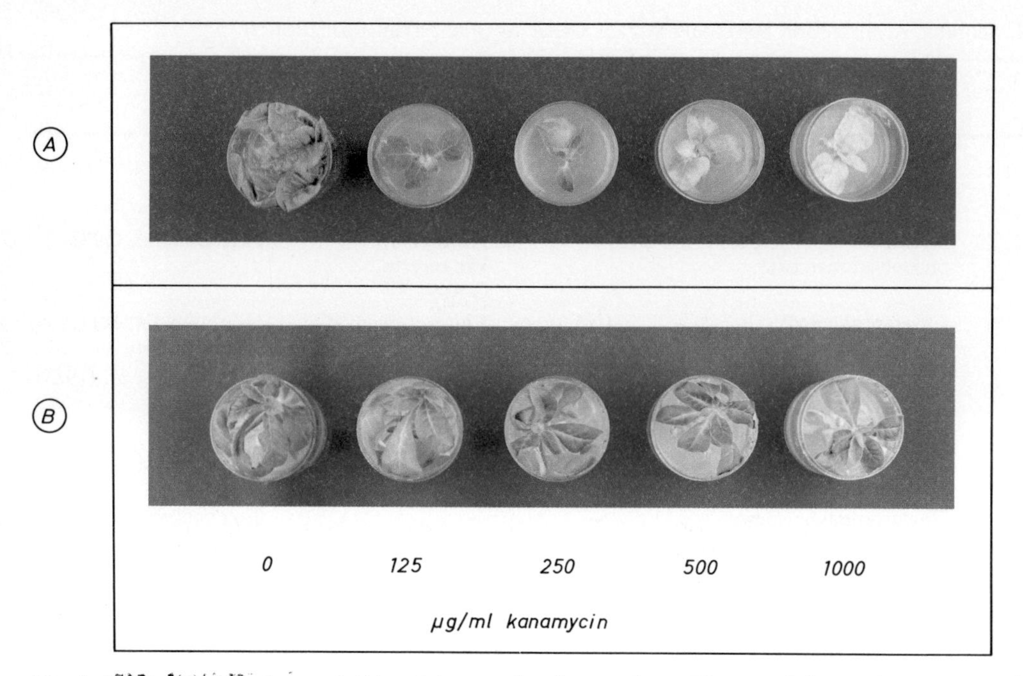

Fig. 3. The sensitivity of normal (A) and transgenic tobacco plants (B) towards increasing concentrations of the aminoglycoside antibiotic kanamycin. The transgenic plants have been engineered with a kanamycin resistance gene derived from transposon 5 (Tn 5), driven by the promoter of the nopaline synthase (nos) gene of the Ti plasmid of *Agrobacterium tumefaciens*. The expression of this gene confers resistance towards kanamycin upon the transgenic plants.

coded vir functions were not *cis*-essential for T-DNA transfer, but could be located on a different plasmid or even on the bacterial chromosome (HOEKEMA et al., 1983; 1984). "Binary" vector systems could therefore be constructed that consisted of a pair of plasmids both of which replicate in *Agrobacterium*. One component is a (non)modified Ti plasmid providing the vir-functions in *trans,* while the second component is the vector itself, carrying a gene construct of interest within T-DNA-borders, and mobilization and replication functions as well as a bacterial selection marker outside the T-DNA borders, allowing the maintenance, replication, and mobilization of the vector between *E. coli* and *Agrobacterium*. Vector "cassettes" may combine several useful elements within the T-DNA insert, depending on the aim of the transformation experiment (see Fig. 4):

(1) selectable marker genes,
(2) reporter genes,
(3) multiple cloning sites to introduce the gene construct of interest,
(4) elements, which facilitate the rescue of transferred DNA from plants to phage or bacterial plasmids by either the inclusion of a lambda cos site or a plasmid replicon plus a bacterial selectable marker within the T-DNA borders (KONCZ and SCHELL, 1986),
(5) elements designed to circumvent position effects (see Sect. 5.3),
(6) transposable elements that allow gene tagging in the foreign host genome.

2.2.4 Vector Design

Various modifications of binary vector systems are now available (BEVAN, 1984; AN et al., 1985; HOEKEMA et al., 1985; KLEE et al., 1985; VAN DEN ELZEN et al., 1990; KONCZ and SCHELL, 1986; KONCZ et al., 1987b;

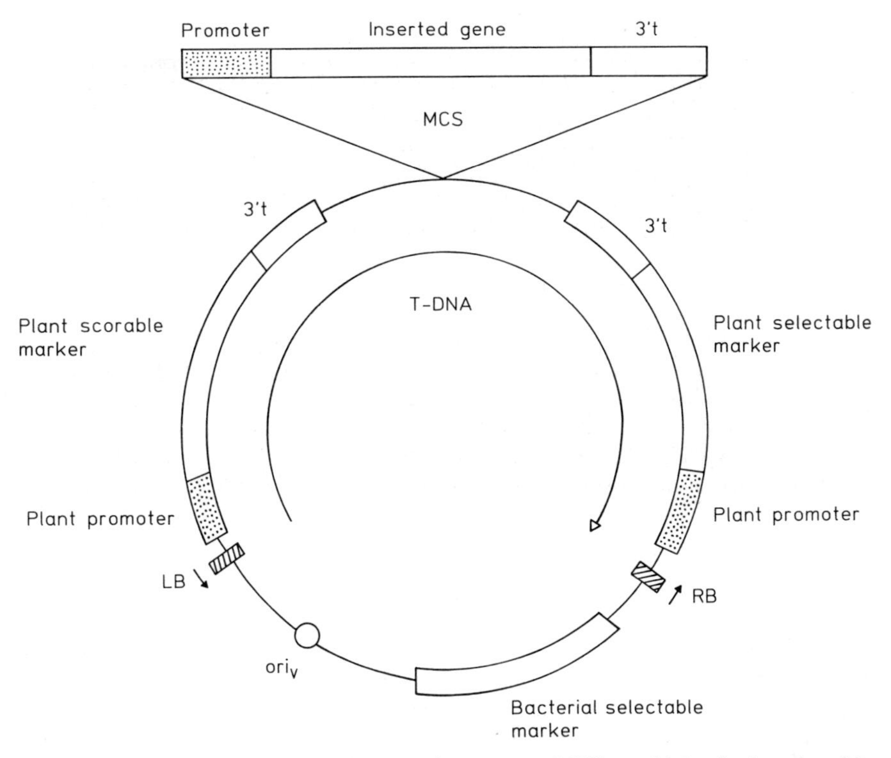

Fig. 4. The prototype of a plant transformation and expression vector. MCS, multiple cloning site; 3′t, 3′-terminating sites including poly(A) site; LB, RB, left and right border sequences of T-DNA; oriV, origin of vegetative replication.

McBride and Summerfelt, 1990; Ott et al., 1990). Dictated by the experimental strategy, several different classes of vectors are constructed: vectors designed for testing regulatory regions of genes, expression vectors that harbor the coding region or a cDNA of a gene of interest fused to its own or a heterologous promoter that confers high level and/or regulated expression. Elements that positively influence transcriptional and post-transcriptional events (e.g., introns and translational enhancers) may be included in such constructs. Modifications were created that allow the coordinate transcription of two genes of choice under the control of a dual promoter element (Velten et al., 1984; Velten and Schell, 1985; Ott et al., 1990). Another vector cassette combined Tn5-mediated gene mutation in bacteria with the Ti plasmid-mediated transfer of these genes into plants (Koncz et al., 1987b). This construct allows the random mutagenesis of any DNA sequence carried by any *E. coli* plasmid and its transfer to the plant genome.

The choice of promoters to direct expression of foreign genes in transgenic plants will depend on the objective, and thus it is important to develop various promoters with well-defined characteristics.

Promoter test vectors, often used in transient expression assays, usually contain transcriptional or translational fusions of (dissected) plant gene control regions and the coding region of a suitable reporter gene. Such control regions can be isolated by a technique known as "promoter-trapping", which is based on the cloning of genomic DNA sequences in front of promoterless reporter genes, followed by selection for those DNA fragments that activate the silent gene. Promoter trap vectors have been designed and used successfully (Herman et al., 1990; Claes et al., 1991;

GOLDSBROUGH and BEVAN, 1991). In a somewhat different approach plant cells are transformed with specially designed T-DNA vectors (see Sect. 2.2) that carry a reporter gene devoid of any *cis*-acting transcriptional and translational expression signals, located at the end of the T-DNA. Upon integration, the initiation codon of the promoterless reporter gene will be juxtaposed to plant sequences. If these sequences contain promoter elements, the reporter gene will be transcribed. The hybrid genes, called T-DNA mediated gene fusions, are then composed of uncharacterized plant promoters residing at their natural location within the genome, and the reporter sequence located on the inserted T-DNA (ANDRÉ et al., 1986; TEERI et al., 1986; FOBERT et al., 1991; KERTBUNDIT et al., 1991; see FELDMANN, 1991).

High levels of foreign gene expression are, for example, desirable after the introduction of (1) genes conferring resistance to insects, fungi and other pests, (2) herbicide resistance genes, (3) genes that code for economically important proteins, and (4) genes encoding enzymes that catalyze important or rate-limiting steps in metabolism (see WILLMITZER, Chapter 16, this volume). Overexpression is also a valuable tool for understanding the function of gene products. *Regulated promoters,* on the other hand, are useful when the expression of a transgene product is only desired in special situations (e.g., chitinase and β-1,3-glucanase expression upon fungal attack) or at distinct stages of development (e.g., antisense genes that inhibit fruit ripening). Moreover, the ability to switch a gene on and off is especially useful for functional studies on gene products and their interactions (e.g., on the coordinated expression of the small and large subunits of ribulose-bisphosphate-carboxylase; RODERMEL et al., 1988).

The promoter of choice for obtaining high levels of constitutive transgene expression in dicots is the cauliflower mosaic virus 35S promoter (CaMV 35S). The promoter and enhancer regions of CaMV 35S are among the best-characterized gene regulatory sequences in plants (reviewed by BENFEY and CHUA, 1991). In a comparative analysis, this promoter was shown to direct npt II gene expression at 30 to 100 fold higher levels than did the nopaline synthase promoter combined with the same reporter gene in transgenic plants (SANDERS et al., 1987). The constitutive transcription of CaMV 35S-controlled genes is mediated by the combined action of various modular elements, each of which shows a tissue- and development-specific expression pattern (BENFEY et al., 1990a, b). The promoter is thus able to function in a variety of tissues.

High-level expression, directed by CaMV promoters, overproduced EPSP synthase in transgenic plants leading to an enhanced resistance against the herbicide glyphosate (SHAH et al., 1986), or to the synthesis of large quantities of a tobacco mosaic virus (TMV) coat protein in transgenic plants resulting in enhanced tolerance to TMV infection (POWELL ABEL et al., 1986). The strategy of coat-protein expression allowed the engineering of a number of virus-resistant plants (see WILLMITZER, Chapter 16, this volume). Overexpression of a yeast ornithine decarboxylase gene in transgenic roots of *Nicotiana rustica* led to the enhanced accumulation of nicotine (HAMILL et al., 1990). In a series of experiments on overexpressed anionic peroxidase genes in transgenic tobacco, LAGRIMINI et al. (1990) observed a variety of unprecedented effects of high peroxidase activity on the phenotype of the transgenic plant.

Several plant- and T-DNA derived promoters have been shown to confer high levels of foreign gene expression. For example, the ST-LS1 promoter from potato (ECKES et al., 1986) overexpressed alfalfa glutamine synthetase genes in transgenic tobacco to levels comparable to those obtained with CaMV 35S (ECKES et al., 1989). The dual 1′-2′-promoter element from octopine-type T-DNA (VELTEN and SCHELL, 1985) directs the expression rate of two genes in a coordinated manner (VELTEN et al., 1984; SAITO et al., 1991a). Thus, selection of transformants expressing high levels of marker gene activity should allow coselection for high expression rates of the gene of interest. A recent study on a large number of individual transgenic tobacco calli showed, however, that cat and gus reporter gene activities driven by the 1′- and 2′-promoter, respectively, were also prone to position effects resulting in non-coordinate transcription (PEACH and VELTEN, 1991). Moreover, like other T-DNA gene promoters (AN et al., 1988; 1990), this promoter is also transcribed in an inducible and tissue-specific fashion rather than being constitutive (LANGRIDGE et al., 1989; TEERI et al., 1989).

Promoter strength is, of course, also determined by the host plant under study, and large phylogenetic distances between donor and acceptor plants might result in loss of function (see Sect. 5.1). According to their biology, T-DNA-derived as well as viral promoters are adapted to work in a variety of hosts and are thus most versatile. However, in spite of its high efficiency in transgenic rice plants (BATTRAW and HALL, 1990), the CaMV 35S promoter is comparably weak in some other monocot species (VASIL et al., 1989; PETERHANS et al., 1990b; HAGIO et al., 1991). The rice actin promoter has recently been shown to be highly expressed in monocot cells (MCELROY et al., 1990; 1991).

Frequently, a distinct promoter has been chosen for plant transformation experiments on the basis of certain features (e.g., sequence elements responding to exogenous parameters), but does not express the transferred gene at levels required for detection or exploitation of the product. Several strategies have been developed to increase promoter strength, such as insertion of enhancer elements, removal of silencer sequences, optimization of spacing between the different promoter elements, and others. For plant promoters one approach has been rewarding: the insertion of intron sequences between the 5′ terminus of the promoter and the 3′ end of the first exon of the adjacent gene. CALLIS and colleagues fused the first intron of the maize alcohol dehydrogenase (Adh-1) gene to the promoters of Adh-1, CaMV 35, or nopalin synthetase. The constructs were electroporated into maize protoplasts where they increased transient reporter gene expression to 50–100 times of the control (CALLIS et al., 1987).

The positioning of the introns seems to be of strategic importance, since they have to be placed immediately downstream of the 5′ terminus of the promoter and upstream of the reporter gene. Constructs containing only the exonic Adh-1 sequences did not enhance reporter-gene expression. This enhancing effect of intron spacing has repeatedly been found in a series of monocotyledonous plants (e.g., wheat, OARD et al., 1989; rice, KYOZUKA et al., 1990; maize, GORDON-KAMM et al., 1990; FROMM et al., 1990), and with introns from other genes (e.g., first intron of the shrunken-1 (Sh-1) gene of maize; VASIL et al., 1989). The enhancing capacity of the Sh-1 intron has been shown to be superior to that of the Adh-1 intron, but both do not seem to exert their activating influence beyond monocotyledonous plants (e.g., tobacco: MAAS et al., 1991). It may well be that different splicing strategies in mono- and dicots are responsible for this bias (KEITH and CHUA, 1986). Though the intron effect was originally thought to be restricted to intron sequences, more recently exon sequences were isolated that also influence gene expression. For example, reporter-gene expression in electroporated maize, rice, and tobacco protoplasts was stimulated by the insertion of the Sh-1 exon 1. If both the Sh-1 intron 1 and Sh-1 exon 1 were combined, the expression of a chimeric 35S-CAT construct could be dramatically enhanced in maize and rice protoplasts (MAAS et al., 1991).

Whatever the precise mechanism of action, the improvement of promoter efficiency by changing its architecture or the structure of its immediate environment certainly has a promising future. To achieve regulated expression, the use of *inducible promoters* that respond to exogenous signals like heat shock (AINLEY and KEY, 1990) or wounding (LOGEMANN et al., 1989) might be envisaged. For some purposes, however, it may be desirable to specifically (in)activate the integrated transgene without influencing other genes of the host plant. In these cases, the installation of a "switch" might be the strategy of choice. Such a switch can be provided by introducing gene regulatory mechanisms not usually present in plants. Using cotransformation of a repressor gene and a test gene, GATZ and QUAIL (1988) introduced the components of the Tn10-encoded bacterial tet repressor system into tobacco cells. Tet repressor was shown to bind to its target sites inserted close to the TATA box, thereby inhibiting expression of the test gene (CaMV 35S-cat). Addition of the specific inducer tetracycline released the inhibition. The system did also work in transgenic plants using a different reporter gene (GATZ et al., 1991), and was shown to be useful for the analysis of interactions between *trans*-acting factors and the transcriptional machinery (FROHBERG et al., 1991). However, its general use as a gene switch is probably limited, since control of target-gene expression is somewhat leaky (GATZ et al., 1991).

An alternative "switching" strategy was recently proposed by SCHENA et al. (1991). The authors cotransfected a rat glucocorticoid receptor cDNA controlled by CaMV 35S, and a test gene carrying glucocorticoid responsive elements linked to a CaMV 35S minimal promoter/cat reporter fusion into tobacco protoplasts. In a transient expression assay, nanomolar concentrations of glucocorticoids were sufficient to induce cat activity more than 150 fold. Since glucocorticoids are taken up by roots, and have no detectable effects on growth rate and metabolism of *Arabidopsis* (SCHENA et al., 1991), this strategy might also be applied to transgenic plants.

2.3 Plant Viruses as Gene Vectors for Plants

2.3.1 Introduction

One of the major motors driving the development of gene technology in bacteria and mammals was the availability of naturally occurring gene vectors such as bacteriophages and viruses, their modification to serve as potent and versatile vehicles for cloning, the design of recombinant derivatives with streamlined genomes to accomodate large amounts of foreign DNA, and the precise knowledge of the infection and replication details. Compared to the immense progress in this field of molecular biology, the exploitation of plant viruses as gene vectors is still in an infantile stage. However, the need for vector systems that complement other techniques (e.g., *Agrobacterium*-mediated gene transfer, see Sects. 2.1 and 2.2) recently attracted attention to plant viruses as potential gene vectors for plant genetic manipulation.

Plant viruses fall into two broad categories, the RNA and DNA viruses (Tab. 3). The majority of the 650 known plant viruses, namely more than 90%, possess an RNA genome and can therefore be classified as RNA viruses, most of which carry a single-stranded RNA genome (ssRNA). This documents the great evolutionary success of these viruses, which are again divided into some 30 groups. Each group harbors viruses with similar physico-chemical properties, but different host ranges. The DNA viruses in contrast are in the minority and can be classified as single-stranded DNA viruses (gemini viruses) and double-stranded DNA viruses (caulimoviruses, see Tab. 3). Members of the dsDNA subgroup (e.g., cauliflower mosaic virus, CaMV) have been exten-

Tab. 3. Plant Viruses

Nucleic Acid	Genome Type	Number	Example
RNA	Positive-strand type	494	
	Monopartite		Tobacco mosaic virus (TMV)
	Bipartite		Cowpea Mosaic virus (PMV)
	Tripartite		Brome mosaic virus (BMV)
	Negative-strand type	85	Rhabdoviruses
	Double-stranded	30	Phytoreoviruses
DNA	Single-stranded	26	Gemini viruses
	Monopartite		Wheat dwarf virus (WDV)
	Bipartite		Cassava latent virus (CLV)
	Double-stranded	16	Caulimoviruses

sively characterized, also because most gene techniques have been evolved with or adopted for double stranded DNA.

The attractiveness of viruses as gene vectors for plants generally is based on a series of specific characters:

1. Viruses adsorb to and infect cells of intact plants.
2. Some viruses spread systemically from the infection site and transport viral nucleic acid into virtually every living cell of the host plant.
3. The virus may replicate up to several thousand times per cell and be present in 10^3-10^6 genome copies per cell.
4. The viral genes are driven by strong and plant-constitutive promoters and are therefore strongly expressed.
5. As a group, plant viruses have an extremely broad host range (e.g., it is suspected that every existing plant is infected by at least one RNA virus).
6. Specific viruses infect monocotyledons (e.g., cereals) that are recalcitrant towards *Agrobacterium*-mediated gene tranfer.
7. Relatively large numbers of viruses accumulate in plants. Consequently, large amounts of viral nucleic acid and also viral proteins are present in infected hosts.

All these, and other, features (e.g., the ease of handling viruses and performing infections) have made plant viruses attractive for genetic engineers. In fact, they bear a potential for amplification of desired genes (if these genes can be inserted into the viral genome without interfering with viral functions) and production of gene products. Moreover, in many cases the purified viral nucleic acid is infectious to plants, which facilitates manipulation of plants. The systemic spread of the virus also circumvents the requirement for laborious and tedious (in many cases yet impossible-to-perform) regeneration protocols (as is the case with most other transformation systems). And even one single plant cell supports the replication of plant viruses *in vitro* for transient expression studies.

This list of advantageous traits is not complete. With all their virtues, however, plant viruses also have inherent disadvantages that stand against their use as potent gene vectors, and it is the purpose of this chapter to illustrate both pros and contras realistically.

2.3.2 RNA Viruses

In view of the large numbers of different RNA viruses it might be that all plants are hosts for at least one, or maybe several such pathogens and so, in principle, it should be possible to develop viral gene vectors for any particular plant species. However, RNA is still difficult to manipulate *in vitro*, though RNA recombinant technology is rapidly improving. The reverse transcription of viral RNA into a first strand cDNA copy, using retroviral reverse transcriptase (RTase), and *E. coli* DNA polymerase I to synthesize a second strand after RNase H digestion of the template RNA (yielding a double-stranded cDNA) has made genetic manipulation of RNA viruses possible. Now the viral genome (as cDNA) can be cloned into bacterial plasmids and be manipulated at will *in vitro*, using the whole repertoire of recombinant DNA technology. These recombinant cDNA clones can then be directly used to infect plants successfully, at least in some cases. These strategies have been applied to a number of positive strand RNA viruses (that encapsidate single-stranded messenger-sense RNA, see Tab. 3). However, since cDNAs are not necessarily infectious as such, infectious RNAs have been transcribed *in vitro* from viral cDNAs (e.g., brome mosaic virus, BMV, AHLQUIST et al., 1984; tobacco mosaic virus, TMV, DAWSON et al., 1986; MESHI et al., 1986; cowpea mosaic virus, CPMV, VOS et al., 1988; cowpea chlorotic mottle virus, CCMV, ALLISON et al., 1988).

We will portray the progress made in this area of virology using two model systems: the monopartite tobacco mosaic virus (TMV) and the tripartite brome mosaic virus (BMV).

Tobacco mosaic virus (TMV)

The monopartite RNA viruses, like TMV, have a single-stranded RNA genome, usually of plus-sense polarity, that is packaged in a single nucleoprotein particle. The TMV plus-strand RNA genome of 6.4 kb is encapsidated in rod-shaped, 300 nm long particles, it serves as messenger RNA, and is itself infectious. It encodes at least four proteins. The 183 kDa (P183) and 126 kDa (P126) polypeptides are translated directly from the same initiation codon on genomic RNA and serve viral replication functions. The two other proteins, the 30 kDa P30 and the 17 kDa P17, are translated from subgenomic RNAs. P30 is essential for TMV propagation in whole plants and represents the movement protein that catalyzes viral spread in some yet unknown way. P17 synthesis is driven by a very effective promoter to high levels (i.e., several mg per g of infected tobacco leaf), it functions in long-distance spread of the virus and represents the coat protein (cp). Since the cp gene is non-essential for viral replication, it is an attractive target site for insertion or replacement strategies.

The infection cycle starts with the uncoating of the virus particles upon entry into the presumptive host cell. An exposed part of the 5′ terminus of the viral RNA binds to ribosomes while the remainder of the molecule is still covered with cp subunits. As translation of the viral messenger RNA proceeds, coat proteins are progressively stripped off by the ribosomes towards the 3′ terminus of the template (WILSON, 1985). After multiple rounds of replication the assembly of a virion is nucleated by coat proteins which form a disc. Progressively more and more discs are added to the extending rod with free viral RNA looped through the central hole of the disc. An origin of assembly has been located in the P30 coding region that allows the formation of three hairpin loops as potential address sites for proteins (TURNER and BUTLER, 1986).

The manipulation of TMV as gene expression vector has been repeatedly successful. In 1987, TAKAMATSU and coworkers placed the bacterial chloramphenicol acetyltransferase (CAT) gene downstream in-frame of the initiation codon of the cp gene and produced *in vitro* transcripts. When these were inoculated into tobacco plants, CAT expression could be detected, but no systemic spread of infection occurred. If inoculated as part of a full-length cDNA clone of TMV-RNA, the CAT gene was also expressed in host plants. In another approach, TMV sequences specifying the origin of assembly of cp subunits were cloned side by side with the CAT gene. Messenger RNA transcribed from this construct *in vitro* was mixed with purified TMV cp subunits, which assembled the RNAs into TMV-like rods. Following inoculation of protoplasts and plants with these pseudovirions, CAT activity increased to significantly higher levels than in plants inoculated with unprotected RNA (GALLIE et al. 1987).

All these, and other reports, simply document that the TMV genome can be manipulated to accomodate a foreign sequence of limited length, that the foreign DNA is transferred into the plant by TMV, and is expressed. The expression levels are usually higher compared with other transfection techniques, most probably a consequence of systemic spread and overreplication of the virus. However, integration of viral (or foreign) sequences into the plant genome was not observed and is most likely impossible. The TMV can therefore be used as a plant expression vector, but not as plant transformation vector.

Brome mosaic virus (BMV)

The BMV, which infects mainly monocotyledonous plants (including Gramineae), contains a tripartite single-stranded plus-sense RNA genome. Two of the RNA segments, RNA 1 (3.2 kb) and RNA 2 (2.9 kb) encode transacting replication factors (KIBERSTIS et al., 1981; FRENCH et al., 1986). RNA 3 (2.1 kb) codes for the viral coat protein, that is, however, transcribed from a subgenomic RNA 4 (produced during infection from RNA 3) via a 120 bases promoter cassette (FRENCH and AHLQUIST, 1988). Shifting this cassette within the RNA 3 produced up-promoter (or, down-promoter) effects, suggesting a potential for gene expression vector design. Yet this RNA 3 is completely dispensable for BMV-RNA replication in plant protoplasts. Considerable deletions, insertions, and rearrangements are tolerated without interfering with replication functions. The observed variability in RNA 3 is possibly the cause for the broad host range of the virus (AHLQUIST and PACHA, 1990). As with TMV, the strategy to modify BMV to become a plant transformation vector included the insertion of the CAT gene from Tn 9, or the replacement of the cp gene by the CAT sequences (FRENCH et al., 1986). Replacement of the cp gene is a potentially dangerous approach, since there will be no packaging of viral RNA without coat proteins (AHLQUIST et al., 1987). Chimeric BMV-RNA 3 together with wild-type RNA 1 and 2 were transfected into barley *(Hordeum vulgare)* protoplasts, where replication and production of subgenomic mRNA ensued, concomitantly with a high-level expression of the CAT gene (AHLQUIST et al., 1987; AHLQUIST and PACHA, 1990). CAT expression is also controllable at the translational level (FRENCH et al., 1986).

Again, there was only expression, but no integration of the foreign gene reported. This leaves us with the conclusion that the potential of RNA viruses as gene vectors is strictly and insurmountably limited to transfer and expression of foreign DNA in plants. The importance of RNA virus vectors will therefore be restricted to the production of economically interesting proteins in plants. However, improvements of the vector RNA are necessary. First, the capacity of RNA viruses is limited. However, many RNA viruses have filamentous morphology, and since the length of a virus particle is determined by the length of the viral nucleic acid, there is probably no strict size limitation for the foreign gene to be packaged. Second, a potentially serious obstacle for successful gene transfer and expression via RNA viruses is the repeatedly observed instability of the foreign insert. DAWSON and colleagues (1989) introduced a 1 kb CAT reporter gene into the TMV genome between the 30 kDa and 17 kDa protein gene reading frames in a cDNA clone. After infection this construct behaved like wild-type virus, and the CAT gene was expressed, but not in all transformants. It turned out that the CAT sequences had been precisely deleted to produce wild-type TMV (DAWSON et al., 1989). This tendency towards *in vivo* stream-lining of the genome is obviously counterproductive and undesirable. The integration of some viral genes into the recipient genome and the supply of their functions in *trans* would expand the RNA virus capacity and could circumvent packaging constraints or other inadequacies. Third, the relatively low fidelity of RNA replication (error frequency during alfalfa mosaic virus replication: higher than 10^{-5}; VAN VLOTEN-DOTING et al., 1985) imposes a restraint upon the versatility of RNA viruses as efficient expression vectors (see also SIEGEL, 1985).

In summary, RNA virus vectors will certainly be improved to accomodate more foreign DNA, to replicate with higher fidelity, and to allow the stable maintenance of the foreign gene. As such these vectors will be very useful for the expression of foreign sequences in plants. However, we regard their potential as vectors for integrative transformation of plants as discouraging low.

2.3.3 DNA Viruses

The smaller group of plant viruses with a DNA genome can be divided into two subgroups:

A. Single-stranded DNA viruses that again fall into two broad categories, the ssDNA viruses with a monopartite genome, transmitted mainly by leafhoppers (example: maize streak virus, MSV), and the ssDNA viruses with a bipartite genome, transmitted by whiteflies (example: cassava latent virus, CLV). The ssDNA-containing viruses altogether are members of the gemini virus group.
B. Double-stranded DNA viruses (caulimoviruses).

Both groups have received much attention as potential vectors for the introduction of foreign DNA into plants, so that their biology is rather well understood, and vector design advanced, because viral DNA is readily accessible for all kinds of recombinant DNA techniques.

2.3.3.1 Gemini Viruses

These viruses derive their name from their paired (twinned, or 'geminate') appearance in the electron microscope. They contain a covalently closed circular (ccc) single-stranded DNA genome in the molecular-weight range between 7 and 9×10^5, which replicates via a double-stranded (ds) DNA intermediate. Gemini viruses can be categorized by their host range, transmission vector, and genome structure (LAZAROWITZ, 1987; DAVIES and STANLEY, 1989). All members infecting monocotyledonous plants are transmitted by leafhoppers (e.g., *Circulifer tenellus, Cicadulina mobila*) and carry a single genomic DNA of 2.7 kb, the smallest infectious DNA known so far. Maize streak virus (MSV), wheat dwarf virus (WDV), digitaria streak virus (DSV), and chlorosis striate mosaic virus (CSMV) belong to this group of viruses that infect such important crops such as maize, wheat, rice, millet, and sorghum. Most members infecting dicotyledonous plants are transmitted by whiteflies (e.g., *Bemisia tabaci*) and have a bipartite genome. Members of this group are cassava latent virus (CLV), tomato golden mosaic virus (TGMV), and bean golden mosaic virus (BGMV). The individual DNAs are designated DNA 1 (or A) and DNA 2 (or B), respectively. Neither DNA alone is infectious. DNA 1 encodes replication functions and the coat protein, and DNA 2 is required for cell-cell and/or long distance, i.e., vascular spread of the viral DNA within the host plant. Systemic infection, or better, viral movement in the infected plant is not fully understood. However, the coat protein gene is not essential for this process (STANLEY and TOWNSEND, 1986; WARD et al., 1988; GARDINER et al., 1988; HAYES et al., 1988a).

The different viral genomes share a high degree of sequence similarity, especially in a common region at which transcription is initiated bidirectionally (Fig. 5). In CLV DNA A three conserved open reading frames (AC 1–AC 3) are read in the complementary sense, one (AV 1) in the virion sense, whereas DNA B contains one of each kind (BC 1 and BV 1). Though both DNAs are necessary to incite systemic infection, DNA A alone directs replication and virion formation, and in addition, the synthesis of coat protein (cp) messenger RNA. Since the coat proteins accumulate to high levels in infected plants, the cp gene promoter obviously is a very strong promoter. Consequently, reporter genes (e.g., CAT gene), if fused in-frame to the amino terminus of the CLV cp gene, are expressed extensively (WARD et al., 1988). Moreover, the cp gene itself can be deleted without interfering with viral replication functions. Coat-protein deletion mutants retain their infectivity, so that the cp sequence can be replaced with foreign sequences (e.g., viral genes, ETESSAMI et al., 1988; CAT gene, WARD et al., 1988). This allows the construction of viral *replacement vectors* (see Fig. 4). Although very attractive, their use is still problematic, because the pathogenicity is not abolished, and the construct is unstable. For example, CLV coat protein mutants still produce severe disease symptoms (ETESSAMI et al., 1988; WARD et al., 1988), though in some cases pathogenicity was attenuated (as e.g. in TGMV; GARDINER et al., 1988; HAYES et al., 1988a). Usually, replacement sequences that restore the size of wild-type DNA A are relatively stable (ETESSAMI et al., 1988; WARD et al., 1988). If the insert size exceeds the wild-type size of the virus, it becomes inherently unstable (e.g., CLV DNA and DNA B, STANLEY and TOWNSEND, 1986). One possibility to circumvent such constraints is to release the insert from the construct. This approach has been successful with TGMV-DNA. Monomeric inserts from integrated multimeric copies of DNA A constructs were released and subsequently replicated to high copy number (ROGERS et al., 1986b; HAYES et al., 1988a). However, the frequency of such release is apparently low, so that improvements are required (SCOTT-ELMER et al., 1988).

Gemini viruses have a very broad host range which makes them attractive as potential vectors for the transfer of genes into plants, particularly monocotyledons. One example is the engineering of the monopartite gemini virus wheat dwarf virus (WDV) as a vector for cereals. Protoplasts, derived from suspension cultures of *Triticum monococcum,* were transfected with dimeric or linearized WDV-DNA. Replication of the WDV genome occurred (MATZEIT, 1987). Obviously a cloned WDV genome present as genomic dimer is superior to a genomic monomer, since the latter would not be replicated, unless released from the bacterial vector (GRONENBORN and MATZEIT, 1989). The insertion of reporter genes such as neomycin phosphotransferase (npt II) or β-galactosidase genes does not affect the replication of the viral genome in *Triticum* protoplasts. This means that the cloning capacity of WDV

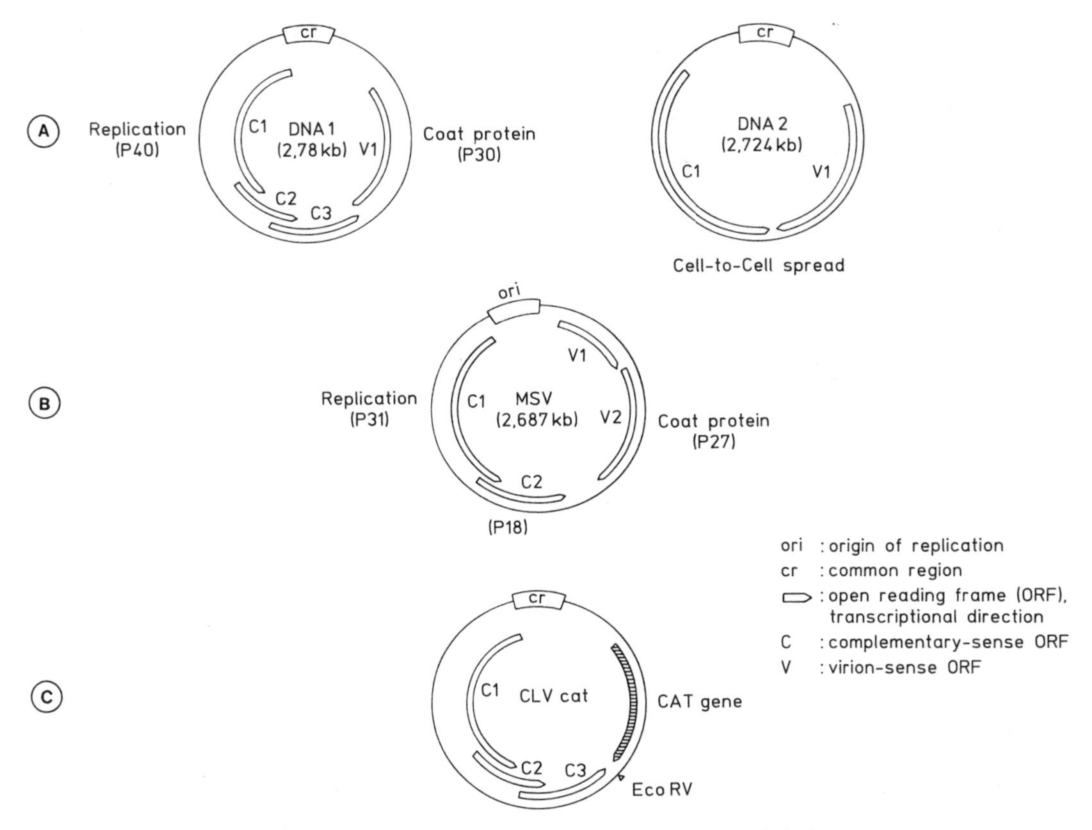

Fig. 5. Functional organization of the genomes of bi- and monopartite gemini viruses.
(A) Bipartite gemini virus genome (vector: whitefly; hosts: dicotyledons; examples: CLV, TGMV, BGMV).
(B) Monopartite gemini virus genome (vector: leafhopper; hosts: monocotyledons; examples: MSV, WDV, DSV, CSMV).
(C) Coat protein replacement and expression vector, based on the CLV genome 1 (or A).

can be expanded to accomodate gene inserts of up to 3 kb. The reporters are also expressed in infected *Triticum monococcum* and *Zea mays* cells. In another approach, an npt II gene has been inserted into the WDV genome, and this construct incubated with dry, seed-derived embryos of wheat. Then seedlings were regenerated that expressed the npt II gene to high levels (in case of vector pWDV neo 2, that allows the replication of the recombinant WDV genome; TÖPFER et al., 1990).

The use of gemini viruses as gene vectors can be expanded by agroinfection (agroinoculation; see Sect. 3.2). The fact that DNA A (1) encodes all functions for replication and virion assembly (SUNTER et al., 1987), could be exploited, if a two-component replication system could be established. The genes involved in replication could be engineered into the host-plant genome and constitutively be expressed, and viral DNA void of the replication functions could be introduced via agroinfection. Replication proteins would then be supplied in *trans*. Experiments towards this goal have already been performed. The cp gene of TGMV was replaced by an npt II gene, and partial dimers of both DNA A and B were constructed and transferred by agroinfection. The construct containing the reporter gene in a partial dimer of DNA A was either inserted into the T-DNA alone (A 1.6 neo) or with dimers of DNA B (A 1.6 neo B 2). In both cases a freely replicating DNA A containing npt II sequences was found in transgenic plants. After agroinfection of plants with A 1.6 neo B 2, replicating DNA A containing the npt II gene was detected. The levels of NPT II activity in the agroinoculated tissue correlated with the copy number of the DNA A (HAYES et al., 1988a).

We have detailed the present state of the art of gemini virus vectors, because we expect improvements of this vector system in the near future. Main obstacles for efficiency have to be overcome (e.g., host specificity, tissue specificity, restriction of viruses to vascular tissues, limited loading capacity, instability of constructs). Whereas the prospects for gemini virus-based high-copy number replicating vectors and expression vectors are indeed good, we do not yet see reliable and effective strategies to develop such viruses for stable transfer (i.e., integration into the host's genome).

2.3.3.2 Caulimoviruses

The 16 members of this group of viruses containing a double-stranded genome have similar particle sizes (isometric, 50 nm in diameter), and infection behavior. For example, all caulimoviruses infect only a very narrow range of plant hosts (e.g., cauliflower mosaic virus, CaMV, only Cruciferae; carnation etched ring virus, CERV, only Caryophyllaceae; mirabilis mosaic virus, MMV, only Nyctaginaceae; to name only a few). Caulimoviruses also induce the formation of dense virion-containing proteinaceous inclusion bodies ("viroplasms") in infected cells, presumed sites of virion assembly. The viruses are mostly transmitted by aphids, and are responsible for a number of economically important diseases in cultivated crops.

As a representative member of this group, CaMV shows all the characteristics, and we will limit our discussion to this prototype. CaMV has a circular double-stranded DNA genome of about 8 kb (different isolates differ in total genome size due to substitutions, deletions, and insertions), that is interrupted by so-called 'gaps', single-stranded discontinuities. Such gaps occur in both strands, usually two in one, and one in the other strand (Fig. 6), and represent regions of sequence overlap produced by strand displacement. Another peculiarity is the presence of ribonucleotides covalently attached to the 5′ termini of the discontinuities.

The sequence of CaMV-DNA reveals eight tightly packed open reading frames that are asymmetrically transcribed from the minus-strand into proteins, six of which have been detected in infected plants, namely, the major constituent of the inclusion body, P62, the product of gene VI; the capsid protein, encoded by gene IV; the viral replicase, encoded by gene V; the cell-to-cell movement protein, encoded by gene I; a DNA-binding protein, encoded by gene III; an aphid transmission protein, encoded by gene II. The actual template for transcription is the supercoiled form of the viral DNA that is read out by host RNA polymerase II into two major CaMV RNA transcripts, the 19S and 35S RNA. The syntheses of both RNAs are driven by relatively strong promoters with sequence elements typical of eukaryotic promoters. These promoters are frequently used for the construction of chimeric genes (see Sect. 2.2.4). The 19S mRNA derived from gene VI is capped and polyadenylated (size: 1850 nucleotides), the 35S mRNA is also polyadenylated, polycistronic (comprising sequences from genes I–III), and functions in CaMV replication in very much the same way as retroviral replication proceeds (HOHN et al., 1985; GRONENBORN et al., 1987; BONNEVILLE et al., 1988).

The design of CaMV as a gene vector for plants has been limited by the capacity of its capsid which does not allow the packaging of DNA substantially larger than the wild-type genome. These physical constraints require substitution of viral genes by foreign DNA. However, *in vitro* mutagenesis proved the majority of coding regions to be absolutely necessary for normal infection and replication, with the exception of ORF II (coding for an insect transmission factor) and ORF VII (unknown function). Both ORFs can be completely deleted without interfering with infectivity of CaMV (mechanical transmission provided). The obvious strategy for vector development then, was substitution.

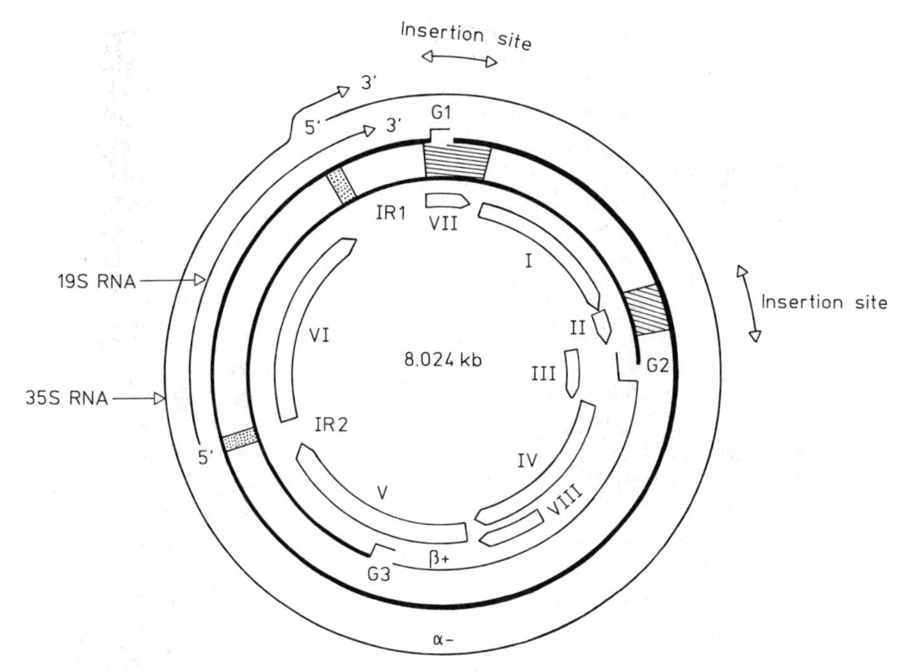

Fig. 6. Genome organization and transcripts of the cauliflower mosaic virus (CaMV). Thick lines represent the two strands ($\alpha-$, $\beta+$) of the viral DNA with the gaps (G1, G2, G3). Open reading frames are numbered I - VIII, the direction of transcription is symbolized by arrow heads. Transcription of 19S and 35S RNAs from the supercoiled form of CaMV-DNA is driven by eukaryotic promoters (dotted boxes). Hatched areas delimit non-essential regions into which foreign DNA can be inserted.

At least three such substitutions with specific coding regions were stable. First, a bacterial dihydrofolate reductase (DHFR) gene of 0.24 kb was cloned into the dispensable ORF II and conferred methotrexate resistance to turnip plants infected with the recombinant CaMV (BRISSON et al., 1984; BRISSON and HOHN, 1986). Second, an npt II gene was transferred to the same host (PASZKOWSKI et al., 1986). Third, a human D interferon (IFN) gene of 0.5 kb was transferred to plants and expressed (DE ZOETEN et al., 1989); and fourth, a chinese hamster metallothionein (CHMT) of 0.2 kb was transfected into plants and also expressed (LEFEBRE et al., 1987). Nevertheless, the capacity of the CaMV genome is severely restricted and does not tolerate more than 1 kb of foreign DNA, substituting for ORF II, ORF VII, and a small part of the large intergenic region. The maximum size of a recombinant viral genome is definitely 8.3 kb (GRONENBORN et al., 1981; DAUBERT et al., 1983). It seems that the foreign sequences can have regulative effects on the polycistronic translation mechanism of the virus, or that the original ORF architecture is perturbed (FÜTTERER et al., 1990). Therefore, an extremely precise replacement of, e.g., ORF II by the insert is necessary for stability of the construct. In addition, there exists the danger that the recombinant virus deletes all or part of the inserted sequences after a few infection cycles (FÜTTERER et al., 1990). Yet another obstacle precludes the complementation of viral functions in *trans* (as for example in other helper virus systems): illegitimate template switches allow a high recombination rate during CaMV replication (GRIMSLEY et al., 1986a, b; DIXON et al., 1986).

All these considerations limit the use of CaMV-based vectors to the transfer of foreign genes into host plants by infection and the expression of the transgenes to high levels with concomitant accumulation of the encoded proteins. At present CaMV vectors cannot be used as plant transformation vectors, ensuring stable integration of the foreign DNA. Aside from the high-level expression of genes in plants, the major contribution of CaMV to gene technology of plants has been the provision of 19S and 35S RNA promoters for the construction of constitutive expression vectors.

2.3.4 Other Potential Vector Systems

A series of other potential gene vector systems for plants can be listed, although their molecular characterization is far from complete. Since the precise knowledge of the structure of such vectors and their replication cycle is absolutely necessary to estimate their usefulness, we will merely summarize some data and leave it to future developments whether the listed molecules will ever become gene vectors for plants.

Viroids

Viroids are probably the smallest plant pathogenic molecules. They are sub-viral particles, consisting of a protein-free, 200 - 400 nucleotides long circular single-stranded (ss) RNA molecule that undergoes extensive intramolecular base pairing (SÄNGER 1982; ROBERTSON et al., 1983). The intracellular base pairing causes viroid RNA to appear as rods in electron microscopic pictures (e.g., the potato spindle tuber viroid consists of double-stranded rods of about 50 nm). Distinct regions of the RNA of different viroids share a high degree of sequence homology. Viroids become associated with the nucleus of an infected cell, where they exploit host cell enzymes for their replication. There is evidence that the circular viroid RNA is copied by a rolling-circle mechanism to generate linear multimers of viroid minus-strand RNA. These undergo self-cleavage into monomeric minus-strand circles which in turn are copied into a multimeric plus-strand linear form. Monomeric progeny plus-strand viroid circles are then excised autocatalytically. Although the exact mechanism of pathogenicity is yet not fully unraveled, it is suggestive that viroid RNA interferes with nuclear RNA processing (e.g., splicing of pre-mRNA), because it has a similar structure as small nuclear RNAs (involved in splicing reactions). The viroids are interpreted as escaped intron sequences.

Viroids cause a series of disease symptoms on mostly tropical or subtropical plants. A well-known example is the so-called Cadang-Cadang disease of coconut-trees that leads to the disrupture of the tree top from its stem. The apparent disadvantages of viroids as gene vectors for plants, such as the interference with nuclear splicing reactions and the pathogenicity, are counterbalanced by a number of advantages. These include mechanical transmissibility, movement from the infection site to other parts of the plant via phloem sap, transmission through seed (in some cases), relative broad host range (though restricted to tropical crops), and the infectivity of full-length cDNA copies of viroid RNAs (ROBERTSON et al., 1983). We expect viroids, even if extensively modified, to be inferior to other gene transfer and expression systems.

Satellite RNAs, satellite viruses, and satelloids

All these entities consist of sub-viral RNA molecules that are either encapsidated together with the genome of their helper virus (satellite RNAs) or encapsidated separately from their helper virus (satellite viruses). Helper virus replication is totally independent of both satellite RNA and satellite virus, but the replication of both is fully dependent on the replication machinery of the helper virus. Both entities influence the pathogenicity of the virus and may therefore either decrease or increase virulence. They also may interfere with helper virus replication.

The satellite RNAs (example: CARNA 5 cucumber mosaic virus, CMV) are distinct from satellite viruses and unrelated to their helper virus. Their size ranges from 270 bases (tobacco ringspot virus satellite) to 1.5 kb (tomato black ring virus satellite). The satellite viruses are known to encode proteins (example: the satellite virus, STNV, of tobacco necrosis virus, TNV, encodes its coat protein). Satelloids are satellite RNAs of the satellite viruses which are encapsidated together with the satellite virus (e.g., STNV).

The molecular biology of all these particles is not very much advanced so that it would be premature to discuss their potential usefulness as gene vectors.

3 Transformation of Plant Cells

3.1 *Agrobacterium*-Mediated Transfer of Foreign Genes into Target Plants

Early experiments relied on the infection of intact plants, or the cocultivation of protoplasts with agrobacteria (see Fig. 7) harboring wild-type or manipulated Ti plasmids (reviewed by WEISING et al., 1988). Since selection is not easy with whole plants, and cocultivation is only feasible for plant species allowing regeneration from protoplasts, alternatives were sought. In 1985, a new generation of *Agrobacterium*-mediated transfer techniques was created by the invention of the leaf disc transformation method (HORSCH et al., 1985; ROGERS et al., 1986a). Sterile leaf explants

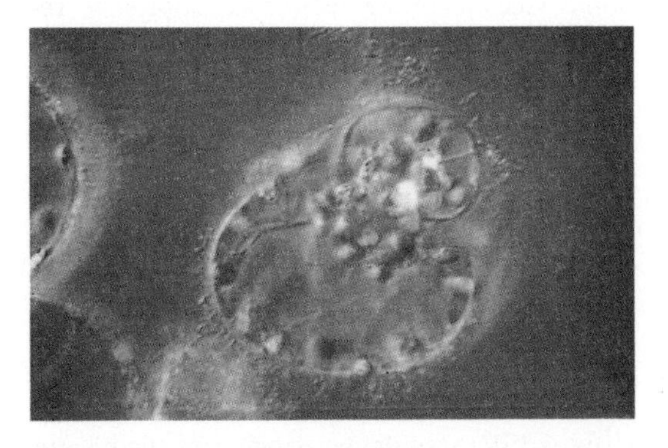

Fig. 7. Cocultivation of mesophyll protoplasts of *Nicotiana tabacum* with *Agrobacterium tumefaciens*. Note the bacteria attached to the cell membrane (courtesy of Dr. H. H. STEINBISS, Max-Planck-Institut für Züchtungsforschung, Köln, FRG).

were inoculated with recombinant *Agrobacterium* and cultured on a medium which favored shoot formation. Subsequently, bacteria were killed by an antibiotic, and transformants were selected by another antibiotic. Surviving shoots were rooted and transferred to soil.

Since the leaf-disc transformation procedure represented an ideal combination of high transformation frequency with easy and fast selection and regeneration of transformants, it was soon adopted and modified by many laboratories, (e.g., acetosyringone treatment of bacteria, which was shown to enhance transformation frequency for leaf discs of *Arabidopsis;* SHEIKOLESLAM and WEEKS, 1987). Co-cultivation of *Agrobacterium* with stem segments (AN et al., 1986), suspension-cultured cells (SCOTT and DRAPER, 1987), microcalli (POLLOCK et al., 1985), epidermal segments (TRINH et al., 1987), thin-layer explants (CHAREST et al., 1988), cotelydon (SCHMIDT and WILLMITZER, 1988) and hypocotyl explants (BARFIELD and PUA, 1991), and germinating seeds (FELDMANN and MARKS, 1987) proved to be similarly successful. More recently, *Agrobacterium*-mediated gene transfer was also obtained by co-cultivation with *Petunia* pollen, as proven by Southern analysis (SÜSSMUTH et al., 1991).

A major challenge that remains with *Agrobacterium*-mediated gene transfer is the recalcitrance of important crop species such as many legumes and monocotyledons, especially the cereals (DE CLEENE and DE LEY, 1976; DE CLEENE, 1985; VASIL, 1988). Though several studies indicate that transformation of monocotyledonous plants is principally possible (SCHÄFER et al., 1987; BYTEBIER et al., 1987; GRIMSLEY et al., 1987; DOMMISSE et al., 1990; RAINERI et al., 1990; GOULD et al., 1991), the process is much less effective as compared with even recalcitrant dicotyledons such as soybean (transgenic soybean plants have been recently obtained: HINCHEE et al., 1988). We can only guess, why this is the case, but should keep in mind that the complex machinery of *Agrobacterium*-mediated transformation presupposes many activities on both the bacterial and the plant side to function optimally. If any of these activities is missing, reduced or going awry, the overall transformation efficiency would be reduced or even abolished. We share the view that wounding does not incite competence in monocot cells, or does so relatively inefficiently (KAHL, 1982; POTRYKUS, 1990, 1991). As a consequence, *Agrobacterium* cannot start the infection process (e.g., cannot attach), or cannot deliver T-DNA or, if this is possible, the T-DNA is integrated, but only into the genomes of the dying cells adjacent to the wound site (as, e.g., in cereals, see POTRYKUS, 1991). Finally, T-DNA might be expressed but does not lead to a transformed phenotype.

Nevertheless, several strategies may be followed to expand the host range of Ti plasmid-derived vectors to more recalcitrant species:
(1) Careful evaluation of the host range of existing strains of *Agrobacterium tumefaciens* as well as of its close relative, *A. rhizogenes.* Both species transform plant cells in a similar way in terms of T-DNA transfer and the production of opines in the target tissue (see Sect. 2.1.4). Hairy roots induced by *A. rhizogenes* have been shown to possess the capacity of regenerating T-DNA-containing plants which exhibit an aberrant phenotype: wrinkled leaves and a reduced apical dominance (TEPFER, 1984; GUERCHE et al., 1987). Normal-looking plants carrying only vector DNA were obtained by the use of binary vectors in combination with the Ri plasmid (SHAHIN et al., 1986; SUKHAPINDA et al., 1987). *A. rhizogenes* strains proved especially useful for the transformation of legumes, e.g. *Lotus corniculatus* (JENSEN et al., 1986), and *Medicago arborea* (DAMIANI and ARCIONI, 1991).
(2) New strains with an extended host range may be generated by mutagenesis of specific vir genes. Several vir genes have been shown to affect the host range (see Sect. 2.1.2). Overexpression of virD has been shown to enhance T-strand formation and plant transformation (WANG et al., 1990). The "supervirulent" phenotype of strain A281 that infects several species, only weakly affected by other *Agrobacterium* strains, has been localized to the virG region of the Ti plasmid pTiBo542 (JIN et al., 1987; PYTHOUD et al., 1987).
(3) The availability of histochemical reporter gene assays (e.g., the gus assay) allows to screen for individual transformed cells on a histochemical basis (JANSSEN and GARDNER, 1990; VANCANNEYT et al., 1990). These techniques might help to investigate the cell-, tissue-, and development-specific aspects of accessibility to *Agrobacterium* in recalcitrant species.
(4) Tissue culture methods may be developed for weakly susceptible species that allow regeneration of single transformed cells from tissues amenable to *Agrobacterium*-mediated transformation.

Even if all these strategies are exploited, it may turn out that the majority of monocot crop plants are no suitable hosts for *Agrobacterium*-mediated transformation. For those species, the application of one of the direct gene transfer techniques outlined in the following sections appears to be more promising.

3.2 Viral Inoculation and Agroinfection (Agroinoculation)

The different strategies for transferring foreign genes into plant cells by means of vectors derived from RNA and DNA viruses are outlined in Sect. 2.3. More recently the combination of the infectivity of a plant virus with the transforming capacity of the Ti plasmid of *Agrobacterium tumefaciens* produced exciting results. This type of *Agrobacterium*-mediated virus infection has been introduced with an engineered cauliflower mosaic virus (CaMV) genome (see Sect. 2.3.3.2). Two copies of the CaMV-DNA in tandem were inserted in-between the borders of the T-DNA and agrobacteria carrying these constructs used to infect turnip *(Brassica rapa)* plants through an artificial wound. Not only was systemic infection of the host detected after use of purified infectious CaMV-DNA, but also if agrobacteria with the recombinant Ti plasmid were inoculated (GRIMSLEY et al., 1986a, b). In subsequent experiments, GRIMSLEY and colleagues constructed plasmids containing a tandemly repeated dimer of the genome of maize streak virus (MSV), a circular single-stranded DNA that replicates to high copy number in nuclei of infected cells. The virus is transmitted by leafhopper vectors and causes stunting and formation of yellow-streaked leaves. The isolated MSV-DNA alone is not infectious on maize plants. The MSV-DNA dimer was cloned in-between T-DNA borders, and the agrobacteria harboring the recombinant Ti plasmid used to infect wounded leaves of maize. Two weeks after inoculation the treated plants developed symptoms of viral infection. Detection of the replicative form (RF) of the viral DNA and Southern analysis proved the presence of MSV sequences in plants, whereas controls, inoculated with naked MSV-DNA, or *Agrobacterium* carrying mutations in the virA gene, or plasmids lacking the T-DNA borders did not show any symptoms (GRIMSLEY et al., 1987). A prerequisite for this type of viral infection is the presence of functional vir A, B, C, D, and G regions, whereas mutations in vir E attenuated the symptoms (GRIMSLEY et al., 1989).

From these and other reports it is very clear that *Agrobacterium*-mediated transfer of virus-carrying T-DNA into wound-adjacent cells of monocotyledonous host cells occurs. The viral DNA is somehow released, replicates, and spreads systemically. Obviously a number of other viruses and viroids, if present between the T-DNA borders in tandem duplications, allow the escape of single genome DNA that initiates infections. The whole process has been named "agroinfection" (GRIMSLEY et al., 1986b), or, more recently, agroinoculation.

A series of reports shows the potential of this technique for the transfer of viral DNA into different tissues of maize (GRIMSLEY et al., 1988), to *Digitaria sanguinalis* and *Avena sativa* (DONSON et al., 1988), to *Triticum* (WOOLSTON et al., 1988; HAYES et al., 1988b), to *Hordeum vulgare, Panicum milaceum,* and *Lolium temulentum* (BOULTON et al., 1989), and to *Aegilops speltoides, Triticum monococcum,* and *T. durum* (DALE et al., 1989; MARKS et al., 1989). Agroinfection encompasses the transfer of potato spindle tuber viroid (PSTV, GARDNER et al., 1986), cauliflower mosaic virus (CaMV, see above), four gemini viruses, maize streak virus (MSV, see above; LAZAROWITZ, 1988); digitaria streak virus (DSV; DONSON et al., 1988), wheat dwarf virus (WDV; HAYES et al., 1988b), tomato golden mosaic virus (TGMV), and cassava latent virus (CLV; SCOTT-ELMER et al., 1988; GARDINER et al., 1988). This is ample evidence that *Agrobacterium tumefaciens* recognizes monocotyledonous wound substances, attaches (at least in many Gramineae) and transfers its T-region into the cells adjacent to the wound. Agroinfection then represents an alternative route to direct gene transfer, in so far as expression of viral sequences (or sequences cloned into these viral sequences) is concerned. Therefore, agroinfection may serve to express foreign sequences in host plants. However, the potential for integrative transformation of target plants seems to be low, though the T-DNA can integrate into the nuclear genome of plant cells. It may very well be that agroinfection leads to the integration of virus-containing T-DNA into the few competent cells at the wound site. In most monocots, however, these cells die. Consequently, agroinfection, in spite of advantages for basic science, cannot lead to transgenic plants.

3.3 Vectorless Gene Transfer: Physical and Chemical Transfer Techniques

3.3.1 Introduction

The evolution of the gene transfer system of *Agrobacterium* most probably took more than one million years. During this time the complex cascade of signal perception, signal transduction, vir induction, T-strand excision, *in vivo* packaging, piloting, and integration of the T-DNA into the nuclear target sites of a plant genome has been optimized to some extent. It is for this reason that *Agrobacterium*-mediated gene transfer is relatively efficient. Transfer efficiency can be roughly calculated from co-cultivation experiments (BINNS, 1990) and usually ranges from 0.1 – 5% (FRALEY et al., 1984; EAPEN et al., 1987) to more than 50% (VAN DEN ELZEN et al., 1985a). The differences in such figures partly reflect the number of competent plant cells in the different co-cultivation mixtures. Competence certainly is a function of the quality of wound response of these cells (KAHL, 1982). As compared to transformation frequencies in other gene transfer systems (see below), the effectiveness of *Agrobacterium* doubtless is remarkably high. This efficiency together with the relative simplicity of many of the *Agrobacterium*-based transformation protocols led to their widespread usage and, initially, to the erroneous belief that each and every plant could be transformed by *Agrobacterium*. This assumption catalyzed the development of numerous Ti-plasmid-based gene vector systems, elegant protocols to use these vectors, and an ever-increasing number of transgenic plants with impressive new properties (see WILLMITZER, Chapter 16, this volume).

However, as outlined in the previous section, *Agrobacterium* did not evolve its transforming capacities equally well with each plant species, nor did it manage to transform monocotyledonous plants. Notwithstanding the actual cause of *Agrobacterium*'s inherent failure to transform monocotyledonous plants, this drawback elicited a wealth of activities around the world, seeking a way out of the dilemma that the economically most interesting crops, being monocots, could not be engineered this way. One possible solution was the development of an alternative technology, called direct or vectorless gene transfer (PASZKOWSKI et al., 1984), that is totally independent of viral or bacterial vectors. The transforming DNA is instead introduced into target protoplasts (i.e., cells whose cell walls have been removed by enzymatic digestion), cells, tissues, organs, or whole plants by a series of chemical treatments, or by electrical or mechanical force.

The ideal system for gene transfer to plants are protoplasts, because the DNA has access to free plasma membranes and all recipients in a population, and each protoplast has undergone an isolation procedure that mimics wounding. Therefore, it is fair to anticipate that more cells have been made competent for integrative transformation than were in the intact tissue. Such protoplasts can also be transformed by *Agrobacterium,* probably because they immediately begin to resynthesize cell-wall material, thereby offering signal molecules and attachment sites. It is, however, easier to transform protoplasts than to regenerate them into plants. Regeneration, then, is the actual bottle-neck in the generation of plants from protoplasts transformed by direct gene transfer (see Chapter 16, this volume). This is, again, especially true for the cereals. Considerable progress has been made in the last few years concerning the regeneration of previously recalcitrant species from protoplasts (reviewed by VASIL, 1988; VASIL et al., 1990). Consequently, transgenic plants have been generated via direct gene transfer to protoplasts of maize (RHODES et al., 1988) and rice (TORIYAMA et al., 1988; SHIMAMOTO et al., 1989; DATTA et al., 1990; HAYASHIMOTO et al., 1990; TADA et al., 1990). However, not all varieties respond to the regeneration procedure.

Since cell-wall-containing plant cells and tissue explants are usually easier to regenerate than protoplasts, the transformation of intact cells would provide a more straight-forward avenue to the routine transformation of many important crop plants. Intact cells, on the other hand, require more drastic gene transfer techniques than protoplasts, because the DNA is efficiently trapped by the cellulose fibers and other macromolecules of the cell wall. Moreover, at least some plants excrete a series of hydrolases (among them exo- and endonucleases) into the extracellular space. Adsorption and destruction of the transforming DNA then is counterproductive for high-efficiency integrative transformation. These difficulties could be circumvented by the development of systems that deliver the DNA directly into the cytoplasm, or preferably the nucleus of the recipient cell. Still, regeneration of transformed cells into plants again remains an obstacle for many species, and these considerations also hold for tissues and organs.

No matter what plant system and what specific gene transfer technique is used, we consider the following remarks as essential for understanding the *potential* of vectorless gene transfer methods.

1. Gene transfer may mean that a particular gene has been taken up by a recipient cell and even entered the cell's nucleus. It is safe to assume that this type of gene transfer will work with almost any plant cell and direct transfer technique. However, the gene – if driven by appropriate promoters – will only be transiently active *(transient expression),* because it is not covalently integrated into the re-

cipient's genome. While transient expression of gene constructs is an elegant method to test regulatory element qualities in a transgenic environment (see Sect. 3.4.2), it is definitely not the ultimate aim of gene transfer. On the contrary, the *stable integration* of the transferred DNA is in the mind of everybody who uses these techniques. However, in comparably few cases has the stable integration of foreign DNA been proven by standard techniques (e.g., Southern-type hybridization with purified high-molecular weight genomic DNA), though "transformation" has enthusiastically been claimed. In our view, the simple demonstration of the activity of the enzyme encoded by the transferred gene is no longer acceptable as indicator for transformation, if not supplemented by reliable data on the integration of that gene into the plant DNA (Southern analysis of flanking regions), its integrity, and its regulated or constitutive expression, depending on the type of promoter used (Northern analysis). Keeping in mind that such proof for integrative transformation has only occasionally been demonstrated in the literature, we consider the potential of direct gene transfer methodology for integrative transformation as probably overestimated.

2. To date, reproducible and reliable transformation protocols work with protoplasts, less so with cells. However, even if the stable transformation of such protoplasts or cells succeeds, the relatively rare event of integrative transformation can only be exploited for science and agroindustry, if the transformants can be regenerated into complete plants which retain the transformed state in a series of progeny. However, regeneration of protoplasts is more an art than a science, notwithstanding recent successes in protoplast culture (ROEST and GILISSEN, 1989). In but few cases is a reliable regeneration protocol for protoplasts available, and the majority of economically important monocot crops still resists regeneration (some do not even divide in culture). We would like to stress that integrative transformation of plant cells is only successful for plant biotechnology on the long run, if it will result in fertile plants that allow exploration of the stability of the transferred DNA over many generations as well as the exploitation of the transferred trait.

For reasons of simplicity we will arbitrarily categorize the different direct gene transfer techniques into techniques that involve membrane-destabilizing chemicals (chemical transfer techniques) and techniques that work with various physical devices (physical transfer techniques).

3.3.2 Gene Transfer Mediated by Membrane-Destabilizing Agents

Probably the easiest, but nevertheless reasonable technique for gene transfer involves plasma-membrane destabilizing and/or precipitating chemical agents. This type of chemical transformation uses mostly protoplasts that are simply incubated with DNA in buffers containing either polyethylene glycol (PEG), poly-L-ornithine, polyvinyl alcohol, or divalent ions (for a review see WEISING et al., 1988). The exact mechanism of action of these diverse agents is not known, but is regarded as destabilizing membrane structures. Generally spoken, the chemical transformation works satisfactorily with protoplasts from a broad spectrum of plants (see, e.g., NEGRUTIU et al., 1987, 1990; JUNKER et al., 1987; MAAS and WERR, 1989; ARMSTRONG et al., 1990), sometimes even better than standard electroporation procedures (see below; NEGRUTIU et al., 1990). Starting from earlier, less effective DNA uptake experiments (see COCKING et al., 1981), the method has considerably been improved with concomitant increase in transformation frequencies. The optimization encompasses the stream-lining of the procedures. For example, earlier protocols demanded a 20 – 30 minutes incubation time of protoplasts with DNA and PEG. However, shorter incubation periods (2 – 5 min) are probably optimal (NEGRUTIU et al., 1990). This suggests that DNA uptake is a passive but instantaneously occurring process, if PEG is used. Other parameters have to be thoroughly optimized for any protoplast type separately. This holds true for the origin of the protoplasts, their isolation, purity, quality and handling, the pH of the transformation mixture, the concentration of the chemicals and divalent ions, the potential benefit of a heat shock, the topology of the transforming DNA (e.g., linear versus circular supercoiled, single-stranded versus double-stranded), and the form of carrier DNA. Optimizing these parameters might yield different results for different species. For example, linear DNA has been frequently found to be superior to supercoiled DNA in terms of transformation efficiency (SHILLITO et al., 1985; NEGRUTIU et al., 1987, 1990) as well as transient expression (BALLAS et al., 1988; BATES et al., 1988). However, the reverse was found in other systems (OKADA et al., 1986a; BOSTON et al., 1987; EBERT et al., 1987), and single-stranded DNA was also shown to be a very effective transforming agent (RODENBURG et al., 1989; FURNER et al., 1989).

Recently, intact plant cells were also shown to be amenable to PEG-mediated transformation (LEE et al., 1991b). Small groups of 50 to 100 cells, prepared from rice suspension cultures, were transformed with a vector harboring gus- and npt II-constructs. According to the gus activity of calli regenerating from these cell groups, the transformation frequency was about 7%, and transgenic gus-positive plantlets were easily regenerated. This method appears quite promising, since it is easy to perform and does not require plant regeneration from protoplasts.

A somewhat different technique to introduce DNA into a target protoplast exploits the precipitation of the DNA in insoluble calcium phosphate complexes directly onto the membranes (calcium phosphate coprecipitation technique). These calcium phosphate precipitates are generated after adding a DNA-$CaCl_2$ solution to an isotonic phosphate buffer. The precipitates form after about 30 minutes and are effectively taken up by endocytosis. The relative success of this comparably simple technique (see, e.g., HAIN et al., 1985) relies on the high local concentration of DNA and its apparent protection in the precipitate (KRENS et al., 1982).

3.3.3 Gene Transfer Mediated by Physical Techniques

3.3.3.1 Electroporation and Electrophoresis

Both techniques work with electrical fields but differ in their mechanism to transfer genes across membranes. *Electroporation* works with a discharge of a capacitor that perforates cell membranes of target cells with a short (range: 1 millisecond) electrical pulse and voltage gradients of about 700 V/cm. The perforation process is only transient (i.e., the generated holes in the membranes are sealed by self-assembly processes), but sufficient to allow the entry of even large DNA molecules. In *electrophoresis* (electrophoretic transfection, electrofection, ETF), the protoplasts or cells are suspended in a microdroplet between the anode and cathode, and the DNA is electrophoresed through the membranes. Whereas electroporation of monocot- and dicot-derived plant protoplasts (SHILLITO et al., 1985; FROMM et al., 1986, 1987; HAUPTMANN et al., 1987; LINDSEY and JONES, 1990) as well as of cells from other organisms (reviewed by SHIGEKAWA and DOWER, 1988) has been proven to be very efficient and is now a routine technique for transformation, electrophoresis has only rarely been employed to drive DNA uptake (e.g., in shoot meristems of barley seeds; AHOKAS, 1989), and results were not indicative for integrative transformation. Electroporation can also be used with cells (LINDSEY and JONES, 1990), germinating tobacco pollen (MATTHEWS et al., 1991), and tissue slices from rice (DEKEYSER et al., 1990). Since in the latter case even cells remote from the surface layer express the genes, the DNA obviously has been pushed across many cell walls, though this conclusion has been challenged (POTRYKUS, 1991). In summary, we think that despite being a routine technique for protoplast transformation, the full potential of electroporation for the transfer of genes to cells or tissues has not yet been fully explored.

3.3.3.2 Particle Bombardment ("Biolistics")

As a relatively young technique, biolistic ("*bio*logical and bal*listic*") plant transformation has already proven to be highly promising. The method is based on the introduction of macromolecules into intact cells and tissues with the aid of high-velocity microprojectiles (KLEIN et al., 1987, 1988a, b, c, 1989, 1992; SANFORD, 1990; JOHNSTON, 1990). The motive force for driving these projectiles, spherical high-density metal bullets of 0.2 – 2 μ in diameter, is generated by gunpowder, air pressure, or electrical discharge. In the latter technique, a 10 μL-water droplet is placed between two electrodes in a spark-discharge chamber. A high voltage capacitor is discharged through the droplet which vaporizes instantly, creating a shock wave. By either technique, macroprojectiles carrying the DNA-coated microbeads on their surface are accelerated towards a stopping plate. Small perforations in this plate allow the DNA-coated tungsten or gold particles to proceed to the target cells or tissues which are penetrated. During penetration the DNA is removed from the surface of the particles and either transiently expressed (see Fig. 8), and/or stably integrated.

Using this technique, intact cells of a variety of important crop plants recalcitrant to *Agrobacterium*-mediated transformation have been transformed (e.g., *Sorghum vulgare:* HAGIO et al., 1991; barley: MENDEL et al., 1989; LEE et al., 1991a; wheat: DANIELL et al., 1991; VASIL et al., 1991; OARD et al., 1990; rice and wheat: WANG et al., 1988; rapeseed: SEKI et al., 1991) and occasionally regenerated to transgenic plants (e.g, soybean: CHRISTOU et al., 1988, 1989; MCCABE et al., 1988; corn: FROMM et al., 1990; GORDON-KAMM et al., 1990; cotton: FINER and MCMULLEN, 1990). Using the particle gun technique, organelle transformation has also become feasible (e.g., mitochondria: JOHNSTON et al., 1988; FOX et al., 1988; chloroplasts: BOYNTON et al., 1988; BLOWERS et al., 1989; DANIELL et al., 1990, 1991). DNA has also been delivered into yeast, *Chlamydomonas,* and animal cells (see SANFORD, 1990; KLEIN et al., 1992). Notwithstanding the type of cells or tissues, biolistic gene transfer with subsequent integrative transformation seems to be generally applicable. Most importantly, regenerable target tissues have biolistically been transformed, such as embryogenic cell suspensions (WANG et al., 1988; DANIELL et al., 1990), meristems (MCCABE et al., 1988), embryos (CAO et al., 1990; LONSDALE et al., 1990), and pollen (MCCABE et al., 1988; TWELL et al., 1989). Biolistics seems to be the only reliable technique for organelle transformation to date, and certainly has its potential for monocot transformation, if a number of parameters will be optimized, e.g., the particle size, shape and coating efficiency, the topology of the delivered DNA, and the physiological state of the target cells. For example, the efficiency of microprojectile-mediated transformation was recently shown to depend on the cell cycle stage of synchronized tobacco cells (IIDA et al., 1991). Cells bombarded at G2 and M phases gave four to six times higher transformation rates than those bombarded at the G1 and S phases. Taken together, the fundamental advantage of the "gene gun" technique, in addition to its versatility and avoidance of protoplast technology, is the possibility to reach cells within, and not only on the surface, of the target tissue, thus increasing the number of hit competent cells.

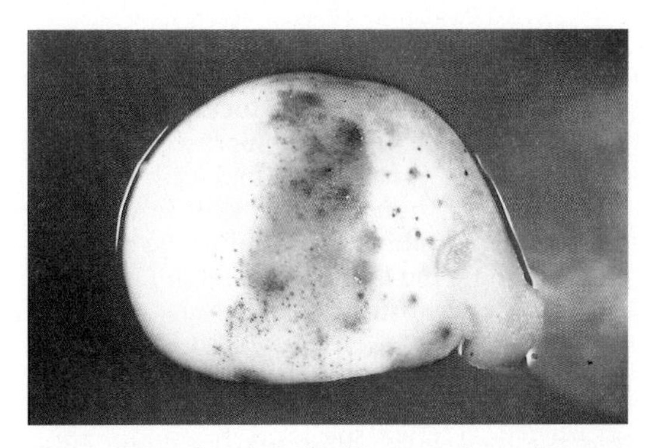

Fig. 8. Histochemical localization of transient gus reporter gene expression in pigeon pea (*Cajanus cajan* L.) cotyledons after bombardment with microprojectiles coated with a vector construct carrying the gus gene controlled by CaMV35S 5′ and 3′ regulatory regions (courtesy of Dr. H. H. STEINBISS, Max-Planck-Institut für Züchtungsforschung, Köln, FRG).

3.3.3.3 Laser Microbeam Technique

A UV-laser microbeam focused into the light path of an inverted microscope allows perforation of membranes and cell walls of target cells. If these cells are kept in a hypertonic, i.e., plasmolyzing medium, the perforations facilitate the uptake of foreign DNA into cells, and even into chloroplasts (WEBER, 1988; WEBER et al., 1988, 1990). The transferred DNA can transiently be expressed, but has not yet been shown to be stably integrated into the nuclear or chloroplast genome (WEBER et al., 1990). Laser delivery of DNA into plant cells, or their organelles, shares common problems with electroporation, but suffers from a very unfavorable cost–effect ratio. Moreover, the technique cannot be as precise as, e.g., microinjection, as far as intracellular targeting is concerned. While deserving more experimental development, this technique will probably never be able to compete with less expensive but more efficient methods for vectorless gene transfer.

3.3.3.4 Silicon Carbide Fiber-Mediated Gene Transfer

As an intermediate between particle gun and microinjection techniques, the silicon carbide fiber-facilitated DNA transfer simply uses fibers that are vortexed with transforming DNA and the target cells. This rather drastic procedure leads to the multiple puncture of cell walls and plasma membranes, and the introduction of the DNA via the fibers (KAEPPLER et al., 1990). The precise mechanism of uptake is not yet clear, but is most probably a simple diffusion through the holes generated by the fibers. Recently, intact cells of the agronomically important grass *Agrostis alba* have been transformed by the silicon carbide fiber technique (ASANO et al., 1991).

Two aspects make the silicon carbide fiber technique an interesting alternative to other methods of vectorless gene transfer: (1) its simplicity, and (2) its applicability to intact cells and thus to plant species that do not regenerate from protoplasts. However, the viability of the punctured cells, the fate of the fibers and the foreign DNA has not yet been investigated, and in view of the probably cancerogenic properties of carbide fibers, we do not regard fiber-mediated transfer as a method of choice.

3.3.3.5 Microinjection

Originally developed to inject macromolecules into animal cells (CAPECCHI, 1980), the technique of microinjection has produced transgenic animals (BRINSTER et al., 1981; RUSCONI and SCHAFFNER, 1981), and has later on been adapted to plant cells (NEUHAUS et al., 1984, 1986). The technique capitalizes on either an injection capillary containing the transforming DNA and a support onto which the protoplasts or cells are transiently immobilized, or the injection capillary, working cooperatively with a so-called holding capillary, serving to fix the target for effective injection of DNA. Usually, immobilization of the target cells is achieved by poly-L-lysine, agarose, or sodium alginate (LAWRENCE and DAVIES, 1985; ALY and OWENS, 1987; TOYODA et al., 1988). The injection process itself can be monitored microscopically, and successful injection is detected by, e.g., coinjected fluorochromes (e.g., lucifer yellow; see Fig. 9; STEINBISS and STABEL, 1983; REICH et al., 1986). The number of microinjected protoplasts, cells, or more complex cell aggregates like proembryos remains limited, even if considera-

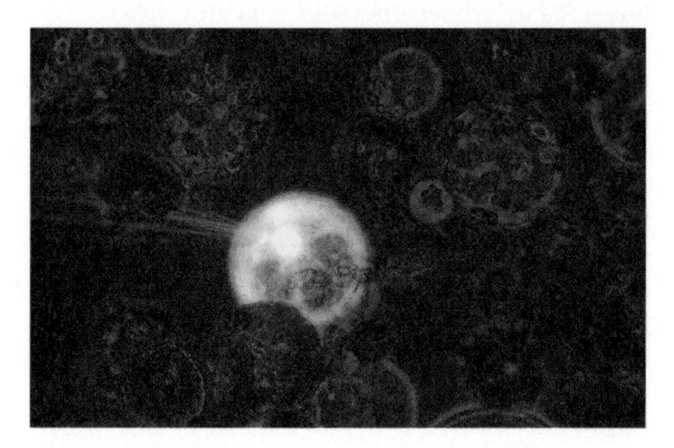

Fig. 9. Microinjection of the fluorochrome lucifer yellow into protoplasts of *Nicotiana tabacum* SR1 (courtesy of Dr. H. H. STEINBISS, Max-Planck-Institut für Züchtungsforschung, Köln, FRG).

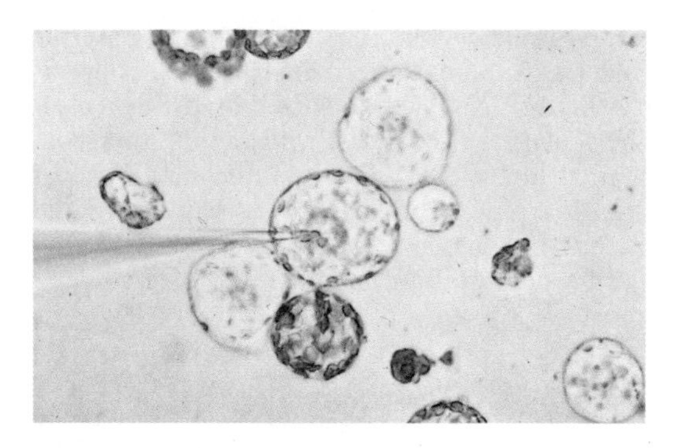

Fig. 10. Intranuclear microinjection in protoplasts of *Nicotiana tabacum* SR1 (courtesy of Dr. H. H. STEINBISS, Max-Planck-Institut für Züchtungsforschung, Köln, FRG).

ble experience and sophisticated technical setups are involved (100 - 200 cells per hour). Moreover, specific culture systems for the injected recipients are needed (e.g., microdroplets: CROSSWAY et al., 1986; individual microcultures: SCHWEIGER et al., 1987; nurse culture: LAWRENCE and DAVIES, 1985; MORIKAWA and YAMADA, 1985; REICH et al., 1986). Notwithstanding these (and other) obstacles, microinjection allows the efficient and precise delivery of DNA into protoplasts, cells, cell aggregates and also subcellular organelles (see Fig. 10), and the generation of transgenic plants (if the targets survive the treatment and can be regenerated into whole plants).

A few examples should illustrate the potential of this technique. First of all, protoplasts have been microinjected successfully, and some of these protoplasts also seemed to have the foreign DNA integrated into their genome (MELNIKOV et al., 1985; CROSSWAY et al., 1986; REICH et al., 1986; NEUHAUS et al., 1987; SPANGENBERG et al., 1990). The transformation frequencies in some of these experiments were high (14 - 16%), but could be exploited for the generation of transgenic plants in only few cases (e.g., *Acetabularia:* NEUHAUS et al., 1984, 1986). In an indirect approach, the stable transformation of the economically more important rapeseed and tobacco cells has been achieved. The foreign DNA (chimeric npt II gene constructions) was injected into both cytoplasts and karyoplasts, and both were reconstituted to viable cells by electrofusion. Such reconstitutes expressed the introduced gene that has been integrated physically into the recipient genome (NEUHAUS and SPANGENBERG, 1990).

Second, intact cells of carrot and tomato have also been microinjected, and microinjection monitored with fluorescent dyes. However, sound evidence for integrative transformation is not yet available (NOMURA and KOMAMINE, 1985, 1986; TOYODA et al., 1988, 1990).

Third, microinjection has been used to transform microspore-derived embryoids (NEUHAUS et al., 1987). Dot blot and Southern-type analyses of the genomic DNA of the regenerated plantlets proved the integration of full-length as well as rearranged reporter npt-II gene copies, and the cells of the primary regenerants also expressed the transgene. However, the regenerants were chimeras (POTRYKUS, 1991). This leaves us with the assumption that the relatively costly microinjection procedure is but one technique for direct gene transfer to protoplasts, cells, and multicellular structures like somatic or zygotic embryos, which still needs further sophistication for a routine introduction.

3.3.3.6 Liposome Fusion and Liposome Injection

One of the major obstacles for an efficient transformation of plant cells is their relatively high endo- and exonuclease content. Any DNA entering the plant cell will inevitably be exposed to these hydrolases, and be degraded to some extent. The protection of the transforming DNA, however, has not been a major concern of experimentors. On the contrary, *Agrobacterium tumefaciens* packages the T-strand with proteins, capping the termini (vir D2) and protecting the single-stranded remainder from endonucleolytic breakdown (vir E). In the early times of gene transfer to animal cells it was found that encapsidation of nucleic acids improved their delivery into recipients (FRALEY and PAPAHADJOPOULOS, 1982). Since this delivery involves passage through membranes, where nuclease concentrations are known to be high, a reasonable protection was expected from encapsulation in liposomes, lipid vesicles of various sizes. Macromolecules can be entrapped in these liposomes, preferably into the large unilamellar vesicles (LUV), though preparation and handling (e.g., the use of sonication) is somewhat tedious. The liposomes then fuse with the plasmalemma of the target cell (although endocytosis is not necessarily excluded, see GAD et al., 1990), and probably pour their content into the cytoplasm. If this is the case, then the liposome-mediated transfer has in common with other direct gene transfer techniques that the delivered DNA is virtually unprotected within the cell.

Nevertheless, liposome-mediated gene transfer has been achieved with protoplasts and been monitored by transient expression of reporter genes (ROSENBERG et al., 1988). In few cases transfer resulted in integrative transformation (e.g., in tobacco: DESHAYES et al., 1985; CABOCHE, 1990). Other targets have been exposed to liposomes (e.g., pollen: AHOKAS, 1987; various cells and tissues: GAD et al., 1990), but in no case was unequivocal proof for stable transformation presented. So we are left with the conclusion that liposomes as protective vehicles for direct gene transfer do not possess convincing advantages over the vectorless techniques.

A variation on the theme is the transformation of protoplasts in the presence of commercially available cationic liposomes in a process called "lipofection" (FELGNER et al., 1987; FELGNER and RINGOLD, 1989; ANTONELLI and STADLER, 1990). In this procedure, the liposomes do not encapsulate the DNA but rather facilitate DNA uptake by reacting with and transiently destabilizing the anionic plasma membrane. SPÖRLEIN and KOOP (1991) showed recently that the efficiency of the lipofection method is further increased by its combination with other techniques such as PEG treatment and electroporation.

Another variant, liposome injection into vacuoles, combines liposome encapsidation and microinjection and tries to exploit the fusion of liposomes with the tonoplast membrane for the safe delivery of DNA into the cell (LUCAS et al., 1990). Whereas liposome–tonoplast fusion was verified in suspension-cultured cells of maize, rice, tobacco, and carrot, the lipofected firefly luciferase reporter gene has in no case been expressed. It is possible that nucleases are localized on the cytoplasmic side of the tonoplast, preventing the transport of intact DNA into the nucleus. We find that microinjection of DNA directly into the nucleus of the target cell is a more straight-forward technique and should be preferred over liposome injection.

3.3.3.7 Protectifer

Protection of the transforming DNA is desirable, if not essential for effective and stable transformation of plant cells. As detailed in the previous section, liposome encapsidation will protect macromolecules as long as the vesicles are intact, i.e., before fusion. After fusion of liposomes with the plasmalemma the DNA is exposed to exo- and endonucleolytic attack. As a consequence, most of the delivered DNA never reaches the nucleus, or does so after exonucleolytic trimming, truncation, recombination, or rearrangements. All these rather drastic operations are reflected in studies on the physical structure of the transgenes (as, e.g., most Southern-type analyses). In order to prevent such dramatic alterations of the transforming DNA, we have packaged it in a way nature does. The nuclear DNA of eukaryotic organisms is associated with a multitude of proteins, and forms with a group of basic proteins, the histones, aggregates called nucleosomes. Such nucleosomes are composed of 1.75 turns of B-DNA wrapped around a histone octamer of two each of histone H2A, H2B, H3, and H4 (reviewed by KAHL et al., 1987). For protection of transforming DNA plant histones and linearized vector DNA have been reconstituted to nucleosomes *in vitro,* using salt gradient dialysis. This procedure permits package of the DNA in very much the same way as it appears *in nucleo* (HOFMANN et al., 1989). Such protected DNA was electroporated into tobacco protoplasts, the protoplasts regenerated to whole plants, and the plants screened for expression and physical structure of the transferred npt II genes. As compared to naked DNA, protected DNA allowed a much higher transformation frequency, an appreciably enhanced expression of the transgene, and resulted in the integration of at least one *intact* copy of the marker gene (HOFMANN et al., in preparation). We do not necessarily recommend to package all transforming DNA, but would like to stress the importance of DNA protection within recipient cells. Moreover, we hypothesize that this technique (protectifer) also guides the nucleosomally organized foreign DNA into the nucleus.

3.3.3.8 Sonication

A novel procedure involves the transient destabilization of the plasma membrane by mild sonication (JOERSBO and BRUNSTEDT, 1990). Uptake and transient expression of reporter genes in protoplasts of sugar beet and tobacco was obtained after a brief exposure of the transformation mixture to 20 kHz ultrasound. Like other direct gene transfer techniques, the procedure is rather simple. However, since overdoses of ultrasound will cause lethal damage to the protoplasts, parameters will have to be carefully optimized in terms of protoplast viability and transformation efficiency.

3.3.3.9 Incubation of Cells, Tissues, and Organs in DNA

Probably the easiest and cheapest way to bring target cells or tissues into contact with DNA is a simple incubation of both target and transforming nucleic acid. Such incubation procedures are expected to channel at least some DNA molecules into cells inspite of the limiting porosity of the cell wall and the plasma membrane with all its nucleases. The efficiency of this approach cannot be high. Unprotected DNA is adsorbed to cellulose fibers, even under very mild conditions. And it probably does so *in planta*. However, in the highly complex architecture of plant cell walls a multitude of interacting macromolecules (polygalacturonates, glucans, proteins, among them hydrolytic enzymes) probably interfere with the passage of DNA. In our view, an even more serious obstacle for free diffusion of DNA is the plasmalemma, the physical border between a cell and its microenvironment. Here one expects an accumulation of various nucleases. Free passage of DNA across cell walls and plasma membranes is therefore unlikely, but at least extremely ineffective. Therefore, the following approaches, all based on the assumption of an unrestrained movement of genes through cellular barriers, cannot be reliable.

(1) Imbibition of cells, tissues, embryos and seeds

The early experiments to imbibe cells, tissues, and organs with DNA in order to transform them have met with no, or only little success, and were accordingly interpreted critically (see e.g. KLEINHOFS et al., 1975; KLEINHOFS and BEHKI, 1977; SOYFER and TITOV, 1981). It is fair to state that imbibition experiments did not result in any proven case of integrative transformation. This also holds for the imbibition of dry seeds and embryos, though defined reporter DNA has been taken up and been expressed (TÖPFER et al., 1989; SENARATNA et al., 1991). The system in itself holds great promise, because eventually transformed mature embryos could give rise to transgenic plants, especially in recalcitrant crops such as grain, legumes or cereals. However, more experimentation is needed to substantially improve the procedure.

(2) Macroinjection

In contrast to microinjection, macroinjection uses normal injection syringes with a needle diameter far wider than cell diameters to deliver the DNA into tissues. The DNA would then have to travel across the cell walls and plasma membranes of the cells adjacent to the wound site, a process which causes an appreciable loss of transforming DNA. Though macroinjection has been exercised in many laboratories, there are only few reports on a successful outcome. One of these (DE LA PENA et al., 1987) used macroinjection of a plasmid carrying an npt II gene into the stem below the immature floral meristem of rye (*Secale cereale*) prior to meiosis, and demonstrated that some of the selected offspring had acquired kanamycin resistance. Southern blot hybridization also indicated integration of the npt II gene. In view of the low transformation frequency obtained with macroinjection, we regard this technique not too promising for integrative transformation of cereals.

(3) Pollen transformation and pollen tube pathway

If mature or germinating pollen is incubated with DNA, the DNA is taken up (HESS, 1987), and may be integrated into the sperm nuclei, or reach the egg cell with the pollen tube. As such an ideal gene transfer technique, it has met with great interest and yielded a series of remarkable results (OHTA, 1986; HESS, 1987). Though integrative transformation has been claimed, definite proof is missing (reviewed by POTRYKUS 1990, 1991). We expect that future experiments will prove the validity of this system, but foresee difficulties because the highly condensed state of pollen cell chromatin almost excludes recombination of foreign DNA, and any freely moving DNA would be exposed to the unknown environment of the synergid-egg-complex, not to speak of the nucleases present in pollen (MATOUSEK and TUPY, 1984, 1985).

An aproach circumventing these difficulties, the *in vitro* maturation of microspores, may be of future use. Special treatment of isolated microspores leads to their maturation *in vitro*. *In vitro* matured pollen can effectively be used to pollinate flowers *in situ* (BENITO MORENO et al., 1988; ALWEN et al., 1990). During *in vitro* maturation, foreign DNA could be transferred into such pollen cells (using imbibition, electroporation, microinjection, or co-cultivation with *Agrobacterium tumefaciens,* see e.g. SÜSSMUTH et al., 1991) and be integrated into the generative nuclear DNA during S-phase replication. Such transformed DNA would then reach the egg apparatus via normal fertilization.

In another approach, DNA has been applied to stigmas after their pollination, or to styles after removing the stigma, and the DNA was expected to be guided to the embryo sac via the *pollen tube pathway* (LUO and WU, 1988). This technique seemed to work with rice (*Oryza sativa*), though the data do not provide unequivocal proof for integrative transformation, nor has any transgenic rice plant been obtained by this approach. It is also difficult to understand how DNA could survive the pollen-tube environment with its nucleases and adsorption sites on exposed cell walls. Even if DNA would reach the micropylus and enter the embryo sac, how should it be targeted to the egg nucleus? So, this technique faces a series of difficulties common with other methods of imbibition. Of course, the DNA could also be taken up by the growing pollen tube, but again the highly condensed chromatin of the pollen nuclei, especially the generative nucleus, probably prevents any covalent integration. Taken together, the obstacles are many, the experiments are few. However, if these difficulties are overcome, this relatively simple system might become attractive, also for the transformation of monocotyledons.

3.4 Characterization of the Transformed State

3.4.1 Integration and Inheritance of Transgenes

Foreign DNA transferred to plant cells by one of the various techniques described above is usually integrated into the nuclear plant genome and inherited in a Mendelian manner (DE BLOCK et al., 1984; HORSCH et al., 1984; POTRYKUS et al., 1985; DEROLES and GARDNER, 1988a, b). Transformation of organellar DNA is, however, also observed (DE BLOCK et al., 1985; see Sect. 4.1). While earlier experiments indicated a random distribution of integration sites (WALLROTH et al., 1986), gene tagging experiments (KONCZ et al., 1989) and the investigation of transgene chromatin structure (WEISING et al., 1990) suggested a preferential integration of T-DNA-derived vectors in active chromatin regions. Target-specific integration based on homologous recombination is a rare event in higher plants as compared to lower eukaryotes (see Sect. 4.2), though recombination of short sequence elements was shown to participate in Ti plasmid-mediated integration events (see Sect. 2.1.4).

Foreign DNA sequences are integrated at single or multiple loci (AMBROS et al., 1986;

WALLROTH et al., 1986; SPIELMANN and SIMPSON, 1986), and usually transmitted to the progeny as dominant Mendelian traits (POTRYKUS et al., 1985; BUDAR et al., 1986; MÜLLER et al., 1987). However, physical or functional inactivation of transgenes may occur in the progeny (HÄNISCH TEN CATE et al., 1990; SAUL and POTRYKUS, 1990; see Sect. 5). Tandem-like arrays of multiple inserts in a head-to-head or head-to-tail arrangement were frequently observed (JORGENSEN et al., 1987). Truncation, tandemerization, and other kinds of rearrangements are especially common with vectorless gene transfer techniques, and are probably caused by the combined action of recombination (see Sect. 4.2), ligation (BATES et al., 1990) and nucleolytic events prior to integration. Recombinational events occurring during the passage of the foreign DNA through the cytoplasm of the recipient cell are also held responsible for the frequently observed co-transformation of physically unlinked genes (reviewed by SAUL and POTRYKUS, 1990). The higher tendency of *Agrobacterium*-mediated transfer techniques to result in low copy, non-tandemerized, and unrearranged integration events is probably a consequence of the protection of the transferred T-DNA-complex by Vir-proteins (see Sects. 2.1.2 and 2.1.3). Simple integration patterns were also observed by the inclusion of a "transformation booster sequence" from the petunia genome (MEYER et al., 1988). This sequence element has some homology to scaffold attachment regions (KARTZKE et al., 1990).

3.4.2 The Analysis of Transgene Expression

Regulated and/or maximal foreign gene expression in target plants is desirable for many areas of applied plant gene technology. However, the availability or reliable genetic transformation systems has also provided a powerful instrument to explore basic mechanisms of gene regulation. The introduction of appropriate constructs, usually consisting of dissected plant gene regulatory regions fused to a minimal promoter and a reporter gene, into homo- or heterologous host genomes allowed the functional characterization of a large variety of enhancer and silencer elements involved in transcriptional control as well as of target "boxes" for trans-acting factors (reviewed by WILLMITZER, 1988; WEISING and KAHL, 1991; KATAGIRI and CHUA, 1992).

The easiest technique to monitor foreign gene expression capitalizes on the fusion of reporter genes to putative control sequences of plant genes, and measuring reporter gene activity in transient (WERR and LÖRZ, 1986) or stable expression systems. Sensitive enzymatic assays have been developed for a variety of proteins encoded by marker and reporter genes, such as GUS (JEFFERSON et al., 1987; JEFFERSON, 1989; SPÖRLEIN et al., 1991a), CAT (GORMAN et al., 1982; GENDLOFF et al., 1990), NPT II (REISS et al., 1984; ROY and SAHASRA BUDHE, 1990; MCDONNELL et al., 1987), luciferase (reviewed by KONCZ et al., 1990), octopine and nopaline synthase (OTTEN and SCHILPEROORT, 1978). In the majority of recent transgene studies, GUS was included as a reporter that allows not only fast quantitation of transgene activity in a non-radioactive assay, but also the histochemical analysis of cell-specific expression patterns (see Fig. 11a, b).

However, several potential obstacles narrow the use of reporter gene strategies. First, some plant species exhibit considerable levels of background reporter activity, or produce compounds that inhibit the expression of the respective reporter genes in the transgenic host. For both of these reasons, the cat reporter is not at all useful in *Brassica* (CHAREST et al., 1989) and *Triticum* (CHIBBAR et al., 1991). Second, the level of reporter gene expression does not necessarily reflect the actual transcription rate. It has been shown in transient expression experiments that reporter enzymes might be extraordinarily stable and persist extended times after transcription ceased (PRÖLS et al., 1988). Third, it might be of interest for some studies to transfer not only the 5′ and 3′ regions, but also the coding region of a gene into the host genome, in order to follow the processing and stability of an encoded protein. In these cases, more classical techniques of transcriptional analysis have to be used, e.g., at the RNA or protein level. If full-length plant genes are transferred to a host that alrea-

a
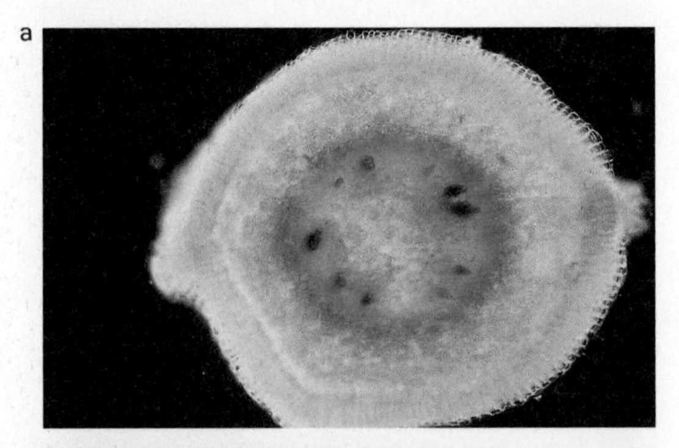
b

Fig. 11. Histochemical localization of gus reporter gene expression in cross-sections of tuber tissues (a) and root tissues (b) of potato (*Solanum tuberosum* L.) transformed with a vector construct carrying the gus gene controlled by patatin 5′ and nopaline synthase 3′ regulatory regions. Note the cell-specific action of the patatin promoter (courtesy of Dr. M. KÖSTER-TÖPFER, see KÖSTER-TÖPFER et al., 1989).

dy contains one or more similar resident genes, care has to be taken to distinguish between the expression of transferred and resident gene copies. This may be achieved by "exon tagging" (ECKES et al., 1986), i.e., the insertion of a tagging sequence useful as gene-specific probe into one of the exons of the transferred gene.

Most techniques for gene transfer into cells introduce much more DNA into the recipient nucleus than is finally integrated into the nuclear genome. This extrachromosomal DNA is usually degraded *in nucleo*, or gradually diluted by cell divisions, so that one or two weeks after transfer the transferred DNA can no longer be detected. However, during its existence the foreign DNA is expressed, and this expression can be easily monitored, provided a scorable reporter gene was included in the transfected DNA *(transient gene expression assay).* Such assays are rapid and rather reliable tools to monitor successful gene transfer and promoter effectivity (WERR and LÖRZ, 1986). However, to exploit the advantages of transient gene expression assays, a number of parameters have to be controlled:

(a) The stability of the transferred DNA. Depending on the specific nuclease complement of the host cell, the incoming DNA will be more or less stable, and hence the transient assay be more or less reliable. This difficulty can be partly overcome by packaging the DNA with proteins (protectifer, see Sect. 3.3.3.7).

(b) The integrity of the transferred DNA. During the transfer process the foreign DNA frequently undergoes tandemerization, truncation, concatenation, and other rearrangements (for plants see CZERNILOFSKY et al., 1986a, b; RIGGS and BATES, 1986; WIRTZ et al., 1987). No effective means to prevent such disintegration of the foreign DNA are at hand.

(c) The topology of the transferred DNA. The structure of the foreign DNA at the moment of entry into the recipient cell strongly influences its subsequent transient expression. Whereas only few experiments with plant cells addressed this problem (e.g., BALLAS et al., 1988; MAAS and WERR, 1989), studies of animal systems suggest that DNA conformation is governing its expression (e.g., WEINTRAUB et al., 1986). However, there are as yet conflicting data as to what specific conformation (e.g., linear relaxed versus supercoiled) is the best prerequisite for efficient transient expression.

(d) The physiological state of the target cells. Generally, actively dividing cells seem to set a favorable environment for transient expression (OKADA et al., 1986a, b; KARTHA et al., 1989), though this has been questioned (see, e.g., JUNKER et al., 1987).

(e) The specific design of the transferred construct. Transient expression vector construction should consider problems like promoter bias (e.g., monocot–dicot) and reporter gene suitability (e.g., background in the host cell).

Transient gene expression is extremely useful for the study of promoter function (i.e., gene expression), which can be tested after only a few hours following transfer. Moreover, the promoter can work without the modulating influences of flanking sequences (which is not the case after its integration, see position effect, Sect. 5.3). The transient assay allows the rapid selection and evaluation of promoters for use in stable transformation systems, and the detection of upstream regulatory elements (e.g., EBERT et al., 1987; ELLIS et al., 1987; HORTH et al., 1987). Additionally, the expression of various coding sequences and the stability of the encoded protein product can be rapidly estimated. Since the introduction of biolistic procedures (see Sect. 3.3.3.2), transient gene expression is no longer restricted to protoplasts or cells, but can also be successfully applied to tissues and whole organs.

4 Directed Genetic Engineering of Plant Cells

4.1 Organelle Transformation

Only recently, several techniques of direct gene transfer have been successfully employed in plastid transformation. Vectors carrying GUS reporter genes flanked by chloroplast promoters of atpA, atpB, and rbcL genes were delivered into the unicellular alga *Chlamydomonas reinhardtii* where they were introduced into chloroplast chromosomes by homologous recombination to produce photosynthetically competent stable transformants (BLOWERS et al., 1989). The gus genes were effectively transcribed, and their transcription rates were identical to those of their endogenous counterparts (BLOWERS et al., 1990). Subsequently it has been proven that the transformed chloroplast of this alga indeed produced the encoded protein (GOLDSCHMIDT-CLERMONT, 1991). The author constructed improved transformation-expression vectors by fusing an *Escherichia coli* aadA sequence (encoding aminoglycoside adenyltransferase) to transcription and translation signals from chloroplast genes. Biolistic transfer of these constructs conferred spectinomycin and streptomycin resistance to the alga. The chloroplasts of higher plants can also be transformed by simple direct gene transfer techniques (SVAB et al., 1990; DANIELL et al., 1990). An improved helium-driven biolistic device allowed the transfer of GUS constructs into chloroplasts of tobacco (YE et al., 1990), avoiding gunpowder effects, and a PEG-induced general transformation of tobacco protoplasts resulted in the transient expression of the transgenes (GUS) in chloroplasts (SPÖRLEIN et al., 1991 b). The term "transplastomic" clones has been suggested for the transformants (SVAB et al., 1990).

Though the efficiency of plastid transformation is less or about the same as nuclear gene transformation (SVAB et al., 1990), chloroplast transgenesis has opened the way for a detailed and systematic analysis of chloroplast gene function and regulation, especially the identification of critical *cis*-acting and *trans*-acting regulatory elements, the exploration of the architecture of chloroplast promoters using the transient expression assay (SPÖRLEIN et al., 1991 b) and the involvement of plastid genes in nucleo-plastid interaction(s). Moreover, site-directed mutagenesis of loci required for photosynthesis (e.g., psaC or tscA) establishes "disruption" mutants that allow fundamental photosynthetic functions to be probed in detail (GOLDSCHMIDT-CLERMONT, 1991).

4.2 Gene Targeting: Site-Specific Integration into the Nuclear Genome

Two different recombination systems may operate in the process of foreign DNA integration into the genome of a host organism: while *illegitimate recombination* is independent of the target sequence and therefore results in random insertion patterns, *homologous recombination* specifically targets an incoming DNA molecule to homologous regions in the host genome. Both systems may coexist, and compete for the same DNA substrate (CAPECCHI, 1990). Homologous recombination ("gene targeting") offers the potential for controlling the location of transgene inserts in host cell DNA (CAPECCHI 1989a, b). This could greatly improve the efficiency of plant genetic engineering by, e.g., the targeting of transgenes to sites conferring high levels of expression. The introduction of precise mutations into the structure of endogenous genes would allow the study of structure-function relationships as well as the targeted inactivation of a particular gene of interest.

The major obstacle for the widespread use of gene targeting vectors is the fact that homologous recombination is a rare event in higher eukaryotes as compared to fungi. The much higher efficiency of illegitimate recombination requires a positive selection procedure for homologous recombination events that is usually based on a two-step gain-of-function procedure. First, a transgenic organism is produced that carries a non-functional deletion derivative of a reporter or selectable marker gene. In the second step, this organism is retransformed with a construct carrying the same gene which is mutated at a different site. A functional marker gene can only be formed by homologous recombination. Using this kind of approach, a considerable number of studies performed with mammalian systems have revealed targeting frequencies between 10^{-2} and 10^{-5} (reviewed by BOLLAG et al., 1989). In contrast, only few studies have been performed in plants. Using either *Agrobacterium*-mediated or direct gene transfer techniques, successful gene targeting has been reported for tobacco (PASZKOWKSI et al., 1988; LEE et al., 1990; OFFRINGA et al., 1990) and *Arabidopsis* (HALFTER et al., 1992). In one study, an endogenous chlorosulfuron-sensitive acetolactate synthase (ALS) gene was successfully converted to a resistant gene by homologous recombination with a non-functional part of a resistant ALS gene (LEE et al., 1990). For all experiments, targeting frequencies were in the same range (between 10^{-4} and 10^{-5}), irrespective of the plant species, the gene under study, and the transformation technique employed. A similar range was also observed for intrachromosomal recombination between a closely linked complementary pair of mutated reporter genes in tobacco (PETERHANS et al., 1990a). Since the *Arabidopsis* genome is about 20 times smaller than the tobacco genome, genome size is obviously not a major determinant of targeting frequencies. This has also been demonstrated for mammalian cells carrying normal versus amplified copy numbers of target sequences (ZHENG and WILSON, 1990).

Much higher recombination frequencies are usually observed when both participating DNAs exist extrachromosomally, i.e., upon cotransformation of two or more plasmids carrying non-overlapping deletions (WIRTZ et al., 1987; BAUR et al., 1990; LYZNIK et al., 1991; PUCHTA and HOHN 1991a, b). Extrachromosomal recombination occurring prior to integration might also be responsible for the frequently observed tandemerization of vector DNA. The mechanisms involved in homologous recombination are still a matter of debate (for a discussion see PUCHTA and HOHN, 1991b). Though different mechanisms might operate in intra- and extrachromosomal recombination, respectively, the higher frequency of the latter event allowed the study of the structural requirements for recombining molecules in plant cells. Extrachromosomal recombination frequencies in plant protoplasts have been found to depend on the length of the homologous overlap (PUCHTA and HOHN, 1991a), on the linear conformation of DNA (BAUR et al., 1990), and the existence of terminal regions of homology (LYZNIK et al., 1991). This combination of requirements distinguishes extrachromosomal recombination in plant cells from similar processes both in fungi (BINNINGER et al., 1991) and mammals (MANSOUR et al., 1988).

A different gene targeting strategy has recently been reported that relied on site-specific rather than homologous recombination (DALE and OW, 1990; 1991; ODELL et al., 1990; MAESER and KAHMANN, 1991). In these studies, recombinases derived from bacteriophages were shown to catalyze site-specific recombination at their target sites present in test genes, when expressed in transgenic *Arabidopsis* protoplasts or tobacco plants. These findings suggest that (1) foreign DNA can be inserted at preselected locations into the plant genome, and (2) undesired vector sequences can be removed when flanked by two recombinase target sites. In a recent study, DALE and OW (1991) reported the precise excision of a selectable marker gene mediated by the CRE/lox recombination system. After removal of the hygromycin phosphotransferase gene construct, plants were obtained that had only incorporated the desired transgene. This strategy could greatly increase the public acceptance of field studies using genetically modified plants.

4.3 Protein Targeting

The compartmentalization of a foreign gene product can be triggered by exploiting transport mechanisms that specifically target translation products of nuclear genes to different organelles, vacuoles, or the extracellular space. The analysis of specific translocation signals coding for transit peptides was greatly facilitated by the availability of transgenic plants and *in vitro* import systems (e.g., for chloroplasts). Chimeric genes were created that consist of a heterologous promoter fused to sequences coding for putative transit peptides and a suitable reporter gene. These constructs were either transferred into isolated organelles or into plants. Their expression resulted in fusion proteins which were specifically transported into the corresponding compartment. Extensive manipulation of the signal elements thus permitted to define amino acid sequence requirements for a specific transmembrane transport.

Early experiments demonstrated that transit peptide sequences derived from the rbcS gene (SCHREIER et al., 1985; VAN DEN BROECK et al., 1985) were able to target selectable marker genes into chloroplasts. Specific targeting of a reporter enzyme into plant mitochondria was then detected using a transit peptide of the β-subunit of a mitochondrial ATPase (BOUTRY et al., 1987), and nuclear targeting was achieved by, e.g., a signal peptide present in the amino-terminal portion of the *Agrobacterium* virD2 protein (HERRERA-ESTRELLA et al., 1990). Since then, a large number of studies has been performed on protein targeting to nuclei (reviewed by HUNT, 1989; GOLDFARB, 1989; BURKE, 1990; SILVER, 1991), mitochondria (reviewed by HARTL and NEUPERT, 1990), chloroplasts (reviewed by KEEGSTRA, 1989; SMEEKENS et al., 1990), vacuoles, and the extracellular space (reviewed by CHRISPEELS, 1991). Some targeting signals are surprisingly conserved. For example, a yeast mitochondrial transit peptide functioned as a dual targeting signal for both chloroplasts and mitochondria in transgenic tobacco (HUANG et al., 1990a).

Taken together, the increasing insight into the sequence requirements for sorting signals as well as their state of modification (e.g., by glycosylation) will allow targeting of transgene products into any compartment of interest in the near future.

4.4 Targeted Inactivation of Resident Genes

For certain experimental purposes it is desirable to inactivate a resident plant (or viral) gene rather than to introduce a new foreign gene. This is especially true if one wishes to establish the causative relationship(s) between a gene of unknown function and its role in plant development and metabolism, to isolate genes via gene tagging strategies, to suppress commercially undesirable genes (e.g., to prevent premature fruit ripening), and to inactivate plant-invading viruses.

Two principal strategies for gene inactivation may be considered:

(1) The *physical disruption* of resident genes, e.g., by T-DNA or transposon tagging (KONCZ et al., 1989; FELDMANN et al., 1989). This procedure is usually applied to the isolation and cloning of genes, or for their replacement by engineered homologs ("gene targeting", see Sect. 4.2).

(2) The *functional inactivation* of resident genes by the introduction of antisense constructs (MOL et al., 1990), antigene oligonucleotides (HÉLÈNE and TOULMÉ, 1990), or gene-specific ribozymes which possess the ability to catalyze cleavage of defined target RNAs (SYMONS, 1991). The potential of both the antisense RNA and ribozyme approaches will be portrayed.

4.4.1 Antisense RNA

Among the strategies for functional gene inactivation, the antisense approach is by far the most advanced. Originally recognized as a natural mechanism for gene regulation in prokaryotes (reviewed by GREEN et al., 1986), the block of the information flow from RNA to protein by a complementary RNA was introduced into gene technology by IZANT and WEINTRAUB (1984). Because antisense RNA acts at the transcript level, it allows the coordinate inactivation of multiple gene copies in *trans*.

Antisense RNA technology was first applied to plant cells by ECKER and DAVIS (1986), who found that CAT activity was inhibited after cotransformation of sense and antisense CAT constructs into carrot cells. Since then, the inactivation by antisense RNA of a large variety of previously introduced reporter genes as well as of resident plant genes has been reported (see Tab. 4). Even strongly transcribed genes such as the small subunit gene for ribulosebisphosphate carboxylase/oxygenase were effectively inhibited (RODERMEL et al., 1988). There is, however, one remarkable exception of the rule: plant RNA virus-specific mRNA levels (e.g., of tobacco mosaic virus, cucumber mosaic virus, and potato virus X) are barely influenced by the introduction of antisense constructs (CUOZZO et al., 1988; HEMENWAY et al., 1988; REZAIAN et al. 1988; POWELL et al., 1989). Attempts to obtain resistance against infection by RNA viruses using antisense RNA were thus rather unsuccessful (reviewed by VAN DEN ELZEN et al., 1989). The reasons for this ineffectiveness possibly reside in the cytoplasmic life cycle of the RNA viruses (potato virus X, cucumber mosaic virus). Infection by plant *DNA viruses*, on the other hand, seems to be more efficiently counteracted by antisense constructs. Increased resistance towards tomato golden mosaic virus infection was recently reported for transgenic tobacco plants which harbored an antisense gene encoding AL-1, a protein necessary for replication of the virus (DAY et al., 1991). Symptom development was strongly reduced in plants expressing high levels of AL-1 antisense RNA.

What is the precise mechanism of antisense action? In prokaryotes, the formation of double-stranded RNA inhibits translation (MIZUNO et al., 1984). In eukaryotes, however, more than one mechanism seems to exist. Inhibition of transcription, enhanced degradation rates of duplex RNA, interference with mRNA processing and transport, and/or translation have been discussed (CORNELISSEN, 1989; MOL et al., 1990).To explain gene inactivation mediated by antisense RNA, several phenomena have to be considered: (1) Antisense RNA is effective in substoichiometric amounts. (2) Only part of the gene in an antisense configuration is sufficient to completely block sense gene expression. (3) In plants harboring active sense and antisense constructs, neither duplex RNA nor free antisense RNA are usually detected by Northern analysis. (4) Antisense-like effects are sometimes also observed upon the introduction of one or more *sense* constructs (NAPOLI et al., 1990; VAN DER KROL et al., 1990a; SMITH et al., 1990b, GORING et al., 1991). One hypothesis to explain this so-called "sense inhibition" or "homologous co-suppression" (see Sect. 5.5) supposes the presence of an antisense RNA which is transcribed from the opposite strand of DNA (GRIERSON et al., 1991; MOL et al., 1991). Further studies will certainly reveal the actual mechanisms of both sense and antisense inhibition of gene activity.

For several applications it might be useful to control antisense transcription by fusing the antisense construct to an inducible or tissue-specific promoter (see Sect. 2.2.4). An even more effective "switch" can be provided by bacterial repressor systems (GATZ and QUAIL, 1988; GATZ et al., 1991). This technique would involve the inclusion of target sequences for repressor molecules in the antisense gene promoter, and cotransfer of the respective repressor genes controlled by a constitutive promoter. In the absence of an inducer, the repressor inhibits the expression of the antisense gene by binding to its target sequence. Specific inducer molecules that can be applied to the plant (tetracycline in case of the tet repressor; GATZ and QUAIL, 1988) remove the repressor, activate the antisense gene, and hence inactivate the target gene(s).

Tab. 4. Gene Inactivation in Transgenic Plant Systems by Antisense and Sense Constructs

Gene	Plant Species	Remarks	References
Antisense Constructs			
cat gene	carrot	Cotransformation of protoplasts	ECKER and DAVIS (1986)
cat gene	Tobacco	Retransformation of transgenic plants	DELAUNEY et al. (1988)
nos gene	Tobacco	Retransformation of transgenic plants	ROTHSTEIN et al. (1987)
nos gene	Tobacco	Retransformation of transgenic plants, constructs from different regions of the nos gene	SANDLER et al. (1988)
pat gene	Tobacco	Cotransformation of protoplasts, retransformation of plants	CORNELISSEN and VANDEWIELE (1989) CORNELISSEN (1989)
gus gene	Tobacco	Retransformation of transgenic plants, 41 bp antisense RNA length is sufficient for inhibition	CANNON et al. (1990)
gus gene	Tobacco	Crossing of plants with gus- and anti-gus gene, or retransformation. Heat shock promoter allows reversible gus inactivation	ROBERT et al. (1989a, 1990)
rbcS gene	Tobacco	Effects of reduced rbcS levels on photosynthesis and rbcL regulation	RODERMEL et al. (1988) STITT et al. (1991) QUICK et al. (1991a, b)
ST-LS1 gene	Potato	Gene function studied	STOCKHAUS et al. (1990)
pTOM5 (cDNA)	Tomato	Gene function studied: pTOM5 is involved in carotenoid biosynthesis	BIRD et al. (1991)
pTOM13 (cDNA)	Tomato	Gene function studied: pTOM13 is involved in ethylene biosynthesis	HAMILTON et al. (1990)
ACC synthase gene	Tomato	Fruit ripening inhibited	OELLER et al. (1991)
Polygalacturonase gene	Tomato	Gene function studied	SHEEHY et al. (1988) SMITH et al. (1988; 1990a) SCHUCH et al. (1989)
Chalcone synthase gene	Petunia	Altered flower pigmentation; antisense RNA is effective as inhibitor when truncated	VAN DER KROL et al. (1988) VAN DER KROL et al. (1990b, c)
CMV coat protein gene	Tobacco	Viral protection not very effective	CUOZZO et al. (1988)
CMV (three different regions)	Tobacco	Viral protection not very effective	REZAIAN et al. (1988)
PVX coat protein gene	Tobacco	Viral protection not very effective	HEMENWAY et al. (1988)
TMV coat protein gene	Tobacco	Viral protection not very effective	POWELL et al. (1989)
TGMV AL-1 gene	Tobacco	Viral protection effective	DAY et al. (1991)
Sense Constructs			
Chalcone synthase gene	Petunia	Altered flower pigmentation	VAN DER KROL et al. (1990a) NAPOLI et al. (1990)
Dihydro-flavonol-4-reductase gene	Petunia	Altered flower pigmentation	VAN DER KROL et al. (1990c)
Polygalacturonase gene	Tomato	Truncated gene is inhibitory	SMITH et al. (1990b)
nos gene	Tobacco	Truncated gene is inhibitory	GORING et al. (1991)

4.4.2 Ribozymes

In recent years, a variety of RNA molecules were shown to possess self-splicing or self-cleaving properties, e.g., a ribosomal intervening sequence of *Tetrahymena*, and the RNA of several viroids (reviewed by CECH, 1987; SYMONS, 1991). The discovery that these cleavage reactions mediated by RNA "enzymes" do also function catalytically in *trans*, prompted the development of a new strategy for functional gene inactivation: the design of sequence-specific *trans*-acting *ribozymes* (HASELOFF and GERLACH, 1988). So-called "hammerhead" ribozymes (Fig. 12) derived from plant viroids seemed especially suitable for this purpose: only about 50 nucleotides were shown to be necessary for the formation of an active structure which consists of three base-paired stems plus 13 conserved nucleotides. Target specificity is mediated by base pairing to the substrate RNA, thereby forming stem I and III. The catalytic activity of the "hammerhead" structure, mainly located in stem II, then introduces a specific cut 3′ to a GUC motif within the substrate. HASELOFF and GERLACH (1988) demonstrated that small hammerhead ribozymes designed for specific cleavage were able to recognize and cut three distinct sites of the cat reporter gene transcript *in vitro*. Since these initial experiments, several groups reported the successful *in vitro*-cleavage of different target RNAs by specific ribozymes. For example, LAMB and HAY (1990) prepared ribozymes which were directed against plasmid-derived transcripts containing sequences of the coat protein gene and the polymerase gene of the potato leaf roll virus. Site-specific cleavage was observed within these transcripts, and also within full-length genomic RNA of this virus.

Though the potential application of hammerhead ribozymes *in vivo* has created much enthusiasm, this approach is still in an early

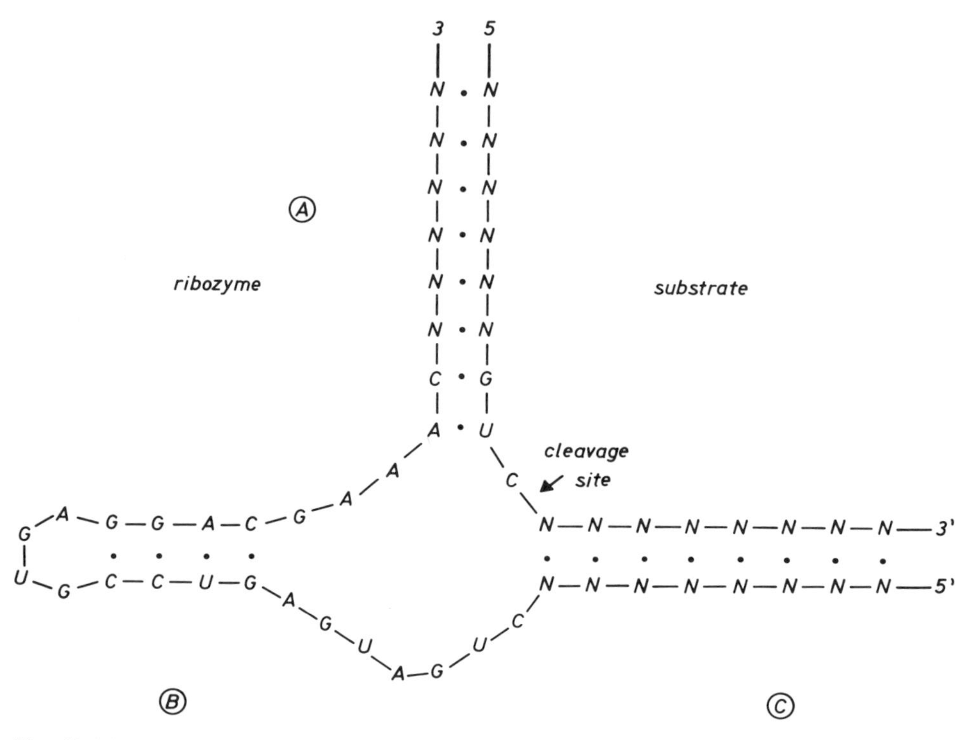

Fig. 12. The hammerhead structure derived from a separate substrate RNA and a ribozyme acting in *trans*. The cleavage site is indicated by an arrow (redrawn according to HASELOFF and GERLACH, 1988, and SYMONS, 1991).

experimental phase. The functional suppression of genes by vector-encoded ribozymes in transgenic mammalian cells as well as in microinjected *Xenopus* oocytes is already feasible (CAMERON and JENNINGS, 1989; SARVER et al., 1990; COTTEN and BIRNSTIEL, 1989; SAXENA and ACKERMANN, 1990). In most cases, suppression was only observed if the ribozyme was present in several-fold molar excess. Using appropriate controls, SAXENA and ACKERMANN (1990) demonstrated that the ribozyme effect was actually an antisense effect: microinjection of active as well as of functionally inactivated ribozymes into *Xenopus* oocytes resulted in similar suppression rates.

Taken together, several problems and questions will have to be solved before ribozyme constructs will complement antisense technology for functional gene inactivation *in vivo*. Are ribozymes acting truly catalytic *in vivo* – or is it merely an antisense effect that was observed in all the experiments reported to date? What amounts of ribozymes will be needed to exert an effect? If high-level expression is desired, the catalytic RNA sequences can be inserted into a vector containing sequences encoding tRNA, which are transcribed to high copy numbers by RNA polymerase III (COTTEN and BIRNSTIEL, 1989). Finally, how do ribozymes find their target RNA? This question will be especially important for cleavage reactions restricting plant pathogenic viruses with a cytoplasmic life cycle.

5 Factors Affecting Foreign Gene Expression

With the increasing wealth of data on the regulated transcription of foreign genes in their transgenic environment, several obstacles to a straight-forward gene expression in genetically engineered plant cells became apparent. While the *quality* of transcription is mainly determined by the nature of cotransferred *cis*-regulatory DNA sequences (i.e., promoters, enhancers, and silencers), it also depends on the presence of suitable transacting factors in the host organism. Problems might be encountered when phylogenetic distances between donor and acceptor organisms are large. Even more importantly, foreign gene expression levels may vary considerably between individual transformants harboring identical constructs. A variety of factors seem to be involved in the *quantitative* control of transgene expression: gene dosage, "position effects", DNA methylation, and homologous co-suppression. The latter factor, also called "sense inhibiton", is especially intriguing since transferred genes were found to be silenced along with their homologous counterparts in the resident genome.

5.1 Phylogenetic Distance

With a few exceptions (e.g., T-DNA genes from *Agrobacterium*), bacterial genes controlled by their own promoters are transcriptionally silent in transgenic plants (reviewed by FRALEY et al., 1986). The same holds true for most but not all animal genes. Though early experiments showed that mammalian promoters are usually not recognized by the transcriptional machinery of plants (SHAW et al., 1983; KONCZ et al., 1984; AN, 1986), exceptions to the rule have been found. Thus, the combination of a *Drosophila* heat-shock promoter with the npt II coding region resulted in heat-inducible expression of the marker gene in transgenic tobacco (SPENA et al., 1985; SPENA and SCHELL, 1987; WING et al., 1989), and a promoter derived from the *Drosophila* copia long terminal repeat fused to the cat reporter gene directed high levels of transient cat expression in electroporated protoplasts of different monocots (OU-LEE et al., 1986). At least in case of the heat-shock gene, the correct transcription in plant cells is probably a consequence of the unusually high conservation of heat-shock regulatory sequences throughout evolution.

If driven by a promoter functioning in plants, animal genes are transcribed to pre-mRNAs which are only inefficiently spliced and polyadenylated (BARTA et al., 1986; VAN SANTEN and SPRITZ, 1987; HUNT et al., 1987; MARTINEZ-ZAPATER et al., 1988). Most recently, an exception has been observed: the SV-40 small t intron was correctly spliced from a CaMV35S-CAT-SV40 transcription unit in tobacco cells (HUNT et al., 1991). The reason for the aberrant behavior of the SV40 small t intron might lie in its unusually high AU-content (80%). According to GOODALL and FILIPOWICZ (1989), the only non-junction requirement for intron removal in plants is the presence of AU-rich sequences close to the splicing junction.

The use of cDNA instead of genomic coding sequences may circumvent problems associated with incorrect processing. If under control of appropriate promoters, cDNA derived from animal genes was transcribed and translated efficiently in plants (OW et al., 1986; EICHHOLTZ et al., 1987; MAITI et al., 1988; HIATT et al., 1989; SAITO et al., 1991b).

Within the plant kingdom, correct transcription of transgenes and splicing of the pre-mRNA are usually observed in homo- as well as in heterologous host backgrounds (reviewed by WEISING et al., 1988). There are, however, several exceptions, and considerable barriers seem to exist between monocots and dicots. For example, seed-specific expression patterns of monocotyledonous storage protein genes was shown to be conserved in transgenic dicots (COLOT et al., 1987; SCHERNTHANER et al., 1988; MARRIS et al., 1988; WILLIAMSON et al., 1988; ROBERT et al., 1989b; MATZKE et al., 1990). However, the observed transcription rates were often low, and translation products sometimes undetectable. In at least two cases, a maize zein promoter in transgenic petunia (UENG et al., 1988) and a rice glutelin promoter in transgenic tobacco (LEISY et al., 1990), transcription even escaped the strict tissue-specific control. Conflicting results were also obtained with light-regulated genes. Whereas 5′ regulatory sequences derived from a monocot chlorophyll a/b binding protein gene retained their function in transgenic dicots (LAMPPA et al., 1985; NAGY et al., 1986, 1987), an oat phytochrome gene promoter (KELLER et al., 1989) and a wheat rbcS promoter (KEITH and CHUA, 1986) were found to be non-functional in transgenic tobacco. If placed under the control of a CaMV 35S promoter, the wheat rbcS gene was transcribed, but introns were only inefficiently removed and multiple novel polyadenylation sites were used (KEITH and CHUA, 1986). In a more detailed analysis, SCHÄFFNER and SHEEN (1991) demonstrated by transient expression assays that rbcS promoters from three different dicot species were non-functional in maize protoplasts. The same was true for the opposite approach: two monocot-derived rbcS promoters were not expressed in tobacco protoplasts. In contrast, the rbcS promoters were efficiently recognized by protoplasts derived from their own kingdom. Promoter deletion and sequence analyses revealed considerable differences between the architecture of regulatory elements of monocot and dicot rbcS promoters (SCHÄFFNER and SHEEN, 1991). Insufficient transcription (ELLIS et al., 1987) and aberrant processing (KEITH and CHUA, 1986) were also observed in the case of the maize Adh1 gene transferred to tobacco. The use of dicot-specific promoters and polyadenylation sites together with cDNA constructs of monocot genes was shown to circumvent problems inherent with the monocot-dicot barrier (e.g., BOYLAN and QUAIL, 1989; KAY et al., 1989). These problems are, however, not always restricted to long-distance gene transfer. BUDELIER et al. (1990) demonstrated that a pistil-specific promoter from tomato was non-functional in the closely related tobacco genome.

In contrast to the above mentioned examples, several genes show a remarkable evolutionary conservation of regulatory patterns involved in transcription and mRNA processing. For example, a set of different angiosperm promoters was found to be active in transiently transformed gymnosperm cells (ELLIS et al., 1991). The maize sucrose synthase-1 promoter directed phloem-specific expression (YANG and RUSSELL, 1990) in transgenic tobacco. The PEP-carboxylase gene of the monocot crop *Sorghum vulgare* was correctly transcribed and efficiently spliced in transgenic tobacco cells (TAGU et al., 1991), and a construct consisting of the CaMV35S promoter and the npt II coding region interrupted by a phaseolin intron was correctly transcribed and spliced in transgenic rice cells (PETERHANS et al., 1990b).

To summarize, these conflicting results indicate that the accurate expression of monocot genes in dicots and *vice versa* may depend on the specific gene transferred and the plant species to be transformed. Transcription, termination, and processing signals appear to have diverged significantly between bacteria, fungi, animals and plants, and to a certain extent also within the plant kingdom. Whether regulatory patterns are retained or not, is at present unpredictable. Thus, if an optimal expression of genes derived from phylogenetically distant organisms in transgenic plants is desired, e.g., for the production of antibodies (HIATT et al., 1989; HIATT, 1990), metallothioneins (MAITI et al., 1988; MISRA and GEDAMU, 1989) or bacterial insecticides (VAECK et al., 1987), vectors have to be constructed in which intronless or processed genes of interest are flanked by host-derived promoter and polyadenylation signals. Even though, the translation efficiency and stability of the resulting protein will probably depend on the particular gene under study.

5.2 Gene Dosage

The impact of copy number on foreign gene expression varies from one system to another. While some authors found the copy number of transgenes to be positively correlated to their expression rate (STOCKHAUS et al., 1987; BÄUMLEIN et al., 1988; GENDLOFF et al., 1990), others found a negative correlation (e.g., HOBBS et al., 1990; LINN et al., 1990; MITTELSTEN SCHEID et al., 1991), or no correlation at all (e.g., CZERNILOFSKY et al., 1986a, b; SANDERS et al., 1987; SHIRSAT et al., 1989). Therefore, copy number is likely to be of secondary importance for foreign gene expression.

5.3 Position Effects

The expression of foreign genes in transgenic animals (reviewed by WILSON et al., 1990) and plants (reviewed by WEISING et al., 1988) is known to be strongly influenced by their chromosomal location. This so-called position effect might be related to the chromatin structure at the integration site (and/or DNA methylation, see below) on one hand, and to the presence of endogenous promoters/enhancers close to the integration site on the other.

Two main strategies have been pursued to circumvent this unpredictable and undesirable effect. First, vectors were designed that carry divergent, co-regulated promoters: one for the selectable marker, and the other for the non-selectable gene of interest (VELTEN and SCHELL, 1985; GIDONI et al., 1988; OTT et al., 1990). Since the chromosomal position was expected to exert a similar effect on both genes, transformants with the desired quality and quantity of transgene expression were thought to be directly selectable. However, the expression level of physically linked genes co-transformed to the host plant genome can vary independently (JONES et al., 1985; DEBLAERE et al., 1985; WEISING et al., 1990), even if they are controlled by dual, normally coregulated promoters (PEACH and VELTEN, 1991).

A second strategy, recently designed for animal systems, capitalizes on the inclusion in vectors of so-called "dominant control regions" (DCRs) that were identified far up- and downstream of the human globin gene cluster (GROSVELD et al., 1987). So-called "miniloci" consisting of a homo- or heterologous gene flanked by the DCRs were transfected into mammalian cells and resulted in high-level, position-independent expression correlating with copy number (BLOM VAN ASSENDELFT et al., 1989; TALBOT et al., 1989). Similar observations were reported by SIPPEL and coworkers who used DNA sequences derived from scaffold attachment regions (SARs) that flank the chicken lysozyme gene for vector construction (STIEF et al., 1989; PHI-VAN et al., 1990). DCRs may relate to domain borders and/or "insulate" the transgene from the surrounding chromatin architecture by creating a functionally distinct, transgene-harboring domain.

Some evidence suggests that DCR-like DNA sequence elements also exist in plants. A sequence element from petunia called "transformation booster sequence" provided enhanced transformation frequency, simple integration patterns, and copy-number-dependent expression of cotransferred genes in transgenic petunia and tobacco (MEYER et al., 1988). This sequence bears some homology to scaffold attachment regions from *Drosophila* (KARTZKE et al., 1990) and possibly relates to the type of control regions described above. On the other hand, attempts to override the between-transformant variability by cotransferring 10 kb of 5′- and 13 kb of 3′-flanking regions of the petunia rbcS gene into transgenic tobacco plants failed (DEAN et al., 1988). The recent development of convenient isolation methods for plant nuclear scaffolds (MORENO DIAZ DE LA ESPINA et al., 1991) and SARs (SLATTER et al., 1991) will now allow to test whether the inclusion of SARs in plant transformation vectors can abolish the position effects in transgene expression.

5.4 DNA Methylation

DNA methylation is inversely correlated to transcriptional activity in a huge variety of plant, fungal, and animal genes (reviewed by CEDAR and RAZIN, 1990; MAGILL and MAGILL, 1989; CHOMET, 1991). In eukaryotes, this modification is mainly limited to cytosine residues, though a few reports on the methylation of adenine residues exist (e.g., in rice: DHAR et al., 1990). Cytosine methylation of plant DNA usually occurs in the sequence context -CG- and -CXG- (GRUENBAUM et al., 1981; BROWN, 1989; BELANGER and HEPBURN, 1990). Modified cytosines were predominantly found in repetitive DNA, but also in the vicinity of genes. In most (but not all) plant genes examined, transcriptional activity was correlated to hypomethylation of promotor regions (e.g., in pea ribosomal DNA: WATSON et al., 1987; FLAVELL et al., 1988; maize zein genes: BIANCHI and VIOTTI, 1988; maize photosynthetic genes: NGERNPRASIRTSIRI et al., 1989). Transcription and mobility of plant transposable elements are probably regulated by DNA methylation, e.g., of the maize elements Mu (CHANDLER and WALBOT, 1986; BENNETZEN, 1987), Ac/Ds (SCHWARTZ and DENNIS, 1986; CHOMET et al., 1987; KUNZE et al., 1988; SCHWARTZ, 1989; BRETTELL and DENNIS, 1991), and Spm (BANKS et al., 1988).

Transferred genes in foreign plant genomes are also subject to cytosine methylation. Methylation of wild-type T-DNA genes was observed in crown gall tumors (GELVIN et al., 1983; VAN LIJSEBETTENS et al., 1986), in revertant plants regenerated from hairy root (SINkar et al., 1988b) and crown gall (AMASINO et al., 1984), and in transgenic plants transformed with individual T-DNA genes (JOHN and AMASINO, 1989; KLAAS et al., 1989). Wherever investigated, gene activity was found to be inversely correlated to the occurrence of methylated cytosines.

Methylation of foreign DNA is not restricted to T-DNA genes. Though infrequently, *de novo*-methylation of transposable elements was also shown to occur in transgenic plants (MARTIN et al., 1989; NELSEN-SALZ and DÖRING, 1990). In a detailed study, HOBBS et al. (1990) investigated the effect of T-DNA copy number, chromosomal position and methylation of cytosine residues on CaMV 35S-GUS gene expression in individual tobacco transformants. The authors found a correlation between the occurrence of multiple insertions of the transgene, low reporter gene expression, and increased levels of methylation.

Similar observations were made by LINN et al. (1990) who analyzed a large number of transgenic petunia plants for the structure and expression of a maize gene involved in anthocyanin biosynthesis. The expression of the foreign gene was dependent on its chromosomal position, and negatively correlated to the extent of DNA methylation within the CaMV 35S promoter sequence. Transgenes were preferentially methylated when present in high copy numbers. Reversible inactivation of the hygromycin resistance gene in transgenic *Arabidopsis thaliana* was reported by MITTELSTEN SCHEID et al. (1991). Fifty percent of the investigated plants failed to transmit the resistant phenotype to the progeny. Low expression rates of the transgene were again correlated with multicopy integration, but in this case *not* with methylation.

MATZKE and coworkers found the expression in transgenic tobacco of one T-DNA insert to be reversibly silenced in *trans* by the introduction of a second, different T-DNA (MATZKE et al., 1989; MATZKE and MATZKE, 1990; 1991). The suppression of gene activity was accompanied by an increase of cytosine methylation within the promoter of the resident gene. The suppressed phenotype was reverted upon segregation of both T-DNAs in the progeny, accompanied by a progressive decrease of transgene methylation in subsequent generations. The mechanism by which one gene is able to exert such a negative epigenetic effect onto the expression of another gene, is not clear. The authors proposed that suppression of the resident T-DNA in the presence of the incoming T-DNA is mediated by competition between homologous regions on each T-DNA for attachment to fixed sites in the nucleus which activate expression. However, it is neither clear how methylation could influence such a process, nor whether T-DNA methylation is a cause or a consequence of transgene suppression.

These few examples show that the (un)specific *de novo*-methylation of foreign genes represents a conceivable possibility of their inactivation. Now what are the determining factors of foreign gene methylation? This question is part of two more general problems: (1) How does methylation modulate transcription, and (2) how do methyltransferases recognize which sequences have to be methylated and which not? Evidence is now emerging that methylation acts on principally two levels. First, the presence of 5mC in regulatory regions might directly interfere with the binding of transcriptional factors. This kind of interference was demonstrated in animals (reviewed by DYNAN, 1989) and plants (STAIGER et al., 1989; INAMDAR et al., 1991). However, since a variety of mammalian factors bind their target sequences irrespective of their methylation status (e.g., Sp1; IGUCHI-ARIGA and SCHAFFNER, 1989), this is probably not the whole story. The second hypothesis to explain methylation-mediated gene inactivation postulates a strong interdependence of cytosine methylation and chromatin structure (KESHET et al., 1986). In a model put forward by SELKER (1990), an open chromatin structure ("form I") is maintained by the binding of sufficient quantities of sequence-specific trans-acting factors. Loss of these proteins accompanied by decreasing transcriptional activity causes the chromatin to switch to an inactive, condensed, and more regular structure ("form II"). DNA methylation, solely occurring with form II, "locks" the inactive state. Reversion to form I is principally possible, but can only be mediated by a few "keysN, i.e., methylation-insensitive trans-acting factors being able to bind their target sites in condensed, methylated chromatin. SELKER's model is consistent with a variety of observations, e.g., the correlation between enhanced DNase I-sensitivity, relaxed configuration and hypomethylated state of actively transcribed chromatin (KESHET et al., 1986; KLAAS and AMASINO, 1989; TAZI and BIRD, 1990), the existence of methyl-sensitive and -insensitive transcriptional factors (DYNAN, 1989), and the recent identification of a mammalian protein that preferentially binds methylated DNA in a non-sequence-specific manner (MEEHAN et al., 1989; BOYES and BIRD, 1992). Functionally similar proteins are also present in plants (ZHANG et al., 1989) and might well serve as the "lock" for methylated form II chromatin. SELKER's model would also explain the preferential methylation of repetitive sequences in plants, i.e., transposable elements, ribosomal DNA, satellite DNA, duplicated sequences in fungi (MAGILL and MAGILL, 1989), and multicopy insertions of T-DNA (HOBBS et al., 1990; LINN et al., 1990): if the chromatin structure is actively maintained in an open and therefore unmethylated state by the binding of transcription factors, multiplication of a given sequence motif may lead to its reduced occupation by factors. Consequently, a switch to the inactive form II takes place that may or may not be locked by DNA methylation.

5.5 Homologous Co-Suppression

A different, but possibly related means of foreign gene inactivation was first observed upon the introduction of genes encoding key enzymes of the anthocyanin biosynthesis pathway into transgenic petunia plants (NAPOLI et al., 1990; VAN DER KROL et al., 1990a). In a considerable percentage of transformants, the introduction of a chalcone synthase gene controlled by a CaMV 35S promoter did not result in overexpression (as was expected), but instead in the inactivation of the transferred gene. Moreover, the resident members of the chalcone synthase gene family were co-suppressed. This effect, called "homologous co-suppression" (NAPOLI et al., 1990), was also observed with a dihydroflavonol-4-reductase gene in transgenic petunia (VAN DER KROL et al., 1990c), a truncated polygalacturonase gene in transgenic tomato (SMITH et al., 1990b), and a fragment of the nopaline synthase gene in retransformed tobacco plants already harboring a wild-type copy of this gene (GORING et al., 1991).

Two main hypotheses have been put forward to explain these unexpected epigenetic effects (reviewed by JORGENSEN, 1990). On one hand, homologous co-suppression might be associated with DNA methylation, and rely on the same or similar mechanisms as discussed in the previous section. Methylation was, however, either not examined in these experiments, or not found associated with suppression. The second hypothesis predicts that sense inhibition is caused by antisense RNA produced by transcriptional readthrough from adjacent promotors located on the opposite DNA strand (GRIERSON et al., 1991; MOL et al., 1991). The promoter responsible for antisense RNA transcription might be either derived from plant sequences in the vicinity of the integration site, or, alternatively, from marker and reporter genes located on the vector itself. Since considerable antisense effects were observed with substoichiometrical amounts of antisense RNA, and with fragments as short as 41 bp (CANNON et al., 1990), antisense transcripts must neither be abundant nor full-length to achieve homologous co-suppression, according to this model. At least two lines of evidence support this explanation. (1) Sense and antisense inhibitions show remarkable similarities, e.g., the inhibitory potential of truncated constructs (SMITH et al., 1990b; GORING et al., 1991), the absolute dependence of inhibition on the homology of the coding regions, and the mutual gene inactivation. (2) Using nuclear run-on assays for the transcriptional analysis of sense-inhibited chalcone synthase genes in petunia, MOL et al. (1991) observed normal levels of sense RNA *and* low levels of antisense RNA transcription. These results strongly suggest that co-suppression is a post-transcriptional event involving antisense RNA formation.

If the readthrough hypothesis is correct, these findings have to be considered for expression vector construction. Vectors designed to overexpress a gene sharing homology with resident genes should then be constructed with divergent promoters (VELTEN and SCHELL, 1985; OTT et al., 1990). Even then, homologous co-suppression cannot be totally excluded. Since T-DNA-derived vectors tend to be preferentially integrated into active chromatin regions (KONCZ et al., 1989; HERMAN et al., 1990; WEISING et al., 1990), a considerable influence of plant promoters and enhancers adjacent to the integration site is to be expected. It remains to be determined whether such influences may be blocked by the inclusion in the vector of, e.g., scaffold attachment regions that provide topological barriers to the surrounding plant DNA.

6 Prospects: The Future Potential of Genetic Engineering of Plants

The past decade has seen a vigorous research using all the tools of plant genetic engineering. As a consequence, our knowledge about how plant genes are constructed, how they function in a homo- and heterologous environment, how they influence each other and how they are regulated has dramatically increased. So has the number of genes that became known in molecular detail. We are confident that this development will continue, and we expect another decade of original research. Two major areas will certainly benefit from any progress in plant genetic engineering. First, basic research. This is amply documented in this chapter. Second, plant breeding. The precision with which plant genetic engineering can modify a plant genome is new in breeding technology. The successful engineering of insect resistance, herbicide resistance, bacterial resistance, virus resistance, and new traits such as improved amino acid content in proteins and the production of economically interesting proteins (to name only a few) by transferring one single dominant gene is proof enough. However, genetic engineering is but one technique more in the whole repertoire of plant breeding, and certainly cannot solve all agronomical problems. Yet the potential benefits of this type of research have to be accepted by the public which is not fully true at present. So we ought to invest a lot more energy in public information and discussion about the exciting advantages, but also the supposed or potential risks of plant genetic engineering.

Acknowledgements

The dedicated help of Mrs. S. Kost during preparation of this article is greatly appreciated. Kurt Weising acknowledges a fellowship from the FAZIT foundation (Frankfurt am Main, Germany), and both authors thank Dr. Steinbiß and Dr. Köster-Töpfer for providing some of the figures. Research of the authors is supported by grants from BMZ (No. 89.7860. 3-01.130) and DFG (grant Ka 332/14-1).

7 References

AHLQUIST, P., PACHA, R. F. (1990), Gene amplification and expression by RNA viruses and potential for further application to plant transfer, *Physiol. Plant.* **79**, 163-167.

AHLQUIST, P., FRENCH, R., JANDA, M., LOESCH-FRIES, L. S. (1984), Multicomponent RNA plant virus infection derived from cloned viral cDNA, *Proc. Natl. Acad. Sci. USA* **81**, 7066-7070.

AHLQUIST, P., FRENCH, R., BUJARSKI, J. J. (1987), Molecular studies of brome mosaic virus using infectius transcripts from cloned cDNA, *Adv. Virus Res.* **32**, 215-242.

AHOKAS, H. (1987), Transfection by DNA-associated liposomes evidenced at pea pollination, *Hereditas* **106**, 129-138.

AHOKAS, H. (1989), Transfection of germinating barley seed electrophoretically with exogenous DNA, *Theor. Appl. Genet.* **77**, 469-472.

AINLEY, W. M., KEY, J. L. (1990), Development of a heat shock inducible expression cassette for plants: characterization of parameters for its use in transient expression assays, *Plant Mol. Biol.* **14**, 949-967.

AKIYOSHI, D. E., KLEE, H., AMASINO, R. M., NESTER, E. W., GORDON, M. P. (1984), T-DNA of *Agrobacterium tumefaciens* encodes an enzyme of cytokinin biosynthesis, *Proc. Natl. Acad. Sci. USA* **81**, 5994-5998.

ALBRIGHT, L. M., YANOFSKY, M. F., LEROUX, B., MA, D., NESTER, E. W. (1987), Processing of the T-DNA of *Agrobacterium tumefaciens* generates border nicks and linear, single-stranded T-DNA, *J. Bacteriol.* **169**, 1046-1055.

ALLISON, R., JANDA, M., AHLQUIST, P. (1988), Infectious *in vitro* transcripts from cowpea chlorotic mottle virus cDNA clones and exchange of individual RNA components with brome mosaic virus, *J. Virol.* **62**, 3581-3588.

ALT-MÖRBE, J., RAK, B., SCHRÖDER, J. (1986), A 3.6.kbp segment from the vir region of Ti plasmids contains genes responsible for border sequence-directed production of T region circles in *E. coli, EMBO J.* **5**, 1129-1135.

ALT-MÖRBE, J., KÜHLMANN, H., SCHRÖDER, J. (1989), Differences in induction of Ti plasmid virulence genes virG and virD, and continued control of virD expression by four external factors, *Mol. Plant-Microbe Interact.* **2**, 301-308.

ALWEN, A., ELLER, N., KASTLER, M., BENITO MORENO, R. M., HEBERLE-BORS, E. (1990), Potential of *in vitro* pollen maturation for gene transfer, *Physiol. Plant.* **79**, 194-196.

ALY, M. A. M., OWENS, L. D. (1987), A simple system for plant cell microinjection and culture, *Plant Cell Tissue Organ Cult.* **10**, 159-174.

AMASINO, R. M., POWELL, A. L. T., GORDON, M. P. (1984), Changes in T-DNA methylation and expression are associated with phenotypic variation and plant regeneration in a crown gall tumor line, *Mol. Gen. Genet.* **197**, 437-446.

AMBROS, P. F., MATZKE, M. A., MATZKE, A. J. M. (1986), Detection of a 17 kb unique sequence (T-DNA) in plant chromosomes by *in situ* hybridization, *Chromosoma* **94**, 11-18.

AN, G. (1986), Development of plant promoter expression vectors and their use for ánalysis of differential activity of nopaline synthase promoter in transformed tobacco cells, *Plant Physiol.* **81**, 86-91.

AN, G., WATSON, B. D., STACHEL, S., GORDON, M. P., NESTER, E. W. (1985), New cloning vehicles for transformation of higher plants, *EMBO J.* **4**. 277-284.

AN, G., WATSON, B. D., CHIANG, C. C. (1986), Transformation of tobacco, tomato, potato and *Arabidopsis thaliana* using a binary Ti vector system, *Plant Physiol.* **81**, 301-305.

AN, G., COSTA, M. A., MITRA, A., HA, S.-B., MARTON, L. (1988), Organ-specific and developmental regulation of the nopaline synthase promoter in transgenic tobacco plants, *Plant Physiol.* **88**, 547-552.

AN, G., COSTA, M. A., HA, S.-B. (1990), Nopaline synthase promoter is wound inducible and auxin inducible, *Plant Cell* **2**, 225-233.

ANDRE, D., COLAU, D., SCHELL, J., VAN MONTAGU, M., HERNALSTEENS, J. P. (1986), Gene tagging in plants by a T-DNA insertion mutagen that generates APH(3′)II-plant gene fusions, *Mol. Gen. Genet.* **204**, 512-518.

ANKENBAUER, R. G., NESTER, E. W. (1990), Sugar-mediated induction of *Agrobacterium tumefaciens* virulence genes: structural specificity and activities of monosaccharides, *J. Bacteriol.* **172,** 6442–6446.

ANTONELLI, N. M., STADLER, J. (1990), Genomic DNA can be used with cationic methods for highly efficient transformation of maize protoplasts, *Theor. Appl. Genet.* **80,** 395–401.

AOYAMA, T., TAKANAMI, M., MAKINO, K., OKA, A. (1991), Cross-talk between the virulence and phosphate regulons of *Agrobacterium tumefaciens* caused by an unusual interaction of the transcriptional activator with a regulatory DNA element, *Mol. Gen. Genet.* **227,** 385–390.

ARMSTRONG, C. L., PETERSEN, W. L., BUCHHOLZ, W. G., BOWEN, B. A., SULC, S. L. (1990), Factors affecting PEG-mediated stable transformation of maize protoplasts, *Plant Cell Rep.* **9,** 335–339.

ASANO, Y., OTSUKI, Y., UGAKI, M. (1991), Electroporation-mediated and silicon carbide fiber-mediated DNA delivery in *Agrostis alba* L. (Redtop), *Plant Sci.* **79,** 247–252.

ASHBY, A. M., WATSON, M. D., SHAW, C. H. (1987), A Ti-plasmid determined function is responsible for chemotaxis of *Agrobacterium tumefaciens* towards the plant wound product acetosyringone, *FEMS Microbiol. Lett.* **41,** 189–192.

ASHBY, A. M., WATSON, M. D., LOAKE, G. J., SHAW, C. H. (1988), Ti plasmid-specified chemotaxis of *Agrobacterium tumefaciens* C58C1 toward vir-inducing phenolic compounds and soluble factors from monocotyledonous and dicotyledonous plants, *J. Bacteriol.* **170,** 4181–4187.

BAKKEREN, G., KOUKOLIKOVA-NICOLA, Z., GRIMSLEY, N., HOHN, B. (1989), Recovery of *Agrobacterium tumefaciens* T-DNA molecules from whole plants early after transfer, *Cell* **57,** 847–857.

BALLAS, N., ZAKAI, N., FRIEDBERG, D., LOYTER, A. (1988), Linear forms of plasmid DNA are superior to supercoiled structures as active templates for gene expression in plant protoplasts, *Plant Mol. Biol.* **11,** 517–527.

BANKS, J.A., MASSON, P., FEDOROFF, N. (1988), Molecular mechanisms in the developmental regulation of the maize suppressor-mutator transposable element, *Genes Dev.* **2,** 1364–1380.

BARFIELD, D. G., PUA, E.-C. (1991), Gene transfer in plants of *Brassica juncea* using *Agrobacterium tumefaciens*-mediated transformation, *Plant Cell Rep.* **10,** 308–314.

BARNES, W. M. (1990), Variable patterns of expression of luciferase in transgenic tobacco leaves, *Proc. Natl. Acad. Sci. USA* **87,** 9183–9187.

BARTA, A., SOMMERGRUBER, K., THOMPSON, D., HARTMUTH, K., MATZKE, M. A., MATZKE, A. J. M. (1986), The expression of a nopaline synthase – human growth hormone chimaeric gene in transformed tobacco and sunflower callus tissue, *Plant Mol. Biol.* **6,** 347–357.

BATES, G. W., CARLE, S. A., PIASTUCH, W. C. (1988), Electroporation: form of plasmid DNA affects the level of transient gene expression, in: *The Second International Congress of Plant Molecular Biology,* Abstract, **567,** Jerusalem, November 13–18.

BATES, G. W., CARLE, S. A., PIASTUCH, W. C. (1990), Linear DNA introduced into carrot protoplasts by electroporation undergoes ligation and recircularization, *Plant Mol. Biol.* **14,** 899–908.

BATTRAW, M., HALL, T. C. (1990), Histochemical analysis of CaMV 35S promoter-β-glucuronidase gene expression in transgenic rice plants, *Plant Mol. Biol.* **15,** 527–538.

BÄUMLEIN, H., MÜLLER, A. J., SCHIEMANN, J., HELBING, D., MANTEUFFEL, R., WOBUS, U. (1988), Expression of a *Vicia faba* legumin B gene in transgenic tobacco plants: gene dosage-dependent protein accumulation, *Biochem. Physiol. Pflanz.* **183,** 205–210.

BAUR, M., POTRYKUS, I., PASZKOWSKI, J. (1990), Intermolecular homologous recombination in plants, *Mol. Cell. Biol.* **10,** 492–500.

BELANGER, F. C., HEPBURN, A. G. (1990), The evolution of CpNpG methylation in plants, *J. Mol. Evol.* **30,** 26–35.

BENFEY, P. N., CHUA, N.-H. (1991), The cauliflower mosaic virus 35S promoter: combinatorial regulation of transcription in plants, *Science* **250,** 959–966.

BENFEY, P. N., REN, L., CHUA, N.-H. (1990a), Combinatorial and synergistic properties of CaMV 35S enhancer subdomains, *EMBO J.* **9,** 1685–1696.

BENFEY, P. N., REN, L., CHUA, N.-H. (1990b), Tissue-specific expression from CaMV 35S enhancer subdomains in early stages of plant development, *EMBO J.* **9,** 1677–1684.

BENITO MORENO, R. M., MACKE, F., ALWEN, A., HEBERLE-BORS, E. (1988), *In situ* seed production after pollination with *in vitro* matured, isolated pollen, *Planta* **176,** 145–148.

BENNETZEN, J. L. (1987), Covalent DNA modification and the regulation of mutator element transposition in maize, *Mol. Gen. Genet.* **208,** 45–51.

BEVAN, M. (1984), Binary *Agrobacterium* vectors for plant transformation, *Nucleid Acids Res.* **12,** 8711–8721.

BEVAN, M., FLAVELL, R. B., CHILTON, M. D. (1983), A chimaeric antibiotic resistance gene as a selectable marker for plant cell transformation, *Nature* **304,** 184–187.

BIANCHI, M. W., VIOTTI, A. (1988), DNA methylation and tissue-specific transcription of the storage protein genes of maize, *Plant Mol. Biol.* **11,** 203–214.

BINNINGER, D. M., LE CHEVANTON, L., SKRZYNIA, C., SHUBKIN, C. D., PUKKILA, P. J. (1991), Targeted transformation in *Coprinus cinereus, Mol. Gen. Genet.* **227,** 245–251.

BINNS, A. N. (1990), *Agrobacterium*-mediated gene delivery and the biology of host range limitations, *Physiol. Plant.* **79,** 135–139.

BIRD, C. R., RAY, J. A., FLETCHER, J. D., BONIWELL, J. M., BIRD, A. S., TEULIERES, C., BLAIN, I., BRAMLEY, P. M., SCHUCH, W. (1991), Using antisense RNA to study gene function: inhibition of carotenoid biosynthesis in transgenic tomatoes, *Bio/Technology* **9,** 635–639.

BLOM VAN ASSENDELFT, G., HANSCOMBE, O., GROSVELD, F., GREAVES, D. R. (1989), The β-globin dominant control region activates homologous and heterologous promoters in a tissue-specific manner, *Cell* **56,** 969–977.

BLOWERS, A. D., BOGORAD, L., SHARK, K. B., SANFORD, J. C. (1989), Studies on *Chlamydomonas* chloroplast transformation: Foreign DNA can be stably maintained in the chromosome, *Plant Cell* **1,** 123–132.

BLOWERS, A. D., ELLMORE, G. S., KLEIN, U., BOGORAD, L. (1990), Transcriptional analysis of endogenous and foreign genes in chloroplast transformants of *Chlamydomonas, Plant Cell* **2,** 1059–1070.

BOLLAG, R. J., WALDMAN, A. S., LISKAY, R. M. (1989), Homologous recombination in mammalian cells, *Annu. Rev. Genet.* **23,** 199–225.

BOLTON, G. W., NESTER, E. W., GORDON, M. P. (1986), Plant phenolic compounds induce expression of the *Agrobacterium tumefaciens* loci needed for virulence, *Science* **232,** 983–985.

BONNEVILLE, J. M., HOHN, T., PFEIFFER, P. (1988), Reverse transcription in the plant virus, cauliflower mosaic virus, in: *RNA Genetics,* Vol. 2, pp. 23–42 (DOMINGO, E., HOLLAND, J. J., AHLQUIST, P., Eds.). Boca Raton, Florida: CRC Press.

BOSTON, R. S., BEVWAR, M. R., RYAN, R. D., GOLDSBROUGH, P. B., LARKINS, B. A., HODGES, T. K. (1987), Expression from heterologous promoters in electroporated carrot protoplasts, *Plant Physiol.* **83,** 742–746.

BOULTON, M. I., BUCHHOLZ, W. G., MARKS, M. S., MARKHAM, P. G., DAVIES, J. W. (1989), Specificity of *Agrobacterium*-mediated delivery of maize streak virus DNA to members of the Gramineae, *Plant Mol. Biol.* **12,** 31–40.

BOUTRY, M., NAGY, F., POULSEN, C., AOYAGI, K., CHUA, N.-H. (1987), Targeting of bacterial chloramphenicol acetyltransferase to mitochondria in transgenic plants, *Nature* **328,** 340–342.

BOYES, J., BIRD, A. (1992), Repression of genes by DNA methylation depends on CpG density and promoter strength: evidence for involvement of a methyl-CpG binding protein, *EMBO J.* **11,** 327–333.

BOYLAN, M. T., QUAIL, P. H. (1989), Oat phytochrome is biologically active in transgenic tomatoes, *Plant Cell* **1,** 765–773.

BOYNTON, J. E., GILLHAM, N. W., HARRIS, E. H., HOSLER, J. P., JOHNSON, A. M., JONES, A. R., RANDOLPH-ANDERSON, B. L., ROBERTSON, D., KLEIN, T. M., SHARK, K. B., SANFORD, J. C. (1988), Chloroplast transformation in *Chlamydomonas* with high velocity microprojectiles, *Science* **240,** 1534–1537.

BRETTELL, R. I. S., DENNIS, E. S. (1991), Reactivation of a silent Ac following tissue culture is associated with heritable alterations in its methylation pattern, *Mol. Gen. Genet.* **229,** 365–372.

BRINSTER, R. L., CHEN, H. Y., TRUMBAUER, M., SENEAR, A. W., WARREN, R., PALMITER, R. D. (1981), Somatic expression of herpes thymidine kinase in mice following injection of a fusion gene into eggs, *Cell* **27,** 223–231.

BRISSON, N., HOHN, T. (1986), Plant virus vectors: Cauliflower mosaic virus, *Methods Enzymol.* **118,** 659–668.

BRISSON, N., PASZKOWSKI, J., PENSWICK, J. R., GRONENBORN, B., POTRYKUS, I., HOHN, T. (1984), Expression of a bacterial gene in plants by using a viral vector, *Nature* **310,** 511–514.

BROWN, D. (1989), DNA methylation in plants and its role in tissue culture, *Genome* **31,** 717–729.

BUCHMANN, I., MARNER, F.-J., SCHRÖDER, G., WAFFENSCHMIDT, S., SCHRÖDER, J. (1985), Tumour genes in plants: T-DNA encoded cytokinin biosynthesis, *EMBO J.* **4,** 853–859.

BUDAR, F., THIA-TOONG, L., VAN MONTAGU, M., HERNALSTEENS, J.-P. (1986) *Agrobacterium*-mediated gene transfer results mainly in transgenic plants transmitting T-DNA as a single Mendelian factor, *Genetics* **114,** 303–313.

BUDELIER, K. A., SMITH, A. G., GASSER, C. S. (1990), Regulation of a stylar transmitting tissue-specific gene in wild-type and transgenic tomato and tobacco, *Mol. Gen. Genet.* **224,** 183–192.

BURKE, B. (1990), The nuclear envelope and nuclear transport, *Curr. Opin. Cell Biol.* **2,** 514–520.

BYTEBIER, B., DEBOECK, F., DE GREVE, H., VAN MONTAGU, M., HERNALSTEENS, J. P. (1987), T-DNA organization in tumor cultures and transgenic plants of the monocotyledon *Asparagus officinalis, Proc. Natl. Acad. Sci. USA* **84,** 5345–5349.

CABOCHE, M. (1990), Liposome-mediated transfer of nucleic acids in plant protoplasts, *Physiol. Plant.* **79,** 193-196.

CALLIS, J., FROMM, M., WALBOT, V. (1987), Introns increase gene expression in cultured maize cells, *Genes Dev.* **1,** 1183–1200.

CAMERON, F. H., JENNINGS, P. A. (1989), Specific gene suppression by engineered ribozymes in monkey cells, *Proc. Natl. Acad. Sci. USA* **86,** 9139–9143.

CAMILLERI, C., JOUANIN, L. (1991), The TR-DNA region carrying the auxin synthesis genes of the *Agrobacterium rhizogenes* agropine-type plasmid pRiA4: nucleotide sequence analysis and introduction into tobacco plants, *Mol. Plant-Microbe Interact.* **4,** 155–162.

CANGELOSI, G. A., ANKENBAUER, R. G., NESTER, E. W. (1990), Sugars induce the *Agrobacterium* virulence genes through a periplasmic binding protein and a transmembrane signal protein, *Proc. Natl. Acad. Sci. USA* **87,** 6708–6712.

CANNON, M., PLATZ, J., O'LEARY, M., SOOKDEO, C., CANNON, F. (1990), Organ-specific modulation of gene expression in transgenic plants using antisense RNA, *Plant Mol. Biol.* **15,** 39–47.

CAO, J., WANG, Y. C., KLEIN, T. M., SANFORD, J., WU, R. (1990), Transformation of rice and maize using the biolistic process, in: *Plant Gene Transfer* (LAMB, C. J., BEACHY, R. N., Eds.), pp. 21–33, New York: Wiley-Liss.

CAPECCHI, M. R. (1980), High efficiency transformation by direct microinjection of DNA into cultured mammalian cells, *Cell* **22,** 479–488.

CAPECCHI, M. R. (1989a), Altering the genome by homologous recombination, *Science* **244,** 1288–1292.

CAPECCHI, M. R. (1989b), The new mouse genetics: altering the genome by gene targeting, *Trends Genet.* **5,** 70–76.

CAPECCHI, M. R. (1990), How efficient can you get? *Nature* **348,** 109.

CAPONE, I., SPANO, L., CARDARELLI, M., BELLINCAMPI, D., PETIT, A., COSTANTINO, P. (1989a), Induction and growth properties of carrot roots with different complements of *Agrobacterium rhizogenes* T-DNA, *Plant Mol. Biol.* **13,** 43–52.

CAPONE, I., CARDARELLI, M., TROVATO, M., COSTANTINO, P. (1989b), Upstream non-coding region which confers polar expression to Ri plasmid root inducing gene rolB, *Mol. Gen. Genet.* **216,** 239–244.

CAPONE, I., CARDARELLI, M., MARIOTTI, D., POMPONI, M., DE PAOLIS, A., COSTANTINO, P. (1991), Different promoter regions control level and tissue specificity of expression of *Agrobacterium rhizogenes* rolB gene in plants, *Plant Mol. Biol.* **16,** 427–436.

CARDARELLI, M., SPANO, L., MARIOTTI, D., MAURO, M. L., VAN SLUYS, M. A., COSTANTINO, P. (1987), The role of auxin in hairy root induction, *Mol. Gen. Genet.* **208,** 457–463.

CARRER, H., STAUB, J. M., MALIGA, P. (1991), Gentamycin resistance in *Nicotiana* conferred by AAC(3)-I, a narrow substrate specificity acetyltransferase, *Plant Mol. Biol.* **17,** 301–303.

CECH, T. R. (1987), The chemistry of self-spicing RNA and RNA enzymes, *Science* **236,** 1532–1539.

CEDAR, H., RAZIN, A. (1990), DNA methylation and development, *Biochim. Biophys. Acta* **1049,** 1–8.

CHANDLER, V. L., WALBOT, V. (1986), DNA modification of a maize transposable element correlates with loss of activity, *Proc. Natl. Acad. Sci. USA* **83,** 1767–1771.

CHAREST, P. J., HOLBROOK, L. A., GABARD, J., IYER, V. N., MIKI, B. L. (1988), *Agrobacterium*-mediated transformation of thin cell layer explants from *Brassica napus* L., *Theor. Appl. Genet.* **75,** 438–445.

CHAREST, P. J., IYER, V. N., MIKI, B. L. (1989), Factors affecting the use of chloramphenicol acetyltransferase as a marker for *Brassica* genetic transformation, *Plant Cell Rep.* **7,** 628–631.

CHIBBAR, R. N., KARTHA, K. K., LEUNG, N., QURESHI, J., CASWELL, K. (1991), Transient expression of marker genes in immature zygotic embryos of spring wheat (*Triticum aestivum*) through microprojectile bombardment, *Genome* **34,** 453–460.

CHOMET, P. S. (1991), Cytosine methylation in gene-silencing mechanisms, *Curr. Opin. Cell Biol.* **3,** 438–443.

CHOMET, P. S., WESSLER, S., DELLAPORTA, S. L. (1987), Inactivation of the maize transposable element activator (Ac) is associated with its DNA modification, *EMBO J.* **6,** 295–302.

CHRISPEELS, M. J. (1991), Sorting of proteins in the secretory system, *Annu. Rev. Plant Physiol. Plant Mol. Biol.* **42,** 21–53.

CHRISTIE, P. J., WARD, J. E., WINANS, S. C., NESTER, E. W. (1988), The *Agrobacterium tumefaciens* virE2 gene product is a single-stranded-DNA-binding protein that associates with T-DNA, *J. Bacteriol.* **170,** 2659–2667.

CHRISTIE, P. J., WARD, J. E., GORDON, M. P., NESTER, E. W. (1989), A gene required for transfer of T-DNA to plants encodes an ATPase with autophosphorylating activity, *Proc. Natl. Acad. Sci. USA* **86,** 9677–9681.

CHRISTOU, P., MCCABE, D. E., SWAIN, W. F. (1988), Stable transformation of soybean callus by DNA-coated gold particles, *Plant Physiol.* **87,** 671–674.

CHRISTOU, P., SWAIN, W. F., YANG, N.-S., MCCABE, D. E. (1989), Inheritance and expression of foreign genes in transgenic soybean plants, *Proc. Natl. Acad. Sci. USA* **86,** 7500–7504.

CHYI, Y.-S., JORGENSEN, R. A., GOLDSTEIN, D., TANKSLEY, ST. D., LOAIZA-FIGUEROA, F. (1986), Locations and stability of *Agrobacterium*-mediated T-DNA insertions in the *Lycopersicon* genome, *Mol. Gen. Genet.* **204,** 64–69.

CITOVSKY, V., DE VOS, G., ZAMBRYSKI, P. (1988), Single-stranded DNA binding protein encoded by the virE locus of *Agrobacterium tumefaciens, Science* **240,** 501–504.

CITOVSKY, V., WONG, M. L., ZAMBRYSKI, P. (1989), Cooperative interaction of *Agrobacterium* virE2 protein with single-stranded DNA: implications for the T-DNA transfer process, *Proc. Natl. Acad. Sci. USA* **86,** 1193–1197.

CLAES, B., SMALLE, J., DEKEYSER, R., VAN MONTAGU, M., CAPLAN, A. (1991), Organ-dependent regulation of a plant promoter isolated from rice by 'promoter-trapping' in tobacco, *Plant J.* **1,** 15–26.

CLOSE, T. J., TAIT, R. C., REMPEL, H. C., HIROOKA, T., KIM, L., KADO, C. I. (1987), Molecular characterization of the virC genes of the Ti plasmid, *J. Bacteriol.* **169,** 2336–2344.

COCKING, E. C., DAVEY, M. R., PENTAL, D., POWER, J. B. (1981), Aspects of plant genetic manipulation, *Nature* **293,** 265–270.

COLOT, V., ROBERT, L. S., KAVANAGH, T., BEVAN, M. W., THOMPSON, R. D. (1987), Localization of sequences in wheat endosperm protein genes which confer tissue-specific expression in tobacco, *EMBO J.* **6,** 3559–3564.

CORNELISSEN, M. (1989), Nuclear and cytoplasmic sites for anti-sense control, *Nucleic Acids Res.* **17,** 7203–7209.

CORNELISSEN, M., VANDEWIELE, M. (1989), Both RNA level and translation efficiency are reduced by anti-sense RNA in transgenic tobacco, *Nucleic Acids Res.* **17,** 833–843.

COTTEN, M., BIRNSTIEL, M. L. (1989), Ribozyme mediated destruction of RNA *in vivo, EMBO J.* **8,** 3861–3866.

CROSSWAY, A., OAKES, J. V., IRVINE, J. M., WARD, B., KNAUF, V. C., SHEWMAKER, C. K. (1986), Integration of foreign DNA following microinjection of tobacco mesophyll protoplasts, *Mol. Gen. Genet.* **202,** 179–185.

CUOZZO, M., O'CONNELL, K. M., KANIEWSKI, W., FANG, R.-X., CHUA, N.-H., TUMER, N. E. (1988), Viral protection in transgenic tobacco plants expressing the cucumber mosaic virus coat protein or its antisense RNA, *Bio/Technology* **6,** 549–557.

CZERNILOFSKY, A. P., HAIN, R., BAKER, B., WIRTZ U. (1986a), Studies of the structure and functional organization of foreign DNA integrated into the genome of *Nicotiana tabacum, DNA* **5,** 473–482.

CZERNILOFSKY, A. P., HAIN, R., HERRERA-ESTRELLA, L., LÖRZ, H., GOYVAERTS, E., BAKER, B., SCHELL, J. (1986b), Fate of selectable marker DNA integrated into the genome of *Nicotiana tabacum, DNA* **5,** 101–113.

DAHL, G. A., TEMPÉ, J. (1983), Studies on the use of toxic precursor analogues of opines to select transformed plant cells, *Theor. Appl. Genet.* **66,** 233–239.

DALE, P. J., MARKS, M. S., BROWN, M. M., WOOLSTON, C. J., GUNN, H. V., MULLINEAUX, P. M., LEWIS, D. M., KEMP, J. M., CHEN, D. F., GILMOUR, D. M., FLAVELL, R. B. (1989), Agroinfection of wheat: inoculation of *in vitro* grown seedlings and embryos, *Plant Sci.* **63,** 237–245.

DALE, E.C., OW., D. W. (1990), Intra- and intermolecular site-specific recombination in plant cells mediated by bacteriophage P1 recombinase, *Gene* **91,** 79–85.

DALE, E. C., OW, D. W. (1991), Gene transfer with subsequent removal of the selection gene from the host genome, *Proc. Natl. Acad. Sci. USA* **88,** 10558–10562.

DAMIANI, F., ARCIONI, S. (1991), Transformation of *Medicago arborea* L. with an *Agrobacterium rhizogenes* binary vector carrying the hygromycin resistance gene, *Plant Cell Rep.* **10,** 300–303.

DANIELL, H., VIVEKANANDA, J., NIELSEN, B. L., YE, G. N., TEWARI, K. K., SANFORD, J. C. (1990), Transient foreign gene expression in chloroplasts of cultured tobacco cells after biolistic delivery of chloroplast vectors, *Proc. Natl. Acad. Sci. USA* **87,** 88–92.

DANIELL, H., KRISHNAN, M., MCFADDEN, B. F. (1991), Transient expression of β-glucuronidase in different cellular compartments following biolistic delivery of foreign DNA into wheat leaves and calli, *Plant Cell Rep.* **9,** 615–619.

DAS, A. (1988), *Agrobacterium tumefaciens* virE operon encodes a single-stranded DNA-binding protein, *Proc. Natl. Acad. Sci. USA* **85,** 2909–2913.

DATTA, S. K., PETERHANS, A., DATTA, K., POTRYKUS, I. (1990), Genetically engineered fertile indica-rice recovered from protoplasts, *Bio/ Technology* **8,** 736–740.

DAUBERT, S., SHEPERD, R., GARDNER, R.C. (1983), Insertional mutagenesis of the cauliflower mosaic virus genome, *Gene* **25,** 201–208.

DAVIES, J. W., STANLEY, J. (1989), Geminivirus genes and vectors, *TIG* **5,** 77–81.

DAWSON, W. O., BECK, D. L., KNORR, D. A., GRANTHAM, G. L. (1986), cDNA cloning of the complete genome of tobacco mosaic virus and production of infectious transcripts, *Proc. Natl. Acad. Sci. USA* **83,** 1832–1836.

DAWSON, W. O., LEWANDOWSKI, D. J., HILF, M. E., BUBRICK, P., RAFFO, A. J., SHAW, J. J., GRANTHAM, G. I., DESJARDINS, P. R. (1989), A tobacco mosaic virus-hybrid expresses and looses an added gene, *Virology* **172,** 285–292.

DAY, A. G., BEJARANO, E. R., BUCK, K. W., BURRELL, M., LICHTENSTEIN, C. P. (1991), Expression of an antisense viral gene in transgenic tobacco confers resistance to the DNA virus tomato golden mosaic virus, *Proc. Natl. Acad. Sci. USA* **88,** 6721–6725.

DEAN, C., JONES, J., FAVREAU, M., DUNSMUIR, P., BEDBROOK, J. (1988), Influence of flanking sequences on variability in expression levels of an introduced gene in transgenic tobacco plants, *Nucleic Acids Res.* **16,** 9267–9283.

DEBLAERE, R., BYTEBIER, B., DE GREVE, H., DEBOECK, F., SCHELL, J., VAN MONTAGU, M., LEEMANS, J. (1985), Efficient octopine Ti plasmid-derived vectors for *Agrobacterium*-mediated gene transfer to plants, *Nucleic Acids Res.* **13,** 4777–4788.

DE BLOCK, M., HERRERA-ESTRELLA, L., VAN MONTAGU, M., SCHELL, J., ZAMBRYSKI, P. (1984), Expression of foreign genes in regenerated plants and in their progeny, *EMBO J.* **3,** 1681–1689.

DE BLOCK, M., SCHELL, J., VAN MONTAGU, M. (1985), Chloroplast transformation by *Agrobacterium tumefaciens, EMBO J.* **4,** 1367–1372.

DE BLOCK, M., BOTTERMAN, J., VANDEWIELE, M., DOCKX, J., THOEN, C., GOSSELÉ, V., RAO MOVVA, N., THOMPSON, C., VAN MONTAGU, M., LEEMANS, J. (1987), Engineering herbicide resistance in plants by expression of a detoxifying enzyme, *EMBO J.* **6,** 2513–2519.

DE CLEENE, M. (1985), The susceptibility of monocotyledons to *Agrobacterium tumefaciens, Phytopathol. Z.* **113,** 81–89.

DE CLEENE, M., DE LEY, J. (1976), The host range of crown gall, *Bot. Rev.* **42,** 389–466.

DE CLEENE, M., DE LEY, J. (1981), The host range of infectious hairy-root, *Bot. Rev.* **47,** 147–194.

DEKEYSER, R. A., CLAES, B., DE RYCKE, R. M. U., HABETS, M. E., VAN MONTAGU, M. C., CAPLAN, A. B. (1990), Transient gene expression in intact and organized rice tissues, *Plant Cell* **2,** 591–602.

DE LA PENA, A., LÖRZ, H., SCHELL, J. (1987), Transgenic rye plants obtained by injecting DNA into young floral tillers, *Nature* **325,** 274–276.

DELAUNEY, A. J., TABEIZADEH, Z., VERMA, D. P. S. (1988), A stable bifunctional antisense transcript inhibiting gene expression in transgenic plants, *Proc. Natl. Acad. Sci. USA* **85,** 4300–4304.

DEROLES, S. C., GARDNER, R. C. (1988a), Expression and inheritance of kanamycin resistance in a large number of transgenic petunias generated by *Agrobacterium*-mediated transformation, *Plant Mol. Biol.* **11,** 355–364.

DEROLES, S. C., GARDNER, R. C. (1988b), Analysis of the T-DNA structure in a large number of transgenic petunias generated by *Agrobacterium*-mediated transformation, *Plant Mol. Biol.* **11,** 365–377.

DESHAYES, A., HERRERA-ESTRELLA, L., CABOCHE, M. (1985), Liposome-mediated transformation of tobacco mesophyll protoplasts by an *Escherichia coli* plasmid, *EMBO J.* **4,** 2731–2737.

DE VOS, G., ZAMBRYSKI, P. (1989), Expression of *Agrobacterium* nopaline-specific VirD1, VirD2, and VirC1 proteins and their requirement for T-strand production in *E. coli, Mol. Plant-Microbe Interact.* **2,** 43–52.

DE ZOETEN, G. A., PENSWICK, J. R., HORISBERGER, M. A., AHL, P., SCHULZE, M., HOHN, T. (1989), The expression, localization and effect of a human interferon in plants, *Virology* **172,** 213–222.

DHAR, M. S., PETHE, V. V., GUPTA, V. S., RANJEKAR, P. K. (1990), Predominance and tissue specificity of adenine methylation in rice, *Theor. Appl. Genet.* **80,** 402–408.

DIXON, L., NYFFENEGGER, T., DELLEY, G., MARTINEZ-IZQUIERDO, J., HOHN, T. (1986), Evidence for replicative recombination in cauliflower mosaic virus, *Virology* **150,** 463–468.

DOMMISSE, E. M., LEUNG, D. W. M., SHAW, M. L., CONNER, A. J. (1990) Onion is a monocotyledonous host for *Agrobacterium, Plant Sci.* **69,** 249–257.

DONSON, J., GUNN, H. V., WOOLSTON, C. J., PINNER, M. S., BOULTON, M. I., MULLINEAUX, P. M., DAVIES, J. W. (1988), *Agrobacterium*-mediated infectivity of cloned digitaria streak virus DNA, *Virology* **162,** 248–250.

DOUGLAS, C. J., STANELONI, R. J., RUBIN, R. A., NESTER, E. W. (1985), Identification and genetic analysis of an *Agrobacterium tumefaciens* chromosomal virulence region, *J. Bacteriol.* **161,** 850–860.

DÜRRENBERGER, F., CRAMERI, A., HOHN, B., KOUKOLIKOVA-NICOLA, Z. (1989), Covalently bound virD2 protein of *Agrobacterium tumefaciens* protects the T-DNA from exonucleolytical degradation, *Proc. Natl. Acad. Sci. USA* **86,** 9154–9158.

DYNAN, W. S. (1989), Understanding the molecular mechanism by which methylation influences gene expression, *Trends Genet.* **5,** 35–36.

EAPEN, S., KÖHLER, F., GERDEMANN, M., SCHIEDER, O. (1987), Cultivar dependence of transformation rates in moth bean after co-cultivation of protoplasts with *Agrobacterium tumefaciens, Theor. Appl. Genet.* **75,** 207–210.

EBERT, P. R., HA, S. B., AN, G. (1987), Identification of an essential upstream element in the nopaline synthase promoter by stable and transient assays, *Proc. Natl. Acad. Sci. USA* **84,** 5745–5749.

ECKER, J. R., DAVIS, R.W. (1986), Inhibition of gene expression in plant cells by expression of antisense RNA, *Proc. Natl. Acad. Sci. USA* **83,** 5372–5376.

ECKES, P., ROSAHL, S., SCHELL, J., WILLMITZER, L. (1986), Isolation and characterization of a light-inducible, organ-specific gene from potato and analysis of its expression after tagging and transfer into tobacco and potato shoots, *Mol. Gen. Genet.* **205,** 14–22.

ECKES, P., SCHMITT, P., DAUB, W., WENGENMAYER, F. (1989), Overproduction of alfalfa glutamine synthetase in transgenic tobacco plants, *Mol. Gen. Genet.* **217,** 263–268.

EICHHOLTZ, D. A., ROGERS, S. G., HORSCH, R. B., KLEE, H. J., HAYFORD, M., HOFFMANN, N. L., BRAFORD, S. B., FINK, C., FLICK, J., O'CONNELL, K. M., FRALEY, R. T. (1987), Expression of mouse dihydrofolate reductase gene confers methotrexate resistance in transgenic *Petunia* plants, *Somat. Cell Mol. Genet.* **13,** 67–76.

ELLIS, D. D., MCCABE, D., RUSSELL, D., MARTINELL, B., MCCOWN, B. H. (1991), Expression of inducible angiosperm promoters in a gymnosperm, *Picea glauca* (white spruce), *Plant Mol. Biol.* **17,** 19–27.

ELLIS, J. G., LLEWELLYN, D. J., DENNIS, E. S., PEACOCK, W. J. (1987), Maize Adh-1 promoter sequences control anaerobic regulation: addition of upstream promoter elements from constitutive genes is necessary for expression in tobacco, *EMBO J.* **6,** 11–16.

ESTRUCH, J. J., CHRIQUI, D., GROSSMANN, K., SCHELL, J., SPENA, A. (1991 a), The plant oncogene rolC is responsible for the release of cytokinins from glucoside conjugates, *EMBO J.* **10,** 2889–2895.

ESTRUCH, J. J., SCHELL, J., SPENA, A. (1991 b), The protein encoded by the plant rolB oncogene hydrolyses indole glucosides, *EMBO J.* **10,** 3125–3128.

ETESSAMI, P., CALLIS, R., ELLWOOD, S., STANLEY, J. (1988), Delimination of essential genes of cassava latent virus DNA 2, *Nucleic Acids Res.* **16,** 4811–4829.

FELDMANN, K. A. (1991), T-DNA insertion mutagenesis in *Arabidopsis:* mutational spectrum, *Plant J.* **1,** 71–82.

FELDMANN, K. A., MARKS, M. D. (1987), *Agrobacterium*-mediated transformation of germinating seeds of *Arabidopsis thaliana:* a non-tissue culture approach, *Mol. Gen. Genet.* **208,** 1–9.

FELDMANN, K. A., MARKS, M. D., CHRISTIANSON, M. L., QUATRANO, R. S. (1989), A dwarf mutant of *Arabidopsis* generated by T-DNA insertion mutagenesis, *Science* **243,** 1351–1354.

FELGNER, P. L., RINGOLD, G. M. (1989), Cationic liposome-mediated transfection, *Nature* **337,** 387–388.

FELGNER, P. L., GADEK, T. R., HOLM, M., ROMAN, R., CHAN, H. V., WENZ, M., NORTHROP, J. P., RINGOLD, G. M., DANIELSON, H. (1987), Lipofection, a highly efficient, lipid-mediated DNA transfection procedure, *Proc. Natl. Acad. Sci. USA* **84,** 7413–7417.

FINER, J. J., MCMULLEN, M. D. (1990), Transformation of cotton (*Gossypium hirsutum* L.) via particle bombardment, *Plant Cell Rep.* **8,** 586–589.

FLAVELL, R. B., O'DELL, M., THOMPSON, W. F. (1988), Regulation of cytosine methylation in ribosomal DNA and nucleolus organizer expression in wheat, *J. Mol. Biol.* **204,** 523–534.

FOBERT, P. R., MIKI, B. L., IYER, V. N. (1991), Detection of gene regulatory signals in plants revealed by T-DNA mediated fusions, *Plant Mol. Biol.* **17,** 837–851.

FOX, T. D., SANFORD, J. C., MCMULLIN, T. W. (1988), Plasmids can stably transform yeast mitochondria lacking endogenous mtDNA, *Proc. Natl. Acad. Sci. USA* **85,** 7288–7292.

FRALEY, R. T., PAPAHADJOPOULOS, D. (1982), Liposomes: The development of a new carrier system for introducing nucleic acids into plant and animal cells, *Curr. Tech. Microbiol.* **96,** 171–191.

FRALEY, R. T., ROGERS, S. G., HORSCH, R. B., SANDERS, P. R., FLICK, J. S., ADAMS, S. P., BITTNER, M. L., BRAND, L. A., FINK, C. L., FRY, J. S., GALLUPPI, G. R., GOLDBERG, S. B., HOFFMANN, N. L., WOO, S. C. (1983), Expression of bacterial genes in plant cells, *Proc. Natl. Acad. Sci. USA* **80**, 4803–4807.

FRALEY, R. T., HORSCH, R. B., MATZKE, A. J. M., CHILTON, M. D., CHILTON, W. S., SANDERS, P. R. (1984), *In vitro* transformation of petunia cells by an improved method of co-cultivation with *A. tumefaciens* strains, *Plant Mol. Biol.* **3**, 371–378.

FRALEY, R. T., ROGERS, S. G., HORSCH, R. B., EICHHOLTZ, D. A., FLICK, J. S., FINK, C. L., HOFFMANN, N. L., SANDERS, P. R. (1985), The SEV system: A new disarmed Ti plasmid vector system for plant transformation, *Bio/Technology* **3**, 629–635.

FRALEY, R. T., ROGERS, S. G., HORSCH, R. B. (1986), Genetic transformation in higher plants, *CRC Crit. Rev. Plant Sci.* **4**, 1–46.

FRENCH, R., AHLQUIST, P. (1988), Characterization and engineering of sequences controlling *in vivo* synthesis of brome mosaic virus subgenomic RNA, *J. Virol.* **62**, 2411–2420.

FRENCH, R., JANDA, M., AHLQUIST, P. (1986), Bacterial gene inserted in an engineered RNA virus: efficient expression in monocotyledonous plant cells, *Science* **231**, 1294–1297.

FROHBERG, C., HEINS, L., GATZ, C. (1991), Characterization of the interaction of plant transcription factors using a bacterial repressor protein, *Proc. Natl. Acad. Sci. USA* **88**, 10470–10474.

FROMM, M. E., TAYLOR, L. P., WALBOT, V. (1986), Stable transformation of maize after gene transfer by electroporation, *Nature* **319**, 791–793.

FROMM, M., CALLIS, J., TAYLOR, L. P., WALBOT, V. (1987), Electroporation of DNA and RNA into plant protoplasts, *Methods Enzymol.* **153**, 351–366.

FROMM, M. E., MORRISH, F., ARMSTRONG, C., WILLIAMS, R., THOMAS, J., KLEIN, T. M. (1990), Inheritance and expression of chimeric genes in the progeny of transgenic maize plants, *Bio/Technology* **8**, 833–839

FURNER, I. J., HIGGINS, E. S., BERRINGTON, A. W. (1989), Single-stranded DNA transforms plant protoplasts, *Mol. Gen. Genet.* **220**, 65–68.

FÜTTERER, J. BONNEVILLE, J. M., HOHN, T. (1990), Cauliflower mosaic virus as a gene expression vector for plants, *Physiol. Plant.* **79**, 154–157.

GAD, A. E., ROSENBERG, N., ALTMAN, A. (1990), Liposome-mediated gene delivery into plant cells, *Physiol. Plant.* **79**, 177–183.

GALLIE, D. R., SLEAT, D. E., WATTS, J. W., TURNER, P. C., WILSON, T. M. A. (1987), *In vivo* uncoating and efficient expression of foreign mRNAs packaged in TMV-like particles, *Science* **236**, 1122–1124.

GARDINER, W. E., SUNTER, G., BRAND, L., ELMER, J. S., ROGERS, S. G., BISARO, D. M. (1988), Genetic analysis of tomato golden mosaic virus: the coat protein is not required for systemic spread or symptom development, *EMBO J.* **7**, 899-904.

GARDNER, R. C., CHONOLES, K. R., OWENS, R. A. (1986), Potato spindle tuber viroid infections mediated by the Ti plasmid of *Agrobacterium tumefaciens, Plant Mol. Biol.* **6**, 221–228.

GATZ, C., QUAIL, P. H. (1988), Tn10-encoded tet repressor can regulate an operator-containing plant promoter, *Proc. Natl. Acad. Sci. USA* **85**, 1394–1397.

GATZ, C., KAISER, A., WENDENBURG, R. (1991), Regulation of a modified CaMV 35S promoter by the Tn10-encoded tet repressor in transgenic tobacco, *Mol. Gen. Genet.* **227**, 229–237.

GELVIN, S. B. (1990), Crown gall disease and hairy root disease. A sledgehammer and a tackhammer, *Plant Physiol.* **92**, 281–285

GELVIN, S. B., KARCHER, S. J., DIRITA, V. J. (1983), Methylation of the T-DNA in *Agrobacterium tumefaciens* and in several crown gall tumors, *Nucleic Acids Res.* **11**, 159–174.

GENDLOFF, E. H., BOWEN, B., BUCHHOLZ, W. G. (1990), Quantitation of chloramphenicol acetyl transferase in transgenic tobacco plants by ELISA and correlation with gene copy number, *Plant Mol.Biol.* **14**, 575–583.

GHAI, J., DAS, A. (1989), The virD operon of *Agrobacterium tumefaciens* Ti plasmid encodes a DNA-relaxing enzyme, *Proc. Natl. Acad. Sci. USA* **86**, 3109–3113.

GHEYSEN, G., VAN MONTAGU M., ZAMBRYSKI, P. (1987), Integration of *Agrobacterium tumefaciens,* transfer DNA (T-DNA) involves rearrangements of target plant DNA sequences, *Proc. Natl. Acad. Sci. USA* **84**, 6169–6173.

GIDONI, D., BOND-NUTTER, D., BROSIO, P., JONES, J., BEDBROOK, J., DUNSMUIR, P. (1988), Coordinated expression between two photosynthetic petunia genes in transgenic plants, *Mol. Gen. Genet.* **211**, 507–514.

GIETL, C., KOUKOLIKOVÁ-NICOLÁ, Z., HOHN, B. (1987), Mobilization of T-DNA from *Agrobacterium* to plant cells involves a protein that binds single-stranded DNA, *Proc. Natl. Acad. Sci. USA* **84**, 9006–9010.

GOLDFARB, D. S. (1989), Nuclear transport, *Curr. Opin. Cell. Biol.* **1**, 441–446.

GOLDSBROUGH, A., BEVAN, M. (1991), New patterns of gene activity in plants detected using an *Agrobacterium* vector, *Plant Mol. Biol.* **16**, 263–269.

GOLDSCHMIDT-CLERMONT, M. (1991), Transgenic expression of aminoglycoside adenine transferase in the chloroplast: a selectable marker for site-directed transformation of *Chlamydomonas, Nucleic Acids Res.* **19,** 4083–4089.

GOODALL, G. J., FILIPOWICZ, W. (1989), The AU-rich sequences present in the introns of plant nuclear pre-RNAs are required for splicing, *Cell* **58,** 473–483.

GORDON-KAMM, W. J., SPENCER, T. M., MANGANO, M. L., ADAMS, T. R., DAINES, R. J., START, W. G., O'BRIEN, J. V., CHAMBERS, S. A., ADAMS, W. R., WILLETTS, N. G., RICE, T. B., MACKEY, C. J., KRUEGER, R. W., KAUSCH, A. P., LEMAUX, P. G. (1990), Transformation of maize cells and regeneration of fertile transgenic plants, *Plant Cell* **2,** 603–618.

GORING, D. R., THOMSON, L., ROTHSTEIN, S. J. (1991), Transformation of a partial nopaline synthase gene into tobacco suppresses the expression of a resident wild-type gene, *Proc. Natl. Acad. Sci. USA* **88,** 1770–1774.

GORMAN, C. M., MOFFAT, L. F., HOWARD, B. H. (1982), Recombinant genomes which express chloramphenicol acetyltransferase in mammalian cells, *Mol. Cell. Biol.* **2,** 1044–1051.

GOULD, J., DEVEY, M., HASEGAWA, O., ULIAN, E. C., PETERSON, G., SMITH, R. H. (1991), Transformation of *Zea mays* L. using *Agrobacterium tumefaciens* and the shoot apex, *Plant Physiol.* **95,** 426–434.

GREEN, P. J., PINES, O., INOUYE, M. (1986), The role of antisense RNA in gene regulation, *Annu. Rev. Biochem.* **55,** 569–597.

GRIERSON, D., FRAY, R. G., HAMILTON, A. J., SMITH, C. J. S., WATSON, C. F. (1991), Does co-suppression of sense genes in transgenic plants involve antisense RNA? *Trends Biotechnol.* **9,** 122–123.

GRIMSLEY, N., HOHN, B., HOHN, T., WALDEN, R. (1986a), "Agroinfection", an alternative route for viral infection of plants by using the Ti plasmid, *Proc. Natl. Acad. Sci. USA* **83,** 3282–3286.

GRIMSLEY, N., HOHN, T., HOHN, B. (1986b), Recombination in a plant virus: template-switching in cauliflower mosaic virus, *EMBO J.* **5,** 641–646.

GRIMSLEY, N., HOHN, T., DAVIES, J. W., HOHN, B. (1987), *Agrobacterium*-mediated delivery of infectious maize streak virus into maize plants, *Nature* **325,** 177–179.

GRIMSLEY, N., RAMOS, C., HEIN, T., HOHN, B. (1988), Meristematic tissues of maize plants are most susceptible to agroinfection with maize streak virus, *Bio/Technology* **6,** 185–189.

GRIMSLEY, N., HOHN, B., RAMOS, C., KADO, C., ROGOWSKY, P. (1989), DNA transfer from *Agrobacterium* to *Zea mays* or *Brassica* by agroinfection is dependent on bacterial virulence functions, *Mol. Gen. Genet.* **217,** 309–316.

GRONENBORN, B., MATZEIT, V. (1989), Plant gene vectors and genetic transformation: Plant viruses as vectors, in: *Cell and Culture and Stomatic Cell Genetics of Plants* (SCHELL, J., VASIL, J. K., Eds.), Vol. 6, pp. 69–100, New York: Academic Press. 715006-4.

GRONENBORN, B., GARDNER, R. C., SCHAEFER, S., SHEPHERD, R. J. (1981), Propagation of foreign DNA in plants using cauliflower mosaic virus as vector, *Nature* **294,** 773–776.

GRONENBORN, B., GARDNER, R. C., SCHAEFER, S., SHEPHERD, R. J. (1987), The molecular biology of cauliflower mosaic virus and its application as plant gene vector, in: *Plant DNA Infectious Agents* (HOHN, T., SCHELL, J., Eds.), pp. 1–29, Wien: Springer-Verlag.

GROSVELD, F., VAN ASSENDELFT, G. B., GREAVES, D. R., KOLLIAS, G. (1987), Position-independent, high-level expression of the human β-globin gene in transgenic mice, *Cell* **51,** 975–985.

GRUENBAUM, Y., NAVEH-MANY, T., CEDAR, H., RAZIN, A. (1981), Sequence specificity of methylation in higher plant DNA, *Nature* **292,** 860–862.

GUERCHE, P., JOUANIN, L., TEPFER, D., PELLETIER, G. (1987), Genetic transformation of oilseed rape *(Brassica napus)* by the Ri T-DNA of *Agrobacterium rhizogenes* and analysis of inheritance of the transformed phenotype, *Mol. Gen. Genet.* **206,** 382–386.

GUERINEAU, F., BROOKS, L., MEADOWS, J., LUCY, A., ROBINSON, C., MULLINEAUX, P. (1990), Sulfonamide resistance for plant gene transformation, *Plant Mol. Biol.* **15,** 127–136.

HAGIO, T., BLOWERS, A. D., EARLE, E. D. (1991), Stable transformation of sorghum cell cultures after bombardment with DNA-coated microprojectiles, *Plant Cell Rep.* **10,** 260–264.

HAIN, R., STABEL, P., CZERNILOFSKY, A. P., STEINBIß, H. H., HERRERA-ESTRELLA, L., SCHELL, J. (1985), Uptake, integration, expression and genetic transmission of a selectable chimaeric gene by plant protoplasts, *Mol. Gen. Genet.* **199,** 161–168.

HALFTER, U., MORRIS, P. C., WILLMITZER, L. (1992), Gene targeting in *Arabidopsis thaliana, Mol. Gen. Genet.* **231,** 186–193.

HAMILL, J. D., ROBINS, R. J., PARR, A. J., EVANS, D. M., FURZE, J. M., RHODES, M. J. C. (1990), Over-expressing a yeast ornithine decarboxylase gene in transgenic roots of *Nicotiana rustica* can lead to enhanced nicotine accumulation, *Plant Mol. Biol.* **15,** 27–38.

HAMILTON, A. J., LYCETT, G. W., GRIERSON, D. (1990), Antisense gene that inhibits synthesis of the hormone ethylene in transgenic plants, *Nature* **346**, 284–287.

HÄNISCH TEN CATE, C. H., LOONEN, A. E. H. M., OTTAVIANI, M. P., ENNIK, L., VAN ELDIK, G., STIEKEMA, W. J. (1990), Frequent spontaneous deletions of Ri T-DNA in *Agrobacterium rhizogenes* transformed potato roots and regenerated plants, *Plant Mol. Biol.* **14**, 735–741.

HARTL, F. U., NEUPERT, W. (1990), Protein sorting to mitochondria, evolutionary conservations of folding and assembly, *Science* **247**, 930–938.

HASELOFF, J., GERLACH, W. L. (1988), Simple RNA enzymes with new and highly specific endoribonuclease activities, *Nature* **334**, 585–591.

HAUPTMANN, R. M., OZIAS-AKINS, P., VASIL, V., TABAEIZADEH, Z., ROGERS, S. G., HORSCH, R. B., VASIL, I. K., FRALEY, R. T. (1987), Transient expression of electroporated DNA in monocotyledonous and dicotyledonous species, *Plant Cell Rep.* **6**, 265–270.

HAYASHIMOTO, A., LI, Z., MURAI, N. (1990), A polyethylene glycol-mediated protoplast transformation system for production of fertile transgenic rice plants, *Plant Physiol.* **93**, 857–863.

HAYES, R. J., PETTY, I. T. D., COUTTS, R. H. A., BUCK, K. W. (1988a), Gene amplification and expression in plants by a replicating geminivirus vector, *Nature* **334**, 179–182.

HAYES, R. J., MACDONALD, H., COUTSS, R. H. A., BUCK, K. W. (1988b), Agroinfection of *Triticum aestivum* with cloned DNA of wheat dwarf virus, *J. Gen. Virol.* **69**, 891–896.

HAYFORD, M. B., MEDFORD, J. I., HOFFMAN, N. L., ROGERS, S. G., KLEE, H. J. (1988), Development of a plant transformation selection system based on expression of genes encoding gentamicin acetyltransferases, *Plant Physiol.* **86**, 1216–1222.

HÉLÈNE, C., TOULMÉ, J.-J. (1990), Specific regulation of gene expression by antisense, sense and antigene nucleic acids, *Biochim. Biophys. Acta* **1049**, 99–125.

HELMER, G., CASADABAN, M., BEVAN, M., KAYES, L., CHILTON, M. D. (1984), A new chimeric gene as a marker for plant transformation: the expression of *Escherichia coli* β-galactosidase in sunflower and tobacco cells, *Bio/Technology* **2**, 520–527.

HEMENWAY, C., FANG, R.-X., KANIEWSKI, W. K., CHUA, N.-H., TUMER, N. E. (1988), Analysis of the mechanism of protection in transgenic plants expressing the potato virus X coat protein or its antisense RNA, *EMBO J.* **7**, 1273–1280.

HERMAN, L., JACOBS, A., VAN MONTAGU, M., DEPICKER, A. (1990), Plant chromosome/marker gene fusion assay for study of normal and truncated T-DNA integration events, *Mol. Gen. Genet.* **224**, 248–256.

HERNALSTEENS, J. P., VAN VLIET, F., DE BEUCKELEER, M., DEPICKER, A., ENGLER, G., LEMMERS, M., HOLSTERS, M., VAN MONTAGU, M., SCHELL, J. (1980), The *Agrobacterium tumefaciens* Ti plasmid as a host vector system for introducing foreign DNA in plant cells, *Nature* **287**, 654–656.

HERRERA-ESTRELLA, A., CHEN, Z.-M., VAN MONTAGU, M., WANG, K. (1988), VirD proteins of *Agrobacterium tumefaciens* are required for the formation of a covalent DNA-protein-complex at the 5′ terminus of T-strand molecules, *EMBO J.* **7**, 4055–4062.

HERRERA-ESTRELLA, A., VAN MONTAGU, M., WANG, K. (1990), A bacterial peptide acting as a plant nuclear targeting signal: the amino-terminal portion of *Agrobacterium* VirD2 protein directs a β-galactosidase fusion protein into tobacco nuclei, *Proc. Natl. Acad. Sci. USA* **87**, 9534–9537.

HERRERA-ESTRELLA, L., DEPICKER, A., VAN MONTAGU, M., SCHELL, J. (1983a), Expression of chimeric genes transferred into plant cells using a Ti-plasmid-derived vector, *Nature* **303**, 209–213.

HERRERA-ESTRELLA, L., DE BLOCK, M., MESSENS, E., HERNALSTEENS, J. P., VAN MONTAGU, M., SCHELL, J. (1983b), Chimeric genes as dominant selectable markers in plant cells, *EMBO J.* **2**, 987–995.

HESS, D. (1987), Pollen based techniques in genetic manipulation, *Int. Rev. Cytol.* **107**, 169–190.

HESS, K. M., DUDLEY, M. W., LYNN, D. G., JOERGER, R. D., BINNS, A. N. (1991), Mechanism of phenolic activation of *Agrobacterium* virulence genes: development of a specific inhibitor of bacterial sensor/response systems, *Proc. Natl. Acad. Sci. USA* **88**, 7854–7858.

HIATT, A. (1990), Antibodies produced in plants, *Nature* **344**, 469–470.

HIATT, A., CAFFERKEY, R., BOWDISH, K. (1989), Production of antibodies in transgenic plants, *Nature* **342**, 76–78.

HILLE, J., VERHEGGEN, F., ROELVINK, P., FRANSSEN, H., VAN KAMMEN, A., ZABEL, P. (1986), Bleomycin resistance: a new dominant selectable marker for plant cell transformation, *Plant Mol. Biol.* **7**, 171–176.

HINCHEE, M. A. W., CONNOR-WARD, D. V., NEWELL, C. A., MCDONNELL, R. E., SATO, S. J., GASSER, C. S., FISCHHOFF, D. A., RE, D. A., FRALEY, R. T., HORSCH, R. B. (1988), Production of transgenic soybean plants using *Agrobacterium*-mediated DNA transfer, *Bio/Technology* **6**, 915–922.

HIRAYAMA, T., MURANAKA, T., OHKAWA, H., OKA, A. (1988), Organization and characterization of the virCD genes from *Agrobacterium rhizogenes, Mol. Gen. Genet.* **213**, 229–237.

HOBBS, S. L. A., KPODAR, P., DELONG, C. M. O. (1990), The effect of T-DNA copy number, position and methylation on reporter gene expression in tobacco transformants, *Plant Mol. Biol.* **15**, 851–864.

HOEKEMA, A., HIRSCH, P. R., HOOYKAAS, P. J. J., SCHILPEROORT, R. A. (1983), A binary plant vector strategy based on separation of vir- and T-region of the *Agrobacterium tumefaciens* Ti-plasmid, *Nature* **303**, 179–180.

HOEKEMA, A., ROELVINK, P.W., HOOYKAAS, P. J. J., SCHILPEROORT, R. A. (1984), Delivery of T-DNA from the *Agrobacterium tumefaciens* chromosome into plant cells, *EMBO J.* **3**, 2485–2490.

HOEKEMA, A., VAN HAAREN, M. J. J., FELLINGER, A. J., HOOYKAAS, P. J. J., SCHILPEROORT, R. A. (1985), Non-oncogenic plant vectors for use in the *Agrobacterium* binary system, *Plant Mol. Biol.* **5**, 85–89.

HOFMANN, D., ZENTGRAF, H., KAHL, G. (1989), *In vitro* nucleosome assembly with plant histones, *FEBS Lett.* **256**, 123–127.

HOHN, T., HOHN, B., PFEIFFER, P. (1985), Reverse transcription in CaMV, *Trends Biochem. Sci.* **10**, 205–209.

HOOYKAAS, P. J. J. (1989), Transformation of plant cells via *Agrobacterium, Plant Mol. Biol.* **13**, 327–336.

HOOYKAAS, P. J. J., HOFKER, M., DEN DULK-RAS, H., SCHILPEROORT, R. A. (1984), A comparison of virulence determinants in an octopine Ti plasmid, a nopaline Ti plasmid, and an Ri plasmid by complementation analysis of *Agrobacterium tumefaciens* mutants, *Plasmid* **11**, 195–205.

HOOYKAAS-VAN SLOGTEREN, G. M. S., HOOYKAAS, P. J. J., SCHILPEROORT, R. A. (1984), Expression of Ti plasmid genes in monocotyledonous plants infected with *Agrobacterium tumefaciens, Nature* **311**, 763–764.

HORSCH, R. B., FRALEY, R. T., ROGERS, S. G., SANDERS, P. R., LLOYD, A., HOFFMANN, N. (1984), Inheritance of functional foreign genes in plants, *Science* **223**, 496–498.

HORSCH, R. B., FRY, J. E., HOFFMANN, N. L., EICHHOLTZ, D. E., ROGERS, S. G., FRALEY, R. T. (1985), A simple and general method for transferring genes into plants, *Science* **227**, 1229–1231.

HORTH, M., NEGRUTIU, I., BURNY, A., VAN MONTAGU, M., HERRERA-ESTRELLA, L. (1987), Cloning of a *Nicotiana plumbaginifolia* protoplast-specific enhancer-like sequence, *EMBO J.* **6**, 2525–2530.

HOWARD, E., CITOVSKY, V. (1990), The emerging structure of the *Agrobacterium* T-DNA transfer complex, *BioEssays* **12**, 103–108.

HOWARD, E. A., WINSOR, B. A., DE VOS, G., ZAMBRYSKI, P. (1989), Activation of the T-DNA transfer process in *Agrobacterium* results in the generation of a T-strand-protein complex: tight association of virD2 with the 5′ ends of T-strands, *Proc. Natl. Acad. Sci. USA* **86**, 4017–4021.

HUANG, J., HACK, E., THORNBURG, R. W., MYERS, A. M. (1990a), A yeast mitochondrial leader peptide functions *in vivo* as a dual targeting signal for both chloroplasts and mitochondria, *Plant Cell* **2**, 1249–1260.

HUANG, Y., MOREL, P., POWELL, B., KADO, C. I. (1990b), VirA, a coregulator of Ti-specified virulence genes, is phosphorylated *in vitro, J. Bacteriol.* **172**, 1142–1144.

HUNT, A. G., CHU, N. M., ODELL, J. T., NAGY, F., CHUA, N. H. (1987), Plant cells do not properly recognize animal gene polyadenylation signals, *Plant Mol. Biol.* **8**, 23–35.

HUNT, A. G., MOGEN, B. D., CHU, N. M., CHUA, N. H. (1991), The SV40 small t intron is accurately and efficiently spliced in tobacco cells, *Plant Mol. Biol.* **16**, 375–379.

HUNT, T. (1989), Cytoplasmic anchoring proteins and the control of nuclear localization, *Cell* **59**, 949–951.

IGUCHI-ARIGA, S. M. M., SCHAFFNER, W. (1989), CpG methylation of the cAMP-responsive enhancer/promoter sequence TGACGTCA abolishes specific factor binding as well as transcriptional activation, *Genes Dev.* **3**, 612–619.

IIDA, A., YAMASHITA, T., YAMADA, Y., MORIKAWA, H. (1991), Efficiency of particle-bombardment mediated transformation is influenced by cell cycle stage in synchronized cultured cells of tobacco, *Plant Physiol.* **97**, 1585–1587.

INAMDAR, N. M., EHRLICH, K. C., EHRLICH, M. (1991), CpG methylation inhibits binding of several sequence-specific DNA-binding proteins, from pea, wheat, soybean and cauliflower, *Plant Mol. Biol.* **17**, 111–123.

INZÉ, D., FOLLIN, A., VAN LIJSEBETTENS, M., SIMOENS, C., GENETELLO, C., VAN MONTAGU, M., SCHELL, J. (1984), Genetic analysis of the individual T DNA genes of *Agrobacterium tumefaciens;* further evidence that two genes are involved in indole-3-acetic acid synthesis, *Mol. Gen. Genet.* **194**, 265–274.

IZANT, J. G., WEINTRAUB, H. (1984), Constitutive and conditional suppression of exogenous and endogenous genes by antisense RNA, *Science* **229**, 345-352.

JANSSEN, B.-J., GARDNER, R. C. (1990), Localized transient expression of GUS in leaf discs following cocultivation with *Agrobacterium, Plant Mol. Biol.* **14**, 61-72.

JARCHOW, E., GRIMSLEY, N. H., HOHN, B. (1991), virF, the host-range-determining virulence gene of *Agrobacterium tumefaciens,* affects T-DNA transfer to *Zea mays, Proc. Natl. Acad. Sci. USA* **88**, 10426-10430.

JAYASWAL, R. K., VELUTHAMBI, K., GELVIN, S. B., SLIGHTOM, J. L. (1987), Double-stranded cleavage of T-DNA and generation of single-stranded T-DNA molecules in *Escherichia coli* by a virD-encoded border-specific endonuclease from *Agrobacterium tumefaciens, J. Bacteriol.* **169**, 5035-5045.

JEFFERSON, R. A. (1989), The GUS reporter gene system, *Nature* **342**, 837-838.

JEFFERSON, R. A., KAVANAGH, T. A., BEVAN, M. W. (1987), GUS fusions: β-glucuronidase as a sensitive and versatile gene fusion marker in higher plants, *EMBO J.* **6**, 3901-3907.

JENSEN, J. S., MARCKER, K. A., OTTEN, L., SCHELL, J. (1986), Nodule-specific expression of a chimaeric soybean leghaemoglobin gene in transgenic *Lotus corniculatus, Nature* **321**, 669-674.

JIN, S., KOMARI, T., GORDON, M. P., NESTER, E. W. (1987), Genes responsible for the supervirulence phenotype of *Agrobacterium tumefaciens* A 281, *J. Bacteriol.* **169**, 4417-4425.

JIN, S., ROITSCH, T., ANKENBAUER, R. G., GORDON, M. P., NESTER, E. W. (1990a), The virA protein of *Agrobacterium tumefaciens* is autophosphorylated and is essential for vir gene regulation, *J. Bacteriol.* **172**, 525-530.

JIN, S., PRUSTI, R. K., ROITSCH, T., ANKENBAUER, R. G., NESTER, E. W. (1990b), Phosphorylation of the virG protein of *Agrobacterium tumefaciens* by the autophosphorylated virA protein: essential role in biological activity of virG, *J. Bacteriol.* **172**, 4945-4950.

JIN, S., ROITSCH, T., CHRISTIE, P. J., NESTER, E. W. (1990c), The regulatory virG protein specifically binds to a cis-acting regulatory sequence involved in transcriptional activation of *Agrobacterium tumefaciens* virulence genes, *J. Bacteriol.* **172**, 531-537.

JOERSBO, M., BRUNSTEDT, J. (1990), Direct gene transfer to plant protoplasts by mild sonication, *Plant Cell Rep.* **9**, 207-210.

JOHN, M. C., AMASINO, R. M. (1989), Extensive changes in DNA methylation patterns accompany activation of a silent T-DNA ipt gene in *Agrobacterium tumefaciens*-transformed plant cells, *Mol. Cell. Biol.* **9**, 4298-4303.

JOHNSTON, S. A. (1990), Biolistic transformation: microbes to mice, *Nature* **346**, 776-777.

JOHNSTON, S. A., BUTOW, R., SHARK, K. B., SANFORD, J. C. (1988), Transformation of yeast mitochondria by bombardment of cells with microprojectiles, *Science* **240**, 1538-1541.

JONES, J. D. G., DUNSMUIR, P., BEDBROOK, J. (1985), High level expression of introduced chimaeric genes in regenerated transformed plants, *EMBO J.* **4**, 2411-2418.

JONES, J. D. G., SVAB, Z., HARPER, E. C., HURWITZ, C. D., MALIGA, P. (1987), A dominant nuclear streptomycin resistance marker for plant cell transformation, *Mol. Gen. Genet.* **210**, 86-91.

JORGENSEN, R. (1990), Altered gene expression in plants due to trans interactions between homologous genes, *Trends Biotechnol.* **8**, 340-344.

JORGENSEN, R., SNYDER, C., JONES, J. D. G. (1987), T-DNA is organized predominantly in inverted repeat structures in plants transformed with *Agrobacterium tumefaciens* C58 derivatives, *Mol. Gen. Genet.* **207**, 471-477.

JOUANIN, L., BOUCHEZ, D., DRONG, R. F., TEPFER, D., SLIGHTOM, J. L. (1989), Analysis of TR-DNA/plant junctions in the genome of a *Convolvulus arvensis* clone transformed by *Agrobacterium rhizogenes* strain A4, *Plant Mol. Biol.* **12**, 75-85.

JUNKER, B., ZIMNY, J., LÜHRS, R., LÖRZ, H. (1987), Transient expression of chimaeric genes in dividing and non-dividing cereal protoplasts after PEG-induced DNA uptake, *Plant Cell Rep.* **6**, 329-332.

KADO, C. I. (1991), Molecular mechanisms of crown gall tumorigenesis, *Crit. Rev. Plant Sci.* **10**, 1-32.

KAEPPLER, H. F., GU, W., SOMERS, D. A., RINES, H. W., COCKBURN, A. F. (1990), Silicon carbide fiber-mediated DNA delivery into plant cells, *Plant Cell Rep.* **9**, 415-418.

KAHL, G. (1982), Molecular biology of wound healing: The conditioning phenomenon, in: *Molecular Biology of Plant Tumors* (KAHL, G., SCHELL, J., Eds.), pp. 211-267, New York: Academic Press.

KAHL, G., WEISING, K., GÖRZ, A., SCHÄFER, W., HIRASAWA, E. (1987), Chromatin structure and plant gene expression, *Dev. Genet.* **8**, 405-434.

KAMAKURA, T., YONEYAMA, K., YAMAGUCHI, I. (1990), Expression of the blasticidin S deaminase gene (bsr) in tobacco: fungicide tolerance and a new selective marker for transgenic plants, *Mol. Gen.Genet.* **223**, 332-334.

KARTHA, K. K., CHIBBAR, R. N., GEORGES, F., LEUNG, N., CASWELL, K., KENDALL, E., QURESHI, J. (1989), Transient expression of chloramphenicol acetyltransferase (CAT) gene in barley cell cultures and immature embryos through microprojectile bombardment, *Plant Cell Rep.* **8,** 429–432.

KARTZKE, S., SAEDLER, H., MEYER, P. (1990), Molecular analysis of transgenic plants derived from transformations of protoplasts at various stages of the cell cycle, *Plant Sci.* **67,** 63–72.

KATAGIRI, F., CHUA, N. H. (1992), Plant transcription factors: present knowledge and future challenges, *TIG* **8,** 22–27.

KAY, S. A., NAGATANI, A., KEITH, B., DEAK, M., FURUYA, M., CHUA, N.-H. (1989), Rice phytochrome is biologically active in transgenic tobacco, *Plant Cell* **1,** 775–782.

KEEGSTRA, K. (1989), Transport and routing of proteins into chloroplasts, *Cell* **56,** 247–253.

KEITH, B., CHUA, N.-H. (1986), Monocot and dicot pre-mRNAs are processed with different efficiencies in transgenic tobacco, *EMBO J.* **5,** 2419–2425.

KELLER, J. M., SHANKLIN, J., VIERSTRA, R. D., HERSHEY, H. P. (1989), Expression of a functional monocotyledonous phytochrome in transgenic tobacco, *EMBO J.* **8,** 1005–1012.

KERTBUNDIT, S., DE GREVE, H., DEBOECK, F., VAN MONTAGU, M., HERNALSTEENS, J.-P. (1991), *In vivo* random β-glucuronidase gene fusions in *Arabidopsis thaliana, Proc. Natl. Acad. Sci. USA* **88,** 5212–5216.

KESHET, I., LIEMAN-HURWITZ, J., CEDAR, H. (1986), DNA methylation affects the formation of active chromatin, *Cell* **44,** 535–543.

KIBERSTIS, P. A., LOESCH-FRIES, L. S., HALL, T. C. (1981) Viral protein synthesis in barley protoplasts inoculated with native and fractionated brome mosaic virus RNA, *Virology* **112,** 804–808.

KLAAS, M., AMASINO, R. M. (1989), DNA methylation is reduced in DNase I-sensitive regions of plant chromatin, *Plant Physiol.* **91,** 451–454.

KLAAS, M., JOHN, M. C., CROWELL, D. N., AMASINO, R. M. (1989), Rapid induction of genomic demethylation and T-DNA expression in plant cells by 5-azacytosine derivatives, *Plant Mol. Biol.* **12,** 413–423.

KLEE, H. J., YANOFSKY, M. F., NESTER, E. W. (1985), Vectors for transformation of higher plants, *Bio/Technology* **3,** 637–642.

KLEIN, T. M., WOLF, E. D., WU, R., SANFORD, J. C. (1987), High-velocity microprojectiles for delivering nucleic acids into living cells, *Nature* **327,** 70–73.

KLEIN, T. M., FROMM, M., WEISSINGER, A., TOMES, D., SCHAAF, S., SLETTEN, M., SANFORD, J. C. (1988a), Transfer of foreign genes into intact maize cells with high-velocity microprojectiles, *Proc. Natl. Acad. Sci. USA* **85,** 4305–4309.

KLEIN, T. M., HARPER, E. C., SVAB, Z., SANFORD, J. C., FROMM, M. E., MALIGA, P. (1988b), Stable genetic transformation of intact Nicotiana cells by the particle bombardment process, *Proc. Natl. Acad. Sci. USA* **85,** 8502–8505.

KLEIN, T. M., GRADZIEL, T., FROMM, M. E., SANFORD, J. C. (1988c), Factors influencing gene delivery into *Zea mays* cells by high-velocity microprojectiles, *Bio/Technology* **6,** 559–563.

KLEIN, T. M., KORNSTEIN, L., SANFORD, J. C., FROMM, M. E. (1989), Genetic transformation of maize cells by particle bombardment, *Plant Physiol.* **91,** 440–444.

KLEIN, T. M., ARENTZEN, R., LEWIS, P. A., FITZPATRICK-MCELLIGOTT, S. (1992), Transformation of microbes, plants, and animals by particle bombardment, *Bio/Technology* **10,** 286–291.

KLEINHOFS, A., BEHKI, R. (1977), Prospects for plant genome modification by non-conventional methods, *Annu. Rev. Genet.* **11,** 79–101.

KLEINHOFS, A., EDEN, F. C., CHILTON, M.-D., BENDICH, A. J. (1975), On the question of the integration of exogenous bacterial DNA into plant DNA, *Proc. Natl. Acad. Sci. USA* **72,** 2748–2752.

KNIGHT, M. R., CAMPBELL, A. K., SMITH, S. M., TREWAVAS, A. J. (1991), Transgenic plant aequorin reports the effects of touch and cold-shock and elicitors on cytoplasmic calcium, *Nature* **352,** 524–526.

KONCZ, C., SCHELL, J. (1986), The promoter of TL-DNA gene 5 controls the tissue-specific expression of chimaeric genes carried by a novel type of *Agrobacterium* binary vector, *Mol. Gen. Genet.* **204,** 383–396.

KONCZ, C., KREUZALER, F., KALMAN, ZS., SCHELL, J. (1984), A simple method to transfer, integrate and study expression of foreign genes, such as chicken ovalbumin and alpha-actin in plant tumors, *EMBO J.* **3,** 1029–1037.

KONCZ, C., OLSSON, O., LANGRIDGE, W. H. R., SCHELL, J., SZALAY, A. A. (1987a), Expression and assembly of functional bacterial luciferase in plants, *Proc. Natl. Acad. Sci. USA* **84,** 131–135.

KONCZ, C., KONCZ-KALMAN, Z., SCHELL, J. (1987b), Transposon Tn5 mediated gene transfer into plants, *Mol. Gen. Genet.* **207,** 99–105.

KONCZ, C., MARTINI, N., MAYERHOFER, R., KONCZ-KALMAN, Z., KÖRBER, H., REDEI, G. P., SCHELL, J. (1989), High-frequency T-DNA-mediated gene tagging in plants, *Proc. Natl. Acad. Sci. USA* **86,** 8467–8471.

KONCZ, C., LANGRIDGE, W. H. R., OLSSON, O., SCHELL, J., SZALAY, A. A. (1990), Bacterial and firefly luciferase genes in transgenic plants: advantages and disadvantages of a reporter gene, *Dev. Genet.* **11,** 224–232.

KÖRBER, H., STRIZHOV, N., STAIGER, D., FELDWISCH, J., OLSSON, O., SANDBERG, G., PALME, K., SCHELL, J., KONCZ, C. (1991), T-DNA gene 5 of *Agrobacterium* modulates auxin response by autoregulated synthesis of a growth hormone antagonist in plants, *EMBO J.* **10,** 3983–3991.

KÖSTER-TÖPFER, M., FROMMER, W. B., ROCHA-SOSA, M., ROSAHL, S. (1989), A class II patatin promoter is under developmental control in both transgenic potato and tobacco plants, *Mol. Gen. Genet.* **219,** 390–396.

KOUKOLÍKOVÁ-NICOLA, Z., SHILLITO, R. D., HOHN, B., WANG, K., VAN MONTAGU, M., ZAMBRYSKI, P. (1985), Involvement of circular intermediates in the transfer of T-DNA from *Agrobacterium tumefaciens* to plant cells, *Nature* **313,** 191–196.

KRENS, F. A., MOLENDIJK, L., WULLEMS, G. J., SCHILPEROORT, R. A. (1982), *In vitro* transformation of plant protoplasts with Ti-plasmid DNA, *Nature* **296,** 72–74.

KULDAU, G. A., DE VOS, G., OWEN, J., MCCAFFREY, G., ZAMBRYSKI, P. (1990), The virB operon of *Agrobacterium tumefaciens* pTi C58 encodes 11 open reading frames, *Mol. Gen. Genet.* **221,** 256–266.

KUNZE, R., STARLINGER, P., SCHWARTZ, D. (1988), DNA methylation of the maize transposable element Ac interferes with its transcription, *Mol. Gen. Genet.* **214,** 325–327.

KYOZUKA, J., IZAWA, T., NAKAJIMA, M., SHIMAMOTO, K. (1990), Effect of the promoter and the first intron of maize Adh 1 on foreign gene expression in rice, *Maydica* **35,** 353–357.

LAGRIMINI, L. M., BRADFORD, S., ROTHSTEIN, S. (1990), Peroxidase-induced wilting in transgenic tobacco plants, *Plant Cell* **2,** 7–18.

LAMB, J., HAY, R. T. (1990), Ribozymes that cleave potato leafroll virus RNA within the coat protein and polymerase genes, *J. Gen. Virol.* **71,** 2257–2263.

LAMPPA, G., NAGY, F., CHUA, N. H. (1985), Light-regulated and organ-specific expression of a wheat cab gene in transgenic tobacco, *Nature* **316,** 750–752.

LANGRIDGE, W. H. R., FITZGERALD, K. J., KONCZ, C., SCHELL, J., SZALAY, A. A. (1989), Dual promoter of *Agrobacterium tumefaciens* mannopine synthase genes is regulated by plant growth hormones, *Proc. Natl. Acad. Sci. USA* **86,** 3219–3223.

LAWRENCE, W. A., DAVIES, D. R. (1985), A method for the microinjection and culture of protoplasts at very flow densities, *Plant Cell Rep.* **4,** 33–35.

LAZAROWITZ, S. G. (1987), The molecular characterization of geminiviruses, *Plant Mol. Biol. Rep.* **4,** 177–192.

LAZAROWITZ, S. G. (1988), Infectivity and complete nucleotide sequence of the genome of a South African isolate of maize streak virus, *Nucleic Acids Res.* **16,** 229–249.

LEE, B. T., MURDOCH, K., TOPPING, J., JONES, M. G. K., KREIS, M. (1991 a), Transient expression of foreign genes introduced into barley endosperm protoplasts by PEG-mediated transfer or into intact endosperm tissue by microprojectile bombardment, *Plant Sci.* **78,** 237–246.

LEE, K. Y., LUND, P., LOWE, K., DUNSMUIR, P. (1990), Homologous recombination in plant cells after *Agrobacterium*-mediated transformation, *Plant Cell* **2,** 415–425.

LEE, N., WANG, Y., YANG, J., GE, K., HUANG, S., TAN, J., TESTA, D. (1991 b), Efficient transformation and regeneration of rice small cell groups, *Proc. Natl. Acad. Sci. USA* **88,** 6389–6393.

LEFEBRE, D. D., MIKI, B. L., LALIBERTÉ, J.-F. (1987), Mammalian metallothionein functions in plants, *Bio/Technology* **5,** 1053–1056.

LEISY, D. J., HNILO, J., ZHAO, Y., OKITA, T. W. (1990), Expression of a rice glutelin promoter in transgenic tobacco, *Plant Mol. Biol.* **14,** 41–50.

LEROUX, B., YANOFSKY, M. F., WINANS, S. C., WARD, J. E., ZIEGLER, S. F., NESTER, E. W. (1987), Characterization of the virA locus of *Agrobacterium tumefaciens:* a transcriptional regulator and host range determinant, *EMBO J.* **6,** 849–856.

LINDSEY, K., JONES, M. G. K. (1990), Electroporation of cells, *Physiol. Plant.* **79,** 168–172.

LINN, F., HEIDMANN, I., SAEDLER, H., MEYER, P. (1990), Epigenetic changes in the expression of the maize A1 gene in *Petunia hybrida:* role of numbers of integrated gene copies and state of methylation, *Mol. Gen. Genet.* **222,** 329–336.

LLOYD, A. M., BARNASON, A. R., ROGERS, S. G., BYRNE, M. C., FRALEY, R. T., HORSCH, R. B. (1986), Transformation of *Arabidopsis thaliana* with *Agrobacterium tumefaciens, Science* **234,** 464–466.

LOGEMANN, J., LIPPHARDT, S., LÖRZ, H., HÄUSER, I., WILLMITZER, L., SCHELL, J. (1989), 5′ Upstream sequences from the wun1 gene are responsible for gene activation by wounding in transgenic plants, *Plant Cell* **1,** 151–158.

LONSDALE, D., ÖNDE, S., CUMING, A. (1990), Transient expression of exogenous DNA in intact, viable wheat embryos following particle bombardment, *J. Exp. Bot.* **41**, 1161–1165.

LUCAS, W. J., LANSING, A., DE WET, J. R., WALBOT, V. (1990), Introduction of foreign DNA into walled plant cells via liposomes injected into the vacuole: a preliminary study, *Physiol. Plant.* **79**, 184–189.

LUO, Z.-X., WU, R. (1988), A simple method for the transformation of rice via the pollen-tube pathway, *Plant Mol. Biol. Rep.* **6**, 165–174.

LYZNIK, L. A., MCGEE, J. D., TUNG, P. Y., BENNETZEN, J. L., HODGES, T. K. (1991), Homologous recombination between plasmid DNA molecules in maize protoplasts, *Mol. Gen. Genet.* **230**, 209–218.

MAAS, C., WERR, W. (1989), Mechanism and optimized conditions for PEG mediated DNA transfection into plant protoplasts, *Plant Cell Rep.* **8**, 148–151.

MAAS, C., LAUFS, J., GRANT, S., KORRHAGE, C., WERR, W. (1990), The combination of a novel stimulatory element in the first exon of the maize Shrunken-1 gene with the following intron 1 enhances reporter gene expression up to 1000-fold, *Plant Mol. Biol.* **16**, 199–207.

MACHIDA, Y., USAMI, S., YAMAMOTO, A., NIWA, Y., TAKEBE, I. (1986), Plant-inducible recombination between the 25 bp border sequences of T-DNA in *Agrobacterium tumefaciens, Mol. Gen. Genet.* **204**, 374–382.

MAESER, S., KAHMANN, R. (1991), The gin recombinase of phage Mu can catalyse site-specific recombination in plant protoplasts, *Mol. Gen. Genet.* **230**, 170–176.

MAGILL, J. M., MAGILL, C. W. (1989), DNA methylation in fungi, *Dev. Genet.* **10**, 63–69.

MAITI, I. B., HUNT, A. G., WAGNER, G. J. (1988), Seed-transmissable expression of mammalian metallothionein in transgenic tobacco, *Biochem. Biophys, Res. Commun.* **150**, 640–647.

MANSOUR, S. L., THOMAS, K. R., CAPECCHI, M. R. (1988), Disruption of the proto-oncogene int-2 in mouse embryo-derived stem cells: a general strategy for targeting mutations to non-selectable genes, *Nature* **336**, 348–352.

MARKS, M. D., FELDMANN, K. A. (1989), Trichome development in *Arabidopsis thaliana*. I. T-DNA tagging of the GLABROUS1 gene, *Plant Cell* **1**, 1043–1050.

MARKS, M. S., KEMP, J. M., WOOLSTON, C. J., DALE, P. J. (1989), Agroinfection of wheat: a comparison of *Agrobacterium strains, Plant Sci.* **63**, 247–256.

MARRIS, C., GALLOIS, P., COPLEY, J., KREIS, M. (1988), The 5′flanking region of a barley B hordein gene controls tissue and developmental specific CAT expression in tobacco plants, *Plant Mol. Biol.* **10**, 359–366.

MARTIN, C., PRESCOTT, A., LISTER, C., MACKAY, S. (1989), Activity of the transposon Tam3 in *Antirrhinum* and tobacco: possible role of DNA methylation, *EMBO J.* **8**, 997–1004.

MARTINEZ-ZAPATER, J. M., FINKELSTEIN, R., SOMERVILLE, C. R. (1988), Drosophila P-element transcripts are incorrectly processed in tobacco, *Plant Mol. Biol.* **11**, 601–607.

MATOUSEK, J., TUPY, J. (1984), Purification and properties of extracellular nuclease from tobacco pollen, *Biol. Plant.* (Praha) **26**, 62–73.

MATOUSEK, J., TUPY, J. (1985), The release and some properties of nuclease from various pollen species, *J. Plant Physiol.* **119**, 169–178.

MATSUMOTO, S., TAKEBE, I., MACHIDA, Y. (1988), *Escherichia coli* lacZ gene as a biochemical and histochemical marker in plant cells, *Gene* **66**, 19–29.

MATSUMOTO, S., ITO, Y., HOSOI, T., TAKAHASHI, Y., MACHIDA, Y. (1990), Integration of *Agrobacterium* T-DNA into a tobacco chromosome: possible involvement of DNA homology between T-DNA and plant DNA, *Mol. Gen. Genet.* **224**, 309–316.

MATTHEWS, B. F., ABDUL-BAKI, A. A., SAUNDERS, J. A. (1991), Expression of a foreign gene in electroporated pollen grains of tobacco, *Sex Plant Reprod.* **3**, 137–141.

MATZEIT, V. (1987), cited in Göbel, E., Lörz, H. (1988), Genetic manipulation of cereals, *Oxford Surv. Plant Mol. Cell Biol.* **5**, 1–22.

MATZKE, A. J. M., MATZKE, M. A. (1986), A set of novel Ti plasmid-derived vectors for the production of transgenic plants, *Plant Mol. Biol.* **7**, 357–365.

MATZKE, A. J. M., STÖGER, E. M., SCHERNTHANER, J. P., MATZKE, M. A. (1990), Deletion analysis of a zein gene promoter in transgenic tobacco plants, *Plant Mol. Biol.* **14**, 323–332.

MATZKE, M. A., MATZKE, A. J. M. (1990), Gene interactions and epigenetic variation in transgenic plants, *Dev. Genet.* **11**, 214–223.

MATZKE, M. A., MATZKE, A. J. M. (1991), Differential inactivation and methylation of a transgene in plants by two suppressor loci containing homologous sequences, *Plant Mol. Biol.* **16**, 821–830.

MATZKE, M. A., PRIMIG, M., TRNOVSKY, J., MATZKE, A. J. M. (1989), Reversible methylation and inactivation of marker genes in sequentially transformed tobacco plants, *EMBO J.* **8**, 643–649.

MAUREL, C., BREVET, J., BARBIER-BRYGOO, H., GUERN, J., TEMPÉ, J. (1990), Auxin regulates the promoter of the root-inducing rolB gene of *Agrobacterium rhizogenes* in transgenic tobacco, *Mol. Gen. Genet.* **223**, 58–64.

MAYERHOFER, R., KONCZ-KALMAN, Z., NAWRATH, C., BAKKEREN, G., CRAMERI, A., ANGELIS, K., REDEI, G. P., SCHELL, J., HOHN, B., KONCZ, C. (1991), T-DNA integration: a mode of illegitimate recombination in plants, *EMBO J.* **10**, 697–704.

MCBRIDE, K. E., SUMMERFELT, K. R. (1990), Improved binary vectors for *Agrobacterium*-mediated plant transformation, *Plant Mol. Biol.* **14**, 269–276.

MCCABE, D. E., SWAIN, W. F., MARTINELL, B. J., CHRISTOU, P. (1988), Stable transformation of soybean *(Glycine max)* by particle acceleration, *Bio/Technology* **6**, 923–926.

MCDONNELL, R. E., CLARK, R. D., SMITH, W. A., HINCHEE, M. A. (1987), A simplified method for the detection of neomycin phosphotransferase II activity in transformed plant tissues, *Plant Mol. Biol. Rep.* **5**, 380–386.

MCELROY, D., ZHANG, W., CAO, J., WU, R. (1990), Isolation of an efficient actin promoter for use in rice transformation, *Plant Cell* **2**, 163–171.

MCELROY, D., BLOWERS, A. D., JENES, B., WU, R. (1991), Construction of expression vectors based on the rice actin 1 (Act1) 5′ region for use in monocot transformation, *Mol. Gen. Genet.* **231**, 150–160.

MEEHAN, R. R., LEWIS, J. D., MCKAY, S., KLEINER, E. L., BIRD, A. P. (1989), Identification of a mammalian protein that binds specifically to DNA containing methylated CpGs, *Cell* **58**, 499–507.

MELCHERS, L. S., THOMPSON, D. V., IDLER, K. B., NEUTEBOOM, S. T. C., DE MAAGD, R. A., SCHILPEROORT, R. A., HOOYKAAS, P. J. J. (1988), Molecular characterization of the virulence gene virA of the *Agrobacterium tumefaciens* octopine Ti plasmid, *Plant Mol. Biol.* **11**, 227–237.

MELCHERS, L. S., REGENSBURG-TUINK, A. J. G., SCHILPEROORT, R. A., HOOYKAAS, P. J. J. (1989a), Specificity of signal molecules in the activation of *Agrobacterium* virulence gene expression, *Mol. Microbiol.* **3**, 969–977.

MELCHERS, L. S., REGENSBURG-TUINK, T. J. G., BOURRET, R. B., SEDEE, N. J. A., SCHILPEROORT, R. A., HOOYKAAS, P. J. J. (1989b), Membrane topology and functional analysis of the sensory protein VirA of *Agrobacterium tumefaciens, EMBO J.* **8**, 1919–1925.

MELCHERS, L. S., MARONEY, M. J., DEN DULK-RAS, A., THOMPSON, D. V., VAN VUUREN, H. A. J., SCHILPEROORT, R. A., HOOYKAAS, P. J. J. (1990), Octopine and nopaline strains of *Agrobacterium tumefaciens* differ in virulence; molecular characterization of the virF locus, *Plant Mol. Biol.* **14**, 249–259.

MELNIKOV, P. V., PASTERNAK, T. P., GLEBA, Y. Y., SYTNIK, K. M. (1985), Microinjection of DNA into cells of higher plants, *Dokl. Akad. Nauk SSR. Geol. Khim. Biol. Nauki* **10**, 69–71.

MENDEL, R. R., MÜLLER, B., SCHULZE, J., KOLESNIKOV, V., ZELENIN, A. (1989), Delivery of foreign genes to intact barley cells by high-velocity microprojectiles, *Theor. Appl. Genet.* **78**, 31–34.

MESHI, T., ISHIWAKA, M., MOTOYOSHI, F., SEMBA, K., OKADA, Y. (1986), *In vitro* transcription of infectious RNAs from full-length cDNAs of tobacco mosaic virus, *Proc. Natl. Acad. Sci. USA* **83**, 5043–5047.

MESSENS, E., LENAERTS, A., VAN MONTAGU, M., HEDGES, R. W. (1985), Genetic basis for opine secretion from crown gall tumor cells, *Mol. Gen. Genet.* **199**, 344–348.

MESSENS, E., DEKEYSER, R., STACHEL, S. E. (1990), A nontransformable *Triticum monococcum* monocotyledonous culture produces the potent *Agrobacterium* vir-inducing compound ethyl ferulate, *Proc. Natl. Acad. Sci. USA* **87**, 4368–4372.

MEYER, P., KARTZKE, S., NIEDENHOF I., HEIDMANN, I., BUSSMANN, K., SAEDLER, H. (1988), A genomic DNA segment from *Petunia hybrida* leads to increased transformation frequencies and simple integration patterns, *Proc. Natl. Acad. Sci. USA* **85**, 8568–8572.

MISRA, S., GEDAMU, L. (1989), Heavy metal tolerant transgenic *Brassica napus* L. and *Nicotiana tabacum* L. plants, *Theor. Appl. Genet.* **78**, 161–168.

MITTELSTEN SCHEID, O., PASZKOWSKI, J., POTRYKUS, I. (1991), Reversible inactivation of a transgene in *Arabidopsis thaliana, Mol. Gen. Genet.* **228**, 104–112.

MIZUNO, T., CHOU, M., INOUYE, M. (1984), A unique mechanism of regulating gene expression: translational inhibition by a complementary RNA transcript (micRNA), *Proc. Acad. Natl. Sci. USA* **81**, 1966–1970.

MOL, J. N. M., VAN DER KROL, A. R., VAN TUNEN, A. J., VAN BLOKLAND, R., DE LANGE, P., STUITJE, A. R. (1990), Regulation of plant gene expression by antisense RNA, *FEBS Lett.* **268**, 427–430.

MOL, J. N. M., VAN BLOKLAND, R., KOOTER, J. (1991), More about co-suppression, *Trends Biotechnol.* **9**, 182–183.

MORENO DIAZ DE LA ESPINA, S., BARTHELLEMY, I., CEREZUELA, M. A. (1991), Isolation and ultrastructural characterization of the residual nuclear matrix in a plant cell system, *Chromosoma* **100**, 110–117.

MORIKAWA, H., YAMADA, Y. (1985), Capillary microinjection into protoplasts and intranuclear localization of injected materials, *Plant Cell Physiol.* **26**, 229–236.

MÜLLER, A. J., MENDEL, R. R., SCHIEMANN, J., SIMOENS, C., INZÉ, D. (1987), High meiotic stability of foreign gene introduced into tobacco by *Agrobacterium*-mediated transformation, *Mol. Gen. Genet.* **207**, 171–175.

NAGY, F., KAY, S. A., BOUTRY, M., HSU, M.-Y., CHUA, N.-H. (1986), Phytochrome-controlled expression of a wheat cab gene in transgenic tobacco seedlings, *EMBO J.* **5**, 1119–1124.

NAGY, F., BOUTRY, M., HSU, M.-Y., WONG, M., CHUA, N.-H. (1987), The 5′-proximal region of the wheat Cab-1 gene contains a 268-bp enhancer-like sequence for phytochrome response, *EMBO J.* **6**, 2537–2542.

NAPOLI, C., LEMIEUX, C., JORGENSEN, R. (1990), Introduction of a chimeric chalcone synthase gene into petunia results in reversible co-suppression of homologous genes in trans, *Plant Cell* **2**, 279–289.

NEGRUTIU, I., SHILLITO, R., POTRYKUS, I., BIASINI, G., SALA, F. (1987), Hybrid genes in the analysis of transformation conditions. I. Setting up a simple method for direct gene transfer in plant protoplasts, *Plant Mol. Biol.* **8**, 363–373.

NEGRUTIU, I., DEWULF, M., PIETRZAK, M., BOTTERMAN, J., RIETVELD, E., WURZER-FIGURELLI, E. M., DE YE, JACOBS, M. (1990), Hybrid genes in the analysis of transformation conditions: II. Transient expression vs stable transformation – analysis of parameters influencing gene expression levels and transformation efficiency, *Physiol. Plant.* **79**, 197–205.

NELSEN-SALZ, B., DÖRING, H.-P. (1990), Rare *de novo* methylation within the transposable element activator (Ac) in transgenic tobacco plants, *Mol. Gen. Gent.* **223**, 87–96.

NEUHAUS, G., SPANGENBERG, G. (1990), Plant transformation by microinjection techniques, *Physiol. Plant.* **79**, 213–217.

NEUHAUS, G., NEUHAUS-URL, G., GRUSS, P., SCHWEIGER, H. G. (1984), Enhancer-controlled expression of the simian virus 40 T-antigen in the green alga *Acetabularia mediterranea, EMBO J.* **3**, 2169–2171.

NEUHAUS, G., NEUHAUS-URL, G., DE GROOT, E. J., SCHWEIGER, H.-G. (1986), High yield and stable transformation of the unicellular green alga *Acetabularia* by microinjection of SV40 DNA and pSV2neo, *EMBO J.* **5**, 1437–1444.

NEUHAUS, G., SPANGENBERG, G., MITTELSTEN SCHEID, O., SCHWEIGER, H.-G. (1987), Transgenic rapeseed plants obtained by the microinjection of DNA into microspore-derived embryoids, *Theor. Appl. Genet.* **75**, 30–36.

NGERNPRASIRTSIRI, J., CHOLLET, R., KOBAYASHI, H., SUGIYAMA, T., AKAZAWA, T. (1989), DNA methylation and the differential expression of C4 photosynthesis genes in mesophyll and bundle sheath cells of greening maize leaves, *J. Biol. Chem.* **264**, 8241–8248.

NOMURA, K., KOMAMINE, A. (1985), Identification and isolation of single cells that produce somatic embryos at high frequency in a carrot suspension culture, *Plant Physiol.* **79**, 988–991.

NOMURA, K., KOMAMINE, A. (1986), Embryogenesis from microinjected single cells in a carrot cell suspension, *Plant Sci.* **44**, 53–58.

OARD, J. H., PAIGE, D., DVORAK, J. (1989), Chimeric gene expression using maize intron in cultured cells of breadwheat, *Plant Cell Rep.* **8**, 156–160.

OARD, J. H., PAIGE, D. F., SIMMONDS, J. A., GRADZIEL, T. M. (1990), Transient gene expression in maize, rice, and wheat cells using an airgun apparatus, *Plant Physiol.* **92**, 334–339.

ODELL, J., CAIMI, P., SAUER, B., RUSSEL, S. (1990), Site-directed recombination in the genome of transgenic tobacco, *Mol. Gen. Genet.* **223**, 369–378.

OELLER, P. W., MIN-WONG, L., TAYLOR, L. P., PIKE, D. A., THEOLOGIS, A. (1991), Reversible inhibition of tomato fruit senescence by antisense RNA, *Science* **254**, 437–439.

OFFRINGA, R., DE GROOT, M. J. A., HAAGSMAN, H. J., DOES, M. P., VAN DEN ELZEN, P. J. M., HOOYKAAS, P. J. J. (1990), Extrachromosomal homologous recombination and gene targeting in plant cells after *Agrobacterium* mediated transformation, *EMBO J.* **9**, 3077–3084.

OHTA, Y. (1986), High-efficiency genetic transformation of maize by a mixture of pollen and exogenous DNA, *Proc. Natl. Acad. Sci. USA* **83**, 715–719.

OKADA, K., TAKEBE, I., NAGATA, T. (1986a), Expression and integration of genes introduced into highly synchronized plant protoplasts, *Mol. Gen. Genet.* **205**, 398–403.

OKADA, K., NAGATA, T., TAKEBE, I. (1986b), Introduction of functional RNA into plant protoplasts by electroporation, *Plant Cell Physiol.* **27**, 619–626.

OKAMOTO, S., TOYODA-YAMAMOTO, A., ITO, K., TAKEBE, I., MACHIDA, Y. (1991), Localization and orientation of the VirD4 protein of *Agrobacterium tumefaciens* in the cell membrane, *Mol. Gen. Genet.* **228**, 24–32.

OLSSON, O., KONCZ, C., SZALAY, A. A. (1988), The use of the luxA gene of the bacterial luciferase operon as a reporter gene, *Mol. Gen. Genet.* **215**, 1-9.

OTT, R. W., REN, L., CHUA, N.-H. (1990), A bidirectional enchancer cloning vehicle for higher plants, *Mol. Gen. Genet.* **221**, 121-124.

OTTEN, L. A. B. M., SCHILPEROORT, R. A. (1978), A rapid microscale method for the detection of lysopine and nopaline dehydrogenase activities, *Biochim. Biophys. Acta* **527**, 497-500.

OU-LEE, T.-M., TURGEON, R., WU, R. (1986), Expression of a foreign gene linked to either a plant-virus or a *Drosophila* promoter, after electroporation of protoplasts of rice, wheat, and sorghum, *Proc. Natl. Acad. Sci. USA* **83**, 6815-6819.

OW, D. W., WOOD, K. V., DELUCA, M., DE WET, J. R., HELINSKI, D. R., HOWELL, S. H. (1986) Transient and stable expression of the firefly luciferase gene in plant cells and transgenic plants, *Science* **234**, 856-859.

PASZKOWSKI, J., SHILLITO, R. D., SAUL, M., MANDAK, V., HOHN, T., HOHN, B., POTRYKUS, I. (1984), Direct gene transfer to plants, *EMBO J.* **3**, 2717-2722.

PASZKOWSKI, J., PISAN, B., SHILLITO, R. D., HOHN, T., HOHN, B., POTRYKUS, I. (1986), Genetic transformation of *Brassica campestris* var. *rapa* protoplasts with an engineered cauliflower mosaic virus genome, *Plant Mol. Biol.* **6**, 303-312.

PASZKOWSKI, J., BAUR, M., BOGUCKI, A., POTRYKUS, I. (1988), Gene targeting in plants, *EMBO J.* **7**, 4021-4026.

PAZOUR, G. J., DAS, A. (1990), VirG, an *Agrobacterium tumefaciens* transcriptional activator, initiates translation at a UUG codon and is a sequence-specific DNA-binding protein, *J. Bacteriol.* **172**, 1241-1249.

PEACH, C., VELTEN, J. (1991), Transgene expression variability (position effect) of CAT and GUS reporter genes driven by linked divergent T-DNA promoters, *Plant Mol. Biol.* **17**, 49-60.

PEERBOLTE, R., LEENHOUTS, K., HOOYKAAS-VAN SLOGTEREN, G. M. S., HOGE, J. H. C., WULLEMS, G. J., SCHILPEROORT, R. A. (1986), Clones from a shooty tobacco crown gall tumor I: deletions, rearrangements and amplifications resulting in irregular T-DNA structures and organizations, *Plant Mol. Biol.* **7**, 265-284.

PERALTA, E. G., REAM, L. W. (1985), T-DNA border sequences required for crown gall tumorigenesis, *Proc. Natl. Acad. Sci. USA* **82**, 5112-5116.

PERALTA, E. G., HELLMISS, R., REAM, W. (1986), Overdrive, a T-DNA transmission enhancer on the *A. tumefaciens* tumour-inducing plasmid, *EMBO J.* **5**, 1137-1142.

PEREZ, P., TIRABY, G., KALLERHOFF, J., PERRET, J. (1989), Phleomycin resistance as a dominant selectable marker for plant cell transformation, *Plant Mol. Biol.* **13**, 365-373.

PETERHANS, A., SCHLÜPMANN, H., BASSE, C., PASZKOWSKI, J. (1990a), Intrachromosomal recombination in plants, *EMBO J.* **9**, 3437-3445.

PETERHANS, A., DATTA, S. K., DATTA, K., GOODALL, G. J., POTRYKUS, I., PASZKOWSKI, J. (1990b), Recognition efficiency of Dicotyledoneae-specific promoter and RNA processing signals in rice, *Mol. Gen. Genet.* **222**, 361-368.

PHI-VAN, L., VON KRIES, J. P., OSTERTAG, W., STRÄTLING, W. H. (1990), The chicken lysozyme 5′ matrix attachment region increases transcription from a heterologous promoter in heterologous cells and dampens position effects on the expression of transfected genes, *Mol. Cell. Biol.* **10**, 2302-2307.

POLLOCK, K., BARFIELD, D. G., ROBINSON, S. J., SHIELDS, R. (1985), Transformation of protoplast-derived cell colonies and suspension cultures by *Agrobacterium tumefaciens, Plant Cell Rep.* **4**, 202-205.

PORTER, J. R. (1991), Host range and implications of plant infection by *Agrobacterium rhizogenes, CRC Crit. Rev. Plant Sci.* **10**, 387-421.

POTRYKUS, I. (1990), Gene transfer to cereals: an assessment, *Bio/Technology* **8**, 535-542.

POTRYKUS, I. (1991), Gene transfer to plants: Assessment of published approaches and results, *Annu. Rev. Plant Physiol. Plant Mol. Biol.* **42**, 205-225.

POTRYKUS, I., PASZKOWSKI, J., SAUL, M. W., PETRUSKA, J., SHILLITO, R. D. (1985), Molecular and general genetics of a hybrid foreign gene introduced into tobacco by direct gene transfer, *Mol. Gen. Genet.* **199**, 169-177.

POWELL, P. A., STARK, D. M., SANDERS, P. R., BEACHY, R. N. (1989), Protection against tobacco mosaic virus in transgenic plants that express tobacco mosaic virus antisense RNA, *Proc. Natl. Acad. Sci. USA* **86**, 6949-6952.

POWELL ABEL, P., NELSON, R. S., DE, B., HOFFMANN, N., ROGERS, S. G., FRALEY, R. T., BEACHY, R. N. (1986), Delay of disease development in transgenic plants that express the tobacco mosaic virus coat protein gene, *Science* **232**, 738-743.

PRÖLS, M., TÖPFER, R., SCHELL, J., STEINBIß, H.-H. (1988), Transient gene expression in tobacco protoplasts: I. Time course of CAT appearance, *Plant Cell Rep.* **7**, 221-224.

PUCHTA, H., HOHN, B. (1991a), A transient assay in plant cells reveals a positive correlation between extrachromosomal recombination rates and length of homologous overlap, *Nucleic Acids Res.* **19,** 2693–2700.

PUCHTA, H., HOHN, B. (1991b), The mechanism of extrachromosomal homologous DNA recombination in plant cells, *Mol. Gen. Genet.* **230,** 1–7.

PYTHOUD, F., SINKAR, V. P., NESTER, E. W., GORDON, M. P. (1987), Increased virulence of *Agrobacterium rhizogenes* conferred by the vir region of pTiBo 542: Application to genetic engineering of poplar, *Bio/Technology* **5,** 1323–1327.

QUICK, W. P., SCHURR, U., FICHTNER, K., SCHULZE, E.-D., RODERMEL, S. R., BOGORAD, L., STITT, M. (1991a), The impact of decreased Rubisco on photosynthesis, growth, allocation and storage in tobacco plants which have been transformed with antisense rbcS, *Plant J.* **1,** 51–58.

QUICK, W. P., SCHURR, U., SCHEIBE, R., SCHULZE, E. D., RODERMEL, S. R., BOGORAD, L., STITT, M. (1991b), Decreased ribulose-1,5-bisphosphate carboxylase-oxygenase in transgenic tobacco transformed with antisense "rbcS". I. Impact on photosynthesis in ambient growth conditions, *Planta* **183,** 542–554.

RAINERI, D. M., BOTTINO, P., GORDON, M. P., NESTER, E. W. (1990), *Agrobacterium*-mediated transformation of rice (*Oryza sativa* L.), *Bio/Technology* **8,** 33–38.

REAM, W. (1989), *Agrobacterium tumefaciens* and interkingdom genetic exchange, *Annu. Rev. Phytopathol.* **27,** 583–618.

REICH, T. J., IYER, V. N., MIKI, B. L. (1986), Efficient transformation of alfalfa protoplasts by the intranuclear microinjection of Ti plasmids, *Bio/Technology* **4,** 1001–1004.

REISS, B., SPRENGEL, R., WILL, H., SCHALLER, H. (1984), A new sensitive method for qualitative and quantitative assay of neomycin phosphotransferase in crude cell extracts, *Gene* **30,** 211–218.

REZAIAN, M. A., SKENE, K. G. M., ELLIS, J. G. (1988), Anti-sense RNAs of cucumber mosaic virus in transgenic plants assessed for control of the virus, *Plant Mol. Biol.* **11,** 463–471.

RHODES, C. A., PIERCE, D. A., METTLER, I. J., MASCARENHAS, D., DETMER, J. J. (1988), Genetically transformed maize plants from protoplasts, *Science* **240,** 204–207.

RIGGS, C. D., BATES, G. W. (1986), Stable transformation of tobacco by electroporation: Evidence for plasmid concatenation, *Proc. Natl. Acad. Sci. USA* **83,** 5602–5606.

ROBERT, L. S., DONALDSON, P. A., LADAIQUE, C., ALTOSAAR, I., ARNISON, P. G., FABIJANSKI, S. F. (1989a), Antisense RNA inhibition of β-glucuronidase gene expression in transgenic tobacco plants, *Plant Mol. Biol.* **13,** 399–409.

ROBERT, L. S., THOMPSON, R. D., FLAVELL, R. B. (1989b), Tissue-specific expression of a wheat high molecular weight glutenin gene in transgenic tobacco, *Plant Cell* **1,** 569–578.

ROBERT, L. S., DONALDSON, P. A., LADAIQUE, C., ALTOSAAR, I., ARNISON, P.G., FABIJANSKI, S. F. (1990), Antisense RNA inhibition of β-glucuronidase gene expression in transgenic tobacco can be transiently overcome using a heat-inducible β-glucuronidase gene construct, *Bio/Technology* **8,** 459–464.

ROBERTSON, H. D., HOWELL, S. H., ZAITLIN, M., MALMBERG, R. L. (1983), *Plant Infectious Agents: Viruses, Viroids, Virusoids and Satellites,* Cold Spring Harbor, NY: Cold Spring Harbor Laboratory Press.

RODENBURG, K. W., DE GROOT, M. J. A., SCHILPEROORT, R. A., HOOYKAAS, P. J. J. (1989), Single-stranded DNA used as an efficient new vehicle for transformation of plant protoplasts, *Plant Mol. Biol.* **13,** 711–719.

RODERMEL, S. R., ABBOTT, M. S., BOGORAD, L. (1988), Nuclear-organelle interactions: nuclear antisense gene inhibits ribulose bisphosphate carboxylase enzyme levels in transformed tobacco plants, *Cell* **55,** 673–681.

ROEST, S., GILISSEN, L. J. W. (1989), Plant regeneration from protoplasts: a literature review. *Acta Bot. Neerl.* **38,** 1–23.

ROGERS, S. G., HORSCH, R. B., FRALEY, R. T. (1986a), Gene transfer in plants: Production of transformed plants using Ti plasmid vectors, *Methods Enzymol.* **118,** 627–640.

ROGERS, S. G., BISARO, D. M., HORSCH, R. B., FRALEY, R. T., HOFFMANN, N. L., BRAND, L., SCOTT ELMER, J., LLOYD, A. M. (1986b), Tomato golden mosaic virus A component DNA replicates autonomously in transgenic plants, *Cell* **45,** 593–600.

ROITSCH, T., WANG, H., JIN, S., NESTER, E. W. (1990), Mutational analysis of the VirG protein, a transcriptional activator of *Agrobacterium tumefaciens* virulence genes, *J. Bacteriol.* **172,** 6054–6060.

ROSENBERG, N., GAD, A. E., ALTMAN, A., NAVOT, N., CZOSNEK, H. (1988), Liposome-mediated introduction of the chloramphenicol acetyl transferase (CAT) gene and its expression in tobacco protoplasts, *Plant Mol. Biol.* **10,** 185–191.

ROTHSTEIN, S. J., DIMAIO, J., STRAND, M., RICE, D. (1987), Stable and heritable inhibition of the expression of nopaline synthase in tobacco expressing antisense RNA, *Proc. Natl. Acad. Sci. USA* **84,** 8439–8443.

ROY, P., SAHASRA BUDHE, N. (1990), A sensitive and simple paper chromatographic procedure for detecting neomycin phosphotransferase II (NPTII) gene expresseion, *Plant Mol. Biol.* **14,** 873–876.

RUSCONI, A., SCHAFFNER, W. (1981), Transformation of frog embryos with a rabbit β-globin gene, *Proc. Natl. Acad. Sci. USA* **78,** 5051–5055.

SAITO, K., YAMAZAKI, M., KANEKO, H., MURAKOSHI, I., FUKUDA, Y., VAN MONTAGU, M. (1991a), Tissue-specific and stress-enhancing expression of the TR promoter for mannopine synthase in transgenic medicinal plants, *Planta* **184,** 40–46.

SAITO, K., NOJI, M., OHMORI, S., IMAI, Y., MURAKOSHI, I. (1991b), Integration and expression of a rabbit liver cytochrome P-450 gene in transgenic *Nicotiana tabacum, Proc. Natl. Acad. Sci. USA* **88,** 7041–7045.

SANDERS, P. R., WINTER, J. A., BARNASON, A. R., ROGERS, S. G., FRALEY, R. T. (1987), Comparison of cauliflower mosaic virus 35S and nopaline synthase promoters in transgenic plants, *Nucleic Acids Res.* **15,** 1543–1558.

SANDLER, S. J., STAYTON, M., TOWNSEND, J. A., RALSTON, M. L., BEDBROOK, J. R., DUNSMUIR, P. (1988), Inhibition of gene expression in transformed plants by antisense RNA, *Plant Mol. Biol.* **11,** 301–310.

SANFORD, J. C. (1990), Biolistic plant transformation, *Physiol. Plant.* **79,** 206–209.

SÄNGER, H. L. (1982), Biology, structure, function and possible origin of viroids, in: *Encyclopedia of Plant Physiology,* New Series, Vol. 14B, pp. 368–454, (PARTHIER, B., BOULTER, D., Eds.), Berlin: Springer-Verlag.

SARVER, N., CANTIN, E. M., CHANG, P. S., ZAIA, J. A., LADNE, P. A., STEPHENS, D. A., ROSSI, J. J. (1990), Ribozymes as potential anti-HIV-1 therapeutic agent, *Science* **247,** 1222–1225.

SAUL, M. W., POTRYKUS, I. (1990), Direct gene transfer to protoplasts: fate of the transferred genes, *Dev. Genet.* **11,** 176–181.

SAXENA, S. K., ACKERMANN, E. J. (1990), Ribozymes correctly cleave a model substrate and endogenous RNA *in vivo, J. Biol. Chem.* **265,** 17106–17112.

SCHÄFER, W., WEISING, K., KAHL, G. (1984), T-DNA of a crown gall tumor is organized in nucleosomes, *EMBO J.* **3,** 373–376.

SCHÄFER, W., GÖRZ, A., KAHL, G. (1987), T-DNA integration and expression in a monocot crop plant after induction of *Agrobacterium, Nature* **327,** 529–532.

SCHÄFFNER, A. R., SHEEN, J. (1991), Maize rbcS promoter activity depends on sequence elements not found in dicot rbcS promoters, *Plant Cell* **3,** 997–1012.

SCHENA, M., LLOYD, M., DAVIS, R. W. (1991), A steroid-inducible gene expression system for plant cells, *Proc. Natl. Acad. Sci. USA* **88,** 10421–10425.

SCHERNTHANER, J. P., MATZKE, M. A., MATZKE, A. J. M. (1988), Endosperm-specific activity of a zein gene promoter in transgenic tobacco plants, *EMBO J.* **7,** 1249–1255.

SCHMIDT, R., WILLMITZER, L. (1988), High efficiency *Agrobacterium tumefaciens*-mediated transformation of *Arabidopsis thaliana* leaf and cotyledon explants, *Plant Cell Rep.* **7,** 583–586.

SCHMÜLLING, T., SCHELL, J., SPENA, A. (1988), Single genes from *Agrobacterium rhizogenes* influence plant development, *EMBO J.* **7,** 2621–2629.

SCHMÜLLING, T., SCHELL, J., SPENA, A. (1989), Promoters of the rolA, B, and C genes of *Agrobacterium rhizogenes* are differentially regulated in transgenic plants, *Plant Cell* **1,** 665–670.

SCHREIER, P. H., SEFTOR, E. A., SCHELL, J., BOHNERT, H. J. (1985), The use of nuclear-encoded sequences to direct the light-regulated synthesis and transport of a foreign protein into plant chloroplasts, *EMBO J.* **4,** 25–32.

SCHUCH, W., BIRD, C. R., RAY, J., SMITH, C. J. S., WATSON, C. F., MORRIS, P. C., GRAY, J. E., ARNOLD, C., SEYMOUR, G. B., TUCKER, G. A., GRIERSON, D. (1989), Control and manipulation of gene expression during tomato fruit ripening, *Plant Mol. Biol.* **13,** 303–311.

SCHWARTZ, D. (1989), Gene-controlled cytosine demethylation in the promoter region of the Ac transposable element in maize, *Proc. Natl. Acad. Sci. USA* **86,** 2789–2793.

SCHWARTZ, D., DENNIS, E. (1986), Transposase activity of the Ac controlling element in maize is regulated by its degree of methylation, *Mol. Gen. Genet.* **205,** 476–482.

SCHWEIGER, H. G., DIRK, J., KOOP, H.-U., KRANZ, E., NEUHAUS, G., SPANGENBERG, G. (1987), Individual selection, culture and manipulation of higher plant cells, *Theor. Appl. Genet.* **73,** 769–783.

SCOTT, R. J., DRAPER, J. (1987), Transformation of carrot tissues derived from proembryogenic suspension cells: A useful model system for gene expression studies in plants, *Plant Mol. Biol.* **8,** 265–274.

SCOTT-ELMER, J., SUNTER, G., GARDINER, W. E., BRAND, L., BROWNING, C. K., BISARO, D. M., ROGERS, S. G. (1988) *Agrobacterium*-mediated inoculation of plants with tomato golden mosaic virus DNAs, *Plant Mol. Biol.* **10,** 225–234.

SEKI, M., KOMEDA, Y., IIDA, A., YAMADA, Y., MORIKAWA, H. (1991), Transient expression of β-glucuronidase in *Arabidopsis thaliana* leaves and roots and *Brassica napus* stems using a pneumatic particle gun, *Plant Mol. Biol.* **17**, 259–263.

SELKER, E. U. (1990), DNA methylation and chromatin structure: a view from below, *Trends Biochem. Sci.* **15**, 103–107.

SEN, P., PAZOUR, G. J., ANDERSON, D., DAS, A. (1989), Cooperative binding of *Agrobacterium tumefaciens* virE2 protein to single-stranded DNA, *J. Bacteriol.* **171**, 2573–2580.

SENARATNA, T., MCKERSIE, B. D., KASHA, K. J., PROCUNIER, J. D. (1991), Direct DNA uptake during the imbibition of dry cells, *Plant Sci.* **79**, 223–228.

SHAH, D. M., HORSCH, R. B., KLEE, H. J., KISHORE, G. M., WINTER, J. A., TUMER, N. E., HIRONAKA, C. M., SANDERS, P. R., GASSER, C. S., AYKENT, S., SIEGEL, N. R., ROGERS, S. G., FRALEY, R. T. (1986), Engineering herbicide tolerance in transgenic plants, *Science* **233**, 478–481.

SHAHIN, E. A., SUKHAPINDA, K., SIMPSON, R. B., SPIVEY, R. (1986), Transformation of cultivated tomato by a binary vector in *Agrobacterium rhizogenes:* transgenic plants with normal phenotypes harbor binary vector T-DNA, but no Ri-plasmid T-DNA, *Theor. Appl. Genet.* **72**, 770–777.

SHAW, C. H. (1991), Swimming against the tide: chemotaxis in *Agrobacterium, BioEssays* **13**, 25–29.

SHAW, C. H., LEEMANS, J., SHAW, C. H., VAN MONTAGU, M., SCHELL, J. (1983), A general method for the transfer of cloned genes to plant cells, *Gene* **23**, 315–330.

SHAW, C. H., ASHBY, A. M., BROWN, A., ROYAL, C., LOAKE, G. J., SHAW, C. H. (1988), virA and virG are the Ti-plasmid functions required for chemotaxis of *Agrobacterium tumefaciens* towards acetosyringone, *Mol. Microbiol.* **2**, 413–417.

SHEEHY, R. E., KRAMER, M., HIATT, W. R. (1988), Reduction of polygalacturonase activity in tomato fruit by antisense RNA, *Proc. Natl. Acad. Sci. USA* **85**, 8805–8809.

SHEIKHOLESLAM, S. N., WEEKS, D. P. (1987), Acetosyringone promotes high efficiency transformation of *Arabidopsis thaliana* explants by *Agrobacterium tumefaciens, Plant Mol. Biol.* **8**, 291–298.

SHEN, W. H., PETIT, A., GUERN, J., TEMPÉ, J. (1988), Hairy roots are more sensitive to auxin than normal roots, *Proc. Natl. Acad. Sci. USA* **85**, 3417–3421.

SHIGEKAWA, K., DOWER, W. J. (1988), Electroporation of eukaryotes and prokaryotes: a general approach to the introduction of macromolecules into cells, *BioTechniques* **6**, 742–751.

SHILLITO, R. D., SAUL, M. W., PASZKOWSKI, J., MÜLLER, M., POTRYKUS, I. (1985), High efficiency direct gene transfer to plants, *Bio/Technology* **3**, 1099–1103.

SHIMAMOTO, K., TERADA, R., IZAWA, T., FUJIMOTO, H. 1989), Fertile transgenic rice plants regenerated from transformed protoplasts, *Nature* **338**, 274–276.

SHIMODA, N., TOYODA-YAMAMOTO, A., NAGAMINE, J., USAMI, S., KATAYAMA, M., SAKAGAMI, Y., MACHIDA, Y. (1990), Control of expression of *Agrobacterium* vir genes by synergistic actions of phenolic signal molecules and monosaccharides, *Proc. Natl. Acad. Sci. USA* **87**, 6684–6688.

SHIRSAT, A. H., WILFORD, N., CROY, R. R. D. (1989), Gene copy number and levels of expression in transgenic plants of a seed specific gene, *Plant Sci.* **61**, 75–80.

SIEGEL, A. (1985), Plant-virus-based vectors for gene transfer may be of considerable use despite a presumed high error frequency during RNA synthesis, *Plant Mol. Biol.* **4**, 327–329.

SILVER, P. A. (1991), How proteins enter the nucleus, *Cell* **64**, 489–497.

SINKAR, V. P., PYTHOUD, F., WHITE, F. F., NESTER, E. W., GORDON, M. P. (1988a), rolA locus of the Ri plasmid directs developmental abnormalities in transgenic tobacco plants, *Genes Dev.* **2**, 688–697.

SINKAR, V. P., WHITE, F. F., FURNER, I. J., ABRAHAMSEN, M., PYTHOUD, F., GORDON, M. P. (1988b), Reversion of aberrant plants transformed with *Agrobacterium rhizogenes* is associated with the transcriptional inactivation of the TL-DNA genes, *Plant Physiol.* **86**, 584–590.

SLATTER, R. E., DUPREE, P., GRAY, J. C. (1991), A scaffold-associated DNA region is located downstream of the pea plastocyanin gene, *Plant Cell* **3**, 1239–1250.

SLIGHTOM, J. L., DURAND-TARDIF, M., JOUANIN, L., TEPFER, D. (1986), Nucleotide sequence analysis of TL-DNA of *Agrobacterium rhizogenes* agropine type plasmid, *J. Biol. Chem.* **261**, 108–121.

SMEEKENS, S., WEISBEEK, P., ROBINSON, C. (1990), Protein transport into and within chloroplasts, *TIBS* **15**, 73–76.

SMITH, C. J. S., WATSON, C. F., RAY, J., BIRD, C. R., MORRIS, P. C., SCHUCH, W., GRIERSON, D. (1988), Antisense RNA inhibition of polygalacturonase gene expression in transgenic tomatoes, *Nature* **334**, 724–726.

SMITH, C. J. S., WATSON, C. F., MORRIS, P. C., BIRD, C. R., SEYMOUR, G. B., GRAY, J. E., ARNOLD, C., TUCKER, G. A., SCHUCH, W., HARDING, S., GRIERSON, D. (1990a), Inheritance and effect on ripening of antisense polygalacturonase genes in transgenic tomatoes, *Plant Mol. Biol.* **14,** 369–379.

SMITH, C. J. S., WATSON, C. F., BIRD, C. R., RAY, J., SCHUCH, W., GRIERSON, D. (1990b), Expression of a truncated tomato polygalacturonase gene inhibits expression of the endogenous gene in transgenic plants, *Mol. Gen. Genet.* **224,** 477–481.

SOYFER, V. N., TITOV, Y. B. (1981), The absence of a definite gene-specific effect in wheat after seed treatment with exogenous DNA, *Mol. Gen. Genet.* **182,** 361–363.

SPANGENBERG, G., NEUHAUS, G., POTRYKUS, I. (1990), Micromanipulation of higher plant cells, in: *Plant Cell Line Selection* (DIX, P. J., Ed.), Weinheim–New York–Basel–Cambridge: VCH.

SPANIER, K., SCHELL, J., SCHREIER, P. H. (1989), A functional analysis of T-DNA gene 6b: The fine tuning of cytokinin effects on shoot development, *Mol. Gen. Genet.* **219,** 209–216.

SPANO, L., MARIOTTI, D., CARDARELLI, M., BRANCA, C., COSTANTINO, P. (1988) Morphogenesis and auxin sensitivity of transgenic tobacco with different complements of Ri T-DNA, *Plant Physiol.* **87,** 479–483.

SPENA, A., SCHELL, J. (1987), The expression of a heat-inducible chimeric gene in transgenic tobacco plants, *Mol. Gen. Genet.* **206,** 436–440.

SPENA, A., HAIN, R., ZIERVOGEL, U., SAEDLER, H., SCHELL, J. (1985), Construction of a heat-inducible gene for plants. Demonstration of heat-inducible activity of the *Drosophila* hsp70 promoter in plants, *EMBO J.* **4,** 2739–2743.

SPENA, A., SCHMÜLLING, T., KONCZ, C., SCHELL, J. S. (1987), Independent and synergistic activity of rol A, B, and C loci in stimulating abnormal growth in plants, *EMBO J.* **6,** 3891–3899.

SPENCER, P. A., TOWERS, G. H. N. (1988), Specificity of signal compounds detected by *Agrobacterium tumefaciens, Phytochemistry* **27,** 2781–2785.

SPIELMANN, A., SIMPSON, R. B. (1986), T-DNA structure in transgenic tobacco plants with multiple independent integration sites, *Mol. Gen. Genet.* **205,** 34–41.

SPÖRLEIN, B., KOOP, H.-U. (1991), Lipofection: direct gene transfer to higher plants using cationic liposomes, *Theor. Appl. Genet.* **83,** 1–5.

SPÖRLEIN, B., MAYER, A., DAHLFELD, G., KOOP, H. U. (1991a), A microassay for quantitative determination of β-glucuronidase reporter gene activities in individually selected single higher plant cells, *Plant Sci.* **78,** 73–80.

SPÖRLEIN, B., STREUBEL, M., DAHLFELD, G., WESTHOFF, P., KOOP, H. U. (1991b), PEG-mediated plastid transformation: a new system for transient gene expression assays in chloroplasts, *Theor. Appl. Genet.* **82,** 717–722.

STACHEL, S. E., NESTER, E. W. (1986), The genetic and transcriptional organization of the vir region of the A6 Ti plasmid of *Agrobacterium tumefaciens, EMBO J.* **7,** 1445–1454.

STACHEL, S. E., ZAMBRYSKI, P. C. (1986a), virA and virG control the plant-induced activation of the T-DNA transfer process of *A. tumefaciens, Cell* **46,** 325–333.

STACHEL, S. E., ZAMBRYSKI, P. C. (1986b), *Agrobacterium tumefaciens* and the susceptible plant cell: a novel adaptation of extracellular recognition and DNA conjugation, *Cell* **47,** 155–157.

STACHEL, S. E., MESSENS, E., VAN MONTAGU, M., ZAMBRYSKI, P. (1985), Identification of the signal molecules produced by wounded plant cells that activate T-DNA transfer in *Agrobacterium tumefaciens, Nature* **318,** 624–629.

STACHEL, S. E., NESTER, E. W., ZAMBRYSKI, P. C. (1986a), A plant cell factor induces *Agrobacterium tumefaciens* vir gene expression, *Proc. Natl. Acad. Sci. USA* **83,** 379–383.

STACHEL, S. E., TIMMERMAN, B., ZAMBRYSKI, P. (1986b), Generation of single-stranded T-DNA molecules during the initial stages of T-DNA transfer from *Agrobacterium tumefaciens* to plant cells, *Nature* **322,** 706–712.

STACHEL, S. E., TIMMERMANN, B., ZAMBRYSKI, P. (1987), Activation of *Agrobacterium tumefaciens* vir gene expression generates multiple single-stranded T-strand molecules from the pTiA6 T-region: requirement for 5′virD gene products, *EMBO J.* **6,** 857–863.

STAIGER, D., KAULEN, H., SCHELL, J. (1989), A CACGTG motif of the *Antirrhinum majus* chalcone synthase promoter is recognized by an evolutionary conserved nuclear protein, *Proc. Natl. Acad. Sci. USA* **86,** 6930–6934.

STANLEY, J., TOWNSEND, R. (1986), Infectious mutants of cassava latent virus generated *in vivo* from intact recombinant DNA clones containing single copies of the genome, *Nucleic Acids Res.* **14,** 5981–5998.

STECK, T. R., CLOSE, T. J., KADO, C. I. (1989), High levels of double-stranded transferred DNA (T-DNA) processing from an intact nopaline Ti plasmid, *Proc. Natl. Acad. Sci. USA* **86,** 2133–2137.

STEINBISS, H. H., STABEL, P. (1983), Protoplast derived tobacco cells can survive capillary microinjection of the fluorescent dye lucifer yellow, *Protoplasma* **116,** 223–227.

STIEF, A., WINTER, D. M., STRÄTLING, W. H., SIPPEL, A. E. (1989), A nuclear attachment element mediates elevated and position-independent gene activity, *Nature* **341,** 343–345.

STITT, M., QUICK, W. P., SCHURR, U., SCHULZE, E.-D., RODERMEL, S. R., BOGORAD, L. (1991), Decreased ribulose-1,5-bisphosphate carboxylase-oxygenase in transgenic tobacco transformed with antisense rbcS. II. Flux-control coefficients for photosynthesis in varying light, CO_2, and air humidity, *Planta* **183,** 555–566.

STOCKHAUS, J., ECKES, P., BLAU, A., SCHELL, J., WILLMITZER, L. (1987), Organ-specific and dosage-dependent expression of a leaf/stem specific gene from potato after tagging and transfer into potato and tobacco plants, *Nucleic Acids Res.* **15,** 3479–3491.

STOCKHAUS, J., HÖFER, M., RENGER, G., WESTHOFF, P., WYDRZYNSKI, T., WILLMITZER, L. (1990), Anti-sense RNA efficiently inhibits formation of the 10 kd polypeptide of photosystem II in transgenic potato plants: analysis of the role of the 10 kd protein, *EMBO J.* **9,** 3013–3021.

SUKHAPINDA, K., SPIVEY, R., SIMPSON, R. B., SHAHIN, E. A. (1987), Transgenic tomato (*Lycopersicon esculentum* L.) transformed with a binary vector in *Agrobacterium rhizogenes:* Nonchimeric origin of callus clone and low copy numbers of integrated vector T-DNA, *Mol. Gen. Genet.* **206,** 491–497.

SUNTER, G., GARDINER, W. E., RUSHING, A. E., ROGERS, S. G., BISARO, D. M. (1987), Independent encapsidation of tomato golden mosaic virus A component DNA in transgenic plants, *Plant Mol. Biol.* **8,** 477–484.

SÜSSMUTH, J., DRESSLER, K., HESS, D. (1991), *Agrobacterium*-mediated transfer of the GUS gene into pollen of *Petunia, Bot. Acta* **104,** 72–76.

SVAB, Z., HAJDUKIEWICZ, P., MALIGA, P. (1990), Stable transformation of plastids in higher plants, *Proc. Natl. Acad. Sci. USA* **87,** 8526–8530.

SYMONS, R. H. (1991), Ribozymes, *Crit. Rev. Plant Sci.* **10,** 189–234.

TADA, Y., SAKAMOTO, M., FUJIMURA, T. (1990), Efficient gene introduction into rice by electroporation and analysis of transgenic plants: use of electroporation buffer lacking chloride ions, *Theor. Appl. Genet.* **80,** 475–480.

TAGU, D., CRETIN, C., BERGOUNIOUX, C., LEPINIEC, L., GADAL, P. (1991), Transcription of a sorghum phosphoenolpyruvate carboxylase gene in transgenic tobacco leaves: maturation of monocot PRE-mRNA by dicot cells, *Plant Cell Rep.* **9,** 688–690.

TAIT, R.C., KADO, C. I. (1988), Regulation of the virC and virD promoters of pTi C58 by the ros chromosomal mutation of *Agrobacterium tumefaciens, Mol. Microbiol.* **2,** 385–392.

TAKAMATSU, N., ISHIKAWA, M., MESHI, T., OKADA, Y. (1987), Expression of bacterial chloramphenicol acetyltransferase gene in tobacco plants mediated by TMV-RNA, *EMBO J.* **6,** 307–311.

TALBOT, D., COLLIS, P., ANTONIOU, M., VIDAL, M., GROSVELD, F., GREAVES, D. R. (1989), A dominant control region from the human β-globin locus conferring integration site-independent gene expression, *Nature* **338,** 352–355.

TAZI, J., BIRD, A. (1990), Alternative chromatin structure at CpG islands, *Cell* **60,** 909–920.

TEERI, T. H., HERRERA-ESTRELLA, L., DEPICKER, A., VAN MONTAGU, M., PALVA, E. T. (1986), Identification of plant promoters *in situ* by T-DNA-mediated transcriptional fusions ot the npt-II gene, *EMBO J.* **5,** 1755–1760.

TEERI, T. H., LEHVÄSLAIHO, H., FRANCK, M., UOTILA, J., HEINO, P., PALVA, E. T., VAN MONTAGU, M., HERRERA-ESTRELLA, L. (1989), Gene fusions to lacZ reveal new expression patterns of chimeric genes in transgenic plants, *EMBO J.* **8,** 343–350.

TEPFER, D. (1984), Transformation of several species of higher plants by *Agrobacterium rhizogenes:* Sexual transmission of the transformed genotype and phenotype, *Cell* **37,** 959–967.

THOMASHOW, M. F., HUGLY, S., BUCHHOLZ, W. G., THOMASHOW, L. S. (1986), Molecular basis for the auxin-independent phenotype of crown gall tumor tissues, *Science* **231,** 616–618.

THOMPSON, C. J., RAO MOVVA, N., TIZARD, R., CRAMERI, R., DAVIES, J. E. (1987), Characterization of the herbicide resistance gene bar from *Streptomyces hygroscopicus, EMBO J.* **6,** 2519–2523.

THOMPSON, D. V., MELCHERS, L. S., IDLER, K. B., SCHILPEROORT, R. A., HOOYKAAS, P. J. J. (1988), Analysis of the complete nucleotide sequence of the *Agrobacterium tumefaciens* virB operon, *Nucleic Acids Res.* **16,** 4621–4636.

TIMMERMAN, B., VAN MONTAGU, M., ZAMBRYSKI, P. (1988), Vir-induced recombination in *Agrobacterium*. Physical characterization of precise and imprecise T-circle formation, *J. Mol. Biol.* **203,** 373–384.

TINLAND, B., HUSS, B., PAULUS, F., BONNARD, G., OTTEN, L. (1989), *Agrobacterium tumefaciens* 6b genes are strain-specific and affect the activity of auxin as well as cytokinin genes, *Mol. Gen. Genet.* **219,** 217–224.

TINLAND, B., ROHFRITSCH, O., MICHLER, P., OTTEN, L. (1990), *Agrobacterium tumefaciens* T-DNA gene 6b stimulates rol-induced root formation, permits growth at high auxin concentrations and increases root size, *Mol. Gen. Genet.* **223**, 1–10.

TÖPFER, R., PRÖLS, M., SCHELL, J., STEINBIß, H.-H. (1988), Transient gene expression in tobacco protoplasts: II. Comparison of the reporter gene systems for CAT, NPT II, and GUS, *Plant Cell Rep.* **7**, 225–228.

TÖPFER, R., GRONENBORN, B., SCHELL, J., STEINBISS, H.-H. (1989), Uptake and transient expression of chimeric genes in seed-derived embryos, *Plant Cell.* **1**, 133–139.

TÖPFER, R., GRONENBORN, B., SCHAEFER, S., SCHELL, J., STEINBIß, H.-H. (1990), Expression of engineered wheat dwarf virus in seed-derived embryos, *Physiol. Plant.* **79**, 158–162.

TORIYAMA, K., ARIMOTO, Y., UCHIMIYA, H., HINATA, K. (1988), Transgenic rice plants after direct gene transfer into protoplasts, *Bio/Technology* **6**, 1072–1074.

TORO, N., DATTA, A., YANOFSKY, M., NESTER, E. (1988), Role of the overdrive sequence in T-DNA border cleavage in *Agrobacterium, Proc. Natl. Acad. Sci. USA* **85**, 8558–8562.

TORO, N., DATTA, A., CARMI, O. A., YOUNG, C., PRUSTI, R. K., NESTER, E. W. (1989), The *Agrobacterium tumefaciens* vir C1 gene product binds to overdrive, the T-DNA transfer enhancer, *J. Bacteriol.* **171**, 6845–6849.

TOYODA, H., MATSUDA, Y., UTSUMI, R., OUCHI, S. (1988), Intranuclear microinjection for transformation of tomato callus cells, *Plant Cell Rep.* **7**, 293–296.

TOYODA, H., YAMAGA, T., MATSUDA, Y., OUCHI, S. (1990), Transient expression of the β-glucuronidase gene introduced into barley cells by microinjection, *Plant Cell Rep.* **9**, 299–302.

TRINH, T. H., MANTE, S., PUA, E.-C., CHUA, N.-H. (1987), Rapid production of transgenic flowering shoots and F1 progeny from *Nicotiana plumbaginifolia* epidermal peels, *Bio/Technology* **5**, 1081–1084.

TURNER, D. R., BUTLER, P. J. G. (1986), Essential features of the assembly origin of tobacco mosaic virus RNA as studied by directed mutagenesis, *Nucleic Acids Res.* **14**, 9229–9242.

TWELL, D., KLEIN, T. M., FROMM, M. E., MCCORMICK, S. (1989), Transient expression of chimeric genes delivered into pollen by microprojectile bombardment, *Plant Physiol.* **91**, 1270–1274.

UENG, P., GALILI, G., SAPANARA, V., GOLDSBROUGH, P. B., DUBE, P., BEACHY, R. N., LARKINS, B. A. (1988), Expression of a maize storage protein gene in *Petunia* plants is not restricted to seeds, *Plant Physiol.* **86**, 1281–1285.

URSIC, D., SLIGHTOM, J. L., KEMP, J. D. (1983), *Agrobacterium tumefaciens* T-DNA integrate into multiple sites of the sunflower crown gall genome, *Mol. Gen. Genet.* **190**, 494–503.

VAECK, M., REYNAERTS, A., HÖFTE, H., JANSENS, S., DE BEUCKELEER, M., DEAN, C., ZABEAU, M., VAN MONTAGU, M., LEEMANS, J. (1987), Transgenic plants protected from insect attack, *Nature* **328**, 33–37.

VANCANNEYT, G., SCHMIDT, R., O'CONNOR-SANCHEZ, A., WILLMITZER, L., ROCHA-SOSA, M. (1990), Construction of an intron-containing marker gene: splicing of the intron in transgenic plants and its use in monitoring early events in *Agrobacterium*-mediated plant transformation, *Mol. Gen. Genet.* **220**, 245–250.

VAN DEN BROECK, G., TIMKO, M. P., KAUSCH, A. P., CASHMORE, A. R., VAN MONTAGU, M., HERRERA-ESTRELLA, L. (1985), Targeting of a foreign protein to chloroplasts by fusion to the transit peptide from the small subunit of ribulose 1,5-bisphosphate carboxylase, *Nature* **313**, 358–363.

VAN DEN ELZEN, P., LEE, K. Y., TOWNSEND, J., BEDBROOK, J. (1985a), Simple binary vectors for DNA transfer to plant cells, *Plant Mol. Biol.* **5**, 149–154.

VAN DEN ELZEN, P. J. M., TOWNSEND, J., LEE, K. Y., BEDBROOK, J. R. (1985b), A chimaeric hygromycin resistance gene as a selectable marker in plant cells, *Plant Mol. Biol.* **5**, 299–302.

VAN DEN ELZEN, P. J. M., HUISMAN, M. J., POSTHUMUS-LUTKE WILLINK, D., JONGEDIJK, E., HOEKEMA, A., CORNELISSEN, B. J. C. (1989), Engineering virus resistance in agricultural crops, *Plant Mol. Biol.* **13**, 337–346.

VAN DER KROL, A. R., LENTIN, P. E., VEENSTRA, J., VAN DER MEER, I. M., KOES, R. E., GERATS, A. G. M., MOL, J. N. M., STUITJE, A. R. (1988), An anti-sense chalcone synthase gene in transgenic plants inhibits flower pigmentation, *Nature* **333**, 866–869.

VAN DER KROL, A. R., MUR, L. A., BELD, M., MOL, J. N. M., STUITJE, A. R. (1990a), Flavonoid genes in petunia: addition of a limited number of gene copies may lead to a suppression of gene expression, *Plant Cell* **2**, 291–299.

VAN DER KROL, A. R., MUR, L. A., DE LANGE, P., MOL, J. N. M., STUITJE, A. R. (1990b), Inhibition of flower pigmentation by antisense CHS genes: promoter and minimal sequence requirements for the antisense effect, *Plant Mol. Biol.* **14**, 457–466.

VAN DER KROL, A. R., MUR, L. A., DE LANGE, P., GERATS, A. G. M., MOL, J. N. M., STUITJE,

A. R. (1990c), Antisense chalcone synthase genes in petunia: visualization of variable transgene expression, *Mol. Gen. Genet.* **220,** 204–212.

Van Lijsebettens, M. Inzé, D., Schell, J., Van Montagu, M. (1986), Transformed cell clones as a tool to study T-DNA integration mediated by *Agrobacterium tumefaciens, J. Mol. Biol.* **188,** 129–145.

Van Santen, V. L., Spritz, R. A. (1987), Splicing of plant pre-mRNAs in animal systems and *vice versa, Gene* **56,** 253–265.

Van Vloten-Doting, L., Bol, J. F., Cornelissen, B. (1985), Plant-virus-based vectors for gene transfer will be of limited use because of the high error frequency during viral RNA synthesis, *Plant Mol. Biol.* **4,** 323–326.

Vasil, I. K. (1988), Progress in the regeneration and genetic manipulation of cereal crops, *Bio/Technology* **6,** 397–402.

Vasil, V., Clancy, M., Ferl, R. J., Vasil, I. K., Hannah, L. C. (1989), Increased gene expression by the first intron of maize shrunken-1 locus in grass species, *Plant Physiol.* **91,** 1575–1579.

Vasil, V., Redway, F., Vasil, I. K. (1990), Regeneration of plants from embyrogenic suspension culture protoplasts of wheat (*Triticum aestivum* L.), *Bio/Technolgy* **8,** 429–434.

Vasil, V., Brown, S. M., Re, D., Fromm, M. E., Vasil, I. K. (1991), Stably transformed callus lines from microprojectile bombardment of cell suspension cultures of wheat, *Bio/Technology* **9,** 743–747.

Velten, J., Schell, J. (1985), Selection-expression plasmid vectors for use in genetic transformation at higher plants, *Nucleic Acids Res.* **13,** 6981–6998.

Velten, J., Velten, R., Hain, R., Schell, J. (1984), Isolation of a dual plant promoter fragment from the Ti plasmid of *Agrobacterium tumefaciens, EMBO J.* **3,** 2723–2730.

Veluthambi, K., Jayaswal, R. K., Gelvin, S. B. (1987), Virulence genes A, G, and D mediate the double-stranded border cleavage of T-DNA from the *Agrobacterium* Ti plasmid, *Proc. Natl. Acad. Sci. USA* **84,** 1881–1885.

Veluthambi, K., Ream, W., Gelvin, S. B. (1988), Virulence genes, borders, and overdrive generate single-stranded T-DNA molecules from the A6 Ti plasmid of *Agrobacterium tumefaciens, J. Bacteriol.* **170,** 1523–1532.

Veluthambi, K., Krishnan, M., Gould, J. H., Smith, R. H., Gelvin, S. B. (1989), Opines stimulate induction of the vir genes of the *Agrobacterium tumefaciens* Ti plasmid, *J. Bacteriol.* **171,** 3696–3703.

Vilaine, F., Casse-Delbart, F. (1987), Independent induction of transformed roots by the TL and TR regions of the Ri plasmid of agropine type *Agrobacterium rhizogenes, Mol. Gen. Genet.* **206,** 17–23.

Vos, P., Jaegle, M., Wellink, J., Verver, J., Eggen, R., van Kammen, A., Goldbach, R. (1988), Infectious RNA transcripts derived from full-length DNA copies of the genomic RNAs of cowpea mosaic virus, *Virology* **165,** 33–41.

Waldron, C., Murphy, E. B., Roberts, J. L., Gustafson, G. D., Armour, S. L., Malcolm, S. K. (1985), Resistance to hygromycin B, *Plant Mol. Biol.* **5,** 103–108.

Wallroth, M., Gerats, A. G. M., Rogers, S. G., Fraley, R. T., Horsch, R. B. (1986), Chromosomal localization of foreign genes in *Petunia hybrida, Mol. Gen. Genet.* **202,** 6–15.

Wang, K., Herrera-Estrella, L., Van Montagu, M., Zambryski, P. (1984), Right 25 bp terminus sequence of the nopaline T-DNA is essential for and determines direction of DNA transfer from *Agrobacterium* to the plant genome, *Cell* **38,** 455–462.

Wang, K., Stachel, S. E., Timmermann, B., Van Montagu, M., Zambryski, P. C. (1987), Site-specific nick in the T-DNA border sequence as a result of *Agrobacterium* vir gene expression, *Science* **235,** 587–591.

Wang, K., Herrera-Estrella, A., Van Montagu, M. (1990), Overexpression of virD1 and virD2 genes in *Agrobacterium tumefaciens* enhances T-complex formation and plant transformation, *J. Bacteriol.* **172,** 4432–4440.

Wang, Y.-C., Klein, T. M., Fromm, M., Cao, J., Sanford, J. C., Wu, R. (1988), Transient expression of foreign genes in rice, wheat and soybean cells following particle bombardment, *Plant Mol. Biol.* **11,** 433–439.

Ward, A., Etessami, P., Stanley, J. (1988), Expression of a bacterial gene in plants mediated by infectious geminivirus DNA, *EMBO J.* **7,** 1583–1587.

Ward, E. R., Barnes, W. M. (1988), VirD2 protein of *Agrobacterium tumefaciens* very tightly linked to the 5′ end of T-strand DNA, *Science* **242,** 927–930.

Ward, J. E., Dale, E. M., Nester, E. W., Binns, A. N. (1990), Identification of a virB10 protein aggregate in the inner membrane of *Agrobacterium tumefaciens, J. Bacteriol.* **172,** 5200–5210.

Watson, J. C., Kaufman, L. S., Thompson, W. F. (1987), Developmental regulation of cytosine methylation in the nuclear ribosomal RNA genes of *Pisum sativum, J. Mol. Biol.* **193,** 15–26.

Weber, G. (1988), Microperforation of plant tissue with a UV laser microbeam and injection of DNA into cells, *Naturwissenschaften* **75,** 35–36.

WEBER, G., MONAJEMBASHI, S., GREULICH, K. O., WOLFRUM, J. (1988), Genetic manipulation of plant cells and organelles with a laser microbeam, *Plant Cell Tissue Organ Cult.* **12,** 219–222.

WEBER, G., MONAJEMBASHI, S., WOLFRUM, J., GREULICH, K.-O. (1990), Genetic changes induced in higher plant cells by a laser microbeam, *Physiol. Plant.* **79,** 190–193.

WEINTRAUB, H., CHANG, P. F., CONRAD, K. (1986), Expression of transfected DNA depends on DNA topology, *Cell* **46,** 115–122.

WEISING, K., KAHL, G. (1991), Towards an understanding of plant gene regulation: the action of nuclear factors, *Z. Naturforsch.* **46c,** 1–11.

WEISING, K., SCHELL, J., KAHL, G. (1988), Foreign genes in plants: transfer, structure, expression, and applications, *Annu. Rev. Genet.* **22,** 421–477.

WEISING, K., BOHN, H., KAHL, G. (1990), Chromatin structure of transferred genes in transgenic plants, *Dev. Genet.* **11,** 233–247.

WERR, W., LÖRZ, H. (1986), Transient gene expression in a Gramineae cell line. A rapid procedure for studying plant promoters, *Mol. Gen. Genet.* **202,** 471–475.

WHITE, F. F., TAYLOR, B. H., HUFFMAN, G. A., GORDON, M. P., NESTER, E. W. (1985), Molecular and genetic analysis of the transferred DNA regions of the root-inducing plasmid of *Agrobacterium rhizogenes, J. Bacteriol.* **164,** 33–44.

WILLIAMSON, J. D., GALILI, G., LARKINS, B. A., GELVIN, S. B. (1988), The synthesis of a 19 kilodalton zein protein in transgenic *Petunia* plants, *Plant Physiol.* **88,** 1002–1007.

WILLMITZER, L. (1988), The use of transgenic plants to study plant gene expression, *Trends Genet.* **4,** 13–18.

WILLMITZER, L., SIMONS, G., SCHELL, J. (1982), The TL-DNA in octopine crown-gall tumours codes for seven well-defined polyadenylated transcripts, *EMBO J.* **1,** 139–146.

WILLMITZER, L., DHAESE, P., SCHREIER, P. H., SCHMALENBACH, W., VAN MONTAGU, M., SCHELL, J. (1983), Size, location and polarity of T-DNA-encoded transcripts in nopaline crown gall tumors, common transcripts in octopine and nopaline tumors, *Cell* **32,** 1045–1056.

WILSON, C., BELLEN, H. J., GEHRING, W. J. (1990), Position effects on eukaryotic gene expression, *Annu. Rev. Cell. Biol.* **6,** 679–714.

WILSON, T. M. A. (1985), Nucleocapsid disassembly and early gene expression by positive-strand RNA viruses, *J. Gen. Virol.* **66,** 1201–1207.

WINANS, S. C. (1990), Transcriptional induction of an *Agrobacterium* regulatory gene at tandem promoters by plant-released phenolic compounds, phosphate starvation, and acidic growth media, *J. Bacteriol.* **172,** 2433–2438.

WINANS, S. C., KERSTETTER, R. A., NESTER, E. W. (1988), Transcriptional regulation of the virA and virG genes of *Agrobacterium tumefaciens, J. Bacteriol.* **170,** 4047–4054.

WINANS, S. C., KERSTETTER, R. A., WARD, J. E., NESTER, E. W. (1989), A protein required for transcriptional regulation of *Agrobacterium* virulence genes spans the cytoplasmic membrane, *J. Bacteriol.* **171,** 1616–1622.

WING, D., KONCZ, C., SCHELL, J. (1989), Conserved function in *Nicotiana tabacum* of a single *Drosophila* hsp70 promoter heat shock element when fused to a minimal T-DNA promoter, *Mol. Gen. Genet.* **219,** 9–16.

WIRTZ, U., SCHELL, J., CZERNILOFSKY, A. P. (1987), Recombination of selectable marker DNA in *Nicotiana tabacum, DNA* **6,** 245–253.

WOHLLEBEN, W., ARNOLD, W., BROER, I., HILLEMANN, D., STRAUCH, E., PÜHLER, A. (1988), Nucleotide sequence of the phosphinothricin N-acetyltransferase gene from *Streptomyces viridochromogenes* Tü494 and its expression in *Nicotiana tabacum, Gene* **70,** 25–37.

WOOLSTON, C. J., BARKER, R., GUNN, H., BOULTON, M. I., MULLINEAUX, P. M. (1988), Agroinfection and nucleotide sequence of cloned wheat dwarf virus DNA, *Plant Mol. Biol.* **11,** 35–43.

YANG, N.-S., RUSSELL, D. (1990), Maize sucrose synthase-1 promoter directs phloem cell-specific expression of Gus gene in transgenic tobacco plants, *Proc. Natl. Acad. Sci. USA* **87,** 4144–4148.

YANOFSKY, M., NESTER, E. W. (1986), Molecular characterization of a host-range-determining locus from *Agrobacterium tumefaciens, J. Bacteriol.* **168,** 244–250.

YANOFSKY, M. F., PORTER, S. G., YOUNG, C., ALBRIGHT L. M., GORDON, M. P., NESTER, E. W. (1986), The virD operon of *Agrobacterium tumefaciens* encodes a site-specific endonuclease, *Cell* **47,** 471–477.

YE, G.-N., DANIELL, H., SANFORD, J. C. (1990), Optimization of delivery of foreign DNA into higher-plant chloroplasts, *Plant Mol. Biol.* **15,** 809–819.

YOUNG, C., NESTER, E. W. (1988), Association of the virD2 protein with the 5′end of T strands in *Agrobacterium tumefaciens, J. Bacteriol.* **170,** 3367–3374.

ZAMBRYSKI, P. (1988), Basic processes underlying *Agrobacterium*-mediated DNA transfer to plant cells, *Annu. Rev. Genet.* **22,** 1–30.

ZAMBRYSKI, P., HOLSTERS, M., KRUGER, K., DEPICKER, A., SCHELL, J., VAN MONTAGU, M., GOODMAN, H. (1980), Tumor DNA structure in plant cells transformed by *A. tumefaciens*, *Science* **209**, 1385–1391.

ZAMBRYSKI, P., LOOS, H., GENETELLO, C., LEEMANS, J., VAN MONTAGU, M., SCHELL, J. (1983), Ti plasmid vector for the introduction of DNA into plant cells without alteration of their normal regeneration capacity, *EMBO J.* **2**, 2143–2150.

ZAMBRYSKI, P., TEMPE, J., SCHELL, J. (1989), Transfer and function of T-DNA genes from *Agrobacterium* Ti and Ri plasmids in plants, *Cell* **56**, 193–201.

ZERBACK, R., DRESSLER, K., HESS, D. (1989), Flavonoid compounds from pollen and stigma of *Petunia hybrida:* inducers of the vir region of the *Agrobacterium tumefaciens* Ti plasmid, *Plant Sci.* **62**, 83–91.

ZHANG, D., EHRLICH, K. C., SUPAKAR, P. C., EHRLICH, M. (1989), A plant DNA-binding protein that recognizes 5-methylcytosine residues, *Mol. Cell. Biol.* **9**, 1351–1356.

ZHENG, H., WILSON, J. H. (1990), Gene targeting in normal and amplified cell lines, *Nature* **344**, 170–173.

16 Transgenic Plants

LOTHAR WILLMITZER

Berlin, Federal Republic of Germany

1 Introduction

Classical plant breeding has by all accounts, been most successful in the last couple of decades. However, the rapid population growth, the resulting food and energy shortages anticipated, and the numerous ecological problems due partly to intensive agricultural practices of our present society, necessitate the major limitations of classical breeding to be overcome.

First, classical plant breeding is limited by sexual compatibility. This results in a very limited gene pool (when compared to the total gene pool of all different organisms) being available for breeding processes involving sexual crosses. Second, inherent to the use of sexual crosses, the offspring consists of a 1:1 mixture of the genes present in the wild relative and in the elite plant. This leads to many undesirable traits from the wild-type relative being incorporated, in addition to the desired trait, into the elite plant. An extensive time- and labor-consuming effort is consequently required to eliminate the undesired characteristics by out-crossing, while still maintaining the desired ones. A typical breeding scheme will involve numerous back-crosses lasting at least 10–15 years.

Both these major limitations of classical plant breeding can be overcome by the advances made in the field of plant biotechnology during the last decade. The major breakthroughs include: the ability of transferring isolated and well-defined genes into plant cells, the identification and characterization of regulatory structures present in plant genes, the regeneration of intact and fertile transgenic plants starting from a single transformed cell, and the generation of the first examples of transgenic crop plants which show improved ugronomic traits.

Since genes are transferred by ways different from that of sexual processes, there is no limitation with respect to the gene pool available. In addition, since only well-defined single genes or small groups of genes encoding the desired trait are transferred, gene technology should enable the starting elite plant to keep all its good characteristics while being specifically improved for the desired traits. Thus, ideally, no back-crossing is needed as no genes exerting a negative effect should be transferred.

This however, represents the ideal situation and it will be shown later that the reality might be different. Nevertheless, gene technology could have a substantial impact on plant breeding, and the first examples which demonstrate the power of this technology, will be discussed in this chapter.

2 Transfer of DNA into Plant Cells and Regeneration of Intact and Fertile Plants

The transfer of isolated DNA into plant cells has become a routine procedure which is applicable to many different plants, including crop species. Several methods exist to transfer DNA into cells. The use of the Gram negative soil bacterium *Agrobacterium tumefaciens* to introduce genetic material into the nucleus of most dicotyledonous plants (cf. ZAMBRYSKI et al., 1989, for a review) remains the method of choice for transforming dicotyledonous plants. Second to this, is the direct transformation of wall-less protoplasts of plants by naked DNA present in the medium, in the presence of polyethyleneglycol or calcium (cf. POTRYKUS, 1991, for a review). Numerous other possibilities of delivering DNA into cells include microinjection (CROSSWAY et al., 1986), liposome fusion (CABOCHE, 1990), and the biolistics approach (KLEIN et al., 1987). While it is facile to obtain single cells or undifferentiated cell clumps that are genetically altered due to gene transfer, obtaining intact and fertile transgenic plants has proved a much greater obstacle. Thus, at present, the number of plant species and genotypes accessible to gene manipulation is less limited by the ability to transfer single, isolated genes into plant cells, but more by the inability to regenerate intact and fertile plants from such cells. The various strategies used to overcome this problem employ different vector and delivery systems for transferring the DNA

Tab. 1. Chimeric Marker Genes Used for Recognizing Transformed Plant Cells

Enzymatic Activity	Resistance to	Reference
A Selectable Marker Genes		
Neomycin phosphotransferase	Kanamycin, G418	HERRERA-ESTRELLA et al., 1983
Hygromycin phosphotransferase	Hygromycin	WALDRON et al., 1985
Dihydrofolate reductase	Methotrexate	HERRERA-ESTRELLA et al., 1983
Gentamycin acetyltransferase	Gentamycin	HAYFORD et al., 1988
Bleomycin resistance	Bleomycin	HILLE et al., 1986
Phosphinotricine acetyltransferase	Phosphinotricine	DE BLOCK et al., 1987
B Detectable Marker Genes		
Nopaline synthase		OTTEN and SCHILPEROORT, 1978
β-Glucuronidase		JEFFERSON, 1987
Streptomycin phosphotransferase		JONES et al., 1987
Firefly luciferase		OW et al., 1987
Bacterial luciferase		KONCZ et al., 1987

into plant cells, and are specific to the plant species and its ability to be cultured.

2.1 Use of *Agrobacterium*-Based Vectors

Both *A. tumefaciens* and less widely used *Agrobacterium rhizogenes*, have the ability to transform multicellular explants as well as protoplasts/single cells. In the case of *A. tumefaciens* containing disarmed vectors (i.e., devoid of oncogenes), transformation events can only be recognized with the help of dominant, selectable markers. A variety of chimeric genes, either encoding enzymes able to detoxify an otherwise toxic compound, or allowing an easy detection in plant cells, have been constructed. Their use has been demonstrated in transgenic plants (Tab. 1). For most plant species, it is easier to regenerate plants starting from a multicellular explant than from a single cell. Hence, *Agrobacterium* vectors are used preferentially for transformation provided the plant is susceptible to *Agrobacterium*, which holds true for most dicotyledonous plants. Various multicellular explants have been used for transformation, the most common ones being leaf discs, stem discs, cotyledon explants, discs from storage tissues such as tuber or tap root, and root pieces. An increasing number of intact, fertile transgenic plants has been obtained by *Agrobacterium*-mediated transformation and an incomplete list is given in Tab. 2. Thus, *Agrobacterium*-based vectors remain the most versatile vectors. Their major disadvantage however, is *Agrobacterium*'s limited host range. Despite several reports claiming the successful use of *Agrobacterium* in transforming cereals (RAINERI et al., 1990; HESS et al., 1990; MOONEY et al., 1991), monocotyledonous plants still seem to be a major barrier to *Agrobacterium*-based vectors (for a critical review see POTRYKUS, 1990). To date, there are no unequivocal reports of fertile, transgenic cereals having been obtained by *Agrobacterium*-mediated transformation.

Tab. 2. Transgenic Plant Species Obtained with *Agrobacterium* Vectors

Species	Reference
Petunia	HORSCH et al., 1985
Tomato	MCCORMICK et al., 1986
Potato	STIEKEMA et al., 1988
Tobacco	HORSCH et al., 1985
Arabidopsis	LLOYD et al., 1986
Lettuce	MICHELMORE et al., 1987
Sunflower	EVERETT et al., 1987
Oilseed rape	FRY et al., 1987
Flax	BASIRAN et al., 1987
Cotton	UMBECK et al., 1987
Sugarbeet	BOTTERMAN and LEEMANS, 1989
Celery	CATLIN et al., 1987
Soybean	MCCABE et al., 1988
Alfalfa	KUCHUK et al., 1990
Cucumber	TRULSON et al., 1986
Cauliflower	DAVID and TEMPE, 1988
Horseradish	NODA et al., 1987
Morning glory	TEPFER, 1984
Pea	PUONTI-KAERLAS et al., 1990
Chrysanthemum	LEDGER et al., 1991
Strawberry	NEHRA et al., 1990
Kalanchoe laciniata	JIA et al., 1989
Stylosanthes humilis	MANNERS and WAY, 1989
Pepino	ATKINSON and GARDNER, 1991
Eggplant	ROTINO and GLEDDIE, 1990
Muskmelon	FANG and GRUMET, 1990
Vicia narbonensis	PICKARDT et al., 1991
Poplar	PHYTHOUD et al., 1987
Walnut	MCGRANAHAN et al., 1988
Apple	JAMES et al., 1989
Allocasarina verticillata Lam.	PHELEP et al., 1991
Asparagus	BYTEBIER et al., 1987

Tab. 3. Transformation of Cereal Species (Transgenic Plants)

Species	Reference
Rice (japonica type)	TORIYAMA et al., 1988
	ZHANG and WU, 1988
	SHIMAMOTO et al., 1989
Rice (indica type)	DATTA et al., 1990
Maize	RHODES et al., 1989
	FROMM et al., 1990
	GORDON-KAMM et al., 1990
	MÜLLNER, 1991

2.2 Direct Delivery of DNA into Protoplasts

Plant protoplasts, like cells and spheroplasts from many pro- and eukaryotic organisms, take up DNA readily in the presence of polyethyleneglycol (PEG) or calcium. This method along with dominant selectable marker genes (see Tab. 1) has been used to transform protoplasts of a wide variety of sources. When compared to the *Agrobacterium*-based delivery system, the direct transfer of DNA has the unique advantage of being applicable to protoplasts of any plant source and thus, is not limited by the host range. The direct delivery of DNA into plant protoplasts has been applied to both dicotyledonous and monocotyledonous plant species, including cereals (cf. POTRYKUS, 1991, for a review).

The major disadvantage of using protoplasts however, is that the conditions necessary for protoplast to plant regeneration have been ascertained for fewer plant species than those for the regeneration of multicellular explants.

Nevertheless, during the last couple of years, protoplast to plant regeneration systems have been established in particular for the most important crop species, i.e., cereals such as rice (ABDULLAH et al., 1986; cf. VASIL, 1988, for a review), maize (SHILLITO et al., 1989; PRIOLI and SÖNDAHL, 1989), wheat (VASIL et al., 1990), and barley (JÄHNE et al., 1991). In addition, transgenic and fertile rice and maize plants have been obtained (cf. VASIL et al., 1990, for a review, and Tab. 3).

A major breakthrough in this area was the establishment of callus and suspension cultures able to undergo somatic embryogenesis. These cultures provided protoplasts which were subsequently used in transformation experiments. Thus, the starting material and not so much the culture conditions, proved to be the most important parameter (cf. VASIL, 1988, for a review of the earlier work). Despite these successes, major problems are anticipated as a result of the differences in the tissue culture ability of various genotypes of a certain species.

2.3 Use of the Particle Gun – the Biolistic Approach

Heavy particles of, for example, gold and tungsten, measuring 1–2 micrometer in diameter and coated with DNA, can be used to transport DNA into plant cells of intact tissues (KLEIN et al., 1987). This technique has been used not only in higher plants, but also in algae enabling organelles, i.e., chloroplasts to be transformed for the first time (BOYNTON et al., 1988; BLOWES et al., 1989). Since this method theoretically combines the advantages of both systems discussed above, i.e., the transformation of a multicellular explant and absence of host limitation, it was expected to solve all problems remaining in transforming plants. However, the success achieved up to date, has been rather limited; transgenic plants with fertile offspring have been obtained in the case of maize (FROMM et al., 1990; GORDON-KAMM et al., 1990) and soybean (MCCABE et al., 1988). Thus, despite the fact that this method provides a vector-independent DNA delivery system which enables many different cell types in all plant species to be reached, it has not proved as successful as promised.

2.4 Injection of DNA into Various Tissues

Both micro-injection into protoplasts and micro-injection into pre-embryos have been used to obtain transgenic plants (cf. NEUHAUS and SPANGENBERG, 1990, for a review). Whereas the micro-injection into protoplasts (MIKI et al., 1987) is not more efficient than the direct DNA uptake by protoplasts with the help of PEG/calcium, micro-injection into pre-embryos has produced transgenic rapeseed (NEUHAUS et al., 1987). This method has the potential of solving the problems inherent in cereal transformation. Thus, by injecting DNA into numerous cells of the embryo (in the hope of also transforming the germ-line cells), the F1-progeny of such a chimeric plant should also include transgenic plants of (by definition) clonal origin. The major advantage of this approach is that, since the embryo is used as the target for the injection, the inherent differentiation program of the embryo should automatically lead to the formation of an intact plant, thus circumventing the problems normally associated with regeneration.

2.5 Transformation of Organellar DNA

With respect to the transformation of organellar DNA, one report described the transformation of tobacco chloroplasts using *Agrobacterium*-based vectors (DE BLOCK et al., 1985). This work has to date, not been reproduced. Recently, the successful transformation of tobacco chloroplasts by the biolistic approach was reported (SVAB et al., 1990). It remains to be ascertained how widely applicable this method will be.

2.6 Conclusion

The transfer of isolated DNA into plant cells is no longer a problematic task. Among the different methods developed, the *Agrobacterium*-based vectors, the direct transfer into plant protoplasts in the presence of polyethyleneglycol/calcium, and to a lesser extent, the biolistic approach, are the most successful. Numerous selectable and, or detectable marker genes are available which enable the transformation event to be monitored. The major obstacle remaining, is the availability of suitable tissue culture techniques which allow the regeneration of intact and fertile plants from multicellular or unicellular explants. This is primarily a problem with cereal and legume plants.

3 Controlled Expression of Foreign Genes in Transgenic Plants

One important aspect of plant gene technology and the production of transgenic plants, is the controlled and predetermined expression of foreign genes. It might be desirable for a certain enzyme/protein to be produced only during certain developmental phases of the plant (such as embryogenesis or germination) or in certain organs (e.g., seeds, leaves, tubers, roots or flowers), tissues (e.g., epidermis, phloem, tapetum, mesophyll cells or meristematic tissues) or only under certain environmental conditions (e.g., in light, darkness, upon pathogen attack, wounding, under drought conditions, etc.). Because the identification and characterization of plant genes has been a major research area of plant molecular biology during the last couple of years, chimeric genes which express their coding sequence in a predetermined manner can now be constructed.

The regulatory structures of plant genes which control their transcription, are very similar to those from other eukaryotic sources (cf., e.g., EDWARDS and CORUZZI, 1990; BENFEY and CHUA, 1989, for a review). It has been observed that in general, the sequences located 5′ of the RNA coding sequence of the gene, contain many if not all, the signals necessary for its correct expression. This has been confirmed by fusing the 5′-upstream region of plant genes to a marker gene which encodes an easily detectable product and, after adding a poly-adenylation signal, transferring the chimeric gene back to the plant, using one of the above-mentioned techniques. By studying derivatives of the 5′-upstream region (e.g., deletions) it has furthermore, been ascertained that the promoter is composed of several elements which are important for either the qualitative and/or quantitative traits of expression. One of the best studies and most commonly used promoters, is that of the 35s RNA of the cauliflower mosaic virus (CaMV). This promoter has been shown to contain different domains. Whereas the intact promoter is active in most tissues and cells and thus, represents an almost constitutive promoter, the fusion of certain domains only, results in a dramatically different, highly specific expression pattern. In this latter case, gene expression might occur for example, in roots and embryonic tissue only, or high expression might be observed in leaf, cotyledon, and stem tissue (BENFEY et al., 1990). In addition, the 35s RNA promoter contains a strong transcriptional enhancer which, when fused to other promoters, can stimulate the level of expression up to tenfold without interfering with the specificity of expression which remains controlled by the basal promoter (NAGY et al., 1987).

These characteristics of plant promoters have made the following procedures possible:

- the identification of a plant gene by the differential screening of cDNA which is expressed specifically under the desired conditions,
- the isolation of the corresponding genomic clone,
- the identification of the regulatory sequences and promoter by comparing the cDNA and the genomic clone,
- proof of the identity of the promoter by fusing it to a reporter gene and monitoring its activity in transgenic plants containing this chimeric gene.

Various types of promoters which have been identified in different plants are listed by EDWARDS and CORUZZI (1990).

As a rule, plant-derived promoters maintain their expression specificity in heterologous plants and can therefore, be widely used. There is no positive evidence that promoters derived from other eukaryotic kingdoms or from prokaryotes, are able to express in plant cells. The only exceptions apart from promoters of plant viral genomes, are the promoters of genes located on the DNA segment which is transferred from *Agrobacterium* species into plant cells (ZAMBRYSKI et al., 1989).

Although many promoters retain their expression specificity in heterologous plants, the level of expression of the same gene in independent transformants, varies widely (WILLMITZER, 1988). This phenomenon, which is not specific to plants, but has also been observed in other transgenic eukaryotes, is probably caused by the sequences which flank the transferred gene after its integration. Surprisingly, although these neighboring sequences greatly influence the quantitative level of expression of the transferred gene, as a rule, they do not influence the qualitative expression pattern at all.

Recently, artificial promoters were constructed which displayed a new specificity.

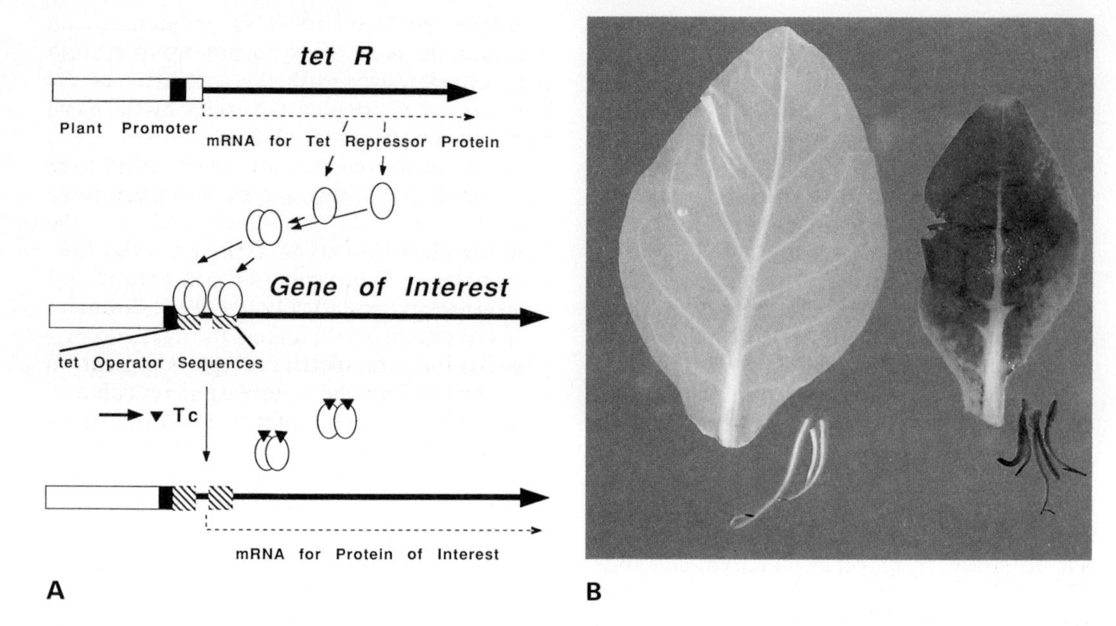

Fig. 1. Regulation of a plant promoter by a bacterial repressor protein and its induction by a small chemical compound (e.g., tetracycline).
(A) The upper part shows the gene coding for a bacterial repressor protein (repressor protein from the *tet* gene of Tn10; GATZ et al., 1991) controlled by a constitutive plant promoter which leads to the synthesis of the repressor protein (0). The middle part shows the gene of interest which is to be expressed in the plant only after addition of the low-molecular weight inducer (tetracycline). In this chimeric gene, the constitutive plant promoter is modified by inserting operator sites for the *tet* repressor protein in such a way that, in the absence of the low-molecular weight compound (tetracycline), the repressor binds to the operator sites, thus preventing the expression of the gene of interest. The lower part shows the dissociation of the *tet* repressor from the operator sites due to the addition of tetracycline, thus allowing the expression of the gene of interest. The dissociation of the repressor from the operator sites is due to an allosteric change induced by the binding of the low-molecular weight compound tetracycline. The advantage of this system is that it permits expression of a gene to be induced very specifically by the addition of a small chemical compound.
(B) Addition of tetracycline leads to the expression of a chimeric gene encoding the β-glucuronidase of *Escherichia coli*. The coding region of the β-glucuronidase gene of *E. coli* was fused to a 35s CaMV promoter which contained the operator sites recognized by the *tet* repressor protein. The chimeric gene was transferred into a transgenic tobacco plant which constitutively expressed the *tet* repressor protein. Due to the binding of the repressor protein to the operator sites, the expression of the β-glucuronidase gene is prevented (left hand side). Upon addition of tetracycline, the repressor disossociates from the operator sites, thus allowing the expression of the β-glucuronidase gene (right hand side). The formation of the enzyme β-glucuronidase is shown by the formation of a brilliant-blue color (JEFFERSON, 1987). This picture was supplied by Dr. C. GATZ, Institut für Genbiologische Forschung Berlin GmbH, Berlin.

One such chimeric promoter is composed of the basic promoter of the 35s RNA of the CaMV, the operator sequences of the *tet* gene of the transposable element Tn10, and the repressor of Tn10 (GATZ et al., 1991). In bacteria, the *tet* repressor attached itself to the operator sequences, thus, preventing the expression of genes situated behind this operator. Upon the addition of tetracycline the repressor protein changed its conformation which lead to its release from the operator site, thus allowing transcription to take place. When these operator sites were placed in certain regions of the 35s RNA promoter and concurrently with the *tet* repressor protein in the same plant cell, the 35s RNA promoter became inactive in transgenic plants. This was due to the binding of the repressor molecule. Most interestingly, however, this binding ceased when tetracycline was added to the plants. Thus, an artificial promoter was constructed which could be activated or inactivated by adding a certain low-molecular weight chemical to the plant (GATZ et al., 1991; Fig. 1). Although the antibiotic activity of tetracycline excludes its application in the field, this concept of a promoter which can be activated or inactivated specifically by a chemical, has enormous potential with respect to the controlled expression of genes in transgenic plants.

3.1 Directing Foreign Proteins into Certain Subcellular Compartments

A characteristic feature of the eukaryotic cells is the presence of various compartments which serve, for example, to separate anabolic from catabolic pathways. In order to change plants in a predetermined way, one may wish not only to control the formation of a protein/enzyme with respect to a certain tissue, developmental or environmental conditions (affected by the promoter used), but also to control its subcellular location within the cell. This could be achieved by targeting proteins which have a predetermined destination within the cell.

The transportation of nuclear-encoded proteins within a eukaryotic cell, occurs in several ways depending on the final destination of the protein. Proteins which are destined for the mitochondria or the chloroplasts are synthesized on free polysomes and given an N-terminal extension (transit peptide). This transit peptide directs them into either the chloroplast or mitochondrium where the transit peptide is cleaved. In the case of peroxisomal proteins, evidence suggests that post-translational sorting by sequences present in the carboxy terminus, also occurs. In order to remain in the cytosol, proteins are synthesized on free polysomes and remain in the cytosol due to the absence of a transit peptide.

A large class of proteins is finally destined to move through the endoplasmic reticulum (ER)/Golgi pathway. They are either secreted into the extracellular space, become part of the plasma membrane, remain in the ER/Golgi complex, or move into the vacuole. This large and complex class of proteins is believed to be initially co-translationally imported into the lumen of the endoplasmic reticulum, whereafter it is sorted in the cis- and trans-Golgi. All these proteins contain an N-terminal extension (signal peptide) which directs the nascent polypeptide chains to specific receptors present on the membrane of the endoplasmic reticulum.

By fusing reporter genes to DNA sequences encoding the N-terminal extension of proteins destined for the chloroplast or the mitochondrium, chimeric genes were constructed which, after adding plant promoters and poly-adenylation signals, resulted in the formation of fusion proteins in transgenic plants. Moreover, these proteins were observed to leave the cytosol and to move into different organelles. The organelle of destination was determined by the origin of the N-terminal sequence added (KEEGSTRA et al., 1989; BOUTRY et al., 1987). As more and more transit peptides of nuclear-encoded proteins located in the chloroplast or mitochondrium have been identified, so the possibilities of creating chimeric proteins have increased. Hence, directing foreign proteins into either chloroplasts or mitochondria of transgenic plants is a fairly straightforward procedure.

The situation is slightly different for proteins which are co-translationally imported into the endoplasmic reticulum and subsequently distributed among the vacuole, plasma membrane, extracellular space and ER-Golgi complex. With respect to directing proteins into the lumen of the endoplasmic reticulum, an experimental procedure, similar to that described above could be followed. The fusion of the N-terminal extension (signal peptide) of vacuolar or secreted proteins to the foreign protein resulted in its assimilation into the ER (ITURRIAGA et al., 1989). Certain aspects of the subsequent sorting procedure of the proteins, however, remain obscure. Thus, only one example exists of a foreign protein being directed into the vacuoles of transgenic plants (SONNEWALD et al., 1991). This was achieved by fusing a very large section (146aa) of a vacuolar plant protein to the N-terminus of the reporter protein. In other cases, it has been speculated that the last 10–12 amino acids located at the C-terminus are responsible for vacuolar targeting (cf. SEBASTIANI et al., 1991, for a review). The secretion of proteins into the extracellular space seems to occur if no signal is present apart from the signal peptide. The secretion of chimeric proteins that were linked only to the signal peptide of a plant protein, was demonstrated in transgenic plants (DOREL et al., 1989; VON SCHAEWEN et al., 1990).

3.2 Inactivation of the Expression of Endogenous Genes in Transgenic Plants

Apart from being able to add new traits to a plant, the ability to inhibit the manifestation of unwanted characteristics would clearly be advantageous. Theoretically this could be achieved by mutating the DNA by gene disruption via homologous recombination. This would create a null mutant and therefore, abolish the activity of the targeted gene. Although certain model experiments demonstrated that gene targeting could be achieved in higher plants, the low frequency of its occurrence precludes it from being carried out routinely (PASZKOWSKI et al., 1988). A different approach involving anti-sense RNA however, proved surprisingly efficient at silencing specific endogenous genes in transgenic plants and eukaryotic organisms (MOFFAT, 1991). The chimeric gene introduced, encodes anti-sense RNA which is complementary to the target mRNA which is to be silenced. It is assumed (although not proven) that the mRNA encoded by the endogenous gene and the anti-sense RNA produced by the introduced gene, form a duplex RNA structure which, by an unknown mechanism, leads to the inactivation and disappearance of both sense- and anti-sense mRNA molecules (Fig. 2).

The anti-sense RNA-mediated inhibition of the expression of endogenous genes has been demonstrated for a number of genes (discussed by VAN DER KROL et al., 1988).

One further advantage of the use of anti-sense RNA is that, as a rule, plants which are independently transformed with the anti-sense RNA-coding gene, display different degrees of inhibition. Thus, not only is it possible to suppress almost entirely the expression of the endogenous gene (as is the case for gene disruption via homologous recombination), but also to obtain transgenic plants which differ widely with respect to the degree of gene expression inhibition.

A depressed expression of endogenous genes in transgenic plants has also been observed to result from the introduction of multiple copies of the gene in its normal (sense) orientation. This phenomenon called co-suppression, is even less understood than anti-sense RNA-mediated inhibition of gene expression (JORGENSEN, 1990). It is at present unclear whether a similar or different mechanism is responsible for this phenomenon. In the case of sense inhibition however, the depression of gene expression seems to be more dependent on environmental conditions than in the case of anti-sense RNA-mediated inhibition.

3.3 Stability of Proteins and RNA Produced from Genes Introduced into Transgenic Plants

The methods discussed above allow the introduction and predetermined expression of a foreign gene in that RNA is synthesized and the resulting foreign protein is directed into a specific subcellular compartment.

The aim of genetic engineering however, is to alter the phenotype of the plant. To this end, the product of the transferred gene must either be very active (i.e., have a high specific activity) or be produced in large amounts. In the first case, a small quantity of protein produced from the transferred gene may be sufficient to obtain the desired phenotype alteration. In the second case, however, it is not sufficient to merely ensure the expression (i.e., transcription) of the foreign gene, but the stability of the RNA and protein formed must in addition be taken into account. This is applicable where the aim is to modify the nutritional

EFFECT OF ANTISENSE INHIBITION

WILD TYPE SITUATION

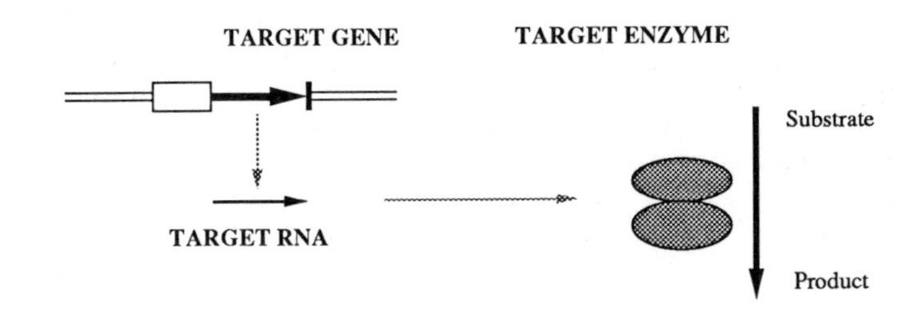

ANTISENSE INHIBITION

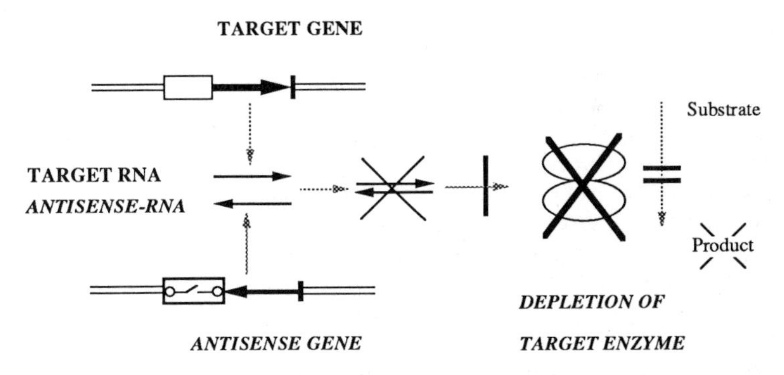

Fig. 2. Schematic diagram of the inhibition of the expression of an endogenous gene by introducing a chimeric gene encoding an anti-sense RNA.

value of a plant seed or to produce certain new metabolites/proteins using the plant as a bioreactor.

With respect to ascertaining the RNA stability, few studies have been undertaken. It is generally thought however, that the RNA stability is not a critical determinant of expression levels. The only example of an altered RNA stability was observed when a premature stop codon was present (VAN CANNEYT et al., 1990). This however, would also abolish the formation of a functional protein.

With respect to protein stability, the situation is more complex. In several cases, little or no protein was produced despite the fact that a large amount of RNA was formed and that this RNA could be translated into protein *in vitro* (GOLDSBROUGH et al., 1986; MATZKE et al., 1984). Thus, it can be concluded that certain proteins are unstable in a plant cell or in a certain tissue. The protein instability can be overcome by either expressing the protein in different tissues or different organs by using specific promoters, or to direct the protein to different subcellular compartments. Few studies have been conducted to address this topic. One investigation however, showed that depending on the protein, the transfer into different subcellular organelles led to either a stabilization of, or no change in the protein (SONNEWALD et al., 1990). Post-translational modifications such as glycosylation were observed to exercise no, or very little, influence on the protein stability (VOELKER et al., 1989; SONNEWALD et al., 1990). Probably the most detailed study aimed at improving the amount of a foreign protein produced in a transgenic plant, was conducted by PERLAK et al. (1991). By eliminating potential poly-A sites and adjusting the GC content of a bacterial gene to account for the plant's codon usage, the level of expression of a certain protein increased by a factor of 100.

3.4 Conclusion

Numerous promoters derived from plant genes have been identified which permit the expression of foreign coding sequences in transgenic plants to occur in a predetermined way with respect to various developmental and environmental conditions. Furthermore, strategies are available which would enable novel promoters which displayed particular new specificities, to be isolated. Techniques have been established for directing foreign proteins into certain subcellular organelles (i.e., chloroplasts, mitochondria, vacuoles, and the extracellular space). The major problem remaining is the stability of foreign proteins in higher plant cells. Since one general solution is probably not available, the expression of each protein requires particular attention, as is the case in other expression systems.

4 Improving Plants with Respect to Agronomic Traits

The ability to create transgenic plants that express a foreign gene in a predetermined manner, has created possibilities of improving plants with respect to a variety of agronomic characteristics. In this section an overview of the progress made in different fields, will be presented with particular emphasis on plants which display improved resistance to abiotic and biotic stresses, including agrochemicals.

4.1 Herbicide Resistance

The creation of transgenic plants which are resistant to herbicides, has been a prime goal of plant genetic engineering. Although this goal is contested publically, herbicide-resistant plants offer an array of economic and ecological advantages.

(1) Transgenic plants which are resistant to a total herbicide would in most cases, alleviate the farmer's use of pre-emergence herbicides as he could delay applying herbicides until a certain threshhold was reached beyond which, the excessive growth of weeds would damage the crop. Ignoring the application parameters pertaining to conventional herbicides, he could use the total herbicide which would kill the weeds, but leave the crop plant untouched. The replacement of pre-emergence herbicides by post-emergence herbicides will without doubt, lead to a reduction in the total quantity of herbicides used in agriculture.

(2) The absence of pre-emergence herbicides would allow the weeds to germinate. The total herbicide would reduce these growing weeds into mulch on the soil surface which would have the ecologically advantageous result of alleviating soil erosion.

(3) Creation of transgenic plants resistant to total herbicides, would be particularly advantageous in cases where a certain weed was very closely related to the crop plants, for example; mustard growing amongst rape seeds, or red rice in a rice paddy. To date, selective herbicides have failed to control weed proliferation because of the high plant similarity.

(4) The fairly wide variety of herbicides which are currently required to exploit the plant-endogenous differences in susceptibility to certain chemicals, could be replaced by a few compounds which have a much broader range of application. Thus, the plethora of molecules each having a different action, toxicity and persistence in the environment could be replaced by few compounds, which could be analyzed more extensively with respect to their influence on the environment because they would probably hold a larger share

of the whole market. Thus in conclusion, the strongest argument in favor of using plants resistant to herbicides, is the resulting ecological benefit.

In order to obtain herbicide-resistant plants via genetic engineering two strategies can be distinguished (MAZUR and FALCO 1989; OXTOBY and HIGHES, 1990):

- Modification of the target of the herbicide. Since in many cases, the target of the herbicide is a protein, altering (increasing) the level of this protein and/or altering (decreasing) its sensitivity to the herbicide are two possible approaches.
- Transformation of the herbicide into a non-toxic component within the transgenic plants. As a rule, this can be achieved by introducing a gene encoding an enzyme which detoxifies the herbicide by a chemical modification or cleavage.

4.1.1 Modification of the Target of the Herbicide

Numerous proteins representing the target for herbicides, have been identified in higher plants. With respect to genetic engineering, work has been performed mainly on two classes of herbicides:

- inhibitors of photosynthesis (i.e., triazines)
- inhibitors of amino acid biosynthesis (i.e., glyophosate, phosphinotricine and sulfonylureas or imidazolinones)

Resistance to triazines

Triazines bind to the Qb protein present in the thylakoid membrane of chloroplasts, thus interfering with the binding of plastoquinone. The Qb protein is encoded by the *psb*A gene present in the chloroplast genome of higher plants and algae. Triazine-resistant mutants which have one altered amino acid and thus, have lost their ability to bind the triazines, have been identified in several microbial and higher plant species. As there is no routine method available to genetically modify the genome of chloroplasts, a *psb*A gene encoding a triazine-insensitive Qb protein was fused to the transit peptide of a nuclear-encoded chloroplast-located protein. After *Agrobacterium*-mediated transformation of tobacco cells, the mutant Qb protein was found to be present in the chloroplasts into which it had been directed via the transit peptide. The transgenic plants expressing the mutant Qb protein, were found to have a somewhat increased tolerance for atrazine which however, was far below any level required to show resistance in the field (CHEUNG et al., 1988).

Inhibitors of amino acid biosynthesis
Sulfonylureas and imidazolinones

A broad spectrum of weeds is sensitive to this class of herbicides. The target enzyme inhibited is the acetolactate synthase (ALS). In both yeast and higher plants, mutant forms of the ALS resistant to either one or both classes of herbicides, have been identified which differ in one amino acid from the sensitive wild-type ALS. The transfer of such mutant genes derived from either *Arabidopsis* or tobacco into sensitive, wild-type plants, has led to increased resistance under greenhouse and field conditions (four-fold above the level normally applied under field conditions) (LEE et al., 1988; MAZUR and FALCO, 1989.

Phosphinotricine

Phosphinotricine (PPT) which is a structural analog of glutamine, inhibits the glutamine synthase (GS) of both plant and bacterial origin. Inactivation of the GS will lead to ammonia accumulation which is toxic to the cell. No mutants of GS which are resistant to PPT have been identified. Thus, it was tested whether the over-expression of the target enzyme would lead to an increased tolerance by out-titrating the herbicide. Transgenic tobacco lines which contained an over-expressing GS gene derived from alfalfa, showed some increased tolerance for PPT (ECKES et al., 1989).

Glyphosate

The most successful attempt at amplifying the sensitivity of the target has been made in the case of the herbicide glyphosate. It specifically inhibits the 5-enol-pyruvylshikimate-3-phosphate synthase (EPSPS). Both strategies have been followed, i.e., increasing the level of the EPSPS protein and identifying mutants of a petunia-derived EPSPS protein by screening in *Escherichia coli.* The combination of both strategies led to the creation of a number of fully resistant plants, including important crops such as tomato, potato, tobacco, soybean, brassica, and sugarbeet (DELLA CIOPPA et al., 1987; SHAH et al., 1986, 1988; KISHORE and SHAH, 1988). Field trials of these plants have been conducted in the United States and Canada with positive results (MAZUR and FALCO, 1989).

The creation of glyphosate-resistant brassica plants is a good example how several modifications to the expression of the gene in the transgenic plant can yield the desired result. Since glyphosate is most active in the apical meristems, it was necessary to increase the level of the EPSP synthase in these areas of the brassica plant. New promoters, specifically and highly expressed in the apical meristem were isolated following the strategy described earlier in order to replace the 35s RNA promoter which proved too inefficient in these tissues. The combination of a specific promoter and the over-expression of a mutated form of the EPSP synthase protein yielded the resistance to glyphosate without any reduction in the total yield of the crop.

4.1.2 Transformation of the Herbicide into a Non-Toxic Compound

The expression of herbicide-detoxifying enzymes in transgenic plants has been a very successful method of counteracting the effect of several herbicides. Since many herbicides are readily degraded in soil, microorganisms able to use the herbicide as the sole carbon and/or nitrogen source enabled those to degrade the herbicide to be identified. This was followed by the isolation of the degrading enzyme, and by the identification of the gene.

An example of the successful application of this strategy is the creation of phosphinotricine-resistant plants, in which the phosphinotricine-acetyltransferase gene (*bar*) was cloned from *Streptomyces hygroscopicus* (DE BLOCK et al., 1987). The enzyme encoded by this gene acetylates PPT, thus converting it to a non-toxic form. After adding suitable promoters, this gene was introduced via *Agrobacterium*-mediated transformation into several plants including crop species such as potato, tomato, sugarbeet, and rapeseed (BOTTERMAN and LEEMANS, 1989) as well as maize (MÜLLNER, 1991) (in maize, protoplasts were transformed). Transgenic plants were fully resistant to the herbicide, and field tests performed with transgenic tobacco and potato plants demonstrated no reduction in the yield (DE GREEF et al., 1989). Thus through the creation of PPT-resistant transgenic plants, this very effective and ecologically very advantageous herbicide can be used as a broad-spectrum, post-emergence herbicide.

Other examples include the expression of a nitrilase gene (*bxn*) from *Klebsiella ozaenae,* which converts the herbicide bromoxynil into the non-toxic compound 3,5-dibromo-4-hydroxybenzoic acid. The *bxn* gene was introduced into transgenic tobacco and tomato plants which became resistant to bromoxynil (STALKER et al., 1988).

Finally the gene *tfda* from the soil bacterium *Alcaligenes eutrophus*, after the addition of suitable plant expression signals, was transferred into tobacco plants where it conferred resistance to the herbicide 2,4-dichlorophenoxyacetic acid (2,4-D) (Fig. 3). The gene *tfda* encodes a mono-oxygenase which converts the 2,4-D into its non-toxic phenol derivative (STREBER and WILLMITZER, 1989; LYON et al., 1989).

4.1.3 Conclusion

Strategies have been developed which permit herbicide-resistant crop plants to be created by genetic engineering methods. Of the two strategies, (1) changing the amount or sensitivity of the target enzyme, and (2) introduction of a

Fig. 3. Expression of the mono-oxygenase *tfda* from *Alcaligenes eutrophus* in transgenic plants leads to resistance to the herbicide 2,4-D. A transgenic tobacco plant expressing the *tfda* encoded mono-oxygenase (left) and a control plant were sprayed with a solution equivalent to 10 kg 2,4-D/ha. The picture was taken one week after spraying.

detoxifying enzyme, the second approach seems more straightforward, provided toxicological tests can prove that the detoxified compounds are no risk for human or animal health.

4.2 Use of Transgenic Plants to Obtain Protection against Viral Infections

In order to obtain transgenic plants which are protected against viral infection, three approaches have been followed:

- Coat protein-mediated protection,
- protection via the expression of satellite RNA,
- protection via expression of anti-sense RNA.

Coat protein-mediated protection has proved to be the most successful approach by far, and will therefore be discussed in more detail.

4.2.1 Coat Protein-Mediated Protection

It is a long-standing observation that the infection of a plant by a mild virus, prior to infection by a serologically related, aggressive virus, leads to a decrease of the symptoms normally evoked by the infection of the aggressive virus. Since the two viruses must be related, it has been speculated that the coat protein (determining the serological response) is responsible for this phenomenon (DE ZOETEN and FULTON, 1975; SHERWOOD and FULTON, 1982; SEQUEIRA, 1984).

It was assumed that the mild virus which infected the cell first, produced excessive amounts of coat protein which remained in a free state, i.e., not bound to the viral RNA genome. The unbound, freely floating coat protein inhibited the uncoating of the RNA of the second, aggressive virus. Consequently, the viral RNA expression, replication and the resulting symptom development were delayed. It was therefore suggested as early as 1980, that the presence of coat proteins in transgenic plants could protect them against viral attack.

This hypothesis has subsequently been demonstrated to be true. The first report of the delay of symptom development, described transgenic tobacco plants which expressed about 1–2 μg/g fresh weight of the tobacco mosaic virus (TMV) coat protein when infected with TMV (POWELL-ABEL et al., 1986). Since then, an ever-increasing list has been published of plants being protected against viral infection by expressing the coat protein gene of the corresponding virus in the transgenic plants (cf. Tab. 4 for a summary). The exact molecular mechanism underlying the coat protein-mediated protection, is unknown.

Tab. 4. Transgenic Plants Protected against Viral Infection by Expression of a Viral Coat Protein

Plant Species	Origin of Coat Protein	Protection against	Reference
Tobacco	TMV	TMS	POWELL-ABEL et al., 1986
Tobacco	AlMV	AlMV	TUMER et al., 1987 LOESCH-FRIES et al., 1987 VAN DUN et al., 1987
Tobacco	CMV	CMV	CUOZZO et al., 1988 QUEMADU et al., 1991
Tobacco	TSR	TSV	VAN DUN et al., 1988
Tobacco	PVX	PVX	HEMENWAY et al., 1988
Tobacco	TRV	TRV	VAN DUN et al., 1987
Tobacco	SMV	TEV, PVY	STARK and BEACHY, 1989
Tobacco	PRV	TEV, PVY, PMV	LING et al., 1991
Potato	PVX	PVX	LAWSON et al., 1990 HOEKEMA et al., 1989
Potato	PVY	PVY	LAWSON et al., 1990
Potato	PLRV	PLRV	KAWCHUK et al., 1990
Tomato	TMV	TMV	POWELL-ABEL et al., 1986 NELSON et al., 1988
Tomato	AlMV	AlMV	TUMER et al., 1987
Alfalfa	AlMV	AlMV	HILL et al., 1991

Abbreviations: AlMV, alfalfa mosaic virus; CMV, cucumber mosaic virus; PEV, pea early browning virus; PVX, potato virus X; PVY, potato virus Y; PLRV, potato leafroll virus; PMV, pepper mottle virus; SMV, soybean mosaic virus; TEV, tobacco etch virus; TRV, tobacco ringspot virus; TMV, tobacco mosaic virus; TSV, tobacco streak virus

Several studies however, suggest that at least two different steps/mechanisms are operative (BEACHY et al., 1990). One of them acts during the very early phase of infection, i.e., during the initial phase after the virus has entered the cell. This phase is characterized by a swelling of the virus particle, the release of parts of the RNA genome, binding of the ribosomes, and the concurrent disassembly of the virus and translation which initiate the infection process. The main evidence indicating that one of these steps during the early phase is actually involved in the protection, is the observation that the coat protein-mediated resistance can be suspended by infecting with the naked viral RNA, as opposed to virus particles consisting of the RNA encapsulated in the coat protein. One possible explanation for this is that the coat protein present in transgenic plants interferes with the disassembly of the virions by re-encapsulating the RNA molecule.

A second process, distinct from this early resistance mechanism must be operative during the spread of the infection. Thus, the concentration of TMV in transgenic plants expressing the coat protein, was found to be similar in the tissue adjacent to the initially infected locus (1–3 mm). However, it was much lower in more distant tissues (5–10 mm from the original site of infection) and in other leaves when compared to non-coat protein expressing plants.

Another interesting feature of coat protein-mediated protection is the observation that in some cases, the plant is not only protected against an infection of the virus from which the coat protein was received, but also against other serologically non-related viruses. STARK and BEACHY (1989) demonstrated that the expression of the coat protein of the potyvirus soybean mosaic virus (a non-pathogen on tobacco) in transgenic tobacco plants, led to resistance to the two serologically unrelated potyviruses, potato virus Y (PVY) and tobacco etch virus. In a similar way, the expression of the coat protein of papaya ringspot virus in transgenic tobacco plants led to significant resistance to the potyviruses; tobacco etch virus, PVY and pepper mottle virus (LING et al., 1991). The expression of the coat protein of tobacco mosaic virus led to considerable resistance to a number of viruses from the same group, i.e., tobamoviruses. This was true only in cases where the homology of tobamovirus and TMV was 60% or higher on the amino acid level, whereas those showing a lower degree of homology, were significantly less affected (NEJIDAT and BEACHY, 1990).

Several field trials of transgenic plants expressing virus-derived coat proteins have been performed in the United States and in Europe. Among these are tomato plants expressing the TMV coat protein gene (NELSON et al., 1988), tobacco plants expressing the AlMV coat protein gene (BEACHY et al., 1990) and potato plants expressing coat proteins of both PVX and PVY (KANIEWSKI et al., 1990). In general, coat protein-mediated protection led to a dramatically reduced infection by the virus in question, which in turn, led to an increase in yield. Thus, in the case of the potato variety Russet Burbank, the expression of the coat proteins of both PVX and PVY, one transgenic line was nearly fully resistant to infection by both PVX and PVY (percentage of infected plants being as low as 8% compared to 80% in non-coat protein expressing controls) which corresponds to no loss in yield. By comparison, the control experiments showed a 30% loss in yield (KANIEWSKI et al., 1990).

4.2.2 Protection of Plants by the Expression of Satellite RNA

Some RNA viruses are associated with small, extragenomic RNA molecules which are not autonomous, but require the presence of an intact viral genome to replicate and propagate within the plant and be transmitted to other hosts. These small RNA species are generally called satellite RNA.

A striking observation with respect to satellite RNA molecules, is that is they are able to modulate the symptoms produced by the corresponding helper virus. Thus, a cucumber mosaic virus associated RNA, attenuates the symptoms evoked by the cucumber mosaic virus on tobacco. In a similar way, a certain satellite RNA of tobacco ringspot (TRV) virus leads to an attenuation of the symptoms produced by the tobacco ringspot virus on tobacco.

This observation of the ability of certain satellite RNA molecules to modulate the symptom development induced by the corresponding virus, has been exploited in two cases, to obtain protection against virus infection. To achieve this goal, the genomes of the satellite RNA of CMV and TRV under the control of constitutive promoters, were incorporated into the genome of transgenic plants (HARRISON et al., 1987; GERLACH et al., 1987). When these transgenic tobacco plants which constitutively expressed the satellite RNA, were infected with the corresponding virus, a significant delay in symptom development was observed when compared to plants devoid of the gene encoding the satellite RNAs. However, although this approach works in theory, it is too hazardous to be used in field experiments, since very small differences exist on the nucleotide level between benign and virulent satellite RNA species.

4.2.3 Protection of Plants against Viral RNA by the Expression of Anti-Sense RNA

A few reports have been made of attempts to obtain resistance against virus infection by introducing a gene into the genome of transgenic plants which encoded anti-sense RNA to the viral RNA. Weak or no protection, not comparable to that obtained by expressing the coat protein, was observed (CUOZZO et al., 1988; HEMENWAY et al., 1988; POWELL et al., 1989; REZIAN et al., 1988). A different result however, was obtained with a DNA virus (geminivirus). The expression of the anti-sense RNA corresponding to part of the viral genome, led to a significant level of resistance (DAY et al., 1991).

4.2.4 Other Ways of Protection against Viral Infections

Transgenic tobacco plants, transformed with the 54 kDa non-structural protein of TMV, showed complete resistance against two closely related strains of TMV (GOLEMBOSKI et al., 1990). This resistance mechanism is not understood.

4.2.5 Conclusion

The expression with the help of constitutive plant promoters, of the coat protein of the corresponding virus in transgenic plants, has proved to be the most effective way to create resistance against viral infection. This strategy seems to be nearly universally applicable. Other strategies of obtaining resistance against viral infections are either too hazardous to be used in the field, or have not yet proved to be generally applicable.

4.3 Transgenic Plants Resistant to Insects

The control of insects in crop plants is largely dependent on synthetic insecticides which are provided by the organic chemical industry. During the last five years, transgenic plants have been created which express insecticidal proteins resulting in the protection of these plants against certain types of insects (cf. VAN RIE, 1991; BRUNKE and MEEUSEN, 1991, for recent reviews). The approach of inducing plants to produce their own insecticide has several advantages over the conventional insect control mechanisms which involve externally applied synthetic chemicals. These advantages include: the absence of non-proteinaceous residues in soil or ground-water, high specificity with respect to the target organism, and protection of parts of the plant such as the roots, which are difficult to reach by conventional methods.

Two types of insect-control agents have been developed for use in transgenic plants:

- the so-called delta-endotoxin proteins obtained from various *Bacillus thuringiensis* species,
- certain classes of proteinaceous proteinase inhibitors.

4.3.1 *Bacillus thuringiensis* Endotoxin Proteins

Insecticidal proteins (delta-endotoxin proteins) in spores of the Gram positive bacterium *Bacillus thuringiensis* have been exploited for more than 30 years. These spores are used as biological control agents for several insects. The delta-endotoxin proteins are present as crystals in the spores of *B. thuringiensis.* Upon ingestion, these proteins are solubilized in the insect's mid-gut due to the alkaline pH and the presence of certain proteases in the mid-gut. The proteases cleave the endotoxin protein resulting in the conversion of the protoxin into the active toxin. The presence of the toxin leads to mid-gut paralysis and eventual disruption of the mid-gut cells (cf. HÖFTE and WHITELEY, 1989, for a review).

One of the major advantages of using the *B. thuringiensis* spores as a biological control agent is its high specificity. In addition to differentiating between lepidopteran, dipteran and coleopteran species, certain spores display a differential activity against different members of the lepidopteran group. The mechanism of this high specificity has been elucidated recently. A correlation has been found between the toxicity of a certain endotoxin protein and its ability to bind to specific receptors on the brush-border membrane of the insect's mid-gut, which is the target site for the endotoxin protein's insecticidal action (HOFMANN et al., 1988; VAN RIE et al., 1989; 1990). For example, seven different lepidoptera-specific endotoxin proteins have been identified which differ slightly in their amino acid sequences. Among these the toxins CryIA(b) and CryIB are both toxic to larvae of *Pieris brassicae,* but only one toxin, CrYIA(b), is toxic to *Manducta sexta,* which is reflected by different receptors present on the mid-gut of the two insect species. An analysis of the toxicity versus the binding capacity to the re-

Fig. 4. Expression of the *Bacillus thuringiensis* endotoxin gene in transgenic tobacco plants leads to protection against insects.
The left and the right sides show plants transformed with a chimeric gene expressing the *B. thuringiensis* var. *kurstaki* endotoxin gene, the middle plant is a control. The left and the middle plants have been infested with larvae of *Manducta sexta.* It is apparent that the expression of the *B. thuringiensis* endotoxin protects the plant against insect attack. These pictures were supplied by Plant Genetic Systems, Ghent/Belgium.

ceptors of all seven lepidoptera-specific endotoxin proteins with respect to a range of insects, revealed a strong correlation existing between the two factors. These studies, therefore, suggest that families of different receptors are present in different lepidopteran insects, each of which binds with high affinity to one or several endotoxin proteins and thus, leads to the insecticidal activity of the endotoxin protein.

Based on the long-standing successful use of *B. thuringiensis* spores for biological insect control, it was therefore, a straightforward experiment to attempt to induce the plants to produce their own specific insecticide by transferring the coding sequence of the *B. thuringiensis* endotoxin gene furnished with appropriate plant regulatory sequences, into transgenic plants. The first successful uses of the *B. thuringiensis* in transgenic plants were reported in 1987 (VAECK et al., 1987; FISCHHOFF et al., 1987; BARTON et al., 1987) (Fig. 4). By expressing the insecticidal protein of the *Bacillus thuringiensis* strain *kurstaki,* in transgenic tobacco and tomato plants, transgenic plants were created which were highly resistant to infestation by larvae of certain lepidoptera species such as *M. sexta. B. thuringiensis* strain *tenebrionis* contains an endotoxin protein which is highly active against certain coleoptera species, mainly the colorado beetle. Transgenic potatoes were shown to be highly resistant against infection by the colorado beetle (BRUNKE and MEEUSEN, 1991). Another example showing the power of the *B. thuringiensis* insecticidal proteins was demonstrated by the expression of the *B. thuringiensis kurstaki* strain endotoxin protein in tubers of transgenic potato. This led to significant prevention of infestation by the tuber moth (BOTTERMAN and LEEMANS, 1989).

In the case of tomato, tobacco, and potato, numerous field experiments which achieved control over the target insects, have been performed in the USA and several countries in Europe (DELANNEY et al., 1989). A further improvement of insect control via transgenic plants expressing *B. thuringiensis* endotoxin proteins was made by massively increasing the amount of protein produced (up to 1% of the total protein). This not only led to a more rapid killing of the susceptible larvae, but also led to a reduction in fertility of the mature insects, thus reducing their progeny. In addition, in the case of transgenic tomato, significant control of the tomato fruitworm (*Heliothis zea*) and the tomato pinworm (*Keiferia lycopericella*), was observed (PERLAK et al., 1991; BRUNKE and MEEUSEN, 1991).

One argument often raised against the intensive use of transgenic crop plants expressing the endotoxin protein, is the increased pressure exerted upon the insect population to develop resistance. Several cases have been reported where insects are resistant to certain formulations of *B. thuringiensis* spores. Thus, a *Plodia interpunctella* strain (McGaughey, 1985) developed resistance against widely used commercial applications of *B. thuringiensis* spores. The examination of the strain showed that the resistance correlated with a 50-fold reduction in the affinity of a certain receptor to the endotoxin protein present in the *B. thuringiensis* formulation used. Interestingly, the resistant insect strain remained sensitive to another enodotoxin-type protein which was not used in the formulation (Van Rie et al., 1990; cf. Fig. 4). Thus, it seems a very plausible strategy to construct transgenic plants which express not only one type of insecticidal protein, but two types acting against the same insect. If the mechanism described above for developing resistance is of a general type, it would imply that two independent mutations would have to occur almost simultaneously in the same insect which is likely to occur at a very low frequency. Further modifications of the use of *B. thuringiensis* endotoxin-expressing plants could include the fine-tuning of expression such that an accumulation of the endotoxin protein would occur only in the case of real threat. This could be achieved by placing the endotoxin protein coding sequences under the control of a wound-inducible promoter, for example.

4.3.2 Use of Proteinase Inhibitors as Insect Control Agents

Proteinase inhibitors are wide-spread among higher plants. Interestingly, they reach their highest concentration in organs or tissues necessary for somatic or sexual propagation, such as seeds or tubers, whereas in other tissues such as leaves, they are often induced under conditions which mimic the attack of a herbivorous insect, e.g., by wounding. This expression pattern, combined with the fact that most protease inhibitors display an activity not against plant endogenous proteinases, but against proteases of microbial and mammalian origin, has led to the interpretation that proteinaceous protease inhibitors represent a defense barrier against certain pathogens (Richardson, 1977).

In contrast to *Bacillus thuringiensis* endotoxins, protease inhibitors have anti-metabolic activity in a wide range of insects. Thus, it is an attractive strategy to make plants resistant to herbivorous insects by introducing genes for certain protease inhibitors which normally would be absent.

Despite the potential of using protease inhibitors, there are only few reports on their expression in transgenic plants. Thus the expression of a plant-derived protease inhibitor, the cowpea trypsin inhibitor in transgenic tobacco plants, led to a significant resistance to tobacco budworm (*Heliothis virescens*) (Hilder et al., 1987). By adding the protease inhibitor to their diets artificially, enzymatic activity in a wide spectrum of insects including *Spodoptera litoralis, Heliothis zea,* and *Diabrotica undecimpunctata,* was observed (Hilder et al., 1990). These initial studies show the potential use of expressing protease inhibitor genes in transgenic plants. Compared to *B. thuringiensis* endotoxin proteins, their main advantage might be their broad spectrum of activity in many different insects, and the fact that such inhibitors are commonly found in the food of humans and animals. Their major disadvantage is the fairly high level of protein needed to kill the insect larvae.

4.3.3 Conclusion

In order to obtain insect resistance in transgenic plants, two strategies have been developed and demonstrated to be functional, i.e., the use of the endotoxin proteins from the various *B. thuringiensis* strains and the use of protease inhibitors. The *B. thuringiensis* endotoxin protein approach has the advantage of combining a high specificity with respect to the target insect, with a high specific activity of the protein, and only small amounts of the protein are required. Since the insecticidal action of the endotoxin proteins is better understood than that of protease inhibitors, strategies might be developed preventing the early formation of resistant insects. The increasing list of *B. thuringiensis* strains exhibiting slightly different insecticidal action (Höfte and Whiteley, 1989) could provide a long-lasting reservoir for insect control.

Protease inhibitors have a similar potential, although more evidence of transgenic plants protected against insect attack by their expression is desirable. Their major disadvantage is the high amount of protein needed for significant protection. Nevertheless, attacking the insect's metabolizing system by the use of protease inhibitors or by the often overlooked, amylase inhibitors, should prove a valuable approach.

4.4 Possibilities of Obtaining Transgenic Plants Resistant to Fungi or Bacteria

In sharp contrast to the progress made in obtaining transgenic plants protected against virus infection or insect attack, no such generally applicable methods have yet been developed to combat infection by fungi or bacteria.

In order to make plants resistant against fungi, three main concepts have been followed. First, genes encoding enzymes involved in the formation of small organic compounds were introduced in the hope of producing a new compound (phytoalexin) which is more aggressive against the fungus. Secondly, genes, which encoded enzymes postulated to degrade or destroy the cell wall of the fungal hyphae, were introduced into transgenic plants. Finally, genes encoding low-molecular weight proteins with a direct antifungal or antibacterial action, were introduced into plants.

Upon infection with a fungal or bacterial pathogen, the plant responds by re-adjusting its biosynthetic machinery (cf. LAMB et al., 1989, for a review). One of the changes observed is the formation of low-molecular weight organic compounds which display an antifungal/antibacterial activity in *in vitro* tests. These low-molecular weight compounds which are thought to play a major role in the defense of the plant against phytopathogenic fungi, have generally been called phytoalexins (cf. BAILEY,1987, for a review).

One approach aimed at modifying the phytoalexin composition in transgenic plants involved transferring a gene encoding stilbene synthase from groundnut into transgenic tobacco plants (HAIN et al., 1990). Stilbenes, a group of secondary plant products with properties of phytoalexin and antifungal compounds, are synthesized in a limited number of plant families only. The biochemical action of stilbene synthase is the conversion of *p*-coumaryl CoA and three molecules of malonyl CoA into 3,4,5-trihydroxystilbene, commonly known as resveratrol. Although tobacco is devoid of stilbene synthase, it contains all the precursors necessary for its synthesis. As stilbenes are known to be potent inhibitors of fungal growth, it was hoped therefore, that resistance would be obtained. The transgenic tobacco plants obtained, were shown to produce resveratrol and displayed an increased level of resistance against phytopathogenic fungi.

Chitinases and β-1,3-glucanases display an antimicrobial activity (SCHLUMBAUM et al., 1986). Several groups have tried therefore, to obtain resistance against fungi by (over)expressing a new chitinase activity in transgenic plants (LUND et al., 1989; NEUHAUS et al., 1991). Although chitinase activity could be shown, no increased resistance due to the expression of these proteins was observed. Recently, however, the expression of a chitinase gene of bean in tobacco, rendered the tobacco resistant to some extent against *Rhizoctonia* (BROGLIE et al., 1991). Other proteins which display a clear antifungal activity *in vitro*, comprise osmotin, zeamatin and thionins (VIGERS et al., 1991; ROBERTS et al., 1990; BOHLMANN and APEL, 1991; WOLOSHUK et al., 1991). Whether or not over-expression and, or ectopic expression of these genes leads to any increased resistance in transgenic plants, remains to be ascertained.

With respect to engineering plants to obtain resistance against bacteria, similar approaches have been applied. Thus, enzymes from heterologous sources which degrade the cell wall of Gram positive and Gram negative bacteria, have been introduced into transgenic plants such as the bacteriophage T4-encoded lysozyme (DÜRING, 1989). It remains to be determined whether this leads to some resistance against bacteria. A special case was exploited by ANZAI et al. (1990) to achieve resistance against *Pseudomonas syringae* pv. *tabaci.* This bacterium produces a toxin, called tabtoxin, which is a phytotoxic dipeptide and acts by inhibiting the glutamine synthetase. This toxin is necessary for the bacterium to invade the tobacco plant successfully. Following a strategy very similar to that to obtain herbicide resistance, ANZAI et al. (1990) isolated an acetyltransferase-encoding gene from *P. syringae* pv. *tabaci*. This activity is used by the *Pseudomonas* strain to protect itself against the action of the tabtoxin which otherwise would inhibit its own glutamine synthase. The transfer of this gene modified by suitable plant regulatory sequences, into tobacco plants resulted in transgenic plants which were resistant against infection by *P. syringae* pv. *tabaci*.

Clearly, this successful strategy, i.e., the detoxification of the toxin, should be applicable to all pathogens where a toxin is an essential part of the infection and/or colonization process.

Conclusion

In comparison to virus and insect resistance, the methods of obtaining resistance against fungi and bacteria are less developed. Nevertheless, the approaches of introducing genes which lead to the synthesis of new phytoalexins or cell-wall degrading enzymes, show great potential.

5 Engineering of Other Traits

5.1 Introducing Male Sterility into Transgenic Plants

The intense use of hybrid seeds obtained by crossing two highly inbred lines, has been the major reason for the yield increase in several crop plants such as maize. Male sterility has been of enormous advantage in the production of hybrid seeds in that self-fertilization of the female partner can be excluded in the male sterile line. Apart from genetic male sterility, the only other way of preventing self-fertilization is mechanical emasculation.

The main crop plant where hybrid seed is of major importance, is maize. Two decades ago, an effective genetic male sterility system which was controlled by the mitochondria of maize and therefore, called cytoplasmic male sterility, was effectively used in the corn seed industry (LAUGHNAN and LAUGHNAN, 1983). Unfortunately, this male sterility trait is linked to a susceptibility to the phytopathogenic fungus *Bipolaris maydis* which, about 19 years ago, led to the eradication of most maize plants due to the massive spread of a new genotype of this fungus. Since then, no reliable male sterility system has been developed and thus, most corn hybrid seed is produced by the hand-emasculation of the male tassel of the corn plant. Obviously, the creation of a new and universally applicable male sterility system would be of major advantage, not only for maize, but also for other crop plants which, due to the very extensive costs of hand-emasculation have not yet been considered for hybrid seed production.

This goal was recently achieved by MARIANI et al. (1990). A highly active RNase of *Bacillus amyloliquefaciens* was expressed in transgenic plants using a specific promoter from a tobacco gene which was expressed exclusively in the tapetum at a particular stage of flower development. The tapetum is a single layer of cells surrounding the pollen sac inside the anthers, and is necessary for pollen development. Due to the high specific activity of the RNase from *B. amyloliquefaciens*, the tapetal cells in the transgenic plants are destroyed, leading to the abortion of pollen development, while the female organs remain unchanged. This is another impressive example of the usefulness of specific promoters which are tightly controlled in either a temporal or spatial way. It is apparent that a leaky expression of the RNase in tissues other than the tapetum, would have deleterious effects on plant development.

The possibility of specifically destroying pollen can be applied directly to crops where seeds do not represent the harvestable part of the plant, such as tuber-forming plants or vegetables. In seed-producing plants, a restorer gene is needed to inhibit the activity of the RNase in the hybrid plant, thus allowing the formation of seeds. This has also been achieved by genetic engineering: a proteinaceous inhibitor of the RNase under the control of the same tapetum-specific promoter, was introduced into plants. When present in the same cell such as in a hybrid seed, the inhibitor suppresses the RNase activity completely, thus leading to the restoration of pollen fertility (DILWORTH, 1991). Clearly, the technique could establish hybrid seeds in many species.

5.2 Changing Characteristics of Fruit: Producing Transgenic Tomato Plants with a Longer Shelf-Life and a Better Taste

According to a US Department of Agriculture estimate, nearly half of the fresh fruit and vegetables harvested annually, are lost due to spoilage. This spoilage is mainly due to the formation of ethylene which triggers fruit ripening.

In order to prevent or delay fruit ripening, sequestrants of ethylene are used, or fruit are harvested well before they ripen on the plant. Both ways have their disadvantages; early harvest as a rule, results in an unpleasant taste and sequestering ethylene may involve the use of chemicals and increase the price of the fruit.

Thus, possibilities have been explored aiming at modifying the ethylene formation or content in plants and fruit.

Ethylene is formed from S-adenosylmethionine via the intermediate 1-aminocyclopropane-1-carboxylic acid (ACC). The formation of ACC is catalyzed by the enzyme ACC synthase. The second step leading to ethylene formation is catalyzed by the ethylene forming enzyme (EFE) or ACC oxidase. The genes encoding the ACC synthase have been cloned from tomato and squash (SATO and THEOLOGIS, 1989; VAN DER STRAETEN et al., 1990), and those encoding ACC oxidase have been cloned from tomato (HAMILTON et al., 1990).

Two approaches have been followed to inhibit or depress the regulation of the formation of ethylene. In one approach, transgenic tomato plants were produced which showed a highly reduced level of the ethylene forming enzyme. This was due to the expression of a chimeric gene which was under the control of the constitutive 35s CaMV promoter and which encoded the anti-sense RNA of the ACC oxidase. This resulted in the reduction of the formation of ethylene. In wounded leaves the maximal decrease of ethylene formation was 68% and in ripening fruit, 97%. No data on the effect of this dramatic reduction of ethylene biosynthesis on ripening, has yet been reported (HAMILTON et al., 1990).

Another approach to inhibit the formation of ethylene, was followed by KLEE and coworkers from the Monsanto Company. They identified a bacterial gene encoding an enzyme able to degrade ACC, the immediate precursor of ethylene synthesis, which also significantly decreased ethylene biosynthesis (DILWORTH, 1991). The reduced level of ethylene was accompanied by a 2-week delay in fruit ripening which clearly, has commerical potential. The appearance of the tomato fruit, apart from the delay in ripening, was completely normal when compared to control experiments.

Whereas experiments aimed at lowering the ethylene level would have immediate effects on fruit ripening, the softening of the tomato fruit is another area where transgenic plants could create new opportunities. The softening of the tomato has been correlated with changes in cell wall structure and the activities of cell wall degrading enzymes (cf. KRAMER et al., 1989, for a review). Several such enzymes are candidates determining the softening process, and to date, the highest correlation to softening has been observed for the enzyme polygalacturonase. In order to understand the precise role of polygalacturonase better and to perhaps alter the process of fruit softening, chimeric gene encoding an anti-sense RNA for the polygalacturonase under the control of the 35s CaMV promoter, have been introduced into tomato plants (SMITH et al., 1988; SHEEHY et al., 1988). As a result, the level of polygalacturonase was reduced by as much as 99%.

In tomato fruit containing a residual 1% polygalacturonase activity, the depolymerization of the pectin was inhibited, whereas other ripening related processes such as production of ethylene or lycopene production, were not affected (SMITH et al., 1990). Most importantly however, the fruits of these tomato plants were more resistant to mechanical stress associated with, for example, packaging and transport, despite the fact that their compressibility was not changed. This could be a significant improvement as it might allow the fruit to be left longer on the plant until they are fully ripe, and yet they would still withstand the mechanical stress associated with transportation. Whether or not this might improve the taste of the tomato fruit, remains to be ascertained.

Conclusion

With respect to changing the quality of vegetables and fruit in transgenic crops, some successful, early attempts have been described. The ability to influence ripening by reducing the ethylene level is not only of importance to the ripening of tomato fruit, but could well influence and thus change, many parameters related to crop physiology.

6 Transgenic Plants as a Means of Manipulating Protein and Carbohydrate Composition – Plants as Bioreactors

One of the potential uses of plant genetic engineering is the construction of plants producing modified or new compounds, i.e., using plants as bioreactors.

Both low-molecular weight (e.g., lipids, sugars, secondary metabolites) and high-molecular weight compounds (proteins, carbohydrate polymers, fibres) are potential targets to be manipulated by using genetically modified plants. Although many different laboratories and companies are actively involved in this area of research, the majority of data published is restricted to producing new or modified proteins and altered starches.

6.1 Production of New or Modified Proteins in Transgenic Plants

Two main lines are followed in this area:

- improvement of the nutritional quality of plants by improving the amino acid composition of their storage proteins,
- production of new, high-value proteins with pharmaceutical applications.

The benefits of plants which produce seed proteins with an improved amino acid composition are numerous. One way to reach this goal is to increase the amount of essential amino acids present in seed material used either for human consumption, or for animal feedstock. Among the plant seeds consumed by animals, the vast majority comes from maize and soybeans. Unfortunately, the content of some essential amino acids is too low in both species: sulfur-containing amino acids such as methionine and cysteine are missing in soybean and lysine and tryptophan are missing in corn. A straightforward way to alter the composition of the seed protein, is to express a foreign seed protein which contains a high proportion of the missing amino acids in seeds of transgenic plants. Many experiments have shown that genes encoding seed proteins retain their developmentally controlled (endosperm-specific) expression in heterologous plants, irrespective of being derived from monocotyledonous or dicotyledonous plants (GOLDBERG et al., 1989). In addition, in most cases, seed proteins are correctly targeted to the protein body in heterologous plants and undergo correct processing.

The most encouraging example of improving the amino acid composition of seed proteins, has been reported by ALTENBACH et al. (1990), who expressed a seed protein coding gene from the Brazil nut in transgenic tobacco plants. Nearly 30% of the amino acids of this seed protein are sulfur amino acids. It is therefore, a good candidate for complementing the deficiency of sulfur amino acids of some seeds. This Brazil nut seed protein contributed up to 8% of the total protein in seeds of transgenic tobacco plants, resulting in a significant increase in the methionine content.

6.1.1 Production of High-Value Proteins in Transgenic Plants

The first reported example of the production of high-value peptides/proteins, was the formation of leuenkephalin, a pentapeptide which has some interesting pharmaceutical applications in seeds of oilseed rape (VANDERKERCKHOEVE et al., 1989). This peptide was produced by inserting its sequence into a seed protein gene from *Arabidopsis thaliana,* the so-called 2s albumin in such a way that, after isolation of the 2s albumin-leuenkephalin fusion protein, it could be released easily by the application of proteases. Authentic leuenkephalin was recovered from seed extracts after protease treatment and HPLC purification. With respect to the production of leuenkephalins in transgenic rape seed, it was estimated that from ten to several hundred grams could be obtained per hectare (KREBBERS and VANDERKERCKHOEVE, 1990).

Other examples of the production of high molecular weight proteins produced in plants are the synthesis of human serum albumin in tubers of transgenic potato (SIJMONS et al., 1990) and the production of immunoglobulins in transgenic tobacco plants (HIATT et al., 1989; DÜRING et al., 1990). Whereas plant-produced human serum albumin could be of interest, due to the demand for a human serum albumin not contaminated by human pathogens, the production of functionally active, monoclonal antibodies in transgenic plants could open completely new perspectives for a wide use of these proteins in fields other than the diagnostic field, i.e., in industrial processes (SWAIN, 1991).

6.1.2 Conclusion

From the examples described, it is apparent that all the tools have been developed to use plants as factories for producing proteins for pharmaceutical and technical/industrial applications. Whether or not this technique will be applied in the future, will depend on economic considerations. The production costs in transgenic plants are lower as compared to those of other production systems, like mammalian tissue cultures or transgenic animals. The main question is whether the (low) production cost will be offset by, for example, higher purification cost. Thus it may be advantageous to produce those compounds in plants which do not need such a sophisticated purification as is required for pharmaceutical agents. This could be true for enzymes used in technical processes. In addition, other disadvantages of plants such as seasonal production have to be considered.

6.2 Production of New or Modified Carbohydrates and Oils in Transgenic Plants

Plants represent the major renewable resource of oils as well as complex carbohydrate polymers such as fibers and starches.

With respect to oils, there are many possibilities of improvement which might permit a broader use of these compounds as food or for industrial uses. However, despite intense efforts in companies and public institutes, no successes have, as yet, been reported (cf. SOMERVILLE and BROWSE, 1991, for a review).

With respect to starch composition, one of the priorities is to produce plants which would only produce one sort of starch, i.e., either amylopectin or amylose. This goal was recently reached in transgenic potatoes. By expressing a gene under the control of the 35s RNA promoter, which encoded the anti-sense RNA of the granule-bound starch synthase (the enzyme responsible for amylose synthesis), transgenic potatoes were created which were devoid of amylose and contained only amylopectin (VISSER et al., 1991; SONNEWALD and WILLMITZER, unpublished observations, Fig. 5).

It is less likely that plants will find application as bioreactors for producing high-value proteins than for producing new varieties of oils or complex carbohydrate polymers. The main reason for this, is that plants have one great advantage over all other organisms which is their mass potential. Only plants are able to produce millions of tons of, for example, carbohydrates which are needed for technical processes. With respect to proteins where

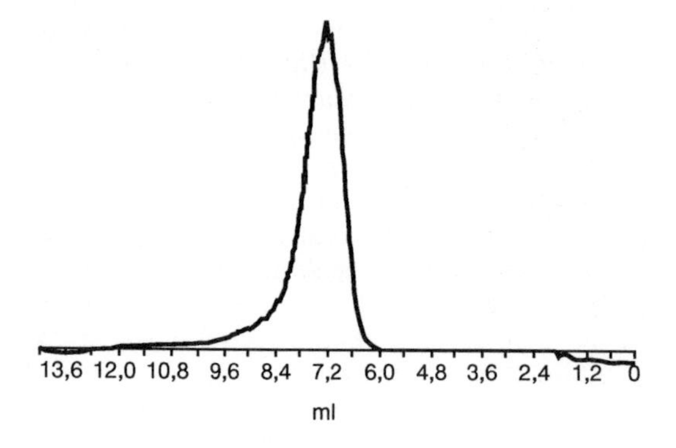

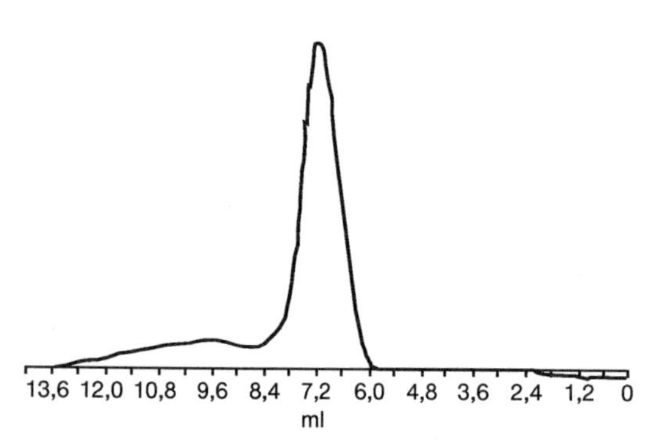

Fig. 5. Inhibition of the granule-bound starch synthase in transgenic potato plants leads to the formation of amylose-free starch. The starch content was analyzed of tubers of transgenic potato plants in which the expression of the granule-bound starch synthase gene was inhibited by the anti-sense RNA approach. These were compared with the starch extracted from control potato plants. The figure shows the elution profile of starch from a transgenic plant (upper part) and a control plant (lower part) separated by gel permeation chromatography. The first peak eluted is amylose, the second, amylopectin. It is evident from the elution profile that the starch of the transgenic plant is nearly devoid of amylose.

the requirement may be just a few kilograms (as for many proteins with a pharmaceutical application), or a few tons worldwide (as for proteins being used in technical processes), plants are not the unique source to produce these compounds, but they have to compete with methods of producing these proteins in other organisms.

7 References

ABDULLAH, R., THOMSON, J., COCKING, E. (1986), Efficient plant regeneration from rice protoplasts through somatic embryogenesis, *Biotechnology* **4,** 1087–1090.

ALTENBACH, S., PEARSON, K., MEEKER, G., STARACI, L., SUN, S. (1990), Enhancement of the methionine content of seed proteins by the expression of a chimeric gene encoding a methionine rich protein in transgenic plants, *Plant Mol. Biol.* **13,** 513–522.

ANZAI, H., YONEYAMA, K., YAMAGUCHI, I. (1989), Transgenic tobacco resistant to a bacterial disease by the detoxification of a pathogenic toxin, *Mol. Gen. Genet.* **219,** 492–494.

ATKINSON, R., GARDNER, R. (1991), *Agrobacterium* mediated transformation of pepino and regeneration of transgenic plants, *Plant Cell. Rep.* **10,** 208–212.

BAILEY, J. A. (1987), Phytoalexins: a genetic view of their significance, in: *Genetics and Plant Pathogenesis* (DAY, P., ELLIS, G. Eds.), Oxford: Blackwell Scientific Publishers, pp. 233–244.

BARTON, K. A., WHITELEY, H., YANG, N. (1987), *Bacillus thuringiensis* delta-endotoxin expressed in transgenic *Nicotiana tabacum* provides resistance to Lepidopteran insects, *Plant Physiol.* **85,** 1103–1109.

BASIRAN, N., ARMITAGE, P., SCOTT, R., DRAPER, J. (1987), Genetic transformation of flax (*Linum usitatissimum*) by *Agrobacterium tumefaciens:* Regeneration of transformed shoots via a callus phase, *Plant Cell Rep.* **6,** 396–399.

BEACHY, R., LOESCH-FRIES, S., TUMER, N. (1990), Coat-protein mediated resistance against virus infection, *Annu. Rev. Phytopathol.* **28,** 451–474.

BENFEY, P., CHUA, N. (1989), Regulated genes in transgenic plants, *Science* **244,** 174–181.

BENFEY, P., REN, L., CHUA, N. (1990), Tissue-specific expression from CaMV 35s enhancer subdomains in early stages of plant development, *EMBO J.* **9,** 1677–1684.

BLOWERS, A., BOGORAD, L., SHARK, K., SANFORD, J. (1989), Studies on *Chlamydomonas* chloroplast transformation: Foreign DNA can be stably maintained in the chromosome, *Plant Cell* **1,** 123–132.

BOHLMANN, H., APEL, K. (1991), Thionins, *Annu. Rev. Plant Physiol. Plant Mol. Biol.* **42,** 227–240.

BOTTERMAN, J., LEEMANS, J. (1989), Field testing of insect and herbicide resistant crops, *Vortr. kflanzenzüchtung* **16,** 455–461.

BOUTRY, M., NAGY, F., POULSON, C., AOYAGO, K., CHUA, N. (1987), Targeting of bacterial chloramphenicol-acetyltransferase to mitochondria in transgenic plants, *Nature* **328,** 340–342.

BOYNTON, J., GILLHAM, N., HARRIS, E., HOSLER, J., JOHNSON, A., JONES, A., RANDOLPH-ANDERSON, B., ROBERTSON, D., KLEIN, T., SHARK, K., SANFORD, J. (1988), Chloroplast transformation in *Chlamydomonas* with high velocity projectiles, *Science* **240,** 1534–1538.

BROGLIE, R., BROGLIE, K., CHET, I., ROBY, D., HOLLIDAY, M. (1991), Chitinase expression in transgenic plants: A molecular approach to fungal disease resistance, *J. Cell Biol., Suppl.* **15A,** 9.

BRUNKE, K., MEEUSEN, R. (1991), Insect control with genetically engineered crops, *Trends Biotechnol.* **9,** 197–200.

BYTEBIER, B., DEBOECK, F., DE GREVE, H., VAN MONTAGU, M., HERNALSTEENS, J.-P. (1987), T-DNA organization in tumor callus and transgenic plants of the monocotyledon *Asparagus officinalis, Proc. Natl. Acad. Sci. USA* **84,** 5345.

CABOCHE, M. (1990), Liposome mediated transfer of nucleic acids into plant cells, *Physiol. Plant.* **79,** 173–176.

CATLIN, D., OCHOA, O., MCCORMICK, S., QUIROS, C. (1988), Celery transformation by *Agrobacterium tumefaciens:* Cytological and genetic analysis of transgenic plants, *Plant Cell Rep.* **7,** 100–103.

CHEUNG, A., BOGORAD, L., VAN MONTAGU, M., SCHELL, J. (1988), Relocating a gene for herbicide tolerance: a chloroplast gene is convered into a nuclear gene, *Proc. Natl. Acad. Sci. USA* **85,** 391–395.

CROSSWAY, A., OAKES, J. V., IRVINE, J. M., WARD, B., KNAUF, V. C., SHEWMAKER, C. K. (1986), Integration of foreign DNA following microinjection of tobacco mesophyll protoplasts, *Mol. Gen. Genet.* **202,** 179–185.

CUOZZO, M., O'CONNELL, K., KANIEWSKI, W., FANG, R., CHUA, N., TUMER, N. (1988), Viral protection in transgenic tobacco plants expressing the cucumber mosaic virus protein or its antisense RNA, *Biotechnology* **6,** 549–557.

DATTA, S., PETERHANS, A., DATTA, K., POTRYKUS, I. (1990), Genetically engineered fertile indica-rice recovered from protoplasts, *Biotechnology* **8,** 736–740.

DAVID, C., TEMPE, J. (1988), Genetic transformation of cauliflower (*Brassica oleracea* L. var. *Botrytis*) by *Agrobacterium rhizogenes, Plant Cell Rep.* **7,** 88–91.

DAY, A., BEJARANO, E., BUCK, K., BURRELL, M., LICHTENSTEIN, C. (1991), Expression of an antisense viral gene in transgenic tobacco confers resistance to the DNA virus golden moscaic virus, *Proc. Natl. Acad. Sci USA* **88,** 6721–6725.

DE BLOCK, M., VAN MONTAGU, M., SCHELL, J. (1985), Chloroplast transformation by *Agrobacterium tumefaciens, EMBO J.* **4,** 1367–1372.

DE BLOCK, M., BOTTERMAN, J., VANDEWIELE, M., DOCKX, J., THOEN, C., GOSSELE, V., MOVVA, N., THOMPSON, C., VAN MONTAGU, M., LEEMANS, J. (1987), Engineering herbicide resistance in plants by expression of a detoxifying enzyme, *EMBO J.* **6,** 2513–2518.

DE GREEF, W., DELON, R., DE BLOCK, M., LEEMANS, J., BOTTERMAN, J. (1989), Evaluation of herbicide resistance in transgenic crops under field conditions, *Biotechnology* **7,** 61–64.

DELANNEY, X., LAVALLEE, B., PROKSCH, R., FUCHS, R., SIMS, S., GREENPLATE, J., MARRONE, P., DODSON, R., AUGUSTINE, J., LAYTON, J., FISCHHOFF, D. (1989), Field performance of transgenic tomato plants expressing the *Bacillus thuringiensis* var. *kurstaki* insect control protein, *Biotechnology* **7,** 1265–1269.

DELLA CIOPPA, G., BAUER, S., TAYLOR, M., ROCHESTER, D., KLEIN, B., SHAH, D., FRALEY, R., KISHORE, G. (1987), Targetting a herbicide resistant enzyme from *E. coli* to chloroplasts of higher plants, *Biotechnology* **5,** 579–584.

DE ZOETEN, G., FULTON, R. W. (1975), Understanding generates possibilities, *Phytopathology* **57,** 1347–1352.

DILWORTH, M. (1991), Molecular Biology comes home, *Plant Cell* **3,** 213–218.

DOREL, C., VOELKER, T., HERMAN, E., CRISPEELS, M. (1989), Transport of proteins to the plant vacuole is not by bulk flow through the secretory system and requires positive sorting information, *J. Cell. Biol.* **108,** 327–337.

DÜRING, K. (1989), Wundinduzierbare Expression und Sekretion von T4-Lysozym und monoklonalen Antikörpern in *Nicotiana tabacum, Dissertation,* Universität Köln, FRG.

DÜRING, K., HIPPE, S., KREUZALER, F., SCHELL, J. (1990), Synthesis and self-assembly of a functional monoclonal antibody in transgenic tobacco, *Plant Mol. Biol.* **15,** 281–293.

ECKES, P., SCHMITT, P., DAUB, W., WENGENMAYER, F. (1989), Overproduction of alfalfa glutamine synthetase in transgenic tobacco plants, *Mol. Gen. Genet.* **217,** 263–268.

EDWARDS, J., CORUZZI, G. (1990), Cell-specific gene expression in plants, *Annu. Rev. Genet.* **24,** 275–303.

EVERETT, N., ROBINSON, K., MASCARENHAS, D. (1987), Genetic engineering of sunflower (*Helianthus annuus* L.), *Biotechnology* **5,** 1201–1204.

FANG, G., GRUMET, R. (1990), *Agrobacterium* mediated transformation and regeneration of muskmelon plants, *Plant Cell Rep.* **9,** 160–164.

FISCHHOFF, D., BOWDISH, K., PERLAK, F., MARRONE, P., MCCORMICK, S., NIEDERMEYER, J., DEAN, D., KUSANO-KRETZMER, K., MAYER, E., ROCHESTER, D., ROGERS, S. FRALEY, R. (1987), *Biotechnology* **5**, 807–813.

FROMM, M., MORRISH, F., ARMSTRONG, C., WILLIAMS, R., THOMA, J., KLEIN, T. (1990), Inheritance and expression of chimeric genes in the progeny of transgenic maize, *Biotechnology* **8**, 833–839.

FRY, J., BARNASON, R., HORSCH, R. (1987), Transformation of *Brassica napus* with *Agrobacterium tumefaciens* based vectors, *Plant Cell Rep.* **6**, 321–325.

GATZ, C., KAISER, A., WENDENBURG, R. (1991), Regulation of a modified CaMV 35s promoter by the Tn10-encoded Tet repressor in transgenic tobacco, *Mol. Gen. Genet.* **227**, 229–237.

GERLACH, W., LLEWELLYN, D., HASELOFF, J. (1987), Construction of a plant disease resistance gene from the satellite RNA of tobacco ringspot virus, *Nature* **328**, 802–805.

GOLDBERG, R., BARKER, S., PEREZ-GRAU, L. (1989), Regulation of gene expression during plant embryogenesis, *Cell* **56**, 149–160.

GOLDSBROUGH, P., GELVIN, S., LARKINS, B. (1986), Expression of maize zein genes in transformed sunflower cells, *Mol. Gen. Genet.* **202**, 374–389.

GOLEMBOSKI, D., LOMONOSSOFF, G., ZAITLIN, M. (1990), Plants transformed with a tobacco mosaic virus nonstructural gene sequence are resistant to the virus, *Proc. Natl. Acad. Sci. USA* **87**, 6311–6315.

GORDON-KAMM, W., SPENCER, T., MANGANO, M., ADAMS, T., DAINES, R., START, W., O'BRIEN, J., KRUEGER, R., KAUSCH, A., LEMAUX, P. (1990), Transformation of maize and regeneration of fertile transgenic plants, *Plant Cell* **2**, 603–618.

HAIN, R., BIESELER, B., KINDL, H., SCHRÖDER, G., STÖCKER, R. (1990), Expression of a stilbene synthase gene in *Nicotiana tabacum* results in synthesis of the phytoalexin resveratrol, *Plant Mol. Biol.* **15**, 325–335.

HAMILTON, A., LYCETT, G., GRIERSON, D. (1990), Antisense gene that inhibits synthesis of the hormone ethylene in transgenic plants, *Nature* **346**, 284–287.

HARRISON, B., MAYO, M., BAULCOMBE, D. (1987), Virus resistance in transgenic plants that express cucumber mosaic virus satellite RNA, *Nature* **328**, 799–802.

HAYFORD, M., MEDFORD, J., HOFFMANN, N., ROGERS, S., KLEE, H. (1988), Development of a plant transformation selection system based on expression of genes encoding gentamycin acetyltransferase, *Plant Physiol.* **86**, 1216–1222.

HEMENWAY, C., FANG, R., KANIEWSKI, W., CHUA, N., TUMER, N. (1988), Analysis of the mechanism of protection in transgenic plants expressing the potato virus x coat protein or its anti-sense RNA, *EMBO J.* **7**, 1273–1280.

HERRERA-ESTRELLA, L., DEBLOCK, M., MESSENS, E., HERNALSTEENS, J., VAN MONTAGU, M., SCHELL, J. (1983), Chimeric genes as dominant selectable markers in plant cells, *EMBO J.* **2**, 987–995.

HESS, D., DRESSLER, K., NIMMRICHTER, R. (1990), Transformation experiments by pipetting *Agrobacterium* into the spikelets of wheat, *Plant Sci.* **72**, 233–244.

HIATT, A., CAFFERKY, R., BOWDISH, K. (1989), Production of antibodies in transgenic plants, *Nature* **342**, 76–78.

HILDER, V., GATEHOUSE, A., SHEERMAN, S., BARKER, R., BOULTER, D. (1987), A novel mechanism of insect resistance engineered into tobacco, *Nature* **330**, 160–163.

HILDER, V., GATEHOUSE, A., BOULTER, D. (1990), Genetic engineering of crops for insect resistance using genes of plant origin, in: *Genetic Engineering of Crop Plants* (LYCETT, G., GRIERSON, D., Eds.), pp. 51–66, London–Boston: Butterworth.

HILL, K., JARVIS-EAGAN, N., HALK, E., KRAHN, K., LIAO, L., MATHEWSON, R., MERLO, D., NELSON, S., RASHKA, K., LOESCH-FRIES, S. (1991), The development of virus-resistant alfalfa, *Medicago sativa* L., *Biotechnology* **9**, 373–379.

HILLE, J., VERHEGGEN, T., ROELVINK, A., FRANSSEN, H., VAN KAMMEN, A., ZABEL, P. (1986), Bleomycin resistance: a new dominant selectable marker for plant cell transformation, *Plant Mol. Biol.* **7**, 171–176.

HOEKEMA, A., HUISMAN, M., MOLENDIJK, L., VAN DEN ELZEN, P., CORNELISSEN, B. (1989), The genetic engineering of two commercial potato cultivars for resistance to potato virus x, *Biotechnology* **7**, 273–278.

HOFMANN, C., VANDERBRUGGEN, H., HÖFTE, H., VAN RIE, J., JANSENS, S., VAN MELLAERT, H. (1988), Specificity of *Bacillus thuringiensis* delta-endotoxins is correlated with the presence of high affinity binding sites in the brush order membrane of target insect midguts, *Proc. Natl. Acad. Sci. USA* **85**, 7844–7848.

HÖFTE, H., WHITELEY, H. (1989), Insecticidal crystal proteins of *Bacillus thuringiensis, Microb. Rev.* **53**, 242–255.

HORSCH, R., FRY, J., HOFFMANN, N., EICHOLTZ, P., ROGERS, S., FRALEY, R. (1985), A simple and general method for transferring genes into plants, *Science* **227,** 1229.

ITURRIAGA, G., JEFFERSON, R., BEVAN, M. (1989), Endoplasmic reticulum targetting and glycosylation of hybrid proteins in transgenic tobacco, *Plant Cell* **1,** 381–390.

JÄHNE, A., LAZZERI, P., LÖRZ, H. (1991), Regeneration of fertile plants from protoplasts derived from embryogenic cell suspensions of barley, *Plant Cell Rep.* **10,** 1–6.

JAMES, D., PASSEY, A., BARBARA, D., BEVAN, M. (1989), Genetic transformation of apple (*Malus pumila* Mill) using a disarmed Ti-binary vector, *Plant Cell Rep.* **7,** 658–661.

JEFFERSON, R. (1987), Assaying chimeric genes in plants: The GUS gene fusion system, *Plant Mol. Biol. Rep.* **5,** 387–405.

JIA, S., YANG, M., OTT, R., CHUA, N. (1989), High frequency transformation of *Kalanchoe laciniata, Plant Cell Rep.* **8,** 336–340.

JONES, J., SVAB, Z., HARPER, E., HORWITZ, C., MALIGA, P. (1987), A dominant nuclear streptomycin resistance marker for plant cell transformation, *Mol. Gen. Genet.* **210,** 86–91.

JORGENSEN, R. (1990), Altered gene expression in plants due to trans-interactions between homologous genes, *Trends Biotechnol.* **8,** 340–344.

KANIEWSKI, W., LAWSON, C., SAMMONS, B., HALEY, L., HART, J., DELANNAY, X., TUMER, N. (1990), Field resistance of transgenic Russet Burbank potato to effects of infection by potato virus x and potato virus y, *Biotechnol.* **8,** 750–754.

KAWCHUK, L, MARTIN, R., MCPHERSON, J. (1990), Resistance in transgenic plants expressing the potato leafroll luteovirus coat protein gene, *Mol. Plant Microbe Interact.* **3.**

KEEGSTRA, K., OLSEN, L., THEG, S. (1989), Chloroplast precursors and their transport across the envelope membranes, *Annu. Rev. Plant Physiol. Mol. Biol.* **40,** 471–501.

KISHORE, G., SHAH, D. (1988), Amino acid biosynthesis inhibitors as herbicides, *Annu. Rev. Biochem.* **57,** 627–663.

KLEIN, T. M., WOLF, E. D., WU, R. D., STANFORD, J. C. (1987), High velocity microprojectiles for delivering nucleic acids into living cells, *Nature* **327,** 70–73.

KONCZ, C., OLSSON, O., LANGRIDGE, W., SCHELL, J., SZALAY, A. (1987), Expression and assembly of functional bacterial luciferase in plants, *Proc. Natl. Acad. Sci. USA* **84,** 131–135.

KRAMER, M., SHEEHY, R., HIATT, W. (1989), Progress towards the genetic engineering of tomato fruit softening, *Trends Biotechnol.* **7,** 191–194.

KREBBERS, E., VANDEKERCKHOEVE, J. (1990), Production of peptides in plant seeds, *Trends Biotechnol.* **8,** 1–3.

KUCHUK, N., KOMARNITSKY, I., SHAKOVSKY, A., GLEBA, Y. (1990), Genetic transformation of *Medicago* species by *Agrobacterium tumefaciens* and electroporation of protoplasts, *Plant Cell Rep.* **8,** 660–663.

LAMB, C., LAWTON, M., DRON, M., DIXON, R. (1989), Signals and transduction mechanisms for activation of plant defenses against microbial attack, *Cell* **56,** 215–224.

LAUGHNAN, J., GABAY-LAUGHNAN, S. (1983), Cytoplasmic male sterility in maize, *Annu. Rev. Genet.* **17,** 27–48.

LAWSON, C., KANIEWSKI, W., HALEY, L., ROZMAN, R., NEWELL, C., SANDERS, P., TUMER, N. (1990), Engineering resistance to mixed virus infection in a commercial potato cultivar: Resistance to potato virus X and potato virus Y in transgenic Russett Burbank, *Biotechnology* **8,** 127–135.

LEDGER, S., DEROLES, S., GIVEN, N., (1991), Regeneration and *Agrobacterium*-mediated transformation of chrysanthenum, *Plant Cell Rep.* **10,** 195–199.

LEE, K., TOWNSEND, J., TEPPERMAN, J., BLACK, M., CHUI, C. (1988), The molecular basis of sulfonylurea herbicide resistance in higher plants, *EMBO J.* **7,** 1241–1248.

LING, K., NAMBA, S., GONSALVES, C., SLIGHTOM, J., GONSALVES, D. (1991), Protection against detrimental effects of potyvirus infection in transgenic tobacco plants expressing the papaya ringspot virus coat protein gene, *Biotechnology* **9,** 752–758.

LLOYD, A., BARNASON, A., ROGERS, S., BYRNE, M., FRALEY, R., HORSCH, R. (1986), Transformation of *Arabidopsis thaliana* with *Agrobacterium tumefaciens, Science* **234,** 464–466.

LOESCH-FRIES, L., MERLO, D., ZINNEN, T., BURHOP, L., HALL, T. (1987), Expression of alfalfa mosaic virus RNA 4 in transgenic plant confers virus resistance, *EMBO J.* **6,** 1845–1851.

LUND, P., LEE, R., DUNSMUIR, P. (1989), Bacterial chitinase is modified and secreted in transgenic tobacco, *Plant Physiol.* **91,** 130–135.

LYON, B., LLEWELLYN, D., HUPPATZ, J., DENNIS, L., PEACOCK, W. (1989), Expression of a bacterial gene in transgenic tobacco confers resistance to the herbicide 2,4-dichlorophenoxyacetic acid, *Plant Mol. Biol.* **13,** 533–540.

MANNERS, J., WAY, H. (1989), Efficient transformation with regeneration of the tropical pasture legume *Stylosanthes humilis* using *Agrobacterium rhizogenes* and a Ti-Plasmid binary vector, system, *Plant Cell Rep.* **8,** 341–345.

MARIANI, C., DE BEUCKELEER, M., TRUETTNER, J., LEEMANS, J., GOLDBERG, R. (1990), Induction of male sterility in plants by a chimeric ribonuclease gene, *Nature* **347**, 737–741.

MATZKE, M., SUSANI, M,. BINNS, A., LEWIS, E., RUBENSTEIN, J., MATZKE, A. (1984), Transcription of a zein gene introducted into sunflower using a Ti-plasmid vector, *EMBO J.* **3**, 1525–1531.

MAZUR, B., FALCO, S. (1989), The development of herbicide resistant crops, *Annu. Rev. Plant Physiol. Plant Mol. Biol.* **40**, 441–470.

MCCABE, D., SWAIN, W., MARINELL, B., CHRISTOU, P. (1988), Stable transformation of soybean *Glycine max* by particle acceleration, *Biotechnology* **6**, 923–926.

MCCORMICK, S., NIEDERMEYER, J., FRY, J., BARNASON, A., HORSCH, R., FRALEY, R. (1986), Leaf disc transformation of cultivated tomato using *Agrobacterium tumefaciens, Plant Cell Rep.* **5**, 81–84.

MCGAUGHEY, W. (1985), Insect resistance to the biological insecticide *Bacillus thuringiensis, Science* **229**, 193–195.

MCGRANAHAN, G., LESLIE, C., URATSU, S., MARTIN, L., DARDEKAR, A. (1988), *Agrobacterium* mediated transformation of walnut somatic embryos and regeneration of transgenic plants, *Biotechnology* **6**, 800–804.

MICHELMORE, R., MARSH, E., SEELY, S., LANDRY, B. (1987), Transformation of lettuce (*Lactuca sativa*) mediated by *Agrobacterium tumefaciens, Plant Cell Rep.* **6**, 439–442.

MIKI, L., REICH, T., IYER, V. (1987), Microinjection: an experimental tool for studying and modifying plant cells, in: *Plant Gene Research: Plant DNA Infectious Agents* (Hohn, T., Schell, J., Eds.), pp. 249–266, Wien–New York: Springer Verlag.

MOFFAT, A. (1991), Making sense of antisense, *Science* **253**, 510–511.

MOONEY, P., GOODWIN, P., DENNIS, E., LLEWELLYN, D. (1991), *Agrobacterium*-mediated gene transfer into wheat tissues, *Plant Cell Tissue Organ Cult.* **25**, 209–218.

MÜLLNER, H. (1991), Herbicide tolerance, a contribution to integrated crop management in: *Pesticide Chemistry: Advances in International Research, Development and Legislation* (FREHSE, H. Ed.) pp. 131–138, Weinheim-New York-Basel-Cambridge: VCH.

NAGY, F., BOUTRY, M., HSU-WONG, M., CHUA, N. (1987), The 5′-proximal region of the wheat Cab-1 gene contains a 286 bp enhancer like sequence for phytochrome response, *EMBO J.* **6**, 2537–2542.

NEHRA, N., CHIBBAR, R., KARTHA, K., DATLA, R., CROSBY, W., STUSHNOFF, C. (1990), *Agrobacterium* mediated transformation of strawberry calli and recovery of transgenic plants, *Plant Cell Rep.* **9**, 10–13.

NEJIDAT, A., BEACHY, R. (1990), Transgenic tobacco plants expressing a tobacco virus coat protein gene are resistant to some tobamoviruses, *Mol. Plant. Microbe Interact.* **3**, 247–251.

NELSON, R., MCCORMICK, S., DELANNAY, X., DUBE, P., LAYTON, J., ANDERSON, E., PROKSCH, R., HORSCH, R., ROGERS, S., FRALEY, R., BEACHY, R. (1988), Virus tolerance, plant growth and field performance of transgenic tomato plants expressing the coat protein gene of tobacco mosaic virus, *Biotechnology* **6**, 403–409.

NEUHAUS, G., SPANGENBERG, G. (1990), Plant transformation by microinjection techniques, *Physiol. Plant.* **79**, 213–217.

NEUHAUS, G., SPANGENBERG, G., SCHEID, O., SCHWEIGER, H. (1987), Transgenic rape seed plants obtained by microinjection of DNA into microspore derived proembryoids, *Theor. Appl. Genet.* **75**,30–36.

NEUHAUS, J., AHL-GOY, P., HINZ, U., FLORES, S., MEINS, F. (1991), High-level expression of a tobacco chitinase gene in *Nicotiana sylvestris.* Susceptibility of transgenic plants to *Cercospora nicotianae* infection, *Plant Mol. Biol.* **16**, 141–151.

NODA, T., TANAKA, N., MANO, Y., NABESHIMA, S., OHKAWA, H., MATSUI, C. (1987), Regeneration of horseradish hairy roots incited by *Agrobacterium rhizogenes* infection, *Plant Cell Rep.* **6**, 283–286.

OTTEN, L., SCHILPEROORT, R. (1978), A rapid microscale method for the detection of lysopine and nopaline dehydrogenase activity, *Biochim. Biophys. Acta* **527**, 497–500.

OW, D., JACOBS, J., HOWELL, S. (1987), Functional regions of the cauliflower mosaic virus 35s RNA promoter determined by use of the firefly luciferase gene as a reporter of promoter activity, *Proc. Natl. Acad. Sci. USA* **84**, 4870–4874.

OXTOBY, E., HUGHES, M. (1990), Engineering herbicide tolerance into crops, *Trends Biotechnol.* **8**, 61–65.

PASZKOWSKI, J., BAUR, M., BOGUCKI, A., POTRYKUS, I. (1988), Gene targeting in plants, *EMBO J.* **7**, 4021–4026.

PERLAK, F., FUCHS, R., DEAN, D., MCPHERSON, S., FISCHHOFF, D. (1991), Modification of the coding sequence enhances plant expression of insect control protein genes, *Proc. Natl. Acad. Sci. USA* **88**, 3324–3328.

PHELEP, M., PETIT, A., MARTIN, L., DUHOUX, E., TEMPE, J. (1991), Transformation and regeneration of a nitrogen-fixing tree, *Allocasuarina verticillata* Lam., *Biotechnology* **9**, 461–466.

PHYTHOUD, F., SINKAR, V., NESTER, E., GORDON, M. (1987), Increased virulence of *Agrobacterium rhizogenes* conferred by the vir-region of pTiBo542: Application to genetic engineering, *Biotechnology* **5**, 1323–1328.

PICKARDT, T., MEIXNER, M., SCHADE,V., SCHIEDER, O. (1991), Transformation of *Vicia narbonensis* via *Agrobacterium* mediated gene transfer, *Plant Cell Rep.* **9**, 535–538.

POTRYKUS, I. (1990), Gene transfer to cereals: an assessment, *Biotechnology* **8**, 535–542.

POTRYKUS, I. (1991), Gene transfer to plants: assessment of published approaches and results, *Annu. Rev. Plant Physiol. Plant Mol. Biol.* **42**, 205–225.

POWELL, P., STARK, D., SANDERS, P., BEACHY, R. (1989), Protection against tobacco mosaic virus in transgenic plants that express tobacco mosaic virus anti-sense RNA, *Proc. Natl. Acad. Sci. USA* **86**, 6949–6952.

POWELL-ABEL, P., NELSON, R., HOFFMAN, N., ROGERS, S., FRALEY, R. (1986), Delay of disease development in transgenic plants that express the tobacco mosaic virus coat protein gene, *Science* **232**, 738–743.

PRIOLI, L., SÖNDAHL, M. (1989), Plant regeneration and recovery of fertile plants from protoplasts of maize, *Biotechnology* **7**, 589–594.

PUONTI-KAERLAS, J., ERIKSSON, T., ENGSTRÖM, P. (1990), Production of transgenic pea plants by *Agrobacterium tumefaciens* mediated gene transfer, *Theor. Appl. Genet.* **80**, 246–252.

QUEMADA, H., GONSALVES, D., SLIGHTOM, J. (1991), Expression of coat protein gene from cucumber mosaic virus strain C in tobacco: Protection against infection by CMV strains transmitted mechanically or by aphids, *Phytopathology* **81**, 794–802.

RAINERI, D., BOTTINO, P., GORDON, M., NESTER, E. (1990), *Agrobacterium* mediated transformation of rice, *Biotechnology* **8**, 33–40.

REZIAN, M., SKENE, K., ELLIS, J. (1988), Antisense RNA of cucumber mosaic virus in transgenic plants asessed for control of the virus, *Plant Mol. Biol.* **11**, 463–471.

RHODES, C. A., PIERCE, D. A., METTLER, I., MASCARENHAS, D., DETMER, J. (1989), Genetically transformed maize plants from protoplasts, *Science* **240**, 204–207.

RICHARDSON, M. (1977), The proteinase inhibitors of plants and microorganisms, *Phytochemistry* **16**, 159–169.

ROBERTS, W., SELITRENNIKOFF, C. (1990), Zeamatin, an antifungal protein from maize with membrane-permeabilizing activity, *J. Gen. Microbiol.* **136**, 1771–1778.

ROTINO, G., GLEDDIE, S. (1990), Transformation of eggplant (*Solanum melongena* L.) using a binary *Agrobacterium tumefaciens* vector, *Plant Cell Rep.* **9**, 26–29.

SATO, T., THEOLOGIS, A. (1989), Cloning the mRNA encoding ACC synthase, the key enzyme for ethylene synthesis in plants, *Proc. Natl. Acad. Sci. USA* **86**, 6621-6625.

SCHLUMBAUM, A., MAUCH, F., VÖGELI, U., BOLLER, T. (1986), Plant chitinases are potent inhibitors of fungal growth, *Nature* **327**, 365-367.

SEBASTIANI, F., FARRELL, L., VASQUEZ, M., BEACHY, R. (1991), Conserved amino acid sequences among plant proteins sorted to protein bodies and plant vacuoles, *Eur. J. Biochem.* **199**, 441–450.

SEQUEIRA, L. (1984), Cross protection and induced resistance: their potential in plant disease control, *Trends Biotechnol.* **2**, 25–29.

SHAH, D., HORSCH, R., KLEE, H., KISHORE, G., WINTER, J., TURNER, N., HIRONAKA, C., SANDERS, P., GASSER, C., AYKENT, S., SIEGEL, N., ROGERS, S., FRALEY, R. (1986), Engineering herbicide tolerance in transgenic plants, *Science* **233**, 478–481.

SHAH, D., GASSER, C., DELLA-CIOPPA,G., KISHORE, G. (1988), Genetic engineering of herbicide resistance genes, in: *Plant Gene Research, Temporal and Spatial Regulation of Plant Genes* (VERMA, P. S., Ed.), Vol. 5, pp. 297–309, Berlin-New York: Springer-Verlag.

SHEEHY, C., KRAMER, M., HIATT, W. (1988), Reduction of polygalacturonase activity in tomato fruit by antisense RNA, *Proc. Natl. Acad. Sci. USA* **85**, 8805–8809.

SHERWOOD, J., FULTON, R. (1982), The specific involvement of coat protein in tobacco mosaic virus protection, *Virology* **119**, 150–158.

SHILLITO, R., CARSWELL, G., JOHNSON, C., DIMAIO, J., HARMS, C. T. (1989), Regeneration of fertile plants from protoplasts of elite inbred maize, *Biotechnology* **7**, 581–589.

SHIMAMOTO, K., TERADA, R., IZAWA, T., FUJIMOTO, H. (1989), Fertile rice plants regenerated from transformed protoplasts, *Nature* **338**, 274–276.

SIJMONS, P., DEKKER, B., SCHRAMMEIJER, B., VERWOERD, T., VAN DEN ELZEN, P., HOEKEMA, A. (1990), Production of correctly processed human serum albumin in transgenic plants, *Biotechnology* **8**, 217–221.

SMITH, C., WATSON, R., RAY, J., BIRD, C., MORRIS, P., SCHUCH, W., GRIERSON, D. (1988), Antisense RNA inhibition of polygalacturonase gene expression in transgenic tomatoes, *Nature* **334**, 724–726.

SMITH, C., WATSON, C., MORRIS, P., BIRD, C., SEYMOUR, G., GRAY, J., ARNOLD, C., TUCKER,

G., Schuch, W., Harding, S., Grierson, D. (1990), Inheritance and effect on ripening of antisense polygalacturonase genes in transgenic tomatoes, *Plant Mol. Biol.* **14**, 369–379.

Somerville, C., Browse, J. (1991), Plant lipids: Metabolism, mutants and membranes, *Science* **252**, 80–87.

Sonnewald, U., Von Schaewen, A., Willmitzer, L. (1990), Expression of mutant patatin protein in transgenic tobacco plants: Role of glycans and intracellular location, *Plant Cell* **2**, 345–355.

Sonnewald, U., Brauer, M., Von Schaewen, A., Stitt, M., Willmitzer, L. (1991), Transgenic tobacco plants expressing yeast-derived invertase in either the cytosol, vacuole or apoplast: a powerful tool for studying sucrose metabolism and sink/source interactions, *Plant J.* **1**, 95–106.

Stalker, D., McBurke, K., Malyji, L. (1988), Herbicide resistance in transgenic plants expressing a bacterial detoxification gene, *Science* **242**, 419–423.

Stark, D., Beachy, R. (1989), Protection against potyvirus infection in transgenic plants: evidence for broad spectrum resistance, *Biotechnology* **7**, 1257–1262.

Stiekema, W., Heidekamp, F., Louwers, J., Verhoeven, H., Dijkhuis, P. (1988), Introduction of foreign genes into potato cultivars Bintje and Desiree using an *Agrobacterium tumefaciens* binary vector, *Plant Cell Rep.* **7**, 47–50.

Streber, W., Willmitzer, L. (1989), Transgenic tobacco plants expressing a bacterial detoxifying gene are resistant to 2,4 D, *Biotechnology* **7**, 811–816.

Svab, Z., Hajdukiewicz, P., Maliga, P. (1990), Stable transformation of plastids in higher plants, *Proc. Natl. Acad. Sci. USA* **87**, 8526–8530.

Swain, W. (1991), Antibodies in plants, *Trends Biotechnol.* **9**, 107–109.

Tepfer, D. (1984), Transformation of several species of higher plants by *Agrobacterium rhizogenes:* Sexual transmission of the transformed genotype and phenotype, *Cell* **37**, 959–967.

Toriyama, K., Arimoto, Y., Uchiyima, H., Hinata, K. (1988), Transgenic rice plants after direct gene transfer into protoplasts, *Biotechnology* **6**, 1072–1074.

Trulson, A., Simpson, R., Shahin, E. (1986), Transformation of cucumber (*Cucumis sativas* L.) plants with *Agrobacterium rhizogenes, Theor. Appl. Genet.* **73**, 11–15.

Tumer, N., O'Connell, K., Nelson, R., Sanders, R., Beachy, R. (1987), Expression of alfalfa mosaic virus coat protein confers cross-protection in transgenic tobacco and tomato plants, *EMBO J.* **6**, 1181–1188.

Umbeck, P., Johnson, G., Barton, K., Swain, W. (1987), Genetically transformed cotton (*Gossypium hirsatum* L.), *Biotechnology* **5**, 263–266.

Vaeck, M., Reynaerts, A., Höfte, H., Janssens, S., De Beuckeleer, M., Dean, C. Zabeau, M., Van Montagu, M., Leemans, J. (1987), Transgenic plants protected from insect attack, *Nature* **328**, 33–37.

Van Canneyt, G., Rosahl, S., Willmitzer, L. (1990), Translatability of a plant-mRNA strongly influences its accumulation in transgenic plants, *Nucleic Acids Res.* **18**, 2917–2921.

Vandekerckhoeve, J., Van Damme, J., Van Lijsbetens, M., Botterman, J., De Block, M., Vandewiele, M., De Clerq, A., Leemans, J., Van Montagu, M., Krebbers, E. (1989), Enkephalins produced in transgenic plants using modified 2s seed storage proteins, *Biotechnology* **7**, 929–933.

van der Krol, A., Mol, J., Stuitje, A. (1988), Modulation of eucaryotic gene expression by complementary RNA or DNA sequences, *BioTechniques* **6**, 958–976.

Van der Straeten, D., Van Wiemeersch, L., Goodman, H., Van Montagu, M. (1990), Cloning and sequence of two different cDNAs encoding 1-aminocyclopropane1-carboxylate synthase in tomato, *Proc. Natl. Acad. Sci. USA* **87**, 4859–4863.

Van Dun, C., Bol, J., Van Vloten-Doting, L. (1987), Expression of alfalfa mosaic virus and tobacco rattle virus protein genes in transgenic plants, *Virology* **159**, 299–305.

Van Rie, J. (1991), Insect control with transgenic plants: resistance proof? *Trends Biotechnol.* **9**, 177–179.

Van Rie, J., Jansens, S., Höfte, H., Degheele, D., Van Mellaert, H. (1989), Specificity of *Bacillus thuringiensis* delta-endotoxins, *Eur. J. Biochem.* **186**, 239–247.

Van Rie, J., Jansens, S., Höfte, H., Degheele, D., Van Mellaert, H. (1990), Receptors on the brush border membrane of the insect midgut as determinants of the specificity of *Bacillus thuringiensis* delta-endotoxins, *Appl. Environ. Microbiol.* **56**, 1378–1385.

Vasil, I. (1988), Progress in the regeneration and genetic manipulation of cereal crops, *Biotechnology* **6**, 397–402.

Vasil, I. (1990), Transgenic cereals become a reality, Biotechnology **8**, 797–794.

Vasil, V., Redway, F., Vasil, I. (1990), Regeneration of plants from embryogenic suspension culture protoplasts of wheat, *Biotechnology* **8**, 429–434.

Vigers, A., Roberts, W., Selitrennikoff, C.

(1991), A new family of antifungal proteins, *Mol. Plant Microbe Int.* **4,** 315–323.

VISSER, R., SOMHORST, I., KUIPERS, G., RUYS, N., FEENSTRA, W., JACOBSEN, E. (1991) Inhibition of the expression of the gene for granule-bound starch synthase in potato by antisense constructs, *Mol. Gen. Genet.* **225,** 289–296.

VOELKER, T., HERMAN, E., CHRISPEELS, M. (1989), *In vitro* mutated phytohemagglutinin genes expressed in tobacco seeds: Role of glycans in protein targeting and stability, *Plant Cell* **1,** 95–104.

VON SCHAEWEN, A., STITT, M., SCHMIDT, R., SONNEWALD, U., WILLMITZER, L. (1990), Expression of a yeast-derived invertase in the cell wall of tobacco and *Arabidopsis* plants leads to accumulation of carbohydrate and inhibition of photosynthesis and strongly influences growth and phenotype of transgenic tobacco plants, *EMBO J.* **9,** 3033–3044.

WALDRON, C., MURPHY, E., ROBERTS, J., GUSTAFSON, G., ARMOUR, S., MALCOLM, S. (1985), Resistance to hygromycin B: A new marker for plant transformation studies, *Plant Mol. Biol.* **5,** 103–108.

WILLMITZER, L. (1988), The use of transgenic plants to study plant gene expression, *Trends Genet.* **4,** 13–18.

WOLOSHUK, C. P., MEULENHOFF, J., SELA-BUURLAGE, M., VAN DEN ELZEN, P., CORNELISSEN, B. (1991), Pathogen-induced proteins with inhibitory activity toward *Phytophtora infestans, Plant Cell* **3,** 619–628.

ZAMBRYSKI, P., TEMPE, J., SCHELL, J. (1989), Transfer and function of T-DNA genes from *Agrobacterium* Ti- and Ri-plasmids in plants, *Cell* **56,** 193–201.

ZHANG, W., WU, R. (1988), Efficient regeneration of transgenic rice protoplasts and correctly regulated expression of foreign genes in the plants, *Theor. Appl. Genet.* **76,** 835–840.

V. Genetic Engineering of Animals

17 Genetic Engineering of Animal Cells

MANFRED WIRTH
HANSJÖRG HAUSER

Braunschweig, Federal Republic of Germany

1 Introduction

The development of gene cloning and recombinant DNA methods provided the feasibility to convert bacteria or eukaryotic cells for the production of valuable proteins. While it has been clear for some time that *Escherichia coli* could be genetically programmed to synthesize virtually any linear peptide, it was uncertain whether all polypeptides produced in these microorganisms would contain the unique structure essential for authentic biological activity. Experience with the bacterial systems has revealed that intracellular expression of cloned eukaryotic proteins is often accompanied by denaturation and/or degradation of the engineered protein. Although there are many examples in which denatured polypeptides are refolded into the correct conformation by sophisticated methods, there are several classes of protein which cannot be produced in bacteria to meet the above mentioned authenticity. Post-translational modifications such as glycosylation, processing and assembly are unlikely to be properly performed by prokaryotic cells and are difficult or even impossible to reconstruct *in vitro*. Some of these modifications are also performed in lower eukaryotes like yeasts, but often do not result in products identical to those of the human gene.

The production of recombinant proteins in animal cells obviously overcomes most of these problems and therefore becomes an increasingly important source for products of commercial interest.

In addition, there are many other uses for animal cells which have been manipulated to express foreign genes (see Tab. 1).

One application is based on the principles of reverse genetics, which involves the introduction of wild-type or mutated gene(s) into the host cells and monitoring expression. Another aim is to construct cells which can be used as test tubes for evaluating external influences on animal cell-specific functions. An example is the test of inhibitors for specific viral functions which are simulated in the recombinant cells.

It is important to note that the use of animal cells as an analytical tool for the investigation of gene function and expression, is not limited to the introduction of additional DNA into the host. The elimination or distortion of host chromosomal gene function can also be studied.

The strength of gene expression, whether regulated or constitutive, is important for most of the purposes summarized in Tab. 1. In many cases, expression should be as high as possible. For other purposes, expression regulated by external stimuli or intracellular signaling of the recombinant gene is important. Eukaryotic expression is regulated at different levels. The basic parameters for expression efficiency of individual genes are summarized in Tab. 2.

In Sect. 2 the essentials of gene expression in higher eukaryotes are reviewed in relation to genetic engineering of cells. Section 3 describes approaches and experiences in the genetic manipulation of animal cells.

Particular emphasis is placed on mammalian cells. This reflects the fact that gene expression and the technology of gene manipulation are best understood for these animal cells, since they are most frequently used for applied biotechnological research related to human studies.

2 Expression of Endogenous and Heterologous Genes in Animal Cells

2.1 Transcriptional Control

In general, the rate of transcription initiation is the rate-limiting step in the production of transcripts. Despite pre-mature termination and pausing in a few genes, the rate of transcriptional elongation is constant until the movement of RNA polymerase is stopped by termination signals. As a consequence, the density of polymerase molecules on a given gene reflects the frequency of initiation events. The conditions which govern transcription ini-

Tab. 1. Purposes for the Use of Animal Cells

1. Confirmation that isolated genes direct the synthesis of a desired protein or mediate defined physiological effects to the cells.
2. Evaluation of the effect of mutations introduced into a gene.
3. Isolation of genes directly based on screening or selection of recipient cells for the production of a particular protein.
4. Analysis of physiological consequences or expression of specific proteins in mammalian cells in order to study biological regulatory control.
5. Test tubes for assaying effectors (drugs) of regulatory processes.
6. Production of large amounts of proteins that are normally available in only limited quantity.

Tab. 2. Expression Parameters for Protein Coding Genes

Transcriptional efficiency	Structure of the chromosomal neighborhood of the gene of interest; presence of transcription modulating chromosomal DNA elements (CpG methylation, positive and negative promoter proximal elements, enhancers, silencers)
RNA processing and transport	Capping, splicing, polyadenylation, transfer to the cytoplasm
mRNA turnover	RNA and derived peptide sequences influencing mRNA stability; shortening of the poly(A) tail
Translation efficiency	Recognition sequences of the mRNA, secondary structures, cap, modification of translation factors
Protein stability	Peptide signals binding specific structures or proteins for transport

tiation are therefore important for the strength of expression.

2.1.1 The Transcription Complex

Transcription is achieved by an interplay of DNA elements and proteins that function through complex interactions to affect and regulate gene expression. There are elements which are able to form higher-order chromatin structures, and by this regulate transcription of genes. While this is a prerequisite for transcription, another set of DNA elements is responsible for the genes transcribed by RNA polymerase II, the polymerase which is responsible for expression of protein coding genes. The sequence elements controlling RNA polymerase II dependent gene expression can be divided into several categories (WASYLYK, 1986). A short characterization of these elements is given in Tab. 3.

2.1.1.1 Signals for Transcription Initiation and Termination

By molecular definition, a gene in animal cells comprises the transcribed portion (transcription unit) as well as 5′ and 3′ flanking control DNA sequences (Fig. 1). The sites which determine the 5′ and 3′ ends of mRNAs are defined as short sequence motifs (Fig. 2; Tab. 3). The initiator motif found in most genes is the TATA- or Goldberg-Hogness box (BREATHNACH and CHAMBON, 1981) which is located 25–35 base pairs upstream of the start of RNA polymerase II transcription. The function of the TATA box is to direct the start of RNA transcription at the so-called cap site. The start is usually an adenosine nucleotide which is flanked by pyrimidines. In addition to the TATA box, another type of initiator motif (INR) which surrounds the transcription start site was recently discovered. This motif is suf-

Tab. 3. DNA Control Elements for RNA Polymerase II Transcribed Genes

Type	Position with Respect to the Cap Site	Properties	Examples
TATA box	−25 to −35	Responsible for positioning of RNA polymerase II and the transcriptional start site. Binding site for the RNA polymerase II cofactor TFII D	TATAAAA
Initiators (INR)	Overlap the cap site	Directs the transcription initiation to the cap site. Several sequence elements with this function are identified.	+1 YAYTCYYY
AATAAA plus a downstream G/T box	Downstream; the distance between AATAAA and the G/T box is usually less than 40 bases	Transcriptional termination signals. AAUAAA serves as a recognition sequence for cleavage and constitutes the polyadenylation signal.	AATAAA ... TGTGTGTTGGTTTTTGTGTGT
Positive promoter proximal elements	Proximal; −40 to −200	Different sequence elements of 10–20 bp length are present in typical promoters. The elements form binding sites for transcription factors.	GGCCAATCT (CAAT box) GGGCGG (GC box) ATTTGCAT (octamer box)
Negative promoter proximal elements	Proximal	Similar to the positive proximal promoter elements but with a negative effect. Binding sites for proteins (e.g. repressors) which upon binding reduce the positive effect of *cis*-acting elements; can also be located within enhancers.	Negative element with tissue-restricted function in the IgH chain enhancer (IMLER et al., 1987)
Enhancer	Proximal or distal; up to 10 kbp, upstream or downstream, both orientations possible	Large sequences containing several often repeated elements that can function independently. The closest promoter is the preferential target.	72 bp-repeat of SV40; IgH chain enhancer in the intron of the structural gene
Silencer	Proximal or distal	Short sequence elements, sometimes repeated. Binding sites for proteins which either block the signaling from positive *cis*-acting sequence elements or directly block the action of RNA polymerase II.	Distal element which acts on the rat insulin 1 gene (LAIMINS et al., 1986)

ficient to direct transcription initiation in the absence of the TATA box (ROEDER, 1991; SETO et al., 1991).

The mRNA start site of a particular gene is usually conserved. However, for a few genes it has been found that alternative mRNA start sites exist. These may be tissue-specific in character.

The corresponding sequence to the initiator motifs, usually preceding the end of the

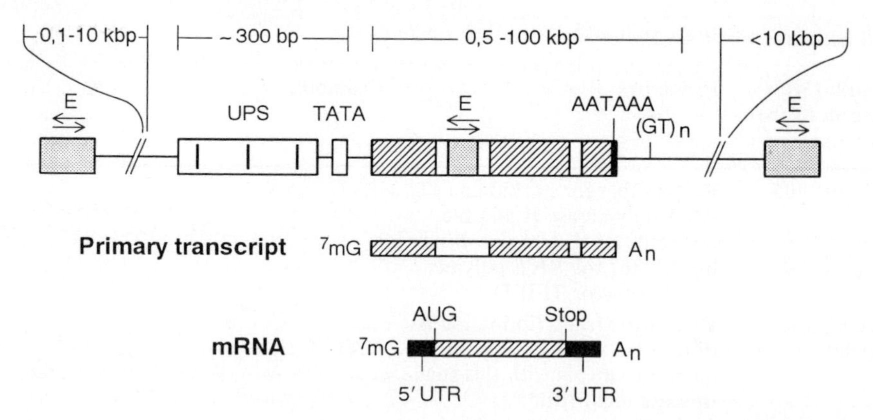

Fig. 1. Structure of a eukaryotic protein coding gene. A complete gene consists of the transcribed region (structural gene) and the flanking DNA plus transcription regulating DNA elements. These transcriptional activating elements are the TATA box, the promoter proximal elements (upstream promoter elements, UPS) and enhancers (E) (dotted). The enhancers can exist in either orientation in the DNA at different positions with respect to the structural gene. The primary transcript is a result of transcription and subsequent cleavage downstream of the polyadenylation site (AATAAA). RNA splicing and modification at the 5′ (capping, ^{7}mG) and 3′ (polyadenylation, A_n) leads to mature mRNA. The protein coding region (hatched) is flanked by the UTRs (solid boxes).

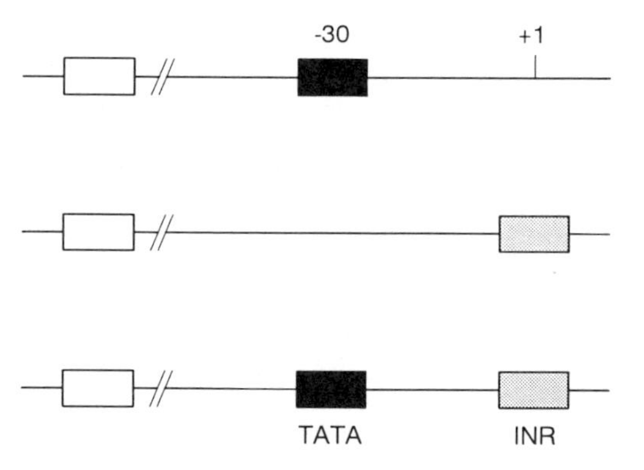

Fig. 2. Transcriptional DNA elements in animal cell promoters. Three types of elements are shown: The TATA box (solid boxes) found about 30 base pairs upstream of the transcriptional start site, the initiator (INR) (dotted boxes) at the start site, and activation sequences (open boxes) which are usually located further upstream. Promoters may contain either a TATA box or the INR or both of these elements.

mRNA, is the poly(A) addition site. This sequence, AATAA, is the most conserved DNA recognition element in eukaryotes. This sequence motif directs the cleavage of primary transcripts approximately 20 base pairs downstream. The cleavage process, also requires another, less conserved, normally G/U-rich, recognition sequence which is located downstream of the poly(A) site. This sequence, in collaboration with the poly(A) site, leads to the endonucleolytic attack (PROUDFOOT, 1991). Most messenger RNAs have a post-transcriptionally added poly(A) tract. Termination of the transcription reaction takes place further downstream. The termination site is not known in most cases. After AATAAA-mediated cleavage, the 3′ part of the transcript is degraded in the nucleus.

Some eukaryotic transcription units contain more than one poly(A) addition site. The differential choice for the site is influenced by adjacent sequences and secondary structures. Several examples of tissue-specific usage of multiple poly(A) sites have been reported. In addition, some poly(A) sites are promiscuous, which leads to a partial read-through and the production of RNAs with different 3′ terminal ends. In specific cases this can even influence transcription of neighboring genes, since transcription running through a 3′ located promoter can lead to occlusion of its activity (EMERMAN and TEMIN, 1986; CULLEN et al., 1984).

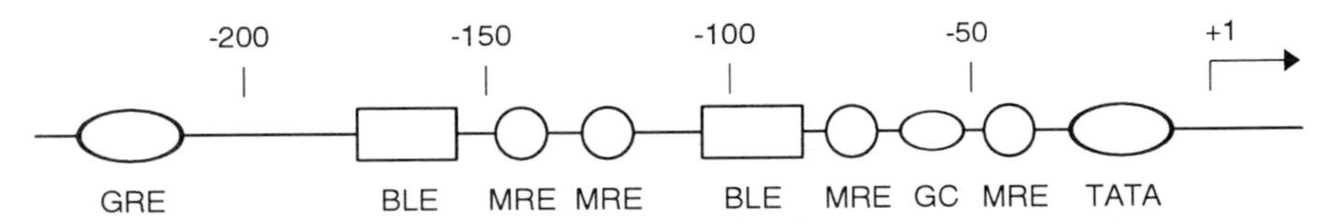

Fig. 3. Transcriptional control elements upstream of the transcriptional start site in the gene encoding a metallothionein. The TATA box, the GC- and BLE-boxes bind factors which are involved in constitutive transcription. Glucocorticoid response elements (GRE) and metal response elements (MRE) bind factors involved in the induction of gene expression in response to specific stimuli.

2.1.1.2 Proximal Promoter Elements

In contrast to the prokaryotic promotors the initiator sequences by themselves are not active in initiating RNA polymerase II transcription *in vivo*. In order to obtain a functional RNA polymerase II promoter, further sequence elements, positively *cis*-acting elements, must be present in the proximity of the TATA box (upstream promoter sequences, UPS) (see Tab. 3). They can occur in different positions around 70 to 80 nucleotides upstream of the mRNA cap site, as in the case of the CCAAT-motif homology in the herpes simplex virus thymidine kinase and human β-globin promoters. Many housekeeping genes contain several structural components in the upstream region of the mRNA cap site, namely G/C-rich regions. In addition to the afore mentioned sequence elements, a series of other consensus sequences has been recognized in the 5′ promoter proximal region (Fig. 3) (LATCHMAN, 1991).

2.1.1.3 Enhancers

A third group of elements is comprised of the enhancer sequences, first identified in the genomes of SV40 and murine retroviruses. Since that time, many cellular enhancers have been identified. Although little is known about their mechanism of action, one of the key properties that makes an expression element an enhancer is the fact that it can act over large distances, up to several thousand base pairs in either orientation (Tab. 3, Fig. 1) (SERFLING et al., 1985). Enhancers are relatively large sequence elements which often contain repeated sequences. They are found proximal to promoter regions of genes, within introns, and far upstream as well as downstream of the transcribed region. Normally, the enhancer stimulates the closest promoter.

Enhancers are composed of multiple functional units, each of which cooperates with the others or with duplicates of itself to enhance transcription. In many cases individual elements can act autonomously when present in multiple tandem copies, exhibiting distinct cell-specific activities.

The elevation of expression by enhancers can be between 2- and 1000-fold, and although often active in a wide variety of cell types, some enhancers show cell- or tissue-specificity. The immunoglobulin genes, for instance, have enhancers which are normally active only in B-lymphocytes. Their enhancer is located in the first intron of the constant part of the gene and can exert its effect only after recombinational rearrangements. B-cell specific recombination brings the promoter from the variable part to the enhancer in the constant part. Other enhancers, e.g., the polyomavirus enhancer, although showing a host-cell preference, still function in cells of different tissues.

In contrast to the chromosomal *cis*-acting sequences as introduced in Sect. 2.1.2, typical enhancers function as transcriptional activators in episomal as well as in chromosomal surroundings.

2.1.1.4 Inducible Transcription Elements

Cellular and viral promoters/enhancers are known to respond to external stimuli by hormones, temperature, growth factors, viruses,

Tab. 4. Inducible Promoters/Enhancers

Gene from which the Promoter/Enhancer is Derived	Inducing Agent	Base Level Transcription	References
Mammalian type I interferon (IFN)	Virus, dsRNA	very low/none	LENGYEL (1986)
Mammalian Mx	Type I interferons (α, β)	very low/none	HUG et al. (1988)
Mammalian or *Drosophila* heat shock proteins	Heat shock	low	PELHAM and BIENZ (1982)
Mammalian metallothionein I or II	Cadmium-, zinc-ions, glucocorticoids[a], IFN	high	PAVLAKIS and HAMER (1983)
Mouse mammary tumor virus (MMTV)	Glucocorticoids[a]	medium	LEE et al. (1981); BEATO (1989)

[a] The expression of the respective hormone receptor in the recipient cells is required.

etc. Tab. 4 gives an overview of some well-known inducible promoters/enhancers. They contain short DNA sequences which are required for response to external stimuli (response elements). Response elements have the same general characteristics as other promoter or enhancer elements. They are located in promoters, as well as enhancers, as single or multiple copies (see Fig. 3).

2.1.1.5 Negatively *cis*-Acting Elements

Apart from positively acting elements the existence of negatively *cis*-acting DNA sequences have been demonstrated (RENKAWITZ, 1990) (Tab. 3). Repressor elements are comparable to the positively *cis*-acting sequences, since they require proximity to the TATA box. Silencers are defined as negatively acting correspondents to enhancers. They act independently of orientation and distance to the TATA box (ALBERTS and STERNGLANZ, 1990). The action of enhancers and silencers is orientated to the TATA box activity either directly, or indirectly by an interaction with a proximal promoter element.

In specific cases, the same DNA sequence can function as a positive as well as a negative promoter element. For example, a virus-responsive element of interferon type I genes when inserted between the TATA box and the SV40 enhancer, serves to silence the promoter. This silencing is fully reversed after virus induction (KUHL et al., 1987). Repeats of the same sequence element make heterologous promoters virus-inducible (FUJITA et al., 1987).

2.1.1.6 Transcription Factors

Two different types of components, usually proteins, are involved in initiation of transcription: the RNA polymerase II with its cofactors and regulatory transacting factors. RNA polymerase II by itself is not able to initiate transcription from a functional promoter. It requires cofactors for transcriptional initiation and factors which participate in the elongation of the transcripts (Fig. 4). These general factors are required for transcription of all protein coding genes. In addition, promoter-specific factors help to initiate transcription of particular genes. A limited number of DNA-binding proteins (a few hundred) have been suggested to interact directly or indirectly with the RNA polymerase II-complex in order to stimulate its binding and initiation (WINGENDER, 1988, 1990). Indirect stimulation most probably requires factors which are capable of bridging between the DNA binding factors and the polymerase II and its accessory proteins. The positively transacting DNA-binding fac-

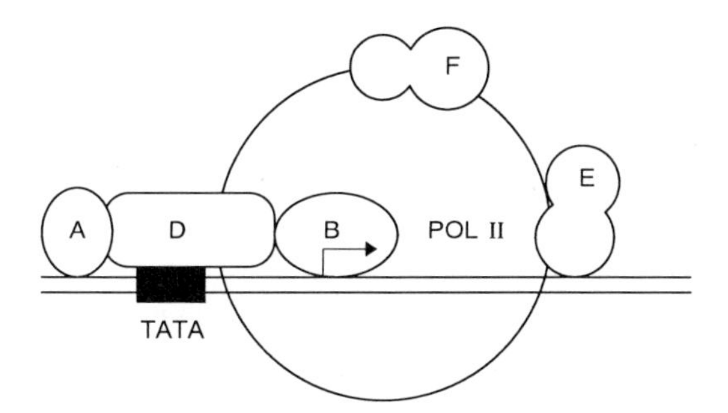

Fig. 4. RNA polymerase II in the preinitiation complex. In TATA-containing promoters general transcription initiation factors, beginning with the binding of TFIID to the TATA box, assemble successively. The final preinitiation complex which contains all effectors (TFIIA, B, D, E, F and Pol II) is able to initiate transcription. Adapted from ROEDER (1991).

tors specifically recognize single *cis*-acting sequence elements in the proximity of the TATA box or in enhancers (Fig. 5a). The concentrations and availabilities of these factors in a certain cell type are crucial for the transcriptional strength of individual genes.

Cell-specific gene control requires both, widely distributed transcription factors and other factors that are limited or totally cell-specific. Thus, the availability of the correct set of transcription factors is an important determinant of which genes are active in a given cell.

Factors which do not directly bind to DNA may be active in the modulation of transcription. Their mechanism of action with the components of the transcriptional complex is not fully understood. Examples are given in Fig. 5b.

Many of the DNA-binding transcription factors belong to one of the three groups of DNA-binding proteins: The helix-turn-helix family, the zinc-finger family, or the amphipathic helix family, including the leucine zipper and the helix-loop-helix motifs. The structural design of members of these families is very similar (Tab. 5). In addition to these three families a number of unclassified binding factors which are active in transcriptional initiation have been described.

Since the classification of the transcription factors not only refers to the structural motif

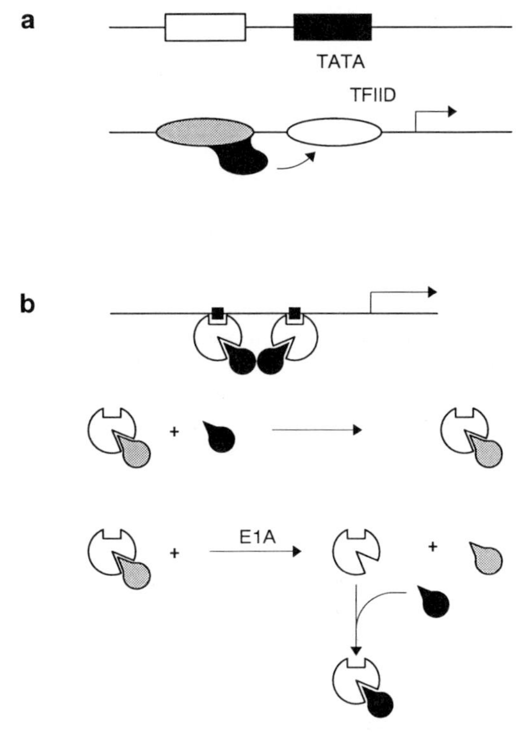

Fig. 5. Action of transcription-modulating factors. (a) The activation of a promoter which contains an activating sequence (open box) and a TATA box may be activated by binding of a sequence-specific binding protein which specifically recognizes the activator sequence and thereby facilitates the interaction of TFIID (open oval) with the TATA box. The transcriptional activating protein contains a DNA-binding domain (dotted) and an activating domain (black). (b) Another promoter which contains two identical activating elements (black squares) is active after association with two DNA-binding factors which form a stable DNA–protein complex after interaction with linking factors (black droplets). The DNA-binding factors and the DNA alone would form an unstable complex. The linking proteins are responsible for the cooperative binding of the DNA-binding factors to both adjacent sites in the promoter resulting in a stable DNA–protein complex. This stable complex cannot be assembled when another protein which is depicted as a dotted droplet competes for the binding of the linking factors. This would result in the inactivity of this promoter (middle). The dissociation of the DNA-binding protein and the competing factors with other components (in this case the adenovirus E1A protein) allows the interaction of the linker factor with the DNA-binding protein and thereby the activation of transcription (lower). Adapted from NEVINS (1991).

Tab. 5. DNA-Binding Proteins

Class	Common Structural Motif and Properties	Examples
Helix-turn-helix according to DNA-binding domains	3 α-Helical regions separated by short turns. The carboxy-terminal helices make contact with DNA. DNA binding is enhanced by dimerization.	Oct 1, Oct 2, Pit; homeotic gene products, Mat locus proteins
Zinc finger according to DNA-binding domains	Array of more than one short protein loop structure. Cysteines and histidines spaced at regular intervals from the base of the loops by complexing a zinc ion each.	Steroid receptors, Sp 1
Amphipathic helix according to dimerization domains	Amphipathic helical domains are responsible for dimerization by forming a coiled-coil structure. A nearby region rich in basic amino acids constitutes the DNA-binding domain.	
a) Leucine zipper	4 Leucines, separated by 6 amino acids each, are on the hydrophobic surfaces of two protein coils interdigitated by zipping up. Some transcription factors have a leucine zipper and a zinc finger or a homeo-domain (helix-turn-helix motif).	Fos, Jun, C/EBP
b) Helix-loop-helix	2 α-Helical regions separated by a loop on each molecule, form the basis for dimerization.	MyoD, E 12

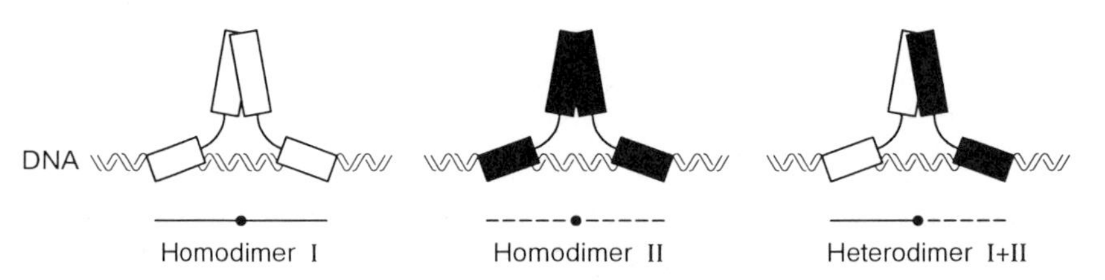

Fig. 6. DNA-binding specificity by heterodimerization. The binding of protein-dimers which are connected by leucine zippers to different DNA sequences is shown. The DNA-binding domains of the zipper proteins (white or black rectangles) recognize specific half sites of regulatory DNA elements. Homodimeric zipper proteins (left, middle) recognize palindromic DNA sequences. Heterodimers of these zipper proteins bind to a hybrid DNA recognition site (right). Adapted from LAMB and MCKNIGHT (1991).

by which they bind to DNA, a certain factor might have both, a structural motif for DNA-binding (e.g., a zinc finger) and another motif for dimerization (e.g., a leucine zipper).

Many factors have a domain which enables interaction with other proteins (e.g., the leucine zipper). This allows them to form homo- or heterodimers. In general, dimerization seems to increase the affinity to DNA. Some DNA-binding factors are related to members of a family (e.g., C/EBP) which are predestinated to form heterodimers with each other. Heterodimer formation by combination of a limited set of proteins can produce a large array of transcription factors with differential activity (Fig. 6) (LAMB and MCKNIGHT, 1991).

All DNA-binding factors, which have been described to stimulate transcription, exhibit at least two features: one is a DNA-binding domain and the other is an activation domain. Other domains include sites for regulatory post-translational modifications and low- or high-molecular weight effector binding sites.

Many viruses encode trans-activating factors that are active in the viral replication cycle as well as in cellular activities (NEVINS, 1991).

A number of DNA-binding factors have been recognized as proto-oncogenes. Mutants

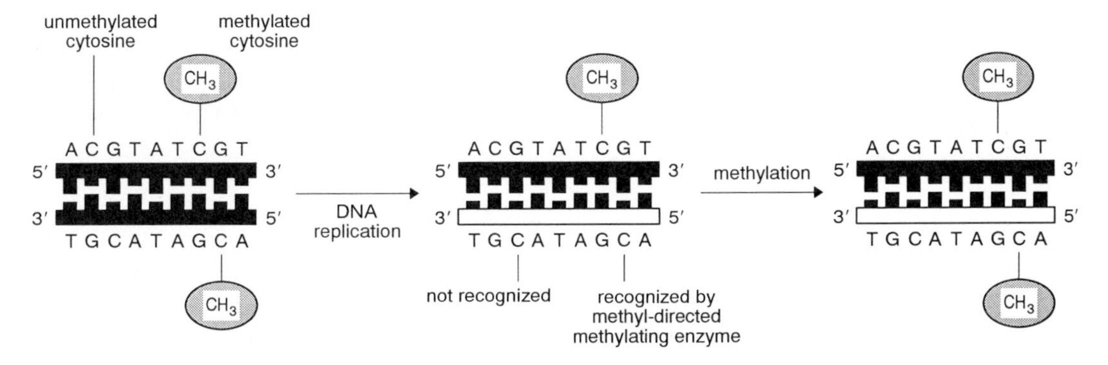

Fig. 7. Maintenance methylation of newly incorporated cytosine residues after DNA replication. DNA replication leads to hybrid DNA double-strands with one parental strand (black) and one daughter strand (white). Only those cytosines are recognized by the methylating enzyme which carry a methylated cytosine on the parental strand.

of these proteins are found in retroviruses as oncogenic agents (LATCHMAN, 1991).

The TAT protein, encoded by the genome of human immunodeficiency virus I, is an unusual transcription factor. This gene product binds to the extreme 5′ terminus of transcripts from the HIV-1 promoter, thereby enhancing transcription. Although it is still disputed whether this transactivation is due to enhanced transcriptional initiation or due to the relief of premature termination, TAT is the first promoter-specific transcription initiation factor recognized to bind a RNA sequence instead of a DNA sequence (ROSEN, 1991).

2.1.1.7 DNA Methylation

In the DNA of higher animals, about 70% of the cytosine residues in the 5′-CpG-3′ sequence are methylated whereas most other cytosine residues are not. During cell culture methylated DNA sequences are inherited conservatively (Fig. 7). Since m^7C is easily mutated to T in vertebrate genomes, the CpG sequence occurs less frequently than statistically expected. Exceptions are islands which are mostly found around the promoter areas of genes transcribed by RNA polymerase II. DNA regions which are active in transcription, in particular housekeeping genes, preferentially lack 5-methylcytidine. A close correlation exists between transcriptional inactivity and methylation (RAZIN and CEDAR, 1991). After *in vitro* CpG methylation of cloned DNA and its introduction into cells, it cannot be expressed to the same extent. Little is known about the mechanism of change in methylation patterns of genes.

2.1.2 Chromosomal Elements

In the previous sections simple eukaryotic genes have been described (Fig. 1). When viewing these genes in their chromosomal environment, they may be described as an assembly of genes, complexed with nuclear compounds and arranged in a highly organized spatial structure. Although the number of chromosomes of a given animal species is fairly constant, the detailed chromosomal structure changes during the course of the cell cycle and as a function of development and differentiation. The fine structure of the chromosomal DNA-protein interaction which differs from one cell type to the other, influences expression of the transcription units therein.

The DNA of animal cells is associated with a number of nuclear proteins. It is packaged on cores of octamers of histone proteins to form the so-called nucleosomes. B-type helical DNA (140 base pairs) is wound in one and three quarter turns of a left-handed superhelix around the outside of a nucleosome (Fig. 8a). DNA between the core particles is a flexible linker of variable length (0–70 base pairs). The association, via the linker, with another his-

a

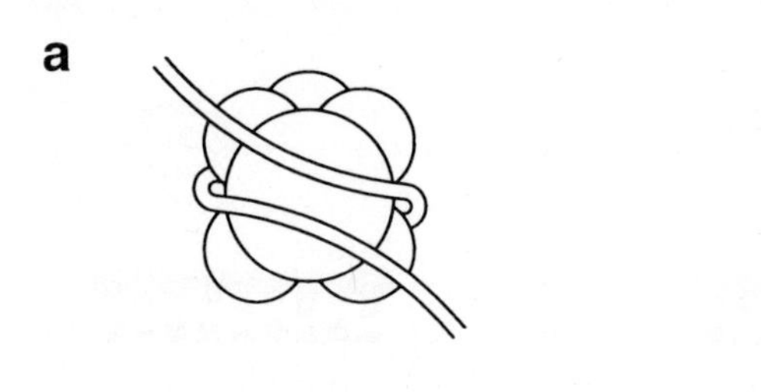

b

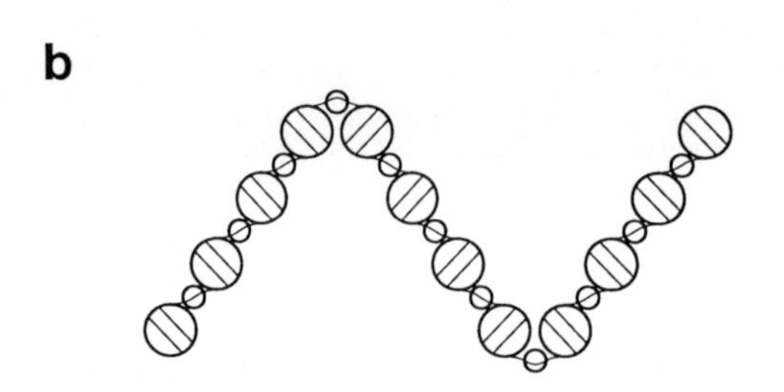

Fig. 8. Packaging of DNA on nucleosomes. (a) shows the structure of a nucleosome. The circles represent monomers of H2A, H2B, H3, and H4, while the tape winding round them represents the DNA double helix.
(b) shows the packaging of nucleosomes. The larger circles represent the nucleosomes with the DNA double helix coiling round them. The smaller circles represent the H1 molecules bound to the flexible linkers of DNA.

tone molecule, H1, leads to a denser packaging of DNA into solenoids. Solenoids are regular structures (30 nm fibers) with a higher degree of order (Fig. 8), which, in turn, are formed into loops or domains of 5 to 200 kbp of DNA.

The histone proteins are known to be posttranslationally modified by phosphorylation and acetylation. These modifications, which decrease the positive charge on the molecules result in a weaker binding to DNA. This results in conformational changes of the nucleosomes, which is believed to alter the transcriptional potential of the associated DNA by changing the helicity of the double strand within chromosomal loops.

Interactions occur between the basic chromosomal elements, the nucleosomes and DNA-binding factors. Usually the promoter regions of actively transcribed genes are free of nucleosomes. For several promoters there is evidence that binding of *trans*-acting transcription factors leads to the removal of nucleosomes rendering the adjacent gene active and the site hypersensitive to DNAse I (CORDINGLEY et al., 1987).

2.1.2.1 Chromatin Domains

Transcriptional activation by regulatory molecules (transcription factors) is dependent on the excess of these factors relative to the target DNA sequences. When DNA is compacted in the form of tightly coiled chromatin, the structure may prevent these molecules from interaction with their cognate binding sites. Active genes are in a less densely packed conformation, and therefore they are more easily digested by DNAse I compared to densely packed inactive genes (GROSS and GARRARD, 1988).

A group of nuclear proteins (scaffold proteins) is involved in the organization of the structural organization of long chromosomal domains. These domains are structurally characterized by long loops (5–200 kbp) which are anchored to the non-histone-protein scaffold (GASSER and LAEMMLI, 1987). The DNA sites which interact with the nuclear scaffold proteins (the matrix) are called scaffold associated regions (SARs) (Fig. 9) or matrix associated regions (MARs).

Topoisomerase II, an enzyme that is critical in DNA replication and RNA transcription because it relieves torsional stress by cutting and religating DNA, is often associated to the nuclear scaffold. However, the SARs are not necessarily identical to the topoisomerase II binding sites.

It appears that the position of the constitutive attachment of DNA to the nuclear scaffold is not developmentally regulated. However, recent work has revealed that SARs organize specific chromosomal domains in a way to facilitate transcription and to fix single origins of replication (EISSENBERG and ELGIN, 1991).

Isolation and delimitation of SAR element DNA revealed that SAR elements promote the activity of enhancers and promoters. As expected, they function in a cell-type unspecific manner and only when the DNA has been inte-

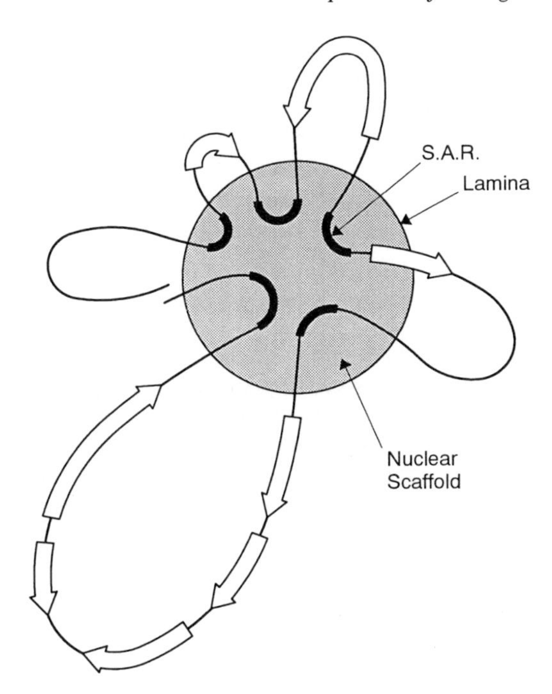

Fig. 9. Attachment of DNA to the nuclear scaffold. The chromosomal DNA is attached to the nuclear scaffold at multiple sites (SAR) resulting in the formation of looped-out DNA. The genes on the DNA are shown as white arrows. Adapted from GASSER and LAEMMLI (1987).

grated into the chromosomal DNA of the host (Tab. 6) (MIELKE et al., 1990; KLEHR et al. 1991). Transcription units flanked by SAR elements may be stimulated by adding butyrate or proprionate to the cells. These agents counteract the deacetylation of histones and lead to unwinding of DNA (negative superhelicity), indicating that SAR elements may stabilize looped-out DNA with open (underwound) DNA strands (Tab. 6).

2.1.2.2 Locus Control Regions

A locus control region (LCR) is another type of chromosomal element which specifically works in the chromosomal context. The LCR adjacent to the β-globin domain confers position-independent expression to the β-globin gene specifically in erythroid cells. As indicated in Tab. 6, the LCR specifically interacts with an activating sequence in the β-globin promoter and thereby enhances transcription. Although this property would be typical for tissue-specific enhancers, the property which distinguishes enhancers from the LCR is the fact that newly discovered LCR elements only function in a chromosomal context. Another LCR element with similar properties has been identified in the flanking DNA of the mammalian CD2 gene, which is specifically expressed in T-cells (GREAVES et al., 1989), and the chicken lysozyme gene, which is expressed in myeloid cells (STIEF et al., 1989; BONIFER et al., 1991). These findings indicate that regions with similar function may flank other tissue-specific regulated genes.

2.2 RNA Processing, Transport, and Degradation

Production of functional mRNA from the primary transcripts (heterologous nuclear RNA; hnRNA) involves four major steps: Capping, splicing, polyadenylation, and transport to the cytoplasm. Since the efficiency of these processes can influence gene expression, they will be briefly described. We would like to mention that RNA during processing is not naked, but associated with specific proteins that form particles which assist in carrying out

Tab. 6. Distal Transcription Influencing DNA Elements

	Activity During Transient Expression	Activity after Stable Integration	Interaction with Other Promoter/ Enhancer Elements	Cell-type Specificity	Affinity to Nuclear Scaffold
Enhancer	+	+	±	+	–
LCR	–	+	+	+	–
SAR/MAR	–	+	±	–	+

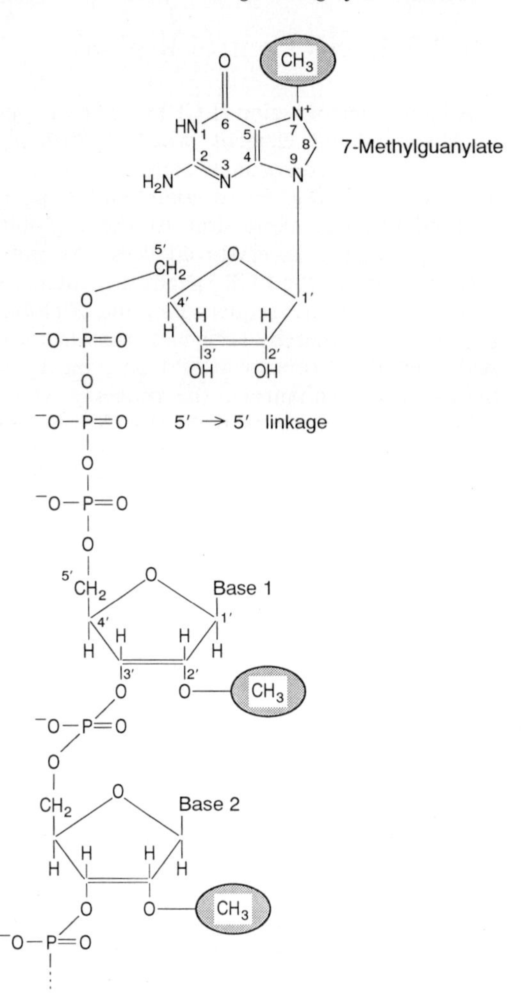

Fig. 10. Cap structure of eukaryotic mRNAs.

post-transcriptional modifications and transport.

2.2.1 Capping

All RNA polymerase II transcribed RNAs formed in normal cells are capped. The cap consists of a terminal nucleotide, 7-methylguanylate (^{7}mG) in a 5-linkage with the initial nucleotide of the mRNA chain (Fig. 10). It is co-transcriptionally added to the primary transcript. Capping is therefore one of the first steps of hnRNA modification. The capping enzyme and its substrate are located in the nucleus. The existence of the cap positively influences translation of cellular mRNAs as well as polyadenylation and splicing.

2.2.2 Splicing

The most obvious step in conversion of the primary transcript to mRNA is splicing, namely the elimination of internal RNA sequences (introns). The remaining sequences (exons) are subsequently ligated. Splicing of the primary transcripts starts on the nascent transcript and is completed after a maximum of 20 minutes after termination of transcription. The unstable intronic sequences are readily eliminated by degradation in the nucleus. The splicing process involves a cleavage at the 5′ exon/intron boundary, followed by a ligation of the 5′ end of the intron to a 2′ OH of an adenylate residue (branch point) at a downstream site of the intron. The next step is the cleavage of the intron/3′ exon junction and the subsequent ligation of the 5′ exon to the 3′ exon.

The exon/intron boundaries as well as the branch point of the sequences are moderately conserved. Two strict consensus nucleotides each are only found at the 5′ and 3′ ends of the intron (Fig. 11) (SHARP, 1987). Splicing signals have been highly conserved during evolution. Primary transcripts from heterologous species in the animal kingdom are readily recognized and processed.

Most higher eukaryotic genes contain introns. When introduced into mammalian cells, efficient expression of several genes has been shown to depend on the presence of an intron (HAMER and LEDER, 1979; BUCHMAN and BERG, 1988). Interestingly, this appears related to the type of promoter/enhancer which drives the expression of the gene, indicating that the transcription and the splicing event are not fully independent. For example, an intron is required for efficient expression from an immunoglobulin heavy chain promoter/enhancer, but the intron need not be specific. In contrast, no intron is required for efficient expression from other strong promoters, including the hCMV or HSP70 promoters (NEUBERGER and WILLIAMS, 1988). Many genes do not require introns for mRNA formation when in-

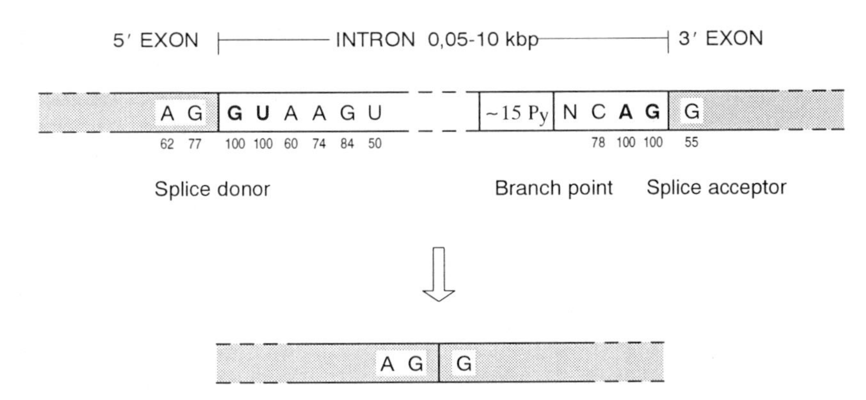

Fig. 11. Conserved sequence elements around intron/exon junctions in eukaryotic pre-mRNAs. Splicing requires consensus nucleotide sequences at the splice donor, the splice acceptor and the branch point. At the 5′ and 3′ intron termini, GU and AG, respectively, are 100% conserved. The frequency of occurrence is indicated by the numbers below the nucleotides. '15 Py' stands for 15 pyrimidine bases. After completion of the splicing reaction, the exon sequences are joined.

troduced into mammalian cells. This follows the observation that some intron-less genes are efficiently expressed. The mechanism by which expression efficiency is influenced by the presence of introns or the splicing process itself, remains to be elucidated.

Surprisingly, the requirements for the presence of introns are even stronger for constructs introduced into transgenic mice. The construct in which the rat growth hormone gene is driven by the mouse metallothionein I promoter, requires an intronic sequence for efficient expression in transgenic mice, but not in tissue culture experiments (Brinster et al., 1988). This observation may be related to the interdependence of promoters and introns as described above. It implies that most genes which have been processed through development, become dependent on mRNA splicing.

Investigations concerning the formation of adenovirus late mRNAs, led to the discovery that multiple RNAs can be produced from one primary transcript by differential splicing. From one transcript, three different RNAs are created by removal of an RNA fragment which starts with the same splice donor site in all three cases, but uses three different splice acceptor sites. Many examples of differential processing of animal RNAs have been described. Alternative splicing may be optional in a cell, or it may be a cell-specific function (Maniatis, 1991). Differential splicing can lead to proteins which share a common N-terminus but differ in the C-terminal region. In the extreme case, the resulting proteins are completely different.

2.2.3 Polyadenylation

After cleavage of the primary transcripts by a nuclease ~20 bp downstream of the AATAAA box (cf. Sect. 2.1.1.1), poly(A) is added to the newly generated 3′ ends. The process is called polyadenylation and is executed by poly(A) polymerase, using ATP as substrate (Munroe and Jacobsen, 1990).

With the exception of histone mRNAs, all known mature mRNAs are polyadenylated. Since only a portion of the hnRNA is polyadenylated, it was suggested that polyadenylation could provide a signal for hnRNA processing. Polyadenylation by itself seems not to be regulated. The length of the synthesized poly(A) tail is usually 250 adenosine residues.

The function of the poly(A) tail of mRNAs is not fully understood, but it is definitely related to the stability and translatability of the mRNA (Munroe and Jacobsen, 1990; Jackson and Standart, 1990). In the cytoplasm, the poly(A) end is successively shortened. Poly(A) ends with less than 27 adenosine residues are unstable in the cytoplasm.

2.2.4 Other Post-Transcriptional Processes

It is still unclear how transport from the nucleus to the cytoplasm is regulated. The discovery of the REV protein from HIV-1 and its interaction with REV-responsive RNA elements (RRE) on HIV-1 mRNAs, has demonstrated that this level can also be influenced. The REV protein seems to influence both, splicing and transport of HIV-1 transcripts to the cytoplasm, resulting in a higher representation of unspliced HIV-1 mRNAs in the cytoplasm (ROSEN and PAVLAKIS, 1990).

2.2.5 Cytoplasmic mRNA Turnover

In contrast to rRNA and tRNA which are synthesized at a rate that allows doubling their amounts with each cell generation, mRNA is synthesized faster than is apparently required for maintenance of a steady-state amount. The reason is that most mRNAs species are constantly degraded.

Half-lives of mRNAs in animal cells differ greatly. The turnover of several housekeeping gene products is very low, with a half-life of more than 24 hours. In contrast, mRNAs for gene products which are only required for short-time periods, like cytokinines, growth factors and some proto-oncogenes (c-myc, c-fos) are short-lived with half-lives of less than 30 minutes. Assuming that translation of a specific mRNA is only a function of the respective mRNA concentration in the cytoplasm, it is easy to understand that specific protein production can vary by orders of magnitude depending on the mRNA turnover rate.

The detailed mechanisms of mRNA turnover are not fully understood. Most probably, distinct mechanisms operate in the degradation of different mRNA species. However, one observation seems to be true for all cases: a poly(A) end which is longer than 30 base pairs protects the mRNA from being rapidly degraded; e.g., histone mRNAs which lack poly(A) at their 3′ end have a very short half-life time. Degradation of the poly(A) tail is dependent on sequence information within the mRNA, in particular in the 3′ untranslated region (3′UTR). The poly(A)-binding factor (PABF) covers 27 base pair units of the poly(A) tract and protects the mRNA from nucleolytic attack (BERNSTEIN and ROSS, 1989). When a poly(A) tail is shorter than 27 nucleotides, the mRNA is converted to a completely nucleolytic sensitive state.

One of the sequence conditions found to be responsible for short half-lives of mRNAs encoding proto-oncogenes and cell growth factors is the repeated occurrence of the sequence AUUUA in the 3′UTR (Tab. 7) (SHAW and KAMEN, 1986). The existence of these sequences in a certain mRNA species have been shown to correlate to enhanced degradation of the poly(A) tract (WILSON and TREISMAN, 1988). However, this mechanism must be more complex; e.g., AUUUA sequences alone are not always sufficient for mRNA destabilization. And, in some cases they act in concert with other RNA sequences.

Rapid degradation of histone mRNAs and others, including those having the AUUUA tract in the 3′UTR, does not occur in the absence of translation. This association suggests

Tab. 7. AT-Rich Sequences in the 3′UTRs of Genes Encoding Short-Lived mRNAs

Hu GM-CSF	TAATATTTATATATTTATATTTTTAAAATATTTATTTATTTATTTATTTAA
Hu G-CSF	TATTTATCTCTATTTAATATTTATGCTATTTAA
Hu IL-2	TATTTATTTAAATATTTAAATTTTATATTTATT
Hu IFN-γ	TATTTATTAATATTTAACATTATTTATAT
Hu TNF	TTATTTATTATTTATTTATTATTTATTTATTTA
Hu c-*fos*	TTTTTAATTTATTTATTAAGATGGATTCTCAGATATTTATATTTTTATTT-TATTTTTT

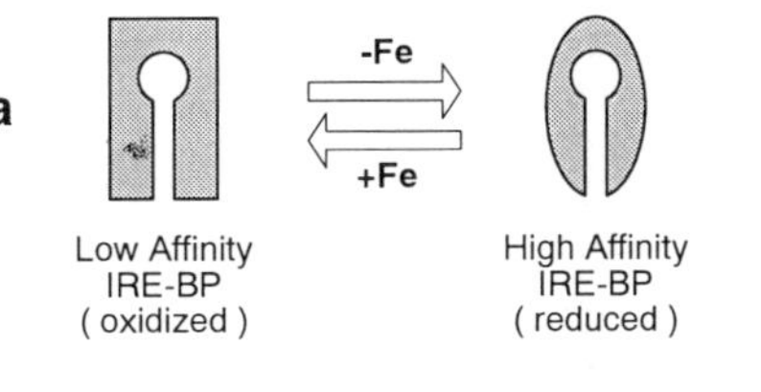

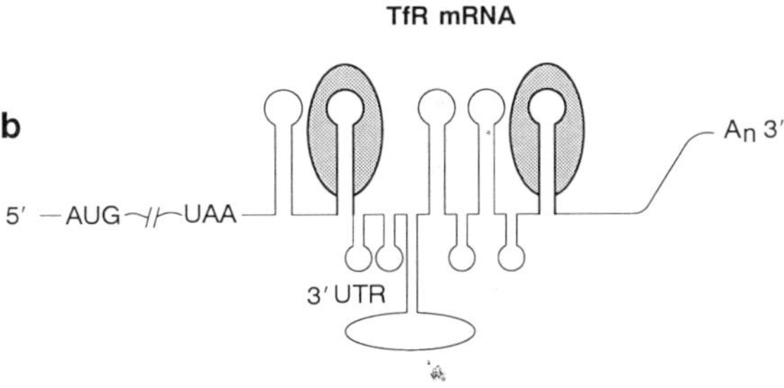

Fig. 12. Regulation of the transferrin receptor mRNA degradation rate. (a) The cytoplasmic iron-response element/binding protein (IRE-BP) exists in two different conformations depending on the intracellular iron concentration.
(b) The stem-loop structures in the transferrin receptor mRNA (IREs) associate with the high affinity form of IRE-BP. At low intracellular iron concentrations the binding of the IRE-BP to the transferrin mRNA prevents its degradation.

that normal mRNA turnover is due to ribosomal associated nucleases (BERNSTEIN and ROSS, 1989). Furthermore, in some cases the signal which is responsible for initiation of mRNA degradation may be the nascent protein chain.

mRNA secondary structures are also involved in regulation of the degradation rate. An interesting example concerns the transferrin receptor mRNA (see also Sect. 2.3.4.1). At high concentrations of intracellular iron ions, the transferrin receptor mRNA is rapidly destroyed. The existence of five stem-loop structures in the 3′UTR of the transferrin receptor mRNA (TfR) is responsible for its degradation (Fig. 12). These stem-loop structures are called iron response elements (IRE), each of which is able to bind a protein called IRE-BP. The binding of the IRE-BP depends on the intracellular iron ion concentration. Iron excess results in low affinity IRE-BP. The affinity change is based on a change in the protein's redox state (termed sulfhydryl switch), which is determined by the iron status of the cell. At high concentrations the IRE-BP does not bind to the stem-loop structures, allowing the respective mRNA to be degraded. At low iron concentrations the IRE-BP binds to the IRE sequences. This leads to a block of the degradation of the TfR mRNA (MÜLLNER et al., 1989; HENTZE, 1991).

Mediation of the TfR mRNA turnover by physiologically active compounds is not the only example of regulation of specific mRNA turnover. Stabilization of specific mRNAs can also be mediated by hormones, such as prolactin, estrogen, glucocorticoids, and intracellular gene products.

2.3 Translational Control

2.3.1 The Translational Machinery

The translational machinery of eukaryotes is complex and involves a variety of factors. A comparison with the translational machinery of prokaryotes is shown in Tab. 8. It is evident that eukaryotes have developed related, but distinct mechanisms of protein synthesis regulation.

Translation occurs on 80S ribosomes, which are composed of a 40S and a 60S subunit. The translational mechanism can be divided into three phases: initiation, elongation, and termination.

Initiation: In most eukaryotic mRNAs the AUG closest to the 5′ end is used for initiation of translation. The choice of the initiator-AUG is thought to occur after binding of a 40S ribosomal subunit complex to the mRNA 5′cap structure (see Fig. 10) and 'scanning' of this complex along the mRNA. The initiation process involves mRNA, a special tRNA (Met-$tRNA_i$), GTP, ATP, a number of initiation factors (eIF's), and the ribosomal subunits.

Elongation requires the aminoacyl-tRNAs, GTP, and diverse elongation factors (EF's). The reaction cycle involves binding of the amino acyl-tRNA to the ribosomal A-site, transpeptidation to form a bond between the new

Tab. 8. Prokaryotic and Eukaryotic Translational Components

Component	Prokaryotes	Eukaryotes
tRNA used for start of translation	Formyl-tRNAMeT$_i$	tRNAMet$_i$
Preferred start codon	AUG, less: GUG, UUG, CUG, even AUU	AUG rarely CUG
Start selection-, scanning-sequences	Internal initiation determined by context around the initiation codon, Shine-Dalgarno	Cap-dependent in 90% of the AUG located nearest to the mRNA's 5′ end, Ribosomal landing pads
Initiation factors	IF-1, IF-2, IF-3	Multiple, called eIFs. Often composed of numerous subunits
Elongation factors	EF-Tu, EF-TS EF-G	EF1-alpha EF1-beta/gamma EF-2
Termination factors	RF1, RF2, RF3	RF

aminoacyl and the peptidyl-moiety of the ribosomal P-site, and translocation of elongated peptidyl-tRNA to the P-site, accompanied by a one-codon movement of mRNA.

The *termination* process requires GTP and the RF-protein factor which recognizes the specific stop codons. Binding of the termination factor to the A-site results in hydrolysis of the peptidyl-tRNA at the P-site and in release of the polypeptide chain and dissociation of the ribosomal subunits.

Although all steps are important for successful translation, most interest has focused on the initiation of translation, as this step is rate-limiting. Translation efficiency of an mRNA is, therefore, mainly determined by the fidelity of the initiation step. The actual concentration of protein factors in the cell system, as well as the mRNA's structure (primary, secondary, and even tertiary), are the components, which define the translation efficiency of a given gene. Therefore, further discussion will be led on the conditions for optimal initiation of translation.

2.3.2 Initiation of Translation

2.3.2.1 Steps in the Initiation Process

The initiation phase of translation can be separated into distinct steps (Fig. 13):

(1) ribosome dissociation,
(2) formation of the ternary complex eIF2 + Met-tRNA$_i$ + GTP,
(3) formation of the 43S preinitiation complex,
(4) binding of mRNA to the preinitiation complex,
(5) association of 60S subunits to the mRNA-40S complex to form an 80S initiation complex.

The dimeric form of the ribosome is not able to initiate translation and therefore must dissociate into the 40S and 60S subunits. Subunits are in equilibrium with the associated form, albeit the 80S form is clearly predominant. The subunit forms are stabilized by the binding of some initiation factors: eIF3 and eIF1A (previously termed eIF4C) attach to the 40S subunit, while eIF3A (previously eIF6) combines with the 60S form. Although their role in inhibiting the association (anti-association) of the

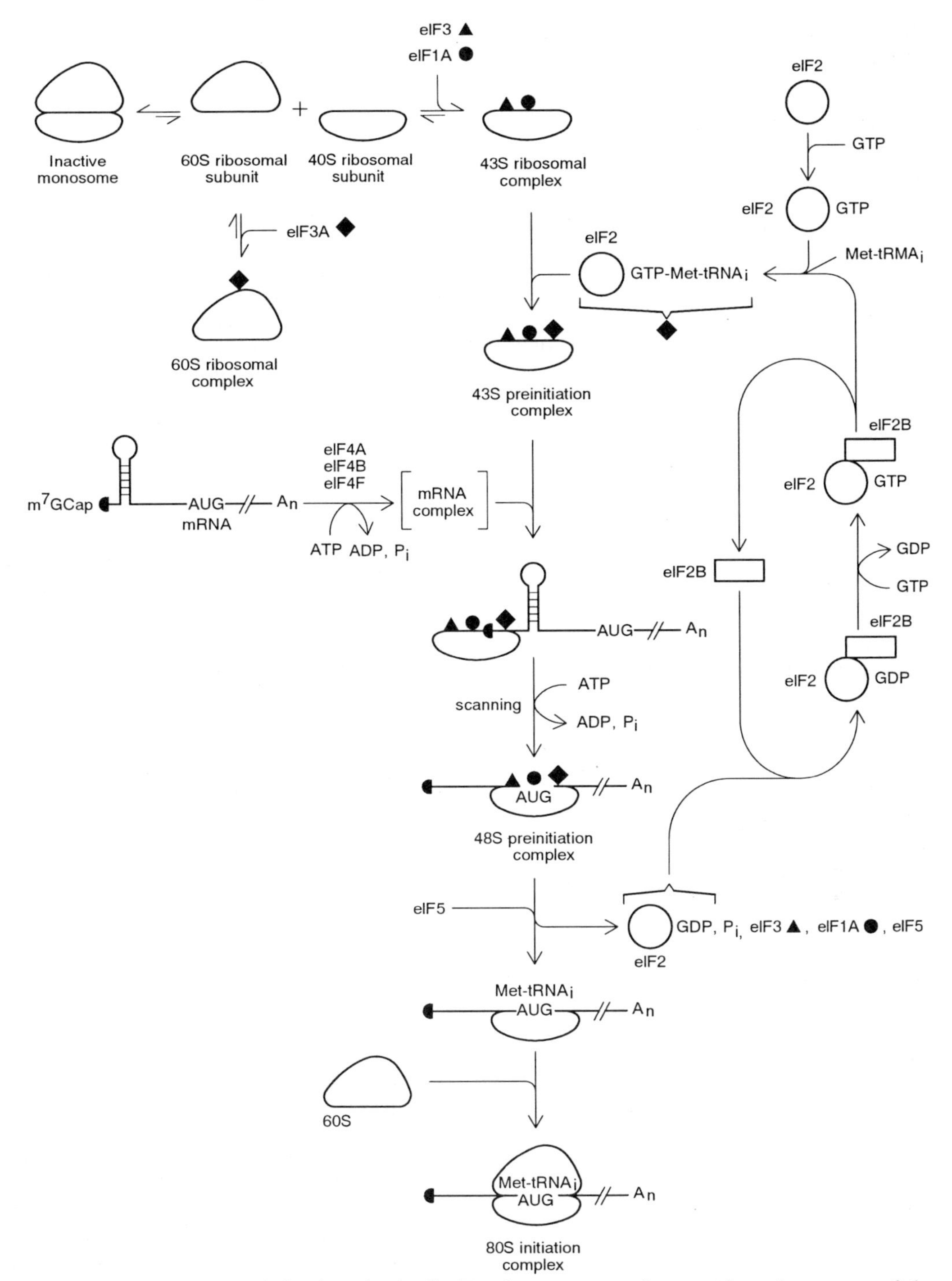

Fig. 13. The initiation of translation in animal cells. For the new nomenclature and previous names of the multiple factors involved in initiation of protein synthesis, see SAFER (1989). A detailed description of the processes is given in the text.

ribosomal subunits has been demonstrated, an active role in the dissociation process has also been discussed for these factors.

The formation of the ternary complex eIF2 + Met-tRNA$_i$ + GTP is a very specific process. The specificity of interaction may depend on distinct features of the special initiator tRNA. eIF2 plays a crucial role in regulation of translation as its affinity to individual mRNAs has been shown to determine the efficiency with which the individual mRNA is translated (KAEMPFER et al., 1981). It is composed of three subunits, alpha, beta, gamma. eIF2α can be phosphorylated by certain factors preventing its further utilization. The ternary complex associates with the 43S ribosomal complex. This preinitiation complex is shown to bind to the mRNA's cap structure. The importance of the cap structure is underscored by experiments which show that the interaction of the individual constituents is severely impaired by cap analogs. Most eukaryotic mRNAs are capped; however, several animal viruses are able to perform their translation in the absence of a cap structure.

Free accessibility of the mRNA has been postulated as crucial in both cases. A complex, formerly termed CBP, consisting of factors eIF4F, A, B, has been shown to resolve hindering secondary structures in an ATP driven process.

After the initiator AUG has been determined, the helping factors are removed from the 40S ribosome by the action of eIF5 accompanied by GTP hydrolyzation and the 60S subunit joining to the 40S counterpart.

2.3.2.2 Selection of the Start Codon

Usually, the AUG codon next to the mRNA cap is used for initiation of translation. However, exceptions accumulated in the last years have led to a revision of initial dogmas.

Several models have been proposed which explain the choice of the initiator AUG. The most common one is the scanning model of translation initiation postulated by MARILYN KOZAK. In an updated version (KOZAK, 1986c, 1989a) a 43S ribosomal complex binds to the

Tab. 9. Translation Efficiency Dependence upon Sequences Around the Initiation Codon[a]

−3	−2	Position −1		+4	Relative Efficiency
T	T	T	ATG	T	0.2
G	T	T	ATG	T	0.7
A	T	T	ATG	T	2.2
A	T	T	ATG	G	5
A	C	C	ATG	G	6
A	C	C	ATG	T	5
T	C	C	ATG	T	5
C	C	C	ATG	T	2

[a] according to KOZAK (1986c)

cap structure and moves along the 5′ untranslated region (5′UTR) until an AUG in a *favorable* sequence context is reached. This is followed by the formation of a functional 80S complex by association of the 60S subunit. The idea that a favorable context may be involved came from compilation of sequences around the AUG of numerous vertebrate mRNAs. Two positions around the initiator AUG were most prominently conserved, position −3 which is often occupied by a purine residue, predominantly A, and position +4 which most often is a G residue. Model experiments confirmed the importance of these positions for efficient translation, as mutations in −3 and +4 severely altered translational efficiency (see Sect. 2.3.3.2 and Tab. 9).

Although many points may argue for a scanning of ribosomes evidence is only indirect. Additionally, the model cannot account for a cap-independent, internal initiation which has been shown for some picornaviruses and special cellular RNAs, such as BiP, *Drosophila* antennapediae and ultrabithorax. These findings have prompted the development of the 'melting model' which accounts for both types of translation initiation. The model is based on the observation that factors involved in unwinding of hindering secondary structures are capable of binding mRNA. In the melting model, scanning of the leader is performed by the melting eIF4A plus eIFB rather than by the 40S ribosome. Since the affinity of free eIF4B to the AUG codon is documented, it is speculated that this factor is involved in the recognition process of the start site. The mechanism is supported by the ability of eIF4A to bind independently to single-stranded RNA (SONENBERG, 1988).

A limited number of internal initiation events has been investigated. Steps in this type of initiation of translation include the binding of eIF4A, B and F to regions in the mRNA devoid of secondary structure, followed by scanning as described above. Additional proteins, which may help in elucidating one of the specific steps, have been identified in picornavirus-infected cells (p52, p57).

eIF2 has also been suggested to participate in the selection of AUG initiation codons. Evidence for this comes from experiments, in which eIF2 was specifically bound to mRNA in an *in vitro* filter binding assay and had affinity to the AUG-initiation codon of Mengo and STNV mRNA (KAEMPFER et al., 1981; PEREZ-BERCOFF and KAEMPFER, 1982). However, there is no consensus in the literature about the participation of eIF2 in AUG selection.

2.3.3 Factors Influencing Translation Efficiency

Two levels of regulation in the translation initiation step must be considered: global regulations influencing the initiation rates of all or many mRNAs, and selective regulations which are due to features of the individual mRNAs. Regulatory steps may involve the concentration and modification of protein factors, and the primary or secondary structure of the mRNA.

2.3.3.1 Modification of Translation Initiation Factors

Modifications of translation factors can lead to changes in their biological activity and therefore may influence the efficiency of translation. This will be demonstrated by three examples.

Phosphorylation of the alpha-subunit of eIF2 is a global mechanism for regulation at the level of initiation. The enzymes responsible for the phosphorylation are p68-kinase (double-stranded RNA activated kinase) or the HCI (heme controlled inhibitor). p68 is a constitutively expressed cellular enzyme which can be induced by IFN and which is activated by double-stranded RNA. This reaction is part of the cellular response to virus infection. The phosphorylated eIF2 has a much greater affinity to GDP than to the unchanged factor. As a result, the GEP (eIF2B) is not able to recycle the eIF2-GDP, the system is depleted for functional eIF2, and a severe reduction of translational efficiency occurs.

eIF2 modification may be prevented by viral products, such as the adenovirus VAI RNA, and the Epstein-Barr virus EBER1/2, or may be impaired by mechanisms which have been accomodated by certain viruses (Fig. 14). In transient expression experiments, KAUFMAN and coworkers showed that coexpression of VAI RNA can increase expression. This observation was interpreted as an inhibition of p68, induced by mRNA from transfected nucleic acids (KAUFMAN and MURTHA, 1987).

Polioviruses favor their own synthesis by inducing the proteolysis of p220, a component of the cap-binding complex (CBP). Although the role of p220 is not fully understood, a p220 shortage results in a reduction of cap-dependent initiation of translation. The viral RNA translation remains unaffected, as it is initiated by a cap-independent mechanism. Negative effects on the efficiency of cap-dependent translation are exerted by competition for cap-binding proteins, specifically with the cap analog m^7G.

eIF4E is the only factor with affinity for the mRNA cap structure in the absence of ATP. It is thought to interact early with the initiating mRNA. eIF4E is one of the least abundant initiation factors and, therefore, cap binding serves as the rate-limiting step in initiation of translation. In HeLa cells and reticulocytes it exists as a mixture of phosphorylated and unphosphorylated forms. Several findings argue for a global regulation of initiation at the level of cap recognition. Heat shock of cells diminishes the initiation of protein synthesis, which is accompanied by a decrease in eIF4E phosphorylation. The effect can be reversed by the addition of the eIF4F complex. eIF4E, which

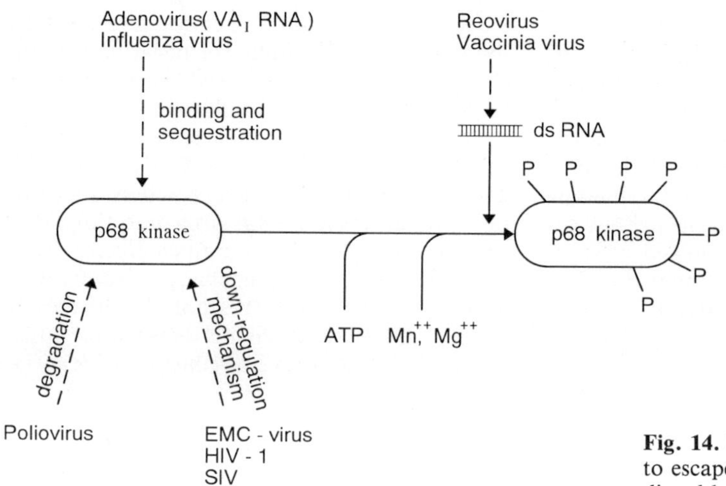

Fig. 14. Mechanisms evolved by viruses to escape inhibition of translation mediated by the p68 kinase.

is a subunit of eIF4F, was shown to be involved in the regulation of proliferation. It has been shown that the state of its phosphorylation in mammalian cells correlates well with decreased protein synthesis during mitosis or as a result of serum starvation. Activation of eIF4E through phosphorylation is stimulated by extracellular agents such as TNF, TPA, or PDGF-BB (BU and HAGEDORN, 1991). The overexpression of eIF4E in NIH3T3 or RAT2 cells, which cause oncogenic transformation, underscores an eIF4E role in growth control (LAZARIS-KARATZAS et al., 1990).

2.3.3.2 Primary and Secondary Structure of mRNA

Translation efficiences are recognized to be due to mRNA specific differences, which are determined by the primary and secondary structure of individual mRNAs.

The cap structure is found on most eukaryotic mRNAs and is generally thought to be necessary for an efficient translation of eukaryotic mRNAs (reviewed, e.g., in BANERJEE, 1980). However, experiments which compared naturally capped mRNAs with their synthetic uncapped counterparts showed that the requirement for a cap structure varied in individual mRNAs and test systems. Excluding the special case of internal initiation of translation, a cap structure at the 5′ end facilitates the entry of the initiation complex and consequently promotes translation.

The accessibility of the cap correlates with the efficiency of initiation. Evidence for this came from the investigation of several viral mRNAs. For example, discrimination of alfalfa mosaic virus (AIMV RNA) occurred at the level of cap binding of translation factors. There binding and translation correlates with the enzymatic accessibility of the cap.

Early work of M. KOZAK has shown that the primary sequence around the AUG determines the fidelity of initiation. Mutations at position -3 (which is often a purine) or at $+4$ (which is often a G) converting the purines to pyrimidines had the most pronounced effect on translation of a preproinsulin-RNA in COS cells (Tab. 9). *In vivo* evidence for context-dependent translation has been found in the case of α-thalassemia, where an A to C substitution at position -3 of the human α-globin gene is the cause for a severe reduction in translation efficiency resulting in the disease. In an additional investigation the sequence GCCGCCACC**AUG**G was postulated to be the optimal sequence for translation in the preproinsulin system (KOZAK, 1986b, 1987a). Converting the context around the AUG of the gene to be expressed in the animal cells to the optimum 'Kozak sequence' should, therefore, increase expression of the respective transfected gene. However, extensive data which support these initial observations are missing.

Data on the influence of the 5′UTR length are rare. JOHANSEN et al. (1984) found no effect on galactokinase K expression by varying the length of a 5′UTR from 40 to 100 bases (provided that the GalK-AUG remained the first AUG in the transcription unit) (JOHANSEN et al., 1984). Under conditions of hypertonic stress, the length of synthetic, unstructured 5′UTR influences translation (KOZAK, 1988).

Stabilization of secondary structures may be responsible for the effect of salt concentration on translation and may account for earlier observations that cellular translation initiation is more sensitive to excess salt than certain viral systems.

The degree of secondary structure in the 5′UTR is important for translation. Secondary structures in the 5′UTR may hinder the binding step, scanning, and AUG selection. Therefore, an unfolded 5′ leader optimally should support these steps. However, most RNA-5′UTRs contain stem loops with considerable stability. To allow binding and scanning of the translation elements, cellular factors, eIF4A and eIF4B, are able to melt existing secondary structures up to a certain stability. The position of hairpins seems to be crucial.

5′ Proximal secondary structures up to 15 bp from the cap (created by hybridization of oligonucleotides to mRNA) impaired translation, while hybrids further downstream did not. This strengthens the theory that the cap structure must be accessible (e.g., GODEFROY-COLBURN et al., 1985). Similarly, a hairpin 7 bases downstream of the cap inhibits translation, while the same hairpin located 34 bases downstream does not (PELLETIER and SONENBERG, 1985). Both results were explained by distortion of a proposed two-step initiation process consisting of an unaffected weak binding of eIF4E to the cap followed by the ATP-dependent formation of a secondary structure sensitive, stable initiation factor-mRNA complex involving eIF4A and eIF4B.

The influence of the stability of hairpin structures further downstream of the cap and upstream of the AUG has also been addressed. To determine the resolving power of the translational machinery, KOZAK introduced stem loops of increasing stability into a preproinsulin leader RNA. A stem-loop structure with a stability of $\Delta G = -126$ J/mol had little effect, while structures of $\Delta G = -210$ J/mol in the leader or around the AUG severely affected the translation.

Interestingly, strong secondary structures in the coding region of a gene exhibited no effect on translation. Significantly, moderate secondary structures 12 – 15 bp downstream of the AUG initiator codon augmented the initiation of AUG in a weak context, possibly by causing the scanning ribosome to pause above the AUG (KOZAK, 1990).

The effect of natural 5′UTR sequences on gene expression was measured after transfection of various coding regions which were fused to specific 5′UTRs. Favorable 5′UTRs were found in the *Xenopus* L. *β*-globin gene, the tobacco mosaic virus RNA, and the alfalfa RNA4 gene (FALCONE and ANDREWS, 1991; GALLIE et al., 1987, 1988; JOBLING and GEHRKE, 1987). If 5′UTRs were fused to the body of reporter genes, the *in vitro* translation efficiencies of these genes were increased 10- to 300-fold compared to the mRNAs with the authentic leader (FALCONE and ANDREWS, 1991) or up to 80-fold in *in vivo* investigations (GALLIE et al., 1987). Increased affinity to a limiting initiation factor or a diminished requirement for a limiting factor (e.g., for the unwinding activity in the case of low secondary structure of the 5′UTR) were discussed as possible reasons for high-level translation. However, when the mechanism for augmentation was investigated in the case of the *Xenopus β*-globin leader, an increase in expression was found to be due to an increased utilization of translation initiation factors.

The influence of 3′UTRs on translation initiation was also investigated. Regions of complementarity in the 3′UTR of mRNA to the 5′UTR of the same mRNA molecule may lead to formation of inhibitory as well as stimulatory secondary structures. Support for such an interaction comes from zein, IFN-β, TMV mRNA, and model systems (SPENA et al., 1985; KRUYS et al., 1987; GALLIE, 1991; KOZAK, 1989b). In the zein example, deletion of the complementary 3′UTR which was assumed to interact with the 5′UTR by imperfect in-

Tab. 10. Translationally Controlled Systems in Animal Cells

mRNA	Mediator
Ferritin	Iron-ion concentration
Heat shock	Heat shock
Picornavirus (Poliovirus, EMC)	Infection via internal initiation
Maternal embryonal mRNA	Fertilization
Reticulocyte mRNA	Hemin
mRNA in virus infected cells	Double-stranded RNA

verted repeats, resulted in a considerable increase in translation of the specific mRNA without affecting the mRNA's stability.

2.3.4 Translationally Regulated Systems

In nature, several eukaryotic systems are known in which expression is regulated at the level of translation. Translational control which is mediated by the 5′ untranslated regions can be conferred to heterologous genes. Some examples are listed in Tab. 10.

Two systems, the iron-controlled ferritin system and the regulation of picornavirus protein synthesis, will be described in more detail.

2.3.4.1 Post-Transcriptional Regulation of Ferritin/Transferrin Receptor Synthesis

In higher eukaryotes two proteins, ferritin and transferrin receptor, mediate iron uptake and detoxification. The synthesis of both proteins is coordinately regulated via the concentration of intracellular iron. The ferritin synthesis is translationally controlled (Fig. 15), while in the case of transferrin, receptor synthesis is regulated at the level of mRNA stability (cf. Sect. 2.2.5). Both systems involve a protein (IRE-BP) capable of binding to a region of secondary structure (IRE) occurring once in the 5′ leader of ferritin mRNA or five times in the 3′ leader of the transferrin receptor mRNA. IRE-BP, a 90 kDa protein, exists in two forms depending on the iron concentration. Under conditions of iron deprivation, the high affinity form of IRE-BP is prevalent. Iron excess results in the low affinity IRE-BP. The affinity change is based on the change in the redox state of the protein (termed sulfhydryl switch) which is determined by the iron status of the cell. The high affinity form binds

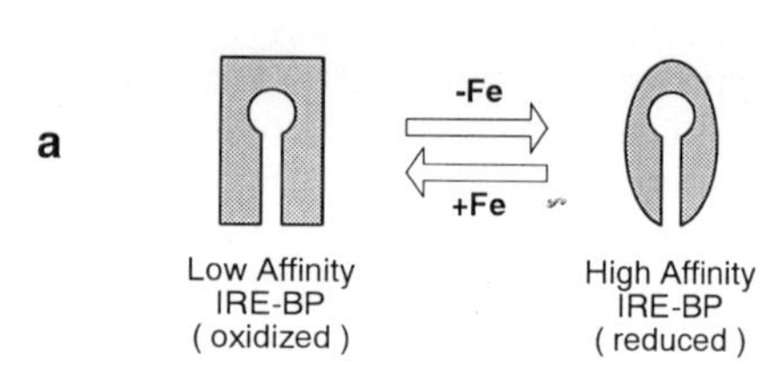

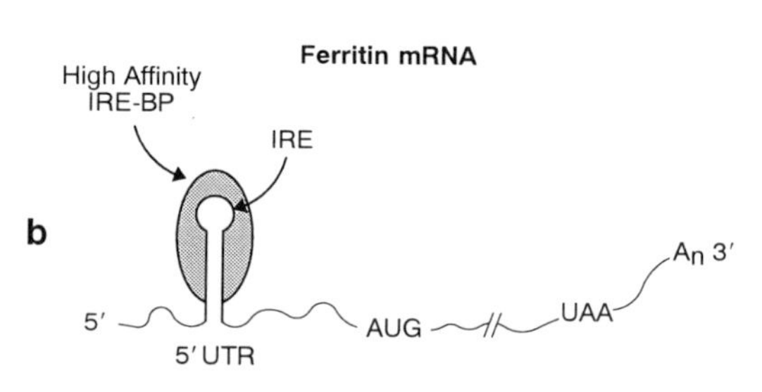

Fig. 15. Post-transcriptional regulation of ferritin synthesis. (a) Depending upon the concentration of iron ions, the IRE-binding protein (IRE-BP) exhibits different affinities to the iron-responsive element (IRE).
(b) Upon iron deprivation the high-affinity form of the IRE-BP is prevalent. Its binding to the IRE inhibits mRNA translation.

to the 5′IRE in ferritin mRNA and prevents its translation, while surplus iron induces the translation of ferritin (reviewed in KLAUSNER and HARFORD, 1989).

2.3.4.2 Cap-Independent Initiation of Translation in Picornaviruses

Picornaviruses are positive-strand RNA viruses. The single mRNA codes for a long polyprotein, which is further processed to viral structural and non-structural proteins. Picornaviruses have unusually long 5′ untranslated leader regions, ranging from 600 to 1300 bp. In these leaders up to 10 AUGs precede the AUG codon which is used for translation. The mechanism responsible for this unusual translation initiation has been elucidated for some members of the picornavirus family. It seems that, in general, translation is performed by cap-independent, internal initiation. The internal entry of ribosomes is mediated by a region termed ribosome landing pad (RLP) or internal ribosomal entry site (IRES) (Fig. 16). The delimitation of *cis*-elements in the leader, which is necessary for efficient internal initiation has been determined (reviewed by JACKSON, 1991; for poliovirus, additionally notice SIMOES and SARNOW, 1991). According to this mapping, picornaviruses can be divided into two groups with respect to the mode of AUG selection. In the enterovirus group (poliovirus), internal initiation is mediated by a 400 bp region of distinct secondary structure more than one hundred bp 5′ of the authentic AUG codon. Initiation is thought to occur after ribosomal landing and subsequent scanning. In contrast, in the cardiovirus group (e.g., encephalomyocarditis virus (EMC)) the respective *cis*-regions are located adjacent to the initiator AUG, and initiation occurs without prior scanning of ribosomes. Several proteins with ability to bind picornaviral 5′UTRs have been identified. A 52 kDa protein, which associates with the poliovirus leader at position 560–620, was isolated from HeLa cell extracts. A 57 kDa protein showed affinity to a region in the EMC leader. The function of these proteins is still unclear, but they are not identical with known cellular initiation factors.

Some picornaviruses favor their own synthesis by inducing the proteolysis of the 220 bp subunit of eIF4F, the cap-binding complex. As a consequence the cap-dependent initiation of cellular mRNAs is severely impaired, but eIF4-independent picornaviral translation can proceed.

Cap-independent initiation has been reported for some cellular RNAs. The synthesis

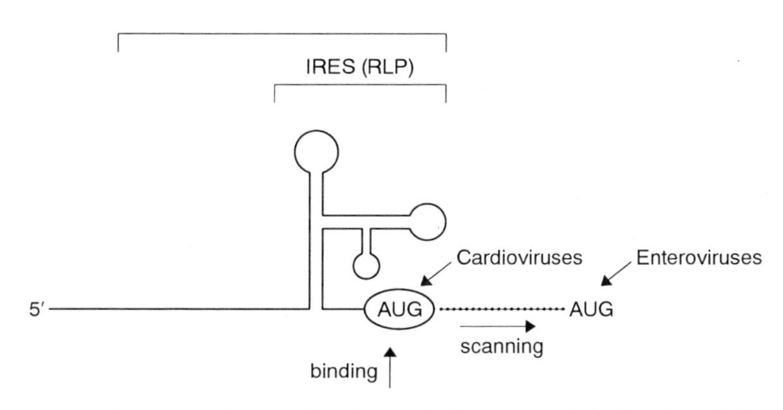

Fig. 16. Internal initiation of translation in picornaviruses. The picornaviral 5′UTR and the location of the initiator codons of cardioviruses and enteroviruses is depicted. The numerous AUGs preceding the actual translation initiation sites are omitted for reasons of simplicity. The upper bracket indicates the region in the 5′UTR necessary for efficient translation. The smaller bracket below delimits the internal ribosomal entry site (IRES, sometimes designated as ribosomal landing pad, RLP). The modes of initiation after internal entry of the ribosome in cardioviruses and enteroviruses (scanning and directed AUG binding, respectively) are indicated.

of BiP protein (heavy chain immunoglobulin binding protein) is not reduced upon poliovirus infection, and therefore a cap-independent initiation could be postulated (MACEJAK and SARNOW, 1991). *Drosophila* developmental genes antennapediae and ultrabithorax contain unusually long 5′UTRs. They are also suspected to mediate translation by cap-independent initiation of translation.

The potential of internal initiation on picornaviral 5′UTRs is exploited in eukaryotic expression vectors carrying polycistronic transcription units. This development promises an improvement in the expression of homo- and heterodimeric proteins (for more details see Sect. 3.2.1.5).

2.3.5 Translation from Polycistronic Transcription Units. Coupled Internal Initiation and Reinitiation

Translation from multiple open reading frames (cistrons) on a single mRNA is a common mechanism of protein synthesis in bacteria. In animal cells, most mRNAs are monocistronic and code for the synthesis of only one protein. The utilization of polycistronic transcription units is exceptional. The study of these exceptions has proven valuable for evaluation of translation mechanisms and in the construction of new, eukaryotic expression vectors. Proteins can be synthesized from overlapping or non-overlapping reading frames by bifunctional mRNAs (for an alternative see Sect. 2.3.6).

The realization of proteins from overlapping reading frames is mainly due to a leaky scanning mechanism where some 40S ribosomes by-pass the first weak context AUG and initiate at the next AUG downstream (for examples see KOZAK, 1986b). However, an interesting mechanism has been found in artificial constructs with a certain codon overlap. If the stop codon of the first cistron is contained in the start codon of the second cistron, such as in the sequence AUGA, initiation of the second open reading frame (ORF) can occur. A special mechanism of 'reach back translation' was postulated (PEABODY and BERG, 1986).

Some examples for bicistronic mRNAs with non-overlapping reading frames are known (reviewed in KOZAK, 1986b). Often the first ORF is terminated early and contains a minicistron which codes for a short polypeptide of unknown function. Several models have emerged to account for a mechanism of translation of these untypical mRNAs. Translation of the second ORF can occur by reinitiation of ribosomes (Fig. 17), coupled internal initiation, or internal initiation as described for picornaviruses. In the reinitiation model the same ribosome is postulated to be responsible for translation remaining associated with the intercistronic region, while in the coupled internal initiation mechanism, dissociation of ribosomes after translation of the upstream ORF may allow a different or the same ribosome to initiate at the downstream ORF.

The requirements for efficient initiation at a second ORF have been elucidated. The frequency of reinitiation is dependent on the length of the intercistronic region. In a model system composed of a minicistron and a preproinsulin gene, KOZAK determined optimal intercistronic lengths of 80 to 150 bp (allowing for greater than 50% translation efficiency). However, investigations which confirm the reinitiation efficiencies when longer 5′ORF's or genes precede the second ORF, have not yet been published. KOZAK explained the 5′UTR length dependence by a time requirement to reload a scanning ribosome with components necessary for starting initiation at the second AUG (KOZAK, 1987b). In another investigation, an excess of eIF4F and eIF4B was shown to promote translation from a second ORF (ANTHONY and MERRICK, 1991).

2.3.6 Ribosomal Frameshifting and Suppression of Termination

Ribosomal frameshifting is an alternative strategy for the expression of two or more genes encoded by one mRNA. In ribosomal frameshifting a reading frame is changed at a specific site (s) resulting in the synthesis of one fusion protein from overlapping genes. Ribosomal frameshifting has been observed in bacteria, yeast, and certain animal viruses, such as

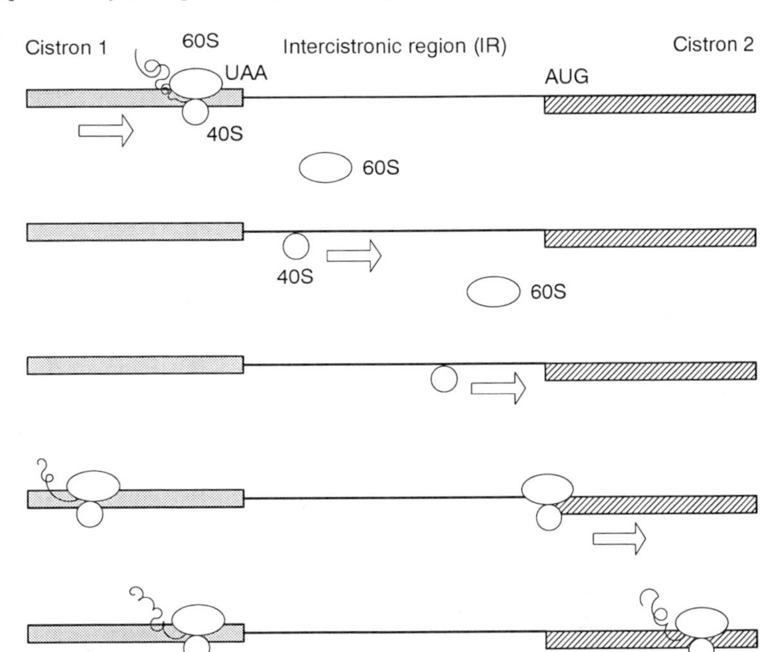

Fig. 17. Reinitiation of translation on a bicistronic mRNA. After synthesis of the first cistron (stippled), the 40S subunit (white circle) is assumed to move along the intercistronic region. Reinitiation may occur on a following AUG after association with a large ribosomal subunit (white oval). Reinitiation frequency increases with the length of the intercistronic region and is thought to reflect time requirements for reloading of the 40S subunit, e.g., with the initiator-tRNA.

coronaviruses or retroviruses. Depending on the species, the reading is changed in the 5′ or 3′ direction by a frameshift of 1 nucleotide (Tab. 11; for a review see, e.g., JACKS, 1990). Some members of the retrovirus family, including MMTV or HTLV-I, use frameshifting twice, rather than once, to synthesize their proteins. The efficiency of synthesis of the frameshift product is in the range of 5–30% compared to the unshifted protein. Mutational analysis of the overlap region showed that the frameshift signal consists of a heptanucleotide sequence followed by a downstream stem-loop structure. For some members of the frameshift family it has been demonstrated that the integrity of the stem structure and its distance from the frameshift signal are important for efficient frameshifting. In some cases, the loop structure contributes by forming a complex tertiary structure (RNA pseudoknot) with a region further downstream (Fig. 18). The precise mechanism of ribosomal frameshifting is not known. However, a good explanation is provided by the 'simultaneous slippage model' (JACKS et al., 1988a).

Several mammalian C-type viruses express their replication/processing enzymes (pro, pol) through suppression of a UAG-termination codon. The specific codon is located at the end of the gag reading frame. Suppression is mediated by a glutamine tRNA, and the reading frame of gag and pol is maintained. Similarly to frameshifting, downstream of the stop codon, a region has been found, which harbors potential stem-loop structures. The exact mechanism of suppression is not known, but the level of suppressor tRNA has been suspected to play a key role.

In both events, frameshifting and suppression of termination, involvement of a ribosome 'pausing' is discussed. In this view the stem loop causes ribosomes to slow down or halt near or at the frameshift site, thereby increasing the time for 'errors' like a tRNA slippage. However, this model needs yet to be confirmed. Suppression of termination and also ribosomal frameshifting gained pharmaceutical interest, as drugs are or may become available which interfere with these processes (e.g., avarol and paramomycin for translational suppression) and therefore inhibit virus replication.

Tab. 11. Frameshifting and Termination Suppression in Animal Retroid Genes

Representatives	Genes Overlap	Type	Efficiency	Bases at and around Frameshift Sequence	Stem Loop Involved	References
IBV	F1–F2	−1 Frameshift	25–30% [a,b]	UAU UUA AACGGG [e] UUU UUA AACGGG	+	BRIERLEY et al. (1987)
MHV	1a-1b	−1 Frameshift	35–40% [a,c]	CUU UUA AACGGA	n.d.	BREDENBEEK et al. (1990) LEE et al. (1991)
HTLV-I	Gag-pro pro-pol	−1 Frameshift	n.d.	UCA AAA AACUAA CCU UUA AACCAG	n.d. +	NAM et al. (1988)
MMTV	gag-pro pro-pol	−1 Frameshift	25% 5%	UCA AAA AACUUG	— [d]	HIZI et al. (1987) JACKS et al. (1987)
HIV-1	gag-pol	−1 Frameshift	10% 3%	AAU UUU UU**A**GGG	—	JACKS et al (1988b) REIL and HAUSER (1990)
RSV	gag-pol	−1 Frameshift	5%	ACA AAU UU**A**UAG	+	JACKS et al. (1988a) JACKS and VARMUS (1985)
Mouse IA	gag-pol	−1 Frameshift	n.d.	CUG GGU UUUCCU	n.d.	MIETZ et al. (1987)
MLV FeLV BAEV	gag-pol	Terminatition Suppression	5–10% [a]	Stop codon	± [f]	JONES et al. (1989); HONIGMAN et al. (1991)

[a] *in vitro*
[b] *in vivo* in oocytes
[c] *in vivo* in animal cell lines
[d] Tested in *E. coli*
[e] Shifty signals are underlined, frameshift occurs at the base shown bold.
[f] Sequences downstream of UAG necessary; yet no consensus on a stem-loop involvement.

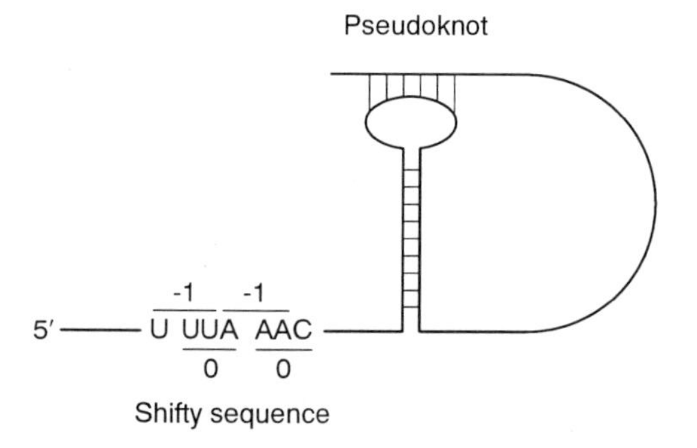

Fig. 18. Structural requirements for ribosomal frameshifting. The frameshift region of the infectious bronchitis virus (IBV) is shown schematically as an example for frameshifting. It encompasses a 7 bp shifty sequence and a tertiary structure termed pseudoknot, which is fixed by two regions of sequence complementarity (stems). In the shifty sequence the different reading frames are underlined and indicated by numbers.

2.4 Co- and Post-Translational Events

2.4.1 Protein Folding and Assembly

Proteins are unfolded when they are synthesized in the endoplasmic reticulum or cytosol. The folding process begins immediately and is completed within minutes. The folding process involves three steps which can be distinguished by means of NMR, circular dichroism, and hydrophobic dye exclusion. In the first step, regions of local secondary structures are formed. This is followed by collapse into a molten globule state, in which a phase of loose tertiary structure exists. Finally, formation of the finite tertiary structure of the native protein occurs. The first two steps are rapid, while the last step takes several minutes (described in

PTITSYN, 1991). The information which is necessary to complete the three-dimensional structure is fully encoded in the amino acid sequence. Nevertheless, folding into the final structure or stabilization of intermediate forms is assisted by a set of proteins. Folding of proteins which are synthesized in the ER involves several ER-resident helpers including BiP (Ig heavy chain binding protein) which is identical with grp78 (glucose regulated protein), protein disulfide isomerase, and proline *cis-trans*-isomerase. All of these accelerate the folding process. On the contrary, proteins synthesized in the cytosol, are somehow partially prevented from folding by so-called chaperons like the heat-shock protein 70 (hsp70) which maintains these proteins in a translocation-competent state, allowing their protrusion through membranes. In the case of multi-subunit enzymes, folding is followed by assembly of the subunits into a quaternary structure. Assembly may occur in the ER but can take place in other compartments as well. Correct folding and complete assembly of subunits (in the ER) is a prerequisite to achieve transport competence for some proteins (e.g., GETHING et al., 1986; GETHING and SAMBROOK, 1992). Misfolded and misassembled proteins formed as a result of mutations or lack of glycosylation are retained within the ER as aggregates, which are often strongly associated with BiP. Overexpression of recombinant proteins can result in a partial retardation of secretion. This was circumvented in one experiment by blocking expression of BiP with antisense RNA (DORNER et al., 1988). The fate of misfolded proteins is degradation within the ER or transport into degradative compartments.

Unassembled, correctly folded subunits may remain in the ER until they have engaged their partners for assembly (a form of transport retention which does not necessarily involve BiP). Although the binding of BiP to IgG heavy chains has been demonstrated, this is not generally true for other multi-subunit proteins (HAAS and WABL, 1983; BOLE et al., 1986). And, although oligomerization may start if one partner is in the nascent state of synthesis, as has been demonstrated for IgG assembly (BERGMAN and KÜHL, 1979), post-translational assembly through fortuitous collision is more common.

2.4.2 Protein Transport and Sorting

Animal cells represent complex factories with a huge number of compartments in which functions are specialized. Transport and sorting of proteins into these compartments must somehow be directed. Protein transport and sorting are functions of the protein structure and are supported by specific cellular factors. The necessary information for delivery of proteins resides in the protein structure (such as amino acid sequence, post-translational modification, three-dimensional structure).

The sequences which determine protein localization can be subdivided into four groups (BLOBEL, 1980). First, signal sequences provide unidirectional transport into or across membranes. Second, stop transfer sequences block complete translocation across membranes which is necessary for transmembrane protein anchoring. Third, membrane insertion sequences mediate insertion of sequences into the membrane without translocation. Finally, sorting signals direct transport of proteins to their final destination.

Newly synthesized proteins are delivered by two routes to their final destination (Fig. 19). Route 1 is the secretory pathway and includes synthesis of the growing peptide chain in the ER and unidirectional, vesicular transport to intermediate organelles and the Golgi apparatus. From there proteins are delivered into secretory vesicles or vesicles with lysosomal destination. Route 2 encompasses synthesis of proteins on free ribosomes into the cytoplasm and delivery for organelles like mitochondria or peroxisomes and the nucleus.

The process of translocation for route 1 proteins is well investigated and has been reviewed previously (e.g., WALTER and LINGAPPA, 1986). In brief, a factor termed signal recognition particle (SRP), which is built of 7S RNA and six proteins, binds to the signal sequence of the growing peptide chain when it protrudes into the cytoplasm, thereby halting further synthesis of the protein until the ribosome SRP complex has bound to the docking protein (SRP receptor) at the ER membrane. Insertion of the signal sequence, then, presumably occurs by adopting a helical hairpin structure with the amino terminus which faces the cytosol (Fig. 20) (ENGELMAN and STEITZ,

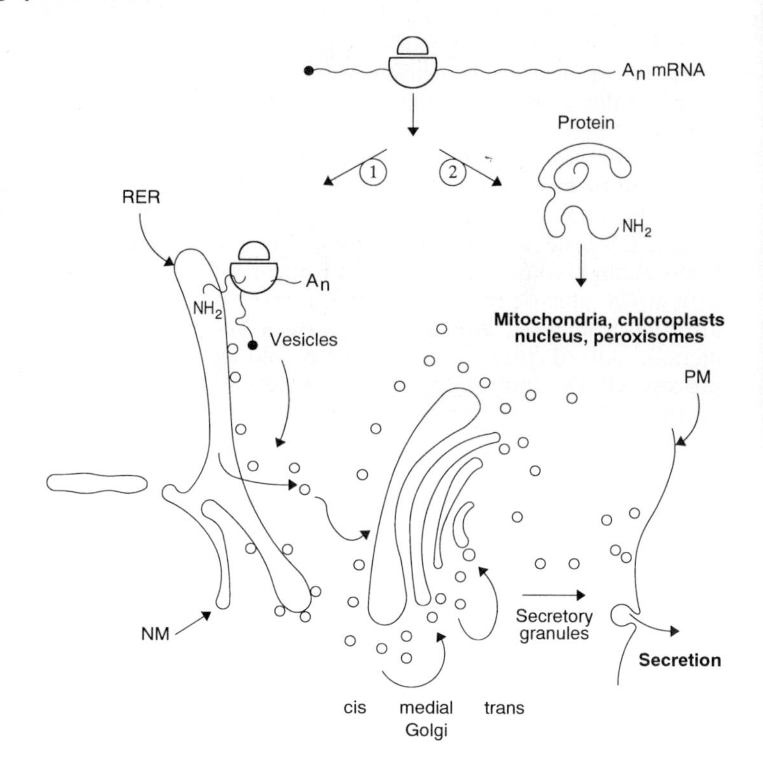

Fig. 19. Protein transport in eukaryotic cells. The two routes of transport are indicated by numbers as described in the text. Abbreviations: NM, nuclear membrane; PM, plasma membrane; RER, rough endoplasmic reticulum.

1981; SHAW et al., 1988). Restarting protein synthesis allows the movement of the growing peptide chain into the lumen of the ER. A membrane resident protease then cleaves off the signal sequence.

Cleavable signal sequences for ER translocation are generally located at the amino terminus of the protein. However, functional internal, uncleavable signal sequences, which remain part of the protein after its translocation, are also documented. Examples comprise ovalbumin, bovine opsin, coronavirus matrix proteins, and influenza virus neuraminidase.

Comparison of numerous eukaryotic signal peptides yielded no obvious homology on the level of amino acid sequence but revealed that a common overall structure prevails (VON HEIJNE 1983, 1984, 1985, 1986b). Signal peptides encompass 13–30 amino acids. They usually have a net positive charge at the amino terminus, a core of at least 9 hydrophobic amino acids, and a region for cleavage by the signal peptidase which is characterized by small-sized residues at positions −1 and −3 (with respect to the cleavage site). Signal sequences are functionally independent as they work in different contexts, e.g., if connected to a heterologous protein body. However, cleavage may be impaired in special cases.

It is possible to deviate proteins, which are normally cytoplasmically localized, to the secretory pathway by addition of a respective signal sequence. However, signal sequences differ in their potential for translocation. The tPA leader peptide has been assumed to be highly efficient for translocation (BURKE et al., 1986). To date, extensive *in vivo* comparisons which addressed this question have not been published. Nevertheless, measurement of the insertion capability of different leader peptides by physical means made possible an estimation for bacterial signal peptides (BRUCH et al., 1989). Astonishingly, substitution of the core region at the DNA level by random human cDNA fragments resulted in hybrid signal peptides which directed a reporter protein into the ER in 20% of the investigated cases (KAISER et al., 1987; KAISER and BOTSTEIN, 1990). Functional hybrid leader peptides often contained a hydrophobic core region.

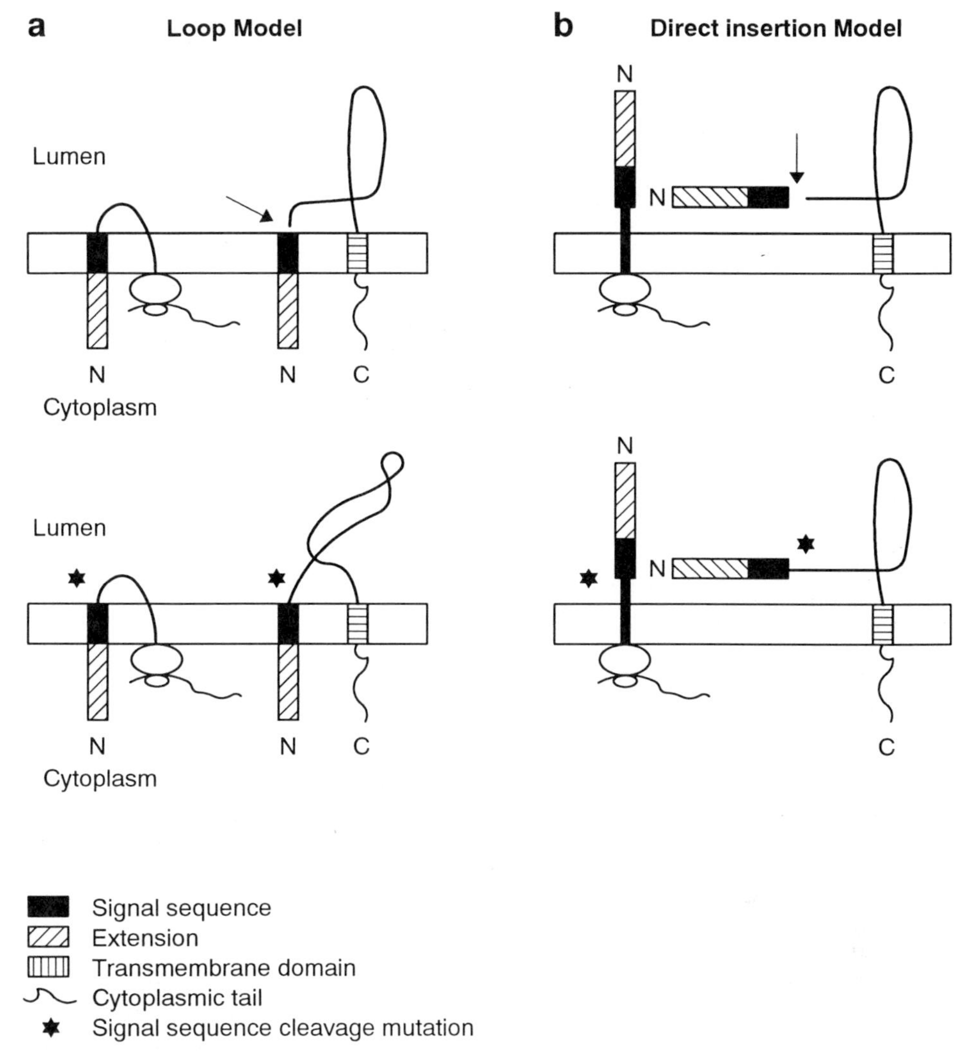

Fig. 20. Evidence for a looped insertion of the signal sequence into the ER membrane. Schematic drawing showing the predicted orientation and localization of the signal sequence according to the loop model (a) (ENGELMAN and STEITZ, 1981) and the direct insertion model (b). To detect the signal sequence SHAW et al. (1988) extended the signal sequence of the transmembranal VSV-G protein amino-terminally by a neutral peptide sequence using a vector for *in vitro* RNA transcription. *In vitro* translation experiments in the presence of dog pancreas microsomes suggested a looped insertion of the signal sequence into those membranes, as only a cytoplasmic orientation was found for the amino-terminal peptide. The arrow indicates cleavage by a cellular protease. Results were confirmed in a second experiment in which a mutation was introduced into the cleavage signal that prevented the separation of the signal sequence and the extension.

In addition to the signal sequence for ER translocation, integral membrane proteins usually contain a membrane anchor domain built of hydrophobic residues which signals a stop transfer and leads to residence of the protein in the membrane. Integral membrane proteins may have one or more transmembrane domains with different or equally orientated

N- and C-termini (for topological nomenclature see GAROFF, 1985). The topology of integral membrane proteins is determined by the cleavability of the signal sequence and the location of stop transfer signals. A well investigated example for an integral membrane protein is influenza hemagglutinin, which has an N-terminal signal peptide and a transmembrane domain near the C-terminus. This configuration determines the N-terminus to face the lumen of the ER (which at a later state becomes the outside of the cell) and the C-terminus to face the cytoplasm (which remains inside of the cell). Another influenza protein, neuraminidase, has a non-cleaved signal sequence and a membrane anchor near the C-terminus which leads to the inverse configuration in the membrane.

Several eukaryotic proteins span the membrane several times (opsin, ion transporters, amino acid permeases, certain signal receptors). The topology of such proteins is explained by sequential utilization of signal and stop transfer signals; e.g., bovine opsin, which spans the membrane seven times, has four signal sequences each followed by a membrane anchor serving as a stop transfer signal (FRIEDLANDER and BLOBEL, 1985). Due to context independence of the membrane anchor, secretory proteins can be genetically engineered into plasma membrane proteins by addition of a membrane anchor. *Vice versa,* membrane anchor proteins might be converted into secreted forms. An example for a membrane protein which has been artificially converted into a secretory protein is the soluble CD4 protein, which is misused in its transmembrane form as a receptor for entry of the HIV virus. The soluble protein inhibits HIV infection.

The IgM molecule is a protein which naturally occurs in membrane- and secreted forms. The synthesis of both forms is developmentally regulated. Alternative splicing determines whether or not an exon coding for the membrane anchor is utilized.

Although translocations are co-translational events, growing evidence is presented from yeast and mammalian experiments that post-translational insertion into the ER is possible, as it has been found for some bacterial membrane proteins.

Signal sequences for route 2 proteins differ from the signal sequences for route 1 proteins. The signal sequence for the outer mitochondrial membrane usually has a high number of basic and hydroxylated amino acid residues and forms amphiphilic alpha-helices (VON HEIJNE, 1986a). Translocation of route 2 proteins is post-translational, and protein unfolding has been shown to be required for transport into the mitochondrial outer membrane (EILERS and SCHATZ, 1986). For the hierarchical order of signals directing proteins into other mitochondrial compartments, consult a recent review (e.g., HARTL and NEUPERT, 1990).

2.4.2.1 Sorting Signals

Constitutive secretion is the 'default route' of the secretory pathway and requires no specific sorting signal for translocation through the ER, except the signal peptide. Most striking evidence for a bulk flow came from a membrane-permeable acyltripeptide Asn-Tyr-Thr, which contains an N-glycosylation recognition site. This peptide entered the ER and was glycosylated and secreted within 10 minutes (WIELAND et al., 1987). These data and other experiments suggested that movement of proteins along the secretory pathway is non-selective and unidirectional. Only deviations from this pathway or retention in a specific compartment are signal-mediated. Transport along the secretory pathway has been investigated extensively, as it can be monitored easily by following the sequential changes of oligosaccharide patterns on transported proteins. A compilation of sorting signals for route 1 and also route 2 proteins is depicted in Tab. 12.

In contrast to constitutive secretion, regulated secretion is observed in a set of specialized cells. Therein, proteins destined for regulated secretion are released from dense secretory granules upon a specific stimulus. The mechanism for delivery in secretory granules is not known, but such secretory proteins accumulate in a late Golgi compartment where a specific receptor is postulated to mediate the transport (reviewed in BURGESS and KELLY, 1987). Expression of genes for constitutively secreted proteins, when cloned into specialized

Tab. 12. Compilation of Sorting Signals for Higher Eukaryotes

Destination	Signal	Example	Literature
Constitutive secretion	Bulk flow, no signal	Interferons, antithrombin III	Reviewed by BURGESS and KELLY (1987)
Storage vesicles for regulated secretion	Not known	Human growth hormone	Reviewed by BURGESS and KELLY (1987)
ER	$KDEL_{COOH}$	BiP	PELHAM (1986)
	$KKXX_{COOH}$ or $KXKXX_{COOH}$	Adenovirus 19K protein HMGCoA	JACKSON et al. (1989)
Golgi	Contained in a specific transmembrane region	Coronavirus matrix protein	MAYER et al. (1988) MACHAMER and ROSE (1987)
Lysosomes	Man-6-phosphate	Lysosomal hydrolases	REITMAN and KORNBERG (1981)
Polarized transport:	Possibly GPI anchoring not known	Thyl and many other GPI-anchored proteins,	Reviewed in LISANTI and RODRIGUEZ-BOULAN (1990)
apical		VSV G protein	
basolateral		HA influenza	
Nucleus	KRPRP	Adenovirus E1A	LYONS et al. (1987)
	PKK^{128a}KRKV	SV40 late T antigen	KALDERON et al. (1984)
	VSRK^{192}RPRP	Polyoma large T	RICHARDSON (1986)
	PPKK^{282}ARED	Polyoma large T	
	RRNRRRRW	Rev	MALIM et al. (1989)
	GRKK R	Tat	COCHRANE et al. (1990)
	KR PAA TKKAGQA KKKKL	Nucleoplasmin (*Xenopus*)	DINGWALL and LASKEY (1991)
	KRKTEEESPLKD KDAKKS	N1 (*Xenopus*)	DINGWALL and LASKEY (1991)
	KR ALPNNTSSS PQ KKK	P53 (human)	DINGWALL and LASKEY (1991)
Peroxisomes	SKL-COOH	Luciferase acyl coenzyme A oxidase	KELLER et al. (1987) MIYAZAWA et al. (1989)
Mitochondria	Leader peptide for mitochondria membrane translocation	CoXIV (cytochrome C oxidase subunit IV)	HURT et al. (1984)

[a] number indicates the position of the most essential amino acid

secretory cells (e.g., pituitary cells AtT20, GH4) which exhibit regulated secretion results in constitutive secretion of the respective protein. However, proteins destined for regulated secretion follow the constitutive pathway in normal fibroblasts. The mechanisms underlying these phenomena have not yet been elucidated. Interestingly, transport vesicles, which are destined for lysosomes and secretory granules, differ in their composition of coat proteins from vesicles for constitutive secretion.

Among the known sorting signals the signal for lysosomal delivery was first discovered. In the Golgi, lysosomal proteins acquire a mannose-6-phosphate residue, which results from addition of a blocked GlcNac to a trimmed mannose-core-glycan and subsequent removal of the GlcNac-moiety. Transport is mediated by a receptor located in an early Golgi compartment.

Polarized transport first was investigated by means of virus infections, as it had been noted that viruses differ in their ability to bud from membranes of polarized cells. Exit of influenza virus was observed from the basolateral membrane of such cells, while VSV used the apical side. Budding is determined by the localization of influenza virus envelope proteins, which exhibit an opposite transport behavior. Despite intensive investigations the mechanism of polarized transport is not understood. It has been proposed that one pathway is the bulk pathway while the other is signal-mediated. Recently, it has been shown that GPI-anchoring of membrane proteins is a signal for apical transport (reviewed in LISANTI and RODRIGUEZ-BOULAN, 1990; see, e.g., BROWN and ROSE, 1992).

Since the constitutive secretion runs by bulk flow, the maintenance of proteins in specific compartments must occur by specific retention signals. Two such signals have been identified for retention in the endoplasmic reticulum, each residing at the carboxy terminus of either soluble or transmembrane ER proteins. The first signal was identified when it was determined that certain soluble ER proteins, such as grp78, protein disulfide isomerase, calreticulin, and grp94, share a common carboxy-terminal sequence KDEL (MUNRO and PELHAM, 1986). Deletion of this signal caused slow secretion of the respective proteins, while addition of KDEL to the carboxy terminus of proteins destined for secretion or transport into lysosomes resulted in their retention (MUNRO and PELHAM, 1986; PELHAM, 1988).

KDEL proteins are abundant in the ER, and it was questionable whether binding by an ER transmembrane receptor could account for the retention mechanism. A post-ER salvage compartment was therefore postulated and has been found to be localized in tubules and vesicles grouped around the Golgi. Escaped KDEL proteins are recycled from this compartment into ER, presumably mediated by divalent cation dependent receptor binding. The KDEL receptor, a 72 kDa protein, has been localized to the salvage compartment by antiidiotype technology (VAUX et al., 1990).

The second ER retention signal was identified by comparison of ER membrane proteins and expression of hybrid constructs. Two lysine residues, 3 and 4 or 5 amino acids apart from the carboxy terminus, were determined to be sufficient for ER localization of these proteins. Examples for this type of signal involve the adenoviral 19K protein and HMG CoA (JACKSON et al., 1990). The mechanism for retention of transmembrane proteins is not clear. Oligomeric assembly and fixation with ER submembrane structures or the matrix are being discussed.

Signals for Golgi residence are not well understood. Initial clues came from investigations of coronaviruses IBV and MHV, which bud into pre-Golgi and early Golgi compartments. Budding is determined by the intracellular localization of their matrix protein, a protein that spans the membrane three times. Transport into the *cis*-Golgi was shown to be directed by the first membrane spanning domain (MACHAMER and ROSE, 1987; MACHAMER et al., 1990; MAYER et al., 1988). In the case of galactosyl-transferase, a *trans*-Golgi membrane protein, it was shown that especially two residues, cysteine and histidine, in the membrane-spanning domain were required for Golgi localization of the protein (AOKI et al., 1992).

Sorting signals for nuclear destination are not homogeneous. One type usually encompasses 5–10 amino acids and has only a clustering of basic residues in common. Representatives include the nuclear localization signal (NLS) of SV40 large T or adenovirus E1A protein. These types of NLS may occur once or twice (e.g., polyomavirus large T) in a protein (reviewed in GARCIA-BUSTOS et al., 1991). The other type of NLS encompasses a bipartite motif of basic residues and was first found in *Xenopus* nuclear proteins N1 and nucleoplasmin (ROBBINS et al., 1991). To date, the motif has been identified in many other nuclear proteins as well (DINGWALL and LASKEY, 1991). The bipartite NLS is comprised of two basic amino acids, followed by a spacer region of about 10 random amino acids, and then a cluster of amino acids in which three out of the next five amino acids must be basic.

The nuclear localization signals are not located at specific sites within the protein. Translocation into the nucleus does not involve a translocation through a membrane, but occurs by a unidirectional transport through nuclear pores. One or more receptors reside at the nuclear pore which recognize proteins with nuclear destination. Internalization is then mediated by a transporter which is part of the constituents of the nuclear pore (nuclear pore complex). The possibility of specific retention in the nucleus could not been verified. The nuclear sorting signal is not processed, a feature that allows nuclear proteins to be relocalized into these compartments after loss of nuclear structures in mitosis.

The elucidation of protein transport signals does not only satisfy a basic research interest of cell biologists, but has an important impact on the biotechnology of animal cells. The knowledge of transport signals allows the tailoring of protein production to specific cellular compartments where specific functions are exhibited. An example is the anchoring of the HIV receptor CD4 in the ER by insertion of the retention signal from the adenovirus 19K protein. The resulting retention of CD4 in the ER leads to capturing of HIV envelope proteins and therefore impairs the HIV replication cycle. The retainment of cell surface molecules was first discovered during the study of adenovirus replication. The 19K protein retains MHC class I antigens and thereby impairs the natural immune response supporting the propagation of adenovirus. A similar application involves the intracellular fixation of specific antibodies in order to retain their antigen within intracellular compartments. Many questions remain, e.g., what determines the rate of constitutive secretion for which $t_{1/2}$ value are known to vary from 30 minutes to as long as 3 hours.

2.4.3 Co- and Post-Translational Modifications

Covalent modifications of amino acids during or after translation are numerous. However, the influence of these modifications on the property of a certain protein is not always evident at the first glance, and some features still

Tab. 13. Some Co- and Post-Translational Modifications of Proteins and their Biological Relevance

Modification	Function/Involvement
N-Glycosylation	Protein structure Biological activity Transport out of ER Clearance (blood) Antigenicity Solubility
O-Glycosylation	Protein conformation Stability Clearance Solubility
Mannose phosphorylation	Signal for transport into lysosomes
Proteolytic processing	Biological activity Protein assembly
Acylation	Membrane association Virus assembly Protein-protein interaction
GPI-addition	Membrane anchoring Transport in polarized cells
Sulfation	
γ-Carboxylation	Calcium binding Biological activity
β-Hydroxylation	Biological activity

Tab. 14. Subcellular Localization of Post-Translational Modifications

Modification Reaction	Organelle	Type (Modification→Target)	Literature
N-Linked glycosylation: Dolichol precursor addition Glucose trimming Removal of single mannose Mannose trimming Terminal sugar transfer (GlcNAc, gal, sialyl, fucosyl)	ER Er ER ER Golgi Golgi	Asn-X-Ser(Thr) ↑ GlcNAc ↕ R	Reviewed in HIRSCHBERG and SNIDER (1987)
O-Linked glycosylation	Golgi, transitional elements	R-Gal-Nac→O-Ser/Thr GlcNac→O-Ser/Thr	Reviewed in HART et al. (1988)
Mannose-6-phosphate signal addition for lysosomal enzymes	Golgi	Phosphate→Man	REITMAN and KRONBERG (1981)
Proteolytic processing	Golgi, PM, granules	Proteins are cleaved into functional subunits at specific sites	Reviewed in NEURATH (1989)
Fatty acid acylation	Post-translational ER, cotranslational Golgi, PM	Palmitoyl (C16) →Cys Myristoyl (C14)→Gly	Reviewed in MCKINNEY (1990)
GPI-addition	ER	Glycophosphatidyl inositol	
Sulfation		Sulfate→O-tyrosine	BAEUERLE and HUTTNER (1987)
β-Hydroxylation		Hydroxyl→Asn, Asp	STENFLO et al. (1989)
γ-Carboxylation	Microsomes	Carboxyl→Glu	SUTTIE (1985)

remain obscure. Modifications may alter a protein by changing, e.g., its tertiary structure, charge, or hydrophilicity. There is no general rule for the effects triggered by a certain modification on an individual protein. Tab. 13 lists some protein modifications and their effects. The effects are often no more than an estimation which results from the evaluation of reported data.

The most prominent modification in mammalian cells concerns glycosylation, which, therefore, will be described in more detail. Features of other modifications such as acylation, sulfation, γ-carboxylation etc. are summarized in Tab. 14.

Acylation is the attachment of fatty acid residues at certain amino acids of a protein. There are distinct classes of acylated proteins distinguishable by the nature of the fatty acid incorporated and its acyl linkage to the protein. Palmitate (C16) is bound by thioester linkage predominantly to a cysteine residue, while myristate (C14) is exclusively linked through an amide bond to an N-terminal glycine (Tab. 14). Both forms can be distinguished after labeling with ^{3}H-fatty acid by the difference in sensitivity of the linkage to alkali or hydroxylamine.

Regarding the biological function, the fatty acylation is considered to provide the protein with a lipophilic handle for membrane anchoring or for association with proteins in membranes. However, the appearance of soluble, acylated proteins and, e.g., the requirement of myristylation of p60^{src} to induce transformation (KAMPS et al., 1985) shed a different light on this modification. The possibility of other functions came from evidence that certain viruses, such as picorna- or retroviruses, require acylation of specific proteins for virus assembly which, in turn, may be specifically inhibited by fatty acid analogs without harmful effects on the cell (BRYANT et al., 1989; HEUCKEROTH and GORDON, 1989).

Several proteins involved in hemostasis, such as factors II, VIII, IX, and protein C, exhibit a set of post-translational modifications including γ-carboxylation, β-hydroxylation, and proteolytic processing. In general, deficiency in one of these modifications severely impairs the biological activity of these proteins. Glutamate residues are γ-carboxylated by a vitamin K-dependent microsomal carboxylase. The enzyme responsible for β-hydroxylation, which modifies aspartate and asparagine residues, has not yet been identified. Liver, kidney, and bone cells are known to harbor γ-carboxyglutamate (gla) containing proteins and β-hydroxyaspartate modified proteins are found in the kidney. The limited distribution of the enzymes responsible for these modifications has impact on the choice of animal cell lines for use in synthesis of the above mentioned blood proteins. It has been shown that upon gene transfer of the respective genes, certain fibroblasts like hamster CHO, mouse C127, and mouse L cells were not able to produce functionally active forms of the vitamin K-dependent proteins in sufficient quantities. In contrast, two kidney cell lines, human 293 and hamster BHK, performed the modifications correctly and substantially.

Among the co- and post-translational modifications, glycosylation of proteins has attracted the most attention. Glycosylation involves the modification of single residues with carbohydrate structures. Two types of glycosylation can be distinguished with respect to carbohydrate moiety, time point of addition, linkage to the protein, and target amino acids.

N-glycosylation of proteins involves the en-bloc, cotranslational transfer of a dolichol-carbohydrate precursor to an asparagine via N-linkage of GlcNAc. The asparagine residue is contained within the discrete recognition motif Asn-X-Ser(Thr), which must be freely accessible for attachment of the sugar moiety. During transport of the protein from ER to Golgi, numerous processing steps are performed involving addition and removal of certain residues in the carbohydrate chain. Processing involves removal of glucose, trimming of the mannose-rich ER-precursor structures and addition of sugars to the core structure. The trimming procedures are carried out by enzymes residing in individual sectors of the ER and Golgi apparatus. Investigation of the glycosylation steps has been simplified, as a huge number of inhibitors of the modifying enzymes are known (Fig. 21) and specific glycosidases are available. Although N-glycans differ considerably in their final structures, all exhibit a common core structure constructed of $Man_3GlcNAc_2$.

In contrast to the N-glycan addition, O-glycosylation of proteins is a post-translational modification, which is carried out sequentially along the late ER (transitional elements) and Golgi. O-glycans are often found as clusters coupled via O-linkage of GalNAc to the hydroxyl group of adjacent serines or threonines. Interestingly, proline often resides in the carboxy-terminal neighborhood of such O-glycosylated serine and threonine residues (Conradt et al., 1990). The O-linked sugars can be distinguished from N-linked by their sensitivity to alkali or O-glycanase. O-glycans differ in their terminal sugar structures, and the only common residue shared by O-glycans in the structure is the GalNAc. Recently, a new type of O-glycosylation has been identified in certain cytosolic and nuclear proteins which consists of a single GlcNAc O-linked to serine or threonine (Hart et al., 1988).

Sequence diversity in both types of sugars results from the addition of repeats of Gal-β(1,4)-GlucNAc-β(1,3) to the initial core structures and decoration of these units with a variety of terminal sugars of which neuraminic acid is the most prominent.

Variation in the sugar moiety of a glycoprotein is clearly cell-type-specific and developmentally regulated, and heterogeneity is known to result from the differential expression of the various glycosyl transferase genes in a given cell type. Data on the potential glycosylation of a limited number of proteins from different cell lines resulting from investigation are summarized in Conradt et al., 1990.

The influences of carbohydrates on the function of a glycoprotein are numerous. However, it is not easy to delineate general rules, and the role of the sugar moiety of a glycoprotein should be considered case-by-case. For pharmaceutical recombinant proteins, which are delivered via the blood, it is important to ensure that a sialic acid constitutes the

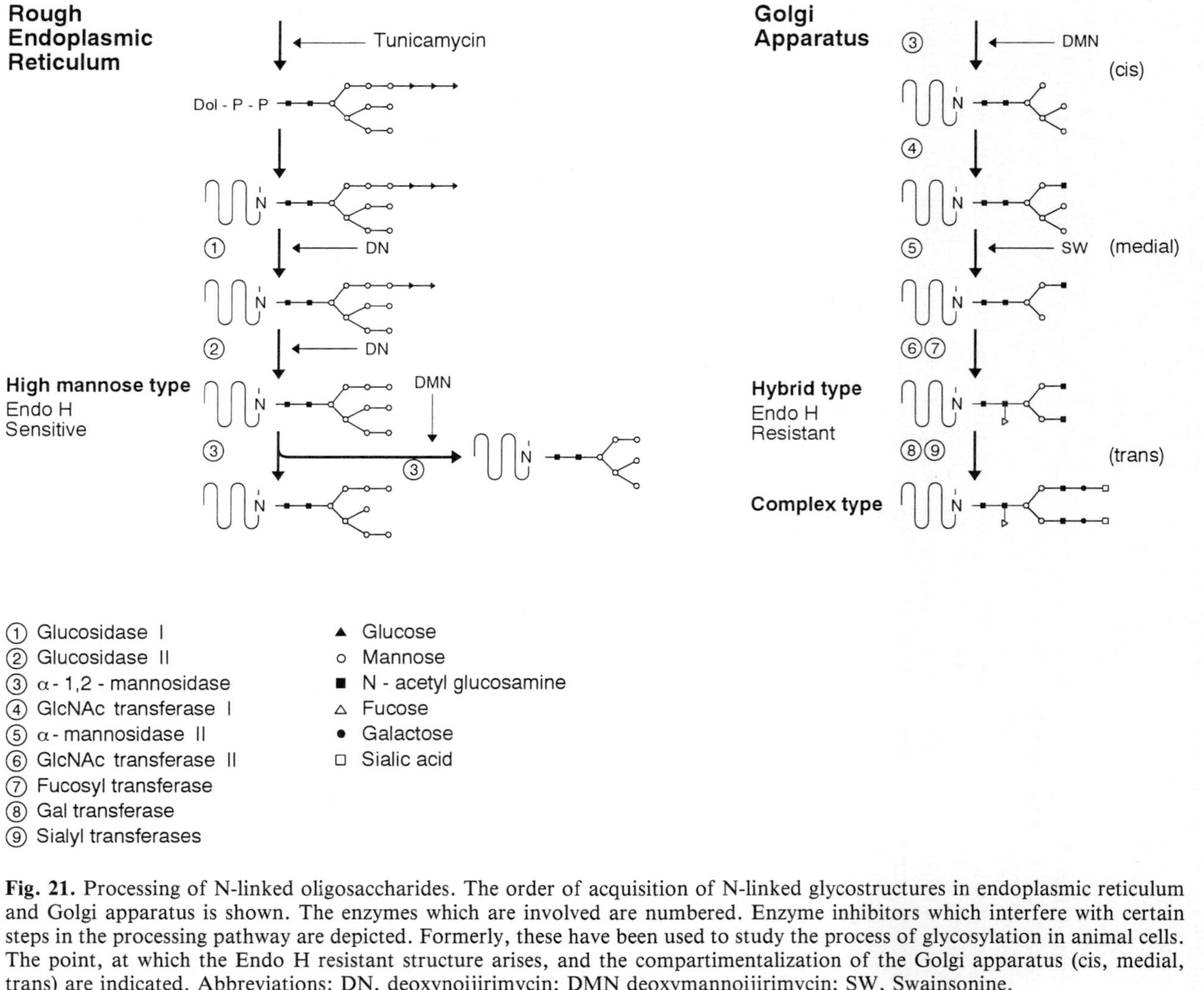

Fig. 21. Processing of N-linked oligosaccharides. The order of acquisition of N-linked glycostructures in endoplasmic reticulum and Golgi apparatus is shown. The enzymes which are involved are numbered. Enzyme inhibitors which interfere with certain steps in the processing pathway are depicted. Formerly, these have been used to study the process of glycosylation in animal cells. The point, at which the Endo H resistant structure arises, and the compartimentalization of the Golgi apparatus (cis, medial, trans) are indicated. Abbreviations: DN, deoxynoijirimycin; DMN deoxymannoijirimycin; SW, Swainsonine.

terminal sugar of the glycan structure. Sialic acid has a masking function, as it prevents clearance from circulation by asialo- and asialogalacto-protein receptors. This favors production of such proteins in mammalian cells, which efficiently perform sialylation, and excludes production in insect cells or yeast production systems, since the sialic acid deprived sugars of proteins generated in these systems would be cleared rapidly *in vivo*.

Glycosylation is known to stabilize the conformation of a protein, and in the case of cotranslational N-glycosylation, it facilitates the folding of a newly synthesized protein. This is demonstrated by the fact that inhibition of N-glycosylation by tunicamycin treatment results in association of newly synthesized protein with BiP, which is indicative for a malfolding.

Glycosylation can affect the biological activity of a protein. This topic has attracted much attention because of the fact that cells differ in their ability to glycolysate a protein, and this results in a frustrating microheterogeneity of a glycoprotein sugar moiety. However, the influence on bioactivity is not distinct, and in many cases there is no effect at all. Nevertheless, changes in solubility or antigenicity may result from even subtle changes in the carbohydrate moiety and should, therefore, not be underestimated.

The role of O-linked carbohydrates in the function of glycoproteins is still suspect. However, O-linked sugars may contribute to the stability of a protein. Expression of an LDL receptor, for example, in a mutant CHO cell line deficient in the enzyme glucose 4-epimerase, which renders O-glycosylation dependent on the addition of N-GalNAc, results in LDL receptors that are properly transported to the plasma membrane, but which are rapidly degraded (KINGSLEY, 1986). O-linked sugars may provide charge and water-binding. Highly O-glycosylated cellular membrane proteins, such as mucins, leukosialin or LDL receptors, are thought to exist as sugar-mediated, stiff, and extended structures, which protrude from the cell's surface and may support their function as recognition signals.

Carbohydrate tailoring is a desirable goal in genetic engineering of animal cells. Certain progress was made in this field when the genes for sugar modifying enzymes became available. This enabled geneticists to equip special cells, which are valuable for production purposes, with modifying enzymes normally not present in these cells (e.g., the enzyme for NeuAc-2,6-Gal modification in CHO cells; LEE et al., 1989). To minimize microheterogeneity, numerous CHO-glycosylation mutants have been generated which exhibit a defect in one or more steps of carbohydrate biosynthesis (STANLEY, 1989). Erythropoietin (EPO) expression in the CHO mutant Lec 3.2.8.1 may serve as an example. EPO is a glycoprotein with three N-linked lactosamine units and one O-linked sugar which are all essential for bioactivity. Lec 3.2.8.1 derived EPO bears Man5GlcNAc2 units at the three N-glycosylation sites and a GalNAc at the single O-glycosylation site, thus eliminating the heterogeneity observed in wild-type CHO cells. Due to the N-glycan moiety, purification by ConA-Sepharose is easily accomplished (STANLEY, 1991).

2.5 Protein Turnover

The overall concentration of a protein in a cell is a function of its rate of synthesis and degradation. In the animal cell several pathways of degradation have been found. Among these, the ubiquitin-dependent pathway has been intensively investigated. Steps in the ubiquitin-mediated type of degradation are marking of proteins by enzymatic covalent attachment of ubiquitin molecules at the protein's lysine residues and degradation of the tagged protein by a multi-subunit protease concomitant with the release of ubiquitin. Other degradative pathways involve calcium-dependent proteases or lysomal proteases. Degradation also is performed by ER- or Golgi-resident proteases. Lysosomal degradation of protein occurs following cell lysis or with turnover of extracellular or membrane bound proteins that enter via receptor-mediated endocytosis, and is also the proposed pathway for proteins that are not digested in an ATP and/or ubiquitin-dependent manner.

At the protein level, several signals have been identified which determine the stability of a protein in a cell. Evidence has accumulated

Tab. 15. Compilation of Aminoterminal Amino Acids Conferring Stability/Instability to Proteins in Mammalian Cells

N-end Rule[a]	
Destabilizing Amino Acid[b]	Stabilizing Amino Acid
Arg	Gly
His	Val
Phe	Met
Leu	Pro
Trp	
Tyr	
Ala	
Ser	
Thr	

[a] as obtained from degradation experiments of ubiquitin-x-β-Gal fusion proteins in rabbit reticulocyte lysates.

[b] only 'primary' destabilizing residues are shown. For more details, see GONDA et al. (1989).

in the last few years that the nature of the N-terminal amino acid influences protein degradations. In a series of experiments with mutant β-galactosidases modified with ubiquitin at the N-terminus, it was found that certain residues at the amino terminus exhibited a destabilizing function in yeast and higher eukaryotes (Tab. 15) (BACHMAIR et al., 1986; BACHMAIR and VARSHAVSKY, 1989; GONDA et al., 1989). In a number of cases, accumulation of the amino acids proline, glutamic acid, serine, and threonine, when flanked by basic residues (so-called PEST sequences) is correlated with a short-life time of a protein (ROGERS et al., 1986; GHODA et al., 1989). Five-residue motifs have been found in other proteins, which assist in transfer of proteins to the lysosome (DICE, 1987). Interestingly, N-terminal protein modifications like N-terminal acylation of a protein may, on the other hand, increase stability.

3 Manipulation of Animal Cells

3.1 Gene Transfer Methods, Selection, and Gene Amplification

3.1.1 Cells and Cell Lines

The choice of host cells very much depends upon the purpose of the study or on questions that need to be answered (see Tab. 1). The following discussion will describe the rationales for the choice of a host cell which will achieve high expression of heterologous gene products with defined biochemical and biological parameters.

Theoretically, any cell which is maintained in tissue culture can be manipulated to express foreign genes. However, depending upon the cell type selected, the efficiencies of gene transfer, expression, and stability will differ considerably.

Transient expression is a method which rapidly yields high concentrations of the expressed products. Transient expression systems are, therefore, convenient to check the integrity of the expression construct and some properties of the product derived thereof. Provided that only small amounts are required, transient expression allows one to analyze the nature of the gene product. Transient expression in COS cells is a commonly used system which produces rapid results (Sect. 3.2.2.2). However, vectors with specific properties are required to obtain optimal expression. Other host cell lines which lead to very efficient transient expression, but do not require specific features of the plasmid expression vector, include human 293 and hamster BHK-21 cells (Tab. 16). In both of these hosts efficient transfer and expression, but not episomal replication, are responsible for high expression of the exogenous DNA.

If more gene product is needed or continuous production is required, a selection of *stably transfected* cells with high-level production has to be performed. For high-level expression of a foreign gene (>> 1 μg/10^6 cells/24 h), only a limited number of suitable cell lines is

Tab. 16. Host Cell Lines

Cell Line	Species Tissue[a]	Adherence	Transfection Method	Specific Features	Prerequisites for High Expression	Long-term Production	References
CHO dhfr$^-$	Hamster	+	Calcium phosphate coprecipitation	dhfr$^-$	Gene amplification	+	URLAUB and CHASIN (1980), KAUFMAN and SHARP (1982)
BHK-21	Hamster	+	Calcium phosphate coprecipitation	Efficient transient and stable expression	Gene amplification is not required	+	WIRTH et al. (1988)
C127	Mouse	+	Calcium phosphate coprecipitation	Replication of BPV-vectors is possible	BPV-based episomaly replicating vectors	+	LOWY et al. (1980a), SARVER et al. (1981)
Myeloma, e.g. SP2/0, 3558L, P3-X63-Ag8.653	Mouse	–	Lipofection, protoplast fusion, electroporation		Ig-promoter/enhancers are favorable	+	OI et al. (1983)
293	Human	+	Calcium phosphate coprecipitation	Constitutive expression of Adenovirus Ela gene; extremely high transient expression; high expression in stable transferrants	Promoters. which are efficiently *trans*-activated by Ela	+	GRAHAM et al. (1977), WALLS et al. (1989)
MEL	Mouse, erythroleukemia	–	Electroporation		Only after differentiation and the combination of β-globin DCRs with suitable promoters	+	COLLIS et al. (1990)
COS	Monkey	+	Calcium phosphate coprecipitation, DEAE-dextran	Constitutive expression of SV40 TAG. Only transient expression is efficient	Circular plasmids containing the SV40 ori of replication	–	GLUZMAN (1981)
Sf9	*Spodoptera frugiperda*	+	Calcium phosphate coprecipitation, infection		P10 or polyhedrin promoters from Baculovirus are favorable (usually used in Baculovirus-infected cells)	–	SMITH et al. (1983), LUCKOW and SUMMERS (1988)
Schneider 2	*Drosophila*	+	Calcium phosphate coprecipitation		Strong *Drosophila* promoters	+	SCHNEIDER (1969), CULP et al. (1991)

[a] If not stated, the cell line is of epithelial or fibroblast origin.

Tab. 17 Properties of Host Cell Lines for Recombinant Protein Production

Property	Examples
Efficient expression of foreign genes	Transcription and post-transcriptional processing Protein synthesis Secretion
Posttranslational modifications	e.g., Protein folding Glycosylation, phosphorylation
Stability of transgene expression	Genotypic (chromosomal stability) Phenotypic (CpG methylation)
Absence of adventitious agents	Viruses Mycoplasms
Cultivation requirements	Serum, growth factors Stabilizing proteins
Fermentation properties	Growth to high cell densities Resistance to shear forces Aeration processes
Production under fermentation conditions	High production at high cell density Reduced cell growth Limited or zero external protein (serum/growth factors) requirements
Cell growth	Rapid growth at low cell density Reduced or zero growth at high cell density

presently available. The economic large-scale production has further requirements and reduces the number of usable cell lines. In this respect mammalian cells, such as CHO, C127, myeloma, and BHK-21, have been successfully used. Alternatively, baculovirus-infected insect cells could be used for large-scale production. However, this system is presently not available for continuous production. Tab. 17 gives an overview of the properties of an ideal host cell system for biotechnological use.

In principle, the establishment of a single host cell system for protein production is desirable. However, the fact that different cell types give rise to specifically modified proteins from the same gene annuls this idea. Although the different features exhibited by the cell lines concentrate on detailed post-translational modifications, these modifications could be important for application of the respective protein (e.g., WASLEY et al., 1991). In certain cases, therefore, different cell lines must be tested in order to find a host that will perform the appropriate post-translational processing of the produced protein. In this context it should be mentioned that the number of mammalian cell lines currently being used is probably not sufficient to cover future requirements in pharmaceutical protein production.

To overcome this problem large numbers of established cell lines are screened to find those which fulfill the requirements stated in Tab. 17. Improvements over the properties of established cell lines can be achieved by cell fusion or genetic engineering. New properties can be transmitted by the transfer of genes which are required for serum- or protein-free cell growth, specific post-translational modifications or the enhancement of gene expression by additional factors. A step towards this goal has been achieved by bio-engineering cells to become independent of certain growth factors (e.g., WATTS and MACDONALD, 1991). Genetic modifications of the biochemical pathways which are responsible for post-translational modifications may be included in this approach (cf. Sect. 2.4.3).

3.1.2 Non-Viral Gene Transfer Methods

A number of methods have been developed to transfer DNA and other macromolecules into mammalian cells. Apart from the viral transfer methods which will be introduced in Sect. 3.2.3, the available methods are based on different principles:

(1) The DNA is transferred as a precipitate or in charged complexes (calcium phosphate precipitation, DEAE-dextran transfection).
(2) The genetic information is transferred to the target cells after attachment to or packaging into membraneous or lipid vesicles which are able to fuse with the plasma membrane of the recipient cells (protoplast fusion, lipofection, erythrocyte ghost transfer).
(3) Transfer of nucleic acids is accomplished by physically opening the cells (microinjection, laser poring, electroporation).

A number of different factors determines the method which is optimal for the individual case: the vector, the type of cells to be transfected, the number of samples, the duration of expression (transient versus stable), the type of analysis for the gene transfer, and the number of DNA molecules which are expected to remain in the target cells.

A list of usable methods is given in Tab. 18. Since the detailed methods and their application to certain cell types are described in many protocols and reviews (SAMBROOK et al., 1989; KRIEGLER, 1990; KAUFMAN, 1990), the following discussion will focus on the principal features of these methods in order to facilitate the choice.

Tab. 18. Non-Viral Gene Transfer Methods

Methods	Notes	References
Calcium phosphate coprecipitation	For adherent cells; large amounts of DNA per cell are delivered. Most frequently used method.	GRAHAM and VAN DER EB (1973), WIGLER et al. (1977), CHEN and OKAYAMA (1987)
DEAE-dextran	Mostly for suspension cells; large amounts of DNA are delivered.	MCCUTCHAN and PAGANO (1968), SOMPAYRAC and DANNA (1981)
Lipofection	Efficient method; RNA and proteins are also delivered; toxicity of the lipids is limiting.	PAPAHADJOPOULOS et al. (1975), WONG et al. (1980), FELGNER et al. (1987)
Electroporation	Low amount of DNA or single copy integrates are stably transferred to individual cells; advantageous for lymphoid cells.	NEUMANN et al. (1982), CHU et al. (1987)
Protoplast fusion	Frequently used for lymphoid cells. Introduction of total protoplast compounds may be disadvantageous.	SCHAFFNER (1980), RASSOULZADEGAN et al. (1982)
Fusion with erythrocyte ghosts	Nucleic acids and proteins can be delivered.	WIBERG et al. (1986)
Laser poring	High efficiency. Complex and expensive apparatus.	KURATA et al. (1986), TAO et al. (1987)
Microinjection	Complex and expensive set-up. A limited amount of cells can be injected, but with 100% efficiency; nuclear or cytoplasmic targeting is possible.	GRAESSMANN and GRAESSMANN (1976), CAPECCHI (1980)

The *calcium phosphate transfection method* is the most commonly used for transient and stable expression. The DNA is delivered as a coprecipitate with calcium phosphate. These precipitates enter the cells by a phagocytotic process. Many cell types, especially adherent cells, are efficiently transfected by this method. Since large amounts of DNA of different length are taken up by the target cells, chromosomal DNA as well as mixtures of different types of cloned plasmid DNA (cotransfer) have been successfully used (WIGLER et al., 1978). Frequently, high-molecular weight chromosomal DNA is used as a "carrier DNA" for increasing the transfer efficiency. Within the nucleus, the DNA assembles to high-molecular weight forms, including homologous recombination which often results in tandem arrays of the transferred plasmids (FOLGER et al., 1982). These "transgenomes" which consist of plasmid as well as carrier DNA, can integrate into chromosomal DNA of the host cell. The amount of the integrated DNA in stable transformants varies considerably (1–10000 kb/cell) (KUCHERLAPATI and SKOULTCHI, 1984). Several factors determine the efficiency of the calcium phosphate transfection method: DNA form (circular DNA is more active than linear DNA), the form of precipitate which depends on the length and the concentration of the DNA, temperature and pH during transfection (CHEN and OKAYAMA, 1987). Chemical shock treatment of cells (DMSO, glycerol, polyethyleneglycol, sucrose) improves transfection efficiency (FROST and WILLIAMS, 1978).

The *DEAE-dextran transfection method* is simpler to perform than the calcium phosphate precipitation technique. DEAE-dextran is therefore the method of choice for transfer of a large number of DNA samples. The method is very efficient for transient expression, but it can also be used for stable gene transfer (HAUSER, unpublished). However, a higher mutation rate of the transferred DNA is observed in comparison to other methods (CALOS et al., 1983). Although the mechanism of DNA uptake is not fully understood, the fact that the treatment of the transfected cells with chloroquin improves the efficiency of DEAE-dextran transfections, is indicative for the involvement of lysosomes in this process. Alternatively, the application of chemical shock as in calcium phosphate transfection, improves transfection efficiency. The DEAE-dextran method is mainly used for non-adherent cells, but is preferred to the calcium phosphate precipitation method in special cases (e.g., for transient expression in adherent monolayers of COS cells).

The *protoplast fusion* is less frequently used, but has specific advantages and disadvantages. Since the method entails the direct fusion of *E. coli* protoplasts to mammalian cells, it does not require plasmid DNA isolation and is therefore convenient for a rapid check of newly created expression constructs or mutants in transient expression experiments. Secondly, a series of non-adherent cells can be efficiently transfected by this method. The method is not restricted to a specific cell type. Protoplast fusion is not convenient for quantitative determinations, cotransfer of different DNAs, or if a large number of DNA samples are to be transfected. Another disadvantage is that the whole sample of *E. coli* protoplasts is introduced into the mammalian cells, and this might result in adverse or side effects.

In *electroporation* the cells are exposed to short, high-voltage electric fields. This treatment creates transient pores in the cell membrane which allows the passage of DNA into the cells. A part of the DNA sample is obviously transported to the nucleus and a subset thereof integrates into the host chromosomes. Efficient transient, as well as stable transfection, can be obtained with the electroporation method. Interestingly, nearly all cell types can be successfully electroporated, even those which are resistant to other transfection methods. A number of parameters including the geometry of the electrodes, field strength, duration and shape of the pulse, buffer composition, cell concentration, as well as concentration and form of the DNA (linear versus circular) affect gene transfer efficiency and the fate of the DNA; e.g. the amount of stably integrated DNA per cell can be roughly determined by the proper adjustment of these parameters (MIELKE et al., 1990). Additionally, the application of appropriate conditions allows the cotransfer of different genes.

Utilization of synthetic cationic lipids has

greatly improved the efficiency of *lipofection.* Usually, unilamellar lipid vesicles (liposomes) which are loaded with DNA are applied to the cells resulting in attachment and subsequent fusion of positively charged vesicles to negatively charged cell membranes. Again, the mechanism of DNA entry into the nucleus is not known. The method strongly depends on type and composition of the lipids used for liposome preparation and also varies considerably with the recipient cell type. This probably reflects the dependence of lipofection on the composition of the membrane constituents. Newly developed liposomes which form a complex with the DNA rather than trapping the DNA inside allow efficient transient and stable transfection (FELGNER et al., 1987; BEHR et al., 1989). Major disadvantages of this method are the toxicity of the lipids and cell type dependence.

The most effective way of inserting foreign DNA into defined compartments of cells is *microinjection.* Efficiencies of up to 100% means that all of the cells injected with DNA express the foreign genes. However, the method requires expensive equipment and considerable training. Apart from the efficiency, the major advantage is the fact that defined amounts of DNA can be directly introduced into the nucleus. One important application of microinjection is the manipulation of embryonal cells, a prerequisite for the production of transgenic animals. Furthermore, the effects of the transferred DNA on individual cells in culture can be followed microscopically. The number of cells which can be injected is a limiting factor. This is the major disadvantage of microinjection. New automated set-ups are available which allow the injection of more than 1000 cells per hour (PEPPERKOK et al., 1988).

The remaining methods which are mentioned in Tab. 18 do not play a significant role in standard gene transfer, but may be advantageous for specific cell types or applications.

3.1.3 Markers for Selection

In most gene transfer methods the rate for obtaining stable transformants is lower than 10^{-3}. To identify the cells which have taken up transferred DNA, a selection system must be used. Selective systems are usually based on the cointegration of the gene of interest with a selectable marker gene which leads to a new phenotype of the transformant upon expression. Selection for this phenotype is then performed.

Markers for selection are distinguished by different principles of selection. Toxic substances are most frequently used against which the expression of the selectable gene provides resistance enabling the transformants to survive under conditions which kill all other cells (Tab. 19).The first example of this category involves the complementation of a defect in the salvage pathway of mammalian cells. Such a defect is found in thymidine kinase negative (TK^-) cell mutants. While cells are grown in the presence of aminopterin, the *de novo* synthesis of the thymidine is blocked and TK^- cells cannot survive. Expression of a transferred external thymidine kinase gene (e.g., from Herpes simplex virus) is able to restore the viability of the cells provided that the cells are supplied with thymidine and hypoxanthin. Other examples of selection systems which are based on complementation include adenosine phosphoribosyltransferase (APRT) or dihydrofolate reductase (DHFR) negative cells (URLAUB and CHASIN, 1980; LOWY et al., 1980b). Upon use of wild-type selectable markers the application of these systems is restricted to cell lines which are deficient in TK or DHFR or APRT.

With the development of dominant selectable markers, this problem has been overcome. Dominant selectable markers render the cells resistant to substances that are not usually found in the cells. Such markers include prokaryotic genes like the aminoglycoside phosphotransferases from transposon Tn5 in *E. coli* (COLBÈRE-GARAPIN et al., 1981; BLOCHLINGER and DIGGELMANN, 1984). The respective toxic drugs, e.g., G418 and hygromycin B, are inactivated by phosphorylation. The number of available dominant selection markers of this type is steadily growing (Tab. 19). Another group of dominant selectable markers is based on the expression of prokaryotic genes involved in the biosynthesis of amino acids. The expression of these genes allows growth in

Tab. 19. Markers for Selection and Gene Amplification

Gene	Selective Drug	Amplification	References
Thymidine kinase (TK)	Hypoxanthine, aminopterin, thymidine (HAT)	±	COLBÈRE-GARAPIN et al. (1981); ROBERTS and AXEL (1982)
Xanthine-guanine phosphoribosyltransferase (XGPRT)	Xanthine, mycophenolic acid	–	MULLIGAN and BERG (1980)
Dihydrofolate reductase (DHFR)	Methotrexate	+	ALT et al. (1978); SUBRAMANI et al. (1981)
Adenosine phosphoribosyltransferase (APRT)	Adenine, aminopterin, thymidine (HAT)	–	LOWY et al. (1980b); ROBERTS and AXEL (1982)
Hypoxanthine-guanine phosphoribosyltransferase (HGPRT)	Hypoxanthine, aminopterin, thymidine (HAT)	±	ROBERTS and AXEL (1982)
Mutant HGPRT or mutant thymidine kinase			
Neomycin phosphotransferase	G418	–	COLBÈRE-GARAPIN et al. (1981)
Hygromycin phosphotransferase	Hygromycin	–	BLOCHINGER and DIGGELMANN, (1984); BERNARD et al. (1985)
Puromycin acetyltransferase	Puromycin	–	VARA et al. (1986); DE LA LUNA et al. (1988)
Bleomycin resistance	Bleomycin	–	GENILLOUD et al. (1984)
Tryptophan synthetase	TRP^- medium, indol	–	HARTMANN and MULLIGAN (1988)
Histidine aldehydrogenase	HIS^- medium, histidinol	–	HARTMANN and MULLIGAN (1988)
Glutamine synthetase	Methionine sulfoximine	+	YOUNG and RINGOLD (1983)
Metallothionein I	Cadmium	+	BEACH and PALMITER (1981)
Carbamoylphosphate synthetase, aspartate transcarbamylase (CAD)	N-phosphonoacetyl-L-aspartate (PALA)	+	WAHL et al. (1979)
UMP synthetase	Pyrazofurin	+	SUTTLE (1985)
Ornithine decarboxylase	Difluoromethyl ornithine	+	CHIANG and MCCONLOGUE (1988)
Adenosine deaminase	Xyl-A-or adenosine and 2′-deoxycoformycin	+	YEUNG et al. (1983)
Adenylate deaminase	Adenine, azaserine, and coformycin	+	DEBATISSE et al. (1988)
UMP synthetase	6-Azauridine, pyrazofuran	+	KANALAS and SUTTLE (1984)
IMP 5′-dehydrogenase	Mycophenolic acid	+	HUBERMAN et al. (1981)
Thymidylate synthetase	5-Fluorodeoxyuridine	+	ROSSANA et al. (1982)
P-glycoprotein 170	Multiple drugs like colchicine, puromycin, vincristin, cytochalasin B	+	RIORDAN et al. (1985)
Asparagine synthetase	Beta-aspartylhydroxamate or Albizziin	+	ANDRULIS et al. (1983)
Arginosuccinate synthetase	Canavanine	+	SU et al. (1981)
HMG-CoA reductase	Compactin	+	LUSKEY et al. (1984) REYNOLDS et al. (1984)
Na^+, K^+-ATPase	Ouabain	+	PAUW et al. (1986)

media which lack certain amino acids (e.g., HARTMANN and MULLIGAN, 1988).

A second category of selectable markers is based on the morphological transformation of the transfectants. Cells which express DNA encoding oncogenes often change their growth properties, resulting in multi-layer growth of cell colonies (foci). These foci can be seen under the microscope and mechanically isolated. The bovine papillomavirus vectors (BPV) prominently exhibit this type of selection system (LOWY et al., 1980a; SARVER et al., 1981).

A series of other cell properties which can be induced by gene transfer have been successfully used as selectable markers. The expression of surface molecules allows the separation of cells by fluorescence activated cell sorting (FACS) or by panning (KAVATHAS and HERZENBERG, 1983; SEED and ARUFFO, 1987). Furthermore, genes encoding secreted proteins which can be detected by their enzymatic properties or by specific antibodies in overlay assays, can serve as selectable markers (WIRTH et al., 1990). An excellent example of such a marker is secreted alkaline phosphatase (SEAP). This enzyme first diffuses, then adheres onto overlayed nitrocellulose paper, after which it is easily identified by a color reaction linked to the decomposition of an appropriate substrate (WIRTH et al., unpublished). In principle, internally expressed proteins could be used as selectable markers. Lysates from the colonies could be tested by immunological or enzymatic means for the existence of these gene products. However, this approach requires the application of replica methods which are usually difficult to perform with animal cells. Therefore, this method is applied only in very specific cases.

Selectable markers are either included in the recombinant vector or cotransferred. Cotransfer offers the possibility to adjust the amounts of DNAs to be expressed. In this way one can regulate the relative expression of both genes by gene dosage. This is important since the overexpression of some selective markers is detrimental to the cells. Differential expression of the gene of interest and the selective marker on the same plasmid can either be achieved by using promoters of different strength, or by expression of the selective marker as a second cistron from one transcript which contains the gene of interest as a first cistron (bicistronic expression vectors). The latter trick will be further discussed in Sect. 3.1.4. Although a large number of these types of vectors are in use, it should also be mentioned that the integration of the neomycin resistance gene into expression vectors is not advisable, as the neomycin resistance gene exerts a negative effect on adjacent promoters (silencing). This may lead to reduced expression of the gene of interest (ARTELT et al., 1991).

3.1.4 Screening Procedures

Cell clones resulting from stably transfected expression plasmids exhibit marked differences in expression levels of the introduced gene. There are two predominant reasons for these differences: (1) Expression plasmids are maintained in each cell clone with different copy numbers. (2) They are assumed to be randomly distributed over the host genome, meaning that integration has occurred into more or less transcriptionally active sites. Good candidates for a subsequent large-scale production are single clones which exhibit a stable and high-level expression of the foreign gene. For isolation of such a candidate a huge number of single cell clones (usually much greater than 100) need to be screened for high-level expression. Working with a characterized single clone is preferred. Clone mixtures are not homogeneous with respect to either the growth properties of the individual cell clones, or product quality. Specifically, non-producing cells may overgrow the high-level producing cell clones, if the product has adverse effects on cell growth. In a low percentage of cells, rearrangements in the protein coding region may have occurred resulting in expression of an altered protein.

Several methods for screening are currently available. They mainly differ in their time requirements. Direct screening methods which measure an immunological, biochemical, or biological property of the product are preferentially used as they achieve the most accurate results. However, this brute force method of 'picking', clone extension, and assay is extremely time-consuming and unattractive to perform. The *filter immunoassay* has become a simple alternative to this method. It allows the rapid screening of a large number of individual transformants (McCracken and Brown, 1984; Wirth et al., 1989; Walls and Grinnell, 1990). The procedure is based on the secretion of the protein of interest and requires antibodies directed against the protein. Cell clones are overlaid with a protein-binding filter, and the secreted protein is allowed to diffuse to the filter. Protein on the filter is detected by antibodies and color reactions as used in Western blotting. The intensity of the colorspot correlates with expression levels of the cell clones.

Often the lack of simple detection procedures for a protein favors the use of *indirect screening procedures,* which involves the stoichiometric coexpression of a gene product for which screening is easily performed. A straightforward application is the screening for overexpression of cotransferred selectable markers with high concentrations of the selective drug. In most cases the cotransfected selectable marker DNA is ligated together with the DNA of the expression plasmid as concatamers, and then integrated into the chromosomal DNA of the host cell. Selection for high level expression of the selectable marker identifies cell clones with high-level expression of the gene of interest.

To avoid selection of unspecific resistant cells, which are induced by the increased concentration of the selective drug, Wirth et al. introduced a second selectable marker which is simultaneously selected for at basal concentrations of the respective drug. This *'combined selection procedure or double selection procedure'* allows access to cell clone mixtures within 3 weeks expressing up to 100-fold the amount that has been achieved by single selection (Wirth et al., 1988; Page and Sydenham, 1991). Instead of selectable genes, other screening marker genes could be cotransferred. Such screening markers are represented by genes with gene products which can be simply quantified. Examples include secreted proteins detected in a filter assay, such as the secreted form of placental alkaline phosphatase (SEAP) and human growth hormone, or firefly luciferase monitored in an *in situ* luminometric assay from cell clones (Wirth et al., 1991a), cell surface markers to be selected for by FACS analysis (Kavathas and Herzenberg, 1983; Beckers et al., 1988; Wirth et al., 1990).

Physical linkage of selectable markers and screening markers, by including them in one plasmid, is preferred by some researchers. Although this method allows equimolar transfer of both genes, some drawbacks have to be considered. (1) The 1:1 expression may be disadvantageous, as high-level expression of the selectable marker may be harmful to the cell. Cotransfer, in contrast, allows one to regulate the expression of the gene and the marker by simply adjusting the ratio of transfected DNAs. (2) The physical linkage of the screening marker might have negative effects on the expression of the gene of interest. Such is the case for the neomycin resistance gene which exerts a silencing effect on several neighboring promoters (Artelt et al., 1991). (3) An additional DNA cloning step is necessary, since many eukaryotic expression plasmids do not contain a selectable marker which functions in animal cells.

In the procedure reviewed above the gene of interest and the screening marker gene are linked at the DNA level but are translated from two separate mRNAs. A tighter linkage of expression is achieved if the genes are located on one single mRNA. Such vectors are referred to as *bicistronic expression vectors.* Expression vectors carrying such bicistronic transcription units have been used to screen for high levels of foreign gene expression. In these constructs the foreign gene is contained in the first cistron, and a selectable marker gene or a screening marker gene represents the second cistron. Screening for high expression of the second cistron also identifies clones expressing high levels of the first cistron (Fig. 22) (Kaufman et al., 1987; Boel et al., 1987; Wirth et al., 1990; 1991a). Insufficient translation reinitiation could be responsible for realization of the second cistron, which therefore is underexpressed. Expression levels for the second cistron are usually 10- to 200-fold lower when compared to a respective monocistronic expression unit. Due to the low expression of the second cistron, strong selection pressure favors the undesired rearrangement of the transcription unit (Adam et al., 1991; Kaufman et al., 1991). Therefore, the use of

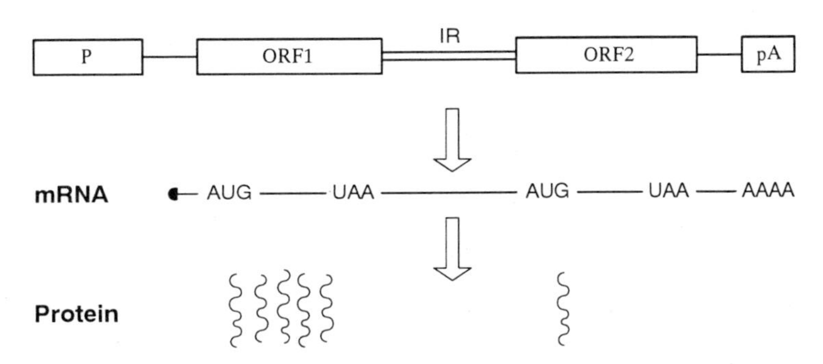

Fig. 22. Expression of two genes from a bicistronic transcription unit. A bicistronic gene construct from an expression vector is shown. The two cistrons (ORF1, ORF2) are flanked by a eukaryotic promoter (P) and a polyadenylation signal (pA). Transcription and processing give rise to one single, capped, and polyadenylated mRNA containing both cistrons. Cap-dependent translation frequently results in a coexpression of both cistrons. Due to reduced translation efficiency of the second cistron, less ORF2 protein is synthesized. Since coexpression occurs in defined molar ratios, expression of the second cistron can be used to monitor the expression level of the first cistron (indirect screening).

a screening marker instead of a selective marker is recommended (WIRTH et al., 1990, 1991a).

3.1.5 Gene Amplification

3.1.5.1 Definition and Examples in Nature

Gene amplification can be defined as an increase in the relative amount of DNA in a given cell line due to multiplication of portions of the genome. The phenomenon of gene amplification is widespread and occurs naturally in many species. Examples of such amplification include the rDNA of *Tetrahymena* and amphibian oocytes, or the chorion genes in *Drosophila*. With these types of gene amplification, the increase in the copy number is developmentally programmed, resulting in the overaccumulation of a desired gene product. However, abnormal amplifications are not rare and are observed with some oncogenes in certain kinds of cancer and in cells resistant to various kinds of antiproliferative or toxic drugs. Gene amplification is considered to be a spontaneous event in mammalian cells, occurring with a frequency of 10^{-4} to 10^{-6}. A few genomic hot spot regions which confer gene amplification to heterologous, linked genes, have been identified and cloned (e.g., DEBATISSE et al., 1988).

3.1.5.2 Application of Gene Amplification to Overexpression of Transfected DNA

The dihydrofolate reductase gene is a well investigated example of an amplification marker which is used for the detection of such events. Overexpression of this gene can be selected for by the antifolate drug methotrexate, which competitively inactivates increasing amounts of the DHFR protein. Some other common amplification marker genes and the corresponding drugs are listed in Tab. 20. Some of these, such as the wild-type DHFR gene, are recommended for use in combination with a cell line defective in the respective gene. Others are dominantly acting and are not restricted to a specific cell line. In a normal selection procedure for an amplification marker gene, first, cell lines are selected which have integrated the respective expression plasmids and, therefore, exhibit a drug-resistant phenotype. Thereafter, selection for clones, which have highly amplified the introduced DNA is performed, by stepwise increases in the concentrations of the selective drug.

Tab. 20. Common Amplification Markers and their Selection

Marker	Cell Type	1. Selection for DNA Uptake	2. Selection for Coamplification[a]	Literature
Adenosine deaminase (ADA)	CHO DUKX B11 ($DHFR^-$)	α^-MEM, HT, Xyl-A	1. plus αCF or Xyl A	YEUNG et al (1983); KAUFMAN et al. (1986)
CAD complex (*de novo* uridine synthesis)	CHO UrdA mutant	Ham's F12	1. plus PALA	DE SAINT VINCENT et al. (1981)
Asparagine synthetase (AS)	CHO N3 (AS^-) any	α^-MEM, minus asparagine	1. plus β-AH 1. plus Albizziin	CARTER et al. (1984); CARTIER and STANNERS (1990)
Dihydrofolate reductase (DHFR)	CHO DUKX B11 ($DHFR^-$)	α^-MEM	1. plus MTX	KAUFMAN and SHARP (1982)
Mutant DHFR	any	α^-MEM, MTX	1. plus MTX	SIMONSEN and LEVINSON (1983)
Glutamine synthetase	any	Glasgow MEM without glutamine plus glutamate	1. plus MSX	BEBBINGTON and HENTSCHEL (1987)
Multiple drug resistance	any	Medium plus colchicine	Medium plus colchicine	KANE et al. (1989)
Ornithine decarboxylase (ODC)	CHO C55.7 (ODC^-)	DMEM minus putrescine plus DFMO	DMEM minus putrescine plus DFMO	CHIANG and MCCONLOGUE (1988)

[a] Stepwise increasing concentrations of selective agent Xyl-A, (9-β-D-xylofuranoxyl); PALA, N-phosphoneacetyl-L-aspartate; MTX, methotrexate; DCF, deoxycoformicine; β-AH, β-aspartyl-hydroxamate; DFMO, difluoromethylornithine; MSX, methioninesulfoximine

Together with the multiplication of a specific gene, large regions in the neighboring DNA are co-amplified. Therefore, DNA which has been cointroduced with an amplification marker is also multiplied. This principle has been widely used to isolate cell lines overexpressing genes which were cointroduced on expression plasmids. The coamplification is not dependent on whether the amplification marker is contained in the expression plasmid, or is cotransfected on a separate expression construct (KAUFMAN et al., 1985).

3.1.5.3 Mechanism of Gene Amplification

The molecular basis of gene amplification is still under investigation. The initial events have become the focus of interest. In several cases palindromic gene duplications, with novel inverted junctions in the intergenic regions, have been observed as intermediates in amplification (Fig. 23) (e.g., FORD and FRIED,

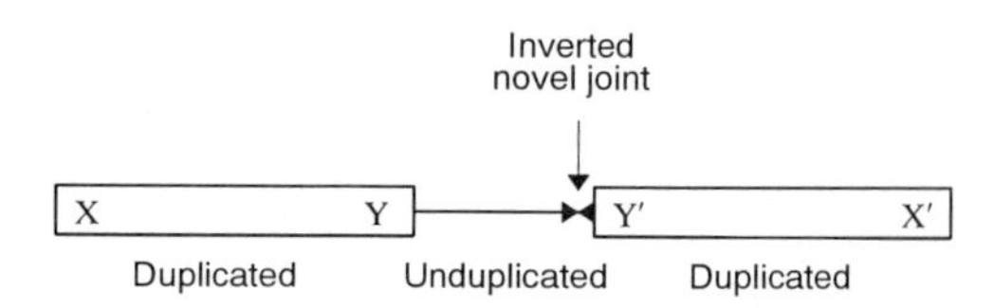

Fig. 23. Structure of amplified DNA. Schematic drawing of a structure which is found as repeat arrays in some amplified cell lines. The long palindromic arms (open bars) and the inverted novel joint are shown. X, Y and X′, Y′ represent a 5′ to 3′ single-stranded DNA and its complement, respectively. The emergence of repeated arrays of such structures can be explained by an extrachromosomal double rolling circle model (PASSANANTI et al., 1987), as it was suggested for the overreplication of yeast 2 μ circles.

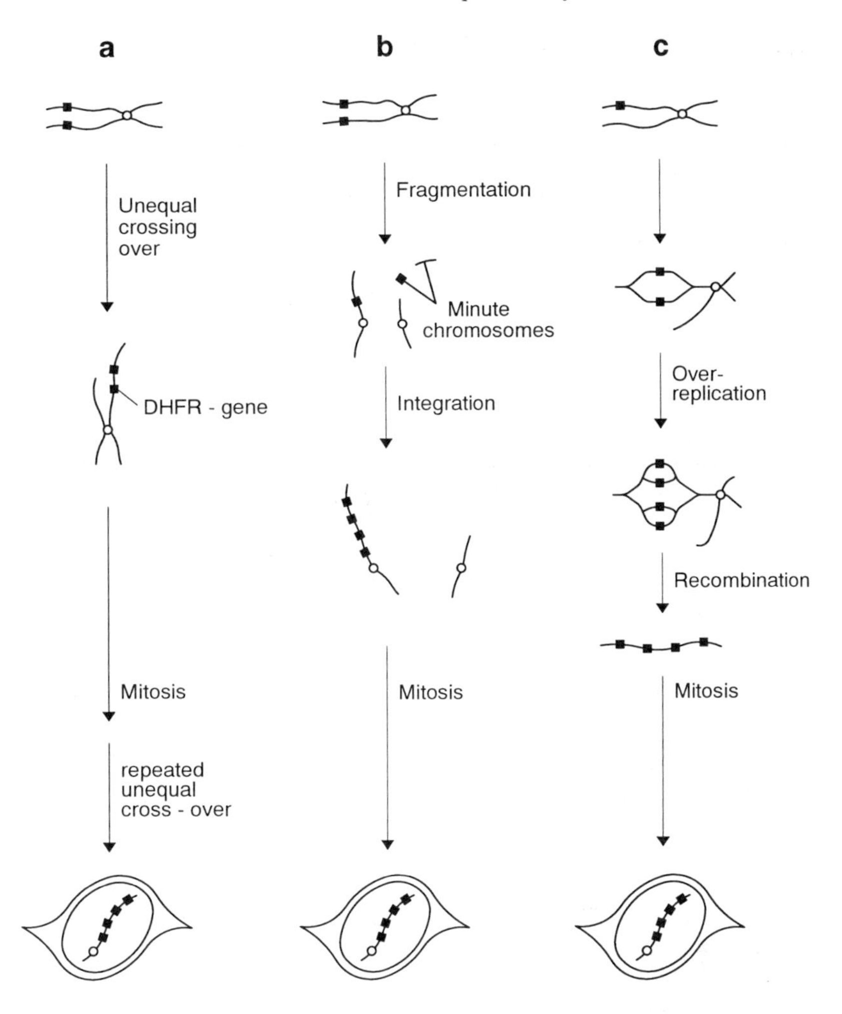

Fig. 24. Some of the mechanisms leading to gene amplification. Some events leading to gene amplification of a DHFR gene and its selection with methotrexate are shown. Filled square: DHFR gene; line: chromosome.
(a) Sister chromatid exchange by unequal crossing-over. (b) Emergence of extrachromosomal double minutes which are able to reintegrate into the chromosome. (c) Onion skin model: multiple initiation at one locus within a single cell cycle and subsequent recombination of the overreplicated region.

1986; Ruiz and Wahl, 1988). Double minutes (DM) and extended chromosome regions (ECR) are observed in amplified cell lines. DMs are extrachromosomal paired DNA fragments of up to 1000 kb in length which are formed in some, but not all cell lines upon selection for amplification. Their reintegration into the genome has been proposed to contribute to the amplification phenotype (Fig. 24; Carroll et al., 1988; Ruiz and Wahl, 1988), ECRs or enlarged chromosomal regions are detected frequently, and their contribution to the stable phenotype of amplification is under investigation.

Several models have been proposed which explain these different findings, e.g., the onion skin model and the sister chromatid exchange model (Fig. 24). In the first model, overreplication of a locus within a single cell cycle is responsible for the appearance of extra copies. In contrast, replication is normal according to the second model, but unequal segregation during mitosis is the causative event. For a more detailed description, the following review is recommended: STARK et al. (1990).

3.1.5.4 Parameters Affecting Gene Amplification

Although spontaneous and uninduced in nature, several agents are known which increase the amplification frequences. Besides certain viruses like EBV, treatment of cells with DNA damaging agents or agents interfering with DNA synthesis can stimulate amplification events. Examples include exposure to UV or ionizing radiation or treatment of cells with hydroxyurea and carcinogens.

3.1.5.5 Stability of Amplified Genomes

Amplified genomes in selected cell lines exhibit different degrees of stability depending on the specific localization of the multiplied copies and on the maintenance of selection pressure. Loss of amplified regions in the absence of selection is often observed if these regions are contained in extrachromosomal DMs. Chromosomal location seems to be a prerequisite for stable maintenance of amplified DNA. Recently, several reports described the loss of chromosomally located amplified DNA in the absence of selection pressure (WEIDLE et al., 1988; PALLAVACINI et al., 1990; SAITO et al. 1989). These data argue in favor of the existence of different types of stable and unstable amplicons. In one investigation the stable phenotypes are represented often by a single, joined amplified structure occurring once on a chromosome and comprising 10% of the chromosome's DNA content (PALLAVACINI et al., 1990).

3.1.6 Gene Targeting

3.1.6.1 Homologous and Non-Homologous Recombination in Mammalian Cells

The mode of integration of exogenously added DNA into the chromosomal DNA of the animal host cell has been the subject of numerous investigations. Generally, transfected DNA may be integrated by two mechanisms, homologous recombination or non-homologous (illegitimate) recombination. Illegitimate recombination does not depend upon regions of homology between the DNA molecules, and the mechanism is still obscure. Homologous recombination involves regions of homology between the host's DNA and the exogenous DNA. Models describing the mechanism of homologous recombination have been proposed for yeast and amimal cells (SZOSTAK et al., 1983; LIN et al., 1984). Integration directed into certain loci of mammalian chromosomes (gene targeting) has been attempted mainly via homologous recombination (SEDIVY and JOYNER, 1992).

3.1.6.2 Selection and Screening for Homologous Recombinants

The gene targeting ratio (the ratio of homologous to illegitimate integrations) has often been found to be lower than 1/1000. Therefore, the use of gene targeting for the manipulation of a cell's genome was dependent on the development of efficient methods for enrichment or screening for the targeted event. To date, several techniques have been elaborated which allow rapid identification of *bona fide* homologous recombinants (Fig. 25).

One method is the positive/negative selection (PNS) system which has been elaborated by MANSOUR et al. (1988) (Fig. 25a). Two selective markers are involved. The first marker, which is positively selected for (e.g., the neomycin resistance gene) is flanked by two regions of homology with the target region and indicates both random or targeted integration events. The second marker which lies outside the homology regions, can be negatively selected for (e.g., the HSV-TK gene). Targeted integration involves exchange of the DNA in and between the regions of homology of exogenous and endogenous DNA (replacement recombination) and will result in the loss of this marker. Loss of the HSV-TK gene renders cells resistant to an exogenously added drug like Gancyclovir (related to Acyclovir) or FIAU (a iodouracil derivative), which are metabolized by the viral TK gene to products toxic to mammalian cells. In random integration events the negative marker is retained, and cells exhibiting this genotype die upon drug addition. In general, PNS selection is not dependent on the presence of transcriptional elements in the integration site. However, it may fail if the chromosomal surrounding exhibits a silencing effect on expression of the markers.

Another method is based on conditional positive selection, which means that only under conditions of homologous recombination of substrate and target DNA will a selectable marker become expressed (Fig. 25b). Such a procedure was successfully used by JASIN and BERG (1988) who transfected constructs carrying an enhancer- and promoter-less SV40 transcription region and the Ecogpt gene as a selectable marker. The homology of the shortened SV40 region to an endogenous complete SV40 transcriptional control region was large enough to direct the transfected construct to this locus in nearly 50% of the transformants. False positive clones arose by illegitimate recombination into loci containing enhancer/ promoter sequences. Refinements and modifications of this procedure have been reported (e.g., ITZHAKI and PORTER, 1991; WOOD et al., 1991). In general, conditional positive selection is restricted to actively transcribed target regions.

Targeting events can be confirmed directly by the loss of function (e.g., if a gene coding region is affected) or gain of a function. Usually, targeting is detected by Southern blotting or PCR techniques (KIM and SMITHIES, 1988).

3.1.6.3 Parameters Affecting Gene Targeting

Several parameters are known to influence the efficiency of gene targeting. (1) The method of gene transfer is important. Microinjection gives the best results, followed by electroporation which is superior to calcium phosphate precipitation. (2) The length of the homology region is correlated with efficient homologous recombination (THOMAS and CAPECCHI, 1987). At least 200 bp of homology are necessary to yield low levels of targeted recombinants. Increasing the length of homology usually improves the targeting ratio. (3) The extent of the non-homologous region (e.g., between the two homology regions in the PNS method) has no influence. (4) A cut in the region of homology usually exhibits a positive effect (KUCHERLAPATI, 1984). (5) Neither the DNA concentration of the incoming DNA nor the copy number of the endogenous target locus, affects the number of homologous recombinants (ZHENG and WILSON, 1990). (6) The accessibility of the target locus (e.g., open chromatin) may exert certain effects on recombination and is thought to be responsible for the variation in gene targeting efficiency which has been observed with different targets.

3.1.6.4 Applications

Gene targeting procedures now have been widely used to elucidate the function of various genes *in vivo* (Tab. 21). This also has been accomplished by knocking out the respective gene in an embryonic stem cell by homologous recombination (gene ablation) and introduction of the cell into mouse blastocytes. The developing heterozygotic animal carrying the mutation, hopefully in the germline, is crossed to produce the desired homozygote with a null mutation. If viable, the progeny can be subject to further investigation. With the accuracy of recombination and the low frequency of mutations during gene targeting, a somatic correction of genetic defects using such methods is envisioned. On the other hand, null mutations in a certain cell line can be achieved by consecutive inactivation of the respective gene on

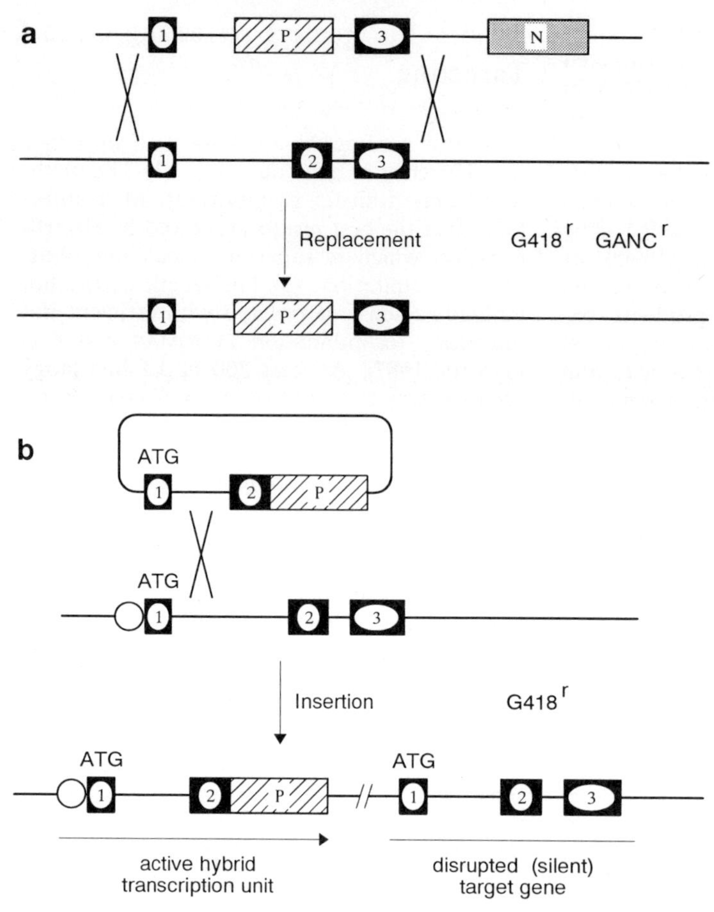

Fig. 25. Gene targeting by homologous recombination in animal cells. (a) Disruption of a gene containing three exons (black boxes 1, 2, 3) by replacement integration. The upper line represents the targeting vector. Double reciprocal recombination into the chromosomal DNA (middle line) results in the structure shown in the lower line. These structures of targeted integration are found after positive-negative selection (PNS selection). The double reciprocal exchange is accompanied by loss of the N gene gut not the P gene. In the original approach, MANSOUR et al. (1988) used a neomycin transcription unit as a positive (P) selectable marker and the HSV-TK transcription unit as a negative selectable marker (N). Targeted integration is indicated by a G418 and Gancyclovir (GANC) resistant phenotype.
(b) Disruption by insertional inactivation. A 'conditional positive selection' approach is indicated. Upon targeted integration the formerly silent positive selectable marker (P) (e.g, the neomycin-resistance gene) is transcribed.
(c) Targeted introduction of mutations into chromosomal genes by a two-step recombination mechanism (insertion/excision). The P and N genes are located outside of the region of homology. In the first step, insertion of the vector creates a duplication of genomic sequences, which creates a G418-resistant and FIAU-sensitive phenotype (if the neomycin and HSV-TK gene are used). Subsequent intrachromosomal recombination within the duplication leads to loss of the duplication, indicated by a FIAU-resistant and G418-sensitive phenotype.
N, negative selectable gene; P, positive selectable gene.

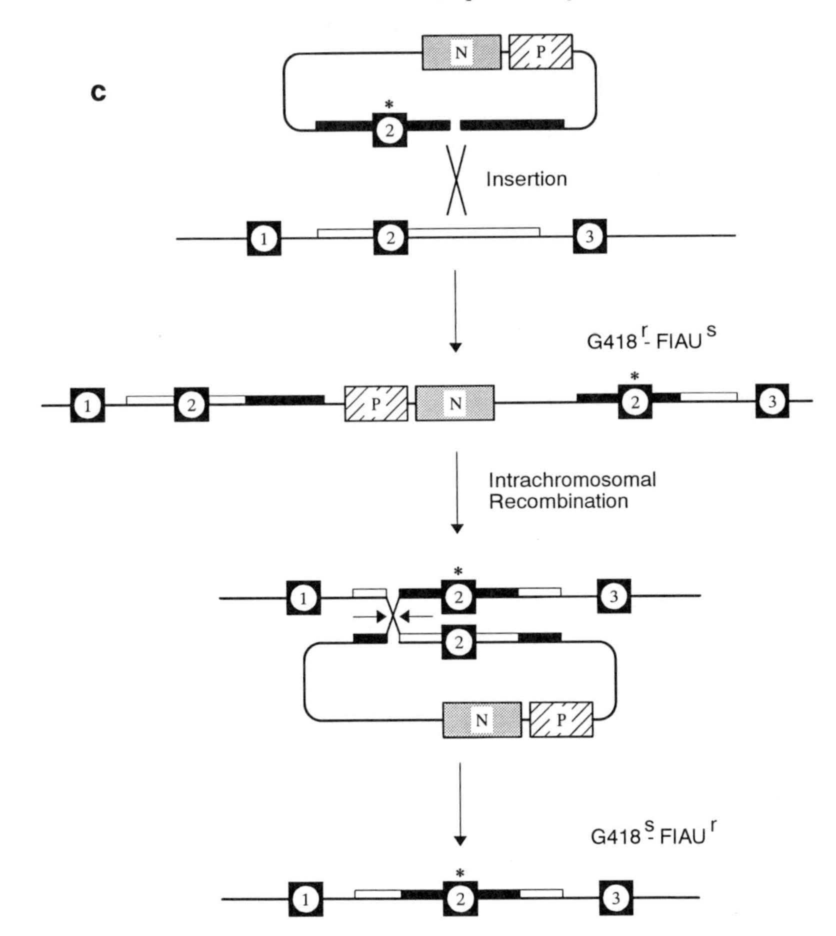

Fig. 25c.

each allele. Gene targeting methods also permit the directed transfer of expression vectors into transcriptionally active regions, such as SAR environments or hot spots for amplification of the genome.

3.2 Animal Expression Vectors

Animal expression vectors are used to introduce and express cloned genes in animal cells. These vectors normally contain the prokaryotic elements which allow replication and selection in *Escherichia coli* as well as permit isolation of pure DNA. They also harbor elements to mediate transcription initiation and termination (polyadenylation) in the host cells. In addition to these basic components animal expression vectors may contain a series of other elements which modulate transcription efficiency, post-transcriptional processes, replication, specific integration, and gene amplification or helper function for delivery of the vector nucleic acids into the target cells. The following section will review these elements which can modify vector function. Currently used animal gene expression systems, vectors, and hosts will be introduced in Sects. 3.2.2 and 3.2.3.

Tab. 21. Gene Targeting in Mammalian Cells

Target	Cells	Selection	Screen	Type of HR[c]	Reference
HPRT int-2	ES	PNS	SB	R	MANSOUR et al. (1988)
en-2	ES	PNS	SB	R	JOYNER et al. (1989)
APRT	CHO	PNS	SB	R	ADAIR et al. (1989)
DHFR	CHO C400[b]	PNS	SB	R	ZHENG and WILSON (1990)
c-fyn	ES	PNS[a]	SB	R	YAGI et al. (1990)
IGF-II	ES	PNS	SB	R	DECHIARA et al. (1990)
Insulin	3T3-L1	PNS	SB	R	ACCILI and TAYLOR (1991)
Ig μ	preB	PNS	SB	R	KITAMURA et al. (1991)
Hox 1.1	ES	–	PCR	R	ZIMMER and GRUSS (1989)
IFN-ind. 6-16 gene	HT1080	–	hGH	I	ITZHAKI and PORTER (1991)
IG μ	J558L	–	AB	I	FELL et al. (1989)
Ig μ	J558L	–	AB	I	SMITH and KALOGERAKIS (1990)
Ig μ	hybridoma	CPS	AB	I	WOOD et al. (1991)
SV40 early	COS1	CPS	SB	I	JASIN and BERG (1988)
c-abl		CPS	SB	I	SCHWARTZBERG et al. (1989)
N-myc	ES pre b	CPS	SB	I	CHARRON et al. (1990)
CD4	JM	CPS	SB	I	JASIN et al. (1990)
Creatine kinase M	ES	CPS	SB	I/R	VAN DEURSEN et al. (1991)
HPRT Hox 2.6	ES	PNS	SB	I/REV	HASTY et al. (1991)

[a] Toxin gene expression was used for negative selection.
[b] CHO cell line with about 400 copies of the DHFR gene.
[c] Homologous recombination.

3.2.1 Control Elements for Construction of Expression Vectors

Control of gene expression occurs mainly at the transcription level, since this is a prerequisite for the subsequent post-transcriptional events. However, post-transcriptional events can influence the strength of expression to the same extent, as observed in the differential transcription of many genes. In consequence, a defined vector/host system and manipulation method to express individual recombinant genes, does not necessarily result in identical levels of expression.

Because all steps of gene expression up to transcription and eventual splicing are independent of the protein coding region, expression can be adjusted by vector construction, the method of gene transfer, and selection. In contrast, control of expression for the subsequent levels is mostly dictated by the individual gene. In the early days of eukaryotic gene expression technology, emphasis had been placed on elements involved in transcription regulation. Translation control in eukaryotes had not been considered. However, now that it is known that certain features of the transcribed mRNA, such as the 5′ or 3′ untranslated region or the context around the initiator AUG of a given gene, may influence its expression efficiency, more attention is placed on determining the most favorable signals for translational control in expression constructs. The principles which are responsible for post-transcriptional control and the consequences for gene expression have been introduced in Sects. 2.2 and 2.3. Often the mechanisms governing this type of regulation either are not known or appear extremely complex. It should be kept in mind that at the RNA level not only linear sequences, but also secondary and tertiary structures play an important functional role. In many cases it is not possible to manipulate gene expression at the post-transcriptional level. Moreover, as coding sequences are involved in many post-transcriptional events, manipulations in the nucleic acid sequence must be made without changing the amino acid sequence of the gene product.

Many events controlling gene expression are partially or completely tissue-specific. Therefore, expression parameters including transcriptional or post-transcriptional events, which may be exhibited in a particular host cell line, cannot be expected to be identical in another cell line, even if the cell line originates from the same species and tissue.

3.2.1.1 General and Tissue-Specific Gene Transcription

Most expression vectors use strong promoter/enhancer combinations to drive transcription. A series of promoters has been shown to issue a strong effect in a variety of cell types (Tab. 22). The strongest of these vectors are of viral origin. A combination of elements from two or more distinct promoters has led to the construction of hybrid promoters with higher activity or an increased spectrum of cell and species specificity, e.g., the fusion of the polyoma or SV40 enhancer to the HSV TK promoter results in a strong promoter for many cell types, including undifferentiated embryonal stem cells (Tab. 22).

In contrast to the more general elements of transcriptional control which may be necessary, expression in a specialized cell type could be required. In Tab. 23 some highly tissue-specific promoters or enhancers are listed. The most prominent is the immunoglobulin heavy chain enhancer which directs high level expression only in B-lymphoid cells.

Polyadenylation and signals for transcription termination are important elements for all expression vectors. cDNAs usually possess an AATAAA element located upstream of the poly(A) tract. The T/G-box which is important for efficient polyadenylation, (Sects. 2.1.1.1 and 2.2.3) is usually not contained in the cDNA sequence because the processing of the mRNA precursor occurs between the AATAAA- and the T/G-box. Therefore, mRNAs derived from expression vectors are often either incompletely or not at all polyadenylated at the authentic site. In these cases, another polyadenylation site and sequences influencing termination further downstream must be supplied by the expression vector. Such DNA fragments are included in most expression vectors. This results in transcripts with different 3′ ends, as poly(A) signals from both the inserted cDNA sequence as well as the

Tab. 22. Strong Universal Mammalian Promoters

Promoter	References
SV40, early	MOREAU et al. (1981), NEUHAUS et al. (1984)
Rous sarcoma virus (RSV), LTR	GORMAN et al. (1982)
Human cytomegalovirus early (hCMV)	BOSHART et al. (1985)
Myeloproliferative sarcoma virus (MPSV)	ARTELT et al. (1988), BOWTELL et al. (1988)
Combined: SV40 early/Ad MLP	WOOD et al. (1984), BERG et al. (1988)
Combined: MPSV/hCMV	WIRTH et al. (1991b)
Combined: Py/TK	THOMAS and CAPECCHI (1987)

Tab. 23. Examples of Tissue-Specific Transcription Elements

Gene	Tissue	References
IgH promoter/enhancer	B-lymphoid	PICARD and SCHAFFNER (1984)
β-Globin DCR	Erythroid cells	BLOM VAN ASSENDELFT et al. (1989)
Elastase enhancer	Pancreatic cells	ORNITZ et al. (1987)
Transthyretin	Hepatocytes, choroid plexus cells	COSTA et al. (1989)
α-1-Antitrypsin	Hepatocytes	XANTHOPOULOS et al. (1989)

Tab. 24. Inducible Expression Systems

Induction	Based on	Restrictions	References
Heavy metals	Mouse MT-1	–	SEARLE et al. (1985)
Glucocorticoids	MMTV LTR	Receptor expression required	ISRAEL and KAUFMAN (1989)
Estrogen	ERE in *Xenopus* l. vitellogenin A2	Receptor expression required	KLEIN-HITPASS et al. (1986)
Heat shock	HSP70	–	WURM et al. (1986), BENDIG (1987)
Virus, dsRNA	Type I interferons	–	KUHL et al. (1987)
Serum	c-fos	–	TREISMAN et al. (1985)
Butyrate, propionate	SAR elements	–	KLEHR et al. (1992)
Transactivation[a]	1. HIV-1 LTR + TAT 2. HBV promoter and many others	–	SODROSKI et al. (1985), RUBEN et al. (1989)
Interferons type I	Mx	–	LLEONART et al. (1990)
Erythroid differentiation	DCR from β-globin	Only in murine erythroleukemia cells (MEL)	COLLIS et al. (1990)
Post-transcriptional control: iron response	1. Ferritin 2. Transferrin receptor (TfR)	–	KLAUSNER and HARFORD (1989)

[a] The transactivator is delivered by protein or gene transfer.

expression vector are used. The consequences of this alteration are not predictable. It seems, however, that certain polyadenylation/termination sequences are advantageous for use in expression vectors (VAN HEUVEL et al., 1986).

Fragments from different genes which mediate transcription termination do not necessarily function very efficiently in heterologous surroundings. Promiscuous termination leads to mRNAs with different 3′ ends. Run-through transcription can result in a series of difficulties, including transcriptional interference on downstream promoters, instability of the oversized mRNA, and hybrids with antisense RNA.

3.2.1.2 Regulated Gene Expression

In certain circumstances, e.g., for the production of a toxic protein, conditional expression is required. The gene of interest can be stably introduced into the host cells without any negative effect from the gene product. After induction of specific gene expression the product can be produced for a certain interval. In Tab. 24 some inducible expression systems are listed. Besides inducible promoters or enhancers, systems are included which are based on the stimulation of expression by changing chromosomal structure, the transactivation of otherwise silent promoters, the differentiation of cells to a new phenotype, and the use of post-transcriptional control elements like the iron responsive element of the human ferritin and transferrin receptor genes (Sects. 2.2.5 and 2.3.4).

Regions responsible for induced transcription have been identified from a number of promoters. They function as enhancer elements. When selecting an inducible vector system for a particular gene, it is important to ensure that the inducing stimulus does not interfere with the properties under study. It is also important to know the extent of induction which is required, including the base level expression and the maximal achievable expression level. By using multiple copies of the induction responsive elements in combination with convenient basic promoter elements, an improvement of the induction ratio of some naturally inducible promoters was achieved (Tab. 24).

For industrial protein production in fermenters, the availability of an expression system in which induction is coupled to the downregulation of cell growth would be very helpful. Unfortunately, systems like this are still not available. The use of dominant control regions from the β-globin locus in differentiating mouse erythroleukemia cells might be a first step in this direction (Sect. 2.1.2.2) (COLLIS et al., 1990).

3.2.1.3 Prokaryotic and Synthetic Gene Expression Systems

The use of prokaryotic elements is an attractive way to maximize gene expression and create inducible expression systems in animal cells. It has been shown that a bacterial control system can function in eukaryotic cells. First, DNA sequences which bind *Escherichia coli* repressor proteins were inserted into functional animal cell promoters (HU and DAVIDSON, 1987). With another construct the same cell was engineered to produce the respective repressor. Expression could be induced by inactivation of the repressor (Fig. 26).

The bacteriophage-derived T7-polymerase actively transcribes T7-promoters upon expression in many different cell types. This system can be used to drive expression independently of cellular RNA polymerases. The gene of interest is delivered on a construct which is under the control of the T7-promoter. The polymerase could either be constitutively or inducibly expressed in the host cell. The prokaryotic T7-promoter is presently used in two different systems:

(1) One expression system was established as a vaccinia virus driven system which initiates cytoplasmic transcription (Sect. 3.2.3.2). This system gives rise to very efficient, however, transient expression. In order to produce an inducible system, the T7-polymerase can be delivered by infection with a recombinant vaccinia virus (FUERST et al., 1987).

(2) Another expression system was developed by LIEBER et al. (1989) who engineered a nuclear targeting sequence into the T7-polymerase reading frame. This polymerase is transported to the nucleus of mammalian cells and is fully functional. The gene of interest is under control of the T7-promoter in normal expression plasmids. This system allows constitutive or inducible expression in cells stably transfected by appropriate methods.

In some cases inducible or constitutive expression systems require the presence of a high level of transacting factors, e.g., the glucocorticoid or estrogen receptor proteins. Cells which do not possess these properties can compensate by transfecting the respective expression constructs separately. As the genes for other transactivators are isolated, they can be used to manipulate cell lines which lack these factors and render dependent promoters highly active. A characterization of the structural features of transcription factors will allow one to engineer novel potent transactivators which elicit the activity of specific promoters. This was successfully approached by fusing the genes from different transcription factors to those encoding the hormone-binding domains of steroid receptors, thus, rendering the activi-

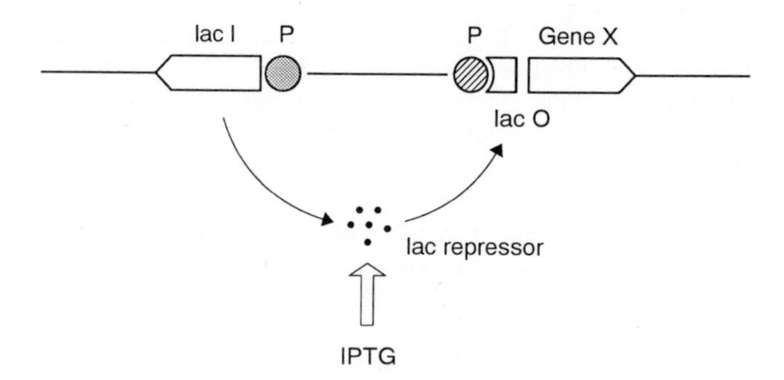

Fig. 26. Inducible gene expression using eukaryotic and prokaryotic control elements. Juxtaposition of the lac operator eliminates the activity of a eukaryotic promoter in the presence of the functional lac repressor. The lac I gene is constitutively expressed due to fusion to a functional eukaryotic promoter. Inactivation of the lac repressor protein by the addition of IPTG to the medium, activates the hybrid promoter and leads to expression of gene X.

ty of these transcription factors hormone-inducible (EILERS et al., 1989).

An independent but complementing approach results from the creation of synthetic promoters which are based on known elements from natural promoters. This requires the detailed functional analysis of natural promoters. The promoters may exhibit high expression-specific regulation or new cell type specificity. One example concerns the construction of a strictly metal-inducible promoter by multimerization of a metal-responsive element from a metallothionein promoter (MCNEALL et al., 1989).

3.2.1.4 Introns in Expression Vectors

Since splicing is a prerequisite for the production of stable cytoplasmic mRNA in some recombinant genes (see Sect. 2.2.2), most vectors for cDNA expression contain introns. Although dependency on introns for efficient mRNA formation is usually not known, it is still recommended to use intron-containing vectors.

However, the use of introns can also create artefacts resulting from aberrant splicing. The SV40 small t-intron is one of the most commonly used introns in expression vectors. This intron is one of the shortest introns to produce a lariat branch site complex. During expression from vectors in which the SV40 small t-intron is present in the 3′ UTR, 5′ splice sites from the preceding cDNA genes are preferentially used. Splicing from these sites together with the 3′ splice site of the SV40 small t-intron result in deletions within the protein coding region (HUANG and GORMAN, 1990). 5′ Splice sites in the cDNA could exist as a consequence of splice junctions from former introns. Due to conservation of the exonic nucleotides in the splice-donor and -acceptor sequences, the probability for a restoration of 5′ splice sites in a cDNA is one of five splice junctions. Furthermore, ordinarily inactive splice consensus sequences (cryptic splice sites) are found in many genes. These may be utilized together with introns from vector sequences. The positioning of small introns in the 3′UTR of cDNA expression vectors should be avoided. If introns are used, they should be located in the 5′UTR of the vector. The probability of aberrant splicing from this position is lower, since the splicing apparatus requires not only the 3′ splice site, but also the adjacent branch point sequence (cf. Sect. 2.2.2).

3.2.1.5 Bicistronic Expression Vectors

Bicistronic expression vectors have been employed for screening of high-level expression clones. The second underexpressed cistron is used as an indicator for the expression of the first cistron (WIRTH et al., 1990). Equimolar expression of both cistrons has not been achieved by using intercistronic regions which mediate reinitiation or coupled initiation. Nevertheless, stoichiometric expression is desirable for proteins built of similar or heterologous subunits (Fig. 27). An increased tendency to

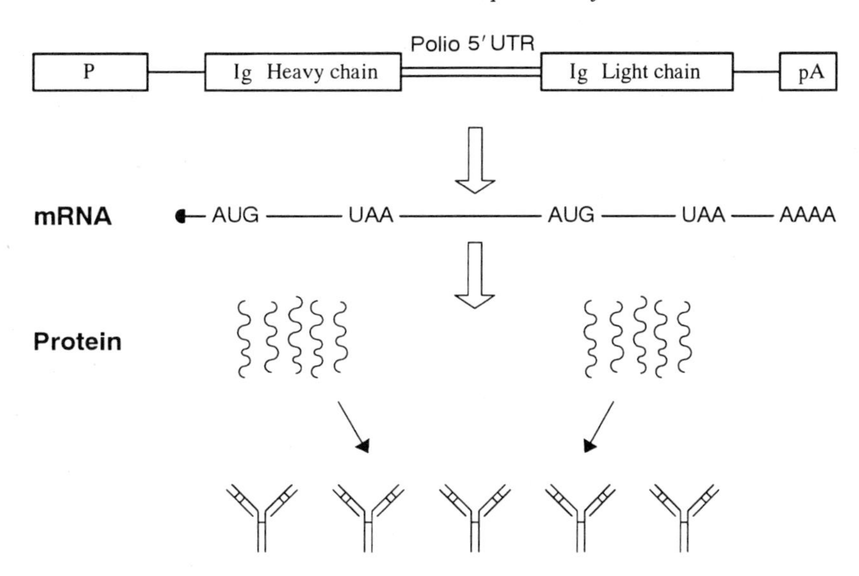

Fig. 27. Coexpression of antibody genes using bicistronic expression vectors. Heavy and light chain genes are expressed from a bicistronic mRNA by a cap-dependent and a cap-independent internal initiation of translation. Internal entry of ribosomes is mediated by the poliovirus 5′UTR as the intercistronic region. Coexpression of both subunits occurs in defined, favorable ratios and in close proximity, which should, therefore, promote association of heavy and light chains.

associate into oligomers is expected, as the close proximity of both subunits during coupled translation should favor their association. To increase translation efficiency of the second cistron, 5′UTRs have been used which promote efficient cap-independent initiation as intercistronic regions (WIRTH et al., 1991a; KAUFMAN et al., 1991; WOOD et al., 1991; ADAM et al., 1991; GHATTAS et al., 1991).

3.2.2 Plasmid Expression Vectors

Generally expression vectors are based on prokaryotic plasmids which are combined with eukaryotic gene transcription elements (Fig. 28). In order to facilitate manipulation and propagation of the DNA in bacteria, the vectors include plasmid components, resistance genes, and bacterial origins of replication. The eukaryotic part consists of a transcriptional initiation element (promoter/enhancer), a multiple cloning site, and a sequence allowing transcriptional termination. Since the vectors are propagated in bacteria, usually *Escherichia*

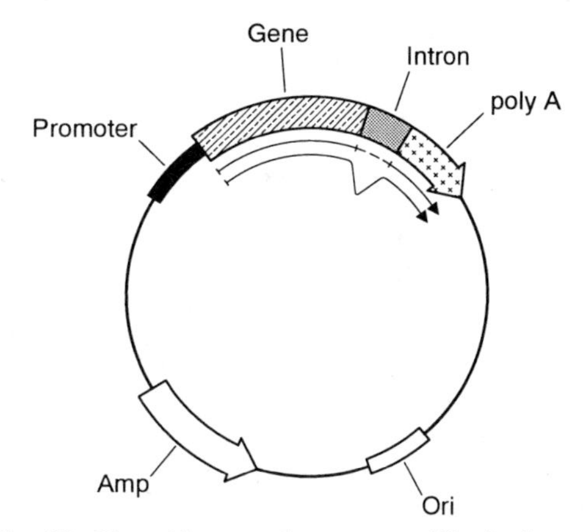

Fig. 28. Plasmid expression vectors. The basic constituents of a plasmidal expression vector are shown. The promoter and the poly(A) addition site are the transcriptional control sequences supplied by the vector. In the empty vector a multiple cloning site for insertion of the gene of interest is located between the transcriptional control elements. The intron may be located 5′ or 3′ to the multiple cloning site. Additional elements, e.g., to improve transcription, are integrated between the bacterial elements (resistance gene, origin of replication) and the basic transcriptional control elements.

coli, and transferred to animal cells they are called shuttle vectors. As discussed in Sect. 3.2.1.4, most plasmidal expression vectors contain an intron with splice consensus sequences 5′ or 3′ to the multiple cloning site.

Apart from these basic elements, expression vectors contain additional elements which facilitate replication in specific cell lines, confer resistance to selectable or amplification markers, and contain further expression cassettes and chromosomal sequences, which are convenient for homologous recombination or favorable exposure of the vector sequences.

3.2.2.1 Plasmid Vectors Based on Viral Replication Elements

Viral elements which enhance the transcription of heterologous genes are frequently used in expression vectors. Furthermore, other properties of viruses, namely, episomal replication or integration can be transferred by nucleic acid sequences to plasmid vectors. The use of viral elements for the improvement of integration into favorable chromosomal sites is an interesting approach for achieving high-level expressing cell lines, but will not be discussed further in this chapter. The use of viral segments for episomal replication, however, is involved in the most efficient vector systems for heterologous gene expression. Prominent examples of these host-vector systems are the SV40-COS system (Sect. 3.2.2.2) and the bovine papillomavirus (BPV)-C127 cell system (Sect. 3.2.2.3). These will be introduced in the following sections and will serve as examples for analogous systems which are not as efficient or remain under development.

3.2.2.2 Transient Expression in COS Cells

This system is based on Simian virus 40 (SV40), which is able to undergo a complete replication cycle in certain primate cells (TOOZE, 1980). For SV40 replication, one of the early viral gene products, that acts on the origin of replication in the SV40 DNA, large T antigen, is required. The minimal requirement for vectors to replicate is the SV40 origin of replication, a fragment of less than 300 bp DNA. All other elements may be identical to those from simple plasmid expression vectors. However, certain sequences appear to interfere with replication and transcription in mammalian cells (LUSKY and BOTCHAN, 1981), the so-called poison sequences.

Optimal expression of SV40 origin containing plasmids is gained in CV1 derived cells, which constitutively express large T antigen (GLUZMAN, 1981). Due to the replication of SV40 origin containing plasmids in these cells, high levels of foreign DNA sequences within the vectors are expressed.

Unfortunately, expression in this system is transient. This most probably results from uncontrolled accumulation of extrachromosomal DNA.

Apart from the COS cells, some other primate cell lines, e.g., HeLa cells, allow replication of SV40 origin containing plasmids, provided that large T antigen is supplied in *trans.*

Efforts have been made to improve the basic COS system by external control of large T-antigen expression. These include the creation of thermolabile T antigen or the expression of T antigen under the control of regulated promoters. However, permanent cell lines with high constitutive expression have not been developed by this approach.

A series of useful SV40 origin vectors have been developed. For review see SAMBROK et al. (1989) and KRIEGLER (1990).

3.2.2.3 Bovine Papillomavirus Vectors

The bovine papillomavirus 1 (BPV-1) is a member of the double-stranded DNA papovavirus family. The virus is able to transform certain rodent cells. The virus and the derived vectors are capable of maintaining permanent episomal replication in some murine cells (BROKER and BOTCHAN, 1986). The vectors which are based on BPV-1, contain either the entire viral genome (7.95 kbp) or the 69% transforming fragment. The other part of the vector contains the components typical for plasmid vectors (DIMAIO et al., 1982). In order to get stable episomal replication in mammalian cells, additional DNA fragments, e.g., from the human β-globin locus, were used.

Two methods can be used to identify stable transfectants: in a few cell lines such as the murine C127 line BPV vectors lead to a morphological transformation and can be isolated by their focus formation character in tissue culture plates (cf. Sect. 3.1.4). Alternatively, the vectors may require dominant selectable markers, since replication of BPV-1 derived plasmids does not manifest a transformed phenotype in many other cell lines.

Stable expression from BPV vectors can be rather high. Some obvious disadvantages, however, are connected with the use of this expression system: a series of genes cannot be expressed in these vectors since they undergo rearrangement. Furthermore, although most DNA remains in the episomal state, integration of the BPV plasmids into the chromosomal DNA of the host takes place. Finally, due to the large size of the vectors, the integration of foreign genes into the vectors is often cumbersome.

3.2.3 Gene Transfer and Expression Based on Viral Infection

Many viruses are able to infect a broad spectrum of cells with high efficiency and thereby introduce their genetic material into the hosts. Replacement of the genes which are non-essential for the simple infection-replication cycle of viruses by heterologous genes has been used to take advantage of specific viral properties. In this technology, the nucleic acids containing the foreign gene are encapsulated into virions. These pseudovirions are infectious. In some systems the pseudovirions even replicate concomitantly with the production of viral particles which kill the host cells. Depending on the type of system, expression of the inserted gene is transient or stable. Further development of these kinds of vectors has led to the establishment of helper cells in which virus particles are produced. These virus particles are able to infect other cells, but do not complete their replication cycle with the production of new virions.

A series of pseudovirion systems has been developed. In this chapter only the most prominent and widely used systems, namely those based on retroviruses, vaccinia virus, and baculovirus are presented. The properties of these expression systems are summarized in Tab. 25.

3.2.3.1 The Retrovirus System

The family of retroviruses has some common features which are important for the use of these viruses as vectors. Viral particles contain two single-stranded RNAs which resemble eukaryotic RNA (cap, poly(A) tail) and some enzymes encoded by the retroviral genome (reverse transcriptase, protease). The viruses are covered by an envelope containing viral glycoprotein and cellular plasma membrane components. The viral RNA is embedded in structural proteins (gag), constituting the viral core. The viral envelope glycoprotein is responsible for recognition of a cellular receptor on the plasma membrane of the cell to be infected. After receptor binding, the virus is internalized and uncoated. The viral core remains intact for some time and allows the reverse transcriptase to drive the synthesis of double-stranded DNA assisted by tRNAs as primers and host cell low-molecular weight components. The resulting proviral DNA is a molecule of linear double-stranded DNA. It is transferred into the nucleus and a part of it is circularized. At this time or some time later, depending on the virus and the host cell, the majority of proviral DNA integrates into the cellular genome. A fraction also remains episomal for some time. The integrated proviral DNA behaves as a cellular gene and is transcribed and processed as such. Some retroviruses (lentiviruses) contain genes which are involved in the regulation of their own proviral gene expression. After retroviral gene expression is activated, which results in the production of full-length viral RNA and partially or fully spliced derivatives, the RNA is translated into the viral proteins (VARMUS, 1988).

In retroviral vectors the internal viral function genes are replaced by foreign genes. Viral proteins are provided by packaging cell lines that are by themselves unable to make packa-

Tab. 25. Animal Expression Systems

Type	Viral Components	Nucleic Acids		Production of Recombinant Infectious Particles	Expression		Target Cell Spectrum	Post-translational Modifications[b]	Time and Skill Requirements	System-born Safety Considerations
		Insert Size (kbp)	Localization		Stable/Transient	Strength[a]				
Nonviral expression vectors	–	up to 50	chromosomal	–	Stable/transient	middle high	Vertebrate and invertebrate cells depending on the transfer method	Cell-type specific	Minimal	–
BPV	Bovine papillomavirus	up to 15	nuclear, episomal and chromosomal	–	Stable	middle high	C127 and some murine cells	Cell-type specific	Cloning selection	(+)
COS	SV40	<5	nuclear, episomal	–	Transient	high	COS cells and some other primate cells with T-antigen expression	COS cell specific	Minimal	–
SV40 Pseudovirions	SV40	<3.5	nuclear, episomal + chromosomal	+	Transient	middle	Some primate cells and organisms	No information available	Helper system	+
Retrovirus	Murine and avian retroviruses (MoMuLV, MPSV, SNV)	<10	chromosomal	+	Stable	low	Many mammalian and avian cells and organisms	Cell-type specific	Helper system	+
Vaccinia	Vaccinia	>20	cytoplasmic (episomal)	+	Transient	high	Vertebrate and some invertebrate cells and organisms	Influences by virus-mediated inactivation of cellular functions	Helper system	+ +
Baculovirus	Baculovirus lepidopteras	>10	nuclear, episomal	+	Transient	middle	Some insect cell lines; e.g. Sf9	Not identical to mammalian cell-derived proteins	Helper system selection	+

[a] low: <0.5 μg/24 h/10^6 cells; middle: <0.5–5 μg/24 h/10^6 cells; high: <5 μg/24 h/10^6 cells
[b] Very high expression may cause exhaustion of cellular modifying systems and thereby lead to incomplete protein modifications

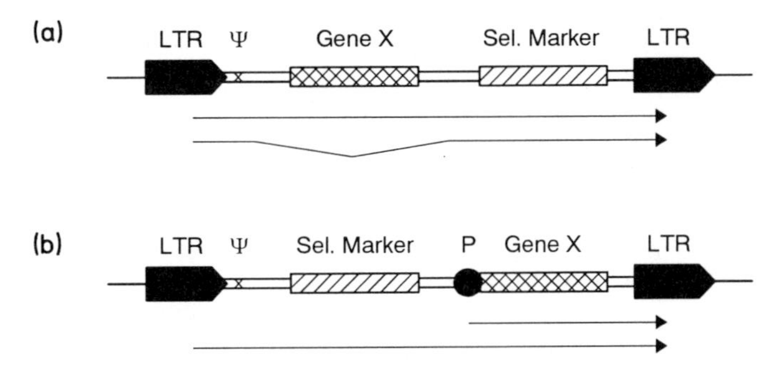

Fig. 29. Retroviral vectors. The genomic structure (provirus) of two typical types of vectors is shown. The black bars with the arrow heads represent LTR sequences, and the filled circle indicates an internal promoter. Thin arrows indicate transcripts derived from the proviral DNA, and Ψ represents the packaging sequence.
For the vector depicted in (a), the selectable marker protein may be either realized after splicing out the first cistron. Alternatively, an internal constitutive promoter to initiate transcription of the right half of the provirus may be positioned 5′ to the selectable marker gene. An example of this type of vector is the pBABE-Puro (MORGENSTERN and LAND, 1990), which gives rise to strong expression of the gene of interest.
The vector in (b) shows a type of virus in which the selectable marker is transcriptionally controlled by the retroviral LTR. This allows the use of any type of promoter (P) to direct the expression of the gene of interest. A frequently used vector of this type is pCIPNEOSV(X) (CEPKO et al., 1984).

ble transcripts (McLAUCHLIN et al., 1990). The vectors carry prokaryotic amplification and selection functions, the retroviral LTRs, the packaging sequence (Ψ), which functions as a selectable marker in animal cells, and short sequences, which represent the tRNA primer binding sites. The gene of interest is either under the control of the 5′LTR or under the control of an internal promoter (Fig. 29). The other promoter is mostly used to drive transcription from the selectable marker gene. After transfection or infection, 3′ end formation of the transcripts from these retroviral vectors occurs in the 3′LTR. In packaging cell lines, the RNA containing the Ψ sequence is predominantly packaged into complete virus particles (Fig. 30).

The retroviral vectors are transfected into the helper cells. Currently available retroviral vectors for mammalian cells are based on the Moloney murine leukemia virus (MoMuLV). In the first engineered helper cell line, Ψ2, a defective mouse Moloney murine leukemia virus retroviral genome provides the proteins for packaging recombinant retroviral RNA (MANN et al., 1983). Recent efforts in vector development have concentrated on minimizing the chances of recombination betwen the vector and the packaging defective helper functions within packaging cells (MILLER and BUTTIMORE, 1986; MILLER and ROSMAN, 1989; DANOS and MULLIGAN, 1988; MARKOWITZ et al., 1988).

In some retroviral vectors, deletions or additional genes are inserted into the U3 region of the 3′LTR. After infection, reverse transcription of the retroviral RNA from these vectors results in the duplication of this modified LTR so that both LTRs are now identical (double-copy vectors). Such double-copy vectors have integrated a dominant selectable marker (DHFR) or a transcribed gene (STUHLMANN et al., 1989; HANTZOPOULOS et al., 1989). Self-inactivating vectors (SIN), in which a deletion in the 3′LTR leads to the inactivation of the proviral transcriptional unit, have also been constructed (YU et al., 1986).

The host range of retroviruses and also the recombinant vectors derived thereof are dependent on the type of envelope protein (env) supplied by the helper cell. The murine ecotropic env-packaged viruses are able to infect murine and some other rodent cells whereas the amphotropic env-derived viruses infect murine as well as non-murine cells. The choice of the helper cell line is therefore important

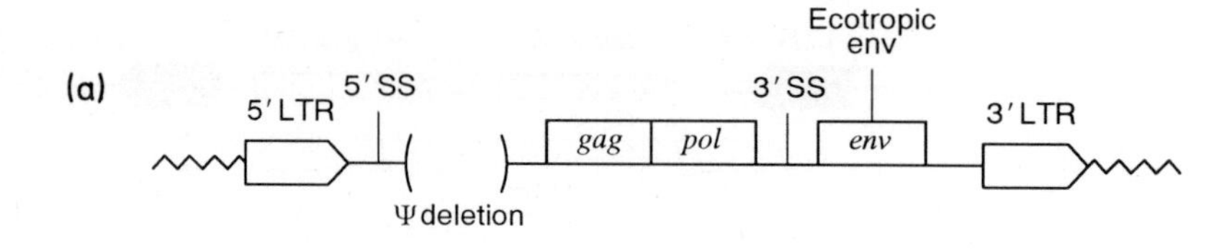

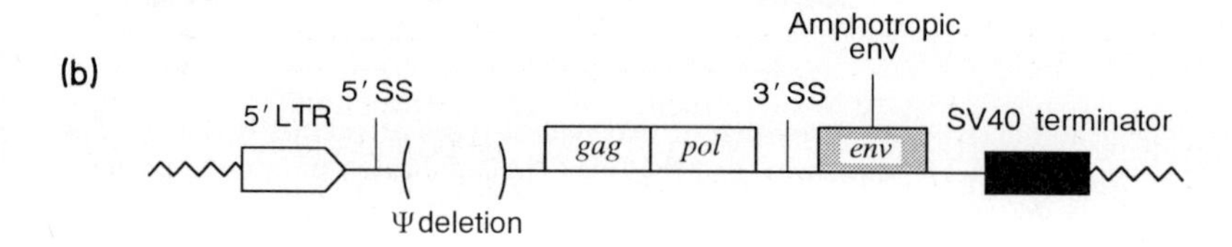

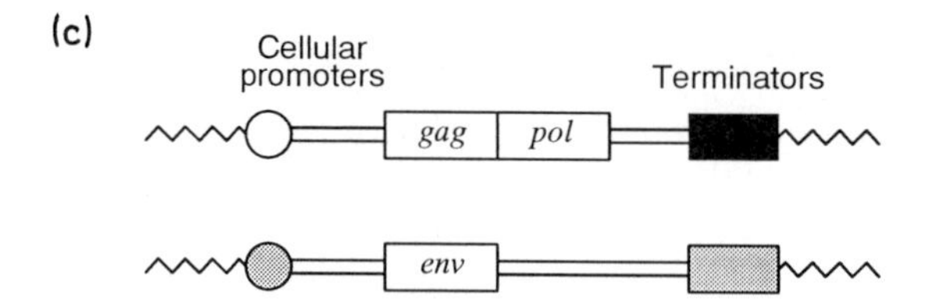

Fig. 30. Recombinant retroviral genomes in packaging cell lines. The factors required to rescue defective viral genomes (retroviral vectors) are supplemented in *trans*. Three typical viral genomes are shown.
(a) The first developed recombinant packaging cell line (Ψ2) contains a wild-type murine leukemia virus genome with an extended deletion covering the packaging signal sequence Ψ (MANN et al., 1983).
(b) In other packaging cell lines, the ecotropic env gene is replaced by the amphotropic env. In the case of PA317 part of the 5′LTR is deleted, the entire 3′LTR is deleted, and an SV40 polyadenylation signal is inserted in place of the 3′LTR (MILLER and BUTTIMORE, 1986).
(c) Recent developments have led to creation of safer packaging cell lines. The helper functions, the reading frames of gag/pol, and env are located on two different genomes. The retroviral sequences are reduced to a minimum in order to minimize homologous recombination with retroviral vectors or with each other.

for the use of the recombinant retrovirus. A series of ecotropic and amphotropic helper cell lines are now available (MCLAUCHLIN et al., 1990).

Of great importance is the titer of virus particles that can be achieved. The titer is dependent on many parameters including the origin of the packaging cell line, the expression of the helper functions, the retroviral vector, the inserted genes, and the mechanism of transfer into the helper cells. Since the strength of expression of the retroviral vector is important for the titer, different approaches to improve its expression or to select for high expression helper cell clones have been undertaken. Another concept includes the cocultivation of two different packaging cell lines (e.g., amphotropic and ecotropic), each producing differently packaged vectors. This approach is called the "ping-pong" amplification method (BESTWICK et al., 1988; BODINE et al., 1990). Titers of up to 10^{10} cfu/mL were reported, but could not be reproduced in several laboratories.

Expression of foreign genes by using retrovirus vectors is usually low. Some vectors have been constructed in which expression is improved (MORGENSTERN and LAND, 1990). The major advantage in using retroviral expression vectors is that one can define the number of gene copies by choice of the number of particles infecting the host cell, and this will achieve stable expression with high efficiency of gene transfer in a broad range of cell types (LEVER, 1991).

3.2.3.2 The Vaccinia System

Vaccinia virus (a pox virus) contains a large double-stranded DNA. It has a very wide host range being able to infect vertebrate and some invertebrate cells. It replicates in the cytoplasm of infected cells and is therefore isolated from the host's nuclear functions. After infection and uncoating of the envelope the early viral genes are activated by a completely functional virus-encoded transcriptional complex contained within the viral core particles and then transcribed. The DNA is released into the cytoplasm of the host cells followed by DNA replication, down-regulation of early gene transcription, and the initiation of late gene expression. The late genes encode structural proteins which function during the formation of mature virions, as well as catalytic proteins which are packaged into the newly arising virus particles. After a complex morphogenetic process, progeny viral particles are produced and released from the cells by cellular lysis.

Vaccinia virus vectors have been primarily used during the generation of experimental vaccines (PANICALI and PAOLETTI, 1982). In the last years, however, it was found that this system is a good candidate to overexpress certain proteins. Vaccinia viral infections shut down the host cell protein synthesis during high-level production of the vaccinia virus derived proteins.

Since the vaccinia virus genome is large (185 kbp), handling would be extremely difficult. Therefore, transfer vectors were constructed which contain vaccinia virus DNA flanking the gene of interest and the selectable marker. The selectable marker for these vectors is usually the thymidine kinase (TK) gene of the vaccinia virus itself. The production of recombinant vaccinia viruses is accomplished by recombination between the transfer vector DNA and vaccinia DNA introduced into the cells by infection (Fig. 31). The transfer vector DNA is transfected into cultured cells. The cell is coinfected with wild-type vaccinia virus. Recombination between homologous DNA of both parts occurs during replication. The resulting recombinant vaccinia viral DNA can be replicated and packaged into infectious vaccinia virus particles. The transfer vectors contain vaccinia virus TK gene sequences which are interrupted by the gene of interest. Upon recombination the vaccinia viral TK gene is interrupted by the gene of interest rendering the recombinant virus TK^-. Such recombinant viruses are selected on the basis of their TK^- phenotype. For expression of the recombinant gene usually vaccinia promoters have been used. Improvement of the expression system was achieved by using the prokaryotic T7 promoter (FUERST et al., 1987; MOSS et al., 1990). Its expression is independent of the vaccinia life cycle. A series of further improvements have led to the generation of excellent systems for very high gene expression. This includes the use of dominant selectable markers which allow the selection for cells infected with recombinant vaccinia virus (FALKNER and MOSS, 1988). Further improvements were obtained by making the gene of interest regulatable using the prokaryotic lac system (MOSS et al., 1990).

The major restrictions for the use of vaccinia virus systems are the following: (1) The host range of vaccinia viruses imposes strict safety considerations for the recombinant vaccinia viruses. (2) The expression system, although very efficient, is transient. (3) Since a series of host cell functions are shut off during vaccinia virus infection, the production or modification of proteins derived from recombinant vaccinia virus infected cells might not be identical to those produced from non-infected cells.

3.2.3.3 The Baculovirus System

Expression systems based on insect cells infected with recombinant baculoviruses have provided a valuable source of protein expression. The natural hosts of baculoviruses are arthropods. Virus replication in other organisms has not been reported. The system, therefore, represents a safe system for laboratory use. Some cell lines from arthropods have been isolated and allow the propagation of wild-type and recombinant baculoviruses. The baculovirus which is most widely used in expression systems is *Autographa californica* nuclear po-

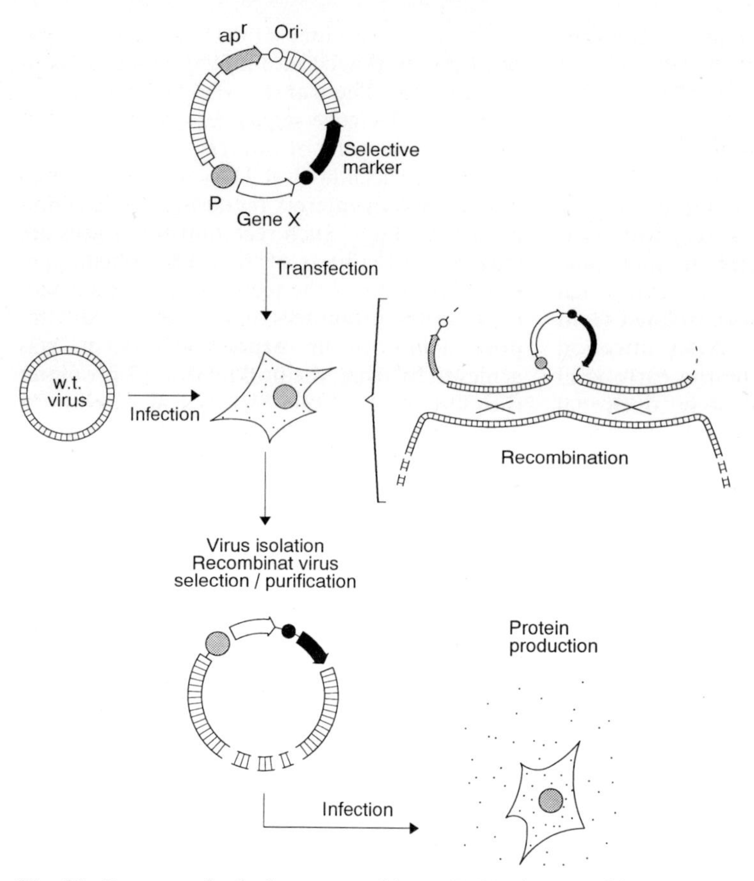

Fig. 31. Gene transfer by large recombinant DNA-viruses. This schematical approach describes the creation of recombinant infectious particles based on vaccinia- and baculoviruses. The gene of interest is integrated into a small vector which contains sequences homologous to the wild-type virus. The vector also contains a marker gene which permits selection of cells which do or do not express this marker. Stable transfectants with this type of vector are selected for and infected with wild-type virus. Double reciprocal recombination of the vector with the wild-type virus leads to hybrids in which a short piece of the virus DNA is replaced by the gene of interest. These cells produce a mixture of wild-type and recombinant viral particles. When these particles are used to infect new cells with a low multiplicity of infection, the hosts which only harbor the recombinant virus are selected for. These cells produce homogeneous recombinant virus particles which can be used to infect the cells in which the gene of interest will be expressed. The selection system depends on the virus chosen, the vector, and the promoter which drives the gene of interest.

lyhedrosis virus (AcMNPV). AcMNP viruses are produced in two different forms during baculoviral infection. Release of extracellular virus takes place throughout the early time of infection and declines in the late phases of infection. During the late phase an occluded form, packaged in polyhedra, is produced. The latter form is extremely resistant and is only important for the lateral transmission of the virus in insects. The polyhedra are highly stable and consist mostly of a single protein, the polyhedrin (MILLER, 1988; BISHOP and POSSEE, 1990).

The polyhedrin protein accumulates to more than half of the total cellular protein in the late phase of virus infection. This protein is not necessary for viral infection or replication of the extracellular virus in tissue culture. In recombinant baculoviruses, the polyhedrin gene is replaced by the gene of interest. This is achieved by homologous recombination. Since the resulting mutant viruses do not produce occlusions, the host cells can be isolated by distinct occlusion-negative plaque morphology. Viruses isolated from cells with the occlusion-negative plaque morphology give rise to overexpression of the gene of interest in secondary infection cycles of new infectants. Another protein, p10, with unknown function is also efficiently expressed in the late phase of infection. Since it is non-essential for virus replication/production in insect cells, it can be replaced by genes of interest as well.

The large size of the baculovirus genomic DNA (130 kbp) allows the insertion of large amounts of heterologous DNA. Single genes as well as multiple genes from one expression vector have been expressed (FRENCH and ROY, 1990).

The following discussion describes expression of foreign genes in baculovirus expression vectors (BISHOP and POSSEE, 1990). The principle is very similar to that described for production of recombinant vaccinia viruses (see Fig. 31). As a first step the gene of interest is cloned into a plasmidal vector, usually under the control of the polyhedrin promoter. In this vector baculoviral DNA sequences flank the recombinant gene. In the second step the vector with the gene of interest is cotransfected together with AcMNPV-DNA into insect cells. During DNA replication in these cells, homologous recombination takes place between the flanking viral sequences of the vector and the wild-type virus DNA. This event, which occurs in a small percentage of cells, results with the excision of the wild-type polyhedrin gene and its replacement by the gene of interest (see Fig. 31). Viruses from cells in which recombination has taken place produce a mixture of wild-type and recombinant viruses. These viruses are then used for a second round of infection. Recombinant viruses are isolated by limited dilution during infection and selection for occlusion-minus plaques. Recombinant virus isolation may be repeated several times until wild-type virus is no longer present. The plaque purified recombinant virus can be propagated and stored for further infections. Infected cells in the late phase of virus production produce large amounts of the gene of interest.

The expression yield is usually no more than 50 μg/10^6 cells. This yield varies according to the gene product that is being expressed. In general, 1–10 μg of heterologous protein per 10^6 cells are obtained.

The production of mammalian proteins in insect cells has only a few limitations. Normally secreted proteins are also secreted in baculovirus-infected insect cells. Furthermore, a series of post-translational modifications including phosphorylation, glycosylation, and myristilation have been demonstrated. However, these modifications do not occur in all cases, as found in mammalian cells, e.g., O-glycosylation seems to be inefficient in insect cells, and the processing of N-glycosylated proteins also differs between insect and mammalian cells (KURODA et al., 1990; LUCKOW, 1991).

Acknowledgement

We thank I. Dortmund for secretarial help, and Drs. J. Bode, H. Conradt, G. Gross, and E. Wingender for critical comments.

4 References

ACCILI, D., TAYLOR, S. I. (1991), Targeted inactivation of the insulin receptor gene in mouse 3T3-L1 fibroblasts via homologous recombination, *Proc. Natl. Acad. Sci. USA* **88**, 4708–4712.

ADAIR, G. M., NAIRN, R. S., WILSON, J. H., SEIDMAN, M. M., BROTHERMAN, K. A., MACKINNON, C., SCHEERER, J. B. (1989), Targeted homologous recombination at the endogenous adenine phosphoribosyltransferase locus in Chinese hamster cells, *Proc. Natl. Acad. Sci. USA* **86**, 4574–4578.

ADAM, M. A., RAMESH, N., MILLER, A. D., OSBORNE, W. R. A. (1991), Internal initiation of translation in retroviral vectors carrying picornavirus 5′ nontranslated regions, *J. Virol.* **65**, 4985–4990.

ALBERTS, B., STERNGLANZ, R. (1990), Gene expression: Chromatin contract to silence, *Nature* **344**, 193–194.

ALT, F., BEATINO, R., SCHIMKE, R. T. (1978), Selective multiplication of dihydrofolate reductase in methotrexate resistant variants of cultured murine cells, *J. Biol. Chem.* **253**, 1357–1370.

ANDRULIS, I. L., DUFF, C., EVANS-BLACKLER, S., WORTON, R., SIMINOVITCH, L. (1983), Chromosomal alterations associated with overproduction of asparagine synthetase in Albizziin-resistant Chinese Hamster ovary cells, *Mol. Cell. Biol.* **3**, 391–398.

ANTHONY, D. D., MERRICK, W. C. (1991), Eukaryotic initiation factor (eIF)-4F, *J. Biol. Chem.* **266**, 10218–10226.

AOKI, D., LEE, N., YAMAGUCHI, N., DUBOIS, C., FUKUDA, M. N. (1992), Golgi retention of a *trans*-Golgi membrane protein, galactosyl-transferase, requires cysteine and histidine residues within the membrane-anchoring domain. *Proc. Natl. Acad. Sci. USA* **89**, 4319–4323.

ARTELT, P., MORELLE, C., AUSMEIER, M., FITZEK, M., HAUSER, H. (1988), Vectors for efficient expression in mammalian fibroblastoid, myeloid and lymphoid cells via transfection or infection, *Gene* **68**, 213–219.

ARTELT, P., GRANNEMANN, R., STOCKING, C., FRIEL, J., BARTSCH, J., HAUSER, H. (1991), The prokaryotic neomycin resistance gene acts as a transcriptional silencer in eukaryotic cells, *Gene* **99**, 249–254.

BACHMAIR, A., VARSHAVSKY, A. (1989), The degradation signal in a short-lived protein, *Cell* **56**, 1019–1032.

BACHMAIR, A., FINLEY, D., VARSHAVSKY, A. (1986), *In vivo* half-life of a protein is a function of its amino-terminal residue, *Science* **234**, 179–186.

BAEUERLE, P. A., HUTTNER, W. B. (1987), Tyrosine sulfation is a trans-Golgi-specific protein modification, *J. Cell. Biol.* **105**, 2655–2664.

BANERJEE, A. K. (1980), 5′-Terminal cap structure in eukaryotic messenger ribonucleic acid, *Microbiol. Rev.* **44**, 175–205.

BEACH, L. R., PALMITER, R. D. (1981), Amplification of the metallothionein-I gene in cadmium-resistant mouse cells, *Proc. Natl. Acad. Sci. USA* **78**, 2110–2114.

BEATO, M. (1989), Gene regulation by steroid hormones, *Cell* **56**, 335–344.

BEBBINGTON, C. R., HENTSCHEL, C. C. G. (1987), The use of vectors based on gene amplification for the expression of cloned genes in mammalian cells, in: *DNA Cloning* (GLOVER, D. M., Ed.), Vol. III, pp. 163–168, Oxford: IRL Press.

BECKERS, T., HAUSER, H., HÜSKEN, D., ENGELS, J. W. (1988), Expression and characterization of a des-methionine mutant interleukin-2 receptor (Taq protein) with interleukin-2 binding affinity, *J. Biol. Chem.* **236**, 8359–8365.

BEHR, J.-P., DEMENEIX, B., LOEFFLER, J.-P., PEREZ-MUTUL, J. (1989), Efficient gene transfer into mammalian primary endocrine cells with lipopolyamine-coated DNA, *Proc. Natl. Acad. Sci. USA* **86**, 6982–6986.

BENDIG, M. M., STEPHENS, P. E., CROCKETT, M. I., HENTSCHEL, C. C. G. (1987), Mouse cell lines that use heat shock promoters to regulate the expression of tissue plasminogen activator, *DNA* **6**, 343–352.

BERG, D. T., Q. MOONEY, P., BAEZ, M., GRINNELL, B. W. (1988), Tandem promoter/enhancer units create a versatile regulatory element for the expression of genes in mammalian cells, *Nucleic Acids Res.* **16**, 1635.

BERMAN, L. W., KUEHL, W. M. (1979), Formation of an intrachain disulfide bond on nascent immunoglobulin licht chains, *J. Biol. Chem.* **254**, 8869–8876.

BERNARD, H.-U., KRÄMMER, G., RÖWEKAMP, W. G. (1985), Construction of a fusion gene that confers resistance against hygromycin B to mammalian cells in culture, *Exp. Cell Res.* **158**, 237–243.

BERNSTEIN, P., ROSS, J. (1989), Poly(A), poly(A) binding protein and the regulation of mRNA stability, *Trends Biochem. Sci.* **14**, 373–377.

BESTWICK, R. K., KOZAK, S. L., KABAT, D. (1988), Overcoming interference to retroviral superinfection results in amplified expression and transmission of cloned genes, *Proc. Natl. Acad. Sci. USA* **85**, 5404–5408.

BISHOP, D. H. L., POSSEE, R. D. (1990), Baculovirus expression vectors, *Gene Technol.* **1**, 55–72.

BLOBEL, G. (1980), Intracellular protein topogenesis, *Proc. Natl. Acad. Sci. USA* **77**, 1496–1500.

BLOCHLINGER, K., DIGGELMANN, H. (1984), Hygromycin B phosphotransferase as a selectable marker for DNA transfer experiments with higher eukaryotic cells, *Mol. Cell. Biol.* **4**, 2929–2931.

BLOM VAN ASSENDELFT, G., HANSCOMBE, O., GROSVELD, F., GREAVES, D. R. (1989), The beta-globin dominant control region activates homologous and heterologous promoters in a tissue-specific manner, *Cell* **56**, 969–977.

BODINE, D. M., MCDONAGH, K. T., BRANDT, S. J., NEY, P. A., AGRICOLA, B., BYRNE, E., NIENHUIS, A. W. (1990), Development of a high-titer retrovirus producer cell line capable of gene transfer into rhesus monkey hematopoietic stem cells, *Proc. Natl. Acad. Sci. USA* **87**, 3738–3742.

BOEL, E., BERKNER, K. L., NEXOE, B. A., SCHWARTZ, T. W. (1987), Expression of a human pancreatic polypeptide precursors from a dicistronic mRNA in mammalian cells, *FEBS Lett.* **219**, 181–188.

BOLE, D. G., HENDERSHOT, L. M., KEARNEY, J. F. (1986), Posttranslational association of immunoglobulin heavy chain binding protein with nascent heavy chains in nonsecreting and secreting hybridomas, *J. Cell. Biol.* **102,** 1558–1566.

BONIFER, C., HECHT, A., SAUERESSIG, H., WINTER, D. M., SIPPEL, A. E. (1991), Dynamic chromatin: The regulatory domain organization of eukaryotic gene loci, *J. Cell. Biochem.* **47,** 99–108.

BOSHART, M., WEBER, F., JAHN, G., DORSCH-HÄSLER, K., FLECKENSTEIN, B., SCHAFFNER, W. (1985), A very strong enhancer is located upstream of an immediate early gene of human cytomegalovirus, *Cell* **41,** 521–530.

BOWTELL, D. D. L., CORY, S., JOHNSON, G. R., GONDA, T. J. (1988), Comparison of expression in hemopoietic cells by retroviral vectors carrying two genes, *J. Virol.* **62** (7), 2464–2473.

BREATHNACH, R., CHAMBON, P. (1981), Organisation and expression of eukaryotic split genes coding for proteins, *Annu. Rev. Biochem.* **50,** 349–383.

BREDENBEEK, P. J., PACHUK, C. J., NOTEN, A. F. H., CHARITE, J., LUYTJES, W., WEISS, S. R., SPAAN, W. J. M. (1990), The primary structure and expression of the second open reading frame of the polymerase gene of the coronavirus MHV-A59. A highly conserved polymerase is expressed by an efficient ribosomal frameshifting mechanism, *Nucleic Acids Res.* **18,** 1825–1832.

BRIERLEY, I., BOURSNELL, M. E. G., BINNS, M. M., BILIMORIA, B., BLOK, V. C., BROWN, T. D. K., INGLIS, S. C. (1987), An efficient ribosomal frame-shifting signal in the polymerase-encoding region of the coronavirus IBV, *EMBO J.* **6,** 3779–3785.

BRINSTER, R. L., ALLEN, J. M., BEHRINGER, R. R., GELINAS, R. E., PALMITER, R. D. (1988), Introns increase transcriptional efficiency in transgenic mice, *Proc. Natl. Acad. Sci. USA* **85,** 836–840.

BROKER, T. R., BOTCHAN, M. (1986), Papillomaviruses: Retrospectives and prospectives, *Cancer Cells* **4,** 17–43.

BROWN, D. A., ROSE, J. K. (1992), Sorting of GPI-anchored proteins to glycolipid-enriched membrane subdomains during transport to the apical cell surface, *Cell* **68,** 533–544.

BRUCH, M.D., MCKNIGHT, C. J., GIERASCH, L. M. (1989), Helix formation and stability in a signal sequence, *Biochemistry* **28,** 8554–8561.

BRYANT, M. L., HEUCKEROTH, R. O., KIMATA, J. T., RATNER, L., GORDON, J. I. (1989), Replication of human immunodeficiency virus 1 and Moloney murine leukemia virus is inhibited by different heteroatom-containing analogs of myristic acid, *Proc. Natl. Acad. Sci. USA* **86,** 8655–8659.

BU, X., HAGEDORN, C. H. (1991), Platelet-derived growth factor stimulates phosphorylation of the 25 kDa mRNA cap binding protein (eIF-4E) in human lung fibroblasts, *FEBS Lett.* **283,** 219–222.

BUCHMAN, A. R., BERG, P. (1988), Comparison of intron-dependent and intron-independent gene expression, *Mol. Cell. Biol.* **8,** 4395–4405.

BURGESS, T. L., KELLY, R. B. (1987), Constitutive and regulated secretion of proteins, *Annu. Rev. Biochem.* **3,** 243–293.

BURKE, R. L., PACHL, C., QUIROGA, M., ROSENBERG, S., HAIGWOOD, N, NORDFANG, O., EZBAN, M. (1986), The functional domains of coagulation factor VIII:C*, *J. Biol. Chem.* **261,** 12574–12578.

CALOS, M. P., LEBKOWSKI, J. S., BOTCHAN, M. R. (1983), High mutation frequency in DNA transfected into mammalian cells, *Proc. Natl. Acad. Sci. USA* **80,** 3015–3019.

CAPECCHI, M. R. (1980), High efficiency transformation by direct microinjection of DNA into cultured mammalian cells, *Cell* **22,** 479–488.

CARROLL, S. M., DEROSE, M. L., GAUDRAY, P., MOORE, C. M., NEEDHAM-VANDEVANTER, D. R., VON HOFF, D. D., WAHL, G. M. (1988), Double minute chromosomes can be produced from precursors derived from a chromosomal deletion, *Mol. Cell. Biol.* **8,** 1525–1532.

CARTER, A. D., FELBER, B. K., WALLING, M. J., JUBIER, M. F., SCHMIDT, C. J., HAMER, D. H. (1984), Duplicated heavy metal control sequences of the mouse metallothionein-I gene, *Proc. Natl. Acad. Sci. USA* **81,** 7392–7396.

CARTIER, M., STANNERS, C. P. (1990), Stable, high-level expression of a carcinoembryonic antigen-encoding cDNA after transfection and amplification with the dominant and selectable asparagine synthetase marker, *Gene* **95,** 223–230.

CEPKO, C. L., ROBERTS, B. E., MULLIGAN, R. C. (1984), Construction and applications of a highly transmissible murine retrovirus shuttle vector, *Cell* **37,** 1053–1062.

CHARRON, J., MALYNN, B. A., ROBERTSON, E. J., GOFF, S. P., ALT, F. W. (1990), High-frequency disruption of the N-myc gene in embryonic stem and pre-B cell lines by homologous recombination, *Mol. Cell. Biol.* **10,** 1799–1804.

CHARTIER, M., CHANG, M., STANNERS, C. (1987), Use of the *Escherichia coli* gene for asparagine synthetase as a selective marker in a shuttle vector capable of dominant transfection and amplification in animal cells, *Mol. Cell. Biol.* **7,** 1623.

CHEN, C., OKAYAMA, H. (1987), High-efficiency transformation of mammalian cells by plasmid DNA, *Mol. Cell. Biol.* **7,** 2745–2752.

CHIANG, T., MCCONLOGUE, L. (1988), Amplification and expression of heterologous ornithine carboxylase in Chinese hamster cells, *Mol. Cell. Biol.* **8**, 764.

CHU, G., HAYAKAWA, H., BERG, P. (1987), Electroporation for the efficient transfection of mammalian cells with DNA, *Nucleic Acids Res.* **15**, 1311-1326.

COCHRANE, A. W., PERKINS, A., ROSEN, C. A. (1990), Identification of sequences important in the nucleolar localization of Human Immunodeficiency Virus Rev: Relevance of nucleolar localization to function, *J. Virol.* **64**, 881-885.

COLBÈRE-GARAPIN, F., HORODNICEANU, F., KHOURILSKY, P., GARAPIN, A. C. (1981), A new dominant hybrid selective marker for higher eukaryotic cells, *J. Mol. Biol.* **150**, 1-13.

COLLIS, P., ANTONIOU, M., GROSVELD, F. (1990), Definition of the minimal requirements within the human beta-globin gene and the dominant control region for high level expression, *EMBO J.* **9**, 233-240.

CONRADT, H. S., HOFER, B., HAUSER, H. (1990), Expression of human glycoproteins in recombinant mammalian cells: Towards genetic engineering of N- and O-glycoproteins, *TIGG* **2**, 168-181.

CORDINGLEY, M. G., RIEGEL, A. T., HAGER, G. L. (1987), Steroid-dependent interaction of transcription factors with the inducible promoter of mouse mammary tumor virus, *Cell* **48**, 261-270.

COSTA, R. H., GRAYSON, D. R., DARNELL, J. E., JR. (1989), Multiple hepatocyte-enriched nuclear factors function in the regulation of transthyretin and alpha 1-antitrypsin genes, *Mol. Cell. Biol.* **9**, 1415-1425.

CULLEN, B. R., LOMEDICO, P. T., JU, G. (1984), Transcriptional interference in avian retroviruses-implications for the promoter insertion model of leukaemogenesis, *Nature* **307**, 241-245.

CULP, J. S., JOHANSEN, H., HELLMIG, B., BECK, J., MATTHEWS, T. J., DELERS, A., ROSENBERG, M. (1991), Regulated expression allows high level production and secretion of HIV-1 gp120 envelope glycoprotein in *Drosophila* Schneider cells, *BioTechnology* **9**, 173-177.

DANOS, O., MULLIGAN, R. C. (1988), Safe and efficient generation of recombinant retroviruses with amphotropic and ecotropic host ranges, *Proc. Natl. Acad. Sci. USA* **85**, 6460-6464.

DEBATISSE, M., SAITO, I., BUTTIN, G., STARK, G. R. (1988), Preferential amplification of rearranged sequences near amplified adenylate deaminase genes, *Mol. Cell. Biol.* **8**, 17-24.

DECHIARA, T. M., EFSTRATIADIS, A., ROBERTSON, E. J. (1990), A growth-deficiency phenotype in heterozygous mice carrying an insulin-like growth factor II gene disrupted by gene targeting, *Nature* **345**, 78-80.

DE LA LUNA, S., SORIA, I., PULIDO, D., ORTIN J., JIMENEZ, A. (1988), Efficient transformation of mammalian cells with constructs containing a puromycin-resistance marker, *Gene* **62**, 121-125.

DE SAINT VINCENT, B. R., DELBRÜCK, S., ECKHART, W., MEINKOTH, J., VITTO, L., WAHL, G. (1991), The cloning and reintroduction into animal cells of a functional CAD gene, a dominant amplifiable genetic marker, *Cell* **27**, 267-277.

DICE, J. F. (1987), Molecular determinants of protein half-lives in eukaryotic cells, *FASEB J.* **1**, 349-357.

DIMAIO, D., TREISMAN, R., MANIATIS, T. (1982), Bovine papillomavirus vector that propagates as a plasmid in both mouse and bacterial cells, *Proc. Natl. Acad. Sci. USA* **79**, 4030-4035.

DINGWALL, C., LASKEY, R. A. (1991), Nuclear targeting sequences - a consensus, *Trends Biochem. Sci.* **16**, 478-481.

DORNER, A. J., KRANE, M. G., KAUFMANN, R. J. (1988), Reduction of endogenous GRP78 levels improves secretion of a heterologous protein in CHO cells, *Mol. Cell. Biol.* **8**, 4063-4070.

EILERS, M., SCHATZ, G. (1986), Binding of a specific ligand inhibits import of a purified precursor protein into mitochondria, *Nature* **322**, 228-232.

EILERS, M., PICARD, D., YAMAMOTO, K. R., BISHOP, J. (1989), Chimaeras of Myc oncoprotein and steroid receptors cause hormone-dependent transformation of cells, *Nature* **340**, 66-68.

EISSENBERG, J. C., ELGIN, S. C. R. (1991), Boundary functions in the control of gene expression, *Trends Genet.* **7**, 335-340.

EMERMAN, M., TEMIN, H. M. (1986), Quantitative analysis of gene suppression in intergrated retrovirus vector, *Mol. Cell. Biol.* **6**, 793-800.

ENGELMAN, D. M., STEITZ, T. A. (1981), The spontaneous insertion of proteins into and across membranes: The helical hairpin hypothesis, *Cell* **23**, 411-422.

FALCONE, D., ANDREWS, D. W. (1991), Both the 5′untranslated region and the sequences surrounding the start site contribute to efficient initiation of translation *in vitro*, *Mol. Cell. Biol.* **11**, 2656-2664.

FALKNER, S. G., MOSS, B. (1988), *Escherichia coli gpt* gene provides dominant selection of vaccinia virus open-reading-frame expression vectors, *J. Virol.* **62**, 1849.

FELGNER, P. L., GADEK, T. R., HOLM, M., ROMAN, R. H., CHAN, H. W., WENZE, M., NORTHROP, J. P., RINGOLD, G. M., DANIELSEN, M. (1987), Lipofection: A highly efficient, lipid mediated DNA-transfection procedure, *Proc. Natl. Acad. Sci. USA* **84,** 7413–7417.

FELL, H. P., YARNOLD, S., HELLSTROEM, I., HELLSTROEM, K. E., FOLGER, K. R. (1989), Homologous recombination in hybridoma cells: Heavy chain chimeric antibody produced by gene targeting, *Proc. Natl. Acad. Sci. USA* **86,** 8507–8511.

FOLGER, K.R., WONG, E. A., WAHL, G., CAPECCHI, M. (1982), Patterns of integration of DNA microinjected into cultured mammalian cells: evidence for homologous recombination between injected plasmid DNA molecules, *Mol.Cell. Biol.* **2,** 1372–1387.

FORD, M., FRIED, M. (1986), Large inverted duplications are associated with gene amplification, *Cell* **45,** 425–434.

FRENCH, T. J., ROY, P. (1990), Synthesis of bluetongue virus (BTV) corelike particles by a recombinant baculovirus expressing the two major structural core proteins of BTV, *J. Virol.* **64,** 1530–1536.

FRIEDLANDER, M., BLOBEL, G. (1985), Bovine opsin has more than one signal sequence, *Nature* **318,** 338–342.

FROST, E., WILLIAMS, J. (1978), Mapping temperature-sensitive and host-range mutations of adenovirus type 5 by marker rescue, *Virology* **91,** 39–50.

FUERST, T. R., EARL, P. L., MOSS, B. (1987), Use of a hybrid vaccinia virus-T7 RNA polymerase system for expression of target genes, *Mol. Cell. Biol.* **7,** 2538–2544.

FUJITA, T., SHIBUYA, H., HOTTA, H., YAMANISHI, K., TANIGUCHI, T. (1987), Interferon-beta gene regulation: tandemly repeated sequences of a synthetic 6 bp oligomer function as a virus-inducible enhancer, *Cell* **49,** 357–367.

GALLIE, D. R. (1991), The cap and poly(A) tail function synergistically to regulate mRNA translational efficiency, *Genes Dev.* **5,** 2108–2116.

GALLIE, D. R., SLEAT, D. E., WATTS, J. W., TURNER, P. C., WILSON, T. M. A. (1987), A comparison of eukaryotic viral 5′-leader sequences as enhancers of mRNA expression *in vivo, Nucleic Acids Res.* **15,** 8693–8711.

GALLIE, D. R., SLEAT, D. E., WATTS, J. W., TURNER, P. C., WILSON, T. M. A. (1988), Mutational analysis of the tobacco mosaic virus 5′-leader for altered ability to enhance translation, *Nucleic Acids Res.* **16,** 883–893.

GARCIA-BUSTOS, J., HEITMAN, J., HALL, M. N. (1991), Nuclear protein localization, *Biochim. Biophys. Acta* **1071,** 83–101.

GAROFF, H. (1985), DNA techniques to study protein targeting in the eukaryotic cell, *Annu. Rev. Cell. Biol.* **1,** 403–445.

GASSER, S., LAEMMLI, U. K. (1987), A glimpse at chromosomal order, *Trends Genet.* **3,** 16–22.

GENILLOUD, O., GARRIDO, M. C., MORENO, F. (1984), The transposon Tn5 carries a bleomycin-resistance determinant, *Gene* **32,** 225–233.

GETHING, M.-J., SAMBROOK, J. (1992), Protein folding in the cell, *Nature* **355,** 33–45.

GETHING, M.-J., MCCAMMON, K., SAMBROOK, J. (1986), Expression of wild-type and mutant forms of influenza hemagglutinin: The role of folding in intracellular transport, *Cell* **46,** 939–950.

GHATTAS, I. R., SANES, J. R., MAJORS, J. E. (1991), The encephalomyocarditis virus internal ribosome entry site allows efficient coexpression of two genes from a recombinant provirus in cultured cells and in embryos, *Mol. Cell. Biol.* **11,** 5848–5859.

GHODA, L., WETTERS, VAN DAALEN, MACRAE, M., ASCHERMAN, D., COFFINO, P. (1989), Prevention of rapid intracellular degradation of ODC by a carboxyterminal degradation, *Science* **243,** 1493–1495.

GLUZMAN, Y. (1981), SV40-transformed simian cells support the replication of early SV40 mutants, *Cell* **23,** 175–182.

GODEFROY-COLBURN, T., RAVELONANDRO, M., PINCK, L. (1985), Cap accessibility correlates with the initiation efficiency of alfalfa mosaic virus RNAs, *Eur. J. Biochem.* **147,** 549–552.

GONDA, D. K., BACHMAIR, A., WÜNNING, I., TOBIAS, J. W., LANE, W. S., VARSHAVSKY, A. (1989), Universality and structure of the N-end rule, *J. Biol. Chem.* **264,** 16700–16712.

GORMAN, C. M., MERLINO, G. T., WILLINGHAM, M. C., PASTAN, I., HOWARD, B. H. (1982), The Rous sarcoma virus long terminal repeat is a strong promoter when introduced into a variety of eukaryotic cells by DNA-mediated transfection, *Proc. Natl. Acad. Sci. USA* **79,** 6777–6781.

GRAESSMANN, M., GRAESSMANN, A. (1976), "Early" simian virus 40 specific RNA contains information for tumor antigen formation and chromatin replication, *Proc. Natl. Acad. Sci. USA* **73,** 366–370.

GRAHAM, F., VAN DER EB, L. (1973), A new technique for the assay of infectivity of human adenovirus DNA, *Virology* **52,** 456–487.

GRAHAM, F. L., SMILEY, J., RUSSELL, W. C., NAIRN, R. (1977), Characteristics of a human cell line transformed by DNA from adenovirus type 5, *J. Gen. Virol.* **36,** 59–72.

GREAVES, D. R., WILSON, F. D., LANG, G., KIOUSSIS, D. (1989), Human CD2 3′ flanking sequence confer high-level, T cell-specific position-independent gene expression in transgenic mice, *Cell* **56**, 979–986.

GROSS, D., GARRARD, W. T. (1988), Nuclease hypersensitive sites in chromatin, *Annu. Rev. Biochem.* **57**, 159–197.

HAAS, I. G., WABL, M. (1983), Immunoglobulin heavy chain binding protein, *Nature* **306**, 387–389.

HAMER, D. H., LEDER, P. (1979), Splicing and the formation of stable RNA, *Cell* **18**, 1299–1302.

HANTZOPOULOS, P. A., SULLENGER, B. A., UNGERS, G., GILBOA, E. (1989), Improved gene expression upon transfer of the adenosine deaminase minigene outside the transcriptional unit of a retroviral vector, *Proc. Natl. Acad. Sci. USA* **86**, 3519–3523.

HART, G. W., HOLT, G. H., HALTIWANGER, R. S. (1988), Nuclear and cytoplasmic glycosylation, *Trends Biochem. Sci.* **13**, 380–384.

HARTL, F.-U., NEUPERT, W. (1990), Protein sorting to mitochondria: Evolutionary conservations of folding and assembly, *Science* **247**, 930–938.

HARTMANN, S. C., MULLIGAN, R. C. (1988), Two dominant-acting selectable markers for gene transfer studies in mammalian cells, *Proc. Natl. Acad. Sci.* **85**, 8047–8051.

HASTY, P., RAMIREZ-SOLIS, R., KRUMLAUF, R., BRADLEY, A. (1991), Introduction of a subtle mutation into the Hox-2.6 locus in embryonic stem cells, *Nature* **350**, 243–246.

HENTZE, M. W. (1991), Translational regulation of ferritin biosynthesis by iron: A review, *Curr. Stud. Hematol. Blood Transfus.* **58**, 115–126.

HEUCKEROTH, R. O., GORDON, J. I. (1989), Altered membrane association of p60 v-src and a murine 63-kDa N-myristoyl protein after incorporation of an oxygen substituted analog of myristic acid, *Proc. Natl. Acad. Sci. USA* **86**, 5262–5266.

HIRSCHBERG, C. B., SNIDER, M. D. (1987), Topography of glycosylation in the rough endoplasmic reticulum and Golgi apparatus, *Annu. Rev. Biochem.* **56**, 63–89.

HIZI, A., HENDERSON, L. E., COPELAND, T. D., SOWDER, R. C., HIXSON, C. V., OROSZLAN, S. (1987), Characterization of mouse mammary tumor virus gag-pro gene products and the ribosomal frameshift site by protein sequencing, *Proc. Natl. Acad. Sci. USA* **84**, 7041–7045.

HONIGMAN, A., WOLF, D., YAISH, S., FALK, H., PANET, A. (1991), Cis-acting RNA sequences control the gag-pol translation readthrough in murine leukemia virus, *Virology* **183**, 313–319.

HU, M. C.-T., DAVIDSON, N. (1990), A combination of derepression of the lac operator-repressor system with positive induction by glucocorticoid and metal ions provides a high-level-inducible gene expression system based on the human metallothionein-IIA promoter, *Mol. Cell. Biol.* **10**, 6141–6151.

HUANG, M. T. F., GORMAN, C. M. (1990), The simian virus 40 small-t intron, present in many common expression vectors, leads to aberrant splicing, *Mol. Cell. Biol.* **10**, 1805–1810.

HUBERMAN, E., MCKEOWN, C. K., FRIEDMAN, J. (1981), Mutagen-induced resistance to mycophenolic acid in hamster cells can be associated with increased inosine 5′-phosphate dehydrogenase activity, *Proc. Natl. Acad. Sci. USA* **78**, 3151–3154.

HUG, H., COSTAS, M., STAEHELI, P., AEBI, M., WEISSMAN, C. (1988), Organization of the murine Mx gene and characterization of its interferon- and virus-inducible promoter, *Mol. Cell. Biol.* **8**, 3065–3079.

HURT, E. C., PESOLD-HURT, B., SCHATZ, G. (1987), The cleavable prepiece of an imported mitochondrial protein is sufficient to direct cytosolic dihydrofolate reductase into the mitochondrial matrix, *FEBS Lett.* **178**, 306–310.

ISRAEL, D., KAUFMAN, R. J. (1989), Highly inducible expression from vectors containing multiple GRE's in CHO cells overexpressing the glucocorticoid receptor, *Nucleic Acids Res.* **17**, 4589–4604.

ITZHAKI, J. E., PORTER, A. C. G. (1991), Targeted disruption of a human interferon-inducible gene detected by secretion of human growth hormone, *Nucleic Acids Res.* **19**, 3835–3842.

JACKS, T. (1990), Translational suppression in gene expression in retroviruses and retrotransposons, in: R. SWANSTROM, P. K. VOGT (Eds.), *Retroviruses – Strategies of Replication,* Vol. **157** of *Curr. Top. Microbiol. Immunol.*, pp. 93–124, Berlin: Springer.

JACKS, T., VARMUS, H. E. (1985), Expression of the Rous Sarcoma Virus pol gene by ribosomal frameshifting, *Science* **230**, 1237–1242.

JACKS, T., MADHANI, H. D., MASIARZ, F. R., VARMUS, H. E. (1988a), Signals for ribosomal frameshifting in the Rous Sarcoma Virus gag-pol region, *Cell* **55**, 447–458.

JACKS, T., POWER, M. D., MASIARZ, F. R., LUCIW, P. A., BARR, P. J., VARMUS, H. E. (1988b), Characterization of ribisomal frameshifting in HIV-1 gag-pol expression, *Nature* **331**, 280–283.

JACKSON, M.R., NILSSON, T., PETERSON, P. A. (1990), Identification of a consensus motif for retention of transmembrane proteins in the endoplasmic reticulum, *EMBO J.* **9**, 3153–3162.

JACKSON, R. J. (1991), mRNA translation: Initiation without an end, *Nature* **353**, 14–15.

Jackson, R. J., Standart, N. (1990), Do the poly(A) tail and 3′ untranslated region control mRNA translation? *Cell* **62**, 15–24.

Jasin, M., Berg, P. (1988) Homologous integration in mammalian cells without target selection, *Genes Dev.* **2**, 1353–1363.

Jasin, M., Elledge, S. J., Davis, R.W., Berg, P. (1990), Gene targeting at the human CD4 locus by epitope addition, *Genes Dev.* **4**, 157–166.

Jobling, S. A., Gehrke, L. (1987), Enhanced translation of chimeric messenger RNAs containing a plant viral untranslated leader sequence, *Nature* **325**, 622–625.

Johansen, H., Schümperli, D., Rosenberg, M. (1984), Affecting gene expression by altering the length and sequence of the 5′ leader, *Proc. Natl. Acad. Sci. USA* **81**, 7698–7702.

Jones, D. S., Nemoto, F., Kuchino, Y., Masuda, M., Yoshikura, H., Nishimura, S. (1989), The effect of specific mutations at and around the gag pol junction of Moloney murine leukemia virus, *Nucleic Acids Res.* **17**, 5933–5945.

Joyner, A. L., Skarnes, W. C., Rossant, J. (1989), Production of a mutation in mouse En-2 gene by homologous recombination in embryonic stem cells, *Nature* **338**, 153–155.

Kaempfer, R., van Emmelo, J., Fiers, W. (1981), Specific binding of eukaryotic initiation factor 2 to satellite tobacco necrosis virus RNA at a 5′ terminal sequence comprising the ribosome binding site, *Proc. Natl. Acad. Sci. USA* **78**, 1542–1546.

Kaiser, C. A., Botstein, D. (1990), Efficiency and diversity of protein localization by random signal sequences, *Mol. Cell. Biol.* **10**, 3163–3173.

Kaiser, C. A., Preuss, D., Grisafi, P., Botstein, D. (1987), Many random sequences functionally replace the secretion signal sequence of yeast invertase, *Science* **235**, 312–317.

Kalderon, D., Roberts, B. L., Richardson, W. D., Smith, A. E. (1984), A short amino acid sequence able to specify nuclear location. *Cell* **39**, 499–509.

Kamps, M. P., Buss, J. E., Sefton, B. M. (1986), Rous sarcoma virus transforming protein lacking myristic acid phosphorylates known polypeptide substrates without inducing transformation, *Cell* **45**, 105–112.

Kanalas, J. J., Suttle, D. P. (1984), Amplification of the UMP synthase gene and enzyme overproduction in pyrazofurin-resistant rat hepatoma cells, *J. Biol. Chem.* **259**, 1848–1853.

Kane, S. E., Reinhard, D. H., Fordis, C. M., Pastan, I., Gottesman, M. M. (1989), A new vector using the human multidrug resistance gene as a selectable marker enables overexpression of foreign genes in eukaryotic cells, *Gene* **84**, 439–446.

Kaufman, R. (1990), Vectors used for expression in mammalian cells, *Methods Enzymol.* **185**, 487–512.

Kaufman, R. J., Sharp, P. A. (1982), Amplification and expression of sequences cotransfected with a modular dihydrofolate reductase complementary DNA gene, *J. Mol. Biol.* **159**, 601–621.

Kaufman, R. J., Murtha, P. (1987), Translational control mediated by eucaryotic initiation factor-2 is restricted to specific mRNAs in transfected cells, *Mol. Cell. Biol.* **7**, 1568–1571.

Kaufman, R. J., Wasley, L. C., Spiliotes, A. J., Gossels, S. D., Latt, S. A., Larsen, G. R., Kay, R. M. (1985), Coamplification and coexpression of human tissue-type plasminogen activator and murine dihydrofolate reductase sequences in Chinese hamster ovary cells, *Mol. Cell. Biol.* **5**, 1750–1759.

Kaufman, R. J., Murtha, P., Ingolia, D. E., Yeung, C.-Y., Kellems, E. R. (1986), Selection and amplification of heterologous genes encoding adenosine deaminase in mammalian cells, *Proc. Natl. Acad. Sci. USA* **80**, 3136–3140.

Kaufman, R. J., Murtha, P., Davies, M. V. (1987), Translational efficiency of polycistronic mRNAs and their utilization to express heterologous genes in mammalian cells, *EMBO J.* **6**, 187–193.

Kaufman, R. J., Davies, M. V., Wasley, L. C., Michnick, D. (1991), Improved vectors for stable expression of foreign genes in mammalian cells by use of the untranslated leader sequence from EMC virus, *Nucleic Acids Res.* **19**, 4485–4490.

Kavathas, P., Herzenberg, L. A. (1983), Amplification of a gene coding for human T-cell differentiation antigen, *Nature* **306**, 385–387.

Keller, G.-A., Gould, S., Deluca, M., Subramani, S. (1987), Firefly luciferase is targeted to peroxisomes in mammalian cells, *Proc. Natl. Acad. Sci. USA* **84**, 3264–3268.

Kim, H.-S., Smithies, O. (1988), Recombinant fragment assay for gene targetting based on the polymerase chain reaction, *Nucleic Acids Res.* **16**, 8887–8903.

Kingsley, D. M., Kozarsky, K. F., Hobbie, L., Krieger, M. (1986), Reversible defects in O-linked glycosylation and LDL-receptor expression in a UDP-Gal/UDP-GalNAc 4-epimerase deficient mutant, *Cell* **44**, 749–759.

Kitamura, D., Roes, J., Kühn, R., Rajewski, K. (1991), A B cell-deficient mouse by targeted disruption of the membrane exon of the immunoglobulin μ chain gene, *Nature* **350**, 423–426.

KLAUSNER, R. D., HARFORD, J. B. (1989), Cis-trans models for posttranscriptional gene regulation, *Science* **246**, 870–872.

KLEHR, D., MAASS, K., BODE, J. (1991), Scaffold-attached regions from the human interferon-beta domain can be used to enhance the stable expression of genes under the control of various promoters, *Biochemistry* **30**, 1264–1270.

KLEHR, D., SCHLAKE, T., MAASS, K., BODE, J. (1992), Scaffold-attached regions (SAR elements) mediate transcriptional effects due to butyrate, *Biochemistry,* in press.

KLEIN-HITPASS, L., SCHORPP, M., WAGNER, U., RYFFEL, G. U. (1986), An estrogen-responsive element derived from the 5′ flanking region of the *Xenopus* vitellogenin A2 gene functions in transfected human cells, *Cell* **46**, 1053–1061.

KOZAK, M. (1986a), Influences of mRNA secondary structure on initiation by eukaryotic ribosomes, *Proc. Natl. Acad. Sci. USA* **83**, 2850–2854.

KOZAK, M. (1986b), Bifunctional messenger RNAs in eukaryotes, *Cell* **47**, 481–483.

KOZAK, M. (1986c), Point mutations define a sequence flanking the AUG initiator codon that modulates translation by eukaryotic ribosomes, *Cell* **44**, 283–292.

KOZAK, M. (1987a), At least six nucleotides preceding the AUG initiator codon enhance translation in mammalian cells, *J. Mol. Biol.* **186**, 947–950.

KOZAK, M. (1987b), Effects of intercistronic length on the efficiency of reinitiation by eucaryotic ribosomes, *Mol. Cell. Biol.* **7**, 3438–3445.

KOZAK, M. (1988), Leader length and secondary structure modulate mRNA function under conditions of stress, *Mol. Cell. Biol.* **8**, 2737–2744.

KOZAK, M. (1989a), The scanning model for translation: an update, *J. Cell. Biol.* **108**, 229–241.

KOZAK, M. (1989b), Circumstances and mechanisms of inhibition of translation by secondary structure in eucaryotic mRNAs, *Mol. Cell. Biol.* **9**, 5134–5142.

KOZAK, M. (1990), Downstream secondary structure facilitates recognition of initiator codons by eukaryotic ribosomes, *Proc. Natl. Acad. Sci. USA* **87**, 8301–8305.

KRIEGLER, M. (1990), *Gene Transfer and Expression. A Laboratory Manual*, Stockton Press.

KRUYS, V., WATHELET, M., POUPART, P., CONTRERAS, R., FIERS, W., CONTENT, J., HUEZ, G. (1987), The 3′ untranslated region of the human interferon-β mRNA has an inhibitory effect on translation, *Proc. Natl. Acad. Sci. USA* **84**, 6030–6034.

KUCHERLAPATI, R., SKOULTCHI, A. I. (1984), Introduction of purified genes into mammalian cells, *CRC Biochem.* **16**, 349–379.

KUHL, D., DE LA FUENTE, J., CHATURVEDI, M., PARIMOO, S., RYALS, J., MEYER, F., WEISSMANN, C. (1987), Reversible silencing of enhancers by sequences derived from the human IFN-alpha promoter, *Cell* **50**, 1057–1069.

KURATA, S. I., TSUKAKOSHI, M., KASUYA, T., IKAWA, Y. (1986), The laser method for efficient introduction of foreign DNA into cultured cells, *Exp. Cell. Res.* **162**, 372–378.

KURODA, K., GEYER, H., GEYER, R., DOERFLER, W., KLENK, H.-D. (1990), The oligosaccharides of influenza virus hemagglutinin expressed in insect cells by a baculovirus vector, *Virology* **174**, 418–429.

LAIMINS, L., HOLMGREN-KÖNIG, M., KHOURY, G. (1986), Transcriptional "silencer" element in rat repetitive sequence associated with the rat insulin 1 locus, *Proc. Natl. Acad. Sci. USA* **83**, 3151–3155.

LAMB, P., MCKNIGHT, S. L. (1991), Diversity and specificity in transcriptional regulation: the benefits of heterotypic dimerization, *TIBS* **16**, 417–422.

LATCHMAN, D. S. (1991), *Eukaryotic Transcription Factors*, London: Academic Press.

LAZARIS-KARATZAS, A., MONTINE, K. S., SONENBERG, N. (1990), Malignant transformation by a eukaryotic initiation factor subunit that binds to mRNA 5′ cap, *Nature* **345**, 544–547.

LEE, F., MULLIGAN, R., BERG, P., RINGOLD, G. (1981), Glucocorticoids regulate expression of dihydrofolate reductase cDNA in mouse mammary tumour virus chimaeric plasmids, *Nature* **294**, 228.

LEE, E. U., ROTH, J., PAULSON, J. C. (1989), Alteration of terminal glycosylation sequences on N-linked oligosaccharides of Chinese hamster ovary cells by expression of β-galactoside α-2,6-sialyltransferase, *J. Biol. Chem.* **264**, 13848–13855.

LEE, H.-J., SHIEH, C.-K., GORBALENYA, A. E., KOONIN, E. V., LA MONICA, N., TULER, J., BAGDZHADZHYAN, A., LAI, M. M. C. (1991), The complete sequence (22 kilobases) of murine coronavirus gene 1 encoding the putative proteases and RNA polymerase, *Virology* **180**, 567–582.

LENGYEL, P. (1982), Biochemistry of interferons and their actions, *Annu. Rev. Biochem.* **51**, 251–282.

LEVER, A. M. L. (1991), Vectors, *Biochem. Soc. Trans.,* 379–383.

LIEBER, A., KIESSLING, U., STRAUSS, M. (1989), High level gene expression in mammalian cells by a nuclear T7-phage RNA polymerase, *Nucleic Acids Res.* **17**, 8485–8493.

LIN, F.-L., SPERLE, K., STERNBERG, N. L. (1984),

Model for homologous recombination during transfer of DNA into mouse L cells: role for DNA ends in the recombination process, *Mol. Cell. Biol.* **4,** 1020–1034.

LISANTI, M. P., RODRIGUEZ-BOULAN, E. R. (1990), Glycophospholipid membrane anchoring provides clues to the mechanism of protein sorting in polarized epithelial cells, *Trends Biochem. Sci.* **15,** 113–118.

LLEONART, R., NAEF, D., BROWNING, H., WEISSMANN, C. (1990), A novel, quantitative bioassay for type I interferon using a recombinant indicator cell line, *BioTechnology* **8,** 1263–1267.

LOWY, D. R., DVORETZKY, I., SHOBER, R., LAW, M.-F., ENGEL, L., HOWLEY, P. M. (1980a), *In vitro* tumorigenic transformation by a defined sub-genomic fragment of bovine papilloma virus DNA, *Nature* **287,** 72–74.

LOWY, I., PELLICER, A., JACKSON, J. F., SIM, G.-K., SILVERSTEIN, P., AXEL, R. (1980b), Isolation of transforming DNA: cloning the hamster aprt gene, *Cell* **22,** 817–823.

LUCKOW, V. A. (1991), Cloning and expression of heterologous genes in insect cells with baculovirus vectors, in: *Recombinant DNA Technology and Applications* (PROKOP, A., BAJPAI, R. K., HO, C. S., Eds.), pp. 97–152, New York: McGraw-Hill.

LUCKOW, V. A., SUMMERS, M. (1988), Trends in the development of baculovirus expression vectors, *Bio/Technology* **6,** 47–55.

LUSKEY, K. L., FAUST, J. R., CHIN, D. J., BROWN, M. S., GOLDSTEIN, J. L. (1983), Amplification of the gene for 3-hydroxy-3-methylglutaryl coenzyme A reductase, but not for the 53-kDa protein, in UT-1 cells, *J. Biol. Chem.* **258,** 8462–8469.

LUSKY, M., BOTCHAN, M. (1981), Inhibition of SV40 replication in simian cells by specific pBR322 DNA sequences, *Nature* **293,** 79–81.

LYONS, R. H., FERGUSON, B. Q., ROSENBERG, M. (1987), Pentapeptide nuclear localization signal in the adenovirus E1A, *Mol. Cell. Biol.* **7,** 2451–2456.

MACEJAK, D. G., SARNOW, P. (1991), Internal initiation of translation mediated by the 5′ leader of a cellular mRNA, *Nature* **353,** 90–94.

MACHAMER, C. E., ROSE, J. K. (1987), A specific transmembrane domain of a coronavirus E1 glycoprotein is required for its retention in the golgi region 1, *J. Cell. Biol.* **105,** 1205–1214.

MACHAMER, C. E., MENTONE, S. A., ROSE, J. K., FARQUHAR, M. G. (1990), The E1 glycoprotein of an avian coronavirus is targeted to the cis Golgi complex, *Proc. Natl. Acad. Sci. USA* **87,** 6944–6948.

MALIM, M. H., BOEHNLEIN, S., HAUBER, J., CULLEN, B. R. (1989), Functional dissection of the HIV-1 Rev trans-activator - derivation of a trans-dominant repressor of Rev function, *Cell* **58,** 205–214.

MANIATIS, T. (1991), Mechanisms of alternative pre-mRNA splicing, *Science* **251,** 33–34.

MANN, R., MULLIGAN, R. C., BALTIMORE, D. (1983), Construction of a retrovirus packaging mutant and its use to produce helper-free defective retrovirus, *Cell* **33,** 153–159.

MANSOUR, S. L., THOMAS, K. R., DENG, C., CAPECCHI, M. R. (1990), Introduction of a lacZ reporter gene into the mouse int-2 locus by homologous recombination, *Proc. Natl. Acad. Sci. USA* **87,** 7688–7692.

MARKOWITZ, D., GOFF, S., BANK, A. (1988), A safe packaging line for gene transfer: Separating viral genes on two different plasmids, *J. Virol.* **62,** 1120–1124.

MAYER, T., TAMURA, T., FALK, M., NIEMANN, H. (1988), Membrane integration and intracellular transport of the coronavirus glycoprotein E1, a class III membrane glycoprotein, *J. Biol. Chem.* **263,** 14956–14963.

MCCRACKEN, A. A., BROWN, J. L. (1984), A filter immunoassay for detection of protein secreting cell colonies, *BioTechniques* **2,** 82–87.

MCCUTCHAN, J. H., PAGANO, J. S. (1968), Enhancement of the infectivity of simian virus 40 deoxyribonucleic acid with diethylaminoethyl-dextran, *J. Natl. Cancer Inst.* **41,** 351–357.

MCILHINNEY, R. A. J. (1990), The fats of life: the importance and function of protein acylation, *Trends Biochem. Sci.* **15,** 387–391.

MCLAUCHLIN, J. R., CORNETTA, K., EGLITIS, M. A., ANDERSON, W. F. (1990), Retroviral-mediated gene transfer, *Prog. Nucleic Acid Res. Mol. Biol.* **18,** 91–135.

MCNEALL, J., SANCHEZ, A., GRAY, P. P., CHESTERMAN, C. N., SLEIGH, M. J. (1989), Hyperinducible gene expression from a metallothionein promoter containing additional metal-responsive elements, *Gene* **76,** 81–88.

MIELKE, C., KOHWI, Y., KOHWI-SHIGEMATSU, T., BODE, J. (1990), Hierarchical binding of DNA fragments derived from scaffold-attached regions: Correlations of properties *in vitro* and function *in vivo*, *Biochemistry* **29,** 7475–7485.

MIETZ, J., GROSSMAN, Z., LUEDERS, K. K., KUFF, E. L. (1987), Nucleotide sequence of a complete mouse intracisternal A-particle genome: relationship to known aspects of particle assembly and function, *J. Virol.* **61,** 3020–3029.

MILLER, A. D., BUTTIMORE, C. (1986), Redesign of retrovirus packaging cell lines to avoid recombination leading to helper virus production, *Mol. Cell. Biol.* **6,** 2895–2902.

MILLER, A. D., ROSMAN, G. J. (1989), Improved retroviral vectors for gene transfer and expression, *BioTechniques* **7**, 980–990.

MILLER, L. K. (1988) Baculoviruses as gene expression vectors, *Annu. Rev. Microbiol.* **42**, 177–199.

MIYAZAWA, S., OSUMI, T., HASHIMOTO, T., OHNO K., MIURA, S., FUJIKI, Y. (1989), Peroxisome targeting signal of rat liver acyl-coenzyme A oxidase resides at the carboxy terminus, *Mol. Cell. Biol.* **9**, 83–91.

MOREAU, P., HEN, R., WASYLYK, B., EVERETT, R., GAUB, M. P., CHAMBON, P. (1981), The SV40 base-pair repeat has a striking effect on gene expression both in SV40 and other chimeric recombinants, *Nucleic Acids Res.* **9**, 6047–6059.

MORGENSTERN, J. P., LAND, H. (1990), Advanced mammalian gene transfer: high titre retroviral vectors with multiple drug selection markers and a complementary helper-free packaging cell line, *Nucleic Acids Res.* **18**, 3587–3596.

MOSS, B., ELROY-STEIN, O., MIZUKAMI, T., FUERST, W. A., ALEXANDER T. R. (1990), New mammalian expression vectors, *Nature* **348**, 91–92.

MULLIGAN, R. C., BERG, P. (1980), Expression of a bacterial gene in mammalian cells, *Science* **209**, 1422–1427.

MÜLLNER, E. W., NEUPERT, B., KÜHN, L. C. (1989), A specific mRNA binding factor regulates the iron-dependent stability of cytoplasmic transferrin receptor mRNA, *Cell* **58**, 373–382.

MUNRO, S., PELHAM, H. R. B. (1986), An Hsp70-like protein in the ER: Identity with the 78 kd glucose-regulated protein and immunoglobulin heavy chain binding protein, *Cell* **46**, 291–302.

MUNROE, D., JACOBSEN, A. (1990), mRNA poly(A) tail, a 3′ enhancer of translational initiation, *Mol. Cell. Biol.* **10**, 3441–3455.

NAM, S. H., KIDOKORO, M., SHIDA, H., HATANAKA, M. (1988), Processing of gag precursor polyprotein of human T-cell leukemia virus type I by virus encoded protease, *J. Virol.* **62**, 3718–3728.

NEUBERGER, M. S., WILLIAMS, G. T. (1988), The intron requirement for immunoglobulin gene expression is dependent upon the promoter, *Nucleic Acids Res.* **16**, 6713–6724.

NEUHAUS, G., NEUHAUS-URI, G., GRUSS, P., SCHWEIGER, H.-G. (1984), Enhancer-controlled expression of the simian virus 40 T-antigen in the green alga *Acetabularia, EMBO J.* **3**, 2169–2172.

NEUMANN, E., SCHAEFER-RIDDER, M., WANG, Y., HOFSCHNEIDER, P. (1982), Gene transfer into mouse lyoma cells by electroporation in high electric fields, *EMBO J.* **1**, 841.

NEURATH, H. (1987), Proteolytic processing and physiological regulation, *Trends Biochem. Sci.* **14**, 268–271.

NEVINS, J. R. (1991), Transcriptional activation by viral regulatroy proteins, *TIBS* **16**, 435–440.

OI, V. T., MORRISON, S. L., HERZENBERG, L. A., BERG, P. (1983), Immunoglobulin gene expression in transformed lymphoid cells, *Proc. Natl. Acad. Sci. USA* **80**, 825–829.

ORNITZ, D. M., HAMMER, R. E., DAVISON, B. L., BRINSTER, R. L., PALMITER, R. D. (1987), Promoter and enhancer elements from the rat elastase I gene function independently of each other and of heterologous enhancers, *Mol. Cell. Biol.* **7**, 3466–3472.

PAGE, M. J., SYDENHAM, M. A. (1991), High level expression of the humanized monoclonal antibody campath-1h in Chinese hamster ovary cells, *Bio/Technology* **9**, 64–68.

PALLAVACINI, M. G., DETERESA, P. S., ROSETTE, C., GRAY, J. W., WURM, F. M. (1990), Effects of methotrexate on transfected DNA stability in mammalian cells, *Mol. Cell. Biol.* **10**, 401–404.

PANICALI, D., PAOLETTI, E. (1982), Construction of poxviruses as cloning vectors: insertion of thymidine kinase gene from herpes simplex virus into the DNA of infectious vaccinia virus, *Proc. Natl. Acad. Sci. USA* **79**, 4927.

PAPAHADJOPOULOS, D., VAIL, W. J., JACOBSON, K., POSTE, G. (1975), Cochleate lipid cylinders: Formation by fusion of unilamellar lipid vesicles, *Biochim. Biophys. Acta* **394**, 483–491.

PASSANANTI, C., DAVIES, B., FORD, M., FRIED, M. (1987), Structure of an inverted duplication formed as a first step in a gene amplification event: Implications for a model of gene amplification, *EMBO J.* **6**, 1697.

PAUW, P. G., JOHNSON, M. D., MOORE, P., FINEMAN, R. M., KALKA, T., ASH, J. F. (1986), Stable gene amplification and overexpression of sodium- and potassium-activated ATPase in HeLa cells, *Mol. Cell. Biol.* **6**, 1164–1171.

PAVLAKIS, G. N., HAMER, D. H. (1983), Regulation of a metallothionein-growth hormone hybrid gene in bovine papilloma virus, *Proc. Natl. Acad. Sci. USA* **80**, 397–401.

PEABODY, D. S., BERG, P. (1986), Termination-reinitiation occurs in the translation of mammalian cell mRNAs, *Mol. Cell. Biol.* **6**. 2695–2703.

PELHAM, H. R. B. (1986), Speculation on the functions of the major heat shock and glucose-regulated proteins, *Cell* **46**, 959–961.

PELHAM, H. R. B. (1988), Evidence that luminal ER proteins are sorted from secreted proteins in a post-ER compartment, *EMBO J.* **7**, 913–918.

PELHAM, H. R. B., BIENZ, M. (1982), A synthetic heat-shock promoter element confers heat-inducibility on the herpes simplex virus thymidine kinase gene, *EMBO J.* **1,** 1473–1477.

PELLETIER, J., SONENBERG, N. (1985), Photochemical cross-linking of cap binding proteins to eucaryotic mRNAs: Effect of mRNA 5′ secondary structure, *Mol. Cell. Biol.* **5,** 3222–3230.

PEPPERKOK, R., SCHNEIDER, C., PHILIPSON, L., ANSORGE, W. (1988), Single cell assay with an automated capillary microinjection system, *Exp. Cell. Res.* **178,** 369–376.

PEREZ-BERCOFF, R., KAEMPFER, R. (1982), Genomic RNA of mengovirus: Recognition of common features by ribosomes and eukaryotic initiation factor 2, *J. Virol.* **41,** 30–41.

PICARD, D., SCHAFFNER, W. (1984), A lymphocyte-specific enhancer in the mouse immunoglobulin kappa gene, *Nature* **307,** 80–82.

PROUDFOOT, N. (1991), Poly(A) signals, *Cell* **64,** 671–674.

PTITSYN, O. B. (1991), How does protein synthesis give rise to the 3D-structure? *FEBS Lett.* **285,** 176–181.

RASSOULZADEGAN, M., BINETRUY, B., CUZIN F. (1982), High frequency of gene transfer after fusion between bacterial and eukaryotic cells, *Nature* **295,** 257–259.

RAZIN, A., CEDAR, H. (1991), DNA methylation and gene expression, *Microbiol. Rev.* **55,** 451–455.

REIL, H., HAUSER, H. (1990), Test system for determination of HIV-1 frameshifting efficiency in animal cells, *Biochim. Biophys. Acta* **1050,** 288–292.

REITMAN, M. L., KORNBERG, S. (1981), Lysosomal enzyme targeting; N-acetylglucosaminyl-phosphotransferase selectively phosphorylates native lysosomal enzymes, *J. Biol. Chem.* **256,** 11977–11980.

RENKAWITZ, R. (1990), Transcriptional repression in eukaryotes, *Trends. Genet.* **6,** 192–196.

REYNOLDS, G. A., BASU, S. K., OSBORNE, T. F., CHIN, D. J., GIL, G., BROWN, M. S., GOLDSTEIN, J. L., LUSKEY, K. L. (1984), HMG CoA reductase: a negatively regulated gene with unusual promoter and 5′untranslated regions, *Cell* **38,** 275–285.

RICHARDSON, W. D., ROBERTS, B. L., SMITH, A. E. (1986), Nuclear location signals in polyoma virus large-T, *Cell* **44,** 77–85.

RIORDAN, J. R., DEUCHARS, K., KARTNER, N., ALON, N., TRENT, J., LING, V. (1985), Amplification of P-glycoprotein genes in multidrug-resistant mammalian cell lines, *Nature* **316,** 817–819.

ROBBINS, J., DILWORTH, S. M., LASKEY, R. A., DINGWALL, C. (1991), Two interdependent basic domains in nucleoplasmin nuclear targeting sequence: Identification of a class of bipartite nuclear targeting sequence, *Cell* **64,** 615–623.

ROBERTS, J. M., AXEL, R. (1982), Gene amplification and gene correction in somatic cells, *Cell* **29,** 109–119.

ROEDER, R. G. (1991), The complexities of eukaryotic transcription initiation: regulation of preinitiation complex assembly, *TIBS* **16,** 402–408.

ROGERS, S., WELLS, R., RECHSTEINER, M. (1986), Amino acid sequences common to rapidly degraded proteins: The PEST sequences. *Science* **234,** 364–368.

ROSEN, C. A. (1991), Regulation of HIV gene expression by RNA-protein interactions, *Trends Genet.* **7,** 9–14.

ROSEN, C. A., PAVLAKIS, G. N. (1990), Tat and Rev: Positive regulators of HIV gene expression, *AIDS* **4,** 499–509.

ROSSANA, C., RAO, L. G., JOHNSON, L. F. (1982), Thymidilate synthetase overproduction in 5-Fluorodeoxyuridine-resistant mouse fibroblasts, *Mol. Cell. Biol.* **2,** 1118–1125.

RUBEN, S., PERKINS, A., PURCELL, R., JOUNG, K., SIA, R., BURGHOFF, R., HASELTINE, W. A., ROSEN, C. A. (1989), Structural and functional characterization of Human Immunodeficiency Virus tat protein, *J. Virol.* **63,** 1–8.

RUIZ, J. C., WAHL, G. M. (1988), Formation of an inverted duplication can be an initial step in gene amplification, *Mol. Cell. Biol.* **8,** 4302.

SAFER, B. (1989), Nomenclature of initiation, elongation and termination factors for translation in eukaryotes, *Eur. J. Biochem.* **186,** 1–3.

SAITO, I, GROVES, R., GIULOTTO, E., ROLFE, M., STARK, G. R. (1989), Evolution and stability of chromosomal DNA coamplified with the CAD gene, *Mol. Cell. Biol.* **9,** 2445–2452.

SAMBROOK, J., FRISCH, E. F., MANIATIS, T. (1989), *Molecular Cloning; A Laboratory Manual,* Cold Spring Harbor, NY: Cold Spring Harbor Laboratory Press.

SARVER, N., GRUSS, P., LAW, M.-F., KHOURY, G., HOWLEY, P. M. (1981), Bovine papilloma virus deoxyribonucleic acid: a novel eukaryotic cloning vector, *Mol. Cell. Biol.* **1,** 486–496.

SCHAFFNER, W. (1980) Direct transfer of cloned genes from bacteria to mammalian cells, *Proc. Natl. Acad. Sci. USA* **77,** 2163–2167.

SCHNEIDER, I. (1969), Establishment of three diploid cell lines of *Anopheles stephensi* (Diptera: Culicidae), *J. Cell. Biol.* **42,** 603–606.

SCHWARTZBERG, P. L., GOFF, S. P., ROBERTSON, E. J. (1989), Germ-line transmission of a c-abl mutation produced by targeted gene disruption in ES cells, *Science* **246,** 799–803.

SEARLE, P. F., STUART, G. W., PALMITER, R. D. (1985), Building a metal-responsive promoter with synthetic regulatory elements, *Mol. Cell. Biol.* **5**, 1480–1489.

SEDIVY, J. M., JOYNER, A. L. (1992), *Gene Targeting*, New York: Freeman and Co.

SEED, B., ARUFFO, A. (1987), Molecular cloning of the CD2 antigen, the T-cell erythrocyte receptor, by a rapid immunoselection procedure, *Proc. Natl. Acad. Sci. USA* **84**, 3365–3369.

SERFLING, E., JASIN, M., SCHAFFNER, W. (1985), Enhancers and eukaryotic gene transcription, *TIG* **1985**, 224–230.

SETO, E., SHI, Y., SHENK, T. (1991), YY1 is an initiator sequence-binding protein that directs and activates transcription *in vitro, Nature* **354**, 241–245.

SHARP, P. A. (1987), Splicing of messenger RNA precursors, *Science* **235**, 766–771.

SHAW, A. S., ROTTIER, P. J. M., ROSE, J. K. (1988), Evidence for the loop model of signal-sequence insertion into the endoplasmic reticulum, *Proc. Natl. Acad. Sci. USA* **85**, 7592–7596.

SHAW, G., KAMEN, R. (1986), A conserved AU sequence from the 3′untranslated region of GM-CSF mRNA mediates selective mRNA degradation, *Cell* **46**, 659–667.

SIMOES, E. A. F., SARNOW, P. (1991), An RNA hairpin at the extreme 5′ end of the poliovirus RNA genome modulates viral translation in human cells, *J. Virol.* **65**, 913–921.

SIMONSEN, C., LEVINSON, A. (1983), Isolation and expression of an altered mouse dihydrofolate reductase cDNA, *Proc. Natl. Acad. Sci. USA* **80**, 2495.

SMITH, A. J. H., KALOGERAKIS, B. (1990), Replacement recombination events targeted at immunoglobulin heavy chain DNA sequences in mouse myeloma cells, *J. Mol. Biol.* **213**, 415–435.

SMITH, G. E., SUMMERS, M. D., FRASER, M. J. (1983), Production of human beta interferon in insect cells infected with a Baculovirus expression vector. *Mol. Cell. Biol.* **3**, 2156–2165.

SODROSKI, J., GOH, W. C., ROSEN, C., DAYTON, A., TERWILLIGER, E., HASELTINE, W. (1986), A second post-transcriptional trans-activator gene required for HTLV-III replication, *Nature* **321**, 412–417.

SOMPAYRAC, L. M., DANNA, K. J. (1981), Efficient infection of monkey cells with DNA of simian virus 40. *Proc. Natl. Acad. Sci. USA* **78**, 7575–7578.

SONENBERG, N. (1988), Cap-Binding Proteins of eukaryotic messenger RNA: Functions in initiation and control of translation. *Prog. Nucleic Acid Res.* **35**, 173–207.

SPENA, A., KRAUSE, E., DOBBERSTEIN, B. (1985), Translation efficiency of zein mRNA is reduced by hybrid formation between the 5′ and 3′ untranslated regions, *EMBO J* **4**, 2153–2158.

STANLEY, P. (1989), Chinese hamster ovary cell mutants with multiple glycosylation defects for production of glycoproteins with minimal carbohydrate heterogeneity, *Mol. Cell. Biol.* **9**, 377–383.

STANLEY, P. (1991), Glycosylation engineering: CHO mutants for the production of glycoproteins with tailored carbohydrates, in: *Protein Glycosylation: Cellular, Biotechnological and Analytical Aspects* (CONRADT, H. S., Ed.), GBF Monographs Vol. 15, pp. 225–234, Weinheim: VCH.

STARK, G. R., DEBATISSE, M., WAHL, G. M., GLOVER, D. M. (1990), DNA amplification in eukaryotes, in: *Gene Rearrangement* (HAMES, B., GLOVER, D. M., Eds.), Vol. *Frontiers in Molecular Biology,* pp. 99–149, Oxford: IRL Press.

STENFLO, J., HOLME, E., LINDSTEDT, S., CHANDRAMULI, N., HUANG, L., TAM, J., MERRIFIELD, R. (1989), Hydroxylation of aspartic acid in domains homologous to the epidermal growth factor precursor is catalyzed by a 2-oxoglutarate-dependent dioxygenase, *Proc. Natl. Acad. Sci. USA* **86**, 444–447.

STIEF, A., WINTER, D. M., STRÄTLING, W. H., SIPPEL, A. E. (1989), A nuclear DNA attachment element mediates elevated and position-independent gene activity, *Nature* **341**, 343–345.

STUHLMANN, H., JAENISCH, R., MULLIGAN, R. C. (1989), Construction and properties of replication-competent murine retroviral vectors encoding methotrexate resistance, *Mol. Cell. Biol.* **9**, 100–108.

SU, T.-S., BOCK, H.-G. O., O'BRIEN, W. E., BEAUDET, A. L. (1981), Cloning of cDNA for argininosuccinate synthetase mRNA and study of enzyme overproduction in a human cell line, *J. Biol. Chem.* **256**, 11826–11831.

SUBRAMANI, S., MULLIGAN, R., BERG, P. (1981), Expression of the mouse dihydrofolate reductase complementary deoxyribonucleic acid in simian virus 40 vectors, *Mol. Cell. Biol.* **1**, 854–864.

SUTTLE, J. W. (1985), Vitamin K-dependent carboxylase, *Annu. Rev. Biochem.* **54**, 459–477.

SZOSTAK, J. W., ORR-WEAVER, T. L., ROTHSTEIN, R. J., STAHL, F. W. (1983), The double-strand-break repair model for recombination, *Cell* **33**, 25–35.

TAO, W., WILKINSON, J., STANBRIDGE, E. J., BERNS, M. W. (1987), Direct gene transfer into human cultured cells facilitated by laser micropuncture of the cell membrane, *Proc. Natl. Acad. Sci. USA* **84**, 4180–4185.

THOMAS, K. R., CAPECCHI, M. R. (1987), Site-directed mutagenesis by gene targeting in mouse embryo-derived stem cells, *Cell,* **51,** 503–512.

TOOZE, J. (1980), *Molecular Biology of Tumor Viruses,* 2nd Ed.: *DNA Tumor Viruses,* Cold Spring Harbor, NY: Cold Spring Harbor Laboratory Press.

TREISMAN, R. (1985), Transient accumulation of c-fos RNA following serum stimulation requires a conserved 5′element and c-fos 3′ sequences, *Cell* **42,** 889–902.

URLAUB, G., CHASIN, L. A. (1980), Isolation of Chinese hamster cell mutants deficient in dihydrofolate reductase activity, *Proc. Natl. Acad. Sci. USA* **77,** 4216–4220.

VAN DEURSEN, J., LOVELL-BADGE, R., OERLEMANS, F., SCHEPENS, J., WIERINGA, B. (1991), Modulation of gene activity by consecutive gene targeting of one creatine kinase M allele in mouse embryonic stem cells, *Nucleic Acids Res.* **19,** 2637–2643.

VAN HEUVEL, M., BOSVELD, I. J., LUYTEN, W., TRAPMAN, J., ZWARTHOFF, E. C. (1986), Transient expression of murine interferon-alpha genes in mouse and monkey cells, *Gene* **45,** 159–165.

VARA, J., PORTELA, A, ORTIN, J., JIMENEZ, A. (1986), Expression in mammalian cells of a gene from *Streptomyces alboniger* conferring puromycin resistance, *Nucleic Acids Res.* **14,** 4617–4624.

VARMUS, H. (1988), Retroviruses, *Science* **240,** 1427–1435.

VAUX, D., TOOZE, J., FULLER, S. (1990), Identification by anti-idiotype antibodies of an intracellular membrane protein that recognizes a mammalian endoplasmatic reticulum retention signal, *Nature* **345,** 495–502.

VON HEIJNE, G. (1983), Patterns of amino acids near signal-sequence cleavage sites, *Eur. J. Biochem.* **133,** 17–21.

VON HEIJNE, G. (1984), Analysis of the distribution of charged residues in the N-terminal region of signal sequences: implications for protein export in prokaryotic and eukaryotic cells, *EMBO J.* **3,** 2315–2318.

VON HEIJNE, G. (1985), Signal sequences. The limits of variation, *J. Mol. Biol.* **184,** 99–105.

VON HEIJNE, G. (1986a), Mitochondrial targeting sequences may form amphiphilic helices, *EMBO J.* **5,** 1335–1342.

VON HEIJNE, G. (1986b), A new method for predicting signal sequence cleavage sites, *Nucleic Acids Res.* **14,** 4683–4690.

WAHL, G. M., PADJETT, R. A., STARK, G. R. (1979), Gene amplification causes overproduction of three enzymes of UMP in N-(phosphonoacetyl)-L-aspartate resistant hamster cells, *J. Biol. Chem.* **254,** 8679–8689.

WALLS, J. D., GRINNELL, B. W. (1990), A rapid and versatile method for the detection and isolation of mammalian cell lines secreting recombinant proteins, *Biotechniques* **8,** 138–142.

WALLS, J. D., BERG, D. T., YAN, S. B., GRINNELL, B. W. (1989), Amplification of multicistronic plasmids in the human 293 cell line and secretion of correctly processed recombinant human protein C, *Gene* **81,** 139–149.

WALTER, P., LINGAPPA, V. R. (1986), Mechanism of protein translocation across the endoplasmic reticulum membrane, *Annu. Rev. Cell. Biol.* **2,** 499–516.

WASLEY, L, C., TIMONY, G., MURTHA, P., STOUDEMIRE, J., DORNER, A. J., CARO, J., KRIEGER, M., KAUFMAN, R. J. (1991), The importance of N- and O-linked oligosaccharides for the biosynthesis and *in vitro* and *in vivo* biologic activities of erythropoietin, *Blood* **77,** 2624–2632.

WASYLYK, B. (1986), Protein coding genes of higher eukaryotes, promoter elements and trans-acting factors, in: *Maximising Gene Expression* (REZNIKOFF, W., GOLDS, L., Eds.), pp. 79–99, Guilford, UK: Butterworth.

WATTS, P. L., MACDONALD, C. (1991), Production of biologically active insulin-like growth factor by cells transfected with a recombinant retrovrial vector encoding IGF-1, in: *Production of Biologicals from Animal Cells in Culture* (SPIER, R. E., GRIFFITHS, J. B., MEINIER, B., Eds.), pp. 726–734, Oxford: Butterworth-Heinemann.

WEIDLE, U., BUCKEL, P., WIENBERG, J. (1988), Amplified expression constructs for human tissue-type plasminogen activator in Chinese hamster ovary cells: instability in the absence of selective pressure, *Gene* **66,** 193–203.

WIBERG, F. C., SUNNERHAGEN, P., BJURSELL, G. (1986), New, small circular DNA in transfected mammalian cells, *Mol. Cell. Biol.* **6,** 653–662.

WIELAND, F. T., GLEASON, M. L., SERAFINI, T. A., ROTHMAN, J. E. (1987), The rate of bulk flow from the endoplasmic reticulum to the cell surface, *Cell* **50,** 289–295.

WIGLER, M., SILVERSTEIN, S., LEE, L.-S., PELLICER, A., CHENG, Y.-C., AXEL, R. (1977), Transfer of purified herpes virus thymidin kinase gene to cultured mouse cells, *Cell* **11,** 223–232.

WIGLER, M., PELLICER, A., SILVERSTEIN, S., AXEL, R. (1978), Biochemical transfer of single-copy eucaryotic genes using total cellular DNA as donor, *Cell* **14,** 725–731.

WILSON, T., TREISMAN, R. (1988), Removal of poly(A) and consequent degradation of c-fos mRNA facilitated by AU-rich sequences, *Nature* **336,** 396–399.

WINGENDER, E. (1988), Compilation of transcription regulating proteins, *Nucleic Acids Res.* **16,** 1879–1902.

WINGENDER, E. (1990), Transcription regulating proteins and their recognition sequences, *Crit. Rev. Eucaryotic Gene Expression* **1,** 11–48.

WIRTH, M., BODE, J., ZETTLMEISSL, G., HAUSER, H. (1988), Isolation of overproducing recombinant mammalian cell lines by a fast and simple selection procedure, *Gene* **73,** 419–426.

WIRTH, M., LI, S.-Y., SCHUMACHER, L., LEHMANN, J., ZETTLMEISSL, G., HAUSER, H. (1989), Screening for and fermentation of high producer cell clones from recombinant BHK cells, in: *Advances in Animal Cell Biology and Technology for Bioprocesses* (SPIER, R. E., GRIFFITHS, J. B., STEPHENNE, J., CROOY, P. J., Eds.), pp. 44-50, London: Butterworth.

WIRTH, M., HÖXTER, M., MORELLE, C., HAUSER, H. (1990), Bicistronic expression vectors facilitate screening for overexpressing mammalian cells, in: *DECHEMA Biotechnology Conferences* (BEHRENS, D., KRÄMER, P., Eds.), Vol. 4, pp. 69–73, Weinheim: VCH.

WIRTH, M., SCHUMACHER, L., HAUSER, H. (1991 a), Use of dicistronic transcription units for the correlated expression of two genes in mammalian cells, in: *Modern Approaches to Animal Cell Technology* (GRIFFITHS, B., SPIER, R., MEIGNER, R., Eds.), pp. 338–343, London: Butterworth.

WIRTH, M., SCHUMACHER, L., HAUSER, H. (1991 b), Construction of new expression vectors for mammalian cells using the immediate early enhancer of the human Cytomegalovirus to increase expression from heterologous enhancer/promoters, in: *Protein Glycosylation: Cellular, Biotechnological, and Analytical Aspects* (CONRADT, H. S., Ed.), GBF Monographs, Vol. 15, pp. 49–52, Weinheim: VCH.

WONG, T.-K., NICOLAU, C., HOFSCHNEIDER, P. H. (1980), Appearance of beta-lactamase activity in animal cells upon liposome-mediated gene transfer, *Gene* **10,** 87–94.

WOOD, W. I., CAPON, D. J., SIMONSEN, C. C., EATON, D. L., GITSCHIER, J., KEYT, B., SEEBURG, P. H., SMITH, D. H., HOLLINGSHED, P., WION, K. L., DELWART, E., TUDDENHAM, E. G. D., VEHAR, G. A., LAWN, R. M. (1984), Expression of active human factor VIII, *Nature* **312,** 330–337.

WOOD, C. R., MORRIS, G. E., ALDERMAN, E. M., FOUSER, L., KAUFMAN, R. J. (1991), An internal ribosome binding site can be used to select for homologous recombinants at an immunoglobulin heavy-chain locus, *Proc. Natl. Acad. Sci. USA* **88,** 8006–8010.

WURM, F. M., GWINN, K. A., KINGSTON, R. E. (1986), Inducible overproduction of the mouse c-myc protein in mammalian cells, *Proc. Natl. Acad. Sci. USA* **83,** 5414–5418.

XANTHOPOULOS, K. G., MIRKOVITCH, J., DEKKER, T., KUO, C. F., DARNELL, J. E. Jr. (1989), Cell-specific transcriptional control of the mouse DNA-binding protein mC/EBP, *Proc. Natl. Acad. Sci. USA* **86,** 4117–4121.

YAGI, T., IKAWA, Y, YOSHIDA, K., SHIGETANI, Y., TAKEDA, N., MABUCHI, I., YAMAMOTO, T., AIZAWA, S. (1990), Homologous recombination at c-fyn locus of mouse embryonic stem cells with use of diphtheria toxin A-fragment gene in negative selection, *Proc. Natl. Acad. Sci. USA* **87,** 9918–9922.

YEUNG, C., INGOLIA, E., BOBONIS, C., DUNBAR, B., RISER, M., SICILIANO, M., KELLEMS, R. (1983), Selective overproduction of adenosine deaminase in cultured mouse cells, *J. Biol. Chem.* **258,** 8338–8346.

YOUNG, A. P., RINGOLD, G. M. (1983), Mouse 3T6 cells that overproduce glutamine synthetase, *J. Biol. Chem.* **258,** 11260–11266.

YU, S.-F., VON RÜDEN, T., KANTOFF, P. W., GARBER, C., SEIBERG, M., RÜTHER, U., FRENCH ANDERSON, W., WAGNER, E. F., GILBOA, E. (1986), Self-inactivating retroviral vectors designed for transfer of whole genes into mammalian cells, *Proc. Natl. Acad. Sci. USA* **83,** 3194–3198.

ZHENG, H., WILSON, J. H. (1990), Gene targeting in normal and amplified cell lines, *Nature* **344,** 170–173.

ZIMMER, A., GRUSS, P. (1989), Production of chimaeric mice containing embryonic stem (ES) cells carrying a homoeobox Hox 1.1 allele mutated by homologous recombination, *Nature* **338,** 150–153.

18 Transgenic Animals

GOTTFRIED BREM

München, Federal Republic of Germany

1 Introduction

Part III of this volume contains detailed discussions of the techniques and possibilities of genetically altering prokaryotes, lower eukaryotes, plant and animal cells. While Chapter 15 has focussed on transgenic plants as genetically engineered systems, this chapter will review the methodology of creating transgenic animals, the establishment of transgenic lines, the expression of the transferred gene constructs, and the possibilities offered by transgenic animals for basic and applied research.

For thousands of years the genetic alteration of animals by man has been the driving force behind the domestication of wild animals. Initially these interventions were carried out on an empirical basis. It should be noted that the spectrum of genetic alterations introduced during the process of domestication is by far much more dramatic than all other alterations introduced since then by purposeful and selective breeding. To an increasing extent animal breeding programs carried out since the last century have become much more consistent in pursuing the selective, purposeful, and noticeable alteration of individual traits. The truly remarkable results obtained so far are also one reason for the very favorable assessment and much higher ratings given to these achievements. The extreme development of certain traits is the result of formulating distinct breeding aims. The improvement of milk yields in dairy cows may serve as an example: while the annual milk yield per cow was approximately 1000 kg in the middle of the 19th century, it reached more than 7000 kg in some populations at the end of the 1980s with herd averages of over 10000 kg. The genetic alterations responsible for these accomplishments are by far less dramatic and less complex than those associated with the domestication of wild cattle.

The most difficult problem in animal breeding is that the (often complex) genotypes responsible for the development of desired phenotypic traits cannot be recognized *per se*. Frequently, information about the genetic causes of certain phenotypic characters can be obtained only by analyzing an extensive body of data obtained from related animals (pedigree analyses and estimation of breeding values). This problem is not a new one and also confronts geneticists engaged in basic research. The most important question regarding polygenically determined characters is the distinction between genetic and environmental influences for the evaluation of cause-and-effect relationships between genotypes and phenotypes. The analysis of genetic defects and their consequences, and the prediction of the results of genetic modifications is of greatest importance regarding mono- or oligogenically determined characters. The developments in molecular biology (see Part II) have revolutionized our knowledge about genetics and have provided some conclusive and competent answers to at least some of these questions.

The past few years have seen the development of methods which, unlike conventional breeding techniques, allow the genetic composition of organisms to be modified directly and selectively rather than indirectly by estimating the breeding values on the basis of phenotypes in order to establish a ranking order for selection. In principle these new techniques of genetic manipulation can be categorized into those which treat the entire genome as a compact unit (genome manipulation; e.g., nuclear transfer) and those in which individual genes are manipulated (gene manipulation; e.g., gene transfer) (Fig. 1). Furthermore, these techniques can be differentiated into those allowing manipulation of somatic cells (e.g., somatic gene therapy; see Sect. 2.4) and those directed at altering the germ line of animals (see Sect. 3). The latter techniques give rise to transgenic lines of animals characterized by the stable transmission of the genetic modification.

The application of gene transfer techniques has furnished us with new insights in developmental biology and the principles underlying tissue-specific gene expression (see Sect. 4). It has also provided oncologists, immunologists, and medical geneticists with a plethora of important and detailed information (see Sect. 5). In addition, gene transfer also allows the development of new production systems (see Sect. 6.3) for pharmaceutically important proteins (drug farming). As far as animal production is concerned, gene transfer appears to be a promising technique for improving disease re-

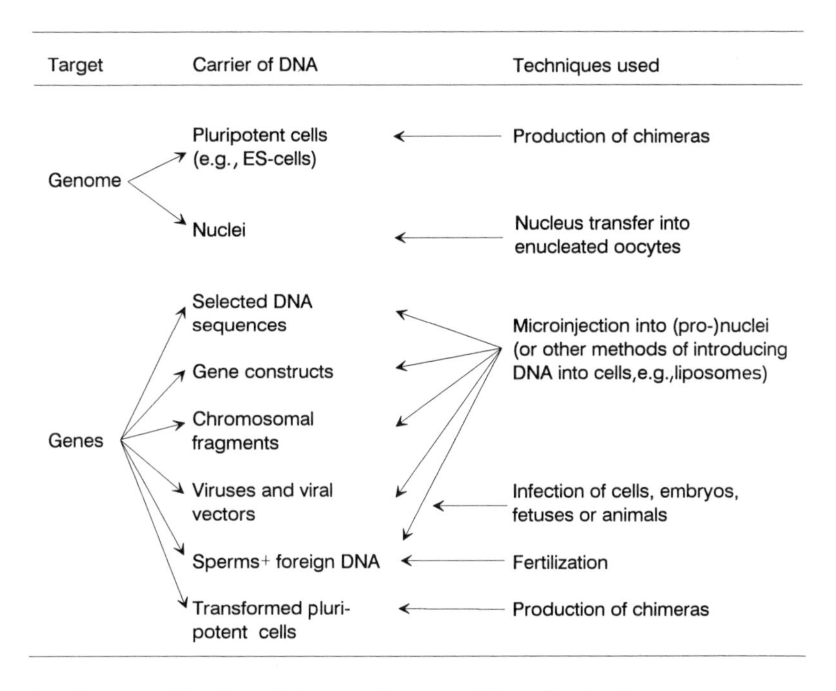

Fig. 1. Means available for the direct modification of the genetic composition of animals.

sistance, the performance, and the quality of animal products by modifying, for example, metabolic pathways and hormone status (see Sect. 6).

2 Transfer of Transgenes

The term "transgenic" has been used for the first time by GORDON and RUDDLE (1982) to describe animals harboring new genes within their genome. This term is now generally applied to the characterization of certain variants of species whose genome has been altered by the transfer of genes.

The first animals containing experimentally introduced foreign DNA were obtained by micro-injection of SV40 DNA into the blastocoel of mice (JAENISCH and MINZ, 1974; JAENISCH et al., 1975). The integration of the foreign DNA into the genome of the resulting animals has not been investigated in these early experiments. Later studies by JAENISCH (1976) have demonstrated that the infection of mouse embryos with Moloney leukemia retrovirus (M-MuLV) indeed yielded the first transgenic mouse line (see also Sect. 2.3).

Apart from retroviral infection of mouse embryos another technique, i.e., the direct micro-injection of DNA into pronuclei, has been used extensively for the generation of transgenic animals. The most significant technical contribution in the development and utilization of this technique has been made by LIN (1966) who was first in being able to inject measurable amounts of liquids into fertilized mouse oocytes. He also demonstrated that these oocytes were capable of developing into normal fetuses and upon further development into normal mice. Amphibian oocytes have been used more or less without any problems for injection experiments. Experiments carried out in the 70s have demonstrated that mRNA and DNA micro-injected into *Xenopus laevis* oocytes remained biologically active for some time before disappearing in the course of further development. The ability of a cloned gene to be replicated as extrachromosomal circular DNA after micro-injection has been demonstrated by using cloned rabbit β-globin genes

(RUSCONI and SCHAFFNER, 1981). GURDON et al. (1974) have shown that mRNA is translated in *Xenopus* oocytes and that injection of nucleic acids can be used for gene transfer in amphibian oocytes (GURDON and MELTON, 1981). Other cell types such as HeLa cells and fibroblasts also translate injected foreign DNA. BRINSTER et al. (1980) have shown that mRNA coding for murine and rabbit globin genes is translated after micro-injection into fertilized mouse oocytes.

A recombinant plasmid composed of segments of herpes simplex virus and SV40 viral DNA inserted into the bacterial plasmid pBR322 has been micro-injected into the pronuclei of fertilized mouse oocytes by GORDON et al. (1980). Two out of 78 mice contained the foreign DNA sequences, demonstrating that genes can be introduced into the mouse genome by direct injection of DNA into the nuclei of early embryos. WAGNER et al. (1981) have introduced foreign cloned genes into mouse oocytes by micro-injection into pronuclei. 15% of the fetuses analyzed ($n=33$) contained 3–50 copies of the human β-globin gene and the HSV tk gene that had been retained and replicated without significant loss or rearrangement of sequences. At least one gene copy was accurately transcribed and translated to produce a functional protein. As shown also by several groups during the early 80s (GORDON and RUDDLE, 1981, 1982; CONSTANTINI and LACY, 1981; BÜRKI and ULLRICH, 1982; BRINSTER et al., 1981, 1982; WAGNER et al., 1981; PALMITER et al., 1982), recombinant molecules of any type can be introduced into embryos of the 1-cell stage (for reviews see GORDON and RUDDLE, 1982, 1985). In 1985 transgenic farm animals were produced for the first time by DNA micro-injection into the pronuclei of rabbit, pig, and sheep embryo cells (HAMMER et al., 1985; BREM et al., 1985).

Another potential strategy for embryo gene transfer has been provided with the successful production of chimeric mice by the introduction of mouse teratocarcinoma cells (BRINSTER, 1974; PAPAIOANNOU et al., 1975; MINTZ and ILLMENSEE, 1975) and totipotent embryonic stem (ES) cells (EVANS and KAUFMAN, 1981; MARTIN, 1981). It has been observed that ES cells injected into blastocysts can colonize all tissues of the developing animal, including the germ line (ROBERTSON et al., 1986). If ES cells are genetically modified before the transfer into blastocysts they can stably transmit a transgene into chimeras and, through offspring, also to several subsequent generations (GOSSLER et al., 1986). This procedure also allows the investigation of effects of insertional mutations in transgenic offspring (BRADLEY and ROBERTSON, 1986). With the recent advances in homologous recombination and gene targeting it is also possible to select ES cells for the production of chimeras and to obtain transgenic mice with ablated genes or homologous recombinated gene loci (see Sect. 2.2).

2.1 Micro-Injection

One aim in the generation of transgenic animals is to guarantee that all somatic cells, and germ line cells in particular, will carry the foreign gene construct. It is therefore mandatory that the transfer of foreign genes is carried out as early as possible during the development of an animal. This usually means that the foreign genes are introduced into fertilized oocytes or into 2-cell stage embryos at the latest. Of course, this can be achieved only if suitable techniques are available for isolating, manipulating, and culturing such early embryonal stages for the species of interest.

The technique that suggests itself for this purpose and that is still one of the most frequently used is the transfer of gene constructs by direct micro-injection of DNA solutions into the pronuclei of zygotes and embryos. By taking into account the expertise gained with the development of transgenic mice it has been possible to obtain transgenic individuals and lines for all economically important farm animals. In principle the methodology is similar to that used for the production of transgenic mice. However, depending on the species in question, it has been necessary to introduce certain modifications mainly because of distinct differences in the morphology of embryos, the early embryonic development, and the handling of donor and recipient animals required for the embryo transfer. First, the essential techniques required for generating

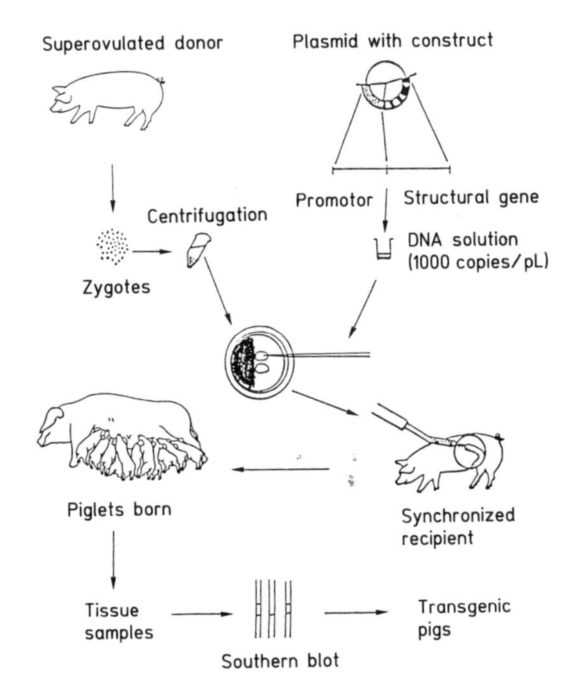

Fig. 2. Scheme depicting the gene transfer by DNA micro-injection in pigs.

transgenic mice will be discussed briefly. The most important alterations of gene transfer in different mammalian species will then be outlined in the following paragraphs. Detailed descriptions of the methodology can be found in the pertinent original publications and reviews dealing with the generation of transgenic mice (GORDON and RUDDLE, 1985; HOGAN et al., 1986; WALTON et al., 1987; PALMITER and BRINSTER, 1985, 1986; WAGNER and STEWART, 1986; DE PAMPHILIS et al., 1988), rabbits (HAMMER et al., 1985; BREM et al., 1985), pigs (HAMMER et al., 1985; BREM et al., 1985), sheep (HAMMER et al., 1985), cattle (CHURCH, 1986; BIERY et al., 1988; ROSCHLAU et al., 1989; ROSCHLAU, 1991; MASSEY, 1990; KRIMPENFORT et al., 1991), and goats (EBERT et al., 1991).

In principle a program aimed at the generation of transgenic mammals consists of the following steps (Fig. 2):

1. cloning of the gene construct
2. preparation of the DNA solution to be used for micro-injection
3. preparation of oocytes and embryos
4. micro-injection of the DNA into the pronuclei
5. transfer of injected embryos into suitable foster animals
6. detection of the transgene in newly born animals

The establishment of transgenic lines by crosses between transgenic founder animals and the analysis of transgene expression as well as the biological activities of the transgene will be discussed in Sections 3 and 4.

Gene constructs

It is a generally known fact that the structure of the DNA double helix is the same for eukaryotes and prokaryotes and that the genetic code is universal with the exception of mitochondrial DNA. In principle any DNA fragment, e.g., chemically synthesized DNA, cloned DNA, or fragments of chromosomes, can be micro-injected and will be integrated into the host genome with more or less the same frequency. The size of the micro-injected DNA does not appear to be subject to limitations. The fact that most of the transferred gene constructs are less than 20 kb in length is mainly due to constraints in the cloning capacities of cloning vectors currently available. The analysis of integrated gene constructs actually suggests that considerably longer inserts with lengths of up to 1 Mbp or even longer can be integrated without difficulties. The analysis of copy numbers of discrete gene constructs at a single chromosomal locus shows that the presence of such long inserts may be explained by concatamerization of the injected DNA fragments before or during the integration process.

The molecular structure of the injected DNA molecules is a significant parameter influencing the frequency of integration (BRINSTER et al., 1985). It has been observed that linearized DNA molecules integrate approximately five-fold better than circular molecules. Under favorable conditions the use of linear DNA constructs leads to an integration frequency of approximately 25% in mouse oocytes. The use of DNA fragments with staggered ends rather than blunt-ended ones also appears to be of

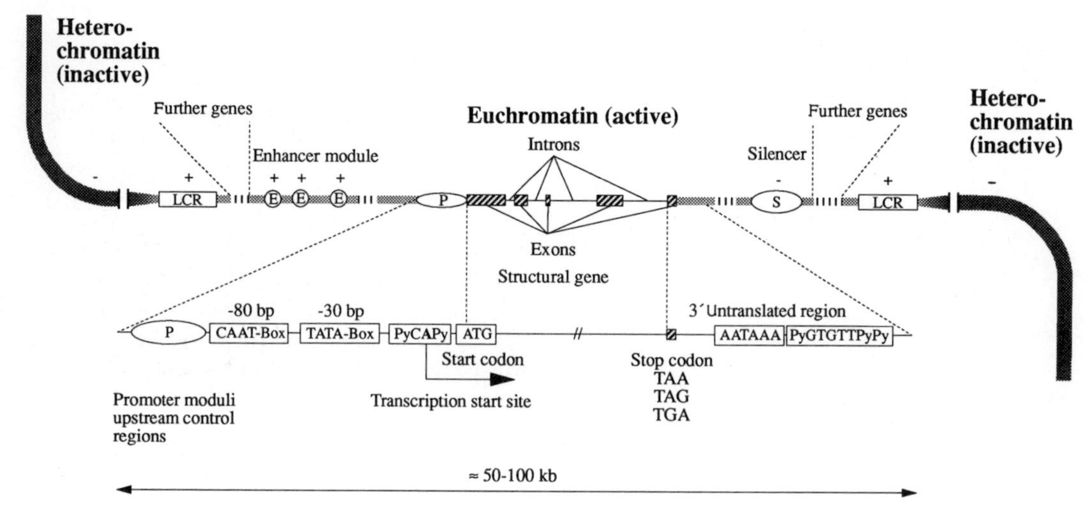

Fig. 3. Structure of a typical eukaryotic transcription unit.

advantage. By analyzing the DNA 24 h after micro-injection BRINSTER et al. (1981) have shown that mouse oocytes contain many different enzyme activities capable of modifying the incoming DNA. It has also been demonstrated that foreign genes are transcribed and degraded to a small extent only during the first 24 h following micro-injection. Transcription of micro-injected plasmids is reduced considerably after the first cell division (CHEN et al., 1986).

In order to ensure that the transgenes are also transcribed it is of advantage to remove any prokaryotic vector sequences from the DNA constructs because they may later inhibit gene activity. It has also been shown that it is more advantageous to use transgenes in their original genomic form rather than using cDNA copies. The correct exon–intron structure appears to favor transcription efficiencies of transgenes (BRINSTER et al., 1985). If the genomic sequences are not available or difficulties arise in preparing the gene construct, such constructs should at least contain one or more shorter intron sequences. In many cases the cDNA coding for the gene to be transferred is therefore furnished at its 3′ end with exon–intron sequences derived from the untranslated region of another available gene. Frequently the polyadenylation region of the gene in question or another gene is used for this purpose.

In a remarkable series of experiments PALMITER at al. (1991) have demonstrated that heterologous introns may also enhance the expression of a transgene and that the position of these introns relative to the cDNA moiety of the construct appears to be of importance.

In principle a functional gene construct must carry regulatory sequences located 5′ to the coding regions. Isolated DNA sequences without promoter/enhancer regions will also be integrated but they will not be expressed. Gene expression may be observed only in those rare cases in which by accident integration has taken place exactly 3′ to an endogenous promoter.

Structural genes can be combined with any regulatory elements even those which are not normally associated with them. The promoter for the gene to be transferred is chosen to maximize tissue specificity, the desired extent and the time of transgene expression (see also Sect. 3). If the protein encoded by the transgene is to be secreted, a suitable sequence encoding a signal sequence must be attached to the 5′ region of the coding sequence.

Apart from regulatory and coding sequences a complete eukaryotic transcription unit (Fig. 3) must also contain sequences in its 3′ region which allow correct 3′ end processing of mRNA and also influence the processing, the translocation of the transcript to the cytoplasm, and the translation of the transcript.

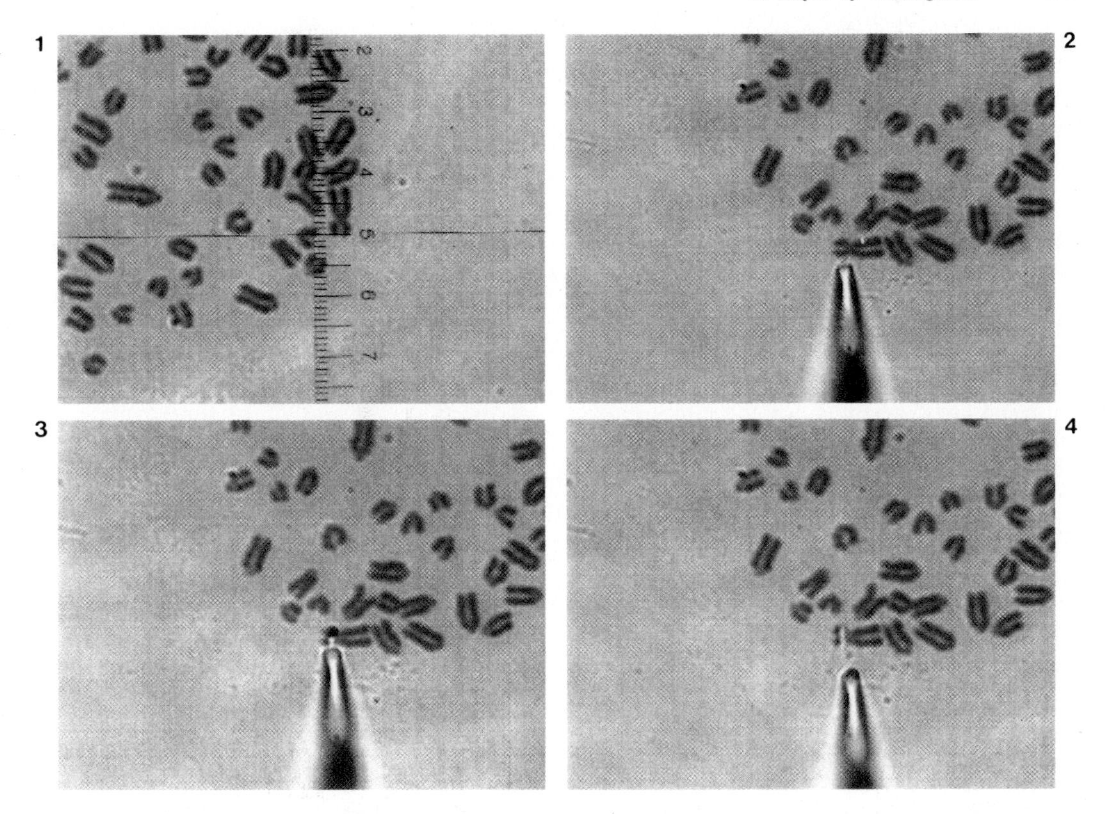

Fig. 4. Microdissection of metaphase chromosomes as exemplified by the X chromosome of cattle. Micrograph: courtesy of M. FÖRSTER (1991)
The bovine X chromosome has a length of approximately 160 centiMorgan (cM). At a total length of 8 μm 1 μm corresponds to ca. 20 cM. The chromatin fragment removed represents ca. 20 Mbp.
(1) Total length of the X chromosome ~8 μm (length unit = 1 μm).
(2) Positioning of the needle at region Xp 1.4–1.5.
(3) Careful dissection of the desired fragment with a length on the order of 1 μm.
(4) Completed dissection and uptake of the chromatin segment with the tip of the needle. The fragment is recognizable as a dark spot (compare with the previous needle).

These sequences comprise the hexanucleotide 5′-AATAAA-3′, which is essential for polyadenylation, and another highly conserved sequence with the consensus sequence 5′-PyGTGTTPyPy-3′ which is required for the formation of translatable mRNA (MCLAUGHLAN et al., 1985).

Designing and cloning of the gene construct is the fundamental step in the generation of transgenic animals, essentially determining success or failure of all subsequent steps and whether the questions asked can in fact be answered by analyzing the transgenic animal. One helpful approach to avoid potential problems associated with the expression of the gene construct in transgenic animals is to test the construct beforehand by *in vitro* transfection of cells. It should be noted, however, that the *in vitro* tests must be regarded with caution because they will not always yield unequivocal and sufficient results for the transgenic animals.

Last but not least another technique permitting the transfer of very large DNA fragments of chromosomal origin should be mentioned. RICHA and LO (1989) have obtained centromeric fragments with a length of approximately 0.5–1.0 μm (corresponding to 15–30 Mbp) by microdissection of metaphase chromosomes (Fig. 4) and introduced them by direct microinjection into the pronuclei of mouse oocytes. The transferred chromosomal fragment was detectable in approximately 50% of the morula/blastocyst stages of embryos developed *in vitro* and also in 13 days old fetuses. By transferring non-centromeric fragments it should be possible to obtain stable lines of transomic animals, i.e., animals carrying chromosomal subfragments of another species.

DNA micro-injection solution

Recombinant plasmids or cosmids containing the gene construct used for the establishment of transgenic animals are isolated as supercoiled DNA molecules by standard procedures from bacterial cultures. They are subsequently cleaved by suitable restriction enzymes to obtain the insert which can then be purified by agarose or polyacrylamide gel electrophoresis. The DNA is subsequently extracted (NaDod-SO_4/phenol/chloroform), precipitated (ethanol), washed (ethanol), and resuspended in injection buffer (TE, pH 7.5) at a suitable dilution. Depending on the length of the construct, a solution of approximately 1 µg/mL will contain several hundred copies of the construct per picoliter. This concentration seems to be optimal for obtaining the highest integration frequencies (BRINSTER et al., 1985). DNA solutions for micro-injection must be absolutely free of particulate matter in order to avoid clogging of the micropipette or damage of the injected cells. All solutions used for the preparation of DNA must be filter-sterilized by using filters of 0.22 µm pore size. Glass ware must be rinsed thoroughly with filter-sterilized water (BRENIG, 1987).

Preparation of embryos

a) Mice
Fertilized mouse oocytes for micro-injection can be obtained from fertile inbred, outbred, or hybrid strains. A maximum number of injectable oocytes from a single donor animal is obtained by treating the female mice with hormones at the age of 6–8 weeks. Three days before the micro-injection these animals receive an intraperitoneal injection of 5–10 IU of PMSG (pregnant mare serum gonadotropin). This is followed 48 h later by an intraperitoneal injection of human chorinonic gonadotropin (HCG). Ovulation induced by this treatment usually takes place approximately 12 hours after the last hormone administration. Each super-ovulated female is then put together with a male in a separate cage. Mating usually takes place in the middle of the dark period and can be assayed on the next morning by vaginal plug control. Successfully mated females are used to obtain embryos while negative females may be used for another superovulation procedure after approximately 2 weeks. Depending on the genetic background of the mice 5–8 out of 10 superovulated females will normally be mated successfully and will each yield between 15–30 oocytes that can be used for the gene transfer experiments.

Plug-positive mice are sacrificed by cervical dislocation in the course of the morning of the next day. The reproductive organs are removed to obtain the oocytes. Upon separation of the oviducts at the uterotubular junction and separation from the ovaries, a fine aspiration needle is inserted into the infundibulum and the oviduct is rinsed with M2 medium (modified Krebs–Ringer solution with HEPES buffer). The rinsing solution is collected in a small Petri dish so that oocytes still surrounded by cumulus cells can be picked upon microscopic observation. The rather sticky cumulus cells can be removed from the oocytes by the addition of hyaluronidase (300 µg/mL). The purified oocytes are then washed again with fresh M2 medium and are transferred to M16 medium (similar to Witten medium) at 37 °C and 5% CO_2 until they are used for micro-injection.

b) Rabbits
Adult animals are housed in individual cages 21 days before superovulation. Four days before the recovery of the embryos the animals receive intramuscular injections of 20 IU PMSG/kg body weight. Ovulation is induced by intravenous injection of HCG (180 IU). Immediately before or after the injection of HCG the donor animals are either artificially inseminated or put together with fertile males. Copulation is repeated 1 hour later to ensure fertilization of a maximum number of oocytes. The activity of PMSG can be neutralized after fertilization has taken place by using anti-PMSG

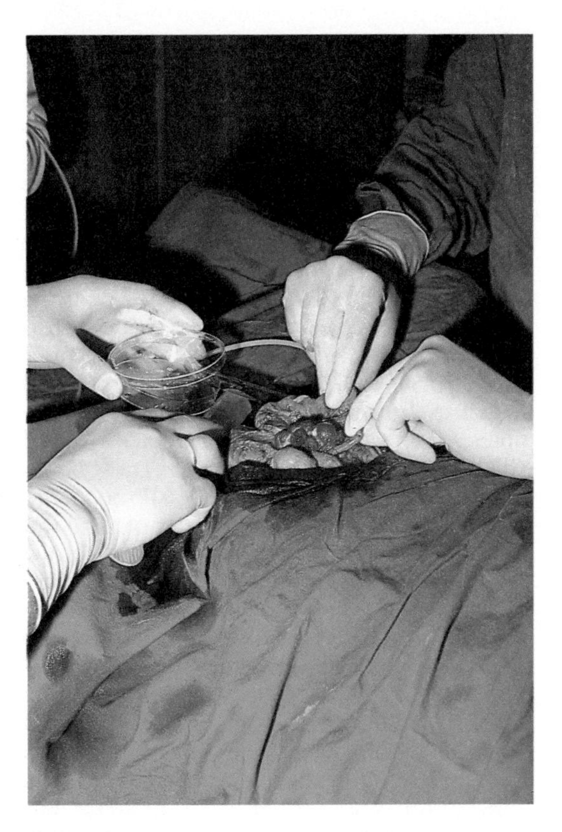

Fig. 5. Surgical recovery of pig embryos by oviduct flushing.

antibodies. The animals are sacrificed 19–21 hours after fertilization to obtain the reproductive organs (ovaries, oviducts, and cranial uterus horns). The mesosalpinx and fat tissue are removed from the oviducts which are then rinsed through the infundibulum. BSM II medium (MAURER, 1978) or PBS (phosphate-buffered saline) supplemented by 20% fetal calf serum are suitable media for storing embryos and short-term cultivation. Fertilized oocytes of good quality are free of cumulus cells and can be used for micro-injection immediately after having been washed with medium. Our own experiments have shown that a single donor animal will yield approximately 25 fertilized oocytes which can be micro-injected.

c) Pigs

Prepuberal gilts are housed in individual cages 132–135 hours before the embryos are to be removed. The animals receive an injection of 1250 IU of PMSG. 72 hours later the ovulation is induced by administration of 750 IU of HCG. The animals are artificially inseminated 24 and 36 hours later. Embryos are collected 24–27 hours after the second insemination by surgery or after sacrificing the animals. For surgery the animals are anesthetized with 160 mg of Azaperon (Stresnil®) and 400 mg of Metomidate hydrochloride (Hypnodil®). Upon fixation of the animals on their backs the site of operation is cleaned and covered with sterile cloth. The skin is opened by a median incision of approximately 10 cm between the last pair of teats. Muscles are severed by a blunt cut and the linea alba is dissected free. Upon opening of the abdominal cavity uterus, oviduct and ovaries are eventerated. A glass canula (diameter 1 cm) is inserted into and immobilized in the uterus horn that has been severed at the uterotubal junction by a blunt cut. A curved eye canula of 8 cm length with a rubber tubing is inserted into the oviduct through the fimbriae of the uterine tube and immobilized. 50 mL of Dulbecco's PBS are used for rinsing the oviduct. The washing fluid is collected in a Petri dish to collect the oocytes (Fig. 5). One donor animal will yield approximately 30–35 oocytes and 50–60% of these will be suitable for micro-injection.

d) Sheep

14–20 days before the insemination the sheep receive an intravaginal progesteron-impregnated sponge (SIMONS et al., 1988; HAMMER et al., 1985; REXROAD and PURSEL, 1988; MURRAY et al., 1989). Three days before the removal of the pessary the animals are injected with 5 mg FSH (follicle stimulating hormone). 2.5 mg FSH are administered in the mornings and evenings of the following three days. In the afternoon of the day on which the pessary is removed most animals will be in heat and can be inseminated surgically. The surgical removal of the embryos takes place on the morning of the second day after the onset of the oestrus. The washing medium is PBS supplemented with 10% FCS, while TCM 199 medium is used for the subsequent culture. On the average one animal will yield approximately 5 embryos, and approximately 50–70% of these can be used for micro-injection.

e) Goats

An experimental protocol for the production of transgenic goats has been reported by EBERT et al. (1991). The timing of oestrus was synchronized in the donors with Norgestomet ear implants. Prostaglandin was administered after the first 7–9 days to remove endogenous sources of progesterone. At day 13 following progesterone administration FSH was given to the animals at concentrations of 18 mg over three days in twice daily injections. 24 hours following implant removal the donor animals were mated several times to fertile males over a 2-day period. The embryos were recovered surgically 72 h after the removal of the implant. Each animal will yield approximately six oocytes/embryos, half of which are suitable for micro-injection.

f) Cattle

Between the 8th and the 12th day after the onset of oestrus the animals receive 2000–3000 IU of PMSG. Two days after the PMSG injection the donor animals receive 2 mL of prostaglandin F2α (cloprostenol) to induce luteolysis of the corpus luteum. The oestrus observed approximately 48 hours later is exploited for artificial insemination. A second insemination follows 12 hours later. The embryos are usually collected upon slaughtering. However, it is also possible to remove the oviducts by castration and to collect the embryos by the very time-consuming and laborious procedure of surgically flushing the oviducts. The most suitable time for the recovery of the embryos is 78 to 82 hours after the prostaglandin treatment. Approximately 12 oocytes can be collected per donor animal, and ca. half of them are suitable for micro-injection. An alternative source of oocytes is the *in vitro* production of cattle embryos through collection, maturation, and fertilization of oocytes form untreated animals obtained from slaughterhouse ovaries (BERG and BREM, 1989). By injecting embryos produced *in vitro* KRIMPENFORT et al. (1991) have established 21 pregnancies from which 19 calves were born. In two cases the micro-injected DNA had been integrated into the host genome.

DNA micro-injection

Micro-injection of embryos requires a stable and preferably shock-proof work bench. The set-up consists of an inverted microscope, two micro-manipulators to handle the holding and the injection pipettes, and an injection apparatus allowing the injection pressure to be regulated. An injection chamber containing the embryos in medium with a top layer of paraffin oil is positioned under the microscope. The embryo to be injected is held by the holding pipette under reduced pressure and positioned so that the pronuclei are visible. The injection pipette has a tip diameter of 1 μm and is filled with DNA solution at the tip. For DNA injection the tip is inserted carefully through the zona pellucida and the cell membrane until the tip is positioned within the pronucleus (Fig. 6). The volume of the pronucleus increases by approximately 50% if 1–2 pL of DNA solution are injected. This visible swelling of the pronucleus is also the only indication for a successful injection of the DNA solution into the nucleus. After the injection the oocyte is released from the holding pipette. All injected embryos are transferred to culture medium and are kept at 37 °C or 39 °C until they are used for the transfer into a recipient animal. The pronuclei of fertilized mouse or rabbit embryos which were collected at the right age are easily discernible and can be injected with ease and precision. In embryos of farm animals the pronuclei are obscured by dark lipid granulae. They can be moved to one side of the oocytes by centrifugation at 13000–15000 *g* for 3–5 min so that the pronuclei which are usually located near the center become visible (Fig. 6; 3). The pronuclei of sheep embryos may be visualized under phase-contrast Nomarski optics. In spite of the centrifugation step the oocytes of farm animals may often be more difficult to micro-inject and cannot be injected always with the same precision as those of mice and rabbits.

Embryo transfer

The embryos will usually be transferred to synchronized recipient animals after a short period of *in vitro* culture (up to several hours). Morphologically defect embryos are, of course, discarded and not used for the embryo trans-

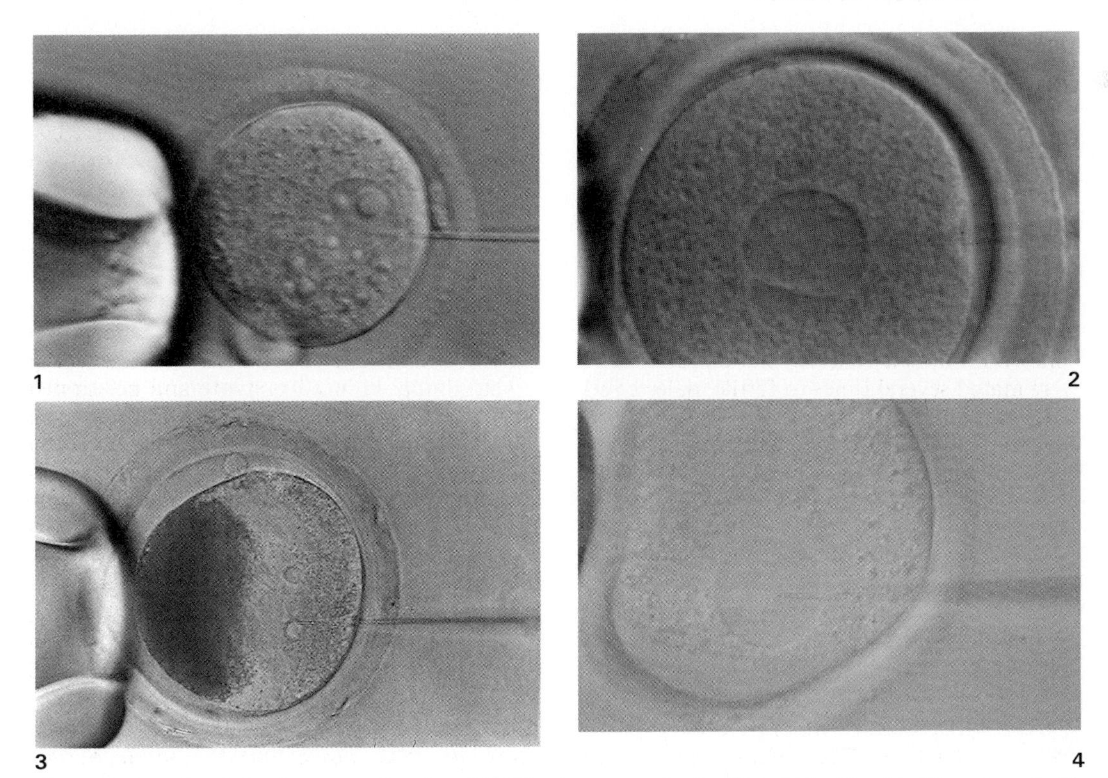

Fig. 6. DNA micro-injection into pronuclei of fertilized oocytes of mice (1), rabbits (2), pigs (3), and cattle (4).

fer. An alternative to an embryo transfer carried out at the same day as the micro-injection is the intermediate cultivation of embryos for 3 days (mice) or up to 7 days (cattle). This intermediate culture has the advantage that embryos are identified which have been damaged in their development as a result of the injection procedure itself. On the other hand, the intermediate culture may also introduce an additional stress factor so that difficulties may be encountered in certain species. Under optimal culture conditions rabbit embryos, for example, will develop into the morula/blastocyst stage. However, these embryos cannot implant themselves properly in the uterus after transfer into the recipient animal because the mucin layer, which is a characteristic of this species, cannot develop *in vitro*. Cattle embryos will only develop well *in vitro* if they are cocultivated with granulosa cells (BERG and BREM, 1989), oviduct cells, or similar cells, or if they are cultivated with appropriately conditioned medium.

The oestrus cycle of the recipient animals must be synchronized with that of the animals used as embryo donors so that ovaries, oviducts, und uterus will be in a condition that will guarantee the physiological development of the transferred embryos. Synchronization is achieved in the species mentioned previously either by mating with vasectomized (sterile) males (mice), induction of ovulation with HCG (rabbit), "mild" superovulation in pigs, sheep, and goats, or luteolysis with prostaglandins in cattle. If embryos must be cultivated for longer periods of time *in vitro*, it is of advantage to use donors and recipients whose oestrous cycle is asynchronous by one day. The chances of embryos to develop normally appear to be much better if the *in vitro* cultivated embryos are in the lead with respect to the stage of development of the recipient animals. This effect is quite pronounced in mice: 3 day old blastocysts will usually develop much more frequently if they are transferred to day 2-recipients (1 day asynchrony) rather than being transferred into exactly synchronous recipients.

The direct transfer of micro-injected oocytes into the oviducts of ultimate recipients will always require surgery. The environment needed for the correct development of the oocytes can be provided only by the oviduct. For anatomical reasons the oviducts cannot be reached without surgical interference (laparatomy) even in large farm animals. Recipient animals are anesthetized, the field of operation is prepared properly, and the abdominal cavity is opened. The embryos are taken up with a suitable transfer catheter which is then used to release them into the oviduct. Before the transfer the reaction of the ovaries towards induction of ovulation is controlled (points of ovulation, development of the corpora rubra resp. lutea) in order to exclude unsynchronized animals which are unsuitable for embryo transfer. Occasionally temporary recipients may also be used for the transfer of all injected oocytes. The oocytes can be stored in the ligated oviduct for 4–7 days before they are re-isolated. Only those oocytes which have developed further into the appropriate developmental stage are then transferred into the ultimate recipient animals. This technique is used especially with large farm animals because the numbers of required recipients can be reduced considerably. Intermediate recipients are frequently employed in particular with cattle embryos in which case the oviducts of sheep or rabbits may be used as intermediate recipients. 7 day old cattle embryos can then be transferred non-surgically to the recipient heifers.

Mouse and rabbit recipients usually receive 20 to 30 injected embryos distributed evenly between the two oviducts. In pigs ca. 30–40 embryos are transferred into one oviduct, because the embryos subsequently distribute themselves evenly between the two uterus horns (spacing). Oviduct transfer in sheep, goats, and cattle usually involves the transfer of 2–4 embryos per recipient animal. A comparison of success rates of gene transfer programs in experimental farm animals is shown in Tab. 1.

Depending upon the species, a successful gravidity may be diagnosed several days or weeks later and may then proceed under controlled conditions until birth. The diagnosis of gravidity at an early stage also provides early data about the survival rates of embryos.

Detection of integration

Tissue or blood samples of animals obtained after the transfer of injected oocytes can be used for the preparation of DNA. In cattle, sheep, goats, and pigs it is therefore possible to obtain data about the integration of the transferred gene construct at the day of birth. In mice and rabbits, the DNA will be analyzed for integrated sequences usually 3 and 6 weeks, respectively, after birth. When tissue or blood samples are taken, the animals must be marked permanently and unmistakably. In rare cases animals which harbor an integrated transferred construct may already show phenotypic alterations at birth. The expression of phenotypes distinguishing transgenic and non-

Tab. 1. Comparison of Reproductive Data Important for Generation of Transgenic Animals in Different Species

Species	Mouse	Rabbit	Pig	Sheep	Cattle
Number of injectable eggs per superovulated donor (D)	15	20	15	4	5
Number of donors per recipient (R)	2	2	2	1,5	1
Pregnancy rate	60%	50%	40 %	40%	20 %
Animals born/injected embryos	10–20%	10%	5 %	15%	10 %
Integration frequency	15%	10%	10–15 %	5–10%	5–10 %
Transgenic animals/injected eggs	2%	1%	0.5%	1%	0.5%
Number of animals (D + R) required per transgenic animals	10	15	20	40	80

transgenic animals will, of course, depend on the specific construct used. Since transgenes are normally not expressed in all individuals, unequivocal proof of the presence of an integrated gene construct must be obtained by analysis of the genomic DNA.

Several different techniques are available to investigate the integration of the transgene in the DNA of animals. It has been shown that DNA with a mean fragment size of 25–35 kb is quite sufficient to carry out the analysis. It is, therefore, not necessary to use specifically high molecular-weight preparations of DNA.

Tissue specimens are usually completely lysed by treatment for several hours with a lysis buffer containing proteinase K at 55 °C. The DNA is extracted with phenol/chloroform and can then be precipitated with sodium acetate and ethanol. A second precipitation step may be carried out after centrifugation of the first precipitate. The concentration of the DNA is measured with a diluted aliquot by spectrophotometry.

Conclusive evidence for the integration of a gene construct can be obtained by Southern blot analyses (SOUTHERN, 1975). The analyses can be carried out with approximately 20–40 μg of high molecular-weight genomic DNA.

A fast protocol suitable for analyzing a large number of different DNA samples has been described by BRENIG et al. (1989). The procedure circumvents standard blotting protocols and essentially involves the transfer of aliquots of crude preparations of denatured DNA by vacuum blotting onto suitable carrier membranes and allows the detection of micro-injected gene constructs in the DNA of animals within 25–30 hours. Hybridization probes for the detection of gene constructs are usually labeled by nick translation or oligopriming utilizing radioactively labeled deoxynucleoside triphosphates. An alternative to the use of radioactive labels is the use of biotinylated dUTP (LANGER et al., 1981). This labeling procedure also allows detection of minute amounts of biotinylated DNA complexes by exploiting the very high affinity of biotin towards avidin which itself can be detected by suitable antibodies, enzymes, or colored dyes.

The correct choice of restriction enzymes used to cleave high molecular-weight genomic DNA is of paramount importance for Southern blot analysis. If the cellular DNA is cleaved with the enzyme used to remove the gene construct from the vector, Southern blot analysis of the genomic DNA will yield hybridizing DNA fragments corresponding to the size of the entire gene construct and frequently may also yield one or two other fragments which are longer than the entire original construct (Fig. 7).

The analysis of such hybridization patterns allows several conclusions to be drawn about the integration of the gene construct in the genome of the transgenic animals. It is known that many transgenic animals do not harbor a single integrated copy of the injected gene construct but rather contain several copies orientated in the form of head-to-tail or tail-to-tail cointegrates, depending on the nature of the terminal cleavage site. It has also been observed that several bases may be lost upon integration of the construct especially if fragments with staggered ends were involved. Under these circumstances cleavage sites at the 5′ and 3′ ends of the integrated DNA fragment may be lacking. If the genomic DNA is then cleaved with an enzyme that cleaves at the ends of the gene construct the integrated concatamers are cleaved so that internal full-length copies of the original gene construct are released. If the terminal cleavage sites are missing, the 5′ and 3′ regions will yield larger fragments consisting of the original gene construct plus a piece of genomic DNA whose length will depend on the position of the appropriate cleavage site in the genomic DNA. The lengths of these terminal gene fragments cannot be predicted beforehand and will depend on the presence of suitable cleavage sites within the abutting genomic DNA for the restriction enzyme in question.

In rare cases the gene construct will be integrated at two or even more sites within the genome of the transgenic animal. It may sometimes be possible to detect the existence of such different integration sites by Southern blot analysis (Fig. 8). Frequently, however, the existence of several integration sites may not be noticeable in primary animals and may be detected only after segregation of the transgenes in the progeny of primary animals.

If the gene transfer involves constructs containing homologous gene sequences, i.e., se-

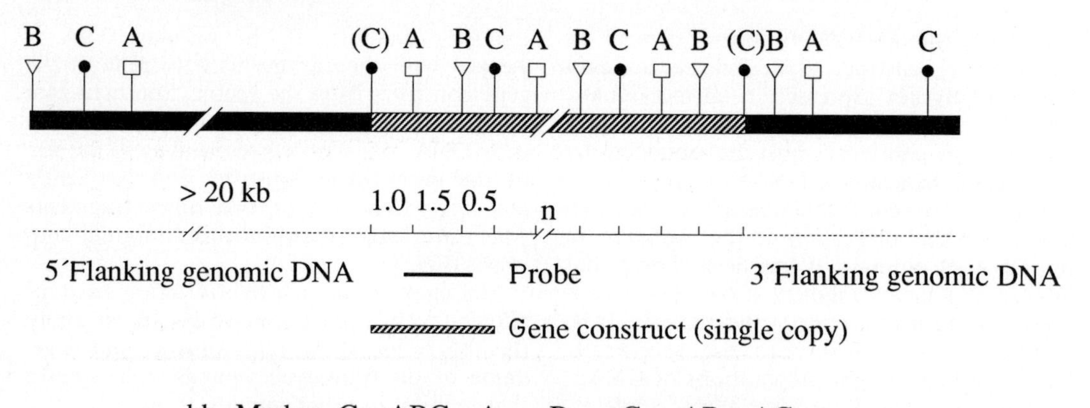

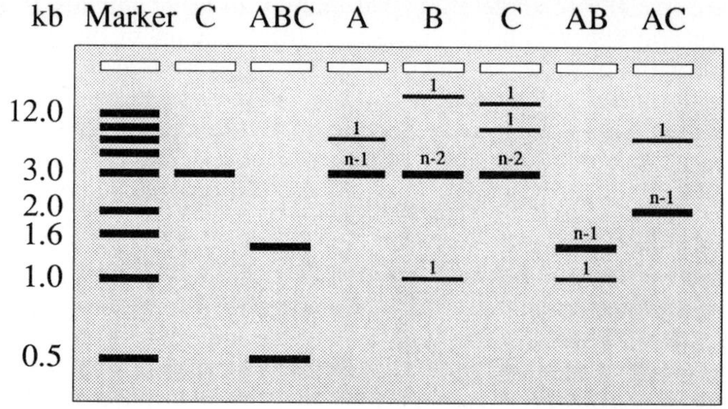

Fig. 7. Theoretical example of a Southern blot analysis of a transgenic animal. Numbers above the bands indicate the calculated copy numbers.

quences derived from the genome of the same species, the Southern blot analysis carried out for detecting integration will have to be planned carefully and will be much more demanding. A specific integration event can be detected only if suitable restriction enzymes and DNA probes for hybridization are available which allow endogenous DNA to be differentiated from the transferred DNA. Since the order of untranslated regions, promoter regions, and structural genes in the gene construct will normally differ from that in the genome, the junctions between these different structural elements will be particularly well suited as hybridization probes. Nevertheless, cross-hybridization of these probes with genomic sequences may still yield signals which are difficult to interpret unequivocally.

Occasionally a Southern blot may yield completely unexpected hybridization patterns which cannot be explained by the restriction pattern of the original construct. In such cases differential restriction of the DNA with several restriction enzymes and subsequent Southern analysis may be helpful to resolve the complex integration pattern. One reason for the occurrence of such complicated integration patterns may be the generation of deletions or mutations within the injected gene constructs. If these altered constructs are amplified before the integration, some or all integrated copies may show the same alterations. It is also conceivable that genomic DNA fragments may have become positioned between integrated copies of the constructs, thus also altering the expected restriction pattern.

Intermolecular ligation usually yields complex concateneric structures. The ligation of the injected DNA constructs usually proceeds faster than the integration of the constructs into the chromosomal DNA. If concateneric structures are integrated within chromosomal

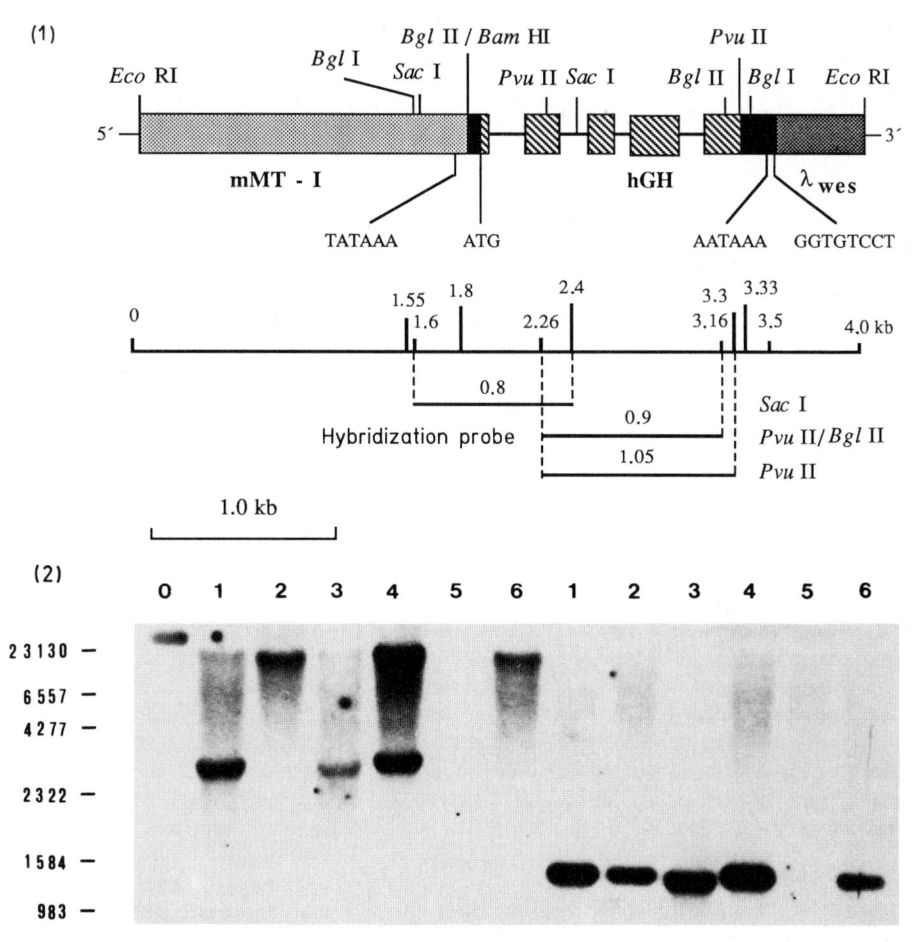

Fig. 8. Structure of the MThGH gene construct (1) and Southern blot analysis (2) of the offspring of a transgenic mouse with two integration sites (BRENIG and BREM, 1988).
Lanes 1–4 and 6 show the DNA of offspring. Lane 5 contains control DNA of a non-transgenic mouse; the left half shows *Eco* RI-cleaved DNA (terminal fragments), while the right half shows *Pvu* II-cleaved DNA (internal fragments).

regions undergoing rearrangements, the positions of transgenes may also change even after the integration event.

The copy numbers of the integrated gene constructs may be determined by dot blot analysis of DNA immobilized on filters. Autoradiographs of the hybridized filters are analyzed by densitometry which yields data that can be used to calculate copy numbers of the transgene (haploid genome size $\sim 3 \times 10^9$ bp). This approach is particularly useful because it does not require electrophoresis of the DNA, and the DNA does not have to be of the same quality as the DNA used in Southern blot analyses.

PCR

The polymerase chain reaction (PCR) is an extremely efficient method to detect transgenes in genomic DNA isolated from minute amounts of tissues within a relatively short time (see Fig. 15). KING and WALL (1988) have demonstrated for the first time the detection of transgenes in pre-implantation embryos by PCR. This technique should also allow identification of potentially transgenic embryos in livestock. Its application may help to clarify one point which still is of considerable concern, namely the very low efficiency of transgenic livestock production.

In situ hybridization

If the site of integration of a transgene is to be determined on the chromosomal level, the method of choice is *in situ* hybridization on metaphase chromosomes which can also be used to detect DNA and RNA in cryostat sections. By *in situ* hybridization experiments of metaphase chromosomes COSTANTINI et al. (1984) have demonstrated that the integration of a transgene (β-globin construct) has occurred at a specific single chromosomal locus in each mouse but not into the homologous β-globin locus in the mouse genome. *In situ* hybridization has been used to examine the spatial and development control of transcription in transgenic mice (KOOPMAN et al., 1989). *In situ* transgenic enzyme markers can be used to monitor migration of cells for the mid-gestation mouse embryos and the visualization of inner cell mass clones during early organogenesis (BEDDINGTON et al., 1989; BEDDINGTON and MARTIN, 1989).

2.2 Use of Embryonic Stem Cells

Undeterminated pluripotent cell lines may be obtained and established either directly from embryos or indirectly from teratocarcinomas. The former are known as ES cells (embryonic stem cells) while the latter have been designated EC cells (embryonic carcinoma cells).

The establishment of pluripotent stem cells has been investigated in detail in the mouse system. The first visible sign of differentiation of mammalian embryos is the formation of the blastocyst in which trophoectodermal cells develop into extraembryonic cell lines. Only a few cells constitute the so-called inner cell mass (ICM) which gives rise to all organs of the growing fetus including the germ cells. In the mouse the ICM is presumably made up of only three cells. It has been known for many years that pluripotent stem cells are an essential constituent of teratocarcinomas. These malignant tumors of ectodermal origin develop in the gonads of male and female mammals and consist of differentiated and undifferentiated cells comprising a variety of different tissues including muscles, brain, skin, glands, blood, bone, hair, teeth and so on, in different stages of development. The benign form of the teratoma may develop into a malignant form, showing unrestricted growth in teratocarcinomas. This process usually takes place if the teratoma contains undifferentiated cells. Teratomas have been observed first in the testes and ovaries of certain strains of mice (STEVENS and VARNUM, 1974).

Several studies involving the transplantation and the transfer of teratomas have shown that teratocarcinomas originate from primordial germ cells. Ectopic transplantation into the kidney or testes of syngenic mice of 5–6 day old embryos, or of cells derived from the embryonic crest of 12 day old mouse fetuses, will result in the development of teratomas or teratocarcinomas. These tumors can be propagated by subcutaneous or intraperitoneal transplantation to other mice. When grown in the ascites fluid these cells develop into so-called embryoid bodies whose morphology shows all features of very early embryos. These bodies contain pluripotent stem cells which are still capable of differentiation. Although the stem cells can be propagated easily by continuous passage of tumors, their *in vitro* culture appears to be very difficult.

The successful isolation of ES cells has been described for the first time by EVANS and KAUFMAN (1981) and MARTIN (1981). The starting material in both cases were murine blastocysts, mainly because such cells were easily obtained, but also due to the fact that some equivalence between EC cells and ICM was already known. The observation that EC cells which had been injected into blastocysts were capable of taking part in the development of somatic tissues has been an important observation indicating the pluripotent nature of such cells (PAPAIOANNOU and ROSSANT, 1983).

In contrast to embryonic carcinoma cells (EC) embryonic stem cells (ES) can be cultivated *in vitro* (EVANS and KAUFMAN, 1981). Such cells are usually obtained by culturing ex-

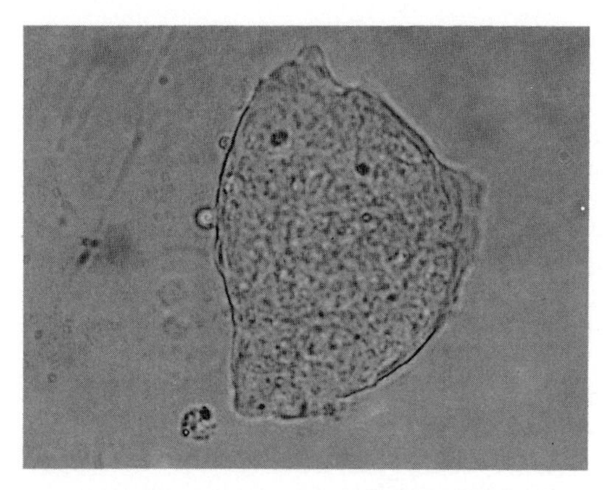

Fig. 9. Colony of embryonic stem cells (ES). Micrograph: courtesy of E. WOLF (1991).

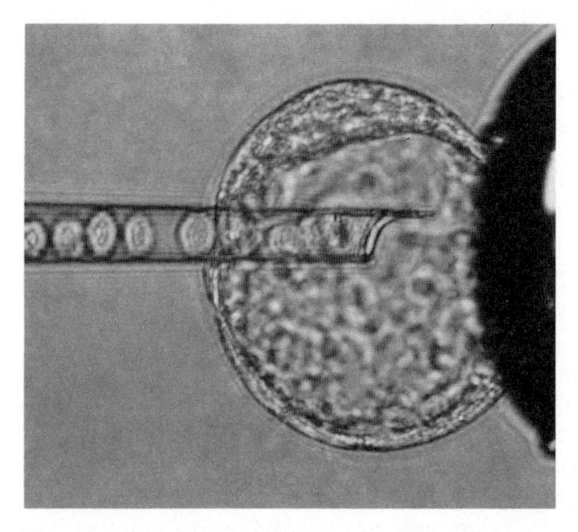

Fig. 10. Blastocyst injection using ES cells. Micrograph: courtesy of E. WOLF (1991).

panded 3.5 day old blastocysts until day 6.5–7 of their development. The ICM moiety is easily discernible because it forms a localized cell mass within the trophoectodermal cell layer. Such cells are collected by mechanical and enzymatical disruption of the tissue and culturing them on a feeder layer of mitotically inactivated fibroblasts (Fig. 9). Apart from a series of differentiated cell lines this procedure also yields lines of stem cells. By continuously subcultivating the cells every 4–5 days it is possible to obtain permanently growing lines of pluripotent stem cells. Approximately 10% of all embryos cultivated in this way have been shown to give rise to continuous cell lines (HOGAN et al., 1986). The crucial step in the establishment of permanent pluripotent stem cell lines is the removal of undifferentiated cells from the lumps of ectodermal cells developing from blastocyst cultures. Three parameters govern the isolation of ES cells (BRADLEY, 1990):

1. The cells must be present in the egg cylinder at the time of the initial disaggregation.
2. The cells must be deprived effectively of the differentiation signals.
3. The cells must be able to continue to proliferate *in vitro*.

Production of chimeras

EC cells may lose their tumorigenic potential once they have been injected into blastocysts (Fig. 10) and grow in the embryonal environment. The cells may then differentiate into normal cells participating in the development of all organs and thus give rise to chimeric animals (BRINSTER, 1974; MINTZ and ILLMENSEE, 1975; ILLMENSEE and MINTZ, 1976).

Chimeras may also be obtained by the aggregation of stem cells with morula stages of embryos. Cells from different EC lines may not show the same potential for generation of chimeras; only 20% of the embryos and approximately 10% of all born animals will be chimeric. Some of these animals may later develop tumors. The presence of EC cells in the germ line is a relatively rare event since at most 10% of the chimeras will transmit the EC-specific genotype to the offspring.

Much better results are obtained if embryonic stem cells are used for the generation of chimeras. Most of the available ES cell lines are capable of participating normally in the development of the embryo. The resulting chimeras are identical in all aspects to those obtained from the transfer of ICM cells. In summary chimeras obtained from ES cells may be characterized as follows (ROBERTSON, 1986; ROBERTSON et al., 1986):

1. The frequency with which chimeras are observed among the animals born is in the order of 35%. Prenatal losses are not significantly higher than those observed with control embryos, since 70% of all injected embryos develop into adult animals.
2. The frequency of ES cells present in the chimeras usually exceeds 50%.
3. All organs are usually chimeric.
4. The frequency of germ line-chimeras is usually relatively high. EVANS et al. (1985) have reported that 31 of 78 phenotypically male chimeras were infertile; however, 15 of 37 fertile males did transmit the genome of *in vitro*-cultivated cells to their offspring.

The generation of chimeras, obtained in particular after DNA transformation or homologous recombination (see later), is usually carried out with XY cell lines for three reasons:

1. XY male karyotypes appear to be more stable *in vitro* than XX female karyotypes.
2. Male chimeras produce functional sperms and usually generate much more offspring than females.
3. It has been observed that large contributions of XY stem cells in developing XX fetuses may cause the development of a male in many instances (sex conversion).

Currently, great efforts are made to establish stem cell lines of commercial livestock. Some results obtained with pigs demonstrate that it is possible to cultivate stem cell-like colonies from embryos (PIEDRAHITA et al., 1988). EVANS et al. (1990) have succeeded in cultivating cells derived from pigs and cattle for up to nine months in continuous *in vitro* culture (NOTARIANNI et al., 1990). The phenotype of these cells appears to be identical to that of embryonic stem cells, but it has not been proven yet that these cells are also pluripotent because no chimeras have been obtained. It can be expected, however, that embryonic stem cells will also be established for commercial livestock and hence may be available for gene transfer.

Gene transfer via stem cell transformation

The considerably improved methods for the transformation of *in vitro*-cultivated cells have been used for several years to transfer gene constructs into embryonic stem cell lines. The use of embryonic stem cells as vehicles for the introduction of novel genes has already been suggested by MINTZ (1977). An experimental model for the establishment of transgenic commercial animals by the production of chimeras with transformed stem cells has also been proposed (BREM, 1986). So far the application of this gene transfer technique has been hampered by the lack of suitable cell lines.

A number of techniques are available which allow the genome of ES cells to be altered *in vitro*. They include the transfer of chromosomes, cell hybridization, microcell fusion, DNA transfer by the calcium phosphate precipitation technique, infection with retroviral vectors, DNA micro-injection, and also electroporation.

In the mouse STEWART et al. (1985) and ROBERTSON et al. (1986) have shown that the gene constructs introduced into embryonic stem cells are expressed in the somatic tissues of the resulting mouse chimeras. The frequency of chimeric animals among the born animals was very high, and in some cases the transmission of the transgene to the offspring has also been demonstrated.

Gene transfer mediated by the use of transformed stem cell lines has several advantages. On the one hand, the integration and possibly also the expression of the transgene can be monitored in the resulting cell lines. On the other hand, the transfer can be effected by manipulation of morula and blastocyst stages. Since these stages can be isolated non-surgically, in particular in cattle, this procedure would greatly facilitate gene transfer programs.

The definitive advantage of the gene transfer involving embryonic stem cells is the possibility of direct manipulation of individual genes by homologous recombination. The DNA transfer into the ES cells is generally carried out by electroporation. This technique has the disadvantage that the frequency of stable transformants is on the order of 10^{-2} to 10^{-3} and therefore relatively low. On the other hand, the technique allows many cells to be treated at the same time. Under suitable conditions it is possible to greatly favor the formation of those types of transformants that primarily contain single copy insertions at single sites. Due to the low efficiency electroporation requires the use of a dominant selectable marker such as neo^R to allow untransformed cells to be selectively removed from the cultures.

Early homologous recombination experiments in ES cells have been carried out with the HPRT gene because this gene is located on the X chromosome and therefore exists only in a single copy in XY cells. Moreover, the activity of the HPRT gene leads to resistance against the purine analog 6-thioguanine. THOMAS and CAPECCHI (1986, 1987) have shown that the use of insertion and replacement vectors which require one and two cross-over events, respectively, leads to homologous recombination with almost equal frequencies. To a large extent the recombination frequency depends on the length of the homologous DNA sequences while it is relatively independent of the length of the non-homologous sequences which are to be inserted into the target gene. The great importance of sequence homology between the vector DNA and the target gene has been pointed out recently by TE RIELE et al. (1991).

Screening for homologous recombination is a difficult and time-consuming task for which the use of PCR is almost indispensable. PCR analysis involves the use of a primer which is specific for the wild-type locus and another primer which is specific for the genomic alteration present on the vector. Under these conditions PCR products are obtained only if homologous recombination occurred.

To avoid the difficulties associated with mere screening strategies several enrichment protocols for homologous recombinants have been developed (for review see MANSOUR, 1990). One protocol applies only to genes that are expressed in ES cells (neo^R, HPRT, c-abl, N-myc). The strategy known as positive/negative selection (PNS) can be used to enrich for genes that are no longer expressed after homologous recombination. The PNS vector used in this procedure carries an independently expressed positive selectable marker located in an exon and a negative selectable marker that

Fig. 11. Four chimeric mice obtained after ES cell injection. Injection of D3 ES cells (agouti) into Balb/c (albino) blastocysts. Courtesy of E. WOLF (1991).

is placed outside of the region of homology used for recombination (MANSOUR et al., 1988).

VALANCIUS and SMITHIES (1991) have introduced a 4 bp insertion into the HPRT gene of an ES cell line by using an „in-out" targeting procedure. After in „in"-step which involves homologous integration by using an insertional vector, recombinants were isolated by direct screening. The cells were then grown to select for revertants resulting from the excision of the integrated vector sequences.

The alteration created by gene targeting can be introduced into the germ line by injecting the ES cells into blastocysts and producing chimeras (Fig. 11). Breeding of homozygous F2 offspring determines the phenotypes that result from the recombination event (Tab. 2). These technologies are very important for generating mouse models for human recessive genetic diseases, and the animals obtained in this way should provide useful systems for testing diagnostics, prophylactic and therapeutic agents and strategies.

2.3 Retroviral Vectors

Retroviruses are viruses with a single-stranded RNA genome which is replicated via a double-stranded DNA intermediate stably integrated into the cellular DNA (=provirus). Most retroviruses do not destroy their host

Tab. 2. Introduction of Targeted Mutations into the Mouse Genome Using Embryonic Stem Cells

Gene	ES Cell Line	Number of Targeted Clones	Blastocyst Donors	Number of Blastocysts Injected and Transferred	Number of Pups Born	Number of Chimeras (Males)	Number of Germ Line Chimeras	Phenotype of Heterozygous Mice	Phenotype of Homozygous Mice	Ref.
Hox-1.1	D3	5	C57BI/6	–	–	4 (–)	–	–	–	(1)
Hox-1.5	CC1.2	1	C57BI/6	–	–	7 (7)	3	normal	Death around birth; athymic, aparathyroid, reduced thyroid and submaxillary tissue, throat abnormalities; defects of heart and arteries	(2)
Hox-1.6	D3	2	C57BI/6	822	96	29 (16)	3	normal	Death at birth; delayed hind brain neural tube closure; absence of certain cranial nerves and ganglia; malformed inner ears and bones of the skull	(3)
En-2	D3	3	CD1 C57BI/6	261 104	162 24	48 (–) 6 (–)	– –	–	–	(4)
β_2-microglobulin	D3	10	C57BI/6	89	24	10 (10)	6	normal	Lack mature $CD4^-8^+$ T cells; are defective in $CD4^-8^+$ T cell-mediated cytotoxicity	(5, 6)
N-myc	D3	7	C57BI/6	–	–	–	4	–	–	(7)
HPRT	E14TG2a	1	C57BI/6JLac x CBA/CaLac	396	266	63 (50)	19	normal	–	(8)
HPRT	CCE	3	MF1	629	384	203 (133)	11	normal	–	(9)
$HPRT^-$	E14TG2a	1	C57BI/6/Ola x CBA/Ca/Ola	93	26	15 (12)	1	–	–	(10)
$HPRT^-$	ES98-12	1	C57BI/6	–	11	9 (6)	2	–	–	(11)
Wnt-1 (int-1)	AB-1	6	C57BI/6	75	27	26 (24)	19	normal	Death within 24 h after birth; absence of midbrain and cerebellum	(12)

Tab. 2. Continued

Gene	ES Cell Line	Number of Targeted Clones	Blastocyst Donors	Number of Blastocysts Injected and Transferred	Number of Pups Born	Number of Chimeras (Males)	Number of Germ Line Chimeras	Phenotype of Heterozygous Mice	Phenotype of Homozygous Mice	Ref.
IGF II	CCE.33	2	CD-1 MF1 C57BI/6	64 58 135	24 24 61	6 (-) 13 (7) 22 (15)	- 3 4	Dwarfs, if inherited by a male, normal, if transmitted by a female	Dwarfs (60% smaller than controls)	(13, 14)
MHC Class II	D3	5	C57BI/6	-	45%	28%	yes[a]	normal	Near complete elimination of $CD4^+$ thymocytes from spleen and lymph nodes; alterations in the B cell compartment	(15)
c-abl	CCE	7	CD-1 MF1 C57BI/6	156 189 331	49 80 106	20 (15) 42 (27) 29 (21)	0 0 6	normal	Perinatal mortality; runtedness; abnormal spleen, head and eye development	(16, 17)
c-abl	CCE	15	C57BI/6	78	29	21 (-)	1	normal	Neonatal lethality, thymic and splenic atrophy	(18)
IL-2	E14	2	C57BI/6	-	40	8 (-)	3	normal	Dysregulation of the immune system, changes in the isotype levels of serum immunoglobulins	(19)
c-src	AB2.1	4	C57BI/6	51	30	27 (22)	14	normal	Death within the first weeks of life, deficient in bone remodeling, impaired osteoclast function, osteopetrosis	(20)

(1) ZIMMER and GRUSS (1989), (2) CHISAKA and CAPECCHI (1991), (3) LUFKIN et al. (1991), (4) JOYNER et al. (1989), (5) ZIJLSTRA et al. (1989), (6) ZIJLSTRA et al. (1990), (7) STANTON et al. (1990), (8) HOOPER et al. (1987), (9) KUEHN et al. (1987), (10) THOMPSON et al. (1989), (11) KOLLER et al. (1989), (12) MCMAHON and BRADLEY (1990), (13) DECHIARA et al. (1990), (14) DECHIARA et al. (1991), (15) COSGROVE et al. (1991), (16) SCHWARTZBERG et al. (1989), (17) SCHWARTZBERG et al. (1991), (18) TYBULEWICZ et al. (1991), (19) SCHORLE et al. (1991), (20) SORIANO et al. (1991)

[a] have been obtained

cells and may even bestow selective growth advantages upon them. The proviruses whose genes are transcribed and translated by the cellular machinery are also stably inherited together with flanking DNA sequences.

Retroviruses possess a number of features which suggest their use as gene transfer vehicles. These viruses can infect cells on their own and are efficiently integrated into the cellular genome. The retroviral genes are also efficiently expressed inside the cells. Many retroviruses already contain cellular sequences (oncogenes) and may be considered natural vectors allowing the transfer of non-viral and cellular DNA sequences.

The advantages of using retroviral vectors as vehicles for the gene transfer into cells are the relatively high efficiency of infection, the single-copy integration, and the ease with which gene transfer in suitable target cells can be effected. A disadvantage of these vectors is the fact that the upper size limit of genes that can be incorporated into the retroviral genomes is in the order of 6–8 kb. It should also be pointed out that several aspects concerning the biohazards of these vehicles have not yet been entirely clarified.

Retroviral life cycle

Retroviruses with a relatively restricted host range, i.e., those which only replicate in one host species or a closely related species are called ecotropic. In order to understand the functional mechanisms of retroviral vectors it is be necessary to summarize (SALMONS et al., 1991) briefly the life cycle of retroviruses (Fig. 12). An example of an ecotropic retrovirus is MLV which replicates in mouse and rat cells. Amphotropic retroviruses have a broad host spectrum and are capable of replicating in several mammalian species. The genome of retroviruses consists of RNA (Fig. 12). The infection is initiated by the interaction of the virus with the cellular membrane and involves the binding of the viral coat protein (env) with a specific receptor protein (Fig. 12; 2). In the case of MLV the receptor has been shown to be a cationic amino acid transporter (KIM and CUNNINGHAM, 1991). The retrovirus is taken up by the cells by micropinocytosis (Fig. 12; 3) which is then followed by the release of the viral core. The core structure consists of the gag proteins, reverse transcriptase, integrase, and also contains the retroviral RNA (Fig. 12; 4). In some retroviruses the core structure is transported into the nucleus (Fig. 12; 5, 6) while in others reverse transcription proceeds in the cytoplasm. Reverse transcriptase which is part of an enzyme complex of the core converts the genomic viral RNA into a double-stranded DNA copy (Fig. 12; 7). Some sequences at the ends of the viral genome are duplicated during the transcription process. In their DNA form these sequences are known as LTR (long terminal repeat) (Fig. 13). LTRs consist of an array of three distinct sequence elements, U3-R-U5, with U3 being derived from the 3′ end, and U5 of the 5′ end of the viral genome. The double-stranded DNA is transported into the nucleus and integrated into the cellular genome

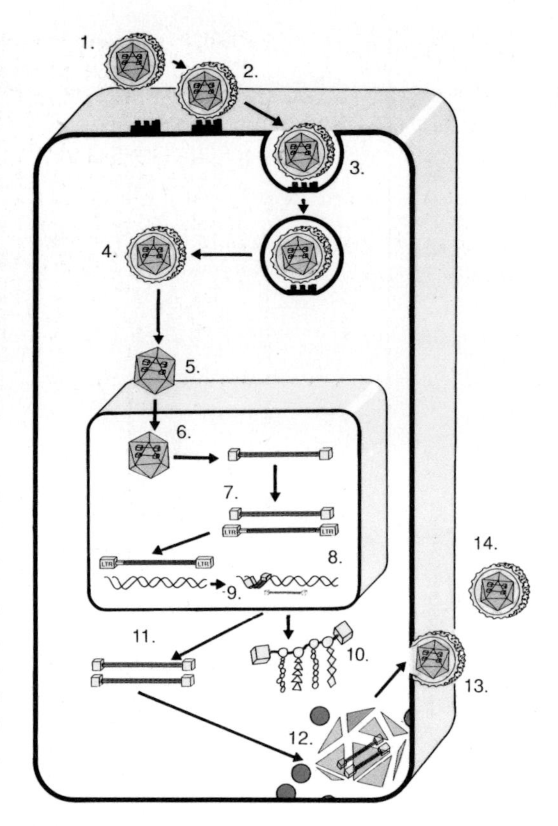

Fig. 12. Life cycle of retroviruses (SALMONS et al., 1991). For details see text.

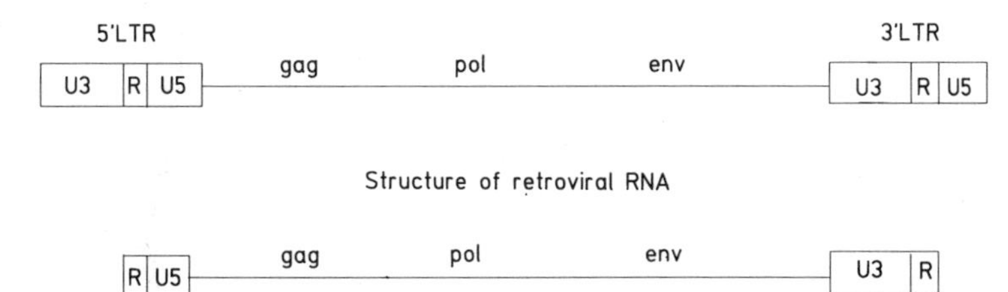

Fig. 13. Structure of an integrated provirus and retroviral RNA (SALMONS et al., 1991).

(Fig. 12; 8) (reviewed by GRANDGENETT and MUMM, 1990).

The integration event requires the presence of 9 base pairs at either end of the DNA (BUSHMAN and CRAIGIE, 1990). These bases are recognized and cleaved by the virus-specific integrase which is also part of the core enzyme complex (SCHWARTZBERG et al., 1984). The integrase must be associated with the proviral DNA, since the integration will not take place if the enzyme is provided in *trans* (KRIEGLER and BOTCHAN, 1983). The cleavage reaction is a specific and essential step for the integration process and involves the successive removal of two base pairs at both ends of the provirus (FUJIWARA and CRAIGIE, 1989). It is still an open question whether the unspecific cleavage of the host DNA is mediated by a nucleophilic attack through the integrase or the viral DNA (MIZUUCHI and CRAIGIE, 1991).

The integration event is always associated with the loss of two base pairs at the two ends of the proviral DNA (MAJORS and VARMUS, 1981; DONEHOWER et al., 1981) and a duplication of 3–6 base pairs of the cellular DNA (4 bp in MLV) (MCCLEMENTS et al., 1980; SHIMOTOHNO and TEMIN, 1981; SHOEMAKER et al., 1981). A number of *in vitro* integration systems have been developed recently which allow the analysis of the integration process in more detail. Nucleoprotein complexes of ca. 160 S which are sufficient for DNA integration have been isolated from MLV-infected cells (BROWN et al., 1987; FUJIWARA and MIZUUCHI, 1988; BOWERMAN et al., 1989). The MLV-specific complex contains at least the p30gag protein, the p46pol integrase, and presumably also the reverse transcriptase (BOWERMAN et al., 1989).

The proviral DNA integrated into the host genome (Fig. 12; 8) is transcribed (Fig. 12; 9) and translated (Fig. 12; 10) like other cellular genes. The translated viral gag and pol proteins are complexed with the genomic viral RNA and form the viral core which then leaves the cell by budding to yield an infectious virus particle (Fig. 12; 12). During the budding process the core is surrounded with part of the host cell membrane which also contains the viral env proteins (Fig. 12; 13).

The growth of cells is usually not affected by the infection with a retrovirus. Dramatic effects may be observed, however, if the provirus has integrated near a cellular gene. Some protooncogenes are activated by such integration events (VARMUS, 1983; PETERS et al., 1983; NUSSE, 1986). The retrovirus may also integrate into a cellular gene, e.g., tumor repressor genes such as p53 or the retinoblastoma gene, leading to an inactivation of these genes (GRIDLEY et al., 1987).

Retroviral vectors

A variety of retroviral vectors based on several different retroviruses has been developed. The most frequently used retrovirus vectors are based on murine leukemia viruses (C type retroviruses) (for reviews see EGLITIS and ANDERSON, 1988; MCLACHLIN et al., 1990). These vectors consist of two components, i.e., the vector construct and a packaging cell line (Fig. 14). The genes encoding the structural proteins gag, pol, and env have been removed from the vector constructs (proviral DNA) and replaced by other genes such as the β-galactosidase gene (gal) or the neomycin resistance gene (neo). Since the structural proteins are essential for the replication cycle of the retroviruses

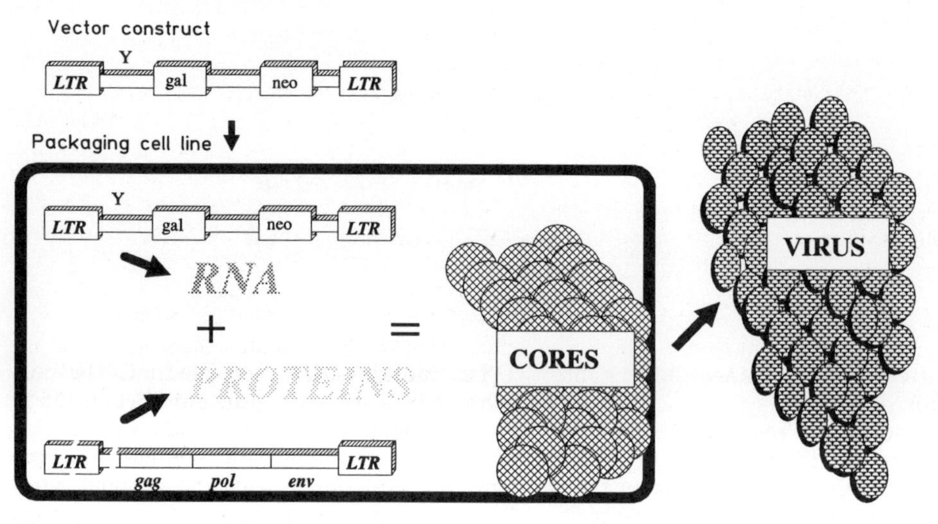

Fig. 14. Packaging cell line for retroviral vectors (SALMONS et al., 1991).

a suitable packaging cell line must be used which provides these proteins in *trans*. Such packaging cell lines contain a provirus which directs the synthesis of large amounts of the structural genes. The provirus has been modified in a way that prevents the package of the retroviral RNA to be packaged into the ribonucleoprotein complex: it lacks the so-called ψ recognition signal which is essential for packaging the retroviral RNA. Since this recognition sequence is present in the vector construct introduced into the packaging cell line, it can be packaged correctly. Therefore, infectious virus particles harboring the genetic information of the retroviral vector can be produced. The target cell infected with such particles is incapable of further producing infectious particles because it lacks the retroviral genes encoding the structural proteins.

The production of transgenic animals by retroviral vectors has many advantages. Most notable are the stable integration of the desired genes, the low copy numbers – usually one – and the colinear integration into the host cell DNA with the heterologous gene always flanked by the LTR sequences (SALMONS et al., 1991). Unfortunately, the use of retrovirus vectors is still associated with a certain risk, albeit a small one, of producing wild-type viruses by recombination between retroviral sequences. This might lead to diseases such as tumors or immunosuppression.

The transfer of foreign genes by means of retroviral vectors is not restricted to the gene transfer into fertilized oocytes. These vectors also allow gene transfer into the germ line during the prenatal development. Recombinant retrovirus vectors are particularly well suited for the transformation of hematopoietic stem cells and hence may offer some possibilities for somatic gene therapy (see also Sect. 2.4). These vectors can also be used for the investigation of insertion mutations and for tagging sites on chromosomes. From a practical point of view the greatest advantage of the retroviral system is the ease with which the gene transfer can be effected because, in principle, it only requires cocultivation of the embryos (without zona pellucida) with the cells producing the vector.

It has been shown for the first time in 1974 that the injection of SV40 DNA into the blastocoel of murine blastocysts may yield adult animals with cells containing this DNA (JAENISCH, 1974; JAENISCH and MINTZ, 1974). Subsequent experiments with Mo-MuLV-provirus DNA have demonstrated that the DNA was integrated into the genome and was stably transmitted to the offspring, thus enabling the establishment of stable lines (JAENISCH, 1976;

STUHLMANN et al., 1981). The next step has been the development of retroviral vectors and the transfer of heterologous genes into mice using recombinant vector systems.

A general strategy for the efficient insertion of recombinant retroviral vector DNA into the mouse germ line is the infection of preimplantation mouse embryos. VAN DER PUTTEN et al. (1985) have generated transgenic mice harboring a replication-competent recombinant retrovirus (ΔMo + Py M-MuLV) that lacks the M-MuLV-type enhancer sequence in the LTR. 16 mice were also generated that harbor proviral DNA of a defective recombinant retrovirus carrying a mutant dihydrofolate reductase gene (VAN DER PUTTEN et al., 1985).

HUSZAR et al. (1985) have been able to integrate the bacterial neo^R gene into the germ line of mice by retrovirus-mediated gene transfer. After a 24 h cocultivation of zona-free 8-celled embryos with Rat-2 cells constitutively expressing the retroviral MLV-NEO.1 vector 65 of 119 (55%) developed until birth. 12% of the animals contained sequences of the neo gene as shown by the analysis of DNA derived from tails. 6 different independent integration sites were detected. However, the neo^R gene which was heavily methylated was not expressed. At least one provirus was inherited like a Mendelian gene into the F3 generation, yielding some homozygous transgenic mice in the process. RUBINSTEIN et al. (1986) have also succeeded in creating mice with integrated proviruses by cocultivation of zona pellucida-free 2-celled embryos with vector-producing cells.

SORIANO et al. (1986) have produced transgenic mouse lines by infection of preimplantation embryos with recombinant retroviruses containing the complete human β-globin gene under the control of its own promoter and the bacterial neomycin phosphotransferase gene under the control of a viral promoter. 3 out of 69 (5%) of the born mice were transgenic. In comparison to the frequencies obtainable with M-MuLV (>50%) this is a relatively low yield, and it may be due to the fact that virus stocks with a 25-fold lower titer had been used. In all transgenic lines the β-globin gene was expressed predominantly in the hematopoietic system. In addition one line expressed the β-globin gene ectopically in the same tissues that also contained a high level of RNA expressed from the viral promoter. It appears that, in contrast to the internal promoter, the LTR-driven gene expression is also influenced by the integration site (SORIANO et al., 1986). As a whole, the expression of the transgene was markedly lower than the level of endogenous murine globin gene expression or the level of human β-globin in the circulation.

STEWART et al. (1987) have infected preimplantation mouse embryos with a retroviral vector containing the v-myc gene and a neomycin resistance gene under the control of the thymidine kinase gene promoter. Of 104 embryos 17 (17%) developed into mice, 6 (35%) animals were positive and 3 mice were used as founder animals which transmitted the transgene to approximately 8% of their offspring. An embryo survival rate of 34% has been observed with another retroviral gene construct containing the cDNA of the human adenosine deaminase gene and an SV40 promoter (M-SAX mice). 19% of the animals developing from the embryos were positive, and 4 out of 7 primary animals transmitted the transgene to 10% of their progeny. In one line (M-TKneo1) the transgene was found to be integrated into the X chromosome. The male founder animal transmitted the transgene only to female progeny (5/11) while one positive F1 female yielded positive males and females. The 35 females obtained from an F2 positive male were all positive while none of the male offspring did inherit the transgene.

Southern blot analysis has revealed that all transgenic lines contained a single intact integrated copy of the transgene. Replication-competent viruses were not detectable in any of the mice. The neo^R gene under the control of the TK promoter was expressed in all three lines whereas the ADA cDNA was not detected in any of the four M-SAX substrains. Mice which expressed genes under the control of the 5′ LTR were not observed.

There are currently only a few reports about the application of retroviral vectors in domestic animals. The use of such vectors in poultry is described in Sect. 7.

Some reports are available on the use of retroviral vectors in sheep and pigs. One group in Glasgow has reported experiments in which concentrated virus solutions containing feline leukemia virus (FeLV) were injected into the perivitteline space of 2- to 4-celled sheep embryos. These embryos were then transferred to suitable recipients and the developing fetuses were analyzed after 50 days. FeLV-specific sequences were detectable in 2 out of 17 fetuses (12%) (HARVEY et al., 1990). Work is in progress to develop a suitable packaging line for FeLV.

PETTERS et al. (1989) have propagated a poultry retrovirus (SNV, spleen necrosis virus) in a dog cell line. Approximately 100–150 infected cells were injected into the blastocoel of 122 pig blastocysts which were subsequently transferred to 12 recipients. 21 normally developed fetuses were recovered from 4 pregnant animals after 6 weeks. In 17 cases (80%) PCR analysis of several tissues, including brain, heart, liver, and muscle, revealed the presence of SNV sequences in at least one of these tissues. The same group has also investigated the survival of early preimplantation porcine embryos after coculture with cells producing an avian retrovirus (JIN et al., 1991). Using D-17 canine cells (see also Sect. 7) 58% of all cocultivated pig embryos developed to blastocyst stages and 12% of them developed to normal fetuses ($n=6$).

One serious disadvantage of the generation of transgenic animals with retrovirus vectors appears to be the lack of expression of the transgene in mice. It appears that some factors suppress the expression of these vectors in early embryos and also in embryonic stem cells and only allow a very limited expression in adult animals. Initially the methylation of the DNA was considered to be responsible for this phenomenon, but this may not be the sole cause because the degree of methylation does not correlate well with the observed levels of gene expression. This disadvantage may be overcome by constructing vectors which contain the desired structural gene and, in addition, also a suitable external promoter sequence.

Another complication may be the rather limited capacity of retroviruses, since 8 kb of foreign DNA constitute the upper limit of DNA that may be incorporated into such vectors. Therefore, this system is not suited for a number of gene transfer experiments requiring the transfer of long genomic DNA fragments.

Moreover, there are three problems associated with the retrovirus-mediated gene transfer which are more or less specific for retroviruses. The first problem is concerned with the spread of the vector virus in infected organisms. This possibility can be ruled out in most instances because the virus vector systems generally in use do not require the use of helper viruses. Secondly, it has been shown that the integration of a retrovirus into the host genome may lead to the activation of cellular oncogenes by viral transcription signals.

The third potential problem associated with the gene transfer via retrovirus vectors has to be discussed in greater detail. It concerns the probability of creating active forms of retroviruses by recombination between the defective retroviruses generally used as vectors with endogenous retroviral sequences of the host. The possibility for such events to occur may not be ruled out altogether but is generally considered to be extremely low. It should also be pointed out that the activation of a cellular retrovirus mediated by the gene transfer process is equivalent to the reactivation of ecotropic viruses independent of the transfer procedure. It appears advisable to carry out suitable long-term experiments involving the transmission of retroviral transgenes over many generations in the mouse in order to find out whether such cases will indeed occur and what consequences this may have. Eventually it should be possible to design retrovirus vector systems which will also exclude this residual risk. DOUGHERTY and TEMIN (1987) have constructed a retrovirus vector, for example, which lacks all viral transcription signals. This vector is much safer for use in gene transfer experiments because it does not contain any U3 sequences apart from those 10 base pairs derived from the right-hand LTR which are absolutely required for integration.

To reduce the risk of generating wild-type viruses we have isolated subretroviral particles (cores) as an efficient integration system for the establishment of transgenic mammalians. These cores only contain those retroviral sequences which are required for a successful colinear integration of the gene construct into the genome of the host cell. This subretroviral vector system therefore contains so little retroviral information that the possibility of a recombination event yielding wild-type virus can be excluded. Of course, due to the lack of infectious virus particles this safety strategy requires that the recombinant gene constructs are transferred into the target cells (e.g., fertilized oocytes) by micro-injection which is much more demanding than the introduction of the foreign gene by retrovirus infection. Nevertheless, this core transfer technique is much more efficient than the injection of naked DNA. It has the advantage in that the use of cores leads to higher integration frequencies and a colinear integration of DNA in low copy numbers, thus guaranteeing a greater stability of the transferred DNA.

2.4 Other Methods of Gene Transfer

The search for alternatives to the conventional techniques of gene transfer by micro-injection has led to the conclusion that appropriate techniques for the transfer of genes into mammalians or embryos might be developed from the plethora of methods based on the transformation of eukaryotic cells. Some of these techniques such as transfection by the calcium phosphate precipitation or chemical procedures involving the use of DEAE dextran can be excluded from the beginning: they are relatively inefficient and also lead to unstable integration events because the foreign DNA frequently becomes mutated and rearranged. The electroporation allows the transfer of DNA into mammalian cells with good efficiency. In most cases, however, the expression of the transferred gene is transient and the gene does not integrate into the DNA of the host cell. Another technique which can be used for gene transfer into blastocysts is the micro-injection of DNA packaged into liposomes. ROTTMANN et al. (1985) have linked the Moloney mouse sarcoma virus LTR to the genomic fragment encoding the surface antigen of hepatitis B virus. The construct has been encapsulated into liposomes which were then injected into the blastocoel of mouse embryos. 80 injected and transferred embryos yielded 24 animals of which 5 were transgenic. REED et al. (1988) have encapsulated recombinant DNA into liposomes and injected them into cattle blastocysts. However, positive data have not yet been reported.

A very simple procedure for producing transgenic mice involving the incubation of sperms with DNA before *in vitro* fertilization has been described by LAVITRANO et al. (1989). Approximately 30% of the animals obtained in this way contained foreign DNA integrated into their genomes and also transmitted this DNA to their offspring. Moreover, the expression of the pSV-CAT gene construct used in this study has also been reported. The same gene construct has been incubated with pig sperms. After surgical insemination of 22 female pigs 16 became pregnant and yielded 48 piglets. 10 (21%) of these piglets were transgenic and expressed the CAT gene (GANDOLFI et al., 1989). Another report published in the same year has described the generation of sea urchin embryos in the blastula stage expressing the CAT gene after treatment of sperms with the CAT construct (AREZZO, 1989). As early as 1971 BRACKETT et al. have described the incubation of rabbit sperms with SV40 DNA and the transport of the foreign DNA into oocytes following artificial insemination. In both cases, however, chromosomal integration of the SV40 DNA has not been observed.

Because of the great importance of the sperm-mediated gene transfer many groups have tried to repeat these experiments immediately after publication. In October 1989, BRINSTER et al. (1989b) published the results of an analysis of 1300 mice born after *in vitro* fertilization with sperms treated with DNA. They were unable to detect a single case of genetic transformation.

However, further reports have now been published that the sperms of several species are permeable for exogenous DNA and that sperm-mediated gene transfer has been repeated successfully. For the work pertaining to poultry the reader is referred to Sect. 7.

DE LA FUENTE et al. (1990) have employed the B6D2F1 strain of mice and the pUC-HBV plasmid. 37 of the born animals were tested for the presence of the foreign gene, and three were found to be positive when the serum was analyzed. The analysis of liver tissues of these animals by Southern blotting also yielded positive signals while DNA from the tails was negative.

The PCR technique can be used to monitor the fate of the exogenous DNA transported into the oocytes by sperms. HOCHI et al. (1990) have incubated mouse sperms with a pSV2-gpt construct and used them for *in vitro* fertilization. Embryos of the 1-, 2-, 4-, 8-cell stage, morulae and blastocysts were analyzed by PCR and the foreign DNA was detected in 46 out of 50 embryos. However, the gene construct was not detectable in 130 mice born after the transfer of embryos from this *in vitro* fertilization program. Our own investigations with an MT-GHRH construct have shown the presence of a positive signal by PCR analysis in mouse embryos obtained with DNA-treated sperms. The foreign DNA constructs were absent, however, in 10 day old fetuses and adult animals.

Some studies have also focused their attention on the fate of the foreign DNA on the sperms. LAVITRANO's team has detected the radioactively labeled plasmid DNA on the surface and also in the nucleus of mouse sperms (LAVITRANO et al., 1991a, b). Sperms of pigs and cattle were also capable of binding and incorporating foreign DNA (Fig. 15).

The foreign DNA is usually found in the equatorial segment and in the postacrosomal region. The highest rates of uptake are observed after an incubation of sperms with DNA for 20–40 minutes. Another investigation on the fate of foreign DNA molecules on and in sperms has been carried out by HORAN et al. (1991) who used motile and non-motile pig sperms treated for 15 min with radioactively labeled DNA. The sperms were washed 5 times, and aliquots were scintillation-counted and prepared for autoradiography. The results showed that approximately 400 DNA molecules were bound by a single sperm. Motile sperms were associated with much higher concentrations of radiolabeled DNA than non-motile sperms.

FRENCH et al. (1990) have suggested that the binding of DNA molecules to sperms is reversible. ATKINSON et al. (1991) have reported experiments in which bull sperms were treated with radioactively labeled OY 11.1 DNA. After several washing procedures one half was treated with DNAse I while the other half remained untreated. 19% of the sperms treated with the nuclease and 39% of the untreated sperms still had retained foreign DNA. Further analysis has shown the existence of four different binding regions, i.e., the acrosomal cap (I), the equatorial segment (II), the caudal part of the sperm head (III), and the end of the head (IV) all of which were occupied in sperms of DNAse I treated and untreated sperms: fraction without DNAse I treatment: 0.28% at I, 2.25% at II, 84.79% at III, and 12.68% at IV; DNAse I-treated fraction: 0.28% at I, 0.56% at II, 11.70% at III, and 87.47% at IV. Sperms that had been subjected to a cold shock and also non-motile sperms had foreign DNA bound exclusively to positions III and IV.

The absorption of DNA by sperms can be increased by electroporation. The electroporated sperms can the be used to transfer the foreign DNA by fertilizing oocytes. An increased fraction of day-5 cattle embryos has revealed a positive reaction for foreign DNA as shown by PCR analysis (GAGNÉ et al., 1991).

LAVITRANO et al. (1990) have reported that approximately 70% of the DNA molecules remain attached to the head of sperms treated with DNAse I. It has been shown that the liposome-mediated gene transfer allows efficient transfer of DNA into the sperms. Although the fertilization frequency remained unaffected, it has been impossible to obtain transgenic mice in this way (BACHILLER et al., 1991).

Similar observations have been made with sperms of mice and carp (CASTRO et al., 1990). Among the sperms treated with DNAse I approximately 4% less had bound foreign DNA than untreated sperms. Sperms of a variety of species which were treated with formaldehyde did take up foreign DNA only marginally. CASTRO et al. have reported that approximately 1000–6000 DNA molecules are bound by one sperm when the input is on the order of 10^4–10^6 molecules of DNA. ATKINSON et al. (1991) have reported that insect sperms (*Apis mellifera, Lucilla cuprina*) can bind DNA over the entire surface and that no DNA is retained after DNAse I treatment.

MILNE et al. (1989) have used sperms of the honey bee (*Apis mellifera*) as a vector for a 1 kb DNA fragment derived from plasmid pSP6/T7-19 and inseminated young queens.

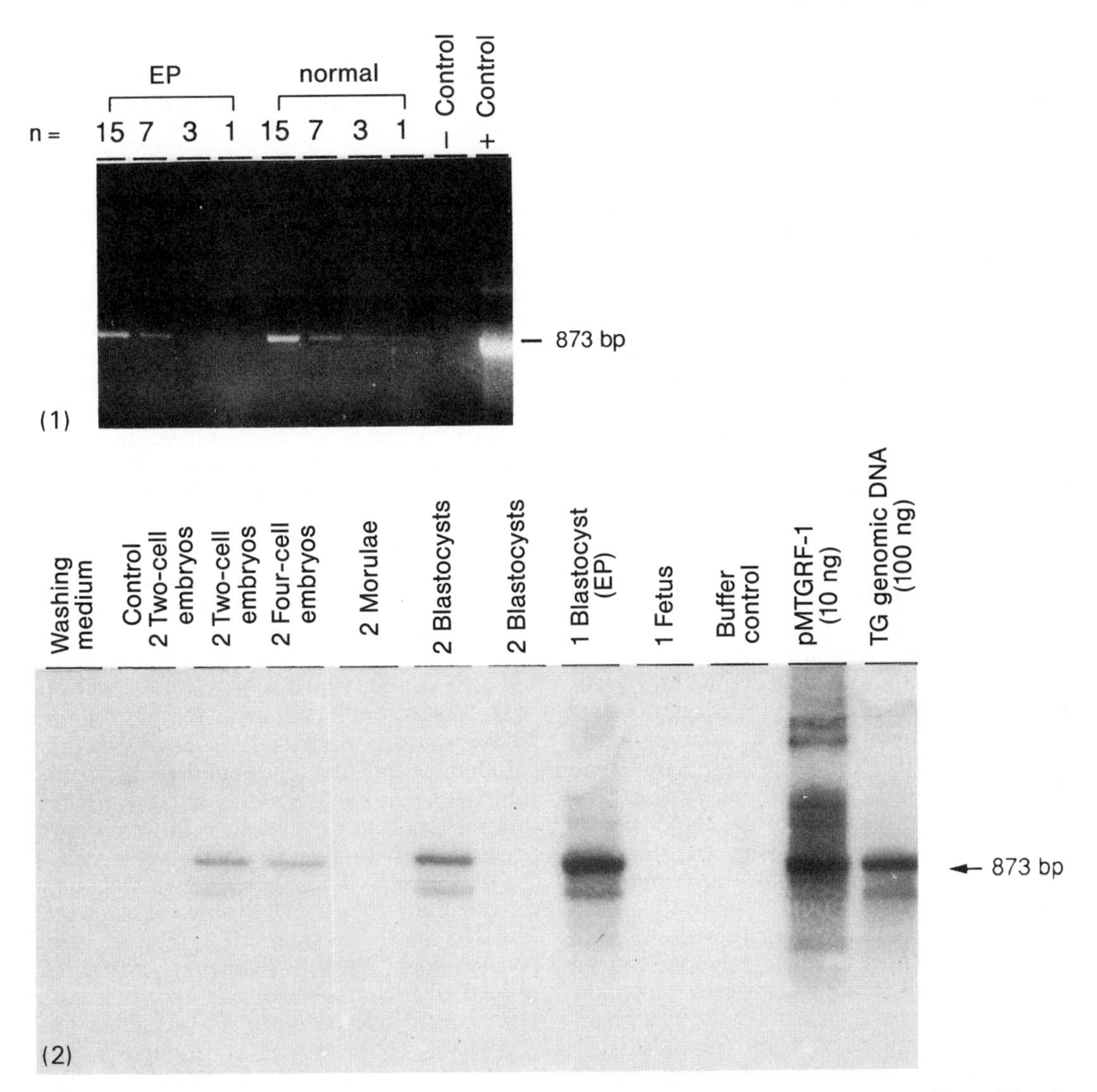

Fig. 15. PCR analysis of embryos obtained by *in vitro* fertilization with DNA-treated sperms (1), and Southern blot analysis (2) after PCR of embryos and fetuses obtained after *in vitro* fertilization with MT GRF-1-treated sperms, negative, and positive controls, and washing fluid.

The foreign gene was found in 29.6% of the larvae of worker bees.

Liposome-mediated gene transfer via spermatozoons into avian egg cells (ROTTMANN et al., 1991) and into mice (BACHILLER et al., 1992) was described recently. Liposomes were used to transfect DNA into the sperm head prior to fertilization. Although DNA transfer into sperms mediated by liposomes was very efficient and no obvious reduction in the fertilization frequency of oocytes was detected, BACHILLER et al. (1991) failed to generate transgenic mice by this method. After insemination of 200 hens with liposome-treated sperms ROTTMANN et al. (1991) detected in 26% of all fetuses banding patterns of the transferred plasmid but the sequences indicated that the foreign DNA was not chromosomally integrated but was present in episomal form.

A positively charged glycoprotein with a molecular mass 35 kD present in the epididymal sperms of sea urchins as well as in sperms of other species, including humans, is assumed to play a role in the interaction of sperms with DNA molecules (LAVITRANO et al., 1991a).

According to LAVITRANO and coworkers (1991b) protection against the undesired introduction of foreign molecules is mediated by two different mechanisms. A protein found in the seminal plasma can prevent binding of DNA to the sperm cell by high affinity interaction with the 35 kD protein. In addition, a negatively charged glycoprotein with a molecular mass of approximately 70 kD has been isolated from sperms. It is completely absent in epididymal sperms and also reduces the permeability of sperms for foreign DNA.

Much more work is required before the gene transfer into the germ line via sperms will become an effective method. Using this technique the efficiency of gene transfer can be increased by a factor of 3–5 in mice while the costs are greatly reduced at the same time (BARINAGA, 1989). For these reasons the sperm-mediated gene transfer could be of enormous importance for gene transfer programs carried out with animals of economical importance.

I will conclude this discussion with the presentation of a rather unusual technique for the establishment of transgenic progeny which has been used successfully in mice: the transplantation of transgenic ovaries into non-transgenic females. This approach might be invaluable if offspring cannot be obtained, for example, if the development and fertility of transgenic mice produced by conventional techniques such as DNA micro-injection has been impaired completely by the expression of the gene construct. By surgically transplanting transgenic ovaries into ovarectomized non-transgenic mice it will then be possible to obtain transgenic offspring (BAUNACK et al., 1988). In our own experiments we have transplanted the ovaries of 8 MTI-hGH-transgenic mice, which are usually infertile, into 15 ovarectomized recipient mice with each mouse receiving one half of the pair of ovaries. 10 recipient mice were fertile and yielded 20 litters with a total of 112 animals. Approximately 50% of these were transgenic (BREM et al., 1990b).

For the sake of completeness I would like to mention briefly another application of gene transfer, i.e., the transfer of genes into somatic cells. Apart from different techniques allowing the transfer of foreign genes into the germ line much effort has been put into the establishment of somatic gene transfer techniques that might provide safe ways for somatic gene therapy. The most frequently used cells are hematopoietic stem cells which can be transformed by means of retroviral vectors. SELDEN et al. (1987) have used transfection of mouse fibroblasts with a hGH fusion gene and transplantation of these cells at different localizations in the mouse as a model of somatic gene therapy. This so-called transkaryotic implantation technique led to the synthesis and secretion of growth hormone in the serum.

It has also been demonstrated that endothelial cells can be transformed *in vivo* and express the recombinant gene (NABEL et al., 1989). Another approach is the genetic modification of endothelial cells obtained *ex vivo* and the subsequent reimplantation into the original donor animals in which these cells were shown to survive up to five weeks (WILSON et al., 1989). Recently it has been shown that even the direct injection of DNA into muscle tissue leads to the expression of a reporter gene (WOLFF et al., 1990; ACSADI et al., 1991). The *in vivo* treatment of liver and skin cells with DNA-coated microprojectiles also allows foreign genes to be expressed in these tissues (WILLIAMS et al., 1991). Analbuminemic rats were injected intravenously with a plasmid containing the structural gene for human serum albumin, driven by mouse albumin enhancer–rat albumin promoter elements. Two weeks post infection the targeted DNA was found to exist primarily in plasmid form in the cells. Circulating human albumin was detected in the serum at a level of 34 mg/mL (WU et al., 1991).

A very interesting and promising approach for transforming somatic cells may be the membrane receptor-mediated gene transfer technique which can be used for a variety of different somatic tissues. WAGNER et al.

(1990) have used the transferrin receptor for this purpose. Apparently the concept of specific receptor-mediated gene transfer can also be extended to T-cell-specific receptors (M. BIRNSTIEL, personal communication, cited in RUSCONI, 1991). An alternative strategy to improve the receptor-mediated gene transfer has been described by CURIEL et al. (1991). This strategy essentially exploits the ability of adenoviruses to destroy endosomes. Infections with adenoviruses increased the efficiency of gene transfer by transferrin-polylysine conjugate in a dosis-dependent manner by a factor of more than 2000. This improvement of gene transfer has been observed for various target cells, some of which could not be transformed without additional adenovirus infection. The treatment of skin and liver tissue with DNA-coated microprojectiles also appears to be suitable for the genetic transformation of cells.

3 Transmission of Transgenes

3.1 Establishment of Transgenic Lines

The transmission of the transferred gene is of decisive importance for the establishment of transgenic lines and the application of gene transfer in breeding animals of economic importance. The transmission of the transgene requires that all or at least some of the germ line cells of the primary animals (those derived from the injected embryos) contain the transgene. Unfortunately little is known about the molecular-biological processes associated with the integration of injected DNA into the genome. For example, the exact time at which the integration takes place is not known, and it is also unknown whether the integrated state remains stable in all cells during the further development of the embryos. The analyis of newly born transgenic animals and also attempts to establish transgenic offspring have demonstrated that genetic mosaics may develop although the DNA had been injected into the pronuclei of fertilized oocytes (WILKIE et al., 1986).

Genetic mosaics are animals that consist of cell lines which are derived from the same zygote but possess different genotypes. Transgenic mosaics contain non-transgenic and transgenic cells. The establishment of transgenic lines may therefore create problems. If, for example, the gonads do not contain transgenic cells the injected gene construct cannot be transmitted to the offspring. Previous experience with transgenic animals has shown that approximately 30% of the primary transgenic animals obtained after micro-injection will not transmit the transgene with the expected frequency of 50% and these animals are therefore of limited value.

In animals which have stably integrated the gene construct within their genomes and transmit it to their offspring, the mode of inheritance is usually Mendelian because integration usually has occurred at a single site within a chromosome. By definition such animals are hemizygous transgenic. The term heterozygous cannot be applied because the corresponding allele does not occur in the homologous chromosome. One should expect that 50% of the offspring of such a transgenic animal will also contain the transgene. If the gonads are mosaics of transgenic and non-transgenic cell lines the frequency of transgenic offspring may vary between 0 and 50%, depending on the fraction of transgenic cells within the gonads (Fig. 16). Mosaics are usually observed only in the F0 generation. Transgenic F1 animals and all subsequent generations will contain the transgene in all somatic and germ line cells. It has been observed, however, that the integration may not be stable and the transgene may therefore not be inherited stably either.

In rare cases one observes more than one integration site in a primary transgenic animal. If the distances between these sites are sufficiently large to allow independent genetic recombination to take place between these two loci, one may obtain a higher frequency of transgenic offspring (Fig. 16). In a transgenic animal with two independent integration sites one may expect that 75% of the offspring will inherit a transgene with 25% of the animals inheriting either of the two transgenes and 25%

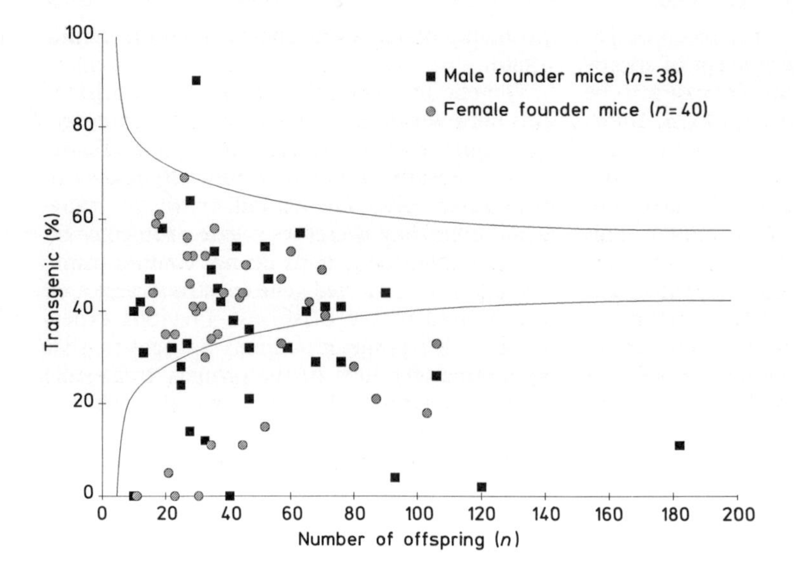

Fig. 16. Pattern of inheritance of various transgenes in primary transgenic mice. Each symbol represents a founder mouse. The position within the scheme depends on the total number of offspring and the frequency of transgenic offspring (significant deviations from the expected value of 50% of transgenic offspring verified by the Dixon & Mood sign test).

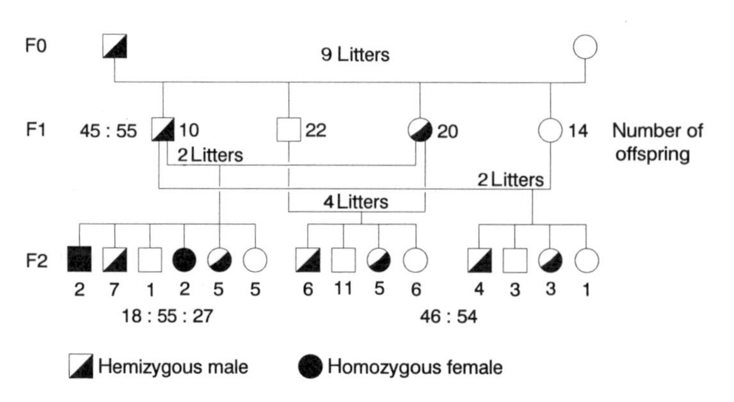

Fig. 17. Pedigree of a transgenic mouse line with mating of hemizygous offspring and breeding of homozygous transgenic mice.

inheriting both integration sites (see Fig. 8). If hemizygous transgenic animals are crossed 50% of the resulting progeny will be hemizygous transgenic, 25% will be homozygous transgenic and 25% of the animals will not be transgenic (Fig. 17).

Since integration of the gene construct in primary animals is essentially a random process, the transgene may also integrate at a locus which is crucial for the normal development of the animal. As long as there is another intact allele of this gene on the homologous chromosome, these insertion mutations may not lead to severe developmental aberrations. However, problems may arise if the effects of the affected gene are additive, i.e., if the integration reduces the expression of the affected additive trait. If the gene affected by the insertion of the transgene is essential for the development of the fetus, it may not be possible to obtain homozygous transgenic animals by crosses between hemizygous transgenic animals. The proportion of hemizygous animals in the offspring obtained from such crosses will be approximately 66% (25% of the fetuses will die in the uterus and/or will be resorbed, and the 50% hemizygous animals will correspond to ⅔ of all animals born).

WAGNER et al. (1983) have observed this phenomenon with hGH-transgenic mice. Although homozygous transgenic animals were not obtainable due to prenatal lethality a continuation of the hemizygous line was rendered possible by mating with normal animals, yielding 50% of normal transgenic offspring.

Such insertional mutations may be of great scientific interest. WOYCHIK et al. (1985) have found insertion mutations in transgenic mice which led to a known defect in the development of the limbs. Such insertion mutations may therefore be useful in isolating and characterizing the gene responsible for the observed defects (see also Sect. 5.1).

3.2 Breeding Programs with Transgenic Animals

The generation of genetic mosaics and insertional mutations as well as the fact that there are differences in the expression levels of transgenes have important and far reaching consequences for breeding. The introduction of a certain gene construct into a breeding population will require more than a single transgenic animal. At least 5 to 10 primary transgenic animals will be needed to ensure that appropriate transgenic lines will be established with an acceptable probability.

In general very high biological effects of transgenes are required for traits which are important as breeding aims in artificial insemination breeding programs in order to obtain the same level of genetic improvement. Experience has shown that extremely high biological gene effects are frequently associated with undesired side reactions. It seems appropriate, therefore, that breeding aims in transgenic lines should involve those traits that cannot be improved by employing conventional methods or can be improved only unsatisfactorily in traditional breeding programs, e.g., due to the low heritability of the trait concerned. Examples are traits which influence the resistance against disease or the quality of animal products.

The use of transgenic lines for breeding in the population does not require any specific breeding measures. Apart from the effect of the transgene, the constitution of the gene pool of the starting population (herdbook or nucleus population) will be another significant factor.

Like in any other breeding program breeding of transgenic lines is governed by the breeding aim and the speed with which genetic alterations will be achieved (genetic improvement). For genetic alterations which cannot be achieved by traditional selection the cost/effect considerations resemble those of traditional breeding programs. If the breeding aim includes traits which are also part of traditional breeding programs it will be necessary to compare the genetic improvement achieved with and without employment of transgenic lines. Such a comparison has to be carried out for the most important production characteristics of cattle and pigs (Tab. 3). It shows that,

Tab. 3. Examples for the Required Increase of Performance in Transgenic Lines for Equal Improvement as in Selected Populations

Trait	Mean	Estimated Improvement per Year in AI Breeding Programs	Years Needed for Establishing Transgenic Lines (F3) (see Tab. 4)	Minimum of Required Increase in Performance by Transgenic Effect
Cattle				
Milk yield	6000 kg	1.5%	11	16.0% (=960 kg)
Weight gain	1200 g	1.5%	11	16.5% (=200 g)
Lean	60 %	0.5%	11	5.5% (= 3.3 %)
Pig				
Backfat	20 mm	2.0%	4	8 % (= 1.6 mm)
Weight gain	700 g	2.5%	4	10 % (= 70 g)
Lean	57 %	1.5%	4	6 % (= 3.4 %)

Tab. 4. Time Schedule (in Months) Required for the Generation of Transgenic Lines in Different Mammalian Species

Step	Mouse	Rabbit	Pig	Sheep	Cattle
Collection, micro-injection and transfer of embryos	1	1	6	6	12
Parturition of FO animals	0.75 (1.75)	1 (2)	4 (10)	5 (11)	9 (21)
Generation interval (F1 offspring born)	2.50 (4.25)	6 (8)	12 (22)	18 (29)	30 (51)
Homozygous transgenic F2 animals available for breeding	3.25 (7.50)	6 (14)	16 (38)	23 (52)	39 (100)
Usage of F3 animals in population	2.50 (10.00)	6 (20)	12 (50)	18 (70)	30 (130)

Values in parentheses are the added sum of months necessary for finishing the program to this step.

depending on the individual trait, performance in transgenic lines must increase by 5.5–16.5% in order to achieve the same level of genic improvement that is achieved in optimal artificial insemination breeding programs in the same time (BREM, 1989).

Considering the rather demanding task of establishing transgenic lines and the high costs associated, the selection of the appropriate animals for the gene transfer will be of paramount importance. A time schedule required for the establishment and evaluation of transgenic lines in different species is given in Tab. 4. Transgenic lines should be especially helpful as paternal lines since the transgene may be propagated relatively fast either by artificial insemination or naturally, especially if the transgenic allele already leads to the desired effect in hemizygous animals. SMITH et al. (1987) have discussed the establishment of a pool of transgenes in nucleus and MOET (multiple ovulation embryo transfer) breeding programs. Transgenes can be considered as unique identified endogenous genes; their advantage is the potentially large effect and the addition of genetic variation to the population, and their disadvantage is some genetic lag once they have been entered into the population (GIBSON, 1991). The current state of our knowledge is still too limited to exactly evaluate the consequences of an accumulation of transgenic alleles in a gene pool.

4 Expression of Transgenes

4.1 Eukaryotic Gene Structure and Function

The crucial problem in gene transfer programs is the prediction of tissue- or cell-specific expression of the transferred genes. In the first experiments involving the establishment of transgenic animals most micro-injected genes were expressed, albeit at low and/or extremely variable levels. Further studies have demonstrated that the chromosomal location of the integrated foreign gene often affects the expression pattern and that prokaryotic vector sequences should be removed before transfer of gene constructs because they frequently inhibit expression.

It is possible to direct the expression of transferred genes to distinct or various cell types in transgenic animals by using gene constructs combining the structural gene with distinct regulatory elements. As shown in Fig. 3, a typical eukaryotic gene consists of various operationally distinguishable portions (for review see RUSCONI, 1991):

1. Coding region (which is only partly preserved in the mature transcripts). The introns are spliced out during the maturation of mRNA while the exons give rise to leader, open reading frame, and trailer sequences of the mature mRNA.

2. Proximal *cis*-acting control elements. The promoter, a region upstream of the transcription start site, specifies quantity, accuracy of initiation, and polarity of transcription.
3. Distal *cis*-acting elements. The rate of transcription is specified by enhancer or silencer elements in cooperation with the promoter.
4. Remote control elements. Dominant control regions and/or other regulatory sequences may specify the overall on-off state of the chromatin containing and surrounding the transcribed region.

The accepted view of transcriptional regulation in mammalian cells holds that genes encoding mRNAs are preceded by a TATA box and an upstream control region located upstream from the transcriptional start site at positions −25 to −30 and −40 to −110, respectively. Some promoters apparently deviate from this consensus organization. These new promoters have a very high G+C content and no apparent TATA box sequence, and they apparently direct transcription of house-keeping genes such as HPRT and DHFR (DYNAN, 1986).

The promoters and enhancers in mammals are composed of multiple genetic elements or modules. The cellular transcriptional machinery is capable of handling and integrating the regulatory information conveyed by each module, thus permitting the evolution of distinct and often quite complex patterns of transcriptional regulation in different genes (DYNAN, 1989). A good example is provided by the metallothionein promoters which contain separate modules responding to glucocorticoids, metals, and other stimuli.

4.2 Tissue Specificity of Transgene Expression

The development of animals and other organisms depends upon the activation and differential expression of many genes in various cell types. Transgenic mice have been instrumental in localizing some *cis*-acting elements responsible for tissue-specific gene regulation. Some genes are expressed exlusively in one cell type, others are expressed in only a few cell types, and some are expressed in most cells (for review see PALMITER and BRINSTER, 1986; JAENISCH, 1988; CUTHBERTSON and KLINTWORTH, 1988; BABINET et al., 1989; RUSCONI, 1991). The tissue specificity of gene expression is usually studied by transferring gene constructs containing regulatory 5′ flanking sequences, proximal or distal to the promoter, introns, and 3′ flanking sequences. The various tissues of transgenic mice then are assayed for the specific mRNA or the protein product of the transferred gene. The functional analysis of gene expression has been greatly facilitated by the use of reporter genes whose products are readily and easily detectable. Several genes, including the SV40 late tumor antigen and SV40 capsid protein, chloramphenicol acetyl transferase gene, firefly luciferase gene, herpes simplex virus thymidine kinase (HSV-TK) gene, *E. coli* β-galactosidase gene, various oncogenes, and growth hormone genes have been used for this purpose. In order to discriminate between the products of injected and endogenous genes the transgene may be derived from a different species, or may be a minigene with some exons deleted or modified by the insertion or deletion of a few nucleotides.

The effects of some regulatory elements are subject to modulation by external stimuli such as metal ions, hormones, heat, cold, food additives, and so forth. The intrinsic properties of such elements are also retained in transgenic mice. Therefore, the *in vivo* expression of any transgene that has been coupled with such elements will also be subject to the control by these external factors. The metallothionein promoter, for example, can be stimulated by supplementing the food or the drinking water of transgenic mice containing this promoter with heavy metals (Fig. 18). Unfortunately, there is no possibility to shut off this promoter once it has been turned on by the external stimulus.

The tissue-specific expression of a desired gene can be determined by the combination of tissue-specific enhancer elements with particular promoter sequences. Recently, regulatory elements designated dominant control regions (DCR) resp. locus control regions (LCR) have

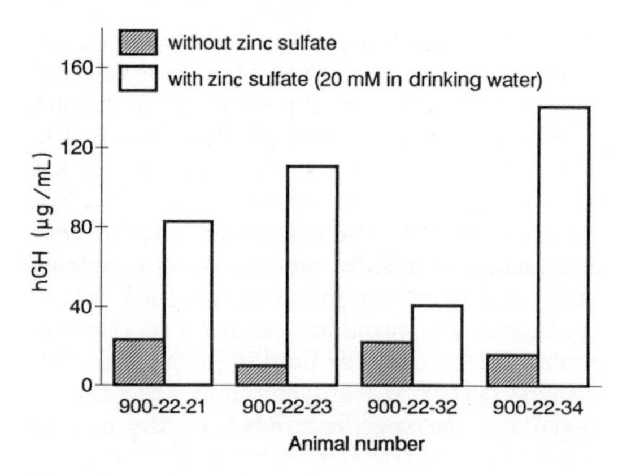

Fig. 18. Concentration of hGH in the serum of MT-hGH-transgenic mice before and after the administration of zinc sulfate with drinking water (WOLF, 1990).

been characterized (GROSVELD et al., 1987; STIEF et al., 1989; VAN ASSENDELFT et al., 1989; BONIFER et al., 1990). These regions, which constitute DNAse I-hypersensitive sites, are located up to 65 kb upstream of the transcription start site of certain genes. They may contain independent elements capable of conferring high-level expression to genes with which they are associated.

4.3 Efficiency of Transgene Expression

A pronounced variability with respect to the expression of transgenes is observed much more frequently than the occurrence of instable integration. All forms of alterations ranging from increased or decreased expression to virtual absence of expression may be observable throughout several generations. Even within groups of full sibs individual animals expressing the transgene may show a more or less pronounced variability in the levels of expression. The mechanisms underlying these differences are largely unknown. It may be surmised, however, that the phenomenon known as genetic imprinting may be one factor influencing transgene expression. Among other mechanisms yet unknown, differential methylation may influence gene expression, depending on whether the gene was inherited from the father or the mother of the animal in question.

So far, no method is available to predict the levels of transcriptional activity of genes that have randomly integrated into the host genome of transgenic animals. Transcription levels may vary with the chromosomal site of integration. This position effect on gene expression may reflect the organization of chromosomes into topologically constrained loops and functional domains. It has been shown that the effect is not observed in transgenic mice carrying the human β-globin gene if a suitable DNAse-hypersensitive site (LCR) has been incorporated into the gene construct (GROSVELD et al., 1987). The correct expression pattern of the β-globin gene may, in fact, not be dependent on competition *per se* but rather on the order of genes relative to the LCR (HANSCOMBE et al., 1991).

Cis-acting regions mediating the attachment of the chromatin to the nuclear scaffold have recently been shown to flank the chicken lysozyme gene. These domains have been designated A elements and map to the 5′ and 3′ boundaries of the region of general DNAse sensitivity in the active chromatin which contains the gene and its *cis*-regulatory elements. STIEF et al. (1989) have demonstrated that the expression in stably transfected cells of a reporter gene flanked by 5′ A elements from the lysozyme gene is significantly elevated and is no longer influenced by the chromosomal position of this gene.

REITMAN et al. (1990) have used a 4.5 kb DNA fragment carrying the β^{A}-globin gene and its downstream enhancer without any further upstream elements and have shown that all transgenic mice expressed globin mRNA at levels proportional to the transgene copy number.

cDNA constructs are frequently used for gene transfer in mice because cDNA sequences are more abundantly available and because many natural genes, due to their sizes, cannot be cloned and inserted into gene constructs. Many of these cDNA constructs are expressed quite well in cell culture but are not or poorly expressed in transgenic mice. BRINSTER et al. (1988) have therefore tested the effect of introns on gene expression. Four different pairs of gene constructs (one member of each pair lacked all introns) were used to generate transgenic mice. On the average 10- to 100-fold more mRNA was produced from constructs containing intron sequences. These experiments demonstrate that introns may facilitate the transcription of transgenes, and this effect may be manifest only on genes exposed to developmental influences.

Using the mMT-rGH gene constructs as a model PALMITER et al. (1991) have shown that the first intron of the rGH gene is essential for high-level expression, whereas the contribution of the other three introns is less pronounced. Heterologous introns placed 3′ of the cDNA of rGH were also ineffective. Insertion of heterologous introns between the MT promoter and the GH gene enhanced the expression. It appears that the expression of genes is influenced by – as yet unknown – events taking place during development which do not occur in *in vitro* culture of cells. LCR elements may be instrumental in establishing active chromosomal domains; in conjunction with enhancer elements and sequences contained within introns these elements may alter nucleosomal DNA interactions and may influence the assembly of functional transcription complexes at the promoter.

5 Application of Transgenic Mice

Within the space allotted to this chapter it is, of course, impossible to present and discuss all the different applications of data obtained from the use of gene transfer techniques in the mouse. The estimated number of publications dealing with transgenes falls short of 2000. The following passages will therefore be a small and admittedly subjective overview of selected topics. For these reasons the sections about disease models, oncogenes, and the immune system will be restricted to discussions of some fundamental findings and exemplary experiments. The reader is referred to special reviews and the pertinent primary literature cited, at least in part, in this chapter for further information.

5.1 Disease Models

The application of gene transfer techniques offers new strategies for the establishment of disease models. In principle there are two different approaches, namely, the directed and specific establishment of disease models, and the chance observations made in transgenic mice or their progeny, which may then be used as disease models. Tab. 5 summarizes the fundamental possibilities of establishing transgenic disease models.

Genes normally integrate at random (see also Sect. 2) and sometimes this process may not be without consequences. If the gene construct used for generating a transgenic animal becomes integrated in the vicinity of an endogenous gene, this may lead to noticeable insertional mutations. Such insertional mutations have been observed to occur in approximately 5% of the transgenic lines established so far. In hemizygous transgenic mice such insertions are usually relatively inconspicuous. They are normally detected only during an attempt to breed homozygous transgenic animals by observing either that no homozygous animals are obtained or that the born animals are characterized by a genetic disease whose nature, of course, depends on the affected gene. With the aid of such mutants it is possible to detect and isolate the gene affected by the mutation. The preferred method of screening for interesting genes is the random insertion mutagenesis by means of retroviral vectors. This technique has the advantage that insertion mutations occur much more frequently due to the relatively high frequency of integration. An additional advantage is that single copy integration facilitates the isolation of the affected gene because proviral sequences can be used as a hybridization probe. This approach can also be employed for the mutagenesis of embryonic stem cells *in vitro*. ES cells carrying the desired insertion mutation, i.e., those in which a certain gene has been disrupted, are then used to generate transgenic mice (see Sect. 2.2).

Insertion mutations in transgenic animals have been described for the first time by WAGNER et al. (1983). Following the injection of an hGH gene they have obtained 6 transgenic mouse lines of which only 4 could be used to produce homozygous transgenic mice

Tab. 5. Possible Disease Models in Transgenic Mice

Mode of Establishment	Mode of Action
Expression-related alterations	Transgenic gene product with additive effects in the organism
Dominant-negative mutations	Mutated gene product results in required alterations in spite of normally expressed endogenous proteins
Insertional mutations	Random interruption (mutation) of an endogenous gene with corresponding consequences predominantly observable in homozygous transgenic mice
Antisense-RNA-transgenic mice	Neutralization of the function of an endogenous gene
Promoter/diphtheria toxin-transgenic mice	Promoter-dependent ablation of a cell line or cell type
Homologous recombination	Homologous replacement of an endogenous gene by a new construct obtained by *in vitro* recombination
Oncogenes	Transgene induces tumor formation

by crossing between transgenic founder animals. In two lines the litter sizes were reduced and no homozygous transgenic mice were identified. In these lines the integration of the foreign DNA had caused a recessive lethal defect which was characterized by the prenatal death of the homozygous fetuses. By now several other lethal mutations caused by the insertion of foreign DNA have been described (HARBERS et al., 1984; STEWART et al., 1985; WOYCHIK et al., 1985; MARX, 1985; MCNEISH et al., 1988).

The insertion of foreign DNA into the β-1 collagen gene has yielded a lethal mutation in which the fetuses degenerate approximately between day 12–15 of gestation (JAENISCH et al., 1983).

Mice from the mutant Mpv 17 line have been generated after insertional mutagenesis. Homozygous mice were shown to develop a nephrotic syndrome and chronic renal failure associated with progressive glomerular sclerosis. This condition resembles that of human patients with progressive renal function deterioration (WEIHER et al., 1990) and the Mpv 17 mutant may therefore provide a useful experimental system for studying the mechanisms leading to this kind of renal disorder in humans.

A special case of an insertion mutation has been described by PALMITER et al. (1984) to occur in a transgenic mouse line in which the transmission of the transgene was restricted to female mice. The offspring of fertile males did not inherit the transgene presumably because the insertion affected a gene expressed during spermatogenesis.

Many lethal mutants are characterized by the death of the embryos occurring very early during embryonal or fetal development, usually beginning with the time of implantation. One may argue that the exceptional transcriptional activity of the cells during this stage may be one reason for this apparent sensitivity (COVARRUBIAS et al., 1986).

Dominant negative mutations

STACEY et al. (1988) have pursued a new approach for the establishment of models for certain human diseases. They have introduced well-known mutations into the pro-alpha (I) collagen gene and used these mutated genes for the production of transgenic mice. They have observed that the normal function of collagen was already disturbed if the level of expression of the mutated transgene was in the order of 10% of that of the endogenous gene. It has been surmised that the intracellular accumulation of the transgene product induces not only the degradation of the mutated but also of the normal collagens thus causing either a reduction in the level of collagen or disrupting the correct alignment of collagen and the formation of fibrils. The transfer of a mutated collagen gene into the embryos of normal mice has therefore yielded a disease model for the perinatal osteogenesis imperfecta type II.

There are many potential applications of introducing dominant negative mutations via gene transfer into mice. This approach appears to be particularly promising for the study of collagen genes and other genes encoding proteins of the cytoskeleton such as actin or filaments (STACEY et al., 1988).

The role of HTLV-I as one of the etiologic agents of chronic arthritis in humans has been elucidated by HTLV-I transgenic mice. These mice develop chronic arthritis resembling rheumatoid arthritis (IWAKURA et al., 1991). Synovial and periarticular inflammation with articular erosion caused by invasion of granulation tissues were quite pronounced.

Transgenic mice expressing mutant prion proteins spontaneously develop neurological dysfunctions and spongiform neuropathology (SCOTT et al., 1989; PRUSINER et al., 1990; HSIAO et al., 1990; WESTAWAY et al., 1991). The transgenic mice studies established that the prion protein gene influences all aspects of scrapie, including the species barrier, the replication of prions, the incubation times, the synthesis of the disease-specific protein and the neuropathological changes (PRUSINER, 1991).

The great advantage of disease models based on the use of dominant negative mutations lies in the fact that the effects of such mutations are additive and that such mutations may be introduced by simple gene transfer of appropriately mutated genes.

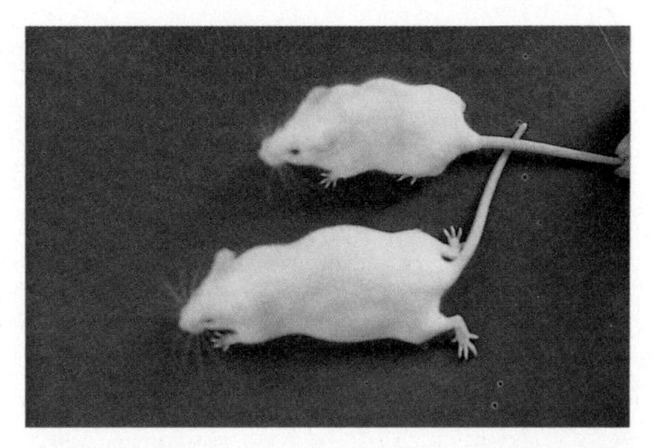

Fig. 19. MT-hGH-transgenic giant mouse in comparison with the non-transgenic full sibling (yellow marked).

Expression-related effects of transgenes

The expression of a transgene mostly depends on the regulatory sequences which have been used in the design of the gene construct. A massive overexpression of the transgene may be obtained by employing strong promoters which at the same time are independent of endogenous control circuits governed by the structural gene in question. A classical example are transgenic giant mice in which the expression of a growth hormone gene (GH) is driven by the metallothionein promoter (mMTI) (PALMITER et al., 1982, 1983).

It has been suggested that by their nature these transgenic giant mice (Fig. 19) may be useful models for the study of human growth deviations such as gigantism and acromegaly. They are excellent models for studying the entire spectrum of activities of growth hormones in the absence of interfering immunological processes (WOLF et al., 1991a, b, c). Our own investigations with mMTI-hGH transgenic mice were aimed at deepening our knowledge of growth processes. The pathological consequences observed in these mice are an example of a fortuitous disease model. We have observed, for example, that the expression of the GH transgene invariably causes nepropathological lesions (BREM et al., 1988, 1989; WANKE et al., 1991) which are excellent models not only for the study of the pathogenesis of renal glomerulosclerosis (Fig. 20) but also for the evaluation of new therapeutic concepts.

A recent example of a transgenic disease model depending on the level of expression of a transgene is Alzheimer's disease for which no suitable experimental animal system has been available so far. WIRAK et al. (1991a, b) have produced transgenic mouse lines which express the human amyloid β-protein under the control of the human amyloid precursor protein promoter in the dendrites of some but not all hippocampal neurons in 1-year old mice. Aggregates of the amyloid β-protein formed amyloid-like fibrils that were similar in appearance to those in the brains of patients suffering from Alzheimer's disease. Formation of the β-amyloid protein deposits in the brains of transgenic mice was also described at the same time by QUON et al. (1991).

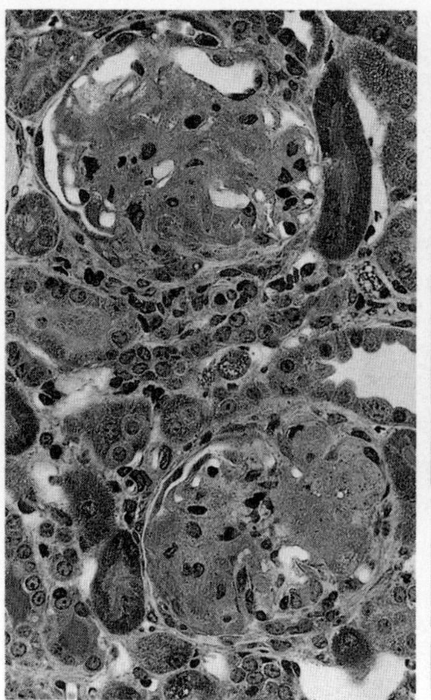

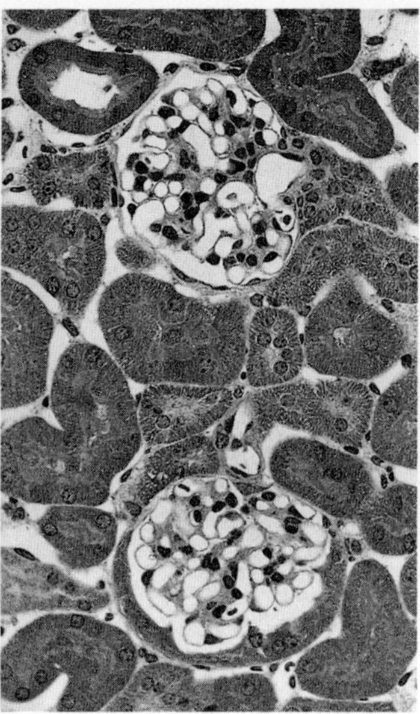

Fig. 20. Global glomerulosclerosis in a 24-week old MT-bGH-transgenic mouse (right); for comparison, glomerulae of a control kidney are shown at the same magnification (left). Glycol/methacrylate sections; hematoxilin/eosin stain (WANKE et al., 1991).

Genetic ablation

The techniques used currently for the generation of transgenic mice by DNA micro-injection essentially expand the normal genome by the addition of multiple copies of a gene. This additive approach does not allow the removal of endogenous sequences. However, two different procedures allow the specific suppression of gene activity.

The activity of an individual gene may be neutralized by the transfer of a gene construct which contains a structural gene encoding an antisense RNA for the corresponding endogenous transcript. Intracellular hybridization of antisense RNA with the mRNA (sense RNA) encoded by the gene under study either effectively inhibits translation or leads to the synthesis of functionally crippled protein fragments. The efficiency of the inhibition of the endogenous gene expression primarily depends on the synchronous or overlapping expression of the antisense transgene and the endogenous gene. It also depends on a relative excess of the antisense transcript because only this will guarantee that more or less all endogenous RNA molecules of the structural gene in question will indeed be complexed by the antisense RNA.

If in a particular transgenic model the aim is to repress the growth of a certain cell population rather than the activity of an individual gene this can be achieved by a process known as genetic ablation or amputation. BREITMANN et al. (1987) have demonstrated the feasibility of such an approach by constructing a fusion gene in which the gene encoding the A chain of diphtheria toxin was coupled to the δ2 crystalline gene promoter. Due to the toxic effect of the gene it has been possible to suppress in part and also completely the growth of a particular cell population of the lens, thus producing microphthalmia in the transgenic mice expressing the construct. PALMITER et al. (1987) have used a similar approach to suppress the growth of a cell population in the pancreas.

In principle the technique of genetic ablation should allow the growth of any desired cell line or cell type to be suppressed specifically, if suitable regulatory sequences were available that are specific for the desired cell type. It should be noted, however, that the expression of the toxin may already exert its effects in early phases of embryonal development, depending on the particular construct used. This may also suppress the further development of the fetuses.

Homologous recombination

There are many instances in which a suitable disease model can be established only if the function of an endogenous gene is inactivated effectively by the introduction of a mutation. The mutations of conventional experimental animals are usually either spontaneous or induced. For a long time directed and specific *in vivo* mutagenesis has not been feasible. SMITHIES et al. (1985) have shown that homologous recombination (Fig. 21) of a foreign gene construct with the endogenous gene locus is indeed possible (see Sect. 2.2). They have transformed carcinoma cells with a gene construct consisting of 4.6 kb of the β-globin locus, the neo and the supF gene. In approximately 0.1% of all cases they were able to demonstrate that this construct had integrated into the endogenous β-globin gene by homologous recombination.

The next step in the development of this technique has been the use of homologous recombination for directed mutagenesis in embryonic stem cells of mice (THOMAS et al., 1986; DOETSCHMAN et al., 1987). The hypoxanthine phosphoribosyl transferase gene (HPRT) has been chosen as a target gene because it is located on the X chromosome and a selection procedure for HPRT-negative mutants is available (see Sect. 2.2). Another reason for choosing this gene was the fact that HPRT-negative transgenic mice might be valuable models for the Lesh Nyhan syndrome. It has been possible to mutate the endogenous HPRT gene in embryonic stem cells by gene targeting (THOMAS et al., 1987) and also to correct the defect in HPRT-negative stem cells by homologous recombination and to produce transgenic mice (DOETSCHMAN et al., 1987). The results of these experiments have been very promising and have suggested that it might be possible to use these embryonic stem cells to produce chimeras and mice carrying a mutated gene at a desired gene locus. Since it is difficult to mutate both alleles of the gene on both homologous chromosomes, the resulting transgenic mice are chimeric and their F1 offspring is hemizygous. By crossing these hemizygous transgenic mice it is possible to obtain homozygous mutants in the next generation, if the mutated gene is not lethal or creates other problems. ZIMMER and GRUSS (1989) have used homologous recombination to mutate the homeobox 1.1 gene and have obtained chimeras carrying the expected mutation after injection of cells of a cloned mutated stem cell line (see Tab. 2).

A mutation has also been introduced into the c-src proto-oncogene by homologous re-

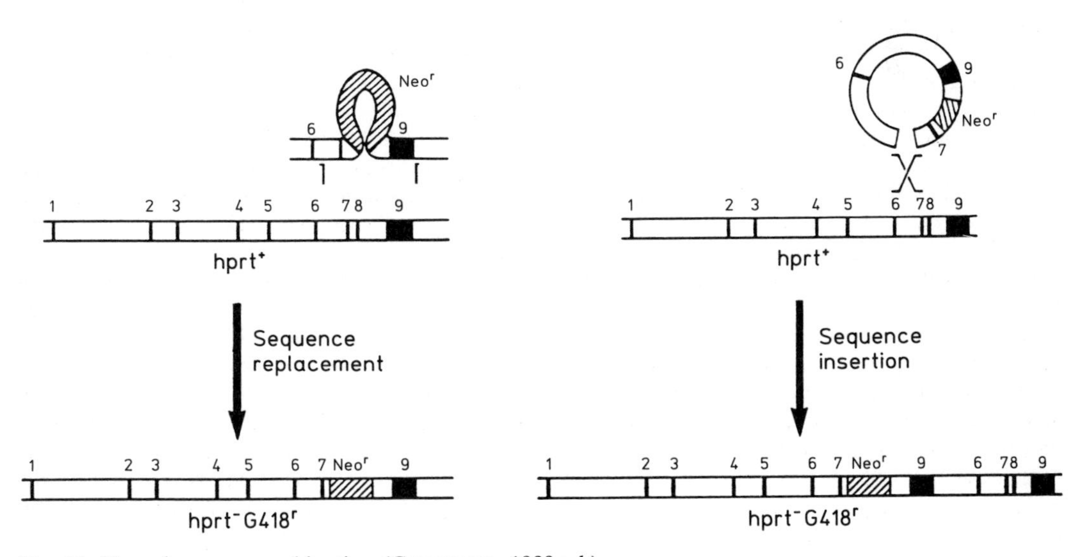

Fig. 21. Homologous recombination (CAPECCHI, 1989a, b).

combination in ES cells. Homozygous F2 offspring from chimeras died within the first week of birth. They were deficient in bone remodeling and suffered from osteopetrosis, indicating osteoclast function deficiencies (SORIANO et al., 1991).

It now appears that the use of the available techniques should allow any gene to be mutated directly and specifically (CAPECCHI, 1989a, b). The expression of the endogenous gene is not a critical factor for homologous recombination in embryonic stem cells although mutation frequencies appear to be higher if the gene in question is also expressed. By producing stem cell chimeras the mutated cells can be used to produce mice carrying the desired mutation in their genomes. A summary of the genes mutated in ES cells and the resulting phenotypes of mice carrying these mutations is shown in Tab. 2.

It has been described recently that the micro-injection of a DNA construct can also be used to correct the deleterious effects of a deleted endogenous gene. BRINSTER et al. (1989a) have injected 5′ sequences of a functional MHC II E gene into mouse cells carrying a 630 bp deletion of the endogenous locus. The injection of over 10000 oocytes yielded 1800 live mice, 500 of which were transgenic. The homologous recombination of the injected gene construct into the mutated locus was observed in one mouse. The recombined allele was characterized by the occurrence of many new point mutations. The resulting mRNA was expressed in a tissue-specific manner although it differed in its length from the wildtype RNA.

At present more than 3500 different genetic diseases of man are known. In principle it should be possible to create a suitable model system for each of these defects by using the approach discussed previously if the genes responsible for each disease were identified, cloned, and mutated accordingly in murine embryonic stem cells. Such models would facilitate the pathogenetic analysis of these diseases. They would also be very useful as models allowing the development and evaluation of new therapeutic approaches, in particular somatic gene therapy.

5.2 Oncogenes

Gene transfer is particularly valuable for studying the effects and consequences of oncogene expression in animals. As proto-oncogenes are important for normal development, the transfer and expression of these genes may be lethal in transgenic mice. Lethal genes can be investigated if the expression is restricted to a specific tissue which can be achieved by choosing suitable regulatory elements. The use of oncogene-transgenic mice allows the spectrum of tissues susceptible to the transforming activity of an oncogene, the relation between multistep oncogenesis and the cooperativity of oncogenes, and the effect of oncogenes on growth and differentiation (JAENISCH, 1988). Both viral and cellular oncogenes have been used for generating transgenic mice. The detailed analysis of these mice may be helpful in elucidating at a molecular level the mechanisms of (proto-)oncogene functions during normal and malignant development.

Viral oncogenes

A variety of viral oncogenes, including, e.g., the large T antigen gene of SV40 virus, polyoma virus large and middle T genes, bovine papilloma virus, human JC and BK viruses, human T cell leukemia virus tat oncogene, and human hepatitis B virus have been used for generating transgenic mice (PALMITER and BRINSTER, 1986; CAMPORE et al., 1988; CUTHBERTSON and KLINTWORTH, 1988; BABINET et al., 1985, 1989; JAENISCH, 1988).

BRINSTER et al. (1984) have used a gene construct of SV40 (Simian virus) genes and the murine metallothionein promoter to produce transgenic mice. These mice developed papillomas and carcinomas of the choroid plexus. Further investigations have demonstrated that a 72 bp element of the SV40 promoter was of crucial importance for the genesis of these tumors (PALMITER et al., 1985). When promoters of the insulin and crystalline genes were used the resulting transgenic mice suffered from tumors of the pancreas (HANAHAN, 1985; TEITELMAN et al., 1988) and the eye lens (OBERBEEK et al., 1985; MAHON et al., 1987), respectively.

WINDLE et al. (1990) have reported that the expression of the viral SV40 T antigen oncogene in the retina of transgenic mice produces heritable ocular tumors with histological ultrastructural and immunohistochemical features identical to those of human retinoblastomas.

Particular attention has been paid to hepatitis B virus (HBV) because there are more than 200 million chronic carriers of this infection. HBV transgenic mice harboring sequences encoding the surface antigen, pre-S, and X-antigen under the control of the normal viral and mMT promoter have been produced to study the mechanisms which are responsible for the acute and chronic hepatocellular damages observed in infected patients (CHISARI et al., 1985). These studies have demonstrated that the liver damage is not a direct consequence of the expression of pre-S or HB regions of the HBV genome. In another model transgenic mice carrying the HBV genome without the core gene have been established to study the regulation of virus expression *in vivo* (BABINET et al., 1985).

HBV envelope transgenic mice have shown that HBV-encoded antigens are expressed at the surface of hepatocytes. The surface antigens are recognized by MHC class I-restricted CD-8 positive cytotoxic T lymphocytes specific for a dominant T cell epitope within the major envelope polypeptide and by envelope-specific antibodies (MORIYAMA et al., 1990). Both interactions led to the death of hepatocytes *in vivo*.

The transfer of the bovine papilloma virus 1 (BPV-1) genome has yielded transgenic mice that develop the typical picture of fibropapillomas of the skin. The tumor tissue contained extrachromosomal copies of BPV-1 DNA while normal tissues only contained integrated copies of the viral DNA. No effects were visible when only 69% of the transforming region of BPV were used for the production of transgenic animals (LACEY et al., 1986).

These BPV-1 transgenic mice are an excellent model for the study of the tissue specificity in the expression of BPV, the activity of the BPV oncogene, and the genetic make-up leading to the development of neoplasias. They may also be helpful in establishing similar models involving the use of human papilloma viruses in mice.

Transgenic mice containing a construct consisting of the LTR sequence of Rous sarcoma virus (RSV) and the chloramphenicol acetyl transferase gene as reporter gene have shown a tissue specificity predominantly for bones and muscles (OBERBEEK et al., 1986). This is another indication of the importance of the LTR region in determining tissue specificity. Insertion mutations have been observed in two transgenic mouse lines. One mutation was dominant and led to embryo lethality while the other was recessive, leading to fused toes in all four extremities of the affected animals.

A transgenic disease model for progressive multifocal leukoencephalopathy has been established by the transfer of the JVC early region gene under the control of its own promoter/enhancer region (SMALL et al., 1986). The transgenic mice were characterized by tonic seizures of 15–30 s duration at the age of approximately 3 weeks which resembled those observed in the mouse mutants "jimpy" and "quaking". The neuropathological analysis of these transgenic mice has revealed the loss of myelin in the central but not in the peripheral nervous system.

Human T-lymphotropic viruses of type I (HTLV-1) are associated with neurological diseases characterized by topical spastic paraparesis and possibly also with multiple sclerosis. In transgenic mice the expression of the tat gene of HTLV under the control of its own LTR region leads to tumors which strongly resemble human neurofibromatosis generalisata (von Recklinghausen disease) (HINRICHS et al., 1987). In three of eight transgenic founder animals the expression of the tat gene of HTLV-1 induced tumors of soft tissues at the age of 12–17 weeks (NERENBERG et al., 1987).

The human immunodeficiency virus (HIV) is a retrovirus whose genome contains at least 6 genes in addition to the normal gag, pol, and env genes. These additional genes are important for the regulation of the expression of viral genes and also influence the replication of the viral genome. The tat gene in particular is a potent transactivator *in vitro*. HIV/LTR-tat2-transgenic mice develop skin alterations which resemble those of Kaposi sarcomas observed in AIDS patients (VOGEL et al., 1988). Transgenic mice which harbor the entire intact provirus did not develop an HIV viremia and remained healthy. Transgenic F1 offspring of one line, however, developed the syndrome and infectious viruses were recovered from spleen, lymph nodes, and the skin of the animals (LEONARD et al., 1988).

Epstein Barr virus (EBV) is a human herpesvirus and the etiologic agent of a lymphoproliferative disorder. The BNLF-1 gene encoding the latent membrane protein (LMP) has been expressed in transgenic mice. LMP expression in the epidermis induces the phenotype of hyperplastic dermatosis and the induction of the expression of a hyperproliferative keratin at aberrant locations (WILSON et al., 1990). This implies that the LMP plays an important role in the development of the acanthotic condition of the tongue epithelium and possibly also in the predisposition of the nasopharyngeal epithelium to carcinogenesis.

The spontaneous occurrence of tumors in myc-transgenic mice has been described for the first time by STEWART et al. (1984) who used gene constructs in which the normal murine myc gene was linked to the MMTV (murine mammary tumor virus) promoter. Although the expression of the construct differed widely in the 13 lines established in this study, two female founder animals developed spontaneous adenocarcinomas of the mammary gland during early pregnancy. The offspring of these animals also developed mammary carcinomas during the second and third pregnancy. These results demonstrate that the deregulated expression of the myc gene can be considered a heritable predisposing factor which is responsible for the development of adenocarcinomas.

Other experiments have revealed that no tumors develop despite the expression of a myc gene under the control of a variety of promoters in a variety of tissues, including pancreas, lungs, brain, and the salivary glands. This finding supports the notion that additional tissue-specific factors may be responsible for tumorigenesis. Transgenic mice carrying a 1 kb fragment of the murine immunoglobulin enhancer for the heavy chain inserted into the first intron of the c-myc gene developed clonal lymphomas predominantly originating from pre B cells (ADAMS et al., 1985; SCHMIDT et al., 1988). When these mice were crossed with other transgenic mice expressing a membrane-bound form of the heavy chain of immunoglobulin, the double-transgenic animals showed a markedly reduced tumor incidence. This suppressive effect is the result of a subtle change in the development of B cells which is induced by the immunoglobulin transgene (NUSSENZWEIG et al., 1988). KNIGHT et al. (1988) have induced lymphocytic leukemia in transgenic rabbits by transfer of a c-myc oncogene fused with the immunoglobulin heavy chain enhancer.

High level expression of c-myc driven by the mouse Thy-1 transcriptional unit has been observed in the thymus of transgenic mice. These mice developed thymic tumors which contained proliferating thymocytes and expanded populations of epithelial cells (SPANOPOULOU et al., 1989). A marked synergy between bcl-2 (a putative oncogene which has not yet shown a propensity for spontaneous tumorigenesis in transgenic mice) and myc in double transgenic mice has been described by STRASSER et al. (1990). Eμ-bcl-2/myc mice showed hyperproliferation of pre-B and B cells and developed tumors much faster than Eμ-myc mice. Surprisingly, the tumors are derived from a cell with the hallmark of primitive hematopoietic cells, perhaps a lymphoid-commited stem cell. In contrast to Eμ-myc Eμ-pim-1 transgenic mice are predisposed to T-cell lymphomas. Double transgenic Eμ-myc/Eμ-pim-1 mice develop pre-B cell leukemia prenatally (VERBEEK et al., 1991).

The allele-specific activation of the myc gene in BL has been studied in transgenic mice carrying either the normal or the translocated and mutated myc allele isolated from the Burkitt lymphoma cells (line BL64). In the constructs used for the establishment of the transgenic animals the functional κ light chain gene from BL64 was linked 3′ to both myc genes in order to provide an active chromatin for the B cell lineage. Mice carrying either the mutated (t-myc-κ) or the normal (n-myc-κ) myc allele expressed both transgenes to a similar extent specifically in the spleen. t-myc-κ and n-myc-κ mice developed B cell lymphomas (Fig. 22) between 3 and 11 months (median 6) with identical low frequencies (13–15%). Control mice (t-myc) carrying the mutated myc allele without the κ chain did not develop any tumors (MÖRITZ et al., 1991). We conclude from this study that the transposition of myc into an ac-

Fig. 22. Myc-transgenic mouse.

tivated chromatin structure conferred by the immunoglobulin loci is sufficient to active its expression. Somatic mutations within the myc gene are not a prerequisite for deregulation of myc in Burkitt's lymphoma.

Transgenic mice carrying the ras oncogene express this gene in a variety of tissues and show various effects. Acinar cells of the pancreas were transformed by the ras oncoprotein when an elastase promoter was used and this led to the rapid development of a pancreatic hypoplasia (QUAIFE et al., 1987) although these cells were later demonstrated to be non-tumorigenic. When linked to a mamma-specific promoter, ras expression induced the formation of mammary tumors with a long latency (ANDRES et al., 1987; SINN et al., 1987). In contrast to ras constructs the corresponding MMTV-and WAP-myc constructs showed a high incidence of tumors. By crossing MMTV-myc and -ras lines SINN et al. (1987) have been able to demonstrate that the double transgenic offspring developed tumors much faster, and that tumors did develop not only in the mammary gland but also in other organs.

Transgenic mice have also been obtained for the fos, fms, and mos oncogenes.

A novel target cell for c-fos-induced oncogenesis has been described recently by WANG et al. (1991). A high frequency of cartilage tumors developed as early as 3–4 weeks of age in embryonic stem cell chimeras apparently independently of the extent of chimerism.

The expression of the majority of oncogenes is generally not sufficient to induce, for example, mammary tumors without additional stimuli. In contrast the expression of the activated c-neu oncogene in transgenic mice has been shown to be the sole factor responsible for causing the induction of mammary adenocarcinomas (MULLER et al., 1988). These transgenic mice do not develop single tumors but rather mammary carcinomas extending over the entire epithelium of each gland. It appears that the activation of the c-neu oncogene is sufficient to induce malignant transformation in the cells of this tissue. In other tissues such as the parotid gland or the epididymis the expression of the c-neu oncogene leads to hypertrophy and hypoplasia without accompanying malignant transformation. The activated oncogene therefore only causes malignant transformation in special tissues.

Transforming growth factor α (TGF-α) is a potent mitogen. Overexpression of TGF-α by the mMT-1 promoter induces epithelial hyperplasia, liver neoplasia, and abnormal development of the mammary gland (carcinomas) and pancreas (metaplasia) (JHAPPAN et al., 1990; SANDGREN et al., 1990).

There can be no doubt that the estalishment of transgenic mice is and will be of paramount and invaluable importance for studying the effects of cellular and viral oncogenes in their normal environment. It has also been shown that the lack of a correlation between oncogene expression and tissue transformation is an indication of a complex interaction between oncogenes and other cellular factors in specific tissues.

5.3 Immune System

The establishment of transgenic mice is also invaluable for the analysis of the immune system (STORB, 1987), in particular for studies of B and T cell development and the mechanism of actions of lymphokines. It is sufficiently known that the immune system of higher vertebrates operates on two lines of defense, namely the humoral and the cellular immune system. B lymphocytes, the precursors of plasma cells which produce and secrete antibodies, are responsible for the humoral branch of the immune system. The immunoglobulins recognize and bind specifically to foreign antigens. The genes encoding the immunoglobulins (Ig) consist of several exons encoding the constant (C) and the variable (V) regions of the antibodies. Igs consist of two identical light (L) and two identical heavy (H) polypeptide chains. Both chains contain variable regions which recognize the antigen and constant regions which are involved in effector functions.

The B cell population of an individual provides an extremely variable and versatile antigen binding spectrum which essentially results from different variable regions of the antibodies. These variable regions are not fixed genetically. Instead, the variable regions are linked during the rearrangement of Ig gene sequences taking place during the maturation of the B cells to yield the final variable region. These rearrangements sequentially join gene segments, known as V (variable), D (diversity), and J (joining), which encode the variable elements of the H chain and form a V_HDJ_H complex. In the second step the corresponding elements for the L chain are rearranged to form the V_LDJ_L complex. Although B cells are diploid and all gene segments involved in these rearrangements therefore exist in two copies, only one Ig allele is rearranged and expressed (allelic exclusion).

B cell development

Transgenic mice expressing the genes for functionally rearranged transgenic H chains are characterized by a reduced expression of the endogenous H genes (RUSCONI and KÖHLER, 1985). This mechanism of transgenic allelic exclusion also operates if the differences between the endogenous and the transgenic isotypes are quite pronounced.

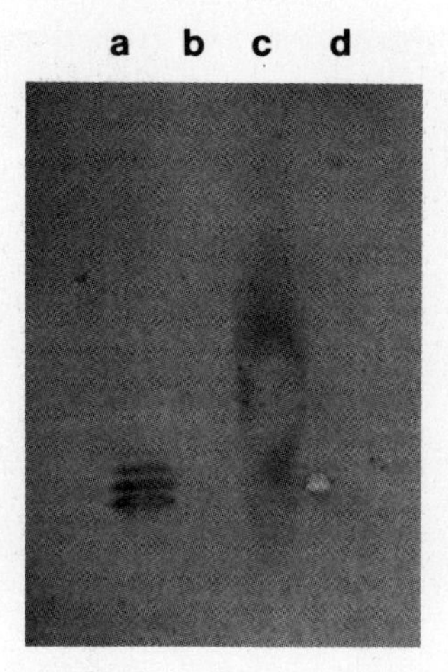

Fig. 23. Isoelectric focussing and subsequent immunofixation of serum (dilution 1:20) of a transgenic pig (c) in comparison with a control animal (b, d), and 20 ng of antibody A20/44 (a) (WEIDLE et al., 1991).

In our own experiments we have produced transgenic mice, rabbits, and pigs carrying the genes for the light and heavy chain of a mouse monoclonal antibody which yield up to 1 g mAb/L serum in one transgenic pig line. Isoelectric focusing (IEF) analysis of sera has revealed that in the transgenic pig only a minority of the antibody molecules correspond exactly to those of purified monoclonal antibodies from ascites (Fig. 23). In the sera of the transgenic rabbits no IEF bands coinciding with those of the monoclonal antibody were observed. The existence of the same discrete bands in the serum fraction and the purified antibody preparation from pig serum argues against an association of heterologous chains with those of the mouse. The expression of the κ chain may therefore be insufficient to effect complete allelic exclusion (WEIDLE et al., 1991; BREM and WEIDLE, 1991).

The fact that the transgenes also cause allelic exclusion suggests that the expression of the transgene in pre-B cells precedes the endogenous rearrangement of the corresponding antibody gene segments (IGLESIAS, 1991). The expression of transgenes for H and L chains and the observed allelic exclusion also suggest that L chains on their own are not sufficient to block rearrangement processes.

Transgenic mice expressing functionally rearranged Ig genes provide an adequate model system for the study of B cell maturation and the formation and internal dynamics of the antibody reservoir (reviewed by IGLESIAS, 1991). Transgenic mice expressing immunoglobulins directed against mouse self components may also be valuable tools for the investigation of B cell tolerance.

T cell development

The T cell receptor (TCR) is another molecule of the immune system which is characterized by an astounding diversity of features that allow the immune system to react to all sorts of different antigens. This diversity results from the somatic rearrangement of certain genetic elements and the addition of discrete N-terminal regions. This maturation takes place after the emigration of the thymocyte precursors from the bone marrow into the thymus and the onset of cell proliferation during fetal development. Most T cells carry TCRs made of α- and β-chains and possess a constant and a variable region. T cells also show the phenomenon of allelic exclusion, which is not a random process but rather a regulated event. In transgenic mice showing low level expression of a transgenic TCR β-chain the same cells expressing the transgene also express the endogenous TCR chain (PULLEN et al., 1988). The introduction of a mutated TCR β-chain led to the arrest of thymocyte maturation with transgenic mice being completely deficient in functional $\alpha\beta$ T cells (KRIMPENFORT et al., 1989). The events taking place during T cell development involve positive and negative selection processes generating functional and selftolerant mature T cells, respectively. The detailed analysis of these processes has been made possible for the first time by the establishment of appropriate transgenic mice (reviewed by BLÜTHMANN, 1991). It has been shown recently that mice transgenic for a mutant MHC class I molecule which cannot interact with CD8 do not delete CD8-dependent T cells reactive with the wildtype molecule (INGOLD et al., 1991). This finding unequivocally establishes that for negative selection in the thymus CD8 must interact with the same MHC class I molecule as the TCR. ALDRICH et al. (1991) conclude from their gene transfer experiments that not only the $\alpha 1/\alpha 2$ domains of MHC class I molecules but also the $\alpha 3$ domain plays an important part in the positive and negative selection of antigen-specific cells. CD8 expression under the T cell specific CD2 regulatory sequences in transgenic mice does not affect thymic selection of CD4-positive cells although the selection of a class I-specific TCR in the CD8 subset is substantially improved (ROBEY et al., 1991). This is consistent with a model of positive selection in which selection occurs at a developmental stage in which both CD4 and CD8 are expressed and positive selection by MHC class I generates an instructive signal that directs the differentiation to a CD8 lineage. By disruption of the MHC class $II^{b}\beta$ gene MHC class II-deficient mice can be depleted of mature CD4-positive T cells and are then deficient in cell-mediated immune response (GRUSBY et al., 1991). These findings provide genetic evidence for the observation that class II molecules are required for the maturation and function of mature CD4-positive T cells.

There is a good chance that the study of transgenic mice will allow elucidation of those processes that are involved in the induction of tolerance (and the lack of it) that leads to autoimmune diseases. To evaluate the tolerance of neutralizing B cell responses, ZINKERNAGEL et al. (1990) have used transgenic mice expressing the cell membrane-associated glycoprotein of vesicular stomatitis virus as a self-antigen. Autoantibodies were not induced by vaccinia virus expressing this glycoprotein but were triggered by an infection with wildtype virus. The data show that helper T cell tolerance is crucial in the maintenance of B cell non-reactivity and that cognate T-B recognition is necessary to break tolerance of self-reactive B cells. Autoimmune diseases may in some instances arise by bypassing the T cell tolerance (ADELSTEIN et al., 1991).

TCR transgenes in SCID (severe combined immunodeficiency) and LPF (lymphoproliferation) mice have provided new data about the development of thymocytes (BLÜTHMANN, 1991). Mice homozygous for a $\beta 2$-microglobulin gene disruption express little if any functional MHC class I antigen on the cell surface yet are fertile and apparently healthy (ZIJL-

STRA et al., 1990). They show no mature CD4-negative CD8-positive cells and are defective in CD4-negative/CD8-positive T cell-mediated cytotoxicity.

Lymphokines

Interleukins play a decisive role in mediating the growth and differentiation of leukocytes and also in eliciting immune responses and inflammatory processes. The overexpression of a transgene encoding granulocyte-macrophage colony-stimulating factor (GM-CSF) leads to pathological alterations in the retina and causes blindness but also muscle deterioration and premature death (LANG et al., 1987). A very pronounced increase of activated macrophages is observed in these mice. In addition, the excess of GM-CSF leads to the activation of mature macrophages which in turn secrete large amounts of interleukin 1 and tumor necrosis factor. Histopathology reveals a pronounced increase in the progenitor cells of the monocyte lineage. Similar results have been obtained in mice that have received transplants of bone marrow cells harboring a recombinant retrovirus and overexpressing GM-CSF. It should be noted that these results may not be relevant for the clinical application of this factor. The long-term treatment of primates and mice with GM-CSF has shown, for example, that life-threatening effects are not observed.

The overexpression of IL-2 and IL-4 also has characteristic pathological consequences (reviewed by IGLESIAS, 1991). Transgenic mice harboring the human gene for IL-2 or the TAC-IL-2 receptor constitutively express IL-2 in the thymus, spleen, bone marrow, lungs, muscle, and skin. Among other things these animals show pronounced growth retardation and die prematurely. Histologically one observes a selective loss of Purkinje cells in the cerebellum and a focal infiltration of lymphocytes, leading to pneumonia. The spleen shows a massive increase of cells with the marker spectrum thy1$^+$/CD3$^-$4$^-$8$^-$. The expression of the human IL-2 gene which has been transferred into murine T cells by means of a retrovirus vector leads to growth autonomy and malignant transformation of the cells (ISHIDA, 1989a, b; NISHI et al., 1988).

The expression of IL-4 in transgenic animals impairs the maturation of T cells and reduces considerably the population of immature CD4/CD8-positive thymocytes and peripheral T cells, while the number of mature CD8-positive thymocytes increases markedly. In CD4-negative mouse strains the development of CD8-positive cells proceeds normally. However, the activity of helper cells is markedly reduced. The proliferation of thymocytes with the marker spectrum CD4$^-$8$^-$, CD4$^+$8$^-$, CD4$^-$8$^+$ is stimulated by interleukin 7 which is therefore an important developmental factor of functionally different subpopulations of T lymphocytes. IL-4-transgenic animals express IL-4 in all of their tissues, and this is accompanied by an excessive increase of serum IgE levels. The levels of IL-4 cannot be increased over a certain level because the uncontrolled overexpression leads to the death of these animals. The introduction of a recombinant IL-4 gene and the resulting overproduction of IL-4 leads to factor-independence in IL-4-dependent cells although it does not cause cell transformation.

Mice overexpressing IL-5 have a characteristic continuous blood and spleen eosinophilia and eosinophil infiltration in lungs and gut, but they remain normal and do not show any signs of tissue damage (DENT et al., 1990). In IL-6-transgenic mice one observes a marked stimulation of proliferation and maturation of B cells without accompanying malignant transformation of plasma cells (KISHIMOTO, 1989; SUEMATSU et al., 1989). Il-6-transgenic mice harboring the human IL-6 gene show a dramatic increase of serum levels of IL-6 and of polyclonal IgG1. In addition to a massive plasmacytosis one also observes all typical signs of a proliferative mesangial glomerulonephritis which may be attributed to the deregulated synthesis of the cytokine. IL-7 preferentially promote the proliferation of B cell precursors without causing a concomitant increase in granulocytes and macrophages (SAMARIDIS et al., 1991).

The expression of IFN-γ in the pancreas of transgenic mice precipitates autoimmune diabetes (SARVETNICK, 1988).

6 Application of Transgenic Livestock

Until now it is possible to alter by gene transfer only those traits of economically important animals that are caused by single or a few genes. A prerequisite for such interventions is a detailed knowledge of the underlying physiological processes on the molecular level because that allows suitable gene constructs to be developed and applied successfully. At present only a few traits are known which are interesting from a breeder's point of view and which are caused by single genes. However, the generation of giant mice by PALMITER et al. (1982, 1983) expressing the human and rat growth hormone gene under the control of the metallothionein gene promoter has illustrated quite well that the boundaries between qualitative and quantitative traits may be rather fluid at times. These experiments have shown that growth which is one of the classical quantitative traits in animal breeding can be converted into a qualitative trait simply by the transfer of a major gene, the growth hormone gene, whose expression makes the synthesis of growth hormone independent of any regulatory feedback mechanisms. A similar effect has also been described for the application of bovine somatotropin in dairy cows. Our knowledge in this field is still rather inadequate. However, the results of these experiments and in particular the great variability of the effects in individual animals suggest that the contribution with respect to increased performance of major genes, whose effects are subject to certain internal standards, may resemble that of additive gene dosis effects.

At present research in transgenic animals of economic importance is focused on features such as growth, new metabolic pathways, quality of animal products, gene farming, and disease resistance.

6.1 Growth and Carcass Composition

Growth is a very complex process which is influenced by the interactions of hormones and autocrine/paracrine factors, nutritional

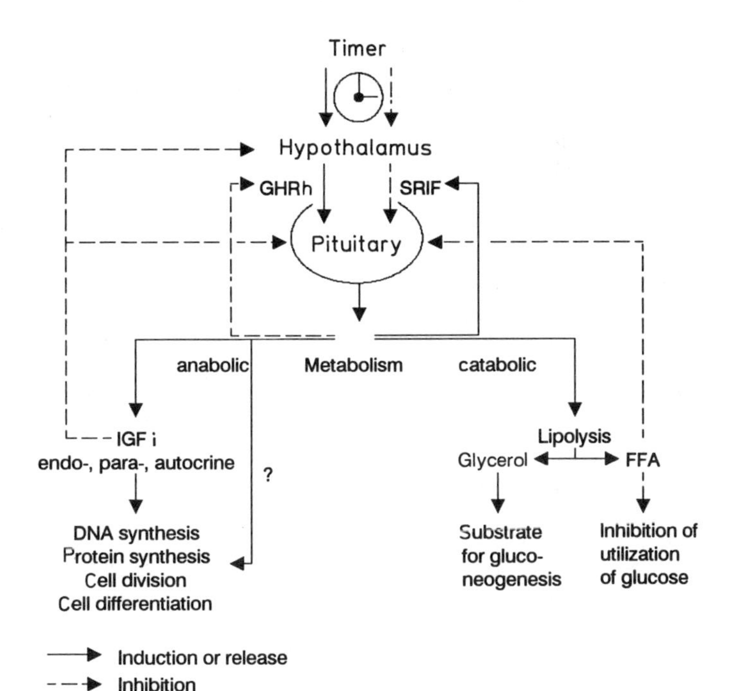

Fig. 24. Modes of action and regulation of the growth hormone cascade (WOLF et al., 1991a).

conditions, and environmental factors set against a discrete genetic background. Among the genetically determined factors the genes encoding proteins of the growth hormone cascade are of particular interest. This cascade is initiated in the hypothalamus and the pituitary, includes the liver, and eventually affects peripheral target organs (Fig. 24).

The hypothalamus is subject to regulation by a number of different factors, in particular the serum concentrations of growth affecting hormones as GH, IGF-I, and others. Among other things it is responsible for the circadian expression of the stimulatory hormone somatoliberin (GHRH, growth hormone releasing hormone), and the inhibitory hormone somatostatin (SRIF, somatotropin release-inhibiting factor).

GHRH is an oligopeptide of 43 or 44 amino acids which is synthesized in the nucleus arcuatus and the nucleus ventromedialis of the hypothalamus. Porcine and human proteins differ in three positions only, while rat and human hormones differ by 14 amino acid positions. SRIF consists of two different molecules, 14 and 28 amino acids in length, whose sequence has been conserved in different species. The hormone is synthesized in the hypothalamus but also in the pancreas and the gut.

Growth hormone (GH, growth hormone, STH, somatotropic hormone, somatotropin) is synthesized in the somatotropic cells of the anterior lobe of the pituitary and consists of 190 or 191 amino acids. The protein contains two disulfide bridges. Apart from a 22 kD predominant form, other forms (20, 45, 80–90 kD) also occur in humans. The signal sequence attached to the primary translation product is cleaved off in the endoplasmic reticulum. The amino acid sequences and genes for these proteohormones of the most important species are known and the genes have been cloned (Tab. 6).

Second to the growth hormone comes somatomedin C (IGF-I, insulin-like growth factor I) which is a mitogenic basic polypeptide of 70 amino acids with a molecular mass of 7.5 kD. It is synthesized in the liver but also in other organs including kidney, lung, heart, testes, mammary gland, and the epiphyses of bones. Several tumor cell lines have also been observed to synthesize the factor. IGF-I acts as an endocrine hormone but also exerts auto- and paracrine effects. IGF-I is produced from two different precursors (prepro IGF-IA and IGF-IB) which result from differential splicing of the corresponding mRNA. They are identical in the aminoterminal and coding regions and differ in their carboxyterminal ends (Tab. 7) (MÜLLER, 1989).

The release of growth hormone depends on the concentrations of GHRH and SRIF and shows marked fluctuations with several peaks occurring over a single day. Growth hormone is distributed into the organism via the circulation and binds to specific GH receptors in particular in the liver but also in fat tissues, muscle, kidney, heart, and epiphyses of bones.

Growth hormone-transgenic mice have been produced for the first time in 1982 (PALMITER et al., 1982). These mice showed a very pronounced growth increase with a fourfold increase in growth rates (Fig. 25) and a twofold increase in the final body weight (see Fig. 19). Subsequently growth hormone genes of other species and also other genes of the growth hormone cascade such as GHRH and IGF-I have also been used to produce transgenic mice (reviewed by WOLF, 1990). A number of reports have been published about transgenic animals of economic importance that harbor and express genes of the growth hormone family.

In contrast to the results obtained with growth hormone-transgenic mice and the administration of growth hormone in economically important animals, the corresponding growth hormone-transgenic animals did not show the same biological effects. Transgenic pigs and rabbits expressing the growth hormone gene did not show an increase in growth rates (HAMMER et al., 1985). Transgenic pigs raised on the usual fattening ration containing 16% of crude protein even appeared to grow less well then control pigs. Injection experiments carried out at the same time, however, have revealed that an increase in growth performance requires a protein-rich diet (18% raw protein) supplemented with lysine (0.25%), minerals, and vitamins. When, as a consequence, the transgenic offspring of the growth hormone-transgenic pigs received this modified diet, the daily growth rates were indeed increased by 15% (PURSEL et al., 1988a,

Tab. 6. Amino Acid Sequences of Different Growth Hormones and Sequence Comparison of Different Species (WOLF, 1990)

```
      Signal Peptide

hGH   [M-ATGSRTSLLLAFGLLCLPWLQEGSA?FPTIPLSRPFDNAMLRAHRLHQLAFDTYQEFEE
bGH   [.M.AGPR.SL..AFA....P.TQVVG.]..AMS..GL.A..V...QH.....A..FK...R
oGH   [.M.AGPR.SL..AFT....P.TQVVG.]..AMS..GL.A..V...QH.....A..FK...R
pGH   [.-.AGPR.SA..AFA....P.TREVG.]..AMP..SL.A..V...QH.....A..YK...R
rGH   [.-.ADSQ.PW..TFS....L.PQEAG.]..AMP..SL.A..V...QH.....A..YK...R
mGH   [.-.TDSR.SW..TVS....L.PQEAS.]..AMP..SL.S..V...QH.....A..YK...R

hGH   AYIPKEQKYSFLQNPQTSLCFSESIPTPSNREETQQKSNLELLRISLLLIQSWLEPVQFLRS
bGH   T...EG.R..-I..T.VAF....T..A.TGKN.A..KSDL....I.........G.L...SR
oGH   T...EG.R..-I..T.VAF....T..A.TGKN.A..KSDL....I.........G.L...SR
pGH   A...EG.R..-I..A.AAF....T..A.TGKD.A..RSDV....F.........G.V...SR
rGH   A...EG.R..-I..A.AAF....T..A.TGKE.A..RTDM....F.........G.V...SR
mGH   A...EG.R..-I..A.AAF....T..A.TGKE.A..RTDM....F.........G.V...SR

hGH   VFANSLVYGASDSNVYDLLKDLEEGIQTLMGRLEDGSPRTGQIFKQTYSKFDTNSHNDDALL
bGH   V.T...VF.T..R-..EK........LA..RE...GT..A...L....D...T.MRS.....
oGH   V.T...VF.T..R-..EK........LA..RE...VT..A...L....D...T.MRS.....
pGH   V.T...VF.T..R-..EK........QA..RE...GS..A...L....D...T.LRS.....
rGH   I.T...MF.T..R-..EK........QA..QE...GS..I...L....D...A.MRS.....
mGH   I.T...MF.T..R-..EK........QA..QE...GS..V...L....D...A.MRS.....

hGH   KNYGLLYCFRKDMDKVETFLRIVQCRS-VEGSCGF
bGH   ......S..R..LH.T..Y..VMK..RFG.A..A.
oGH   ......S..R..LH.T..Y..VMK..RFG.A..A.
pGH   ......S..K..LH.A..Y..VMK..RFV.S..A.
rGH   ......S..K..LH.A..Y..VMK..RFA.S..A.
mGH   ......S..K..LH.A..Y..VMK..RFV.S..A.
```

Comparison of the primary structures of human (h), bovine (b), ovine (o), porcine (p), rat (r), and mouse (m) growth hormone (GH)

	hGH	bGH	oGH	pGH	rGH	mGH	
hGH	*	66.3%	65.8%	67.9%	65.8%	66.3%	
bGH	64	*	99.5%	90.5%	87.9%	87.4%	Homology
oGH	65	1	*	90.0%	87.4%	86.8%	
pGH	61	18	19	*	94.7%	94.7%	
rGH	65	23	24	10	*	98.4%	
mGH	64	24	25	10	3	*	

Number of different amino acids

1989b), and the food utilization was also increased (2.46 kg of food per kg weight increase in transgenes compared to 3.12 kg in control animals).

A transgenic pig harboring a MLV-rGH gene construct and expressing biologically active rat growth hormone did not show an increased weight gain in the main growth period

Tab. 7. Comparison of IGF-I Polypeptide Sequences of Different Species: Pig (pIGF-I), Human (hIGF-IA), Rat (rIGF-IA), Mouse (muIGF-IA), Cattle (bIGF-I), Sheep (ovIGF-I), Salmon (sIGF-I). Sequence of IGF-I between (1) and (70) (MÜLLER, 1989)

pIGF-I	MGKISSLPTQLFKCCFCDFLK VKMHITSSSHLFYLALCLL	40
hIGF-IA	....TM.............	
rIGF-IA	.... I.I..M.............	
muIGF-IA	MTAPA I.I..M.............................	
bIGF-I	....................Q...P................	
ovIGF-I		
sIGF-I	M.SGHLFQWHLCDV.KS AMCC.SCTHT.SL.LCV.T	
	(1)	
pIGF-I	SFTSSATAGPETLCGAELVDALQFVCGDRGFYFNKPTGYG	80
hIGF-IA	T.......................................	
rIGF-IA	T..........................P............	
muIGF-IA	T....T.....................P............	
bIGF-I	A.......................................	
ovIGF-I	....	
sIGF-I	LTSAATG.............T......E.....S......	
	(70)	
pIGF-I	SSSRRAPQTGIVDECCFRSCDLRRLEMYCAPLKPAKSARS	120
hIGF-IA	....	
rIGF-IA	..I..............................T......	
muIGF-IA	..I..............................T.A....	
bIGF-I	....	
ovIGF-I	..................................A.....	
sIGF-I	P....SHNR.......Q..E..........V.SG.A....	
porcine IGF-I	VRAQRHTDMPKAQKEVHLKNTSRGSSGNKNYRM	153
human IGF-IA	...........T........A....A.......	
rat IGF-IA	I..........T.............A...T...	
murine IGF-IA	I..........T.............A...T...	
bovine IGF-I	.........................A.......	
ovine IGF-I		
salmon IGF-I	..........RTP....Q..S...NT.GR....	

between the 2nd and 6th month compared to control animals. It appeared, however, that the growth period of this pig has been prolonged so that at the age of 8 months it was 28% heavier than a non-transgenic male sibling (EBERT et al., 1988). In a single MT-pGH transgenic pig daily weight gains of 1273 g have been observed by VIZE et al. (1988) which compares favorably to control animals which showed a weight gain of 781 g. So far this observation has not been corroborated by similar observations in other transgenic pigs.

Only 3 out of 11 transgenic sheep harboring the bovine growth hormone under the control of a mouse transferrin enhancer/promoter, and 2 out of 4 animals harboring the human growth hormone-releasing factor gene under the control of the murine albumin enhancer/promoter expressed the corresponding genes (REXROAD et al., 1991). An increase in growth performance has not been observed, probably, because these animals showed pathological alterations and fertility disturbances and developed a pronounced diabetic condition.

Another problem associated with the use of growth hormones in pigs is the depression of food intake which may be as high as 20% and which is, of course, counterproductive as far as further increases in weight gains are concerned. However, GH-transgenic pigs show a marked improvement in food utilization of up

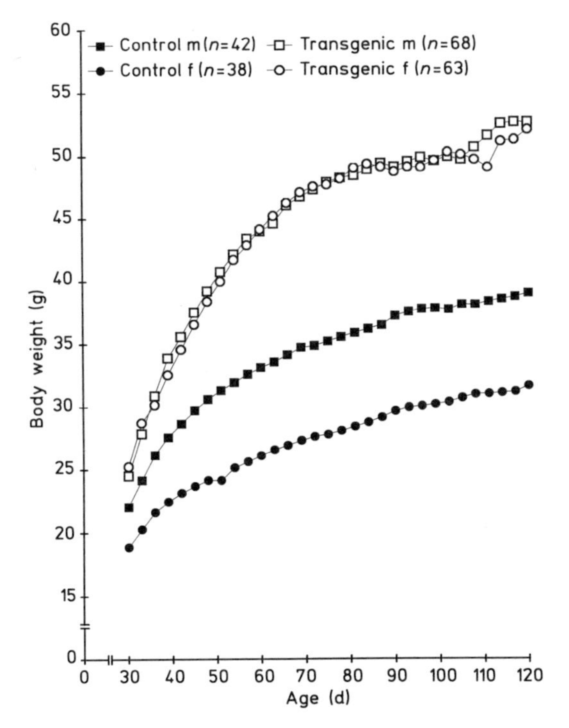

Fig. 25. Growth of MT-hGH-transgenic mice and control animals.

to 18%. The most pronounced alteration observable in transgenic pigs is the massive reduction of the fat content which is documented by a reduction of the back fat thickness from 18–20 mm to 7–8 mm (HAMMER et al., 1986; PURSEL et al., 1989, 1990a).

The most interesting positive feature of GH-transgenic pigs may be the phenomenon known as nutrient partitioning. Growth is the result of an initial hyperplasia of differentiated cells which is then followed by hypertrophy. If tissues have become terminally differentiated and have reached a relatively static state with respect to the DNA content, growth essentially is the result of hypertrophy of different tissue types competing for extracellular nutrients. Even under conditions of scarce nutrition neural and bone tissues have priority, and muscle and fat tissues will show signs of hypertrophy when nutrients become abundant with the degree of hypertrophy depending upon the genetic potential (STEELE and PURSEL, 1990).

Because of the problems associated with the overexpression of growth hormones discussed in the previous section attempts have been made to utilize other regulatory elements in order to tailor hormone gene expression and adjust it to physiological requirements. It has been possible to alleviate but not to reduce completely the adverse effects of growth hormone overexpression by using a liver-specific 460 bp 5′ flanking sequence of the rat PEPCK gene (phosphoenolpyruvate carboxykinase). Expression of these constructs in transgenic pigs reduced the back fat thickness by 40–50% (WIEGHART et al., 1990).

In another experiment the bovine prolactin gene promoter has been used to drive the expression of the bovine growth hormone. The resulting transgenic pigs expressed bGH in physiological concentrations of 20 ng/mL and their growth rates were normal. Treatment with sulpiride, a dopamine antagonist, or TRH (thyreotropin releasing hormone) resulted in the episodical release of bGH, suggesting that the expression of the transgene is subject to the normal feedback mechanisms associated with prolactin secretion control (POLGE et al., 1989).

GH- and GHRH-transgenic sheep show elevated levels of growth hormone although they do not display increased growth performance (REXROAD et al., 1990). The use of the metallothionein Ia promoter region and the growth hormone gene of sheep also increased the level of growth hormone (NANCARROW et al., 1991). Although the growth performance of these animals were not altered, the transgenic sheep only had 5–7% body fat in contrast to control animals (25–30%). A chicken cSKI gene has been transferred into pigs to determine whether it might enhance muscle development as it did in mice (PURSEL et al., 1992). Expression of cSKI in transgenic pigs resulted in phenotypic responses varying from hypertrophy of skeletal muscle in some pigs to muscular hypotony in others.

The growth hormone experiments carried out with mice and livestock differ fundamentally in that no selection with respect to growth performance has ever been exerted in the mouse lines whereas the pigs used in these experiments were derived from populations in which selection for the optimization of growth parameters had been carried out for several decades.

Experiments with mice selected for an increased weight at the age of 8 weeks have led almost to a duplication of the body weight after 30 generations (von Butler-Wemken et al., 1984; von Butler-Wemken and Pirchner, 1985). The growth performance and the final body weights of these "giant mice" are only slightly lower than those of GH-transgenic mice. They differ from transgenic mice in several important aspects, namely the fat content and the fertility. However, these experiments demonstrate that in principle the growth potential of normal experimental mice has not yet been stretched to the limit. One is tempted to speculate that the application of gene transfer in mice has led to a high-level growth plateau within a single generation, because these mice never were subject to discrete selecting efforts aimed at improving their growth capacity. In pigs, on the other hand, the plateau of the genetic growth potential appears to have been reached almost entirely by the continued breeding efforts of the past.

6.2 Biochemical Pathways and Quality of Products

The alteration of biochemical and metabolic pathways is a very interesting approach to improve the productivity of economically important animals. Ward et al. (1986) have pointed out that this approach might involve the re-establishment of certain metabolic pathways which have been lost in certain species and also the introduction of novel pathways that have not been observed so far. The genes responsible for these pathways would have to come from other sources and would require a tailoring process streamlining them for the expression in mammalians.

It is obvious that the main interest in altering metabolic pathways lies in the introduction of biosynthetic processes for essential factors. It has been known for some time that the availability of the amino acid cysteine is a factor limiting the synthesis of sheep wool. Moreover, the serum level of cysteine cannot be raised by simply adding cysteine to the food because it is degraded in the ruminant stomach of sheep. It would be of considerable advantage if transgenic sheep were capable of synthesizing this amino acid. This would require the introduction of genes encoding the enzymes serine transacetylase and the O-acetyl serine sulfhydrolase (Fig. 26) of *Escherichia coli* after providing them with suitable regulatory sequences allowing their expression in mammalians. Cells of the gastric epithelium of such transgenic sheep should be capable of synthesizing cysteine by utilizing the H_2S present in the stomach.

As a first step Ward et al. (1990, 1991) have linked these genes with the MT-promoter of sheep and with GH sequences at the 3′ regions and used these constructs to produce transgenic mice. Surprisingly they have found that constructs free of introns are also expressed quite satisfactorily. After the transfer of a construct containing the coding sequences of the genes, the SV40 late promoter, and the SV40 polyadenylation signal, a constitutive expression of both genes was observed in transgenic mice and also in sheep (Rogers, 1990). This finding suggests that the bacterial genes are transcribed and translated correctly in some sheep tissues.

Another example of the establishment of new metabolic pathways is the introduction of the glyoxylate pathway which would allow sheep to synthesize glucose from acetate. This approach is of particular interest because ruminants require glucose as a source of energy for certain tissues such as the brain and the fetus although they lack a sufficient source of glucose. Glucose therefore has to be synthesized via gluconeogenesis from free fatty acids such as propionic acid and also from amino acids. Acetate is usually available in amounts

$$\text{serine} + \text{acetyl-CoA} \xrightarrow[\text{serine transacetylase}]{} \text{O-acetylserine} + \text{CoA-SH}$$

$$\text{O-acetylserine} + H_2S \xrightarrow[\text{O-acetylserine sulfhydrylase}]{} \text{cysteine} + \text{acetate}$$

Fig. 26. Reactions involved in the biosynthesis of cysteine (Ward and Nancarrow, 1991).

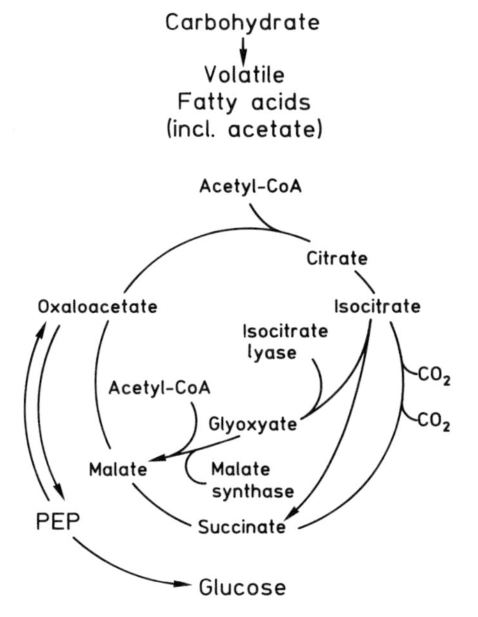

Fig. 27. Biochemical reactions of the glyoxylate cycle (WARD and NANCARROW, 1991). PEP, phosphoenol pyruvate.

exceeding those of metabolites required for gluconeogenesis. The new metabolic pathways that must be introduced to achieve this goal are depicted in Fig. 27. The isocitrate is synthesized from acetate and oxalacetate with isocitrate lyase yielding glyoxylate and succinate. In each cycle two molecules of acetate will yield one molecule of succinate. The presence of active isocitrate lyase and malate synthase has been demonstrated by using transformed cells and also transgenic mice (BYRNE, 1990, cited in WARD and NANCARROW, 1991). Among other things it has also been suggested to use the biosynthetic pathways for threonine and lysine in transgenic animals of economic importance (REES et al., 1990).

It should be pointed out that the introduction of such novel biosynthetic pathways is not a trivial task. The process requires that a number of parameters and conditions be observed and fulfilled, e.g., the availability of the corresponding enzymes in cells which require them, and the necessity to utilize preexisting cellular substrates and co-enzymes. There is also another restriction, namely, the use of pathways involving a limited number of biosynthetic steps which are encoded by a few genes only.

A very interesting observation concerning the establishment of new biosynthetic pathways has been made in one transgenic mouse line harboring 250 copies of the gene encoding the intermediate filament of keratin of sheep (POWELL and ROGERS, 1990). This line was characterized by the cyclic loss and the subsequent new growth of hairs which is due to the imbalance created by an increased level of intermediate keratin filaments and the resulting reduced level of filament-associated proteins.

Modifications of the milk composition (Tab. 8) that might be brought about by the transfer of suitable gene constructs (see also Sect. 6.3) are frequently mentioned discussing the improvement the quality of animal products. A model proposed by MERCIER (1987) suggests a reduction of the lactose content of milk. By establishing transgenic sheep or cattle containing and expressing a lactase gene under the control of an udder-specific promoter it should be possible to cleave lactose into glucose and galactose. The milk of such animals should be tolerated by those patients suffering from lactose intolerance due to the lack of the enzyme lactase. This would be interesting for developing countries.

Another approach is concerned with the partial inhibition of lipid synthesis in the mammary gland. Since no suitable embryonic stem cells of economically important animals are available as yet for carrying out homologous recombination experiments, a succesful solu-

Tab. 8. Endogenous Milk Proteins

Encoded Protein	Concentration (μg/mL)
Caseins (ruminants)	
αS1	10000–12000
αS2	3400– 3800
β	10000–16000
K	3900– 4600
Major whey (ruminants) proteins	
α-Lactalbumin	800– 1000
β-Lactoglobulin	2800– 3000
Whey acidic protein (WAP) (rodents)	2000

Tab. 9. Potential Changes in Milk through Genetic Engineering (BREMEL et al., 1989)

Gene	Change
Casein	Increase of protein
Engineered casein	Manufacturing properties
Anti-sense β-lactoglobulin	Reduction/removal
Anti-sense acetyl CoA carboxylase	Reduction/removal of fat
β-Galactosidase, lactase	Increase of solids content
Antibodies of pathogens	Safer food, mastitis prevention

tion would require other measures to suppress gene activity. One possible approach is the expression of antisense RNA which has already been used successfully in transgenic mice to reduce the translation of proteins. In tissue culture the efficiency of the antisense approach has been increased by combination with a self-catalytic RNA cleavage structure known as the ribozyme. Some conceivable ways of allowing milk components to be altered specifically are summarized in Tab. 9 (BREMEL et al., 1989).

The alteration of the milk composition *per se* has also been discussed. The aim would be an alteration of cow milk so that it would resemble more closely the composition of human milk. Such a product might be of advantage in feeding human babies and infants.

6.3 Gene Farming

Many different proteins are currently produced by recombinant DNA technology involving the large-scale expression of genes in bacteria, yeasts, and tissue cultures. However, there are several good reasons to look for alternative sources of, for example, certain pharmaceuticals, diagnostics, and food components. It has been suggested (for obvious reasons) to use the mammary glands of transgenic animals (CLARK et al., 1987) for the production of heterologous proteins:

- The biological activity of many proteins frequently depends on a variety of posttranslational modifications such as glycosylation, hydroxylation, and carboxylation. In many cases the "simple" recombinant production systems including *Escherichia coli* and also *Saccharomyces cerevisiae* do not allow such modifications to be introduced at all or with a satisfactory precision. Proteins expressed in these organisms are frequently altered; they may be antigenic or show no activities or altered activities. In contrast to prokaryotic production systems mammalian cells are normally quite capable of carrying out the posttranslational modification of heterologous proteins. It is still an open question, however, whether the posttranslation modifications of these proteins carried out in cells grown *in vitro* and also in mammary gland cells will be exactly identical to those carried out in the original cells (e.g., liver cells).
- The expression of foreign proteins in the milk is a system which would allow the recovery of the product in a conventional way (i.e., by milking) without exerting any adverse effects on the animals.
- The efficiency with which mammary glands synthesize proteins is enormous. The concentration of endogenous milk proteins is on the order of 4–6%, depending on the species. Even if gene farming in the forseeable future would not allow recombinant proteins to be expressed in quantities approaching 60 g per liter, concentrations of 1–2 g/L would be sufficient to yield considerable amounts of important recombinant proteins.
- Milk is a pure and hygienic product. The purification of foreign proteins from milk should not create any unsurmountable problems. It should also be

pointed out that the purified proteins would be entirely free of prokaryotic impurities.

The mammary gland-specific expression of transgenes in mice has been described for constructs involving the use of mouse mammary tumor virus promoter (STEWART et al., 1984) or rather its LTR sequences. The activity of this promoter is not strictly specific for mammary tissues, and an active search has been made for other regulatory sequences allowing the transferred gene constructs to be expressed exclusively in mammary glands. It is obvious that the most likely candidates for such sequences are the genes of proteins which are found under physiological conditions in milk. As shown in Tab. 8, there are several milk proteins which are found in concentrations of above 1 g per liter in the milk. The genes encoding α_{s1}- and β-casein, α-lactalbumin, β-lactoglobulin, and whey acidic protein (WAP) have been isolated and cloned (reviewed by HENNIGHAUSEN, 1990; WILMUT et al., 1990).

Several gene constructs used for the generation of transgenic mice contain an activated human Ha-ras oncogene under the control of the WAP promoter. WAP-ras expression has been observed in the mammary glands in 2 out of 5 transgenic mouse lines. After a long latency tumors arose in tissues expressing WAP-ras, i.e., in mammary and salivary glands (ANDRES et al., 1987). If a truncated mouse c-myc gene driven by the WAP promoter was used the WAP-myc transgene was abundantly expressed in the mammary gland analogous to the expression pattern of the endogenous WAP gene. As early as two months after the onset of WAP-myc expression, tumors occurred in the mammary glands of the transgenic mice (SCHÖNENBERGER et al., 1988). Coexpression of both oncogenes in double transgenic Ha-ras and c-myc mice synergistically affected the differentiation and resulted in a high number of neoplastic foci. Palpable tumors were observed only after a latency of 3–5 months (ANDRES et al., 1988).

The transfer of a genomic clone encoding sheep β-lactoglobulin into mice has demonstrated for the first time that the mammary gland-specific expression of heterologous milk protein genes in transgenic animals may be very efficient. In 3 out of 5 female β-LacβLG transgenic mice SIMONS et al. (1987) have observed concentrations of β-lactoglobulin of up to 23 mg/mL (Tab. 10). This concentration is a factor of 5 over the expression level observed is sheep's milk, and it should be noted that this protein is not a normal constituent of mouse milk. A 14 kb fragment of the 3′ region of the β-casein gene has been cloned and transferred into mice, leading to the expression of the protein corresponding to 0.01–1% of the endogenous mouse β-casein (LEE et al., 1988). The expression of a cattle α-lactalbumin transcription unit comprising 750 bp of the 5′ and 336 bp of the 3′ flanking region has been described by VILOTTE et al. (1989) who observed concentrations of the protein corresponding to endogenous levels. The guineapig α-lactalbumin

Tab. 10. Expression of Milk Proteins in Transgenic Mammals

Gen Construct	Encoded Protein	Concentration (µg/mL)	Reference
		Mouse	
βLac-βLG	Lactoglobulin (ovine)	23000	SIMONS et al. (1987)
βCas-βCas	β-Casein (rat)	2–160	LEE et al. (1988)
αLac-bLac	Lactalbumin (bovine)	2.5–450	VILOTTE et al. (1989)
αLac-gpLac	Lactalbumin guineapig	~1000	MASCHIO et al. (1991)
WAP WAP	Whey acidic protein (mouse)	20–1900	BAYNA and ROSEN (1990)
WAP-WAP	Whey acidic protein	60–1080	BURDON et al. (1991)
		Pig	
WAP-WAP	Whey acidic protein	1000–2000	PURSEL et al. (1990), WALL et al. (1991)

Tab. 11. Expression of Foreign Proteins in the Milk of Transgenic Mammals

	Encoded Protein	Concentration (μg/mL)	Reference
		Mouse	
WAP-tPA	Plasminogen/activator	<0.4	GORDON et al. (1987)
WAP-CD4	Cd4 receptor	0.01–0.2	PITTIUS et al. (1988b)
WAP-hGH	Growth hormone	0.006–0.2	BREM et al. (1991a)
WAP-hGH	Growth hormone	0.065–410	REDDY et al. (1991)
WAP-PS 2-1	PS 2	1.5	TOMASETTO et al. (1989)
WAP-hPC	Protein C	0.2	VELLANDER (1991)
βLac-αA1AT2	α1-Antitrypsin	7000	ARCHIBALD et al. (1990)
βCas-hUK	Urokinase	1000–2000	MEADE et al. (1990)
αLac-oTP	Trophoblast interferon	1000	STINNAKRE et al. (1991)
		Rabbit	
βCas-hIL2	Interleukin 2	0.051–0.430	BÜHLER et al. (1990)
WAP-hGH	Growth hormone	<3.7	BREM et al. (1991b)
α_{S1}Cas-chym	Prochymosin	<10000	BREM and HARTL (1991)
α_{S1} IGF-I	IGF-I	2500	HARTL and BREM (1992)
		Pig	
WAP-hGH	Growth hormone	0.04	BREM et al. (1991b)
		Sheep	
βLac-α_1AT1	α1-Antitrypsin	5	CLARK et al. (1989a)
βLAc-FIX	Clotting factor IX	0.025	CLARK et al. (1989b)
	α1-Antitrypsin	35000	WRIGHT et al. (1991)
		Goat	
WAP-LAtPA	Plasminogen/activator	3	EBERT et al. (1991)

gene is also correctly regulated in the mammary gland of transgenic mice, and the protein is secreted into the milk (MASCHIO et al., 1991). Surprisingly, *in situ* hybridization analysis has shown that the gene was also expressed in the undifferentiated cells in the basal layer of the sebaceous glands.

The whey acidic protein (WAP) is an abundant milk protein in rodents but is not present in the milk of pigs or other farm livestock (CAMPBELL and ROSEN, 1984). PURSEL et al. (1990b) have transferred several hundred copies of a 7 kb fragment containing the mouse WAP gene into five transgenic pigs. Three lines were analyzed, and mouse WAP was detected in the milk from all lactating females at concentrations of about 1 g/L (WALL et al., 1991). Similar levels are observed in mice (Tab. 10).

GORDON et al. (1987) have demonstrated for the first time that proteins usually not found in milk can also be expressed in the mammary gland of transgenic mice. By using the WAP promoter and the cDNA of human tissue plasminogen activator (tPA) with its cognate secretion signal sequence, they have been able to obtain transgenic mice whose milk contains up to 460 ng/mL of active tPA (Tab. 11). The amount of tPA found in the milk of one line including the same construct was about 50 μg/mL as judged by an ELISA (PITTIUS et al., 1988a, b). Mice harboring the fusion construct WAP-PS2-1 in which the PS2-coding sequence was inserted into the 5′ untranslated region of the complete WAP gene have been observed to express the transgene at a level of approximately 1.5 μg/mL (TOMASETTO et al., 1989). The appearance of the PS2 protein in milk is approximately 10-fold reduced from the expected values, suggesting a defect in the translation, stability, and secretion of the PS2 protein.

YU et al. (1989) have used the WAP promoter to express the CD4 receptor in milk of

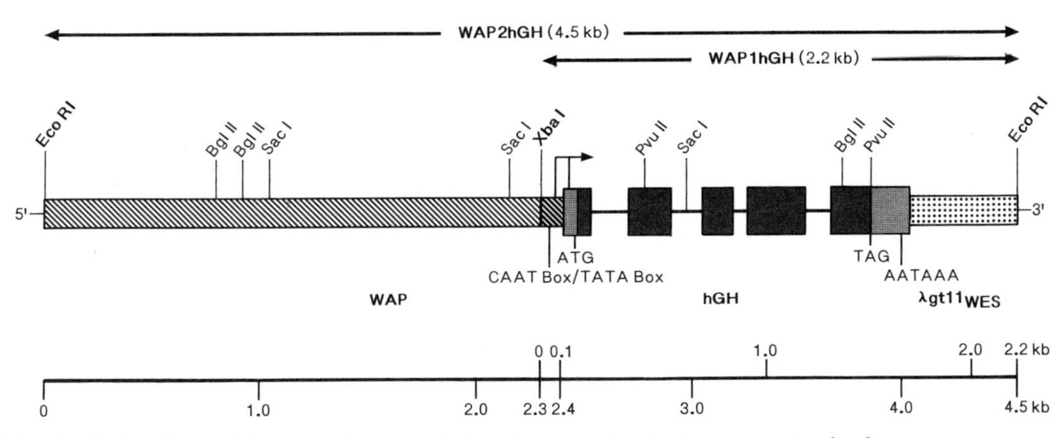

Fig. 28. WAP (whey acidic protein) constructs for the expression in the mammary glands.

transgenic mice (0.2 µg/mL). VELANDER (cited in HENNINGHAUSEN, 1990) has gained the expression of protein C at a concentration of 0.2 µg/mL in the milk of transgenic mice. A rat genomic clone which carried 949 bp of the 5′ flanking DNA and 1.4 kb of 3′ sequences of WAP was present in the milk of 8 out of 9 mouse lines at an average concentration of 27% of the endogenous mouse protein.

The group headed by John CLARK in Edinburgh has reported for the first time in 1990 the high-level expression of a biologically active heterologous protein in the milk of transgenic mice. A hybrid construct using sequences from ovine β-lactoglobulin fused to an α1 AT minigene was used. 5 out of 13 transgenic lines expressed the hybrid gene in the mammary gland; 5 in the salivary glands and 2 in both tissues. 4 out of 7 lines produced concentrations of at least 0.5 mg of α1 AT per mL in their milk (Tab. 11), and one line produced 7 mg (ARCHIBALD et al., 1990).

High-level expression in transgenic mouse milk has also been observed using a hybrid bovine αS1 casein/human urokinase gene. Urokinase is normally synthesized in the kidney. However, the transgenic mice secreted the active human enzyme in their milk at concentrations of 1–2 mg/mL (MEADE et al., 1990). In 1 out of 8 transgenic mouse lines carrying an ovine trophoblastin cDNA the bovine α-lactalbumin gene expression has been found to be restricted to the mammary glands (STINNAKRE et al., 1991).

Several studies have been carried out to determine if human growth hormone (hGH) can also be produced in the milk of transgenic mammals without adverse effects on growth and reproduction. Our own experiments have involved the use of two constructs in which WAP sequences (either 110 bp = WAP1, or 2.4 kb = WAP2) from the 5′ flanking region of the WAP gene were fused to the structural gene encoding hGH (Fig. 28). We have been able to establish a total of 31 transgenic lines. Pathological alterations and also reproductive problems have not been observed. Although RNA analyses revealed that the expression rates were in some cases comparable to that of the endogenous WAP (Fig. 29), the concentrations of GH in the milk were only on the order of 6–200 ng in 10 out of 14 lines studied at days 3–7 of lactation (BREM et al., 1991a). Transgenic mice which contained the short WAP fragment usually expressed less hGH-specific mRNA. These findings demonstrate that these constructs did not contain important regulatory elements which control the physiological expression of the protein during pregnancy and lactation.

When organs from transgenic mice were examined, high-level expression of hGH was unexpectedly also observed in the brain. Using *in situ* hybridization or immunohistochemistry hGH expression from the WAP-2-hGH-transgene was seen to occur specifically in Bergmann glia cells (GÜNZBURG et al., 1991). Thus we propose that the combination of the WAP

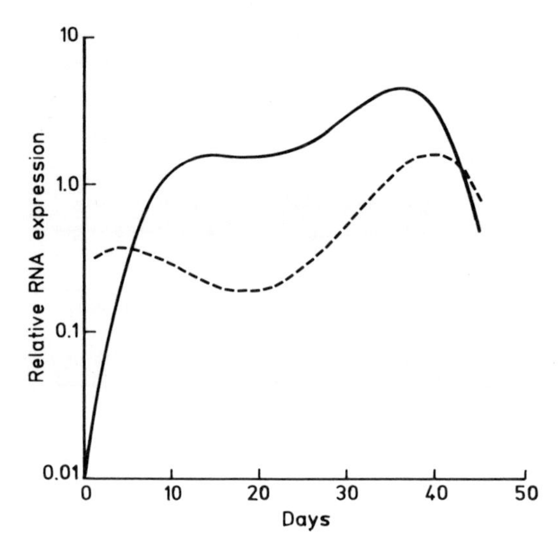

Fig. 29. Relative expression levels of endogenous WAP and WAP-hGH transgenes (GÜNZBURG et al., 1991).
The calculated level of expression during gravidity (parturition on day 19) and lactation of the endogenous WAP (solid line) is shown in comparison to the WAP2-hGH (dotted line).

promoter and the hGH structural gene results in novel tissue specificity in the Bergmann glia.

REDDY et al. (1991) have also used a WAP-hGH construct. Of 24 transgenic founder mice 17 lines were established. In four lines hGH levels in the milk were 65, 525, 970, and 410000 ng/mL, respectively. Homozygous offspring obtained from the high-level expressing animal had up to 1 mg/mL of hGH in their milk. Human GH was also detectable in the serum of 19 animals, indicating that the expression of the hGH by the WAP promoter was not tissue-specific. The serum levels observed by REDDY et al. (1991) varied from 16–9500 ng/mL which is 1- to 18-fold higher than the endogenous growth hormone level normally observed in mice. A greater weight gain (40%) was only seen in one lineage. Blood serum levels of growth hormone in our WAP-hGH transgenic mice (BREM et al., 1991b) were at the limits of detection (<0.7 ng/mL). Differences in the growth behavior of these animals were also not observed.

Transgenic mice containing a chimeric gene comprising the hydroxymethylglutaryl coenzyme A reductase promoter and the hGH gene had mammary glands at an age of 8 weeks that were morphologically and functionally comparable to those normally reached after 14–15 days of gestation. Precocious development correlated with local expression of hGH in the mammary gland (BCHINI et al., 1991).

To study the developmental regulation of the mouse WAP gene, a 7.2 kb WAP transgene has been used for generating transgenic mice. 6 out of 13 lines expressed the transgene during lactation at levels between 3–54% of the endogenous gene. Expression was observed to depend on the site of integration. However, within a given locus the levels of expression were entirely dependent on the copy numbers. The induction of the WAP transgenes during pregnancy preceded that of the endogenous gene. During lactation two lines increased expression levels coordinately with the endogenous WAP and in three lines expression decreased to basal levels. These data indicate that the 7.2 kb gene contains some but not all of the elements necessary for the correct developmental regulation and that a repressor and an induction element may be lacking in the transgene (BURDON et al., 1991).

Eventually the most prominent aim of the research described in the previous paragraphs about the mamma-specific expression of genes in transgenic animals is to evaluate whether transgenic farm animals might be suitable alternatives for the production of proteins whose commercial production has not been possible so far. CLARK et al. (1987, 1989a, b) have used three different gene constructs (pMK, BLG-FIX, BLG-α1AT) to create six transgenic sheep. Two animals which contained approximately 10 copies of the β-lactoglobulin promoter and the human hemophilic factor IX (BLG-FIX) only expressed 25 ng/mL, which is approximately $1/250$ of the levels found in normal human plasma and only $1/10000$ of the β-lactoglobulin concentration in normal sheep milk (CLARK et al., 1989a, b). Using a minigene containing the sheep β-lactoglobulin promoter fused to hα_1AT (antitrypsin) sequences which comprise part of exon 1 and the remaining downstream introns and exons, excluding intron 1, transgenic sheep

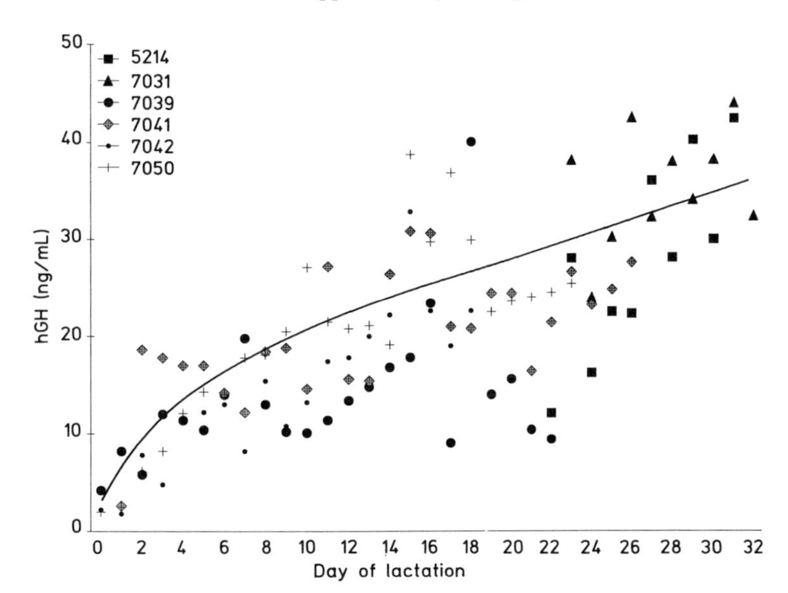

Fig. 30. hGH expression in the milk during lactation of one line of WAP-hGH-transgenic pigs.

were generated. Two of the animals produced 1–5 grams per liter while the third produced approximately 35 grams of hα_1AT/liter milk (WRIGHT et al., 1991).

By using the rabbit β-casein promoter BÜHLER et al. (1990) have produced human interleukin-2 in the milk of transgenic rabbits. In 4 female rabbits tested expression levels reached only 50–430 ng/mL. Our own investigations with WAP-hGH constructs (see Fig. 28) have shown expression levels in transgenic rabbits to be on the order of up to 3.6 µg/mL in the offspring of the transgenic rabbits. In the six offspring of one WAP2-hGH-transgenic pig the expression of the transgene increased with the progression of lactation (Fig. 30) but reached only 10 (1st week) to 40 (5th week) ng/mL (BREM et al., 1991b). Preliminary results of RNA and Western blot analyses indicate, however, that transcription levels are higher and that apart from the physiological (21 kD) hGH protein other forms are also present which are not recognized by monoclonal antibodies directed against hGH.

The mammary gland-specific expression of genes following the gene transfer has also been demonstrated with goats. EBERT et al. (1991) have examined the production of a glycosylation variant of human tPA (longer acting tissue plasminogen activator) from an expression vector containing the murine WAP promoter in dairy goats. Milk from one female transgenic goat was shown by ELISA to contain LA tPA at a concentration of 3 µg/mL.

In our own approch for the large-scale production of proteins in rabbits we have generated 14 lines of transgenic rabbits carrying 3 different hybrid gene constructs comprising bovine α_{s1}-casein gene regulatory sequences and the bovine prochymosin gene. The milk of females of 9 transgenic lines contained concentrations of up to 10 g/L of bovine prochymosin. We have found this high level of protein in two different lines. We have also carried out Northern blotting and protein analyses and quite surprisingly observed different products, namely, the prochymosin and shorter pseudoprochymosins, which may have been created by alternative splicing of the prochymosin primary transcript (HARTL and BREM, 1991; BREM and HARTL, 1991).

Despite the fact that a commercial application of these transgenic animals is quite a long way off, we may expect that these systems in the future will make an important contribution towards the production of proteins which are of medical importance or which are required for other purposes.

6.4 Disease Resistance

Genetic variations in the defense of animals against infectious microorganisms can be detected at all levels. In most cases the susceptibility against infection has a polygenic hereditary basis. It is therefore not astonishing that only a few cases are known in which a specific gene locus has been shown to be involved in disease resistance. A particular example is the well-studied system of the Mx-1 gene product found in certain strains of mice which selectively protects against infections of influenza virus.

Gene transfer is of particular interest for the manipulation of disease resistance because it has been impossible so far to reduce the susceptibility against diseases in farm animals by conventional breeding programs. Apart from the possibilities offered by gene diagnosis the improvement of the disease resistance of farm animals has been discussed in particular with a view towards strategies involving the use of gene transfer techniques. A survey of attempts to increase the disease resistance of farm animals by the employment of modern molecular biological techniques has been presented by MÜLLER and BREM (1991).

Five different classes of mammalian genes are currently discussed as likely candidates for gene transfer experiments because these genes appear to be associated with the regulation of disease resistance:

- MHC genes
- T cell receptor genes
- immunoglobulin genes
- genes encoding lymphokines
- specific disease resistance genes.

Most of these genes and their actions are currently under investigation in mice. The following sections will specifically address some aspects of strategies involving the transfer of genes to improve the disease resistance of farm animals.

Transfer of specific resistance genes

One of the very few examples of a specific resistance against viral infections is the gene known as the Mx gene (LINDENMANN, 1962). Genetic analysis has revealed that the resistance of certain strains of mice against infection with influenza A and B viruses is caused by a single autosomal dominant $Mx1^+$ allele. The expression of this gene is controlled by type I interferons (IFN α/β). Mice which contain the $Mx1^-$ allele are not protected against influenza virus infections. The transfection of the $Mx1^+$ allele into cells that are susceptible to influenza virus has revealed that this gene is necessary and sufficient to protect against influenza virus infections (NOTEBORN et al., 1987; STAEHELI et al., 1986). It has been suggested that the inhibition of either primary transcription of the virus genome or the translation of the viral RNA may be involved (KRUG et al., 1985; MEYER and HORISBERGER, 1984).

Mx-homologous genes have also been found in all other eukaryotes investigated so far (reviewed by MÜLLER and BREM, 1991). The mammalians studied so far express at least two Mx-related genes. Our own investigations have demonstrated that pigs also contain at least two Mx genes which show homology with Mx sequences of other mammalians (MÜLLER et al., 1991).

Orthomyxovirus infections in particular are the causative agents of epidemic diseases in pigs. The Mx1 gene of mice was cloned in the mid 80s (STAEHELI et al., 1986). In our own investigations we have tried to integrate several different Mx gene constructs into the genome of pigs by gene transfer. The gene constructs, obtained through the courtesy of WEISSMANN and NOTEBORN (1986) contained the murine Mx1 cDNA driven by three different promoters, respectively, namely the human metallothionein IIA promoter, the SV40 promoter, and the cognate endogenous Mx promoter of mice. The establishment of Mx-transgenic pigs has revealed that the expression of the Mx gene during embryogenesis may prevent further development of the fetuses expressing the transgene. We have been able to produce transgenic pigs with the normally observed frequencies which did not show extensive rearrangements of the transferred gene constructs only if the endogenous Mx promoter of mice was used. Following the injection of the mx-Mx construct we have obtained eight transgen-

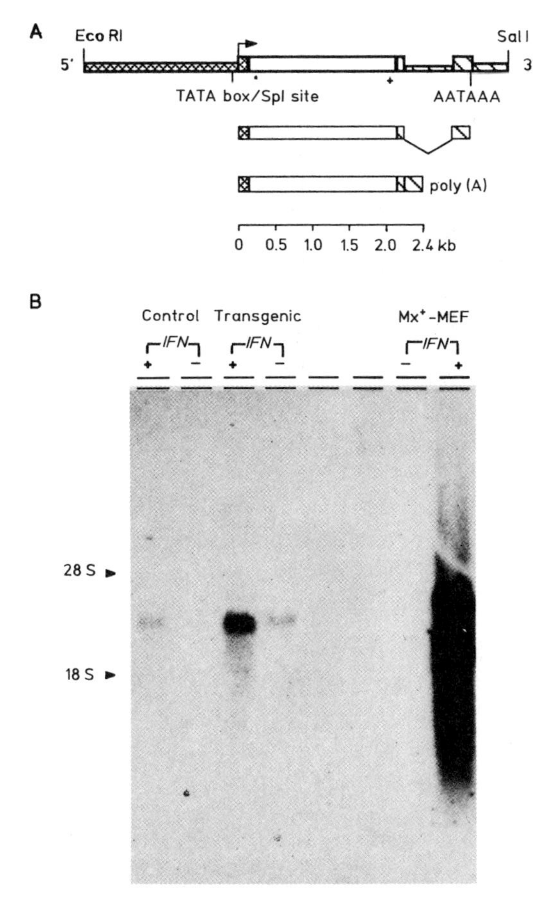

Fig. 31. Northern blot analysis of Mx-transgene expression in peripheral blood lymphocytes.
(A) gene construct (mMxMx) and mRNA with correct transciptional start and correct splicing pattern.
(B) 25 μg of total RNA each. Time of exposition: 24 h. (Mx^+-MEF = Mx^+ mouse fibroblasts as positive control).

ic piglets all of which contained correctly integrated gene constructs. In five lines the gene construct was stably transmitted to the offspring (MÜLLER et al., 1992).

We have also observed that the expression of the transgene was inducible with native pig interferon or double-stranded RNA in 2 out of 5 lines of cultivated peripheral blood lymphocytes (Fig. 31). The *in vivo* administration of pig interferon also induced the expression of the Mx genes in these two transgenic lines. However, an increase of the murine Mx1 protein in cells of these transgenic pigs was not detectable. One possible explanation may be that the response of the transgenes to interferon stimulation is so weak as to be practically not detectable. In contrast to murine $Mx1^+$ lines in which approximately 0.1 of the total polyadenylated mRNA is Mx-specific, levels of Mx mRNA in transgenic pigs are markedly reduced. ARNHEITER et al. (1990) have also shown that Mx1 transgenic mice may either not express the transgene at all or may express the transgenes at low or high levels. While the levels of RNA and proteins in the high-level expression animals reached 60–70% of those in $Mx1^+$ mice, the corresponding values in low-level expressing animals were on the order of 10–20% or lower. This indicates that the expression of the transgene is influenced to a large extent by the genomic localization of the transgene. The levels to which the transgenes can be induced are of critical importance if the aim is to obtain animals resistant against influenza viruses. High-level expression animals are protected against the consequences of an infection with influenza virus, while low-level expression animals are not protected or protected only if they are infected with high doses of viruses. All homozygous transgenic mice of the high responder line 979 survived the virus challenge (virus doses between 10^{-1} and 10^{-6}), while all of the control mice challenged with the same amounts of virus died within several days. Only control mice challenged with a virus dose of 10^{-6} survived (ARNHEITER et al., 1991, personal communication).

Expression of genes for monoclonal antibodies in transgenic animals

The process known as genetic immunization appears to be a realistic possibility to protect animals against infectious agents by the application of gene transfer techniques. This approach essentially involves the expression of genes directing the synthesis of defined antibodies in order to induce a protective state. The results of a variety of studies have demonstrated that it is possible to transfer gene constructs encoding monoclonal antibodies into mice and to express large amounts of the gene products (BRINSTER et al., 1984). The most in-

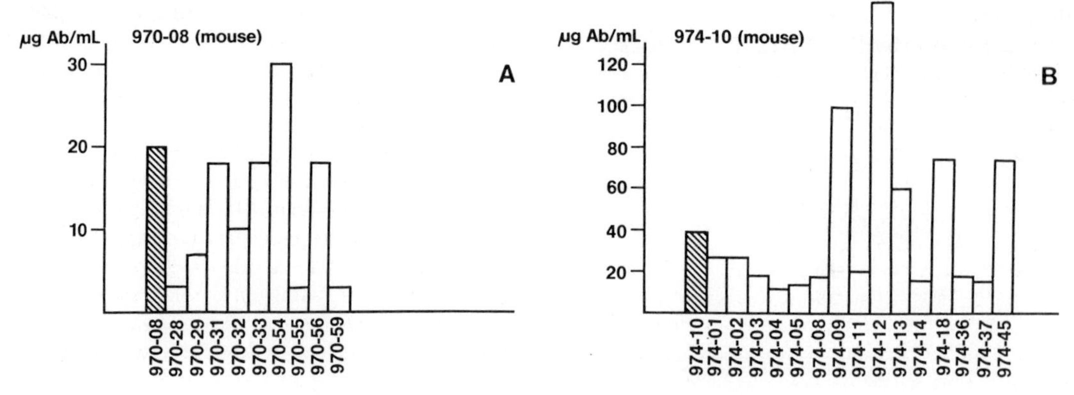

Fig. 32. Expression of antibodies in transgenic mice (WEIDLE et al., 1991).

teresting aspect of this approach is the fact that these transgenic animals produce antibodies against specific antigens without ever having been challenged with these antigens during their life. In the majority of these experiments involving transgenic mice, the main interest focused on the expression of genes encoding heavy and light chains of antibodies and on the consequences brought about on the genomic level by the gene transfer itself. In our own gene transfer experiments we have therefore addressed the question of whether it will be possible to transfer the genes encoding the light and heavy chains of a murine monoclonal antibody and to express a functional and correctly assembled antibody from the transgene. We have used the genes for an anti-idiotypic antibody directed against 4-hydroxy-3-nitrophenylacetate and transferred them into mice, rabbits, and pigs. In several transgenic lines we have been able to demonstrate that the transferred antibody genes lead to the synthesis of up to 150 µg/mL in transgenic mice (Fig. 32), approximately 300 µg/mL of antibodies in rabbits, and of over 1 mg/mL in pigs (WEIDLE et al., 1991; BREM and WEIDLE, 1991). The offspring of the pig also expressed at least 1 g/L of the antibody. The subsequent analysis of the antibodies by isoelectric focusing has shown, however, that only a small fraction was identical to the genuine murine antibody (see also Sect. 5.3). Further experiments should reveal whether the inclusion of enhancer elements of the light chains will improve the low level of expression.

LO et al. (1991) have generated transgenic mice, sheep, and pigs carrying genes encoding the mouse α and κ chains against phosphorylcholine (PC) to determine whether the transgenic antibody might be used to influence susceptibility to disease. In transgenic mice and pigs high serum levels of IgA were detected without suppression of endogenous IgM levels with the exception of one case in a transgenic mouse line. The secreted antibody in the pig presumably included pig light chains and little, if any, of the mouse IgA showed binding specificity for PC.

In the future the strategy of genetic immunization will be of particular interest in those cases in which it is unlawful, difficult or impossible to protect farm animals by vaccination. In principle it should be possible in the future to develop gene constructs whose transfer leads to a sufficiently high expression of antibodies that will protect transgenic animals against certain microorganisms.

Intracellular immunization

A different approach of obtaining resistance against viral infections in mammals is known as intracellular immunization (BALTIMORE, 1988). This process is based on the notion that the endogenous expression of viral proteins or mutated forms of these proteins should provide protection against infections with the genuine virus. According to BALTIMORE this approach will be of particular interest in those cases in which conventional immunization does not lead to protection against viral diseases. SALTER and CRITTENDEN (1989) have shown that transgenic chicken expressing the surface glycoprotein of the avian leukosis virus were resistant against infections with the virus (see also Sect. 7).

Expression of antisense RNA

A particularly interesting strategy to elicit the protection against viral infections is the expression of antisense polynucleotides (MELTON, 1988; WALDER, 1988). A large number of natural antisense RNAs have been found in prokaryotes but there are also examples of eukaryotic cells which contain antisense RNA (asRNA) molecules in their nuclear or cytoplasmic extracts (INOUYE, 1988). Naturally occurring or artificial antisense polynucleotides can be used to suppress specifically the expression of the genes encoding the corresponding sense RNA. These antisense molecules either inhibit the translation of the mRNA or they interrupt splicing processes or the transport of the RNA into the cytoplasm.

A first example of the establishment of transgenic animals expressing asRNA has been described by ERNST et al. (1991) who produced transgenic rabbits expressing adenovirus H5 (Ad5) antisense RNA. 90–98% of the cells in primary kidney cell cultures were resistant against infections with the virus.

7 Transgenic Chicken

Gene transfer in poultry is usually aimed at improving disease resistance or increasing food utilization and growth performance. Although DNA micro-injection has been applied very successfully in mammalians, it has been of little value in poultry because of differences in the reproduction and in the embryonal development (SHUMAN, 1991). If the fertilized eggs are laid they already contain a blastoderm of more than 50000 cells which are organized in several cell layers. A complete *in vitro* culture system for avian embryos has been developed by PERRY (1987, 1988). Although it allows easy access to the embryo, the successes of nuclear injection are still very limited (PERRY and SANG, 1990). Until now gene transfer in poultry has been carried out predominantly by using retroviral vectors (see also Sect. 2.3), sperms treated with DNA (see Sect. 2.4), and irradiated sperms. Totipotent embryonic stem cells which would also allow homologous recombination are currently not available.

The method of choice for the gene transfer in poultry has been the use of retroviral vectors. Retroviruses are natural constituents of the genomes of birds and are also found as endogenous viruses. Because of their life cycle (see Fig. 12) it is possible to transform embryos in the blastoderm stage by the injection of a retrovirus stock solution. The injection procedure does not create any technical problems because the egg shells are easily fenestrated and can be incubated after having been re-sealed. The use of replication-competent retroviruses has the advantage that they can pass through several cycles of replication, thus allowing an entire cell layer to be transformed. However, the use of such viruses is restricted to special cases for safety reasons as well as for possible adverse effects on the embryos and the resulting birds.

These problems are circumvented to a large extent by the use of replication-deficient retroviral vectors and so-called suicide vectors which also reduce the danger of unwanted recombination events.

SALTER et al. (1986) have injected the chicken syncytial strain of reticuloendotheliosis (CS-REV) and wildtype and recombinant avian leukosis virus (ALV) near the blastoderm of inoculated fertilized embryos and CS-REV intraabdominally at the day of hatch. A number of positive animals in the progeny were identified, suggesting that retroviral genetic information had been inserted into the germ line of chicken (SALTER et al., 1987). Of 23 animals with proviral inserts 21 produced virus particles while 2 did not give rise to infectious viruses (CRITTENDEN and SALTER, 1990). Chicken containing the insert called alv6 were highly resistant to infections by the pathogenic subgroup A of ALV (SALTER and CRITTENDEN, 1989). These chicken expressed viral RNA in all tissues tested regardless of the stage of development. The mechanism responsible for the increased resistance is presumed to in-

Tab. 12. Production of Transgenic Birds by Retrovirus Infection (SHUMAN, 1991)

Species	Gene Transferred	Virus	Vector Type	No. FO Tested	No. FO Transmitting	No. F1 Transgenic	No. F1 Negative	No. Expressing	Reference
Chicken Line O	n.a.	RAV-O rec.	r.c.	14	4	21	794	15	(1)
White Leghorn	n.a.	RAV-O rec.	r.c.	9	4	5	532	4	(1)
White Leghorn	n.a.	RAV 1	r.c.	14	1	2	551	1	(1)
White Leghorn	n.a.	CSV	r.c.	8	0–5%	3			(2)
Chicken co.br.	neo-tk	SNV	r.d.	4	4	34	720	yes	(3, 4)
Chicken co.LH	none	RSV	r.c.		0–40%			yes	(5)
Japanese Quail	cat	SNN	r.d.	29	1	1	1594	1	(6)
Japanese Quail	none	SNN	Helper virus	29	1	2	1593	1	(6)

n.a., not attended; r.c., replication-competent; r.d., replication-defective; rec., recombinant; co.br., commercial broiler; co.LH, commercial Leghorn
(1) SALTER and CRITTENDEN (1989), (2) SALTER et al. (1986), (3) BOSSELMAN et al. (1989b), (4) BRISKIN et al. (1991), (5) BUMSTEAD et al. (1987), (6) LEE and SHUMAN (1990)

volve subgroup-specific receptor interference. The results clearly demonstrate that experimentally introduced endogenous proviruses can be expressed at high levels in the avian system (FEDERSPIEL et al., 1991).

In an effort to introduce the bacterial neomycin phosphotransferase (NPT II) the gene was cloned into an infectious avian retrovirus vector derived from the Schmidt-Ruppin strain of RSV. After inoculation of fertilized chicken embryos on day 0 approximately 12% of the embryos were positive for the NPTII gene on day 20 (HIPPENMEYER et al., 1988).

CHEN et al. (1990a, b) have used a modified replication-competent Schmidt-Ruppin A strain of Rous sarcoma virus (RSV) for the infection of embryos and infected early commercial Leghorn embryos with a non-transforming vector containing the bovine GH gene. RSV proviral inserts were transferred by males to offspring at frequencies of 0–40%. These birds did not contain the bovine GH gene. They were, however, protected against A RSV infections.

LEE and SHUMAN (1990) have found two birds carrying spleen necrosis virus (SNV) proviral inserts after infection of Japanese quail embryos with SNV. A summary of the production of transgenic birds by retrovirus infection is given in Tab. 12. It has been shown that replication-defective retroviral vectors can also be transferred into the germ line of birds (BOSSELMAN et al., 1989a, b). 23% of the birds that hatched from eggs previously inoculated with a defective SNV showed evidence of vector sequences in DNA extracted from blood cells (see Tab. 12).

SOUZA et al. (1984), BOSSELMAN et al. (1989a, b), and CHEN et al. (1990) have injected RSV, SNV, and ALV retroviral vectors containing GH genes into embryos. Growth hormone levels were increased in a relatively high fraction of growing embryos and adults developing from the inoculated embryos. Bacterial marker genes have also been transferred into embryos by infection with replication-defective SNV vectors (LEE and SHUMAN, 1987, 1990).

An improvement of the frequency of germline transmission has been attempted by the isolation of helper cell lines that produce high titer virus stocks. HIGHKIN et al. (1991) have engineered the quail cell line QT6 and canine D17 cells for the production of reticuloendotheliosis virus vectors. A high producer QT6 helper cell line (105 cfu/mL) was found to be relatively free of helper virus. An amphotropic murine cell line produced a 6- to 10-fold higher amount of virus and had a comparably high titer on chicken cells.

Gene expression form heterologous promoters in a replication-defective reticuloendotheliosis avian retrovirus vector in quail cells has been investigated by HIPPENMEYER and KRIVI (1991). Under transient conditions in QT6 cells the human cytomegalovirus immediate early promoter was very strong but after integration into the quail genome expression dropped greatly, presumably because it had become methylated after transfection.

It has been shown that genetic transformation of eggs and birds can also be achieved by the use of irradiated sperms (PANDEY and PATCHELL, 1982; BUMSTEAD et al., 1987; TOMITA, 1987; TOMITA et al., 1988). Hens are usually inseminated twice within 24 h, the first time with irradiated semen carrying marker genes and the second time with normal sperms of the same strain as the hen. Feathers, egg color, and MHC haplotypes of a small percentage of the progeny have been shown to be characteristic for the strain from which the irradiated inactivated sperms had been taken.

Recently GRUENBAUM et al. (1991a, b) have introduced a new approach for the generation of transgenic birds which uses transfected sperm cells as vectors. Using a special buffer containing marker genes sperm cells were incubated before the insemination of hens. As shown by PCR and Southern blot analyses, 30–60% of the resulting embryos and/or chicken contained the transferred genes which were also transmitted through the germ line to offspring.

Another approach of great interest for gene transfer technology is the production of germ line chimeras (SHUMAN, 1991) which may be produced by the transfer of primordial germ cells, blastoderm cells, and embryonic stem cells. ES cell lines of chicken are as yet not available (for ES cells see also Sect. 2.2), but primordial germ cells have been shown to be capable of colonizing the host gonads after transfer to an embryo (GUIDO et al., 1991) and to be passed through the germ line (WENTWORTH et al., 1989). Furthermore, it has also been demonstrated that DNA can be introduced into blastodermal cells (BRAZOLOT et al., 1990; VERRINDER-GIBBINS et al., 1990). It is possible to transfer blastodermal cells from an embryo to another where they participate in the development of tissues, including germ cells.

8 Transgenic Fish

The production of transgenic fish has been described recently (ZHU et al., 1985; ROKKONES et al., 1985; for review see CHEN and POWERS, 1990). This is surprising because fish are quite well suited for gene transfer. There is a large variety of fish species, it is possible to obtain large numbers of eggs in all stages which are also easy to cultivate, and a number of fish cell lines are also available. Moreover fish is an important foodstuff for human consumption, and for such important candidates in aquaculture like carps, salmon, and trout great expertise in fish raising and cultivation has been gained. Fish represent the largest and most diverse group of vertebrates, and because of the evolutionary position and their ability to adapt to a wide variety of environments and other characteristics they are excellent experimental models for studying various aspects of the embryology, neurobiology, endocrinology, and environmental biology of organisms (POWERS, 1989).

One reason for gene transfer techniques gaining slow momentum in this field may be the fact that the experimental practicalities are associated at times with some difficulties. For example, DNA micro-injection is hampered in many cases by the specific property of the chorion which becomes very hard especially after fertilization. These problems have been overcome recently by several innovations (culture in suitable media, perforation of the chorion before micro-injection by chorion drilling, mechanical and enzymatic removal of the chorion, injection through the micropyle (Fig. 33) (HALLERMAN et al., 1988; BREM et al., 1988; ROKKONES et al., 1989; CHOURROUT et al., 1986; MACLEAN et al., 1987; ZHU et al., 1985; YOON et al., 1990a, b). Although there are some exceptions it is usually not possible to inject the DNA into the nuclei because they cannot be visualized easily. Instead the DNA is normally injected into the cytoplasm.

Fortunately it has turned out that in most cases the injection of DNA into the cytoplasm is sufficient to guarantee its transfer. Techniques such as the use of retroviral vectors and totipotent embryonic stem cells (see Sect. 2) and also the liposome-mediated gene transfer

Fig. 33. DNA injection through the micropyle of an oocyte from tilapia (BREM et al., 1988).

which are successfully employed in mammals are either not available for fish or are only of limited value. The use of sperms as vehicles (see also Sect. 2.4) for the transfer of DNA has been described at least for the sea urchin (AREZZO, 1989) but failed to yield positive results in trout (HOUDEBINE and CHOURROUT, 1991). So far the injection of DNA into the nuclei has been successful only in medakas (OZATO et al., 1986). However, it has been shown that electroporation may be successfully employed as a new technique for the generation of transgenic fish (INOUE et al., 1990). BUONO and LINSER (1991) have used the plasmid construct RSVCAT which was linearized and electroporated into fertilized zebrafish embryos (*Brachydanio rerio*) at the 2- and 4-cell stages. 60% of 200 treated embryos survived three low-field-strength pulses in a 0.4 cm cuvette. They were tested 6 days after spawning, and more than 40% of the animals tested were carrying the transgene. 10 days post spawning the transgene was transcribed and translated into protein.

Several specific gene transfer conditions have been worked out for a number of different fish species. It is quite remarkable that in contrast to mammals the exogenous DNA transferred into early embryonal stages is rapidly replicated to that an increased number of DNA molecules are available for integration. It should be noted, however, that most of the primary transgenic fish are mosaics. Micro-injection usually involves the transfer of 10^6–10^8 linearized DNA molecules in 1- to 4-celled fish eggs. 35–80% of the eggs survive this treatment and, depending upon the developmental stage analyzed, 50–70% of the embryos, 25% of the fetuses, and 5–10% of the adult fishes possess integrated DNA (CHEN and POWERS, 1990). Tab. 13 summarizes the data available for gene transfer experiments in fish.

Inheritance of the transgene in the offspring is, of course, also observed in transgenic fish. However, since most of the founder animals are genetic mosaics, markedly less than 50% of the progeny also inherit the transgene (STUART et al., 1988, 1990; GUYOMARD et al., 1989; CHEN et al., 1990a, b; MACLEAN et al., 1987; HOUDEBINE and CHOURROUT, 1991).

The transgenes are usually expressed unsatisfactorily in fish (Tab. 13). Some of the transferred DNA constructs of mammals and higher vertebrates are, however, transcribed very efficiently (MOAV et al., 1990; CHEN et al., 1990a, b; HALLERMAN et al., 1990; YOON et al., 1990a, b; HOUDEBINE and CHOURROUT, 1991).

There are some examples of the expression of marker genes in transgenic fish. The CAT gene under the control of the RSV enhancer, the SV40 promoter, and also the human cytomegalovirus early gene promoter have been reported to be expressed in liver, muscle, gut, skin, and blood cells (HOUDEBINE and CHOURROUT, 1991). The chicken δ-crystalline gene was expressed in transgenic medaka. The flounder antifreeze protein gene (FLETCHER et al., 1988, DAVIES et al., 1989; OZATO et al., 1986; INOUE et al., 1989), and the bacterial β-galactosidase gene under the control of the murine MT promoter (MCEVOY et al., 1988) have been expressed in transgenic salmon. The cDNA of trout GH (ZHANG et al., 1990) and bovine GH have been expressed in transgenic carp and the Northern pike (SCHNEIDER et al., 1989). mMT-hGH has been expressed in tilapia (BREM et al., 1988), trout (ROKKONES et al., 1989), and carp (CHEN et al., 1990a, b, 1991). BGH and chinook salmon GH have been expressed in the Northern pike as well (GROSS et al., 1991).

Parallel to the transfer experiments involving the use of growth hormone genes the corresponding genes of fishes (rainbow trout, salmon) have also been analyzed structurally and have been cloned and sequenced (AGELLON et

Tab. 13. Summary of Transgenic Fish Studies Conducted by Various Laboratories, 1985–1990 (CHEN and POWERS, 1990)

Species	Promoter/Gene	I[a]	E[a]	T[a]	Author(s)
Common carp	RSV/rtGHcDNA	+	+	+	CHEN et al. (1989), ZHANG et al. (1990)
Chinese carp	mMT/hGH	+	+	+	CHEN et al. (1989)
Catfish	RSV/rtGHcDNA	+			POWERS (1989)
	mMT/hGH	+			DUNHAM et al. (1987)
	mMT/hGH	+			ZHU et al. (1985)
	SV40/*E. coli* neo	+	+		YOON et al. (1988)
Loach	mMT/hGH	+			ZHU et al. (1986)
Medaka	SV40/c CR	+	+		OZATO et al. (1986)
	mMT/rtGHcDNA	+			INOUE et al. (1989)
	fLuc/flus	+	+		TAMIYA et al. (1989)
Salmon	mMT/*E. coli* β-gal	+	+		MCEVOY et al. (1988)
	mMT/hGH	+	+		ROKKONES et al. (1989)
	fAFP/fAFP	+	+		FLETCHER et al. (1988)
Tilapia	mMT/fGH	+			BREM et al. (1988)
Trout	mMT/hGHcDNA	+	+		CHOURROUT et al. (1986, 1988)
	mMT/rGH	+			MACLEAN et al. (1987)
	cG/cG	+			OSHIRO et al. (1989)
Zebrafish	mMT/*E. coli* hygro	+			STUART et al. (1988)

[a] I, integration; E, expression; T, transmission

al., 1988a, b; GONZÁLEZ-VILLASENOR et al., 1988).

HOUDEBINE and CHOURROUT (1991) have reviewed experiments which show that fish cell lines can be used to evaluate the efficiency of a promoter. The correlation between the activity of a promoter in the cells and in transgenic animals was good, indicating that there are no fundamental differences between fish and mammalian transcription systems. It is more or less impossible to predict the activities of new gene constructs in fish cells, and it is therefore suggested that fish DNA sequences, if available, be used for maximum resutls.

LIU et al. (1990a, b, c) have developed two fish expression vectors which contain the proximal promoter and enhancer regulatory elements of the carp β-actin gene and the polyadenylation signal from the salmon GH gene which should be active in all species of fish. We have fused the promoter of the rainbow trout metallothionein B gene to the bacterial chloramphenicol acetyltransferase gene in an expression vector (tMTb-CAT). This promoter exhibited an extremely low basal expression in all cell lines tested and was zinc- and cadmium-inducible except for the melanoma cell line. In a transient assay it was functional in developing embryos of the medaka fish. These properties make this promoter suitable for inducible, tissue-specific expression of transgenes (HONG et al., 1991).

The practical application of transgenic fish in aquaculture is still a long way off. Specifically it will be necessary to develop cost-effective protocols for gene transfer on a large scale, suitable techniques for the identification of transgenic animals, and selective breeding regimes before further progress can be made. Modifications by gene transfer are aimed at substantial changes of performance characters, extension of environmental tolerance, or expression of novel proteins (KAPUSCINSKI, 1990). Aspects of transgenic fish and public policy anticipating environmental and ecological impacts (KAPUSCINSKI and HALLERMAN, 1990a, b), regulatory concerns (KAPUSCINSKI and HALLERMAN, 1990b), and patenting of transgenic fish (HALLERMAN and KAPUSCINSKI, 1990a, b, 1991; HALLERMAN et al., 1988, 1990) are the main area of future discussions and will have to be solved as soon as possible.

9 Problems and Future Possibilities with Transgenic Animals

Some possible future developments concerning transgenic animals have already been mentioned in several chapters. At the end of Part IV I would like to mention some projects that have been initiated and some promising and long-term goals and developments. It is to be hoped that the techniques for generating transgenic animals will be perfected in the course of the next decade (see also Sect. 2) by improving gene transfer methodologies, especially since it appears that the efficiency of DNA micro-injections cannot be improved any further. New vistas will be opened by the development of new regulatory and transcription units. These developments should also facilitate the analysis of the mechanisms of actions of genes.

There is no doubt that the establishment of new transgenic models will provide an enormous increase in our knowledge of human diseases. The manipulation of genes in ES cells by homologous recombination in particular is still in its infancy and will also open new ways in the utilization of transgenic animals. The same holds true for the elucidation of the complex network of interactions between different genes, which is a fascinating field in its own right. The reader is reminded of studies aimed at the elucidation of the control of embryonal development, the study of additive polygenic actions of genes, and the genetic influences on ageing.

So far the use of gene transfer in the breeding of farm animals, including fishes and poultry, has not led yet to concrete and useful applications. It is to be expected, however, that with our increasing knowledge of molecular basics gleaned from the study of transgenic mice progress will also be made in this field, although it should be noted that progress will be slower due to the long generation times. A very advanced field which is almost at the verge of practical application is gene farming which will allow a number of interesting proteins to be produced in high quality and in a cost-effective way.

It is my opinion that the greatest challenge of gene transfer techniques in farm animals, in particular in pigs, will be the attempt to genetically modify organs and tissues to be used in xenotransplantations. Unfortunately the technical biological basics have not yet been worked out but some first steps in this direction have already been made at least theoretically.

A big problem in the gene transfer in mice and farm animals appears to be the intensive and frequently adverse coverage in the media and the aversion of the public towards gene transfer. One reason may be that this field combines two very sensitive issues to which the public reacts with distrust if not aversion, i.e., genetic engineering and animal experiments. This combination is very likely to increase hidden (and overt) anxieties and ideological aversions. The reader may be aware of the fact that particularly in Germany severe (and often destructive) criticism is widespread (to say the least). Under these circumstances it will be of paramount importance to continue to inform the public objectively, to discuss issues in detail, and to provide intelligible information in order to prevent extremists and fanatic activists from gaining acceptance caused by wrong assumptions.

The importance of gene transfer techniques for biomedical and molecular biological basic research cannot be overemphasized. These techniques have revolutionized our knowledge about fundamental biological and genetic processes to an extent that almost seemed unthinkable a decade earlier. They will also allow us to address many open questions in the future.

10 References

ACSADI, G., JIAO, S., JANI, A., DUKE, D., WILLIAMS, P., CHONG, W., WOLFF, J. A. (1991), Direct gene transfer and expression into rat heart *in vivo, New Biologist* **3**, 71–81.

ADAMS, J. M., HARRIS, A. W., PINKERT, C. A., CORCORAN, L. M., ALEXANDER, W. S., CORY, S., PALMITER, R. D., BRINSTER, R. L. (1985), The c-myc oncogene driven by immunoglobulin enhancers induces lymphoid malignancy in transgenic mice, *Nature* **318**, 533–538.

ADELSTEIN, S., PRITCHARD-BRISCOE, H., ANDERSON, T. A., CROSBIE, J., GAMMON, G., LOBLAY, R. H., BAASTEN, A., GOODNOW, C. C. (1991), Induction of self-tolerance in T cells but not B cells of transgenic mice expressing little self antigen, *Science* **251**, 1223–1225.

AGELLON, L. B., DAVIES, S. L., CHEN, T. T., POWERS, D. A. (1988 a), Structure of a fish (rainbow trout) growth hormone gene and its evolutionary Implications, *Proc. Natl. Acad. Sci. USA* **85**, 5136–5140.

AGELLON, L. B., DAVIES, S. L., LIN, C.-M., CHEN, T. T., POWERS, D. A. (1988 b), Rainbow trout has two genes for growth hormone, *Mol. Reprod. Dev.* **1**, 11–17.

ALDRICH, C. J., HAMMER, R. E., JONES-YOUNGBLOOD, S., KOSZINOWSKI, U., HOOD, L., STROYNOWSKI, I., FORMAN, J. (1991), Negative and positive selection of antigen-specific cytotoxic T lymphocytes affected by the $\alpha 3$ domain of MHC I molecules, *Nature* **352**, 718–721.

ANDRES, A.-C., SCHÖNENBERGER, C.-A., GRONER, B., HENNIGHAUSEN, L., LEMEUR, M., GERLINGER, P. (1987), Ha-ras oncogene expression directed by a milk protein gene promoter: tissue specifity, hormonal regulation, and tumor induction in transgenic mice, *Proc. Natl. Acad. Sci. USA* **84**, 1299–1303.

ANDRES, A.-C., VAN DER VALK, M. A., SCHÖNENBERGER, C. A., FLÜCKIGER, F., LEMEUR, M., GERLINGER, P., GRONER, B. (1988), Ha-ras and c-myc oncogene expression interferes with morphological and functional differentiation of mammary epithelial cells in single and double transgenic mice, *Genes Dev.* **2**, 1486–1495.

ARCHIBALD, A. L., MCCLENAGHAN, M., HORNSEY, V., SIMONS, P., CLARK, A. J. (1990), High-level expression of biologically active human $\alpha 1$-antitrypsin in the milk of transgenic mice, *Proc. Natl. Acad. Sci. USA* **87**, 5178–5182.

AREZZO, F. (1989), Sea urchin sperm as a vector of foreign genetic information, *Cell Biol. Int. Rep.* **13**, 391–404.

ARNHEITER, H., SKUNTZ, S., NOTEBORN, M., CHANG, S., MEIER, E. (1990), Transgenic mice with intracellular immunity to influenza virus, *Cell* **62**, 51–61.

ATKINSON, P. W., HINES, E. R., BEATON, S., MATTHAEI, K. I., REED, K. C., BRADLEY, M. P. (1991), Association of exogenous DNA with cattle and insect spermatozoa *in vitro, Mol. Reprod. Dev.* **29**, 1–5.

BABINET, C., FARZA, H., MORELLO, D., HADCHOUEL, M., POURCEL, C. (1985), Specific expression of hepatitis B surface antigen (HBsAg) in transgenic mice, *Science* **230**, 1160–1163.

BABINET, C., MORELLO, D., RENARD, J. P. (1989), Transgenic mice, *Genome* **31**, 938–949.

BACHILLER, D., SCHELLANDER, K., PELI, J., RÜTHER, U. (1992), Liposome mediated DNA uptake by sperm cells, *Mol. Reprod. Dev.,* in press.

BALTIMORE, D. (1988), Intracellular immunization, *Nature* **335**, 395–396.

BARINAGA, M. (1989), Gene-transfer method fails test, *Science* **246**, 446.

BAUNACK, E., GÄRTNER, K., WERNER, I. (1988), Ovary transplantation, a way to estimate prenatal maternal effects in mice, *Fertilität* **4**, 235–239.

BAYNA, E. M., ROSEN, J. M. (1990), Tissue specific, high level expression of the rat whey acidic protein gene in transgenic mice, *Nucleic Acids Res.* **18**, 2977–2985.

BCHINI, O., ANDRES, A. C., SCHUBAUR, B., MEHTALI, M., LEMEUR, M., LATHE, R., GERLINGER, P. (1991), Precocious mammary gland development and milk protein synthesis in transgenic mice ubiquitously expressing human growth hormone, *Endocrinology* **128** (1), 539–546.

BEDDINGTON, R. S. P., MARTIN, P. (1989), An *in situ* transgenic enzyme marker to monitor migration of cells in the mid-gestation mouse embryo, *Mol. Biol. Med.* **6**, 263–274.

BEDDINGTON, R. S. P., MORGENSTERN, J., LAND, H., HOGAN A. (1989), An *in situ* transgenic enzyme marker for the midgestation mouse embryo and the visualization of inner cell mass clones during early organogenesis, *Development* **106**, 37–46.

BERG, U., BREM, G. (1989), *In vitro* production of bovine blastocysts by *in vitro* maturation and fertilization of oocytes and subsequent *in vitro* culture, *Zuchthygiene* **24**, 134–139.

BIERY, K. A., BONDIOLI, K. R., DE MAYO, F. J. (1988), Gene transfer by pronuclear injection in the bovine, *Theriogenology* **29**, 224 (abstract).

BIRNSTIEL, M. L., CHIPCHASE, M. (1977), Current work on the histone operon, *Trends Biochem. Sci.* **2**, 149–152.

BLÜTHMANN, H. (1991), Analysis of the immune system with transgenic mice: T cell development, *Experientia* **47**, 884–890.

BONIFER, C., VIDAL, M., GROSVELD, F., SIPPL, A. (1990), Tissue specific and position independent expression of the complete gene domain for chicken lysozyme in transgenic mice; *EMBO J.* **9**, 2843–2848.

BOSSELMAN, R. A., HSU, R.-Y., BOGGS, T., HU, S., BRUSZEWSKI, J., OU, S., KOZAR, L., MARTIN, F., GREEN, C., JACOBSEN, F., NICHOLSON, M., SCHULTZ, J. A., SEMON, K. M., RISHELL, W., STEWART, R. G. (1989 a), Germline transmission of exogenous genes in chickens, *Science* **243**, 533–535.

Bosselman, R. A., Hsu, R.-Y., Boggs, T., Hu, S., Bruszewski, J., Ou, S., Souza, L., Kozar, L., Martin, F., Nicolson, M., Rishell, W., Schultz, J. A., Semon, K. M., Stewart, R. G. (1989b), Replication-defective vectors of reticuloendotheliosis virus transduce exogenous genes into somatic stem cells of the unincubated chicken embryo, *J. Virol.* **63**, 2680–2689.

Bowerman, B., Brown, P. O., Bishop, J. M., Varmus, H. E. (1989), A nuclear protein complex mediates the integration of retroviral DNA, *Genes Dev.* **3**, 469–478.

Brackett, B. G., Baranska, W., Sawicki, W., Koprowski, H. (1971), Uptake of heterologous genome by mammalian spermatozoa and its transfer to ova through fertilization, *Proc. Natl. Acad. Sci. USA* **68**, 353–357.

Bradley, A. (1990), Embryonic stem cells: Proliferation and differentiation, *Cell Biol.* **2**, 1013–1017.

Bradley, A., Robertson, E. J. (1986), Embryonic stem cells: a tool for elucidating the development genetics of the mouse, in: *Current Topics in Developmental Biology* (Okada, T. S., Moscona, A. A., Eds.), Vol. 2, New York: Academic Press.

Brazolot, C. L., Petitte, J. N., Clark, M. E., Etches, R. J., Verrinder-Gibbins, A. M. (1990), Introduction of lipofected chicken blastodermal cells into the early chicken embryo, *J. Cell. Biochem. Suppl.* **15E**, 200.

Breitman, M. L., Clapoff, S., Rossant, J., Tsui, L.-C., Glode, M., Maxwell, I. H., Bernstein, A. (1987), Genetic ablation: targeted expression of a toxin gene causes microphthalmia in transgenic mice, *Science* **238**, 1563–1565.

Brem, G. (1986), *Mikromanipulation an Rinderembryonen und deren Anwendungsmöglichkeiten in der Tierzucht*, Stuttgart: Enke.

Brem, G. (1989), Aspects of the application of gene transfer as a breeding technique for farm animals, *Biol. Zentralbl.* **108**, 1–8.

Brem, G. (1990), Transgene Mäuse als Krankheitsmodelle, *Arzneim. Forsch./Drug Res.* **40** (3), 335–343.

Brem, G., Hartl, P. (1991), High-level expression of prochymosin in the milk of transgenic rabbits, in: *Frontiers of Biotechnology in Agriculture*, Conference 1–4 August, Sea of Galilee, Israel.

Brem, G., Wanke, R. (1988), Phenotypic and patho-morphological characteristics in a half-sib-family of transgenic mice carrying foreign MT-hGH genes, in: *New Developments in Biosciences; Their Implications for Laboratory Animal Science*, (Beynen, A. C., Solleveld, H. A., Eds.), pp. 93–98, Dordrecht: Martinus Nijhoff Publishers.

Brem, G., Weidle, U. (1991), Production of proteins with antibody activity in transgenic animals, in: *Frontiers of Biotechnology in Agriculture*, Conference 1–4 August, 1991, Sea of Galilee, Israel.

Brem, G., Brenig, B., Goodman, H. M., Selden, R. C., Graf, F., Kruff, B., Springmann, K., Hondele, J., Meyer, J. Winnacker, E.-L., Kräußlich, H. (1985), Production of transgenic mice, rabbits and pig by microinjection into pronuclei, *Zuchthygiene* **20**, 251–252.

Brem, G., Brenig, B., Hörstgen-Schwark, G., Winnacker, E.-L. (1988), Gene transfer in tilapia (*Oreochromis niloticus*), *Aquaculture* **68**, 209–219.

Brem, G., Wanke, R., Wolf, E., Buchmüller, T., Müller, M., Brenig, B., Hermanns, W. (1989), Multiple consequences of human growth hormone expression in transgenic mice. *Mol. Biol. Med.* **6**, 531–547.

Brem, G., Baunack, E., Müller, M., Winnacker, E.-L. (1990a), Transgenic offspring by transcaryotic implantation of transgenic ovaries into normal mice, *Mol. Reprod. Dev.* **25**, 42–44.

Brem, G., Springmann, K., Meier, e., Kräußlich, H., Brenig, B., Müller, M., Winnacker, E.-L. (1990b), Factors in the success of transgenic pig programs, in: *Transgenic Models in Medicine and Agriculture*, (Church, R. B., Ed.), pp. 61–72, New York: Wiley-Liss.

Brem, G., Brenig, B., Salmons, B., Wolf, E., Müller, M., Erfle, V., Günzburg, H. W., Dahme, E. (1991a), Unerwartete transgene Expression eines gesäugespezifischen Wachstumshormon-Genkonstruktes in den Bergmann-Gliazellen der Maus, *Tierärztl. Prax.* **19**, 1–6.

Brem, G., Wolf, E., Besenfelder, U., Reichenbach, H., Voss, W., Hofmeister, B., Brenig, B., Müller, M., Winnacker, E.-L. (1991b), Expression of hGH into the milk of transgenic rabbits and pigs by the WAP-promoter, unpublished results.

Bremel, R. D., Yom, H. C., Blech, G. T. (1989), Alteration of milk composition using molecular genetics, *J. Dairy Sci.* **72**, 2826–2833.

Brenig, B. (1987), Klonierung mikroinjizierbarer Genkonstrukte und deren Nachweis in transgenen Tieren – Ein Beitrag zur Nutzung transgener Tiere in der Tierzucht, *Dissertation med. vet.*, Universität München.

Brenig, B., Brem, G. (1988), Integration of hGH gene in transgenic mice and transmission to next generation, in: *New Developments in Biosciences: Their Implications for Laboratory Ani-*

mals Science, (BEYNEN, A. C., SOLLEVELD, H. A., Eds.), pp. 331–336.

BRENIG, B., BREM G. (1989), Aufbau eukaryotischer Gene, *Tierärztl. Prax. Suppl.* **4**, 26–30.

BRENIG, B., MÜLLER, M., BREM G. (1989), A fast detection protocol for screening large numbers of transgenic animals, *Nucleic Acids Res.* **17**, 6422.

BRINSTER, R. L. (1974), The effect of cells transferred into the mouse blastocyst on subsequent development, *J. Exp. Med.* **140**, 1049–1056.

BRINSTER, R. L., CHEN, H. Y., TRUMBAUER, M. E., AVARBOCK, M. R. (1980), Translation of globin messenger RNA by the mouse ovum, *Nature* **283**, 499–501.

BRINSTER, R. L., CHEN, H. Y., TRUMBAUER, M., SENEAR, A. W., WARREN, R., PALMITER, R. D. (1981), Somatic expression of herpes thymidine kinase in mice following injection of a fusion gene into eggs, *Cell* **27**, 223–231.

BRINSTER, R. L., CHEN, H. Y., WARREN, R., SARTHY, A., PALMITER, R. D. (1982), Regulation of metallothionein-thymidine kinase fusion plasmids injected into mouse eggs, *Nature* **296**, 39–43.

BRINSTER, L. B., CHEN, H. Y., MESSING, A., VAN DYKE, T., LEVINE, A. J., PALMITER, R. D. (1984), Transgenic mice harboring SV40 T-antigen genes develop characteristic brain tumors, *Cell* **37**, 367–379.

BRINSTER, R. L., CHEN, H. Y., TRUMBAUER, M. E., YAGLE, M. K., PALMITER, R. D. (1985), Factors affecting the efficiency of introducing foreign DNA into mice by microinjecting eggs, *Proc. Natl. Acad. Sci. USA* **82**, 4438–4442.

BRINSTER, R. L., ALLEN, J. M., BEHRINGER, R. R., GELINAS, R. E., PALMITER, R. D. (1988), Introns increase transcriptional efficiency in transgenic mice, *Proc. Natl. Acad. Sci. USA* **85**, 836–840.

BRINSTER, R. L., BRAUN, R. E., LO, D., AVARBOCK, M., ORAM, F., PALMITER, R. D. (1989a), Targeted correction of a major histocompatibility class II Eα gene by DNA microinjected into mouse eggs, *Proc. Natl. Acad. Sci. USA* **86**, 7087–7091.

BRINSTER, R. L., SANDGREN, E. P., BEHRINGER, R. R., PALMITER, R. D. (1989b), No simple solution for making transgenic mice, *Cell* **59**, 239–241.

BRISKIN, J. J., HSU, R.-Y., BOGGS, T., SCHULTZ, J. A., RISHELL, W., BOSSELMAN, R. A. (1991), Heritable retroviral transgenes are highly expressed in chickens, *Proc. Natl. Acad. Sci. USA* **88**, 1736–1740.

BROWN, P. O., BOWERMAN, B., VARMUS, H. E., BISHOP, J. M. (1987), Correct integration of retroviral DNA *in vitro, Cell* **49**, 347–356.

BÜHLER, T. A., BRUY'RE, T., WENT, D. F., STRANZINGER, G., BÜRKI, K. (1990), Rabbit β-casein promoter directs secretion of human interleukin-2 into the milk of transgenic rabbits, *Biotechnology* **8**, 140–143.

BÜRKI, K., ULLRICH, A. (1982), Transplantation of the human insulin gene into fertilized mouse egg, *EMBO* **1**, 127–131.

BUMSTEAD, N., MESSER, L. I., FREEMAN, B. M., MANNING, A. C. C. (1987), Genetic transformation of chickens using irradiated male gametes, *Heredity* **58**, 25–30.

BUONO, J. R., LINSER, P. J. (1991), Transgenic animals by electroporation, *Bio-Rad* **79**, 1–2.

BURDON, T., SANKARAN, L., WALL, R. J., SPENCER, M., HENNIGHAUSEN, L. (1991), Expression of a whey acidic protein transgene during mammary development, *J. Biol. Chem.* **266** (11), 6909–6914.

BUSHMAN, F. D., CRAIGIE, R. (1990), Sequence requirements for integration of Moloney murine leukemia virus DNA *in vitro, J. Virol.* **64**, 5645–5648.

CAMPBELL, S. M., ROSEN, J. M. (1984), Comparison of the whey acidic protein genes of the rat and mouse, *Nucleic Acids Res.* **12** (22), 8685–8697.

CAPECCHI, M. R. (1989a), Altering the genome by homologous recombination, *Science* **244**, 1288–1292.

CAPECCHI, M. R. (1989b), The new mouse genetics: Altering the genome by gene targeting, *Trends Genet.* **5**, 70–76.

CASTRO, F. O., HERNANDEZ, O., ULIVER, C., SOLANO, R., MILANES, C., AQUILAR, A., PEREZ, A., DE ARMAS, R., HERRERA, L., DE LA FUENTE, J. (1990), Introduction of foreign DNA into the spermatozoa of farm animals, *Theriogenology* **34**, 1099–1110.

CHEN, T. T., POWERS, D. A. (1990), Transgenic fish, *Trends Biotechnol.* **8**, 209–215.

CHEN, H. Y., TRUMBAUER, M. E., EBERT, K. M., PALMITER, R. D., BRINSTER, R. L. (1986), Developmental changes in the response of mouse eggs to injected genes, in: *Molecular Development of Biology*, (BOGORAD, L., Ed.), pp. 145–159, 43rd Symp. Soc. Biol. New York.

CHEN, T. T., ZHU, Z., LIN, C. M., GONZALEZ-VILLASENOR, L. I., DUNHAM, R., POWERS, D. A. (1989), Fish genetic engineering: a novel approach in aquaculture, in: *Proc. National Shellfish Association Aquaculture*, Los Angeles.

CHEN, H. Y., GARBER, E. A., MILLS, E., SMITH, J., KOPCHIK, J. J., DILELLA, A. G., SMITH, R. G. (1990a), Vectors, promoters, and expression of genes in chick embryos, *J. Reprod. Fert. Suppl.* **41**, 173–182.

CHEN, T. T., LIN, C. M., ZHU, Z., GONZALEZ-VILLASENOR, L. I., DUNHAM, R. A., POWERS, D. A. (1990b), Gene transfer, expression and inheritance of rainbow trout growth hormone genes in carp and loach, in: *UCLA Symposium on Transgenic Models in Medicine and Agriculture*, pp. 127–139, New York: Wiley-Liss.

CHISAKA, O., CAPECCHI, M. R. (1991), Regionally restricted developmental defects resulting from targeted disruption of the mouse homebox gene hox-1.5, *Nature* **350**, 473–479.

CHISARI, F. V., PINKERT, C. A., MILICH, D. R., FILIPPI, P., MCLACHLAN, A., PALMITER, R. D., BRINSTER, R. L. (1985), A transgenic mouse model of the chronic hepatitis B surface antigen carrier state, *Science* **230**, 1157–1160.

CHISARI, F. V., KLOPCHIN, K., MORIYAMA, T., PASQUINELLI, C., DUNSFORD, H. A., SELL, S., PINKERT, C. A., BRINSTER, R. L., PALMITER, R. D. (1989), Molecular pathogenesis of hepatocellular carcinoma in hepatitis B virus transgenic mice, *Cell* **59**, 1145–1156.

CHOURROUT, D., GUYOMARD, R., HOUDEBINE, L. M. (1986), High efficiency gene transfer in rainbow trout (*Salmo gairdneri* Rich.) by microinjection into egg cytoplasm, *Aquaculture* **51**, 143–150.

CHOURROUT, D., GUYOMARD, R., LEROUX, C., POURRAIN, F., HOUDEBINE, L. M. (1988), Improvement of salmoids: selective breeding or genetic manipulations, *J. Cell. Biochem. Suppl.* **12B**, 188.

CHURCH, R. B. (1986), Embryo manipulation and gene transfer in domestic animals, *Tibtech* **5**, 13–19.

CLARK, A. J., ALI, S., ARCHIBALD, A. L., BESSOS, H., BROWN, P., HARRIS, S., MCCLENAGHAN, M., PROWSE, C., SIMONS, J. P., WHITELAW, C. B. A., WILMUT, I. (1989a), The molecular manipulation of milk composition, *Genome* **31**, 950–955.

CLARK, A. J., BESSOS, H., BISHOP, J. O., BROWN, P., HARRIS, S., LATHE, R., MCCLENAGHAN, M., PROWSE, C., SIMONS, J. P., WHITELAW, C. B. A., WILMUT, I. (1989b), Expression of human anti-hemophilic factor IX in the milk of transgenic sheep, *Biotechnology* **7**, 487–492.

CLARK, A. J., SIMONS, P., WILMUT, I., LATHE, R. (1987), Pharmaceuticals from transgenic livestock, *Tibtech* **5**, 20–24.

COMPERE, S. J., BALDACCI, P., JAENISCH, R. (1988), Oncogenes in transgenic mice, *Biochim. Biophys. Acta* **948**, 129–149.

COSGROVE, D., GRAY, D., DIERICH, A., KAUFMAN, J., LEMEUR, M., BENOIST, C., MATHIS, D. (1991), Mice lacking MHC class II molecules, *Cell* **66**, 1051–1066.

COSTANTINI, F., LACY, E. (1981), Introduction of a rabbit β-globin gene into the mouse germ line, *Nature* **294**, 92–94.

COSTANTINI, F. D., ROBERTS, S., EVANS, E. P., BURTENSHAW, M. D., LACY, E. (1984), Position effects and gene expression in the transgenic mouse, in: *Transfer and Expression of Eukaryotic Genes* (GINSBERG, H. S., VOGEL, H. J., Eds.), pp. 123–134, New York: Academic Press.

COVARRUBIAS, L., NISHIDA, Y., MINTZ, B. (1986), Early postimplantation embryo lethality due to DNA rearrangements in a transgenic mouse strain, *Proc. Natl. Acad. Sci. USA* **83**, 6020–6024.

CRITTENDEN, L. B., SALTER, D. W. (1990), Expression and mobility of retroviral inserts in the chicken germ line, in: *Transgenic Models in Medicine and Agriculture* (CHURCH, R. B., Ed.), pp. 73–87, UCLA Symposia on Molecular and Cellular Biology, New Series, New York: Wiley-Liss.

CURIEL, D. T., AGARWAL, S., WAGNER, E., COTTEN, M. (1991), Adenovirus enhancement of transferrin-polylysine mediated gene delivery, *personal communication*.

CUTHBERTSON, R. A., KLINTWORTH, G. K. (1988), Biology of disease - transgenic mice - a gold mine for furthering knowledge in pathobiology, *Lab. Invest.* **58** (5), 484–502.

DAVIES, P. L., HEW, C. L., SHEARS, M. A., FLETCHER, G. L. (1989), Antifreeze protein expression in transgenic salmon, *J. Cell Biochem.*, Suppl. **13** B, 169.

DE LA FUENTE, J., CASTRO, F. O., HERNANDEZ, O., GUILLEN, I. ULLVER, C., SOLANO, R., MILANES, C., AGULLAR, A., LIEONART, R., MARTINEZ, R., PEREZ, A., DE ARMAS, R., HERRERA, L., LIMONTA, J., CABRERA, E., HERRERA, F. (1990), Sperm mediated foreign DNA transfer experiments in different species, *personal communication*.

DECHIARA, T. M., EFSTRATIADIS, A., ROBERTSON, E. J. (1990), A growth-deficiency phenotype in heterozygous mice carrying an insulin-like growth factor II gene disrupted by targeting, *Nature* **345**, 78–80.

DECHIARA, T. M., ROBERTSON, E. J., EFSTRATIADIS, A. (1991), Parental imprinting of the mouse insulin-like growth factor II gene, *Cell* **64**, 849–859.

DENT, L. A., STRATH, M., MELLOR, A. L., SANDERSON, C. J. (1990), Eosinophilia in transgenic mice expressing interleukin 5, *J. Exp. Med.* **172**, 1425–1431.

DEPAMPHILIS, M. L., HERMAN, S. A., MARTÍNEZ-SALAS, E., CHALIFOUR, L. E., WIRAK, D. O., CUPO, D. Y., MIRANDA, M. (1988), Microinjecting DNA into mouse ova to study DNA replication and gene expression and to produce transgenic animals, *BioTechniques* **6,** 662–680.

DOETSCHMAN, T., GREGG, R. G., MAEDA, N., HOOPER, M. L., MELTON, D. W., THOMPSON, S., SMITHIES, O. (1987), Targetted correction of a mutant HPRT gene in mouse embryonic stem cells, *Nature* **330,** 576–578.

DONEHOWER, L. A., HUANG, A. L., HAGER, G. L. (1981), Regulatory and coding potential of the mouse mammary tumor virus long terminal redundancy, *J. Virol.* **37,** 226–238.

DOUGHERTY, J. P., TEMIN, H. M. (1987), A promoterless retroviral vector indicates that there are sequences in U3 required for 3′RNA processing, *Proc. Natl. Acad. Sci. USA* **84,** 1197–1201.

DUNHAM, R. A., EASH, J., ASKINS, J., TOWNES, T. M. (1987), Transfer of metallothionein human growth hormone fusion gene into channel catfish, *Trans. Am. Fish. Soc.* **116,** 87–91.

DYNAN, W. S. (1986), Promoters for housekeeping genes, *Trends Genet.* **2,** 196–197.

DYNAN, W. S. (1989), Modularity in promoters and enhancers, *Cell* **58,** 1–4.

EBERT, K. M., LOW, M. J., OVERSTROM, E. W., BUONOMO, F. C., BAILE, C. A., ROBERTS, T. M., LEE, A., MANDEL, G., GOODMANN, R. H. (1988), A moloney MLV-rat somatotropin fusion gene produces biologically active somatotropin in transgenic pig, *Mol. Endocrinol.* **2,** 227–283.

EBERT, K. M., SELGRATH, J. P., DITULLIO, P., DENMAN, J., SMITH, T. E., MEMON, M. A., SCHINDLER, J. E., MONASTERSKY, G. M., VITALE, J. A., GORDON, K. (1991), Transgenic production of a variant of human tissue-type plasminogen activator in goat milk: generation of transgenic goats and analysis of expression, *Biotechnology* **9,** 835–838.

EGLITIS, M. A., ANDERSON, W. F. (1988), Retroviral vectors for introduction of genes into mammalian cells, *BioTechniques* **6,** 608–614.

ERNST, L. K., ZAKCHARCHENKO, V. I., SURAEVA, N. M., PONOMAREVA, T. I., MIROSHNICHENKO, O. I., PROKOF'EV, M. I., TIKCHONENKO, T. I. (1991), Transgenic rabbits with antisense RNA gene targeted at adenovirus H5, *Theriogenology* **35,** 1257–1271.

EVANS, M. J., KAUFMANN, M. H. (1981), Establishment in culture of pluripotential cells from mouse embryos, *Nature* **292,** 154–156.

EVANS, M., BRADLEY, A., ROBERTSON, E. J. (1985), EK contribution to chimaeric mice: from tissue culture to sperm, in: *Genetic Manipulation of the Mammalian Ovum and Early Embryos, Banbury Report*. Cold Spring Harbor, NY: Cold Spring Harbor Laboratory Press.

EVANS, M. J., NOTARIANNI, E., LAURIE, S., MOOR, R. M. (1990), Derivation and preliminary characterization of pluripotent cell lines from porcine and bovine blastocysts, *Theriogenology* **33,** 125–128.

FEDERSPIEL, M. J., CRITTENDEN, L. B., PROVENCHER, L. P., HUGHES, S. H. (1991), Experimentally introduced defective endogenous proviruses are highly expressed in chickens, *J. Virol.* **65,** 313–319.

FLETCHER, G. L., SHEARS, M. A., KING, M. J., DAVIES, P. L., HEW, C. L. (1988), Evidence for antifreeze protein gene transfer in Atlantic salmon (*Salmo salar*), *Can. J. Fish. Aquat. Sci.* **45,** 352–357.

FRENCH, D., CAMAIONI, A., ZANI, M., MARIANI-COSTANTINI, R., FRATI, L., SPADAFORA, C., LAVITRANO, M. (1990), Permeability of sperm cells to foreign DNA molecules, in: *Proc. International Symposium on Endocrinology*, Siena, Italy, 8–10 October.

FUJIWARA, T., CRAIGIE, R. (1989), Integration of mini retroviral DNA – a cellfree reaction for biochemical analysis of retroviral integration, *Proc. Natl. Acad. Sci. USA* **86,** 3065–3069.

FUJIWARA, T., MIZUUCHI, K. (1988), Retroviral DNA integration: structure of an integration intermediate, *Cell* **54,** 497–504.

GAGNE, M. B., POTHIER, F., SIRARD, M.-A. (1991), Electroporation of bovine spermatozoa to carry foreign DNA in oocytes, *Mol. Reprod. Dev.* **29,** 6–15.

GANDOLFI, F., LAVITRANO, M., CAMAIONI, A., SPADAFORA, C., SIRACUSA, G., LAURIA, A. (1989), The use of sperm-mediated gene transfer for the generation of transgenic pigs, *J. Reprod. Fert. Abstr.* **4,** 21.

GIBSON, J. P. (1991), Using information on individual genes. Dept. Animal and Poultry Science. Polycopy.

GONZÁLEZ-VILLASENOR, L. I., ZHANG, P., CHEN, T. T., POWERS, D. A. (1988), Molecular cloning and sequencing of coho salmon growth hormone cDNA, *Gene* **65,** 239–246.

GORDON, J. W., RUDDLE, R. H. (1981), Integration and stable germ line transmission of genes injected into mouse pronuclei, *Science* **214,** 1244–1246.

GORDON, J. W., RUDDLE, R. H. (1982), Germ line transmission in transgenic mice, in: *Embryonic Development, Part B: Cellular Aspects*, (BURGER, M. B., WEBER, R., Eds.), pp. 112–124, New York: Liss.

GORDON, J. W., RUDDLE, F. H. (1985), DNA-mediated genetic transformation of mouse embryos and bone marrow – a review, *Gene* **33,** 121–136.

GORDON, J. W., SCANGOS, G. A., PLOTKIN, D. J., BARBOSA, A., RUDDLE, F. H. (1980), Genetic transformation of mouse embryos by microinjection, *Proc. Natl. Acad. Sci. USA* **77**, 7380–7384.

GORDON, K., LEE, E., VITALE, J. A., SMITH, A. E., WESTPHAL, H., HENNIGHAUSEN, L. (1987), Production of human tissue plasminogen activator in transgenic mouse milk, *Biotechnology* **5**, 1183–1187.

GOSSLER, A., DOETSCHMAN, T., KORN, R., SERFLING, E., KEMLER, R. (1986), Transgenesis by means of blastocyst-derived embryonic stem cell lines, *Proc. Natl. Acad. Sci. USA* **83**, 9065–9069.

GRANDGENETT, D. P., MUMM, S. R. (1990), Unraveling retrovirus integration, *Cell* **60**, 3–4.

GRIDLEY, T., SORIANO, P., JAENISCH, R. (1987), Insertional mutagenesis in mice, *Trends Genet.* **3**, 162–167.

GROSS, M. L., SCHNEIDER, J. F., MOAV, N., ALVAREZ, C., MYSTER, S. H., LIU, Z., HALLRMAN, E. M., HACKETT, P. B., GUISE, K. S., FARAS, A. J., KAPUSCINSKI, A. R. (1991), Molecular analysis and growth evaluation of northern pike (*Esox lucius*) microinjected with growth hormone genes, *personal communication*.

GROSVELD, F., VAN ASSENDELFT, G. B., GREAVES, D. R., KOLLIAS, G. (1987), Position-independent, high-level expression of the human beta globin gene in transgenic mice, *Cell* **5**, 975–985.

GRUENBAUM, Y., REVEL, E., YARUS, S., FAINSOD, A. (1991a), Sperm cells as vectors for the generation of transgenic chickens, *J. Cell. Biochem. Suppl.* **15E**, 194.

GRUENBAUM, Y., REVEL, E., YARUS, S., FAINSOD, A. (1991b), Semen preservation and sperm cells as vectors for the generation of transgenic chickens. in: *Frontiers of Biotechnology in Agriculture*, Conference 1–4 August, 1991, Sea of Galilee, Israel.

GRUSBY, M. J., JOHNSON, R. S., PAPAIOANNOU, V. E., GLIMCHER, L. H. (1991), Depletion of CD4+ T cells in major histocompatibility complex class II-deficient mice, *Science* **253**, 1417–1420.

GUIDO, T. C., ABBOTT, U. K., MCCARREY, J. R. (1991), The interspecific transfer of avian primordial germ cells, *J. Cell. Biochem. Suppl.* **15E**, 204.

GÜNZBURG, W. H., SALMONS, B., ZIMMERMANN, B., MÜLLER, M., ERFLE, V., BREM, G. (1991), A mammary-specific promoter directs expression of growth hormone not only to the mammary gland, but also to Bergman glia cells in transgenic mice, *Mol. Endocrinol.* **5**, 123–133.

GURDON, G. B., MELTON, D. A. (1981), Gene transfer in amphibian eggs and oocytes, *Annu. Rev. Genet.* **15**, 180–218.

GURDON, G. B., WOODLAND, H. R., LINGREL, J. B. (1974), The translation of mammalian globin mRNA injected into fertilized eggs of *Xenopus laevis, Dev. Biol.* **39**, 125–133.

GUYOMARD, R., CHOURROUT, D., LEROUX, C., HOUDEBINE, L. M., POURRAIN, F. (1989), Integration and germ line transmission of foreign genes microinjected into fertilized trout eggs, *Biochemie* **71**, 857–883.

HALLERMAN, E. M., KAPUSCINSKI, A. R. (1990a), Transgenic fish and public policy: regulatory concerns, *Fisheries* **15** (1), 12–20.

HALLERMAN, E. M., KAPUSCINSKI, A. R. (1990b), Transgenic fish and public policy: Patenting of transgenic fish, *Fisheries* **15** (1), 21-24.

HALLERMAN, E. M., KAPUSCINSKI, A. R. (1991), Ecological implications of using transgenic fishes in aquaculture, in: *Proc. Int. Symp. on the Effects of Introductions and Transfers of Aquatic Species on Resources and Ecosystems*, (SINOERMANN, C. J., STEINMETZ, B., HERSCHBERGER, W. K., Eds.), Halifax: Nova Scotia.

HALLERMAN, E. M., SCHNEIDER, J. F., GROSS, M. L., FARAS, A. J., HACKETT, P. B., GUISE, K. S., KAPUSCINSKI, A. R. (1988), Enzymatic dechorionation of goldfish, walleye, and Northern pike eggs, *Trans. Am. Fish. Soc.* **117**, 456–460.

HALLERMAN, E. M., SCHNEIDER, J. F., GROSS, M., LIU, Z., YOON, S. J., HE, L., HACKETT, P. B., FARAS, A. J., KAPUSCINSKI, A. R., GUISE, K. S. (1990), Gene expression promoted by the RSV long terminal repeat element in transgenic goldfish, *Anim. Biotechnol.* **1**, 79–93.

HAMMER, R. E., PURSEL, V. G., REXROAD, C. E., WALL, R. J., BOLT, D. J., EBERT, K. M., PALMITER, R. D., BRINSTER, R. L. (1985), Production of transgenic rabbits, sheep and pigs by microinjection, *Nature* **315**, 680–683.

HAMMER, R. E., PURSEL, V. G., REXROAD, C. E., WALL, R. J., BOLT, D. J., PALMITER, R. D., BRINSTER, R. L. (1986), Genetic engineering of mammalian embryos, *J. Anim. Sci.* **63**, 269–278.

HANAHAN, D. (1985), Heritable formation of pancreatic β-cell tumors in transgenic mice expressing recombinant insulin/simian virus 40 oncogenes, *Nature* **315**, 115–122.

HANSCOMBE, O., WHYATT, D., FRASER, P., YANNOUTSOS, N., GREAVES, D., GROSVELD, F. (1991), Globin gene order is important for correct developmental expression, *personal communication*.

HARBERS, K., KUEHN, M., DELIUS, H., JAENISCH, R. (1984), Insertion of retrovirus into the first intron of an α collagen gene leads to embryonic lethal mutation in mice, *Proc. Natl. Acad. Sci. USA* **81**, 1504–1508.

HARTL, P., BREM, G. (1992), Production of human IGF-I and the analogue (Glu 58) IGF-I in the mammary gland of transgenic rabbits, submitted.

HARVEY, M. J. A., HETTLE, S. J. H., CAMERON, E. R., JOHNSTON, C. S., ONIONS, D. E. (1990), Production of transgenic lamb foetuses by subzonal injection of feline leukaemia virus, in: *Transgenic Models in Medicine and Agriculture*, UCLA Symposia on Molecular and Cellular Biology New Series, pp. 11–19, New York: Wiley-Liss.

HENNIGHAUSEN, L. (1990), The mammary gland as a bioreactor: production of foreign proteins in milk, *Protein Expression Purif.* **1**, 3–8.

HIGHKIN, M. K., KRIVI, G. G., HIPPENMEYER, P. J. (1991), Characterization and comparison of avian and murine helper cell lines for production of replication-defective retroviruses for avian transformation, *Poultry Sci.* **70**, 970–981.

HINRICHS, S. H., NERENBERG, M., REYNOLDS, R. K., KHOURY, G., JAY, A. (1987), A transgenic mouse model for human neurofibromatosis, *Science* **237**, 1340–1343.

HIPPENMEYER, P. J., KRIVI, G. G. (1991), Gene expression from heterologous promoters in a replication-defective avian retrovirus vector in quail cells, *Poultry Sci.* **70**, 982–992.

HIPPENMEYER, P. J., KRIVI, G. G., HIGHKIN, M. K. (1988), Transfer and expression of the bacterial NPT-II gene in chick embryos using a Schmidt-Ruppin retrovirus vector, *Nucleic Acids Res.* **16** (15), 7619–7632.

HOCHI, S.-I., NINOMIYA, T., MIZUNO, A., HONMA, M., YUKI, A. (1990), Fate of exogenous DNA carried into mouse eggs by spermatozoa, *Anim. Biotechnol.* **1**, 21–31.

HOGAN, B., COSTANTINI, F., LACY, E. (1986), Manipulation of the mouse embryo, Cold Spring Harbor, NY: Cold Spring Harbor Laboratory Press.

HONG, Y., WINKLER, C., BREM, G., SCHARTL, M. (1991), Development of a heavy metal-inducible fish specific expression vector for gene transfer *in vitro* and *in vivo*, in: *Proc. Aquaculture Meeting*, Wuhan, China.

HOOPER, M., HARDY, K., HANDYSIDE, A. HUNTER, S., MONK, M. (1987), HPRT-deficient (Lesch-Nyhan) mouse embryos derived from germline colonization by cultured cells, *Nature* **326**, 292–295.

HORAN, R., POWELL, R., MCQUAID, S., GANNON, F., HOUGHTON, J. A. (1991), Association of foreign DNA with porcine spermatozoa, *Arch. Androl.* **26**, 83–92.

HOUDEBINE, L. M., CHOURROUT, D. (1991), Transgenesis in fish, *Experientia* **47** (9), 891–897.

HSIAO, K. K., SCOTT, M., FOSTER, D., GROTH, D. F., DEARMOND, S. J., PRUSINER, S. B. (1990), Spontaneous neurodegeneration in transgenic mice with mutant prion protein, *Science* **250**, 1587–1590.

HUSZAR, D., BALLING, R., KOTHARY, R., MAGLI, M. C., HOZUMI, N., ROSSANT, J., BERNSTEIN, A. (1985), Insertion of a bacterial gene into the mouse germ line using an infections retrovirus vector, *Proc. Natl. Acad. Sci. USA* **82**, 8587–8591.

IGLESIAS, A. (1991), Analysis of the immune system with transgenic mice: B cell development and lymphokines, *Experientia* **47**, 878–884.

ILLMENSEE, K., MINTZ, B. (1976), Totipotency and normal differentiation of single teratocarcinoma cells cloned by injection into blastocysts, *Proc. Natl. Acad. Sci. USA* **73**, 549–553.

INGOLD, A. L., LANDEL, C., KNALL, C., EVANS, G. A., POTTER, T. A. (1991), Co-engagement of CD8 with the T cell receptor is required for negative selection, *Nature* **352**, 721–723.

INOUE, K. (1988), Antisense RNA: its function and applications in gene regulation - a review, *Gene* **72**, 25–34.

INOUE, K., OZATO, K., KONDOH, H., IWAMATSU, T., WAKAMATSU, Y., FUJITA, T., OKADA, T. S. (1989), Stage-dependent expression of the chicken-crystalline gene in transgenic fish embryo, *Cell Differ.* **27**, 57–68.

INOUE, K., YAMASHITA, S., HATA, J. I., KABENO, S., ASADA, S., NAGAHISA, E., FRUJITA, T. (1990), Electroporation as a new technique for producing transgenic fish, *Cell Differ.* **29**, 123–128.

ISHIDA, Y. (1989a), Effects of the deregulated expression of human interleukin 2 in transgenic mice, *Int. Immunol.* **1**, 113–120.

ISHIDA, Y. (1989b), Expansion of natural killer cells but not T cells in human interleukin 2/interleukin 2 receptor (Tac) transgenic mice, *J. Exp. Med.* **170**, 1103–1115.

IWAKURA, Y., TOSU, M., YOSHIDA, E., TAKIGUCHI, M., SATO, K., KITAJIMA, I., NISHIOKA, K., YAMAMOTO, K., TAKEDA, T., HATANAKA, M., YAMAMOTO, H., SEKIGUCHI, T. (1991), Induction of inflammatory arthropathy resembling rheumatoid arthritis in mice transgenic for HTLV-I, *Science* **253**, 1026–1028.

JAENISCH, R. (1974), Infection of mouse blastocysts with SV 40 DNA: Normal development of infected embryos and persistance SV 40-specific DNA sequences in the adult animals, *Cold Spring Harbor Symp.* **39**, 375–380.

JAENISCH, R. (1976), Germ line integration and Mendelian transmission of the exogenous Moloney leukemia virus, *Proc. Natl. Acad. Sci. USA* **73,** 1260.

JAENISCH, R. (1988), Transgenic animals, *Science* **240,** 1468–1474.

JAENISCH, R., MINTZ, B. (1974), Simian virus 40 DNA sequences in DNA of healthy adult mice derived from preimplantation blastocysts injected with viral DNA, *Proc. Natl. Acad. Sci. USA* **71,** 1250–1254.

JAENISCH, R., FAN, H., CROKER, B. (1975), Infection of preimplantation mouse embryos and of newborn mice with leukemia virus: tissue distribution of viral DNA and RNA leukemogenesis in the adult animal, *Proc. Natl. Acad. Sci. USA* **72,** 4008–4012.

JAENISCH, R., HARBERS, K., SCHNIEKE, A., LÖHLER, J., CHUMAKOV, I., JÄHNER, D., GROTKOPP, D., HOFFMANN, E. (1983), Germline integration of Moloney murine leukemia virus at the Mov13 locus leads to recessive mutation and early embryonic death, *Cell* **32,** 209–216.

JHAPPAN, C., STAHLE, C., HARKINS, R. N., FAUSTO, N., SMITH, G. H., MERLINO, G. T. (1990), TGFα overexpression in transgenic mice induces liver neoplasia and abnormal development of the mammary gland and pancreas, *Cell* **61,** 1137–1146.

JIN, D. I., PETTERS, R. M., JOHNSON, B. H., SHUMAN, R. M. (1991), Survival of early preimplantation porcine embryos after co-culture with cells producing an avian retrovirus, *Theriogenology* **35** (3), 521–526.

JOYNER, A. L., SKARNES, W. C., ROSSANT, J. (1989), Production of a mutation in mouse En-2 gene by homologous recombination in embryonic stem cells, *Nature* **338,** 153–156.

KAPUSCINSKI, A. R. (1990), Integration of transgenic fish into aquaculture, *Food Rev. Int.* **6,** 373–388.

KAPUSCINSKI, A. R., HALLERMAN, E. M. (1990a), Transgenic fish and public policy: Anticipating environmental impacts of transgenic fish, *Fisheries* **15** (1), 2–11.

KAPUSCINSKI, A. R., HALLERMAN, E. M. (1990b), Transgenic fishes, *Fisheries* **15** (4), 2–5.

KIM, J. W., CUNNINGHAM, J. M. (1991), The mouse ecotropic retrovirus receptor is a transporter of cationic amino acids, *RNA Tumor Virus Meeting* **5,** 5.

KING, D., WALL, R. J. (1988), Identification of specific gene sequences in preimplantation embryos by genome amplification: detection of a transgene, *Mol. Reprod. Dev.* **1,** 57–62.

KISHIMOTO, T. (1989), The biology of interleukin-6. *Blood* **74,** 1–10.

KNIGHT, K. L., SPIEKER-POLET, H., KAZDIN, D. S., OI, V. T. (1988), Transgenic rabbits with lymphocytic leukemia induced by the c-myc oncogene fused with the immunoglobulin heavy chain enhancer, *Proc. Natl. Acad. Sci. USA* **85,** 3130–3134.

KOLLER, B. H., HAGEMANN, L. J., DOETSCHMAN, T. (1989), Germ-line transmission of a planned alteration made in a hypoxanthine phosphoribosyltransferase gene by homologous recombination in embryonic stem cells, *Proc. Natl. Acad. Sci. USA* **86,** 8927–8931.

KOOPMAN, P., POVEY, S., LOVELL-BADGE, R. H. (1989), Widespread expression of human α_1-antitrypsin in transgenic mice revealed by *in situ* hybridization, *Genes Dev.* **3,** 16–25.

KRIEGLER, M., BOTCHAN, M. (1983), Enhanced transformation by a simian virus 40 recombinant containing a Harvey murine sarcoma virus long-terminal repeat, *Mol. Cell. Biol.* **3,** 325–339.

KRIMPENFORT, P., OSSENDORP, F., BORST, J. MELIEF, C., BERNS, A. (1989), T cell depletion in transgenic mice carrying a mutant gene for TCR-β, *Nature* **341,** 742–746.

KRIMPENFORT, P., RADEMAKERS, A., EYESTONE, W., VAN DER SCHANS, A., VAN DEN BROEK, S., KOOIMAN, P., KOOTWIJK, E., PLATENBURG, G., PIEPER, F., STRIJKER, R., DER BOER, H. (1991), Generation of transgenic dairy cattle using "*in vitro*" embryo production, *Biotechnology* **9,** 844–847.

KRUG, R. M., SHAW, M., BRONI, B., SHAPIRO, G., HALLER, O. (1985), Inhibition of influenza viral mRNA synthesis in cells expressing the interferon-induced Mx gene product, *J. Virol.* **56,** 201–206.

KUEHN, M. R., BRADLEY, A., ROBERTSON, E. J., EVANS, M. J. (1987), A potential animal model for Lesch-Nyhan syndrome through introduction of HPRT mutations into mice, *Nature* **326,** 295–298.

LACEY, M., ALPERT, S., HANAHAN, D. (1986), Bovine papillomavirus genome elicits skin tumours in transgenic mice, *Nature* **322,** 609–612.

LANG, R., METCALF, D., CUTHBERTSON, R. A., LYONS, I., STANLEY, E., KELSO, A., KANNOURAKIS, G., WILLIAMSON, D. J., KLINTWORTH, G. K., GONDA, T. J., DUNN, A. R. (1987), Transgenic mice expressing a hemopoietic growth factor gene (GM-CSF) develop accumulations of macrophages, blindness, and a fatal syndrome of tissue damage, *Cell* **51,** 675–686.

LANGER, P. R., WALDROP, A. A., WARD, D. C. (1981), Enzymatic synthesis of biotin-labelled polynucleotides. Novel nucleic acid affinity probes, *Proc. Natl. Acad. Sci. USA* **78,** 6633–6637.

LAVITRANO, M., CAMAIONI, A., FAZIO, V. M., DOLCI, S., FARACE, M. G., SPADAFORA, C. (1989), Sperm cells as vectors for introducing foreign DNA into eggs: genetic transformation of mice, *Cell* **57**, 717–723.

LAVITRANO, M., FRENCH, D., CAMAIONI, A., ZANI, M., MARIANI-CONSTANTINI, R., FRATI, L., SPADAFORA, C. (1990), Uptake of foreign molecules by sperm cells. Factors affecting sperm permeability, in: *VI. International Congress on Spermatology, Comparative Spermatology, 20 years after*, Siena, Italy, August 30–September 1, 1990, in press.

LAVITRANO, M., FRENCH, D., CAMAIONI, A., ZANI, M., FRATI, L., SPADAFORA, C. (1991a), The interaction between exogenous DNA and sperm cells, *Mol. Reprod. Dev.*, in press.

LAVITRANO, M., LULLI, V., MAIONE, B., SPERANDIO, S., FRENCH, D., FRATI, L., FRANCOLINI, M., LORA LAMIA, C., COTELLI, F., SPADAFORA, C. (1991b), The interaction between sperm cells and exogenous DNA: factors controlling DNA uptake, in: *Workshop on Animal Models for Duchenne Muscula Dystrophy and Genetic Manipulation*, August 7–9, 1991, Perth, Australia, in press.

LEE, K.-F., DEMAYO, F. J., ATIEE, S. H. (1988), Tissue-specific expression of the rat β-casein gene in transgenic mice, *Nucleic Acids Res.* **16**, 1027–1041.

LEE, M.-R., SHUMAN, R. M. (1987), Introduction of a bacterial gene into avian embryo by recombinant retrovirus vector, *Poultry Sci. Suppl.* **66**, 24.

LEE, M.-R., SHUMAN, R. M. (1990), Transgenic quail produced by retrovirus vector infection transmit and express a foreign marker gene, in: *Proc. 4th World Congress on Genetics Applied to Livestock Production* (HILL, W. G., THOMPSON, R., WOLLIAMS, J. A., Eds.), Edinburgh, Vol. XVI, pp. 107–110.

LEONARD, J. M., ABRAMCZUK, J. W., PEZEN, D. S., RUTLEDGE, R., BELCHER, J. H., HAKIM, F., SHEARER, G., LAMPERTH, L., TRAVIS, W., FREDRICKSON, T., NOTKINS, A. L., MARTIN, M. A. (1988), Development of disease and virus recovery in transgenic mice containing HIV proviral DNA, *Science* **242**, 1665–1670.

LIN, T. P. (1966), Microinjection of mouse eggs, *Science* **151**, 333–337.

LINDENMANN, J. (1962), Resistance of mice to mouse adapted influenza A virus, *Virology* **16**, 203–204.

LIU, Z., MOAV, B., FARAS, A. J., GUISE, K. S., KAPUSCINSKI, A. R., HACKETT, P. B. (1990a), Development of expression vectors for transgenic fish, *Bio/Technology* **8**, 1268–1272.

LIU, Z., MOAV, B., FARAS, A. J., GUISE, K. S., KAPUSCINSKI, A. R., HACKETT, P. B. (1990b), Functional analysis of elements affecting expression of the β-actin gene of carp, *Mol. Cell. Biol.* **10**, 3432–3440.

LIU, Z., ZHU, Z., ROBERG, K., FARAS, A., GUISE, K., KAPUSCINSKI, A. R., HACKETT, P. B. (1990c), Isolation and characterization of β-actin gene of carp (*Cyprinus carpio*), *J. DNA Sequencing Mapping* **1**, 125–136.

LO, D., PURSEL, V., LINTON, P. J., SANDGREN, E., BEHRINGER, R., REXROAD, C., PALMITER, R. D., BRINSTER, R. L. (1991), Expression of mouse IgA by transgenic mice, pigs and sheep, *Eur. J. Immunol.* **21**, 1001–1006.

LUFKIN, T., DIERICH, A., LEMEUR, M., MARK, M., CHAMBON, P. (1991), Disruption of the Hox-1.6 homeobox gene results in defects in a region corresponding to its rostral domain of expression, *Cell* **66**, 1105–1119.

MACLEAN, N., PENMAN, D., ZHU, Z. (1987), Introduction of novel genes into fish, *Biotechnology* **5**, 257–281.

MAHON, K. A., CHEPELINSKY, A. B., KHILLAN, J. S., OVERBREEK, P. A., PLATIGORSKY, J., WESTPHAL, H. (1987), Oncogenesis of the lens in transgenic mice, *Science* **235**, 1622–1628.

MAJORS, J. E., VARMUS, H. E. (1981), Nucleotide sequences at host proviral junctions for mouse mammary tumor virus, *Nature* **289**, 253–258.

MANSOUR, S. L. (1990), Gene targeting in murine embryonic stem cells: Introduction of specific alterations into the mammalian genome, GATA **7** (8), 219–227.

MANSOUR, S. L., THOMAS, K. R., CAPECCHI, M. R. (1988), Disruption of the proto-oncogene int-2 in mouse embryo-derived stem cells: a general strategy for targeting mutations to non-selectable genes, *Nature* **336**, 348–352.

MARTIN, G. R. (1981), Isolation of a pluripotent cell line from early mouse embryos cultured in medium conditioned by teratocarcinoma stem cells, *Proc. Natl. Acad. Sci. USA* **78**, 7634–7638.

MARX, J. L. (1985), Making mutant mice by gene transfer, *Science* **228**, 1516–1517.

MASCHIO, A., BRICKELL, P. M., KIOUSSIS, D., MELLOR, A. L., KATZ, D., CRAIG, R. K. (1991), Transgenic mice carrying the guinea-pig α-lactalbumin gene transcribe milk protein genes in their sebaceous glands during lactation, *Biochem. J.* **275**, 459–467.

MASSEY, J. M. (1990), Animal production industry in the year 2000, *J. Reprod. Fert. Suppl.* **41**, 199–208.

MAURER, R. R. (1978), Advances in rabbit embryo culture, in: *Methods in Mammalian Reproduc-*

tion (DANIEL, J. C., Jr., Ed.), pp. 259–272, New York, San Francisco, London: Academic Press.

MCCLEMENTS, W. L., ENQUIST, L. W., OSKARSSON, M., SULLIVAN, M., VANDE WOUDE, G. F. (1980), Frequent site-specific deletion of coliphage lambda murine sarcoma virus recombinants and its use in the identification of a retrovirus integration site, *J. Virol.* **35,** 488–497.

MCEVOY, T., STACK, M., KEANE, B., BARRY, T., SREENAN, J., GANNON, F. (1988), The expression of a foreign gene in salmon embryos, *Aquaculture* **68,** 27–37.

MCLACHLIN, J. R., EGLITIS, M. A., UEDA, K. KANTOFF, P. W., PASTAN, I. H., ANDERSON, W. F., GOTTESMAN, M. M. (1990), Expression of a human complementary DNA for the multidrug resistance gene in murine hematopoietic presursor cells with the use of retroviral gene transfer, *J. Natl. Cancer Inst. USA* **82,** 1260–1263.

MCLAUGHLAN, J., GAFFNEY, G., WHITTON, J. L., CLEMENTS, J. B. (1985), The consensus sequence YGTGTTYY located downstream from the AATAAA signal is required for efficient formation of mRNA 3′termini, *Nucleic Acids Res.* **13,** 1347–1368.

MCMAHON, A. P., BRADLEY, A. (1990), The Wnt-1 (int-1) proto-oncogene is required for development of a large region of the mouse brain, *Cell* **62,** 1073–1085.

MCNEISH, J. D., SCOTT, W. J., POTTER, S. S. (1988), Legless, a novel mutation found in PHT1-1 transgenic mice, *Science* **241,** 837–839.

MEADE, H., GATES, L., LACY, E., LONBERG, N. (1990), Bovine alphas1-casein gene sequences direct high level expression of active human urokinase in mouse milk, *Biotechnology* **8,** 443–446.

MELTON, D. A. (1988), Antisense RNA and DNA, in: *Current Commun. Molec. Biol.* (MELTON, D. A., Ed.), Cold Spring Harbor NY: Cold Spring Harbor Laboratory Press.

MERCIER, J. C. (1987), Genetic engineering applied to milk producing animals: some expectations, in: *Exploiting new Technologies in Animal Breeding* (SMITH, C., KING, J. W., MCKAY, J. C., Eds.), pp. 122–131, Oxford: Oxford University Press.

MEYER, T., HORISBERGER, M. A. (1984), Combined action of mouse and β interferons in influenza virus infected macrophages containing the resistance gene Mx, *J. Virol.* **49,** 709–716.

MILNE, JR., C. P., ELSCHEN, F. A., COLLIS, J. E., JENSEN, T. L. (1989), Preliminary evidence for honey bee sperm-mediated DNA transfer, in: *Symposium on Molecular Insect Science*, October 24, 1989, Tucson, Arizona.

MINTZ, B. (1977), Teratocarcinoma cells as vehicles for mutant and foreign genes, in: *Genetic Interaction and Genes Transfer* (ANDERSON, C. W., Ed.), p. 82, Brookhaven Symposia in Biology, Upton.

MINTZ, B., ILLMENSEE, K. (1975), Normal genetically mosaic mice produced from malignant teratocarcinoma cells, *Proc. Natl. Acad. Sci. USA* **72,** 3585–3589.

MIZUUCHI, K., CRAIGIE, R. (1991), Stereochemical course of the HIV DNA strand transfer reaction: evidence for a one step transesterification mechanism, RNA Tumor Virus Meeting **23,** 23.

MOAV, B., LIU, Z., MOAV, N. L., GROSS, M. L., KAPUSCINSKI, A. R., FARAS, A. J., GUISE, K. S., HACKETT, P. B. (1990), Expression of heterologous genes in transgenic fish, in: *Transgenic Fish* (HEW, C., FLETCHER, G. L., Eds.), Singapore: World Science Publishing.

MÖRITZ, H., ZEIDLER, R., LIPP, M., BREM, G.: Transgenic mice bearing either the translocated and mutated or the normal myc-allele from burkitt's lymphoma develop lymphoid malignancies with identical frequencies, in: *Proc. 4th FELASA Symposium Session III*, 144 (abstract).

MORIYAMA, T., GUILHOT, S., KLOPCHIN, K., MOSS, B., PINKERT, C. A., PALMITER, R. D., BINSTER, R. L., KANAGAWA, O., CHISARI, F. V. (1990), Immunobiology and pathogenesis of hepatocellular injury in hepatitis B virus transgenic mice, *Science* **248,** 361–364.

MÜLLER, M. (1989), Molekulargenetische Charakterisierung der porcinen Mx- und IGF-I-Gene und funktionelle Untersuchung transgener Mx-Schweine, *Dissertation med. vet.*, Universität München.

MÜLLER, M., BREM, G. (1991), Disease resistance in farm animals, *Experientia* **47,** 923–934.

MÜLLER, M., WINNACKER, E.-L., BREM, G. (1992a), Molecular cloning of porcine Mx cDNAs: new members of a family of interferon-inducible proteins with antiviral activities and homology to GTP-binding proteins, *J. Interferon Res.* **12,** 119–129.

MÜLLER, M., BRENIG, B., WINNACKER, E.-L., BREM, G. (1992b), Transgenic pigs carrying cDNA copies encoding the murine Mx1 protein which confers resistance to influenza virus infection, *Gene 99*, in press.

MULLER, W. J., SINN, E., PATTENGALE, P. K., WALLACE, R., LEDER, P. (1988), Single-step induction of mammary adenocarcinoma in transgenic mice bearing the activated c-neu oncogene, *Cell* **54,** 105–115.

MURRAY, J. D., NANCARROW, C. D., MARSHALL, J. T., HAZELTON, I. G., WARD, K. A. (1989), Production of transgenic merino sheep by microinjection of ovine metallothionein-ovine growth hormone fusion genes, *Reprod. Fertil. Dev.* **1,** 147–155.

NABEL, E. G., PLAUTZ, G., BOYCE, F. M., STANLEY, J. C., NABEL, G. J. (1989), Recombinant gene expression *in vivo* within endothelial cells of the arterial wall, *Science* **244,** 1342–1344.

NANCARROW, C. D., MARSHALL, J. T. A., CLARKSON, J. L., MURRAY, J. D., MILLARD, R. M., SHANAHAN, C. M., WYNN, P. C., WARD, K. A. (1991), Expression and physiology of performance regulating genes in transgenic sheep, *J. Reprod. Fert. Suppl.* **43,** 277–291.

NERENBERG, M., HINRICHS, S. H., REYNOLDS, R. K., KHOURY, G., JAY, G. (1987), The tat gene of human t-lymphotropic virus type I induces mesenchymal tumors in transgenic mice, *Science* **237,** 1324–1329.

NISHI, M., ISHIDA, Y., HONJO, T. (1988), Expression of functional interleukin-2 receptors in human light chain/Tac transgenic mice, *Nature* **331,** 267–269.

NOTARIANNI, E., LAURIE, S., MOOR, R. M., EVANS, M. J. (1990), Maintenance and differentiation in culture of pluripotential embryonic cell ines from pig blastocysts, in: *Genetic Engineering of Animals* (HANSEL, W., WEIR, B. J., Eds.), pp. 51–56, The Journals of Reproduction and Fertility Ltd., Essex, U.K.

NOTEBORN, M., ARNHEITER, H., RICHTER-MANN, H., BROWNING, H., WEISMANN, C. (1987), Transport of the murine Mx protein into the nucleus is dependent on basic carboxy-terminal sequence, *J. Interferon Res.* **7,** 657–669.

NUSSE, R. (1986), The activation of cellular oncogenes by retroviral insertion, *Trends Genet.* **2,** 244–248.

NUSSENZWEIG, M. C., SCHMIDT, E. V., SHAW, A. C., SINN, E., CAMPOS-TORRES, J., MATHEY-PREVOT, B., PATTENGALE, P. K., LEDER, P. (1988), A human immunoglobulin gene reduces the incidence of lymphomas in c-Myc-bearing transgenic mice, *Nature* **336,** 446–450.

OSHIRO, T., YOSHIZAKI, G., TAKASHINA, F. (1989), in: *Proc. First International Marine Biotechnology Conference*, abstracts Nos. 4–10.

OVERBEEK, P. A., CHEPELINSKY, A. B., KHILLAN, J. S., PIATIGORSKY, J., WESTPHAL, H. (1985), Lens-specific expression and developmental regulation of the bacterial chloramphenicol acetyltransferase gene driven by the murine αA-crystalline promoter in transgenic mice, *Proc. Natl. Acad. Sci. USA* **82,** 7815–7819.

OVERBEEK, P. A., LAI, S.-P., VAN QUILL, K. R., WESTPHAL, H. (1986), Tissue-specific expression in transgenic mice of a fused gene containing RSV terminal sequences, *Science* **231,** 1574–1577.

OZATO, K., KONDOH, H., INOHARA, H., IWAMATSU, T., WAKAMATSU, Y., OKADA, T. S. (1986), Production of transgenic fish: introduction and expression of chicken-crystalline gene im medaka embryos. *Cell. Differ.* **19,** 237–244.

PALMITER, R. D., BRINSTER, R. L. (1985), Transgenic mice. *Cell* **41,** 343–345.

PALMITER, R. D., BRINSTER, R. L. (1986), Germline transformation of mice, *Annu. Rev. Genet.* **20,** 465–499.

PALMITER, R. D., BRINSTER, R. L., HAMMER, R. E., TRUMBAUER, M. E., ROSENFELD, M. G., BIRNBERG, N. C., EVANS, R. M. (1982), Dramatic growth of mice that develop from eggs microinjected with metallothionein-growth hormone fusion genes, *Nature* **300,** 611–615.

PALMITER, R. D., NORSTEDT, G., GELINAS, R. E., HAMMER, R. E., BRINSTER, R. L. (1983), Metallothionein-human GH fusion genes stimulate growth of mice, *Science* **222,** 809–814.

PALMITER, R. D., WILKIE, T. M., CHEN, H. Y., BRINSTER, R. L. (1984), Transmission distortion and mosaicism in an unusual transgenic mouse pedigree, *Cell* **36,** 869–877.

PALMITER, R. D., CHEN, H. Y., MESSING, A., BRINSTER, R. L. (1985), SV40 enhancer and large-T antigen are instrumental in development of choroid plexus tumours in transgenic mice, *Nature* **316,** 457–460.

PALMITER, R. D., BEHRINGER, R. R., QUAIFE, C. J., MAXWELL, F., MAXWELL, I. H., BRINSTER, R. L. (1987), Cell lineage ablation in transgenic mice by cell-specific expression of a toxin gene, *Cell* **50,** 435–443.

PALMITER, R. D., SANDGREN, E. P., AVARBOCK, M. R., ALLEN, D. D., BRINSTER, R. L. (1991), Heterologous introns can enhance expression of transgenes in mice, *Proc. Natl. Acad. Sci. USA* **88,** 478–482.

PANDEY, K. K., PATCHELL, M. R. (1982), Genetic transformation in chicken by the use of irradiated male gametes, *Mol. Gen. Genet.* **186,** 305–308.

PAPAIOANNOU, V. E., ROSSANT, J. (1983), Effects of the embryonic environment on proliferation and differentiation of embryonal carcinoma cells, *Cancer Surveys* **2,** 165–183.

PAPAIOANNOU, V. E., MCBURNEY, M. E., GARDNER, R. L., EVANS, M. J. (1975), Fate of teratocarcinoma cells injected into early mouse embryos, *Nature* **258,** 70–73.

PERRY, M. M. (1987), Nuclear events from fertilisation to the early cleavage stages in the domestic fowl (*Gallus domesticus*), *J. Anat.* **150,** 99–109.

PERRY, M. M. (1988), A complete culture system for the chick embryo, *Nature* **331,** 70–72.

PERRY, M. M., SANG, H. M. (1990), *In vitro* culture and approaches for DNA transfer in the chick embryo, in: *Proc. 4th World Congress on Genetics Applied to Livestock Production* (HILL, W. G., THOMAS, R., WOLLIAMS, J. A., Eds.), Edinburgh, Vol. XVI, pp. 115–118.

PETERS, G., BROOKES, S., SMITH, R., DICKSON, C. (1983), Tumorigenesis by mouse mammary tumor virus: evidence for a common region for provirus integration in mammary tumors, *Cell* **33**, 369–377.

PETTERS, R. M., JOHNSON, B. H., SHUMAN, R. M. (1989), Gene transfer to swine embryos using an avian retrovirus, in: *Proc. UCLA Symposium*, Taos, New Mexico, New York: Wiley-Liss.

PIEDRAHITA, J. A., ANDERSON, G. B., MARTIN, G. R., BONDURANT, R. H., PASHEN, R. L. (1988), Isolation of embryonic stem cell-like colonies from porcine embryos, *Theriogenology* **29**, 286.

PITTIUS, C. W., HENNIGHAUSEN, L., LEE, E., WESTPHAL, H., NICOLS, E., VITALE, J., GORDON, K. (1988a), A milk protein gene protomer directs the expression of human tissue plasminogen activator cDNA to the mammary gland in transgenic mice, *Proc. Natl. Acad. Sci. USA* **85**, 5874–5878.

PITTIUS, C. W., SANKARAN, L., TOPPER, Y. J., HENNIGHAUSEN, L. (1988b), Comparison of the regulation of the whey acidic protein gene with that of a hybrid gene containing the whey acidic protein gene promoter in transgenic mice, *Mol. Endocrinol.* **2**, 1027–1032.

POLGE, E. J. C., BARTON, S. C., SURANI, M. A. H., MILLER, J. R., WAGNER, T., ROTTMAN, F., CAMPER, S. A., ELSONE, K., DAVIS, A. J., GOODE, J. A., FOXCROFT, G. R., HEAP, R. B. (1989), Induced expression of a bovine growth hormone construct in transgenic pigs, in: *Biotechnology in Growth Regulation* (HEAP, R. B., PROSSER, C. G., LAMMING, G. E., Eds.), pp. 189–199, London: Butterworth.

POWELL, B. C., ROGERS, G. E. (1990), Cyclic hairloss and regrowth in transgenic mice overexpressing an intermediate filament gene *EMBO* **9** (5), 1485–1493.

POWERS, D. A. (1989), Fish as model systems, *Science* **246**, 352–358.

POWERS, D. A., GONZALES-VILLASENOR, L. I., ZHANG, P., CHEN, T. T., DANHAM, R. A. (1991), Studies on Transgenic fish: Gene transfer, expression and inheritance, in: *Proc. Conference on Transgenic Technology in Medicine and Agriculture*, in press.

PRUSINER, S. B. (1991), Molecular biology of prion diseases, *Science* **252**, 1515–1522.

PRUSINER, S. B., SCOTT, M., FOSTER, D., PAN, K.-M., GROTH, D., MIRENDA, C., TORCHIA, M., YANG, S.-L., SERBAN, D., CARLSON, G. A., HOPPE, P. C., WESTAWAY, D., DEARMOND, S. J. (1990), Transgenetic studies implicate interactions between homologous PrP isoforms in scrapie prion replication, *Cell* **63**, 673–686.

PULLEN, A. M., MARRACK, P., KAPPLER, J. W. (1988), The T cell repertoir is heavily influenced by tolerance to polymorphic self antigens, *Nature* **335**, 796–801.

PURSEL, V. G., CAMPBELL, R. G., MILLER, K. F., BEHRINGER, R. B., PALMITER, R. D., BRINSTER, R. L. (1988), Growth potential of transgenic pigs expressing a bovine growth hormone gene, *J. Anim. Sci.* **66**, 267.

PURSEL, V. G., MILLER, K. F., BOLT, D. J., PINKERT, C. A., HAMMER, R. E., PALMITER, R. D., BRINSTER, R. L. (1989b), Insertion of growth hormone genes into pig embryos, in: *Biotechnology in Growth Regulation* (HEAP, R. B., PROSSER, C. G., LAMMING, G. E., Eds.), pp. 181–188, London: Butterworth.

PURSEL, V. G., PINKERT, C. A., MÜLLER, K. F., BOLT, D. J., CAMPBELL, R. G., PALMITER, R. D., BRINSTER, R. L., HAMMER, R. E. (1989a), Genetic engineering of livestock, *Science* **244**, 1281–1288.

PURSEL, V. G., BOLT, D. J., MILLER, K. F., PINKERT, C. A., HAMMER, R. E., PALMITER, R. D., BRINSTER, R. L. (1990a), Expression and performance in transgenic pigs, *J. Reprod. Fert. Suppl.* **40**, 235–245.

PURSEL, V. G., WALL, R. J., HENNIGHAUSEN, L., PITTIUS, C. W., KING, D. (1990b), Regulated expression of the mouse whey acidic protein gene in transgenic swine, *Theriogenology* **33** (1), 302.

PURSEL, V. G., SUTRAVE, P., WALL, R. J., KELLY, A. M., HUGHES, S. H. (1992), Transfer of cSKI gene into swine to enhance muscle development, *Theriogenology* **37**, in press.

QUAIFE, C. J., PINKERT, C. A., ORNITZ, D. M., PALMITER, R. D., BRINSTER, R. L. (1987), Pancreatic neoplasia induced by ras expression in acinar cells of transgenic mice, *Cell* **48**, 1023–1034.

QUAIFE, C. J., MATHEWS, L. S., PINKERT, C. A., HAMMER, R. E., BRINSTER, R. L., PALMITER, R. D. (1989), Histopathology associated with elevated levels of growth hormone and insulin-like growth factor I in transgenic mice, *Endocrinology* **124** (1), 40–48.

QUON, D., WANG, Y., CATALANO, R., SCARINA, J. M., MURAKAMI, K., CORDELL, B. (1991), Formation of β-amyloid protein deposits in brains of transgenic mice, *Nature* **352**, 239–241.

REDDY, V. B., VITALE, J. A., WEI, C., MONTOYA-ZAVALA, M., STICE, S. L., BALISE, J., ROBL, J. M. (1991), Expression of human growth hormone in the milk of transgenic mice, *Anim. Biotechnol.* **2**, 15–29.

REED, M. L., ROESSNER, C. A., WOMACK, J. E., DORN, C. G., KRAEMER, D. C. (1988), Microinjection of liposome-encapsulated DNA into murine and bovine blastocysts, *Theriogenology* **29**, 293.

REES, W. D., FLINT, H. J., FULLER, M. F. (1990), A molecular biological approach to reducing dietary amino acid needs, *Biotechnology* **8**, 629–633.

REITMAN, M., LEE, E., WESTPHAL, H., FELSENFELD, G. (1990), Site-independent expression of the chicken βA-globin gene in transgenic mice, *Nature* **348**, 749–752.

REXROAD, C. E., PURSEL, V. G. (1988), Status of gene transfer in domestic animals, in: *Proc. 13th Int. Congr. Animal Production and Artificial Insemination Dublin*, Vol. 5, pp. 29–35, University College, Dublin.

REXROAD, C. E., HAMMER, R. E., BEHRINGER, R. R., PALMITER, R. D., BRINSTER, R. L. (1990), Insertion, expression and physiology of growth-regulating genes in ruminants, *J. Reprod. Fert. Suppl.* **41**, 119–124.

REXROAD, C. E., JR., MAYO, K., BOLT, D. J., ELSASSER, T. H., MILLER, K. F., BEHRINGER, R. R., PALMITER, R. D., BRINSTER, R. L. (1991), Transferrin- and albumin-directed expression of growth-related peptides in transgenic sheep, *J. Anim. Sci.* **69**, 2995–3004.

RICHA, J., LO, C. W. (1989), Introduction of human DNA into mouse eggs by injection of dissected chromosome fragments, *Science* **245**, 175–177.

ROBERTSON, E., BRADLEY, A., KUEHN, M., EVANS, M. (1986), Germ-line transmission of genes introduced into cultured pluripotential cells by retroviral vector, *Nature* **323**, 445–448.

ROBERTSON, E. J. (1986), Pluripotential stem cell lines as a route into the mouse germ line, *Trends Genet.* **2**, 9–13.

ROBEY, E. A., FOWLKES, B. J., GORDON, J. W., KIOUSSIS, D., VON BOEHMER, H., RAMSDELL, F., AXEL, R. (1991), Thymic selection in CD8 transgenic mice supports an instructive model for commitment to a CD4 or CD8 lineage, *Cell* **64**, 99–107.

ROGERS, G. E. (1990), Improvement of wool production through genetic engineering, *Trends Biotechnol.* **8**, 6–11.

ROKKONES, E., ALESTROM, P., SKJERVOLD, H., GAUTVIK, K. M. (1985), Development of a technique for microinjection of DNA into salmonid eggs, *Acta Physiol. Scand.* **124**, *Suppl.* **542**, 417.

ROKKONES, E., ALLESTROM, P., SKJERVOLD, D. H., GAUTVIK, K. M. (1989), Microinjection and expression of a mouse metallothionein human growth hormone gene in fertilized salmon eggs, *J. Comp. Physiol.* **B158**, 751–578.

ROSCHLAU, K. (1991), Gene transfer studies in cattle, *J. Reprod. Fert. Suppl.* **43**, 293–295.

ROSCHLAU, K., ROMMEL, P., ANDREEWA, L., ZACKEL, M., ROSCHLAU, D., ZACKEL, B., SCHWERIN, M., HÜHN, R., GAZARJAN, K. G. (1989), Gene transfer experiments in cattle, *J. Reprod. Fert. Suppl.* **39**, 153–160.

ROTTMANN, O. J., STRATOWA, C., HORNSTEIN, M., HUGHES, J. (1985), Tissue specific expression of hepatitis B surface antigen in mice following liposome – mediated gene transfer into blastocysts, *Zentralbl. Vet. Med.* **A32**, 676–682.

ROTTMANN, O. J., ANTES, R., HÖFER, P., MAIERHOFER, G. (1991), Liposome mediated gene transfer via spermatozoa into avian egg cells, *Züchtungsbiologie* **109**, 64–70.

RUBINSTEIN, J. L., NICOLAS, J.-F., JACOB, F. (1986), Introduction of genes into preimplantation mouse embryos by use of a defective recombinant retrovirus, *Proc. Natl. Acad. Sci. USA* **83**, 366–368.

RUSCONI, S. (1991), Transgenic regulation in laboratory animals, *Experientia* **47**, 866–877.

RUSCONI, S., KÖHLER, G. (1985), Transmission and expression of a specific pair of rearranged immunoglobulin u and k genes in a transgenic mouse line, *Nature* **314**, 330–334.

RUSCONI, S., SCHAFFNER, W. (1981), Transformation of frog embryos with a rabbit beta-globin gene, *Proc. Natl. Acad. Sci. USA* **78**, 5050–5055.

SALMONS, B., GÜNZBURG, W. H., JANKA, I., ERFLE, V., BREM, G. (1991), Verwendung rekombinanter subretroviraler Partikel zur effizienten und sicheren Erstellung transgener Säuger, in: *Fortschritte in der Tierzüchtung – Symposium zu Ehren von Professor Kräußlich* (BREM, G., Ed.), pp. 437–453, Stuttgart: Ulmer.

SALTER, D. W., CRITTENDEN, L. B. (1989), Artificial insertion of a dominant gene for resistance to avian leukosis virus into the germ line of the chicken, *Theor. Appl. Genet.* **77**, 457–461.

SALTER, D. W., SMITH, E. J., HUGHES, S. H., WRIGHT, S. E., FADLY, A. M., WITTER, R. L., CRITTENDEN, L. B. (1986), Gene insertion into the chicken germ line by retroviruses, *Poultry Sci.* **65**, 1445–1458.

SALTER, D. W., SMITH, E. J., HUGHES, S. H., WRIGHT, S. E., CRITTENDEN, L. B. (1987), Transgenic chickens: insertion of retroviral genes into the chicken germ line, *Virology* **157**, 235–240.

SAMARIDIS, J., CASORATI, G., TRAUNECKER, A., IGLESIAS, A., GUTI'RREZ, J. C., MÜLLER, U.,

PALACIOS, R. (1991), Development of lymphocytes in interleukin-7 transgenic mice, *Eur. J. Immunol.* **21**, 453–460.

SANDGREN, E. P., LUETTEKE, N. C., PALMITER, R. D., BRINSTER, R. L., LEE, D. C. (1990), Overexpression of TGFα in transgenic mice: Induction of epithelial hyperplasia, pancreatic metaplasia, and carcinoma of the breast, *Cell* **61**, 1121–1135.

SARVETNICK, N. (1988), Insulin-dependent diabetes mellitus induced in transgenic mice by ectopic expression of class II MHC and interferon gamma, *Cell* **52**, 773–782.

SCHÖNENBERGER, C.-A., ANDRES, A.-C., GRONER, B., VAN DER VALK, M., LEMEUR, M., GERLINGER, P. (1988), Targeted c-myc gene expression in mammary glands of transgenic mice induces mammary tumors with constitutive milk protein gene transcription, *EMBO J.* **7** (1), 169–175.

SCHMIDT, E. V., PATTENGALE, P. K., WEIR, L., LEDER, P. (1988), Transgenic mice bearing the human c-myc gene activated by an immunoglobulin enhancer: A pre-B-cell lymphoma model, *Proc. Natl. Acad. Sci. USA* **85**, 6047–6051.

SCHNEIDER, J. F., HALLERMAN, E. M., YOON, S. J., HE, L., MYSTER, S. A., GROSS, M., LIU, Z., ZHU, Z., HACKETT, P. B., GUISE, K. S., KAPUSCINSKI, A. R., FARAS, A. J. (1989), Microinjection and successful transfer of the bovine growth hormone gene into the Northern pike *Esox lucius, J. Cell. Biochem. Suppl.* **13B**, 173.

SCHORLE, H., HOLTSCHKE, T., HÜNIG, T., SCHIMPL, A., HORAK, I. (1991), Development and function of T cells in mice rendered interleukin-2 deficient by gene targeting, *Nature* **352**, 621–624.

SCHWARTZBERG, P., COLICELLI, J., GORDON, M. L., GOFF, S. P. (1984), Mutations in the gag gene of Moloney murine leukemia virus: Effects on production of virions and reverse transcriptase, *J. Virol.* **49**, 918–924.

SCHWARTZBERG, P. L., GOFF, S. P., ROBERTSON, E. J. (1989), Germ-line transmission of a c-abl mutation produced by targeted gene disruption in ES cells, *Science* **246**, 799–803.

SCHWARTZBERG, P. L., STALL, A. M., HARDIN, J. D., BOWDISH, K. S., HUMARAN, T., BOAST, S., HARBISON, M. L., ROBERTSON, E. J., GOFF, S. P. (1991), Mice homozygous for the abl^{m1} mutation show poor viability and depletion of selected B and T cell populations, *Cell* **65**, 1165–1175.

SCOTT, M., FOSTER, D., MIRENDA, C., SERBAN, D., COUFAL, F., WÄLCHLI, M., TORCHIA, M., GROTH, D., CARLSON, G., DEARMOND, S. J., WESTAWAY, D., PRUSINER, S. B. (1989), Transgenic mice expressing hamster prion protein produce species-specific scrapie infectivity and amyloid plaques, *Cell* **59**, 847–857.

SELDEN, R. F., SKOSKIEWICZ, M. J., BURKE HOWIE, K., RUSSELL, P. S., GOODMAN, H. M. (1987), Implantation of genetically engineered fibroblasts into mice: implications for gene therapy, *Science* **236**, 714–718.

SHIMOTOHNO, K., TEMIN, H. M. (1981), Evolution of retroviruses from cellular movable genetic elements, *Quant. Biol.* **45**, 719–732.

SHOEMAKER, C., GOFF, S., GILBOA, E., PASKIND, M., MITRA, S. W., BALTIMORE, D. (1981), Structure of cloned retroviral circular DNAs: implications for virus integration, *Quant. Biol.* **45**, 711–717.

SHU HUA, Y., DEEN, K. C., LEE, E., HENNIGHAUSEN, L., SWEET, R. W., ROSENBERG, M., WESTPHAL, H. (1989), Functional human CD4 protein produced in milk of transgenic mice, *Mol. Biol. Med.* **6**, 255–261.

SHUMAN, R. M. (1991), Production of transgenic birds, *Experientia* **47** (9), 897–905.

SIMONS, J. P., MCCLENAGHAN, M., CLARK, A. J. (1987), Alteration of the quality of milk by expression of sheep β-lactoglobulin in transgenic mice, *Nature* **328**, 530–532.

SIMONS, J. P., WILMUT, I., CLARK, A. J., ARCHIBALD, A. L., BISHOP, J. O., LATHE, R. (1988), Gene transfer into sheep, *Biotechnology* **6**, 179–183.

SINN, E., MULLER, W., PATTENGALE, P., TEPLER, I., WALLACE, R., LEDER, P. (1987), Coexpression of MMTV/v-Ha-ras and MMTV/c-myc genes in transgenic mice: synergistic action of oncogenes *in vivo, Cell* **49**, 465–475.

SMALL, J. A., KHOURY, G., JAY, G., HOWLEY, P. M., SCANGOS, G. A. (1986), Early regions of JC virus and BK virus induce distinct and tissue-specific tumors in transgenic mice, *Proc. Natl. Acad. Sci. USA* **83**, 8288–8292.

SMITH, C., MEUWISSEN, T. H. E., GIBSON, J. P. (1987), On the use of transgenes in livestock, *Anim. Breed. Abstr.* **55**, 1–6.

SMITHIES, O., GREGG, R. G., BOGGS, S. S., KORALEWSKI, M. A., KUCHERLAPATI, R. S. (1985), Insertion of DNA sequences into the human chromosomal β-globin locus by homologous recombination, *Nature* **317**, 230–234.

SORIANO, P., CONE, R. D., MULLIGAN, R. C., JAENISCH, R. (1986), Tissue-specific and extopic expression of genes introduced into transgenic mice by retroviruses, *Science* **234**, 1409–1413.

SORIANO, P., MONTGOMERY, C., GESKE, R., BRADLEY, A. (1991), Targeted disruption of the c-src proto-oncogene leads to osteopetrosis in mice, *Cell* **64**, 693–702.

SOUTHERN, E. M. (1975), Detection of specific sequences among DNA fragments separated by gel electrophoresis, *J. Mol. Biol.* **98**, 503–517.

SOUZA, L. M., BOONE, T. C., MURDOCK, D., LANGLEY, K., WYPYCH, J., FENTON, D., JOHNSON, S., LAI, P. H., EVERETT, R., HSU, R.-Y., BOSSELMAN, R. (1984), Applications of recombinant DNA technologies to studies on chicken growth hormone, *J. Exp. Zool.* **232**, 465–473.

SPANOPOULOU, E., EARLY, A., ELLIOTT, J., CRISPE, N., LADYMAN, H., RITTER, M., WATT, S., GROSVELD, F., KIOUSSIS, D. (1989), Complex lymphoid and epithelial thymic tumours in Thy1-myc transgenic mice, *Nature* **342**, 185–189.

STACEY, A., BATEMAN, J., CHOI, T., MASCARA, T., COLE, W., JAENISCH, R. (1988), Perinatal lethal osteogenesis imperfecta in transgenic mice bearing an engineered mutant pro-α1 (I) collagen gene, *Nature* **332**, 131–136.

STAEHELI, P., HALLER, O., BOLL, W., LINDENMANN, J., WEISSMANN, C. (1986), Mx protein: constitutive expression in 3T3 cells transformed with cloned Mx cDNA confers selective resistance to influenza virus, *Cell* **44**, 147–158.

STANTON, B. R., REID, S. W., PARADA, L. F.: (1990), Germ line transmission of an inactive N-myc allele generated by homologous recombination in mouse embryonic stem cells, *Mol. Cell Biol.* **10**, 6755–6758.

STEELE, N. C., PURSEL, V. G. (1990), Nutrient partitioning by transgenic animals, *Annu. Rev. Nutrit.* **10**, 213–232.

STEVENS, L. C., VARNUM, D. S. (1974), The development of teratomas from parthenogenetically activated ovarian mouse eggs, *Dev. Biol.* **37**, 369–380.

STEWART, C. L., VANEK, M., WAGNER, E. F. (1985), Expression of foreign genes from retroviral vectors in mouse teratocarcinoma chimaeras, *EMBO J.* **4**, 3701–3709.

STEWART, C., SCHUETZE, S., VANEK, M., WAGNER, E. (1987), Expression of retroviral vectors in transgenic mice obtained by embryo infection, *EMBO J.* **6**, 383–388.

STEWART, T. A., PATTENGALE, P. K., LEDER, P. (1984), Spontaneous mammary adenocarcinomas in transgenic mice that carry and express MTV/myc fusion genes, *Cell* **38**, 627–637.

STIEF, A., WINTER, D. M., STRÄTLING, W. H., SIPPEL, A. E. (1989), A nuclear DNA attachment element mediates elevated and position-independent gene activity, *Nature* **341**, 343–345.

STINNAKRE, M. G., VILOTTE, J. L., SOULIER, S., HARIDON, R. L., CHARLIER, M., GAYE, P., MERCIER, J. C. (1991), The bovine α-lactalbumin promoter directs expression of ovine trophoblast interferon in the mammary gland of transgenic mice, *Fed. Eur. Biochem. Soc.* **284** (1), 19–22.

STORB, U. (1987), Immunoglobulin transgenic mice, *Annu. rev. Immunol.* **5**, 151–174.

STRASSER, A., HARRIS, A. W., BATH, L., CORY, S. (1990), Novel primitive lymphoid tumours induced in transgenic mice by cooperation between myc and bcl-2, *Nature* **348**, 331–333 (abstract).

STUART, G. W., MCMURRAY, J. V., WESTERFIELD, M. (1988), Replication, integration and stable germ-line transmission of foreign sequences injected into early zebrafish embryos, *Dev. Biol.* **103**, 403–412.

STUART, G. W., VIELKIND, J. R., MCMURRAY, J. V., WESTERFIELD, M. (1990), Stable lines of transgenic zebrafish exhibit reproducible patterns of transgene expression, *Dev. Biol.* **109**, 577–584.

STUHLMANN, H., JÄHNER, D., JAENISCH, R. (1981), Infectivity and methylation of retroviral genomes is correlated with expression in the animal, *Cell* **26**, 221–232.

SUEMATSU, S., MATSUDA, T., AOSAZA, K., AKIRA, S., NAKANO, N., OHNO, S., MIYAZAKI, J. YAMAMURA, K., HIRANO, T., KISHIMOTO, T. (1989), IgG1 plasmacytosis in interleukin 6 transgenic mice, *Proc. Natl. Acad. Sci. USA* **86**, 7547–7551.

TAMIYA, E., SUGIYAMA, T., MASAKI, K. H., HIROSE, A., OKOSHI, T., KARUBE, I. (1990), *Nucleic Acids Res.* **18**, 1072.

TE RIELE, H., MAANDAG, E. R., BERNS, A. (1991a), Gene targeting via homologous recombination in embryonic stem cells is strongly stimulated by the use of isogenic DNA constructs, *personal communication*, EMBL, Heidelberg.

TE RIELE, H., MAANDAG, E. R., BERNS, A. (1991b), *personal communication*, EMBL (abstract).

TEITELMAN, G., ALPERT, S., HANAHAN, D. (1988), Proliferation, senescence and neoplastic progression of β cells in hyperplasic pancreatic islets. *Cell* **52**, 97–105.

THOMAS, K. R., CAPECCHI, M. R. (1986), Introduction of homologous DNA sequences into mammalian cells induces mutations in the cognate gene, *Nature* **324**, 34–38.

THOMAS, K. R., CAPECCHI, M. R. (1987), Site-directed mutagenesis by gene targeting in Mouse embryo-derived stem cells, *Cell* **51**, 503–512.

THOMAS, K. R., FOLGER, K. R., CAPECCHI, M. R. (1986), High frequency targeting of genes to specific sites in the mammalian genome, *Cell* **44**, 419–428.

THOMPSON, S., CLARKE, A. R., POW, A. M., HOOPER, M. L., MELTON, D. W. (1989), Germ line transmission and expression of a corrected

HPRT gene produced by gene targeting in embryonic stem cells, *Cell* **56**, 313–321.

TOMASETTO, C., WOLF, C., RIO, M.-C., MEHTALI, M., LEMEUR, M., GERLINGER, P., CHAMBON, P., LATHE, R. (1989), Breast cancer protein PS2 synthesis in mammary gland of transgenic mice and secretion into milk, *Mol. Endocrinol.* **3**, 1579–1584.

TOMITA, T. (1987), Genetic transformation in chickens and mice by the use of irradiated sperm, in: *Symposium on Biotechnology in Animal Breeding*, 11–14 November, pp. 124–128, Berlin, Technische Universität.

TOMITA, T., YAMAMOTO, N., OTSUKA, K., OHTA, M., HIROSE, K. (1988), Genetic transformation of egg shell colour in chicken by the use of irradiated sperm, in: *Proc. 18th World's Poultry Congress*, Nagoya, pp. 515–516.

TYBULEWICZ, V. L. J., CRAWFORD, C. E., JACKSON, P. K., BRONSON, R. T., MULLIGAN, R. C. (1991), Neonatal lethality and lymphopenia in mice with a homozygous disruption of the c-abl proto-oncogene, *Cell* **65**, 1153–1163.

VALANCIUS, V., SMITHIES, O. (1991), Testing an "In-Out" tareting procedure for making subtle genomic modifications in mouse embryonic stem cells, *Mol. Cell. Biol.* **11**, 1402–1408.

VAN ASSENDELFT, G. B., HANSCOMBE, O., GROSVELD, F., GREAVES, D. R. (1989), The beta-globin dominant control region activates homologous and heterologous promoters in a tissue-specific manner, *Cell* **56**, 969–977.

VAN DER PUTTEN, H., BOTTERI, F. M., MILLER, A. D., ROSENFELD, M. G., FAN, H., EVANS, R. M., VERMA, I. M. (1985), Efficient insertion of genes into the mouse germ line via retroviral vectors, *Proc. Natl. Acad. Sci. USA* **82**, 6148–6152.

VARMUS, H. E. (1983), Using retroviruses as insertional mutagens to identify cellular oncogenes, *Prog. Clin. Biol. Res.* **119**, 23–35.

VELANDER, W. (1991), *personal communication*, cited in HENNIGHAUSEN (1990).

VERBEEK, S., VAN LOHUIZEN, M., VAN DER VALK, M., DOMEN, J., KRAAL, G., BERNS, A. (1991), Mice bearing the eu-myc and Eu-pim-1 transgenes develop re-B-cell leukemia prenatally, *Mol. Cell. Biol.* **11** (2), 1176–1179.

VERRINDER-GIBBINS, A. M., BRAZOLOT, C. L., PETITTE, J. N., LIU, G., ETCHES, R. J. (1990), Efficient transfection of chicken blastodermal cells and their incorporation into recipient embryos to produce chimeric chicks, in: *4th World Congress on Genetics Applied to Livestock Production* (HILL, W. G., THOMAS, R., WOLLIAMS, J. A., Eds.), Edinburgh, Vol. XVI, pp. 119–126.

VILOTTE, J.-L., SOULIER, S., STINNARKRE, M.-G., MASSOUD, M., MERCIER, J.-C. (1989), Efficient expression of bovine-lactalbumin in transgenic mice, *Eur. J. Biochem.* **186**, 43–48.

VIZE, P. D., MICHALSKA, A. E., ASHMAN, R., LLOYD, B., STONE, B. A., QUINN, P., WELLS, J. R. E., SEAMARK, R. F. (1988), Introduction of a growth hormone fusion gene into transgenic pigs promotes growth, *J. Cell Sci.* **90**, 295–300.

VOGEL, J., HINRICHS, S. H., REYNOLDS, R. K., LUCIW, P. A., JAY, G. (1988), The HIV tat gene induces dermal lesions resembling Kaposi's sarcoma in transgenic mice, *Nature* **335**, 606–611.

VON BUTLER-WEMKEN, I., PIRCHNER, F. (1985), Growth and reproduction in NMRI-mice after long-term divergent selection for 8 week body weight, *Z. Versuchstierkd.* **27**, 80.

VON BUTLER-WEMKEN, I., WILLEKE, H., PIRCHNER, F. (1984), Two-way within-family and mass selection of 8-week body weight in different mouse populations, *Genet. Res.*, Camb. **43**, 191–200.

WAGNER, E. F., STEWART, C. L. (1986), Integration and expression of genes introduced into mouse embryos, in: *Experimental Approaches to Mammalian Embryonic Development* (ROSSANT, J., PEDERSEN, R. A., Eds.), pp. 509–549, Cambridge: Cambridge University Press.

WAGNER, E. F., COVARRUBIAS, L., STEWART, R. A., MINTZ, B. (1983), Prenatal lethalities in mice homozygous for human growth hormone gene sequences integrated in the germ line, *Cell* **35**, 647–655.

WAGNER, E., ZENKE, M., COTTEN, M., BEUG, H., BIRNSTIEL, M. L. (1990), Transferrin-polycation conjugates as carrier for DNA uptake into cells, *Proc. Natl. Acad. Sci. USA* **87**, 3410–3414.

WAGNER, T. E., HOPPE, P. C., JOLLICK, J. D., SCHOLL, D. R., HODINKA, R. L., GAULT, J. B. (1981), Microinjection of a rabbit β-globin gene into zygotes and its subsequent expression in adult mice and their offspring, *Proc. Natl. Acad. Sci. USA* **78**, 6376–6380.

WALDER, J. (1988), Antisense DNA and RNA progress and prospects, *Genes Dev.* **2**, 502–505.

WALL, R. J., PURSEL, V. G., SHAMAY, A. MCKNIGHT, R. A., PITTIUS, C. W., HENNIGHAUSEN, L. (1991), High-level synthesis of a heterologous milk protein in the mammary glands of transgenic swine, *Proc. Natl. Acad.. Sci. USA* **88**, 1696–1700.

WALTON, J. R., MURRAY, J. D., MARSHALL, J. T., NANCARROW, C. D. (1987), Zygote viability in gene transfer experiments, *Biol. Reprod.* **37**, 957–967.

WANG, Z.-Q., GRIGORIADIS, A. E., MÖHLE-STEINLEIN, U., WAGNER, E. F. (1991), A novel target cell for c-fos-induced oncogenesis: development of chondrogenic tumours in embryonic stem cell chimeras, *EMBO* **10** (9), 2437–2450.

WANKE, R., HERMANNS, W., FOLGER, S., WOLF, E., BREM, G. (1991), Accelerated growth and visceral lesion in transgenic mice expressing foreign genes of the growth hormone family: an overview, *Pediatr. Nephrol.* **5**, 513–521.

WARD, K. A., NANCARROW, C. D. (1991), The genetic engineering of production traits in domestic animals, *Experientia* **47**, 913–922.

WARD, K. A., FRANKLIN, I. R., MURRAY, J. D., NANCARROW, C. D., RAPHAEL, K. A., RIGBY, N. W., BYRNE, C. R., WILSON, B. W., HUNT, C. L. (1986), The direct transfer of DNA by embryo microinjection, in: *Proc. 3rd World Congress Genetics Applied to Livestock Production* (DICKERSON, G. E., JOHNSON, R. K., Eds.), Lincoln, Nebraska, Vol. XII, pp. 6–12, University of Nebraska.

WARD, K. A., NANCARROW, C. D., BYRNE, C. R., SHANAHAN, C. M., MURRAY, J. D., LEISH, Z., TOWNROW, C., RIGBY, N. W., WILSON, B. W., HUNT, C. L. (1990), The potential of transgenic animals for improved agricultural productivity, *Rev. Sci. Tech. Off. Int. Epiz.* **9**, 847–864.

WARD, K. A., BYRNE, C. R., WILSON, B. W., LEISH, Z., RIGBY, N. W., TOWNROW, C. R., HUNT, C. L., MURRAY, J. D., NANCARROW, C. D. (1991), The regulation of wool growth in transgenic animals, *Adv. Dermatol.* **1**, 70–76.

WEIDLE, U. H., LENZ, H., BREM, G. (1991), Genes encoding a mouse monoclonal antibody are expressed in transgenic mice, rabbits and pigs, *Gene* **98**, 185–191.

WEIHER, H., NODA, T., GRAY, D. A., SHARPE, A. H., JAENISCH, R. (1990), Transgenic mouse model of kidney disease: insertional inactivation of ubiquitously expressed gene leads to nephrotic syndrome, *Cell* **62**, 425–434.

WENTWORTH, B. C., TSAI, H., HALLETT, J. H., GONZALES, D. S., RAJCIC-SPASOJEVIC, G. (1989), Manipulation of avian primordial germ cells and gonadal differentiation, *Poultry Sci* **68**, 999–1010.

WESTAWAY, D., MIRENDA, C. A., FOSTER, D., ZEBARJADIAN, Y., SCOTT, M., TORCHIA, M., YANG, S.-L., SERBAN, H., DEARMOND, S. J., EBELING, C., PRUISNER, S. B., CARLSON, G. A. (1991), Paradoxical shortening of scrapie incubation times by expression of prion protein transgenes derived from long incubation period mice, *Neuron* **7**, 1–20.

WIEGHART, M., HOOVER, J. L., MCGRANE, M. M., HANSON, R. W., ROTTMAN, F. M., HOLTZMAN, S. H., WAGNER, T. E., PINKERT, C. A. (1990), Production of transgenic pigs harbouring a rat phosphoenolpyruvate carboxykinase-bovine growth hormone fusion gene, *J. Reprod. Fert. Suppl.* **41**, 89–96.

WILKIE, T. M., BRINSTER, R. L., PALMITER, R. D. (1986), Germline and somatic mosaicism in transgenic mice, *Dev. Biol.* **118**, 9–18.

WILLIAMS, R. S., JOHNSTON, S. A., RIEDY, M., DEVIT, M. J., MCELLIGOTT, S. G., SANFORD, J. C. (1991), Introduction of foreign genes into tissues of living mice by DNA-coated microprojectiles, *Proc. Natl. Acad. Sci. USA* **88**, 2726–2730.

WILMUT, I., ARCHIBALD, A. L., MCCLENAGHAN, M., SIMONS, J. P., WHITELAW, C. B. A., CLARK, A. J. (1990), Modification of milk composition, *J. Reprod. Fert. Suppl.* **41**, 135–146.

WILSON, J. B., WEINBERG, W., JOHNSON, R., YUSPA, S., LEVINE, A. J. (1990), Expression of the BNLF-1 oncogene of Epstein-Barr virus in the skin of transgenic mice induces hyperplasia and aberrant expression of keratin 6, *Cell* **61**, 1315–1327.

WILSON, J. M., BIRINYI, L. K., SALOMON, R. N., LIBBY, P., CALLOW, A. D., MULLIGAN, R. C. (1989), Implantation of vascular grafts lined with genetically modified endothelial cells, *Science* **244**, 1344–1346.

WINDLE, J. J., ALBERT, D. M., O'BRIEN, J. M., MARCUS, D. M., DISTECHE, C. M., BERNARDS, R., MELLON, P. L. (1990), Retinoblastoma in transgenic mice, *Nature* **343**, 665–669.

WIRAK, D. O., BAYNEY, R., KUNDEL, C. A., LEE, A., SCANGOS, G. A., TRAPP, B. D., UNTERBECK, A. J. (1991a), Regulatory region of human amyloid precursor protein (APP) gene promotes neuron-specific gene expression in the CNS of transgenic mice, *EMBO* **10** (2), 289–296.

WIRAK, D. O., BAYNEY, R., RAMABHADRAN, V., FRACASSO, R. P., HART, J. T., HAUER, P. E., HSIAU, P., PEKAR, S. K., SCANGOS, G. A., TRAPP, B. D., UNTERBECK, A. J. (1991b), Deposits of amyloid β protein in the central nervous system of transgenic mice, *Science* **253**, 323–325.

WOLF, E. (1990), Expressionsbedingte Veränderungen bei Wachstumshormon-transgenen Mäusen, *Dissertation med. vet.*, Universität München.

WOLFF, J. A., MALONE, R. W., WILLIAMS, P., CHONG, W., ACSADI, G., JANI, A., FELGNER, P. L. (1990), Direct gene transfer into mouse muscle *in vivo, Science* **247**, 1465–1468.

WOLF, E., RAPP, K., BREM, G. (1991a), Expression of metallothionein-human growth hormone fusion genes in transgenic mice results in disproportionate skeletal gigantism, *Growth* **55**, 117–127.

WOLF, E., WANKE, R., HERMANNS, W., BREM, G., PIRCHNER, F., VON BUTLER-WEMKEN, I.

(1991b), Growth characteristics of metallothionein-human growth hormone transgenic mice as compared to mice selected for high eight-week body weight and unselected controls; I. Body weight gain and external body dimensions, *Growth* **55**, 225–235.

WOLF, E., RAPP, K., WANKE, R., HERMANNS, W., PIRCHNER, F., VON BUTLER-WEMKEN, I., BREM, G. (1991c), Growth characteristics of metallothionein-human growth hormone transgenic mice as compared to mice selected for high eight-week body weight and unselected controls; II. Skeleton, *Growth* **55**, 237–248.

WOYCHIK, R. P., STEWART, C. L., DAVIS, L. G., D'EUSTACHIO, P., LEDER, P. (1985), An inherited limb deformity created by insertional mutagenesis in a transgenic mouse, *Nature* **318**, 36–40.

WRIGHT, G., CARVER, A., COTTOM, D., REEVES, D., SCOTT, A., SIMONS, P., WILMUT, I., GARNER, I., COLMAN, A. (1991), High level expression of active human alpha-1-antitrypsin in the milk of transgenic sheep, *Biotechnology* **9**, 830–834.

WU, G. Y., WILSON, J. M., SHALABY, F., GROSSMAN, M., SHAFRITZ, D. A., WU, C. H. (1991), Receptor-mediated gene delivery *in vivo, J. Biol. Chem.* **286** (22), 14338–14342.

YOON, S. J., LIU, Z., KAPUSCINSKI, A. R., HACKETT, P. B., FARAS, A., GUISE, K. S. (1988), Successful gene transfer in fish, *J. Cell. Biochem. Suppl.* **12B**, 190.

YOON, S. J., HALLERMAN, E. M., GROSS, M. L., LIU, Z., SCHNEIDER, J. F., FARAS, A. J., HACKETT, P. B., KAPUSCINSKI, A. R., GUISE, K. S. (1990a), Transfer of the gene for neomycin resistance into goldfish, *Carassius auratus, Aquaculture* **85**, 21–33.

YOON, S. J., LIU, Z., KAPUSCINSKI, A. R., HACKETT, P. B., FARAS, A., GUISE, K. S. (1990b), Successful gene transfer in fish, *Genet. Aquacult.*, 39–44.

YU, S.-H., DEEN, K. C., LEE, E., HENNIGHAUSEN, L., SWEET, R. W., ROSENBERG, M., WESTPHAL, H. (1989), Functional human CD4 protein produced in milk of transgenic mice, *Mol. Biol. Med.* **6**, 255–261.

ZHANG, P., HAYAT, M., JOYCE, C., GONZALES-VILLASENOR, L. I., LIN, C. M., DUNHAM, R. A., CHEN, T. T., POWERS, D. A. (1990). Gene transfer, expression and inheritance of pRSV-rainbow trout-GH cDNA in the common carp, *Cyprinus carpio* (Linnaeus). *Mol. Reprod. Dev.* **25**, 13–25.

ZHU, Z., LI, G., HE, L., CHEN, S. (1985), Novel gene transfer into the fertilized eggs of goldfish. *Z. Angew. Ichthyol.* **1**, 31–34.

ZHU, Z., XU, K., LI, G., XIE, Y., HE, L. (1986), Biological effects of human growth hormone gene microinjected into the fertilized eggs of loach, *Kexue Tongbao Acad. Sin.* **31**, 988–990.

ZIJLSTRA, M., LI, E., SAJJADI, F., SUBRAMANI, S., JAENISCH, R. (1989), Germ-line transmission of a disrupted β_2-microglobulin gene produced by homologous recombination in embryonic stem cells, *Nature* **342**, 435–438.

ZIJLSTRA, M., BIX, M., SIMISTER, N. E., LORING, J. M., RAULET, D. H., JAENISCH, R. (1990), β2-Microglobulin deficient mice lack CD4-8+cytolytic T cells, *Nature* **344**, 742–746.

ZIMMER, A., GRUSS, P. (1989), Production of chimaeric mice containing embryonic stem (ES) cells carrying a homoeobox Hox 1.1 allele mutated by homologous recombination, *Nature* **338**, 150–153.

ZINKERNAGEL, R. M., COOPER, S., CHAMBERS, J., LAZZARINI, R. A., HENGARTNER, H., ARNHEITER, H. (1990), Virus-induced autoantibody response to a transgenic viral antigen, *Nature* **344**, 68–71.

VI. Biosafety Concepts

19 Safety Aspects in Biotechnology

REINHARD SIMON

Freiburg i. Br., Federal Republic of Germany

WERNER FROMMER

Wuppertal, Federal Republic of Germany

1 Biotechnology and Genetic Engineering

Biotechnology may be broadly defined as "the industrial use of living microorganisms, cell- or tissue cultures for technical or industrial purposes" (KÜENZI et al., 1985), "the integration of natural sciences and engineering sciences in order to achieve the application of organisms, cells, parts thereof and molecular analogues for products and services (FROMMER and KRÄMER, 1990), or simply as "the controlled use of biological information" (HOUWINK, 1989).

Many disciplines such as microbiology, physiology, molecular biology, classical and molecular genetics, biochemistry, chemistry, physics, microelectronics, engineering sciences, process technology, and various others are integral parts of biotechnology. The interdisciplinarity and the wide and different range of applications of biotechnology is exceptional among industrial sections.

Typical products of the more traditional biotechnology are a huge variety of foods and feeds, e.g., cheese, bread, alcoholic beverages, silage etc., but also defined biologically active substances like vitamins, antibiotics etc. as well as rather simple chemicals like citric acid, alcohols and many others.

The capacity or productivity of any living cell used in biotechnological processes ultimately depends on its genetic constitution. Thus, genetics has long been of fundamental interest in biotechnology. Mutagenesis, crosses, and selection methods have been used widely to increase productivity. But by mutagenesis, crosses, and selection it would have been simply impossible to isolate *Escherichia coli* strains able to synthesize, for example, human insulin.

When such possibilities became available through the techniques of genetic engineering, which in turn were logical consequences of the dramatic developments in molecular biology, it was quite obvious that the methodology of *in vitro* DNA recombination would soon become a central part of biotechnology.

The first practical use of genetic engineering in biotechnology was the introduction of single foreign genes into a production strain (e.g., the human insulin- or interferon gene in bacteria) to manufacture the product of the particular gene. Today, recombinant proteins such as interferons are typical products of the large-scale fermentation of genetically engineered microorganisms. The broad spectrum of products of modern biotechnology will increase rapidly, since the manipulation of a wide range of microorganisms through genetic engineering offers enormous commercial potential to the biotechnological industry.

With the rapid increase of basic knowledge in molecular genetics, complex biosynthetic pathways will also become more and more utilizable for biotechnological applications. Genetic engineering also permits strain improvement by directly manipulating the circuits of gene regulation or by designing completely novel genes.

With the refinement of recombinant-DNA technology and all the other disciplines involved in biotechnology, in the future potentially pathogenic organisms or particular genes from them may also become of increasing interest and value for industrial purposes, not only for improved vaccine production but also for other large-scale processes.

Although the use of gene technology will further expand in biological, medical, and other laboratory work as well as in industrial applications, it is generally agreed that the existing high standards of safety in biotechnology can compensate for the risks, provided the already proven safety principles and measures are properly applied and further developed where necessary.

2 Safety Regimentations in Modern Biotechnology

The various disciplines and fields of application of biotechnology have long been areas of regulations and safety measures, since biotechnological activities and processes may affect human health and welfare. For obvious reasons, e.g., the food or pharmaceutical industries have to follow very stringent regulations. In addition, there are a number of more general laws like those concerning environmental protection that biotechnologists have to obey.

As indicated above, biotechnology is used in many industries; thus it cannot be overseen and regulated in a uniform manner which may be possible for other disciplines. Because the handling of pathogenic microorganisms in the laboratory and - in a few cases - in industrial processes is also a part of biotechnology, there is, since the days of Robert Koch, a long history of the development of safety measures for handling them.

Today's persistent debate on biotechnological safety in general mainly results from the controversial discussion about potential hazards of genetic engineering during the 1970s, although genetic engineering is only one, but probably the best publicized discipline in biotechnology.

In the very early days of gene technology, there was a great deal of uncertainty about the risks possibly associated with the new dimension of genetic manipulation of organisms.

The recombinant-DNA debate was started in 1974 by scientists themselves who suggested that certain types of recombinant-DNA experiments should be deferred, voluntarily, until their potential risk could be assessed. As a first global action, an international conference took place in Asilomar in 1975. An immediate consequence of this conference was the drafting of preliminary guidelines which resulted in the first recombinant-DNA research guidelines of NIH (National Institutes of Health) published in 1976. This and the following measures that have been taken primarily by U.S. governmental agencies, again in the form of guidelines, had an immense impact upon the development of guidelines and legal regimentations in other countries. At the beginning, recombinant-DNA organisms had to be handled under conditions that far exceeded those used for safe handling of non-recombinant organisms, including pathogens. Thanks to the practical experience, better scientific understanding, and scientifically based risk assessments, the safety precautions and accordingly the guidelines could be relaxed step by step primarily for laboratory work. Usually, NIH guidelines were relaxed considerably before national governments introduced similar modifications.

There is no doubt that the new biotechnology will contribute to solve severe problems of mankind such as hunger, disease, destruction of the environment, and the like. On the other hand, modern biotechnology is considered to be a new threat to humans and the environment.

Although such discussions are not yet finished, today almost all scientists involved in biotechnology and related fields as well as regulators and an increasing part of the general public estimate the opportunities of the new technology to be greater than their potential risks.

In a report of a WHO working group, it was concluded that "biotechnology in general is regarded as a safe industry". The report further assumed that many countries will develop adequate guidelines to ensure that future developments will be equally safe (WHO, 1982/1984).

Firstly, the American NIH guidelines, although binding only for those institutions receiving financial support from the NIH, constituted a valuable basis for the development of rules for work in genetic engineering laboratories in many countries.

The first major international report with considerable influence on the development of worldwide regulations was published by the Organisation for Economic Co-operation and Development (OECD) in 1986. In this paper a group of national experts compiled a general scientific framework for risk assessment and management for the use of recombinant-DNA organisms in industry, agriculture, and the environment. The report is based on two principles:

a) the classification of experiments into four classes according to the risk potential, and,
b) taking into account the existing experience in medical microbiology, e.g., the use of four safety levels in the laboratory according to the four risk groups.

Especially the GILSP concept (Good Industrial Large Scale Practice) established a number of fundamental principles that have been adopted by all major industrialized countries active in modern biotechnology (see Sect. 4.3.3).

Beside the above mentioned worldwide working organizations WHO and OECD, there exists another transnationally active organization, the European Federation of Biotechnology (EFB), which, by its "Working Party on Safety in Biotechnology", has been engaged in the elaboration of generally acceptable recommendations. – In fact, this chapter is largely based on the publications of these organizations.

The need for internationally accepted codes of practice and regulations is obvious, especially when considering the environmental release of genetically modified organisms which do not respect national borders. In order not to lose public acceptance, it is a vital interest of any biotechnological company and every individual scientist involved in research or production to demonstrate clearly that the commonly used statement is and will stay true, that, "biotechnology if properly practiced is safe".

As mentioned above there is a well established safety system developed during hundred years of experience in handling hazardous microorganisms. This system is being used in the now existing regulatory framework. It is the intention of this chapter to explain what are the fundamentals and the basic principles of the codes of good practice, the guidelines and the legal regulations, and what is meant with the term "properly practiced".

3 Objectives and Consequences of Safety Assessment

When discussing safety aspects in biotechnology it must be distinguished between

- measures ensuring the health of persons engaged directly in the biotechnological work in question;
- protection of the general public, animals, and plants from possible adverse effects that might be caused by particular biotechnological processes;
- measures mainly taken to guarantee high standards of quality of biotechnological products, e.g., Good Manufacturing Practice (GMP).

3.1 Product Integrity

In most biotechnological large-scale processes, e.g., in food production, the only matter of safety concern is the design of equipment and the installation of a procedure which minimize or prevent microbial contamination of the particular process. This issue is regulated in different industries by many laws and regulations. It will not be considered in this chapter, although it is important not only for economical reasons but also for the safety of the consumer of the biotechnological product.

3.2 Occupational Health

Usually measures assuring product integrity also help to increase the safety of the workers. The vast majority of microorganisms used in process biotechnology is harmless. In these cases no special containment conditions are needed to protect the workers. Of course, for every handling of organisms efficient hygienic measures have to be used.

Organisms of potential or real risk to laboratory or plant personnel have to be handled under special conditions which in turn depend on the risk class. In this most important sector of biosafety, microbiologists in general and persons involved in recombinant-DNA technology in particular can utilize the long-term experience of successful prevention of infection among workers in clinical and medical research (Collins, 1990).

3.3 Environmental Protection

Most organisms used for biotechnological purposes are not only harmless to workers who handle them daily but also to the environment including the general public, animals, plants, and ecosystems.

For pathogenic microorganisms with medium to high risk to humans, appropriate phy-

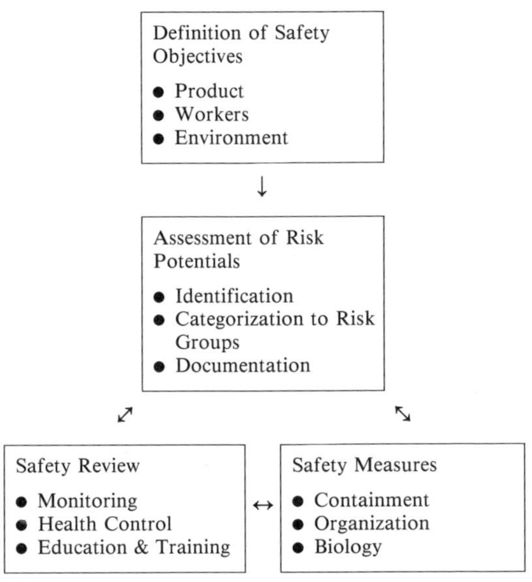

Fig. 1. Basis concept of safety in biotechnology.

sical containment measures are available to minimize or even prevent their spread to the environment. For those organisms that are harmless to man but pathogenic to animals or plants, a subset of safety precautions has to be considered that ensures their containment to an extent that minimizes their environmental risks.

Genetically modified organisms which exhibit reduced biological fitness under natural conditions or wild-type strains with an extended history of safe use can usually be handled without special containment devices (see Sect. 4.3.3).

3.4 Basic Considerations of Biosafety

The actions required to establish and maintain safety in biotechnology cover the four elements shown in the boxes of Fig. 1.

1. Definition of safety objectives
   As discussed above, there are three safety objectives of equal importance. In the case of a new process, a new organism, or a new product, all three points must be considered.
2. Assessment of risk potential
   Before beginning work at any scale, from initial experimental work in the laboratory up to large-scale production in process biotechnology, appropriate risk assessment has to be made based on the identification of the organism. The most important criterion for the clear-cut categorization of an organism into one of the four risk groups is its pathogenic potential to humans, animals, or plants. The process of risk assessment or safety assurance must be documented in a way that the relevant facts are described with clarity and precision. Adequate information must be available not only to satisfy local or national authorities but also to maintain public confidence in applications of biotechnology.
3. Safety measures
   Appropriate risk management should be performed by considering physical containment, biological containment, and organization on the basis of the risk class. The technical containment measures and organizational safety precautions have a long and successful history in microbiology and biotechnology and are not specific to gene technology. The concept of "biological containment" was introduced for work with recombinant DNA organisms. It comprises essentially the use of certified host–vector systems that minimize both, the survival of the genetically modified organism outside the laboratory or the fermenter, and the transmission of recombinant DNA to other hosts. In addition, strains with a long history of safe use are considered to exhibit "biological containment".
4. Safety review
   In order to be effective in the long term, safety measures need to be appropriately scrutinized. Depending on the risk group of the organism used and the corresponding containment level, monitoring programs should be installed that, on the one hand, ensure high hygienic standards in laboratory settings or production facilities and, on the other hand, control possible environmental impacts.

Workers handling biological agents which are or may be a hazard to man should undergo appropriate health surveillance prior to exposure and at regular intervals thereafter. The objectives of medical surveillance are (WHO, 1983):

- to provide a means of preventing occupationally acquired disease by the exclusion of highly suceptible individuals as well as by regularly reviewing those accepted for employment;
- to provide a means for the early detection of laboratory-acquired infection;
- to assess the efficacy of protective equipment and procedures.

Education and training of personnel is the most important element to maintain biosafety at all levels, since human error and poor laboratory practice may compromize even the most sophisticated laboratory safeguards and equipment designed to protect workers and the environment. However, the objectives of occupational safety and avoidance of adverse environmental effects are hardly achieved by mere routine information on sources of danger. Laboratory and plant personnel must be motivated to understand safety at the workplace as an obvious necessity and an essential feature of good professional practice and successful work. Moreover, the best available guideline cannot anticipate every possible situation. Thus, education, motivation, and good judgement are key essentials to safety in biotechnology.

As indicated in Fig. 1, there must be a continuous feedback between the results of safety reviews and both the risk assessment and the safety measures. As a result of this interaction, especially in the case of a new process for which little or no previous experience exists, safety measures may be reduced after some time. More importantly, appropriate scrutiny is necessary in order to be able to react immediately in any case of recognized and perhaps unexpected hazard.

The topics classification of organisms and corresponding biosafety levels mentioned briefly here will be discussed in detail in the following sections.

4 The Classification System in the Contained Use of Microorganisms

4.1 Biological and Physical Containment

In principle the risks that may be associated with biotechnological works can be managed in two ways. On the one hand, there is a strong demand to utilize whenever possible organisms that are known to be harmless, either as wild types or mutant derivatives thereof. In these cases, which have been designated as "biological containment" and which cover 90% or more of biotechnological uses of microorganisms, no special precautionary systems need to be installed that exceed the usual hygienic measures (see Sect. 4.3.2).

On the other hand, there are organisms in use or planned to be used which are assumed or definitely known to be not harmless. The long history of safe handling of pathogenic organisms has demonstrated that appropriate physical containment techniques, i.e., installation and equipment, together with supportive organizational and educational measures are available to handle them safely in the laboratory or in process biotechnology. Not only to ensure the safety of workers, the public or the environment, but also for economical reasons the physical containment and other measures have to be related to the existing or perceived risk of the organism or its products.

It may be of interest to note that as early as 1947 a scheme for the classification of pathogenic microorganisms with four classes was introduced in the U.S. with the aim to define according safety precaution ("containment") levels (COLLINS, 1990).

4.2 Classification of Microorganisms According to Their Risk Potential

The classification system has been established mainly to distinguish clearly between the small proportion of organisms potentially pathogenic to humans or other higher life forms and the huge majority of harmless organisms.

The first prerequisite for the classification is the proper identification of an organism to be used either in research or, much more importantly, in a large-scale biotechnological process. The determination of the genus and species of an organism allows an initial assessment of its probable behavior as a pathogen, based on existing knowledge of the organism itself and of known species closely related to it.

In order to be able to handle a new organism properly, the assessment procedure must result in conclusive information on the safety of the organism to be used or must categorize it according to its potential risk.

Experience has shown that microorganisms isolated from the environment, e.g., from soil samples, can be assumed to belong mainly to risk group I (see Tab. 1). Thus, the process of isolation and identification of particular isolates can be performed under minimal containment conditions but applying basic hygienic measures (see Sect. 4.3.2). For identification purposes clinical material is usually handled as if it consisted of group II organisms. In any case, isolates which are processed further must be identified as soon as possible to define the appropriate safety measures for handling them.

4.2.1 Classification of Wild-Type Strains

The classification system used today is based on a proposal of the World Health Organization, in which naturally occurring microorganisms have been classified into four risk groups as summarized in Tab. 1 (WHO, 1983).

According to the basic consideration that risk is defined as the product of hazard and exposure, the classification of pathogenic microorganisms to risk groups II, III, or IV uses criteria such as

- the history and the known infectivity of the organism,
- the incidence of infection in the community and the presence of vectors and reservoirs,
- the virulence of the pathogen, infection dose, route of attack,
- the feasibility of immunization and the effectiveness of therapy.

The term "pathogenicity" cannot be recorded quantitatively since its manifestation depends

Tab. 1. Classification of Infective Microorganisms by Risk Groups (WHO, *Laboratory Biosafety Manual*, 1983)

Risk Group I (low individual and community risk)

A microorganism that is unlikely to cause human disease or animal disease of veterinary importance.

Risk Group II (moderate individual risk, limited community risk)

A pathogen that can cause human or animal disease but is unlikely to be a serious hazard to laboratory workers, the community, livestock, or the environment. Laboratory exposure may cause serious infection, but effective treatment and preventive measures are available and the risk of spread is limited.

Risk Group III (high individual, low community risk)

A pathogen that usually produces serious human disease but does not ordinarily spread from one infected individual to another.

Risk Group IV (high individual and community risk)

A pathogen that usually produces serious human or animal disease and may be readily transmitted from one individual to another, directly or indirectly.

Tab. 2. EFB Classification of Microorganisms According to Pathogenicity[a]

Class I: Harmless
This class contains those microorganisms that have never been identified as causative agents of disease in man and that offer no threat to the environment. They are not listed in higher classes or in Group E.

Class II: Low Risk
This class contains those microorganisms that may cause disease in man and which might therefore offer hazard to laboratory workers. They are unlikely to spread in the environment. Prophylactics are available and treatment is effective.

Class III: Medium Risk
This class contains those microorganisms that offer severe threat to the health of laboratory workers but comparatively small risk to the population at large. Prophylactics are available and treatment is effective.

Class IV: High Risk
This class contains those microorganisms that cause severe illness in man and offer a serious hazard to laboratory workers and people at large. In general effective prophylactics are not available and no effective treatment is known.

Group E: Environmental Risk
This group contains microorganisms that offer a more severe threat to the environment than to man. They may be responsible for heavy economic losses. National and international lists and regulations concerning these microorganisms are already in existence in contexts other than biotechnology (e.g., for phytosanitary purposes).

[a] from KÜENZI et al., 1985, and FROMMER et al., 1989 (modified).

on many factors that vary broadly. Such factors are, for example, the actual genetic constitution or physiological state of both the infectious agent and the target organism. Furthermore, the resistance or susceptibility of a microorganism to adverse conditions such as temperature, pH, moisture and the like, influence the chances of possible infections.

Thus, a particular risk class contains a spectrum of organisms, the pathogenic potentials of which vary considerably depending on a number of factors.

When comparing various national lists, some unanimities about the allocation of organisms to a category are recognizable (KÜENZI et al., 1985). As far as human pathogens are concerned, this situation reflects some uncertainties about the risks of handling particular organisms. However, the vast majority of classifications of strains and species performed by various national committees of experts showed a uniform result (FROMMER et al., 1989).

Although there is a worldwide agreement that the four-class classification system is sufficient to meet the intended objectives, there have been differences in the description of risk classes in various national lists. In an attempt to harmonize the situation, the Working Party on Safety in Biotechnology of the EFB proposed a "consensus" description for the classification of microorganisms according to their pathogenicity (KÜENZI et al., 1985) (Tab. 2). Tab. 2 contains also another proposal of the above mentioned Working Party, namely to use descriptive adjectives instead of or in addition to numbers to avoid possible confusions with different numbering systems (FROMMER et al., 1989). The special group E mentioned in Tab. 2 will be discussed below.

It should be clearly noted that class I is described as "harmless", which does not mean "no risk at all". Some of these microorganisms may induce diseases under special circumstances, e.g., in immunodeficient individuals. For obvious reasons, in large-scale processes microorganisms of intrinsic low risk, i.e., the harmless ones, should be used whenever possible. In fact, a number of class I organisms have a history of extended safe use in biotechnological industry as well as in human consumption or in agricultural applications.

Class II organisms are all classified as pathogens in the context of hospital, clinical, and diagnostic laboratories. The scale of operation

Tab. 3. EFB Classification of Microorganisms Pathogenic for Plants (KÜENZI et al., 1987)

Class Ep 1
This class contains those microorganism which may cause disease in plants but have local significance only. They may be mentioned in a list of pathogens for individual countries concerned.
Very often they are endemic pathogens for plants and do not require any special physical containment. However, it may be advisable to employ GMT.

Class Ep 2
Microorganisms known to cause outbreaks of disease in crops as well as in ornamental plants. These pathogens are subject to regulations for species listed by authorities of the country concerned.

Class Ep 3
Microorganisms mentioned in quarantine lists. Importation and handling of these organisms are generally forbidden.
The regulatory authorities must be consulted by prospective users.

is rather small compared to most other industrial biotechnological processes.

Some of class II organisms are normally present in the environment and even in food at relatively high concentrations without offering any threat to normal persons (COLLINS, 1990).

Medium risk microorganisms (class III) are rarely used in industry, although established and tested containment systems are available to handle them safely, e.g., in vaccine production.

There is no industrial use of class IV, the high-risk organisms. Worldwide only a few specialized laboratories are properly installed and equipped for their small-scale use.

For obvious reasons, there is a strong tendency to replace medium- or high-risk operations by harmless techniques using the possibilities offered by the methodology of genetic engineering; the well known example of vaccine production using cloned antigens instead of wild-type pathogens demonstrates this clearly.

The classification system described so far has also been applied without problems to wild-type microorganisms that have been modified using the traditional genetic techniques of mutagenesis and selection. To our knowledge, in no case had an organism to be transferred to a higher risk class than the wild type after mutagenesis. However, there is a very popular example for the contrary: *Escherichia coli* wild type is in risk class II as a potential human pathogen, whereas its derivative *E. coli* K12 is classified as harmless in all national lists.

With a few exceptions, human pathogens are relatively easy to categorize in such lists. The situation is much more complex with plant or animal pathogens, the potential risks of which may greatly depend on local situations.

As a possible step towards a solution of the problem of classification of microorganisms that are harmless to man but may be pathogenic for other higher life forms, in Tab. 2 a special group E was established that contains microorganisms of environmental risk (KÜENZI et al., 1985). As far as plant pathogens are concerned, this group has been defined further (Tab. 3) (KÜENZI et al., 1987). Adequate containment conditions have also been elaborated describing three safety levels, one for each of the classes shown in Tab. 3, aimed at minimizing any adverse effects that might result from the use of plant pathogens in laboratories and in large-scale processing facilities (FROMMER et al., 1992a).

4.2.2 Classification of Recombinant-DNA Organisms

There has been a long and sometimes controversial debate about the risk potentials and the classification of organisms modified by recombinant DNA techniques. The discussions resulted in many countries in the elaboration and implementation of laws or other legal regulations specifically dealing with genetic engineering. Nevertheless, it is now generally accepted that genetically modified or engineered organisms must not at all be regarded as hazardous *per se*, but can be classified into the same four-class system used for wild-type organisms. Provided there is sufficient knowledge of the recipient, the origin and nature of the introduced gene(s) and the vector used, a reliable risk assessment can be carried out for the organism modified by recombinant-DNA techniques.

This is one of the major conclusions that can be drawn from almost 20 years of experience with the recombinant-DNA methodology in which no basically unexpected or quantitative novel risks have been realized.

The basic statement of the WHO Biosafety Manual is still valid, and there is no reason to assume that this will change: "There are no unique or specific risks associated with recombinant DNA work (genetic engineering); the risks are no greater than those associated with work with known pathogens and do not necessitate special laboratory design or practice" (WHO, 1983). That means that all sorts of guidelines and regulations set for work with microorganisms and other biological agents can be transferred accordingly to work with recombinant-DNA organisms.

This general statement has been broadly accepted worldwide, also in those countries in which genetic engineering is regulated by specific laws.

For example, the OECD Report (1986), which deals with safety considerations for the use of recombinant-DNA organisms in particular, contains a table of criteria applicable to classify genetically modified organisms into the group of harmless microorganisms (see Sect. 4.3.3). The scientific principles presented in this report greatly influenced the development of a consensus for the safe industrial use of genetically modified organisms.

For this scheme to be applicable in practice it is necessary, of course, to categorize the particular product of recombinant-DNA work, i.e., usually a genetically modified organism, into one of the four risk classes. To assist this theoretical risk assessment, long catalogs of useful criteria exist. As an example, in Tab. 4 the Annex III of the EC COUNCIL DIRECTIVE 90/219/EEC is reproduced, which can be used as a checklist of points to consider when assessing the risk potential of a genetically modified microorganism.

4.2.3 Cell Cultures

Cell cultures of various origins have been used for decades in basic diagnostic, therapeutic, and many other types of research as well as in vaccine production. More recently a growing number of products, e.g., of pharmacological relevance, have been manufactured using permanent mammalian cell lines.

The long history of safe use has clearly demonstrated that cell cultures themselves do not constitute a hazard either to healthy humans or to the environment. Apparently, the immune system normally provides a sufficient level of protection; on the other hand, unlike most prokaryotic organisms, cell cultures depend on very complex growth media and are therefore unable to survive under normal environmental conditions.

The only matter of concern are the possible contaminations of cell cultures with viruses, mycoplasma, bacteria, yeast, or fungi. Thus, the assessment of risks possibly associated with cell cultures or their products are primarily based on the probability of presence of pathogenic adventitious agents. On this basis the safety precautions proven for work with microorganisms can equally well be applied to work with cell cultures (FROMMER et al., 1992b).

4.3 Principles for Handling Microorganisms

For every handling of microorganisms hygienic measures have to be used as they are described in Sect. 4.3.1 (mainly for laboratories) and in Sect. 4.3.2 (mainly for process biotechnology).

More than 30 years of vaccine production have shown that even highly pathogenic organisms can be handled on a large scale without threat to staff or facility environment, provided adequate precautions are applied.

4.3.1 Good Microbiological Techniques

A large number of official and institutional codes of practice and guidelines have been

Tab. 4. Safety Assessment Parameters to be Taken into Account for Genetically Modified Microorganisms[a]

A. Characteristics of the Donor, Recipient or (where appropriate) Parental Organism(s)
- Names and designation
- Degree of relatedness
- Sources of the organism(s)
- Information on reproductive cycles (sexual/asexual) of the parental organism(s) or, where applicable, of the recipient microorganism
- History of prior genetic manipulations
- Stability of parental or of recipient organism in terms of relevant genetic traits
- Nature of pathogenicity and virulence, infectivity, toxicity, and vectors of disease transmission
- Nature of indigenous vectors:
  sequence
  frequency of mobilization
  specificity
  presence of genes which confer resistance
- Host range
- Other potentially significant physiological traits
- Stability of these traits
- Natural habitat and geographic distribution. Climatic characteristics of original habitats
- Significant involvement in environmental processes (such as nitrogen fixation or pH regulation)
- Interaction with, and effects on, other organisms in the environment (including likely competitive or symbiotic properties)
- Ability to form survival structures (such as spores or sclerotia)

B. Characteristics of the Modified Microorganism
- The description of the modification including the method for introducing the vector-insert into the recipient organism or the method used for achieving the genetic modification involved
- The function of the genetic manipulation and/or of the new nucleic acid
- Nature and source of the vector
- Structure and amount of any vector and/or donor nucleic acid remaining in the final construction of the modified microorganism
- Stability of the microorganism in terms of genetic traits
- Frequency of mobilization of inserted vector and/or genetic transfer capability
- Rate and level of expression of the new genetic material. Method and sensitivity of measurement
- Activity of the expressed protein

C. Health Considerations
- Toxic or allergenic effects of non-viable organisms and/or their metabolic products
- Product hazards
- Comparison of the modified microorganism to the donor, recipient, or (where appropriate) parental organism regarding pathogenicity
- Capacity for colonization
- If the microorganism is pathogenic to humans who are immunocompetent:
  a) diseases caused and mechanism of pathogenicity including invasiveness and virulence
  b) communicability
  c) infective dose
  d) host range, possibility of alteration
  e) possibility of survival outside of human host
  f) presence of vectors of means of dissemination
  g) biological stability
  h) antibiotic-resistance patterns
  i) allergenicity
  j) availability of appropriate therapies

Tab. 4. Continued

D. Environmental Considerations
- Factors affecting survival, multiplication, and dissemination of the modified microorganism in the environment
- Available techniques for detection, identification and monitoring of the modified microorganism
- Available techniques for detecting transfer of the new genetic material to other organisms
- Known and predicted habitats of the modified microorganism
- Description of ecosystems to which the microorganism could be accidentally disseminated
- Anticipated mechanism and result of interaction between the modified microorganism and the organisms or microorganisms which might be exposed in case of release into the environment
- Known or predicted effects on plants and animals such as pathogenicity, infectivity, toxicity, virulence, vector of pathogen, allergenicity, colonization
- Known or predicted involvement in biogeochemical processes
- Availability of methods for decontamination of the area in case of release to the environment

[a] Annex III of EC COUNCIL DIRECTIVE of 23 April 1990 on the contained use of genetically modified microorganisms (90/219/EEC)

published. Summarizing much of this experience, the WHO published in its Laboratory Biosafety Manual (1983) a comprehensive and detailed list of guidelines fundamental to all classes of laboratories. These guidelines were the basis for many nationally implemented and internationally accepted rules that ensure occupational safety and hygiene. The principles can be summarized as follows.

- For every laboratory or industrial facility appropriate codes of practice must be formulated, implemented, and its realization adequately controlled.
- Written instructions and training of personnel are essential: good laboratory practice is fundamental to safety and cannot be replaced, but only supported by specialized equipment.
- Exposure to biological agents must be kept at the lowest level which is reasonably practicable, e.g., by decontamination of the working places, protective clothing, minimizing the creation of aerosols, use of washing facilities and so on.
- The ingestion of infectious agents must be prevented: no eating, drinking, smoking during work, no mouth pipetting.
- The use of hypodermic needles and syringes should be restricted to applications where they cannot be replaced by less dangerous instruments.

From the organizational point of view, special attention should be focused on conditions generally known to pose problems; these include:

- creation of aerosols
- work with large volumes and/or high concentrations of microorganisms
- overcrowded, overequipped laboratories
- unauthorized entrance

In a more elaborated form, the principles listed above describe the Good Microbiological Techniques (GMT) for safe handling of microorganisms with risk potential. A summary of rules for GMT has been compiled, for example, by the EFB (FROMMER et al., 1989) (see Tab. 5).

4.3.2 Good Occupational Safety and Hygiene

The term Good Occupational Safety and Hygiene summarizes the normal hygiene procedures applied in large-scale biotechnological processes with microorganisms of group I in which both the contamination of the process itself by microorganisms or other substances must be avoided, as well as their transfer to the personnel and the workplace; or where such transfer can be minimized.

Article 7 of the EC Council Directive on the contained use of genetically modified organ-

Tab. 5. Good Microbiological Techniques (GMT) for the Safe Handling of Microorganisms with Risk Potential (FROMMER et al., 1989)

These techniques are intended to protect both the operator and the product.

1. The operator should have basic knowledge of microbiology. Spreading of pathogens should not occur, e.g., via contaminated surfaces, hands, or clothes.
   All workers should be aware of the risks of cultivated pathogens to people in the vicinity. Entry to the working place should be confined to persons who are aware of these risks.
2. There should be no contact between materials or tools in the working place and the mouth of the operator. Eating, drinking, and mouth pipetting are not allowed.
3. No activities which may produce aerosols are permitted at the working place: centrifuging, blending, filling of bottles or tubes should be carried out in a biosafety cabinet.
4. Infected waste should be placed in sealable containers, the outside of which should be disinfected before transport to the autoclave or incinerator.
5. Heat or chemical sterilization processes should be investigated beforehand to ensure that the required killing rate is obtained.
6. Reliable equipment should be used.
7. Working surfaces, tables, and hands should be disinfected after normal working.
8. Working surfaces, tables, floors, and hands should be disinfected after spillage of infectious material.
9. In case of accidents, an emergency scheme with details of first aid, cleaning and disinfection should be available, and the staff should be trained accordingly.

isms (90/219/EEC), which reproduces the OECD recommendations of 1986 literally, specifies the following principles for working with group I organisms (in the new version of the OECD Report "Safety Considerations for the Use of Genetically Modified Organisms" of 1992, some points are further elaborated, and the concept has been renamed "Fundamental Principles of Good Occupational and Environmental Safety"; major supplements are given in *parentheses*):

1. to keep workplace and environmental exposure to any physical, chemical, or biological agent (*including cellular products and debris*) to the lowest practicable level (*to a level appropriate to the characteristics of the organism, the product and the process*);
2. to exercise engineering control measures at source and to supplement these with appropriate personal protective clothing and equipment when necessary;
3. to test adequately and maintain control measures and equipment. (*The frequency of examination and testing will depend on the nature of the modified organism, the product and the process*);
4. to test, when necessary, for the presence of viable process organisms outside the primary physical containment (*outside the process equipment, both in the workplace and in the environment*);
5. to provide training of personnel;
6. to establish biological safety committees or subcommittees as required (*and/or to consult with worker representatives and to consult with regulatory authorities*);
7. to formulate and implement local codes of practice for the safety of personnel (*and for the protection of the environment*).

4.3.3 Good Industrial Large-Scale Practice: The GILSP Concept

In the above mentioned Report "Recombinant DNA Safety Considerations" (OECD, 1986), a catalog of certain criteria has been suggested to classify a genetically modified organism into the lowest risk group. According to these recommendations, a recombinant-DNA organism of intrinsically low risk can be handled under conditions of "good industrial large scale practice" (GILSP), i.e., under the same conditions of minimal control and containment procedures as they would be used for the harmless host strain from which it is derived.

Tab. 6. Criteria for Classifying Genetically Modified Microorganisms in Group I[a]

A. Recipient or Parenteral Organism
- Non-pathogenic
- No adventitious agents
- Proven and extended history of safe use or built-in biological barriers, which, without interfering with optimal growth in the reactor or fermentor, confer limited survivability and replicability, without adverse consequences in the environment

B. Vector/Insert
- Well characterized and free from known harmful sequences
- Limited in size as much as possible to the genetic sequences required to perform the intended function
- Should not increase the stability of the construct in the environment (unless that is a requirement of intended function)
- Should be poorly mobilizable
- Should not transfer any resistance markers to microorganisms not known to acquire them naturally (if such acquisition could compromise use of drug to control disease agents)

C. Genetically Modified Microorganisms
- Non-pathogenic
- As safe in the reactor or fermentor as recipient or parenteral organism, but with limited survivability and/or replicability without adverse consequences in the environment

D. Other Genetically Modified Microorganisms that could be Included in Group I if they Meet the Conditions in C above
- Those constructed entirely from a single prokaryotic recipient (including its indigenous plasmids and viruses) or from a single eukaryotic recipient (including its chloroplasts, mitochondria, plasmids, but excluding viruses)
- Those that consist entirely of genetic sequences from different species that exchange these sequences by known physiological processes

[a] from Annex II of the EC Council Directive of 23 April 1990 on the contained use of genetically modified microorganisms (90/219/EEC)

Tab. 7. Relationship between Risk Class, Containment Category, and Objectives of Safety Precautions[a]

Risk Class	Containment Category (OECD)	Safety Objectives
Class 1	GILSP	Safeguard hygiene for work with harmless microorganisms which do not require containment. Hygienic process and equipment are used to prevent the contamination of culture or product.
Class 2	C 1	Minimize the release of low-risk microorganisms from primary containment, no secondary containment.
Class 3	C 2	Prevent release of medium-risk microorganisms from primary containment during regular operations; no strict secondary containment.
Class 4	C 3	Absolute containment for high-risk microorganisms; secondary containment strictly required to prevent release in case of breach of primary containment.

[a] Frommer and Krämer, 1990 (modified)

Tab. 8. Relationship between Risk Class, Containment Category, and Safety Precautions (FROMMER et al., 1989)

Harmless Microorganisms (Class 1)
GILSP
Good occupational safety and hygiene principles are to be applied (see Sect. 4.3.2)

Low Risk Microorganisms (Class 2)
Containment category 1
- GMT are to be applied (see Tab. 5)
- Surfaces within the facility should be easily cleaned and disinfected
- Contaminated materials are to be autoclaved or disinfected before cleaning
- Aerosol creating procedures must be controlled and contained
- Access to the facility must be restricted

Medium Risk Microorganisms (Class 3)
Containment category 2
Measure in addition to containment 1 conditions:
- Only authorized personnel is admitted to the facility
- Personnel is vaccinated if possible
- Exhaust air is HEPA filtered
- Effluents from the facility must be decontaminated or sterilized
- An autoclave should be within the facility
- All processes involving medium-risk microorganisms must be carried out in hermetically sealed equipment or in biosafety cabinets
- Protective suits, closing at the back, have to be worn by personnel
- Hands and forearms should be washed and disinfected at regular intervals

High Risk Microorganisms (Class 4)
Containment category 3
Measures in addition to containment 2 conditions:
- No visitors should be admitted
- Facility must be completely isolated
- The rooms for complete change of clothes must include an air-lock facility with compulsory shower
- Negative pressure must be maintained in the facility and the air ducts must be protected by double HEPA filters
- All effluents must be sterilized
- The autoclave must be within the facility
- Materials containing high-risk microorganisms must be absolutely separated from workers
- Protective suits for single use must be worn

Taking into account accumulating experience in the use of genetically modified organisms, an OECD expert committee further developed the GILSP concept by elaborating in more detail the criteria for recombinant-DNA GILSP microorganisms and cell cultures (OECD, 1992).

The Annex II of the EC COUNCIL DIRECTIVE "On the Contained Use of Genetically Modified Microorganisms" (90/219/EEC), that follows these recommendations almost literally is reproduced in Tab. 6. For further interpretation of this Annex detailed guidelines have been elaborated in the Commission Decision of 29 July, 1991 (91/448/EEC; *Official Journal of the European Communities* No. L 239/23).

It is anticipated that the vast majority of biotechnological processes will be designed to use organisms that merit the designation GILSP. This is not only to ensure the safety of workers and of the environment but, of course, also to avoid the need for costly physical containment measures.

Although zero risk is not realistic even for GILSP organisms, experience has clearly demonstrated that the well established principles of good occupational (and environmental) sa-

Tab. 9. Safety Precautions for Biotechnological Operations (FROMMER et al., 1989)

Measures, Operations, Equipment Design, Facilities	GILSP	Containment Category 1	2	3
Procedure				
Written instructions and code of practice	+	+	+	+
Biosafety manual	–	+	+	+
Good occupational hygiene	+	+	+	+
Good microbiological techniques	–	+	+	+
Biohazard sign	–	+	+	+
Restricted access	–	+	+	+
Accident reporting	+	+	+	+
Medical surveillance	–	+	+	+
Primary Containment: Operation and Equipment				
Closed system (CS) designed to minimize (m) or prevent (p) the release of viable microorganisms	–	m	p	p
Treatment of exhaust air or gas from CS	–	m	p	p
Sampling from CS	–	m	p	p
Addition of materials to CS	–	m	p	p
Removal of materials, products, effluents from CS	–	m	p	p
Penetration of CS by agitator shaft and measuring devices	–	m	p	p
Foam-out control	–	m	p	p
Secondary Containment: Facilities				
Protective clothing according to risk category	+	+	+	+
Changing/washing facility	+	+	+	+
Disinfection facility	–	+	+	+
Emergency shower facility	–	–	+	+
Airlock and compulsory shower facility	–	–	–	+
Effluents decontaminated	–	–	+	+
Controlled negative pressure	–	–	–	+
HEPA filters in air ducts	–	–	+	+
Tank for spilled fluids	–	–	+	+
Area hermetically sealable	–	–	–	+

+ required, – not required; CS, closed system; m, minimize release; p, prevent release

fety and hygiene described above are sufficient to handle them safely (OECD, 1992).

4.4 Classification of Safety Measures

4.4.1 The Four-Class Safety Precaution System

If the relevant information about the properties of an organism is available, the procedures of handling this organism together with appropriate containment facilities, if necessary, can be categorized. The internationally accepted four-class risk classification system has been logically related to a four-part safety precaution system both for biotechnological laboratories and industrial processes.

Despite the still existing differences in the numbering system, there is a broad conformity in the contents of the various nationally or internationally used lists of containment levels. In the following three tables (Tabs. 7 through 9), the designations proposed by the OECD (1986) are used. In this classification system the harmless group I organisms are handled under GILSP conditions, whereas for the low to high risk organisms the containment categories C1, C2, and C3, respectively, are defined. A relationship between risk classification and

Tab. 10. Containment Measures for Industrial Processes with Biological Agents[a]

Containment Measures	Containment Levels 2	3	4
1. Viable organisms should be handled in a system which physically separates the process from the environment	Yes	Yes	Yes
2. Exhaust gases from the closed system should be treated so as to:	Minimize release	Prevent release	Prevent release
3. Sample collection, addition of materials to a closed system and transfer of viable organisms to another closed system, should be performed so as to:	Minimize release	Prevent release	Prevent release
4. Bulk culture fluids should not be removed from the closed system unless the viable organisms have been:	Inactivated by validated means	Inactivated by validated chemical or physical means	Inactivated by validated chemical or physical means
5. Seals should be designed so as to:	Minimize release	Prevent release	Prevent release
6. Closed systems should be located within a controlled area	Optional	Optional	Yes, and purpose-built
a) Biohazard signs should be posted	Optional	Yes	Yes
b) Access should be restricted to nominated personnel only	Optional	Yes	Yes, via an airlock
c) Personnel should wear protective clothing	Yes, work clothing	Yes	A complete change
d) Decontamination and washing facilities should be provided for personnel	Yes	Yes	Yes
e) Personnel should shower before leaving the controlled area	No	Optional	Yes
f) Effluent from sinks and showers should be collected and inactivated before release	No	Optional	Yes
g) The controlled area should be adequately ventilated to minimize air contamination	Optional	Optional	Yes
h) The controlled area should be maintained at an air pressure negative to atmosphere	No	Optional	Yes
i) Input air to and exit air from the controlled area should be HEPA filtered	No	Optional	Yes
j) The controlled area should be designed to contain spillage of the entire contents of the closed system	No	Optional	Yes
k) The controlled area should be sealable to permit fumigation	No	Optional	Yes
l) Effluent treatment before final discharge	Inactivated by validated means	Inactivated by validated chemical or physical means	Inactivated by validated chemical or physical means

[a] Annex VI of the EC Council Directive of 26 November 1990 on the protection of workers from risks related to exposure to biological agents at work (90/679/EEC)

the corresponding objectives of safety precautions is summarized in Tab. 7 (FROMMER and KRÄMER, 1990).

4.4.2 Safety Precautions in the Laboratory and in Process Biotechnology

In a review based mainly on the OECD recommendations and in agreement with the basic principles of various national lists, the EFB Working Party on Safety in Biotechnology defined a detailed relationship between the risk classes and the safety precautions to be employed in the laboratory as well as in process biotechnology (FROMMER et al., 1989). This review intended not only to ensure safety by correlating risk classes to various types of laboratory guidelines and physical containment measures, but also to assist in designing new facilities, to enable checking of existing facilities, and to give the manufacturers of equipment appropriate advice for the specification of their products.

A summary of precautionary measures for the four risk classes and the corresponding containment categories is shown in Tab. 8 (FROMMER et al., 1989). A detailed list of risk class related recommendations is compiled in Tab. 9 (FROMMER et al., 1989) which is divided into the following three parts:

a) Procedures: general operational instructions
b) Primary containment: operation and equipment designed to protect the personnel and the immediate processing facility from exposure to microorganisms
c) Secondary containment: facilities available to protect the external laboratory or factory environment from exposure to microorganisms

For laboratories a clear relationship between risk class and safety precaution measures exists.

For large-scale use of microorganisms the overall process may be difficult to be categorized in general. In such cases, it is advisable to divide a manufacturing process into its individual steps such as fermentation, cell harvesting, purification of the product, and so on, and to assess the risks for each of these operation units. This procedure allows the necessary flexibility in selecting appropriate containment measures on the basis of a risk assessment related to any particular process or part of a process.

Finally, it should be emphasized again that the nature of a biological agent is the basis for its risk classification and the conditions appropriate for its use. Thus, there is no reason to design any particular safety precaution measure specifically for recombinant-DNA organisms.

To underline this statement, in Tab. 10 the Annex VI of the EC COUNCIL DIRECTIVE 90/679/EEC is reproduced. The title of this directive ("On the Protection of Workers from Risks Related to Exposure to Biological Agents at Work") shows that it is valid for all biological agents. It contains exactly the same rules concerning containment conditions for industrial processes as the previous EC COUNCIL DIRECTIVE 90/219/EEC, regulating the "Contained Use of Genetically Modified Microorganisms".

5 References

COLLINS, C. H. (1990), Safety in industrial microbiology and biotechnology: UK and European classifications of microorganisms and laboratories, *TIBTECH* **8**, 345–348.

EC COUNCIL DIRECTIVE of 23 April 1990 "On the Contained Use of Genetically Modified Microorganisms" (90/219/EEC), *Off. J. Eur. Commun.* **L 117**/1, May 8, 1990.

EC COUNCIL DIRECTIVE of 26 November 1990 "On the Protection of Workers from Risks Related to Exposure to Biological Agents at Work" (90/679/EEC), *Off. J. Eur. Commun.* **L 374**/1, December 31, 1990.

FROMMER, W., KRÄMER, P. (1990), Safety aspects in biotechnology, *Drug Res.* **40**, 837–842.

FROMMER, W., AGER, B., ARCHER, L., BRUNIUS, G., COLLINS, C. H., DONIKIAN, R., FRONTALI, C., HAMP, S., HOUWINK, E. H., KÜENZI, M. T., KRÄMER, P., LAGAST, H., LUND, S., MAHLER, J. L., NORMAND-PLESSIER, F., SARGEANT, K., TUIJNENBURG MUIJS, G., VRANCH, S. P.,

WERNER, R. G. (1989), Safe Biotechnology III: Safety precautions for handling microorganisms of different risk classes, *Appl. Microbiol. Biotechnol.* **30**, 541–552.

FROMMER, W., AGER, B., ARCHER, L., BOON, B., BRUNIUS, G., COLLINS, C. H., CROOY, P., DONIKIAN, R., DROZD, J., ECONOMIDES, J., FRONTALI, C., HAMP, S., HAYMERLE, H., HUSSEY, C., KÜENZI, M. T., KRÄMER, P., LAGAST, H., LELIEVELD, H. L. M., LOGTENBERG, M. TH., LEMATTRE, M., LUND, S., MAHLER, J. L., NORMAND-PLESSIER, F., RUDAN, H. F. S., SIMON, R., SMITH, I. M., TACHMINTZIS, J., TUIJNENBURG MUIJS, G., VRANCH, S. P., WERNER, R. G., WATERSCHOOT, E. M. A. (1992a), Safe Biotechnology IV: Recommendations for safety levels for biotechnological operations with plant pathogens, *Appl. Microbiol. Biotechnol.*, in press.

FROMMER, W., AGER, B., ARCHER, L., BOON, B., BRUNIUS, G., COLLINS, C. H., CROOY, P., DOBLHOFF-DIER, O., DONIKIAN, R., ECONOMIDES, J., FRONTALI, C., HAMP, S., HAYMERLE, H., HOUWINK, E. H., KÜENZI, M. T., KRÄMER, P., LAGAST, H., LELIEVELD, H. L. M., LOGTENBERG, M. TH., LUND, S., MAHLER, J. L., NORMAND-PLESSIER, F., SIMON, R., TACHMINTZIS, J., TUIJNENBURG MUIJS, G., VRANCH, S. P., WERNER, R. G. (1992b), Safe Biotechnology V: Recommendations for safe work with animal cell culture concerning potential human pathogens, *Appl. Microbiol. Biotechnol.*, in press.

HOUWINK, E. H. (1989), *Biotechnology*, Dordrecht–Boston–London: Kluwer Academic Publishers.

KÜENZI, M., ASSI, F., CHMIEL, A., COLLINS, C. H., DONIKIAN, R., DOMINGUEZ, J. B., FINANCSEK, I., FOGARTY, L. M., FROMMER, W., HASKO, F., HOVLAND, J., HOUWINK, E. H., MAHLER, J. L., SANDKVIST, A., SARGEANT, K., SLOOVER, C., TUIJNENBURG MUIJS, G. (1985), Safe Biotechnology: General considerations, *Appl. Microbiol. Biotechnol.* **21**, 1–6.

KÜENZI, M., ARCHER, L., ASSI, F., BRUNIUS, G., CHMIEL, A., COLLINS, C. H., DEAK, T., DONIKIAN, R., FINANCSEK, I., FOGARTY, L. M., FROMMER, W., HAMP, S., HOUWINK, E. H., HOVLAND, J., LAGAST, H., MAHLER, J. L., SARGEANT, K., TUIJNENBURG MUIJS, G., VRANCH, S. P., OOSTENDORP, J. G., TREUR, A. (1987), Safe Biotechnology II: The classification of microorganisms causing diseases in plants, *Appl. Microbiol. Biotechnol.* **27**, 105.

OECD (Organisation for Economic Co-operation and Development) (1986), Recombinant DNA safety considerations: Safety considerations for industrial, agricultural and environmental applications derived by recombinant DNA techniques, Paris: *OECD Publications*, ISBN 92-64-12857-3.

OECD (Organisation for Economic Co-operation and Development) (1992), Safety considerations for the use of genetically modified organisms, Paris: *OECD Publications*, ISBN 92-64-13641-X.

WHO (World Health Organisation) (1982/1984), *Summary Report on a Working Group on the Health Implications of Biotechnology*, Copenhagen: WHO Regional Office for Europe (1982); published in: *Swiss Biotech.* (1984) **5**, 7–32.

WHO (World Health Organisation) (1983), *Laboratory Biosafety Manual*, Geneva: WHO.

Index

E

M

S

T